WORLD REGIONAL GEOGRAPHY

Second Edition

Global Patterns, Local Lives

Lydia Mihelič Pulsipher
University of Tennessee at Knoxville

Alex A. Pulsipher
Ph.D. Student
Graduate School of Geography
Clark University

with the assistance of

Conrad "Mac" Goodwin
University of Tennessee at Knoxville

W. H. Freeman and Company
New York

Acquisitions Editor: Jason Noe

Developmental Editor: Susan Moran

Marketing Manager: Jeffrey Rucker

Project Editor: Mary Louise Byrd

Cover and Text Designer: Blake Logan

Illustration Coordinator: Shawn Churchman

Maps: University of Tennessee,
Cartographic Services Laboratory,
Will Fontanez, Director

Illustrations: Fine Line Illustrations

Photo Editor: Meg Kuhta

Production Coordinator: Susan Wein

New Media and Supplements Editor: Bridget O'Lavin

Composition: Sheridan Sellers
W. H. Freeman and Company,
Electronic Publishing Center

Manufacturing: RR Donnelley and Sons

Library of Congress Cataloging-in-Publication Data

Pulsipher, Lydia M. (Lydia Mihelič)
  World regional geography : global patterns, local lives / Lydia Mihelič Pulsipher,
Alex A. Pulsipher; with the assistance of Conrad (Mac) Goodwin. — 2nd ed.
    p. cm.
  ISBN 0-7167-3841-4 (pbk.)
  1. Geography. I. Pulsipher, Alex. II. Goodwin, Conrad M., 1940–
  III. Title.
G128 .P85 2002
910 — dc21   2002000274

Printed in the United States of America

First printing 2002

To the people of the Caribbean island of Montserrat who
gave us our first experiences with the local imprints
of globalization

LYDIA MIHELIČ PULSIPHER is a cultural-historical geographer who studies the landscapes of ordinary people through the lens of archaeology and historical geography. She has contributed to several geography-related exhibits at the Smithsonian Museum of Natural History in Washington, D.C., including "Seeds of Change," which featured her research in the eastern Caribbean. Professor Pulsipher has taught cultural and Meso-american geography at the University of Tennessee at Knoxville since 1980, and, through her research, she has given many students their first experience in foreign fieldwork. Previously she taught at Hunter College and Dartmouth College. She received her B.A. from Macalester College, her M.A. from Tulane University, and her Ph.D. from Southern Illinois University.

ALEX A. PULSIPHER is a Ph.D. student in geography at Clark University, where he is studying the vulnerability of human–environment systems to global climate change. In the early 1990s, Alex spent some time in South Asia working for a development research center and then went on to do an undergraduate thesis on the history of Hindu nationalism at Wesleyan University. Beginning in 1995, Alex worked full time on the research and writing of the first edition of this textbook. In 1999 and 2000, he traveled to South America, Southeast Asia, and South Asia, where he collected information for the second edition of the text and for the Web site. In 2000 and 2001, he returned to writing material and designing maps for the second edition.

While writing *World Regional Geography*, Lydia Pulsipher was assisted by her husband, **CONRAD "MAC" GOODWIN**, a historical archaeologist who specializes in sites created during the European colonial era in North America, the Caribbean, and the Pacific. He has particular expertise in the archaeology of agricultural systems, formal gardens, domestic landscapes, and urban spaces. He holds a research appointment in the Department of Anthropology at the University of Tennessee.

BRIEF CONTENTS

CONTENTS

A family and their possessions in the United States (p. 15)

A Guatemalan family takes a break from fieldwork (p. 146)

CHAPTER 5 *Russia and the Newly
Independent States   236*

MAKING GLOBAL CONNECTIONS   238

Members of a Kazakh family do some needlework (p. 284)

**CHAPTER 6 North Africa
and Southwest Asia   288**

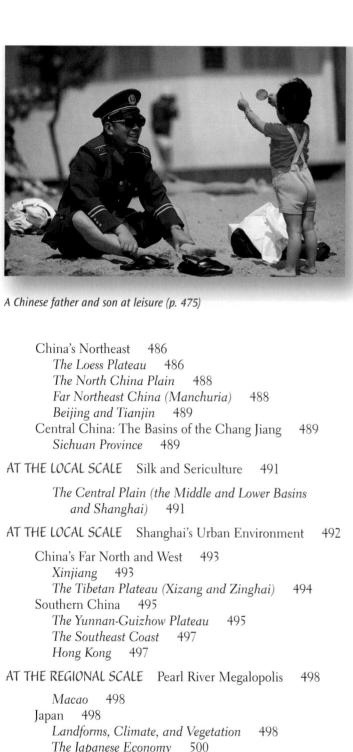

A Chinese father and son at leisure (p. 475)

A Mongolian family with their house and possessions (p. 510)

The motivation to write this text came out of nearly two decades of teaching, research, and travel. When teaching at the introductory level, I (Lydia Pulsipher) found myself longing for a text that conveyed to students more of the current insights from geographic research. I wanted a text that better conveyed the diversity of human culture throughout the world, covering the lives of women, men, and children. I wanted a book that made global patterns of trade and consumption come alive for students by showing the effects of these systems region to region and on the daily lives of ordinary people. In summary, I set out to write a new text that explained and illustrated global patterns while helping readers engage with the way these patterns affect local lives.

THE ENDURING VISION: THE GLOBAL AND THE LOCAL

The Global View

For the second edition we have further emphasized global trends and the interregional linkages that are changing lives throughout the world. The following linkages are explored in every chapter as appropriate:

■ The **multifaceted economic linkages** among world regions. These include (1) the effects of colonialism; (2) trade and particularly the growing number of regional trade organizations such as NAFTA and the European Union; (3) the role of transnational corporations in the world economy; (4) the global debt crisis; and (5) the influence exercised by the World Bank and the IMF in the form of structural adjustment programs (SAPs). For the second edition, we have extensively enhanced discussions of NAFTA, the European Union, and other regional trade organizations, and we have included additional graphics and discussions illustrating trade relationships.

■ The **Internet and computer technology.** In discussions new to this edition, the text explores the variety of roles played by the Internet in economic development, forging cultural identities, and encouraging democratic participation.

■ **Migration.** Migrants are changing economic and social relationships in virtually every part of the globe. People are leaving home and seeking opportunities elsewhere for a variety of reasons, but everywhere the departures of young adults especially and the presence of strangers are having local effects. Discussions, many new to this edition, explore the Indian diaspora; foreign workers in such places as Japan, Europe, the Americas, and Southwest Asia; and internal migration in the United States, Europe, Russia, China, Southeast Asia, and Oceania.

Figure 2.1 Interregional linkages: immigration to the United States.

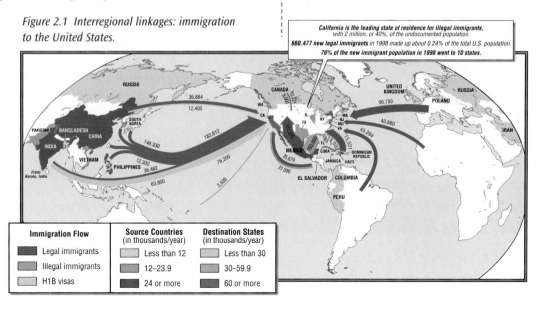

■ **World popular culture.** Mass communications and marketing techniques are promoting world popular culture. Some people across regions, especially young adults, are able to share such things as food, clothing, music, and patterns of socializing (such as watching TV or the frequent use of cell phones). Most of this popular world culture appears benign, perhaps even furthering understanding across cultural divisions. Yet some aspects cause worries about exploitation, consumerism, environmental damage, and loss of values. Like the first edition, the second edition integrates coverage of popular culture and its effects throughout in discussions of topics such as TV viewing in North Africa and Southwest Asia, the blending of Western and traditional culture in Japan, and the blossoming role of the Internet in Africa.

The Local View

Our approach pays special attention to local scales, whether a town, a village, a single household, or a person. Our hope is, first, that stories of individuals and families will make geography interesting and real to students, and, second, that seeing the effects of abstract processes and trends on ordinary lives will make these developments clearer to students. Reviewers have told us that students particularly appreciated the personal vignettes, which are usually stories of real people (with names disguised). We have taken the opportunity in the second edition to ensure we include a balance of vignettes on women, men, and children.

The following local responses are examined for each region as appropriate:

MEASURES OF HUMAN WELL-BEING

VIGNETTE The Parvanov family lives in a small apartment outside Sofia, Bulgaria. The family appears to have the material wealth of the middle class: adequate food, a nice home, a car in the garage. But looks are deceiving: these are largely leftover benefits of the communist economy. As the country adjusts to the free-market economy, inflation is high and Mr. Parvanov's salary is now worth only U.S. $30 a month, so he moonlights on a construction job. Mrs. Parvanov holds three part-time jobs: she works as a clerk and as an assistant in a building company, and at night she machine-knits clothes she sells in the local market. Theodora, the Parvanov's 19-year-old daughter, attends university by day and helps her mother knit at night. Yet with all this effort (six jobs in all), the family earns only about U.S. $100 a month. With this tiny amount, they must also support two grandparents whose pensions will no longer cover food, let alone medicine. The car hasn't been driven in months because gasoline is too expensive. They are able to survive only because the Bulgarian government still heavily subsidizes the costs of food, rent, and other basic necessities, a situation that will not last much longer. [Adapted from Robert Frank, "Price of isolation: impoverished Bulgaria is ready for reform," *Wall Street Journal* (February 28, 1997): 1.]

■ **Cultural change:** changes in the family, gender roles, and social organization in response to urbanization, modernization, and the global economy. The emphasis on cultural change in the first edition continues in the second.

■ **Impacts on well-being:** ideas of what constitutes "well-being" differ from culture to culture yet, broadly speaking, people everywhere try to provide a healthy life for themselves in a community of their choosing. Their success in doing so is affected by local conditions, global forces, and their own ingenuity.

■ **Grassroots development:** extending a feature of the first edition, the revision includes additional examples of projects initiated by local people to improve their own lives.

■ **Issues of identity:** paradoxically, as the world becomes more tightly knit through global communications and media, ethnic and regional identities often become stronger. Several new sections examine how modern developments such as the Internet are used to reinforce particular cultural identities.

■ **Local attitudes toward globalization:** many times people have ambivalent reactions to global forces: they are repelled by the seeming power of these forces, fearing effects on their own jobs and on local traditional cultural values, but they are also attracted by the opportunities, such as new jobs, that may emerge from greater global integration and are pleased when free trade makes consumer goods available at low prices. New sections in several chapters look at what a region's people say, in favor of or against, cultural or economic globalization.

Coffin art in Ghana (p. 370)

THE AUTHOR TEAM

The senior author of this textbook is **Lydia M. Pulsipher,** a professor at the University of Tennessee with research interests in human adaptation to the New World tropics in the post-Columbian era and in the cultural responses to globalization in North America, the Caribbean, Central Europe, and Asia. The junior author is her son **Alex Pulsipher,** a Ph.D. student at Clark University, whose recent travels in South America and Asia have interested him in the local responses to globalization and to physical hazards such as those connected to global warming. He has co-written portions of all chapters, especially those concerned with political, economic, and environmental issues, as well as numerous vignettes drawn from his travels, and he has designed numerous maps.

Another scholar has contributed full time to the project: **Conrad "Mac" Goodwin,** an anthropologist and colonial sites archaeologist, also affiliated with the University of Tennessee, did the bibliographic research and managed the production of the manuscript for the author team, the choice and design of graphics, and the selection of photos.

AUTHORITATIVE UPDATING: THE REGIONAL CONSULTANTS

To provide this text with the authority gained from years of study and travel, we have turned to a group of expert scholars who have contributed to this edition as regional consultants. Every chapter now incorporates additional expert commentary and analysis pertaining to that specific world region, including an interview with the consultant in a special feature entitled **"Geographer in the Field."** We worked closely with each consultant in carefully reviewing the chapters and identifying opportunities to deepen and enhance the regional discussion, so every student now has the opportunity to learn about the world's regions from those who know the regions best.

Our expert regional consultants include:

Chapter 2, North America: **Stanley Brunn, University of Kentucky.** Professor Brunn's teaching and research interests are in political, social, and urban geography, the human geographies of the twenty-first century, and the geographies of knowledge.

Chapter 3, Middle and South America: **Dennis Conway, Indiana University.** Professor Conway's research has centered on the development processes and prospects of Caribbean small islands, in general, and on the complex relationships between patterns of development, urbanization, and migration, in particular.

Chapter 4, Europe: **Stanley Brunn, University of Kentucky.** Professor Brunn's broad-ranging research interests in Europe include the images, identities, and worldviews of new states, landscape and community changes associated with open borders, voting patterns, diaspora networks, environmental quality and health issues, and the personal stories of those affected by war.

GEOGRAPHER IN THE FIELD

Interview with Regional Consultant for Chapter 9, Roman Cybriwsky, Temple University

What led you to become interested in East Asia?
My interest in East Asia was accidental. In 1984, my employer, Temple University, in Philadelphia, opened a campus in Tokyo and asked me to be one of the first faculty. I have since spent a number of years at the Tokyo branch and have traveled widely throughout East and Southeast Asia. As a geographer, I am always interested in place. My new place fascinated me and I made a point of learning as much as I could about it. It turned out to be a good mid-career transformation from earlier interests in U.S. urban geography.

Where have you done fieldwork in East Asia, and what type of fieldwork have you done?
I specialize in urban geography, and I focused on the neighborhoods of Tokyo. I am particularly interested in studying how places and the lives of people change. I do a lot of my work by walking, observing, starting conversations and asking questions, taking notes, and shooting photographs.
   I have also done fieldwork in Batam, a small island in Indonesia. In the last decade or so the island has grown tremendously, as new factories and a budding tourism economy have attracted migrants looking for work. There are enormous social and environmental impacts. For me, Batam is a window on the third world. Research there complements work in the United States and Japan.

How has that experience helped you to understand East Asia's place in the world?
Both sides of the Pacific are closely linked in a global economy. There are economic linkages, too, between Japan and developing countries in Pacific Asia like Indonesia. It has been interesting to see these linkages in person: to see the home base of a big corporation in New York or Tokyo and to see its factories in Batam, Indonesia; to see and compare the life rhythms of workers on both sides of the divide; and, as a result, to understand better my own place in the world.

What are some topics that you think are important to study about your region?
There are lots of critical issues: the spread of global commercial culture, preservation of local cultures, inequality between and within countries, the true meaning of "quality of life," and environmental impacts.

Chapter 5, Russia and the Newly Independent States: **Gregory Ioffe, Radford University.** A native of Moscow, Professor Ioffe has traveled extensively throughout the former USSR and studies the rural population and agriculture of Russia.

Chapter 6, North Africa and Southwest Asia: **Hussein A. Amery, Colorado School of Mines.** Professor Amery has done fieldwork in Lebanon, his place of birth, as well as in Syria, Jordan, Egypt, Israel, and Palestine. He has studied the effects of immigrants' remittances on rural economic development and the relationship between water stress and political conflict, and is now researching and writing about Islamic perspectives on the natural environment.

Chapter 7, Sub-Saharan Africa: **Barbara E. McDade, University of Florida.** Professor McDade's primary areas of research are entrepreneurship and economic development and the economic viability of small- and medium-size businesses in Ghana, West Africa. "Stateside," her research focus is black-owned businesses.

Chapter 8, South Asia: **Carolyn V. Prorok, Slippery Rock University.** Most of Professor Prorok's work focuses on religious material culture in the landscapes of the Hindu diaspora. She also works in the fields of geography of religion in general—especially pilgrimage studies and sacred places—geographic education, and women in development issues.

Chapter 9, East Asia: **Roman Cybriwsky, Temple University.** Professor Cybriwsky's current research interests include urban-social geography, urban planning and development, and neighborhood social change. He also has a special interest in planning and urban design in big cities, and has worked extensively on related issues in central New York City, Tokyo, Philadelphia, and elsewhere.

Chapter 10, Southeast Asia: **Richard Ulack, University of Kentucky.** Professor Ulack's research and teaching interests are in the broad area of development, with emphasis on Southeast Asia and the South Pacific regions. His fieldwork has focused on population geography (internal migration and fertility), urbanization (especially of middle-sized cities), and, most recently, tourism.

Chapter 11, Oceania: **Robert Kuhlken, Central Washington University.** A former regional planner and now a cultural geographer, Professor Kuhlken studies the traditional agricultural practices of Pacific Islanders.

Eroded karst (limestone) deposits form jagged peaks in the Yunnan Plateau of China (p. 496)

GEOGRAPHIC PRINCIPLES IN BALANCE

The core of each chapter combines a portrait of the region's basic economic, political, sociocultural, and environmental geography. In this edition, we have added more material on the economic and political aspects of the region. Moreover, our board of regional consultants ensures that the most modern and authoritative perspectives are explored.

The structure of each chapter follows a consistent model.

■ An introductory section, **"Making Global Connections,"** presents key themes and regional trends, emphasizing **interregional linkages** that connect the region to the rest of the world.

■ The section **"The Geographic Setting"** covers the basic factors of **physical geography, history,** and **population patterns** that contribute to a region's present-day geography.

■ The core of each chapter is the section **"Current Geographical Issues."** This section combines a portrait of the region's basic economic, political, and sociocultural geography with in-depth exploration of contemporary issues.

1. A subsection on **economic and political issues** emphasizes the connection of the region to large-scale or global trends and looks at the distribution of power within a region.

2. A subsection on **sociocultural issues** addresses such topics as family structure, gender roles, the lives of children, and religion in contemporary life.

3. A subsection on **environmental issues** shows how economic and political trends and societal values affect the use of resources and are changing the environment.

■ A special section, **"Measures of Human Well-Being,"** evaluates **human well-being** in a region using a variety of measurements, including gross domestic product (GDP) and the United Nations Human Development Index (HDI) and Gender Empowerment Measure (GEM). The cornerstone of this section is a table that presents these indices, plus male and female literacy figures in most cases, so that students and instructors may compare levels of well-being within and across regions.

■ The final third of each regional chapter provides a more detailed level of analysis. **Each region is divided into five to eight subregions,** covered in turn in separate sections that point out the distinctive human and physical geographic features. Often there is an opportunity to revisit issues explored on the regional scale earlier in the chapter to see the more localized effects on particular subregions.

■ A brief section at the end of the chapter called **"Reflections"** offers a perspective on the region's future.

ENHANCED MAP PROGRAM

New maps reinforce the second edition's emphasis on interregional linkages and core topics. Redesigned maps are more informative, more pedagogically useful, and visually stunning. We have been fortunate to be able to work closely with Will Fontanez and his capable staff at the University of Tennessee Cartographic Services Laboratory in the maps' planning and design.

■ The second edition contains new thematic maps that appear in every regional chapter. In addition to **climate** and **population density maps,** each chapter includes a new, standardized **environmental issues map** that maps land use, pollution, acid rain, and threatened fisheries. A new **economic issues map** in each chapter focuses on one or more key aspects of the region's economy, such as trade, debt, or GDP, and usually illuminates economic linkages with other regions.

■ New maps in several chapters describe access to the Internet and telecommunications, indicating a region's access to **information flows,** an important channel for forces of globalization.

■ Additional maps on topics such as history, political conflict, trade groups, and migration reinforce the core topics of the text.

■ Subregion maps highlighting political divisions, transportation networks, and cities have been redesigned to maximize clarity and visual appeal.

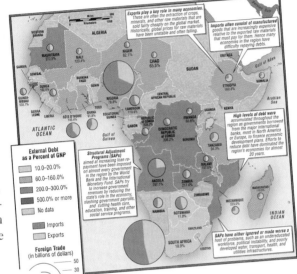

Figure 7.11
Economic Issues: Sub-Saharan Africa

Figure 10.19
Mainland Southeast Asia: Vietnam, Laos, and Cambodia

ENHANCED PEDAGOGICAL FEATURES

Assisting Critical Thinking

Features designed to encourage critical thinking about geographic issues appear in the text, the Web site, the *Instructor's Resource Manual*, and the *Student Study Guide*. These features encourage students to apply geographic principles in interpreting data, analyzing causes and consequences, and exploring the linkages between regions.

We have added discussion questions to the end of each chapter under the title "Thinking Critically About Selected Concepts." These questions ask students to reflect on consequences of a region's geographic features and to compare regions. We have chosen to combine the discussion questions with the end-of-chapter summary of a region's important features, because instructors told us that they liked the immediate reinforcement of those features provided by the discussion questions.

The student Web site offers several types of critical thinking features that are seamlessly woven with the text through the use of media icons in the text margins. These icons refer students to Web exercises that call for student research and analysis related to text discussions. The exercises, written by Tim Oakes and Chris McMorran of the University of Colorado, Boulder, are titled "Thinking Globally: On the Web," "Thinking Geographically: On the Web," and "Thinking Critically: On the Web."

Critical thinking exercises are also to be found in the *Instructor's Resource Manual* and the *Student Study Guide*, written by Helen Ruth Aspaas and Jennifer Rogalsky.

Additional Learning Tools

Several other features of the text also help students get the most from their reading:

- **"Themes to Look For,"** at the beginning of each regional chapter, alerts students to some broad trends in a region that are particularly important in the chapter.

- **"Terms to Be Aware Of,"** in the introduction to each regional chapter, points out some of the biases and unconscious assumptions in our use of particular place names.

- Important terms and concepts are set in boldfaced type the first time they appear in the chapter. They are gathered, with their definitions, in a list of **"Key Terms"** at the end of each chapter.

- All of the chapters' Key Terms and their definitions are also listed alphabetically in a **Glossary** at the back of the book.

- A **Pronunciation Guide** for less familiar place names and foreign terms appears at the back of each chapter. The sounds of vowels are represented by the following system:

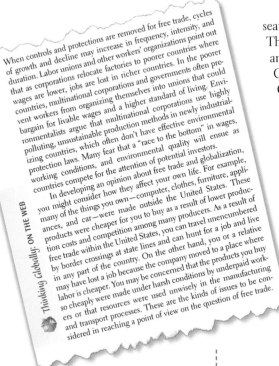

Thinking Globally: ON THE WEB

When controls and protections are removed for free trade, cycles of growth and decline may increase in frequency, intensity, and duration. Labor unions and other workers' organizations point out that as corporations relocate factories to poorer countries where wages are lower, jobs are lost in richer countries. In the poorer countries, multinational corporations and governments often prevent workers from organizing themselves into unions that could bargain for livable wages and a higher standard of living. Environmentalists argue that multinational corporations use highly polluting, unsustainable production methods in newly industrializing countries, which often don't have effective environmental protection laws. Many fear that a "race to the bottom" in wages, working conditions, and environmental quality will ensue as countries compete for the attention of potential investors.

In developing an opinion about free trade and globalization, you might consider how they affect your own life. For example, many of the things you own—computer, clothes, furniture, appliances, and car—were made outside the United States. These products were cheaper for you to buy as a result of lower production costs and competition among many producers. As a result of free trade within the United States, you can travel unencumbered by border crossings at state lines and can hunt for a job and live in any part of the country. On the other hand, you or a relative may have lost a job because the company moved to a place where labor is cheaper. You may be concerned that the products you buy so cheaply were made under harsh conditions by underpaid workers or that resources were used unwisely in the manufacturing and transport processes. These are the kinds of issues to be considered in reaching a point of view on the question of free trade.

A	PAT	p*at*	OO	POOT	p*u*t
AH	PAH-PUH	p*a*p*a*		["oo" as in book]	
AU	PAUT	p*ou*t	OO	POOCH	p*oo*ch
AW	PAW	p*aw*	UH	PUHT	p*u*tt
AY	PAY	p*a*te	AHR	PAHRT	p*ar*t
EE	PEET	p*ea*t	AIR	PAIR	p*air*
EH	PEHT	p*e*t	OHR	POHR	p*our*
EI	PIE	p*ye*	OOR	POOR	p*oor*
IH	PIHT	p*i*t	URR	PURR	p*urr*
OH	POHP	p*o*pe			

MEDIA AND SUPPLEMENTS PACKAGE

The second edition is accompanied by a superior media and supplements package that facilitates student learning and enhances the teaching experience. For students, we have seamlessly integrated topics from the text with the companion Web site; thus the latest technology is being used to reinforce concepts from the text. For instructors, we have created a full-service ancillary package that will help in the preparation of lectures and exams, particularly in regard to electronic classroom presentations.

AIDS FOR STUDENT LEARNING

On the Web

The companion Web site at www.whfreeman.com/pulsipher serves as an online study guide. The authors are Tim Oakes and Chris McMorran, both at the University of Colorado, Boulder. The core of the Web site is a range of features to encourage critical thinking and assist in study and review.

Critical Thinking Features

Thinking Globally: ON THE WEB. How is each world region linked to daily life in the United States? This feature includes a variety of Web site links that students can use to explore such linkages as trade, finance, tourism, and political movements. The links are matched with a series of questions or brief activities or both to give students an opportunity to think about the ways they are connected to the places and people they read about in the text.

Thinking Geographically: ON THE WEB. These exercises allow students to explore a set of issues such as deforestation, human rights, or free trade and examine how geography helps clarify our understanding of them. Students focus on key geography concepts—such as scale, region, place, interaction—as they use these concepts to drive analysis of compelling issues.

Thinking Critically: ON THE WEB. Students will find two types of Web exercises in this category:

■ *Thematic Map Overlays:* Students can overlay various maps from the text to compare and contrast data. Associated **Thinking Critically** questions accompany each overlay option.

■ *Animated Population Maps:* Animated maps show how regional populations have fluctuated over time. Related **Thinking Critically** questions ask students how and why the changes may have occurred.

Cultural Enrichment

World Recipes and Cuisines: From *International Home Cooking,* the cookbook of the United Nations International School.

For Study and Review

Map Learning Exercises: Students use this feature to identify and learn countries, cities, and major geographic features of each region.

Blank Outline Maps: Printable maps of every world region for note-taking or exam review.

Q & A Online: This self-quizzing feature enables students to review key text concepts and sharpen their ability to analyze geographic material for exam preparation. Answers (correct or incorrect) prompt feedback that refers students to the specific section in the text where the question is covered.

Flashcards: Matching exercises to teach vocabulary and definitions.

Audio Pronunciation Guide: Spoken guide of place names, regional terms, and historical figures.

Selected Readings: A selected readings list for each chapter facilitates student projects and more advanced learning when the textbook is used in upper-level courses.

Student CD-ROM

Can be packaged with the text for an additional $3.00 net. Use ISBN 0-7167-3884-8.

For students without convenient access to the Internet, the CD provides all the material from the Web site accessible through the same user-friendly navigation.

Atlas

Rand McNally's Atlas of World Geography, paperback, 176 pages. Available packaged with the text (0-7167-4904-1) or the text and *Student Workbook* (0-7167-4903-3).

Study Guide

Student Workbook for World Regional Geography, Jennifer Rogalsky, University of Tennessee, Knoxville, and Helen Ruth Aspaas, Virginia Commonwealth University, 0-7167-4692-1. The *Student Workbook* functions as both a study guide and book of mapping exercises designed to hone student skills in geographic analysis. The content has been carefully composed in conjunction with Lydia Pulsipher to complement the text.

AIDS FOR THE INSTRUCTOR

Presentation

Instructor's CD-ROM (0-7167-4691-3)

Contains all text images available as PowerPoint slides for use in class presentation. It also contains the *Test Bank* as chapter-by-chapter Word files that can be modified easily by the instructor.

The *Instructor's Resource Manual* is also available on the CD-ROM as chapter-by-chapter Word files.

Instructor's Resource Manual

Jennifer Rogalsky, University of Tennessee, Knoxville, and Helen Ruth Aspaas, Virginia Commonwealth University.

The manual includes suggested lecture outlines, points to ponder for class discussion, and ideas for exercises and class projects. The *Instructor's Resource Manual* is available as chapter-by-chapter Word files on the Web site under "For Instructors" for easy editing and printing, as well as on the *Instructor's Resource CD-ROM*.

Slide Set with Lecture Notes (set of 100), 0-7167-4696-4

A set of more than 100 *National Geographic* images and accompanying explanatory lecture notes for course lectures and presentations.

Overhead Transparencies (set of 100), 0-7167-4695-6

A handy set of key maps and figures from the text for class lecture and presentation.

Instructor's Web Site
For instructors, the Web site offers specific ideas written by Tim Oakes and Chris McMorran on how to incorporate *World Regional Geography* media into their course, as well as the following instructional aids:

Instructor's Resource Manual (Jennifer Rogalsky and Helen Ruth Aspaas) provided as chapter-by-chapter Word files

Password-Protected Test Bank (Andy Walter and Kim Crider) provided as chapter-by-chapter Word files

Syllabus-posting service

All text images available for download; also provided as PowerPoint slides

A *complete citation bibliography* containing an extensive list of the sources of information used in writing each chapter.

Assessment

Test Bank, accessed via the book's Web site under the password-protected "For Instructors" section, and available from the *Instructor's Resource* CD-ROM. Created by Andy Walter, Florida State University, Tallahassee, and Kim Crider, University of Tennessee, Knoxville.

The *Test Bank* is carefully designed to match the pedagogical intent of the text. It contains more than 2000 test questions (multiple choice, short answer, matching, true/false, and essay). The Web site and *Instructor's Resource* CD-ROM versions provide instructors with chapter-by-chapter Word files to download, edit and print.

Online Quizzing, powered by Questionmark, accessed at the book's Web site. Questionmark's Perception is becoming the industry standard for Web-based quizzing and assessment with full multimedia capabilities. With it, instructors can easily and securely quiz students online using prewritten, multiple-choice questions for each text chapter (not from the *Test Bank*). Students receive instant feedback and can take the quizzes multiple times. Instructors can go into a protected Web site to view results by quiz, student, or question, or can receive weekly results via e-mail.

ACKNOWLEDGMENTS

Many of our colleagues in geography have reviewed and commented on various drafts of individual chapters or groups of chapters for the first and second editions, offering advice on content and perspective. Not only have they greatly improved the quality of the book, but many have offered words of encouragement that were especially needed during this long process. Those colleagues who reviewed and commented on at least one chapter of the first edition are listed first, those who reviewed at least one chapter of the second edition are listed second.

FIRST EDITION

Helen Ruth Aspaas
Virginia Commonwealth University

Brad Bays
Oklahoma State University

Stanley Brunn
University of Kentucky

Altha Cravey
University of North Carolina at Chapel Hill

David Daniels
Central Missouri State University

Dydia DeLyser
Louisiana State University

James Doerner
University of Northern Colorado

Bryan Dorsey
Weber State University

Lorraine Dowler
Penn State University

Hari Garbharran
Middle Tennessee State University

Baher Ghosheh
Edinboro University of Pennsylvania

Janet Halpin
Chicago State University

Peter Halvorson
University of Connecticut

Michael Handley
Emporia State University

Robert Hoffpauir
California State University, Northridge

Glenn G. Hyman
International Center for Tropical Agriculture

David Keeling
Western Kentucky University

Thomas Klak
Miami University of Ohio

Darrell Kruger
Northeast Louisiana University

David Lanegran
Macalester College

David Lee
Florida Atlantic University

Calvin Masilela
West Virginia University

Janice Monk
University of Arizona

Heidi Nast
De Paul University

Katherine Nashleanas
University of Nebraska

Tim Oakes
University of Colorado, Boulder

Darren Purcell
Florida State University

Susan Roberts
University of Kentucky

Dennis Satterlee
Northeast Louisiana University

Kathleen Schroeder
Appalachian State University

Dona Stewart
Georgia State University

Ingolf Vogeler
University of Wisconsin, Eau Claire

Susan Walcott
Georgia State University

SECOND EDITION

Helen Ruth Aspaas
Virginia Commonwealth University

Cynthia F. Atkins
Hopkinsville Community College

Robert Maxwell Beavers
University of Northern Colorado

James E. Bell
University of Colorado, Boulder

Richard W. Benfield
Central Connecticut State University

John T. Bowen, Jr.
University of Wisconsin, Oshkosh

Stanley Brunn
University of Kentucky

Donald W. Buckwalter
Indiana University of Pennsylvania

Roman Cybriwsky
Temple University

Cary W. de Wit
University of Alaska, Fairbanks

Ramesh Dhussa
Drake University

David M. Diggs
University of Northern Colorado

Jane H. Ehemann
Shippensburg University

Kim Elmore
University of North Carolina at Chapel Hill

Thomas Fogarty
University of Northern Iowa

James F. Fryman
University of Northern Iowa

Heidi Glaesel
Elon College

Ellen R. Hansen
Emporia State University

John E. Harmon
Central Connecticut State University

Michael Harrison
University of Southern Mississippi

Douglas Heffington
Middle Tennessee State University

Robert Hoffpauir
California State University, Northridge

Doc Horsley
Southern Illinois University, Carbondale

David J. Keeling
Western Kentucky University

James Keese
California Polytechnic State University

Debra D. Kreitzer
Western Kentucky University

Jim LeBeau
Southern Illinois University, Carbondale

Howell C. Lloyd
Miami University of Ohio

Judith L. Meyer
Southwest Missouri State University

Judith C. Mimbs
University of Tennessee, Chattanooga

Monica Nyamwange
William Paterson University

Thomas Paradis
Northern Arizona University

Firooza Pauri
Emporia State University

Timothy C. Pitts
Edinboro University of Pennsylvania

William Preston
California Polytechnic State University

Gordon M. Riedesel
Syracuse University

Joella Robinson
Houston Community College

Steven M. Schnell
Northwest Missouri State University

Kathleen Schroeder
Appalachian State University

Dean Sinclair
Northwestern State University

Robert A. Sirk
Austin Peay State University

William D. Solecki
Montclair State University

Wei Song
University of Wisconsin, Parkside

William Reese Strong
University of North Alabama

Selima Sultana
Auburn University

Suzanne Traub-Metlay
Front Range Community College

David J. Truly
Central Connecticut State University

Alice L. Tym
University of Tennessee, Chattanooga

This book has been a family project many years in the making. I (Lydia) came to the discipline of geography at the age of five, when my immigrant father, Joe Mihelič, hung a world map over the breakfast table in our home in Coal City, Illinois, where he was pastor of the New Hope Presbyterian church (see my childhood map, page 4). We soon moved to the Mississippi Valley of eastern Iowa, where my father, then a professor in the Presbyterian theological seminary in Dubuque, continued his geography lessons on the passing landscapes whenever I accompanied him on Sunday trips to small country churches. Once we had an especially long lesson when the flooding Mississippi River trapped us for hours. Alex and his older brother, Anthony, got their first doses of geography in the bedtime stories I used to tell them. For plots and settings, I drew on Caribbean colonial documents I was then reading for my dissertation. They first learned about the hard labor of field geography when as mid-sized children they were expected to help with the archaeological and ethnographic research my colleagues and I were conducting on the eastern Caribbean island of Montserrat. Mac Goodwin, my husband, has given up several years of his career as a colonial sites archaeologist to help with the research, writing, and production of this book. It was my brother John Mihelič who first suggested that we write a book like this one, after he too came to appreciate geography. He has been a loyal cheerleader during the process, as have our extended family and friends in Knoxville, Montserrat, Slovenia, and beyond.

My graduate students and faculty colleagues in the Geography Department at the University of Tennessee have been generous in their support, serving as helpful impromptu sounding boards for ideas. Ken Orvis, especially, has advised us on the physical geography sections of both editions. Will Fontanez and the cartography shop staff unfailingly pursued the goal of beautiful informative maps, often cheerfully making adjustments when it was definitely not convenient. Special thanks are owed to Will and to Tom Wallin, Susan Carney, Jennifer Barnes, Hilary Martin, Brad Kreps, and Will Albaugh.

Sara Tenney and Liz Widdicombe at W. H. Freeman and Company were the first to persuade us that together we could develop a new direction for *World Regional Geography*, one that included the latest thinking in geography written in an accessible style. In accomplishing this goal, we are especially indebted to our developmental editor, Susan Moran, ably assisted by Marjorie Anderson, and to the W. H. Freeman staff for all they have done to ensure that this book is well written, beautifully designed, and well presented to the public. Over the seven years that have passed since we first began working on this book, Susan has been unfailingly wise, temperate, insightful, and probing, yet also kind and patient. She was always a loyal advocate for the reader—without her help, we could not have produced this book.

We would also like to gratefully acknowledge the efforts of the following people at W. H. Freeman: Melissa Wallerstein and Susan Brennan, acquisitions editors for the first edition, Jason Noe, for the second edition; Mary Louise Byrd, project editor for both editions; Diana Siemens, copy editor; Blake Logan, designer; Bill Page and Shawn Churchman, illustration coordinators; Susan Wein, production coordinator; Sheridan Sellers, W. H. Freeman and Company Electronic Publishing Center; Inge King, photo researcher for the first edition, Meg Kuhta for the second edition; Bridget O'Lavin, supplements and new media editor; and Eleanor Wedge and Martha Solonche, proofreaders for the text and maps, respectively. We are also grateful to Jennifer Rogalsky and Helen Ruth Aspaas, authors of the *Student Workbook* and *Instructor's Resource Manual*; Tim Oakes and Chris McMorran, authors of the Web site; and Andy Walter and Kim Crider, authors of the *Test Bank*.

WORLD REGIONAL GEOGRAPHY

1 | GEOGRAPHY: An Exploration of Connections

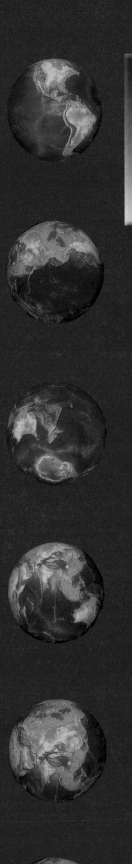

A series of images running vertically at the left gives different views of the earth as seen from outer space. These images are generated by computer from a digital data base containing measurements of ocean depth and land surface elevation. As on the large map of world regions, color is used to indicate elevation. [NOAA.]

North America

Seattle
Portland
San Francisco
Los Angeles
Denver
Phoenix
Dallas
Houston
Chicago
Detroit
Atlanta
Washington, DC
Ottawa
Toronto
Montréal
Boston
New York
Philadelphia
Miami
Havana
Guadalajara
Mexico City

NORTH PACIFIC OCEAN

NORTH ATLANTIC OCEAN

Bogotá
Lima

Middle and South America

Fortaleza
Salvador
Rio de Janeiro
São Paulo
Santiago

SOUTH PACIFIC OCEAN

Buenos Aires

Land Elevations

meters	feet
Ice cap	Ice cap
5183 and above	17,000 and above
3353–5182	11,000–16,999
2134–3352	7000–10,999
914–2133	3000–6999
305–913	1000–2999
0–304	0–999
Below sea level	Below sea level

WHERE IS IT? WHY IS IT THERE?

Where are you? You may be in a house or a library or sitting under a tree on a fine fall afternoon. You are probably in a community (perhaps a college or university); and you are in a country (perhaps the United States) and a region of the world (perhaps North America, Southeast Asia, or the Pacific). Why are you where you are? There are immediate answers, such as "I have an assignment to read." But there are also larger explanations, such as your belief in the value of an education, your career plans, or someone's willingness to sacrifice to pay your tuition. Even past social movements in Europe and America that opened up higher education to more than a fortunate few may help to explain why you are where you are. It's even possible that your desire to learn more about world geography has brought you to your present location.

The questions of *where* and *why* are central to geography. Like anyone who has had to find the site of a party on a Saturday night, the location of the best grocery store, or the fastest and safest route home, geographers are interested in location and spatial relationships. Different places have different sights, sounds, and smells, and these differences are significant. Understanding everything that has contributed to the look and feel of a place, to the standard of living and customs of its people, and to the way people from different places relate to one another helps to answer the question of why places are as they are. Seeking these answers is the special endeavor of geographers.

To make it easier to understand a geographer's many interests, please try this exercise. On a piece of blank typing paper, draw a map of your favorite childhood landscape. Relax, and let your mind recall the objects and experiences that were most important

to you in that place. If the place was your neighborhood, you might start by drawing and labeling your home, then fill in other places you encountered regularly as you went about your life—such as your backyard and the objects in it, your best friend's home, your school. Don't worry about creating a work of art—just make a map that reflects your experience as you remember it.

When you are finished, think about how the map reveals the ways your life was structured by space. Were there places you were not supposed to go, but did anyway? Why were some places off-limits? (And why were they so intriguing to you?) Does your map reveal, perhaps in a very subtle way, such emotions as fear, pleasure, or longing? Does it indicate your sex, your ethnicity, or the makeup of your family? Did you draw some places in greater detail than others? How much space did you decide to cover on the typing paper? Did you draw just a house and yard, a neighborhood, or a square mile or more? Did you give the map reader some clue about how much space is represented on the map? The amount of space you covered with your map may represent the degree of freedom you had as a child to go about on your own, or it may represent something entirely different.

In making your map and analyzing it, you have engaged in the practice of geography:

- Landscape observation

- Description of the earth's surface

- Historical reconstruction of bygone places

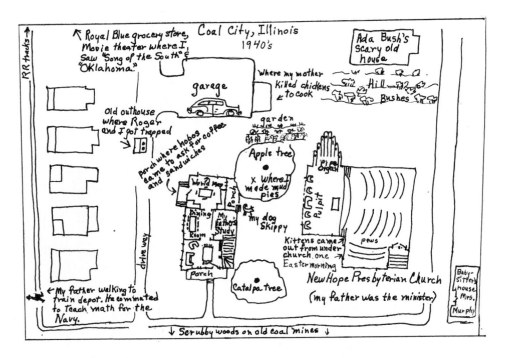

Lydia Pulsipher's childhood landscape map.

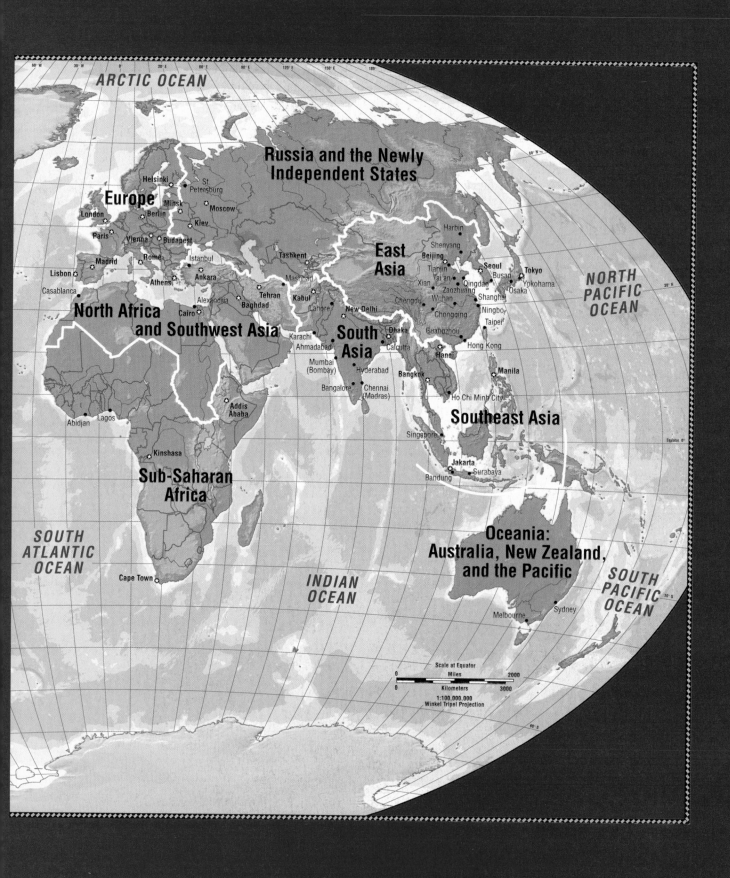

- Spatial analysis (the study of how people, objects, or ideas are, or are not, related to one another across space)

- Cartography (the making of maps)

As you progress through this book, you will acquire geographic information and skills. Whether you are planning a road trip, looking at a photograph of an intriguing landscape, thinking about investing in East Asian stocks, searching for a good place to market an idea, or trying to understand current events, knowing how to practice geography will make the task easier and more engaging.

WHAT IS GEOGRAPHY?

A brief formal definition of geography might be that it is the study of our planet's surface and the processes that shape it. Yet such a succinct statement does not begin to convey the fascinating interactions of physical and human forces that have given the earth its diversity of landscapes and ways of life. Geography, as an academic discipline, links the physical sciences—such as geology, physics, chemistry, biology, and botany—with the social sciences—such as anthropology, sociology, history, economics, and political science. **Physical geography** is the study of earth's physical processes to learn how they work, how they affect humans, and how they are affected by humans in return. **Human geography** is the study of various aspects of human life that create the distinctive landscapes and regions of the world. Physical and human geography are often tightly linked. For example, geographers might aim to understand

- How and why people came to occupy a particular place

- How they assess the physical aspects of that place (climate, landforms, resources) and then set about using and modifying them to suit their particular needs

- How they interact with other places

Yet people are not always tightly bound by their physical surroundings. Transport technology and the Internet link people in one physical place to distant locations and resources, so people no longer need to design their lives around what is available locally. Thus, the investigations of some geographers may seem to have little direct connection to the physical environment. For example, geographers study such issues as political activism among homeowners, the spatial allocation of tax dollars, and socioeconomic variables in housing choice.

Geographers are unified by a common interest in understanding the world in spatial terms. Just as historians study change over time, geographers study variation over space. Most of all, geographers are interested in the explanations for those variations. Among geographers' most important tools are maps, which they use to record and analyze spatial relationships, just as you recorded and analyzed the features that made up your favorite childhood landscape.

Geographers usually specialize in one or more fields of study, or subdisciplines. Table 1.1 gives examples of the different kinds of geographers who might be found in any college or university geography department or working for the government or a private firm.

As you read through the descriptions, it is easy to see that cooperation among the various kinds of geographers would be useful in studying the interactions between people and places. Individual geographers often read and write in several subdisciplines, and geography journals typically contain articles on a wide array of topics.

Many geographers are led by their interest in spatial relationships, and by the difficulties of analyzing any phenomenon on a global scale, to specialize in a particular place or region of the world. They may study their chosen region from several different perspectives. **Regional geography** is the analysis of the geographic characteristics of a particular place (the size of that place can vary radically). The study of a particular place from a geographic perspective can reveal previously unappreciated connections among physical features and ways of life. These connections are a key to understanding the present (and the past) and are essential in planning for the future.

This book follows a "world regional" approach because experience shows that people new to geography find general knowledge about specific regions of the world to be the most interesting and useful introduction to the subject. Such knowledge is produced by the types of geographers described in Table 1.1, and in this book we rely on all of them to provide a holistic perspective on each of the world's regions. Just what geographers mean by *region* is discussed next.

THE REGION AS A CONCEPT

A **region** is a unit of the earth's surface that contains distinct patterns of physical features or of human development. We could speak of a desert region, a region that produces table wine, or a region that is characterized by grassland vegetation or by ethnic violence. Geographers disagree about the concept of region, and a precise definition of particular regions is elusive. For one thing, it is rare for any two regions to be defined by the same set of indicators. One assemblage of places might be considered a region because of its distinctive vegetative landscape and such cultural features as food, accents, and music: the Southwest region of the United States, for example. Another region might be defined primarily by its cold climate and sparse population density, such as Siberia or Alaska. Still another group of places might be considered a region primarily because of its particular political system, such as Eastern Europe before the political and social changes of the 1990s.

Another problem in defining regions is that their boundaries are rarely crisp. The more closely we look at the border zones, the fuzzier the divisions appear. Take the case of the boundary between the United States and Mexico (see Figure 1.1 on page 7). Just where is the true regional boundary? What criteria shall we use to establish this boundary? Although there is a clearly delineated political border between the two countries, it is not really a marker of separation between cultures or even between economies. In a wide band extending over both sides of the border one can find a blend of Native American, Spanish colonial, Mexican, and Anglo-American cultural features: languages, place names, food customs, music, and family organization, to name but a few. And the economy of the border zone—which is based on agriculture, manufacturing, trade, and tourism—depends on interactions across a b·

TABLE 1.1 *The subdisciplines of geography*

Geomorphologists further our understanding of how landforms such as mountains, plains, river valleys, and the ocean floor developed in the past and continue to change. Geomorphologists in academic geography departments (as opposed to geology departments) are often particularly interested in how people alter landforms through deforestation, cultivation, and the manipulation of watersheds and other environments.

Climatologists look at long-term weather patterns (climates), studying the interaction of climate, vegetation, and landforms. Recently, climatologists have begun to study the possibility that humans are altering climate on a worldwide scale by releasing carbon dioxide and other gases into the atmosphere, which results in global warming. The cause is thought to be the burning of wood and fossil fuels (oil, gas, and coal).

Biogeographers study the geography of life on earth: where plants, animals, and other living organisms are found today; where they were found in the past; and the circumstances that affected these distributions. They also examine how the distribution of ecosystems is shaped by human activity, climate change, plate tectonics, and other environmental and evolutionary processes.

Cultural geographers describe and seek to explain the spatial patterns and ecological relationships of culturally distinct groups of people. A cultural geographer might study how the religious or philosophical beliefs of a particular cultural group influence the ways the members of the group practice agriculture, see themselves in relation to nature, organize their settlements, or set down behavior rules for males and females from childhood through old age.

Human (or cultural or political) ecologists combine the interests of biogeographers and cultural geographers. They examine the specific ways in which humans and physical environments (including all types of life-forms) interact.

Historical geographers study long-term spatial processes, such as migrations; changes over time in human habitats; and the spread of technology, ideas, and other aspects of culture. They often use the techniques of biogeographers and cultural, political, and economic geographers in their analyses of past circumstances.

Economic geographers look at the spatial aspects of economic activity, such as how resources are allocated and exchanged from place to place. They also examine how people interact with their environment as they go about earning a living. For example, an economic geographer might focus on where women tend to work within a city, why they work there, and how they get there. They might also study the amount and type of space occupied by particular economic enterprises, such as popcorn vending, shopping malls, or the international cocaine trade.

Political geographers are interested in the spatial expressions of power and in the institutions people have devised to channel that power, such as representative democracies or authoritarian religious states. Often, political power struggles are signified by landscape features such as the Great Wall of China, the Berlin Wall, or the burned villages of Bosnia.

Urban geographers study spatial patterns and processes within cities and the ways in which urban areas interact with surrounding suburban and rural areas.

Gender geographers offer a new perspective on the world by examining gender relations and the roles and status of men and women in society. They also explore how activities that are linked in some way to a person's sex are expressed spatially; and they analyze the meanings of these spatial patterns.

Cartographers specialize in depicting geographic knowledge graphically in maps. They are interested in both the science and the aesthetics of portraying visual information to the map reader. Although maps are often still drawn by hand, computers are increasingly the cartographer's chief tool.

Remote sensing analysts discern patterns on the earth's surface by examining high-altitude photographs and other images collected from space by military and space programs.

Geographic Information Systems (GIS) analysts use computers and software to capture, store, integrate, analyze, and display large bodies of statistical information about spatial relationships. An example of their work would be to analyze the changing patterns of agricultural crop transportation, as well as the location, capacity, and routing of the trucks, trains, and ships required. Recently, GIS analysts have begun to look at complex cultural and social problems, such as welfare reform in the United States. By analyzing the distribution of welfare recipient residences and the location of training, transport, and child-care facilities, as well as the location of potential jobs, GIS specialists can help government officials to understand the problems welfare recipients must resolve if they are to become self-supporting.

---

swath of territory. In fact, the Mexican national economy is more closely connected to the economy of the United States than to its close neighbors in Middle America (the countries in Central America and the island countries of the Caribbean).

Why, then, does this book place Mexico in Middle America? Because at this point in time, Mexico has more in common with Middle America than with North America. Its colonial history and the resulting use of Spanish as the official language tie Mexico to Middle America. The dominant language of Middle and South America is Spanish (in Brazil, it is Portuguese), whereas the dominant language of the United States and Canada is English (except in Québec, where French is primary and English secondary). These language patterns are symbolic of the larger cultural and historical differences between the two regions, which will be dis-

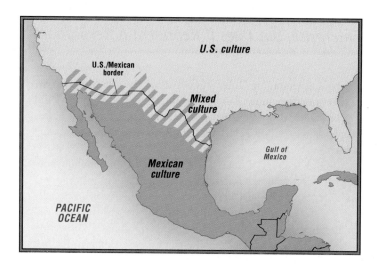

Figure 1.1 Language patterns along the border country of the United States and Mexico. The map reveals that the legal boundary is not a cultural boundary. Along the border there is a wide band of cultural blending. In this blended zone, language, food, religion, and architecture show influences of both U.S. and Mexican culture. [Adapted from *The World Book Atlas* (Chicago: World Book, 1996), p. 89.]

cussed in Chapters 2 and 3. Even though Mexico has much in common with parts of the southwestern United States, on the whole the similarities do not yet override the differences. Note, however, that this situation is changing as the economic connections among the United States, Canada, and Mexico grow stronger and as more Mexicans emigrate to the United States. Indeed, in the not too distant future it might make sense to include Mexico in the same world region as the United States and Canada.

If regions are so difficult to define and describe and are so changeable, why do geographers use them? The obvious answer is that it is impossible to discuss the whole world at once, and so we seek a reasonable way to divide it into manageable parts. There is nothing sacred about the criteria or the boundaries we use; they simply need to be practical. In defining the world regions for this book, we have considered such factors as physical features, political boundaries, cultural characteristics, and what the future may hold. Geographers find that using multiple factors enables them to arrive at regional definitions with which most people can agree.

We have organized the material in this book into four levels, or scales: the global scale, the world regional scale, the subregional scale (Figure 1.2), and the local scale. At the **global scale**, explored in this chapter, the entire world is treated as a single

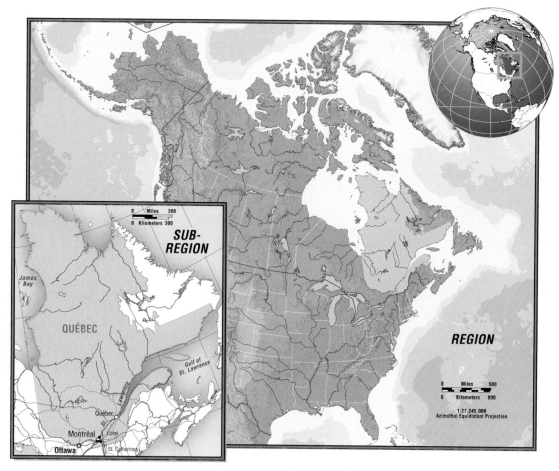

Figure 1.2 Three regional scales: the world, a region (North America), and a subregion (Québec) of North America.

AT THE GLOBAL SCALE   *Cisco Systems: A Global Enterprise, but Does It Represent a Revolution?*

Cisco Systems, Inc., is a global company that provides Internet networking at the local, regional, and global scales. Cisco sells the hardware and services that make it possible to link a company's computers together and then link the company to the World Wide Web. Cisco is young; it shipped its first product in 1986 and just 14 years later (fiscal 2000) had revenues of $18.9 billion. Cisco has sales and service offices in more than 60 countries and sells its products in 115 (Figure 1.3). It trains thousands of technicians in its "Network Academies" and then hires them to provide service to the systems it sells worldwide. The corporate headquarters are in San Jose, California, and the company has operational centers in North Carolina, Texas, Massachusetts, the United Kingdom, and Australia. Cisco employs nearly 40,000 people worldwide.

Those who study cultural change have observed that the present revolution in information technology may rank with the two previous major technological revolutions in human history: the agricultural revolution and the industrial revolution. To be a revolution of the same magnitude, information technology would have to transform the ways in which people relate to their environments and to one another. Is that happening?

John Chambers, president and CEO of Cisco, says it is. He describes what he sees as a fundamental change in human society enabled by the open communication possible over the Internet.

There is an emerging "Internet ecosystem," he says, that encourages cooperation and collaboration along lines that blur social and power differences. To businesspeople, this trend could mean less emphasis on cutthroat competition and more on prospering through interdependency.

John Chambers is not the first to claim that information technology will fundamentally change the ways people relate to one another. But to what extent are these claims believable, especially in the world of commerce? Couldn't this just be business hype, calculated to position a company to make more money? Answering these questions will take time and careful observation to see just where the advantages of and profits from information technology are allocated. The following Web sites are a good starting point for such an assessment.

- http://www.cisco.com/public/countries_languages.shtml

- http://www.cisco.com/warp/public/756/gnb/gnb_wp.htm

- http://www.netaid.org/AboutNetaid/who.html#faq

- http://www.cisco.com/warp/public/779/edu/academy/

- http://www.cisco.com/warp/public/779/edu/media/pdf/cnap_brochure_1200.pdf

- http://www.undp.org/hdr2001

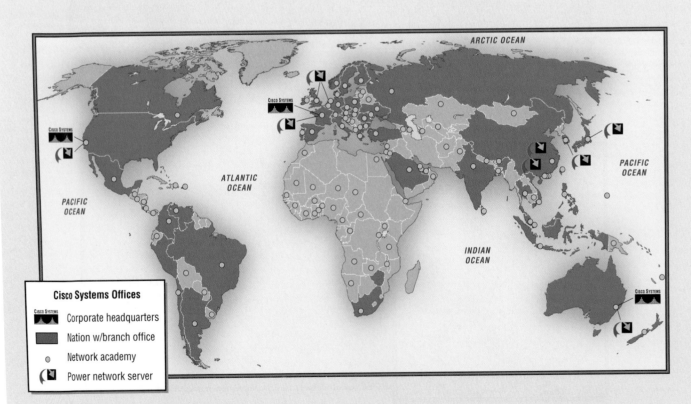

Figure 1.3   Interregional linkages: Cisco's global network.

area—a unity that is still new to all of us but is more and more relevant as we become used to thinking of our planet as a global system. We use the term *world region* for the largest divisions of the globe, such as East Asia, Southeast Asia, and North America. There are 10 world regions, each of which is covered in a separate chapter. The world regions are then divided into subregions, which may be independent countries, groups of countries, or even parts of a single country. For example, the world region of North America contains several subregions. In Canada, the province of Québec is considered a subregion because of its distinct French heritage. In the United States, no one state stands out so distinctively, and subregions tend to be groups of states, such as the Southwest or the Northeast. In some cases, subregions include territory in both the United States and Canada.

The regions, subregions, and local places discussed in this book vary dramatically in size and complexity. A subregion can be as large as a continent: Australia, for example, is part of the world region of Oceania and has an area of nearly 3 million square miles (7.8 million square kilometers). On the other hand, a subregion can be as small as the group of tiny islands known as Micronesia, also in Oceania, which together cover only 270 square miles (700 square kilometers). At the local scale, a place can be a backyard, a neighborhood, or a town.

INTERREGIONAL LINKAGES

Regions and subregions often share some similarities, and they may have overlapping borders. Consequently, **contiguous regions**—those that lie next to or near each other—often interact in various ways. They may trade with each other, share labor pools of available workers, experience each other's economic recessions and booms, borrow new ideas from each other, and help each other in times of crisis.

Our present era is remarkable in that regions widely separated in space can have interdependent economic relationships that used to be possible only between close neighbors. For example, much of the clothing and tropical hardwoods used by North Americans comes from Southeast Asia. South Americans and Africans produce cash crops that end up on European tables. These connections between distant regions began to attain their present reach during the early stages of *European colonialism*: the practice of taking over the human and natural resources of often distant places in order to produce wealth for Europe. The European voyages to America in 1492 led Spain and Portugal to establish colonies in Middle and South America. The British, Portuguese, Dutch, and French founded colonies in North America and Asia in the sixteenth and seventeenth centuries and in the Pacific islands in the eighteenth century. By the end of the nineteenth century, most of Africa was divided into European colonies. Although most colonies are now independent countries, colonization by European countries led to transglobal ties that still remain.

Wherever it took hold, European colonization changed local economies and landscapes. In the Caribbean and the islands of Southeast Asia, for example, forest cover was lost through plantation agriculture, logging, and mining. Sometimes valuable cash crops were transferred to entirely new places; for example, nutmeg trees were sent from the Molucca Islands of Southeast Asia to the British colony of Grenada in the Caribbean. There, nutmeg plantations became an important source of income for Britain. Similarly, sugarcane, originally from Southeast Asia, became a widely grown cash crop in tropical America, and cacao, from Middle America, became a cash crop in several colonies in West Africa. Many of the broad regional linkages we see today evolved from linkages that began hundreds of years ago, in large part under the influence of European colonialism.

Today, because of rapid transportation and the speedy flow of electronic information, widely separated places can be so intimately linked that the details of their relationships can change from day to day. For example, fluctuations in the market price of cacao (chocolate) in London can alter overnight the food that a small cacao grower in West Africa can afford for his family. Construction workers in Saudi Arabia can be sending substantial amounts of their wages home to their families in Pakistan and Malaysia. Then, in a matter of days, hostilities in the Persian Gulf can send hundreds of thousands of workers hurrying home, disrupting not only construction in Saudi Arabia but also tens of thousands of family budgets in Pakistan and Malaysia and the fortunes of shopkeepers in their villages.

The term **globalization** encompasses the many types of interregional linkages and flows and the changes they are bringing about. Here are some trends that have contributed to the globalization of societies around the world.

- As the volume of international trade grows, localities are becoming increasingly connected to this trade.

- Businesses and corporations in rich countries have established manufacturing, sales, and service branches in poorer countries to be closer to their markets and to take advantage of low-cost resources, land, labor, and professional expertise. These transnational or multinational organizations also find lower taxes and less rigorous environmental and trade regulations than in their home countries. See, for example, the map in the box "Cisco Systems: A Global Enterprise."

- As manufacturing and some information processing jobs move to places where labor is cheaper, workers in the old locations are left unemployed.

- International migration is surging as people seek opportunities in better locations.

- The rising volumes of people, goods, and capital flowing across regional and national borders are causing the world's nation-states to become less powerful, as international and supranational organizations evolve to control the new flows.

- Increased cross-cultural contact is leading to greater homogenization of culture, especially through the spread of modern pop culture. At the same time, fast communication is reinforcing cultural differences and enabling the development of distinct cultural identities.

...ology now facilitates the rapid transfer of money, ...g it easier for legitimate firms to do business and for migrant workers to send money to their families. Such illegal activities as drug trading and theft, however, also benefit from rapid international money transactions, in part because identities can be disguised more easily.

■ Local environments increasingly show the effects of worldwide trade and of resource consumption. Resources are moved from one region to serve the needs of another, environmental degradation in one place is often the result of demands for resources or goods in other places, and pollution can flow across national borders.

Because globalization has multiple facets, we will discuss it at many points in this book. Individuals can be harmed, or empowered, by its effects—and sometimes both at once. As you read the many references to globalization in the following sections, reflect on how you as an individual are part of the picture.

Our emphasis in this book is on present-day issues in human geography, especially how human well-being is faring in different world regions, given the trends toward globalization. **Human well-being** is defined broadly as the ability of people to provide a healthy life for themselves in a place and community of their choosing. Human well-being is a general concept, and just what it constitutes will change from culture to culture. Yet some aspects of human well-being can be measured and, therefore, regions of the world can be compared with one another. Each chapter will include a section on the human well-being achievements (as measured by the *United Nations Human Development Report 2000*) for the region discussed.

The rest of this chapter introduces several subfields of human geography emphasized in this book: cultural/social, economic, political, and environmental. Basic concepts in physical geography are also included. This background will help you to understand how people in the various regions of the world have worked out their relationships with the local physical environment, with one another, and with the wider world. The geographic subfields presented in this chapter will guide the description and discussion of each world region. Each chapter is organized along the same lines and covers the same general topics. The order of these topics changes somewhat from chapter to chapter, simply because circumstances in a particular region may lend themselves to a particular arrangement of topics. If you wish to compare general economic or demographic conditions in two or more world regions, however, you will be able to do so.

CULTURAL AND SOCIAL GEOGRAPHIC ISSUES

Culture is an important distinguishing characteristic of human societies. It comprises everything we use to live on earth that is not directly part of our biological inheritance. Culture is represented by the ideas, materials, and institutions that people have invented and passed on to subsequent generations. Among other things, culture includes language, music, belief systems, gender roles, and moral codes (for example, those prescribed in Confucianism, Islam, or Christianity). **Material culture** comprises all the things, living or not, that people use: clothing, houses and office buildings, axes, guns, computers, earthmoving equipment (from hoes to work animals to bulldozers), books, musical instruments, domesticated plants and animals, and agricultural and food processing equipment. **Institutions** are formal or informal associations among people,. The family, in its many different forms, is an **informal institution.** So is a community, which can be made up of the people who share a physical space (a village or a neighborhood) or those who share a belief system. The Roman Catholic community, for example, stretches across countries and continents (Figure 1.4). **Formal institutions** include official religious organizations such as the Presbyterian Church U.S.A. or the Vatican; local, state, and national governments; nongovernmental organizations that provide philanthropic services, such as the Red Cross and Red Crescent; and specific businesses and corporations.

In the next section, we explore the concept of culture groups and some of the cultural attributes, such as value systems and language, that help define them. We will examine gender roles and perceptions about race as cultural phenomena that vary over space and play an important role in human relationships.

CULTURE GROUPS

A group of people who share a set of beliefs, a way of life, a technology, and usually a place form a **culture group.** For example, the Tarahumara of northern Mexico and the Hui people of the Loess Plateau in China are culture groups. The term *culture group* is often used interchangeably with **ethnic group,** but these terms can become problematic. Like the concept of region, the concepts of culture and ethnicity are imprecise, especially as they are popularly used. In the modern era, for instance, as part of the globalizing process, migrating people often move well beyond their customary cultural or ethnic boundaries to cities or to distant countries. In these new places, where they are often a minority, they take on new ways of life or even new beliefs, yet still identify with their culture of origin. The formerly nomadic Kurds in Southwest Asia, for example, are asserting their right to create a nation-state (see Figure 6.20) in the territory in which they once lived but that is now claimed by Syria, Iraq, Iran, and Turkey. We might picture a typical Kurd as a woman living in a certain type of tent and weaving the wool from her family's herd of sheep, and then learn that many Kurds who actively support the cause of the herders are now urban dwellers in Turkey, Iraq, or even London.

They live in apartments and may work as clerks, housekeepers, taxi drivers, rug dealers, or engineers. Although these people think of themselves as ethnic Kurds and are so regarded in the larger society, they do not follow the traditional Kurdish way of life. Hence, we could argue that these urban Kurds both are and are not part of the Kurdish culture or ethnic group.

Another problem with the concept of culture is that it is often applied to a very large group that shares only the most general of characteristics. For example, one often hears the terms *American culture*, *African-American culture*, or *Asian culture*. In each case, the group referred to is far too large to share more than a few broad characteristics.

It might fairly be said, for example, that U.S. culture is characterized by beliefs that promote individual rights, autonomy, and responsibility; by high levels of consumption; by dependence on modern technology; and by a market economy based on credit and widely accessible middle and higher education. If we try to be more specific, we quickly run into disagreements. Can we name a common religious heritage, typical food preferences, or precise tastes in television programming? And although we might all agree that we value individualism, people in the United States constantly debate just how independent the individual should be, at what stages in life, and at whose expense. Think for a moment about current debates over public school dress codes, if the terminally ill have the right to choose "managed death," or how much control women should have over their reproductive systems. In fact, U.S. culture encompasses many subcultures that share some of the core set of beliefs with the majority but disagree over parts of the core and over a host of other matters. The same is true, to some degree, of all other regions of the world.

GLOBALIZATION AND CULTURE CHANGE

There are indications that the diversity of culture is fading as trends and fads circle the globe via the many types of instant communication now available. American fast food, popular music, and clothing styles can now be found from Mongolia to Mozambique. At the same time, a wide variety of ethnic art, textiles, cuisines, and even modes of dress from distant places now grace the lives of consumers in all the world's large cities. Thus, as globalization proceeds, especially as people migrate and ideas are diffused, some measure of **cultural homogeneity**—more overall similarity between culture groups—will occur.

Are we all drifting toward a common repertoire of material culture? Possibly, but there are also countervailing trends. It is now possible for people to reinforce their feeling of **cultural identity** with a particular group through the same channels that are encouraging homogenization. We cite just two examples: the people of Aceh in Southeast Asia and the Nunavut in Canada.

Aceh is a province with a distinctive cultural identity located in northern Sumatra, an island within the country of Indonesia in Southeast Asia. The people of Aceh desire greater autonomy—the right to control their own affairs and especially to retain control of recently discovered oil deposits that the central Indonesian government views as a national resource, not a local one. The Indonesian government has sent military troops to enforce its

interests. The Acehese have established several Web sites and Internet chat groups to build awareness among Aceh migrants worldwide of armed conflict in their homeland.

The Nunavut are a Native American people in northern Canada. They have set up several Web sites through which they teach those who have moved away about Nunavut tribal history and technology. These sites also inform tribal people about efforts to reestablish control over ancestral lands lost hundreds of years ago to British and French colonists.

These two examples show us that the ability to communicate easily over the Internet and to travel quickly may reaffirm cultural diversity. These capabilities enhance the conditions necessary for **multiculturalism**: the state of relating to, reflecting, or being adapted to several cultures.

CULTURAL MARKERS

Members of a particular culture group share features, such as language and common values, that help to define the group. These shared features are called **cultural markers**. The following sections examine the roles of values, religion, language, and material culture as cultural markers. Notice how colonialization and modern communications are causing some markers to disappear and others to become more dominant.

Values and Ways of Knowing

All cultures have ways of establishing, preserving, and passing on ideas and knowledge about how things are. All knowledge systems are grounded in a set of values. For example, knowledge based on scientific rules of observation, experimentation, and proof is now widely accepted in many parts of the world. But although many of us following a way of life based on modern technology—whether in Asia, Africa, Europe, or the Americas—may think that science-based knowledge is the only valid kind of knowledge, many others would disagree. It is wise to be aware of this fact when embarking on the study of places around the world. Many people, including many eminent scientists who grew up in cultures with beliefs based on folk knowledge, are convinced that there are several ways of knowing and that people can gain valuable insights from spiritual as well as scientific experience. Even medical researchers in industrialized societies are now investigating biofeedback, the power of mental processes to influence the body's physical condition. This phenomenon has been recognized by traditional healers around the world for a long time. In some parts of Latin America and Asia, folk healers and practitioners of modern scientific medicine are cooperating in treating patients. As you read about different cultures, you will undoubtedly encounter some perspectives that are alien to your way of thinking or your values.

Occasionally you will hear someone say, "After all is said and done, people are all alike" or "People ultimately all want the same thing." It is a heartwarming sentiment, but it is often wrong. We would be wise not to expect or even to want other people to be like us. People are not all alike, and that is in large part what makes the study of geography interesting. It is often more fruitful to look for the reasons behind differences among people than to search hungrily for similarities. **Cultural diversity** is one reason that humans are such successful and adaptable animals. The various cultures serve as a bank of possible strategies for responding to the social and physical challenges faced by the human species. The reasons for differences in behavior from one culture to the next are usually complex, but they are often related to differences in values.

Let us look at an example that contrasts the values held by modern urban individualistic culture with those held by rural community-oriented culture. In urban areas in many parts of the world, it is often acceptable, even desirable, for a woman to choose clothing that allows her to make a strong statement about who she is as an individual and how well off she is. One rainy afternoon, I saw a beautiful middle-aged Asian woman walking alone down a fashionable street in Honolulu, Hawaii. She wore a white suit with matching white shoes, hat, and bag, and over it all an elegant, clear raincoat. Everyone noticed and admired her because she exemplified an ideal Honolulu woman: beautiful,

self-assured, and rich enough to keep herself outstandingly well dressed. I guessed that she was a businesswoman.

But in her grandmother's village—whether it be in Japan, Korea, Taiwan, or rural Hawaii—people would regard such dress as outrageously immodest and dangerously antisocial. In a rural Asian community, her clothes would breach a widespread traditional ethic that no individual should stand out from the group. Furthermore, her costume exposed her body to open assessment and admiration by strangers of both sexes. In a village context, such behavior would signal that she lacked modesty. The fact that she walked alone down a public street—unaccompanied by her father, husband, or female relatives—would indicate that she was not a respectable woman. Thus, a particular behavior may be admired when evaluated by one set of values, yet considered questionable or even despicable when evaluated by another.

If groups have different sets of values and standards, does that mean that there are no overarching human values or standards? This question increasingly worries geographers, who try to be sensitive both to the particularities of place and to larger issues of

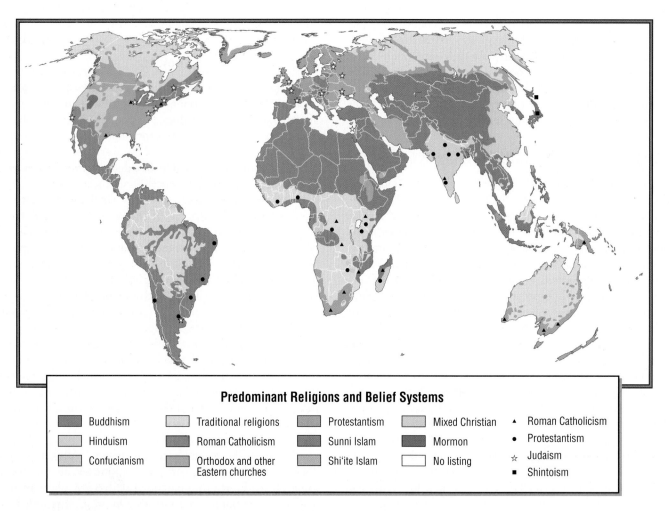

Predominant Religions and Belief Systems

Buddhism	Traditional religions	Protestantism	Mixed Christian	▲ Roman Catholicism
Hinduism	Roman Catholicism	Sunni Islam	Mormon	● Protestantism
Confucianism	Orthodox and other Eastern churches	Shi'ite Islam	No listing	☆ Judaism
				■ Shintoism

Figure 1.4   Major religions around the world. [Adapted from *Oxford Atlas of the World* (New York: Oxford University Press, 1996), p. 27.]

human rights. Those who lean too far in the direction of appreciating difference could be led to the tacit acceptance of inhumane behavior, such as the oppression of minorities and women or even torture and genocide. Acceptance of difference does not mean that we cannot make judgments about the value of certain extreme customs or points of view. Nonetheless, many geographers and others interested in human rights have observed that, although it is important to take a stand against cruelty, deciding when and where to take that stand is rarely easy.

Religion

The religions of the world are formal and informal institutions that embody value systems. Most have roots deep in history, and many include a spiritual belief in a higher power (God, Yahweh, Allah) as the underpinning for their value systems. These days, religions often focus on reinterpreting age-old values for the modern world. Some—such as Islam, Buddhism, and Christianity—*proselytize*; that is, they try to extend their influence by seeking converts. Others, such as Judaism and Hinduism, accept converts only reluctantly.

Religious beliefs are often reflected in the landscape. For example, Hindu religious traditions venerate cows as sacred. Cattle can walk through city streets in India with virtual impunity (see the photo on this page). Settlement patterns often demonstrate the central role of religion in community life: village buildings may be grouped around a mosque or synagogue or an urban neighborhood organized around a Catholic church. In some places, religious rivalry is a major feature of the landscape. Certain spaces may be clearly delineated for the use of one group or another, as in Northern Ireland's Protestant and Catholic neighborhoods. Other examples are Bosnia and Serbia, with their Muslim and Christian villages.

Religion has also been used to wield power. For example, religion sometimes has been an instrument of colonization—a way to impose a quick change of attitude on conquered people. Figure 1.4, which shows the distribution of the major religious traditions on earth today, demonstrates some of the religious consequences of colonization. Note the distribution of Catholicism in the parts of the Americas colonized by France, Spain, and Portugal. Religion can also spread through trade contacts. In the seventh and eighth centuries, Islamic peoples used trade and, occasionally, conquest to spread their religion across North Africa, throughout Central Asia, and eventually into South and Southeast Asia. The distribution of major religions has changed many times over the course of history; a world map is too small in scale to convey the complexity of actual spatial patterns, such as where two or more religious traditions intersect at the local level. Also, as the world's cultural traditions become increasingly mixed and urban life spreads, **secularism,** a way of life informed by values that do not derive from any one religious tradition, is spreading.

Language

Language is one of the most important criteria in delineating cultural regions. The modern global pattern of languages (shown

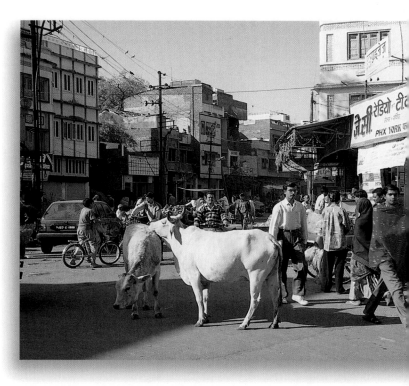

This is a street corner in downtown Bombay, where cattle are allowed to wander freely. According to an ancient Hindu verse, one who kills, eats, or permits the slaughter of a cow will "rot in hell for as many years as there are hairs on the body of the cow so slain." Nonetheless, the resources provided by cattle (especially draft work and dung) are economically important in India. [Michele Burgess/Stock, Boston.]

in Figure 1.5) reflects the complexities of human interaction and isolation over several hundred thousand years. In reality, the map does not begin to depict the actual details of the distribution: between 2500 and 3500 languages are spoken on earth today.

The geographic pattern of languages has continually shifted over time as people have interacted through trade and migration. The pattern changed most dramatically, however, after the age of exploration and colonization began around 1500. From then on, the languages of European colonists often replaced the languages of the colonized people. This is why we find large patches of English, Spanish, Portuguese, and French in the Americas and English in Australia and New Zealand. Pockets of these and several other European languages are found in Africa and Asia as well. In the Americas, the European languages largely replaced Native American languages, but in Africa and Asia, the European tongues coexisted with native tongues. Many people became bilingual or trilingual.

Today, with increasing trade and instantaneous global communication, a few languages are becoming dominant. At the same time, other languages are becoming extinct because children no longer learn them. The important languages of international trade (called **lingua francas**) are English, Spanish, Chinese, and Arabic. Of these, English dominates, largely because the British colonial empire introduced English as a second language to many places

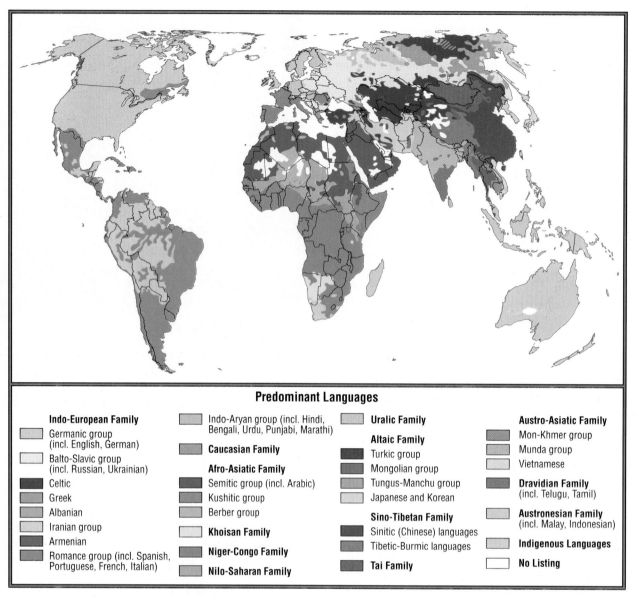

Predominant Languages

Indo-European Family
- Germanic group (incl. English, German)
- Balto-Slavic group (incl. Russian, Ukrainian)
- Celtic
- Greek
- Albanian
- Iranian group
- Armenian
- Romance group (incl. Spanish, Portuguese, French, Italian)

- Indo-Aryan group (incl. Hindi, Bengali, Urdu, Punjabi, Marathi)

Caucasian Family

Afro-Asiatic Family
- Semitic group (incl. Arabic)
- Kushitic group
- Berber group

Khoisan Family

Niger-Congo Family

Nilo-Saharan Family

Uralic Family

Altaic Family
- Turkic group
- Mongolian group
- Tungus-Manchu group
- Japanese and Korean

Sino-Tibetan Family
- Sinitic (Chinese) languages
- Tibetic-Burmic languages

Tai Family

Austro-Asiatic Family
- Mon-Khmer group
- Munda group
- Vietnamese

Dravidian Family (incl. Telugu, Tamil)

Austronesian Family (incl. Malay, Indonesian)

Indigenous Languages

No Listing

Figure 1.5 The world's major language families. Distinct languages (Spanish and Portuguese, for example) are part of a larger language group (Romance languages), which in turn is part of the Indo-European language family. Does the category "indigenous languages" or its distribution raise any questions in your mind about how the mapmaker defined the term *indigenous?* [Adapted from *Oxford Atlas of the World* (New York: Oxford University Press, 1996), p. 27.]

around the globe. Now the need for a common world language in the computerized information age is rapidly pushing the world community to accept English as the primary lingua franca.

Special efforts may be required to preserve the language diversity that still exists today. For many people, a particular language represents a cultural and geographic identity they wish to retain. For example, the newly independent country of Slovenia (once part of Yugoslavia in southeastern Europe) jealously guards its language, Slovene. This small country of just 2 million people is surrounded by more populous countries (Austria, Hungary, Italy, and Croatia), each using a different language. Slovenia's language is one of only a few remaining cultural markers. Slovenes worry that their language will be used less and less as the country becomes integrated into Europe, where French, German, and English are the lingua francas. Indeed, Slovene children study all of these languages in school. As a result, Slovenes devote a great deal of energy to preserving their language. They have one of the highest rates of book publishing on earth and republish many foreign texts in Slovene. They even attempt to extend the use of Slovene beyond the borders of Slovenia by inviting the children and grandchildren of emigrants to return from the Americas and Western Europe for heavily subsidized summer language courses. The Slovenes are an example of a culture group that is intensifying its distinct identity even as its connections to the wider world increase.

Material Culture and Technology

Material culture refers to the tangible items that the members of a specific culture group produce or use. A group's material culture reflects its **technology,** which is the integrated system of knowledge, skills, tools, and methods upon which a culture group bases its way of life. Material culture and technology help to define a culture—its resource base, standard of living, trading patterns, and belief systems. Archaeologists use the interconnectedness of material artifacts, technology, and culture to deduce—from an excavated assemblage of potsherds (bits of broken pottery), tools, animal bones, and other material remains—what was going on long ago in a particular place.

Housing provides an example of how material culture reveals a particular culture's way of life, especially its values, technology, and resource use (see the photo on this page). The typical American suburban ranch house has a living room, a dining room, a kitchen, two baths, three bedrooms, a laundry room, a two-car garage, a front lawn, and a backyard. With its distinctive architecture, electrical and plumbing systems, and landscape, it silently reveals a great deal about American family values. It reflects a nuclear family structure (mother, father, children)—a type that remains an ideal in American society but now constitutes less than 30 percent of American families. This sort of house is less suited to extended families that include aunts, uncles, and grandparents or to single-parent families; in fact, it often has to be modified to accommodate these types of families. The ranch house reflects certain ideas about privacy (separate bedrooms) and gender roles (Mom's special spaces may be the kitchen and laundry; Dad's, the TV room, the garage, and the tool shed). This house bespeaks a certain level of affluence and leisure, set apart as it is on a green lawn requiring constant maintenance and equipped with labor-saving devices and conveniences. It also symbolizes American ideas about private property, polite neighborliness, and mobility.

Because we move so often in our quest for a better job and standard of living, yet insist on privately owned homes, a vast system of interchangeable dwellings has evolved across America, supported by an institution: the American home real estate market. Families moving across the country can find similar houses with nearly the same floor plan and yard in just about any community.

An astute foreign observer could learn a great deal about Americans simply by "reading" the domestic material culture of their homes and surrounding landscapes. The same observer could look at housing in Japan, Mongolia, London, or the West Indies and learn much about each particular culture's notions of proper family structure, gender roles, intimacy rules, aesthetic values, property rights, building technology, use of resources, and trading patterns. Such an observer might even learn how modifications of existing housing reflect cultural change.

GENDER ISSUES

Geographers have begun to pay more attention to gender roles in culture groups. In virtually all parts of the world, and for at least tens of thousands of years, the biological fact of maleness and femaleness has been translated into specific roles for each sex.

This American home in Canyon, California, belongs to the Cavin family. They are pictured (in the foreground) with all of their possessions. After looking over the assemblage, to what extent do you think the house and possessions are typical for an American family? [Peter Menzel/ Material World.]

Although the activities assigned to men and to women can vary greatly from culture to culture and from era to era, there are some rather startling consistencies around the globe and over time. Men are usually expected to fulfill public roles, working outside the home—whether as traveling executives, animal herders, hunters, or government workers. Women are usually expected to fulfill private roles. They keep house; bear and rear the children; tend the elderly; plant the gardens; prepare the meals; fetch the water and firewood; and, in some cultures, run the errands. In nearly all cultures women are defined as dependent on men—their fathers, husbands, or adult sons—even when the women may produce most of the family sustenance. Because their activities are focused on the home, women typically have less access to education and paid employment, and hence less access to wealth and political power. When they do work outside the home (as is the case increasingly in every region), women tend to fill lower paid positions—whether as laborers, service workers, or professionals—and to retain their household duties.

Gender is both a biological and a cultural phenomenon. Men and women have different reproductive roles, and there are certain other physical differences as well. Men have larger muscles, can lift heavier weights, and can run faster than women (but not necessarily for longer periods). In some physical exercises, average women have more endurance and are capable of more precise movements than average men; and in populations that enjoy overall good health, women tend to outlive men an average of three to five years. Women's physical capabilities are somewhat limited during pregnancy and nursing—and, for some, during menstruation—but their susceptibility to pregnancy is limited to about 30 years. Beyond about age 45, women are no longer subject to the physical limits imposed by reproduction.

Although these average physical differences exist, they have been amplified in most cultures to carry greater social significance than the biological facts would warrant. The culturally defined differences between the sexes have enormous impact on the everyday lives of men, women, and children. Customary ideas about masculinity and femininity, proper gender roles, and sexual orientation are handed down from generation to generation within a particular culture group or society. Perhaps more than for any other culturally defined human characteristic, there is significant agreement from place to place and over time that gender is important.

The historical and modern global picture is a puzzlingly negative one for women. In nearly every culture, in every region of the world, and for a great deal of recorded history, women have been (and are) defined as second class. It is hard to find exceptions, although the intensity of this second-class designation varies considerably. In Europe, Asia, the Americas, and Africa; on tiny islands in the Pacific; in cities and villages, people of both sexes routinely accept the idea that males are more productive and intelligent than females. In nearly all cultures, families prefer

boys over girls because, as adults, boys have greater earning capacity (in large part as a result of discrimination against females) and will perpetuate the family name (because of patrilineal naming customs). Around the globe, females are more likely to die in infancy, to start work earlier in their young lives than males, to be compensated less for their work, to work longer hours, and to eat less well. Females have less access to medical care and wield less political power in the larger society. In the United States, when domestic violence occurs, statistics show that females are the victims 96 percent of the time. The actual conditions of women vary, of course; yet, in all regions, the evidence is overwhelming that women have an inferior status.

The obvious question is, "How and why did women become subordinate to men?" Researchers are now pursuing the answers to this question, but serious inquiry began only recently because, oddly enough, few people thought the question significant.

There are growing challenges to traditional notions of femininity and ideas of female capabilities. Female Olympic athletes, for example, have demonstrated superlative strength, speed, and endurance. They make up more than a third of Olympic competitors, and in some countries (such as China), more than half (Figure 1.6).

Yet looking only at women's woes misses half the story. Men are also disadvantaged by strict gender expectations that may ride

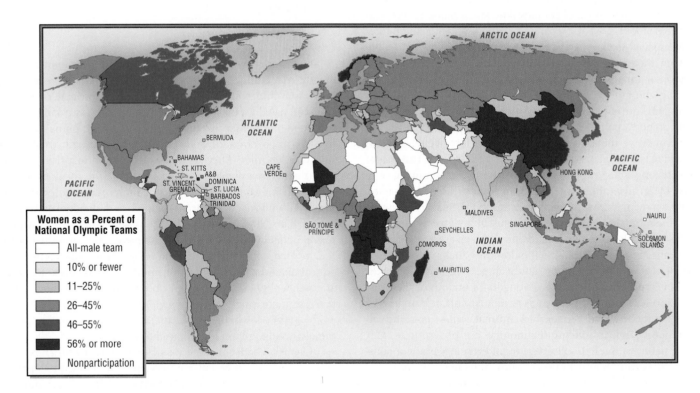

Figure 1.6 Women's participation in the 1996 Olympics. Traditional notions of femininity; female roles; and female strength, speed, and endurance are being strongly challenged by Olympic athletes. In the first modern Olympic Games in 1896, women were barred from participating. In 1996, women made up one-third of the Olympic competitors, and in some countries (China, for instance), more than half of the athletes were women. What might account for the varying representation of women in different parts of the world? [Adapted from Joni Seager, *The State of Women in the World Atlas* (New York: Penguin, 1997), pp. 48–49.]

roughshod over personal preferences. Probably for most of human history, young men have borne the lion's share of onerous physical tasks and dangerous undertakings. Until recently it was usually young men who left home to migrate to distant, low-paying jobs. Overwhelmingly, it is young men who have died alone in faraway wars. It is boys who are taught to repress their feelings, although some cultures emphasize this more than others. Men who grow up with negative attitudes toward women have been taught those attitudes by their families (often female kin) and the larger society.

This book will return repeatedly to the question of gender disparities in an effort to investigate this most perplexing cross-cultural phenomenon. Our examination will also reveal that many societies are addressing gender inequalities and that in every country on earth, what it means to be a man or a woman is being renegotiated, at least in small ways.

RACE

Just as ideas about gender roles affect life in all world regions, ideas about race affect human relationships everywhere on earth. All people now alive on earth are members of one species, *Homo sapiens sapiens*. Biologists do not regard the popular markers of race (skin color, hair texture, face and body shapes) as significant. For any supposed racial trait, such as skin color, there are wide variations. In addition, many invisible biological characteristics, such as blood types and DNA patterns, actually cut across skin color distributions and are shared by what are popularly viewed as different races. Over the last several thousand years, in fact, there has been such massive gene flow among human populations that no modern group presents a discrete set of characteristics. Biologically speaking, we are all related.

It is likely that some of the easily visible features of particular human groups evolved to adapt to an environmental condition, but precise explanations of just how these physical changes occurred are hard to come by. For example, it appears that darker skin evolved in regions close to the equator, where sunlight is most intense. One possible explanation lies in the body's need for the important nutrient vitamin D. Sunlight striking the skin helps the body to absorb vitamin D; but too much of it can result in improper kidney functioning. Perhaps dark skin protects against too much vitamin D absorption in equatorial zones. In high latitudes, where the sun's rays are more dispersed, light skin may allow for the absorption of sufficient vitamin D. Many physical characteristics do not serve any apparent adaptive purpose, however. They are probably the result of random chance and ancient inbreeding within isolated groups.

It may be comforting to learn that, biologically speaking, race is meaningless. But we cannot overlook the significant political and social import that race (often paired with culture or ethnicity) carries in some parts of the world. Race seems to have acquired unprecedented significance over the last 500 years, since the colonization of much of the globe by Europeans. Some researchers have suggested that European colonizers adopted **racism**—the negative assessment of unfamiliar, often darker skinned people—

In all major cities of the world, one will encounter a great diversity of people. Here, in Detroit, Michigan, a group of people from several places around the world are being sworn in as U.S. citizens. [Jim West/Impact Visuals.]

to justify taking land and resources away from supposedly inferior beings. Still, it would be wrong to suggest that disparaging appraisals and exploitation of others are recent cultural innovations, exclusively European traits, or necessarily connected to race or ethnicity. Such inhumane behavior has existed for a long time in all parts of the globe. For more than a thousand years, the Huns and like-minded Central Asian nomadic warriors swept back and forth between Europe in the west and China in the east, killing thousands of people from horseback. The Mongols raided China, the Manchurians raided Mongolia and China, the Japanese raided Korea, and Muslims sacked parts of Central Africa. The Moors of North Africa conquered and colonized Spain, Spain conquered Native Americans, the Turks conquered southeastern Europeans, and the ancient Mexicans harassed the Mayans. As we will see repeatedly in this book, the human animal has committed atrocities against its own kind, often in the name of race.

We should not infer, however, that human history has been marked primarily by conflict and exploitation. Actually, the opposite is probably true: humans have been so successful because of a strong inclination toward *altruism*, the willingness to sacrifice one's own well-being for the sake of others. Writ small, this altruism can be found in the sacrifices individuals make to help family, neighbors, and community. Writ large, it includes charitable giving to help anonymous people in need. It is probably our capacity for altruism that causes us such deep consternation over the relatively infrequent occurrences of inhumane behavior.

PHYSICAL GEOGRAPHY: PERSPECTIVES ON THE EARTH

Humans have always had to adapt to the physical environment, although the nature of the interactions between humans and their environments has changed over time and varied from culture to culture. In this section, we look at two components of the physical environment that are of particular importance to physical geographers: landfoms and climate. We finish by examining the origins of agriculture as an example of how the interactions between humans and the environment can profoundly alter both the physical environment and human society.

LANDFORMS: THE SCULPTING OF THE EARTH

Have you ever marveled at the landforms around you and wondered what awesome forces produced them? The processes that create mountain ranges, continents, and the deep ocean floor are some of the most powerful and slow-moving forces on earth. Originating deep beneath the earth's surface, these forces can move entire continents and often take hundreds of millions of years to do their work. Many of the earth's features, however, such as a beautiful waterfall or a dramatic rock formation, are formed by more rapid and delicate processes that take place on the surface of the earth. All of these forces are studied by *geomorphologists*, geographers who focus on the processes that constantly shape and reshape the earth's surface.

Plate Tectonics

Two key ideas in physical geography are the Pangaea hypothesis and plate tectonics (Figure 1.7). The **Pangaea hypothesis**, first suggested by geophysicist Alfred Wegener in 1912, proposes that

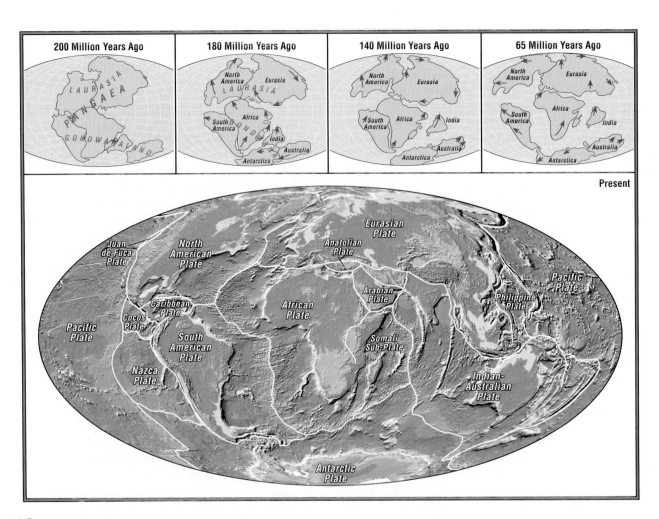

Figure 1.7 **The breakup of Pangaea and resulting continental drift (top) and the modern boundaries of the major tectonic plates (bottom).** [*Top:* Adapted from *Goode's World Atlas*, 19th ed. (Chicago: Rand McNally, 1995), p. 8. *Bottom:* Adapted from Frank Press and Raymond Siever, *Understanding Earth*, 2d ed. (New York: W. H. Freeman, 1998), p. 509.]

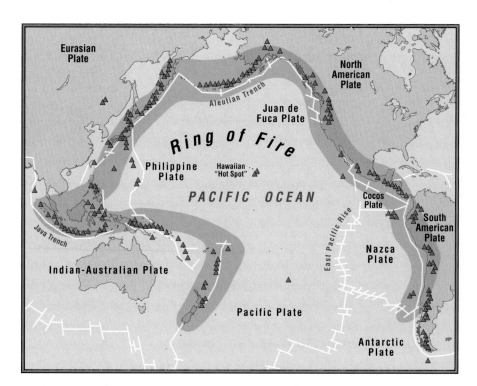

Figure 1.8 **Ring of Fire.** Volcanic formations encircling the Pacific Basin form the so-called Ring of Fire, a zone of frequent earthquakes and volcanic eruptions. [Adapted from http://vulcan.wr.usgs.gov/Glossary/PlateTectonics/Maps/map_plate_tectonics_world.html.]

all the continents were joined in a single vast continent about two hundred million years ago. This supercontinent, called Pangaea, fragmented into continents that drifted apart, reaching their present positions after millions of years. As one piece of evidence for his theory, Wegener pointed to the neat fit between the west coast of Africa and the east coast of South America. For decades, most scientists rejected the Pangaea hypothesis because they could conceive of no natural process that could move the continents across the earth's surface. Wegener's hypothesis finally gained acceptance in the 1960s after evidence for such a process, called plate tectonics, accumulated.

The theory of **plate tectonics** proposes that the earth's surface is composed of large plates that float on top of an underlying layer of molten rock. The plates are of two types. *Oceanic plates* are dense and relatively thin, and they form the floor beneath the oceans. *Continental plates* are thicker and less dense. Much of their surface rises above the oceans, forming continents. These massive plates move slowly, driven by the circulation of the underlying molten rock flowing from hotter deep regions inside the earth to cooler surface regions and back. The creeping movement of tectonic plates is believed to have fragmented Pangaea and separated the continents (see Figure 1.7).

Plate movements influence the shape of major landforms, such as continental shorelines and mountain ranges. Mountain ranges arise mostly from the folding and warping of plates as they collide. The continents have piled up huge mountains on their leading edges as the plates carrying them collided with other plates. Hence, the theory of plate tectonics accounts for the long, linear mountain ranges that run from Alaska to Chile in the Western Hemisphere and from Southeast Asia to the European

Alps in the Eastern Hemisphere. The highest mountain range in the world, the Himalayas of South Asia, was created when what is now India, at the northern end of the Indian-Australian Plate, ground into Eurasia, pushing up huge sections of thick continental crust. The only continent that lacks these long, linear mountain ranges is Africa. Often called the plateau continent, Africa is believed to have been at the center of Pangaea and to have moved relatively little since its breakup.

Humans encounter tectonic forces most directly as earthquakes and volcanoes. Plates slipping past each other create the catastrophic shaking of the landscape we know as an **earthquake.** **Volcanoes** arise at plate boundaries or at weak points in the middle of a plate, where gases and molten rock (called magma) can rise to the earth's surface through fissures and holes in the plate. Volcanoes and earthquakes are particularly common around the edges of the Pacific Ocean, an area known as the **Ring of Fire** (Figure 1.8). In the Philippines, for example, the eruption of the Mount Pinatubo volcano in 1991 killed 550 people, ruined the livelihoods of 650,000 more, and may have influenced global climate patterns.

Landscape Processes

The processes of plate tectonics are **internal processes,** driven by forces that originate deep beneath the surface of the earth. The landforms thus created have been further shaped by **external processes** that are more familiar because we can observe them daily. One such process is **weathering,** the breakdown of rock exposed to the onslaught of sun, wind, rain, snow, ice, and the effects of lifeforms. Exposure to these elements fractures rock and decomposes

In the United States, people living along the Mississippi River floodplain are at risk nearly every year. When winter snows and early summer rains are particularly heavy, floodwaters rise over the banks and drown fields of soybeans and corn, destroy houses, and damage farm buildings—as happened at Wapello, Iowa, in 1993. People who live along the Yangtze and Yellow rivers in China and on the Brahmaputra Delta in Bangladesh are at even greater risk: in some years, thousands die as a result of summer floods. [Chris Stewart/Black Star.]

it into tiny pieces. These particles then become subject to another external process, **erosion.** In erosion, wind and water carry rock particles to new locations and deposit them there. The **deposition** of eroded material can raise and flatten the land around a river, where huge quantities of silt are spread about by periodic flooding. As small valleys between hills are filled in by silt, a **floodplain** is created. Where rivers meet the sea, floodplains often fan out roughly in the shape of a triangle, creating a **delta.** External processes tend to smooth out the dramatic mountains and valleys created by tectonic processes.

Humans often contribute to external landscape processes through building, agriculture, and forestry. By altering the vegetative cover, agriculture and forestry expose the earth's surface to sunlight, wind, and rain—agents that increase weathering and erosion. Flooding becomes more common because the removal of vegetation limits the ability of the earth's surface to absorb rain. As erosion increases, rivers may fill with silt and deltas extend into the oceans. Building with concrete, asphalt, and steel often covers formerly wooded land with impervious surfaces. Again, flooding is the result, because rainwater runs over the surface to the lowest point instead of being absorbed into the ground. The physical effects of human activities vary in intensity from one culture to another depending on the tools used: mechanized earthmovers used to build roads change the earth's surface more quickly and deeply than machetes used to clear a path.

CLIMATE

The processes associated with climate are much more rapid than those that shape landforms. **Weather,** the short-term expression of climate, can change in a matter of minutes. **Climate** is the long-term balance of temperature and precipitation that keeps weather patterns fairly consistent from year to year. By this definition, the last major global change of climate took place 15,000 years ago when the glaciers of the last ice age began to melt.

Energy from the sun gives the earth a temperature range hospitable to life. The atmosphere, oceans, and earth's surface absorb huge amounts of solar energy, insulating the earth from the deep cold of space. Solar energy is also the engine of climate. The most intense, direct sunlight falls in a broad band stretching about 30° north and south of the equator. The highest average temperatures on earth occur within this band. Moving away from the equator, sunlight becomes less intense, and average temperatures drop.

Temperature and Air Pressure

The wind and weather patterns we experience daily are largely a product of complex patterns of air temperature and air pressure. **Air pressure** can best be understood by thinking of air as existing in a particular unit of space; for example, a column of air above a square foot of the earth's surface. Air pressure is the amount of force exerted by that column on that square foot of surface. Air pressure and temperature are related:

1. The gas molecules in warm air are relatively far apart and are associated with low air pressure.

2. The gas molecules in cool air are relatively close together (dense) and are associated with high air pressure.

As a unit of cool air is warmed by the sun, the molecules move farther apart. The air becomes less dense and exerts less pressure. Air tends to move from areas of higher pressure to areas of lower pressure, creating the wind. If you have ever been to the beach on a hot day, you may have noticed a cool breeze blowing in off the water. Land heats up (and cools down) faster than water, and so on a hot day the air over the land warms, rises, and becomes less dense than the air over the water, causing the cooler, denser air to flow inland. Often at night the breeze reverses direction, blowing from the now cooling land onto the now relatively warmer water.

These air movements have a continuous and important influence on global weather patterns. Over the course of a year, continents will heat up and cool off much more rapidly than the oceans that surround them. Hence, the wind tends to blow from the ocean to the land during summer and from the land to the ocean during winter. It is almost as if the continents were breathing once a year, inhaling in summer and exhaling in winter. These yearly shifts in pressure can be seen in maps of air pressure that compare the months of January and July (Figure 1.9).

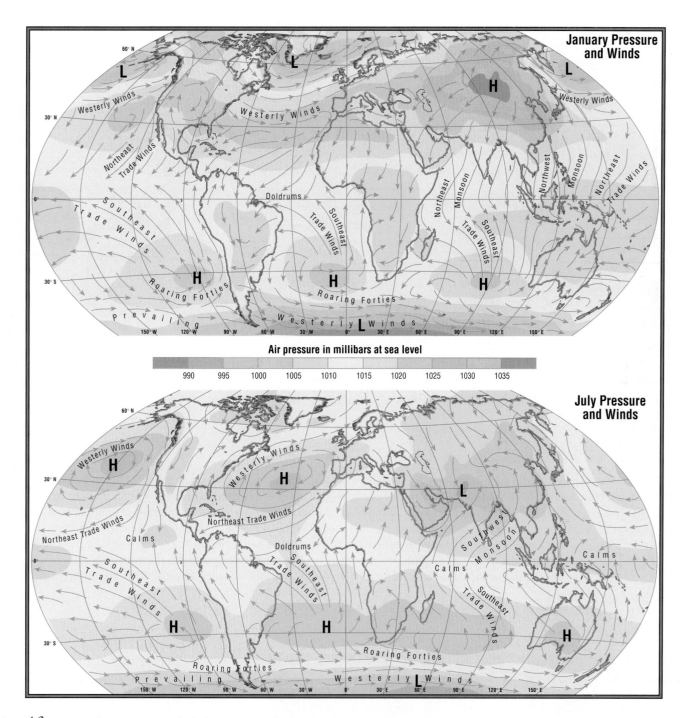

Figure 1.9 **Mean surface pressure and winds, January and July.** [Adapted from *Goode's World Atlas*, 19th ed. (Chicago: Rand McNally, 1995), pp. 14–15.]

Precipitation

Perhaps the most tangible way we experience changes in air temperature and density is through the falling of rain or snow. Precipitation occurs primarily because warm, low-density air holds more moisture than cool, high-density air. Warm air holds water vapor in the form of tiny invisible droplets. When this moist air is pushed up to a higher altitude, the lower temperature reduces the air's ability to hold moisture. The water condenses into drops to form clouds and may eventually fall as rain or snow.

Several conditions that encourage cloud-laden air to rise influence the pattern of precipitation observed around the globe

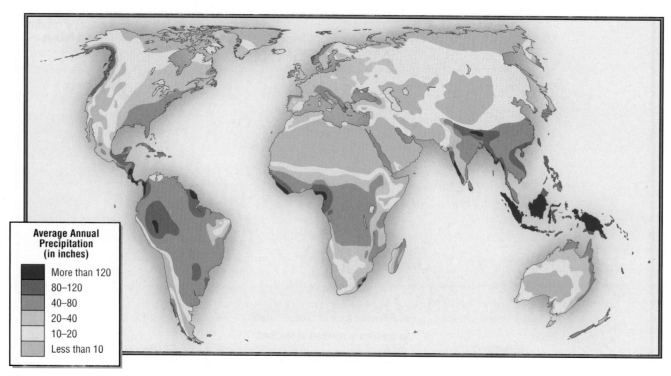

Figure 1.10 World average annual precipitation. [Adapted from *Goode's World Atlas,* 19th ed. (Chicago: Rand McNally, 1995), pp. 16–17.]

(Figure 1.10). When moisture-bearing air is forced to rise over mountain ranges, the air cools and the moisture condenses to produce rainfall. This process, known as **orographic rainfall**, is most common in coastal areas where wind blows moist air from above the ocean onto the land and up the side of coastal mountain ranges (Figure 1.11). Most rain falls as the air is rising on the coastal side of the range. On the inland side, the descending air warms and ceases to drop its moisture. The dry side of a mountain range is called the **rain shadow.** Rain shadows may extend for hundreds of miles across the interiors of continents, as they do on the Mexican Plateau, east of California's Pacific coast, or north of the Himalayas of Eurasia.

A rain belt associated with the equator is primarily the result of warm, moisture-laden tropical air rising to the point where it

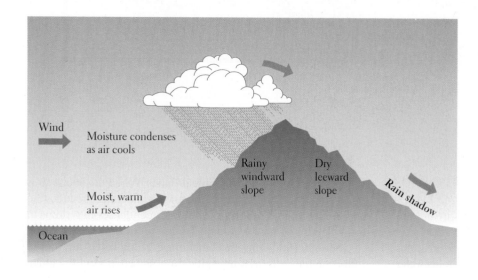

Figure 1.11 Orographic rainfall. [Adapted from Frank Press and Raymond Siever, *Understanding Earth,* 2d ed. (New York: W. H. Freeman, 1998), p. 291.]

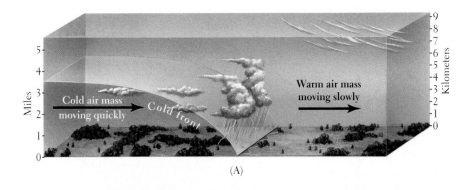

Figure 1.12 **Cold fronts and warm fronts.** Warm air usually overlies colder air along fronts, but the reverse can occur. A cold front (A) encountering warm air slopes backward, and a warm front (B) encountering cold air slopes forward.

releases its moisture as rain. Equatorial areas in Africa, Southeast Asia, and South America are watered in this way. Neighboring nonequatorial areas also receive some of this moisture when seasonally shifting winds blow the rain belt north and south of the equator. The huge downpours of the Asian summer **monsoon** are an example. The Eurasian continental landmass heats up during the summer, causing the overlying air to expand, become less dense, and rise. The somewhat cooler, yet moist, air of the Indian Ocean is drawn inland. The effect is so powerful that the equatorial rain belt is sucked onto the land, resulting in tremendous, sometimes catastrophic, rains throughout virtually all of South and Southeast Asia and much of coastal and interior East Asia (see Figures 8.2 and 8.4). Similar forces pull the equatorial rain belt south during the Southern Hemisphere's summer.

Much of the moisture that falls on North America and Eurasia is **frontal precipitation** caused by the interaction of large air masses of different temperatures and densities. These masses develop when air stays over a particular area long enough to take on the temperature of the land or sea beneath it. Often when we listen to a weather forecast we hear about warm fronts or cold fronts. As shown in Figure 1.12, a front is the zone where warm and cold air masses come into contact, and it is always named after the air mass whose leading edge is moving into an area. At a front, the warm air tends to rise over the cold air, carrying its clouds to a higher altitude, and rain or snow may follow. This kind of precipitation is common across North America and northern Eurasia.

Climate Regions

Geographers have several systems for classifying the world's climates that are based on the patterns of temperature and precipitation just examined. This book uses a modification of the widely known Köppen classification system, which divides the world into several types of climate regions, labeled A, B, C, D, and E on the climate map in Figure 1.13 (see pages 24–25). As we look at each of the regions on this map and read the climate descriptions that accompany it, the importance of climate to vegetation becomes evident. Each regional chapter will contain a climate map; when reading these maps, you can refer back to the verbal descriptions on pages 24–25, if necessary. Keep in mind that the sharp boundaries on climate maps are in reality much more gradual transitions.

THE ORIGINS OF AGRICULTURE

The development of agriculture provides a compelling illustration of how human interactions with the physical environment can transform human society and ways of life. Agriculture includes animal husbandry, or the raising of animals, as well as the cultivation of plants. The practice of agriculture has had long-term effects on human population growth and on rates of natural resource use. It has increased disparities in wealth, created hierarchies of power, introduced the formation of cities, and contributed to a reliance on trade. Agriculture has had to adapt to

Types of Climate Regions

Tropical Humid Climates (A). These climates occupy a wide band within 15° to 20° on either side of the equator. Here we have simplified the variations to just two distinct **A** climates: tropical wet and wet/dry.

In the **tropical wet climate,** rain falls predictably every afternoon and usually just before dawn. The natural vegetation is the tropical rain forest, a broad-leafed evergreen forest consisting of hundreds of species of trees that form a several-layered canopy above the soil.

The **wet/dry tropical climate,** also called **tropical savanna,** experiences a wider range of temperatures than the tropical wet climate and may actually receive more total rainfall, but rain comes seasonally and in great downpours, during the heat of the summer. The tropical forest in these regions is less vigorous and may show signs of having to survive long dry periods that occur unpredictably.

Arid (Desert) and Semiarid (Steppe) Climates (B). Arid and semiarid climates (**B** climates) may be either desert or steppe. **Deserts** generally receive very little rainfall, and most of that comes in downpours that are extremely rare and unreliable but are capable of bringing a brief, beautiful flourishing of desert life. Usually, deserts have little vegetation and almost no cloud cover, which leads to wide swings in temperature between day and night. Life is a battle for both plants and animals because they must be able to survive heat exhaustion during the day and freezing at night. **Steppes** have climates similar to those of deserts, but they are more moderate, usually receiving about 10 inches more rain per year and being covered with grass.

Arid and semiarid climates are found primarily in two locations: the subtropics (meaning slightly poleward of the tropics) and the midlatitudes. Subtropical deserts and steppes are found between 20° and 30° north and south latitudes, where high-pressure air descends in a belt around the planet. They are generally much warmer than the midlatitudinal deserts and steppes, which are found farther toward the poles in the interiors of continents, often in the rain shadows of high mountains. Although soils are generally thin and unproductive in most deserts and steppes, the midlatitude steppes can have some of the thickest and richest soils in the world. The slightly colder temperatures in these steppes keep down rates of decay and hence encourage the accumulation of organic matter in the soil over time. The Great Plains of North America are an example of steppe lands with rich soils.

Temperate Climates (C). In this book, we distinguish among just three **temperate climates.**

Midlatitude C climates, such as those in southeastern North America and China, are moist all year and have short, mild winters and long, hot summers. A variant of the midlatitude **C** climate is the **marine west coast climate,** such as that of western Europe, which is noted for monotonous, drizzling rain.

Subtropical C climates differ from midlatitude **C** climates in that winters are dry.

Mediterranean C climates have moderate temperatures but are dry in summer and wet in winter. Plants do not get moisture

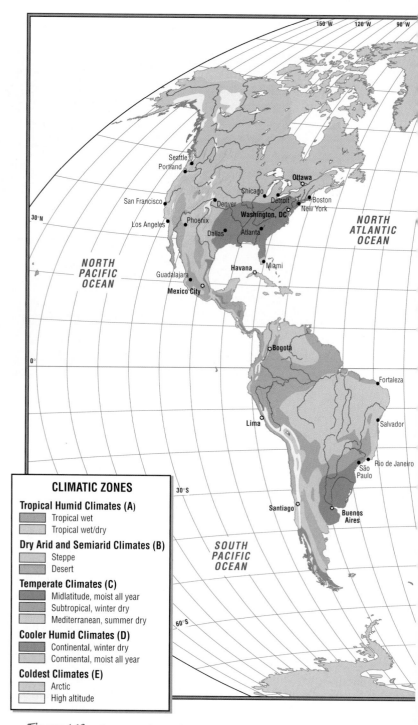

Figure 1.13 Climate regions of the world.

CLIMATIC ZONES

Tropical Humid Climates (A)
- Tropical wet
- Tropical wet/dry

Dry Arid and Semiarid Climates (B)
- Steppe
- Desert

Temperate Climates (C)
- Midlatitude, moist all year
- Subtropical, winter dry
- Mediterranean, summer dry

Cooler Humid Climates (D)
- Continental, winter dry
- Continental, moist all year

Coldest Climates (E)
- Arctic
- High altitude

when temperature and evaporation rates are highest, so the plant species that live in this zone tend to be xerophytic (adapted to dry conditions), with scrubby, shiny leaves capable of storing moisture. California, Portugal, northwestern Africa, southern Italy, Greece, and Turkey are examples of places with this climate type.

Cool Humid Climates (D). Stretching across the broad interiors of Eurasia and North America are continental climates, either **moist all year** (North America and north-central Eurasia) or **dry winters**

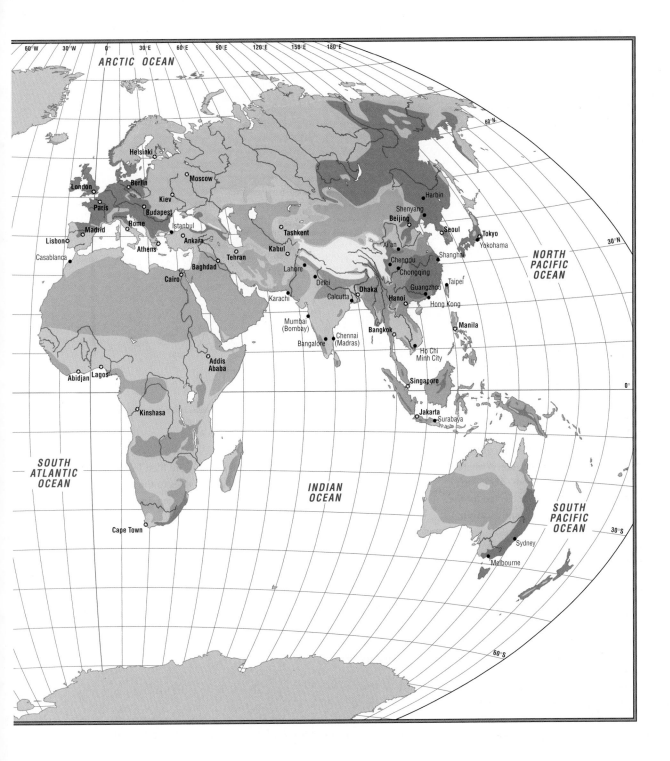

(northwestern Eurasia). Summer in **D** climate zones are short but can have very warm days. The natural vegetation of southern **D** climates is broad-leafed deciduous and evergreen forests. Here the soil is deep and rich, a result of low temperatures that inhibit decay, as in the midlatitude steppes. In the more northerly areas, winters are long and cold. Vast needle-leafed evergreen forests called taiga stretch across the cold interior. Soils here can be deep but are not as rich as they are farther south; growing seasons are so short that cultivation is minimal.

Arctic and High-Altitude Climates (E). These climates are by far the coldest and are also among the driest. Although moisture is present, there is little evaporation because of the low temperatures. The climate is often called tundra, after the low-lying vegetation that covers the ground; the dwarfed vegetation is a response to the 7 to 11 months of below freezing temperatures. What little precipitation there is usually comes during the warmer months, and even this may fall as snow. The high-altitude version of the **E** climate is found in the Rockies and Himalayas and in a thin strip down the high Andes in South America.

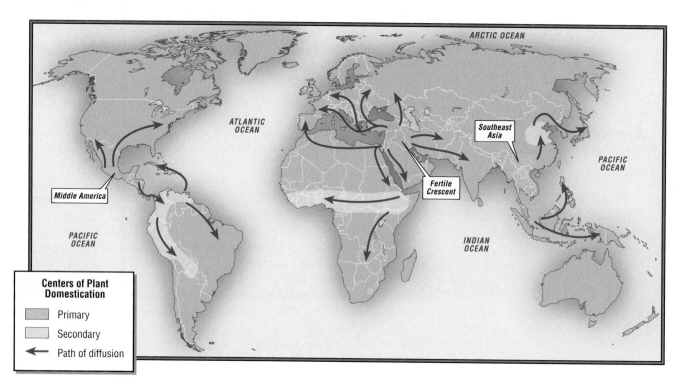

Figure 1.14 **The origins of agriculture.** Scientists have identified six main areas in the world (three primary and three secondary) where agriculture emerged independently. For lengthy periods, people in these different places tended plants and animals and selected for the genetic characteristics they valued, a process called **domestication.** This knowledge then spread around the world in a process known as **diffusion.** Domesticated plants and animals were then further adapted to new locations. This selection process continues to the present. [Adapted from Terry G. Jordan-Bychkov and Mona Domosh, *The Human Mosaic* (New York: Longman, 1999), pp. 108–109.]

climatic and soil conditions and, in turn, has had dramatic effects on the physical environment.

Where and when did plant cultivation and animal husbandry first develop? Very early ancestors of humans hunted animals and gathered plants and plant products (seeds, fruits, roots, fibers) for their food, shelter, and clothing. Even at this early stage, humans were manipulating the environment of wild plants and animals by reducing their numbers and changing their habitats. It is very likely that the transition from hunting and gathering to tending pastures and gardens was a gradual process arising from a long familiarity with the growth cycles and reproductive mechanisms of the plants and animals that humans liked to use. Genetic studies support the view that between 8000 and 20,000 years ago, people in many different places around the globe (the Americas, Southwest Asia, Central Africa, Southeast Asia, and East Asia) independently learned to develop especially useful plants and animals through selective breeding, a process known as **domestication.** The map in Figure 1.14 shows several well-known centers of domestication. A continuing process of agricultural innovation also occurred in many places outside these centers. Sometimes the farmers or herders moved into new areas, taking their tools, methods, or plants and animals with them

Why did agriculture and animal husbandry develop in the first place? Richard MacNeish, an archaeologist who has studied plant domestication in Mexico and Central America, suggests that the chance to trade was a major motivation and may have taken precedence over the simple desire to increase or improve food supplies. Many of the known locations of agricultural innovation lie near early trade centers. People in such places would have had at least two reasons to pursue plant cultivation and animal husbandry:

1.   They would have had access to new information and new plants and animals brought by traders.

2.   They would have had a need for something to trade with the people passing through.

Perhaps, then, agriculture was at first just a profitable hobby for hunters and gatherers that eventually, because of market demand, grew into a "day job"—the primary source of sustenance.

We often think of agriculture as a major human achievement that made it possible to amass surplus stores of food for lean times and allowed some people to specialize in activities other than food procurement. While agriculture did eventually produce these effects, the picture is actually more complex than that. Anthropologist E. N. Anderson, writing about the beginnings of agriculture in China, suggests that agricultural production may have been the impetus for several developments now regarded as problems: rapid population growth, social inequalities, environmental degradation, and famine. He suggests that as groups turned to raising animals and plants for their own use or for trade, more labor was needed. As the population expanded to meet this need and more resources were used to produce food, natural habitats were destroyed, and hunting and gathering was gradually abandoned. The quality of human diets may have declined at first, as people abandoned foraging among diverse wild plants and began to eat primarily corn, wheat, or rice. Moreover, the wealth provided by trade was not shared equally by everyone, and some of those who specialized in nonagricultural activities began to amass wealth and power. Gradually, a small elite emerged amid a mass of much poorer people. Agriculture also increased the vulnerability to drought and other natural disasters that could wipe out an entire harvest. Thus, as ever larger populations depended solely on agriculture, famine became more common.

Anderson's theory illustrates some points that are important for understanding modern geographic issues, especially general human well-being. The potential for human impact on the world's environments increased markedly as fields of cultivated plants and pastures for cattle, sheep, and goats replaced forests and grasslands. This trend of increasingly intense human impact has become even more pronounced over the past few centuries as the human population has doubled and redoubled. We apply ever larger amounts of chemicals (hormones, fertilizers, pesticides) and irrigation water to keep our fields and pastures productive.

ECONOMIC ISSUES IN GEOGRAPHY

The invention of agriculture was an early step in the development of economic systems based on interactions between humans and their environments. Geographers have long been interested in the ways in which economies in different parts of the world interact and in how they affect the resources people use and the ways in which people arrange themselves across the land. In recent decades, those who study economic geography have focused increasingly on the economic aspects of globalization, or the **global economy**—the ways in which goods, capital, labor, and resources are exchanged among distant and very different places. Here are a few examples of people who are part of that exchange and whose day-to-day lives are strongly affected by the global economy.

WORKERS IN THE GLOBAL ECONOMY

VIGNETTE Olivia lives in Soufrière on St. Lucia, an island in the Caribbean. Soufrière was once a quiet fishing village, but it now hosts cruise ship passengers several times a week. Olivia is 60. She, her daughter Anna, and three grandchildren live in a wooden house surrounded by a leafy green garden dotted with fruit and banana trees. Anna has a tiny shop at the side of the house, from which she sells matches, cigarettes, toilet paper, sugar, flour, tinned food, soft drinks, locally grown spices, and preserves she and her mother make from the garden fruits. On days when

From a distance, cruise ship passengers still see the village of Soufrière, St. Lucia, as a quaint place. When carried ashore by lighter, they encounter a small but vibrant entrepreneurial community. [Lydia Pulsipher.]

the cruise ships dock, Olivia strolls down to the market shed on the beach with a basket of papaya, rolls of cocoa paste made from beans picked in a neighbor's yard, homegrown spices, and roasted peanuts packed up in tiny paper bags. She calls out to the passengers as they are rowed to shore, offering her spices and snacks for sale. In a good week she makes U.S. $40, and her daughter about U.S. $100 in the shop. Most days, Olivia tends the garden and some chickens, and her daughter constantly looks for other ways to earn a few dollars, making necklaces for tourists or taking in washing, so usually the family of five makes do with about U.S. $160 a week (U.S. $8000 per year). From this they pay rent on the house, the electric bill, and school fees for the granddaughter who will go to high school in the capital next year. They also buy clothes for the children and whatever food they can't grow themselves.

Henry works in the galley on a luxurious, four-masted German cruise ship that plies the Caribbean from December to June. Every year he leaves the rural Philippine village where he and his wife, two daughters, and a son live. He rides a crowded bus to Manila and buys the cheapest ticket he can find on a freighter to Panama, where he meets the cruise ship on December 1. All winter, he works a 12-hour shift in the galley, preparing food for the chefs and then cleaning up after the three sumptuous meals served every day to the 60 passengers. He receives U.S. $40 a day, or U.S. $3.33 an hour, plus meals, medical care if he needs it, and a hammock in the crew cabin to sleep in. He gets one day off every two weeks. He has little chance to spend his money, so by June he will have sent home U.S. $6000 to his wife. (Crew in the world's merchant marine are paid according to the pay scale of their home country.) While Henry is gone, his wife tends their garden and takes care of the children and her elderly parents, who live downstairs in the new concrete home they finished building themselves two years ago. Occasionally, she clerks in the local general store where they buy food, kerosene, and most of their other goods, except clothes. Their total family cash income is U.S. $7700 a year, which makes them exceptionally well off for their village. There is money in the bank, and all the children plan to attend technical college for at least two years.

Like Reyhan Purwanto (see vignette), these construction workers have fled the economic crisis in Indonesia to work illegally in Malaysia. They are now under arrest and will be questioned by Malaysian immigration officials before being deported. [S. Thinakaran/AP Photo.]

Thirty-year-old Reyhan Purwanto sits on the dock in Riau, Indonesia, anxiously waiting with several other male and female Indonesians for the boat that will take them across the Straits of Malacca to Malaysia, where they plan to work illegally. They will join more than a million fellow Indonesians in Malaysia, some working legally, some not. All are attracted by Malaysia's booming economy, where wages are four times higher than at home. This is Reyhan's second trip to Malaysia. His first was in 1996, when—after many years of working as a farm laborer on the island of Java, Indonesia, earning $600 per year—he signed up for an overseas employment program of the government's Ministry of Manpower. Although he was promised a two-year work visa and a contract to work legally on a Malaysian oil palm plantation, upon arrival Reyhan found that his first three months' wages would go toward paying off his boat fare and that his visa was valid for only two months. Not one to give in easily, he soon escaped to the city of Malacca, where he secured a construction job earning U.S. $10 a day (about U.S. $2600 a year). On this wage, he was able to send enough money to his wife and two children in Indonesia to pay for their food, school fees, and a new roof for their house. In1998, however, Reyhan was one of thousands of Indonesians deported by Malaysia, which was suffering growing unemployment due to the Asian financial crisis and wanted to create more jobs for locals by expelling foreign workers. The crisis affected Indonesia even more severely, and after several years of only part-time farm work, Reyhan is willing to give Malaysia another try.

Tanya is a 50-year-old grandmother who works at a fast-food store in North Carolina making less than U.S. $6 an hour. She had been earning U.S. $8 an hour sewing shirts at a textile plant until it closed and moved abroad. Her husband is a delivery truck driver for a snack-food company and has made enough over the years to build and improve a small home for them along a ridge outside Greenville. Between them, they make $25,000 a year; but from this they must cover their mortgage and car loan, meet regular monthly expenses for food and utilities, and help their daughter, Rayna. She quit school after 11th grade and married a man who is now out of work and can't afford a car to look for a job. They and a baby live in the old mobile home at the back of the lot that Tanya and her family used to live in. With Tanya's new lower wage, there may not be enough money to pay the college tuition for her son, who wants to be an engineer and would be the first in the family to go past high school. For now, he is working at the gas station.

These people, living worlds apart, are all part of the global economy. Workers are paid at startlingly different rates for jobs that require about the same skill level. But the varying costs of living and varying local standards of wealth make the difference between the relative affluence enjoyed by Henry's family and the near poverty and threatened hopes felt by Tanya's family, which actually has the highest income by far.

Most people want a better life than their parents had, but some have a better chance of achieving that goal than others. Henry's children may have the best chance because their parents have some savings. Also, their father's worldwide travel has given him many valuable connections and the experience to know where and how to push his children into the best opportunities. Reyhan, by far the poorest, seems trapped by his status as an illegal worker, which robs him of many of his rights. Still, the higher pay he can earn in Malaysia offers him a way out of poverty.

Olivia and her family, though not well off, do not think of themselves as poor because they have what they need and because others around them live in similar circumstances. The tourist trade promises increased income, but it also means dependence on circumstances beyond their control—in an instant, the tour companies can choose another port of call. Olivia's granddaughter dreams of a well-paid managerial job in tourism, but that requires a college degree available only at the University of the West Indies in faraway Barbados. How will her family pay for that? With no degree, she will be competing with many other undertrained young people for low-paying waitress, kitchen, and maid jobs in resort hotels.

WHAT IS THE ECONOMY?

Earlier in this chapter, we observed that economic geography is the study of how people interact with their environment as they go about earning a living. The **economy** is the forum where people make their living, and **resources** are what they use to do so. Some resources are tangible materials such as mineral ores, timber, plants, and soil. Because they must be mined from the earth's surface or grown from its soil, they are called **extractive resources.** There are also nonmaterial resources, such as skills and brainpower. Often, resources must be transformed to produce new commodities (such as refrigerators or sugar) or bodies of knowledge (such as books or computer software). **Extraction** and **production** are two economic activities; a third is **exchange:** the bartering and trading of resources, products, or people's services.

The **formal economy** includes all the activities that are recorded as part of a country's official production. Examples from the vignettes at the beginning of this section include Olivia's daughter in her shopkeeper's role; Henry as a ship employee and his wife when she clerks in the store; and Tanya, her husband, and their son. All are registered workers who earn recorded wages and presumably pay taxes to their governments. The activity of formal economies is measured by the **gross domestic product,** a number that gives the total value in monetary terms of all goods and services officially recognized as produced in a country during a given year.

Many goods and services are produced outside formal markets, in the **informal economy.** Here, work is often done for no cash payment or for payment that is not reported to the government as taxable income. Only recently have economists begun paying attention to informal economies, and it is estimated that one-third or more of the world's work falls into this category. Examples of workers in the informal economy include Olivia when she sells her goods to the tourists; Reyhan when he works illegally in Malaysia; and any members of the four families when they contribute to their own or someone else's well-being through such unpaid services as housework, gardening, and child care.

WHAT IS THE GLOBAL ECONOMY?

The global economy includes the parts of a country's economy that are involved in global flows of resources: extracted materials, manufactured products, money, and people. Most of us participate in the global economy every day. For example, this book was manufactured using paper made from trees cut down in Southeast Asia, North America, or Siberia and shipped to a paper mill in Oregon. This long-distance movement of resources and products has grown tremendously in the past 200 years, but it existed at least 2000 years ago when silk fabrics and other goods were traded along land routes across Central Asia that connected Rome with China or along sea routes stretching from East Africa to Arabia to East Asia.

Starting in about 1500, long-distance trade entered a period of expansion and change, as Europeans began to take control of distant lands and peoples for economic gain. These acquired lands became known as **colonies.** Europeans began extracting resources from them and organizing systems that processed the resources into higher value goods to be traded back in Europe or wherever else there was a market. Sugarcane, for example, was grown on Caribbean, Brazilian, and Asian plantations; made into sugar and rum locally; and then sold in Europe and North America. The global economy grew as each region produced goods for export, rather than just for local consumption, and became increasingly dependent on imported food, clothing, machinery, and knowledge.

Their new wealth and ready access to global resources allowed Europeans to fund the industrial revolution, a series of innovations and ideas that changed the way goods were produced. One such innovation was the orderly timing and spacing of production steps and the use of specialized labor. Formerly, for example, one woman would produce the fiber (cotton or wool) for cloth, clean it, spin thread, weave the thread into cloth, and sew a shirt. The industrial production of shirts, by contrast, was organized as a set of tasks performed by different workers in separate spaces. Some workers specialized in producing the fiber, others in spinning, weaving, or sewing. Then, perhaps because workers were specialized, they and their supervisors began to see ways to improve the efficiency and speed of their efforts. These innovations led to labor-saving improvements in tools and eventually to mechanized reaping, spinning, weaving, and sewing.

Mechanized production methods that made possible mass-produced goods of all sorts created a demand for many different kinds of raw materials. These were provided at low cost by the various European colonies in the Americas, Africa, and Asia. The colonies also became important markets for European manu-

factured goods. For example, sugar plantation managers in the British Caribbean bought British-made cast iron industrial equipment for crushing sugarcane and boiling cane juice. They also bought cotton cloth woven in England (from cotton grown in colonies in America, Africa, and Asia) to clothe hundreds of thousands of Caribbean slaves. The sugar they produced was transported to European markets in ships made in the British Isles of trees and minerals from various parts of the world. From this one example, we can see that it was during the European colonial and industrial eras (which overlapped in time) that the extension of trade and manufacturing networks to global dimensions increased markedly.

Until the early twentieth century, much of the activity of the global economy took place within the huge colonial empires amassed by a few European nations. However, as we will discuss later, economic and political changes brought an end to these empires by the 1960s, and almost all colonial territories have become independent countries. Nevertheless, the global economy continues to grow as the flows of resources and manufactured goods linking now-independent countries are sustained by private companies, many of which first developed during colonial times. **Multinational corporations** such as De Beers, Nabisco, Nike, Microsoft, Toyota, Cisco, and Bechtel operate across international borders, extracting resources from many places, producing products in factories carefully located to take advantage of cheap labor and transport facilities, and marketing their products wherever they can make an acceptable profit. One of the key characteristics of multinationals is that their global influence, wealth, and importance to local economies enable them to influence the economic and political affairs of the countries in which they operate.

Although the vast majority of multinationals are still headquartered in the industrialized world, some are emerging in less wealthy former colonies. Moreover, while many multinationals still focus on a particular product or set of products, they are also increasingly important as conduits for the flow of **capital,** or investment money. For example, consider what was once a family-owned Mexico City cement company we will call MEXCEM. It recently borrowed heavily from international banks to acquire controlling interests in cement industries in Texas, Mexico, the Philippines, and Thailand. The focus of the company is increasingly global, and MEXCEM now hopes to outmaneuver its rivals by using borrowed money to locate new cement plants that use cutting-edge technology in strategic places.

Such international investment has advantages and disadvantages. In this case, high-quality cement may be delivered more efficiently and cheaply to all the markets served. Competition may spur technological advancement. Jobs will be created in Southeast Asia, Mexico, and Texas. And the diverse new markets may vastly improve MEXCEM's profits. But to maximize its profits, and if local laws allow, MEXCEM may not pay its workers a decent wage or provide a healthy workplace, and its cement plants may cause pollution. Competing smaller cement providers may fail, creating local unemployment. By taking the profits home to Mexico, MEXCEM is reducing the amount of capital

Among the first global economic institutions were Caribbean plantations like Old North Sound on Antigua. In the eighteenth century, thousands of sugar plantations in the British West Indies, subsidized by slaves who were forced to work without wages, provided huge sums of money to England and helped fund the industrial revolution. [Museum of Antigua and Barbuda.]

available elsewhere. Moreover, MEXCEM itself may fail, because its expansion resulted in a huge debt that requires regular payments. If it misses its payments, MEXCEM's creditors could foreclose, putting thousands of jobs from Texas to Southeast Asia at risk.

THE DEBATE OVER FREE TRADE AND GLOBALIZATION

Free trade is the unrestricted international exchange of goods, services, and capital without protections being imposed. There are differing views over the value of free trade and the globalization of markets. Most governments presently impose some restrictions on trade to protect their own national economies from foreign competition. Restrictions take two main forms: tariffs and quotas. **Tariffs** are taxes imposed on imported goods that have the effect of increasing the cost of these goods to the consumer. **Import quotas** set limits on the amount of a given good that may be imported over a period of time. These limits increase the price of the imported good by restricting the amount available, and they are intended to place the good at a price disadvantage when competing with domestic goods.

Other strategies to protect a country's trade include requiring the use of its own ships or other carriers to transport products being imported from abroad. Some countries also try to guard against unpredictable and potentially destructive movements of investment capital and the jobs, resources, and products that come or go with it. To do this, countries may impose **capital controls,** which are restrictions that require investment capital and the profits it earns to stay in a country for a certain amount of time.

Such controls may also require that companies funded by foreign capital be partially controlled by local investors.

Such protections are losing favor. Proponents of free trade have successfully argued that the unrestricted movement of goods and services across national borders encourages efficiency and the production of higher quality goods and services and gives consumers more product choices at lower prices. With free trade, companies can sell to larger markets and take advantage of mass production systems that lower costs further. If local economies obtain access to large markets, they can grow faster, thereby providing people with jobs and opportunities to raise their standard of living.

Restrictions on trade imposed by individual countries are now being reduced through the formation of regional trade blocs and the support of global institutions. **Regional trade blocs** are associations of neighboring countries that agree to lower trade barriers for one another. Examples are the North American Free Trade Agreement (NAFTA), the European Union (EU), the Southern Common Market (Mercosur) in South America, and the Association of Southeast Asian Nations (ASEAN). Examples of global institutions that support free trade are the World Trade Organization (WTO) and the World Bank (officially named the International Bank for Reconstruction and Development). The stated mission of the **World Trade Organization** is to lower trade barriers and establish ground rules for international trade. The **World Bank** is a lending institution that gives loans to countries that need money to pay for development projects. It may require a borrowing country to reorganize its national economy to achieve freer trade. For example, before approving a loan, the World Bank often requires a country to improve conditions for private enterprise by reducing and eventually removing government support for domestic industries and agriculture and by lessening taxes by reducing government services. Such economic reorganization requirements are called **Structural Adjustment Policies, or SAPs.**

Groups that oppose free trade argue that its gains are offset by the instability, even chaos, that comes with a less regulated global economy. They emphasize that even in more regulated times, the global economy has been prone to rapid cycles of growth and decline that can wreak havoc on smaller national economies. When controls and protections are removed for free trade, cycles of growth and decline may increase in frequency, intensity, and duration. Labor unions and other workers' organizations point out that as corporations relocate factories to poorer countries where wages are lower, jobs are lost in richer countries. In the poorer countries, multinational corporations and governments often prevent workers from organizing themselves into unions that could bargain for livable wages and a higher standard of living. Environmentalists argue that multinational corporations use highly polluting, unsustainable production methods in newly industrializing countries, which often don't have effective environmental protection laws. Many fear that a "race to the bottom" in wages, working conditions, and environmental quality will ensue as countries compete for the attention of potential investors.

In developing an opinion about free trade and globalization, you might consider how they affect your own life. For example, many of the things you own—computer, clothes, furniture, appliances, and car—were made outside the United States. These products were cheaper for you to buy as a result of lower production costs and competition among many producers. As a result of free trade within the United States, you can travel unencumbered by border crossings at state lines and can hunt for a job and live in any part of the country. On the other hand, you or a relative may have lost a job because the company moved to a place where labor is cheaper. You may be concerned that the products you buy so cheaply were made under harsh conditions by underpaid workers or that resources were used unwisely in the manufacturing and transport processes. These are the kinds of issues to be considered in reaching a point of view on the question of free trade.

The air that envelops Cairo, Egypt, is by some measures the most polluted of any urban area in the world. Environmental regulations either don't exist or are not enforced. Here, smoke and dust from cement factories cover newly cleaned surfaces within seconds. Much of the funding for Cairo's factories has come from international loans or from internationally controlled companies. The factories provide needed jobs for desperately poor local residents. [Mohamed El Dakhakhny/AP/Wide World Photos.]

DEVELOPMENT AND WELL-BEING

Until recently, the term *development* was used to describe economic changes that lead to better standards of living. These changes often accompany the greater productivity in agriculture and industry that comes from such technological advances as mechanization and computerization. Increasingly, the question "Development for whom?" is being asked, as it becomes clear that merely raising national productivity will not necessarily improve standards of living for most people. Often, the benefits of increased productivity have gone primarily to those who are already economically well off, leaving the majority in circumstances that are little improved, or even worsened. Furthermore, development measured by average economic gains often results in environmental side effects, such as air and water pollution, that reduce the quality of life for everyone. Some development experts (notably, for example, Nobel Prize winner and economist Amartya Sen) are advocating a broader definition of development that will include measures of human well-being (a healthy and socially rewarding standard of living) as well as measures of environmental quality.

MEASURES OF WELL-BEING

The most popular economic measure of development is **gross domestic product (GDP) per capita,** which is the total value of all goods and services produced in a country in a given year,

TABLE 1.2 *Sample human well-being table*

Selected countries (1)	GDP per capita, adjusted for PPP[a] in 1998 $U.S. (GDP of 174 countries ranked from highest) (2)	Human Development Index (HDI) rankings of 174 countries[b] (3)	Gender Development Index (GDI) rankings of 143 countries (4)	Gender Empowerment Measure (GEM) rankings of 70 countries (5)
Japan	23,257 (10)	9 (high)	9	38
United States	29,605 (2)	3 (high)	4	11
Kuwait	25,314 (1996 est.) (5)	36 (high)	34	75 (1999)[c]
Barbados	12,001 (37)	30 (high)	16 (1999)[c]	17

[a]PPP = purchasing power parity.
[b]The high, medium, and low designations indicate where the country ranks among the 174 countries classified into just three categories.
[c]Data not available in 2000.

Source: United Nations Human Development Report 2000.

divided by the number of people in the country. This figure (see column 2 of Table 1.2) is often used as a crude indicator of how well people are living in a given country. There are several problems with GDP per capita. First, because it is an average figure, it can hide the fact that a country has a few fabulously rich people and a mass of the abjectly poor. For example, a GDP per capita of U.S. $20,000 would be meaningless if a few lived on millions per year and most lived on $5000 per year. Second, the purchasing power of currency varies widely around the globe, so a GDP of U.S. $5000 per capita in Jamaica might represent a middle-class standard of living, whereas that same amount in New York City could not buy even basic shelter. Because of these purchasing power variations, in this book we use GDP figures that have been adjusted for **purchasing power parity (PPP)**. PPP is the amount that the local currency equivalent of U.S. $1 will purchase in a given country. For example, according to *The Economist*, on April 17, 2001, a Big Mac at McDonald's in the United States cost U.S. $2.54, in Australia it cost the equivalent of just U.S. $1.52, and in China it cost U.S. $1.20.

A third problem with GDP per capita is that it measures only what goes on in the formal economy, whereas the informal economy may actually account for more activity. Recently, researchers who examined all types of societies and cultures have shown that, on average, women perform about 60 percent of all the work done on a daily basis, much of it unpaid. Nonetheless, only their paid work performed in the formal economy appears in the statistics. Statistics also neglect the contributions of millions of men and children who work in the informal economy as subsistence farmers or as seasonal laborers. For example, a recent worldwide study conducted by the United Nations uncovered the fact that 250 million children under the age of 14 are employed in the informal economy, half of them full time! Most of these working children live in Asia.

The most important failing of GDP per capita as a measure for comparing countries is that it ignores all aspects of development other than economic. There is no way to tell from GDP figures, for example, how fast a country is consuming its natural resources or how well it is educating its young, treating women, caring for the sick, or maintaining its environment. There are a number of movements to refine the definition of development to include these factors and others. Therefore, along with the traditional GDP per capita figure, we have chosen several others to use in this book: the United Nations Human Development Index (HDI), the United Nations Gender Development Index (GDI), and the United Nations Gender Empowerment Measure (GEM). Together, these measures reveal some of the subtleties and nuances of well-being and make comparisons between countries somewhat more valid. Because these more sensitive indexes are also more complex than the purely economic index, they are all still being refined by the United Nations.

The United Nations Human Development Index (HDI). The HDI considers adjusted real income, which takes into account what people can buy with what they earn, as well as data on health and education. Countries are measured on the HDI factors and ranked from highest (1) to lowest (174). The ranks of countries based on HDI (see Table 1.2, column 3) are quite different from those based solely on GDP per capita. The HDI provides no way to score a country directly on the equality of its distribution of income or purchasing power. These factors are indicated only indirectly by the information about health and education. Hence, it is assumed that a country that provides widely available health and education services has a more equitable distribution of wealth than a country that provides low levels of access to health care and education.

The United Nations Gender Development Index (GDI). The Gender Development Index looks at whether countries make basic literacy, health care, and access to income available to both men and women (see Table 1.2, column 4). Because of lack of data, only 143 countries of the 174 ranked by HDI are in this index. GDI does not measure general social acceptance of the idea of gender equality, which is better measured by GEM.

The United Nations Gender Empowerment Measure (GEM). The United Nations devised the GEM to score and rank countries according to how well they enable participation by women in the political and economic life of the country (see Table 1.2, column 5). The indicators used include percentage of women holding parliamentary seats; percentage of administrators, managers, and professional and technical workers who are women; and women's GDP per capita. Although the GEM helps us to learn something about the relative power of men and women in a society, the indicators now available are unsatisfactory. In most countries, very little data is collected separately on men and women, and in many cases data on women is missing altogether, making comparisons impossible. Most important, a high rank does not indicate that a country is treating women and men equally, but only that the country is doing better than those ranked lower. Although women are half the world's population, nowhere on earth do women hold 50 percent of the parliamentary seats or have an average earning power equal with that of men. Also, because sufficient data was available to rank only 70 of the possible 174 countries in the year 2000, a GEM rank higher than an HDI rank may be merely the result of the small number of countries ranked on GEM relative to HDI. This text uses GEM rankings from 1999 where those for 2000 are not available.

The Human Well-Being Table. Table 1.2 shows GDP, HDI, GDI, and GEM rankings for four countries: Japan, the United States, Kuwait, and Barbados. Notice how the rankings change in the four columns. Kuwait's rank drops significantly from 5th in the GDP per capita column to 36th in the HDI column; Japan and Kuwait fall drastically in the GEM ranking, compared to their other rankings; and Barbados rises steadily from the HDI column across the other columns of the table. Perhaps you can already suggest some explanations for why these rankings differ so radically.

Each chapter in this book is designed to provide historical, demographic, cultural, social, economic, and political information that will help you to interpret the rankings for the countries in each world region. A human well-being table similar to this one, with a short discussion of the rankings, will be included in each chapter.

POPULATION PATTERNS

The changing levels of well-being associated with economic development have had dramatic effects on population growth. **Demography,** the study of population patterns and changes, is an important part of geographic analysis. Because geographers are concerned with the interaction between people and their environments, it is essential to know how many people there are on earth and how they are distributed, how fast their numbers are growing, how they make a living, and what their patterns of migration and consumption are.

GLOBAL PATTERNS OF POPULATION GROWTH

It took between 1 and 2 million years, or at least 10,000 generations, for humans to evolve and to reach a population of 2 billion, which happened in about 1945. By the year 2000, in just 55 years, the world's population more than tripled to 6.1 billion. What happened to make the population grow so very quickly in such a short time?

The explanation lies in changing relationships between humans and the environment. For most of human history, fluctuating food availability, natural hazards, and disease kept human death rates high, especially for infants. Periodically, there would even be crashes in human population—as happened in the 1300s throughout Europe and Asia during the pandemic known as the Black Death (Figure 1.15). The astonishing upsurge in human population began about 1500, a time when the technological, industrial, and scientific revolutions were beginning in some parts of the world. These revolutions helped people to exploit land and resources, providing better diets, and improved health, and better disease control. Human life expectancy increased dramatically, and more and more people lived long enough to reproduce successfully, often many times over. The result was an exponential pattern of growth that is often called a J curve (see Figure 1.15), because the ever shorter periods between doubling and redoubling cause an abrupt upward swing in growth, as depicted on a graph.

Today, the human population is growing in virtually all parts of the world—but more rapidly in some places than in others. Even if all couples agreed to have only one child, world population would probably grow to at least 8 billion before zero growth would set in. Growth would continue until then because there are presently so many people who have not yet reached the age of reproduction. Nevertheless, there are indications that the rate of growth is slowing globally. In 1993, the world growth rate was 1.7 percent. In 2000, it had decreased to 1.4 percent, but even this decreased rate still results in a doubling time of only 51 years. If present slower growth trends continue, world population may level off at somewhat less than 11 billion before 2050. Eleven billion people will tax the earth's resources beyond imagining, especially if most people have lifestyles based on mass consumption, as is increasingly the case.

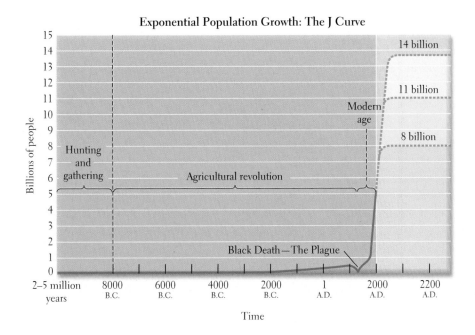

Exponential Population Growth: The J Curve

Figure 1.15 **The J curve depicts the exponential growth of human population.** The curve's shape is a result of successive doublings of the population: it starts out nearly flat, but as doubling time shortens the curve bends ever more sharply upward. Why do you think the plague is visible in the curve but the millions killed in recent wars do not show up? [Adapted from G. Tyler Miller, Jr., *Living in the Environment,* 8th ed. (Belmont, Calif.: Wadsworth, 1994), p. 4.]

LOCAL VARIATIONS IN DENSITY AND GROWTH

If the more than 6 billion people on earth today were evenly distributed across the land surface, they would produce an **average population density** of about 117 people per square mile (45 per square kilometer). As you can see from the population density map (Figure 1.16, pages 36–37), people are not evenly distributed across the face of the earth. We can make certain general statements about the distribution. First, nearly 90 percent of all people live north of the equator, and most of them live between 20° and 60° north latitude. Even within that limited territory, people are concentrated on about 20 percent of the available land. For the most part, people choose to live in lowland regions—nearly 80 percent live below 1650 feet (500 meters)—in zones that have climates warm and wet enough to support agriculture. Many people live along rivers, and most people live fairly close to the sea, within about 300 miles (500 kilometers). In general, people are located where resources are available.

Nevertheless, many places with meager resources contain a great many people because the resources to support them can be garnered from far and near. If people have the means to pay for them, food, water, and clothing; raw materials for building and manufacturing and for producing electricity; and other material support can all be imported. An extreme example is Macao, the former Portuguese trading enclave in South China. The population density in Macao is the highest on earth: 57,400 per square mile (22,000 per square kilometer). Life there is sustained not by local physical resources but by trading connections enhanced by Macao's location on the edge of the huge and populous Chinese mainland. For many centuries, cities around the world have relied on distant resources, and this is increasingly true for more and more of the world's people, no matter where they live.

Just as there is no easy correlation between population density and richness of resources, there is no easy correlation between density and poverty or wealth. Some rather densely occupied places, such as parts of Europe and Japan, are very wealthy. Other densely populated places, such as parts of India and Bangladesh, are desperately poor. To explain population density patterns today, we must look to cultural, social, and economic factors and to such events as past experiences with colonialism. We also must identify present circumstances that attract migrants to some places and cause them to flee others.

Usually, the variable that is most important for understanding population growth in a region is the **rate of natural increase** (often the term is shortened to just **growth rate**). The rate of natural increase is the relationship between the number of people being born (**birth rate**) and the number dying (**death rate**) in a given population, without regard to the effects of migration. The rate of natural increase is usually expressed as a percentage. Take the example of Austria in Europe, which has 8,100,000 people. The annual birth rate is 10 per 1000 people and the death rate 10 per 1000 people. Therefore, the annual rate of natural increase is 0 per 1000 (10 − 10 = 0), or zero percent. (Actually, Austria is probably growing a tiny bit per year.) Just 17 percent of Austrians are under 15 years of age, and 15 percent are over 65 years of age. It is estimated that, under current conditions, Austria will take 2310 years to double its population. For the sake of comparison, consider Jordan in Southwest Asia, which has 5,100,000 people. The birth rate is 33 per 1000 and the death rate is 5 per 1000. The growth rate is thus 28 per 1000 (33 − 5 = 28), or 2.8 percent. At

Figure 1.16 **World population density.** [Adapted from *Hammond Citation World Atlas* (Maplewood, N.J.: Hammond, 1966); and *World Population Data Sheet, 2000* (Washington, D.C.: Population Reference Bureau).]

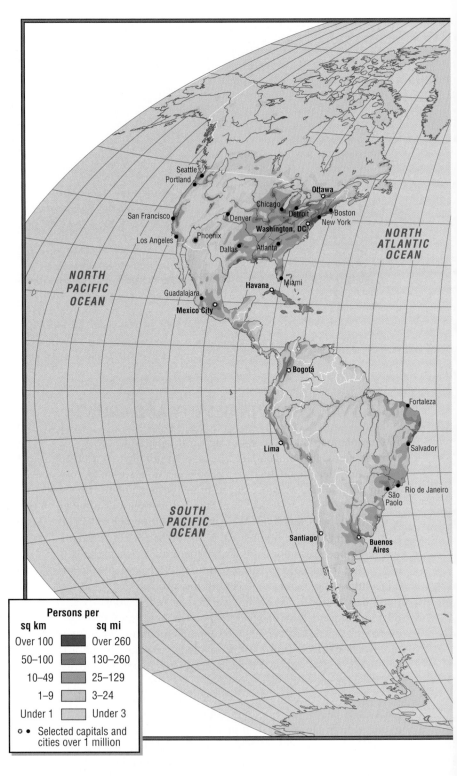

Persons per	
sq km	**sq mi**
Over 100	Over 260
50–100	130–260
10–49	25–129
1–9	3–24
Under 1	Under 3
✪ ●	Selected capitals and cities over 1 million

this rate, Jordan will double its population in just 24 years. As you might expect, the population of Jordan is very young: 42 percent are under 15 years of age, and just 3 percent are over 65.

AGE AND GENDER STRUCTURES

The **age distribution,** also known as the **age structure,** of a population is the proportion of the total population in each age group;

the **gender structure** is the proportion of males and females in each age group. The age and gender structures reflect past and present social conditions, and knowing these structures helps us to predict future population trends.

The **population pyramid** is a graph that depicts age and gender structures. Careful study of the pyramids of places such as Austria and Jordan (Figure 1.17, page 38) reveals the age and gender distribution of the population. As we have noted, most people

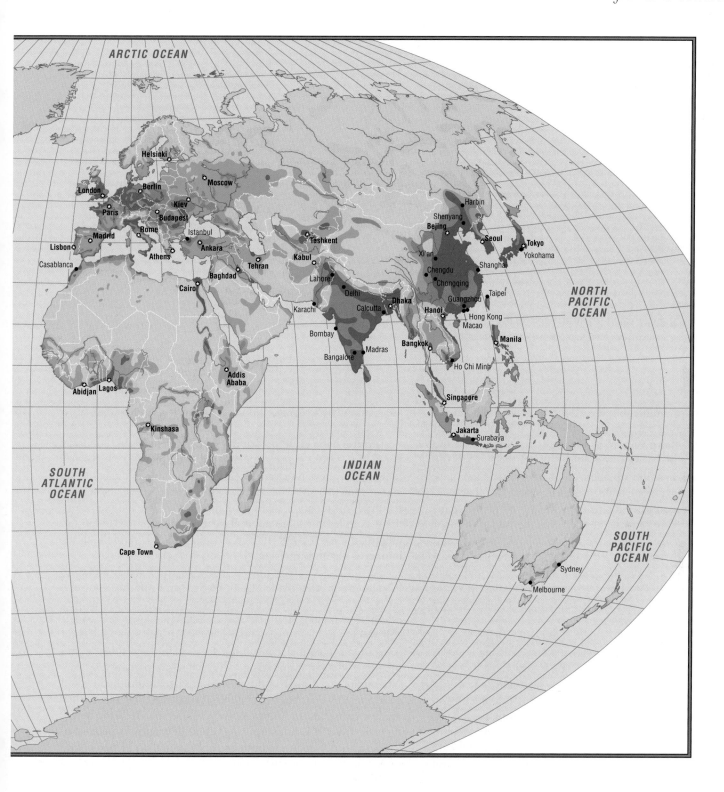

in Jordan are very young, so they are clustered toward the bottom of the pyramid, with the largest groups in the age categories 0 through 9. The pyramid tapers off quickly as it rises above age 34, showing that in Jordan most people die before they reach old age.

On the other hand, Austria's pyramid has an irregular shape, indicating that Austria has had experiences that alternately increased or decreased growth. The narrow base indicates that there are now fewer people in the younger age categories than in young adulthood or middle age, and those over 70 greatly outnumber the youngest (ages 0 to 4). This age distribution tells us that many Austrians now live to an old age and that in the last several decades Austrian couples have been choosing to have only one child or none. Austrians worry that their population will begin to decline and be weighted with elderly who will need care and financial support from an ever declining group of working-age people.

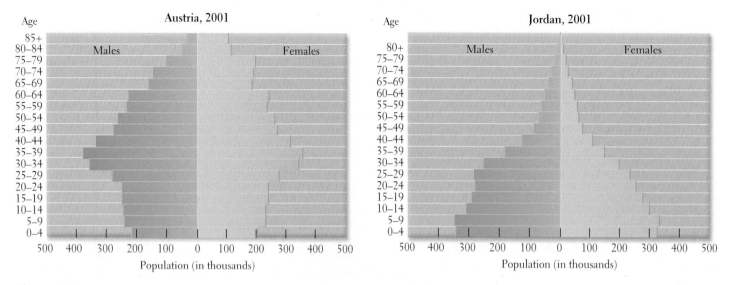

Figure 1.17 **Population pyramids for Austria and Jordan.** [Adapted from U.S. Bureau of the Census, International Data Base, at http://www.census.gov/ipc/www/idbpyr.html.]

Population pyramids also reveal subtle gender differences within populations. Look closely at the right (female) and left (male) halves of the pyramids in Figure 1.17. Notice that the sexes are not evenly balanced on either side of the line. In the Austrian pyramid, near the top, there are more women than men, reflecting the deaths among male soldiers in World War II and the trend in wealthy countries for women to live about five years longer than men. The gender imbalances in the Jordan pyramid occur at younger ages. There are about 230,000 women and about 290,000 men aged 25 to 29. These gender imbalances exist in other age groups as well; for example, among those under age 4, there are about 10,000 fewer females than males. Because gender-based research is so new, explanations for the imbalances are only now being proposed.

One possible explanation is that males and females may be treated differently throughout life (see the box "Missing Females in Population Statistics," and Figure 1.18). In societies afflicted by poverty, girls are sometimes fed less than boys and may die in early childhood more often. Because poor women frequently do not receive needed nutrition and medical care during pregnancy, more tend to die during childbirth. In some places, rules requiring female seclusion from males make professional health care less accessible. All of these factors probably affect the survival rates of women and girls in Jordan.

GROWTH RATES AND WEALTH

Although there is a wide range of variation, regions with slow population growth rates usually tend to be affluent, and regions with fast growth rates tend to have widespread poverty. The reasons that poor people tend to have more children than the rich

are complicated. Again, the cases of Austria and Jordan serve as examples.

Austria has a GDP per capita of $23,166. Its highly educated population is 100 percent literate and is employed in high-tech industry, business, research, teaching and writing, manufacturing, and upscale tourism. The large amounts of time, effort, and money required to prepare a child to compete in this economy cause many couples to choose not to have any children. Jordan, on the other hand, has a GDP (PPP) per capita of $3347, and the distribution of wealth is uneven.

In Jordan, many people still live in rural areas (although 97 percent of the recorded GDP comes from industry and services). Many live a **subsistence lifestyle**—that is, the family produces its own food, clothing, and shelter. Much everyday work is still done in the informal economy by hand, and each new child is seen as a potential contributor to the family well-being and income and as an eventual caregiver when the parents grow old. Not producing enough children who will live to adulthood is still a significant worry, because 34 children per 1000 die before they are five years old. (In Austria, only 5 children per 1000 die before they are five.) It is not surprising, then, that only 53 percent of the women in Jordan use birth control (more than 71 percent do in Austria), in part because in Jordan the chance of losing a child to illness is much greater than in Austria. But the situation is changing in Jordan. Just 25 years ago, the GDP per capita was $993; infant mortality was 77 per 1000; and the average woman had eight children, whereas now she has four.

The sorts of changes evident in Jordan are taking place in many countries. As agricultural production is mechanized, there is less need for agricultural labor and, instead, a demand for fewer, well-trained specialists. Furthermore, subsistence lifestyles

AT THE GLOBAL SCALE   *Missing Females in Population Statistics*

There is now considerable evidence that between 60 million and 100 million females are missing from the world population. The normal ratio of females to males at birth is about 100 to 105. Boy babies are somewhat weaker than girls, so within the first five years of life the ratio evens out naturally. Since 1900, however, the ratio of females to males has been declining; in 1990, it was about 97.6 to 105. Just why the ratio started to decline around 1900 is not yet explained satisfactorily, but the likely reason is the nearly global preference, especially in some developing countries, for boys over girls. Some of the missing females were conceived but were never born because their parents chose to abort them (UNFPA report, "State of the World Population 2000," Sept. 20, 2000, p. 383). Others died at a young age because their parents committed female infanticide or because the girls received inadequate health care and poorer nutrition than boys.

Even surviving females may be invisible in statistics. Much data is collected without distinguishing between the sexes, thus obscuring important statistical differences. For example, research has shown that around the world, in every country from Sweden to Swaziland, females benefit less from development than do males. But development statistics rarely show gender differences, so the lesser well-being of females is not apparent. Another aspect of the "missing female" problem is found in global statistics related to work. Almost universally, women's work has been so undervalued that it is virtually missing from national work statistics. Women's contributions as subsistence farmers, as homemakers, as domestic servants, as child-care providers, as volunteers, and as workers elsewhere in the informal economy have been considered mere pastimes, not real productive labor. Nor have women's earnings in the informal market been calculated into national income statistics. Even though the oversight has been documented for more than 20 years, the customs of statistics gathering are only beginning to change, and the missing female phenomenon continues.

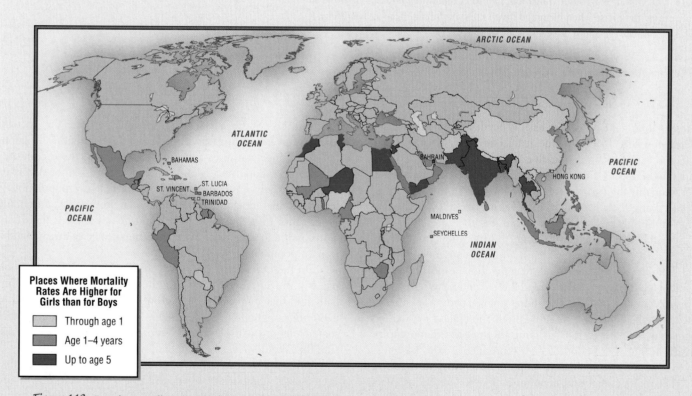

Places Where Mortality Rates Are Higher for Girls than for Boys

- Through age 1
- Age 1–4 years
- Up to age 5

Figure 1.18 **Female mortality rates.** In the countries shaded yellow, orange, and rust, the mortality rate for girls is abnormally high. The darker the color, the longer the risk to girls lasts. How might the explanations for these patterns vary from country to country? [Adapted from Joni Seager, *The State of Women in the World Atlas* (New York: Penguin, 1997), p. 35.]

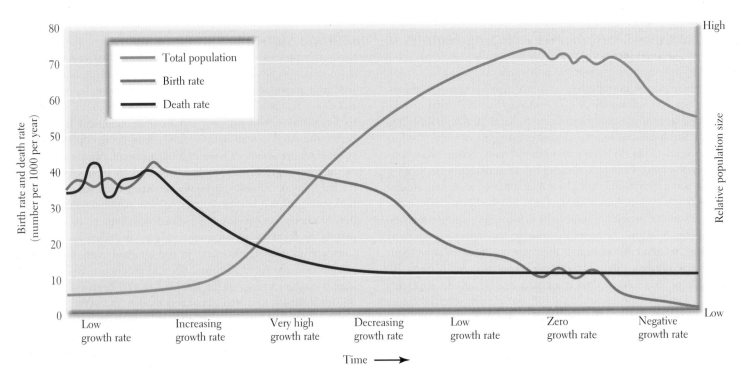

Figure 1.19 The demographic transition. Can you explain why population grows so rapidly during the phase when birth and death rates are falling? [From G. Tyler Miller, Jr., *Living in the Environment*, 8th ed. (Belmont, Calif.: Wadsworth, 1994), p. 218.]

are losing their appeal because cash is needed to buy such goods as television sets, bicycles, blue jeans, sneakers, T-shirts, meat, sugar, and canned goods. In cash economies, children become a drain on the family income. Each child must be educated in order to qualify for a good cash-paying job. Children in school all day and doing their lessons in the evening are not able to earn money until they are 18 or 20 years old.

Societies that are going through the transition from subsistence to cash economies must learn all these consequences first-hand. Until recently, it took one full generation or more to see the benefit of having fewer children and then to choose to have fewer. Now growth rates are dropping more quickly, perhaps as a result of better public education. When reproduction rates slow, demographers say that the region has gone through the **demographic transition:** that is, a period of high reproductive rates has given way to a period of much lower reproductive rates, at the same time that major social and economic changes are taking place within the society (Figure 1.19).

IS THE WORLD OVERPOPULATED?

At the beginning of the twentieth-first century, there is controversy over the question of whether worldwide human population growth should be curtailed. Those who believe so offer several justifications. Some social scientists worry about the societal effects of overcrowding and poverty and believe that if the populations of poor countries can grow more slowly, those countries can develop economically and thereby achieve higher standards of living. Many environmentalists think that overpopulation is leading to the general deterioration of environments and quality of life. Biologists, biogeographers, and botanists focus their concerns on the thousands of plant and animal species threatened by human population expansion.

On the other hand, there are those who think that technology will provide new ways to use the earth's resources more frugally and that markedly slower growth or no growth will lead to economic stagnation. Some leaders can be found in all of the major religions who see human reproduction as a religious mission that should never be tampered with.

Most world leaders advocate slower growth rates so that increases in standards of living will not be canceled out by the need to support new people. Recently, however, sharply declining natural growth rates and aging populations in some industrialized countries (Austria is one example) have led to a realization that ways must be found in these countries to maintain a sufficient workforce and taxpayer base to support those who are too old or too young to work. Such issues will be discussed in several chapters that focus on world regions affected by these trends.

Thinking Geographically: ON THE WEB

HUMANS AND THE ENVIRONMENT

The long, continuous give-and-take between humans and the environment has resulted in many improvements in the circumstances of human life. At the same time, humans have had an enormous impact on physical environments in their efforts to obtain food and shelter; to travel; and to protect themselves from storms, droughts, flooding, and extreme cold and heat. Some of these effects on the environment are now regarded as harmful (and perhaps irreversible). They include ozone depletion, erosion, global warming, acid rain, deforestation, and chemical pollution of groundwater. High standards of living and mass consumption tend to alter the environment most profoundly, but all human ways of life—whether on small cultivation plots in the tropics or in huge industrial cities—have some environmental effects.

Mounting awareness of environmental problems has prompted many proposals to limit the damage. Many of these plans are based on the use of advanced technology—such as catalytic converters to decrease harmful auto emissions and scrubbers on factory chimneys. Finding solutions that will halt or even reverse the damage may be the greatest challenge our species has yet faced. There are many strategies, originating in various world regions, that aim to create more sustainable ways of life, but all of them will require us to change our habits of mass resource consumption.

SUSTAINABLE DEVELOPMENT

The United Nations defines **sustainable development** as the effort to improve present living standards in ways that will not jeopardize those of future generations. The concern is that by destroying resources (as in deforestation) or poisoning them (as in pollution of water and air supplies), we may be depriving future generations of the resources they need for their well-being. Those who support the idea of sustainability hope to promote more environmentally and socially friendly development for all, but especially for the vast majority of earth's people who do not yet enjoy an acceptable level of well-being. Although sustainable development is considered desirable by many people, it is still a concept that no country or other geopolitical entity has yet translated into effective policy.

Geographers who study the interactions among development, human well-being, and the environment are called **political ecologists.** They examine how the power relationships in a society affect the way development proceeds, whose needs it addresses, and how success is measured. Political ecologists have noted, for example, that a country may appear to be developing because its gross domestic product is rapidly increasing—but to achieve that increase, it may be using its resources in ways that cannot be maintained. For instance, the clearing of forests to grow oil palm trees may raise a Southeast Asian country's GDP through the sale of palm oil, but GDP does not take into account the loss of forest resources, the unsustainable use of soils that do not maintain fertility when a single plant species replaces a multispecies forest, or the destruction of a way of life when forest dwellers are forced into urban life where their skills are useless. The incomes of landowners may rise dramatically at the same time as quality of life for the forest dwellers sinks abysmally. Average national production and income may increase while losses in other categories may go unrecognized.

On the other hand, this same country could choose to measure its development by improvements in average human well-being and in present environmental quality. It could consider the potential for maintaining that human well-being and environmental quality into the future. A country using such measures would soon see that the oil palm basis for development wouldn't meet both its development goals and its sustainability goals. The answer to the question "Development for whom?" would be that only a few would benefit, whereas the majority of citizens, the environment, and future generations would lose.

A young man in Yunnan Province, China, cultivates onions and cabbages early in the morning before heading off to a nearby school. Most families in China use intensive agricultural techniques to maximize garden production on small plots that have been sustainable for generations. [Leong Ka Tai/Material World.]

Sustainable Agriculture

Farming that meets human needs without poisoning the environment or using up water and soil resources is called **sustainable agriculture.** Food production on earth has increased remarkably, especially over the last several decades as many types of technology were applied to agricultural production. Figure 1.20 shows that in the 25 years between 1965 and 1990, total food production rose between 70 and 135 percent depending on the region. But population also rose quickly during this period, so the gains were much less per capita. In Africa, per capita production actually decreased. Some people think that we are about to reach the limit of our productive capacity, because as populations boom, environmental problems proliferate. Others say that technological advances will make present land more productive and unused land useful.

Just how sustainable are the world's agricultural systems? The answer is unclear, but previously unrecognized side effects of development are just coming to light. The global map of soil degradation in Figure 1.21 shows that many of the most agriculturally productive parts of North America, Europe, and Asia have already suffered moderate to serious loss of soil through erosion. Moreover, many of the areas labeled as stable on this map may be unusable for agriculture because of extremes in temperature and moisture. Globally, soil degradation and other problems related to food production affect about 7 million square miles (2000 million hectares), putting the livelihoods of a billion people at risk.

The main causes of soil degradation are overgrazing, deforestation, and mismanagement of farmland. Irrigation often makes soil salty and infertile over time. It also can deplete water resources, because it depends on diverting water from rivers and streams or pumping it up from natural underground reservoirs, often at a rate that causes the sources to shrink or even dry up completely. Modern agricultural techniques pioneered in the United States and now spreading throughout the world—such as the use of fertilizers, pesticides, and herbicides—have caused massive die-offs of birds, insect pollinators, fish, and other life-forms.

The definition of sustainable agriculture requires that food production must meet the needs of everyone on the planet. Yet about one-fifth of humanity subsists on a diet too low in total calories and vital nutrients to sustain adequate health and normal physical and mental development. The problem of hunger, so far, is not that the world's agricultural systems cannot produce enough but that food often does not reach hungry people. In the wealthy countries of the world, where for years there have been surpluses of such basic foods as grain, land has been deliberately taken out of production to encourage prices to rise. On the other hand, when rich countries do send surplus food to areas suffering famine, local farmers in the famine zone are often forced out of business because the markets where they previously sold their produce are flooded with free food. Many experts now agree that the most promising solution to hunger seems to be for individual countries and regions to develop their own plans for sustainable agricultural development. In general, it is safe to say that up to now, there have been few truly sustainable agricultural solutions. On the one hand, mass production almost always results in environmental problems and in misallocations, and, on the other hand, small-scale production that is more environmentally sound cannot meet the huge food demands of growing urban populations. The agricultural successes and failures of various countries and agencies will be examined further in the chapters on world regions.

Sustainability and Urbanization

The world is fast becoming urbanized. In 1700, less than 10 percent of the world's total population, about 7 million people, lived

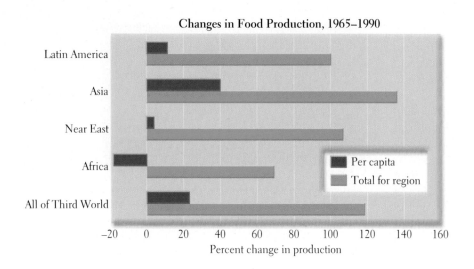

Changes in Food Production, 1965–1990

Figure 1.20 **Changes in food production between 1965 and 1990, by region.** [Adapted from John Bongaarts, "Can the growing human population feed Itself?" in *Global Issues, 1995–96* (Guilford, Conn.: Dushkin/Brown & Benchmark, 1995), p. 119.]

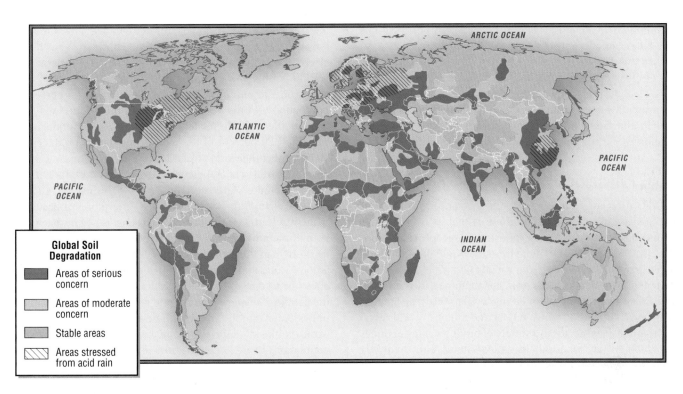

Figure 1.21 **Global soil degradation.** Why might some of the areas labeled "stable" actually be unusable for agriculture? [Adapted from John L. Allen, *Student Atlas of Environmental Issues* (Guilford, Conn.: Dushkin/McGraw-Hill, 1997), p. 109; and Tom McKnight, *Physical Geography* (Upper Saddle River, N.J.: Prentice Hall, 1996), p. 159.]

in cities. Only 5 of those cities had populations as high as several hundred thousand people. By 1995, there were more than 400 cities of over 1 million and more than 10 cities with 10 million. By 2000, 45 percent of the world's population lived in cities. Although many people enjoy city life, it is the relatively idyllic urban existence of a modern, wealthy city that they cherish. Most urban dwellers today cope with far more difficult realities.

Water and sanitation are cases in point. In rapidly growing cities in Asia, Africa, and Latin America, housing is often self-built of scavenged materials, and sanitation systems are absent. A majority of the people in many cities do not have access to working toilets, and desperate citizens relieve themselves in poorly drained pit toilets or on city streets. Sewage and other wastewater is often pumped into a nearby river, swamp, or ocean, even from modern high-rise apartments and luxury hotels. This method of waste disposal causes serious health hazards and widespread ecological damage.

Increasingly, people must boil all the water they use, even for bathing, or must buy drinking water that has been purified. But the poorest urban inhabitants, who are often a majority of the population, lack the money to buy clean water or the fuel to boil it. Adults are often chronically ill, and many babies and young children die of water-related diseases before the age of five.

In sections of Calcutta, India, the only available public water comes from small pipes located at curbsides. Here a woman (*left center*) fills buckets for washing dishes and clothes from a public spigot (under her left hand). [Dilip Mehta, Contact/The Stock Market.]

Building adequate wastewater collection and purifying systems in cities already housing several million inhabitants is so costly that this option is rarely considered, yet new affordable sources of clean water are not available. As we will see in later chapters, clean water is only one of several problems faced by the world's urban dwellers. Technological and other advances will no doubt help alleviate some problems, but for the time being, we are far short of the sustainable ideal for the world's cities.

Changing Patterns of Resource Consumption

One of the most interesting revelations to come from the geographic analysis of rates of resource use is that as people move from rural agricultural work to industry or service sector jobs in cities, they begin to use more resources per capita. Moreover, they draw their resources from a wider and wider area. Water once fetched from nearby wells may now be piped in from hundreds of miles away. Clothing once made laboriously by hand at home is now purchased from manufacturers half a world away. Consumers may have access to a variety of products at lower prices than they would pay for locally produced items. But there is a downside to accessing a wide net of resources: the tendency to overconsume. When water is piped into the house or yard, its very convenience encourages waste. When clothes can be purchased at

low prices, closets become crowded. In fact, resource use has become so skewed toward the affluent with cash to pay that in any given year the relatively rich minority (about 20 percent of the world's population) consumes more than 80 percent of the available world resources. Meanwhile, the poorest 80 percent of the population uses the other 20 percent of the resources.

Another statistic related to resource consumption is that the wealthiest 20 percent of the world's population produces close to 90 percent of the world's hazardous waste, and this same group also consumes well over 50 percent of the world's fossil fuel, metal, and paper resources. The economist E. F. Schumacher, in his book *Small Is Beautiful*, commented that "the problem passengers on spaceship Earth are the first class passengers." Schumacher's point is an important one: consumers in industrialized societies consume far more than their share of resources and produce most of the hazardous wastes and environmental degradation. But this observation does not recognize the full complexity of the situation. Although the people of the developing world presently consume much less as individuals than do those of the developed world, the developing world is already suffering from environmental deterioration. As populations in these countries increase and as their consumption per capita grows closer to consumption rates in the developed world, negative environmental effects will also increase proportionately. A case in point is global warming.

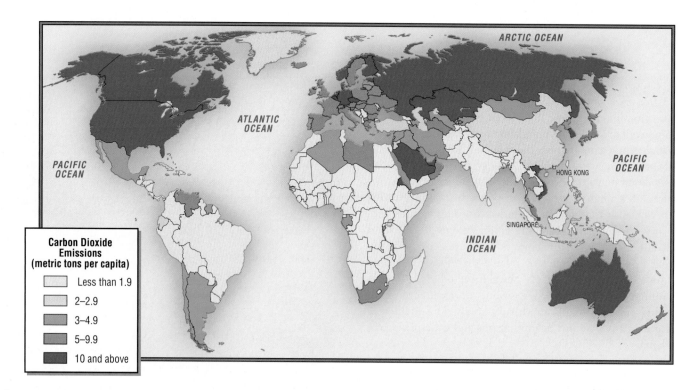

Figure 1.22 Carbon dioxide emissions around the world. [Adapted from UN Framework Convention on Climate Change and UN Environment Programme/GRID Arendal @ http://maps.grida.no/kyoto/ 2000; and *Scientific American* (May 1996): 24.]

GLOBAL WARMING

The theory of **global warming** proposes that the earth's climate is becoming warmer as atmospheric levels of carbon dioxide (CO_2), methane, water vapor, and other gases drastically increase. These gases are collectively known as "greenhouse gases" because their presence allows large amounts of heat from sunlight to be trapped in the earth's atmosphere in much the same way that heat is trapped in a greenhouse or a car parked in the sun. Greenhouse gases exist naturally in the atmosphere. In fact, it is their heat-trapping ability that makes the earth warm enough for life to exist. These gases are also released by such human activities as the burning of coal, oil, and other fossil fuels; the growing of certain crops such as paddy rice; and the use of nitrogen fertilizers.

Over the last several hundred years, human activities that release greenhouse gases have intensified dramatically. Industrial processes and transport vehicles burn large amounts of fossil fuels, and cash-crop agriculture uses nitrogen-based fertilizers. The large-scale raising of grazing animals contributes methane through flatulence. Unusually large quantities of greenhouse gases from these sources have begun to accumulate in earth's atmosphere, and many scientists think that their presence has already led to a warming of the planet's climate.

At the same time, deforestation has removed growing plants (especially trees) from landscapes around the world. Plants absorb CO_2 in the process of photosynthesis. Scientists at the World Resources Institute in Washington, D.C., say that as much as 30 percent of the buildup of CO_2 in the atmosphere results from the loss of trees and plants that would absorb the gas. Most of this loss can be attributed to forest clearing in developing countries.

Climatologists and other scientists are documenting global warming by examining evidence in tree rings, fossilized pollen and marine creatures, and glacial ice. This data indicates that the twentieth century was the warmest in 600 years and that the decade of the 1990s was the hottest since the late nineteenth century. It is estimated that average global temperatures will rise between 3°F and 9°F (about 5°C) by 2030, but it is not clear just what the consequences of such a rise in temperature would be. Certainly the effects will not be uniform across the earth. One prediction is that glaciers and the polar ice caps will melt, causing a corresponding rise in sea level. Potentially, hundreds of millions of people in coastal areas and on low-lying islands could be displaced. Other scientists forecast a shift of warmer climate zones northward in the Northern Hemisphere and southward in the Southern Hemisphere. Such shifts might also displace huge numbers of people, as the zones where specific crops can grow would change dramatically. Animal and plant species that could not adapt to the change rapidly would disappear. Another effect of global warming could be more chaotic and severe weather.

Most of the responsibility for CO_2 emissions rests on the shoulders of the industrialized countries in North America, Europe, northern Eurasia, and Australia, as well as the oil-producing countries in Southwest Asia (Figure 1.22). From 1859 to 1995, developed countries produced roughly 80 percent of the greenhouse gases from industrial sources, and developing countries produced 20 percent. By 1995, however, the developing countries accounted for nearly 40 percent of total CO_2 emissions. Figure 1.23 predicts that as developing nations industrialize over the next century and continue to cut down their CO_2-absorbing forests, they will release more and more greenhouse gases every year. If present patterns hold, the cumulative historical contributions of total greenhouse gases by the developing countries will exceed those of the developed world by about 2040. Members of the Organization for Economic Cooperation and Development (OECD) have drafted an agreement known as the Kyoto Protocol calling for scheduled reductions in CO_2 emissions by developed countries. The agreement encourages OECD cooperation with developing countries to help them curtail their emissions as well. The United States has not signed the Kyoto agreement; opponents in the United States argue that developing countries should be required to reduce their emissions more stringently.

Global warming will probably be a source of disagreement for some time. The Kyoto Protocol contains few provisions for enforcement, and many industries—especially oil and gas—argue that global warming is an unproved theory. In contrast, those concerned about environmental pollution see the currently agreed-upon reductions as far too low. They call for stepped-up energy conservation and for more research into such alternatives as solar, wind, and geothermal energy. Developing countries have steadfastly refused to reduce their emissions until the large emitters do so first.

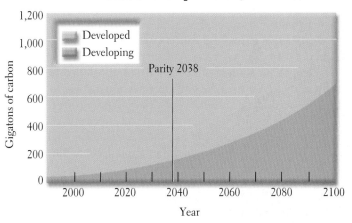

Production of CO₂ Emissions, 1990–2100

Figure 1.23 **Contributions to atmospheric carbon dioxide by developed and developing countries.** The quantity of carbon dioxide releases shown in this graph includes both industrial emissions and amounts released as a result of deforestation. When both these sources are taken into account, the developing countries will exceed the developed countries in CO₂ production after 2038. How could developed countries best encourage lower emission rates in developing countries? [Adapted from Duncan Austin, José Goldemberg, and Gwen Parker, "Contributions to climate change: are conventional metrics misleading the debate?" in *Climate Notes* (World Resource Institute Climate Protection Initiative, Oct. 1998), at http://www.wri.org/cpi/notes/metrics.html.]

POLITICAL ISSUES IN GEOGRAPHY

Political issues in geography revolve primarily around power: its exercise, its allocation to different segments of society, and its spatial distribution within and among world regions. A political issue commonly discussed in geography is geopolitics—the jockeying among countries for territory, resources, or influence. Other political geography topics include how public and private decisions made at the local level affect the allocation of space for various uses; the types of government that exist at local, state, and national levels and how these types affect social organization, resource allocation, and land use; the ramifications of changing borders between countries; and international organizations focused on development, human well-being, and peacekeeping.

GEOPOLITICS

Geopolitics encompasses the strategies that countries use to ensure that their best interests are served in their relations with other countries. For example, if a state perceives a threat to a border from a neighboring state, it may move troops and even settlers (in the case of a lightly populated area) to the border region to make a show of force. An example is provided by the northernmost border zone between India and Pakistan. A dispute over who should control parts of the province of Kashmir erupted when India and Pakistan were separated into two countries at the end of British colonial rule in 1949. The precise border has never been agreed upon, and India keeps 500,000 troops in Kashmir to enforce its interests. For its part, Pakistan funds and supports guerrilla bands that periodically cross the border. There is widespread concern over this dispute because both countries have nuclear weapons and have threatened to use them.

Geopolitics typified the **cold war era,** the period from 1946 to the early 1990s when the United States and its allies in Western Europe faced off against the Union of Soviet Socialist Republics (USSR) and its allies in Eastern Europe. Ideologically, the United States promoted a version of free-market capitalism and democracy, whereas the USSR and its allies favored a centrally planned socialism in which citizens participated in planning and government indirectly, through members of the Communist Party. The cold war grew into a contest to determine which system would prevail throughout the world. It eventually influenced the internal and external policies of virtually every country on earth. In the post–cold war period, geopolitics is shifting: each country is jockeying for a better position in what may be a new era of trade and prosperity, rather than war.

NATIONS AND BORDERS

In this book, you will often see regions defined by the countries within them. Countries are a common unit of geographic analysis, and their boundaries are a major feature of the maps in this book. Why a country has its particular boundaries is a complex subject, but the following discussion addresses some of the reasons.

The eighteenth-century French writer Jean-Jacques Rousseau believed that people realize their true potential by joining together to create a nation-state anchored to a particular territory. According to this view, a **nation** per se is not a political unit or an official country but a group of people who share language, culture, and political identity. Once those people formally establish themselves as occupying a particular territory and become a country, they are a **nation-state.** Japan is a nation-state because it is a country and its people, with only minor exceptions, share a common language, culture, and political identity. The United States and Australia are not nation-states because, although they are countries, their diverse populations originated from virtually every corner of the globe and to the extent that political identity is shared, in both countries it is based on a loosely defined recognition and affirmation of diversity.

Ideas about the nature of national identity spread from France to the rest of Europe, the Americas, and beyond. Although nation-states were seldom feasible because of cultural diversity, nearly all former colonies of Europe eventually became independent states. In some cases, new countries were **pluralistic states** in which power was shared among several groups, each defined by common language, ethnicity, culture, or other characteristics. More typically, only one group, which had managed to gain the advantage over several others, achieved power when the state was created. In some cases, groups found that the territory they had customarily occupied was claimed by several new states, leaving them without a homeland. For example, the once nomadic Kurds, who ranged through parts of what are now Turkey, Syria, Iraq, and Iran, still do not have a recognized territory and are unlikely to get one because all the land has been claimed already. Here and in many other places around the world, groups are in conflict with national governments over rights to territory. The nation-state is theoretically a powerful way to bind together a large group of people in a particular place, but rarely is a substantial territory occupied by only a single group. With no clear way to divide once shared territory equitably among groups, the idea of the nation-state has often led to conflict.

The emergence of the idea of the nation-state was linked with the concept of sovereignty, which means that a country can conduct its internal affairs as it sees fit without interference from outside. Sovereignty increased the importance of precise legal boundaries as demarcations of power and control. A country's borders marked the extent of its territory and hence its sovereignty. When we look at a map, we are viewing a spatial representation of the outcomes of struggles over control of territory. Some such struggles have been resolved only recently or are still under way. In the 1990s, for example, the map of Yugoslavia changed markedly as parts of that country became independent states. Then various ethnic groups within these new states began to struggle over territory they had previously occupied in common.

Thinking Globally: ON THE WEB

Some 200,000 Serbs, expelled from the Krajina region of Croatia, cross through Bosnia to reach Serbia during one of the many "ethnic cleansing" episodes of the conflict in former Yugoslavia. [Peter Turnley/Corbis.]

INTERNATIONAL COOPERATION

The idea of the nation-state has been recognized as deeply flawed, and there is reason to think that the idea of national sovereignty is also becoming less viable. Increasingly, people, goods, and capital are flowing freely across national borders, and this trend favors interdependence over national sovereignty. It also creates a need for some way to enforce laws governing business, trade, and human rights at the international level. When extreme human rights abuses became painfully obvious in the former Yugoslavia, first in

Bosnia and then in Serbia and Kosovo, eventually the United States and Europe reluctantly decided that it was appropriate to intervene and thus breach Yugoslav sovereignty.

There is some evidence that the role of the individual state may decline as international organizations play an increasing role in world affairs. The prime example of government-to-government international cooperation is the **United Nations,** an assembly of 185 member states that sponsors programs and agencies focusing on scientific research, humanitarian aid, planning for development, fostering general health, and peacekeeping assistance in hot spots around the world—such as Angola, Bosnia-Herzegovina, Cambodia, East Timor, Israel, Kosovo, Serbia, Lebanon, Rwanda, Sierra Leone, and Haiti. The United Nations has little legal power and can often enforce its rulings only through the power of persuasion. Even in its peacekeeping mission, there are no true UN forces but rather troops from member states who wear UN designations on their uniforms and temporarily take orders from UN commanders.

The **World Bank,** mentioned earlier, and the **International Monetary Fund (IMF)** are financial institutions funded by the developed nations to help developing countries reorganize and formalize their economies toward a free-market model. These two institutions wield enormous power in global economic development through their lending activities. The **World Trade Organization (WTO),** also mentioned earlier, oversees world trade practices, enforces trade regulations, and settles trade disputes among member countries.

Nongovernmental organizations (NGOs) are an increasingly important embodiment of globalization. They range from groups concerned primarily with local issues, such as East Tennessee's "Save Our Cumberland Mountains," to globally active transnational groups, such as the World Conservation Union and Médecins Sans Frontières (Doctors Without Borders). In such associations, individuals from widely differing backgrounds and locations agree about a political, economic, or environmental issue, such as saving endangered species or dispensing medical care to those who need it most. Sometimes the educational efforts of these organizations succeed in raising awareness of an issue among the general public. NGOs have prepared the way for formal agreements in the United Nations. Although these organizations often operate on a global scale (especially since the advent of the Internet), some, despite their global reach, resemble community-level social groups, in that they solicit input from a wide range of individuals and groups —a feature that political scientists consider essential to participatory democracy.

Thinking Critically About Selected Concepts

1. Geography is the study of the earth's surface and the physical and human processes that shape it. *As you look around your neighborhood with a geographer's eye, can you distinguish the physical and human processes that have made it what it is today?*

2. A region is a unit of the earth's surface that contains distinct patterns of physical features or of human development. Regions exist at different

scales, and their borders are negotiable. *Describe the different scales of regions in which you live, beginning with the neighborhood or campus. What criteria are you using to define each successively larger region?*

3. Culture is represented by the ideas, materials, and institutions that people have invented and passed on to subsequent generations. A culture group is a group of people who share a set of beliefs, a way of life, a

technology, and usually a place. Cultural markers are characteristics, such as language and religion, that help to define a culture group. *When assigning yourself and your classmates to a culture group, what questions are raised in your mind?*

4.   According to the theory of plate tectonics, the earth's surface is composed of large, slow-moving plates that float on top of an underlying layer of molten rock. The movement and interaction of the plates create many of the large features of the earth's surface, particularly mountains, and also short-term events. *Where in the last several months have tectonic events taken place?*

5.   Development is a hazy concept that used to mean modernization and living standard increases measured by income. Now the definition of development is expanding to include human well-being, gender equity, and environmental sustainability. *Does development in your community reveal how its leaders define the concept?*

6.   Demography is the study of population patterns and changes. It is important in geography because it allows us to compare, over time and space, such factors as numbers of people in a given place, their rate of natural increase, their income and spending patterns, and their overall human well-being. *How have the demographics of your community changed over the past decade?*

7.   The global economy is the worldwide system by which capital, labor, goods, and services are exchanged. Key institutions in the global economy are multinational corporations; national governments; and international agencies that manage financial dealings, trade, and development aid. The effects of this worldwide exchange, however, are often experienced at the local level. *In what ways do you and your family experience the global economy?*

8.   Geopolitics encompasses the strategies that countries use to ensure that their best interests are served. Much political contention today centers around the efforts of some culture groups who share a language, culture, or political philosophy to establish nation-states or countries. There is evidence that supranational entities such as trade blocs and international aid institutions are gaining influence because many issues in a global economy cannot be addressed at the country level. *What are some geopolitical issues covered in the news media today where you live?*

Key Terms

age distribution or **age structure** (p. 36)   the proportion of the total population in each age group

air pressure (p. 20)   the force exerted by a column of air on a square foot of surface

average population density (p. 35)   the average number of people per unit area (for example, square mile or square kilometer)

birth rate (p. 35)   the number of births in a given population per unit time, especially per year

capital (p. 30)   wealth in the form of money or property used to produce more wealth

capital controls (p. 31)   restrictions on business investment by foreigners requiring that investment capital stay in the host country for a minimum length of time or that companies be partially controlled by local investors

climate (p. 20)   the long-term balance of temperature and precipitation that characteristically prevails in a particular region

cold war era (p. 46)   the period from 1946 to the early 1990s, when the United States and its allies in Western Europe faced off against the Union of Soviet Socialist Republics and its allies in Eastern Europe

colony (p. 30)   a (usually) distant land acquired by a more powerful country for economic gain

contiguous regions (p. 9)   regions that lie next to or near each other

cultural diversity (p. 12)   differences in ideas, values, technologies, and institutions among culture groups

cultural homogeneity (p. 11)   uniformity of ideas, values, technologies, and institutions among culture groups

cultural identity (p. 11)   a sense of personal affinity with a particular culture group

cultural marker (p. 11)   a characteristic that helps to define a certain culture group

culture (p. 10)   everything we use to live on earth that is not directly part of our biological inheritance

culture group (p. 10)   a group of people who share a set of beliefs, a way of life, a technology, and usually a place

death rate (p. 35)   the ratio of total deaths to total population in a specified community

delta (p. 20)   the triangular-shaped plain of sediment that forms where a river meets the sea

demographic transition (p. 40)   the change from high birth and death rates to low birth and death rates that usually accompanies a cluster of other changes, such as change from a subsistence to a cash economy, increasing education rates, and urbanization

demography (p. 34)   the study of population patterns and changes

deposition (p. 20)   the settling out of rock and soil particles from wind or water as it slows

diffusion (p. 26)   the process by which agriculture and the domestication of animals spread around the world from a few places

domestication (p. 26)   the process of developing plants and animals through selective breeding to live with and be of use to humans

earthquake (p. 19)   a catastrophic shaking of the landscape, often caused by the shifting and friction of tectonic plates

economy (p. 29)   the forum where people make their living, including the spatial, social, and political aspects of how resources are recognized, extracted, exchanged, transformed, and reallocated

erosion (p. 20)   the process by which fragmented rock and soil are moved over a distance, primarily by wind and water

ethnic group (p. 10)   see **culture group**

exchange (p. 29)   barter or trade for money, goods, or services

external processes (geophysical) (p. 19)   landform-shaping processes that originate at the surface of the earth, such as weathering, mass wasting, and erosion

extraction (p. 29)   the acquisition of a material resource through mining, logging, agriculture, or other means

extractive resource (p. 29)   a resource such as mineral ores, timber, or plants that must be mined from the earth's surface or grown from its soil

floodplain (p. 20)   the flat land around a river where sediment is deposited during flooding

formal economy (p. 29)   all aspects of the economy that take place in official channels

formal institutions (p. 10)   associations such as official religious organizations; local, state, and national governments; nongovernmental organizations; and specific businesses and corporations

free trade (p. 31)   the movement of goods and capital without government restrictions

frontal precipitation (p. 23)   rainfall caused by the interaction of large air masses of different temperatures and densities

gender structure (p. 36)   the proportion of males and females in each age group of a population

geopolitics (p. 46)   the use of strategies by countries to ensure that their best interests are served

global economy (p. 27)   the worldwide system in which goods, services, and labor are exchanged

global scale (pp. 7, 9)   the level of geography that encompasses the entire world as a single unified area

global warming (p. 45)   the predicted warming of the earth's climate as atmospheric levels of greenhouse gases increase

globalization (p. 8)   the growth of interregional and worldwide linkages and the changes they are bringing about

gross domestic product (GDP) per capita (pp. 29, 32–33)   the market value of all goods and services produced by workers and capital within a particular nation's borders and within a given year; dividing the value by the number of people in the country results in the per capita value

growth rate (p. 35)   see **rate of natural increase**

human geography (p. 5)   the study of various aspects of human life that create the distinctive landscapes and regions of the world

human well-being (p. 10)   the ability of people to obtain for themselves a healthy life in a place and community of their choosing

import quota (p. 31)   a limit on the amount of a given item that may be imported into a country over a period of time

informal economy (p. 29)   all aspects of the economy that take place outside official channels

informal institutions (p. 10)   ordinary or casual associations, such as the family or a community

institutions (p. 10)   all of the associations, formal and informal, that help people get along together

internal processes (geophysical) (p. 19)   processes, like plate tectonics, that originate deep beneath the surface of the earth

International Monetary Fund (IMF) (p. 47)   a financial institution funded by the developed nations to help developing countries reorganize, formalize, and develop their economies

lingua franca (p. 13)   a language used to communicate by people who don't speak one another's native languages

material culture (pp. 10, 15)   all the things, living or not, that humans use

monsoon (p. 23)   opposing winter and summer patterns of atmospheric and moisture movement between continents and oceans, in which warm, wet air coming in from the ocean brings copious rainfall during the summer months; in winter, cool, dry air moves from the continental interior toward the ocean

multiculturalism (p. 11)   the state of relating to, reflecting, or being adapted to diverse cultures

multinational corporation (p. 30)   a business organization that operates extraction, production, and/or distribution facilities in multiple countries

nation (p. 46)   a group of people who share a language, culture, political philosophy, and usually a territory

nation-state (p. 46)   a political unit, or country, formed by people who share a language, a culture, and a political philosophy

nongovernmental organization (NGO) (p. 47)   an association outside the formal institutions of government, in which individuals from widely differing backgrounds and locations share views and activism on political, economic, social, or environmental issues

orographic rainfall (p. 22)   rainfall produced when a moving moist air mass encounters a mountain range, rises, cools, and releases condensed moisture that falls as rain

Pangaea hypothesis (pp. 18–19)   the proposal that about 200 million years ago all continents were joined in a single vast continent, called Pangaea

physical geography (p. 5)   the study of the earth's physical processes to learn how they work, how they affect humans, and how they are affected by humans in return

plate tectonics (p. 19)   a theory proposing that the earth's surface is composed of large plates that float on top of an underlying layer of molten rock; the movement and interaction of the plates create many of the large features of the earth's surface, particularly mountains

pluralistic state (p. 46)   a country in which power is shared among groups, each defined by common language, ethnicity, culture, or other characteristics

political ecologist (p. 41)   a geographer who studies the interactions among development, human well-being, and the environment

population pyramid (p. 36)   a graph that depicts age and gender structures of a country

production (p. 29)   a form of economic activity in which resources are transformed to produce new commodities or new bodies of knowledge

purchasing power parity (PPP) (p. 33)   the amount of goods or services that U.S. $1 will purchase in a given country

racism (p. 17)   the negative assessment of people, often those who look different, primarily on the basis of skin color and other physical features

rain shadow (p. 22)   the dry side of a mountain range, facing away from the prevailing winds

rate of natural increase (growth rate) (p. 35)   the rate of population growth measured as the excess of births over deaths per 1000 individuals per year, without regard for the effects of migration

region (p. 5)   a unit of the earth's surface that contains distinct patterns of physical features or of human development

regional geography (p. 5)   the analysis of the geographic characteristics of particular places

regional trade bloc (p. 31)   an association of neighboring countries that have agreed to lower trade barriers for one another

resource (p. 29)   anything that is recognized as useful, such as mineral ores, forest products, skills, or brainpower

Ring of Fire (p. 19)   the tectonic plate junctures around the edges of the Pacific Ocean, characterized by volcanoes and earthquakes

secularism (p. 13)   a way of life informed by values that do not derive from any one religious tradition

Structural Adjustment Policies (SAPs) (pp. 31–32)   requirements for economic reorganization toward less government involvement in industry, agriculture, and social services, sometimes made by the World Bank and IMF as conditions for giving loans to borrowing countries

subsistence lifestyle (p. 38)   a way of life in which each family group produces its own food, clothing, and shelter

sustainable agriculture (p. 41)   farming that meets human needs without poisoning the environment or using up water and soil resources

sustainable development (p. 41)   efforts to improve standards of living in ways that will not jeopardize those of future generations

tariff (p. 31)   a tax imposed by a country on imported goods, usually intended to protect industries within that country

technology (p. 15)   an integrated system of knowledge, skills, tools, and methods upon which a culture group's way of life is based

United Nations (UN) (p. 47)   an assembly of 185 member states that sponsors programs and agencies that focus on scientific research, humanitarian aid, planning for development, fostering general health, and peacekeeping assistance

volcano (p. 19)   an area between plates or a weak point in the middle of a plate where gases and molten rock, called magma, can come to the earth's surfaces through fissures and holes in the plate

weather (p. 20)   the short-term (day-to-day) expression of climate

weathering (p. 19)   the process by which rocks are physically or chemically broken up

World Bank (pp. 31, 47)   a global lending institution that makes loans to countries that need money to pay for development projects

World Trade Organization (WTO) (pp. 31, 47)   a global institution whose stated mission is the lowering of trade barriers and the establishment of ground rules for international trade

Pronunciation Guide

Croatia (kroh-AY-shuh)

Krajina (kreye-EE-nuh)

Marie Galante (muh-REE guh-LAHNT)

Molucca Islands (maw-LOOK-uh ["oo" as in "book"])

Pangaea (pan-JEE-uh)

Riau (REE-aw)

Yangtze River (YAHNG-TSUH)

Yunnan (YOO-NAHN)

Selected Readings

A set of Selected Readings for Chapter 1, providing ideas for student research, appears on the *World Regional Geography* Web site at

www.whfreeman.com/pulsipher.

Appendix: Reading Maps

To make the best and fullest use of any map, you should master a few basic map-reading skills. Here we discuss the various aspects of a map: the legend, devices for indicating location and distance on the earth's surface (scale and lines of longitude and latitude), and map projections.

Start by reading any *title*, *caption*, and *legend* that appear on a map. The title tells you the subject of the map, and the caption usually points out some features of the map that the author wants you to notice. The legend is the box that explains what the symbols and colors on the map represent (see Figure 1.13, page 24). In or near the legend usually will be the map's scale, unless it is a world map.

SCALE

The *scale* of a map represents the relationship between the distances shown on the map and the actual distances on the earth's surface. Sometimes, a numerical ratio shows what one unit of measure on the map equals in the same units on the face of the earth. For example, 1:1,000,000, or the fraction 1/1,000,000, means that one inch or one centimeter on the map equals one million inches or one million centimeters on the face of the earth. Other maps may express scale using a phrase such as "One centimeter equals ten kilometers." Alternatively, a simple bar may express the information visually:

Notice that each of the maps of the eastern Caribbean shown in Figure 1.24 is drawn using a different scale. You will see that as the amount of area shown on a map becomes larger, the amount of detail that can be displayed decreases.

LONGITUDE AND LATITUDE

Most maps contain *lines of latitude and longitude*, which enable us to establish a position on the map relative to other points on the globe. European cartographers developed latitude and longitude lines so that navigators far out at sea with no visible landmarks and only the stars and the passage of time to orient them could more easily locate themselves on a map. Lines of longitude run from pole to pole; lines of latitude run parallel to the equator. Lines of longitude (also called *meridians*) come closer together as they move away from the equator, and they all meet at the poles. Lines of longitude are all the same length (half the circumference of the globe). Lines of latitude, also called *parallels*, circle the globe parallel to the equator; they do not meet. The only latitude line that spans the full circumference of the globe is the equator; the rest of the latitude lines describe ever smaller circles of the globe to the north and south of the equator, and the poles are marked with just a point (Figure 1.25).

Both latitude and longitude lines describe circles, so there are 360° in each circle of latitude and in each circle of longitude, 180° for each hemisphere. Each degree spans 60 minutes (designated with the symbol '), and each minute has 60 seconds (designated with the symbol "). Keep in

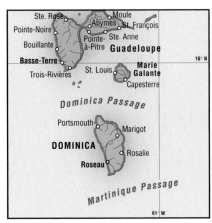

A. The scale of this map is 1:3,000,000.

B. The scale of this map is 1:15,000,000.

C. The scale of this map is 1:45,000,000.

Figure 1.24 **Examples of scale from the eastern Caribbean.** A is a map of Guadeloupe and Dominica in the eastern Caribbean. The scale of 1:3,000,000 makes it possible to show towns, a few roads, and a few landforms, but not much else. Map B is at a scale of 1:13,000,000. You can see much more of the eastern Caribbean, but the only detail that can be shown is the shape of the islands and the location of the capital cities. Map C, at a scale of 1:45,000,000, shows most of the Caribbean Sea and its general location between Central and South America, but now the eastern Caribbean islands are too small to identify clearly. [Adapted from *The Longman Atlas for Caribbean Examinations*, 2d ed. (Essex, UK: Addison Wesley Longman, 1998), p. 4.]

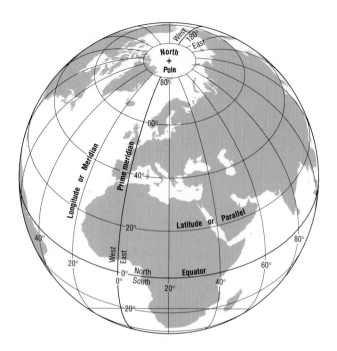

Figure 1.25 Summary of latitude and longitude. Lines of longitude (meridians) extend from pole to pole. The distance between them decreases steadily toward the poles, where they all meet. Lines of latitude (parallels) are equally spaced north and south of the equator and intersect the longitude lines at right angles. The only line of latitude that spans the complete circumference of the earth is the equator; all other lines of latitude describe ever smaller circles heading away from the equator. For example, the 60th lines of latitude (parallels) north and south are one-half the circumference of the equator. [Adapted from *The New Comparative World Atlas* (Maplewood, N.J.: Hammond, 1997), p. 6.]

mind that these are measures of relative space on a circle, not time, nor even real distance (because the circles are different sizes).

The *prime meridian*, which is labeled 0° longitude, runs from the North Pole through Greenwich, England, to the South Pole. The half of the globe's surface west of the prime meridian is called the Western Hemisphere; the half to the east, the Eastern Hemisphere. The longitude lines both east and west of the prime meridian are labeled by their distance in degrees from the prime meridian, from 1° to 180°. Longitude lines lying east of the prime meridian are labeled "east," and those to the west of the prime meridian are labeled "west." For example, 20 degrees east longitude would be written as 20°E. The longitude line at 180° runs through the Pacific and is used as the *international date line*, where the calendar day officially begins. The equator is 0° latitude; the poles are at 90° north and south.

Lines of longitude and latitude form a grid that can be used to designate the location of a place. In Figure 1.24A, you can see that the island of Marie Galante lies just south of the parallel (line of latitude) at 16°N and just about 0.5° west of the 61st meridian (line of longitude). Hence, the position of Marie Galante's northernmost coast is 16°N, 61.5°W.

MAP PROJECTIONS

All maps must solve the problem of showing the spherically shaped earth on a flat piece of paper. When reading a map, it is important to understand the limitations of the different strategies devised to do so. You can appreciate the problems of distortion by drawing a map of the earth on an orange, then peeling the orange and trying to flatten out the orange-peel map and transfer it exactly to a flat piece of paper.

The various ways of showing the spherical surface of the earth on flat paper are called *projections*. All projections entail some distortion. For maps of small parts of the earth's surface, the distortion is minimal. Developing a projection for the whole surface of the earth that minimizes distortion is much more challenging. The *Mercator projection* (Figure 1.26A) is popular, but it is now rarely used by geographers because of its gross distortion near the poles. To make a flat map, the Flemish cartogra-

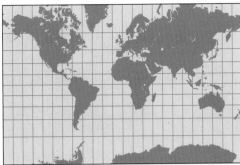

A. Mercator Projection

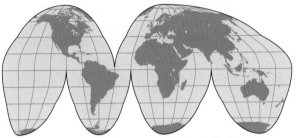

B. Goode's Interrupted Homolosine Projection

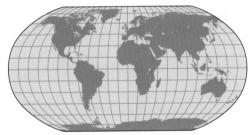

C. Robinson Projection

Figure 1.26 **Three common map projections.** [Mercator and Robinson projections adapted from *The New Comparative World Atlas* (Maplewood, N.J.: Hammond, 1997), pp. 6–7. Goode's interrupted homolosine projection adapted from *Goode's World Atlas,* 19th ed. (Chicago: Rand McNally, 1995), p. x.]

pher Gerhardus Mercator (1512–1594) stretched out the poles, depicting them as lines equal in length to the equator! Greenland, for example, appears about as large as Africa, even though it is only about one-sixth Africa's size. Nevertheless, Mercator's projection is still useful for navigation because it portrays the shapes of landmasses more or less accurately, and a straight line between two points on this map gives the compass direction between them. Also, this type of projection can be safely used for parts of the globe that are within 15 degrees south or north of the equator because the distortion in this range is minimal.

Goode's interrupted homolosine projection (Figure 1.26B) flattens the earth rather like an orange peel, thus preserving some of the size and shape of the landmasses. In this projection, the oceans get snipped up. The *Robinson projection* (Figure 1.26C) sacrifices accuracy at the poles for an uninterrupted view of land and ocean. In this book we often use the Robinson projection for world maps.

Most currently popular world projections reflect the European origin of modern cartography. Europe or North America is usually placed near the center of the map, where distortion is minimal; other population centers, such as East Asia, are placed at the highly distorted periphery. To study the modern world, we need world maps that center on different parts of the globe. For example, much of the world's economic activity is now taking place in and around Japan, Korea, China, Taiwan, and Southeast Asia. Discussions of this activity require maps that focus on these regions but still include the rest of the world in the periphery. Similarly, political developments in the Middle East or in Southwest Asia can best be depicted graphically on world maps that center on such places as Palestine or Iraq.

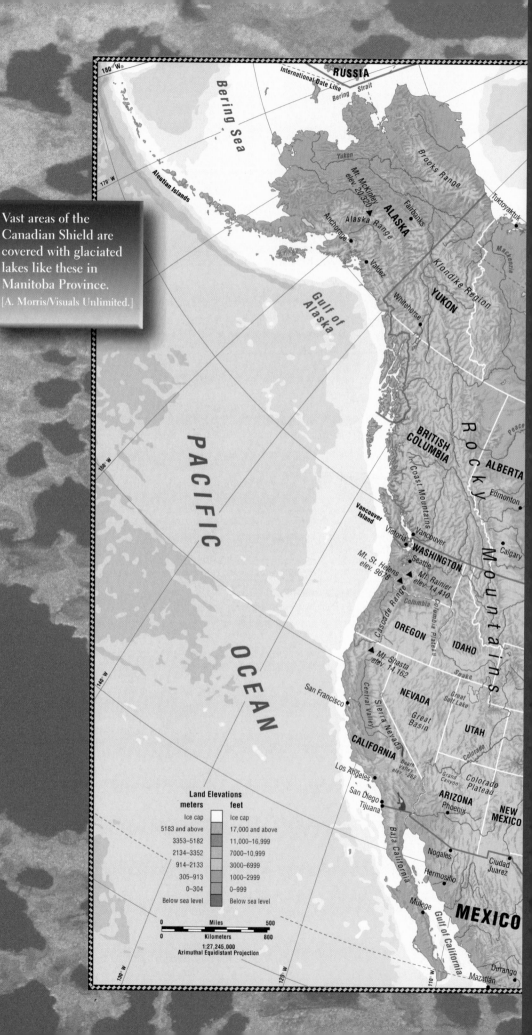

Vast areas of the Canadian Shield are covered with glaciated lakes like these in Manitoba Province.
[A. Morris/Visuals Unlimited.]

Land Elevations

meters	feet
Ice cap	Ice cap
5183 and above	17,000 and above
3353–5182	11,000–16,999
2134–3352	7000–10,999
914–2133	3000–6999
305–913	1000–2999
0–304	0–999
Below sea level	Below sea level

Miles 500
Kilometers 800
1:27,245,000
Azimuthal Equidistant Projection

Chapter 2: North America

MAKING GLOBAL CONNECTIONS

VIGNETTE Mist curls off the Tennessee River just outside of Knoxville as Jesse and Iqbal settle in on a dock for some Saturday morning fishing. After a few minutes, Iqbal quietly puts down his pole and bows to the southeast, toward Mecca, observing his duty as a Muslim.

Jesse keeps fishing. He finds Iqbal's practice a bit strange but is grateful for his company. The two men met along the river early one morning as Jesse was getting ready to fish and Iqbal was finishing his prayers. Iqbal joined in the fishing, and they have been meeting several times a week ever since.

When the Levi Strauss plant moved to Mexico two years ago, Jesse lost his job. Now 56 years old, he works part time at a McDonald's, and even though the pay is low, he feels he has to stay until something better comes along. His social security pension won't begin for years, and even then it won't meet all of his and his wife's expenses. His son has started a job in New Mexico as a schoolteacher and won't be available to help him financially in emergen-

cies. All of these complications make these relaxing mornings with Iqbal even more special.

Iqbal finds his companion and everything else in the United States to be strange. Why, for example, does Jesse insist that they throw the fish they catch back into the river? His assertion that pollution in the river makes the fish unsafe to eat is hard to believe, especially because they look so big and healthy. Iqbal is amazed by the surrounding forested hills and the almost total absence of people on a river so close to a large city. In his home state of Kerala in India, waterways are crowded with boats and fishers, and with people doing laundry. He is amazed by the smooth, clean streets and by the fact that Jesse, a mere restaurant "wallah," as he would be called in India, can afford a house and a lawn that even a well-paid doctor could scarcely afford in Kerala.

All this makes Iqbal determined to make the most of his special temporary visa for high-tech workers. Although his new job as a network administrator at the university pays him less than fellow

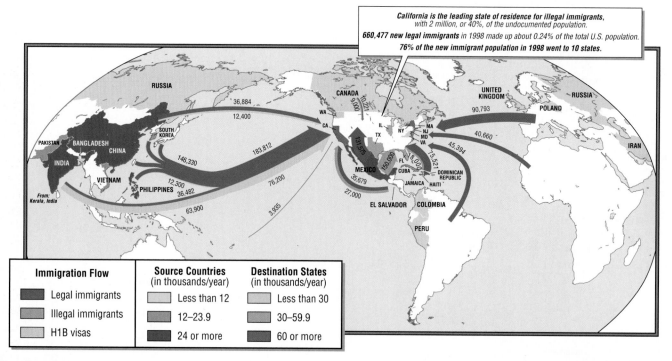

Figure 2.1 **Interregional linkages: immigration to the United States.** The United States is being transformed by the immigrants it attracts from around the globe. Companies and universities recruit highly skilled workers to fill high-tech jobs. At the same time, a considerable flow of less-skilled illegal migrants from Middle and South America take jobs at the lower end of the pay scale. Immigration is also leading to an ethnic transformation. According to data from the U.S. 2000 census, ethnic "minorities" together constitute majorities in half the 100 largest U.S. cities. Census Bureau projections show that all minorities combined will constitute a majority of the U.S. population by 2060.

workers with less technical competence, he works diligently. His visa is good for six years, and if he loses his current job in the meantime, he may be forced to return to India, where the pay is one-tenth what he makes now.

Just as he is finishing his prayer, Iqbal's line goes taut. A long struggle produces an enormous carp. He turns to Jesse and asks, "But . . . must I throw this one back too?" " 'Fraid so," Jesse answers. "If you're hungry, you could come on into work with me and get a filet-o-fish sandwich." [Adapted from author's field notes and experiences, Knoxville, Tennessee, 1998–2001.]

Friendships like that of Iqbal and Jesse are forming more and more often across North America, as ordinary Americans meet and spend time with some of the many migrants from distant countries coming to learn or work, often only temporarily. Immigrants from Asia, Africa, and Middle and South America are bringing new customs and new international sensibilities to even the most remote communities. Rural villages in the mountains of North Carolina have Mexican grocery stores catering to farm laborers from Michoacan in central Mexico. Physicians from India live in eastern Kentucky hill-country towns and treat predominantly Appalachian patients. In Minneapolis, Somali women in Muslim veils lead labor protests for better working conditions in Minnesota's meat-packing plants.

Americans are ambivalent about these new immigrants and guest workers and the major social and economic changes they represent. The family farm is no longer the basic unit of American society, and rural farm communities are shrinking. Manufacturing jobs that once provided a high standard of living are moving out of North America, leaving those who are ill-equipped to take the new high-tech jobs with little but low-wage service employment. A new but still unstable economy is emerging based on sophisticated technology and the high-speed exchange of information, goods, and money at a global scale. There are rapid changes in employment levels, in the location of jobs across the continent, and in the availability of investment funds. Immigrants have a central role in these changes, some entering at the low skill end to take up jobs that Americans no longer wish to fill, others entering at the high skill end to take jobs for which there are too few qualified Americans. These immigrants are changing the face of America, increasing its multiethnic hue.

It is likely that you are a resident of North America. You already have some impressions of this land and of the changes it is undergoing. Please, bring your own experience to the task of defining the evolving character of North America. Think of this chapter as a guide to help you enhance your understanding of what you already know and to construct, over a lifetime, an ever more complex understanding of the geography of North America. Perhaps your experiences with new American residents from faraway places will be part of what you bring to this task. If you grew up in North America, you might want to glance at the map of your childhood landscape that you drew at the beginning of Chapter 1. As you read this chapter, you can place that map in its broader context. You may even see how features of your map—things you noticed as a young child—are related to larger patterns across North America, both physical and cultural.

Themes to Look For

Some central patterns in North American life emerge in this chapter.

1. Mobility. North Americans value the ability to move and to observe different parts of their continent. This behavior is related to the individualism that characterizes North American lifestyles. Links to family and community are important; but migrating to a new geographic location is accepted as a way to achieve individual fulfillment and financial success.

2. Increasing multiculturalism. Many new residents of North America come from Middle and South America and from Asia. According to the 2000 census, Anglo-Americans are minorities in many North American cities, and African-Americans will soon be outnumbered by Hispanic Americans.

3. Changing population distribution. The middle of the country is losing population as older Americans and African-Americans move south and Hispanic Americans and Asian-Americans move into the middle, south, and northeast, as well as other parts of the continent.

4. Similar landscapes; regional differences. All major cities in Canada and the United States have similar suburbs, office towers, malls, ethnic restaurants, and home improvement centers. Yet, despite this sameness, there are physical differences among regions and often stark cultural disparities within regions. Over the past few decades, North Americans have become more, rather than less, culturally and economically diverse.

5. Wealth and poverty. Although Americans as a whole are prospering more than ever, the gap between rich and poor is widening. The number of millionaires has risen sharply, but the 2000 census shows that even though the number below the poverty line has decreased, there are still 37 million poor (13 percent of the population). Children are more likely to be poor than adults.

6. Increasing connections to faraway places and events. On a personal level, Americans can maintain daily relationships with individuals in virtually any country on earth by using e-mail. On a more abstract level, industries contract and expand and jobs are rapidly lost and created as the global economy shifts gears.

7. Growing relationship with Latin America. Two-way trade with Middle and South America and one-way migration into North America, especially from Mexico, are growing.

Terms to Be Aware Of

The region discussed in this chapter consists of Canada and the United States. The term *North America* is used to refer to both countries. Even though it is common on both sides of the border to call the people of Canada "Canadians" and people in the United States "Americans," this text will use the term *United States* rather than America for the United States. The term *American*, then, describes citizens of, or patterns in, both countries, not just the United States. In Chapter 1, we point out that Mexico also might logically be included in North America, and perhaps it will be in the future.

Other terms relate to the growing cultural diversity in this region. For example, the text uses both the terms **Hispanic** and **Latino**. *Hispanic* is a loose ethnic (not racial) term that refers to all Spanish-speaking people from Latin America and Spain. Hispanics may be black, white, Asian, or Native American. *Latino* is an alternative term for the same group. In Canada, the **Québecois,** or French Canadians, are an ethnic group that is becoming assertively distinct from the rest of Canada. They are the largest of an increasingly complex mix of minorities in that country, most of whom are still content to be called simply Canadians.

Diversity is reflected in this group of children as they enjoy a birthday party in the Bunker Hill development, Boston's largest public housing facility. [Joel Sartore/National Geographic Image Collection.]

THE GEOGRAPHIC SETTING

PHYSICAL PATTERNS

The continent of North America begins near the Arctic Circle and stretches southward to the Isthmus of Tehuantepec in southern Mexico. You will find it easier to learn the physical geography of this large and complex territory if you break it into a few smaller segments, as described in the following section on landforms.

Landforms

The most dramatic and complex North American landform is the wide mountainous mass in the west known as the *North American Cordillera* (see the map at the beginning of this chapter). It sweeps down from the Bering Strait in the far north, through Alaska, and all the way to the Isthmus of Tehuantepec in Mexico. (See the map at the beginning of this chapter.) The *Rocky Mountain zone* of Canada and the United States is a major part of this broad swath of mountains and basins. Along the eastern edge of the continent stretches a less rugged formation called the *Appalachians*. Both mountain ranges resulted from the collision of tectonic plates. About 200 million years ago, when the supercontinent Pangaea began to pull apart (see Figure 1.7), the broad band of mountains along the western edge of the continent was created as the Pacific Plate pushed against the North American Plate. The Pacific Plate continues to slip to the northwest, the North American Plate to the southeast. These plate movements cause the earthquakes that are common along the Pacific coast of North America. The much older

and hence more eroded and lower Appalachians resulted from earlier collisions of the North American Plate with North Africa.

Between these two mountain ranges lies a huge, irregularly shaped *central lowland* of undulating plains that stretches through the middle of the continent from the Arctic to the Gulf of Mexico. The central lowland was created by the deposition of material eroded from the mountains. This material was carried to the region by wind and rain and by the rivers that flow from east and west into the Mississippi. These deposits can be several kilometers deep.

Over the last 2 million years, glaciers that formed during periodic ice ages covered the northern portion of the lowlands (as well as adjoining mountain ranges to the west and east) and extended south well into what is now the central United States. The most recent ice age ended between 15,000 and 10,000 years ago. The ice, sometimes as much as 2 miles (about 3 kilometers) thick, scoured the very old exposed rock surface of the Canadian Shield in the far north and picked up sediment from lands to the south, then dumped the sediment as it melted. The Great Lakes are the depressions left by glacial scouring, as are the smaller lakes, ponds, and wetlands that dot Minnesota, Wisconsin, Michigan, and much of central Canada. In the upper Mississippi Valley, some of the sediment dumped by the melting glaciers has been picked up and redeposited by the wind. This wind-deposited layer of soil, called **loess,** is often many yards deep; it has proved particularly suitable for large-scale mechanized agriculture.

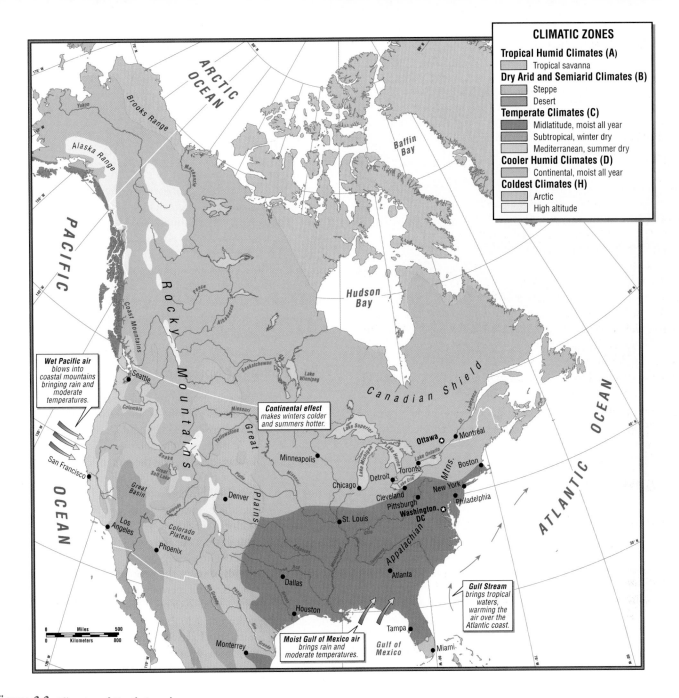

Figure 2.2 Climates of North America.

East of the Appalachians, a *coastal lowland* that is narrow in the north but widens into a broad band toward the south stretches from Virginia to Florida (see the chapter opening map). From Florida, the coastal lowland sweeps to the west, joining the southern reaches of the central lowland along the Gulf of Mexico. In most of this area the slope is nearly flat, and there is a transition zone between land and sea characterized by swamps, lagoons, and sandbars. Much of this coastal lowland was formed from huge loads of silt deposited by North America's rivers during frequent floods. The Mississippi Delta has been growing for more than 150 million years; it began at what is now the junction of the Mississippi and Ohio rivers and has advanced 1000 miles (1600 kilometers) south, filling in much of the lower reaches of the central lowland and creating most of the land of the states of Louisiana and Mississippi.

Human activities such as deforestation, deep plowing, and heavy grazing have added to the silt load of the rivers.

Climate

Every major type of climate on earth, except for the truly tropical, exists in North America. This huge expanse of land experiences widely different temperature patterns from north to south, and different amounts of humidity from the dry continental interior to the moist marine fringes. North America's landforms contribute to its climatic variety by influencing the movement and interaction of air masses.

Look at the climate map (Figure 2.2) and observe along the Pacific coast a thin band of climates that are cool and moist in the

59

Flood control upstream has decreased deposition rates in the Mississippi Delta wetlands. One result is that salt water is introduced into bogs along the delta and kills the cedar trees. (See also Figure 2.20, page 90.) [Annie Griffiths Belt/Corbis.]

north and warmer and drier in the south. Moist air passes onto land from the Pacific Ocean. When this moist air encounters the coastal mountain ranges, it rises, cools, and dumps frequent rain along the windward (western) slopes. Climates are much drier to the east because air masses tend to hold what moisture they have left as they move eastward across the Great Basin and Rocky Mountains. In this arid region, many dams and irrigation projects have been built at tremendous expense to make agriculture and large-scale human habitation possible.

On the eastern side of the Rocky Mountains, the main source of moisture is the Gulf of Mexico. When the continent is warming in the spring and summer, moist air is pulled in from the Gulf. This air interacts with cooler, drier air masses moving from the north and west, often creating violent rainstorms. Generally, the central part of North America is wettest in the east and south and driest in the north and west. Along the Atlantic coast, moisture is supplied by the Gulf Stream, a warm ocean current that flows up the eastern seaboard from the tropics. It brings a supply of warm, wet air that often creates high humidity and rain, and sometimes hurricanes.

The huge size of the North American continent creates large temperature variations in the continental interior. Because land heats up and cools off more rapidly than water, temperatures in the continental interior are hot in summer and cold in winter. Proximity to water generally modifies air temperature extremes, so temperatures are moderated on wide fringes of the continent by the adjacent oceans.

The geographic distribution of the landforms and climates of North America has influenced the settlement patterns of the continent and the behavior, occupations, and problems of the

people living there. We now discuss the historical geography of North America.

HUMAN PATTERNS OVER TIME

We can view the human history of North America as successive arrivals and dispersals of people across the vast continent. These movements began with the original peopling of the continent from Eurasia, and they continued with waves of European immigrants (and, in some cases, enslaved Africans and their descendants) who settled the country from east to west from the 1600s on. Today, immigrants are arriving from Asia and Latin America. The patterns of these movements have resulted in a set of distinct subregions. Today, internal movement is also a defining characteristic of life for North Americans, who are among the most mobile people in the world.

The Peopling of North America

Humans first came to North America from Siberia in northwestern Asia. Small bands of hunters crossed over the Bering land bridge into Alaska between 25,000 and 14,000 years ago, during the last ice age. At that time, the polar ice caps were considerably thicker; sea levels were lower; and the land bridge was a huge, low landmass more than 1000 miles (1600 kilometers) wide. Temperatures rose about 10,000 years ago, and as the ice caps melted, the land bridge sank below the rising sea. By this time, hunter-gatherers had fanned out across the North American continent, throughout Central America, and well into South America, occupying virtually every climate region.

Over thousands of years, the peoples settling in the Americas domesticated plants, built permanent shelters, and sometimes created elaborate social systems. The cultural traditions they developed were distinctive to the regions in which they lived. One group living around the Lake Superior basin about 5000 years ago fashioned copper tools, weapons, and ornaments that eventually were traded up and down the Mississippi River. About 3700 years ago, another group constructed what is believed to be the first large North American settlement, at Poverty Point in Louisiana— a place that became home to perhaps 5000 inhabitants. The domestication of corn is thought to be closely linked to settled life and population growth in North America. Corn was introduced from Mexico into the southwestern desert about 3000 years ago, along with other Mexican domesticates, particularly squash and beans. Even before corn arrived in the eastern woodlands, people had domesticated many native plants that produced nutritious seeds, among them *Cucurbita pepo* (a gourd) and *Helianthus annus* (sunflowers).

These foods provided surpluses that allowed some community members to specialize in activities other than agriculture, hunting, and gathering. The impressive harvests made possible large, city-like regional settlements. For example, by A.D. 200, such distinctive culture groups as the Anasazi, Hohokam, and Mongollon had emerged in the Southwest. These people built villages of attached adobe houses and irrigated their crops. By A.D. 1000, the urban settlement of Cahokia (in what is now central Illinois) covered

5 square miles (12 square kilometers) and was home to an estimated 30,000 people.

The Arrival of the Europeans. North America today bears little resemblance to the land that the Native Americans knew in A.D. 1500, after hundreds of generations of occupation. The climate and landforms are roughly the same, but virtually every other aspect of the continent's geography, from the vegetation to the cultural orientation of the population, has been transformed. The major reason for this transformation was the sweeping occupation of North America by Europeans, beginning in the seventeenth century.

From the early 1600s to the present, settlements have rolled over North America like an unstoppable tide, fed by masses of immigrants from Europe and smaller numbers from Africa and Asia. In the early seventeenth century, the British established colonies along the Atlantic coast in what is now Virginia (1607) and Massachusetts (1620). After the early 1600s, English-speaking colonists built villages, towns, port cities, and plantations up and down the east coast of North America. By the mid-1800s, they had occupied most Native American lands throughout the Appalachian Mountains and into the central part of the continent.

The Disappearance of Native Americans. The rapid expansion of European settlement was assisted by the long biological and technological isolation of Native American populations from the rest of the world. Native Americans had little resistance to diseases, such as measles and smallpox, carried by Europeans. By 1500, these diseases were no longer common killers in Europe, Africa, and Asia because the people there had built up immunity to them. European diseases were particularly deadly to Native Americans, however, killing an average of 90 percent of any given group within 100 years of contact. Conflict between encroaching Europeans and resisting Native Americans also took a large toll. The Europeans were aided by technologically advanced weapons, including primitive firearms, steel swords, and lances, and by trained dogs and horses. Often the Native Americans had only bows, poisoned arrows, and throwing stones, although by the late 1500s Native Americans in the Southwest had acquired horses from the Spanish and begun to breed them and use them in warfare against the Europeans.

Simple numbers tell the general effect of European settlement on Native American populations. There were roughly 18 million Native Americans on the continent in 1492. By 1542, after only a few incursions into North America by Spanish expeditions such as that led by Hernando de Soto, there were half that many. By 1907, slightly more than 400,000 remained.

The European Transformation

European settlement erased many of the landscapes familiar to Native Americans and imposed new ones that fit the needs of the new occupants. Those needs varied according to the opportunities offered by a particular physical environment and the cultural backgrounds of the settlers in that area. Hence distinct regions, which still exist today, developed as settlement proceeded.

The Southern Colonies. European settlement began in Virginia in 1607 with the establishment of the British colony of Jamestown. By the late 1600s, the southern colonies of Virginia, the Carolinas, and Georgia were dependent on the cultivation of cash crops such as tobacco and rice, much of it grown on large holdings known as plantations. Because land was readily available even to the poorer settlers, large plantation owners could not easily secure the large, stable labor force necessary for their huge operations. The unfortunate solution was the use of enslaved African labor. African workers were first brought to North America at Jamestown in 1619. Although slavery was not widely accepted at first, within 50 years enslaved Africans were becoming the dominant labor force on some of the larger southern plantations. By the start of the Civil War in 1861, slaves made up about one-third of the population throughout the southern states.

In many ways, the plantation system was detrimental to the economic development of the South. Much of its wealth was concentrated in the hands of a small class of plantation owners, who made up just 12 percent of Southerners in 1860. The planter elites invested their money in Europe and in the more prosperous northern colonies instead of at home. More than half of Southerners were poor white subsistence and small cash-crop farmers who consumed most of what they could produce and had little money to invest. Because both slaves and poor whites were forced to live simply, their consumption needs did not provide a market for goods, and so there were few trading towns. Plantations tended to be self-sufficient. Some food was produced, a few staple commodities were imported, and planters used off-season labor to produce or repair equipment such as wagons and tools, breed draft animals, and sew garments for the plantation's residents. As a result, plantations generated few of the independent spin-off enterprises—such as transport and repair services, shops, garment making, and small manufacturing—that could have invigorated the regional economy. The plantation economy declined after the Civil War (1861–1865), and the South sank deeper into poverty, remaining economically underdeveloped well into the twentieth century.

There are many lingering effects of the southern plantation era; such cultural relics as place names, architecture, and landholdings are examples. Among the most enduring and significant effects is the fact that in 2001, the largest concentrations of African-Americans are still found in the southeastern states where the plantation economy, which used the labor of enslaved Africans, was centered (see Figure 2.14, page 84).

The Northern Colonies. Throughout the seventeenth and eighteenth centuries, relatively poor subsistence farmers dominated agriculture in the northern colonies of New England and southeastern Canada. In contrast to the southern colonies, there were no plantations and few slaves, and not many cash crops were exported. Instead, farmers often augmented their incomes with exports of timber and animal pelts. Some communities were completely dependent on the rich fishing grounds of the Grand Banks, located off the shores of Newfoundland and Maine. By the late seventeenth century, a few industries were already producing goods both for local consumption and for export. By the eighteenth century, metalworks and pottery, glass, and textile factories were turning out products in Massachusetts, Connecticut, and Rhode Island. They supplied markets in North America but were also beginning to export to British plantations in the Caribbean.

Industry began to flourish in the early nineteenth century, when Francis Lowell designed a water-powered loom. By 1822, "factory girls" were working in the textile mills of Lowell, Massachusetts, and living in company-owned housing. Southern New England, especially the region around Boston, became the center of manufacturing in North America, drawing immigrant labor from French Canada and Europe. It retained that status into the second half of the nineteenth century.

The Economic Core. New England and southeastern Canada were eventually surpassed in numbers of people and in wealth by the mid-Atlantic colonies of New York, New Jersey, and Pennsylvania. This region benefited from more fertile soils, a slightly warmer climate, multiple deepwater harbors, and better access to the resources of the interior. By the end of the Revolutionary War (in 1783), the mid-Atlantic region was on its way to becoming the **economic core,** or the dominant economic region, of North America.

Both agriculture and manufacturing boomed in the early nineteenth century, drawing immigrants from much of northwestern Europe. Their farms produced grain in such vast quantities that the region had been called the "bread colonies" in pre-revolutionary times. The tremendous success of agriculture laid the groundwork for industrial development. Food-processing industries packaged meat and turned grain into cereal, flour, and bread. As farmers prospered, they bought mechanized equipment, appliances, and consumer goods, much of it made in nearby cities. The richest farmers put up money themselves to build the factories. Port cities such as New York, Philadelphia, and Baltimore prospered because they were well positioned to trade industrial products to the vast continental interior. This relatively small, early core region yielded almost a third of U.S. output on the eve of the Civil War (1860).

Throughout the nineteenth century, the borders of the economic core expanded as connections with interior areas deepened. Rivers, canals, railways, and roads connected the big cities of the mid-Atlantic coast with the entire Great Lakes area, as well as with the area that became the states of Ohio, Michigan, Indiana, Illinois, and Wisconsin (Figure 2.3). In these more western states of the economic core, the early settlers were farmers

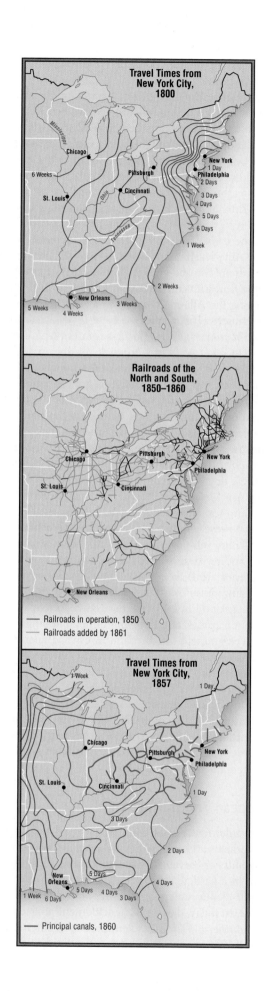

Figure 2.3 **Nineteenth-century transportation.** With the building of the railroads in the decade before the Civil War, the mobility of people and goods increased dramatically across the eastern half of the United States. In 1800, it took a day to travel from New York City to Philadelphia and a week to go to Pittsburgh. By 1857, the travel time between New York and Philadelphia was only two or three hours and to Pittsburgh was less than a day. People could travel farther, easily reaching principal cities along the Mississippi River, and the travel was less expensive and onerous. Because of their ability to move troops and materials quickly, the railroads played a significant role in the North's victory in the Civil War. [Adapted from James A. Henretta, W. Elliot Brownlee, David Brody, and Susan Ware, *America's History,* 2d ed. (New York: Worth, 1993), pp. 400–401.]

who supplied food to growing urban populations. By the mid-nineteenth century, the economy was increasingly based on the steel industry, which diffused from the eastern end of the economic core to the Great Lakes industrial cities of Cleveland, Lorraine, Gary, and Chicago. Local deposits of coal, iron ore, and (later) oil provided power and raw materials for the steel industry. That industry stimulated the mining of coal and iron ore, as well as the manufacture of a wide range of goods needed on farms and in urban households: tools, plows, carriages, pottery, iron cooking and heating stoves, and many others. As mechanization proceeded, heavy farm and railroad equipment, including steam engines, became important products of this region.

By the late nineteenth century, the economic core stretched from the Atlantic to Chicago and from Toronto to Cincinnati. This rapidly industrializing region dominated the other parts of North America economically and politically well into the twentieth century. Most other areas produced food and raw materials for the core's markets and depended on the core's factories for manufactured goods. See pages 96–97, 100 for a detailed discussion of the rise and decline of this region.

Expansion West of the Mississippi

By the mid-nineteenth century, the land in the densely settled eastern parts of the continent had become too expensive for new immigrants. By the 1840s, immigrant farmers and older settlers were pushing their way west across the Mississippi River.

The Great Plains. Much of the land west of the Mississippi was dry grassland, or prairie, which seemed desolate and strange to people whose cultural roots were in the well-watered fields and forests of northwestern Europe and the eastern United States. Some settlers interested in crop farming founded homesteads on the Great Plains and adapted to the lack of trees and water (average annual rainfall is 20 inches or less). The soil usually proved amazingly fertile in wet years. But the naturally arid character of this land eventually created an ecological disaster for Great Plains farmers. After 10 dry years, a series of devastating dust storms in the 1930s blew away topsoil by the ton; animals died, and entire crops were lost. Many farm families packed up what they could and headed west to California and other states on the Pacific coast.

Settlement of the Western Great Plains and Beyond. While the Great Plains filled up, other settlers were alerted to the possibilities farther west. The Louisiana Purchase in 1803 had doubled the size of the United States, and the Lewis and Clark expedition of 1804–1806 reached the Pacific coast along the Columbia River in the Oregon country. Settlers were drawn to the valleys of the Rocky Mountains (especially the Great Basin) and to the well-watered and fertile coastal zones of the Oregon country and California. The great rush to Oregon began in the 1840s and lasted until the 1860s. Over the 20-year period, more than 350,000 people walked the Oregon Trail, their possessions carried in wagons that wore 3-foot-deep ruts still visible today. Although farming was the major occupation in the Oregon country, settlers also turned their attention to the abundance of trees, which filled the rising demand for building materials. Soon logging became the dominant industry in the Pacific Northwest, as

vast stands of enormous redwoods, Douglas firs, spruce, and many other species were clear-cut throughout the region. (**Clear-cutting** is the cutting down of all trees on a given plot of land, regardless of age, health, or species.)

Among the most successful new settlers in the West were the Mormons, a group of utopian Christians also known as the Church of Christ of the Latter-Day Saints. The Mormons began a trek across the continent in the 1830s from western New York. Their leader was Joseph Smith, a visionary who espoused hard work, saving, risk taking, and strict patriarchal rule by the church elders. In Illinois, he received the revelation that polygamy—the taking of more than one wife—was advisable. Many church members dissented from this opinion, and the group split up. One faction of 10,000 that favored polygamy, led by Brigham Young, decided to leave the limits of the United States, in part because polygamy was illegal there. They crossed the Great Plains and settled around the Great Salt Lake in Utah. There, using communal labor, communal water rights, and a system of planned agricultural communities, they demonstrated that the semiarid Great Basin region could be farmed with the use of irrigation. The Mormons abolished polygamy in 1890, six years before Utah became a state, but the practice persists among a few Mormons.

Settlement in the Southwest. The Southwest was first colonized by Spaniards, who at the end of the sixteenth century established sheep ranches and missions in the area that today forms central and southern California, southern Arizona, New Mexico, and Texas. Settlements were few and poorly connected; therefore, as settlers from the United States expanded into the region, Mexico found it increasingly difficult to retain its claimed territory. By 1850, nearly all of the Southwest was under U.S. control, and the new state of California in particular was attracting farmers and laborers. By the twentieth century, a vibrant agricultural economy was developing in California, where a mild Mediterranean climate made it possible to grow vegetables almost year-round. With the advent of refrigerated railroad cars in the early twentieth century, these vegetables could be sent to the major population centers of the East. Massive government-sponsored irrigation schemes greatly increased the amount of cultivable land. Whole rivers were diverted to supply fields and urban developments hundreds of miles away, totally transforming the natural course of water drainage in Southern California and much of the West.

European Settlement and Native Americans

Those Native Americans living in the eastern part of the continent who survived early European incursions often occupied land that Europeans wished to use. These Native Americans were continuously pushed westward. Eventually those who remained were settled on relatively small reservations with few resources on the western side of the Mississippi River. The best known organized relocation event was the Trail of Tears. In the late 1830s, European settlers pushing into Georgia and across the Appalachian Mountains hoped to acquire land at low cost. The Creek and Cherokee, whose lands the settlers wanted, were by then assimilated to European ways. The Cherokee had built roads, schools, and churches, and most earned a living by farming or raising cattle. Some owned ferries and grain mills and lived in substantial brick homes of European style. The European settlers prevailed upon President Andrew Jackson to sign the Indian Removal Act passed by Congress in 1830. The Cherokee resisted by filing suits that reached the U.S. Supreme Court twice, but in May of 1838, General Winfield Scott and 7000 soldiers of the U.S. Army began the removal of the Cherokee to Oklahoma.

Many different reservations were created both east and west of the Mississippi for Native Americans, usually with insufficient attention paid to their needs and customs. Then, especially in the West, as European settlers occupied the Great Plains and prairies, many of the reservations set aside for Native Americans were further shrunk or relocated on ever less desirable land.

Native American populations are now expanding, after plunging from 18 million to less than 400,000 between 1500 and 1900. In 2000, there were almost 2.5 million Native Americans in North America, most living in the United States, where they constitute less than 1 percent of the total population. Ironically, North Americans today often identify with the Native peoples: many now claim some Native American blood, making it difficult to determine who should be called a Native American and just how many there are.

Reservations cover just over 2 percent of the United States, but more than 20 percent of Canada, counting the creation in 1999 of the Nunavut territory in Canada's far north. The Nunavut have won the right to home rule, including an elected legislative assembly, a public service system, and a court. (Most Native groups in Canada do not have legal control over their territory.) Nearly every reservation has insufficient resources to support its population at a standard of living enjoyed by other North Americans. After centuries of mistreatment, many Native Americans are more familiar with poverty than with true Native ways. Many have become severely demoralized. Some have turned to alcohol, and many live as impoverished wards of the state, totally dependent on meager welfare money. Alcohol consumption often leads to family violence, and in the United States, Native American suicide and homicide rates are almost twice the national average.

For many years, Native young people faced a difficult choice: they could remain in poverty with family and community on the reservation, or they could seek a more affluent way of life far from home and kin but would lose touch with traditional ways and might encounter considerable prejudice. In the 1990s, several tribes found avenues to greater affluence on the reservations. Some tribes have sought and received substantial monetary compensation from the government for past losses. Others have opened gambling casinos designed to lure non-Native gamblers. Loopholes in the old treaties allow casinos on reservations in states where local law would otherwise forbid them. Although these few roads to prosperity (especially casinos) can produce enormous income, the stress on weakened tribal structure can be intense. In some cases, graft has diverted funds to a few elites. In others, economic development has been accompanied by environmental pollution, as in the case of the many oil and gas wells and coal and uranium mines located on reservations.

Some tribes have been successful in taking creative approaches to development. The Choctaw in Mississippi have devel-

In the early 1990s, Canadian Mohawks who viewed gambling as cultural pollution strongly opposed the operation of a Mohawk casino across the border in New York State. The protesting Mohawks destroyed some of the U.S. machines, causing a bitter dispute within the Mohawk Nation. [Sarah Leen/Matrix.]

oped factories that produce plastic utensils for McDonald's restaurants, electrical wiring for automobiles, and greeting cards. They employ not only their own people but also 1000 non–Native Americans who come onto the reservation to work. The Salish and Kootenai tribes of Montana have reduced their unemployment to levels below those in nearby communities by creating jobs for themselves in farming, tourism, and recreation. In Canada, Native Americans have a higher standard of living than their U.S. counterparts, and the government spends much more per capita on them. But with the exception of the new Nunavut territory in northern Canada, Canadian tribes have less control over the resources on their lands, which are often developed for use by non–Native Americans (see pages 96 and 104).

The Changing Social Composition of North America

The distinctive regions created by European-led settlement still remain. The boundaries and distinctive characteristics of these regions, however, are increasingly blurred as the process of change continues. In the coming section on geographic issues, we will see that the economic core no longer dominates, as industry has spread to other parts of the country. Some regions that were once dependent on agriculture, logging, or mineral extraction now have high-tech industries as well. In cultural terms, once-distinctive ethnic groups have become assimilated as succeeding generations married outside their groups and moved to new locations. In Los Angeles, for example, one-third of all U.S.-born middle-class Latinos and more than one-quarter of all U.S.-born middle-class Asians marry someone of another ethnic group. Non-European immigrants from

Middle and South America, India, Pakistan, Southeast Asia, East Asia, and Africa are finding a wide variety of social and economic niches as they add to the cultural diversity of their new homeland. Before we look more closely at these emerging patterns, some information about the changing population of North America is useful.

POPULATION PATTERNS

The population map of North America (Figure 2.4) shows the uneven distribution of the more than 316 million people who live

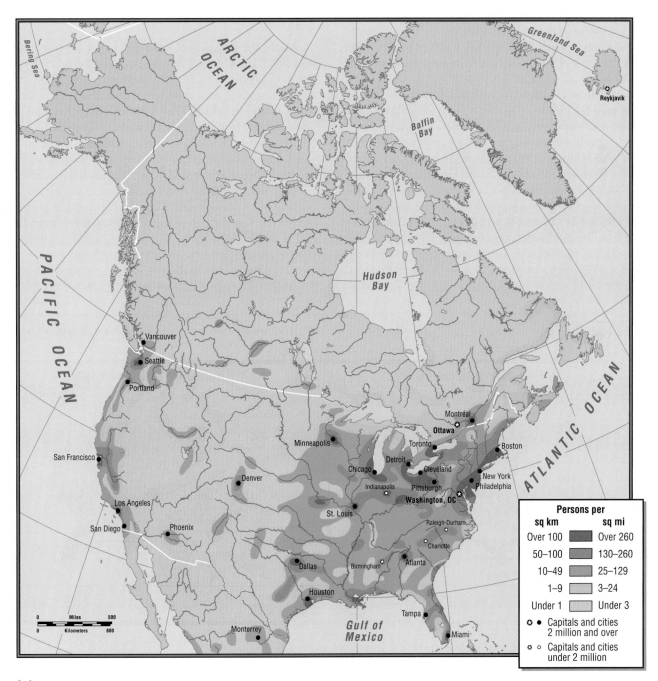

Persons per

sq km		sq mi
Over 100		Over 260
50–100		130–260
10–49		25–129
1–9		3–24
Under 1		Under 3

✿ ● Capitals and cities 2 million and over

✧ ○ Capitals and cities under 2 million

Figure 2.4 **Population density in North America.** [Adapted from *Hammond Citation World Atlas* (Maplewood, N.J.: Hammond, 1996).]

on the continent. Canadians, who make up approximately one-tenth of the population of North America (31 million), live primarily along the southeastern border with the United States (population 285 million). Sixty percent of Canadians live in southern Ontario in the Great Lakes region and in Québec along the St. Lawrence River. The greatest concentration of people in the United States is not far south of the border with Canada, in the northeastern part of the United States. The quadrant marked by Boston and Washington, D.C., along the Atlantic seaboard; St. Louis on the Mississippi River; and Chicago on Lake Michigan contains 7 of the 12 most populous states in the country and several of the largest cities. Population is concentrated in this part of North America because historically this was the economic core. However, the 2000 census shows that other regions of the country are now growing much faster than this one. The Northeast grew by just 5.5 percent in the 1990s, while the West grew by 20 percent and the Southeast by 17 percent.

The Northeast population core is flanked to the south and west by a zone of less dense settlement on rich agricultural land that is dotted with many medium-sized cities, most with populations under 2 million (see Figure 2.4). On the average, rural densities are 3 to 129 people per square mile (1 to 50 per square kilometer) in this landscape of large fields, woodlands, and grasslands. The southern portion of this less dense area contains a number of cities that are growing rapidly and becoming markedly more ethnically diverse: Atlanta; Birmingham; and Charlotte and Raleigh-Durham, North Carolina, for example. Midwestern cities in this zone are also becoming more ethnically diverse with rising populations of Hispanics and Asians in such places as Indianapolis, St. Louis, and Chicago.

To the west and north of this zone, in the continental interior, settlement is very light. In fact, roughly two-thirds of this subregion, including the territory in northern Canada and Alaska, have fewer than three people per square mile (fewer than one per square kilometer). This low population density is mostly a result of mountainous topography; lack of rain; and, in northern or high-altitude zones, a growing season that is too short to sustain agriculture. Nevertheless, there are some population clusters in irrigated agricultural areas such as in the Utah Valley and near rich mineral deposits. Many states within this region grew only very slowly in the 1990s. North Dakota grew the least, by less than 1 percent. At the other extreme, at the southern end of the region, Nevada grew 66 percent, the biggest growth spurt in the United States (see page 109).

Along the Pacific coast, a band of population centers stretches north from San Diego to Vancouver and includes Los Angeles, San Francisco, Portland, and Seattle (see Figure 2.4). These are all port cities engaged in air and maritime trade around the **Pacific Rim** (all the countries that border the Pacific Ocean on the west and east) and in trade with other countries via the Panama Canal. These cities are flanked by agricultural zones, many of them irrigated, that supply a major portion of the fresh produce for North America. Over the past several decades, these cities also have become centers of innovation in computer technology.

Figure 2.5 shows that during the twentieth century, there have been clear shifts in U.S. population patterns from the northeast to the south and west. Since Americans are among the most mobile people on earth, this kind of redistribution of people is common. Every year, in the United States, almost one-fifth of the population relocates. Some are changing jobs, others are attending

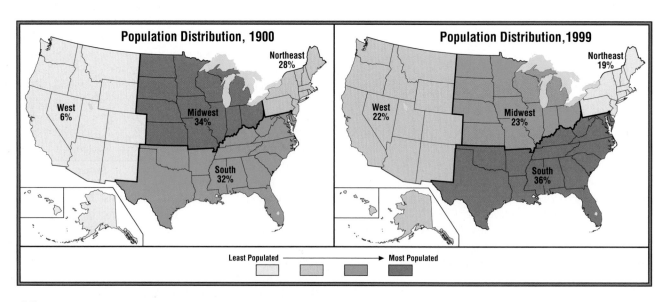

Figure 2.5 **Percentage of population by region in the United States, 1900 and 1999.**
The population of the United States has shifted to the south and west over the past 100 years.
[Adapted from "America's diversity and growth," *Population Bulletin* (June 2000): 12.]

school or retiring to a warmer climate, others are merely moving across town or to the suburbs and countryside. Still others are arriving from outside the region: immigrants enter North America at the rate of about 3000 per day. New residents usually contribute to the local economy by finding work; by buying or renting homes; by becoming taxpayers and consumers; and by increasing the number of representatives in state, provincial, and federal legislatures. They also demand more services and increase the costs of city administration, so gaining and losing residents has important consequences for cities.

People's perceptions of which regions are desirable places to live change over time, so the movement of people ebbs and flows. Why have the South and West of the United States been attracting new residents over the past two decades? There are many factors at work. Migrants are often lured by opportunities for employment. For example, many towns in the Middle West, once magnets for European immigrants, are now losing population as farms consolidate and young people choose careers in cities. Some towns in the Middle West are attempting to recruit new migrants from Middle and South America and central Europe to work in farm-related industries. Southern cities such as Atlanta, Georgia, and Birmingham, Alabama, are growing more rapidly than cities in the old economic core. Atlanta; Birmingham; and Raleigh-Durham, North Carolina, have sprouted satellite or "edge" cities around their peripheries based on technology and international trade. A warm climate may attract the recently retired. You will find additional explanations for these population movements as you read the following sections on geographic issues and individual subregions.

The rate of natural increase in North America (0.6 percent per year) is low, just one-third the rate of the rest of the Americas (1.8 percent). Still, North Americans are adding to their numbers fast enough to double the population in 124 years. Many of the important social issues being debated in North America, and discussed below, are related to population patterns—such issues as legal and illegal immigration, family structure, gender, birth control, the aging of the population, and ethnic diversity.

CURRENT GEOGRAPHIC ISSUES

The citizens of North America live in what is often considered a privileged part of the world. Yet this region faces complex problems posed by a rapidly changing economy and population make-up. Some people fear that the region may not be able to find solutions that maintain the high standard of living currently enjoyed by the vast majority of residents. North America's greater integration with the global economy is visible in nearly every landscape of the region, yet just how these new connections will affect jobs, incomes, and daily life is far from clear. We will examine a series of social, economic, political, and environmental issues as they affect both Canada and the United States.

RELATIONSHIPS BETWEEN CANADA AND THE UNITED STATES

Citizens of Canada and the United States share many characteristics and concerns. Indeed, in the minds of many people (especially those in the United States), the two countries are one. Yet that is hardly the case. In fact, Canadians often make the point that their national identity is focused on how different Canada is from the United States. In this section, we develop a perspective on relations between these two countries. Three key words characterize the interaction between Canada and the United States: *asymmetries*, *similarities*, and *interdependencies*.

Asymmetries

We start with the concept of asymmetry, which means "lack of balance." The two countries occupy about the same amount of space, but much of Canada's territory is sparsely inhabited cold country to the north and west. In about every other way, the United States is much larger than Canada. For example, the U.S. population (285 million) is nearly 10 times the Canadian population (31 million). Although Canada's economy is one of the largest and most productive in the world (U.S. $722 billion in goods and services produced in 1999), the economy of the United States is more than 10 times larger ($9.2 trillion in 1999).

There is also asymmetry in international affairs. The United States is an economic, military, and political superpower with a world leadership role that preoccupies it. Since 1990, U.S. foreign policy has focused primarily on relations with Russia, Europe, China, Japan, Southeast Asia, Southwest Asia (the Middle East), and Middle and South America. Canada is only an afterthought in U.S. foreign policy debates, in part because the country is so close and so secure an ally. But, for Canada, managing its relationship with the United States *is* the foreign policy priority. As a result, Canadians focus more on events and circumstances in the United States and how they should react to them than Americans focus on Canada. These asymmetries are cause for discontent in Canada.

Similarities

Despite the asymmetries, the two countries have much in common. Both are former British colonies, and from this experience they have developed comparable political traditions. Both are federations (of states or provinces) and both are representative democracies; their legal systems are also alike. Not the least of the features they share is the longest undefended border in the world.

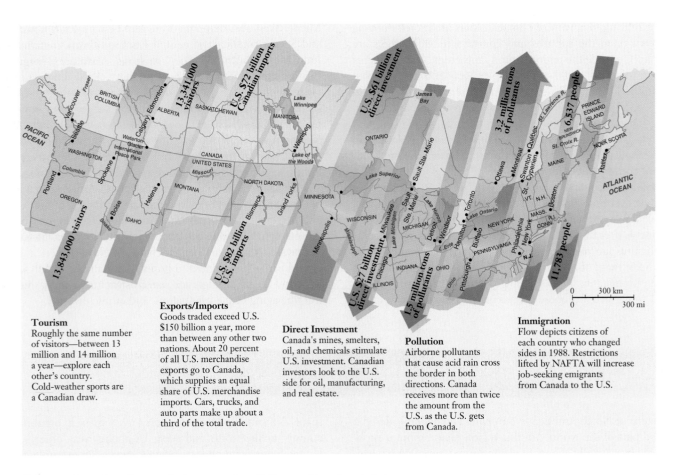

Figure 2.6 Transfers of tourists, goods, investment, pollution, and immigrants between the United States and Canada. [Adapted from *National Geographic* (February 1990): 106–107.]

There are official checkpoints on highways; but for thousands of miles there is not even a fence. In some locations the international boundary runs through bedrooms and backyards. Thousands of residents pass out of one country and into the other many times in the course of a single day simply by going about their usual business of farming, hunting, or housekeeping.

Well beyond the border country, Canada and the United States share many other landscape similarities. Their cities and suburbs look much the same. The billboards that line highways and freeways advertise the same brand names. Shopping malls have followed suburbia into the countryside, drawing millions of affluent buyers away from old urban centers. The same fast-food franchises dispense the same meals of questionable nutrition. And the two countries share similar patterns of ethnic diversity that developed, as we shall see shortly, in nearly identical stages of immigration from abroad.

Interdependencies

Canada and the United States are perhaps most intimately connected by their long-standing economic relationship (Figure 2.6). By 1998, that relationship had evolved into a two-way trade flow of U.S. $307.2 billion annually. Each country is the other's largest

trading partner. Canada sells 80 percent of its exports to the United States and buys 70 percent of its imports from the United States. The United States, in turn, sells 22 percent of its exports to Canada and buys 19 percent of its imports from Canada. Notice that asymmetry exists even in the realm of interdependencies: Canada's smaller economy is much more dependent on the United States than the reverse. Nonetheless, if Canada were to disappear tomorrow, as many as a million American jobs would be threatened. The lopsidedness of the relationships will undoubtedly persist. Because of its size, large population, and giant economy, the United States is likely to remain the dominant partner. As former Canadian Prime Minister Pierre Trudeau once told the U.S. Congress, "Living next to you is in some ways like sleeping with an elephant: No matter how friendly and even-tempered the beast, one is affected by every twitch and grunt."

ECONOMIC AND POLITICAL ISSUES

The economic and political systems of Canada and the United States have much in common. Both evolved from societies based primarily on family farms. After an era of industrialization, both are now increasingly service-based economies with an increasingly important technology sector. Politically, the two countries have similar governments and face similar issues. Yet they often take different approaches to such issues as unemployment and social services.

North America's Changing Agricultural Economy

North America benefits from an abundant supply of food, and exports of agricultural products were once the backbone of the American economy (Figure 2.7). Today, North America's importance as a producer of food for the continent and for foreign consumers remains high, and agricultural exports account for 20 to 30 percent of American farm income. But the overall economic role of agriculture within North America has been shrinking for many years; it now accounts for just 1.7 percent of the region's gross domestic product—this despite the fact that American agriculture is highly productive per worker and per acre.

Agriculture employed 90 percent of the American workforce in 1790 and 50 percent in 1880. European immigrants set up thousands of highly productive, family-owned farms, spreading over much of the United States and southern Canada. These farms provided the majority of domestic consumption of food and fiber and the majority of all exports until 1910. However, agriculture currently employs less than 2 percent of the North American workforce because today's highly mechanized farms need far fewer workers. Instead, farms require heavy capital investment in land and machinery, often a half-million dollars or more per farm. As the farms mechanized, they became substantially more productive, as measured in yield per acre and per worker. But one result of increased yields is declining prices for farm products, making it increasingly difficult for small farms to break even. A successful farm now requires so much investment that large corporations, with their ready access to cash, have an advantage over individual farmers.

Corporate agriculture provides a wide variety of food at low prices for North Americans; nonetheless, in many rural areas,

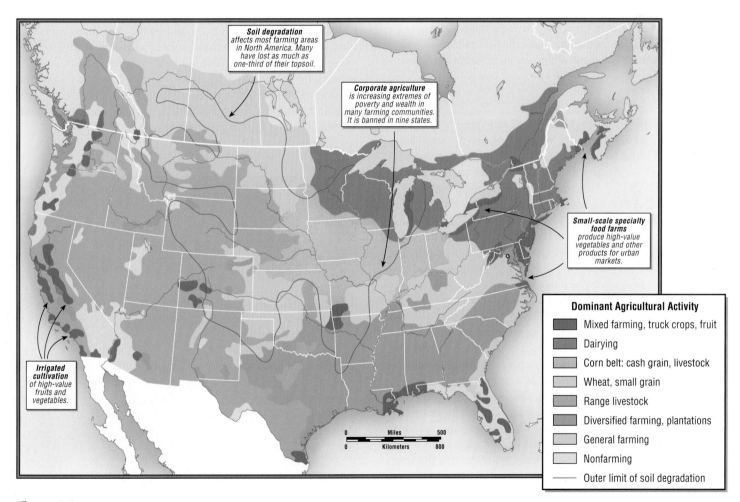

Soil degradation affects most farming areas in North America. Many have lost as much as one-third of their topsoil.

Corporate agriculture is increasing extremes of poverty and wealth in many farming communities. It is banned in nine states.

Small-scale specialty food farms produce high-value vegetables and other products for urban markets.

Irrigated cultivation of high-value fruits and vegetables.

Dominant Agricultural Activity

- Mixed farming, truck crops, fruit
- Dairying
- Corn belt: cash grain, livestock
- Wheat, small grain
- Range livestock
- Diversified farming, plantations
- General farming
- Nonfarming
- Outer limit of soil degradation

Miles 500
Kilometers 800

Figure 2.7 **Agriculture in the United States.** North America is remarkable in that some type of agriculture is possible throughout much of the continent. The major exceptions are the northern reaches of Canada and Alaska and the dry mountain and basin region (the continental interior) lying between the Great Plains and the Pacific coastal zone. Even in the Pacific and continental interior subregions, however, peripheral areas such as southern California, southern Arizona, and the Utah Valley are cultivated with the help of irrigation.
[Adapted from Arthur Getis and Judith Getis, eds., *The United States and Canada: The Land and the People* (Dubuque, Iowa: Wm. C. Brown, 1995), p. 165.]

The rapid industrialization of the South is evident in this recent air photo of the Saturn plant in Springhill, Tennessee. It was laid down over the rural landscape south of Nashville. Old field patterns are still clearly visible within the confines of the new plant. [Saturn Corporation.]

corporate agriculture has depressed local economies and created social problems. Communities in such places as rural Iowa, Nebraska, and Kansas were once made up of farming families with similar incomes, similar social standing, and a commitment to the region. By 2000, farm communities were increasingly composed of a few wealthy farmer-managers amid a majority of poor, often migrant, laborers who work for low wages on corporate farms and in food-processing plants (see the box "Meat Packing in the Great Plains" on page 104). The result is increasing class disparity, low levels of education for many people, and less functional local government. Under the assumption that the family farm has inherent value, the U.S. federal government and some states are increasing efforts to protect family farms and the rural communities of which they are a part. The states of Iowa, Kansas, Minnesota, Missouri, Nebraska, North and South Dakota, Oklahoma, and Wisconsin all have laws that ban corporate agriculture. Family-owned corporations are excepted if family members live on the farm.

Corporate agriculture relies on the use of highly modified seeds and chemical fertilizers, pesticides, and herbicides to increase crop yields, potentially contaminating food and polluting environments. Compared to owner-operators, the small, often salaried labor force of today's farms has a lesser stake in the long-term sustainability of farming. It is easier for them to ignore such environmental effects as erosion and stream pollution. Many North American farming areas have lost as much as one-third of their topsoil. The use of the "low-till" cultivation method, which entails plowing only once every few years, is one strategy for reducing the loss of topsoil. However, low-till methods require

increased use of chemicals to control weeds and pests. An equally controversial effort to reduce dependence on chemicals is to use genetic engineering to create crops that are resistant to pests and diseases. This technique is seen as possibly unsafe in Europe and other parts of the world to which North American farms have traditionally exported their crops.

Changing Transport Networks and the North American Economy

North America is going through dramatic transformations, many of them related to changes in technology and trade. Traditional manufacturing is fading in economic importance, just as agriculture did earlier. A vibrant new economy based on technology is emerging. The old economic core no longer dominates, as new centers of production spring up across the continent. The diffusion of production has been aided in part by improved networks for transportation and information exchange.

From the early years on, an ever-evolving system of transportation and communication has made the movement of goods and people easier and faster. The improvement in transport has had some profound effects on the North American economy. First, increases in the speed and volume of transport have steadily enlarged the number of consumers able to purchase from a given manufacturer, encouraging the growth of industry. Second, changes in transportation have encouraged the gradual dispersion of manufacturing throughout North America rather than its concentration in the former economic core.

This market garden farm in Santa Barbara, California, caters to suburban buyers who pay a seasonal fee for a weekly supply of organically grown fresh vegetables. They come to the farm to pick up their share. [Jim Richardson/Richardson Photography.]

The Manske cousins, who own adjoining farms in Wisconsin, employ contour furrows to curb erosion. They also grow alternating rows of corn (the cash crop) and alfalfa (providing hay for dairy cows) to help control runoff. [Jim Richardson/Richardson Photography.]

These trends have been apparent for at least the past 200 years. By the early 1800s, canals—and the inland lakes and rivers that they connected—replaced roads as the major means for moving goods and people. Between 1817 and 1825, New York State built the Erie Canal, a 40-foot-wide (12-meter-wide) channel of water 364 miles (586 kilometers) long from Albany on the Hudson River to Buffalo on Lake Erie. The canal greatly increased the flow of passengers and cargo. By road, it took four horses one full day to pull a 1-ton load 12 miles. By canal, two horses could pull 100 tons 24 miles in a day. These spectacular savings led to a boom in canal building. Eventually, one could travel by waterway from the Atlantic coast inland to the Mississippi River. By the 1820s, boats powered by steam engines were plying the inland waterways; this improvement in technology again increased the tonnage that could be moved in a given time. Towns located along the waterways, such as Buffalo and Cincinnati, became inland centers of trade and industry, and manufacturers increased their output to reach newly accessible regions.

By 1859, although river traffic remained important, a rapidly growing railroad network had replaced canals as the major carrier of freight and people to the interior (see Figure 2.3, page 62). Railroads could be built more quickly and could reach places to which it was impractical to build canals. This network stimulated a variety of economic activities. As new settlers reached the fresh and remote western lands, they established farms and founded trading towns, using the rails to ship their produce. The railroads were located to connect major cities—Pittsburgh, Cleveland, and Cincinnati, and later Davenport, St. Louis, Kansas City, Omaha, and Denver.

Some towns—such as Davenport, Dubuque, and Cedar Rapids, all in Iowa—became magnets for food-processing operations and manufacturing of farm implements and household goods. Food was sent by water and rail to the eastern cities, and manufactured goods were sent to buyers in the agricultural interior. Specialty services such as grain mills, meat-packing plants, docks, warehouses, and eventually telegraph and telephone services also opened up near rail lines. Thus, in the East and the Middle West, major railroad routes linked a series of industrial and commercial cities. The densest network of rail lines connected cities with their energy and resource suppliers in the economic core in the northeast of the United States and the southeast of Canada. By 1930, rail coverage had spread over most of the more populated eastern half of the continent, though coverage was much less dense to the west, especially beyond the Mississippi River.

The development of the mass-produced and inexpensive automobile in the 1920s changed North American transport once again. Soon trucks were delivering cargo more quickly and conveniently than the railroads could, and the miles of railroad track decreased between 1930 and 1980. The **interstate highway system,** a 45,000-mile (72,000-kilometer) network of high-speed, multilane roads, was begun in the 1950s and completed in 1990. Because this network connected to the vast system of local roads, it introduced previously unimagined flexibility, speed, and low cost to the delivery of manufactured products. Thus, the highway system made possible the dispersal of industry into nonurban locales across the country, away from the old industrial core in the Northeast. The availability of quick delivery virtually anywhere in the country stimulated service and manufacturing industries in the hinterland where labor, land, and living costs were cheaper. By the 1990s, a personal computer could be assembled in Sioux Falls, South Dakota, a thousand miles from primary markets, yet be delivered to its new owner in Sacramento, California, in only 24 hours. Thus, Sioux Falls was not considered unduly remote to be a large-scale national manufacturing center.

Air transport, which got its boost after World War II, also served economic growth in North America, but its special role is business travel. Despite the increasing use in business of instant communication via the Internet, face-to-face contact is still seen as essential to American business culture. To address North America's pattern of many widely dispersed industries in numerous medium-sized cities and to facilitate one-day travel between any two points, air service is organized as a **hub-and-spoke network.** Hubs are strategically located airports used as collection and transfer points for passengers and cargo. Airlines schedule many flights to arrive at a hub within a few minutes of one another. Passengers and cargo then are shifted to flights departing to other hubs or intended final destinations. Although delays can affect the synchronization of the system, causing passengers to miss a connection, other connecting flights are often only a few hours away. In 1999, North American airlines carried nearly 640 million passengers, more than double the continental population. A Gallup survey in 2000 revealed that 80 percent of the U.S. adult population had flown at least once, more than one-third of them in the previous 12 months.

The New Service and Technology Economy

The economic base of North America has recently shifted from manufacturing to a broad and varied **service sector** (economic activity that amounts to doing services for others).

Decline in Manufacturing Employment. As labor unions won higher pay and better working conditions for most workers in the industrial Northeast (economic core), these benefits added to the costs of production. For this reason and others such as the cost of keeping technologically current, the region began to lose its advantageous position. By the 1960s, many companies were moving their factories elsewhere to take advantage of cheaper labor that was less unionized. At first, the southeastern United States was a major destination for these industries, because wages and energy costs were significantly lower there than elsewhere in the country. By the 1980s, however, and especially after the passage of the North American Free Trade Agreement (NAFTA) in 1994, certain industries were moving farther south to Mexico or overseas. In these places labor was vastly cheaper, and laws mandating environmental protection and safe and healthy workplaces were absent or enforced less vigorously.

Another factor in the decline of manufacturing employment is automation. The steel industry provides an illustration. In 1980, huge steel plants, most of them in the economic core, employed more than 500,000 workers. At that time, it took about 10 man-hours and cost about $1000 to produce 1 ton of steel. Spurred by more efficient foreign competitors, the North American steel industry applied new technology in the 1980s and 1990s to lower production costs, improve efficiency, and increase production.

The steel industry as a whole now employs less than half the workers it did in 1980. Revamped plants now produce 1 ton of steel with just 3 man-hours of labor (a 60 percent reduction) at a cost of just $200 per ton. In addition, new technology has allowed the use of cheap scrap metal from many sources (such as junked cars and refrigerators) in smaller operations (minimills) that are located in less urbanized areas (principally the Southeast). These minimills produce steel at the rate of just 0.44 man-hour per ton (a more than 90 percent reduction since 1980) and at a cost of about $165 per ton.

These figures help us to understand the changing employment picture in North America. Far fewer people are now producing more of a given product at far lower cost than was the case 20 years ago. So although the share of the gross domestic product (GDP) produced by manufacturing has declined, the actual level and value of production has not, at least in certain very important sectors, such as steel. It should be noted, however, that in the new era, workers generally are less unionized; often work for lower wages; must put in more overtime (often followed by seasonal layoffs); and, at least in the reorganized steel industry, have somewhat higher accident rates than was the case before reorganization.

Growth of the Service Sector. As factories moved outside the United States and manufacturing employment declined, the economic base of North America shifted increasingly to a broad and varied service sector. As of the year 2000, in both Canada and the United States, about three-quarters of the GDP and a majority of the jobs were in some form of services. In Canada, close to 70 percent of workers—and in the United States, 80 percent—now work in transportation and utilities, wholesale and retail trade, health, leisure, maintenance, government, and education. Many of these activities are connected in some way to international trade and often entail the processing, transport, and trade of agricultural and manufactured products that originate outside North America.

It is customary to subdivide the service sector into two categories, one for those activities just described and the other for the creation, processing, and communication of information. The second category includes such workers as those employed in finance, publishing, higher education, research and development, and professional aspects of health care. The industries that rely on the use of computers and the Internet to process and transport information—banks, software companies, medical technology companies, publishing houses—are increasingly called high-tech or **information technology (IT)** industries (Figure 2.8). They are freer to locate where they wish than were the manufacturing industries of the old economic core, which depended on massive amounts of steel and energy, especially coal. By contrast, technology industries depend mostly on highly skilled thinkers and technicians (as well as on electrical energy), and hence they are often located near major universities and research institutions. Examples of such IT centers include Toronto's Spadina Bus district, near Ryerson Polytechnic; California's Silicon Valley, near Stanford University; the Route 128 corridor near Boston's Massachusetts Institute of Technology (MIT) and Harvard; and Research Triangle near the University of North Carolina. There are IT centers in hundreds of other towns and cities near institutions of higher learning. Patches of IT activity exist around government research labs such as Oak Ridge in Tennessee and Brookhaven on Long Island, New York.

The evolving economy based on information technology relies on a continuous stream of innovation and is dependent on the establishment of networks of individuals, communities, corporate organizations, and government. By the late 1990s, yet another technological change, the Internet, was expanding these networks. The Internet is a computer network that allows for the electronic transfer of all kinds of information virtually instantaneously to any place in the world—information that used to be carried via mail over roads, rails, water, and air. The Internet is emerging as an economic force more rapidly in North America than in any other world region. It was here that the Internet was first widely available, and, for the time being, this is where the largest block of its users live (36 percent of the world's Internet users are in the United States). The *United Nations Human Development Report 2001* indicates that there are 179 Internet hosts per 1000 people in the United States and 108 per 1000 in Canada. In comparison, Europe as a whole has 53 hosts per 1000 people, an increase of 250 percent from the year before (see Figure 4.9, page 199). Finland, a leader in high-tech manufacturing, has one of the highest ratios at 200 hosts per 1000 people; Albania has the lowest ratio at 0.1 host per 1000 people.

The "new economy" of information technology bears some striking resemblances to the "old economy" of industrial production. As we have just observed, over the past several hundred years, North America laid down a network of roads, canals, railroads,

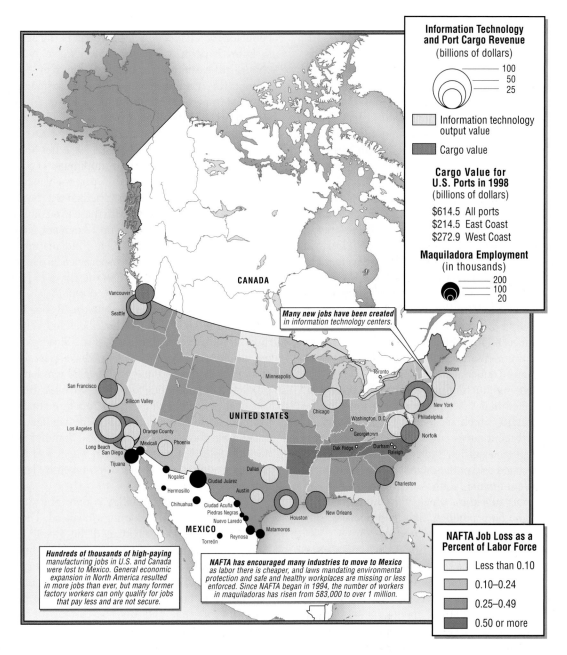

Information Technology and Port Cargo Revenue
(billions of dollars)

100
50
25

Information technology output value

Cargo value

Cargo Value for U.S. Ports in 1998
(billions of dollars)

$614.5 All ports
$214.5 East Coast
$272.9 West Coast

Maquiladora Employment
(in thousands)

200
100
20

Many new jobs have been created in information technology centers.

Hundreds of thousands of high-paying manufacturing jobs in U.S. and Canada were lost to Mexico. General economic expansion in North America resulted in more jobs than ever, but many former factory workers can only qualify for jobs that pay less and are not secure.

NAFTA has encouraged many industries to move to Mexico as labor there is cheaper, and laws mandating environmental protection and safe and healthy workplaces are missing or less enforced. Since NAFTA began in 1994, the number of workers in maquiladoras has risen from 583,000 to over 1 million.

NAFTA Job Loss as a Percent of Labor Force

Less than 0.10
0.10–0.24
0.25–0.49
0.50 or more

Figure 2.8 Economic issues: North America. Trading and industrial patterns for the United States changed during the 1990s. For the first time, U.S. trade with Asia exceeded that with Europe (see Figure 2.9), and New York dropped to the third largest port in value of cargo handled, behind Los Angeles and Long Beach, California. Information technology increased dramatically in value of output, while manufacturing and related jobs decreased—in part due to the effects of NAFTA.

highways, telegraph and telephone lines, and airports to move resources, goods, people, and information. There was a hierarchy of places connected by transport and communications systems. Usually the large cities and highly productive industries were connected early and well. Less influential outlying towns and enterprises were connected later, and rural areas later still. Today, those places first "wired" for information technology tend to be large and wealthy cities, major government research laboratories, and universities on the cutting edge of technology innovations. A spatial

digital divide has developed that leaves small, rural, and poor places out of the loop. Some may wait years for Internet connections. Once the information technology infrastructure is in place, however, distance is much less of a hindrance to participation in the new economy than it used to be in the industrial economy.

New-Technology-Inspired Industry. Across the old industrial region, many outmoded factories are undergoing facelifts so that they can use their skilled manufacturing labor force for new-technology activity. For example "Polymer Valley," near Akron,

Ohio, is switching from manufacturing tires to producing a wide range of chemically based products, from plastic furniture to prosthetics. These are products developed in nearby research universities and private laboratories, and they signal an economic turnaround for the region. Similar industrial facelifts and efforts to develop IT business communities are to be found in Omaha, Nebraska; Huntsville, Alabama; Tulsa, Oklahoma; Cincinnati, Ohio; Louisville, Kentucky; Raleigh-Durham, North Carolina; and Toronto, Ontario, and Québec City, Québec, in Canada.

Globalization and the Economy

Another major force for economic change in North America is globalization: the process of worldwide economic linkages achieved in part through the global reduction of barriers to trade.

NAFTA and Free Trade. In the past, trade barriers have been important aids to North American development. For example, upon achieving independence from Britain in 1776, the new U.S. government imposed tariffs and quotas on imports and gave subsidies to domestic producers to protect fledgling domestic industries and commercial agriculture. As a result, the economic core region flourished. But if tariffs and quotas are kept in place too long, they can result in stagnation—products of poor quality and prices that are too high. The United States is now an active partner in the World Trade Organization, which seeks to lower trade barriers. Nonetheless, some tariffs and quotas against low-priced manufactured goods of developing countries (such as textiles and shoes) remain in effect. The stated role of these tariffs is to prevent these less expensive goods from driving North American producers out of business. **Government subsidies,** which still cover part of the production costs of some farm products, also help domestic producers to sell their goods for less than foreign competitors.

Within the confines of the United States, free trade unrestricted by tariffs or quotas has long been valued. In recent years, Canada and the United States have promoted the idea of free trade in the whole of North America, because tariffs and quotas are now seen as constraints on economic growth that raise prices for consumers and discourage competition among producers. The process formally began with the U.S.–Canada free trade agreement of 1989. The creation in 1994 of the North American Free Trade Agreement (NAFTA) brought in Mexico as well. Today, NAFTA is the world's second largest trading bloc both in size of population and in dollar value of trade (the European Union is the largest). A major long-term goal is to increase the level of trade among Canada, the United States, and Mexico, not necessarily to balance that trade. By the time of its full implementation in 2009, NAFTA will have largely integrated the economies of the United States, Canada, and Mexico by removing or reducing most trade barriers. Part of the process is the establishment of uniform standards. For example, Mexican, Canadian, and U.S. trucks eventually will be equipped with brakes and pollution control devices that meet agreed-upon standards so that they can move freely in all three countries.

It is not possible at this time to assess the overall effects of NAFTA with any accuracy, partly because data has not yet been collected for a long enough time, and partly because it is difficult to know if perceived changes are due to NAFTA or to other changes in world trade. The formation of an economic union in Europe (the European Union) and new trading blocs in South America (especially Mercosur; see Chapter 3), as well as a recession in Asia, changed the dynamics of the North American economy during the 1990s at the same time that NAFTA came into effect. Also, the effects of NAFTA have not been uniform across North America. For example, although U.S. trade with both Canada and Mexico increased 15 percent between 1994 and 1999, the benefit to individual states varied greatly. Some 20 states significantly increased their trade with these two countries, four by more than 10 percent: New Mexico up 37 percent, Kentucky up 35 percent, Colorado up 29 percent, and Indiana up 13 percent. Thirteen states, often poorer ones, experienced a decline, three by more than 10 percent: District of Columbia down 15 percent, Mississippi down 14 percent, and Delaware down 13 percent. The effect of NAFTA on employment is also obscure. By one informed estimate, more than 440,000 U.S. jobs have been lost because of the movement of industries to Mexico and the closing of firms that could not compete with companies in Mexico (see Figure 2.8). At the same time, perhaps as many as 660,000 jobs have been created by economic opportunities forged by NAFTA. But these jobs were usually in very different locations from those that were lost, and the people who took them tended to be younger and more highly skilled than those who had lost jobs. Some of the new jobs were low-skill jobs that paid minimum wages and carried no benefit packages, such as the fast-food job taken by Jesse in the vignette that opens this chapter.

The Asian Link. NAFTA is only one way that the North American economy is becoming globalized. The general lowering of trade barriers has encouraged the growth of trade between North America and Asia (Figure 2.9). During the 1990s, trade with Asia surpassed trade with Europe, long North America's chief trading partner. A graphic example of the increasing links with Asia are the many Japanese automotive plants that dot the landscape of rural mid-America east of the Mississippi River. Japanese companies sought plant locations in North America to be near their most important pool of car buyers: commuting North Americans. They chose rural locations to save money on land, taxes, and utility costs but sought property that was close to the interstate highway system and not far from major cities. In such places, Japanese companies find a ready, inexpensive labor force among those who value life in rural America, yet do not wish to farm. Many are willing to commute 20 miles or more for secure jobs that pay reasonably well and include health and retirement benefit packages.

The Toyota Camry plant constructed during the 1980s in Georgetown, Kentucky, is an example of such an enterprise. Georgetown lies north of Lexington in the heart of the Kentucky Bluegrass country. The Toyota site is adjacent to I-75 and I-64, which form a major interstate highway juncture in the nation's midsection. The plant is Toyota's largest outside Japan and employs 7500 people, who turn out nearly 500,000 vehicles and engines per year (not all Camrys). Several dozen smaller Japanese

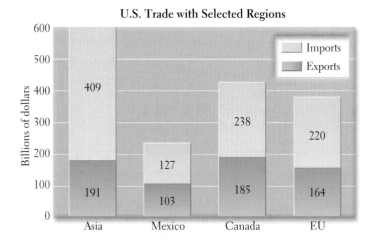

Figure 2.9 U.S. trade with selected regions, 1996. In the 1990s, U.S. trade with Asia surpassed that with Europe for the first time in history.

plants in central and western Kentucky produce parts for the main plant.

The Toyota plant has brought social and physical change to Georgetown. The arrival of Japanese executives and their families brought new ethnic diversity and required adjustments in the public schools, such as the teaching of Japanese and of English as a second language. New religious institutions were built for those Japanese who practice Buddhism. The utility supply (water, electricity) had to be increased greatly to accommodate the new plant and labor force, and new businesses were established to cater to those workers.

Toyota—which also has plants in Princeton, Indiana; Buffalo, West Virginia; Cambridge, Ontario; and other locations—is only one of the Japanese auto companies located in mid-America. Honda, Nissan, Subaru, Mazda, and Mitsubishi all have plants in the province of Ontario, Canada, and in such places as Normal, Illinois (Mitsubishi), and East Liberty, Ohio (Honda), east of the Mississippi.

Foreign manufacturers locate their plants in North America when they are in the business of creating expensive and difficult to transport products that are intended for the American market. American labor is expensive, so it would not be economically sound for foreign companies to manufacture goods in North America for resale abroad.

Canadian and U.S. Responses to Economic Change

The U.S. and Canadian governments have responded differently to the displacement of workers by economic change. For many decades, Canada has strongly protected its working population from job loss or declines in income by enacting strict unemployment insurance laws. It also has spent more than the United States on social programs that lessen the financial burdens of working people in times of economic crisis. Although these poli-

cies have made the financial lives of working people more secure, they have also made Canada slightly less attractive than the United States to new businesses. For example, Canadian employers must pay somewhat higher taxes and sometimes higher wages, and they are required to be more restrained in firing workers. Consequently, they are more hesitant to create new jobs, and as a result Canada's unemployment rate is usually several percentage points higher than that of the United States and its rate of economic growth slightly lower. On the other hand, there are significant social problems associated with the U.S. approach, which provides little job protection or government unemployment assistance. Many poor rural and urban areas in the United States that have lost agricultural and manufacturing jobs have experienced increases in ill health, violent crime, drug abuse, and family disintegration—all of which are social problems connected with declining incomes and the lack of a social safety net.

One important part of a strong social safety net is health care. The Canadian health-care system covers 100 percent of the population. It is heavily subsidized by the government and is less expensive per capita and more effective in preventing illness than the largely private health-care system in the United States. Although the United States spends 13 percent of its gross domestic product on health care, Canada spends only 10 percent. Nevertheless, Canada does better than the United States on most indicators of overall health, such as infant and maternal mortality and overall life expectancy (Table 2.1). Having guaranteed health care removes considerable pressure from those who lose their jobs, even temporarily.

Systems of Government: Shared Ideals, Different Trajectories

Canada and the United States have similar democratic systems of government, but there are differences in the way power is divided between the federal governments and provincial or state governments. There are also differences in the way the division of power has changed since each country achieved independence. Both countries have a federal government with an executive branch, a legislature, and a judiciary. In Canada, however, the executive branch is more closely bound to the legislature, and the Canadian federal government has more and stronger powers (at least constitutionally) than does the U.S. federal government. Over the years, both the Canadian and U.S. federal governments have moved away from the original intentions of their constitutions. Canada's originally strong federal government has become somewhat weaker, whereas the initially more limited federal government in the United States has extended its powers. The main motivation for the weakening of Canada's central government has been demands by provinces for greater autonomy over local affairs. The most assertive effort for provincial control has come from the French-speaking province of Québec, where many people feel overwhelmed by the rest of Canada. In the past decade, several referenda on political independence for Québec have narrowly failed (see the discussion of the Québec separatist movement on pages 95–96).

TABLE 2.1 *Health-related indexes for Canada and the United States*

Country	Health care costs as percent of GDP, 1998	Percent fully insured	Deaths per 1000	Infant mortality per 1000 live births	Maternal mortality per 100,000 live births	Life expectancy at birth (years)
Canada: publicly funded health care	10	100	7	5.5	4	79
United States: privately funded health care	13	66	9	7.0	8	77

Sources: 1998 World Population Data Sheet and Statistical Abstract of the United States 1997, Population Reference Bureau; from http://www.census.gov/prod/www/abs/cc97stab.html and http://www.thirdworldtraveler.com/Health/O_Canada.html.

The U.S. federal government's original source of power was its mandate to regulate trade between states. Over time, this mandate has been interpreted ever more broadly. By the late twentieth century, the federal government was affecting life at the state and local levels primarily through its ability to dispense federal tax monies. The federal government has been able to influence local policy through grants to school systems, federally assisted housing and mortgage programs, the establishment of military bases, anticrime legislation, and the building of interstate highways. The interstate highway system was 90 percent funded by the federal government in accordance with its mandate to facilitate interstate commerce. Ten percent of the funds came from the states linked by the system. Money for these various federal programs is withheld if state and local governments do not conform to federal standards. This practice has made some poorer states dependent on the federal government, but it also has encouraged some state and local governments to enact more enlightened laws than they might have done otherwise. In the 1960s, federally funded programs promoted civil rights for southern African-American citizens by providing enriched educational programs for minority children. Eventually, federal money was made available for programs promoting adult literacy, employment training, job opportunities, and community development for disadvantaged African-American and Hispanic communities. Several U.S. Supreme Court decisions also promoted racial integration in the schools and the workplace.

One example of how broadly federal powers have been interpreted is the case of the Tennessee Valley Authority (TVA) in the southeastern United States. In the 1930s, the federal government attempted to guide the economic revival of a seven-state region bordering and including Tennessee by establishing the TVA. This agency was to promote social and economic development in an area that was plagued by frequent flooding, widespread poverty, and highly eroded agricultural and mined lands. The TVA built a large network of hydroelectric dams and, later, a heavily subsidized coal-fired and nuclear power system. The new power sources attracted industry, and social programs helped to change farming and mining practices. The extension of federal power to undertake such a large project was (and remains) controversial even within the region. However, the seven-state region now enjoys security from floods, inexpensive energy, and relative prosperity, although pockets of persistent poverty can be found throughout the area, especially in Appalachia and the Cumberland Plateau.

Gender in National Politics and Economics

There are some powerful political contradictions in North America with regard to gender. Women apparently decided the U.S. presidential election of 1996, voting overwhelmingly for Bill Clinton; and such women's issues as family leave, equal pay, day care, and reproductive rights have been forced into the national debates. However, men still hold 70 percent of the top executive positions in business and government, and they earn about 30 percent more, on average, than women in those positions. Of the 535 people elected to the U.S. Congress in 2000, just 72, or 13.4 percent, are women (up from 12.5 percent in 1998). At the state level, 22 percent of legislators are women, and the six states with the highest percentage of women legislators all lie in the West and Southwest. In the Canadian Houses of Parliament, 22.7 percent of the members are women, and about 10 percent of provincial elected officials are women. A joint study in 1998 by U.S. and Canadian political scientists shows that in the United States, women are most likely to be elected in rural districts where politics are conservative and the competition for the position is weak. In Canada, on the other hand, women tend to be elected in urban districts where the electorate is multicultural and liberal on gender issues. Consequently, urban women are underrepresented at the state level in the United States, and rural women are underrepresented at the provincial level in Canada. Both the United States and Canada are well behind several other countries in percentages of women in national legislatures: Cuba (27.6), South Africa (28), Mozambique (30), Norway (36.4), and Sweden (42).

Within the broader workforce, women earn only 75 percent of what men earn, but this is up from 59 percent in 1970. Women tend to work more hours than men, however, and their longer working hours contribute to the gain. In the United States, it is estimated that if women and men earned equally, the poverty rate would be cut in half. Although women are now about half the labor force and own more businesses than ever before, they still represent only 38 percent of total business owners in the United States; and their businesses tend to be smaller and less secure, in part because it is harder for women to obtain loans. In education, North America sends more of its women to college than does any other developed region, and Canadian women have attained virtually the same level of education as men in most categories. In the United States, women are still behind: only 19.2 percent of U.S. women hold an undergraduate degree, while 24.8 of men do. This imbalance is likely to disappear, because by the late 1990s women were earning slightly more undergraduate degrees than were men.

SOCIOCULTURAL ISSUES

North America is undergoing rapid sociocultural change, and much of this change has geographic aspects. In this section, we discuss some especially widespread changes that are likely to persist into the future. First we examine a set of topics—urbanization, immigration, race and ethnicity, and religion—that are related to the fact that North Americans are an increasingly diverse people living in concentrated settlements where their differences can make it difficult to find common ground. Then we discuss two issues—the American family and aging—that illustrate changing social organization in this large, complex, and prosperous region.

Urbanization

More than 80 percent of North Americans now live in **metropolitan areas,** meaning cities of 50,000 or more plus their surrounding suburbs. In the nineteenth and early twentieth centuries, North American cities consisted of dense inner cities and less dense suburban peripheries. For many decades, central cities have been losing population and investment while particular kinds of zones in urban peripheries have been growing rapidly. When suburban development began in the early 1900s, workers were drawn out of crowded cities to the suburbs by the opportunity to raise their families in single-family homes with large lots, in secure and pleasant surroundings. Some continued to travel to the inner city to work, usually on streetcars, then increasingly in private cars. After World War II, factories and firms seeking less expensive land on which to develop, as well as better access to transport facilities (highways and airports), followed their workers to the suburbs. They have been joined in the past two or three decades by the high-tech businesses of the new IT economy. Eventually, huge tracts of inner-city land that once held manufacturing industries were abandoned (with an accompanying loss of tax income to the city). These old industrial sites, often contaminated with chemicals and covered with obsolete structures, are called **brown fields** because their degraded conditions pose obstacles to redevelopment.

Also left behind were the least skilled and least educated citizens, often racial and ethnic minorities. In the early 1990s, 70 of the largest U.S. cities had white majorities, but by 2000 almost half of the largest U.S. cities had nonwhite majorities. The majority population in these cities is a mixture of African-Americans, Asians, Hispanics who are nonwhite, and other groups who identify themselves as nonwhite. As whites and some African-Americans moved to the suburbs and urban fringe, immigrants (especially Hispanics) took their places in the inner city. Often those people who remain in the inner cities are in great need of the very services—medical, social support (including churches and synagogues), and educational—that have moved with more affluent people to the suburbs.

Geographers James Fonseca and David Wong have observed that urban and suburban densities continually fluctuate. Dense **nodes** (small regions with extremely dense populations) may appear in urban peripheries where high-tech industries attract thousands of skilled workers. These workers occupy suburban high-rise office and apartment buildings as well as condominiums and single-family homes. Nodes can also reappear in old urban centers and then regenerate into larger dense settlements when businesses and new industry locate on old industrial brown fields that have been rehabilitated. In this way, portions of the old industrial Northeast (Baltimore and Hackensack, New Jersey, for example) are becoming densely populated again, not so much from population growth as from population redistribution. People are moving to the city to take newly available jobs and because of new preferences for urban living. Often, nodes of density follow streets and avenues extending out of the inner-city hub, resembling spokes on a wheel. The rim of the wheel may be formed by interstate beltways that girdle many American cities.

During the 1990s, a significant number of people moved from densely occupied cities to lightly settled areas beyond the urban fringe (farther out than the suburbs). Most of these new rural residents remain employed in urban areas; others work at home via computers or as skilled craftspeople. Many have chosen to lower their standards of living for the simpler pleasures of rural life. Very few have taken up farming, although some raise a small cash crop such as strawberries or tobacco or raise a few animals as a sideline.

The growth of cities and the spread of their suburbs has resulted in two common urban phenomena: the megalopolis and urban sprawl. **Urban sprawl** refers to the encroachment of suburbs on agricultural land. A **megalopolis** is created when several cities grow to the extent that their edges meet and coalesce.

Megalopolis. Megalopolis is the name chosen by French geographer Jean Gottman in the 1960s to describe the 500-mile-long (800-kilometer-long) band of urbanization that stretches across portions of 10 states from Boston, through New York City, Philadelphia, and Baltimore, terminating south of Washington, D.C. Although Megalopolis covers less than 2 percent of the area of the United States, it contains nearly 20 percent of the U.S. population.

There is a continuing battle between urban developers and those who want to preserve landscapes such as wetlands. In 1969, the Hackensack (New Jersey) Meadowlands Development Commission was charged with overseeing development while protecting nearly 20,000 acres of wetlands. Today railroad yards, toxic waste sites, highways, power plants, and other industrial developments have eaten up more than 11,000 of those wetland acres. Hackensack itself is part of the Boston-to-Richmond megalopolis in the Northeast. [Melissa Farlow/National Geographic Image Collection.]

Megalopolis is an informal region in that it has no official standing. It is not a supercity but rather a collection of independent towns and cities whose outer fringes meet. There is no center, no centralized network of services, no shared governmental institutions. In fact, the region crosses literally tens of thousands of administrative units: federal, state, county, township, municipal, and neighborhood. The residents of Megalopolis share a common experience of high population densities, traffic congestion, and high costs of living, especially for housing. But they also enjoy many choices of occupation, housing, neighborhoods, social activities, restaurants, shopping, entertainment, and education. Megalopolis also contains rural landscapes where farmers produce high-quality specialty foods for the cities. In the absence of unifying government, Megalopolis is tied together mainly by a network of interstate highways, railways, and airways superimposed on a complex web of local streets and roads.

Why did Megalopolis develop where it did? One reason was the convoluted coastline of the Atlantic seaboard with its many protective peninsulas and deepwater inlets that are useful seaports. Another reason was topography. Some of the smaller cities developed along rivers just where the foothills (the piedmont) of the Appalachian Mountains meet the rather narrow coastal plain. The abrupt change in elevation created a line of waterfalls and rapids that could not be crossed by boats—commonly known as the **fall line.** Goods had to be off-loaded to other means of transportation, creating a demand for warehouses and labor to shift cargo. The fall line also created a natural waterpower source for milling grain and sawing lumber. Also, the situation of Megalopolis on the edge of a massive and rapidly developing continent provided an unusually rich milieu for long-term growth. The farms and settlements in the vast interior of the continent created a market for all manner of manufactured products shipped through Megalopolis, first from Europe and later from other points in North America. Agricultural products and other goods from the interior found their way into the world market through the coastal cities. The cities of Megalopolis also produced their own products and ideas for global exchange. Even today, 30 percent of U.S. export trade passes through the main ports of Megalopolis: Boston, New York, Newark, Philadelphia, and Baltimore.

Megalopolis is contending both with the results of its own long-term success and with changing patterns of world production and trade. Because it developed early, much of its built environment is now in decline, including its infrastructure of industrial plants, power and other utilities, waste treatment plants, port facilities, and interurban rail and interstate highway systems. Global industrial competition has reduced the earnings of companies located in the cities of Megalopolis. Tax revenues have shrunk just when governments need those revenues to rebuild and update the infrastructure that will keep them competitive. Ironically, some of this competition is coming from other megalopolises that have appeared at Tokyo-Kobe-Osaka in Japan, São Paulo and its satellite cities in Brazil, and Mexico City–Puebla in Mexico.

Urban Sprawl. Among the many consequences of industrialization and urbanization in the economic core and elsewhere is the invasion of farmland by bulldozers preparing the way for suburban development: residences, malls, and discount outlets. Typically, real estate developers find rural land cheaper to develop than city properties, even when inner-city property is occupied only by defunct factories and abandoned houses. The rural land is taxed at a lower rate (at least at the beginning of the development process), is less polluted, and requires less preparation for development than inner-city land. However, the huge suburban housing and mall tracts gobble up farm and wilderness areas around the peripheries of North American cities. This trend is especially problematic for farmers. Even though good fertile land is abundant in North America, some of the best is located close to urban areas because cities and towns were often intentionally founded near rich farmland.

VIGNETTE   Mark Greene's family has been farming in Pittsford, New York (Figure 2.10), near the Canadian border, since 1812, but until 1997 the prospects that his 400-acre farm would be in business for another generation looked dim. As land prices rose, so did property taxes, and the Greene family found that it could not meet its tax payments. New suburban homes were sprouting up on what had been neighboring farms, and

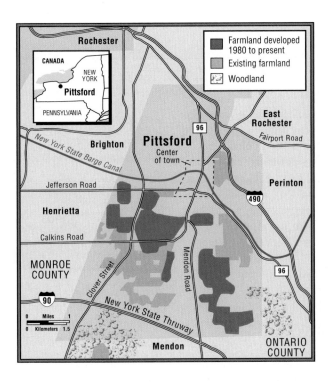

Figure 2.10 The Pittsford, New York, area. [Adapted from *The New York Times* (March 20, 1997): D19.]

the homeowners sometimes pushed local officials to halt normal farm practices, such as noisy nighttime harvesting or planting, spreading smelly manure, or importing bees to pollinate fruit trees. In 1997, though, Pittsford issued $410 million in bonds so that it could pay Greene and six other farmers for promises that they would not sell their 1200 acres to developers but would continue to farm.

Advocates of farmland preservation argue that farms provide more than food and fiber. They provide environmental benefits; soul-soothing scenery; economic diversity; and, especially, tax savings. Saving prime farmland from suburban development helps the nation by reducing food imports and by protecting against volatility in food prices. Urban sprawl eats up 2 acres a minute, a million acres a year, including 400,000 acres of land that is especially well suited for high-quality specialty crops.

Immigration

Most Americans have roots in some other part of the globe. The ancestors of many immigrants entered the North American continent along the eastern seaboard in the seventeenth and eighteenth centuries, most coming from the British Isles and France. Those who settled in what became the United States also arrived in large numbers from Germany, Italy, and Spain, and from many parts of West and Central Africa via the slave trade. After 1880, more and more people migrated from central Europe (Austria-Hungary, Poland, Ukraine, Russia) and southern Europe (Italy, the Balkans,

Greece). Some went to Canada, but most went to the United States. In the nineteenth and early twentieth centuries, a few thousand Asians came to the Pacific coast of North America as low-paid laborers on farms and in mines and to build the new western railroads.

Immigration declined between the two world wars (1921–1939) and then rose dramatically after 1960, with increasing immigration from Latin America, the Caribbean, and Asia. Figure 2.11 shows the proportion of immigrants to the United States from different areas of the world and how the distribution changed over time. The data groups all Asian immigrants together, so we cannot identify how many came from specific Asian countries, such as India, Pakistan, Indonesia, Vietnam, or the Philippines.

Once in North America, the various immigrant groups tended to settle in particular parts of the continent. Early arrivers were particularly attracted to rural agricultural locations that offered affordable land. Because immigrants from the British Isles and France already occupied eastern Canada and the northeastern United States, later waves of immigrants from Europe tended to settle

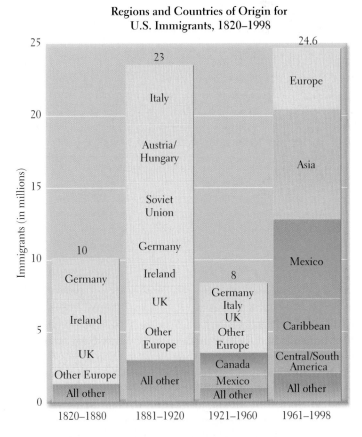

Figure 2.11 National origins of U.S. immigrants, 1820–1998.
[Adapted from Philip Martin and Elizabeth Midgley, "Immigration to the United States: Journey to an uncertain destination," *Population Bulletin* 49 (September 1994): 25; Martha Farnsworth Riche, "America's diversity and growth," *Population Bulletin* 55 (June 2000): 11; and Table 2, "Immigration by region and selected country of last residence: Fiscal years 1820–1998," *1998 Statistical Yearbook of the Immigration and Naturalization Service*, p. 10.]

GEOGRAPHER IN THE FIELD

Interview with Regional Consultant for Chapter 2,
Stanley Brunn, University of Kentucky

What led you to become interested in North America?
I became interested in the region having grown up in different parts of a half-dozen states in the rural Middle West. Observing the many interesting variations in rural and urban economies, ethnic population, political culture, and physical environments stimulated my curiosity about other parts of the United States and Canada.

Where have you done fieldwork in North America, and what kind of work have you done?
Fieldwork to me includes analyzing current or archival records found in libraries, survey data from individuals, and Internet resources, all of which are sources of insight when studying topics or problems within a region. The region under study may be a rural area, a city, a state, or a portion of a much larger region. I have studied the economic functions of small towns in Ohio, African-American population and economic issues in Michigan, political conflicts aroused by controversial land uses in Michigan, the impact of the 1979 Three Mile Island nuclear power reactor accident in Pennsylvania, marriage-mate selection patterns in Appalachia, and the worldviews of southern seminaries. I have also investigated, often with others, regions of growth and no-growth, sparsely populated areas, and patterns of ethnic diversity in U.S. cities.

What is a topic that you think is important to study in your region?
Transborder geographies. Here I am thinking of the border regions between the United States and Mexico and between the United States and Canada. NAFTA is changing the economic face of rural areas, small towns, and cities along the U.S.–Mexico border, espe-

cially with the arrival of new investments and workplace settings. Population growth, especially of youthful populations, presents new challenges to those providing health care, education (including language instruction), and various social services, not only to the legal, but also illegal, residents. The political influence of growing border populations in the United States and Mexico is going to increase, and issues of citizenship, endangered-species protection, and water quality and availability will take center stage. The U.S.–Canada transborder geographies will likely take some different turns. The Pacific Coast region (from California to British Columbia to Alaska) is becoming more closely tied together both economically and culturally. Other developing transborder regions are Canada's Atlantic Provinces and the New England states, Ontario and the Upper Midwest states, and the Prairie Provinces and the northern Great Plains.

The European heritage of both the United States and Canada is becoming less significant; Latin populations are growing in the Sunbelt, and Asian populations on the Pacific Coast. Might one result be some sort of political realignment? Will transborder regional political parties and agendas emerge that promote closer economic and political ties among those along the U.S.–Mexico border, the Pacific Northwest, the northern Atlantic, Upper Great Lakes, and Great Plains–Rocky Mountains? It will be worth observing to see if support grows for transborder environmental, economic, and social agencies. Will transborder attempts to address these common issues diminish the importance of national decision making and increase the significance of the transborder regions that are in the making?

---

farther west. Scandinavians went to the U.S. upper Midwest—to Minnesota and the surrounding states. Many Germans settled in a wedge across the central Midwest, from Ohio to Missouri and in Nebraska and the Dakotas. Other Germans went to southern Appalachia and to south Texas. When early immigrants from Europe sent for friends and relatives, word of an attractive American location would spread to adjacent villages in the homeland. Immigrants from other parts of the world still do the same thing, which explains why people from particular places tend to be concentrated in certain urban neighborhoods or rural communities. For example, shrimp fishermen from coastal Vietnam have congregated in south Texas, where they can ply their trade.

Canada shares the same general immigration pattern as the United States, but specific settlement patterns may vary from

those in the United States. The English, Irish, and Scots tended to settle in the Maritime Provinces bordering the Atlantic coast or around the Great Lakes and on farms in the Prairie Provinces. The French settled primarily along the lower St. Lawrence River, and their descendants remain concentrated in and near Québec. Canada never had a slave-based economy, so Canadians with some African background have roots among those who escaped to Canada from the United States during slavery or who later came voluntarily from the United States, the Caribbean, or Africa. They tend to live in the cities of Ontario and Québec. During most of the twentieth century, Canada encouraged immigration to fill up its empty lands. After farmers from the British Isles came, others from eastern Europe, including Russia and Ukraine, settled mostly on the prairies. By the twenty-first century, as in the United

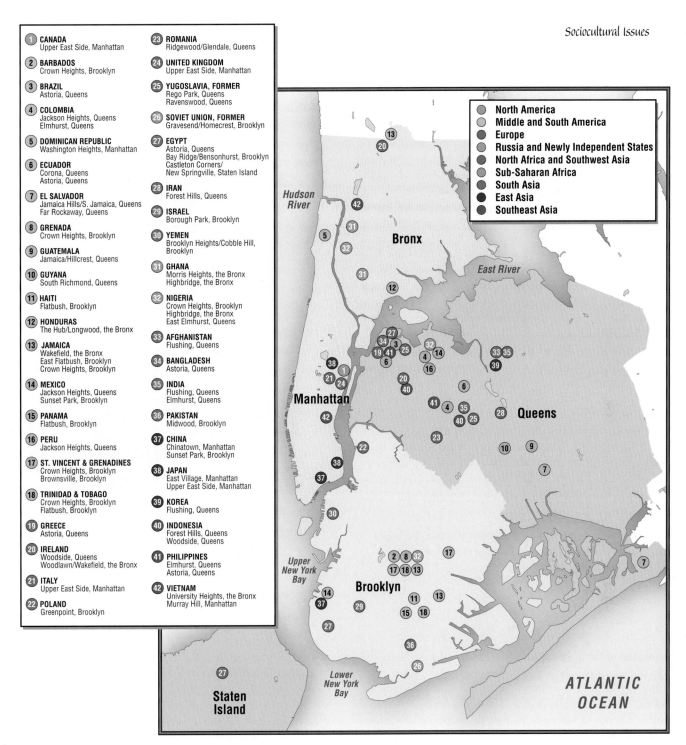

1 CANADA
Upper East Side, Manhattan
2 BARBADOS
Crown Heights, Brooklyn
3 BRAZIL
Astoria, Queens
4 COLOMBIA
Jackson Heights, Queens
Elmhurst, Queens
5 DOMINICAN REPUBLIC
Washington Heights, Manhattan
6 ECUADOR
Corona, Queens
Astoria, Queens
7 EL SALVADOR
Jamaica Hills/S. Jamaica, Queens
Far Rockaway, Queens
8 GRENADA
Crown Heights, Brooklyn
9 GUATEMALA
Jamaica/Hillcrest, Queens
10 GUYANA
South Richmond, Queens
11 HAITI
Flatbush, Brooklyn
12 HONDURAS
The Hub/Longwood, the Bronx
13 JAMAICA
Wakefield, the Bronx
East Flatbush, Brooklyn
Crown Heights, Brooklyn
14 MEXICO
Jackson Heights, Queens
Sunset Park, Brooklyn
15 PANAMA
Flatbush, Brooklyn
16 PERU
Jackson Heights, Queens
17 ST. VINCENT & GRENADINES
Crown Heights, Brooklyn
Brownsville, Brooklyn
18 TRINIDAD & TOBAGO
Crown Heights, Brooklyn
Flatbush, Brooklyn
19 GREECE
Astoria, Queens
20 IRELAND
Woodside, Queens
Woodlawn/Wakefield, the Bronx
21 ITALY
Upper East Side, Manhattan
22 POLAND
Greenpoint, Brooklyn
23 ROMANIA
Ridgewood/Glendale, Queens
24 UNITED KINGDOM
Upper East Side, Manhattan
25 YUGOSLAVIA, FORMER
Rego Park, Queens
Ravenswood, Queens
26 SOVIET UNION, FORMER
Gravesend/Homecrest, Brooklyn
27 EGYPT
Astoria, Queens
Bay Ridge/Bensonhurst, Brooklyn
Castleton Corners/
New Springville, Staten Island
28 IRAN
Forest Hills, Queens
29 ISRAEL
Borough Park, Brooklyn
30 YEMEN
Brooklyn Heights/Cobble Hill, Brooklyn
31 GHANA
Morris Heights, the Bronx
Highbridge, the Bronx
32 NIGERIA
Crown Heights, Brooklyn
Highbridge, the Bronx
East Elmhurst, Queens
33 AFGHANISTAN
Flushing, Queens
34 BANGLADESH
Astoria, Queens
35 INDIA
Flushing, Queens
Elmhurst, Queens
36 PAKISTAN
Midwood, Brooklyn
37 CHINA
Chinatown, Manhattan
Sunset Park, Brooklyn
38 JAPAN
East Village, Manhattan
Upper East Side, Manhattan
39 KOREA
Flushing, Queens
40 INDONESIA
Forest Hills, Queens
Woodside, Queens
41 PHILIPPINES
Elmhurst, Queens
Astoria, Queens
42 VIETNAM
University Heights, the Bronx
Murray Hill, Manhattan
North America
Middle and South America
Europe
Russia and Newly Independent States
North Africa and Southwest Asia
Sub-Saharan Africa
South Asia
East Asia
Southeast Asia
Hudson River
Bronx
East River
Manhattan
Queens
Upper New York Bay
Brooklyn
Lower New York Bay
Staten Island
ATLANTIC OCEAN

Figure 2.12 **Ethnic neighborhoods of New York City.** New York City's newest residents come from around the world and have created discrete neighborhoods within the city. [Adapted from *The New York Times Magazine* (September 17, 2000): 44.]

States, the origin of the immigrants had shifted away from Europe to Latin America, the Caribbean, and Asia. By the 1990s, 55 percent of Canada's new immigrants came from Asia. Many of these Asian immigrants are educated and affluent. They are particularly attracted to Canada's Pacific coast.

Diversity and Immigration. For both Canada and the United States, the relatively recent influx of large numbers of immigrants from areas other than Europe has challenged long-held assumptions that the dominant culture of North America would forever be derived from Europe, with but a few small minorities of Native Americans, African-Americans, Hispanics, and Asians. Figure 2.12 shows the home countries of the 794,000 legal immigrants who came to New York City from 1990 to 1996. Like most new immigrants, they chose neighborhoods where earlier arrivals from their

AT THE LOCAL SCALE   *Ethnicity in Toronto*

The lines dividing Toronto's ethnic neighborhoods aren't strict or exclusionary, but when a city bus moves through town, Italians get off at particular stops, Chinese at others, and Greeks at yet others. Many of the neighborhoods have been rehabilitated and upgraded, or gentrified, to attract the middle-class children and grandchildren of much poorer laborers who arrived earlier in the twentieth century. There are about 400,000 Torontonians of Italian descent, many of whom live in Corso Italia (the Little Italy of Toronto), an area of shops, sidewalk cafés, and elegant homes and apartments. The Chinese community of 350,000 is the fastest growing. Many Chinese who arrived in the 1990s are prosperous businesspeople from Hong Kong who feared the consequences of China's resumption of control of Hong Kong in June 1997. Koreans own many of the fresh produce and grocery stores throughout the city, and Koreatown is home to herb shops and import emporiums. The Greek community revolves around the National Bank of Greece and the Greek Orthodox Church. Jews and people from the Caribbean are sprinkled in small enclaves throughout the city, while Polish people often settle together and specialize in owning bakeries and butcher shops. All in all, there are said to be at least 80 different ethnic neighborhoods in the city. [Adapted from Richard Conniff, "Toronto," *National Geographic* 189(6) (June 1996): 120–139.]

Throughout Toronto, city street signs are written in English, the lingua franca, as well as in the predominant language of the neighborhoods—in this case, Vietnamese. [David R. Frazier/Tony Stone Images.]

homelands had settled. The result is an ethnic patchwork across the boroughs of New York City.

In the United States, immigration and cultural diversity are topics of considerable public debate. The census of 2000 revealed that whites are now a minority in half the nation's 100 largest cities and that immigrants from Middle and South America and Asia have moved into the small cities and towns of the Middle West, the South, and the West Coast. For a variety of reasons, which we explore below, there is a growing sentiment that future immigration should be controlled. Of the two countries, Canada seems to be more receptive to the arrival of newcomers. In recent decades, Canadians have accepted with equanimity many refugees from Asia and Africa. More than in the United States, emigrants to Canada are allowed, even encouraged, to retain their culture, use their native languages in schools, and maintain fairly strict (and usually voluntary) residential segregation. Toronto, Ontario, provides a model of how Canadians accommodate ethnicity (see the box "Ethnicity in Toronto").

Do New Immigrants Cost U.S. Citizens Too Much Money?
Many people believe that immigrants cost taxpayers money be-

cause they use public services such as food stamps and welfare. Several studies have shown that in the long run, immigrants contribute more to the economy than they cost. Most immigrants start to work and pay taxes soon after their arrival. Even immigrants who draw on public services tend to do so only in the very first few years after they arrive. For example, in Los Angeles County, new immigrants make up 25 percent of the population (most are from Mexico and various parts of Asia). These 2.3 million people paid $4.3 billion in taxes in 1991–1992. They received $947 million in county services, less than one-quarter of the taxes they paid.

In fact, immigrants play an important role as taxpayers. As the North American population ages, more people will be drawing social security, and the money to pay their benefits will come out of the pockets of young and middle-aged workers. Yet the base of young workers is shrinking because those now reaching retirement age had relatively few children. The social security contributions of new immigrant workers, who tend to be young adults, will be essential to older Americans.

Do Immigrants Take Jobs Away from U.S. Citizens? Some people in the United States, particularly professional people, are occasionally in competition with highly trained immigrants for jobs. Usually, however, this competition is not a big problem, because there are not enough trained native-born people to fill these positions. The computer technology industry, for example, regularly recruits abroad in such places as Bangalore, India, where there is a surplus of high-tech workers. Immigrants may compete for jobs with the least educated semiskilled American workers, and it is possible that the large pool of immigrant labor drives down the wages of some Americans. Usually, however, immigrants with little education fill the very lowest paid service and agricultural jobs that U.S. workers have already rejected. Although such immigrants are often willing workers, the wages they earn may be too low to support a decent way of life.

Are Too Many Immigrants Being Admitted to the United States? Several circumstances have caused some people to call for curtailing immigration into the United States. First, after a lull at midcentury, the immigration stream has picked up again. In the decade of the 1990s, the rate of immigration appeared to increase faster than ever before (this increase may be partly a result of changes in census coverage); 13.3 million immigrants entered the United States, and 6.4 million children were born to them. These 19.7 million immigrants and children accounted for 78 percent of population growth in the 10-year period. At present rates, by the year 2050, the U.S. population will be over 400 million, nearly one-quarter larger than it would be if all immigration stopped now. Second, illegal immigration reached unprecedented levels during the 1980s and 1990s. This trend is worrisome, because illegal immigrants tend to lack skills, their numbers are uncontrolled, and they are not screened for criminal background. On the other hand, because they must hide from the authorities, they do not partake of public programs, so their drain on the economy is probably not great. Finally, for some people, an added concern is that so many of the new immigrants are from Latin America and Asia. Figure 2.13 shows the projected change in ethnic composition of the U.S. population from 1996 to 2050. Will Euro-Americans and African-Americans lose their sense of home in a United States of America where the largest minority is Hispanic, where Spanish is a prominent second language, and where another large minority is from Asia? Will the U.S. national identity be diluted by so many newcomers from different backgrounds? Or can a new national identity be constructed that will be more useful in the emerging global society? One answer to these questions is that the new immigrants from non-European places are adding talent, entrepreneurial vigor, and creativity to the U.S. cultural mix. Ethnic diversity is already contributing significantly to U.S. culture by enriching friendships, entertainment, business opportunities at home and abroad, and political perspectives. U.S. leadership in international affairs is enhanced by a less monolithic Euro-American presence at conferences and negotiating sessions.

Race and Ethnicity in North America

Although North America is ethnically and racially diverse, the term *race* has most often been applied to the situation and life experience of those who trace their ancestry in whole or in part to Africa. Deeply embedded discrimination and lack of opportunity have clearly hampered the ability of African-Americans to reach social and economic parity with Americans of other ethnic groups, despite the legal removal of many of the formal barriers. African-Americans as a group experience lower life expectancies, higher infant mortality rates, lower levels of academic achievement, higher poverty rates, and greater unemployment than do other Americans, although Hispanics of all racial backgrounds have similar difficulties. In recent years, there has been a tendency to include Hispanics, Asians, Native Americans, and Pacific Islanders in discussions of race and ethnicity, because these other minority groups are growing as a proportion of the population. In the next few years, African-Americans, whose numbers are growing only slowly, will no longer be the largest minority group in the United States. They will be overtaken by Hispanics, who, because of a higher birth rate and a high immigration rate, increased by 58 percent in the 1990s. Asians, who in 2000 made up only 3.6 percent of the U.S. population, nonetheless increased by 48 percent. Figure 2.14 shows the distribution of ethnic minorities in the United States. Figure 2.15 shows the discrepancies in income among Asians, whites, Native Americans, Hispanics, and African-Americans.

There is considerable public debate over the causes and significance of the racial and ethnic differences in income. Over the past few decades, many African-Americans and Hispanics have joined the middle class and achieved success in the highest ranks of government and business. In particular, African-Caribbean emigrants to North America, especially Afro-Cubans and English-speaking West Indians, are outstandingly successful, yet their roots in slavery and past discrimination are very similar to those of African-Americans. Numerous surveys show that Americans of all backgrounds favor equal opportunities for groups that have experienced past discrimination. Nonetheless, many middle-class African-Americans, Hispanics, and Asians report experiencing both overt and covert prejudice that affects them economically.

Is there anything besides the prejudice of others that may be holding back African-Americans, Hispanics, and Native Americans?

U.S. Population by Race and Ethnicity, 1996 and 2050

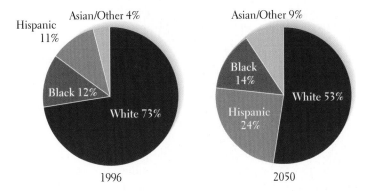

Figure 2.13 **The changing U.S. ethnic composition.** [Adapted from Jorge del Pinal and Audrey Singer, "Generations of diversity: Latinos in the United States," *Population Bulletin* 52 (October 1997): 14.]

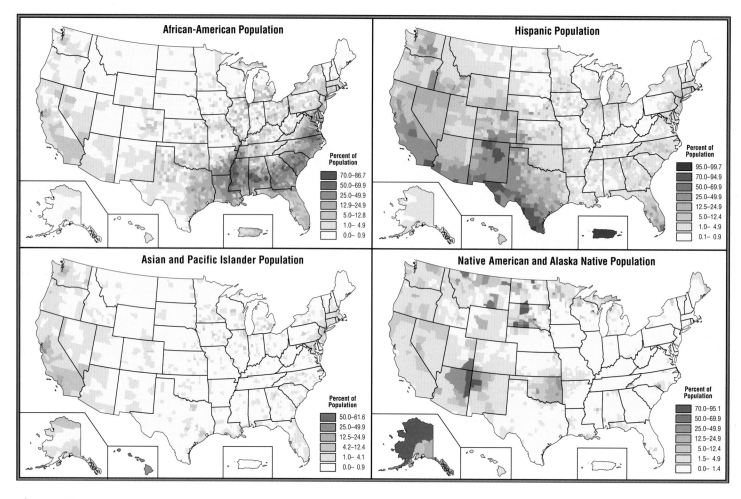

Figure 2.14 Minority population distribution, 2000. Population distribution of four minority groups in the United States in 2000. [U.S. Bureau of the Census, Mapping Census 2000: The Geography of U.S. Diversity.]

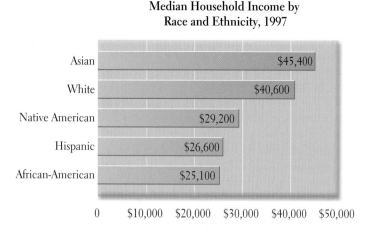

Median Household Income by Race and Ethnicity, 1997

Figure 2.15 Median household income by race and ethnicity, 1997. "Asian" includes Pacific Islanders and excludes Hispanics; "White" and "African-American" exclude Hispanics; and "Native American" includes Eskimos and Aleuts but excludes Hispanics. [Adapted from Kelvin M. Pollard and William P. O'Hare, "America's racial and ethnic minorities," *Population Bulletin* 54 (September 1999): 22, 24–27, 36.]

The experts disagree, but there is a consensus that poverty and low social status breed the perception among the victims that there is no hope for them and hence no point in trying to succeed. Furthermore, the basic skills for success are not being imparted at home or in school to the poorest Americans of whatever background. The increasing number of poor, single-parent families is also thought to play a role. For example, 50 years ago, African-American families were primarily two-parent units or extended families, and fathers and grandfathers influenced the development of their children. In the 1990s, nearly 75 percent of African-American children were born to single mothers, and often the fathers of these children were not active in their support and upbringing. The causes of this change are complex and not yet well understood. They are undoubtedly partly related to employment discrimination against African-Americans and to a welfare system that issues funds only to single parents. Thus, fathers are encouraged to be absent from their families, leaving enormous responsibilities of both child rearing and breadwinning in the hands of often undereducated young mothers.

There is growing evidence that the pattern of poverty, under-achievement, and poor health among African-Americans in particular is part of a larger class problem rather than simply a race problem. For example, there are many poor white and Hispanic Americans of all races in the same predicament. They also experience a growing number of single-parent families, absent fathers, low achievement, and poor health. The crux of the problem may well be the growing economic disparity and spatial separation between poor and rich Americans. The richest 20 percent of U.S. citizens now have an average income that is nine times larger than that of the poorest 20 percent. In Canada, the disparity is less dramatic: the richest are just five times better off than the poorest. In both countries, the increasingly prosperous middle class of whatever race or ethnicity has fled to the suburbs. As geographer John B. Strait found in his study of the concentration of poverty in particular neighborhoods of Atlanta, Georgia, the poor are increasingly isolated in very poor neighborhoods; hence, rich and poor rarely even encounter each other in daily life. This isolation makes it difficult for the affluent of all racial and ethnic backgrounds to see how they might help individuals move out of poverty, and it is difficult for the poor to witness models of success. Evidence for this class-based explanation is the increasing opportunity that seems to exist for middle- and upper-class African, Hispanic, and Asian Americans with education and skills. Privileged Americans of whatever racial background are beginning to share middle- and upper-class neighborhoods, workplaces, places of worship, and marriages, with markedly decreasing attention paid to matters of skin color.

Nonetheless, it would be wrong to conclude that blocks to success have been completely removed from middle-class minority Americans. Robert Fallows, who studies poverty amid wealth in North America, writes that in the new very highly paid IT professions of the new economy, Hispanics and African-Americans are conspicuously absent, despite a broadly multicultural (but overwhelmingly male) workplace. People in the information technology world often have office mates with black, brown, or white skin tones, but they tend to be highly trained immigrants from such places as China, Malaysia, Poland, Colombia, and India.

Religion

Because so many early immigrants were Christian in their home countries, Christianity is the predominant religious affiliation claimed by North Americans. Nonetheless, legally, one is free to accept whatever creed one likes, or none at all; virtually every medium-sized city has at least one synagogue, mosque, and Buddhist temple. In some localities, adherents of Islam, Judaism, or Buddhism are numerous enough to constitute a prominent cultural influence. Religion in North America has some interesting and unusual geographic patterns (Figure 2.16). Two of these are the distribution of particular faiths and the regional differences in the role of religion in daily life.

There are many versions of Christianity in North America, and their distributions are closely linked to the settlement pattern

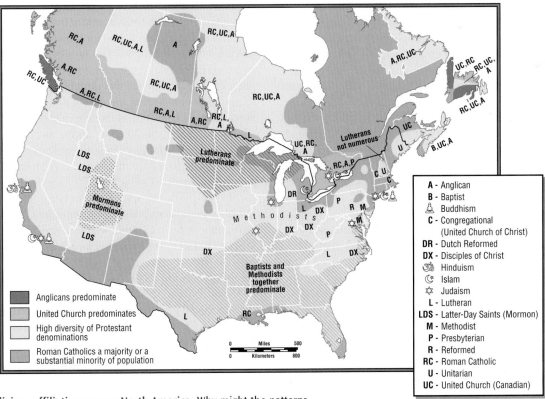

Figure 2.16 Religious affiliations across North America. Why might the patterns show the U.S.–Canadian border so prominently? [Adapted from Jerome Fellmann, Arthur Getis, and Judith Getis, *Human Geography* (Dubuque, Iowa: Brown & Benchmark, 1997), p. 164.]

of the immigrants who brought them. The map in Figure 2.16 shows the dominance of Roman Catholicism in regions where Hispanic and French people have settled—in southern Louisiana, the Southwest, and the far Northeast in the United States; and in Québec and other parts of Canada. Lutheranism is dominant where Scandinavian people have settled, primarily in Minnesota and the eastern Dakotas. The Baptists dominate the religious landscapes of the South. Generally speaking, Southern Baptists tend to take the teachings of the Bible more literally than many other Christian denominations, so this part of the United States has come to be called the Bible Belt. Christianity, and especially the Baptist version, is such an important part of community life in the South that newcomers to the region are asked almost daily what church they attend. (This question is rarely asked in most other parts of the continent.) Those who reply "none" or something other than "Baptist" may be left with the feeling that they have disappointed the questioner.

The proper relationship of religion and politics has long been a controversial issue in the United States, in large part because the framers of the Constitution supported the idea that church and state should remain separate. Even now, most people try to avoid mixing the two in daily life and conversation. In the past decade or two, however, some of the more conservative Protestants have pushed for prayer in the public schools, the teaching of the biblical version of creation as an explanation for the origin of life on earth, and the banning of abortion as a method of birth control. Although policies of conservative Christians have met with the most success in the South, the goals are shared by people scattered across the country.

New immigrants have brought their own faiths and belief systems, and they are contributing to the debate about religion and public life in America. Muslims, for example, are challenging the policies of public schools and local governments regarding public holidays, food served in school cafeterias, and the content of public school texts on such topics as sex education and the religious heritage of America. The long-term outcome of these conflicting views of American religious and political life is not yet apparent, but recent national surveys indicate that a majority of Americans favor the continued separation of church and state and favor choice in personal reproductive matters.

Gender and the American Family

The family has repeatedly been identified as the institution most in need of shoring up in today's fast-changing and ever more impersonal world. Earlier in this century, most people lived with members of several generations in extended families. Family members pooled their incomes and shared chores. Aunts, uncles, cousins, siblings, and grandparents were almost as likely to provide daily care for a child as its mother and father. Humans have long preferred to reside in communities of kinfolk, but in industrialized countries today the extended family has been replaced by the **nuclear family.** The nuclear family, consisting of a married father and mother and their children, is actually a rather recent invention of the industrial age.

Beginning after World War I and especially after World War II, many young people left their large kin groups on the farm and

Many North American suburbs, like this one in Maryland, contain dozens of houses built according to just two or three plans. Paint color, trim, the position of the house on the lot, and landscaping detail are used to give some variety to the neighborhood. There are many other types of residential communities, too. How is (or was) your community similar or different? [Frank Fisher/Gamma-Liaison.]

migrated to distant cities. There they established new families composed of just the basic reproductive unit. Soon suburbia, with its similar single-family homes, provided the perfect domestic space for the emerging nuclear family. This small, compact family suited industry and business, too, because it had no firm ties to other relatives and hence was portable. An industry could draw on a large labor pool willing and able to move. Many North Americans born since 1950 moved as often as 5 to 10 times before reaching adulthood. The grandparents, aunts, and uncles who were left behind missed helping to raise the younger generation, and they had no one to look after them in old age. Institutional care for the elderly proliferated.

In the 1970s, the whole system began to come apart. It was a hardship to move so often. Suburban sprawl meant onerous commutes to jobs for men and long, lonely days at home for women. Women began to want their own careers, and rising consumption patterns made their incomes increasingly essential to family economies. By the 1980s, 70 percent of the females born between 1947 and 1964 were in the workforce, compared to 30 percent of their mothers' generation. But once employed, women could no longer move to a new location with an upwardly mobile husband. Nor could working women manage all the housework and child care as well as a job. Some married men began to handle part of the household management and child care, and the demand for commercial child care grew sharply. With both husband and wife spending long hours at work, many people of both sexes found that their social life increasingly revolved around work, while family life receded in importance. With kinfolk no longer around to shore up the marital bond, and with new options of self-support available to women in unhappy marriages, divorce rates rose drastically. In 1998, half of the marriages in the United States ended in divorce; in Canada, the divorce rate was about two in five.

By the late 1980s, the nuclear family that was dominant for only a few decades was less and less representative of the American family. As Figure 2.17 shows, the percentage of nuclear families continually shrank. By 1997, only about 25 percent of American households were nuclear, while the fastest growing household unit, a single person living alone, had reached 24 percent. More Americans than ever before are living alone; the majority are over the age of 45 and were once part of a nuclear family that dissolved due to divorce or death. In 2000, U.S. Census Bureau analysts reported that there is no longer a typical American household but only an increasing diversity of forms.

Some of these new forms do not necessarily represent an improvement over the nuclear family. By the late 1990s, one-quarter of U.S. children of all races were born to unmarried women, and more than one-quarter lived in single-parent households. Although most single parents do a good job of rearing their children, the responsibilities can be overwhelming, especially because single-parent families tend to be hampered by poverty and lack of education. The vast majority of single-parent households are headed by women, and the incomes of single female-headed households are on average 33 percent lower than those of families headed by single males. In 2000, the U.S. Census Bureau reported that children are more likely than adults to be poor: in 1999, 17 percent of American children lived in poverty, whereas only 11.4 percent of people aged 18 to 64 did.

The changes in family structure, especially the increase in single-parent families and the movement of women into the workplace, have created many social consequences. One often overlooked consequence is that volunteerism in America, long primarily the purview of women, is declining. Over the past 30 years, participation in voluntary activities is down as much as 50 percent, as reported by parent-teacher associations, the League of Women Voters, and the American Red Cross. What happens when the women who formerly provided important services to hospitals, schools, and individuals (such as elderly neighbors) are no longer available because they are working outside the home full time? What are the consequences to society when fewer citizens have the time to help their neighbors? Is the answer that these services should now be paid for by government or individuals, or done without?

Aging

The populations of Canada and the United States are aging, meaning that the numbers of those over 65 are growing more rapidly than the numbers of those under 15. In most societies, people between youth and old age (15 and 65 years) must support those who are younger or older. The age structure of a population tells us how great the burden is likely to be. Wealthier, developed societies tend to have lower birth rates, and their members have longer life spans. Over time, the proportion of those younger than 15 becomes smaller, while the proportion of those over 65 becomes larger. During the twentieth century, the number of older Americans grew rapidly. In 1900, 1 in 25 individuals was over the age of 65; by 1994, 1 in 8 was. By 2050, when most of the current readers of this book will be over 65, 1 in 5 Americans will be elderly.

In North America, the number of elderly people will shoot up especially fast between the years 2010 and 2030. This future spurt is the result of a marked jump in birth rate that took place in the years after World War II, from 1947 to 1964. The so-called **baby boomers** born in those years are the largest age group in North America, as indicated by the wide band through the middle of the population pyramids shown in Figure 2.18. In the years from 2010 to 2030, this group will reach age 65 and retire. During this period, payments from social security and private pensions will be high, and medical costs will leap upward. The boomers had fewer children than their parents had, so there will be fewer people of working age to pay the taxes and pension fund contributions necessary to meet these costs (although the contributions of new young adult immigrants will help). Moreover, people in this smaller population will have fewer brothers and sisters with whom to share the daily personal care and companionship needed by their elderly kin. Most families will not be able to afford assisted living and residential care, which already (in 2000) costs from $30,000 to $60,000 per year for one person. We might expect that once the boomers begin to retire in large numbers, the makeup of many households and the spaces in which they live will reflect people's efforts to find humane and economical means to care for elderly family members at home.

Countries with aging populations need to find ways to lessen the strain on the younger generation. Would it help for healthy people to keep working as long as possible so that they could support themselves and save more private funds for later retirement? Or might this strategy deprive younger adults, in their 40s and 50s, of timely promotions to better paying jobs and leave them

Household Composition, 1970–1994 (percent)

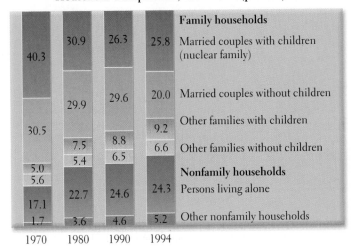

Figure 2.17 **Changing U.S. family compositions, 1970–1994.** [Adapted from Steve W. Rawlings, *U.S. Census Briefs: Households and Families* (Washington, D.C.: U.S. Bureau of the Census, 1995), p. 22.]

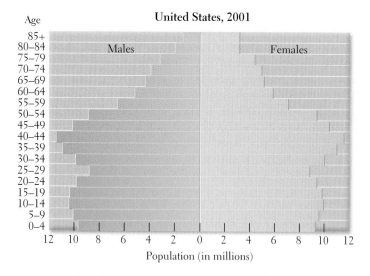

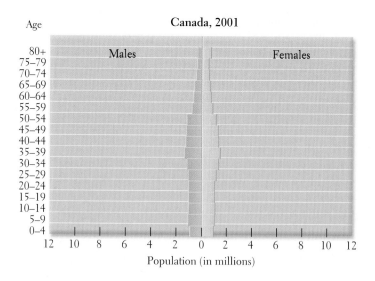

Figure 2.18 **Population pyramids for the United States and Canada.** [Adapted from "Population pyramids of the United States, 2001" and "Population pyramids of Canada, 2001" (Washington, D.C.: U.S. Bureau of the Census), International Data Base, at http://www.census.gov/cgi-bin/ipc/idbpyry.pl.]

less able to save for their own retirement? In many places across North America, local economies depend on spending by retirees. If people retire later, these places will lose income and jobs. And, because retirees often volunteer, agencies that now depend on elderly volunteers will have to pay for labor.

The maps in Figure 2.19 show where the elderly were concentrated in the 1990s and how the patterns will look by 2020. In rural areas (through the Middle West and Northeast), the young have departed in large numbers for other parts of the country, leaving aging parents behind. In other areas—Florida, especially, but also all across the southern United States—the elderly are the

new residents, attracted by the warm, pleasant climate of the Sunbelt. California has the largest number of elderly, but their percentage of the total population is small because there are so many young people.

The problems presented by aging populations reveal a paradox. On the one hand, it is widely agreed that population growth should be reduced to lessen the environmental impact of human life on earth, especially the impact of the societies that consume the most. On the other hand, slower population growth means that as more of earth's citizens grow old, there will be fewer young, working-age people to provide the financial and physical help the elderly require.

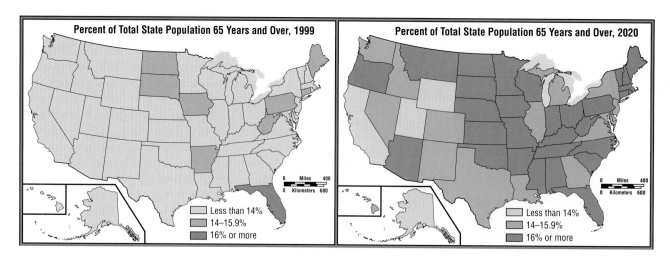

Figure 2.19 **Changing distribution of the elderly in the United States.** The proportion of elderly will increase across the country in the coming decades. Notice that in 1999 the highest density of elderly people was in Florida, but by 2020 more than half of the states will have similar densities. Also, the actual numbers of the elderly in California and Florida are masked on these maps by the fact that these states also have many young people. How might daily life for aging Americans differ between a retirement community in Arizona and a small, depopulated town in South Dakota? Who cares for the very old retired migrants once they can no longer live on their own? [Data from U.S. Bureau of the Census, 2000 census.]

ENVIRONMENTAL ISSUES

VIGNETTE Lisa Crawford lived in Fernald, a rural community in southern Ohio. In the mid-1980s she and her family occupied a rented farmhouse across the road from a "Feed Materials Production Center." Lisa assumed it was an animal feed factory and thought the rural environment would be an excellent place to raise her son—fresh well water, fresh air, clean living. This bucolic prospect changed when Lisa learned that she lived across the road from a U.S. Department of Energy (DoE) military nuclear facility producing nuclear materials for weapons. One day her landlord called to say that the well water was contaminated, and soon Lisa learned that Fernald was also a major storage facility for "hot" nuclear waste. For a while she was at a loss about how to respond but eventually joined with a friend to found a community action group called Fernald Residents for Environmental Safety and Health (FRESH), one of the first grassroots groups that brought public scrutiny to environmental conditions at U.S. military facilities. After years of investigation, the government disclosed that the Fernald facility had released 394,000 pounds (174,000 kilograms) of uranium and 14,300 pounds (6500 kilograms) of thorium into the air. Another 167,000 pounds (85,000 kilograms) of uranium were discharged into the Great Miami (Ohio) River and entered the Great Miami aquifer. In 1989, the DoE closed the facility, which is now waiting for cleanup under a special government program called Superfund. The DoE estimates it will take 50 years and $50 billion to clean up the Fernald site. [Adapted from Joni Seager, "Hysterical housewives and other mad women," in *Feminist Political Ecology*, ed. Dianne Rocheleau, Barbara Thomas-Slayter, and Esther Wangari (London: Routledge, 1996).]

All human activity has environmental consequences, but some actions of North Americans, and especially those in the United States, are particularly damaging. The production of hazardous waste is an example. Hazardous waste is produced by nuclear power generation and weapons manufacture; by mineral mining and drilling; by waste incinerators; and by industry in general, which produces 80 percent of all liquid hazardous waste. Many other harmful wastes are not yet universally defined or regulated by law, such as chemical wastes from small businesses and private homes. These include oven cleaners; drain, toilet, and window cleaners; weed killers; and gasoline. With a population of roughly 280 million, each year the United States generates five times the amount of hazardous waste generated by the entire European Union (population more than 300 million). Canada generates much less hazardous waste per capita each year than the United States, although its citizens generate several times the global average.

The disposal of hazardous waste within the United States has a geographic pattern. Sociologist Robert D. Bullard has shown that a disproportionate amount of hazardous waste is disposed of in the South and in locations inhabited by poor Native American, African-American, and Hispanic people. Bullard writes that "nationally, 60 percent of African-Americans and 50 percent of Hispanics live in communities with at least one uncontrolled toxic-waste site." Military bases are also major sources of hazardous wastes, generating more wastes than the top five U.S. chemical companies combined.

Dumping hazardous waste is only one way that human actions can damage the environment. Consider the many consequences of building and living in a typical North American suburban home. Habitat for wild creatures and plants is lost on the site of the home and on the landfill where the solid waste is disposed of. Lawn fertilizer runoff pollutes nearby streams. Resources for building and maintaining the home and lawn, such as wood and petroleum products, may be drawn from very distant places where their extraction will have caused environmental damage. The rest of this section focuses on a few environmental consequences of the North American lifestyle: air pollution, loss of habitat for plants and animals, and depletion of water resources.

Air Pollution

With only 5 percent of the world's population, North America produces 26 percent of the greenhouse gases released globally by human activity. This large share can be traced to the high North American consumption of fossil fuels, which is in turn related to oil-dependent industrial and agricultural processes, the heating of homes and offices, and a high dependency on automobiles (see Chapter 1, page 45).

Smog is a combination of industrial air pollution and car exhaust that frequently hovers over Los Angeles and other cities, causing a variety of health problems for their inhabitants. The intensity of Los Angeles's smog is due in large part to the city's warm land temperatures and west coast seaside location, which often results in a **thermal inversion**—a warm mass of stagnant air that is trapped beneath cooler air blowing in off the ocean. This inversion is held in place, often for days, by the mountains that surround the city.

Fossil fuels also contribute to **acid rain** because they release sulfur dioxide and nitrogen oxides into the air. Acid rain is created when these gases dissolve in rainwater and make it acidic. Acid rain can kill certain forest trees and, when concentrated in lakes and streams and snow cover, can destroy fish and wildlife. The eastern half of the continent, from the Gulf Coast to Newfoundland and including the entire eastern seaboard, is greatly affected by acid rain. The United States, with its larger population and more extensive industry, is responsible for the vast majority of acid rain. Due to continental weather patterns, however, the area most affected by acid rain is along the U.S.–Canadian border (Figure 2.20).

Loss of Habitat for Plants and Animals

Throughout North America, many plants and animals are in danger of becoming extinct because human activity is destroying the environments in which they live. Of all countries in the world, as of 1998, the United States had both the *largest number* (4669) of

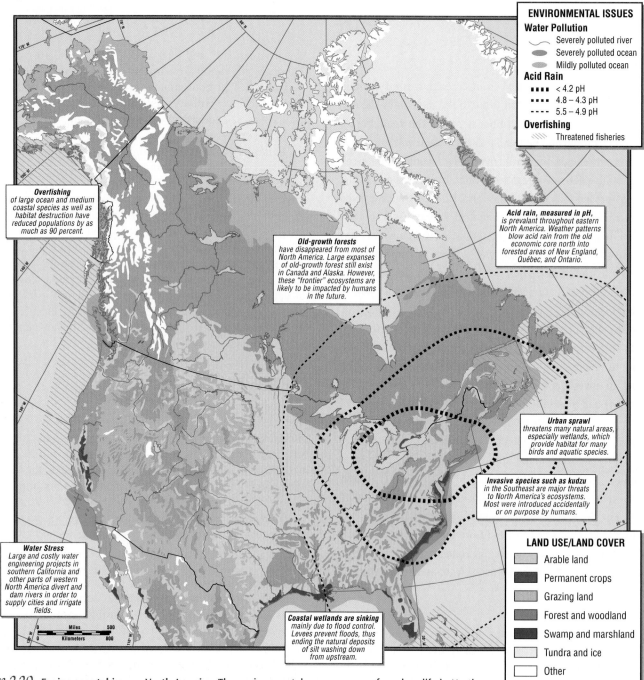

ENVIRONMENTAL ISSUES

Water Pollution
- Severely polluted river
- Severely polluted ocean
- Mildly polluted ocean

Acid Rain
- < 4.2 pH
- 4.8 – 4.3 pH
- 5.5 – 4.9 pH

Overfishing
- Threatened fisheries

Overfishing
of large ocean and medium coastal species as well as habitat destruction have reduced populations by as much as 90 percent.

Acid rain, measured in pH, is prevalent throughout eastern North America. Weather patterns blow acid rain from the old economic core north into forested areas of New England, Québec, and Ontario.

Old-growth forests have disappeared from most of North America. Large expanses of old-growth forest still exist in Canada and Alaska. However, these "frontier" ecosystems are likely to be impacted by humans in the future.

Urban sprawl threatens many natural areas, especially wetlands, which provide habitat for many birds and aquatic species.

Invasive species such as kudzu in the Southeast are major threats to North America's ecosystems. Most were introduced accidentally or on purpose by humans.

Water Stress Large and costly water engineering projects in southern California and other parts of western North America divert and dam rivers in order to supply cities and irrigate fields.

Coastal wetlands are sinking mainly due to flood control. Levees prevent floods, thus ending the natural deposits of silt washing down from upstream.

LAND USE/LAND COVER
- Arable land
- Permanent crops
- Grazing land
- Forest and woodland
- Swamp and marshland
- Tundra and ice
- Other

Miles 500
Kilometers 800

Figure 2.20 Environmental issues: North America. The environmental consequences of modern life in North America are varied and overlapping. Coastal zones are especially affected by pollution and other negative environmental impacts generated by urban and agricultural activities. Note that one result of air pollution is acid rain. Acidity is measured in terms of pH, with 7 representing neutral. A pH less than 7 is acidic: a pH of 5 is approximately the acidity of black coffee; a pH of 4, that of tomatoes. As you can see, eastern North America experiences acid rain with a pH of 5.5 to 4.2, the approximate acidity of tomato juice—sufficient to kill fish, snails, and crustaceans.

plant species and the *highest percentage* (29 percent) of plants threatened by extinction. Earlier in North American history, millions of acres of forests and grasslands were cleared to make way for farms. Now the last bits of natural land near cities are disappearing to make way for residential developments, freeways, golf courses, office complexes, and shopping centers. The destruction of wetlands is especially significant because they are important reproductive zones for many bird and aquatic species.

Some plant and animal species have successfully adapted to living close to humans. Crows, raccoons and opossums raid the garbage cans and refuse dumps of suburban neighborhoods, where there is often enough vegetative cover to provide them with living and breeding spaces. However, human development usually alters environments too radically to support most of the plants and animals that once lived in the vicinity. The animals that can survive face predation by human pets, especially cats,

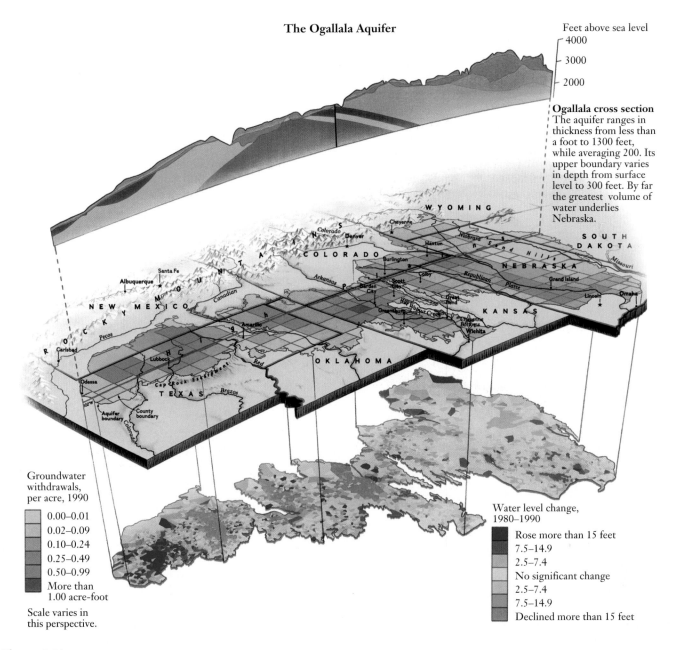

Feet above sea level
- 4000
- 3000
- 2000

Ogallala cross section The aquifer ranges in thickness from less than a foot to 1300 feet, while averaging 200. Its upper boundary varies in depth from surface level to 300 feet. By far the greatest volume of water underlies Nebraska.

Groundwater withdrawals, per acre, 1990

- 0.00–0.01
- 0.02–0.09
- 0.10–0.24
- 0.25–0.49
- 0.50–0.99
- More than 1.00 acre-foot

Scale varies in this perspective.

Water level change, 1980–1990

- Rose more than 15 feet
- 7.5–14.9
- 2.5–7.4
- No significant change
- 2.5–7.4
- 7.5–14.9
- Declined more than 15 feet

Figure 2.21 The Ogallala aquifer. From the 1940s through the 1980s, the aquifer lost an average of 10 feet (3 meters) of water overall, and more than 100 feet (30 meters) of water in some parts of Texas. During the 1980s, the decline was less. Thanks to abundant rain and snow, water management, and new technologies, the water level declined only a foot. [Adapted from *National Geographic* (March 1993): 84–85.]

which are responsible for most of the loss of songbird species in North America. Meanwhile, native North American plants face competition from exotic plant species brought in by humans, sometimes intentionally as landscape plants, sometimes unintentionally as seeds and pollen brought home on clothing from distant places.

Water Resource Depletion

Many people who live in the humid eastern part of North America find it difficult to believe that water is becoming scarce. But as populations grow and per capita water usage increases, it has become necessary to look farther and farther afield for suffi-

cient water resources. New York City, for example, obtains most of its water from the distant Adirondack Mountains in upstate New York. Water becomes increasingly precious the farther west one goes. On the Great Plains, rainfall in any given year may be too sparse to support healthy crops and animals. To make farming more secure and predictable, taxpayers across the continent have subsidized the building of stock tanks and reservoirs. Irrigation has also increased in recent decades, drawing on water that has been stored over the millennia in natural underground reservoirs called **aquifers.** The **Ogallala aquifer** (Figure 2.21) underlying the Great Plains is the largest such reservoir. In parts of the aquifer, water is being pumped out at rates far exceeding natural replenishment.

| TABLE 2.2   Human well-being rankings of Canada, the United States, and selected countries, 2000 |

Country (1)	GDP per capita, adjusted for PPP[a] in 1998 $U.S. (GDP of 174 countries ranked from highest) (2)	Human Development Index (HDI) global rankings, 1998 (3)	Gender Empowerment Measure (GEM) global rankings, 1998 (4)	Gender Development Index (GDI) global rankings, 1998 (5)	Women's average GDP per capita as a percentage of men's average GDP per capita (PPP 1998 $U.S.) (6)
United States	29,605 (2)	3	13	4	61
Canada	23,582 (9)	1	8	1	61
Japan	23,257 (10)	9	41	9	42
Germany	22,169 (15)	14	6	15	52
Sweden	20,659 (20)	6	3	6	82
France	21,175 (17)	12	(31 in 1996)[b]	11	63
Kuwait	25,314 (5) (1996)[b]	36	(75 in 1996)[b]	34	36
Australia	22,452 (13)	4	11	3	67
World	6,526	N/A[c]	N/A	N/A	51

[a]PPP = purchasing power parity.
[b]Data not available in 2000.
[c]N/A = data not available.

Source: United Nations Human Development Report 2000.

Irrigated fields in Southern California supply much of the fresh fruit and vegetables consumed in the United States, but at a high cost to taxpayers. Billions of dollars of federal and state funds have paid for massive water engineering projects that bring enormous quantities of water from hundreds of miles away in Washington and Oregon, Northern California, Colorado, and Arizona, at times pumping it up and over mountain ranges.

Increasingly, citizens in much of western North America are recognizing that the use of scarce water for irrigated agriculture is unsustainable and uneconomical. Conflict over moving water from wet regions to dry ones, or from sparsely inhabited to urban areas, has halted some new water projects and raised awareness of the need to conserve water. Nevertheless, government subsidies keep water artificially low in price, and past successes in harnessing new water supplies provide a disincentive to change.

MEASURES OF HUMAN WELL-BEING

Both Canada and the United States consistently rank high on global scales of well-being, yet a comparison of these two wealthy countries makes it clear why old ways of making such comparisons are misleading. If we looked only at gross domestic product (GDP) per capita for the two countries, we would see that Canada's GDP is U.S. $23,582 and that of the United States is U.S. $29,605. From these numbers, we might conclude that Canadians are not doing as well as people in the United States. But remember that GDP per capita ignores all measures of well-being other than income and is only an average for the entire country.

Table 2.2 compares Canada, the United States, and selected other developed countries on GDP per capita and several other measures developed by the United Nations (see Chapter 1). On all the UN measures, Canada does better than the United States. The United Nations Human Development Index (HDI), column 3 in the table, combines three components—life expectancy at birth, educational attainment, and adjusted real income—to rank 174 countries on how well their citizens achieve basic human capabilities. On this scale, Canada ranks first and the United States third. On the Gender Empowerment Measure (GEM, column 4), which measures the opportunities for women to participate as equals with men, Canada drops to eighth (Norway is first in this category), and the United States drops to thirteenth place. On the Gender Development Index (GDI, column 5), which measures women's access to health care, income, and education, Canada ranks first in the world and the United States fourth. These two gender measures show that women in the United States are not doing as well as men in ways that affect their ability to participate in society. The same is true in Canada, but the disparity between women and men is somewhat less. Many statistics illustrate that in both the United States and

Canada, as well as in other developed countries, women, on average, are poorer than men (column 6), are paid less for the same work, are less the focus of medical research studies (hence female illnesses tend to be less understood), and occupy fewer supervisory positions. As discussed on pages 76–77, women are also severely underrepresented in elected positions in government.

SUBREGIONS OF NORTH AMERICA

The standard way to appreciate the cultural geography of a region as large and varied as North America is to impose some sort of subregional order on the whole. Here we divide North America into subregions, and then we sketch the features that give each subregion its "character of place." But, as we observed in Chapter 1, geographers often have trouble reaching consensus on just where regional boundaries should be drawn.

Joel Garreau, a *Washington Post* editor, designed what he calls "an emotional map" of North America. Garreau paid special attention to the ways in which the continent is changing, especially along cultural or ethnic lines. He noted that state and provincial boundaries are particularly useless for sketching the regional characteristics of the continent. In his successful book, *The Nine Nations of North America* (1981), he proposed a set of regions that cut not only across state boundaries but also across national boundaries (Figure 2.22). Garreau's scheme offers many creative insights, and here we use a modified version in our discussion of the subregions of North America.

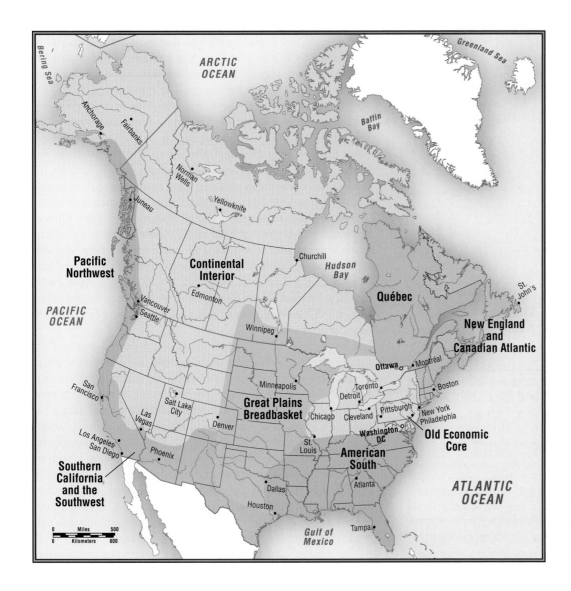

Figure 2.22 Subregions of North America. [Adapted from Joel Garreau, *The Nine Nations of North America* (Boston: Houghton Mifflin, 1981).]

NEW ENGLAND AND CANADA'S ATLANTIC PROVINCES

Among the earliest parts of North America to be settled were New England and Canada's Atlantic Provinces (Figure 2.23). Of all the regions of North America, these maintain perhaps the strongest connection with the past and hold a certain cultural prestige and reputation as North America's cultural hearth. They are especially noted as the source of such early American house styles as Cape Cod and other wood-framed houses, as well as of interior furnishings. Copies of New England–designed furniture are found in all classes of American homes.

Economically, northern New England and the Atlantic Provinces of Canada remain relatively poor. They still depend on fishing, which has become severely depressed, and on timber, vegetable crops, dairying, and other extractive industries. Meanwhile, southern New England is using its celebrated "Yankee ingenuity" to reinvent itself as an information technology center reliant on the intellectual skills of those attracted by its many universities. Although New England has long been the site of several large cities (Boston, New Haven, Hartford), it is becoming both more urban and more ethnically diverse as it shifts its focus, at least partially, to the new IT economy.

Many of the continuities of New England and the Atlantic Provinces with the past derive from the region's geography. During the last ice age, glaciers scraped away much of the topsoil, leaving behind picturesque mountains with rounded, bare summits—the source of some of the finest building stone in North America. But without much topsoil, the land is only marginally productive. Although many farmers settled here, they struggled merely to survive. Today, although much of the landscape is rural, farming is not particularly productive, and many areas try to capitalize on their rural and village ambiance and historic heritage by enticing tourists, especially winter skiers, as well as retirees who will build homes and stay for a decade or more.

After being cleared in the days of early settlement, New England's evergreen and hardwood deciduous forests have slowly filled in the fields abandoned by struggling farmers. These second-growth forests are being clear-cut by logging companies—a practice driven in part by the fear that most of the trees will be lost eventually to a severe, spreading infestation of the northeast budworm. In rural areas, paper milling from wood is still supplying jobs as other blue-collar occupations die out.

Abundant fish were the major attraction that drew the first wave of Europeans to New England in modern times. Throughout the 1500s, hundreds of fishermen from Europe's Atlantic coast came to the Grand Banks, off the shores of Newfoundland and Maine, to take huge catches of cod and other fish. Recently, the Grand Banks have been badly depleted by modern fishing vessels (some from outside North America), which use enormous mechanized nets able to bring in valuable species. These nets damage the fragile sea bottom and kill many unmarketable sea creatures that are nonetheless essential to the marine ecosystem. The fishing industry in this region has also been hurt by competition from fish farming, a rapidly growing industry in many parts of North America. Farmed fish are fed with millions of tons per year (10 million in 1997) of wild oceanic fish and hence pose a further threat to oceanic resources.

In the early 2000s, many parts of southern New England thrive on industries that demand its skilled and educated workers: insurance, banking, high-tech engineering, genetics, and medical research, to name but a few. New England's considerable human resources derive in large part from the strong emphasis placed on education, hard work, and philanthropy by the earliest Puritan settlers, who established many high-quality schools, colleges, and universities. The city of Boston has some of the nation's foremost institutions of higher learning (Harvard, the Massachusetts Institute of Technology, Boston University, Boston College, Northeastern University). Boston has capitalized on its supply of university graduates to become North America's second most important high-technology center after California's Silicon Valley.

At present, New England finds itself confronted by some startling discontinuities with its past. As its remaining agricultural

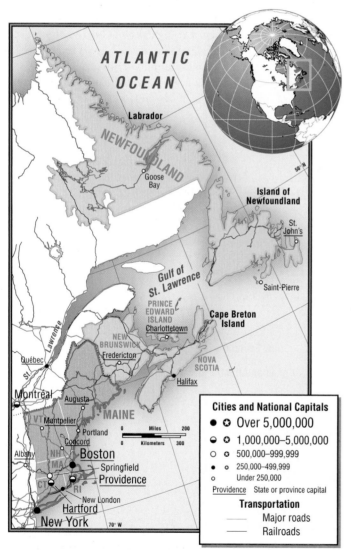

Figure 2.23 **The New England and Atlantic Provinces subregion.**
[Adapted from Joel Garreau, *The Nine Nations of North America* (Boston: Houghton Mifflin, 1981).]

enterprises and rural industries become increasingly mechanized and larger in scale, many rural New Englanders are flocking to the cities for work. Once in the city, they become aware that New England is not the Anglo-American stronghold they had thought it to be. Native Americans, African-Americans, and immigrants from Portugal and southern Europe have long shared New England with those from the British Isles and elsewhere. But today, in such places as Middletown, Connecticut, and the suburbs of Boston, corner groceries are owned by Koreans and Mexicans; Jamaicans sell meat patties and hot-wings; restaurants serve food from Thailand; and schoolteachers are Filipino, Brazilian, and West Indian. The blossoming cultural diversity of New England and the entrepreneurial skills of new migrants are helping New England to keep pace with change across the continent. These trends are less obvious in the Atlantic Provinces of Canada.

QUÉBEC

Québec is the most culturally distinctive formal region in Canada or the United States (Figure 2.24). For more than 300 years, a substantial portion of the population has been French speaking. The French Canadians are now in the majority and are struggling to resolve Québec's relationship with the rest of Canada.

In the seventeenth century, France encouraged its citizens to settle in Canada. By 1760, there were 65,000 French settlers in Canada. Most lived along the St. Lawrence River, on long, narrow strips of land stretching back from the river's edge. This **long-lot system** gave access to resources that extended inland from the river: fishing and river-borne trade; the fertile soil of the river's floodplain; and, on higher ground, the interior forest where the settlers could hunt. Early French colonists commented that one could travel along the St. Lawrence and see every house in Canada. Later, the long-lot system was repeated inland so that today narrow farms also stretch back from roads that parallel the river, forming a second tier of long lots.

Through the first half of the twentieth century, Québec remained a relatively poor land of farmers who grew only enough food to feed their families. After World War II, Québec's economy grew steadily, propelled by increasing demand for the natural resources from northern Québec, such as timber, iron ore, and hydroelectric power. Many of the cities of the St. Lawrence River valley have prospered from the processing and transport of these resources, although most of these enterprises were in the hands of Anglo-Canadians (those of British Isles heritage). Québec's new prosperity was most visible in the rapid growth of Montréal. This city's interior location, near the confluence of several rivers, had always made it an attractive site for British entrepreneurs interested in exporting the natural resources of Québec.

As many French Canadians moved into the cities, the so-called Quiet Revolution occurred. With increasingly better access to

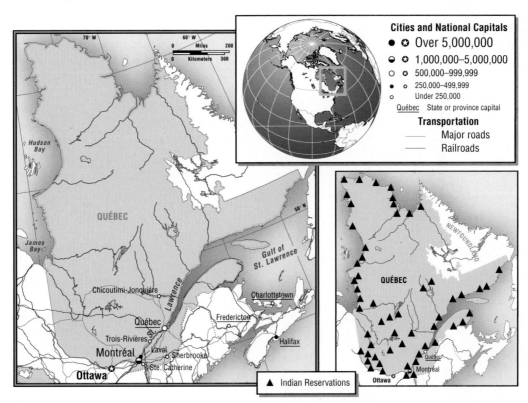

Figure 2.24 **The Québec subregion.** Population in Québec is concentrated along the St. Lawrence River, and a majority of Québecois live in cities. Native Americans from various ethnic groups live throughout Québec, but one-quarter of them live in northern and eastern rural Québec (see insert at right). [Adapted from Joel Garreau, *The Nine Nations of North America* (Boston: Houghton Mifflin, 1981; and "Map of Indian Nations of Québec," Indian and Northern Affairs Canada @http://www.ainc-inac.gc.ca/qc/map/index_e.html.]

education and training, the Québecois were for the first time able to challenge English speakers for higher paying jobs and power. Gradually, the conservative Catholic agricultural society gave way to a more cosmopolitan one that was increasingly resentful of discrimination at the hands of English speakers: social denigration of French culture, blocked access to education, and economic participation. In the 1970s, support for increased autonomy and outright national independence grew. At that time, Québec passed laws that heavily favored the French language in education, government, and business. In response, many English-speaking Québecois relocated to Ontario and elsewhere. Québecois themselves began to fear for Québec's economic survival if the province left Canada. Nevertheless, a referendum on independence failed only narrowly in 1996.

Québec's northern portions extend into areas around Hudson Bay that are rich in timber and mineral deposits (iron ore, copper, and oil). These resources are hard to reach in the remote, difficult terrain of the Canadian Shield, a vast expanse of undulating coniferous forests and tablelands, dotted with small lakes and wetlands.

Although these are the native lands of the Algonquin-speaking Cree people, the Québec government has legal control of resources. In the 1960s, that government became interested in developing hydroelectric power in the vicinity of James Bay (part of Hudson Bay) in order to run mineral-processing plants, sawmills, and paper mills. The Cree protested the clear-cutting of their forests and the changes to community life brought by outsiders. In the 1990s, further hydroelectric projects were put on hold in response to Cree protests that the enormous shallow lakes created by the dams flood sacred ancestral hunting and burial grounds. Nonetheless, development of resources in Cree lands by Euro-Québecois continues; and some electric power from Cree country is sold to the northeastern United States.

THE OLD ECONOMIC CORE

Southern Ontario and the north central part of the United States (Figure 2.25) were once the heart of North American iron- and steel-based heavy industry. This region is still industrial, but it no

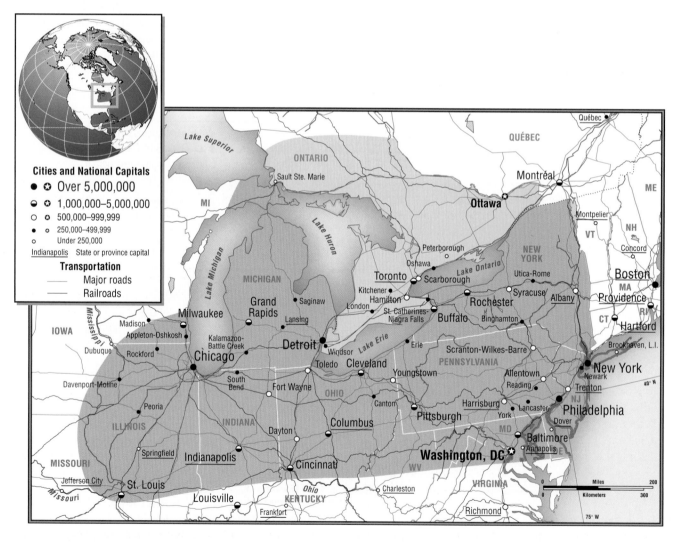

Figure 2.25 **The old economic core subregion.** [Adapted from Joel Garreau, *The Nine Nations of North America* (Boston: Houghton Mifflin, 1981).]

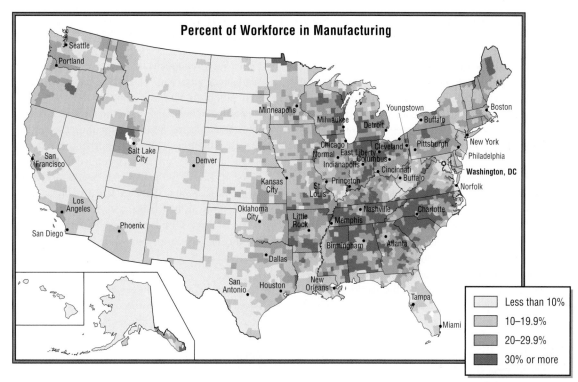

Figure 2.26 **Percentage of the workforce in the manufacturing sector.** [Adapted from Roger Doyle, *1994 Atlas of Contemporary America* (New York: Facts on File, 1994), p. 131.]

longer dominates economically as it once did. Since 1970, many communities have struggled to redefine themselves in the aftermath of plant closings.

For our purposes, the southern boundary of the economic core runs west from the northern tip of Chesapeake Bay and Washington, D.C., along the Ohio River south of Cincinnati and Indianapolis (see Figure 2.25). It then bends to encompass St. Louis on the Mississippi River and follows the river north to Dubuque, Iowa, where it curves back to the northeast. It extends north to Sault Ste. Marie, Ontario, and east to the southern border of Québec. Keep in mind that regional boundaries are often indeterminate. Some of the cities we include in the old economic core have strong ties to the Great Plains and others to the industrial Northeast. And from Washington, D.C., to St. Louis there are many ties to the South.

The economic core is less than 5 percent of the total land area of the United States and Canada; yet as recently as 1975, its industries produced more than 70 percent of the continent's steel and a similar percentage of its motor vehicles and parts. This concentrated output was made possible by the availability of energy and mineral resources in the region, or just beyond its boundaries, to supply the great steel mills and automotive plants. Coal came from Appalachia and southern Illinois; oil and gas came from Pennsylvania; and iron ore came from the great Mesabi Range in Minnesota and deposits like that at Red Lake in Ontario (see Figure 2.28, page 102).

By the 1990s, much of the region was in decline or undergoing economic reorganization. As the map in Figure 2.26 shows, industrial and manufacturing jobs can now be found in the South,

Middle West, and Pacific Northwest, well beyond the old industrial heartland. Brand-new state-of-the-art automobile factories built by Japanese (Honda) and American (Saturn) firms are situated in rural landscapes in Kentucky and Tennessee (see the discussion on pages 74–75). In addition, industrial resources are coming from different places. Much coal now comes from Wyoming, and Canada's coal comes from British Columbia and Alberta, although Appalachian production has remained steady. Steel, the mainstay of the automotive and construction industries, can be more cheaply derived from scrap metal or purchased abroad from Brazil, Japan, or Europe. Hence, huge outdated factories sit empty or underused throughout much of the economic core, and thousands of factory workers have retrained or have entered an early, impoverished retirement. In recent times, some people have taken to calling this obsolete industrialized region the Rust Belt.

It would be wrong to think of the economic core, even during its heyday, as solely a region of factories and mines. In between the great industrial cities of this region are thousands of acres of some of the best farmland in North America. Some of the appliance and food-related industries that grew as offshoots of the farming industry remain in the old economic core. Many, like the meat-packing industry, have consolidated or moved to the South, where labor unions scarcely exist and safety regulations are more lax.

The loss of millions of manufacturing jobs in the old industrial core has had serious consequences. The sociologist William Julius Wilson, in his book *When Work Disappears: The World of the New Urban Poor*, takes special note of the demoralizing effect of plant closings on working-class families in the old industrial

Thinking About the Aftermath of the September 11 Attacks

On September 11, 2001, shortly before 9 on a bright and sunny morning, two passenger jets, carrying 147 people, were purposely flown into the World Trade Center towers in New York City. A third passenger jet struck the Pentagon, and a fourth plane crashed in rural Pennsylvania. All four planes had been commandeered by suicide-hijackers who steered the three craft into the intended targets. The passengers of the Pennsylvania plane apparently were able to force it down before it reached its unknown target in Washington, D.C. Within the hour, the World Trade Center towers collapsed, killing about 3000 people. At the Pentagon, 184 were killed, including those on board the plane, and 40 died in Pennsylvania. Altogether, almost 3500 people died. The September 11 terrorist attacks in New York and Washington have had unprecedented geographic effects on the United States internally and on social and economic relationships around the world, potentially modifying much of what we have written in this book. But at this writing, in the late fall of 2001, it is too early to assess the true depth or duration of the impacts. Here we discuss a handful of patterns that may become better defined in the months and years to come.

A wave of fear swept the United States during and after the attacks. That fear grew first out of immediate worries about the safety of loved ones, but also out of the sudden realization that the United States was not invincible. In the nearly 60 years since the Japanese had bombed Pearl Harbor in December 1941, Americans had come to believe that on the North American continent they were safe from attack. On September 11, it became clear that the very openness of U.S. society and its physically unguarded borders left multiple opportunities for intrusion by those who might do harm.

Redefining America's National Identity. The hijackers were quickly identified as Muslims—most were from Saudi Arabia and Egypt—and it was learned that they probably had Muslim extremist motives, perhaps to undermine the power of the West (both Europe and the United States) and stop the diffusion of Western culture and economic influence. While preparing their attacks, many of the hijackers had lived, worked, and studied among Americans for several years. This realization at first led to anti-Muslim sentiment and some acts of violence against innocent Muslim Americans and other people with obviously foreign origins. But across the country, leaders and ordinary citizens quickly stepped forward to stop such unwarranted reprisals. By October, a sea change in America's national identity seemed to be under way. In part because of frank discussions on radio and television and in print about the injustices of racial profiling and the root causes of hatred toward the United States in distant places, and also because of many public affirmations of the positive features of American life beyond materialism, in a few short weeks Americans of all ethnic and racial backgrounds were reevaluating what it means to be an American.

This redefinition of ethnicity in the United States had already begun during the summer of 2001, when the 2000 census revealed that the United States was no longer primarily a nation of blacks, whites, Christians, and Jews. Hispanic and Asian immigrants were changing the cultural palette of communities across the country. In September, Americans learned to expand their definition of diversity once again when it became clear that there were up to 7 million American Muslims, many of them native-born. Many found themselves reflecting on how the United States might more amicably adjust to being a truly multicultural society. In countless community gatherings, called to further cross-ethnic understanding, Anglo- and black Americans extended tentative hands of friendship to recent arrivals from all over the world. And new residents and citizens realized anew that they had an obligation to show they were assimilating, at least in part, to American ways of life. Americans began to express solidarity across cultural and religious lines by flying the flag or wearing red, white, and blue ribbons and by singing patriotic songs.

A Shaken Economy. The September 11 attacks had immediate negative effects on the American economy. The economy appeared to be going into recession as investors lost confidence, and hundreds of thousands of people lost jobs: in New York City alone 79,000 people lost their jobs as a direct result of the destruction of the World Trade Center. As the map in this box illustrates, nearly all the nation's cities experienced a precipitous decline in economic prospects. Some of the decline was due simply to uncertainty about the future, but some was due to direct loss of business. People afraid of losing jobs in the near future stopped spending. But perhaps most important was a reduction in air travel. Because the hijackers had been able to easily breach airport security and use fully loaded commercial jets as missiles, Americans suddenly cancelled planned trips and appeared to be reevaluating the role air travel would play for them in the future.

Within weeks, several airlines were threatened with bankruptcy. Those cities and businesses dependent on a freely traveling public began to suffer. As the map shows, Las Vegas, San Francisco, Miami, and Honolulu, all dependent on vacation tourism for income, experienced drastic reductions in tourist arrivals. And there were other ripple effects. Companies that had placed a premium on face-to-face relationships with customers stopped flying staff to visit clients and began to use conference calls instead. They reported, however, that deals were lost because of the lack of personal contact. As air transport business shifted to ground systems, interstate highways and trains became, at least temporarily, overloaded. Industrial managers dependent on on-time delivery of parts, resources, and services had to stop production as they waited for deliveries. Drastically increased security measures raised business costs in time and money.

The attacks raised questions about the wisdom of dispersing industrial production into the hinterland, as has been the pattern for several years (see page 71). Might not the dependency of these industries on fast, efficient air and ground transportation place them at a disadvantage if availability of oil from the Arab countries decreased and oil prices rose? Would dispersed factories lose their competitive edge if the highway-based delivery system became overloaded or if air service became more costly and less frequent? Or, conversely, might dispersed industry provide security against attack and hence be a pattern of industrial distribution worth keeping?

Strained External Relations. The United States found reassessing its international role in light of the September 11 attacks particularly perturbing. It was learned that the attacks probably had been planned and funded by the Al Qaeda network headed by Osama bin Laden, who was already implicated in an earlier attack on the World Trade Center in 1993 and in the bombings of U.S. embassies in Kenya and Tanzania in 1998. It seemed logical to take action against bin Laden and Al Qaeda. But their hosts, the Afghan Taliban regime, refused to turn them over. Quickly, U.S. leaders decided to attack Afghanistan. Close to 90 percent of the U.S. population agreed that some type of retribution was necessary for such a diabolical series of acts, but there was less agreement on what form that retribution should take. And it was quickly apparent that elsewhere around the world those who grieved with the United States over the loss of life in the attacks did not necessarily agree that military attack was the answer. In fact, governments in the region of Afghanistan that were expected to support the United States, such as Pakistan, India, Russia, and Saudi Arabia, had first to answer to their own citizens many of whom, like conservative Muslims elsewhere, objected to Western economic and cultural domination. All these governments ultimately supported the U.S. effort in Afghanistan, but with varying degrees of enthusiasm. Were the war to drag on, their continued support would be doubtful, simply because the domestic political costs would be too high.

There were global economic repercussions of the September 11 attacks and of the U.S. attack on Afghanistan. The general slowing of the U.S. economy meant that unemployment quickly spread to distant countries that supplied products to U.S. consumers. For example, a decrease in garment purchases in November in the United States led to a decrease in orders to suppliers in Pakistan and elsewhere. Pakistani workers were laid off just as their government agreed to help the United States with the war it was waging in neighboring Afghanistan.

The politics of oil also played a role, as once again the September 11 attacks perpetrated by at least some Arabs reminded Americans of how dependent they are on oil from Arab countries. As of 2000, the mostly Arab OPEC countries supplied 46 percent of the oil used in the United States. Were that supply to be interrupted, oil production by the United States and its allies would not be sufficient to quickly take up the slack. Nor are Alaskan oil reserves substantial enough to last for long. Some speculated that the time was ripe to develop alternative energy resources, such as solar and thermal. But high oil prices are necessary to give sufficient impetus to the developers of these technologies, and, with shrinking economies after September 11, oil prices fell drastically in the following months.

Those who say that the events of September 11 have changed the world forever are probably right. It would seem, just a few months later, that the effects are widespread, indeed. Because these effects touch prices, investment prospects, jobs, political fortunes, transport costs, food supplies, and much more, they reaffirm the need for continuing geographic analysis.

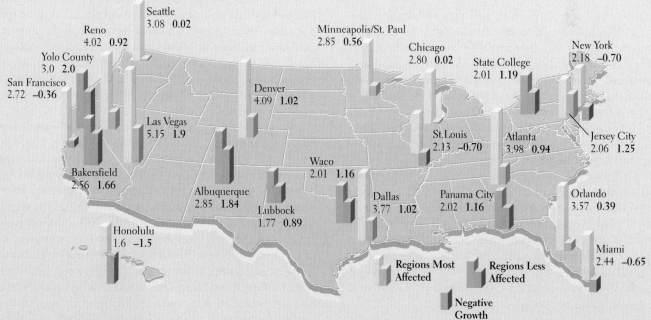

One-Year Economic Growth Forecasts (percent), Before and After September 11, 2001

Seattle 3.08 0.02
Reno 4.02 0.92
Yolo County 3.0 2.0
San Francisco 2.72 −0.36
Minneapolis/St. Paul 2.85 0.56
Chicago 2.80 0.02
State College 2.01 1.19
New York 2.18 −0.70
Denver 4.09 1.02
Las Vegas 5.15 1.9
St. Louis 2.13 −0.70
Atlanta 3.98 0.94
Jersey City 2.06 1.25
Bakersfield 2.56 1.66
Waco 2.01 1.16
Albuquerque 2.85 1.84
Lubbock 1.77 0.89
Dallas 3.77 1.02
Panama City 2.02 1.16
Orlando 3.57 0.39
Honolulu 1.6 −1.5
Miami 2.44 −0.65

Regions Most Affected
Regions Less Affected
Negative Growth

Economic aftermath of the September 11 attacks. The columns show 12-month economic growth forecasts, in percentage points, for some U.S. cities. For each city the left column and number show forecasts made before September 11; the right column and number show forecasts made afterward. Columns for the most affected cities are colored orange; those for the less affected are colored purple; and those for the cities whose economies shrank are colored green. [Adapted from *The New York Times,* September 30, 2001.]

A working-class neighborhood of Youngstown, Ohio, as viewed from the Republic Steel plant, now closed, and which once employed 10,000 people. North Star Steel now operates minimill facilities on the Republic Steel site and employs 250 workers. (See the discussion of the changing steel industry on page 72.) Early twentieth-century frame houses surround a large Roman Catholic church. [Richard Kalvar/Magnum Photos.]

cities. "Neighborhoods that are poor and jobless are entirely different from neighborhoods that are poor and working," he says. "Work is not simply a way to make a living and support a family. It also constitutes a framework for daily behavior because it imposes discipline."

Industrial jobs drew millions of rural men and women, both black and white, from the South to the industrial core after World War II. It is their sons and daughters who are now without work and without funds to retrain and relocate. They constitute some of the more than 35 million U.S. citizens living in poverty. Consider the case of one inner-city black neighborhood on the west side of Chicago, called North Lawndale. In 1960, two large factories employed 57,000 workers, and a large retail chain employed thousands of secretaries and office workers in its corporate headquarters nearby. One factory closed in the late 1960s, removing 14,000 jobs; by 1974, the retail headquarters had moved downtown; and by 1984 the other large factory had closed, eliminating a breathtaking 43,000 jobs. North Lawndale began to disintegrate with the first loss of jobs. Because no one had money to spend anymore, thousands of service jobs disappeared and, with them, many middle-class families. By the early 1990s, the housing stock had deteriorated. As families and businesses left to look for opportunities elsewhere, landlords abandoned buildings, and financial institutions saw little reason to support reinvestment.

THE AMERICAN SOUTH (THE SOUTHEAST)

The regional boundaries of the American South are perhaps fuzzier and based more on a perceived state of mind and way of life than is the case for most other U.S. regions. This region, in fact, covers only the southeastern part of the country rather than the whole of the southern United States (Figure 2.27). Somewhere in east Texas, the American South grades into the Southwest, a region with noticeably different environmental and cultural features. But what characteristics best define the South? And where should we draw its borders?

Within this region there is a complex of features that many people would identify as Southern: food; music; the open friendliness of the people; a variety of dialects; the country Baptist churches; the rolling hills and crooked roads; the early onset of spring; the field patterns and crops such as tobacco and cotton; and the rural settlement patterns, such as those that keep large kin groups together in clusters or lines of mobile homes and newly built houses. But Southern cultural features are hard to measure, there are few clear spatial distributions, and the patterns are not contiguous—nor are all these characteristics ever present in one place. Furthermore, arguments ensue if one tries to define just which accents, or what recipes, or which styles of music, or what sort of friendliness are "Southern." Some places located in the South have few recognizable Southern qualities. Eastern Missouri, Oklahoma, and Texas have strong ties to the Great Plains; western Kentucky has the look of the Midwest; and Miami, on the far southern tip of Florida, seems to have lost all vestiges of the Old South. Miami, with its cosmopolitan Latino culture and trade and immigration ties to the Caribbean and South America, is redefining what it means to be Southern.

North Americans and foreign visitors hold many outdated images of the South. The region today is a different place from the South during the civil rights efforts of the 1960s, let alone the Civil War. It is true that significant racial segregation still persists by custom in that, generally, black people and white people tend to live and worship separately. But the workplace is now integrated, and many African-Americans are in supervisory and administrative positions, especially in government and educational institutions. Black officials have been elected by white constituencies across the South; many schools in both rural and urban settings are integrated, and many more whites and blacks share neighborhoods in the South than in the North.

Poverty is still a problem: the South has the nation's highest concentration of families living below the poverty line. On the other hand, most Southerners, black or white, are able to maintain a substantially better standard of living than their parents did. That older generation toiled as illiterate laborers on plantations or on poor, eroded, hilly farms in Appalachia, or fled the region for jobs in the industrial North. In the early 2000s, the vast majority of Southerners, both black and white, work in the service sector of the economy. Those who left for factory jobs in the North are being attracted back to the region by high-paying jobs; business opportunities; lower taxes; lower costs of living; a milder climate; and safer, more spacious, and friendlier neighborhoods than they could afford in such places as New York, Illinois, or California. The Population Reference Bureau reports (1998) that in the decade of the 1990s, black migration to the South grew dramatically, with people moving there from the Northeast, the Midwest, and the West. Seven of the 10 metropolitan areas that gained the

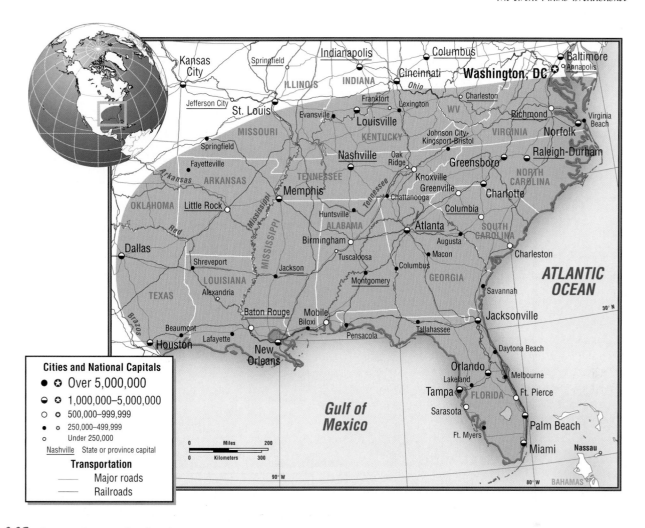

Figure 2.27 **The American South subregion.** [Adapted from Joel Garreau, *The Nine Nations of North America* (Boston: Houghton Mifflin, 1981).]

most black migrants between 1990 and 1996 were in the South, with Atlanta leading in gains: 160,000 for the period.

The South is steadily improving its position as a growth region. The federally funded interstate highway system opened the region to auto and truck transport. Inexpensive industrial locations, close to arterial highways, have drawn many businesses to the South, including automobile and modular home manufacturing, food processing, forest-based building products, light-metals processing, high-tech electronic assembly, furniture manufacturing, and high-end crafts production (for example, the making of fine quilts or stringed instruments such as the dulcimer). More recently, tourists and retirees by the hundreds of thousands have been driving south on the interstate highway system, many attracted by the bucolic rural landscapes and warm temperatures, the scenery of national parks, outlet mall shopping, and recreational theme parks.

Southern agriculture is now mechanized, and at once more diversified and specialized, than ever before. Many cash crops other than tobacco, cotton, and rice are now grown in the South. Strawberries, blueberries, mushrooms, herbs, and vegetables are

often produced for urban consumers on small holdings by part-time farmers who may also have jobs in nearby factories. Most of the country's broiler chickens are now produced on huge operations throughout the South. Interestingly, the laborers on these factory-like chicken farms are often not local. Rather, they are refugee immigrants from Russia, Vietnam, Haiti, Honduras, the Philippines, and Ukraine, who are willing to take the low-wage jobs, at least for a while. They live in communities of prefab houses, another major product of the South.

THE GREAT PLAINS BREADBASKET

The Great Plains (Figure 2.28) receives its nickname, the Breadbasket, from the immense quantities of grain: wheat, corn, sorghum, barley, and oats it produces, much of it exported to Europe and other parts of the world. The gently undulating prairies give the region a certain visual regularity; its weather and climate, in contrast, can be extremely unpredictable, making life precarious at times. The environmental challenges of farming are so great in some places that the land is being abandoned to nature.

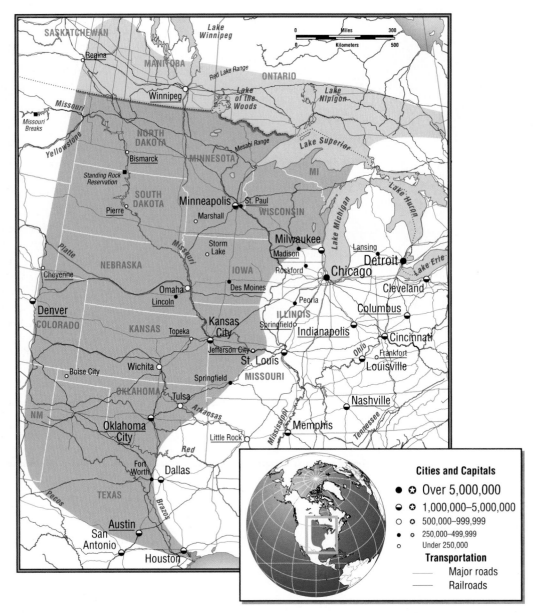

Figure 2.28 **The Great Plains Breadbasket subregion.** [Adapted from Joel Garreau, *The Nine Nations of North America* (Boston: Houghton Mifflin, 1981).]

European settlers began to stake claims on the Great Plains in the 1860s, and for the most part they had to adjust to harsher climates than they were used to. Summers in the middle of the continent, even as far north as the Dakotas, can be oppressively hot and humid. Winters can be terribly cold and dry or so snowy that it may be difficult to get from the house to the barns. Yearly unpredictability is also demonstrated by variable rainfall patterns. Summer rains in some years are plentiful; in other years, hardly a drop falls and a year's labor may come to naught as the crops wither.

The people of the plains have learned to adapt to the harsh weather and unpredictable rainfall. As in yet drier regions to the west, Great Plains farmers have taken to irrigating their crops regularly with water pumped electrically from deep aquifers so that

they won't have to worry about the unpredictability of rain (see the photograph on page 103). They have chosen to grow crops—wheat, corn, oats, barley, sorghum, sugar beets, sunflowers—that suit the environment and world market demand. To select the most profitable crops and find the best time to sell, plains farmers keep close tabs on the global commodities markets, usually with computers installed in barns, homes, or even the cabs of the huge farm machines they use to work their land. The primary crop, however, remains wheat, much of it genetically selected to resist frost damage. Winter wheat, sown in the fall and reaped in the early summer, is grown on the southern and central plains. Fast-growing summer wheat is grown in the north, where the winters are too severe for young wheat plants to survive. Most wheat is

each pound of steak produced in a feedlot results in 35 pounds of eroded soil.

Population patterns on the Great Plains are changing. The few cities around the periphery of the region (Denver, Minneapolis, St. Louis, Dallas, Kansas City) are growing as young people leave the small towns on the prairies. Mechanization has reduced the number of jobs in agriculture and encouraged the consolidation of ownership. Now the men and women who farm may own several large farms in different locales. They often choose to live in cities, traveling to their farms seasonally. As a result of this depopulation, thousands of small prairie towns are dying out. The 2000 census revealed that 60 percent of the counties in the Great Plains lost population. An area equal to the Louisiana Purchase (900,000 square miles, or 2.3 million square kilometers) now has fewer than 6 people per square mile. Those left are struggling to continue to provide schools and opportunities for their own young and, in some towns, for the children of poor Hispanic or Asian immigrants attracted by low-paying packing house or crop-processing jobs.

A countertrend in some parts of the Great Plains is that Native Americans are coming home. Although still a fraction of the total plains population, overall Native American population grew 12 to 23 percent between 1990 and 2000 in the Great Plains states of North and South Dakota, Montana, Nebraska, and Kansas. Myron Gutmann, a University of Texas professor, says these numbers suggest that European agricultural settlement on the Great Plains may have been an experiment that is ending. In hundreds of counties, native grasses and wildflowers, prairie dogs, and bison have made

Cottonwood trees line the Judith River where it joins the Missouri in northern Montana in an area known as the Missouri Breaks. Most of the cottonwoods along the Missouri disappeared over 150 years ago as they fueled the riverboats that helped open the West to European settlement. [William Alber Allard/National Geographic Image Collection.]

harvested by traveling teams of combines that start harvesting in the south in June and move north over the course of the summer.

Women and men have worked equally hard on the Great Plains since the first settlers arrived. For generations, women have borne and raised large families; grown, preserved, and cooked most of the food; and often managed the bookkeeping. They have helped their male family members care for farm stock, milk cows, drive farm equipment, and fix motors. Many women taught in the country schools and organized the church-related social functions that drew widely dispersed farm families together once or twice a week. It is common now to find farm families where several members of both sexes have college degrees. Increasingly, farms are jointly managed by several family members, while others work in a nearby town, not only to earn salaries but also to gain health and retirement benefits for their families.

Cattle raising is another important activity on the Great Plains. Rather than being turned out onto the open range as in the past, cattle now are raised in fenced pastures and then shipped to feedlots where they are fattened for market on a diet rich in sorghum and corn. Cattle (and other livestock such as hogs and turkeys) are slaughtered and processed for market in small plants across the plains. (see the box "Meat Packing in the Great Plains"). Erosion remains a serious problem on the plains: soil is disappearing more than 16 times faster than it can form. Grain fields and pasture grasses do not hold the soil as the original dense prairie grasses once did. The sharp hooves of the cattle loosen the soil, so it is more easily carried away by wind and water erosion. Experts estimate that

Central pivot irrigators create great green circles in the landscapes of the Great Plains and farther west, as in this semiarid environment near Tuscarora, Nevada. The water supports crops that would not grow otherwise. Often, a well at the center of each circle taps an underground aquifer that is then depleted at an unsustainable rate. [Alex S. MacLean/Landslides.]

AT THE LOCAL SCALE *Meat Packing in the Great Plains*

In the 1970s, a meat-packing job in the Great Plains provided a stable annual income of $30,000 or more. The workforce was unionized, and virtually all workers were descendants of German, Slavic, or Scandinavian immigrants. But in the 1980s, a number of highly unionized meat-packing companies closed their doors. Others have opened under nonunion rules, often in isolated small towns in Iowa, Nebraska, or Minnesota. The labor is supplied not by local residents but by immigrants from Mexico, Central America, Laos, and Vietnam. The companies pay $6.00 an hour, which yields an income, after taxes, of less than $12,000 per year. And union-won work rules are gone. The hours are long or short at the convenience of the packing house manager, and those who protest may be summarily fired.

Many of the Latino and Asian workers are refugees from war in their home countries, and most of the Asians have spent a decade or more in refugee camps in Southeast Asia before coming to the United States. These immigrant workers have difficulty affording housing on their wages, and Midwesterners are reluctant to rent to them. Hence, many live in makeshift housing, like the Laotian families in Storm Lake, Iowa, who occupy a series of old railroad cars and shanties. Down the road, Latino workers live in two dilapidated trailer parks.

A few hours away from Storm Lake by car, in Marshall, Minnesota, Roberto Trevino is the personnel director at the Heartland Company, a turkey-processing plant. He supervises 500 workers, 70 percent of whom are from Latin America and Asia, along with a few from Somalia (also war refugees). All wear white smocks and caps and labor with dangerous machinery in icy temperatures. They slaughter, carve, trim, and package 32,000 turkeys per day and ship them under more than 60 different brand names.

Trevino himself is the college-educated son of Hispanic farm workers and feels that his company is providing the immigrants with a stepping-stone. "If you are new in this country, . . . you take the jobs Americans don't want and you may not get ahead. But you do it for your kids," he says. Yet he admits that these jobs are far from stable and do not provide an income sufficient to raise a family.

The Reverend Tom Lo Van, a Laotian Lutheran pastor, sees little chance that the Laotian youth will prosper from their parents' toil. "This new generation is worse off," he says. "Our kids have no self-identity, no sense of belonging . . . no role models. Eighty percent of [them] drop out of high school." [Adapted from Marc Cooper, "The heartland's raw deal: how meatpacking is creating a new immigrant underclass," *The Nation* (February 3, 1997): 1–18.]

---

comebacks. Cattle ranches and grain farms still prevail in the middle and southern plains, but in the northern plains, there are now 300,000 bison—which, unlike cattle, require little management. Mike Faith, a Sioux trained at an accredited Native American college, manages a bison herd at Standing Rock Reservation in North Dakota, not far from where Sitting Bull was killed in 1890.

THE CONTINENTAL INTERIOR

Among the most striking features of the continental interior (Figure 2.29) are its huge size, its physical diversity, and its very low population density. This is a land of extreme physical environments: rugged terrain, frigid temperatures, and lack of water. These physical features restrict many economic enterprises and account for the low population density: fewer than two persons per square mile (one person per square kilometer) in most parts of the region.

Increasing numbers of people are migrating to this region, some to take advantage of its open spaces and often dramatic scenery, others to exploit its considerable natural resources. The two groups often find themselves in conflict with each other and with the indigenous people of the continental interior. In fact, this region is one of the most intense battlegrounds in North America between environmentalists and resource developers.

Physically, there are four distinct zones within the continental interior: the Canadian Shield; the frigid, rugged lands of Alaska; the

Rocky Mountains; and the Great Basin. The Canadian Shield is a vast glaciated territory lying north of the Great Plains that is characterized by thin or nonexistent soils, innumerable lakes, and large meandering rivers. The rugged lands of Alaska lie to the northwest of the shield. Northern coniferous (**boreal**) and subarctic (**taiga**) forests stretch along the southern portion of the shield and Alaska. Farther north, the forest gives way to the **tundra,** a region of winters so long and cold that the ground is permanently frozen to several feet below the surface. Shallow-rooted, ground-hugging plants such as mosses, lichens, dwarf trees, and some grasses are the only vegetation. The Rocky Mountains stretch in a wide belt from southeastern Alaska to New Mexico. The highest areas are generally treeless, with glaciers or tundralike vegetation, while forests line the rock-strewn slopes on the lower elevations. Between the Rockies and the Pacific coastal zone is the Great Basin, a dry region of widely spaced mountains covered mainly by desert scrub and a few woodlands.

The continental interior has the greatest concentration of Native Americans in North America. Most of them live on reservations; in the United States and southern Canada, the reservations are usually not part of their original native lands. Many of those who live in the vast tundra and northern forests of the Canadian Shield, however, still occupy their original territory. For example, the Nunavut recently won rights to their territory after 30 years of negotiation. They are now able to hunt and fish and

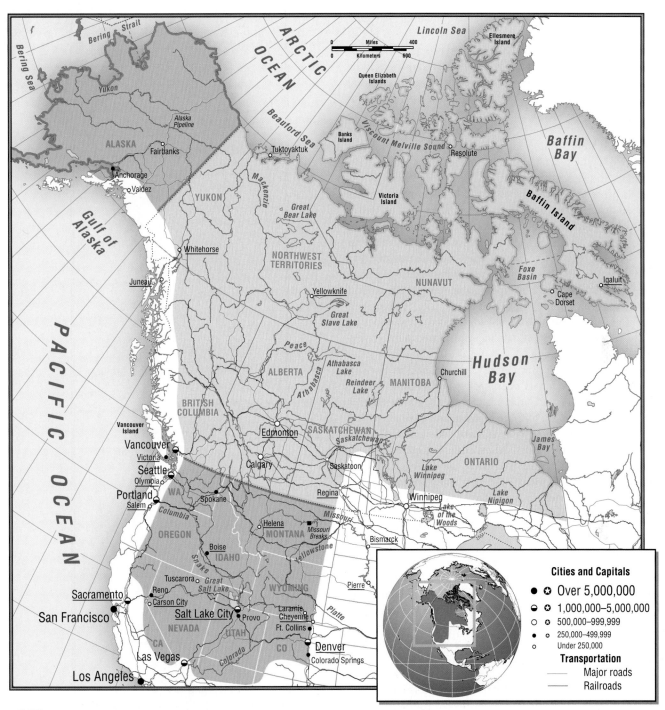

Figure 2.29 The continental interior subregion. [Adapted from Joel Garreau, *The Nine Nations of North America* (Boston: Houghton Mifflin, 1981).]

generally maintain the ways of their ancestors, although most of them now use snowmobiles and modern rifles.

Although much of the continental interior remains sparsely inhabited, nonindigenous people have settled in considerable density in a few places. Where irrigation is possible, agriculture has been expanding—as in the Utah Valley; in the lowlands along the Snake and Columbia rivers of Idaho, Oregon, and Washington;

and as far north as the Peace River district in the Canadian province of Alberta. There are many cattle and sheep ranches throughout much of the Great Basin. Overgrazing and erosion are problems; more serious problems include groundwater pollution from chemical fertilizers and the malodorous effluent from huge feedlots where large numbers of beef cattle are fattened for market. Efforts by the U.S. government to curb abuses on federal land, which is often leased to ranchers in extensive holdings, have not been very successful.

More than half the land in this subregion is federally owned. In Utah, Arizona, and Nevada, 75 percent of the land is held by the federal or state government, and much of this land is leased to mining or logging operators. Mining and oil drilling are by far the largest industries. Since the mid-nineteenth century, the wide range of minerals in the region has supported most of its major permanent settlements. But even in these remote towns, life has been governed by the global economy—which has brought alternating booms and busts, depending on the world market prices for minerals. In recent times, the most stable mineral enterprises have been oil-drilling operations along the northern coast of Alaska. The Trans-Alaska Pipeline runs southward for thousands of miles through mountains and tundra, terminating at the port of Valdez in southern Alaska. The pipeline, which often runs aboveground to avoid shifting as the earth above the permafrost freezes and thaws, constitutes a major ecological disruption, interfering with caribou migrations and always posing the threat of oil spills. A giant spill from the ship *Exxon Valdez* in 1989 devastated 1100 miles of Alaskan shoreline, killed much wildlife, and ruined native livelihoods and commercial fishing.

Increasing numbers of immigrants and vacationers flock to the continental interior for its open spaces and natural beauty. The many national parks in Canada and the United States attract millions of North American tourists; seasonal and permanent residents come to work in such parks as Rocky Mountain, Yellowstone, Glacier, and Grand Teton. Others come to work in mineral mining. Towns such as Laramie, Wyoming, and Calgary, Alberta, have swelled with both seasonal and permanent residents who require services of all kinds. In the United States, environmental groups have pressured the federal government to set aside more land for parks and wilderness preserves. These groups also want to limit or eliminate activities like mining and logging, which they see as damaging to the environment and scenery. A switch to more recreational and preservation-oriented uses would change, and probably lessen, employment opportunities throughout the region. The increasing numbers of visitors would continue to place stress on natural areas, particularly on water resources.

THE PACIFIC NORTHWEST

Once a fairly isolated region, the Pacific Northwest (Figure 2.30) is now at the center of debates about how North America should deal with environmental and development issues. The economy in this region is shifting from logging, fishing, and farming to information technology industries. As this happens, people's attitudes about their environment are also changing. Forests that were once valued primarily for their timber are now also valued for their natural beauty and wildlife, and energy is now more than ever a crucial resource.

The physical geography of this long coastal strip consists of mountains and valleys. Most of the agriculture, as well as the largest cities, is located in the southern part, in a series of valleys and lowlands lying between two long, rugged mountain zones

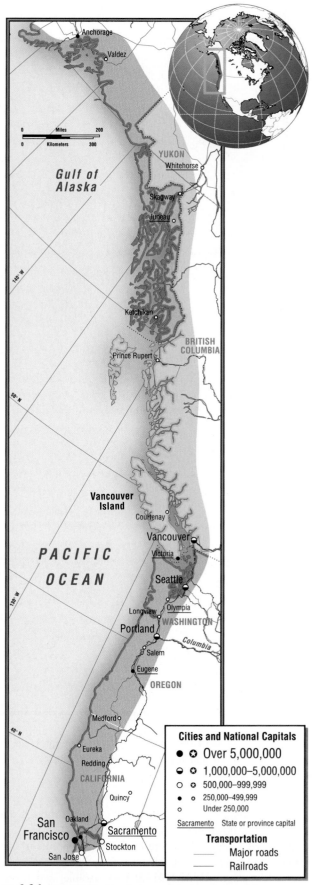

Figure 2.30 **The Pacific Northwest subregion.** [Adapted from Joel Garreau, *The Nine Nations of North America* (Boston: Houghton Mifflin, 1981).]

extending north and south. Throughout the region, the climate is wet. Winds blowing in from the Pacific Ocean bring moist and relatively warm air inland, where it is pushed up over the mountains, resulting in copious orographic rain- and snowfall. The close proximity of the ocean gives this region a milder climate than is found at similar latitudes farther inland. In this rainy, temperate zone, enormous trees and vast forests have flourished for eons. The balmy climate, along with the spectacular scenery, attracts many vacationing or relocating Canadians and U.S. citizens.

Pacific Northwest logging provides most of the construction lumber and an increasing amount of the paper used in North America. Lumber and wood products are important exports to Asia, and the lumber industry is responsible directly and indirectly for hundreds of thousands of jobs in the region. As the forests have shrunk, environmentalists have harshly criticized the logging industry for such practices as clear-cutting, the cheapest and most widely used method of harvesting timber. Clear-cutting destroys wild animal and plant habitats and leaves the land susceptible to erosion and the adjacent forest susceptible to pests. The trees that grow back after clear-cutting tend to be of a single species, allowing harmful insects and infections to spread from tree to tree more easily than in a multispecies forest. There are many reasons to criticize clear-cutting, but an alternative method that will preserve forest diversity and not be unduly costly in time and money is not easy to come by. The battle lines have been drawn between those who make a living from logging or related activities and those who make a living from occupations that tout the beauty of Pacific Northwest forests—people who are often advocates of strict environmental protection.

VIGNETTE In the early 1990s, in the Northern California town of Quincy (slightly east of the border of the Pacific Northwest), several advocates of logging and supporters of environmental preservation decided to meet in neutral territory, the public library, for regular discussions of their differences. Every week, burly loggers sat down with a collection of local environmental advocates. After some years of discussion and conflict with the U.S. Forest Service, which favored clear-cutting, the Quincy Library Group was able to draw up a plan aimed at protecting forest habitats. The plan endorsed managing rather than eliminating fires (fires are needed to clear out highly combustible underbrush) and controlling logging to some extent. The Library Group has also sponsored successful federal legislation to protect old-growth timber. More important, they have established the precedent of citizen input into planning by the U.S. Forest Service, a policy that has changed the dynamics of forestry management in several adjacent subregions.

Several large hydroelectric dams have attracted industries in need of cheap electricity: aluminum smelting and associated manufacturing industries in aerospace and defense. These industries are major employers, but their demand for labor is erratic and periodic layoffs are common. In addition, the hydroelectric

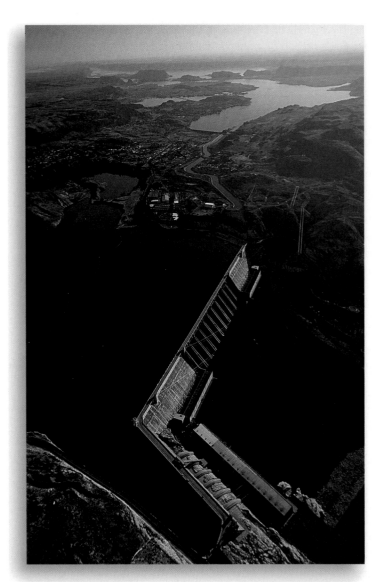

The Grand Coulee Dam is one of 264 dams on the Columbia River and its tributaries. Dams such as the Grand Coulee have almost destroyed the Pacific Northwest's salmon industry while simultaneously providing irrigation water to what was a high desert and generating substantial amounts of affordable electric power. [Jim Richardson/National Geographic Image Collection.]

dams have been criticized by another major employer in the region, the fishing industry. The dams block the seasonal migrations of salmon to and from their spawning grounds, so some of the region's most valuable fish cannot produce young.

Given the environmental impacts of the logging and hydroelectric industries, it is not surprising that many people are enthusiastic about the increasing role of information technology districts within the major urban areas of the Pacific Northwest: San Francisco, Portland, Seattle, and Vancouver. IT companies, for the most part, are clean and efficient, and they produce high-priced finished goods that are easy to transport. With the Silicon Valley just outside San Francisco and with the growing number of companies around Seattle, the Pacific Northwest is the world leader in IT enterprise. But IT firms are particularly dependent on secure and steady sources of electricity. Hence, although IT employees are often vociferous supporters of environmental preservation and restoration, their jobs depend on electric power sources that—whether water, gas, oil, coal, or nuclear fuels—have significant negative environmental impacts.

The leadership of the Pacific Northwest in helping the rest of the United States and Canada to adjust to emerging realities is illustrated by the region's growing connections to Asia. The adjustment of a predominantly Euro-American society to large numbers of Asian immigrants has been well under way along the Pacific coast for many years. This transition has been accompanied by the growing importance of the many countries around the Pacific Rim as trading partners. Asia's economic downturn in the last half of the 1990s deeply hindered the ability of the

Pacific Northwest to market forest products there, particularly to Japan, the largest market. West Coast ports became clogged with timber shipments, and people lost jobs. Japan had been the largest market, but by 1998 it was buying much less. With some measure of economic recovery in Asia, the market rebounded somewhat by 2000. But by then competitors from Europe and South America were also selling in Asia, taking over part of the market and bringing overall prices down. In the short term, the North American economic boom absorbed most of the excess timber for the flourishing housing market. The timber trade with Asia is now recovering, even though many parts of Asia also produce wood.

SOUTHERN CALIFORNIA AND THE SOUTHWEST

Southern California and the Southwest (Figure 2.31) are united primarily by their warm, dry climate, which has attracted many migrants from across America; by their long and deep ties to

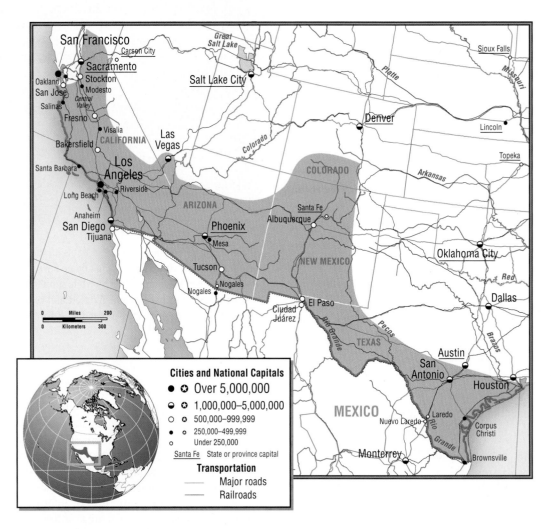

Figure 2.31 **The Southern California and Southwest subregion.** [Adapted from Joel Garreau, *The Nine Nations of North America* (Boston: Houghton Mifflin, 1981).]

José Madrid, 11, picks green chilies in New Mexico. "I'm not good at math, but I'm good at money," he says. Like many child migrant workers in the United States, he goes to school only intermittently. [Eric Draper/AP/Wide World Photos.]

Thinking Geographically: ON THE WEB

Mexico; and by the promise that the North American Free Trade Agreement will make the subregion a nexus of development.

The varied landscapes of the Southwest include the Pacific coastal zone, the hills and interior valleys of California, the widely spaced mountains and dramatic mesas and canyons of southern Arizona and New Mexico, and the gentle coastal plain of south Texas. But the common physical characteristics are their warm average year-round temperatures and arid climate. Most areas receive less than 20 inches of annual rainfall, and the dominant vegetation is scrub, bunch grass, and widely spaced trees. Because of the aridity, ranching is the most widespread form of land use, but other forms are more important economically and in numbers of employees.

Where irrigation is possible, farmers take advantage of the nearly year-round growing season to cultivate fruits and vegetables. California's Central Valley is the leading producer of fruits, vegetables, and wine in the United States and is one of the most valuable agricultural districts in the world. Most of this land is devoted to such crops as grapes, tomatoes, peppers, lettuce, and other vegetables, which are shipped in refrigerated containers throughout the United States and Canada. Many of these crops are produced on vast, plantationlike farms. Migrant Mexican workers, often illegal immigrants, supply most of the labor. These workers, sometimes mere children, lower the cost of North American food by working for very low wages.

Although agriculture is important, the bulk of the region's economy is nonagricultural. The coastal zones of Southern California and Texas nurture oil drilling and refining and associated chemical industries, as well as other industries that need the cheap transport provided by the ocean.

The Southwest's access to the Pacific Ocean and Caribbean Sea, its mild climate, its sunny weather, and its spectacular scenery have made the region a popular location for a wide range of service industries, and it is these activities that employ the vast majority of workers. For example, the U.S. film and television industry has a strong presence in Southern California, drawing writers, actors, producers, investors, and support staff from all over the world. Los Angeles—with its major trade, transport, media, and finance industries, as well as important research facilities—recently surpassed New York as the most populous city in the United States. Like San Francisco, Seattle, and other Pacific coastal cities, Los Angeles is strategically located for trade with Pacific Rim countries, especially in Asia. Los Angeles, rather than New York, is now the largest port in the United States.

On the eastern flank of this region, cities such as Austin, Texas, have also attracted information technology industries, lured by research activities connected to the University of Texas, by the warm climate, and by the ambience of the laid-back, folksy lifestyle of central Texas. Austin has close to 3 million people and grew by more than 30 percent from 1990 to 2000. Across the region, the climate draws large numbers of vacationers and retirees; entire semi-planned communities of 10,000 or 20,000 people have grown up in the desert in the space of just a few years.

The influx of migrants from the rest of North America, from Mexico, and, more recently, from Asia has changed the ambience of many of the major cities of the Southwest. The density of people and businesses, the complex cultural mix, and the gridlock traffic trouble some older residents, but others find the mix exciting and stimulating. The larger, more concentrated population has made the infamous smog of Los Angeles worse. People fleeing pollution and congestion in Los Angeles often head for the cities of southern Nevada, Arizona, and New Mexico. According to the 2000 census, Las Vegas, several hours' drive into the desert from Los Angeles, is the fastest growing city in the United States: its population almost doubled between 1990 (258,295) and 2000 (478,434). The Las Vegas economy is based primarily on casino gambling, glitzy entertainment, and the many services that support these activities. Some IT firms are relocating to Las Vegas from more northerly cities because of lower land, labor, and tax costs.

In the early 2000s, residents of Southern California and the Southwest worry about energy costs along with congestion and smog. Questions about the nation's rate of energy consumption were raised when prices suddenly escalated and supplies became precarious in California. In that state, costs to consumers of electricity had been held artificially low by state regulation of, and subsidies to, utility companies. The low cost of electricity encouraged its ever-increasing use for air conditioning, nighttime commercial lighting, produce refrigeration, and computer-related equipment. State attempts at deregulation failed to take into account the fact that much of the region's power was purchased from wholesale

Los Angeles keeps expanding farther and farther into dry, desertlike foothills. Such development destroys the natural vegetation that holds the soil in place, so large landslides and mud slides may occur when the snowcap melts in the spring or during the occasional heavy rains coming in off the Pacific. Here, the lots have been graded to impede runoff. [Alex S. MacLean/Landslides.]

producers outside the region that had been selling energy to the state-regulated utility companies at inflated prices. By early 2001, costs of electricity in some cases rose to 10 times what they had been a year earlier. At the same time, supplies of power decreased, forcing *rolling blackouts:* sequenced reductions in power to ensure that there is not a general blackout. A wide range of businesses and industries simply could not continue under these conditions, and worries about the region's economic future escalated. In the short term, the electricity-supply problem has been alleviated by a range of strategies, including controls on prices and reduction of nonessential use. It would be wrong to conclude that energy is in critically short supply; the problem is more one of distribution. But like the rest of the nation, California and the Southwest have been consuming energy at rates that are not ultimately sustainable, affordable, or conserving of the environment.

Part of the region was originally a colony of Spain, and this area has maintained the Spanish language, a distinctive Hispanic culture, and other connections with Mexico. Today, Hispanic culture is gaining prominence in the Southwest and spreading beyond it. Large numbers of immigrants are arriving from Mexico, and the economies of the United States and Mexico are becoming increasingly interdependent as NAFTA fosters increased trade (see the discussion on page 74).

In the 1970s, U.S., Canadian, European, and Asian companies began to set up factories, or **maquiladoras,** in Mexican towns just across the border from American towns, in such places as Ciudad Juárez (El Paso), Nuevo Laredo (Laredo), Nogales (Nogales), and Tijuana (San Diego). Maquiladoras produce manufactured goods for sale primarily in the United States and Canada, taking advantage of cheaper Mexican wages, cheaper land and resources, lower taxes,

and weaker environmental regulations. (The maquiladora phenomenon is illustrated in Figure 2.8, page 73, and discussed in greater detail in Chapter 3, pages 161–163.) These factories are a key part of a larger transborder economic network that stretches across North America. In fact, the U.S.–Mexican border is one of the world's most **permeable national borders,** meaning that people and goods flow across it easily (though not without controversy). This permeability has greatly increased since the North American Free Trade Agreement of 1994. There are as many as 200 million legal border crossings each year. According to a U.S. Immigration and Naturalization Service study in 1998, illegal border crossings are less than 10 percent of this figure, at about 150,000 per year.

Several contentious issues surround the estimated 8 to 9 million illegal immigrants currently residing in the United States, 4.5 to 5 million of whom are Mexicans. Many of these people remain in California and the Southwest, where — in addition to contributing vitality to the local economy — they keep down the wages of low-skilled workers. Inadvertently, they have also increased tension over the status of the English language in the United States. Although English speakers in the Southwest far outnumber Spanish speakers, many people fear that English could be challenged in much the same way as it has been challenged by French in Québec. Accordingly, Arizona and California recently became the first states to make English the official language of government. Some bilingual programs, installed in the 1970s to help children make the transition from Spanish to English, are being abolished by those who feel it is better for migrant children to learn English as quickly as possible. Other educators think that migrant children do better in math and science and enjoy higher self-esteem if they can study for a period in their native language.

REFLECTIONS ON NORTH AMERICA

It is not hard to rhapsodize about North America: the sheer size of the continent, its incredible wealth in natural and human resources, its superior productive capacities, and its powerful political position in the world are attributes enjoyed by no other world region. North America enjoys such prosperity, privilege, and power as a result of both fortunate circumstances and the hard work of its inhabitants. Perhaps the most important factor in North America's success has been its democratic governments and supporting institutions, which allow for flexibility as times and circumstances change.

Yet life has not been good to everyone in North America, nor has the influence of North America on other parts of the world always been benign. Native Americans, enslaved Africans and their descendants, and other ethnic minorities suffered as Canada and the United States were being created out of what had been lands long occupied by indigenous peoples. Settlers of all origins and their descendants had, and continue to have, a significant negative impact on the environment; and the standard of living expected by all residents promises to increase the strain on North America's environments. Furthermore, as Canada and the United States developed into wealthy world powers, their impact on environments and people elsewhere around the world, through trade and cultural diffusion, has increased.

There is no guarantee that North America will continue its leadership role into the future. North American models for development—based on democracy and on assumptions of rich and inexhaustible resources—are being challenged as inappropriate for much of the rest of the world. As we shall see, societies elsewhere are beginning to prosper without following North American examples, and sometimes without first installing democratic institutions. Environmental concerns, increasingly a part of the North American consciousness, are not central in many developing countries. Material prosperity is often the chief goal, just as it still is in North America. Many people around the world are eager to bring the North American material miracle to their lands, regardless of negative environmental impacts.

The closer formal economic association of Canada, the United States, and Mexico through NAFTA has strengthened the already strong economic relationships among these three countries. NAFTA may turn out to be just the beginning of new alignments between North America and countries in Middle and South America as well as in Europe and Asia. As we shall see in Chapter 3, over the past few centuries Middle and South America have had experiences very different from those of North America, yet recent social and economic changes in that region have been dramatic and have brought closer ties to North America. Perhaps all of the Americas will eventually be much more integrated economically and, possibly, socially and politically.

Thinking Critically About Selected Concepts

1. Settlement in North America has been characterized by successive movements of people across the vast continent from ancient times to the present. *How have these various movements of people contributed to the present pattern of subregions? What are some other factors contributing to those patterns?*

2. Native Americans experienced enormous losses of people and territory and suffered a decline in their standard of living during European settlement. *Discuss the various innovative ways out of pervasive poverty that are being tried by various Native American groups.*

3. Although Canada and the United States are approximately the same size, Canada has a population only one-tenth that of the United States. *How is population distributed in each country? How does this distribution affect relationships between Canada and the United States?*

4. North Americans are now primarily urban and physically and socially mobile. *How do physical and social mobility fit into the North American version of the market economy? To what extent is social mobility linked to income?*

5. Although Canada and the United States are each other's main trading partners and enjoy many similarities in standard of living and general culture, their relationship is asymmetrical. *Discuss the several ways in which this asymmetry emerges. What are some of its consequences?*

6. Canada and the United States were settled by European immigrants and by people from Africa and their descendants, and both countries continue to accept hundreds of thousands of immigrants each year. *How have migration patterns changed recently? How do the new migrants differ from the old ones in their patterns of assimilation? What are some important reasons that North America needs immigrants? What are some of the disadvantages of large numbers of immigrants?*

7. The pattern of economic and social dominance by particular subregions has changed over time. *Which regions are becoming more dominant, and why? Do you think formerly dominant regions are likely to continue to lose influence?*

8. The nuclear family has long been the norm, but it is now becoming less prevalent as other family types become more prevalent. *Discuss the most important emerging family types and pinpoint some ways in which these types could pose problems for North American societies.*

9. Canada and the United States are economically and politically powerful, are leaders in international matters, and are active participants in the global economy. *What are some of the important ways in which both countries are also facing challenges from the global economy to their prosperity and leadership?*

10. Both Canada and the United States are still striving for gender equality in pay, political representation, and opportunities for jobs and advancement. *What are some important ways in which gender equality has eluded both countries?*

11. North America's high standard of living and growing populations are impinging on the natural habitats of both plants and animals. *Discuss the activities that have the greatest impact on the region's natural environment and the specific environmental effects of each such activity.*

Key Terms

acid rain (p. 89)   acidic precipitation that has formed through the interaction of rainwater or moisture in the air with sulfur dioxide and nitrogen oxides emitted during the burning of fossil fuels

aquifer (p. 91)   a natural underground reservoir

baby boomer (p. 87)   a member of the largest age group in North America, the generation born in the years after World War II, from 1947 to 1964, in which a marked jump in the birth rate occurred

boreal forests (p. 104)   northern coniferous forests

brown field (p. 77)   an old industrial site whose degraded conditions pose obstacles to redevelopment

clear-cutting (p. 63)   the cutting down of all trees on a given plot of land, regardless of age, health, or species

digital divide (p. 73)   the discrepancy in access to information technology between small, rural, and poor areas and large, wealthy cities; major government research laboratories; and universities

economic core (p. 62)   the dominant economic region within a larger region; specifically, southern Ontario and the north central part of the United States

fall line (p. 78)   a line of waterfalls and rapids that cannot be crossed by boats

government subsidies (p. 74)   amounts paid by the government to cover part of the production costs of some products and to help domestic producers to sell their goods for less than foreign competitors

Hispanic (p. 58)   a loose ethnic term that refers to all Spanish-speaking people from Latin America and Spain; equivalent to **Latino**

hub-and-spoke network (p. 71)   the organization of air service in North America around hubs, or strategically located airports used as collection and transfer points for passengers and cargo traveling from one place to another

information technology (IT) (p. 72)   the part of the service sector that relies on the use of computers and the Internet to process and transport information, including banks, software companies, medical technology companies, and publishing houses

interstate highway system (p. 71)   the federally subsidized network of highways in the United States

Latino (p. 58)   a loose ethnic term that refers to all Spanish-speaking people from Middle and South America and Spain; equivalent to **Hispanic**

loess (p. 58)   windblown dust that forms deep soils in North America, central Europe, and China

long-lot system (p. 95)   a system of long, narrow plots of land stretching back from the edge of the St. Lawrence River, which gave French Canadian settlers access to resources extending inland from the river

maquiladoras (p. 110)   foreign-owned factories, often located in Mexican towns just over the border from U.S. towns, that hire people at low wages to assemble manufactured goods that are then sent elsewhere for sale; now also used for similar firms in other parts of Mexico and Middle America

megalopolis (p. 77)   an area formed when several cities grow to the extent that their edges meet and coalesce

metropolitan areas (p. 77)   cities with a population of 50,000 or more and their surrounding suburbs

nodes (p. 77)   small regions with extremely dense populations

nuclear family (p. 86)   a family group consisting of a father and mother and their children

Ogallala aquifer (p. 91)   the largest North American natural aquifer, which underlies the Great Plains

Pacific Rim (p. 66)   all of the countries that border the Pacific Ocean on the west and east

permeable national borders (p. 110)   borders subject to easy flow of people and goods

Québecois (p. 58)   French Canadian ethnic group or members of that group; also, all citizens of Québec, regardless of ethnicity

service sector (p. 72)   economic activity that amounts to doing services for others

smog (p. 89)   a combination of industrial air pollution and car exhaust (*smoke* + *fog*)

taiga forests (p. 104)   subarctic forests

thermal inversion (p. 89)   a warm mass of stagnant air that is temporarily trapped beneath heavy cooler air

tundra (p. 104)   a treeless area between the ice cap and the tree line of arctic regions, which has a permanently frozen subsoil

urban sprawl (p. 77)   the encroachment of suburbs on agricultural land

Pronunciation Guide

Adirondack Mountains (ad-uh-RAHN-dak)

Algonquin (al-GAHN-kwuhn)

Anasazi (ahn-uh-SAH-zee)

Appalachia (ap-uh-LACH-uh)

Appalachian Mountains (ap-uh-LACH-uhn)

Bering Strait (BAIR-ing)

Cahokia (ku-HOH-kee-uh)

Cherokee (CHAIR-uh-kee)

Choctaw (CHAWK-taw)

Ciudad Juaréz (see-oo-DAHD HWAHR-ehss)

Dubuque (duh-BYOOK)

El Paso (el-PAH-soh)

Hohokam (HOH-HOH-kahm)

Kerala (KAIR-uh-luh)

Kootenai (KOOT-uh-nee)

Laredo (luh-RAY-doh)

Latino (lah-TEE-noh)

Louisville (LOO-uh-vuhl/LOO-ee-vihl)

Megalopolis (meh-guh-LAW-puh-lihss)

Mesabi (muh-SAH-bee)

Nogales (no-GAH-layss)

Nuevo Laredo (NWAY-voh lah-RAH-doh)

Ogallala (oh-gu-LAHL-uh)

Ottawa (AW-tuh-waw)

Québec (kay-BEHK)

Québecois (kay-behk-WAH)

Salish (SAY-lihsh)

San Diego (san dee-AY-goh)

Sault Ste. Marie (SOO SAYNT muh-REE)

Seattle (see-AT-uhl)

Sioux Falls (SOO)

taiga (TYE-guh)

Tehuantepec (teh-WAHN-te-pehk)

Tijuana (tee-HWAH-nah)

Toronto (tuh-RAWN-toh)

Valdez (val-DEEZ)

Vancouver (van-KOO-vurr)

Selected Readings

A set of Selected Readings for Chapter 2, providing ideas for student research, appears on the *World Regional Geography* Web site at www.whfreeman.com/pulsipher.

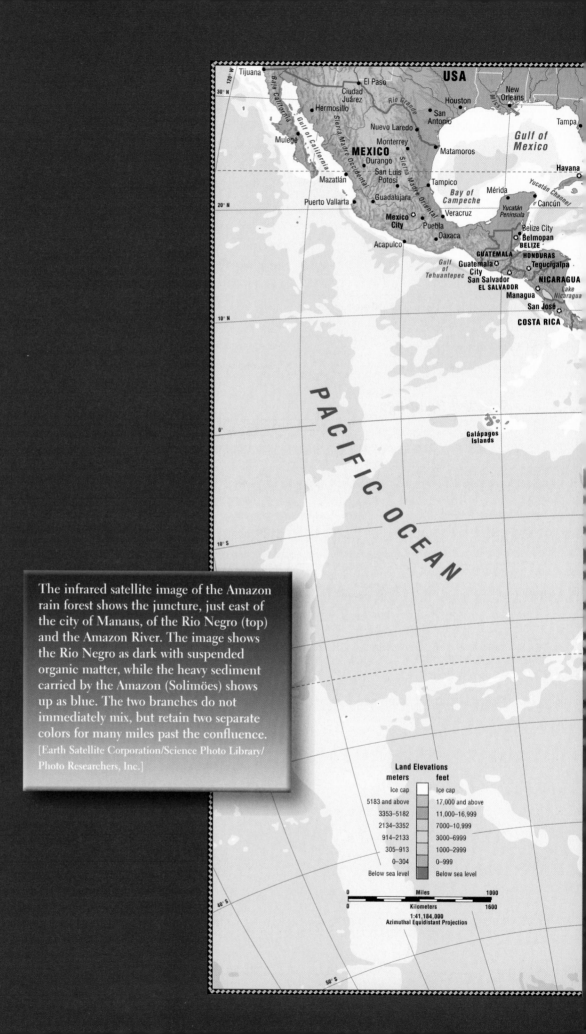

3 | Middle and South America

The infrared satellite image of the Amazon rain forest shows the juncture, just east of the city of Manaus, of the Rio Negro (top) and the Amazon River. The image shows the Rio Negro as dark with suspended organic matter, while the heavy sediment carried by the Amazon (Solimões) shows up as blue. The two branches do not immediately mix, but retain two separate colors for many miles past the confluence. [Earth Satellite Corporation/Science Photo Library/ Photo Researchers, Inc.]

Land Elevations

meters	feet
Ice cap	Ice cap
5183 and above	17,000 and above
3353–5182	11,000–16,999
2134–3352	7000–10,999
914–2133	3000–6999
305–913	1000–2999
0–304	0–999
Below sea level	Below sea level

Miles 0 — 1000
Kilometers 0 — 1600
1:41,184,000
Azimuthal Equidistant Projection

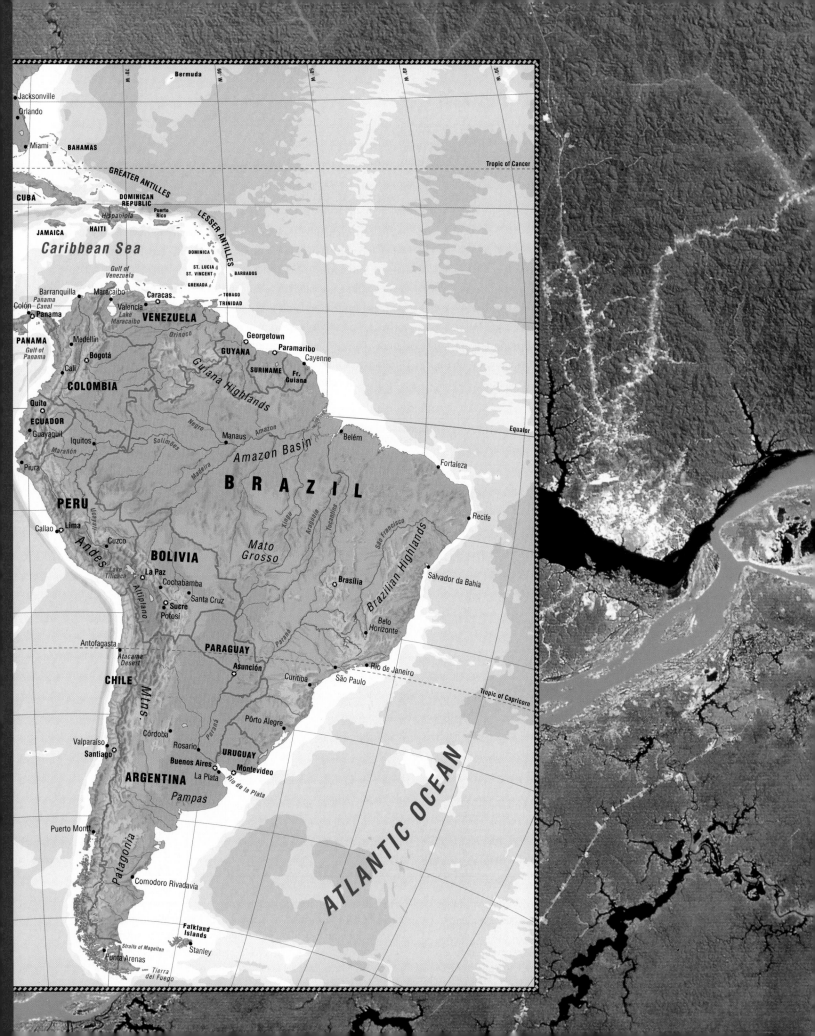

MAKING GLOBAL CONNECTIONS

On a recent trip through the Ecuadorian Amazon, Alex Pulsipher looked into the geography of the region's rapidly expanding oil industry and its effects on local populations.

VIGNETTE The boat trip down the Aguarico River in Ecuador takes one into a world of magnificent trees, river canoes, and houses built high up on stilts to avoid floods and pests and to catch the breeze. I was there to visit the Secoya (Figure 3.1, inset), a group of 350 indigenous people in the Ecuadorian Amazon who are currently negotiating with the U.S. oil company Occidental Petroleum over its plans to drill for oil on Secoya lands. Ecuador's government sees the revenues from oil production as essential to paying off its huge national debt. The Secoya wish to protect themselves from the pollution and cultural disruption that come with oil development. As Colon Piaguaje, chief of the Secoya, puts it, "A slow death will occur. Water will be poorer. Trees will be cut. We will lose our culture, our language, alcoholism will increase, as will marriages to outsiders, and eventually we will disperse to other areas. There is no other way. We will negotiate, but these things will happen." Colon Piaguaje is asking Occidental to use the highest environmental standards in the industry and to establish a fund to pay for the educational and health needs of the Secoya people.

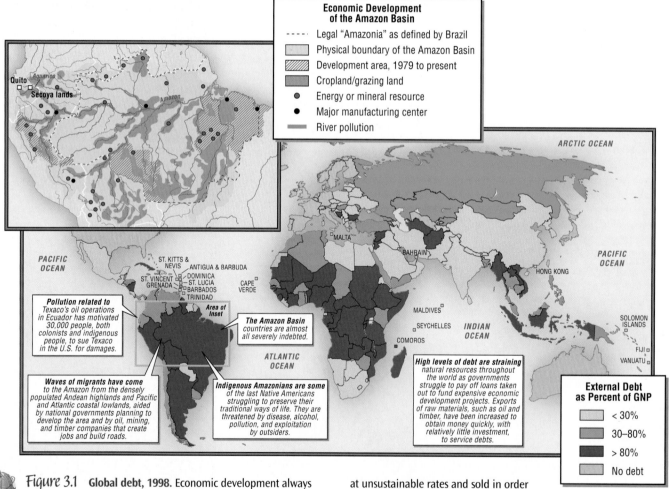

Figure 3.1 **Global debt, 1998.** Economic development always has environmental effects, but environmental damage often increases when countries have moderate to high levels of external debt (that is, debt owed by a country to nonresidents that is repayable in foreign currency). When external debt is high, natural resources may be extracted at unsustainable rates and sold in order to make debt payments. Similar patterns of high debt and unsustainable resource use can be found in the countries that share the Amazon Basin, as well as in Africa, Russia, and Asia. [Debt data from *United Nations HDI Report 2000;* Amazon insert adapted from *Student Atlas of Environmental Issues*, 1997, p. 82.]

Chief Piaguaje bases his assertions on what has happened in parts of the Ecuadorian Amazon that have already experienced several decades of oil development. The U.S. company Texaco was the first major oil developer to establish operations in Ecuador. From 1964 to 1992, its pipelines and waste ponds leaked almost 17 million gallons of oil into the Amazon Basin, 50 percent more than the 11 million gallons the *Exxon Valdez* spilled. Although Texaco sold its operations to the government and left Ecuador in 1992, its oil wastes continue to leak into the environment from hundreds of open pits (*Seattle Times*, June 26, 1996). In addition, many people have come from the densely populated highlands to establish farms along the roads Texaco built to service its oil wells. Some 30,000 people, both colonists and indigenous people, are suing Texaco in the United States for damages from the pollution. They complain that oil contamination has contributed to their skin problems, respiratory ailments, stomach diseases, and dying crops and animals. As an alternative to oil development, many inhabitants of the Ecuadorian Amazon argue for locally based "sustainable development" that won't damage the land. Such development might include programs to help the Secoya market their agricultural produce in the cities or abroad and to start small, environmentally friendly businesses. Others believe that oil development would not be as problematic if effective environmental and cultural protections were in place and if a share of the revenue went to the local community. Still, for the time being, nothing seems likely to stop the "slow death" of the Secoya that Chief Piaguaje describes.

The rich resources of the Americas have attracted outsiders since the first voyage of Columbus in 1492. Europe's encounter with the Americas marked a major expansion of the global econ- omy. Within a few years, the Americas were contributing gold and silver that eventually financed much of the industrial revolution in Europe. Some crops domesticated by Native Americans became cash crops mass-produced for foreign markets with low-paid or slave labor. For several hundred years, Middle and South America occupied a peripheral position in global trade, supplying primarily raw materials. Late in the twentieth century, however, this region began to forge stronger connections with other parts of the world by forming trade blocs within the region, by sending migrants abroad who retained attachments to home, and by attracting outside investment.

Middle and South America differ from North America in several ways. Physically, the region is larger and exhibits a more varied mosaic of environments. Culturally, it is more complex than the continent to the north. Large Native American populations influence daily life just by their presence and also through their legacies of plants, languages, religious beliefs, material culture, and customs. Immigrant groups from Europe, Asia, and Africa have also left distinctive marks, often because circumstances and isolation encouraged a certain cultural conservatism. And the ethnic variety is made yet more complex by highly stratified social systems based on class, sex, and race. Politically, too, the region is more complex; it includes more than three dozen independent countries that implement different models of self-government with different degrees of success. Economically, levels of well-being within and between countries vary more markedly than they do in North America.

Despite the distinctive features of the many parts of Middle and South America, there are regional commonalities. Most of these originate from the various countries' shared experience as colonies of Spain and Portugal (and of Britain, France, and the Netherlands in the Caribbean and Guianas). Colonialism introduced a version of capitalism that did not provide for development in the region but instead made it a producer of raw materials

Themes to Look For

You will encounter a few major themes repeatedly in this chapter:

1. Cultural diversity. The region is notable for its cultural variety and richness.

2. Increasing regional integration of trade. The countries of the region are beginning to form economic links with one another by lowering old colonial trade barriers.

3. Raw materials production. Many countries still rely on income from the export of agricultural products and extracted resources.

4. Highly stratified social systems. Political and economic power remains concentrated in the hands of a small, rich minority, often of European descent. The large Native American, mestizo, and African-American majority is poor and has relatively little political influence. The spread between rich and poor in this region is the widest on earth and may be increasing.

5. Rural-to-urban and international migration. Large numbers of people are migrating from the countryside to the city, where traditional patterns of life are being transformed. Migration to the United States, often for just a few years, is becoming an alternative strategy for the young and adventurous. These young emigrants are most likely to come from Mexico, the Caribbean, or Central America.

6. The extended family. Despite declining birth rates, large multigeneration families, with defined roles for men and women, continue to predominate in the region.

7. Outside influence. For many years, the dominant outside influence in the region has been the United States. Multinational corporations from several world regions are increasing their presence and competing with U.S. corporations for shares in regional trade and commerce.

A worker samples one the several hundred open waste pits that Texaco left behind in the Ecuadorian Amazon. Wildlife and livestock trying to drink from the pits often are poisoned or fall into the pits. After heavy rains, the pits overflow, polluting nearby streams and wells. [Alex Pulsipher.]

for development elsewhere, chiefly Europe. This focus on production for export greatly altered local landscapes, just as oil extraction is altering the homelands of the Secoya people in the Ecuadorian Amazon (see the vignette that opens this chapter). Complex mosaics of land once devoted to multiple uses were transformed into vast stretches of a single crop such as sugarcane or cotton. Sometimes the surface of the land was stripped away to mine silver, copper, or gold. All this was possible because colonialism toppled Native Americans from power in their homelands and instituted a highly stratified class system that placed them at the bottom. This social system still prevails.

Nonetheless, in every country of the region, people like Chief Colon Piaguaje in Ecuador are making efforts to change the patterns of the colonial era, and successes are not hard to find. Self-help projects and efforts to democratize everyday life are evident in countless villages and city neighborhoods. The military, which in the past was often called upon to maintain order (for good or ill), is now much less evident. The economies of most countries are being reorganized to encourage economic growth in markets free of government controls. Such **economic restructuring**, made necessary by national debts incurred to finance large development projects, has been hardest on the poor, who have lost jobs and been greatly affected by rising prices. Yet many have managed to survive using informal networks and communal self-help strategies. In confronting the hardships wrought by restructuring, many people of the region are realizing that development is not just a matter of economic growth, sleek skyscrapers, and massive dam projects. To be judged successful, development must change the lives of the majority for the better.

Terms to Be Aware Of

In this book, **Middle America** refers to Mexico, the narrow ribbon of land south of Mexico that makes up Central America, and the islands of the Caribbean. **South America** refers to the continent south of Central America. For several reasons, we don't usually use the term *Latin America* in this book, even though it is the term most often used by others for what we call Middle and South America. *Latin* refers to the culture of ancient Rome and to the languages (Italian, Spanish, French, Portuguese) that developed from Latin. Latin America is so called because it was colonized, for the most part, by Spain and Portugal, which are thought of as having Latin origins. (Actually, the cultures of Spain and Portugal are the products of many non-Latin influences as well.) When *Latin* is used to refer to the Americas, it serves as a permanent reminder of the region's former colonial status. The designation ignores the other cultures present in the region, most notably the various Native American groups, but also non-Latin people who arrived during and after the colonial period: Africans, Dutch, Germans, British, Chinese, Japanese, and others from elsewhere in Asia. Nor does the term acknowledge the new, distinctly American culture—often called Creole culture(s)—that has been created from these many strands.

The terms *New World* and *Old World* are problematic. Is the New World so designated because it is actually newer in some way than the Old World or only because it was new to Europeans in 1492? In this book, we occasionally use the term *Old World* to mean an entity other than the Americas, which includes virtually the entire rest of the world. This designation is useful at times, because Eurasia, Africa, and Oceania have interacted in many ways over the last 20,000 years and more, exchanging cultural attributes, and it is not always clear where an idea or item of material culture originated.

The recently created colonist town of Joya de las Sachas, in Ecuador, developed along the old Texaco oil pipeline (shown here), which is now owned and operated by the government. Rates of cancer are significantly higher here than in other parts of the Amazon. [Alex Pulsipher.]

THE GEOGRAPHIC SETTING

PHYSICAL PATTERNS

You can see from the map opening this chapter that the region of Middle and South America extends south from the midlatitudes of the Northern Hemisphere all the way across the equator nearly to Antarctica in the Southern Hemisphere. This expanse is part of the reason for the wide range of climates in the region; another factor is the variation in altitude. Tectonic forces have shaped the primary landforms of this huge territory to form an overall pattern of highlands to the west and lowlands to the east.

Landforms

Highlands. A grandly curving and nearly continuous chain of mountains stretches along the western edge of the American continents for more than 10,000 miles (16,000 kilometers) from Alaska to Tierra del Fuego, at the southern tip of South America. This mountain chain is known as the Sierra Madre in Mexico, by various local names in Central America, and as the Andes in South America. This long chain was formed by a process called **subduction** in which the edge of one tectonic plate descends under the edge of another. In this case, the eastward-moving Pacific plates—the Pacific Plate, the Cocos Plate, and the Nazca Plate—are plunging beneath the three American plates—the North American Plate, the Caribbean Plate, and the South American Plate—at a lengthy subduction zone running thousands of miles along the Pacific (western) coast (see Figure 1.8). The overriding plate crumples to create mountain chains, often developing fissures that allow molten rock from beneath the earth's crust to ascend to the surface and form volcanoes.

The entire chain of low and high mountainous islands in the eastern Caribbean is similarly volcanic in origin, created as the Atlantic Plate thrusts under the eastern edge of the Caribbean Plate. It is not unusual for volcanoes to erupt in this active tectonic zone. On the island of Montserrat, for example, people have been coping with an active volcano since July 1995. The eruptions have been violent pyroclastic flows—blasts of superheated rocks, ash, and gas that move with great speed and force down the side of a mountain. Twenty people have been killed and the capital of Plymouth destroyed, along with 20 settlements in the southern two-thirds of the island. Although about 7000 people have left the area, 4000 remain in the north part of the island, struggling to adapt to life in a much changed place.

Lowlands. Lowlands extend over most of the land to the east of the western mountains. In Mexico, however, the Sierra Madre are divided into a broad **V**. Within the **V** sits a plateau, and to the east of the Sierra Madre, a coastal plain borders the Gulf of Mexico. Farther south in Central America, wide aprons of sloping land descend to the Caribbean coast. In South America, a huge wedge of variable lowlands, widest in the north, stretches from the Andes east to the Atlantic Ocean. These South American lowlands are interrupted in the northeast and the southeast by two modest highland zones: the Guiana Highlands and the Brazilian Highlands (Figure 3.2). Elsewhere in the lowlands, grasslands cover huge, flat expanses including the llanos of Venezuela and the pampas of Argentina. The latter has become one of the region's most productive agricultural zones.

The largest feature in the South American lowlands is the Amazon Basin, drained by the Amazon River (see Figure 3.2). This basin lies within Brazil and its western neighbors, but it has global biological significance: it contains the earth's largest expanse of tropical rain forest, 20 percent of the earth's fresh water, and more than 100,000 species of plants and animals. The basin's rivers are so deep that ocean liners can steam 2300 miles (3700 kilometers) upriver, all the way to Iquitos, jokingly referred to as Peru's "Atlantic seaport." The vast Amazon River system starts as streams high in the Andes. These streams eventually unite and flow eastward across the Amazon Plain toward the Atlantic, joined along the way by rivers from the Guiana Highlands to the northeast and the Brazilian Highlands to the southeast. Once these rivers reach the flat land of the Amazon Plain, their velocity slows abruptly and fine soil particles, or **silt**, sink to the riverbed. When the river floods, silt and organic material transported by the floodwaters

On this day in July 1999, the Soufrière volcano on Montserrat was quiet, but you can see the ash deposits left from the pyroclastic flows on the left and right sides of the mountain. Virtually all the houses in the foreground are in the current restricted zone and cannot be occupied. If you are wondering why people reside in volcanic zones, despite the danger, consider that these zones have many attractive features: rich soils; plenty of orographic rainfall; lush vegetation; moderate temperatures; and beautiful, dramatic landscapes. Volcanologists predict this volcano will remain moderately active for several years. [Mac Goodwin.]

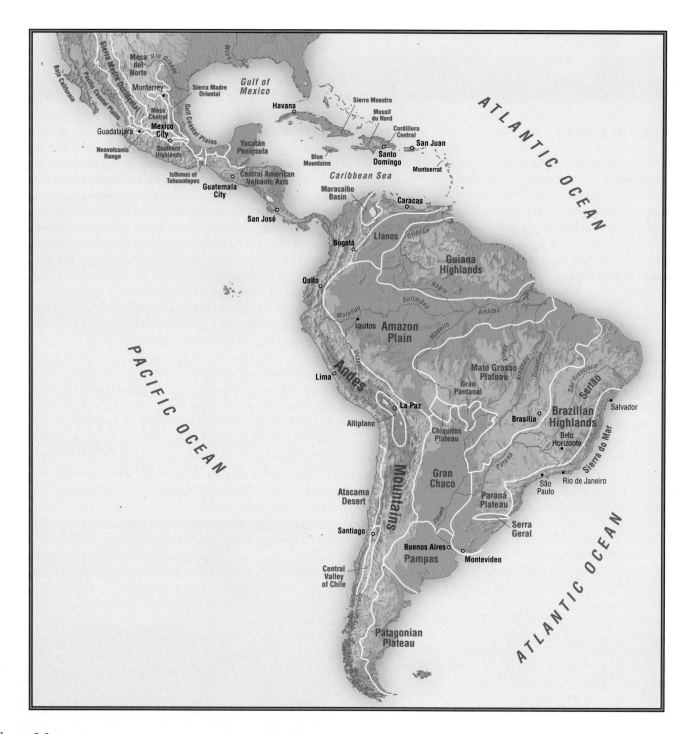

Figure 3.2 Principal landforms and rivers of Middle and South America. [Adapted from David Clawson, *Latin America and the Caribbean* (Dubuque, Iowa: Wm. C. Brown, 1997), pp. 17, 18, 34.]

renew the soil of surrounding areas, nourishing millions of acres of tropical forest. Not all of the Amazon Basin is rain forest, however. Variations in rainfall, cloud cover, wind patterns, and soil types, as well as human activity, can create the conditions for seasonally dry deciduous tropical forest or even grassland.

The interior reaches of the Amazon Basin are home to some of the last remaining relatively undisturbed Native American cultures. When Europeans first came to these river basins, they

were home to perhaps 2 million people living from hunting and gathering and various forms of agriculture. In 1900, there were 230 known ethnic groups in the Amazon Basin, often referred to as tribes. Since then, 87 groups have become extinct, and in some isolated groups only a few hundred members survive. Many have died as the result of introduced diseases or mistreatment while laboring on rubber plantations or in mines. Others have been forced to leave by encroachment on their lands and loss of envi-

CLIMATIC ZONES

Tropical Humid Climates (A)
- Tropical monsoon
- Tropical wet/dry

Dry Arid and Semiarid Climates (B)
- Steppe
- Desert

Temperate Climates (C)
- Midlatitude, moist all year
- Subtropical, winter dry
- Mediterranean, summer dry

Cooler Humid Climates (D)
- Continental, moist all year

Coldest Climates (H)
- High altitude

Northeast trade winds bring heavy seasonal rains to Central America, parts of the Caribbean, and the Amazon Basin of South America.

Peru Current. brings cold surface waters. Air above is very dry, contributing to arid coastal climate. *El Niño* brings warm water instead of cold every few years, impacting climates worldwide.

Rain shadow Andes block the northeast and southeast trade winds off the Atlantic.

Southeast trade winds bring rain to the eastern midlatitudes of South America.

Globe-encircling eastward-blowing winds (also called westerlies) bring steady cold rains. The rainfall supports forests much like those of the Pacific Northwest of North America.

Rain shadow Andes block rains coming from the west.

Figure 3.3 **Climates of Middle and South America.** Can you offer an explanation for the existence of the Atacama and Patagonian desert regions?

ronmental quality caused by logging, ranching, oil exploration, agriculture, and new settlements.

Climate

From the steaming jungles of the Amazon and the Caribbean, to the high, glacier-capped peaks of the Andes, to the parched moonscape of the Atacama Desert, the climatic variety of Middle and South America is astounding (Figure 3.3). This variety results from several factors. The wide range of temperatures reflects the great distance the landmass spans on either side of the equator: northern Mexico lies at 33°N latitude; Tierra del Fuego, the southern tip of South America, lies at 55°S latitude. The region's long mountainous spine creates tremendous changes in altitude that also contribute to the wide range of temperatures. In addition, global patterns of wind and ocean currents result in a distinct pattern of

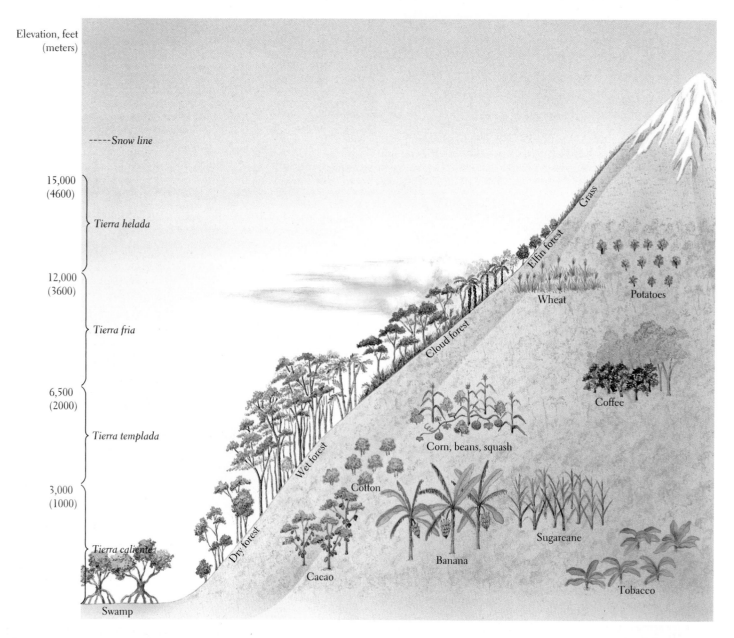

Elevation, feet (meters)

- - - - - *Snow line*

15,000 (4600)

Tierra helada

12,000 (3600)

Tierra fria

6,500 (2000)

Tierra templada

3,000 (1000)

Tierra caliente

Swamp

Cacao

Cotton

Dry forest

Wet forest

Banana

Corn, beans, squash

Sugarcane

Tobacco

Coffee

Cloud forest

Wheat

Potatoes

Elfin forest

Grass

Figure 3.4 **Temperature-altitude zones of Middle and South America.** Each zone is suited for growing specific crops, and the natural vegetation changes with altitude as well. As altitude increases, temperatures usually decrease and cloud cover may increase. The rain forest that flourishes at lower elevations may give way in higher elevations to vegetation adapted to less direct sunlight and somewhat lower temperatures: tree ferns; ground ferns; herbaceous plants such as heliconia; philodendron; grasses; and trees of moderate, or even stunted, height. [Illustration by Tomo Narashima.]

precipitation. People living in mountainous zones use altitude to conceptualize the variety of climates they encounter.

Temperature-Altitude Zones. Four main temperature-altitude zones are commonly recognized by Middle and South Americans (Figure 3.4). As altitude rises, the temperature of the air decreases about 1°F per 300 feet (1°C per 165 meters) of elevation. Thus, temperatures are warmest in the lowlands, which are known in Spanish as the **tierra caliente,** or "hot lands," and extend up to about 3000 feet (1000 meters). In some parts of the region,

lowlands cover wide expanses. Where moisture is adequate, tropical rain forests thrive, as well as a wide range of tropical crops, such as bananas, sugarcane, cacao, and pineapples. Many coastal areas of the hot lands, such as northeastern Brazil, have become zones of plantation agriculture that support populations of considerable size. From 3000 to 6500 feet (1000 to 2000 meters) are the cooler **tierra templada** ("temperate lands"). The year-round springlike climate of this zone drew large numbers of Native Americans in the distant past and, later, Europeans. Here such crops as corn, beans, squash, various green vegetables, wheat, and coffee are grown. From 6500 to 12,000 feet (2000 to 3600 meters) are the *tierra fria* ("cool lands"). Many midlatitude crops—such as wheat, fruit trees, and root vegetables (potatoes, onions, and carrots)— and cool-weather vegetables—such as cabbage and broccoli—do very well at this altitude. Several modern population centers are in this zone, including Mexico City and Quito, Ecuador. Above 12,000 feet (3600 meters) are the *tierra helada* ("frozen lands"). At the lowest reaches of this zone, some grains and root vegetables are cultivated, and some fur-bearing animals—such as llamas, sheep, and guinea pigs—are kept for food and fiber. Higher up, vegetation is almost absent, and mountaintops emerge from under snow and glaciers. The remarkable feature of tropical mountain zones is that in a day of strenuous hiking, one can encounter most of the climate types known on earth.

Precipitation. The pattern of precipitation throughout the region (see Figure 1.11) is influenced by the interaction of global wind patterns, topographic barriers such as mountains, and nearby ocean currents. The northeast trade winds sweep off the Atlantic (see Figures 1.9 and 3.3), bringing heavy seasonal rains to eastern Central America, parts of the Caribbean, and the huge Amazon Basin of South America. The southeast trade winds bring rain to the eastern midlatitudes of South America. Winds blowing in from the Pacific bring seasonal rain to the west coast of Central America, but mountains block the rain from reaching the Caribbean side.

The Andes are a major influence on precipitation in South America (see Figures 1.10 and 3.3). Rains borne by the northeast and southeast trade winds off the Atlantic are blocked, creating a rain shadow on the western side of the Andes in northern Chile and western Peru. Southern Chile is in the path of eastward-trending winds that sweep north from Antarctica, bringing steady cold rains that support forests similar to those of the Pacific Northwest of North America. The Andes block the flow of wet, cool air and divert it to the north, thus creating an extensive rain shadow on the eastern side of the mountains along the southeastern coast of Argentina (Patagonia). The pattern of precipitation is also influenced by the adjacent oceans and their currents. Along the west coasts of Peru and Chile, the cold surface waters of the Peru Current bring cold air that is unable to carry much moisture. The combined effects of the Peru Current and the central Andes rain shadow have created possibly the world's driest desert, the Atacama of northern Chile.

El Niño. An interesting and only partly understood aspect of the Peru Current is its tendency to change significantly every few years. Instead of bringing cold water and drought, it brings warm water and rain to parts of the west coast of South America. Among other consequences, this change in the current causes fish catches to fall drastically as nutrient-poor warm water replaces the nutrient-rich cold water of the Peru Current. The event was named **El Niño** or "the Christ Child" by Peruvian fishermen, who noticed that it reaches its peak sometime in December. El Niño has a worldwide impact, periodically bringing cold air and drought to normally warm and humid western Oceania, torrential rains to the usually dry Peruvian coast, unpredictable weather patterns to Mexico and the southwestern United States, droughts to the Amazon, and perhaps fewer hurricanes to the Caribbean. The El Niño phenomenon in the western Pacific is covered in

El Niño (in red) flares across the Pacific in this image created from satellite data. [Rudolf B. Husar, *National Geographic* (October 2000):90–91.]

Chapter 11, Oceania, where Figure 11.3 illustrates the basic transpacific effects.

HUMAN PATTERNS OVER TIME

The conquest of the Americas by Europe set in motion a series of changes that helped create the ways of life in Middle and South America today. That conquest wiped out much of Native American civilization and set up colonial regimes in its place. It introduced many new cultural influences to the Americas and led to the lopsided distribution of power and wealth that continues to this day.

The Peopling of Middle and South America

Between 13,000 and perhaps 40,000 years ago, groups of hunters and gatherers from Asia spread throughout North America after crossing the Bering land bridge that once connected northeast Asia to Alaska. Many of these groups ventured south through Mexico and Central America. Anthropologists now think that in a period of only several thousand years, these people managed to adapt to a wide range of ecosystems: the dry mountains and valleys of northern Mexico; the tropical rain forests and upland zones of Central America; the vast Amazon Basin; the Andes Mountains; and the cold, windy grasslands near the tip of South America. Recent archaeological evidence from the Monte Verde site in the southern Andes indicates occupation perhaps as much as 30,000 years ago.

By 1492, the North American population numbered at least 18 million; but in Middle and South America there were 50 million to 100 million people. In some places, even rural population densities were high enough to threaten sustainability. These people altered the landscape in many ways. They modified drainage to irrigate crops, constructed raised fields in lowlands, terraced hillsides, built paved walkways across swamps and even mountains, constructed cities with sewer systems and freshwater aqueducts, and raised up huge earthen and stone ceremonial structures that rivaled the Pyramids of Egypt. They also perfected a productive system of **shifting cultivation** that is still common in wet, hot regions in Central America and the Amazon Basin. Small plots were cleared in forestlands, the dried brush was burned to release nutrients, and the clearings were planted with multiple species. Each plot was used for only two or three years and then abandoned for many years of regrowth. This system was (and is) highly productive per unit of land and labor if there is sufficient land for long periods of regrowth. Shifting cultivation produced abundant food for the cultivators but little surplus for marketing. If population pressure increased to the point that the same plot had to be used before it had fully regrown, yields decreased drastically and forest resources were depleted.

These various preconquest activities significantly changed the habitats of land and water species, some of which became extinct; also, some animals were hunted to extinction. Then the trauma of the Spanish conquest and of the ensuing 500 years of European dominance further affected physical environments and obliterated many remarkable accomplishments of Native American cultures, some of which are only now coming to light through archaeological research.

Although they lacked the wheel and gunpowder, the **Aztecs** of the central valley of Mexico had some technologies (such as urban water and sewage systems) and levels of social organization (such as highly organized marketing systems) that rivaled or surpassed civilizations of the time in Asia and Europe. Recent historians have concluded that, on the whole, all social classes of Aztecs lived better and more comfortably than did their contemporaries in Europe.

The largest pre-Columbian state in the Americas was that of the **Incas,** whose domain stretched from southern Colombia to northern Chile and Argentina. The main population clusters were in the Andean Highlands, The cooler temperatures at high altitude eliminated the diseases of the tropical lowlands, yet proximity to the equator guaranteed mild winters and long growing seasons. For several hundred years, until the Spanish arrived in the 1500s, the Incan Empire was one of the most efficiently managed in the history of the world. The ruling class divided the population into a hierarchy of family groups: communities of 10, 50, 100, and all the way up to 40,000 families. Each unit was strictly controlled by a leader who reported to his superior. Highly organized cooperative and reciprocal labor was used to construct paved road systems linking high Andean communities and to build great stone centers in the highlands. Incan agriculture was advanced, particularly in the development of staple crops such as the numerous varieties of potatoes that were adapted to grow in different upland environments.

The Franciscan priest Bernardino de Sahagún interviewed Aztecs who had survived the conquest of 1519–1521. They reported thinking at the time that each armored Spanish soldier on horseback was actually a single, huge, otherworldly creature. This is one of the pictures they drew for Sahagún to illustrate their recollections. [Arthur J. O. Anderson and Charles E. Dibble, *The War of the Conquest* (Salt Lake City: University of Utah Press, 1978), p. 31.]

The Conquest

The European conquest of Middle and South America was one of the most significant events in human history, rapidly altering landscapes and cultures and ending the lives of millions of people. Europeans instigated the conquest after learning of the Americas from Christopher Columbus following his first voyage in 1492. Most early colonizers came from Spain and Portugal on Europe's Iberian Peninsula. By the 1530s, a mere 40 years after Columbus's arrival, all major population centers in the Americas had been conquered and were rapidly being transformed by Iberian colonial policies (Figure 3.5). The superior military technology of the Iberians speeded the process considerably, but the major factor that explains the swiftness and completeness of the conquest was the vulnerability of Native Americans to diseases carried by the Europeans. Epidemics of such diseases as smallpox and measles killed 9 out of 10 Native Americans, who lacked the immunity to these diseases that most Europeans had developed. In about 150 years, the total population of the Americas, estimated at 50 million to 100 million when Columbus arrived, was reduced by more than 90 percent to just 5.6 million.

Columbus established the first Spanish colony in 1492 on the Caribbean island of Hispaniola, now occupied by Haiti and the Dominican Republic. This initial seat of empire expanded to include the rest of the Greater Antilles—Cuba, Puerto Rico, and Jamaica. At first, Native Americans were forced to work on plantations and in mines. Their populations soon plummeted, however, as a result of disease, malnutrition, and brutality. To obtain a new supply of labor, the Spanish initiated the first shipments of enslaved Africans to the Americas in the early 1500s.

The first part of the mainland to be conquered was Mexico, home to several advanced Native American civilizations, most notably the Aztecs. The Spanish, unsuccessful in their first attempt to capture the Aztec capital of Tenochtitlán, succeeded a few months later when a smallpox epidemic they inadvertently brought with them was in full sway. The Spanish demolished the Aztec capital in 1521, including its grand temples, public spaces, causeways, residences, and aqueducts. Mexico City was built on its ruins, to become one of two main seats of part of the Spanish Empire in the Americas. Called the Viceroyalty of New Spain, this part of the empire extended from Panama in the south all the way to what is now San Francisco on the northern California coast. Wealth from the gold and silver mines of Mexico flowed through the port of Veracruz on the Caribbean and from there on to Spain.

The conquest of the Incas bore a remarkable resemblance to that of the Aztecs in that a tiny band of Spaniards, led by Francisco Pizarro in this case, was able to capture the capital city of an extensive empire. Pizarro himself admitted that his campaign received its greatest assistance from a smallpox epidemic (brought unintentionally by earlier Spanish scouts) that preceded his arrival.

Out of the ruins of the Incan Empire in South America, the Spanish created the Viceroyalty of Peru, which originally encompassed all of South America except Portuguese Brazil. The newly constructed capital of Lima flourished on the trade of the huge silver mines established in the highlands of Bolivia at Potosí.

Roman Catholic diplomacy prevented conflict between Spain and Portugal over the lands of the Americas. The Treaty of Tordesillas of 1494 divided the Americas at approximately 46°W longitude. Portugal took all lands to the east and eventually acquired much of what is today Brazil; Spain took all lands to the west.

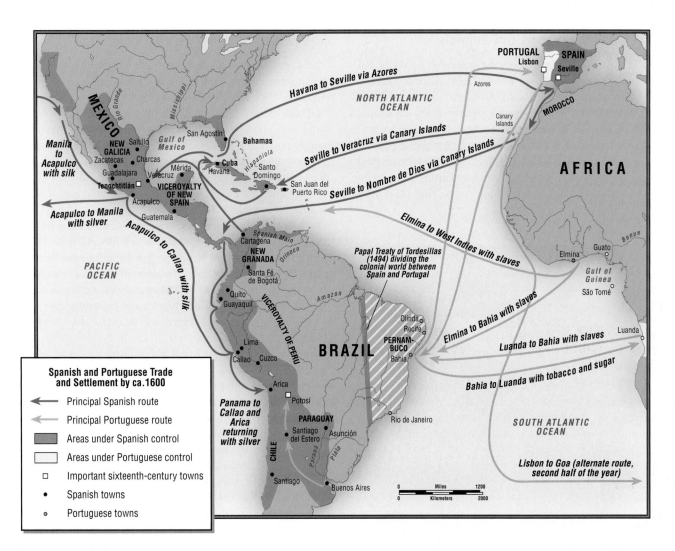

Figure 3.5 **Spanish and Portuguese trade routes and territories in the Americas circa 1600.** The two major trade routes from Spain to its colonies led to the two major centers of its empire, Mexico and Peru. The Spanish colonies could trade only with Spain, but there were direct trade routes from Portuguese colonies in Brazil to Portuguese outposts in Africa. Many millions of Africans were enslaved and traded to Brazilian plantation and mine owners. [Adapted from *Hammond Times Concise Atlas of World History* (Maplewood, N.J.: Hammond, 1994), pp. 66–67.]

The conquest of Brazil by the Portuguese was similar to the Spanish conquest of other areas, in that land was seized and people were killed or enslaved; and yet it differed in some key respects. Brazil was apparently only lightly populated, with people who lived in small villages that were not permanent. There were no highly organized urban cultures as in parts of Middle America and the Andes. Most Atlantic coastal cultures were annihilated early on, and the populations of the huge Amazon Basin declined sharply as contagious diseases spread through trading. It proved difficult to extract quick wealth from tropical forests, however, so most of the Amazon was left alone for several centuries. Instead, the Portuguese focused on extracting mineral wealth, especially gold and precious gems, from the Brazilian Highlands (see Figure 3.2, page 120), and on establishing plantations along the Atlantic coast. Only in the nineteenth and twentieth centuries did commercial interest in the Amazon Basin increase to the point that serious colonization took place. Today, indigenous Amazonians are some of the last Native Americans still struggling to preserve their own ways of life.

A Global Exchange of Crops and Animals

Beginning in the earliest days of conquest, a number of plants and animals were exchanged among the Americas, Europe, Africa, and Asia. Many plants from the Old World are essential to agriculture in Middle and South America today: rice, sugarcane, bananas, citrus, melons, onions, apples, wheat, barley, and oats— to name just a few. When disease decimated the native populations of the Americas, the colonists turned the abandoned land into pasture for herd animals imported from Europe, including sheep, oxen, cattle, donkeys, horses, and mules. European draft

Tomatoes, pineapples, and green peppers are among the many plants domesticated in the Americas that rest beside Old World foods (oranges, grapes, bananas, and eggplant) on this central European vegetable stand. [Lydia Pulsipher.]

animals helped fill in Native American irrigation canals, drain lakes, and plow the relics of complex Indian gardens into huge one-crop fields of sugarcane or wheat. Surviving Native Americans living on the plains of Mexico and Argentina adopted horses, using them to hunt large game. Others learned to herd sheep and used the fleece in ancient spinning and weaving technology.

Plants first domesticated by Native Americans have changed diets everywhere and have become essential components of agricultural economies around the globe. In Europe, for example, the potato had so improved the diet of the poor by 1750 that it fueled a population explosion. Manioc, though less nutritious, played a similar role in West Africa. Corn is a widely grown garden crop in Africa, and peanuts and cacao (chocolate) are essential cash crops. Peppers are a cash crop in China, pineapples in the Pacific; and tomatoes have transformed Mediterranean cuisine. Table 3.1 lists some of the more common globally used plants from Middle and South America and their sites of probable domestication in the Americas.

The Legacy of Underdevelopment

Today, despite Middle and South America's potentially productive environment, perhaps 40 percent of the people are poor and lack access to such basic resources as land, adequate food and shelter, and basic education. A small elite class, however, enjoys levels of affluence equivalent to those of the wealthy in the United States. In part, these inequalities are the lingering effects of colonization: economic policies that still favor the export of raw materials and foster the dominance of outside investors who spend their profits elsewhere rather than reinvesting them within the region. Other factors leading to inequality are official corruption at all levels of government and social attitudes and policies that stifle local entrepreneurial development and upward social mobility. All these factors have resulted in an unskilled populace and infrastructures that are inadequate by modern standards. It is difficult for the region to compete in the global marketplace, which increasingly demands an educated labor force, good transport networks, and technological capabilities.

The first colonial enterprises to be established were based on the extraction of raw materials, including gold, silver, and other minerals; timber; and various agricultural products, such as cotton, indigo, tobacco, and sugar. These activities were part of an emerging policy of **mercantilism** in which the rulers of Spain and Portugal and, later, England and Holland sought to increase the power and wealth of their realms by managing all aspects of production, transport, and trade. Resources were mined or grown in the colonies at the lowest possible cost, brought home to Europe, and manufactured into trade goods (cotton and wool into cloth, indigo into dye, sugar into rum). These finished products were then sold to the colonies and elsewhere in the world markets. Money flowed into Europe as payment for the manufactured goods, enriching the merchants, and taxes on the trade brought revenues to the national treasuries. A merchant marine (fleet of trading ships) was maintained to transport resources; manufactured goods; and, once the slave trade began in earnest in the 1600s, people. Nearly all commercial activity was focused solely on enriching the **colonizing,** or **mother, country.** The colonies of

TABLE 3.1	Major domesticated plants originating in the Americas and now used commercially around the world

Type	Common name	Scientific name	Place of origin
Seeds	Amaranth	*Amaranthus cruentus*	Southern Mexico, Guatemala
	Beans	*Phaseolus* (4 species)	Southern Mexico
	Maize (corn)	*Zea mays*	Valleys of Mexico
	Peanut	*Arachis hypogaea*	Central lowlands of South America
	Quinoa	*Chenopodium quinoa*	Andes of Chile and Peru
	Sunflower	Helianthus annus	Southwestern and southeastern North America
Tubers	Manioc (cassava)	*Manihot esculenta*	Lowlands of Middle and South America
	Potato (numerous varieties)	*Solanum tuberosum*	Lake Titicaca region of Andes
	Sweet potato	*Ipomoea batatas*	South America
	Tannia	*Xanthosoma sagittifolium*	Lowland tropical America
Vegetables	Chayote (christophene)	*Sechium edule*	Southern Mexico, Guatemala
	Peppers (sweet and hot)	*Capsicum* (various species)	Many parts of Middle and South America
	Squash (includes pumpkin)	*Cucurbita* (4 species)	Tropical and subtropical America
	Tomatillo (husk tomato)	*Physalis ixocarpa*	Mexico, Guatemala
	Tomato (numerous varieties)	*Lycopersicon esculentum*	Highland South America
Fruit	Avocado	*Persea americanas*	Southern Mexico, Guatemala
	Cacao (chocolate)	*Theobroma cacao*	Southern Mexico, Guatemala
	Papaya	*Carica papaya*	Southern Mexico, Guatemala
	Passion fruit	*Passiflora edulis*	Central South America
	Pineapple	*Ananas comosus*	Central South America
	Prickly pear cactus (tuna)	*Opuntia* (several species)	Tropical and subtropical America
	Strawberry (commercial berry)	*Fragaria* (various species)	Genetic cross of Chile berry + wild berry from North America
	Vanilla	*Vanilla planifolia*	Southern Mexico, Guatemala, perhaps Caribbean
Ceremonial and drug plants	Coca (cocaine)	*Erythroxylon coca*	Eastern Andes of Ecuador, Peru, and Bolivia
	Tobacco	*Nicotiana tabacum*	Tropical America

Source: B. Kermath and L. Pulsipher, "Guide to Food Plants Now Used in the Americas," unpublished manuscript.

Spain in the Americas, for example, were allowed to trade only with Spain—not with one another. If businesspeople in Peru wanted to trade legally with their counterparts in Mexico, for instance, the goods first had to cross the Atlantic in Spanish ships, be taxed in Spain, and then be reshipped back to Mexico in Spanish ships. These cumbersome and expensive restrictions nurtured a huge underground economy and institutionalized dishonesty as merchants and customers often tried to bribe their way around the trade laws. The trade restrictions also led to considerable resentment of the Spanish authorities.

Figure 3.6 Colonies in the Americas. Except in Panama and the Caribbean, all of Spain's colonies in the New World achieved independence in the 30-year span from 1810 to 1840. [Adapted from *Hammond Times Concise Atlas of World History* (Maplewood, N.J.: Hammond, 1994), p. 69.]

In the early nineteenth century, wars of independence transformed the region. The modern countries of Middle and South America emerged, and Spain was left with only a few colonies in the Caribbean (Figure 3.6). Many supporters of these revolutions were people who had been shut out of the profits of the colonial system due to their race (many were **mestizos,** of mixed European and Native American descent). Once the revolutionary leaders came to power, however, they emulated their colonial predecessors. They became a new elite that controlled the state and monopolized economic opportunity, doing little to expand economic development or the people's access to political power.

During the twentieth century, some countries in the region began to experiment with radical ways of fostering development, as we shall see in the section on economic and political issues. Today, the economies of Middle and South America are much more complex and technologically sophisticated than they once were. Nevertheless, the colonial pattern of dominance by an elite and dependence on extraction of raw materials remain, contributing to a persistent pattern of underdevelopment and unequal distribution of wealth.

POPULATION PATTERNS

The events of the last 500 years have greatly affected the present distribution and character of human settlement in Middle and South America. Conquest and colonization brought the swift demise of Native Americans; soon they were more than replaced by millions of people from Europe, Africa, and Asia. It would be difficult to find a case elsewhere in human history that rivaled this massive shift in human population. Today, the patterns of human settlement continue to change, but these changes are being generated from within the region rather than coming from outside. The population continues to climb, but primarily because of high birth rates rather than migration. At present, the major migration trend is internal: rural-to-urban migration is taking place everywhere in the region and transforming traditional ways of life. Though smaller in numbers, a second major migration trend is international: many people from the Caribbean, Mexico, and Central America are leaving their countries, temporarily or permanently, to seek opportunities elsewhere. In the last 20 years, the United States has been the preferred destination for many of these transnational migrants.

Figure 3.7 Population density in Middle and South America. [Adapted from *Hammond Citation World Atlas* (Maplewood, N.J.: Hammond, 1996.]

Population Numbers and Distributions

As of the year 2000, 518 million people were living in Middle and South America, close to 10 times the population of the region in 1492. This number is about 200 million more than presently live in North America.

Population Distribution. The population density map in Figure 3.7 reveals a very unequal distribution of people in Middle and South America, just as in most places on earth. It is often thought that people concentrate where certain physical landforms are common, but that is only partly true. If you compare the population density map with the map at the beginning of this chapter, you will see that there is no obvious, consistent relationship between density patterns and landforms. Some of the highest densities, such as those around Mexico City and in Colombia and Ecuador, are in highland areas. Elsewhere, high concentrations are found in lowland zones along the Pacific coast of Central America and especially along the Atlantic coast of South America. In tropical and subtropical zones, the cool uplands (*tierra templada*) are particularly pleasant and healthful and were comparatively densely occupied even before the conquest began 500 years ago. Most of the coastal lowland concentrations, in *tierra caliente*, are near important seaports. People are attracted to port cities by the possibility of jobs in the vibrant and varied economy and by the interesting social life. Living near the sea is also attractive because the water modifies the hot, humid climate, making it relatively cooler and breezier than lowlands in the interior.

The map also reveals the relatively empty lands in the region, especially in South America. Cold and windy Patagonia at the far south end of South America is nearly unoccupied, and so are the desert regions of Chile and Peru along the Pacific coast. The vast Amazon River basin is also only lightly settled. Despite recent efforts to develop and populate this wet tropical zone, it seems able to sustain only hunting and gathering, shifting cultivation, and light forestry. Some people think, though, that advancements in technology will make it possible to use and settle the Amazon environments more intensively in the future.

Population Growth. Twentieth-century rates of natural population increase have been high in Middle and South America. (Recall from Chapter 1 that the rate of natural increase is population growth resulting from births alone, not counting growth from immigration.) Although these rates are now declining (Figure 3.8), the time needed for the region's population to double is still just 39 years (in North America, it is 120 years). Such a quick doubling of the population is a disturbing prospect for a region in which the majority of people already suffer from a low standard of living. Any gains that might have gone toward improving their lives will be checked by the costs of supporting more and more new people. In just 39 years, twice as many houses, schools, and hospitals will be needed merely to maintain the present inadequate standard of living.

There are a number of reasons that rates of natural increase remain high when compared with those of North America, Europe, and East Asia, or even with world averages. One is that until recently, most Latin Americans lived in agricultural areas. Children were seen as sources of wealth because they could do useful work at a young age and eventually would care for their aging elders. Also, high infant death rates encouraged parents to have many children to be sure of raising at least a few to adulthood; and the Catholic church has discouraged systematic family planning. Further, as will be discussed later, the cultural mores of *machismo* and *marianismo* have reinforced the idea that both men and women validate themselves as adults by reproducing prolifically. Finally, when medical care improved modestly beginning in the 1930s, death rates began to decline rapidly. By 1975, death rates were about one-third what they had been in 1900. Because birth rates did not decrease as quickly as death rates, population growth was especially rapid between 1940 and 1975.

By the 1980s, for a number of reasons, Latin Americans were beginning to choose contraception to limit family size, and population growth rates started to fall. Now the region is beginning to undergo the demographic transition (see Figure 1.19). Between 1963 and 2000, the rate of natural increase for the entire region fell from about 2.5 to 1.8 percent, still a high rate of growth when compared with those of North America, Europe, and East Asia (all at 1.0 percent or less in 2000). The bar graph in Figure 3.8 compares the natural increase rates of selected countries from 1975–1998 to the projected rates for 1998–2015.

As of the year 2000, the Caribbean as a whole has the lowest rate of natural increase in the region (1.3 percent), with Cuba (0.7 percent), Trinidad and Tobago (0.7 percent), and Barbados (0.5 percent) having the very lowest rates, about the same as North America (0.6 percent). Several factors explain these low rates. Cuba, Trinidad and Tobago, and Barbados provide both women and men with education, meaningful work, and basic health care. Infant mortality rates are low, so people can expect to see their one or two children grow to adulthood. Barbados prospers from manufacturing, computerized information processing, tourism, **remittances** (money sent home) from migrants, and agriculture. Its skilled citizens have

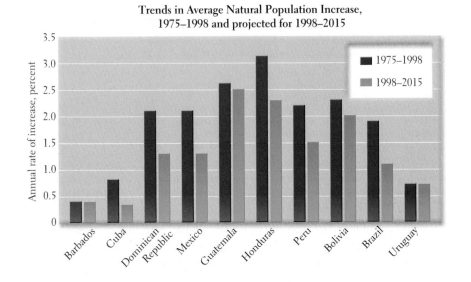

Figure 3.8 Trends in natural population increases. Rates of natural increase have declined throughout the region, but in many countries they remain high enough to outstrip efforts at improving standards of living. [Adapted from *United Nations Human Development Report 2000*, United Nations Development Programme.]

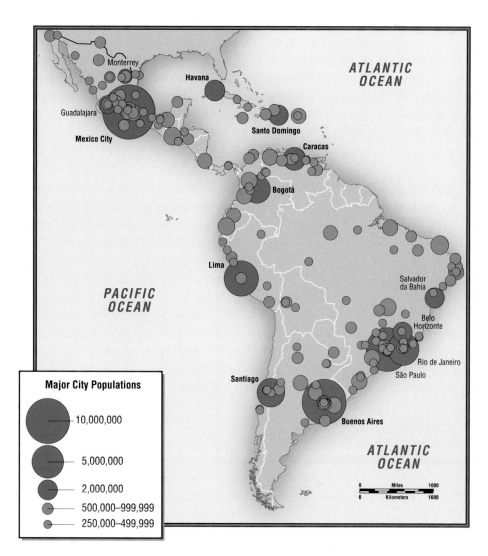

Figure 3.9 Urban concentrations in Middle and South America.

Major City Populations

10,000,000

5,000,000

2,000,000

500,000–999,999

250,000–499,999

many economic and social options for constructing interesting lives that make large families less attractive. Trinidad and Tobago, while less prosperous, has oil-related industries and a thriving business community, and Cuba has made a strong effort to improve social welfare and encourage smaller families over the past 35 years.

In mainland Middle America, poverty is more widespread and women have less access to education and employment; both factors are associated with high rates of natural increase (2.1 percent for the region). South America, also afflicted with high rates of poverty, has a natural increase rate of just 1.7 percent. Within South America, however, there is considerable variation: in the year 2000, the rate of natural increase in Bolivia was 2.0 percent, in Argentina 1.1 percent, in Paraguay 2.7 percent, and in Uruguay 0.7 percent. Again, this variability can be partially explained by differing standards of living and access to education and jobs, especially for women.

Migration and Urbanization

Migration is perhaps the most important social force at work in the world today. Why? Because people are migrating at unprecedented rates and because migration often abruptly introduces large numbers of people from rural areas into new, usually crowded, cities. Residence in urban places may or may not provide more opportunity, but it will surely expose the immigrants to values and ways of life that contrast sharply with those they have known. Since the early 1970s, the region of Middle and South America has led the world in rates of migration.

Geographers recognize the importance of economic and social factors in initiating and sustaining migration. People are drawn to migrate by the possibility of better job opportunities. Often they hear of these opportunities from networks of friends and family who have migrated earlier and report increased prosperity. Communication between the new migrants and their home communities, including the sending of remittances, stimulates still more rural-to-urban migration. As a result, cities throughout the region have grown remarkably quickly. More than 70 percent of the people in the region now live in towns with populations of at least 2000; but increasingly, in Middle and South America and elsewhere, one city in a country is vastly larger than all the others, sometimes accounting for one-quarter or more of the country's total population. Such cities are called **primate cities;** examples are Mexico City; Managua, Nicaragua; Bridgetown, Barbados; San José, Costa Rica; and Buenos Aires, Argentina. Figure 3.9 shows the pattern of urban concentrations at the regional scale.

Thinking Geographically: ON THE WEB

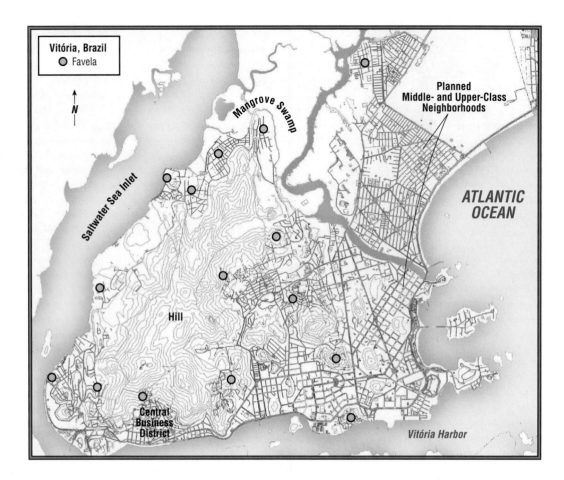

Figure 3.10 The pattern of squatter settlements in Vitória, Brazil. [Adapted from fieldwork map by Eric Spears.]

This converging of people into just one or two cities in a country leads to **lopsided spatial development.** Wealth and power are concentrated in one place, while distant rural areas and even other towns and cities have difficulty competing for talent, investment funds, and government services. Many provincial cities languish, also losing people to migration. Primate cities are usually found in underindustrialized countries. They attract people even though they do not provide enough jobs, housing, or services for the flood of hopeful migrants.

Because the rush to the cities has not been accompanied by a general rise in prosperity, no city in the region is prepared for the influx. Spending on urban infrastructure (such as roads and sanitation) and on social services (such as schools and hospitals) has failed to keep pace with the inflow of people, and the signs of urban decay are everywhere. (For an exception to this pattern, see the box "Curitiba, Brazil: A Model of Urban Planning.") Although there are wealthy neighborhoods in the cities, they are often heavily guarded because of the proximity of unsightly, uncontrolled shantytowns built by the poor. This self-help housing—called **colonias, barrios, favelas,** or **barriadas**—can arise spontaneously. Because most recent urban migrants can't afford to buy or lease land, they become squatters who invade a piece of property after dark and set up simple dwellings by morning. Once established, these communities are extremely difficult to dislodge: the impoverished are such a huge portion of the urban popula-

tion that even those in positions of power will not challenge them directly. The result is a city landscape that is remarkably different from the common U.S. pattern of a poor inner city surrounded by affluent suburbs, with relatively clear spatial separation of classes, races, and family styles. Rather, housing types are intermingled. Inner-city neighborhoods of vastly different socioeconomic classes nestle close to one another, and the suburban fringe is also a mixture of wealthy and poor neighborhoods. The wealthy neighborhoods are similar in architectural style and formality to those of North America but are usually ringed with high fences or walls. The poor neighborhoods are often built of unconventional materials and laid out in an uncontrolled and seemingly chaotic pattern. Figure 3.10 shows the distribution of favelas in Vitória, Brazil, a city about 300 miles northeast of Rio.

The squatters are often enterprising and admirable people who have simply made the most of the bad hand dealt them. Such settlers sometimes organize themselves to pressure for social services: schools, day care, part-time jobs for young mothers, sewer systems. Some far-sighted urban governments, such as that in Fortaleza (Ceará) in northeastern Brazil, contribute building materials so that the squatters can build more permanent structures with at least crude indoor plumbing. Expanses of shacks and lean-tos have thereby been transformed through self-help into modest but livable suburban places. Moreover, these communities can be centers of pride and support for the residents, where community work, music, folk belief systems, and crafts flourish. For example, many of the most prestigious steel bands of Port of Spain, Trinidad, have their homes in such shantytowns. In Brazil, the favelas are home to samba clubs (musical ensembles) and folk religious movements such as Umbanda, Batuque, and Condomble (see the discussion and photo on pages 175–176).

Squatter communities, like this one in Fortaleza, Brazil, sometimes become permanently established with the help of government funding for cinder blocks, roofing tiles, and plumbing. [Lydia Pulsipher.]

VIGNETTE One squatter settlement in Fortaleza, Brazil (shown in the photograph), lies along the Atlantic coast just north of a line of high-rise tourist hotels. One day, I took a stroll past the settlement and was invited by a resident to join him on his porch, just visible in the picture behind the red Coca Cola chest. He maintained a small refreshment stand, and his wife ran a beauty parlor that catered to women from the area. The man explained that they had come five years before, after being forced to leave the drought-plagued interior when they lost the right to cultivate their rented land. The landowner had sold it to the provincial government, which was building a dam to create an irrigation reservoir. Here in Fortaleza, they had slowly constructed the building they used for home and work from resources they collected along the beach, and eventually they were able to purchase the roofing tiles that gave it an air of permanency. [Lydia M. Pulsipher, personal interview, August 1987.]

Interestingly, rural women are just as likely to migrate to the city as rural men. Rural development projects often end up decreasing available jobs, due to mechanization, and many agricultural laborers who lose their jobs may consider migration. Women may be disproportionately affected: despite their experience working in agriculture, they are rarely considered for training as farm equipment drivers and mechanics. In urban areas, unskilled women migrants can usually find work as domestic servants. Low wages, however, force them to live in the households where they work, where they are often subject to sexual overtures, yet are themselves blamed if they fail to adhere to rigid standards of behavior. Male urban migrants tend to depend on

AT THE LOCAL SCALE *Curitiba, Brazil: A Model of Urban Planning*

The southern Brazilian city of Curitiba, capital of the grain-producing state of Paraná, is gaining international renown for its environmentally friendly urban planning. Like most other Brazilian cities, Curitiba has mushroomed, doubling its population in just 20 years to more than 1.3 million people. But Curitiba is unusual in that it has carefully oriented its expansion around an integrated public transport system. Minibuses bring people from their neighborhoods to terminals where they meet express buses to all parts of the city. Being able to get to work quickly and cheaply has helped Curitiba's poor to find and keep jobs.

The city's streets and its many parks and green spaces are kept spotlessly clean and decorated with flowers. The city also has a decentralized public health service. A trash recycling program encourages people in the informal economy to collect specific kinds of trash and sell them to recycling companies.

One goal is to keep migrants from swamping Curitiba's well-designed, environmentally sensitive urban environment with shantytowns. The city first tries to stem the flow by offering free bus tickets back home to new migrants. Twenty-five thousand people have used the tickets. To accommodate those migrants who come to take what are often short-term jobs in Curitiba's industrial sector, the city is building rural satellite towns, *vilas rural*, where people can live and maintain their agricultural skills by farming small plots when they are between industrial jobs. In this way, they will be able to feed themselves and possibly some urbanites as well. Financed by the World Bank and the Inter-American Development Bank, 5 *vilas rural* have been built, 15 are under construction, and 60 more are planned. The strategy is to accommodate a significant proportion of Paraná's landless farmers in these urban fringe communities, where the advantages of both rural and urban life can be enjoyed by those of meager means.

short-term day work, most often in low-skill jobs in construction, maintenance, small-scale manufacturing, and petty commerce. They must compete with a large throng of eager workers, and most are periodically unemployed. Many make some cash in the informal economy, doing such work as street vending, running errands, cleaning, washing cars, and recycling trash; some may engage in crime. For both men and women, the loss of family ties and village life is sorely felt and the chances for re-creating normal family life in the urban context are extremely low. Opportunities to meet and marry mates of their own background are limited, while the chances of having a child as a single parent are very high. Usually, the burden of single parenthood falls to the woman.

It always takes some resourcefulness to move from one place to another. Hence, it is usually those who already have some advantages of education and unusual ambition who migrate to cities (where their talents may, however, go to waste). Thus, the sending communities are deprived of young adults in whom they have invested years of nurture and education. This loss, often referred to as **brain drain,** happens at several scales in Middle and South America. There is rural-to-urban migration from villages to regional towns and from towns to the main cities, as well as international migration from the many countries in the region to North America and Europe. On the other hand, families often encourage their youth to migrate so that they can benefit from the remittances the migrants send home. Often, migrants retain ties to home and return after reaching their financial goals, perhaps with their skills enhanced by the overseas sojourn. The brain drain is therefore offset somewhat by returning migrants and the remittances, goods, and services that migrants provide to their home communities. Thus, there is a mix of consequences from these migrations—some positive, some negative.

The receiving societies, whether cities such as Rio de Janeiro or Mexico City or countries such as the United States, garner considerable benefit from immigration. For example, in 2000, the American Medical Association reported that one-quarter of the doctors practicing in the United States were immigrants, as were about 50 percent of medical students. Of the practicing doctors who were immigrants, more than 5 percent were from Middle and South America. The United States saves money when other countries supply the early education for these physicians (and sometimes even provide government scholarships for higher education). Then these physicians may spend their most productive years treating patients in the United States. And so it is with immigrants generally: most of those who migrate are the ones with education above the average and with the initiative and creativity to take a long, risky journey into unfamiliar territory—all characteristics badly needed by the places they leave

CURRENT GEOGRAPHIC ISSUES

Historically, power and wealth in Middle and South America have been concentrated in the hands of a few, and the last century of economic modernization and the transformation of most countries from rural to urban societies haven't changed that reality. The rewards of urbanization and industrialization have spread only thinly to the masses of poor and not-so-poor, who instead have had to absorb the effects of large government debts incurred in the 1970s and 1980s.

ECONOMIC AND POLITICAL ISSUES

Although this region is not as poor on average as sub-Saharan Africa, South Asia, or Southeast Asia, it has the widest gap between rich and poor. The richest 20 percent of the population is between 10 and 30 times richer than the poorest 20 percent (Table 3.2), and nearly 40 percent of the population is living in poverty. This **income disparity** is troubling in many ways. From a moral standpoint, many people argue that it is wrong for wealth and resources to be concentrated among a tiny number of people while so many have so little. From an economic perspective, poverty prevents people from contributing to economies with their own purchasing power or with their skills, which tend to remain at a low level. Thus, wherever large numbers of people sink into poverty, the potential for a nation's growth diminishes. The danger of political instability also increases because there is the possibility of violent rebellion by impoverished people and equally violent repression by governments. In the not so distant past—in Bolivia, Peru, Chile, Brazil, Argentina, Colombia, Guatemala, Honduras, El Salvador, Mexico, and Haiti—hundreds of thousands have died in repeated waves of government repression, revolution, counter-revolution, and coups d'état.

It has been argued that globalization is reducing the importance of income disparity in the region, as increases in international trade provide more jobs that will raise the income of the poorest. The impact of globalization is seen especially in lowered trade barriers and growing investment by multinational corporations as well as in aid and assistance from the United States, Europe, and some Asian countries. This trend toward globalization is certainly expanding opportunity for an affluent minority. These are the highly educated, urban, technologically proficient businesspeople (sometimes called technocrats) whose new wealth has given them prestige and power. Many poor people have also benefited from jobs created by globalization. However, many have also lost jobs and experienced dramatic declines in welfare support provided by governments. In the political realm, recent advances in democratic participation have been credited with giving the poor a more influential political voice. Nevertheless, many of the region's democratically elected governments are reluctant to

TABLE 3.2 *Share of wealth, 1987–1998*

Country	Ratio of wealth of richest 20% to wealth of poorest 20%[a]
Bolivia	9:1
Brazil	26:1
Chile	17:1
Colombia	20:1
Guatemala	30:1
Mexico	16:1
Peru	12:1
China	8:1
France	6:1
Jordan	6:1
Philippines	10:1
Thailand	8:1
Turkey	8:1
Canada	5.2:1
United States	9:1

[a]Decimals rounded up or down.

Source: United Nations Human Development Report 2000, Tables 4 and 5. The UN used data from 1987 to 1998 derived from available data on either income or consumption. The figures are an approximate representation of how much richer the wealthiest 20 percent of the population is than the poorest 20 percent in selected countries.

embrace egalitarian principles, and they continue to maintain much closer ties to wealthy elites, the military, and foreign corporations and governments than they do to their own poorer citizens, who number in the millions. Thus, many unresolved and difficult challenges remain.

Phases of Economic Development: Globalization and Income Disparity

The current economic and political situation in Middle and South America is deeply rooted in the history of its dependent relationships with external forces: colonial mother countries, Europe's and America's economic interests, and foreign and multinational corporations. There have been three major economic phases: **the early extractive phase,** the **import substitution industrialization phase,** and the **structural adjustment phase** of the present. All three phases have been characterized by various levels of external influence and all have helped entrench wide income disparities.

Early Extractive Phase. This phase, beginning with the conquest and lasting until the early twentieth century, was characterized by colonial policies such as mercantilism that resulted in unequal trade. A relatively small flow of foreign investment and manufactured goods came into the region, and a vast flow of resources left for Europe and beyond. Investment capital (money to fund businesses) came first from Europeans—the Spanish and Portuguese, then other western Europeans—and later from North Americans and other international sources. The capital was used to establish farms, plantations, mines, sawmills, roads and, more recently, industries and railways, generally under the ownership of the foreign investors. The products of these enterprises were exported to Europe, often at bargain prices. The region as a whole was at a global economic disadvantage because the essential manufactured items that its people needed, such as farm tools and household utensils, had to be purchased from Europe and North America at relatively high prices. The same was true for guns and machinery purchased by governments. Moreover, the vast majority of the population received little benefit from the extraction of the region's resources because wages remained low (or, in the case of slave economies, nonexistent) and employment opportunities scarce.

A look at the key economic institutions of the day gives further insight into the dynamics of this colonial mercantile dependency. From the early days of conquest until the nineteenth century, agricultural development was centered on **haciendas,** large rural estates first granted to colonists as a reward for claiming territory and people for Spain and then passed down through families. Laborers on the haciendas were virtually bound to the land like slaves or feudal serfs, while the owner often lived in a distant city or in Europe. Haciendas were inefficient, often producing several types of crops but using only a fraction of their potential agricultural land. Cultivation techniques were labor intensive and outdated. Inefficient management and the lack of specialization meant that haciendas produced goods at relatively high costs.

By the late nineteenth century, as transportation to ports improved, growers gained greater access to global markets. Many haciendas shifted to more efficient styles of production. Some became **plantations,** which are like large factory farms growing a single crop such as sugar, coffee, cotton, or (more recently) bananas. British, French, and Dutch colonizers had already established the cash-crop plantation in the 1600s on Caribbean islands. To maximize production, owners of plantations made much larger investments in equipment and labor than did hacienda owners. Unlike haciendas, which were often established in the continental interior in a variety of climates, plantations were for the most part situated in tropical coastal areas with year-round growing seasons. Their coastal location also gave easier access to global markets via ocean transport.

Another increasingly important institution was the **livestock ranch.** As markets for meat, hides, and wool grew in Europe and North America, many haciendas specialized in raising cattle and sheep. Today, commercial ranches are found in the drier grasslands and savannas of South America, Central America, and northern Mexico and even in the wet tropics on formerly forested land.

The other main industry of the early extractive period was mining. Mines had to be located wherever the desired minerals (at first primarily gold and silver) were to be found. Labor had to be moved to the mine, and the minerals had to be partially refined on site to reduce the costs of transport. Important mines were located on the island of Hispaniola, in north central Mexico, in the Andes Mountains, in the Brazilian Highlands, and in many other locations as well. Today oil and gas have been added to mineral extraction, but there remain rich mineral mines throughout the region, producing gold, silver, copper, tin, precious gems, titanium, bauxite, and tungsten.

The mines, ranches, plantations, and haciendas of the early extractive phase did not contribute to the development of the region because their owners generally invested their earnings outside the region. Owners, who often lived in Europe while receiving income from interests in Middle or South America, preferred to invest their money in Europe, where the more technologically advanced manufacturing enterprises of the industrial revolution were bringing unprecedented opportunities for profit. It is useful to compare this pattern with the economic development of North America at the time. In North America after the American Revolution, the profits from industries based on raw materials went to owners who tended to live in the region and who wanted to bring the industrial revolution to their homeland. Hence, they invested their money locally in industries that processed North America's raw materials into more valuable finished goods. This investment provided the foundation for North America's present greater economic stability and higher living standards relative to Middle and South America. In contrast, the export of raw materials remained the basis of Middle and South America's economy. By the late nineteenth and early twentieth centuries, when nearly all the countries of Middle and South America were independent, wealthy European and North American private investors had purchased many of the extractive industries in the region, thus reinforcing the pattern of profits leaving the region. These investments came to be known as neocolonial because the power these foreign investors wielded resembled that of colonial officials. (**Neocolonialism** refers to modern efforts by dominant countries to control economic and political affairs in other countries to further their own aims.)

Import Substitution Industrialization Phase. The import substitution phase was a reaction against the early extractive phase in that great efforts were made to keep money and resources within the region, in the hope of achieving economic self-sufficiency. These efforts began in the early twentieth century in response to waves of popular political protest against the dominance of the economy and society by foreign businesses and a few wealthy citizens. Many governments—Mexico and Argentina were the most prominent—proclaimed themselves socialist democracies and enacted a set of policies that came to be known as **import substitution industrialization (ISI)**. First, national governments replaced foreigners as the main source of investment money in the region. States either bought or seized (usually with some payment to owners) the most profitable extractive industries, then used their profits to create manufacturing industries that could supply the

The precipitation tanks for the separation of aluminum are part of the C.V.G. Venalum Company holdings in Venezuela. It is the largest aluminum smelter in Latin America. The company began operation in 1973, primarily to make products for export. The Venezuelan government owns about 80 percent of the company (10 to 20 percent of this 80 percent is owned by C.V.G. workers), and the remaining 20 percent is foreign-owned, primarily by Japanese companies. Since 1999, C.V.G. has offered additional shares to foreign investors, and companies from the United States, France, and the United Kingdom have expressed interest. [Mario Corvetto/Evergreen Photo Alliance.]

goods once purchased from Europe and North America. To encourage the local population to buy manufactured goods from local suppliers, governments placed high tariffs on imported manufactured goods. The money and resources kept within each country were expected to provide the basis for further industrial development that would create numerous well-paying jobs and raise living standards for the majority of people. National governments further hoped that these new manufacturing industries would grow large enough to replace the extractive raw materials industries as the backbone of the economy.

Although ISI strategies still survive in a few countries, they failed to bring about a thorough transformation of the industrial sector. In particular, countries remained dependent on imported goods and technology. Some progress was made in reducing disparities in income as governments increased spending on public health, education, and infrastructure and enacted reforms that broke up some large haciendas and distributed them among landless farmers. But the goal of lifting large numbers of people out of poverty through jobs and general economic growth was never realized. The state-owned manufacturing sectors upon which the success of ISI depended were never able to expand sufficiently to absorb the rapid influx of migrants to the cities. Too often, corrupt politicians and bureaucrats manipulated state-owned projects to enrich themselves and their friends. Positions in these industries were awarded to political followers. State-owned manufacturing industries couldn't compete in price or quality with foreign-made products, and tariff barriers were still required to protect locally made goods. Not all state-owned corporations were losing propositions, however: Brazil—with its aircraft, armament, and auto industries—and Mexico—with its oil and gas industries—both experienced success in their ISI development.

Contributing to the overall low success of ISI was the fact that the number of people able to afford manufactured goods did not increase significantly, either within any country or in the region as a whole. The region might have had enough consumers to sustain manufacturing industries if trade and transport barriers between countries had not restricted trade and raised prices to the consumer. The high tariffs applied to European and North American manufactured goods also applied to goods produced by neighboring countries. Even goods that could be traded profitably between countries had to be shipped via an outdated, poorly integrated, and unreliable transportation system. Consider how trade and prosperity in the United States would be constrained if each state had trade barriers against all others and interstate transport consisted of rudimentary dirt or tar-covered roads and a few railroads that did not easily connect at the borders.

For all these reasons, state-owned manufacturing industries—often unprofitable and inefficient—failed as incubators of innovation and expansion or as suppliers of employment on a wide scale. As a result, the region remained ever more dependent on the export of raw materials.

The Debt Crisis. Ultimately, a global financial crisis brought the ISI phase to an end. Starting in the early 1950s, the global economy had enjoyed a period of economic expansion. During this period, the prices for raw materials held steady. That period of prosperity ended in the 1970s. Oil price increases in 1975 and 1978 undermined growth plans in most countries. The prices offered for raw materials (other than oil) in the global market dropped, including prices for exports that were particularly important to Middle and South America, such as various minerals, coffee, and sugar. Exports of raw materials could maintain the region's economies when prices for those materials were high, but not when they were low. Price drops shook national economies repeatedly in the 1970s, putting increasing strain on the finances of the region's governments. At the same time, governments and private interests pursued expansive plans to modernize and industrialize their national economies. They paid for these projects by borrowing millions of dollars from major international banks, most of which were in North America or Europe. The beginning of a global recession in 1980 put a halt to all such ambitious plans. Dragged down by the recession, the Middle and South American economies could not meet their targets for growth. Thus, the governments were also unable to repay their loans (Figure 3.11). The damage to Middle and South America was made worse by the fact that the biggest borrowers—such as Mexico, Brazil, and Argentina—also had the largest economies in the region. Hence, huge debts now burdened the countries that had been the most likely to grow.

Current Structural Adjustment Phase. The advanced industrial economies of the United States, Europe, and Japan rebounded from the global recession of 1980–1983. Influential experts recommended that developing countries adopt the same economic policies that seemed to be so successful in advanced economies. For example, they advocated export-oriented manufacturing as a more robust way to achieve development than import substitution industry, and the success of Japanese and Pacific Rim export economies in the 1980s reinforced this view. Free trade and privatization rather than government-funded industry were considered the soundest ways to achieve economic expansion.

In 1982, a drop in the price of oil left Mexico unable to make its repayments on a huge national debt. Following Mexico's default, international banks refused to lend to the debtor nations of Middle and South America unless they first agreed to a series of policy changes known as **structural adjustment programs (SAPs)**, which are designed to free up money for loan repayment through a number of methods. SAPs were developed and enforced by a global financial institution known as the International Monetary Fund (IMF), acting on behalf of the major international banks. The goal of the IMF was to promote free-market policies through structural adjustment programs. That is, in order to obtain further loans, governments were required to begin removing tariffs on imported goods of all types and to sell state-owned industries, both in raw materials and in manufacturing, to private investors—often multinational corporations located in Asia, North America, and Europe. SAPs also reversed a trend during the ISI era toward the expansion of government programs. Now governments were required to fire many civil servants and drastically reduce spending on public health, education, and infrastructure. Structural adjustment is often associated with *neoliberalism*, the idea that social justice—basic nutritional, shelter, and educational needs met for everyone—can best be achieved through free-market economic development. New policies encouraged the expansion of industries that produced goods for export. Most of these industries were still based on the extraction of raw materials, because they were seen as

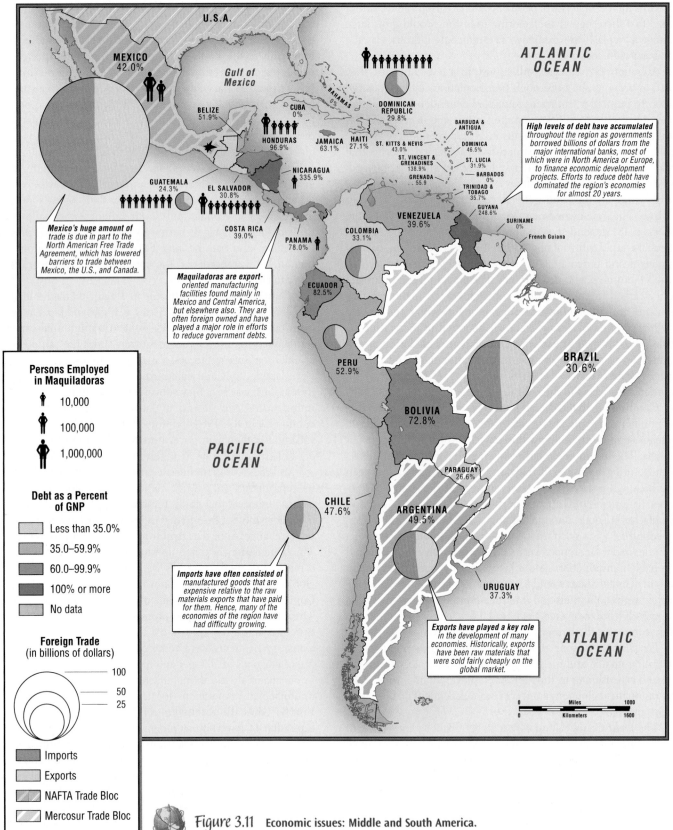

Persons Employed in Maquiladoras

👤 10,000

👤 100,000

👤 1,000,000

Debt as a Percent of GNP

Less than 35.0%

35.0–59.9%

60.0–99.9%

100% or more

No data

Foreign Trade
(in billions of dollars)

100

50

25

Imports

Exports

NAFTA Trade Bloc

Mercosur Trade Bloc

U.S.A.

ATLANTIC OCEAN

MEXICO 42.0%

Gulf of Mexico

BELIZE 51.9%

CUBA 0%

BAHAMAS 0%

DOMINICAN REPUBLIC 29.8%

HONDURAS 96.9%

JAMAICA 63.1%

HAITI 27.1%

BARBUDA & ANTIGUA 0%

ST. KITTS & NEVIS 43.0%

DOMINICA 46.5%

ST. VINCENT & GRENADINES 138.9%

ST. LUCIA 31.9%

BARBADOS 0%

GRENADA .. 55.9

TRINIDAD & TOBAGO 35.7%

GUATEMALA 24.3%

EL SALVADOR 30.8%

NICARAGUA 335.9%

VENEZUELA 39.6%

GUYANA 248.6%

SURINAME 0%

French Guiana

COSTA RICA 39.0%

PANAMA 78.0%

COLOMBIA 33.1%

ECUADOR 82.5%

PERU 52.9%

BRAZIL 30.6%

BOLIVIA 72.8%

PARAGUAY 26.6%

CHILE 47.6%

ARGENTINA 49.5%

URUGUAY 37.3%

PACIFIC OCEAN

ATLANTIC OCEAN

Mexico's huge amount of trade is due in part to the North American Free Trade Agreement, which has lowered barriers to trade between Mexico, the U.S., and Canada.

Maquiladoras are export-oriented manufacturing facilities found mainly in Mexico and Central America, but elsewhere also. They are often foreign owned and have played a major role in efforts to reduce government debts.

High levels of debt have accumulated throughout the region as governments borrowed billions of dollars from the major international banks, most of which were in North America or Europe, to finance economic development projects. Efforts to reduce debt have dominated the region's economies for almost 20 years.

Imports have often consisted of manufactured goods that are expensive relative to the raw materials exports that have paid for them. Hence, many of the economies of the region have had difficulty growing.

Exports have played a key role in the development of many economies. Historically, exports have been raw materials that were sold fairly cheaply on the global market.

Miles 0 — 1000

Kilometers 0 — 1600

Figure 3.11 Economic issues: Middle and South America.

the only reliable way to earn the money necessary to repay national debts quickly.

Outcomes of SAPs. SAPs have increased economic activity in Middle and South America. Multinational corporations have driven an expansion in agriculture, forestry, and the fishing industry. The U.S. corporations Cargill, Del Monte, and Dole are major players in Latin America, producing fresh fruits, flowers, and vegetables for U.S. supermarkets regardless of season. These corporations provide pesticides, fertilizers, and agricultural technologies for Latin American agribusinesses. Asian corporations are moving into timber extraction in Brazil and Guyana. Partnerships between foreign companies and local businesspeople operate the largely unregulated gold mining and forestry industries of the Amazon territories of Brazil, Ecuador, Guyana, and Bolivia.

Some countries are attempting to move beyond extractive industries by establishing **Export Processing Zones (EPZs),** also known as **free-trade zones.** EPZs are specially created legal spaces or industrial parks within a country where duties and taxes are not charged, in order to attract foreign-owned factories. These factories are allowed to assemble products, often textiles or electronics, strictly for export back to the owner's country, such as the United States. By far the largest of the EPZs is the conglomeration of assembly factories, called maquiladoras, located along the Mexican side of the U.S.-Mexican border (see Table 3.4 and the discussion on pages 161–163). Because taxes paid by these industries are low, the main benefit to the host country is the employment of local labor, which helps to ease the country's high rate of unemployment. A few Central American and Caribbean countries—Panama, the Dominican Republic, Jamaica—have also established EPZs, but competition by Mexico's maquiladoras has limited their expansion and success.

Have SAPs Resolved the Debt Crisis and Income Disparity?
The answer is yes and no. SAPs seem to have played a role in the slow economic recovery of the region since the start of the debt crisis in 1982. The burden of debt has fallen throughout the region, but it is still very high and will continue to be so for many years. SAPs have encouraged greater business and industrial efficiency, and some countries (Mexico, Brazil, Chile) have experienced high rates of economic growth recently. Exports have increased dramatically, especially in countries, like Mexico, that have been host to maquiladora-like assembly factories. However, poor working conditions in the maquiladoras and their general disregard for workers' well-being and the environment have also created problems for the communities in which they are located (see pages 161–162). Moreover, except for Mexico, most of the region's exports are still raw materials, and so many economies remain vulnerable to rapid price changes. Thus, while the debt crisis seems to be easing, it could reemerge.

The fact is, SAPs were never intended to decrease income disparity directly, but there was the implication that rising employment would do so. The reality is that SAPs have contributed to increases in the gap between rich and poor. For example, Chile's reforms have produced national economic growth rates as high as 7 percent in recent years—and also some of the highest rates of income disparity in the world. Between 1994 and 1996,

Thinking Globally and Critically: ON THE WEB

More than 2.5 million people live in the shantytowns that ring Lima, Peru. Here, in the Cerro San Cristóbal slum area, ducks are raised on the roof of one of the buildings. [Victor Englebert.]

the average yearly income for the 40 percent of the Chilean population classified as poor was $1440 per capita, while the average yearly income for the top 10 percent of the population was $48,000 a year. For the region as a whole, the number of people living in poverty has increased since SAPs were introduced: between 1980 and 1990, virtually every country in the region experienced a drop in per capita GDP. In most countries, minimum wages are now half, or less than half, of what they were in 1980. Although levels of public funding for education and health care have recently increased in many countries, the years of SAP-mandated cuts in education leave the region poorly equipped to move into the technological age. In most of the region, fewer than

70 percent of those eligible are enrolled in high school; in Ecuador and Venezuela, that number is less than 50 percent. Finally, incomes have declined significantly for those in the middle class, especially former government employees who lost their jobs as a result of SAPs. People who have lost income have increasingly turned to the informal economy to make a living. By the year 2000, the IMF and experts at the United Nations and in North America and Europe were acknowledging that IMF loan programs and subsequent structural adjustment programs had worked undue hardship on the poorest. Moves were initiated to alleviate the hardships by forgiving the overwhelming debt of the poorest countries worldwide. In 2001, Harvard economist Jeffrey Sachs, an architect of SAPs in the 1990s, advocated increasing foreign aid to poor countries, especially those where structural adjustment had increased poverty.

The Informal Economy

VIGNETTE Maria del Rosario Valdez lost her government job as an elevator operator in Mexico City in the 1980s. Unable to find formal employment, she created her own "informal" business: selling secondhand clothing to poor people. Every Thursday evening, she boards a bus for Texas that is filled with people doing similar business—buying and selling toys, auto parts, cosmetics, kitchenware. All along the way, wherever the bus stops, the passengers themselves interact with others in the informal economy—people selling tacos, pastries, cassettes, ice cream. For centuries, such low-profile businesspeople have operated outside the law throughout the Americas, paying no taxes but supporting their families.

By early morning, the bus reaches the border. Maria takes a cab to a huge warehouse where she chooses items she knows will sell from the towers of clothing collected from all over the United States. Some clothes are new—unsalable overstock. She fills boxes and suitcases with about 500 pounds of clothing, for which she pays about $150. Later she will pay another $150 in bribes to Mexican customs officials. Once back in Mexico City, early Saturday, Maria heads immediately to her stall in the market. There she bundles leftover unsold clothing (to be hawked door-to-door in the poor barrios by another woman) and lays out her new purchases. She will sell for 10 or more hours each day until Thursday, when she leaves for the border once again. [Adapted from Randall Hansis, *The Latin Americans: Understanding Their Legacy* (New York: McGraw-Hill, 1997), pp. 178–179.]

The **informal economy** consists of any and all economic activity that is not legal in the sense that taxes are not paid to the government and the work does not appear in official statistics. People working in the informal economy are often small-scale entrepreneurs who see an opportunity to earn some money in the market economy (see the box "The Informal Economy in Peru").

As in the case of Maria in the vignette, the entrepreneurship often entails recycling used items and, in some instances, can serve as an incubator for businesses that might eventually provide legitimate jobs for family and friends. Many people argue, however, that informal workers are just treading water, making too little to ever expand their businesses significantly. Moreover the bribes they have to pay (some have labeled these bribes "informal taxes") are much less beneficial to the economy as a whole than the taxes paid by legitimate businesses. Work in the informal economy is also risky because there is no protection of health and safety and no retirement or disability benefits. In many ways, the growth of the informal economy is a symptom of the increasing disparity between those with substantial access to wealth and power and those with very little.

Regional Trade and Trade Agreements

The growth in international trade and foreign investment encouraged by SAPs has been joined by a growth in regional free-trade agreements, which reduce tariffs and other barriers to trade among a group of neighboring countries. The two most successful trade agreements within the region are the North American Free Trade Agreement (NAFTA) and Mercosur (see Figure 3.11). **NAFTA,** created in 1994, is by far the larger of the two. It links the economies of Mexico, the United States, and Canada in a free-trade zone containing more than 400 million people, with a total annual economy worth almost $10 trillion. **Mercosur** is older, having been created in 1991, and comparatively smaller. It links the economies of Brazil, Argentina, Uruguay, and Paraguay (Chile is an associate member) to create a common market with 220 million potential consumers and an economy worth more than $1.3 trillion per year. There are also a growing number of other free-trade agreements within the region, some of them spurred in part by competition between the European Union (EU) and the United States to expand their influence in the region. Traditionally, the United States and Canada have traded and invested more heavily in the northern parts of Middle and South America—as reflected by Mexico's position within NAFTA and by the many U.S. economic ties to Central America and the Caribbean. Two reasons for North America's greater interest in trade with the northernmost countries of Middle and South America are that these countries are closest, so transport costs are lower, and that this part of the region is seen as strategically important, especially to the United States. The European Union has traded and invested more in the south, as reflected by recent negotiations aimed at linking the EU and Mercosur in a free-trade deal. However, in 2001, the EU and Mexico signed a free-trade agreement, giving Mexico access to 375 million consumers in Europe.

The record of regional free-trade agreements so far is mixed. NAFTA and related agreements that preceded it increased exports from Mexico to the United States by 137 percent over the period from 1990 to 1998, and trade among the Mercosur countries grew by 400 percent in the period from 1990 to 1998—but then underwent a decline that continues in 2001. When trade grows, the benefits are not spread evenly among regions or among

Thinking Globally: ON THE WEB

AT THE LOCAL SCALE *The Informal Economy in Peru*

Street vending is the most visible part of the informal economy in most cities of Middle and South America. The vendors sell useful items such as vegetables, spices, cooked snack foods, sunglasses, and umbrellas. Some sell luxury goods at prices lower than those in shops—such products as perfume, cosmetics, and handwoven rugs. Street vendors use public streets and sell to the public, so the governments in most countries try to formalize and control this activity to protect public health, ensure orderly urban spaces, and collect taxes.

The informal economy is a lifesaver in Peru, which has been hit by unprecedented economic recession, losses in real wages, and underemployment for up to 70 percent of urban workers. It is thought to employ a majority of the working urban population. By the year 2000, perhaps as many as 68 percent of urban workers were in the informal sector, and they generated an estimated 42 percent of the country's total gross domestic product. Most of them work as street vendors; up to one-quarter of the working population is so employed.

The geographer Maureen Hays-Mitchell, who studies street vending in Peruvian cities, found a vibrant, highly organized system in which the vendors assess the market and risks entailed in this semi-illegal activity. She found street vending in Peru to be a highly rational sector of the informal economy, fraught with intense competition for particular urban spaces and requiring continuous strategizing for market advantages. She also found that not only was street vending an important source of sustenance for the vendors, but it improved the overall quality of urban life by making goods available in convenient places and at affordable prices, something the inefficient formal retailing system seemed unable to do.

Street vendors specialize in particular products and carefully place their stands where they can attract the most customers for the products they sell. Vendors sell food and small gifts in front of hospitals during visiting hours; candy, games, and toys in front of schools; shoeshines and newspapers near hotels and restaurants; lotion, hats, and bikinis near beaches. Few opportunities are left unexploited: movie patrons can buy comics and magazines while waiting in line; outside jails, visitors, guards, and even prisoners can buy food, souvenirs, and handicrafts produced by inmates. The most popular locations, though, are along the edges of streets surrounding a city's central retailing district, the entrances to specific buildings and stores, and the intersections of major streets. A single city block may contain, on average, from 15 to 30 street vendors.

A desirable location has itself become a commodity for sale, though the vendors have no legal right to any particular space. Hays-Mitchell observed vendors painting the outlines of their stands on the street itself. The operators have sometimes sold the rights to use such spaces to new vendors for several hundred dollars.

---

sectors of society. In Mexico, the benefits of NAFTA have been concentrated in the northern states that border the United States, where the growth of maquiladoras has been the most intense. Most of those who work in the maquiladoras are paid marginally above the going national wage rate, so their standard of living rises only slightly, if at all (see the later section on the Mexico subregion).

Producers within the Mercosur sphere differ in their views of the trade group. Chilean grain farmers fear competition from cheaper Argentine and Uruguayan wheat and corn, while Chilean fruit growers and winemakers see greater access to buyers in neighboring Brazil as beneficial. In Argentina, Mercosur has been a double-edged sword: it brought greater efficiency and productivity as a result of increased competition, but it also brought high unemployment because excess labor was fired. Brazil favors a customs union: although there would be no tariffs between Mercosur countries, there would be high external tariffs on goods entering from outside Mercosur. These external tariffs are intended to secure regional markets for manufactured goods. However, external tariffs would hamper the ability of Argentina and Uruguay to attract foreign industrial capital. Benefits to all members would grow if improved roads and railroads could be developed. There are now few links across the Andes, and individual countries within Mercosur have only modest transport networks, including air service.

Mercosur meetings provide occasions for discussing regional issues, such as improvements in transport, as well as in South America's infrastructure of power grids, waterways, and pipelines. On the whole, Mercosur seems likely to help economies achieve more independence from unpredictable global markets and greater stability through more competitive and profitable industries, but so far Mercosur accounts for less than one-fifth of the members' total trade.

In April 2001, government leaders from all countries in the Western Hemisphere, except Cuba, signed a new trade agreement for the entire hemisphere, the Free Trade Agreement of the Americas (FTAA). This agreement calls for the removal of trade barriers among members by the year 2005. The current administrations of the United States, Canada, and Mexico view FTAA as a logical expansion of their NAFTA agreement. However, the actual implementation of such an ambitious free-trade plan is a long way away, because it faces considerable opposition throughout the Americas. The meeting, which took place in Québec, drew thousands of protesters. Some support the agreement—if labor and environmental protections can be built into it. Large countries such as Brazil favor the FTAA if it means freer access to U.S. markets, but small countries such as those in the Caribbean fear open competition with such giants as the United States, Canada, Mexico, and Brazil. They are concerned that the highly efficient, well-established corporations of North America will overwhelm their just-developing industries. The burden of economic instability and job loss would then fall mainly on the poor.

Agriculture and Contested Space

VIGNETTE Every day except Sunday, Aguilar Busto Rosalino rises well before dawn and goes to work on a banana plantation. From 5:00 A.M. to 6:00 P.M., stopping only for a half-hour lunch break, he spends his day placing plastic bags containing pesticide around bunches of young bananas. He prefers this work to his former job of applying a more powerful pesticide, work that left him and 10,000 other plantation workers sterile. He works very hard because he is paid according to how many bananas he treats. Usually he earns between $5.00 and $14.50 a day. Right now he is working for a plantation in Costa Rica that supplies bananas to Del Monte, but he thinks that in a few months he will be working for another plantation nearby. It is a common practice for these operations to fire their workers every three months so that they can avoid paying the employee benefits that Costa Rican law mandates. Although Aguilar makes barely enough to live on, he has no plans to protest for higher wages, because he knows that he would be put on a blacklist of people that the plantations agree they will not hire. [Adapted from Andrew Wheat, "Toxic bananas," *Multinational Monitor* 17, no. 9 (September 1996): 6–7.]

Throughout Middle and South America, increasing numbers of rural people find themselves in situations not unlike that of Aguilar Busto Rosalino. For centuries, while people labored on haciendas, the owner would allow them a bit of land to grow a small garden or some cash crops. In recent years, however, governments have encouraged a shift to large-scale, absentee-owned, export-oriented farms and plantations like the one where Aguilar works. Large-scale, primarily export-oriented agriculture is favored because it has the potential to earn large amounts of the foreign currency needed to pay off debts, and it produces food for urban populations. SAPs have encouraged governments to relax their restrictions on foreign ownership, so it is easier for foreign multinationals like Del Monte and others to dominate the most profitable agro-industries. These companies are usually highly efficient, state-of-the-art operations. No longer are subsistence plots or sharecropping parcels available for seasonal laborers. Meanwhile, smaller farmers are squeezed out because they cannot afford the latest production techniques and hence can't sell their produce at the lower prices of the more modern operations. Governments subject to SAPs have cut subsidies and other policies that once supported small farmers. Increasingly, the only new land available to farmers of limited income is in forested areas where soil fertility lasts only a few years or in hilly areas where soil erosion is intense. If they cannot find new land, farmers migrate to cities or work as wage laborers on plantations under conditions similar to those experienced by Aguilar Busto Rosalino.

These trends in rural agriculture have sparked resistance in a number of places. In Brazil, 65 percent of the land is owned by wealthy farmers who make up just 2 percent of the population. Since 1985, more than 2 million small-scale farmers have been pressed to sell their land to larger farms specializing in the production of cattle and other agricultural products for export. These large farms earn money that helps pay off the country's debt, but the now landless farmers have been forced to migrate. To help these farmers, organizations such as the National Movement of Landless Rural Workers began taking over unused portions of some large farms, saying in effect that it is wrong for agricultural land to lie unused while there are people living in extreme poverty. Since the mid-1980s, the landless movements have coordinated the occupation of more than 51 million acres (21 million hectares) of land (an area about the size of Kansas) by 151,427 families. Public opinion polls show that 77 percent of Brazilians support the movement. Nevertheless, rural landowners have used private armies or local military police to remove and, in some cases, massacre landless families. The national government has intervened to avert more violence and has started its own program to settle landless families on farms.

In the southern Mexican state of Chiapas, agricultural activists have mobilized opposition in an armed civil conflict. There, the Zapatista rebellion by Native American agricultural workers (named for the Mexican revolutionary hero Emiliano Zapata) began on the day the North American Free Trade Agreement took effect in 1994. Although the Mexican government redistributed many hacienda lands to poor farmers earlier in the century in land reform programs, most land in Chiapas is still held by a wealthy few who grow cash crops for export. The rest of the population farms tiny plots on unfertile hillsides. In the year 2000, about three-quarters of the rural population was malnourished, and one-third of the children did not attend school. In Chiapas, NAFTA is seen as a threat because it has diverted the government from land reform to the support of large-scale export agriculture. Unlike Brazil, the Mexican government used the army to repress the rebellion rather than to avert violence. Tensions lessened, but the standoff remained unresolved. As of the year 2001, the newly elected government of Vincente Fox is negotiating with the Zapatistas in pursuit of a peaceful solution.

The two instances of agricultural land disputes just discussed have both taken place in areas where subsistence agriculture competes with commercial plantation agriculture or ranching. There are other types of agriculture in the region, as a look at the map of agricultural zones (Figure 3.12) will confirm. A wide belt of commercial grain production, similar to that found in the midwestern United States, stretches through the Argentine pampas. Zones of mixed farming, including the production of vegetables and specialty foods for sale in urban areas, surround the major urban centers of southern Brazil, Argentina, and Uruguay, as well as south central Chile. Irrigated agriculture is practiced along the dry, narrow coastal desert plain of Peru and northern Chile, as well as on the other side of the Andes in the Argentine pampas. These various types of market-based agriculture on large holdings are essential to the trade of many countries and often help to feed local urbanites, thus decreasing the amount of food that must be imported.

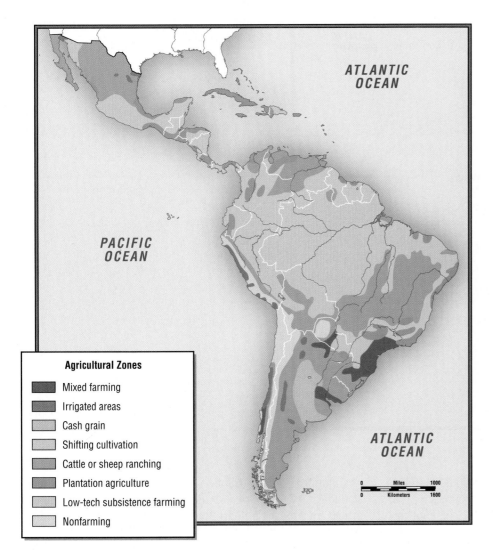

Figure 3.12 **Agricultural zones in Middle and South America.** [Adapted from Edward F. Bergman, *Human Geography—Cultures, Connections, and Landscapes* (Englewood Cliffs, N.J.: Prentice Hall, 1995), p. 194.]

Agricultural Zones

- Mixed farming
- Irrigated areas
- Cash grain
- Shifting cultivation
- Cattle or sheep ranching
- Plantation agriculture
- Low-tech subsistence farming
- Nonfarming

Is Democracy Rising?

For the first time in a long while, all countries in the region except Cuba have multiparty political systems and democratically elected governments. In the last 25 years, there have been repeated peaceful democratic transfers of power in countries once dominated by monolithic alliances composed of the military, wealthy rural landowners, wealthy urban entrepreneurs, foreign corporations, and even foreign governments such as the United States. Change has been especially dramatic in Mexico, where the huge entrenched political machine of the Institutional Revolutionary Party (PRI), which had ruled the country for 71 years, was unseated in a 2000 election by a coalition led by the maverick businessman and reformer, Vincente Fox. The popular election of Hugo Chavez as president of Venezuela could be a harbinger of change, if his call for a fundamental restructuring of the governmental system along more egalitarian lines is heeded.

In much of the region, however, democracy is still fragile and not particularly *transparent*—meaning that official decisions are not open to public input and review. Policy is formulated without sufficient attention to potential effects on the poor or those who lack strong political voices. Officials may remain in power not because they have broad public support but because they are backed by powerful people, who may even be drug czars or otherwise corrupt.

Corruption. High-level corruption undermines democratic institutions by setting a damaging example for the rest of society. Virtually every country in the region has had a serious scandal in the past few years involving high-level officials performing million-dollar favors for friends and families or outright stealing money from taxpayers. The international banks in Miami, Florida, are well known as depositories for these flights of capital, so indirectly, North American financial institutions have benefited, while Middle and South American countries have been robbed of much needed investment capital. For years, in Mexico, banks regularly and openly recorded massive transfers of money into foreign bank accounts at the end of every six-year presidential term. Former president Carlos Salinas fled the country at the end of his term in 1996 with millions of dollars. Some people argue that SAPs are increasing pressure on leaders to be financially accountable, but noticeable effects have yet to appear.

The international drug trade is a major factor contributing to corruption, violence, and the subversion of the democratic

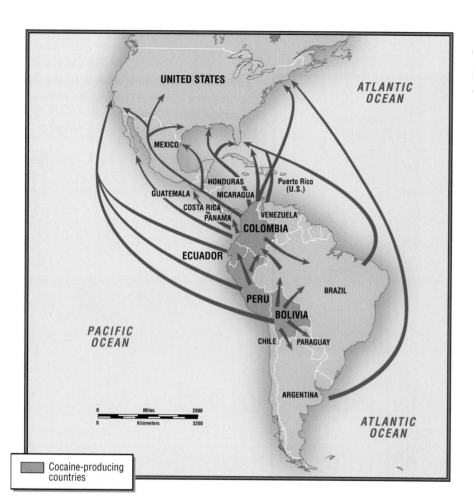

Figure 3.13 Interregional linkages: cocaine sources and traffic in the Americas.
[Adapted from George F. Rengert, *The Geography of Illegal Drugs* (Boulder, Colo.: Westview Press, 1996), p. 16.]

Cocaine-producing countries

process throughout Middle and South America. Middle America and northwestern South America are the primary sources for the raw materials of the drug trade—coca for cocaine, poppies for heroin, and marijuana. Many growers are remote small-scale farmers of Native American or mestizo origin, who can make a better income for their families from these plants than from other cash crops. Although drugs are illegal in all of Middle and South America, in countries such as Colombia, Venezuela, Bolivia, and Mexico, public figures from the local police on up to the highest officials are paid to turn a blind eye to the industry. Because law enforcement is lax in drug-producing areas, drug lords can fund their own private armies to protect drug transport routes and force the compliance of local citizens.

The United States is sponsoring a "war on drugs" that has become controversial in the region. This effort is intended both to stem the flow of cocaine, heroin, and other narcotics into the United States by stopping production abroad and to prosecute users within the United States. Figure 3.13 shows the flow of cocaine to North America from Middle and South America.

An important part of the strategy is to destroy drug crops using chemical defoliants. The effort to stop the production of drugs in Middle and South America has led to the largest U.S. military presence in the region in history. This presence is focused on supplying intelligence, military equipment, and training, but not actual troops. Many leaders in Middle and South America question the wisdom of the U.S. war on drugs. For example, President Andres Pastrana of Colombia has expressed fear that just as countries such

as Colombia, Peru, and Bolivia are trying to turn away from long-standing tendencies toward military rule, U.S. aid to national armies encourages military leaders to maintain a high political profile. Former Bolivian president Gonzalo Sanchez de Lozada referred to pressure by the U.S. Embassy to allow U.S. antinarcotics training for the Bolivian military as "a moment of immense dependency and even embarrassment for the country."

Destroying enough crops to stop the flow of drugs into the United States will be a challenging task. By the U.S. government's own estimates (1993), drug production worldwide is so abundant that the entire U.S. consumption could be supplied by 1 percent of the worldwide drug crop. Oscar Arias, former president of Costa Rica and winner of the 1987 Nobel Peace Prize, asked rhetorically in a letter to the *New York Times* (August 29, 2000), "Haven't the last 20 years shown us that as long as there is a market for drugs in the United States, somebody will find a way to get them there?" The United States takes the position that it is making the security forces of legitimately elected governments more professional and equipping them to combat guerrillas who protect drug-producing underworld figures and force local cultivators to grow illegal drug plants.

Foreign Involvement in the Region's Politics. Intervention by powers outside the region has compromised democracy, even though sometimes the stated aim was to enhance it. Although the former Soviet Union, Britain, France, and other European countries have wielded influence during the twentieth century, the most active foreign power has been the United States. The country proclaimed the Monroe Doctrine in 1823 to warn Europeans that the

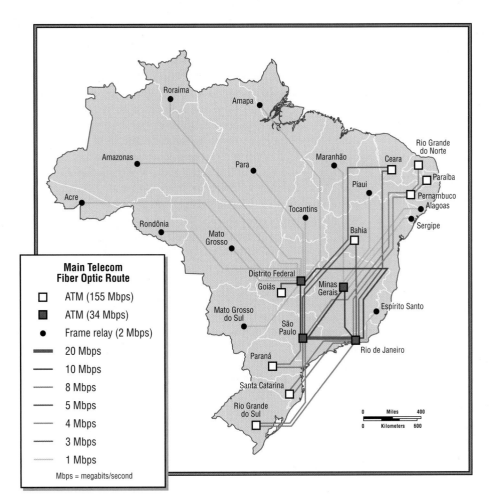

Figure 3.14 **Telecom routes in Brazil.** Brazil is leapfrogging old-fashioned copper wire telephone systems by installing high-speed, broadband fiber optic cable networks for telecommunications and Internet connectivity that are on a par with systems in the United States and Europe. The country is facing a major problem, however: disparity of distribution and access to the routes. The routes connect only the major cities and business districts; 95 percent of Brazil's population cannot gain access to the system. ATM in the key stands for asynchronous transfer mode, a means of very high speed digital (voice, video, image, and data) communications. [Adapted from *Red Herring* (December 19, 2000): 81.]

United States would allow no further colonization in the Americas. Successive U.S. administrations have interpreted the doctrine more broadly to mean that the United States could itself intervene in the affairs of countries in the Americas, ostensibly to make them safe for democracy but in reality to protect U.S. national security interests. One result was the installation of U.S.-backed dictators in the 1930s through the 1950s: the Somozas in Nicaragua, Batista in Cuba, Papa Doc and Baby Doc Duvalier in Haiti, for example. Later, the United States supported armed insurgencies for the purpose of preventing or reversing communist takeovers in Cuba (1961), the Dominican Republic (1965), and Nicaragua (1980s). Corruption linked to the Panama Canal and to the international drug trade led the United States to invade Panama in 1989 in order to put a halt to the activities of the renegade Panamanian general, and then president, Manuel Noriega, who had shadowy connections with both drug dealing and U.S. covert operations in the region. Since 1989 and the end of the cold war, the United States has curtailed direct foreign intervention in the region's politics. The United States is still highly influential but is more likely to wield its influence by enacting economic policies.

How Can Democracy Be Furthered If Not by Outside Intervention? Democracy is strongest where people are educated, healthy, and economically secure. Thus the process of deepening democracy goes hand in hand with efforts to create more socially and economically equal societies. But ideas about how to reach a more egalitarian status quo differ. Supporters of free markets, sometimes called neoliberals, see free-market economic development

and globalization as ultimately furthering social justice and democratic participation by creating jobs, opening up entrepreneurial opportunities, and putting people in touch with one another via the Internet. Critics of globalization and free-market economics think that these macro phenomena are likely to increase income disparities and undermine democratic participation by decreasing environmental quality, lowering wages, and pitting the poorest laborers against one another. They say that egalitarian conditions are more likely to arise out of local control and government participation in social services and supervision of the economy.

If the use of the Internet is to enhance economic and democratic participation, then access to this new technology is crucial. A recent United Nations study of global Internet use showed that users in Middle and South America were just 3.2 percent of the population in 2000, but that rate was four times the 1998 rate. Overall, this region is the most advanced in technology achievement of those outside the economically developed regions (North America, western Europe, Japan and Korea in East Asia, and Oceania). A 2000 study by *Wired* magazine identified those countries worldwide that were leaders in technology innovation, potential leaders, and dynamic adopters. Although no countries in Middle or South America were leaders, nearly half of the dynamic adopters, and one-fifth of the potential leaders (Mexico, Argentina, Costa Rica, and Chile), were in this region. Brazil, a dynamic adopter, has two world-class technology hubs in the environs of São Paulo and is attempting to cover the whole country with broadband fiber optic cable networks for telecommunications and Internet service (Figure 3.14).

SOCIOCULTURAL ISSUES

Under colonialism, a series of social structures evolved that guided daily life—standard ways of organizing the family, the community, and the economy. These included rules for gender roles, rules for race relations, and ways of religious observance. These social structures are still widely in place, but they are changing in response to urbanization, economic development, and the diffusion of ideas from outside the region. The results are varied. In the worst cases, change is leading to the breakdown of family life and the abandonment of children; in the best cases, to a new sense of initiative on the part of women, men, and the poor.

Cultural Diversity

The region of Middle and South America is one of the most culturally complex on earth. Contributing to this diversity are the many indigenous groups present when the Europeans came and the many cultures that were introduced during and after the colonial period. In the Caribbean, the Guianas, and Brazil, the arrival of a relatively small number of European newcomers from Spain, England, Ireland, France, Holland, and Portugal resulted in the annihilation of indigenous cultures, and these regions then became populated almost entirely by various people from the Old World. From 1500 to the early 1800s, some 10 million Africans were brought to plantations on the islands and coastal zones of Middle and South America. After the emancipation of African slaves in the Caribbean in the 1830s, more than half a million Asian indentured servants were brought there from India, Pakistan, and China as agricultural workers. Their cultural impact remains most visible in Trinidad and Tobago, Jamaica, and the Guianas. In some parts of Mexico and in Guatemala, Ecuador, and other Andean zones, indigenous people have remained numerous, and to the unpracticed eye they may appear somewhat unaffected by colonization.

Mestizo is the name for those who are racially and culturally a blend of Native American and European roots. They are now the majority in Mexico, Central America, and much of South America. In some places, such as Argentina, Chile, and southern Brazil, Euro-Americans are now so numerous that they dominate the landscapes and ways of life. The Japanese, though a tiny minority everywhere in the region, are increasingly influential in innovative agriculture and industry in Brazil, the Caribbean, and Peru. In some ways, diversity is increasing as the media and trade introduce new influences into the region. Yet groups may be becoming less distinct in urban areas, where various culture groups live in close proximity. The increasing contact between people of widely different backgrounds is accelerating the rate of **acculturation** (cultural borrowing) and **assimilation** (loss of old ways and adoption of new ones). In the big cities, such as Mexico City, Lima, and Rio de Janeiro, there is a blend of many different strains, so that no one cultural component remains unaffected by the others.

Race and the Social Significance of Skin Color

People from Latin America, and especially those from Brazil, often proudly make the public claim that race and color are of less consequence there than in North America, and they are right in certain ways. It is possible to "erase" the color of one's skin, in effect, through assimilation. For example, a Native American or a dark-skinned person of any background may, by acquiring an education, a good job, a substantial income, the right accent, and a high-status mate, become recognized as not being any particular color at all. Because of this possibility, over the generations race and class designations became more and more complex, and social class was no longer assigned on the basis of race alone. Today, the terms *clear-skinned* for "light" and *black, mulatto,* or *pardo* for "brown" tend to be merely descriptive and do not necessarily carry racial meaning. The dark skin of an upper-class person would not be thought of as racially significant, but family, wealth, education, and occupation would be very significant.

Nevertheless, being able to erase the significance of skin color through accomplishments is not quite the same as color having no significance at all. Although there are poor light-skinned people throughout the region (often the descendants of migrants from central Europe over the last century), overall, those who are poor, less well educated, and of lower social standing tend to have darker skins than those who are educated and well-off. This fact seems to indicate that race and color have not yet disappeared as social factors in Middle and South America.

The Family and Gender Roles

The **extended family** is the basic social institution in all the societies of Middle and South America. Throughout this region, it is generally accepted that the individual should subvert his or her

The Paco Burgos family in the Petén, a region in Guatemala, work together on their milpa, a cleared cultivation plot in the forest that serves as the main source of food. The children do much of the work in cultivating and harvesting the primary crops of corn, beans, and squash. [Miguel Luis Fairbanks/National Geographic Image Collection.]

interests to those of the family and local community and that, by so doing, individual well-being is best secured.

The spatial arrangement of domestic life illustrates the strong family ties. Families of adult siblings, their mates and children, and elderly grandparents frequently live together in domestic compounds of several houses surrounded by walls. Such compounds are rare in the United States and Canada, where the single-family home is the norm. In Middle and South America, social groups out together in public are most likely to be family members of several generations rather than unrelated groups of single young adults or of married couples, as would be the case in Europe or the United States A woman's best friends are likely to be her female relatives. A man's business or social circle will almost certainly include male family members or long-standing family friends.

Gender roles have their roots in the extended family, as well. Throughout Middle and South America, the Virgin Mary, the mother of Jesus, is held up as the model for women to follow. Through their adoration of the Virgin, they absorb a set of values, known as **marianismo,** that puts a priority on chastity and service to the family. The ideal woman is the day-to-day manager of the house and of the family's well-being, training her sons to enter the wider world and her daughters to serve within the home. Over the course of her life, a woman's power increases as her skills and sacrifices for the good of all are recognized and enshrined in family lore.

Her husband, the titular family head, is expected to provide most of his income to his family, but he usually operates in the outside world, working and tending to his social network, which is deemed just as essential to the family's prosperity and status in the community as his work. The rules for how men are to contribute to family stability and well-being are less clearly spelled out than are the rules for women. Men have a good bit more autonomy and freedom to shape their lives than women do, simply because they are expected to move about the community and establish relationships, both economic and personal. In addition, as is the case in most societies, there is an overt double sexual standard for males and females. While expecting strict faithfulness in mind and body from his wife, a man is freer to associate with the opposite sex. Males measure themselves by the model of **machismo,** which considers manliness to consist of honor, respectability, and the ability to father children and be attractive to women, to be an engaging raconteur in social situations, and to be the master of his own household. Under traditional rules of machismo, the ability to acquire money was secondary to the other symbols of maleness. Now a new market-oriented culture also prizes visible affluence as a desirable male attribute.

Many factors are transforming the family and gender roles. For one thing, couples are now having two or three children, instead of five or more. As discussed earlier, with infant mortality declining steeply, it is no longer necessary to have many children simply to see a few survive to adulthood. Because people still marry early, most parents are free of child-raising responsibilities by the time they are 40. For men, the change may require an adjustment of the machismo idea that manliness is defined in part by high fertility. For women, there is the empty-nest syndrome: after the children are grown, 30 or more years of active life loom, to be filled in some way. As it happens, economic change

throughout the region has provided a solution. Increasingly, women, despite little formal education, have been able to find factory jobs or other employment that puts to use the skills they perfected while supervising a family: time and people management, maintenance of equipment, organizational skills, the ability to anticipate problems, and long-range planning. For more and more women of all ages, employment outside the home (often in a distant city) is a way to gain a measure of independence and also contribute to the family's needs. Many families simply bend to accommodate these new situations, others lose many members to migration, and some disintegrate.

Children in Poverty

It is estimated that one-third of the children in Middle and South America are economically active, most because their families need their income. In the countryside, children work in agriculture, logging and mining. But 74 percent of this region's people live in urban areas, and city children also work, often on the street selling items such as chewing gum, flowers, and flip-flops (rubber sandals); cleaning cars; and performing small services. It is understandable that children are asked to help alleviate a family's poverty by doing some work, paid and unpaid. But in societies that traditionally place such high value on the family and children, it is difficult to understand recent news stories about homeless and abandoned children in such countries as Mexico, Guatemala, Colombia, and Brazil. The United Nations reports that in Mexico alone, there are 15,000 homeless children; in Guatemala City, 5000. Some of these reports may be overblown: a study in the year 2000 in São Paulo, South America's largest city, revealed that just 609 children were homeless there. Nonetheless, throughout the region, many children live on the streets of cities, and thousands have been pushed out of the home by overburdened parents. These children must fend for themselves, sleeping in doorways, hustling for money or food. Travelers to the region may experience the extreme distress of eating in an outdoor restaurant ringed with sturdy iron fences through which stretch the skinny arms of children begging to share part of their dinner. Reports of abandoned children are increasing around the world, and the explanations for this phenomenon are numerous.

In Middle and South America, as discussed previously, rural-to-urban migration by young adults is common. Often, inexperienced young women migrants are equipped for only the most menial jobs and rarely earn a living wage. They may soon ally themselves with a man to gain some measure of security or even just a place to live. Soon, children result. The man may already have several informal mating relationships, so his unpredictable income is stretched too thin to support more children adequately. The young mothers' extended families are not around to help them with child care, yet they must keep working. With no older, solid role models to enforce such traditional values as chastity, sobriety, and good parenting, children may grow up neglected, malnourished, and unruly, with lonely, dysfunctional mothers who may turn to drugs or alcohol for solace. Recent reports tell of children who turn to brain-damaging glue sniffing to ease their anxiety (see the photo on page 148).

A Nicaraguan boy sniffs glue from a baby-food jar in the Oriental Market in Managua while his friend pulls his own jar out of his shirt. [Richard Sennott/*Minneapolis–St. Paul Star Tribune.*]

Despite these very real problems, it is important to remember that the vast majority of parents in Middle and South America, even those in the worst of urban slums, heap loving care on their children, providing all that they possibly can and sacrificing so that their children might have a better life.

Religion in Contemporary Life

The Roman Catholic church remains one of the most influential institutions throughout Middle and South America. From the beginning of the colonial era, the church was the major partner of the Spanish and Portuguese colonial governments. It received extensive lands and resources and sent thousands of missionary priests to convert Native Americans. For many centuries, the Catholic church encouraged working people to accept their low status, obedience to authority, and postponement of rewards until heaven. Thus, while serving to unify the region and spread European culture, the church also reinforced class differences. Furthermore, the church ignored those teachings of Christ that admonish the privileged to share their wealth and attend to the needs of the poor. Nonetheless, poor people throughout Spanish and Portuguese America embraced the faith and still make up the majority of church members. They have put their own ethnic spin on Catholicism, creating multiple folk versions of the religion, with music, liturgy, and interpretations of Scripture that vary greatly from the European.

The Catholic church began to see its power erode in the nineteenth century in places such as Mexico. **Populist movements**— popularly based efforts seeking relief for the poor—seized and redistributed church lands and canceled the high fees that the clergy

had been charging for simple rites of passage such as baptisms, weddings, and funerals. Over the years, the church became less obviously connected to the elite and more attentive to the needs of the poor. In the mid-twentieth century, the church began ordaining Native American and African-American clergy as a matter of course. Common people gained the right to read the Bible in their own languages and to replace Gregorian chants with ethnic music. After Vatican II in the 1960s, women were allowed to perform certain rites during Mass.

Just as the Catholic church began to change its stance vis-à-vis the common people, a more radical movement known as **liberation theology,** which promised to reform the church from the ground up, emerged in the 1970s. Begun by a small group of priests and activists, the movement portrays Jesus Christ as a revolutionary who symbolically spoke out for the redistribution of wealth when he divided the loaves and fishes among the multitude. Poverty is viewed as a problem of an entire society rather than as a personal failing. The perpetuation of gross inequality and political repression is viewed as sinful, and social and economic reform is promoted as liberation from evil.

At its height, the liberation theology movement had more than 3 million adherents in Brazil alone and was the most articulate movement for region-wide social change. Today, the movement is somewhat reduced in strength. On the one hand, some of its positions have been adopted by the Vatican; on the other, in such places as Guatemala and El Salvador, it became the target of state-sponsored attacks that vilified participants as communist conspirators. The liberation theology movement has also had to compete with newly emerging evangelical Protestant movements that have many attractions for the poor.

Evangelical Protestantism has diffused into Middle and South America from North America. It is now the fastest growing religious movement in the region, and already about 10 percent of the population, or more than 40 million people, are adherents. Like liberation theology, it appeals to the rural and urban poor— that segment of society most in need of hope for a better future. Evangelical Protestants teach a "gospel of success," stressing that those who are true believers and give themselves to Christ and to a new life of hard work and clean living will experience prosperity of the body as well as of the soul. The movement is *charismatic*, meaning that it focuses on personal salvation and empowerment of the individual through miraculous healing and transformation. People are thought to become "reborn." The movement has no central authority, consisting of a host of small, independent congregations. Leadership is passed around so that both men and women in the congregations develop organizational and public speaking skills. A number of studies have shown a real change in the lives of individuals formerly beset with dejection, lethargy, and vice. After a few years of active church membership, many individuals seize educational and entrepreneurial opportunities and do indeed achieve success aided by their belief in the gospel of Christ. There is considerable evidence that the gospel of success may be an important impetus to the emergence of a middle class.

The geographer David Clawson has provided a map showing the percentages of people throughout the region who are Protestant

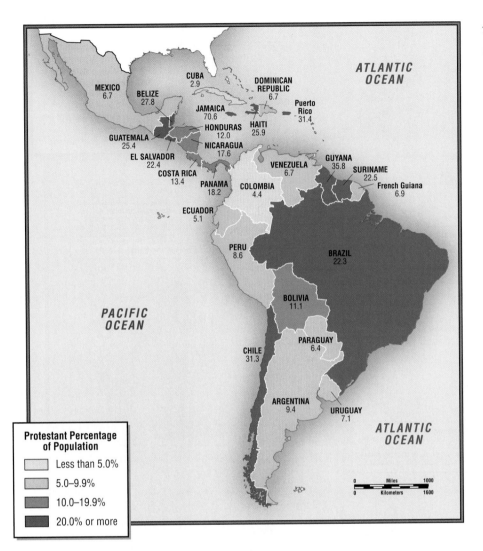

Figure 3.15 **Percentage of Protestants in the populations of Middle and South America.**
[Adapted from David Clawson, *Latin America and the Caribbean* (Dubuque, Iowa: Wm. C. Brown, 1997), p. 210.]

(Figure 3.15). The extent to which these geographic patterns reflect the spread of the evangelical version of Protestantism is not yet well understood. Evangelical Protestantism is growing in such formerly Catholic countries as Brazil and Chile. Brazil now has 34 million Protestants (75 percent of the total for the region), and in Chile nearly a third of the more wealthy and literate people are Protestants. The evangelical movement is also strong in the Caribbean, Mexico, and Middle America, among both the poor and middle classes. The revival tents of North American evangelical pastors are a common sight in the urban fringe of cities throughout the region.

ENVIRONMENTAL ISSUES

Environments in Middle and South America were among the first to inspire concern about the use and misuse of earth's resources. In the 1970s, scholars such as anthropologist Emilio Moran warned of an impending crisis in the tropical rain forests of the Americas. Beginning in that decade, construction of the Amazon Highway provided access to the rain forest to small farmers, who settled in the Brazilian Amazon and began clearing its

forests to grow crops. Scholars now know that human settlement has had consequences for American environments since prehistory. Today's impacts are particularly severe because population has doubled and redoubled at the same time as per capita consumption of resources within the region and exports out of the region have increased.

The Vanishing Forests of the Amazon Basin

During the 1980s and early 1990s, forests were cleared in Middle and South America at the rate of 30.4 million acres (12.3 million hectares) per year, an area about the size of Ohio. The clearing continues today, and close to 75 percent of the forest lost annually from this region is cleared from the huge forestlands of Brazil. These lie mostly within the Amazon Basin (Figure 3.16).

Amazon environments, though vast, face multiple threats. Primary among them is the clearing of land to raise cattle for the export market and to grow tropical cash crops. Governments have offered surplus urban populations cheap land along newly built roads. As a result, landless settlers from southern and eastern

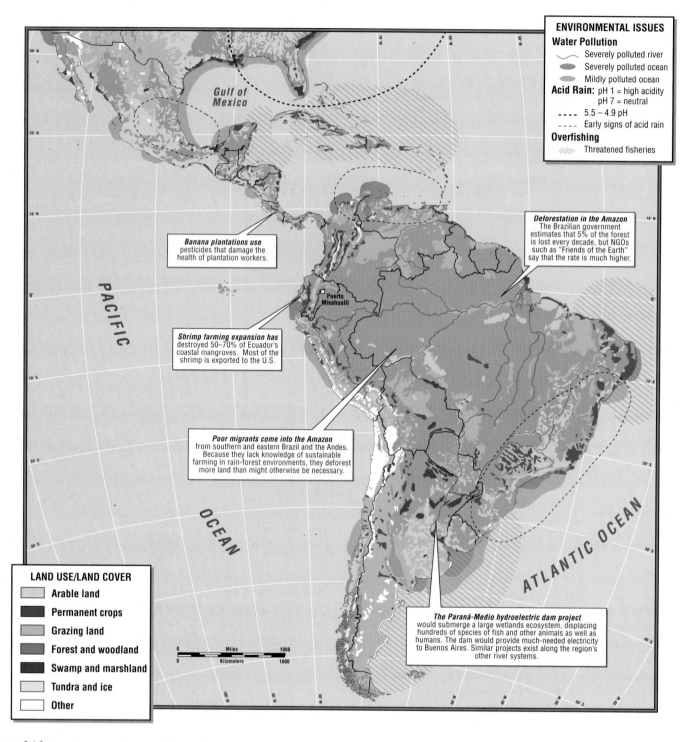

ENVIRONMENTAL ISSUES
Water Pollution
— Severely polluted river
⬮ Severely polluted ocean
⬮ Mildly polluted ocean
Acid Rain: pH 1 = high acidity
pH 7 = neutral
- - - 5.5 – 4.9 pH
- - - - Early signs of acid rain
Overfishing
▨ Threatened fisheries

Deforestation in the Amazon
The Brazilian government estimates that 5% of the forest is lost every decade, but NGOs such as "Friends of the Earth" say that the rate is much higher.

Banana plantations use pesticides that damage the health of plantation workers.

Shrimp farming expansion has destroyed 50–70% of Ecuador's coastal mangroves. Most of the shrimp is exported to the U.S.

Poor migrants come into the Amazon from southern and eastern Brazil and the Andes. Because they lack knowledge of sustainable farming in rain-forest environments, they deforest more land than might otherwise be necessary.

The Paraná-Medio hydroelectric dam project would submerge a large wetlands ecosystem, displacing hundreds of species of fish and other animals as well as humans. The dam would provide much-needed electricity to Buenos Aires. Similar projects exist along the region's other river systems.

Puerto Misahualli

LAND USE/LAND COVER
▨ Arable land
■ Permanent crops
▨ Grazing land
▨ Forest and woodland
■ Swamp and marshland
▨ Tundra and ice
□ Other

Miles 0 — 1000
Kilometers 0 — 1600

Figure 3.16 **Environmental issues: Middle and South America.** Land cover in the region is a mosaic of natural landscapes and human land uses. Superimposed on this mosaic is another pattern of environmental impacts, which are the result of urbanization and industrialization, global market demands, and population pressure. (An explanation of pH is given in the caption to Figure 2.20, page 90.)

Brazil and from the highlands of the Andes are clearing land for use as subsistence cultivation plots. After a few years of farming the poor rain-forest soils, the new settlers often abandon the land, now eroded and depleted of nutrients. Ranchers may buy the worn-out land from failed small farmers to use for cattle pastures.

Another threat to the Amazon forest is the logging of valuable hardwoods for export, which is carried out by local entrepreneurs, landless loggers, large landowners, and Asian multinational companies. The Asian companies have turned to the Amazon forests after having logged up to 50 percent of the tropical forests in Southeast

Asia. Firms from Burma, Indonesia, Malaysia, the Philippines, and Korea are purchasing forestlands in Brazil after already having secured rights to Amazon Basin forests in neighboring countries. Also, oil and gas extraction by multinational petroleum corporations is driving away Native Americans and polluting the water and soil on their lands, as discussed in the opening vignette.

The days of unopposed exploitation may be waning, however. Until the mid-1980s, Brazilian governmental policies encouraged forest clearing, considering it development of unused land. In particular, those who developed forestland did not have to pay taxes on it. Eventually, the Brazilian government changed its tax-relief policies in response to environmental concerns. Although they come late, environmental regulations are now being developed in all the countries of the Amazon. The pace of clear-cutting for pasture and small farming is slowing, but there are new threats to the forest: the rapid growth of urban settlements and suburbs, the increased use of forest timber as fuel wood, and the clearing of forest tracts for the large-scale production of cash crops such as soybeans. In addition, illegal logging has been on the rise since the 1990s.

The destruction of the Amazon may have global implications. Scientists now understand that the absorption of carbon dioxide and the release of free oxygen is one of the most important functions of the world's forests. Many scientists think that the huge tracts of forested land in the Amazon are essential to this process. The loss of these forests may be contributing to global warming through the greenhouse effect. One positive note is that reforestation is more widespread than previously thought. The tropical climate encourages fairly rapid rates of regrowth: abandoned pastures and farmland can regain a mature, although less diverse, forest in 15 to 20 years. Planting fast-growing trees in forest plantations is one way to maintain the exchange of carbon dioxide and oxygen, and the use of wood from such plantations may save some natural forests from cutting in the future. However, forest plantations fall far short of replicating the complex biodiversity of the rain forest.

The forests in other parts of this region are also at risk. About 65 percent of the forest clearing in Central America is intended to create pastures for beef cattle grown primarily for the U.S. fast-food industry. The forests of Central America (Costa Rica, El Salvador, Nicaragua, and Panama) and the Caribbean (Cuba, the Dominican Republic, and Jamaica) have been among the main targets of clearing for ranching and export crops such as bananas. If deforestation continues in these areas at present rates, the natural forest cover will be entirely gone in 20 years.

The forests of Middle and South America are examples of what geographers call **contested space.** Various groups—some new arrivals, some long-time inhabitants—are in conflict over the right to use the space as they see fit. Here are some of the conflicting ideas being developed for the contested space of the Amazon. Deep in the interior, a few Native American leaders hope to defy the odds by stopping development on their ancestral lands. To do so, they are learning litigation and negotiating skills that they will use to fend off the oil-developing conglomerates that want to set up production facilities on their lands. Governments lure surplus urban populations with the offer of cheap land along

In the Brazilian state of Rondônia, colonists from other areas clear the rain forest for settlement and large-scale agriculture. New roads, like the one at upper left, provide the settlers with access. Some of the wood is harvested for local use and to be sold, but much is wasted. [Randall Hyman.]

newly built roads. Then unemployed factory workers, unskilled at agriculture, barely eke out a living on the small homesteads they carve out of the forests. Cattle ranchers buy up the failed plots and clear more land for pastures. One well-financed government research agency (IMPA, in Manaus) promotes the use of forest products, while another less well financed government agency (FUNAE, also in Manaus) works to undo the resulting damage and protect the indigenous tribes from the encroachment of development. Meanwhile, ecotourism entrepreneurs take their clients to resorts in the forest, where luxury and spectacle often take priority over environmental sustainability (see the box "Ecotourism in the Ecuadorian Amazon," page 152).

AT THE LOCAL SCALE   *Ecotourism in the Ecuadorian Amazon*

In a world where human efforts to prosper, or merely survive, continually threaten natural environments, one new strategy for both using and respecting nature is **ecotourism.** Ideally, ecotourism offers both natural and cultural travel experiences in unfamiliar environments. It encourages sustainable use and the conservation of resources while providing a livelihood to local people and the broader host community. Ecotourism is now the most rapidly growing segment of the global tourism and travel industry, which itself is the world's largest industry, with $3.5 trillion spent anually. On a trip through the Ecuadorian Amazon in 1998, Alex Pulsipher saw several sides of the growing ecotourism industry.

VIGNETTE   Puerto Misahualli is a small river town that is currently enjoying an economic boom due to the many European, North American, and other foreign travelers who come there, much as I did, looking for experiences that will bring them closer to the now legendary rain forests of the Amazon. The array of ecotourism offerings is impressive. One capable-looking indigenous man offered to be my personal guide for as long as I wanted as we traveled by boat and on foot, camping out in "untouched forest teeming with wildlife." His guarantee that we would eat monkeys and birds did not seem to promise the nonintrusive, sustainable experience I was after. A well-known "eco-lodge" nearby offered plush hotel-style rooms with a river view, a chlorinated swimming pool, a fancy restaurant serving "international cuisine," and a private 740-acre (300-hectare) nature reserve. A wall topped with broken glass sepa-

rated the lodge from the surrounding community. It seemed more like a fortified beach resort than an eco-lodge.

The solar-powered Yachana Lodge, in contrast, had simple rooms, local cuisine, and a knowledgable resident naturalist who was a veteran of many campaigns to preserve Ecuador's wilderness. Profits from the lodge funded a local clinic and various programs that teach agricultural methods to protect the fragile Amazon soils while increasing earnings from surplus produce. The "local" cultivators were actually poor migrants from Ecuador's densely populated cool highlands who had been given free land in the Amazon by the government—but no training for how to farm in the lowland rain forest. Many of their attempts have resulted in extensive soil erosion and deforestation. I soon learned from people working with the nonprofit group running the Yachana Lodge—the Foundation for Integrated Education and Development (FUNEDESIN)—that the lodge is earning just barely enough to sustain the clinic and agriculture programs.

This story touches on two themes related to environmental issues in Middle and South America: the tensions between economic development and environmental preservation, and the ways in which poverty is linked to environmental degradation. Ecotourism often turns out to be little different from other kinds of tourism that damage the environment and give little back to the surrounding community. Still, there is the potential to use the profits of ecotourism to benefit local communities and the environment.

The Environment and Economic Development

In the past, governments in the region argued that economic development was so desperately needed that environmental regulations were an unaffordable luxury. Now there are increasing attempts to take a middle ground: embracing economic development as necessary to raise standards of living, yet hoping to minimize negative effects on the environment. Grassroots environmental movements are active throughout the region, from the neighborhood on up to the national level. Most such organizations monitor, evaluate, and often challenge economic development projects via public protest, political negotiating, legal action, or public education. They may work with international nongovernmental organizations. Two cases are discussed here; elsewhere, the text describes two locally designed projects that are sensitive to environmental concerns—one fostering urban quality of life in Curitiba, Brazil (see the box on page 133), and one reviving ancient cultivation systems in the Peruvian Andes (see the box on page 171).

Case 1. Hydroelectric dams provide energy that can be used for industrialization and to provide electricity to urban areas. In the swampy central Paraná River basin in Argentina, the Paraná-Medio hydroelectric dam project is to be privately funded by a Metairie, Louisiana, firm that will receive a 50-year concession for electricity sales to Buenos Aires. The dam reservoir would turn a large wetlands ecosystem into the world's second-largest artificial lake. An assessment firm based in the United States has said that the environmental effects would be negligible, but some Brazilian and Argentine scientists say that the lake would inundate the habitat of more than 1100 animal species. This disruption could harm 300 species of fish, fishing tourism, and the livelihood of fishers. Upstream flooding would eliminate seasonal grazing and rice farming and kill many valuable trees.

The public debate has persuaded the government to reevaluate the project, but energy is crucial to most kinds of economic development and is presently in short supply in much of Latin America. Figure 3.17 compares commercial energy use per capita

Thinking Geographically: ON THE WEB

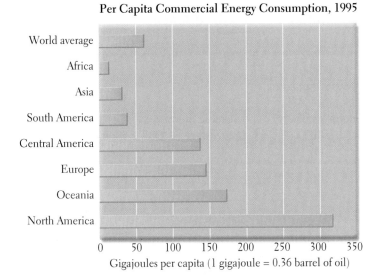

Per Capita Commercial Energy Consumption, 1995

Gigajoules per capita (1 gigajoule = 0.36 barrel of oil)

Figure 3.17 **Commercial energy consumption per capita by world region in 1995.** [*1997 United Nations Energy Statistics Yearbook.*]

for several world regions. South America uses less energy per capita than even Central America, and only about one-tenth that used in North America. Hydroelectric power is favored because it does not produce air pollution as coal or gas does. The Paraná dam is only one of a score of privately funded hydroelectric dam projects being proposed for South America's major river systems.

Case 2. The debt crisis that swept through the region beginning in the 1980s has created tremendous pressure for countries to increase their exports so that they can pay off their debts. Environmental degradation is often one result. During the 1980s and 1990s, Ecuador became a major exporter of shrimp, but in the process it lost 80 to 90 percent of the mangrove swamp forests that once covered much of its coast on the Pacific Ocean. Mangroves are unique ecosystems that provide shelter to birds, fish, mollusks, crustaceans, and other species while also protecting coastal areas from erosion by sea waves. People along Ecuador's coast have traditionally used the mangrove as a source of firewood, clams, fish, and crabs without significantly altering the ecosystem. Because the mangroves flood periodically, those who use them move around constantly to take advantage of resources that shift depending on how much water is present. Shrimp farming requires that the mangrove be cut to make way for large ponds in which shrimp are raised. The industry has spread rapidly both because it is highly profitable and because traditional subsistence users viewed the mangroves as commonly held grounds and never laid legal claim to them. Local and international environmental advocates attempted to halt the encroachment of shrimp farms on mangroves, but bribes to officials by shrimp farmers made it possible for them to claim yet more mangrove land, and now it is nearly gone.

These two cases illustrate how people in Middle and South America are beginning to learn how to be their own environmental advocates in the tension-filled process of economic develop-

ment. So far, success in halting forest exploitation has been elusive. Collaboration is just beginning between local indigenous groups and international conservation groups to pressure for stronger environmental legislation and to negotiate with multinational firms and international lending agencies for more sustainable policies.

MEASURES OF HUMAN WELL-BEING

VIGNETTE Miss Eliza arises at six in the morning and prepares tea, bread, and fruit for her two daughters and her elderly Aunt Hettie, whose two-room wooden house she shares. Auntie helps ready the children for school, and Miss Eliza rushes off with a "head-load" of vegetables and spices to catch the jitney to Montego Bay, Jamaica. There, in the central market, she hustles all day, finding just the right breadfruit for a good customer, putting together a basket of spices and condiments (nutmeg, pimento, cinnamon, fiery hot sauce) for some tourists off a cruise ship, negotiating with a fellow higgler (a person who buys small surpluses to resell) for some special cuttings of dasheen to plant in her home garden. During all this time, she doesn't think once about the fact that her yearly income for herself, Aunt Hettie, and the two children is just $4000. Rather, every single moment of her life is spent dealing with the realities of having very little. [Lydia Pulsipher, fieldwork, 1992.]

As geographers, we need ways to compare the well-being of people like Miss Eliza with that of people in other places. Table 3.3 provides you the opportunity to do just that, according to several different indices.

You will remember from the explanation in Chapter 1 that gross domestic product (GDP) per capita figures are often used as a crude indicator of well-being. In column 2 of Table 3.3, the countries in Middle and South America are compared with Japan, the United States, and Kuwait in per capita GDP. You can see immediately that they all lag far behind the first three. These figures have been adjusted for differences in prices and other factors to obtain internationally comparable indications of purchasing power parity (PPP) in U.S. dollars. Nonetheless, even these adjusted GDP figures can be misleading, because they are averages that mask a very wide disparity of wealth in Middle and South America and because they ignore aspects of well-being other than income. From these figures, we can learn that purchasing power is very likely limited throughout the region. But purchasing power is not the best measure of well-being.

In column 3 of Table 3.3, the United Nations Human Development Index (HDI) adds two other components to GDP per capita (PPP)—life expectancy at birth and educational attainment—to arrive at a ranking of 174 countries that is sensitive to more factors than just income. You can see that some countries in Middle and South America rank close to Kuwait (36) on the

TABLE 3.3   *Human well-being rankings of countries in Middle and South America and selected countries, 2000*

Country (1)	GDP per capita, adjusted for PPP[a] (in 1998 $U.S.) (2)	Human Development Index (HDI) global rankings, 2000[b] (3)	Gender Empowerment Measure (GEM) global rankings, 2000 (4)	Female literacy (percentage), 1998 (5)	Male literacy (percentage), 1998 (6)
Selected countries for comparison					
Japan	23,257	9 (high)	41	99	99
United States	29,605	3 (high)	13	99	99
Kuwait	25,314	36 (high)	75[c]	79	83
Caribbean					
Antigua and Barbuda	9,277	37 (high)	N/A[d]	88[e]	90[e]
Bahamas	14,614	33 (high)	16	96	95
Barbados	12,001	30 (high)	17	97[e]	98[e]
Cuba	3,967	56 (medium)	25[c]	96	97
Dominica	5,102	51 (medium)	N/A	94[e]	94[e]
Dominican Republic	4,598	87 (medium)	39	83	83
Grenada	5,838	54 (medium	N/A	98[e]	98[e]
Guadeloupe	9,200 (1995 est.)[c]	N/A	N/A	90[e]	90[e]
Haiti	1,383	150 (low)	71[c]	46	50
Jamaica	3,389	83 (medium)	N/A	90	82
Martinique	10,000 (1995 est.)[c]	N/A	N/A	93[e]	92[e]
Netherlands Antilles	11,600 (1997 est.)[c]	N/A	N/A	93[e]	98[e]
Puerto Rico	8,600 (1997 est.)[c]	N/A	N/A	88[e]	90[e]
St. Kitts and Nevis	10,672	47 (medium)	N/A	98[e]	97[e]
St. Lucia	5,183	88 (medium)	N/A	N/A	N/A
St. Vincent and the Grenadines	4,692	79 (medium)	N/A	N/A	N/A
Trinidad and Tobago	7,485	50 (medium)	22	92	95

[a]PPP = purchasing power parity.

[b]The high, medium, and low designations indicate where the country ranks among the 174 countries classified by the United Nations; the only country in the Western Hemisphere to rank low is Haiti.

[c]Data from *United Nations Human Development Report 1998* (data not available in 2000).

Country (1)	GDP per capita, adjusted for PPP[a] (in 1998 $U.S.) (2)	Human Development Index (HDI) global rankings, 2000[b] (3)	Gender Empowerment Measure (GEM) global rankings, 2000 (4)	Female literacy (percentage), 1998 (5)	Male literacy (percentage), 1998 (6)
Mexico and Central America					
Mexico	7,704	55 (medium)	35	89	93
Belize	4,566	58 (medium)	40	93	93
Costa Rica	5,987	48 (medium)	24	95	95
El Salvador	4,036	104 (medium)	30	75	81
Guatemala	3,505	120 (medium)	35[c]	60	75
Honduras	2,433	113 (medium)	48	74	73
Nicaragua	2,142	116 (medium)	N/A	69	66
Panama	5,249	59 (medium)	46	91	92
South America					
Argentina	12,013	35 (high)	N/A	97	97
Bolivia	2,269	114 (medium)	54	78	91
Brazil	6,625	74 (medium)	68[c]	85	85
Chile	8,787	38 (high)	51	95	96
Colombia	6,000	68 (medium)	37	91	91
Ecuador	3,003	91 (medium)	43	89	93
French Guiana	6,000[c]	N/A	N/A	82[e]	84[e]
Guyana	3,403	96 (medium)	39	98	99
Paraguay	4,288	81 (medium)	57	92	94
Peru	4,282	80 (medium)	50	84	94
Suriname	5,161	67 (medium)	52	91	95
Uruguay	8,623	39 (high)	45	98	97
Venezuela	5,808	65 (medium)	20	91	93

[d]N/A = data not available.
[e]Data from *CIA World Factbook*, 2000.

Source: United Nations Human Development Report 2000.

HDI index despite having much lower GDP per capita figures: Antigua and Barbuda (37), the Bahamas (33), Barbados (30), Argentina (35), Chile (38), and Uruguay (39). Their rankings are close to or higher than Kuwait's partly because education is more available to both sexes and across classes in Middle and South America than in Southwest Asia, where Kuwait is located and where women are secluded. Also, life expectancy in the Caribbean can be as high as 82 for women, whereas in Kuwait it is 73. Nonetheless, investment in basic and secondary education in Middle and South America is not sufficient to prepare most people for skilled jobs, and poor health care is holding down life expectancy to just age 69 for the region as a whole. The HDI rankings also indicate that poverty is especially deep in Haiti (150), Guatemala (120), Honduras (113), Nicaragua (116), and Bolivia (114), and in these countries the burden of a low income is not eased by government social programs as it is in most low-income Caribbean countries.

The United Nations Gender Empowerment Measure (GEM) (Table 3.3, column 4) measures the extent to which females have opportunities to participate in economic and political life in specific countries and takes into account female per capita GDP. GEM figures are missing for a number of Caribbean countries, but those available show that four Caribbean countries plus Costa Rica and Venezuela rank relatively high on this female empowerment scale: the Bahamas (16), Barbados (17), Venezuela (20), Trinidad and Tobago (22), Costa Rica (24), and Cuba (25). In all cases, their GEM ranks are higher than their HDI ranks. When a country's GEM rank is significantly higher than its HDI rank, it can mean that despite economic problems, the society is comparatively open to female participation in public life: education, jobs outside the home, entrepreneurship, leadership in business, and government. In the Caribbean, women have prominent positions in families, increasingly serve in government and in community organizations, and often have higher educational attainment than men. Many countries in Middle and South America, however, rank relatively low on GEM: Bolivia (54), Brazil (68), Chile (51), Paraguay (57), Peru (50), and others, although Mexico, El Salvador, and Guatemala rank in the 30s (of only 70 countries ranked on GEM). Still, all do better than does relatively rich but gender-stratified Kuwait, where education and health opportunities for women and opportunities for public life have been limited.

To return to the case of Miss Eliza in Jamaica, we can now see that her income of $4000 for a family of four is considerably lower than the average per capita income for that country. However, we also know that social support customs compensate for her low income—exemplified by Miss Eliza's looking after and feeding her aged aunt, who provides a house and child care for Miss Eliza in return. It is through such informal reciprocal exchange that people manage to survive and even thrive despite low incomes and declining tax-supported public services.

SUBREGIONS OF MIDDLE AND SOUTH AMERICA

This tour of the subregions of Middle and South America focuses primarily on how the themes of cultural diversity, economic disparity, and environmental issues are reflected somewhat differently from place to place. In every subregion, examples of connections to the global economy are given.

THE CARIBBEAN

Visiting tourists often misinterpret the Caribbean as being a desperately poor place. Typically, short-term visitors are isolated in enclaves or on cruise ships where they glimpse only tiny swatches of the island landscapes. From that vantage point, it is hard to see beyond the tiny houses and garden plots or quaint, narrow streets to the statistical facts and social relationships that make life in the Caribbean so different from what it first appears to be.

Only in the past 50 years have most of the island societies (Figure 3.18) emerged from colonial status to become independent, self-governing states. These islands, with the exception of Haiti and parts of the Dominican Republic, are no longer the poverty-stricken places they were 30 years ago. Rather, they are managing to provide a reasonably high quality of life for their people (see Table 3.3). Children go to school; literacy rates for people under 70 years of age average close to 99 percent. There is basic health care: mothers receive prenatal care, nearly all babies are born in hospitals, and infant mortality rates are low (much lower than what they were even in the 1970s). Life expectancy is in the 70s for most islands. And people are choosing to have fewer children; the overall rate of population increase for the Caribbean is the lowest in Middle and South America. A number of Caribbean islands are in the most prosperous cohort on the Human Development Index and do particularly well on the Gender Empowerment Measure. Returned emigrants often say that the quality of life on many Caribbean islands actually exceeds that of far more materially endowed societies, because life is enhanced by strong community and family support. And then, of course, there is the healthful and beautiful physical environment.

Beyond the beaches and quaint villages there are local civic organizations such as the Rotary and Lions clubs, libraries, chambers of commerce, gourmet cooking clubs, garden societies, community clean-up associations, and active churches. These organizations practice participatory democracy daily: citizens educate one another about social and environmental issues, and they continuously design and implement solutions to local problems. The progress indicated by the demographic statistics (see Table 3.3) is the result of hard work and civic responsibility, as well as aid from the old colonial powers and Canada (and in Puerto Rico from the United States).

Island ministers of government continuously search for ways to turn plantation economies, formerly managed from Europe and

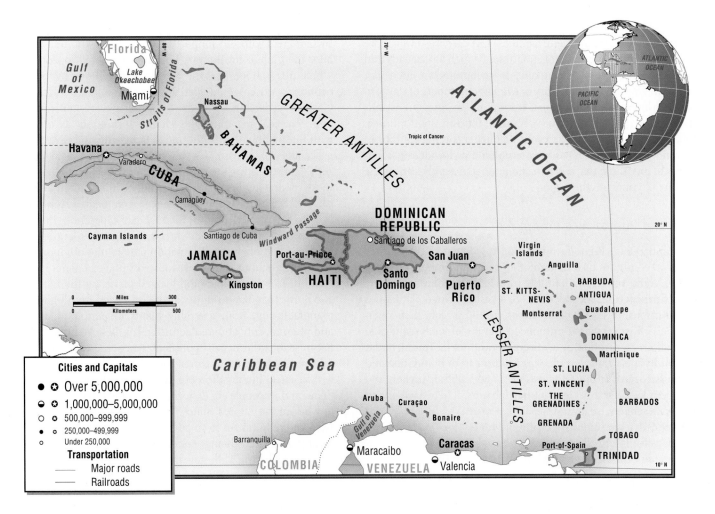

Figure 3.18 The Caribbean subregion.

North America, into more self-directed, self-sufficient, and flexible entities that can adapt quickly to the perpetually changing markets of the global economy. On the one hand, they are tempted to specialize; on the other, they are wary of the dependency and vulnerability that too much specialization can bring. When plantation cultivation of sugar, cotton, and copra died out in the 1950s and 1960s, some islands turned to producing "dessert" crops such as bananas and coffee that they sold for high prices under special agreements with the countries that once held them as colonies. Now these protections are disappearing as the European Union makes such agreements illegal in Europe. Other island countries turned to the processing of special resources (petroleum in Trinidad and Tobago, bauxite in Jamaica, for instance), the assembly and finishing of such high-tech products as computer chips and pharmaceuticals (in St. Kitts and Nevis), or the processing of computerized data (in Barbados). A number of islands combined one or more of these strategies with tourism development.

For some islands, tourism and related activities contribute as much as 80 percent of the gross national product (GNP), as was the case in Antigua and Barbuda, Barbados, and the Bahamas in the1990s. Heavy borrowing to build hotels, airports, water systems, and shopping centers for tourists has left some islands with huge debts to pay off. Furthermore, there is a special stress that comes from dealing perpetually with hordes of strangers, especially when they dress too skimpily, act in ways that violate local mores, and make unflattering assumptions about one's state of well-being—concluding erroneously that poverty and illiteracy are widespread. Every year in the mid-1990s, Jamaica (population 2.6 million) received visitors numbering twice its population; Antigua and Barbuda (population 66,000) hosted more than 4 times its population, and St. Martin (population 20,000), 20 times its population.

Host countries consider tourism difficult to regulate, in part because it is a complex multinational industry. Tourism markets are controlled by external interests such as travel agencies, airlines, and the outside investors needed to fund the construction of hotels, resorts, restaurants, and golf courses. The rapid growth of the industry in an unregulated environment has led to overbuilding and to poorly developed hotel accommodations in such places as St. Croix, St. Thomas, Antigua, and St. Martin. The recent rapid growth of the cruise line industry, operated mostly out of Puerto Rico and Miami, has caused overcrowding, too. Caribbean cruise ships now routinely bring several thousand passengers into small port cities. Plans by such corporations as Disney Enterprises to build even larger cruise ships equipped with elaborate entertainment facilities and carrying even more passengers are troublesome for small islands such as Grenada and St. Lucia, which would be unable to accommodate either the enormous ships in their harbors or the large numbers of day visitors such liners would discharge. Entrepreneurs on some islands are targeting new types of tourists. Nature tourism, sport tourism, and ecotourism are three variations that better suit the Caribbean environment and that have been promoted as low-density alternatives to the "sand, sea, and sun" menu that presently attracts the majority of visitors to the region.

Cuba and Puerto Rico Compared

Cuba and Puerto Rico are interesting to compare because they shared a common history until the 1950s and then diverged starkly. North American interests dominated both islands after the end of Spanish rule, about 1900. U.S. investors owned plantations and other businesses on both islands, and to protect their interests, the U.S. government influenced dictatorial local regimes to keep labor organization at a minimum and social unrest under control. By the 1950s, poverty was widespread in both Cuba and Puerto Rico. Infant mortality was at 50 per 1000 or higher, and most people worked as agricultural laborers. Then each island took a different course toward social transformation. In 1959, Cuba experienced a communist revolution under the leadership of Fidel Castro. Puerto Rico has experienced a more gradual capitalist metamorphosis into a high-tech manufacturing center.

Since Castro seized control, Cuba has dramatically improved the well-being of its general population in all categories of measurement (life expectancy, literacy, and infant mortality). By 1990, it had a solid human well-being record despite its relatively low GDP per capita (see Table 3.3). Cuba thus proved that, once the requisite investment in human capital is made, poverty and social problems in Latin America are not as intractable as some people had thought. Unfortunately, Cuba has been politically repressive in the extreme, forcing out the aristocracy and jailing dissidents. Economically, Cuba has been inordinately dependent on outside help to achieve its social revolution. Until 1991, the Soviet Union was Cuba's chief sponsor. It provided cheap fuel and generous foreign aid and bought Cuba's main export crop, sugar, at preferential prices. With the demise of the Soviet Union, Cuba's economy declined sharply during what became known as the Special Period, when all Cubans were asked to make sacrifices and cut back their consumer spending. To survive this economic crisis, Cuba opened its economy to outside investment, especially European capital, to help redevelop its tourism industry. Canada has invested in joint biotech research on hepatitis and cancer drugs with Cuban medical scientists, who receive particularly good training in Cuba, among the best in Latin America. Biotech exports accounted for $130 million in 2000 and are expected to grow. Health and spa tourism is also a growing industry, with more than 5000 customers per year arriving for face-lifts and medical therapy.

Political relations between the Castro regime and successive U.S. administrations have remained on cold war terms since 1961. A recent example is the Helms-Burton Act adopted in 1996 by Congress with the support of the Cubans who fled to Florida after the Cuban revolution. This law attempted to stop all trade with Cuba in hopes that the hardships brought on by the ensuing economic crisis would lead to popular support for an end to Castro's leadership. Nonetheless, other governments in Europe and America, regarding the Helms-Burton Act as inconsistent with international law, continue to trade with and invest in Cuba, especially in tourism. Cuban tourism grossed $1.5 billion in 1997, mostly from the European market. Nonetheless, Cuba's most natural trading partner would be the United States, and many U.S. businesspeople are eager for a repeal of Helms-Burton.

In the 1950s, Puerto Rico began Operation Bootstrap, a government-aided program to transform the island's economy from its traditional sugar plantation base to modern industrialism. Many international industries took advantage of generous tax-free guarantees and subsidies and located plants on the island. Since 1965, this manufacturing sector has changed from light to heavy manufacturing, from assembly plants to petroleum processing and pharmaceutical manufacturing. It is believed that the Puerto Rican seaboard is heavily polluted as a consequence of chemicals released by these more recent industries.

Because Puerto Rico is a commonwealth within the United States and its people are U.S. citizens, many Puerto Ricans migrate to work in the United States. The manufacturing jobs in Puerto Rico and remittances from people working on the mainland have greatly improved living conditions on the island, but social investment by the U.S. government has also upgraded the standard of living. Many Puerto Ricans receive some sort of support payment from the federal government, including retirement benefits and health care. The infant mortality rate is now 11.5 (Cuba's is just 7.2), and life expectancy is 74 (Cuba's is 75). The connection with the United States helps Puerto Rico's tourism and light-industry economy, but outside of San Juan, with its skyscraper tourist hotels, Puerto Rico's landscape reflects stagnation: few interesting or well-paid jobs, an inadequate transport system, little development at the community level, mediocre schools, and few opportunities for advanced training.

Statehood for Puerto Rico, which some people desire, would mean the end of tax holidays for U.S. companies and hence the end of Puerto Rico's ability to attract assembly plants, some of which are already moving to cheaper labor markets. Also, many fear that statehood would eventually bring cultural assimilation and the loss of the island's language and Spanish heritage.

Haiti and Barbados Compared

Haiti and Barbados are a study in contrasts. During the colonial era, both were European possessions with plantation economies, Haiti a colony of France and Barbados a colony of Britain. Yet they have had very different experiences and today are far apart in economic and social well-being. Haiti, though not without useful resources, is the poorest nation in the Americas, with a ranking on the United Nations Human Development Index of 150 out of 174. Barbadians have the highest HDI ranking in the entire region (30) even though the island has much less space and few resources other than its limestone soil and its people.

Varadero, one of Cuba's new tourism areas, is on the island's north coast, about 100 miles (160 kilometers) east of Havana. [David Alan Harvey/Magnum Photos.]

Haiti was the richest plantation economy in the Caribbean by the end of the eighteenth century. When the Haitian slaves revolted against the brutality of the French planters in 1804, Haiti became the first colony in Middle and South America to achieve independence. Haiti's early promise was lost, however, when the former slave reformist leaders were overthrown by violent and corrupt militarists, who neither reformed the exploitative plantation economy nor sought a new economic base. Under a long series of unstable authoritarian governments, the people sank into abject poverty while the land was badly damaged by particularly wasteful and unprofitable plantation cultivation. In the mid-twentieth century when other Caribbean islands, such as Barbados, were reorganizing their economies and achieving various levels of prosperity, in Haiti a class-based reign of terror under François "Papa Doc" Duvalier (1957–1986) pitted the mulatto elite against the black poor. Today, Haiti remains overwhelmingly rural, with widespread illiteracy and high infant mortality. Its lands are deforested and eroded. Efforts to establish democracy have repeatedly devolved into violence. Since the early 1990s, the United Nations has maintained peacekeeping troops in Haiti; and several humanitarian aid organizations run programs there. Multinational corporations opened maquiladora-like assembly plants employing primarily young women, but the plants are not yet flourishing, and although minerals such as bauxite, copper, and tin exist, cost-effective development is not possible.

Far more prosperous Barbados has fewer natural resources and is more than twice as crowded as Haiti. Barbados has 166 square miles (430 square kilometers) and 1600 people per square mile (618 per square kilometer), whereas Haiti has 10,000 square miles (25,900 square kilometers) and 600 people per square mile (230 per square kilometer). Barbados's present prosperity is explained by the fact that its citizens successfully pressured the British to invest in the people and infrastructure of its colony before giving Barbados

independence in 1966. Although both Haiti and Barbados began the twentieth century with large, illiterate, agricultural populations, their development paths have diverged. In 2001, Haiti remains in this situation: two-thirds of Haitians are still agricultural workers, and less than half are able to read and write. Barbados, on the other hand, now has a diversified economy that includes tourism, sugar production, remittances from migrants, information processing, and modern industries that sell products throughout the Caribbean. Barbadians hold jobs that demand skilled, literate employees; they are well educated and well fed; and most are homeowners. Britain maintains an interest in Barbados and helps it financially in small ways. Meanwhile, the Barbadian government and private businesspeople constantly seek new employment options for the citizens and occupy a central role in Caribbean economic and social development.

MEXICO

Mexico today is working toward becoming a reasonably well managed, middle-income democracy (see the earlier discussion on page 143). Most of the efforts to achieve this goal, both private and governmental, are focused in some way on Mexico's relationship with the United States, its main trade partner.

Mexico's physical geography is dominated by mountains. The Sierra Madre forms a wide **V** that extends through the northern two-thirds of the country (see Figure 3.2). Between the two mountain ranges lies a high northward-tilting plateau, the Mesa Central. To the south, in the crook of the **V**, is a high valley surrounded on three sides by lofty snowcapped volcanoes reaching to nearly 18,000 feet (5500 meters). This high valley once held the famed Aztec city of Tenochtitlán and now is the site of Mexico City (Figure 3.19). With a total metropolitan population

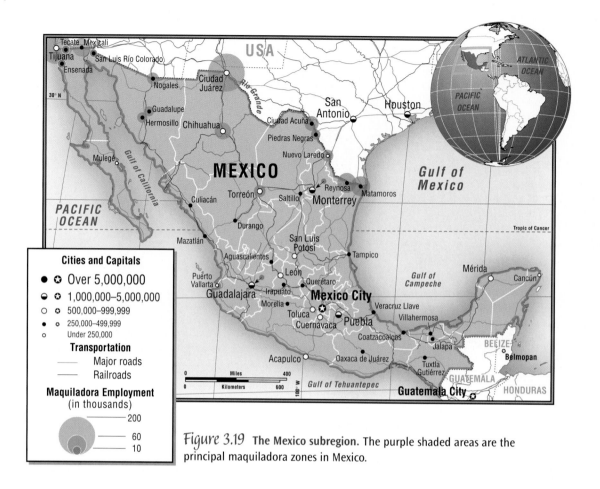

Figure 3.19 The Mexico subregion. The purple shaded areas are the principal maquiladora zones in Mexico.

(including suburbs) of 25 million people—one-quarter of Mexico's total population—Mexico City is not only a primate city but one of the world's largest. East of the volcanoes, the land descends to a wide stretch of humid tropical lowlands along the Caribbean. To the west is the narrow Pacific coast, and to the south more mountains extend beyond the narrow Isthmus of Tehuantepec to Guatemala.

Mexico's formal economy has modernized rapidly in recent decades. Its components are mechanized agriculture, a vastly expanded manufacturing sector, petroleum extraction and refining, tourism, a service and information sector, and the remittances of millions of migrants working abroad. Mexico's large and varied informal economy employs an estimated 16 million people at least part-time, or about half of the total working population of 33 million. They contribute at least 13 percent of the GDP in such activities as subsistence agriculture; fine craft making; and many types of services, such as hairdressing, running errands, tutoring, trash recycling, and street vending. The informal economy also includes a flourishing black market and the crime sector.

The Geographic Distribution of Economic Sectors

Mexico's economic sectors have a geographic distribution pattern. Just 12 percent of Mexico is good agricultural land, and much of that is on hillsides. The biggest agricultural units are large corporate farms aided by government marketing, research, and irrigation programs; these have replaced the old, inefficient

haciendas and ranches. The corporate farms produce high-quality meat and produce—beef, chicken, peaches, nuts, asparagus, chickpeas, citrus, melons, peppers, winter vegetables, oil seeds, strawberries, and raspberries—and sorghum for animal feed. Much of this produce is sent to the North American market, and many farms are located in such places as Sonora, in the arid north close to the U.S. border. In the tropical coastal lowlands farther south, subsistence cultivation on small plots is much more common, but here too tropical fruits and products such as jute, sisal, and tequila are produced for export as well as for internal consumption. Mexico's convoluted upland landscapes provide many small and specialized niches where a wide variety of crops are grown: sesame seeds, hot peppers, and flowers, among others. About 24 percent of Mexico's population is employed in agriculture, though it produces only 5 percent of Mexico's GDP.

Petroleum refining is located along the Gulf coastal plain, and tourism facilities are found primarily along the Caribbean and Pacific coastal plains. Manufacturing employs about 21 percent of the registered workforce and accounts for 27 percent of GDP. It is concentrated along the U.S. border in such cities as Tijuana, Mexicali, Ciudad Juárez, and Reynosa, where numerous U.S. firms operate maquiladoras; some of these factories are also found in the general vicinity of Mexico City and in Yucatán. The service sector employs 55 percent of the population and produces 68 percent of the GDP. Although it includes restaurants, hotels, financial institutions, consulting firms, and museums, this sector is dominated by small operations that cater to everyday needs:

hair salons, tortilla vendors, tailoring shops, shoe shining and repairing services, Internet cafés, cleaning businesses, and so forth. Most such service jobs are concentrated in cities. The informal economy, which can include agriculture, manufacturing, and service activities, is everywhere, but it is most varied and noticeable in the big cities.

The Maquiladora Phenomenon

Maquiladoras are foreign-owned plants that hire people to assemble manufactured goods, usually from imported parts. The materials are imported duty-free and the goods returned to the country of origin for sale there or elsewhere. There are thousands of maquiladoras along the Mexican side of the U.S.-Mexican border, as well as elsewhere in the country (see Figures 3.11 and 3.19). Many maquiladoras are branches of American corporations, though Mexicans hold the majority interest in about half of them. Overall, the number is growing quickly, as Table 3.4 shows. The Mexican side of the U.S. border is a desirable location for factories because there is inexpensive labor within easy reach of the United States. Mexican workers earn an average of about $5.30 a day—and Mexico has far fewer regulations than the United States covering worker safety, environmental protection, and fringe benefits. Mexico also charges much lower taxes on industries. These conditions were attracting a considerable number of maquiladoras even before NAFTA took effect in 1994, although total maquiladora employment has increased dramatically since then (see Table 3.4). Studies in the late 1990s showed that more than half the laborers are women, many of them young and unmarried, but men are gaining in maquiladora employment. The laborers migrate from rural locations across Mexico, and the money they send home is crucial to their families' welfare. Because of their financial contributions, female workers

In Nuevo Laredo, Mexico, workers in this sewing factory, leased by R. G. Barry Corporation of Pickerington, Ohio, are making brand-name slippers for markets in North America and Europe. [Paul S. Howell/Gamma-Liaison.]

enjoy status in the family that earlier generations of women did not enjoy.

Although many Mexicans are pleased to have the maquiladoras and the jobs they bring, the country is increasingly facing negative side effects. The border area has developed serious groundwater and air pollution problems because of the high concentration of poorly regulated factories and unplanned worker shantytowns numbering in the thousands. Social problems are also developing. Workers are crowded into unsanitary living conditions. There are not adequate schools for the increasing numbers of children. Family violence is a growing problem, brought on by the tensions of migration, overcrowded conditions, and changing gender roles. Women often can find work more easily than men because they are compliant and will work for less pay. But men, displaced from agriculture and local factory work, are increasingly filling traditional female jobs, such as sewing garments in maquiladoras.

Mexico and the United States are working to solve the environmental problems related to maquiladora concentrations along the border. The U.S.-Mexico Border XXI Project has joint task forces focusing on water and air pollution mitigation, hazardous waste cleanup and prevention, environmental studies, and emergency preparedness. Early in the twenty-first century, Mexico hopes to provide 93 percent of the border population with clean drinking water, and 16 water-related projects are currently under way along the border, some in Mexico, some in the United States.

VIGNETTE Ciudad Juárez Mayor Ramon Galindo and Ana Serratos, a maquiladora worker, see different sides of rapid industrialization in this Mexican border city, but they reach similar conclusions about the overall situation. Sometimes Ms. Serratos pauses outside a McDonald's restaurant in Ciudad Juárez,

	Rapid growth rate of the maquiladora sector	
TABLE 3.4		

Year	Number of maquiladoras	Number of employees
1968	79	17,000
1974	455	75,974
1981	605	130,973
1988	1400	369,489
1990	2042	486,000
1997	2676	897,354
1998	2983	1,008,021
1999	3500	1,096,000

GEOGRAPHER IN THE FIELD

Interview with Regional Consultant for Chapter 3, Dennis Conway, Indiana University

What led you to become interested in Middle and South America?
I taught geography at Harrison College, a boys' grammar school in Barbados, in the late 1960s. Living in this newly independent island country and observing firsthand the island peoples' struggles to improve their societies and livelihoods gave me a deep interest in the Caribbean. Extensive travels throughout the region and a prolonged visit to Mexico broadened my interest in the region.

Where have you done fieldwork in Middle and South America, and what kind of work have you done?
I conducted my Ph.D. dissertation research on Port of Spain, Trinidad, in the mid-1970s. Between 1983 and 1984, I spent a year in Barbados as a field associate of University Field Staff International (UFSI), traveling to neighboring islands and writing reports on current issues, including the invasion of Grenada. Other trips to Mexico and Guatemala fostered a deep interest in the Maya and Olmec civilizations. My fieldwork has centered on Caribbean migration patterns and processes. I study how migration

affects both the development of home societies and the livelihoods of migrants and their families. My interest in migration has recently extended to Oaxacan (Mexico) migration. The household survival strategies in Oaxaca, like those in the Caribbean, rely heavily on migration both within the country and to and from the United States.

What are some topics that you think are important to study in your region?
One is modern transnational patterns of migration and the strategies used by the dependents of these movers to invest remittances. Another is the difficulties faced by young returning migrants trying to adjust to life back home, as well as their contributions to local economies and their efforts to set up small businesses.

---

just across the border from El Paso, Texas; but she resists the temptation to stop in. Fast food is far beyond her means. A Big Mac, medium Coke, and fries cost $3.05, and for Ms. Serratos this is nearly a full day's wages. She earns 45 cents an hour wrapping tape around bundles of electrical wires in one of Juárez's assembly plants.

"What people in your country make in an hour, we work a whole day for," says Ms. Serratos, 32. She lives with six siblings in a dirt-street suburb of Ciudad Juárez and commutes to work two hours each way on ramshackle buses. "These companies from the United States and Japan don't pay people what they ought to. . . . But we have to recognize that they're important for Juárez."

The vast majority of wage earners in Juárez make little more than Ms. Serratos and are unfamiliar with labor unions, which are not yet allowed by the managers of the maquiladoras. Such low wage earners don't pay much in taxes; hence, as Mayor Galindo notes, the tax revenue of this city of 1.2 million is woefully small. Although drinking water has run out, there is no money for new wells, and it may be 15 years before the streets are blacktopped. The mayor is keenly aware that the factories themselves also pay very low taxes. Although he is hoping to attract yet more factories, the mayor concludes that without change the maquiladora economy may only create "an enormous mass of wretchedly poor people." During the time of rapid maquiladora growth, despite a

minimum wage hike of 14 percent, the consumer price index rose 18.6 percent. The effects on daily life are illustrated by the fact that in 1987 one worker had to work 8 hours to buy the weekly food for a family of four; by 1998, the same food supply required 34 hours of work. [Adapted from Sam Dillon, "Big Mac? Not for maquiladoras workers," *New York Times News Service* (December 5,1995); "Living on Earth," National Public Radio (November 24, 1995); and Coalition for Justice in the Maquiladoras, *Annual Report* (1998), and *Newsletter* 9(1) (Spring 1999): 13.]

NAFTA and Foreign Trade

Even before NAFTA was signed in 1994, Mexico traded predominantly with the United States and Canada. Today, the United States supplies 70 percent of Mexico's imports (industrial and agricultural machinery, electrical products, car parts for assembly and repair) and buys 83 percent of its exports (fresh food, corn, coffee, cotton, oil and oil products, manufactured goods, silver). For many Mexican businesses, NAFTA represents a chance to attract North American investment and to sell their products with fewer restrictions in the rich North American market. NAFTA is expected to raise standards for general manufacturing and business practices in Mexico, and it is also promoted as a way to create jobs for the increasing population of young people. NAFTA is just one of a set of important new free-trade partnerships between Mexico and several other countries. In November 2000, Mexico signed agreements with Iceland, Norway, and Switzerland, and in 2001 with the European Union. Talks are underway with Korea, Singapore, and Japan. Within the region, Costa Rica, Venezuela, Colombia, and Chile have recently entered into free-trade agreements with Mexico. To the countries of Middle and South America, Mexico is a large and relatively prosperous market for their goods. Mexicans, in turn, will receive a greater choice of products. So far, trade with Europe and Asia is equal to only a tiny portion of Mexico's level of trade with the United States and Canada.

Not all Mexicans view NAFTA in a positive light, however. Many small farmers feel that they will not be able to compete with large operations, because they will not enjoy economies of scale—their very smallness will mean that their expenses per unit produced will be higher. Moreover, small farmers will find it difficult to meet new international quality standards for agricultural products. With the loss of protective tariffs and price supports, Mexican corn producers, for example, will have to compete with duty-free mass-produced corn coming in from the United States. Yet switching to new crops is difficult for small farmers, because they cannot afford loans for new equipment and seed. Small local businesses face similar challenges under NAFTA: loans to expand and modernize are expensive, and they will find it difficult to compete with mass-produced imported products. In addition, consumers often choose modern styles and media-promoted fashions over local, artisan-made products.

Population in Mexico

Mexico's many efforts to industrialize, find jobs for its people, and improve trade relationships are all focused on raising standards of living. Yet these efforts are endangered by relentless population growth. Mexico now has almost 100 million people, and at present reproduction rates that number will double in 36 years—a rate of increase that is alarmingly higher than past estimates. And because the Mexican population is very young, with nearly half the people under the age of 21 and more than one-third under the age of 15 (Figure 3.20), even faster growth is possible.

The young people of Mexico could develop into energetic and highly productive workers, filling jobs, serving on committees, running local governments, addressing environmental problems, and consuming products and services. But to do so, they will need healthy childhoods and advanced education. Even today, nearly half of Mexico's people do without adequate sanitation facilities. The infant mortality rate is high, at 32 per 1000 births. Only 66 percent of the eligible students are in secondary school. Mexico needs to be able to spend the coming years improving life for the citizens it already has; but without miraculously high economic growth and the dedication of substantial tax money to human services (especially education and health care), living standards for the majority in Mexico could sink, not rise.

Migration

Migration is a strategy often used to alleviate the problems of poverty. Rapid rural-to-urban migration in Mexico is legendary:

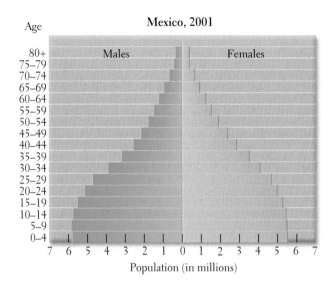

Figure 3.20 Mexico population pyramid, 2001. [Adapted from "Population Pyramids of Mexico" (Washington, D.C.: U.S. Bureau of the Census, International Data Base).]

74 percent of Mexicans (73 million people) live in cities, the majority of them in large cities. The emigration of Mexicans into North America also has an enormous impact on Mexico as a nation. On the one hand, the skills of the best and brightest are lost; on the other, some benefits from their industry are gained.

Unlike many Europeans, who usually cut their ties with family and native land when they came to North America, Mexicans often remain loyal to their home villages. Their migrations are usually undertaken with the express purpose of helping out their own families and communities. A recent paper (1998) by the geographers Dennis Conway and Jeffrey H. Cohen reports that couples who migrate from Native American villages in Mexico often work just long enough at menial jobs in the United States to save a substantial nest egg. Then they return home to build a house (usually a family self-help project) and buy appliances. Afterward, the family may live off that nest egg for several years, while one or both members of the couple renders volunteer community service to the home village—perhaps building schools, serving on the town council, improving the town water system, or refurbishing the town's public spaces. When the money runs out, the couple, or just the husband, may migrate again to save another nest egg. Often these civic-minded migrants cycle other members of their family through a menial U.S. job, such as dishwashing, thus keeping the job open for themselves when they need it again.

Their bosses and coworkers in the United States never guess that such unprepossessing, minimum-wage workers are actually influential and public-spirited citizens in their home villages, who use migration and hard work to enhance community living standards and participatory democracy. All this is possible because Mexican communities tend to be self-reliant and because the Mexican economy is much less affluent than that of the United States. Hence, a few thousand dollars saved in the United States will accomplish a great deal in rural Mexico.

CENTRAL AMERICA

In Central America, wealth is in the soil. Industry accounts for only about 20 percent of the gross domestic product, and the region is not rich in mineral resources. Thus, the seven countries of Central America (Figure 3.21) remain dependent on the produce of their plantations, ranches, and small farms. Here, the disparity of income seen in the region as a whole takes the form of disparity in the distribution of usable land.

Most of the Central American isthmus consists of three physical zones that are not well connected with one another: the narrow Pacific coast; the highland interior; and the long, sloping, rain-washed Caribbean coastal region (see the map at the beginning of this chapter). Along the narrow Pacific coast, mestizo (*ladino* is the local term) laborers work on large plantations that grow sugar, cotton, and bananas and other tropical fruits; coffee is grown in the hills in back of the coast. In the highland interior of Guatemala, Honduras, and Nicaragua, where Native American people have traditionally made their livings from subsistence agriculture, cattle ranching and commercial agriculture have recently displaced subsistence farmers. The humid Caribbean coastal region, long sparsely populated with Native American and African-Caribbean subsistence farmers, is increasingly dominated by commercial agriculture, forestry, tourism development, and resettlement projects for displaced small farmers from the highlands.

In Central America, the majority of people are Native Americans or ladinos, and about half still live in rural villages. In the villages, most people make a spare living by cultivating their own food and cash crops on land they own or rent or as share-croppers and by working seasonally as laborers on large farms and plantations. The people of this region have experienced centuries of hardship, including long hours of labor at low wages and the loss of most of their farmlands to large landholders. In both rural and urban areas, infrastructure development has lagged. Roads are primitive and few. Most people lack clean water, sanitation, health care, protection from poisoning by agricultural chemicals, and basic education. Often, they do not have access to enough cultivable land to meet their basic needs. The majority of the land is held in huge tracts—ranches, plantations, and haciendas—owned by a few families. As a result, the small percentage of agricultural land accessible to the poor for their own cultivation may have densities of 1000 people or more per square mile.

Although there is a general hunger for land among the poor majority, only tiny El Salvador is truly densely populated, with an average of 773 people per square mile (286 per square kilometer). In fact, cultivable land there is so scarce that people desperate for land have slipped over the border to cultivate land in Honduras.

The populations of most Central American countries are growing rapidly, as is common where dire conditions prevail (Table 3.5). Children are still seen as a labor asset in rural settings, and women have very little education for work outside the home. Minimal improvements in diet and health care have decreased deaths, so, although births have declined, they still outnumber deaths by 4 or 5 to 1. In all countries except El Salvador, overpopulation is an issue chiefly because growth rates are outstripping efforts to improve standards of living. Most countries, other than El Salvador, could probably support somewhat larger populations at decent standards of living, if those populations had equitable access to

Figure 3.21 **The Central America subregion.**

TABLE 3.5 *Population data on Central America*

Country	Population	Population density per square mile	Rate of natural increase (percentage)	Literacy rate (percentage)
Belize	300,000	29	2.7	93
Guatemala	12,700,000	301	2.9	67.3
Honduras	6,100,000	142	2.8	73.4
El Salvador	6,300,000	773	2.4	78
Nicaragua	5,100,000	101	3.0	68
Costa Rica	3,600,000	181	2.2	95
Panama	2,700,000	90	1.8	90

employment at living wages and to resources (land, clean water, sanitation) and social services (health and education). But given present realities, continued growth can only cause more hardships.

Costa Rica (and, to some extent, Belize and Panama) is an exception to extreme patterns of elite monopoly, mass poverty, and rapid population growth (see the box "Panama and the Panama Canal"). The huge disparities in wealth between colonists and laborers did not develop in Costa Rica, chiefly because the fairly small native population died out soon after the conquest. Without a captive labor supply, the European immigrants to Costa Rica set up small but productive family farms that they worked themselves, not unlike early North American family farms. Costa Rica's democratic traditions stretch back to the nineteenth century, and it has unusually enlightened elected officials and one of the region's soundest economies. Population growth is low for the region at 1.8 percent per year, and investment in schools and education has been high. As a result, Costa Rica has Central America's highest literacy rates, and on many scales of comparison the country stands out for its high living standards (see Table 3.3). Thus, Costa Rica has often been hailed as a beacon for the more troubled nations of Central America.

Environmental Concerns

Although poor farmers may use unwise practices in wrenching enough to feed their families from the tiny plots they are allowed, and although they are often blamed by the news media for environmental degradation, most environmental problems in Central America are caused by large-scale agriculture and cattle ranching. In some places, commercial farmers have overused chemicals to control weeds and pests. Each year, hundreds of children die from allergies and other effects of chemical poisoning. Land management by the large landowners is also a problem. Clearing enough land for large-scale agriculture and cattle grazing leads to tremendous amounts of erosion. In Honduras, the reservoir for a large hydroelectric dam built only a few years ago has been nearly filled with silt because of clearing on the surrounding land. Electricity output must now be supplemented with generators run on imported oil.

Costa Rica has been a leader in the environmental movement in Central America. The country has established wetland parks

The pattern of tiny farm plots (minifundios) in El Salvador is visible in this photo. Some are the result of land redistribution efforts, but because of high population density, the minifundios are often too small to support a family. [Tomasz Tomaszewski/National Geographic Image Collection.]

AT THE LOCAL SCALE   *Panama and the Panama Canal*

Since early Spanish colonial days, Panama has served as a conduit for moving goods between the Caribbean and the Pacific. In the early days, much of the trip across the isthmus was made by mules and other draft animals. Originally part of Colombia, Panama was created as an independent state under pressure from the United States, which, by the 1890s, wanted to finance the construction of a canal that made use of Gatun Lake for part of the route. The Panama Canal was built primarily with West Indian labor, and many West Indians stayed, making Panama's north coast a distinctly Caribbean place. The country remained a virtual colony of the United States, which managed the canal and maintained a large military presence there. For many years, those Panamanians employed in the Canal Zone or in canal-related occupations lived a way of life that was heavily influenced by U.S.

culture and the U.S. economy; Panama is relatively wealthy for Central America (see Table 3.3), and the canal is the reason. Away from the canal, however, poverty is much more common.

In 1977, under pressure from Panama, the United States agreed to give the canal back to Panama in 1999 and remove itself as a dominant presence. Interestingly, the turnover of the canal to Panama came at a time when the canal was becoming obsolete. It is no longer large enough to accommodate the huge cargo container or cruise ships of the modern era. Thus, traffic through the canal is no longer growing as quickly as it had been. Plans for updating the canal have been delayed by a number of considerations, one of which is that there is no alternative route to use during repairs, except around the tip of South America. Panama's hope for earning revenue from the canal may not be fulfilled over the long term.

---

along the Caribbean coast and encouraged ecotourism, while at the same time paying some attention to the potentially negative environmental side effects of tourism. Costa Rica supports scientific research through several international study centers in its central highlands and lowland rain forests, where students from throughout the hemisphere study tropical environments.

Civil Conflict

Frustrated with governments unresponsive to their plight, Native Americans and other rural people in Guatemala, Honduras, Nicaragua, and El Salvador began organizing protest movements in the 1970s. In some cases, they had the help of liberation theology advocates and Marxist revolutionaries from outside the region. They met with stiff resistance from wealthy elites assisted by national military forces. The consequence has been destructive civil wars that have plagued Guatemala, Honduras, El Salvador, and Nicaragua in recent decades. In the 1980s, protests were particularly strong in Guatemala, which has the largest Native American population in Central America. The military government, armed through aid programs from the United States that were intended to fend off communist insurgencies, killed thousands of Native protesters and drove 150,000 into exile in Mexico. In time, the protesters responded with guerrilla tactics. Rigoberta Menchu is a Native American woman who won the 1992 Nobel Peace Prize for her efforts to stop government violence against her people. Her autobiographical account attracted public attention to the carnage and was important in awakening worldwide concern. Eventually, after a number of regional peacemakers joined Menchu in bringing international pressure to bear on the Guatemalan government, the Guatemalan Peace Accord was signed in September 1996.

The primary actors in civil conflicts are usually men, but recent conflicts in Central America gave rise to the unusual phe-

nomenon of women engaging in public protest and activist politics. The vocal stance of these women, especially in Nicaragua and Guatemala, gained notice for their cause, simply because women protesters were new. Their actions also brought attention to the fact that women head many rural households in Central America and are essential in the economy. They do many of the most labor-intensive tasks involved in animal tending, soil management, farm product marketing, firewood procurement, and reforestation. Leaders of reform efforts—to change age-old economic relationships, to increase entrepreneurial activity, and to initiate the environmentally sensitive management of forests and agricultural resources—are now recognizing that training women as well as men may yield much better results than the former strategy of training men only.

A Case Study of Civil Conflict: Nicaragua

Until recently, Nicaragua showed landownership patterns characteristic of the region: a tiny elite held the usual monopoly of land, while the mass of laborers lived in poverty. Then a socialist revolution attempted to change the distribution of land and reorient society along more egalitarian lines. The effort was thwarted, partly because of internal turmoil but also because Nicaragua became a focus of U.S. worry over communist expansion in the Americas.

North American investors have had an interest in the coffee and fruit plantations of the Nicaraguan Pacific coastal plain since early in the twentieth century. Several times in the early 1900s, the United States sent in the marines to quell labor protests that threatened the export economy. This interference helped the wealthy Somoza family to establish its members as dictators in Nicaragua in the 1930s. The particularly brutal Somoza regime was finally ousted by the Marxist-leaning Sandinista revolution of 1979. The Sandinistas, who eventually won several national elections, embarked on a program of land and agricultural reform, redistributing thousands of acres and downplaying ranching. They also improved basic education and health services. Soon, however, a debilitating war with right-wing counterinsurgents (contras) undid most of the social progress. The contras were a mixed group of dissidents, outlaws, and adventurers from outside the region; they were backed by local elites and by the communist-wary United States during the Reagan administration. A trade embargo imposed by the United States further contributed to the ruin of the economy. By the end of the 1980s, Nicaragua was one of the poorest nations in the Western Hemisphere. In national elections in 1990, a coalition led by Violetta Chamorro defeated the Sandinistas, who had lost popularity as a result of their own incomplete reform programs and public weariness with Sandinista-contra violence. Chamorro intended to privatize both rural and urban economies but was forced by public pressure to continue some redistribution of land. Since 1997, several free elections in which 75 percent of the eligible citizens voted resulted in moderate governments that continue to find it difficult to bring Nicaragua some measure of prosperity.

THE NORTHERN ANDES AND CARIBBEAN COAST

The five countries in the northernmost part of South America share a Caribbean coastline and extend into a remote interior of wide river basins and humid uplands (Figure 3.22). The countries of Guyana, Suriname, and French Guiana resemble the Caribbean countries in that they were once traditional plantation colonies worked with slave and indentured labor, and today their multicultural societies are made up of descendants of African, East Indian, Pakistani, Dutch, French, and English settlers. Venezuela and Colombia share a Spanish colonial past and are predominantly mestizo, with a small upper class of primarily European heritage and a small population of African derivation in the western and Caribbean lowlands. In all the countries of this subregion, small indigenous populations survive, mainly in the lowland Orinoco and Amazon interior where they hunt, gather, and grow subsistence crops.

The Guianas

To the north of Brazil lie three small countries known collectively as the Guianas: Guyana gained independence from the British in 1966, and Suriname from the Dutch in 1975. French Guiana, on the other hand, is not independent, but rather is considered part of France. Today, the common colonial heritage of these three countries is still visible in both the economy and the people. Sugar, rice, and banana plantations established by the Europeans in the coastal areas continue to be economically important, but logging and the mining of gold, diamonds, and bauxite in the resource-rich highlands are now the leading activities. The population descends mainly from laborers who once worked the plantations. These laborers formed two major cultural groups: Africans, brought in as

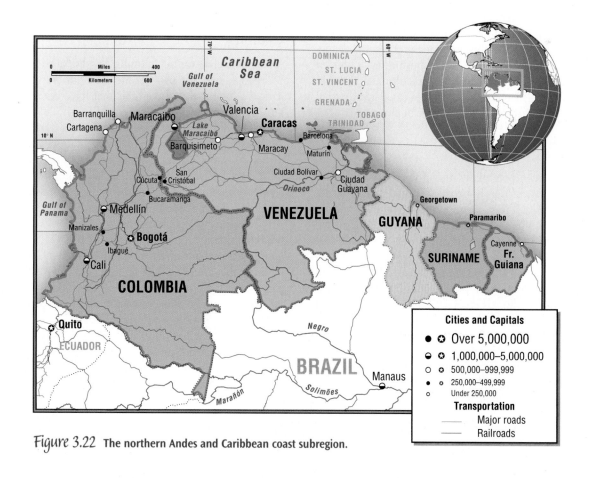

Figure 3.22 **The northern Andes and Caribbean coast subregion.**

In Suriname, Javanese Muslim women, descendants of indentured servants brought by the Dutch in the 1850s, mark the end of Ramadan, a major religious observance. [Robert Caputo/Aurora.]

slaves from 1620 to the early 1800s, and South and Southeast Asians, brought in as indentured servants after the abolition of slavery. The descendants of Asian indentured servants are Hindus and Muslims (see the photograph on this page), many of whom became small-plot rice farmers or owners of small businesses. Those of African descent are Christian and are both agricultural and urban workers. Politics in the Guianas has been complicated by the social differences between citizens of Asian and African derivation.

Venezuela

Venezuela has long been potentially one of the wealthier countries in South America, primarily because it holds large oil deposits and is an active member of the international cartel of oil-producing countries (OPEC). Oil has been the backbone of the economy since the mid-twentieth century. Venezuela is not only among the top suppliers of oil to the United States but is also a major supplier to Japan and Europe. Nonetheless, oil has not generated widespread prosperity even during times of relatively high oil prices. Most of the wealth has been retained by the 20 percent of the population that forms the middle and upper classes—those mainly of European descent. Taxes on this wealth have been kept low, so oil has not generated enough government revenue to fund badly needed improvements in education, transport, and communications. As a result, Venezuela is not building a base for general economic and social advancement. Although the capital city of Caracas is bedecked with gleaming skyscrapers, modern freeways, and universities, it is also surrounded by shantytowns that

lack access to clean drinking water and sanitation and are served by substandard schools.

Despite periodically high oil prices, Venezuela accumulated a large debt in the 1980s, which it planned to pay off with future oil profits. But a slump in oil prices during the early 1990s left the country still deeply in debt. By the year 2000, nearly one-third of the population lived below Venezuela's self-declared poverty line. Fourteen percent of the population gets by on just one dollar a day. Structural adjustment programs (SAPs) that were intended to free money for debt repayment have reduced the role of government in industry and in social programs. Many government jobs disappeared, and the poor lost rent and food subsidies and access to education. Living standards fell even further, and violent riots were one result. Since the mid-1990s, the leftist populist president, Hugo Chavez, has encouraged the poor with suggestions that social welfare programs will return and the rich will lose some of their advantages through higher taxes (and perhaps even through land expropriation). His populist statements, however, have discouraged outsiders contemplating investments in industrialization and mechanized agriculture, thereby hindering economic expansion. Nevertheless, Chavez's populist policies have survived, in part because a rebound in oil prices in the late 1990s and early 2000s brought much-needed capital back into the Venezuelan government coffers.

The drug-related civil war in neighboring Colombia is spilling over into Venezuela. The discovery of large cocaine caches in several raids by U.S. and international police casts doubt on Venezuela's commitment to curtail corruption.

Colombia

An ongoing civil war in Colombia has displaced more than 1.5 million people over the past decade. The current wave of violence is part of a long string of conflicts that have arisen out of inequalities in the country's social order. Although Colombia is today the world's second largest exporter of coffee and a major exporter of oil and coal, a small proportion of the population, mostly of European descent, has received most of the income by keeping wages low and resisting the payment of taxes. In addition, the wealthy have not cooperated with efforts to redistribute some of their extensive landholdings. On one side of the civil war are revolutionary guerrilla bands seeking government-sponsored reforms for the poor. On the other side are the private armies of the wealthy, who have little faith in the will and ability of the ill-equipped Colombian military to defeat the guerrillas. The guerrilla group FARC controls a large part of the Colombian interior south of Bogotá, where they are said to encourage the cultivation of more than 100,000 acres (40,000 hectares) of coca (Figure 3.23).

The current hostilities are still basically driven by unequal access to wealth, power, and land, but they are complicated by the fact that to raise money, all the warring parties participate in the drug trade to varying degrees. This trade depends on the ancient Andean coca plant, whose leaves have traditionally been chewed as a fatigue and hunger suppressant and as a mildly invigorating tonic and mood enhancer. Today, coca leaves are processed into the

addition, there are many efforts to find sources of income for rural people other than coca. In the mountains of southwest Colombia, 1200 farm families have formed a cooperative that produces a line of 20 products—preserved fruits, sauces, and candies marketed especially to Hispanics in North America and Europe. The cooperative also focuses on increasing child and adult education and on enhancing such skills as running an effective business meeting and improving the sales of their products through market research. Elsewhere in the region, rural people working with the food scientists at the International Center for Tropical Agriculture have developed new varieties of corn that will be more productive and nutritious. These farmers now have enough corn to feed their families, raise chickens, and sell corn in the marketplace. These activities diversify diets for rural families, bring in cash, and make the new varieties of corn available to other farmers.

THE CENTRAL ANDES

The central Andes, which includes the countries of Ecuador, Peru, and Bolivia, is the poorest region in Middle and South America (Figure 3.24). On the eve of the Spanish conquest, this was the home of the Incas, one of the two great civilizations of the region. Their legacy is reflected in the roughly one-half of the population that is Native American, the largest proportion in South America.

After its fall to the Spanish, the central Andes, like other parts of the Americas, went into a long decline. During this time, a tiny group of landowners prospered while the vast majority lived in poverty. Most of the twentieth century has been marked by failed efforts at social change and the violence resulting from those failures, but the growing political involvement of the large indigenous population may help achieve greater social equity. In exploring the central Andes, we will look at three main subregions: the economically dynamic coastal lowlands, the poor and largely Native American highlands, and the resource-rich and environmentally threatened Amazonian lowlands.

Environments in the Central Andes

Only Ecuador and Peru have coastal lowlands; Bolivia is landlocked. The coastal population is mainly mestizo and African. Along the Pacific coast are located the largest, most modern and cosmopolitan cities: Peru's capital and commercial center, Lima, and Ecuador's leading industrial center, Guayaquil. Although the climate is often dry, especially in Peru, plantations and other agricultural enterprises that produce crops for export thrive along the coast with the help of irrigation. The production of export crops such as bananas, cotton, tobacco, grapes, citrus, apples, and sugarcane has increased dramatically, and shrimp farming and ocean fishing grounds nourished by the nutrient-rich Peru Current sustain a vital export-oriented fishing industry. Food production for local consumption can be precarious, however, and actually declined 14 percent in Peru between 1985 and 1990.

Like other parts of Middle and South America, this area has had to tighten its belt in response to government debts acquired

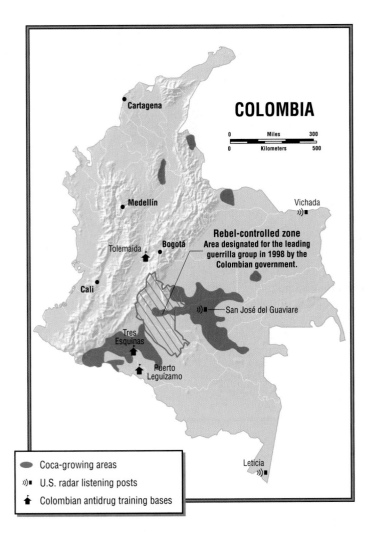

Figure 3.23 Cocaine trade and Colombia. Colombia is reportedly the source for the vast majority of cocaine and most of the heroin intercepted by law enforcement officials in the United States. Antigovernment guerrillas and other paramilitary groups control the cultivation of the coca and poppy crops and the processing and sale of the drug products.
[Adapted from *The New York Times* (International) (April 21, 2000): A12.]

much stronger cocaine (a crystalline extract of coca leaves noted for its euphoric, stimulating, and addicting effects). The coca growers are usually very poor farmers in remote areas, who make a somewhat better income from coca than they would from other crops. However, most of the profits are reaped by competing drug smuggling rings. These drug rings pay the farmers relatively little, and they bribe, intimidate, and murder Colombian government officials who seek to reduce or eradicate the drug trade.

There are some signs that this era of drug-related violence in Colombia will abate, due in part, perhaps, to training and equipment supplied by the United States, but also to local initiatives. Significantly, ordinary Colombians have organized the "No Mas" movement, a civic foundation that works to stop the killing. Five million demonstrators marched in Bogotá in October 1999. In

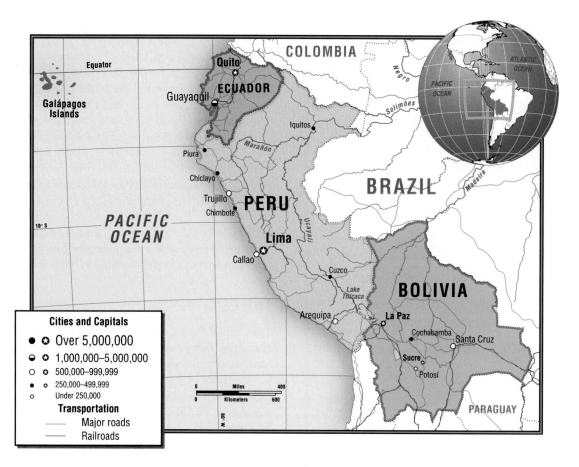

Figure 3.24 **The central Andes subregion. Bolivia has two capitals. La Paz is the seat of the government and Sucre is the seat of the judiciary.**

through borrowing to fund development. Structural adjustment programs have mandated the privatization of state-run industries and the streamlining of government. The SAP reforms resulted in job losses and social turmoil along with stronger economic growth. The government dramatically increased prices for gasoline, electricity, and transport in an effort to raise funds to pay off debts. Such policies hurt the urban poor, whose low wages cannot cover the increases. In rural coastal areas, SAPs have removed government protections for small-scale farmers (low rents, price supports for staple crops, and subsidies for tools). Small farms are now being replaced by corporate farms and high-tech operations such as shrimp farms, most of which, ironically, receive government assistance, at least in the start-up phase. In Ecuador, protests by working people brought down the government in 1997, and the same thing might have happened in Peru had it not been for suppression by the military. Ecuador declared bankruptcy in 1998 because it couldn't repay its debts to the international financial community, and the economy and society are still in a state of crisis and uncertainty. Peru is plagued by the drug trade that has moved across the border from Colombia.

The highlands are the home of the bulk of the region's indigenous people, who for centuries worked on the large haciendas and in the rich mines (copper, lead, and zinc in Peru; tin, bauxite, lead, and zinc in Bolivia) owned by a small group of people of European descent. Most of the Native Americans today obtain their incomes from subsistence cultivation of a wide array of vegetables acclimated to the cool climate: more than 50 varieties of potatoes, for example, are cultivated. Surpluses are traded on market days in towns throughout the highlands. Shepherding, wool processing and weaving, and (for some men) mining work also contribute to family incomes. Recently, foreign and local scientists have explored imaginative new (and ancient) cultivation techniques and crops that may improve standards of living (see the box "Learning from the Incas"). In Peru, agronomist Angel Valledolid oversees 7500 acres (3000 hectares) of *nuña*, "popping beans," an ancient crop that grows without pesticides. Roasted, seasoned, and salted or covered with chocolate, the popped beans are being marketed as a cocktail snack in Europe and North America. If successful in the global marketplace, the popping beans could improve the incomes of 200,000 Peruvian cultivators and their families. In the last few decades, highland Peruvians who cannot get access to land or other means of support have been moving east into lowland regions.

The Amazonian lowlands of Ecuador, Peru, and Bolivia have traditionally been the home of scattered groups of indigenous people. In recent years, this area has seen rapid and often destructive development of natural resources. National governments have encouraged private extraction first of trees and then minerals and agricultural products, and these activities have severely damaged the home territories of Native Americans. At the forefront of this process are oil and mining companies (see the

AT THE LOCAL SCALE   *Learning from the Incas: Agricultural Restoration in the Peruvian Highlands*

Around the ancient Incan capital of Cuzco, in the Peruvian highlands, Incan terraces and canals once supported crops that fed hundreds of thousands of people on what would otherwise have been barren mountain slopes. Much of this infrastructure still remains, but it was not being used until recently. Some of the ancient infrastructure is now being restored, thanks mainly to more than two decades of careful research by the British rural development specialist and archaeologist Ann Kendall. Kendall recognized that the Incan agricultural technologies were more advanced than archaeologists had thought. The construction of terraces, for exam-ple, allows for efficient recycling of nutrients, making high-yield organic cultivation possible. In particular, Kendall found that the Incan use of clay as a flexible, semiwet sealant on both terraces and canals was extremely well adapted to the earthquake-prone environment of the Peruvian highlands because it doesn't break the way concrete does and because cracks tend to reseal themselves. So far, reviving these traditional methods has proved highly productive, allowing some highland peoples who had migrated to lowland cities to return home and take up farming. One canal and its associated terrace system can irrigate enough land to support more than 2000 people.

vignette that opens this chapter and the accompanying illustrations). Roads—such as the one cut across the Andes in Peru from Lima to Pucallpa on the headwaters of the Amazon—are built to gain access to lumber and minerals. Such roads have also opened up this region to massive waves of migrants pushed from the highlands and coastal lowlands by other market forces. Accompanying the migrants are diseases that decimate lowland indigenous people in the Amazon Basin, who lack resistance.

Social Inequalities and Social Unrest

In the twentieth century, various strategies for lessening gross social inequalities were tried, with varying degrees of success. In the 1950s, Bolivia was the first to attempt reform when the government took over the tin mines in the **altiplano,** or highlands, after a revolution in which the miners' unions played a key role. Wages were raised and conditions improved for several decades, but then world prices for tin declined because plastic packaging was replacing tin cans just as tin mining was intensifying around the globe. As tin prices fell, the drug trade replaced tin as Bolivia's chief export. As in Colombia, the military and the government publicly condemn the drug trade, yet members of both are extensively involved, as are local cultivators who grow the poppies for heroin or the coca for cocaine.

In the 1970s, Peru's military dictatorship sought to improve the lives of indigenous peoples by transforming the haciendas of the elite into state-run communes. The experiment was not popular and soon inspired a resistance movement among indigenous small farmers that was taken over in 1980 by a communist guerrilla movement called the Shining Path. Followers of the movement forced cooperation by killing farmers if they participated in the capitalist economy in any way or were associated with the government, whether as traders, community officials, participants in development projects, or even just voters. Eventually, the farmers banded together to fight the Shining Path with the aid of the military. The amount of violence has decreased for several years now, but issues of affordable access to good agricultural land and of democratic participation for the majority remain. The Peruvian

government, hoping to avoid reigniting past grievances, supplies highland communities with subsidies and protections from the harsher aspects of SAP reforms that are straining the lowlands.

The Native Americans of Peru have visibly increased their participation in national politics, having played a part in forcing the resignation of President Alberto Fujimori in the autumn of 2000 after a series of corruption scandals. In Ecuador, a stronger challenge to the status quo may be emerging from the indigenous peoples of the lowlands (see pages 114–115 and Figure 3.15 on page 147) and those of the highlands. CONAIE, a federation representing all of Ecuador's indigenous groups, has been a strong force fighting for the land rights of Native Americans. CONAIE's support was central to the success of the protests against SAPs that brought down Ecuador's government in 1997. Native American demands for an increased voice in national policy continued into 2000, as members of the federation blocked off roads to many remote highland areas in an effort to gain national and international attention for their protest.

THE SOUTHERN CONE

The countries of Chile, Paraguay, Uruguay, and Argentina (Figure 3.25) have diverse physical environments but remarkably similar histories. The southern cone received little European settlement during the Spanish Empire, but in the late nineteenth and early twentieth centuries, European immigrants—mainly Germans, Italians, and Irish—were drawn by the temperate climates, the economic opportunities, and the prospect of landownership. The immigrants swamped the surviving Native American populations throughout most of the region. Paraguay is the only country of the four that has a predominantly mestizo population.

Economics in the Southern Cone

Agriculture has long been a leading economic sector in the southern cone. The primary agricultural zone is the pampas, an area of extensive grasslands and highly fertile soils in northern Argentina and Uruguay. The region is famous for its grain and cattle. Sheep

Figure 3.25 The southern cone subregion.

raising dominates in Argentina's drier, less fertile southern zone, Patagonia. On the other side of the Andes, in Chile's central zone, a Mediterranean climate supports large-scale fruit production that caters to the winter markets of the Northern Hemisphere. Chile also benefits from considerable mineral wealth, especially copper.

Although the agricultural and mineral exports of the southern cone have created considerable wealth, fluctuating prices for raw materials on the global market have periodically sent the economies of this subregion into a sudden downturn. The desire for economic stability was a major impetus for industrialization and urbanization in the mid-twentieth century. At first, the new industries were based on processing agricultural and mineral raw materials. Later, state policies supported diversification into import substitution industries. However, as discussed on page 136, inefficiencies, corruption, and small local markets prevented industry from becoming a leading sector for the region.

Economic policy in the southern cone has been a source of regional conflict for decades (see the discussion of the Mercosur trade association on pages 140–141). Despite their considerable resources, these countries have always had substantial impoverished populations that were, nonetheless, well enough educated to demand attention to their grievances. In response, each country developed mechanisms for redistributing wealth—government subsidies for jobs, food, rent, basic health care, and transport, for example—that did alleviate poverty to some degree. However, when global prices for raw materials fell in the 1970s and produced an economic downturn for the region, people clamored for more fundamental changes. In response, throughout the 1970s and into the early 1980s, military leaders, with the support of the elite, took control of governments and waged bloody campaigns of repression against workers and the socialist and communist opposition. In the years of state-sanctioned violence that followed, known as the "dirty wars," 3000 people were killed in Chile and 14,000 to 18,000 in Argentina, and tens of thousands more were jailed and tortured for their political beliefs.

The International Monetary Fund has consistently applied pressure for repayment of the massive debts accumulated under both civilian and military governments, from the 1960s through the 1990s. These financial pressures effectively shifted countries away from wealth redistribution policies and toward free-market economic reform, which it was hoped would bring economic growth and expansion sufficient to make debt repayment possible. The record of success is mixed. Chile's national economy has witnessed growth of its export-oriented industry, mining and agriculture, whereas Argentina has not been able to rid itself of its debt burden and defaulted in December 2001. An International Monetary Fund package of structural reforms and loans to the Argentine government held off the crisis but did not resolve it. Argentina's debt-induced social disarray and disquiet mirrors Ecuador's (see page 170).

Buenos Aires: A Primate City

The primary urban center in the southern cone is Buenos Aires, the capital of Argentina and one of the world's largest cities. Forty percent of the country's people live in this city. Buenos Aires boasts premier shopping streets, elegant urban landscapes, and dozens of international banks, yet it has weathered six decades of decline that have left it with environmental degradation, empty factories, social conflict, severe poverty, and declining human well-being. Some people argue that the downward slide in quality of life in Buenos Aires is a result of the restructuring required for globalization; others contend that greater integration with the global economy will help to reverse decades of malaise. Argentine leaders are in the latter camp. They want Buenos Aires to be seen as a world city—a center with pools of skilled labor that attract

Modeled after the Champs Élysées in Paris, the Avenida 9 de Julio is the widest avenue in Buenos Aires and is reputed to be the widest in the world. [David Keeling.]

major investment capital, a sophisticated city with a beautiful skyline and a powerful sense of place. An impatient local population, however, may give more importance to obtaining such basics as decent housing and modernized transportation.

BRAZIL

The observant visitor to Brazil (Figure 3.26) is quickly caught up in the country's physical complexity and in the richly exuberant, multicultural quality of Brazilian society. But its landscapes also plainly show, more obviously than those of North America, the environmental effects of both colonialism and recent under-planned economic development. Brazil's people live in a highly stratified society made up of a small, very wealthy elite, a modest but rising middle class, and a huge majority that live in poverty or just barely above the poverty line. In Brazil's megacities of São Paulo and Rio de Janeiro, vast squatter housing tracts, high crime rates, and homeless street children are signs of the gross inequities in opportunity and well-being. Brazil's richest 20 percent has 26 times more wealth than the poorest 20 percent—one of the widest disparities on earth (see Table 3.2). Travelers find themselves caught up and delighted by the flamboyant creativity and elegance of the people, yet sobered by the obvious hardships under which both the land and the people labor.

Brazil's two main physical features are the Amazon Basin and the Brazilian Highlands (see Figures 3.2 and 3.3). The huge Amazon Basin covers the northern two-thirds of the country. The southern third of the country is mostly occupied by the Brazilian Highlands, a variegated plateau that rises abruptly just behind the Atlantic seaboard 500 miles (800 kilometers) south of the mouth of

Figure 3.26 The Brazil subregion.

the Amazon. The northern portion of the plateau is arid; the southern part receives considerably more rainfall. Settlement in Brazil is concentrated in a narrow band along the Atlantic seaboard in the northeast and south through Salvador (see Figure 3.7). Settlement extends deeper inland in Brazil's temperate climate zone, near Rio de Janeiro, São Paulo, Curitiba, and Pôrto Alegre.

Brazil is about the same size as the United States; and the Brazilian economy is the largest in Latin America, the eighth largest in the world. Resources available for development in Brazil are the envy of most nations. Local oil now provides for 60 percent of Brazil's needs, and more has recently been discovered west of Manaus in the Amazon. Gold, silver, and precious gems have been important since colonial days, but it is industrial minerals—titanium, manganese, chromite, tungsten, and especially iron ore—that are the most valuable today. They are found in many parts of the Brazilian Highlands. Hydroelectric power is widely available for development because of the many rivers and natural waterfalls that descend from the plateau. There has been large-scale agriculture in Brazil for 400 years: in the south, European-style farming of fruits and vegetables; in the northeast, plantations, primarily of sugar, tobacco, cotton, cassava, and oil palm; in the interior dry zones, ranching. Agriculture has some potential to expand, but most of the available land is in the tropical north where soils are fragile. Using the land for agriculture there requires special and sustained attention, and more investment of time, research, labor, and skill than has been anticipated by recent entrepreneurs.

Brazil is the most highly industrialized country in South America. Most of the industry—steel, motor vehicles, aeronautics, appliances, chemicals, textiles, and shoes—is concentrated in a triangle formed by the huge southeastern cities of São Paulo, Rio de Janiero, and Belo Horizonte. Until the 1990s, the vast majority of Brazil's mining and industrial operations were developed using government funding, and many are still owned and run by the government. Recently elected governments have chosen to privatize many government industries and businesses (television stations, power plants, mines, and agricultural land) in an effort to make them more efficient and profitable. Many of them have been sold to foreign investors at bargain-basement prices. In 1997 alone, direct foreign investors spent $23 billion on Brazilian properties. Privatizing industry often results in greater productivity, but the low price of sale may be hurting Brazil in the long run. Now any profits that might materialize will not go to the public, which paid for the development of these industries, but to the new private owners, who may well invest the money elsewhere.

Urbanization

Brazil has a number of large and well-known cities: Rio de Janeiro, São Paulo, Belo Horizonte, Salvador, Recife, Fortaleza, Belém, Manaus, and Brasília. All but Manaus, Brasília, and Belo Horizonte are on the Atlantic perimeter of the country. During the global depression of the 1930s, farm workers throughout the country began migrating into urban areas as world prices for Brazil's agricultural exports (jute, sugar, coffee, cacao, rubber) fell. By 1960, Brazil was 22 percent urban. Since then, efforts to modernize and mechanize agriculture have reduced labor requirements further. Agricultural change pushed people off the land; during and after the 1960s, the chance of employment in huge factories being built with government money pulled them into the cities. Brazil's competitive edge globally was its cheap labor; and the military governments of the time thought that the country's then mostly government-owned industries could continue to pay very low wages for some time because there was such a surplus of workers drawn from the low-wage countryside. They were right; but they had to quell many protests.

By 1995, 77 percent of Brazil's population was urban, and at least one-third of the urban population, many of them unemployed, were living in favelas, the Brazilian urban shantytowns. The poverty in the very hardest hit cities in the northeast rivals that of Haiti, the poorest country in the Americas. Brazilians, however, are intrepid at finding ways to make life worth living. Favela dwellers are famous for their efforts to create strong community life and support for those in distress. An example is a religious movement, known as Umbanda, that thrives in all the Atlantic coastal cities from Belém to São Paulo (see the photograph of a similar movement, Batuque). Umbanda groups are centered around a male or female spiritual leader. The leader welcomes celebrants into a neighborhood center and invokes the spirits to help people cope with health problems and the ordinary stresses and strains of a life of urban poverty. Drumming, dancing, spirit possession, and psychological support are the central focus. Umbanda grew out of an older African-Brazilian belief system called Condomble, and similar movements (Voodoo, Santaria, Obeah) are to be found elsewhere in the Americas, including the Caribbean, the United States, and Canada. In Brazil, Umbanda appeals to an increasingly wide spectrum of the populace. There are many centers with adherents of various African, European, and Native American backgrounds.

Rio de Janiero and other cities in Brazil's industrial heartland, the southeast, have districts that are elegant and futuristic showplaces, resplendent with the very latest technology and graced with buildings that would put New York, Singapore, or even Tokyo to shame. The stores are filled with artistically designed consumer goods at attractive prices for those with desirable foreign currency.

The two women in lace shawls are in a trance, having been possessed by the spirits of Oxala and Yemanja during a Batuque ritual. They are greeted and supported by mediums called *filhas de santos*. Batuque ceremonies, like Umbanda and Condomble ceremonies, use mesmerizing drum rhythms to invoke various spirits, including Christian saints. These ceremonies regularly unite large groups of people in a common uplifting experience. [Jacques Jangoux, Brazil.]

Brazilian planners now acknowledge that, in the rush to develop the most modern of urban landscapes, developers neglected to underwrite the parallel development of a sufficient urban infrastructure (for an exception, see the box "Curitiba, Brazil: A Model of Urban Planning" on page 133). In short supply are sanitation and water systems, up-to-date electrical wiring, transportation facilities, schools, housing, and medical facilities—all necessary to sustain modern business, industry, and a socially healthy urban population. But the Brazilians should not be overly criticized for what in hindsight seems like an obvious error. During the 1970s and 1980s, development theory set forth by the World Bank and other financial institutions held that urban-centered investment (growth poles), with its accompanying industrialization and modernization, would deliver development and stimulate economic growth. Social transformation and infrastructure development would be naturally occurring side effects. The debt crisis of the 1980s, which ate up profits with *skyrocketing inflation* (a rapid rise in consumer prices due to the falling value on the international market of the Brazilian currency), put an end to such optimistic plans and policies. Only very recently are all parties coming to recognize that planned development of human capital through education, health services, and community development, whether privately or publicly funded, is a primary part of building strong economies.

Brasília

Brasília, the new capital, is an intriguing example of the effort to lead development by constructing a spectacular urban environment. Built in the state of Goiás in just three years, beginning in 1957, Brasília lies about 600 miles (1000 kilometers) inland from the glamorous old administrative center of Rio de Janiero. The city was built in this location to serve as a **forward capital**—a city built for the avowed purpose of drawing migrants and investment for the development of the western highland territories and then,

eventually, of Amazonia. The scholar William Schurz suggested an alternative to the official explanation for Brasília's location: moving the capital so far away from the centers of Brazilian society was the most efficient way to trim the badly swollen and highly inefficient government bureaucracy. In any case, symbolism figured more prominently than practicality in the design. The city was laid out to look from the air like a swept-wing jet plane. There was to be no central business district, but rather shopping zones in each residential area and one large mall. Pedestrian traffic was limited to a few grand promenades; people were expected to move about in cars and taxis. Public buildings were designed for maximum visual and ceremonial drama.

After people had spent several decades actually using this urban landscape, they made all sorts of interesting changes to the formal design. At the parliament, legislative staff and messengers created footpaths where they needed them: through flower beds and—with little steps notched in the dirt—up and over landscaped banks. Thus, they efficiently connected the administration buildings, bypassing the sweeping esplanades. Commercial districts were retrofitted in and around hotel complexes. Urban workers who couldn't afford taxi fares wore direct-route footpaths across great green swards of grass. Little hints of the informal economy that characterizes life in the old cities of Brazil began to show up—a fruit vendor here, a manicurist there. And the shantytowns, which the planners had tried hard to eliminate, began to rise relentlessly around the perimeter. Today, life in the shantytowns is so much more interesting than life in the sterile environment of Brasília that tour buses take visitors to see them and shop there. Overall, in the 40 years since its inception, Brasília's success as a forward capital has been limited. Although Brasília has drawn poor laborers, the entire province of Goiás still has only 4.5 million people, just 6 percent of the country's total, and the average income is only half that of the country as a whole. Population and investment remain centered in the Atlantic coastal cities.

REFLECTIONS ON MIDDLE AND SOUTH AMERICA

Middle and South America were the first regions in the world to be colonized by Europeans. In a number of ways, European colonialism in the Americas launched the modern global economic system. It was in the Americas that large-scale extractive practices were inaugurated. Raw materials were shipped at low prices to distant locales where they were turned into high-priced products, while the profits went to Europe. Local people were diverted from producing food and other necessities for themselves to working as low-paid or enslaved labor in the fields and mines. The rules governing private property were set down in societies that had long practiced communal land rights. In the process, the landscapes and settlement patterns of the Americas were reorganized as the focus shifted from producing for local and regional economies to producing for global markets.

After the colonial era, this region continued to serve as a testing ground for economic theories, both capitalist and socialist. Middle and South America tried out such capitalist ideas as mechanized agriculture for export, rural-to-urban migration as a solution to rural poverty, and government borrowing for large-scale development projects. More socialist experiments included government-sponsored industrialization to create jobs and produce substitutes for imports, and broad government subsidies to address the basic needs of the poor. External factors, however, intervened to stymie any chance that such optimistic plans for economic expansion and societal transformation would succeed. The global recession of the early 1980s left the region's governments in debt, and the indebtedness forced free-market, structural transformations on governments so that wealth inequities widened and social disparities increased. The rapidly growing numbers of urban poor had to resort to self-help solutions for their shelter; they sought meager livings in the informal sector, and with the reduction of public health and welfare services, many resorted to community activism.

Despite the hardships of the last 20 years—the increased harshness of everyday life for the rural and urban poor, especially women, children, and the elderly—there are signs that the situation is not beyond retrieval. There is increasing recognition that to be judged successful, development must first change the lives of the majority for the better. Middle and South Americans are beginning to build on their strengths and to invent solutions to problems common in many places around the world. Curitiba, Brazil, seeks simple but humane solutions to rapid urbanization and to the pollution and social disruption it brings. Degraded agricultural lands are being rehabilitated in the Peruvian highlands by the revival of ancient indigenous practices. Environmental groups are addressing the root causes of deforestation: inequitable domestic distribution of lands and resources, unsustainable uses for forestlands such as ranching and cash-crop agri-

culture, and demand for forest products in distant markets. Increasingly, regional trade organizations are emerging with sufficient strength and motivation to negotiate for the good of the region rather than for parochial interests.

As you read other chapters of this book, it might be useful to reflect again on the geographic issues of Middle and South America we have discussed. For example, notice how Europe's situation today—its wealth, its position as a world leader, and its emerging commitment to help its former colonies—is related in part to its colonial experiences, which began in the Americas. You will see that Africa and South Asia under colonial rule experienced some conditions similar to those in the Americas; more recently, they also have felt the sting of SAPs. Southeast Asia, Australia, New Zealand, and Oceania, too, have had comparable experiences as colonies; but in recent decades their "colonizers" have included wealthy Asian capitalists. In the former Soviet Union and Central Asia, outside investors intent on exploiting oil and forest resources are reminiscent of the conquistadors and their successors in the Americas.

Thinking Critically About Selected Concepts

1. Mountains and volcanoes often form at subduction zones, where the edge of one tectonic plate plunges below the edge of another. *What are some of the ways in which tectonic activity affects the lives of people in Middle and South America? What role does unpredictability play?*

2. The conquest of Middle and South America by the Spanish and Portuguese transformed life in the region. *Identify some of the lingering effects (social, economic, environmental, and political) of the conquest and colonization.*

3. Influences from outside the region continue into the present. *Do you see any continuity between European influences of the past and current efforts by foreign investors, planners, and lending agencies? How might tourists from abroad influence present societies?*

4. Rural-to-urban migration has fed the rapid growth of cities in the twentieth century. *Do you see any parallels to rural-to-urban migration in the United States? For both regions, identify several environmental and social impacts of this migration on both the places the migrants leave and the places to which they move.*

5. Disparity in the distribution of income and wealth is more severe in this region than elsewhere on earth, even though it is not the poorest region overall. *How do you account for the persistence of this disparity?*

6. The extended family is the basic social institution of Middle and South America. *Discuss the likelihood that this type of family structure*

will persist into the future. What forces are pushing for change? What are the strengths of the extended family that have helped it persist so far?

7. Religion has played a role in this region since precolonial times and was of special importance during the conquest. Often, religion reinforces the status quo in a society. *Would you say that the present rise in various kinds of religious activity in Middle and South America is likely to bring social stability or change?*

8. Since early colonial times, the economies of Middle and South America have been extractive, and trade between parts of the region has been underdeveloped. *What do you see as the important factors likely to change this situation in the next several decades?*

9. In the twentieth century, the countries of the region have experimented with various economic development strategies. *Describe these efforts and explain some of the unforeseen side effects of each.*

10. After a history of authoritarian and military rule, democracy is much more widely practiced now than previously. *What do you see as the chief reasons that democratic institutions remain fragile?*

11. The region of Middle and South America was one of the first to alert the world to the dangers of environmental deterioration. *What types of strategies now being tried in this region to address environmental issues do you see as the most effective? Might these strategies be transferable to other world regions?*

Key Terms

acculturation (p. 146) adaptation of a minority culture to the host culture enough to function effectively and be self-supporting; cultural borrowing

altiplano (p. 171) an area of high plains in the central Andes of South America

assimilation (p. 146) the loss of old ways of life and the adoption of the lifestyle of another culture

Aztecs (p. 124) native people of central Mexico noted for advanced civilization before the Spanish conquest

barriadas (p. 132) see **favelas**

barrios (p. 132) see **favelas**

brain drain (p. 134) the migration of educated and ambitious young adults to cities, depriving the sending communities of talented youth in whom they have invested years of nurturing and education

colonias (p. 132) see **favelas**

colonizing (or **mother**) **country** (p. 126) the country from which the people of a colony or former colony derive their origin

contested space (p. 151) any area that several groups claim or want to use in different ways, such as the Amazon or Palestine

early extractive phase (p. 135) a phase in Central and South American history, beginning with the Spanish conquest and lasting until the early twentieth century, characterized by colonial policies such as mercantilism that resulted in unequal trade

economic restructuring (p. 118) reorganization of an economy to encourage economic growth in markets free of government controls

ecotourism (p. 152) nature-oriented vacations often taken in endangered and remote landscapes, usually by travelers from industrialized nations

El Niño (p. 123) periodic climate-altering changes, especially in the circulation of the Pacific Ocean, now understood to operate on a global scale

evangelical Protestantism (p. 148) a Christian movement that focuses on personal salvation and empowerment of the individual through miraculous healing and transformation; some practitioners preach the "gospel of success" to the poor: that a life dedicated to Christ will result in prosperity of the body and soul

Export Processing Zones (EPZs) (or **free-trade zones**) (p. 139) specially created legal spaces or industrial parks within a country where, to attract foreign-owned factories, duties and taxes are not charged

extended family (p. 146) a family consisting of related individuals beyond the nuclear family of parents and children

favelas (p. 132) Brazilian urban slums and shantytowns built by the poor; called **colonias, barrios,** or **barriadas** in other countries

forward capital (p. 176) a city built to draw migrants and investment for economic development

free-trade zones (p. 139) see **Export Processing Zones**

hacienda (p. 135) a large agricultural estate in Middle or South America, more common in the past; usually not specialized by crop and not focused on market production

import substitution industrialization (ISI) (p. 136) a form of industrialization involving the use of public funds to set up factories to produce goods that previously had been imported

Incas (p. 124) Native American people who ruled the largest pre-Columbian state in the Americas, with a domain stretching from southern Colombia to northern Chile and Argentina

income disparity (p. 134) a dramatic gap in wealth and resources between the rich elite and the poor majority of a country or region

informal economy (p. 140) all aspects of the economy that take place outside official channels

liberation theology (p. 148) a movement within the Roman Catholic church that uses the teachings of Christ to encourage the poor to organize to change their own lives and the rich to promote social and economic reform

livestock ranch (p. 135) a farm that specializes in raising cattle and sheep

lopsided spatial development (p. 132) the concentration of wealth and power in one place

machismo (p. 147) a set of values that defines manliness in Middle and South America

maquiladoras (p. 161) foreign-owned factories, often located in free-trade zones just over the border from the United States, that hire people at low wages to assemble goods that are then sent back to the United States for sale

marianismo (p. 147) a set of values based on the life of the Virgin Mary that defines the proper social roles for women in Middle and South America

mercantilism (p. 126) the policy by which the rulers of Spain and Portugal, and later of England and Holland, sought to increase the power and wealth of their realms by managing all aspects of production, transport, and trade in their colonies

Mercosur (p. 140) a free-trade zone created in 1991 that links the economies of Brazil, Argentina, Uruguay, and Paraguay to create a common market

mestizo (pp. 128, 146) a person of mixed European and Native American descent

migration (p. 131) the movement of people from one place to another

neocolonialism (p. 136) modern efforts by dominant countries to control economic and political affairs in other countries to further their own aims

North American Free Trade Agreement (NAFTA) (p. 140) a free-trade agreement created in 1994 that added Mexico to the 1989 agreement between the United States and Canada

plantation (p. 135) a large estate or farm on which a large group of resident laborers grow (and partially process) a single cash crop

populist movements (p. 148) popularly based efforts seeking relief for the poor

primate city (p. 131) a city that is vastly larger than all others in a country and in which economic and political activity is centered

remittances (p. 130) earnings sent home by immigrant workers

shifting cultivation (p. 124) a productive system of agriculture in which small plots are cleared in forestlands, the dried brush is burned to release nutrients, and the clearings are planted with multiple species; each plot is used for only two or three years and then abandoned for many years of regrowth

silt (p. 119) fine soil particles

structural adjustment programs (SAPs) (p. 137) policy changes designed to free up money for loan repayment to international banks

subduction (p. 119) the sliding of one lithospheric (tectonic) plate under another

tierra caliente (p. 122) low-lying "hot lands" in Middle and South America

tierra fria (p. 123) "cool lands" at relatively high elevations in Middle and South America

tierra helada (p. 123) very high elevation "frozen lands" in Middle and South America

tierra templada (p. 122) "temperate lands" with year-round springlike climates at moderate elevations in Middle and South America

Pronunciation Guide

altiplano (ahl-tee-PLAH-noh)

Amazon River (AM-uh-zahn)

Amazonia (am-uh-ZOH-nee-uh)

Andes (AN-deez)

Antigua (an-TEE-guh)

Argentina (ahr-juhn-TEE-nuh)

Arias, Oscar (AH-ree-ahs)

Atacama Desert (ah-tuh-KAH-muh)

Aztec (AZ-tehk)

Bahamas (buh-HAH-muhz)

Barbados (bahr-BAY-dohss)

Barbuda (bahr-BOO-duh)

Belém (buh-LAYM)

Belize (buh-LEEZ)

Belo Horizonte (BEH-loo aw-rih-ZAWN-teh ["oo" as in "book"])

Bogotá (boh-goh-TAH)

Bolivia (boh-LIHV-ee-ah)

Brasília (bruh-ZIHL-yuh)

Buenos Aires (BWAY-nohss IE-rehss)

Caracas (kah-RAH-kahs)

Caribbean (kuh-RIH-bee-uhn)

Castro, Fidel (KAH-stroh, fee-DEHL)

Cerro San Cristóbal (SEHR-roh, sahn, krees-TOH-vahl)

Chamorro, Violetta (chah-MOHR-oh, vee-oh-LEHT-ah)

Chiapas (chee-AH-pahss)

Chile (CHEE-leh)

Ciudad Juárez (see-oo-DAHD HWAHR-ehss)

Colombia (koh-LOHM-bee-ah)

Costa Rica (KOHSS-tah REE-kah)

Curitiba (koor-ee-TEE-bah)

Cuzco (KOOZ-koh)

Duvallier, François (doo-VAHL-yay, frahnts-WAH)

Ecuador (EHK-wah-dohr)

El Niño (el NEEN-yoh)

El Salvador (ehl sahl-vah-DOHR)

Fortaleza (fohr-tah-LAY-zah)

French Guiana (gee-AHN-ah)

Gatun Lake (gah-TOON)

Goiás (goy-AHSS)

Greater Antilles (an-TIL-eez)

Grenada (gruh-NAY-duh)

Grenadines (GREHN-uh-deenz)

Guadeloupe (gwah-duh-LOOP)

Guatemala (gwah-teh-MAH-lah)

Guayaquil (gwah-yah-KEEL)

Guiana Highlands (gee-AHN-ah)

Guyana (gye-AHN-uh)

hacienda (hah-see-EN-dah)

Haiti (HAY-tee)

Havana (ah-VAHN-ah/huh-VAN-uh)

Hispaniola (hihs-puhn-YOH-luh)

Honduras (ohn-DOO-rahss/hawn-DOOR-uhss)

Inca (ING-kuh)

Iquitos (ih-KEE-tohss)

Lake Titicaca (tee-tee-KAH-kah)

Lima (LEE-muh)

llanos (YAH-nohs)

machismo (mah-CHEEZ-moh)

Managua (mah-NAH-gwah)

Manaus (mah-NAUS)

maquiladora (mah-kee-lah-DOH-rah)

marianismo (mah-ree-ah-NEEZ-moh)

Martinique (mahr-tih-NEEK)

Mesa Central (MAY-zah sehn-TRAHL)

mestizo (mehs-TEE-zoh)

Mexicali (meh-hee-KAH-lee)

Mexico (MEH-hee-koh/MEHK-sih-koh)

minifundios (mih-nee-FOON-dee-ohss ["oo" as in "book"])

Montego Bay (mawn-TEE-goh)

Montserrat (mohnt-srat)

mulatto (moo-LAH-toh)

Nazca Plate (NAHZ-kuh)

Netherlands Antilles (an-TIL-eez)

Nicaragua (nee-kah-RAH-gwah)

Noriega, Manuel (nohr-ee-AY-gah, mahn-WEHL)

Nuevo Laredo (NWAY-voh lah-RAY-doh)

Orinoco River (ohr-ee-NOH-koh)

pampas (PAHM-pahss)

Panama (PA-nuh-mah)

Paraguay (PAHR-uh-gwaiy)

Paraná River (pah-rah-NAH)

pardo (PAHR-doh)

Patagonia (pa-tuh-GOHN-yuh)

Peru (peh-ROO)

Petén (peh-TEHN)

Pizarro, Francisco (pee-ZAHR-oh, frahn-SEESS-koh)

Pôrto Alegre (POHR-too ah-LEHG-reh)

Potosí (poh-toh-SEE)

Puerto Rico (PWAIR-toh REE-koh/ POHR-toh REE-koh)

Quito (KEE-toh)

Recife (reh-SEE-feh)

Reynosa (ray-NOH-sah)

Rio de Janeiro (REE-oh dih zhuh-NAY-roh)

Rondônia (rawn-DOH-nyuh)

Sahagún, Bernardino de (sah-hah-GOON, bair-nahr-DEE-noh deh)

Salinas, Carlos (sah-LEE-nahss, KAHR-lohss)

Salvador (sahl-vah-DOHR)

San Juan (sahn HWAHN)

San José (sahn hoh-SAY)

Sandinista (sahn-dee-NEESS-tah)

São Paulo (sau PAU-loh)

Sierra Madre (see-AIR-ah MAHD-ray)

Somoza, Anastacio (sah-MOH-zah, ah-nah-STAH-see-oh)

St. Croix (KROY)

St. Kitts and Nevis (KIHTSS; NEH-vihss)

St. Lucia (LOO-shuh)

St. Martin (MAHR-tuhn)

Suriname (SOO-rih-NAHM)

Tehuantepec (teh-WAHN-teh-pehk)

Tenochtitlán (teh-nawch-teet-LAHN)

tierra caliente (tee-AIR-ah cah-lee-EHN-tay)

Tierra del Fuego (tee-AIR-ah dehl FWAY-goh)

tierra fria (tee-AIR-ah FREE-uh)

tierra helada (tee-AIR-ah ay-LAH-dah)

tierra templada (tee-AIR-ah temp-LAH-dah)

Tijuana (tee-HWAH-nah)

Tordesillas (tohr-day-SEE-yahss)

Trinidad and Tobago (TRIH-nih-dad, toh-BAY-goh)

Umbanda (oom-BAHN-dah)

Uruguay (OO-roo-gwaiy)

Venezuela (veh-neh-ZWAY-lah)

Veracruz (va-rah-CROOSS)

vilas rural (VEE-lahss roo-RAHL)

Vitória (vee-TOHR-ee-ah)

Selected Readings

A set of Selected Readings for Chapter 3, providing ideas for student research, appears on the *World Regional Geography* Web site at www.whfreeman.com/pulsipher.

The snow-covered peaks of the Alps, the highest part of Europe's central mountain chain, provide income for Austria, Switzerland, France, and Italy by drawing tourists for hiking, winter sports, and spectacular scenery. Below the snow line is a treeless zone of grasslands that for thousands of years have served as summer grazing lands. [NASA/Tom Stack & Associates.]

Chapter 4: Europe

MAKING GLOBAL CONNECTIONS

On a recent visit with his extended family in Ribnica, Slovenia, Alex Pulsipher had an experience that illustrates several current issues in Europe. Slovenia is a small Slavic country in southeastern Europe that was once the northernmost province of the country of Yugoslavia. Slovenia became independent in 1991.

VIGNETTE I was out one night with my cousin Maya when we passed some young men blasting unusual music on their car radio. It sounded almost Middle Eastern, like folk music, and very different from the U.S./Euro-Pop playing in the local bars. "What's that?" I asked. "South music," Maya said, "Those men are from the south (by which she meant Croatia, Bosnia, Serbia, and Macedonia—also once provinces of Yugoslavia). They come here to take the jobs on farms and construction projects that pay too little for Slovenes."

We next stopped at Ribnica's only disco, where the DJ was playing typical U.S./Euro-Pop. When I asked him to play some South music a look of haughty disapproval came over his face. "No! Only Slovene music here!" A few moments later, however, he must have had a change of heart, because the mournful yet rhythmic sounds of the south filled the room. The place erupted! The men who had been quietly drinking and chatting at the bar were suddenly throwing their arms up, spilling beer on one another, and happily singing along with the lyrics. A group raced to the dance floor, where they joined hands to form a semicircle, performing intricate footwork as their formation slowly rotated. But then, just as abruptly, the DJ put on Britney Spears, and the emotion-filled moment ended.

In the days that followed, I asked a number of Slovenes why, if the South music was so popular, the DJ did not play more of it. The answers I received helped me see how Slovenes are motivated on the one hand by love of Slavic culture and on the other by their desire to seem cosmopolitan enough to gain acceptance in the European Union—the primary transnational institution taking shape in Europe.

Slovenes typically have mixed feelings about the countries to the south, even though Slovenia shared the former Yugoslavia with this region until 1991. Slovenia, as the wealthiest and most educated province, and the closest to western Europe, felt both affinity and distaste for the people of the south. Since Slovenia may join the European Union by 2004, Slovenes seem to feel the need to distance themselves from their southern neighbors. Some Slovenes view them as unworthy of living in Slovenia, yet also as possessors of a rich and ancient kindred culture.

Slovene sentiments toward the south echo similar attitudes of countries in North and West Europe about the prospect of countries such as Slovenia joining the **European Union (EU)**, a supra-

---

Themes to Look For The following ideas are developed in this chapter.

1. Economic union. The lure of greater prosperity is drawing the countries of western Europe into economic union and the countries of eastern Europe into closer association with western Europe.

2. Tendencies toward disunion. Despite or even because of increasing economic integration, some Europeans feel an increasing desire to establish a sense of national identity. These people may be wary of immigrants—either from neighboring countries or from outside Europe—or opposed to aspects of European union.

3. Large role for government. European governments play an active role in shaping the urban and rural environments by providing housing, education, health care, transportation, and other services to increase overall human well-being. There are, however, regional variations in the degree to which national governments provide services.

4. Regional differences in well-being. One goal of the European Union is to foster economic growth in the EU subregions that are poorer and less developed, primarily those in southern Europe. The non-EU countries of eastern Europe are trying to improve well-being and enter the free market after emerging from a half-century of control by the Soviet Union, which has left most of them the poorest countries in Europe.

5. Effects of globalization. Europe's overall prosperity arises in part from the many benefits it has obtained—and still obtains—from its access to peoples and resources around the globe, in many cases from its former colonies. At the same time, the globalizing economy has placed some of Europe's workers and products in competition with those in developing parts of the world.

6. Environmental activism. Europe's dense population and its high rates of consumption are contributing to air and water pollution. Environmental activism is growing in Europe.

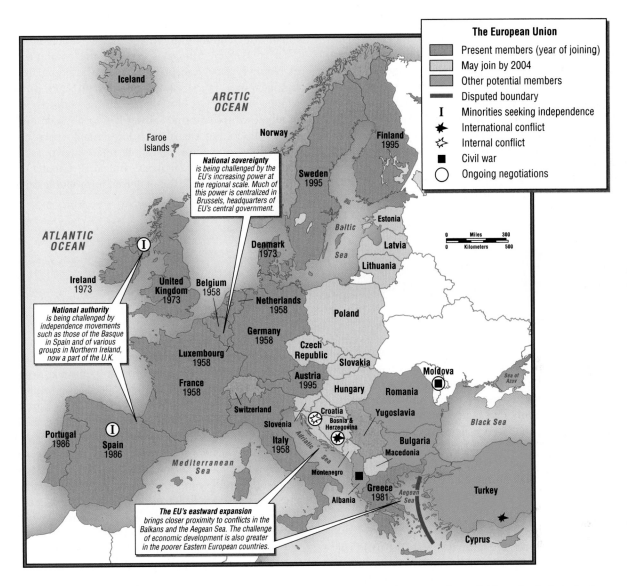

Figure 4.1 **The European Union (EU).** The EU has led the global movement toward greater regional cooperation. It is older and more deeply integrated than its closest competitor, NAFTA, which was largely a response to the EU and is concerned mainly with eliminating trade restrictions. The EU goes beyond this goal to include the free movement of EU citizens across international borders within the EU, a common EU currency (the Euro), and increasing levels of foreign policy and military coordination. [*Source:* http://europa.eu.int/comm/enlargement/report2001/index.htm.]

national organization including most of the countries of West, South, and North Europe (Figure 4.1). The European Union unites member countries in a single economy within which people, goods, and money can move freely. Many western Europeans fear that expansion of the EU into eastern and central Europe will bring new political tensions and a throng of alien immigrants. Similarly, many people who support the idea of European unity harbor the fear that, if the EU succeeds, regional culture of the kind that produced the South music so enjoyed by the Slovenes will disappear.

Terms to Be Aware Of

The end of the cold war changed political alignments, and even some borders, across Europe. For this reason, special care is needed in designating the various parts of the region. In this book, Europe is divided into the following subregions. *North Europe* includes Iceland, Denmark (including Greenland and the Faroe Islands), Sweden, Norway, Finland, Estonia, Latvia, and Lithuania. These last three, formerly part of the Soviet Union, share many economic and cultural characteristics with North Europe. *West Europe* includes the United Kingdom, the Republic of Ireland, France, Belgium, Luxembourg, the Netherlands, Germany, Switzerland, and Austria. *South Europe* includes Portugal, Spain, Italy, and Greece. *East and Central Europe*, the largest subregion, contains (1) those countries of central Europe that were formerly within the Soviet Russian sphere of influence: Poland, the Czech Republic, Slovakia, Hungary, Albania, Romania, and Bulgaria; and (2) all the countries once known as Yugoslavia: Slovenia, Croatia, Bosnia-Herzegovina, Macedonia, Montenegro, and Serbia (the last two now call themselves Yugoslavia).

The actual physical location—east, west, north, south—is less important in this subregional scheme than such other characteristics as historical, economic, political, and cultural alignments. For example, the countries lying between the Adriatic and the Black Sea (once called the *Balkans* but now usually referred to as *southeastern Europe*) are not considered part of South Europe, despite their southerly location. Their cultural history and recent political experience as communist countries have linked them more to East and Central Europe, which is the designation they prefer.

For convenience, we occasionally use the term *Western Europe* to refer to all the countries that were not part of the exper-

iment with communism in the Soviet Union and Yugoslavia. That is, *Western Europe* is used for the combined subregions of North Europe (except Estonia, Latvia, and Lithuania), West Europe (except former East Germany), and South Europe. *Eastern Europe* will be used to designate all the countries once part of, or allied with, the former Soviet Union, including those once a part of the former Yugoslavia. Occasionally, when it is necessary to identify only those countries once known as the *Balkans* (Albania, Bosnia-Herzegovina, Bulgaria, Croatia, Macedonia, Romania, and Serbia), the preferred term *southeastern Europe* is used instead.

THE GEOGRAPHIC SETTING

PHYSICAL PATTERNS

Europe is a region of peninsulas upon peninsulas. The entire European region is one giant peninsula extending off the Eurasian continent, and the whole of its very long coastline is itself festooned with peninsular appendages, large and small. Norway and Sweden share one of the larger appendages. The Iberian Peninsula (shared by Portugal and Spain), Italy, and Greece are other examples of large peninsulas. One result of these many fingers jutting into the oceans and seas is that much of Europe feels the climate-moderating effect of the large bodies of water that surround it.

Landforms

Although European landforms are fairly complex, the basic pattern is not hard to learn. The three basic landforms are mountains, uplands, and lowlands, all of which stretch roughly east-west in wide bands. Look at the map that opens this chapter and notice first the gold color representing mountains. Europe's largest mountain chain stretches west to east through the middle of the continent, from southern France through Austria and curving southeast into Romania. The Alps are the highest and most central part of this formation. This network of mountains is mainly the result of pressure from the collision of the African Plate, which is moving northward, and the Eurasian Plate, which is moving to the southeast. Europe lies on the westernmost extension of the Eurasian Plate. South of the main formation, mountains extend into the peninsulas of Iberia and Italy and along the Adriatic Sea through Greece. The northernmost mountainous formation is shared by Scotland, Norway, and Sweden. These mountains are old (about the age of the Appalachians in North America) and worn down by glaciers.

A band of low-lying hills and plateaus extending northward descends from the central mountain zone to the North European Plain. These uplands form a transition zone between mountains and lowlands (or plains). Crossed by many rivers and holding considerable mineral deposits, this is an area of large industrial cities and densely occupied rural areas.

The lowlands of the North European Plain form the most extensive landform in Europe. The plain begins along the Atlantic coast in western France and stretches in a wide band around the northern flank of the main European peninsula, reaching across the English Channel and North Sea to take in southern England in the British Isles, southern Sweden, and a part of southern Finland. The plain continues east through Poland and then broadens to the south and north to include all the land east to the Ural Mountains. The coastal zones of this plain are densely occupied all the way east through Poland. Much of the coast is low-lying. Over the past thousand years, in countries such as the Netherlands, people have transformed the natural seaside marshes and vast river deltas into farmland, pastures, and urban areas by building dikes and draining the land with wind-powered pumps.

The rivers of Europe link its interior to the surrounding seas (see the chapter opener map). Several of these rivers are naviga-

The Pannonian (or Hungarian) Plain, which is surrounded by mountain ranges, is one of the lowlands of central Europe. [Mac Goodwin.]

ble well into the upland zone, and Europeans have built large industrial cities on their banks. The Rhine carries more traffic than any other European river, and the course it has cut through the Alps and uplands to the North Sea also serves as the route for railways and motorways. The area where the Rhine flows into the North Sea is considered the economic core of Europe, and it is here that Rotterdam, Europe's largest port, is located. The larger and much longer Danube River flows from Germany to the southeast, connecting the center of Europe with the Black Sea and passing the important and ancient cities of Vienna, Budapest, and Belgrade.

Vegetation

Europe was once covered by forests and grasslands. Today, forests exist only in scattered areas, primarily on the more rugged moun-

tain slopes, in the most northern parts of Scandinavia, and in a few places where reforestation is under way. The dominant vegetation in Europe is now crops and pasture grass. Vast areas are covered with industrial sites, railways, roadways, parking lots, canals, cities and their suburbs, and parks.

Climate

Europe has three main climate types: temperate midlatitude, Mediterranean, and humid continental (Figure 4.2). The temperate midlatitude climate dominates in northwestern Europe, where the influence of the Atlantic Ocean is very strong. A broad warm-water current called the North Atlantic Drift, which is really just the easternmost end of the Gulf Stream (see Chapter 2, page 60), brings large amounts of warm water that has traveled from the Gulf of Mexico along the eastern coast of North

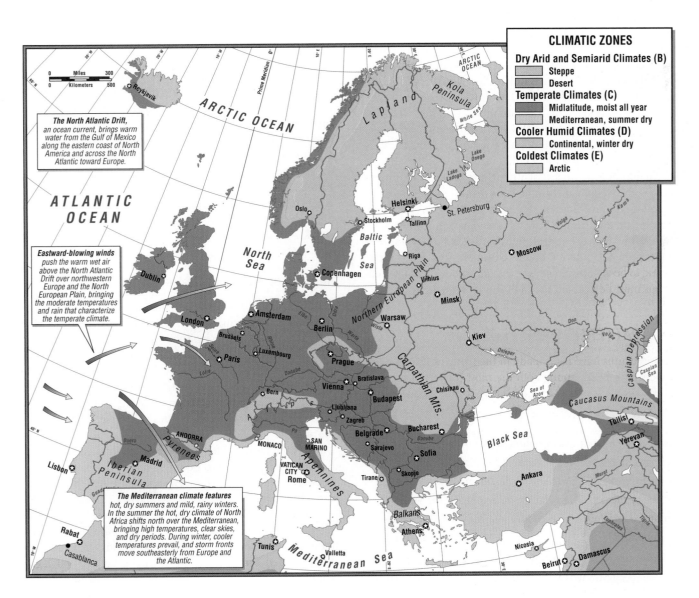

Figure 4.2 Climates of Europe.

America and across the North Atlantic toward Europe. The air above the North Atlantic Drift is quite warm and wet, and eastward-blowing winds push this air over northwestern Europe and the North European Plain, bringing moderate temperatures and rain deep into the Eurasian continent. These factors create a climate that, though still fairly cool, is much warmer than elsewhere in the world at similar latitudes. To minimize the effects of precipitation runoff, people in these areas have developed elaborate drainage systems for houses and communities (steep roofs, rain gutters, storm sewers, drain fields, and canals), and they grow crops such as potatoes, beets, turnips, and cabbages that thrive in cool, wet conditions.

Farther to the south, the Mediterranean climate of warm, dry summers and mild, rainy winters prevails. In the summer, the warm, dry climate of North Africa shifts north over the Mediterranean, bringing high temperatures, clear skies, and dry periods as far north as the Alps. Major crops, such as olives, citrus fruits, and wheat, must be drought resistant or irrigated. In the fall, this warm, dry air shifts to the south and is replaced by cooler temperatures and by thunderstorms sweeping in off the Atlantic. For short periods in winter, the northern Mediterranean climate can resemble the temperate midlatitude climate of northwestern Europe; but overall the climate is mild. Along the Mediterranean, houses by the sea are often open and airy to afford comfort in the hot, sunny, rainless days of Mediterranean summers. Tourists escaping the drab climate of northwestern Europe crowd scenic areas and especially the beaches of Mediterranean shores.

In eastern Europe, the moderating influences of the Atlantic and the Mediterranean are less or absent, and the climate is more extreme. In this region of humid continental climate (eastern Poland and Germany, Slovakia, Hungary, and most of Romania, as well as inland and eastern parts of Scandinavia), summers are fairly hot and winters are longer and colder the farther north or the deeper into the interior of the continent one goes. Here, houses tend to be well insulated, with small windows and low ceilings. Crops must be adapted to much shorter growing seasons.

HUMAN PATTERNS OVER TIME

Some grand claims have been made in the name of Europe, and geographers have been among both the most vociferous supporters and the strongest critics of these ideas, as the following two quotations attest.

Europe has been a great teacher of the world. Almost every vital political principle active in the world today had its origin in Europe or its offspring, European North America. . . . [T]he same is true of the arts. Even though other parts of the world have produced rich folk arts, the culture of the West has become dominant.

[GEORGE F. HEPNER AND JESSE O. MCKEE (EDS.), *WORLD REGIONAL GEOGRAPHY: A GLOBAL APPROACH* (EAGAN, MINN.: WEST PUBLISHING CO., 1992), P. 144.]

[There is] the [unfortunate] notion that European civilization—"The West"—has had some unique historical advantage, some special quality of race or culture or environment or mind or spirit, which gives this [particular] human community a permanent superiority over all other communities, at all times in history and down to the present.

[JAMES BLAUT, *THE COLONIZER'S MODEL OF THE WORLD* (NEW YORK: GUILFORD PRESS, 1993), P. 1.]

We are so used to hearing the praises of Western (European) culture that for many people it is hard to spot the realities that they omit. You may have read the passages above idly and said to yourself, "So?" The first quote fails to recognize that Europe has been as much a learner from the world as a teacher of it, and that many "European" ideas and technologies were adopted from non-European sources. For example, the concept of the peace treaty, so vital to current global stability, was first documented not in Europe but in ancient Egypt. Many people also criticize the assumption that the ability to dominate militarily, politically, economically, or culturally is evidence of overall superiority.

It is true that over the last 500 years Europe *has* influenced how much of the rest of the world trades, fights, thinks, and governs itself. Attempts to explain this influence have ranged from arguments that Europeans are somehow a superior breed of humans to assertions that Europe's many bays, peninsulas, and navigable rivers have promoted commerce to a greater extent there than elsewhere. We may never have a single satisfying answer to the question of why Europe gained the leading role it continues to play. Still, it is worth taking a look at the broad history of this area and considering a few of the developments that have made Europe so influential over the past five centuries.

Sources of European Culture

Starting about 10,000 years ago, the practice of agriculture and animal husbandry gradually spread into Europe from the Tigris and Euphrates river valleys in Southwest Asia (also known as Mesopotamia) and from farther east in Central Asia and beyond. The cultivation of wheat, barley, and numerous vegetables and fruits, and the keeping of cattle, pigs, sheep, and goats, came to Europe from the east, as did pottery making; weaving; mining; metalworking; and, most important, mathematics. All these innovations opened the way for a wider range of economic activity, especially trade. Many of the ideas and techniques that contributed to European civilization had their origins outside Europe, in places that are today known as Turkey, Iraq, Iran, Arabia, and Egypt. Some influences came from as far away as India, Central Asia, Mongolia, and China.

The first European civilizations were ancient Greece (800–86 B.C.) and Rome (100 B.C.–A.D. 500). The innovations of these societies and their extensive borrowings from other culture areas formed some of the most important cultural legacies of Europe. Located in southern Europe, Greece and Rome interacted more with the Mediterranean rim, Southwest Asia, and

North Africa than with the rest of modern Europe, which was then relatively poor and thinly populated. Greek artists, philosophers, and mathematicians were fascinated with the workings of both the natural world and human societies. Later European traditions of science, art, and literature were heavily based on Greek ideas. The Romans, perhaps the greatest borrowers of Greek culture, also left important legacies to Europe, such as the notion of individual ownership of private property and the practices used in colonizing new lands. After a military conquest, Romans would secure control by establishing large plantation-like farms and communities of settlers transplanted from the heartland. Roman systems of colonization were used when European states laid claim to territory in the Americas and other places after 1500. Many Europeans (and Middle and South Americans) today speak Romance languages, such as Spanish, Portuguese, Italian, and French, which are largely derived from Latin, the language of the Roman Empire.

The influence of Islamic civilization on Europe is often overlooked. North African Muslims, called Moors in Europe, had a profound influence on language, music, food customs, architecture, and belief systems in Spain, which they ruled for 700 years, starting in A.D. 711. Similarly, the Muslims of the Ottoman Empire left a deep imprint on various parts of southeastern Europe (the Balkans) and Greece, which they ruled from the sixteenth through the early twentieth centuries. These Muslim rulers brought to Europe many textile and tempered-metal trade goods, food crops, architectural principles, and technologies from Arabia, China, India, and Africa. Muslim scholars preserved much learning from Greece, Egypt, and other ancient civilizations and brought Europe its numbering system as well as significant advances in medicine. Arab mathematicians, perhaps building on ideas picked up in India, brought algebra and algorithms to Europe, both essential elements of modern engineering and architecture. Among the modern applications of these mathematical techniques are the design of computer hardware and software programs.

Growth and Expansion of Continental Europe

Continental Europe became increasingly influential as the Roman Empire declined in the fifth century A.D. and European settlement expanded northward. By about A.D. 900, a social system known as **feudalism** had evolved out of the need for defense against Viking raiders from Scandinavia and nomadic raiders from the Eurasian steppes. The objective of feudalism was to have a sufficient number of heavily armed, professional fighting men, or knights, ready to defend the farmers, or serfs, who cultivated plots of land for them. Over time, some of these knights became an elite class of warrior-aristocrats, called the nobility, who were bound together by a complex web of allegiances obligating them to assist one another in times of war. The often lavish lifestyles of the wealthier knights were supported by the labors of the serfs, who were legally barred from leaving the lands they cultivated for their protectors. Most serfs lived in poverty. Aspects of feudalism were brought to the Americas, where the Spanish crown expropriated land from Native Americans and granted it to colonists, along with the right to treat the native inhabitants as serfs.

While rural life followed established feudal patterns, Europe's towns were developing new institutions that would influence settlement and commerce in modern Europe. Often founded as trading posts by former raiders, towns were able to maintain independence from the feudal knights of the surrounding countryside. They sheltered artisans and merchants who, through innovation, created the main institutions of modern European capitalism: banks, insurance companies, and corporations. Documents called town charters set forth the rights of town citizens and provided the basis for the European notion of civil rights to be enjoyed by all citizens. These strong new social institutions allowed Europe's towns to avoid the extreme divisions of status and wealth of the feudal system and to establish a pace of technological and social change that left the feudal hinterland literally in the Dark Ages. This division in European thought between the "exciting, creative urban" and the "behind-the-times rural" still shapes many of our attitudes toward economic development. In former European colonies, for example, urban development often receives more support than rural development, even though the majority of the population may be rural.

An outgrowth of European town life was the **Renaissance** (French for "rebirth"), a broad cultural movement that began in Italy in the fourteenth century (1300s) and was inspired by the older Greek, Roman, and Islamic civilizations. Renaissance thinkers turned their attention to science, politics, commerce, and the arts. By developing ideas about humanism, a philosophy that emphasizes the dignity and worth of the individual, they provided the foundation of modern European culture.

The liberating effects of the new urban institutions and the Renaissance led to further transformation of life in Europe. One aspect of life so affected was religion. Since Roman times, the Catholic church had dominated religion, politics, and daily life throughout much of Europe. In the sixteenth century, however, a reform movement known as the **Protestant Reformation** arose in the towns of the North European Plain, Scandinavia, and the British Isles. The reformers challenged Catholic practices that stifled public participation in religious discussions, such as the Catholic tradition of printing the Bible and holding church services only in Latin, a language that none but a tiny educated minority understood. The Reformation coincided with the invention of the European version of the printing press, which facilitated widespread diffusion of reformist ideas and stimulated the development of written versions of local languages. In addition to translating the Bible and holding services in the languages of the people, the Protestants also promoted public literacy, individual responsibility, and more open public debate of social issues.

Europe's **Age of Exploration** was a direct outgrowth of the greater openness of the Renaissance and the Reformation, and it began a period of accelerated global commerce and cultural exchange. The globalization of modern international trade had its origins in the Age of Exploration. In the fifteenth and sixteenth centuries, Portugal took advantage of Renaissance advances in navigation, shipbuilding, and commerce to round the Horn of Africa, to set up a trading empire in Asia, and eventually to establish a colony in Brazil. Spain, beginning with the first voyage of Columbus in

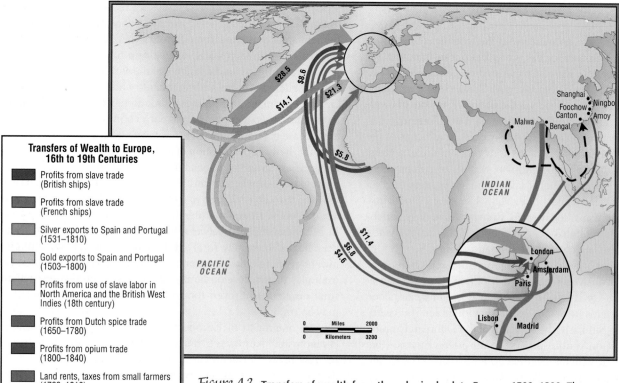

Transfers of Wealth to Europe, 16th to 19th Centuries

- Profits from slave trade (British ships)
- Profits from slave trade (French ships)
- Silver exports to Spain and Portugal (1531–1810)
- Gold exports to Spain and Portugal (1503–1800)
- Profits from use of slave labor in North America and the British West Indies (18th century)
- Profits from Dutch spice trade (1650–1780)
- Profits from opium trade (1800–1840)
- Land rents, taxes from small farmers (1760–1810)
- - - India–China opium trade

Figure 4.3 Transfers of wealth from the colonies back to Europe, 1500–1800. The amounts are in billions of current U.S. dollars (total = U.S. $101 billion). For European colonial holdings in various regions of the world, see Figures 3.5, 6.5, and 7.3. [Adapted from Alan Thomas, *Third World Atlas* (Washington, D.C.: Taylor & Francis, 1994), p. 29.]

1492, founded a vast and profitable American-Pacific mercantile empire. Countries that practiced **mercantilism** attempted to increase their power and wealth by acquiring colonies with human and natural resources and by managing all aspects of production, transport, and trade for their own benefit. Mercantilism supported the industrial revolution in Europe by supplying cheap resources for new factories and markets for European manufactured goods. By the nineteenth century, however, the Spanish and Portuguese empires had weakened, partly because they still relied on archaic Greco-Roman colonial institutions such as the plantation (and the hacienda). In addition, Spain and Portugal maintained a strong alliance with the Catholic church, which discouraged social and technological change. These empires were soon overshadowed by those of the Dutch and the British. These countries, influenced by the Protestant emphasis on individualism and innovation, used somewhat more efficient methods of colonial expansion and control, and extended their influence into Asia and Africa. Regardless of which European powers dominated, by the twentieth century European colonial systems had strongly influenced nearly every part of the world. Colonialism produced great wealth for Europe, often at the expense of people and resources in the overseas colonies (Figure 4.3).

An Age of Revolutions

Unlike other civilizations, Europe did not slow its pace of transformation after achieving a relatively high level of power and prosperity. The Chinese civilization, for example, had been arguably the world's most dynamic and elaborate since the end of the Roman Empire, but by about 1100 its dynamism had slowed. In contrast, as Europe became richer and more influential, the pace of change accelerated. Scholars have developed many theories to explain why this happened; one notion is that Europe's fragmentation into many competing kingdoms pushed it into a continuing race with itself. Whatever the ultimate causes, Europe's ongoing transformation took two main forms: the industrial and the democratic revolutions.

The Industrial Revolution. Europe's industrial revolution was intimately connected with colonial expansion and the Age of Exploration. In particular, Britain's ascendancy as the leading industrial power of the nineteenth century had its origins partly in the expansion of empire. In the sixteenth century, Britain was an island of only modest wealth and resources. In the seventeenth century, however, Britain developed a small but growing trading empire in the Caribbean, North America, and South Asia, which provided access to a wide range of raw materials and to markets for British goods. Sugar, produced by British colonies in the Caribbean, was an especially important crop. Sugar production was a complex process requiring major investments in equipment and labor and long-term management of planting, harvesting, processing, storage, transport, and marketing—all of which were skills needed for industrialization. As demand for sweet foods grew rapidly, sugar produced enormous wealth that, in turn, helped fund industrialization in Britain. By the late eighteenth century, the

Women and young girls in long, full dresses do heavy lifting in a massive cotton factory hall with dangerous open belts and presses. The scene depicts the nineteenth-century outgrowth of an old association between women and large-scale textile production. In medieval times, lords required women laborers to provide them with a certain amount of cloth every year. Women often wove this cloth in a central location within the castle grounds. [Bettmann/Corbis.]

country was straining to meet a demand for more goods than it could produce. Britain met the challenge by introducing mechanization into its industries, first in textile weaving and then in the production of coal and steel. Eventually, cities with factories producing tools, textiles, ceramics, and machinery grew up around Britain's coal and iron fields. By the nineteenth century, industrial technologies were spreading throughout Europe, North America, and elsewhere. Britain was a global economic power with a huge and growing empire, expanding industrial capabilities, and the world's most powerful navy.

The Democratic Revolution. As Europe was industrializing, it experienced political and social transformations that redistributed power more evenly throughout society. Eventually, there was also a more equitable distribution of wealth. For centuries, Europe's power structure had been feudal. But by the eighteenth century, especially in the kingdoms of western Europe, the political elite was expanding to include private property owners as well as nobles. Soon, kings and churches were forced to accept constitutions that restricted their power. But the working classes, both rural and urban, had no vote or other formal role in the political system.

In 1789, the French Revolution arose out of the conflicts created by extreme disparities of wealth in French society. Inspired in part by news of the popular revolution in North America, the French Revolution led to the first major inclusion of the common people into the political process in Europe. As the populace became more involved in governing, the idea of the nation was born. A nation was conceived of as a large group of people, both rich and poor, living in a specific territory and united by certain common cultural traits, such as a common language and shared beliefs and values. Above all, people would be bound by their allegiance to a single country called a **nation-state** that would encompass all people of the same culturally unified group. This new ideology of **nationalism** spread throughout much of Europe, transforming the political structure. Individual kingdoms, in the case of France, or collections of kingdoms, in the case of Germany and Italy, tried to establish the notion that all their people together formed a nation.

Later, the common people gained a formal role in the political life of the nation with the introduction of **democratic institutions** such as constitutions, elected parliaments, and impartial courts. All adults—first men and, after considerable political agitation, women as well—obtained the right to elect their leaders. This trend eventually spread throughout Europe, although Eastern Europe did not adopt democratic institutions until the late twentieth century.

The new industrial production techniques created a demand for skilled and unskilled factory workers that triggered steady migrations from the countryside to the cities. In the cities, workers experienced unsafe working conditions, low wages, and dirty and crowded housing. So, although the common people played an increasingly important role in European society, the industrial revolution failed at first to raise living standards for the vast majority of Europeans. Periodically, popular discontent erupted in violent protests and revolutionary movements that threatened the civic order. As prosperity became more widespread in the mid-twentieth century, the protesters convinced some governments to guarantee to all citizens such basic necessities of life as education, employment, and health care. A social system in which the state accepts responsibility for the well-being of its citizens is known as a **welfare state.** In time, mechanisms were set in place that established more harmonious relations between workers and employers, a reduction in wealth disparities, and an increase in overall civic peace and prosperity. Today, the proper role of the welfare state has been resolved in different ways in various parts of Europe; we will discuss this topic on pages 206–208.

Europe's Difficult Twentieth Century

World War, Cold War, and Decolonization. Despite Europe's many advances, between 1914 and 1945 two horribly destructive world wars removed Europe from its position as the dominant region of the world. By the mid-twentieth century, Europe lay in ruins, millions had died, and Europe still lacked a system of collective security that could prevent war between its rival nations. The defeat of Germany, seen as the instigator of both wars, resulted in a number of enduring changes in Europe. After the end of World War II, in 1945, Germany was divided into two parts. West Germany became an independent democracy allied with the rest of Western Europe, especially Britain and France, and with the United States. Through the Marshall Plan, the United States provided financial assistance to rebuild Western Europe's basic facilities, such as its roads, housing, and schools. Western

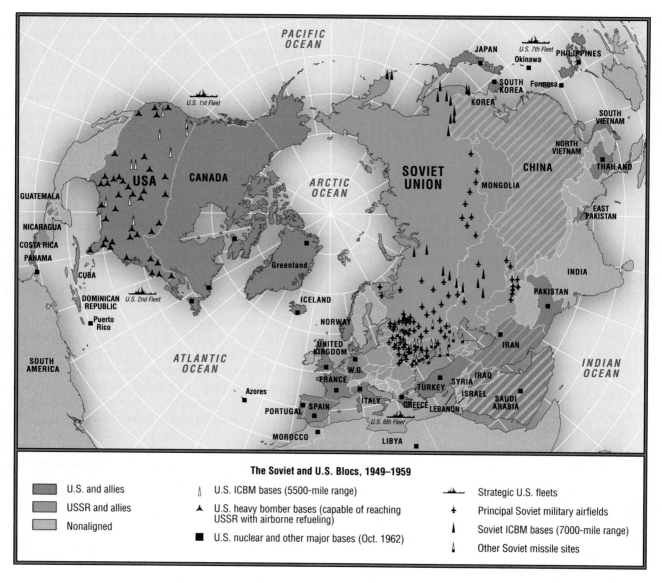

The Soviet and U.S. Blocs, 1949–1959

U.S. and allies	
USSR and allies	
Nonaligned	

- U.S. ICBM bases (5500-mile range)
- U.S. heavy bomber bases (capable of reaching USSR with airborne refueling)
- U.S. nuclear and other major bases (Oct. 1962)
- Strategic U.S. fleets
- Principal Soviet military airfields
- Soviet ICBM bases (7000-mile range)
- Other Soviet missile sites

Figure 4.4 **The Soviet and American blocs, 1949–1959.** The United States and the Soviet Union covered the Northern Hemisphere with military installations. Notice how the United States and its allies nearly managed to encircle the Soviet Union and China. [Adapted from *Hammond Times Concise Atlas of World History* (Maplewood, N.J.: Hammond, 1994), pp. 148–149.]

European countries continued their free-market economic system of privately owned businesses that adjusted prices and output to match the demands of the market.

East Germany and the rest of Eastern Europe (Poland, Czechoslovakia, Hungary, part of Austria, Romania, and Bulgaria) fell under the control of the Soviet Union (Union of Soviet Socialist Republics, or USSR), a revolutionary communist state that had emerged out of the old Russian Empire during World War I. The line between East and West Germany was part of the **iron curtain,** a fortified border zone that separated Western Europe from Eastern Europe. Eastern Europe was integrated into the Soviet Union's sphere of so-called communist states that employed a form of socialism in which the state owned all farms and industry. In contrast to the market economies of Western Europe, these economies were "centrally planned": a central bureaucracy dictated prices and output with the stated aim of allocating goods equitably across society according to need.

The division of Europe laid the foundation for the **cold war** between the United States and the Soviet Union, an era lasting from 1945 to 1991 during which the entire world became a stage on which these two superpowers competed for dominance (Figure 4.4). Once-powerful Europe became subject to the geopolitical manipulations of the two superpowers. Yet another manifestation of Europe's decline was that it could no longer control its colonial empires. By the 1960s, most former colonies were independent and had entered a difficult period of finding ways to thrive on their own.

Europe's Rebirth and Integration.   In the decades after World War II, part of Europe reemerged as a dominant power. Economic reconstruction and industrial growth proceeded rapidly in the free-market democracies of Western Europe, but much more slowly in socialist Eastern Europe, which was saddled with inefficient and highly polluting state-run industries and political systems that did not encourage public debate and citizen participation. In the early 1980s, political and economic reforms in Eastern Europe began with labor protests in Poland. By the end of the 1980s, much of Eastern Europe had abandoned socialism. This change hastened the economic and political collapse of the USSR in 1991, because it had depended to a considerable extent on the resources and skilled labor of Eastern Europe. Meanwhile, some of the free-market democracies of Western Europe had been lowering barriers to trade among themselves and were becoming linked economically. This process, which began in the 1950s, eventually led to the establishment of the European Union in 1993. The EU encourages **economic integration:** the free movement of people, goods, money, and ideas among member countries. EU nations also exhibit some degree of political integration; a constitution for the EU is being developed. Because it is acting more and more as a single large unit, the EU is challenging the dominance of the

United States in world affairs. Later in this chapter, however, we shall see that as the EU expands into the countries of East and Central Europe—the old Soviet Union's sphere of influence—it is encountering many challenges of its own.

POPULATION PATTERNS

Enumerating just how many people live in Europe is complex because, as a result of recent political changes, there is no universally agreed upon eastern border of the region. According to the way we define Europe in this book (see the opening map for Chapter 1), there are about 530 million Europeans, and the population is distributed unevenly. On the map (Figure 4.5), the densest settlement in Europe stretches in a disconnected band from the United Kingdom and north coastal France east through the Netherlands and central Germany all the way to Warsaw and Bucharest (and continues into Ukraine; see Chapter 5). Northern Italy is another zone of density, and pockets along the coasts of Portugal,

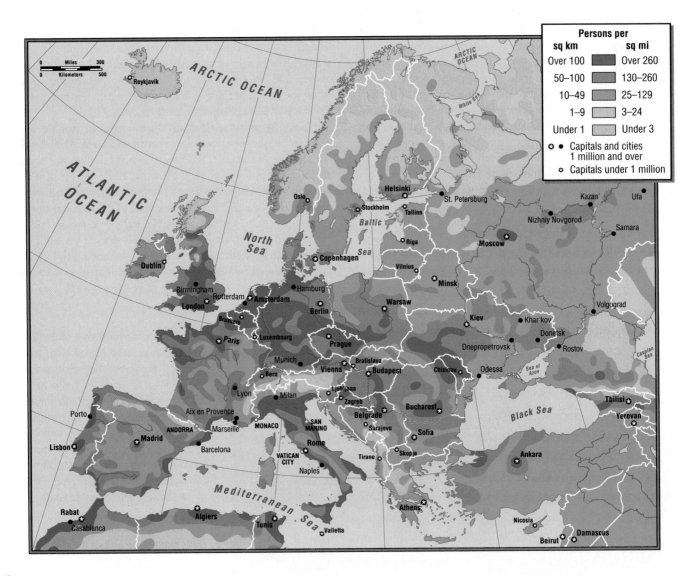

Figure 4.5 **Population density of Europe.** [Adapted from *Hammond Citation World Atlas* (Maplewood, N.J.: Hammond, 1996).]

Spain, southern France, Sicily, southern Italy, and in southeastern Europe are also densely populated. A glance at the global population map in Figure 1.16 (pages 36–37) will show that, overall, Europe is one of the most densely occupied regions on earth. You might assume that at such high population densities there would be insufficient resources to go around, so an interesting question is how Europeans maintain their high standards of living.

Population Density and Access to Resources

Population density tells us little about how well people are doing in a particular place. If people have access to adequate resources and if these resources are allocated fairly among the population, they may live well even at high densities. A region's **resource base** is the selection of raw materials available for domestic use and industrial development: coal, petroleum, iron ore, cotton and wool fiber, food, soil and water, for example. Europe depleted many of its own resources, especially forests and minerals, during the early industrial revolution. Nonetheless, Europe is now both densely populated and wealthy. This is possible because for hundreds of years it has had access to a resource base that is global in extent, not confined to Europe. The map in Figure 4.3 shows transfers of wealth to Europe from the sixteenth to the nineteenth centuries. Since about 1500, colonies in the Americas, Africa, and Asia have contributed tremendous wealth to Europe in free or very low paid labor, in forest and mineral resources, and in agricultural products extracted at low costs. This wealth has added considerably to the overall standard of living of Europeans and to the ability of Europe to develop an industrial economy. Even ordinary laborers in Europe have benefited. In contrast, those far away in colonies or former colonies who have helped to create wealth for Europeans often have not fared as well.

Consider the case of the European chocolate industry. The raw material for fine chocolates is cacao beans, which are grown in the tropics of the Americas, Africa, and Asia. Most cacao farmers are small producers who barely eke out a living and have little or no power to negotiate the prices they are paid. Meanwhile, European cacao traders are wealthy from the earnings they make in the world cacao market; European workers in chocolate factories also live well in comparison to cacao farmers. Some observers would argue that the European cacao traders and chocolate factory workers are rewarded with high incomes for their higher levels of skill and technology and for developing the chocolate market in the first place. Critics of global market systems might counter that the cacao farmers are also quite skilled, despite little access to schooling. They might also question whether it is right that education and technology should provide such overwhelming benefits to those who trade and process a product, in comparison with those who produce the raw materials. (See page 215 for a further discussion of the global role of chocolate.)

Urbanization in Europe

Europe is a region of cities surrounded by well-developed rural hinterlands. Even in sparsely settled North Europe, 84 percent of the people live in urban areas. South Europe is least urbanized, with 61 percent living in cities. Many European cities began as trading centers more than a thousand years ago and still bear the architectural marks of medieval life in their historic centers. Most of these old cities are located on navigable rivers in the interior or along the coasts, because water transport figured prominently (and still does) in Europe's trading patterns.

Since World War II, nearly all the cities in Europe have expanded in concentric circles of apartment blocks. Well-developed rail and bus lines link these blocks to one another and to the old central city. Land is scarce and expensive in Europe, so only a small percentage of Europeans live in single-family homes, although the number is growing. These homes tend to be attached or densely arranged on small lots surrounded by walls or fences to ensure privacy and to protect small gardens. Rarely does one see the sweeping lawns that North Americans spend so much effort grooming. Publicly funded transportation is widely available, so many people

Strasbourg, at the confluence of the Ill River with the Rhine at the edge of Alsace wine country, is the seventh largest city in France. The medieval towers of the Ponts Couverts are all that remain of fortifications that surrounded the old "free" city, originally founded in 12 B.C. as a Roman camp. Today Strasbourg, a major attraction for tourists from around the world, is also home to the European Parliament, where representatives of the EU meet; the Council of Europe, which is concerned with social, cultural, educational, environmental, and human rights matters throughout the region; and the European Court of Human Rights, which is responsible for individuals' rights in all member countries. [Mac Goodwin.]

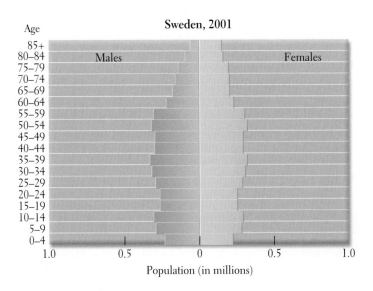
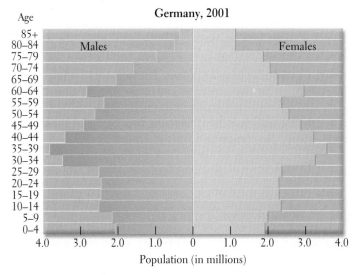

Figure 4.6 **Population pyramids for Germany and Sweden. Notice that the population scales are different for the two countries.** [Adapted from "Population Pyramids of Germany" and "Population Pyramids of Sweden" (Washington, D.C.: U.S. Bureau of the Census), International Data Base.]

prefer to live in apartments near city centers. They can either walk or take public transport to their jobs and to shopping centers, and they can easily do without a car. Public transport also links the countryside to urban centers, and many Europeans commute daily from their ancestral villages to work in nearby cities. The income of these urban workers is often invested in maintaining and landscaping European villages.

Europe contains a number of large cities: London has 11 million people in its metropolitan area, Paris 10 million, Madrid 5 million, Essen 5 million, and Berlin 4 million. Yet, even in these cities, everyday life can be intimate and personal. Although deteriorating housing and slums do exist, European city dwellers are very well off compared to the residents of Mexico City or Calcutta, where the majority are very poor and live in shantytowns. Substantial public spending on social welfare systems, sanitation, water, utilities, education, housing, and public transportation accounts for the high standard of urban living in Europe. Tourists from around the world are drawn to Europe's large cities by their art, music, architectural heritage, and generally pleasant urban ambience, and to Europe's small villages for their picturesque charm.

Population Growth Patterns

Though Europe is densely occupied, the population is aging and is no longer growing rapidly. In fact, in 2001, birth rates in Europe are the lowest on earth, and for the region as a whole there is actually a negative rate of natural increase (–0.1). Death rates are higher than birth rates in part because the average age of the population is high. This circumstance is most marked in countries that were once part of, or allied with, the former Soviet Union. In these countries, people have had few children for years, and death rates are fairly high because health care has deteriorated and because there are many elderly. The one-child family is

increasingly common throughout Europe, and in western Europe immigrants are a major source of population growth.

The declining birth rate is illustrated in the population pyramids of European countries, which look more like lumpy towers than pyramids. The population diagram of Germany (reunited in 1990) is an example (Figure 4.6). The diagram's narrow bottom indicates that since about 1978 very few babies have been born, compared with the number born in the 1950s, 1960s, and 1970s when there was a baby boom across Europe. By 2000, in Germany, an estimated 25 percent of women and men were choosing to remain unmarried well into their 30s (unmarried parenthood is increasingly common, with one European child in four born to unmarried parents); 35 to 40 percent of the people were choosing to have no children at all. The reasons for these trends are complex. For one thing, more and more women desire professional careers. The need for advanced education alone could account for late marriage and low birth rates; in addition, the German government makes very few provisions for working mothers. There is little day care available and school days are short, with no lunches provided, so mothers must be home at noon. Hence, even many married women are choosing not to become mothers. Two results of the small number of births are that the vast majority of Germans are over the age of 20 and that the population pyramid continues to narrow at the bottom. If these trends continue, the population will eventually settle into a stable age structure. The population pyramid for Sweden (see Figure 4.6) depicts such an age structure. It resembles an elongated box, with nearly equal numbers from birth through middle age, tapering at the top after age 64, especially for males. The more rapid death rate for males than females after age 64 is a common pattern throughout the developed world, although it is not yet well understood.

A stable population with a low birth rate has several consequences. Although families with few children have extra money

Thinking Critically: ON THE WEB

for luxury spending, new consumers are not being produced at the rate that elderly consumers are dying. Hence, in time, markets will contract unless immigrants can be attracted. Demands for new laborers, especially highly skilled ones, may go unmet, again unless immigration supplies a solution. The number of younger people available to provide expensive and time-consuming health care to the elderly, either personally or through tax payments, is small; currently, for example, there are just two German workers for every retiree. By comparison, in the United States, there were three workers for every retiree in 2001, but by 2020 there will also be just two. There are other consequences of stable or declining and aging populations: as we have noted, Europe has long been a center of creative achievement in the arts and sciences, but as the pool of young people shrinks, leadership in scientific and technological innovation may pass to newer, more youthful states in Asia and the Americas. For centuries, economic growth has been the hallmark of progress, and the success of states has been measured by the ever-increasing wealth of their economies. But with declining and aging populations, economic growth in Europe may be difficult to maintain.

CURRENT GEOGRAPHIC ISSUES

At the start of the twenty-first century, the social, economic, and political geography of Europe is in a state of flux. This circumstance is the result of three major changes that occurred during the 1990s: the demise of the Soviet Union, the end of the cold war, and the effort toward economic and political integration that has produced the European Union. These developments, especially the emergence of the European Union, could ultimately bring greater peace and prosperity to Europe, but many problems and tensions still have to be resolved.

ECONOMIC AND POLITICAL ISSUES

Almost all economic and political issues in Europe today are linked to the European Union in one way or another. At the end of World War II, European leaders felt that closer economic ties would prevent the hostilities that had led to two world wars. The first major step in achieving this economic unity took place in 1958, when Belgium, Luxembourg, the Netherlands, France, Italy, and West Germany formed the **European Economic Community (EEC)**. The members of the EEC agreed to eliminate certain tariffs against one another and to promote mutual trade and cooperation. Four episodes of expansion followed: Denmark, Ireland, and the United Kingdom joined in 1973; Greece joined in 1981; Portugal and Spain joined in 1986; and Austria, Finland, and Sweden joined in 1995. In 1992, a treaty enlarging the concept of the EEC to that of a European Union was signed at Maastricht in the Netherlands. In 2001, the EU consisted of the 15 countries shown in Figure 4.7.

Goals of the European Union

The original plan at the founding of the EEC, and still the core idea of the European Union, was to work toward a level of economic and social integration that would make possible the free flow of goods and people across national borders. European countries have smaller populations than their competitors in North America and Asia. With smaller markets for their products, companies earn lower profits. European businesses can sell their products outside their home countries, as is common in the global economy, but the extra costs incurred by tariffs, currency exchanges, and border regulations sap their earnings. The EU tries to solve this problem by joining national economies into a common market, thus providing the companies in any one country with access to a much larger market and the potential for larger profits through economies of scale. **Economies of scale** are reductions in the unit costs of production that occur when goods or services are produced in large amounts, resulting in a rise in profits per unit.

There are now close to 370 million people in the EU (out of a total of 530 million in the whole of Europe), which is roughly a hundred million more than the population of the United States and almost three times that of Japan. Overall, the countries of the EU are rich. The combined exports of the EU are 40 percent of the world's total, and their average gross domestic product (GDP) per capita is comparable to that of the United States (see Table 4.2, page 212). Yet some countries are notably wealthier than others (see Figure 4.7), and one aim of the European Union is to promote the equitable distribution of economic activity, opportunity, and environmental quality across Europe, while at the same time building institutions that respect the many different regional identities within Europe.

The European Regional Development Fund (ERDF) is one program that aims to reduce disparities and inequities among regions and social groups within the EU. ERDF monies (U.S. $195 billion for the six-year period 2000–2006) are allocated first to those countries that have internal regions with serious unemployment problems (Spain, Italy, Greece, Germany, Portugal, and the United Kingdom). Spain, Italy, Greece, and Portugal all have poor, underdeveloped rural regions. Germany and the United Kingdom, both rich countries overall, are included because Germany has recently absorbed poor and underdeveloped East Germany, and the U.K. has old industrial regions that have exceedingly high unemployment.

A significant share of ERDF funds goes to improve the **infrastructure.** The infrastructure consists of the skeleton of basic facilities that a society needs to operate, including such things as transportation systems, communication systems, power and water, schools, and banks. For example, EU funds are assigned to improve roads, railroads, and airports in Portugal, Spain, Greece, and Ireland. In these areas defined as underdeveloped, infrastructures need modernization to foster economic growth. Other EU funds

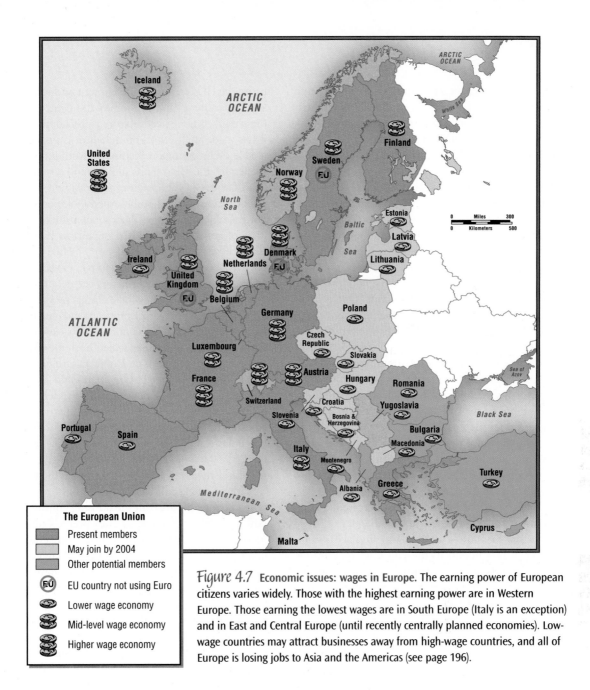

Figure 4.7 **Economic issues: wages in Europe.** The earning power of European citizens varies widely. Those with the highest earning power are in Western Europe. Those earning the lowest wages are in South Europe (Italy is an exception) and in East and Central Europe (until recently centrally planned economies). Low-wage countries may attract businesses away from high-wage countries, and all of Europe is losing jobs to Asia and the Americas (see page 196).

are allocated to environmental programs in EU countries to ensure water quality, solid waste disposal, erosion control, and reforestation. EU funds are raised through an annual 1.27 percent tax on the gross national product (GNP) of all members. Although financial allotments change from year to year, some member countries—such as Germany, the Netherlands, Austria, Sweden, the United Kingdom, France, and Italy—contribute more than they receive back in grants, while others—Greece, Ireland, Portugal, Spain, and to a lesser extent Finland and Denmark—tend to receive more than they contribute.

The majority of EU members are committed to enlarging the organization, adding those countries in Western Europe that have declined membership so far (Norway, Iceland, and Switzerland). The EU is interested in adding the countries of Eastern Europe, including those formerly under the control of the USSR as well as those formerly part of Yugoslavia, which disintegrated in 1991. Turkey, long a European ally and a source of many immigrants, is

also a candidate for eventual membership, as are the Mediterranean islands of Malta and Cyprus. Increasing the size of the European Union will increase its influence in global political and economic affairs and enlarge the market for the products of current members.

Since 1993, the EU's agenda has expanded to include the creation of a common European currency, the defense of Europe's interests in international forums, and negotiation of EU-wide agreements on human rights and social justice. The European Union is also experimenting with various forms of political unification (including the creation of a common European military). One day, the people of Europe may become citizens of the European Union and live under a common EU constitution. Drafts of such a constitution are already being discussed.

A Common European Currency. On January 1, 2002, the **Euro** became the official currency in 12 of the 15 EU countries. Three EU countries—the United Kingdom, Denmark, and Sweden—have elected not to use the Euro because they fear losing control over their own national economies and prefer to keep their national currencies alive as a patriotic symbol. Although the viability of the Euro is not yet established, it is expected that a single currency will bring gains in efficiency. For example, the common currency will eliminate the fees charged when money is exchanged while traveling from country to country. Countries that use the Euro have a greater voice in the creation of EU economic policies; those that don't use it face the economic uncertainty of using a currency whose value fluctuates relative to the value of the Euro. In the U.K., for example, when the pound rises in value relative to the Euro, British exporters are put at a disadvantage when selling in Europe, because their goods become more expensive to consumers using the Euro.

The European Union in the Global Economy. One goal of European economic unity is to provide member nations with a competitive edge in the global economy. Although western Europe was the birthplace of the industrial revolution, in the last three decades some of its facilities—its factories and mines, roads and railroads—and industries such as steel and textiles became technologically obsolete. The outdated technology and high wages paid to its workforce have made it more expensive to produce goods in Europe than in competing countries such as the United States and Japan, where technologies in some industries are more advanced and labor is often cheaper. An important reason for establishing the common market was to allow companies in the EU to keep pace technologically with their competitors in the United States and Japan. In the larger EU market, their potential for profits has increased, giving them the capital to invest in technological improvements.

Managing industrial location is another way the EU tries to reduce the high cost of production in the countries that are its economic leaders. If factories in the wealthiest countries can be relocated to the EU's relatively poorer, low-wage countries, the costs of production will fall. The EU's economic leaders are Germany, France, the U.K., the Netherlands, Belgium, Austria, and Luxembourg in West Europe, as well as Denmark, Sweden, and Finland in North Europe. Italy is the only such leader in South Europe. In all these countries, workers and managers are highly paid by world standards, which (along with related expenses) drives up production costs. The EU has encouraged its economic leaders to locate new facilities in the poorer, lower wage countries of Portugal, Spain, Ireland, and Greece. Promoters of the EU hope that these efforts will help poorer European countries to prosper and that the removal of tariffs, the elimination of border regulations, and the use of the Euro as the common currency will keep the costs of doing business in the EU low enough to restrain European companies from moving to Mexico, Southeast Asia, or China, where wages are much lower still.

Like the United States, the EU exerts a powerful influence in the global trading system, often negotiating privileged access to world markets for European companies. However, the EU also occasionally considers the needs of less developed countries. For example, the EU has supported the inclusion of worker and environmental protection standards in global free-trade negotiations within the World Trade Organization (WTO). The desire to include these standards reflects Europe's own emphasis on a strong role for government in human welfare and environmental issues. In some

cases, the EU has also pushed to remove tariffs protecting the markets of wealthy countries (but not the United States) so as to make these markets more accessible to poorer countries. For the most part, it has made such efforts on behalf of Europe's former colonies to enable them to support themselves with decent standards of living. Generally, the EU retains protectionist measures that favor European farmers at the expense of consumers, who must pay higher prices as a result, and at the expense of producers outside Europe. For example, the EU has had a major impact on the international trade of **genetically modified organisms (GMOs)**—plants or animals whose genetic code has been modified to make them more attractive as commodities. For example, GMOs may produce greater yields, have a longer shelf life, or be more nutritious. The EU has generally opposed the free trade of such products, largely in response to the sentiments of many Europeans and their national governments that these products may contain drawbacks that are not yet understood. This opposition has scaled back the global trade of GMOs and curtailed the profits of American farmers who produce them.

Future Paths of EU Organizational Development. At present, the EU governing body in Brussels, Belgium, makes decisions on the basis of consensus: issues are discussed and proposals are adjusted until all members agree on them. Supporters of a strong EU believe it can be strengthened by giving the central government greater power. Under their proposals, decisions would be made in Brussels by elected delegates on the basis of majority rule.

Many citizens of EU member countries are wary of this proposed change because they fear losing control of decisions important to their own countries. They prefer a more flexible arrangement in which groups of countries with common interests can band together to work out individual agreements. Under this plan, a single country would no longer have the power to block projects or policies, as they do now. France and Germany, for example, could purchase arms for a European army that would be staffed by soldiers only from the countries that agreed to this program. The other member countries would not have to participate, but they could not block the arrangement either.

This discussion of the European Union has made it clear that the region is far more integrated than ever before and that there is considerable momentum for further cooperation both among existing members and among those countries that are likely to join the EU in the near future. The possibility of a supranational political union grows as the benefits of economic union become more apparent. Still, it is equally evident that smaller countries, certain ethnic groups, and special interest groups are likely to resist overarching political union because they fear the loss of autonomy to the EU governing body in Brussels.

Eastern Europe and the EU

Membership in the EU is attractive to countries in East and Central Europe because the demise of the Soviet Union and the move toward more open and competitive market systems has placed many of their citizens in jeopardy. Industries that once received considerable support from the government are now forced to survive on their own revenues; hence, operations have been streamlined and many workers have lost their jobs. Increasing scar-

CULTURAL INSIGHT *The Hungarian Sausage Experience and the EU*

The Hungarian writer, humorist, and self-professed anthropologist Dork Zygotian (probably a pseudonym) assesses the prospects of European union from the perspective of a sausage lover. Zygotian writes that when you bite into a Hungarian sausage (*kolbasz*), "great torrents of paprika colored grease and juice should explode into the atmosphere around you. If you eat more than two, you should expect to bite on some piece of bone or possibly find a tooth or hair sometime during your meal. There will be a large yellow gelatinous bit somewhere in your sausage that you should not be able to identify." This is all part of the tasty Hungarian *kolbasz* experience.

But now there is talk that Hungary, long a part of the East, will eventually join the EU, and Zygotian worries about the effect of membership on his country's sausages, which he rates as Europe's best. He fears that overzealous EU standards of cleanliness and purity will kill the special flavors of Hungarian sausages. Worse, "Eurofication" (his term) may well leave Hungarians unable to afford their own beloved *kolbasz* because prices in the small, poorer countries will rise to match those of wealthy Germany, France, the Netherlands, and Switzerland.

cities of food and other necessities are leading to social turmoil, including a rising crime rate. Although the social programs and funds that benefit EU members are appealing, it is the possibility of attracting foreign investment as EU members that is seen as a source of economic salvation. Foreign investors provide funds for upgrading technology and for creating new industries (and jobs) so that the EU countries can begin to compete in the global marketplace. Also, it is already commonplace for businesses from EU member countries to invest in the economies of East Europe. For example, the Germans and the French are active investors in Poland, the Czech Republic, and Hungary. Austrians and Italians invest in Slovenia, Croatia, and Hungary. Swedes and Finns invest in Latvia, Lithuania, and Estonia.

Standards for EU membership, however, are exacting. To be considered for membership, a country must have achieved political stability and have a democratically elected government. It must have a constitution that guarantees the rule of law, human rights, and respect for minorities. It must also have a functioning market economy and the capacity to cope with competition from within the EU, and it must be able to take on the financial and administrative obligations of membership. The EU is committed to helping countries achieve these standards, and in all countries aspiring to membership, the EU maintains commissions to help them adjust administrative structures, modify legal systems, and create banks and other financial institutions that will support a market economy. Finally, no country may become an EU member without the agreement of all 15 present members.

Poland, the Czech Republic, Hungary, Estonia, and Slovenia are scheduled to join the EU in 2004, and all are now exempt from tariffs the EU applies to non-EU countries (see Figure 4.7). Others are expected to join later in the decade. But the timely acceptance of even these five is not guaranteed. Many people in EU member countries fear that more jobs will leave the wealthier countries and go to these poorer, lower wage countries, or that skilled and educated migrants from the new Eastern European EU member states will move throughout Europe, taking jobs from locals. Some also fear that standards of all sorts will be lowered and that cheap products made in Eastern Europe will flood EU markets, competing unfairly and putting a further strain on economies. Indeed, these

and similar worries motivated the citizens of Ireland, in a plebiscite held in June 2001, to vote against letting East and Central European countries into the EU. Some East and Central Europeans also have their reservations about joining the EU, as exemplified by Hungarian sausage lover Dork Zygotian. (See the box "The Hungarian Sausage Experience and the EU.")

The idea that European unification will erase distinguishing cultural features, encourage boring homogenization, and even diminish well-being comes up again and again throughout Europe. One result has been the rise of conservative political movements, as in France, that foster a sort of "circling the wagons" mentality. In such countries as Hungary and Poland, this attitude has led to the election of some former communist leaders to positions of power.

NATO's Role in the EU

From World War II to the early 1990s, at the same time that Europe was moving toward economic union, the cold war was an overarching fact of life. To counterbalance the Soviet bloc—the USSR and its Eastern European allies—the United States, Canada, Western Europe, and Turkey formed a military alliance called the North Atlantic Treaty Organization (NATO). NATO nations cooperated militarily and learned to share authority, and this experience was an important precursor to European economic union. Now, with the breakup of the USSR, NATO has become a stabilizing institution that fosters the cooperation needed to expand the EU. Indeed, membership in NATO is considered a stepping-stone to membership in the EU, especially for countries once allied with the USSR. Three such countries—Poland, the Czech Republic, and Hungary—joined NATO in 1999 and hope to join the EU in 2004. As cold war tensions fade and as strengthening global economic and political forces bring the world closer together, NATO's role will doubtless change further. Some observers see it expanding its role as a peacekeeper to hot spots in adjacent regions, such as southeastern Europe (the Balkans), Russia (Chechnya), Southwest Asia (Israel and Palestine), and Central Asia (Afghanistan). Others see NATO's importance declining as the EU slowly develops its own military establishment to address security issues in Europe and beyond, an idea that is increasingly discussed in the EU today.

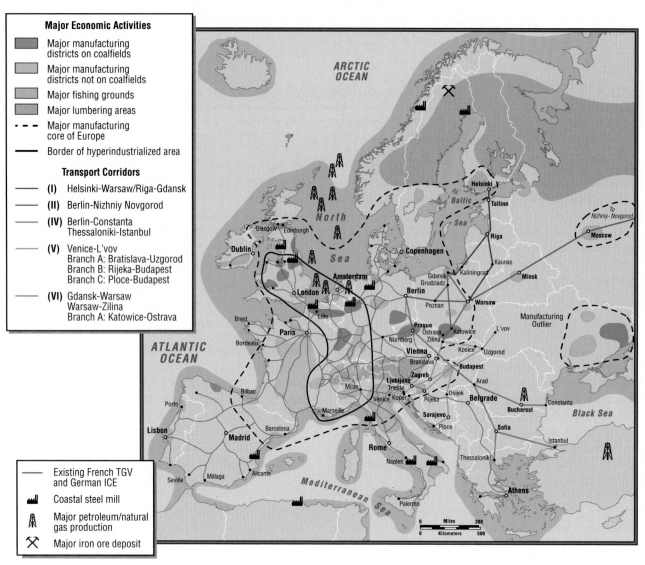

Major Economic Activities

- Major manufacturing districts on coalfields
- Major manufacturing districts not on coalfields
- Major fishing grounds
- Major lumbering areas
- - - Major manufacturing core of Europe
- ⎯ Border of hyperindustrialized area

Transport Corridors

- **(I)** Helsinki-Warsaw/Riga-Gdansk
- **(II)** Berlin-Nizhniy Novgorod
- **(IV)** Berlin-Constanta Thessaloniki-Istanbul
- **(V)** Venice-L'vov
  Branch A: Bratislava-Uzgorod
  Branch B: Rijeka-Budapest
  Branch C: Ploce-Budapest
- **(VI)** Gdansk-Warsaw
  Warsaw-Zilina
  Branch A: Katowice-Ostrava

- ⎯ Existing French TGV and German ICE
- Coastal steel mill
- Major petroleum/natural gas production
- Major iron ore deposit

Figure 4.8 **Europe's principal industrial centers. Industry is widely distributed across North and West Europe, and industrial centers are linked by transport corridors made up of roads, railroads, rivers, and canals.** [Adapted from Terry Jordan, *The European Culture Area* (New York: Harper & Row, 1988), p. 338; William H. Berentsen, *Contemporary Europe: A Geographic Analysis* (New York: Wiley, 1997), p. 164; and Terry G. Jordan-Bychov and Bella Bychova Jordan, *The European Culture Area: A Systematic Geography* (Lanham, Md.: Rowman & Littlefield, 2000), p. 300.]

The Expanding Distribution of Economic Activity in Europe

The EU powerfully influences all aspects of life in Europe as it works for progressive changes in European societies. In this section, we have included several examples that explore aspects of economic activity in Europe.

Heavy industry based on coal and steel production grew increasingly important In Europe throughout the nineteenth century. Industrial activity was concentrated in the British Midlands and along the Rhine and other major rivers of the North European Plain, close to Europe's main coal fields. In the twentieth century, industry began to diversify and spread into North and South Europe. Figure 4.8 shows these patterns. Over the past several decades, new service industries, many of which rely on advanced technology, have developed in all of Europe's major metropolitan areas. Although the EU's easing of trade and migration restrictions has done much to make industrial location more flexible, the switch away from coal to oil, gas, and nuclear power as sources of energy, and the expansion of transport systems, especially highways, have also been factors.

Europe's Service Economy. Most Europeans (about 70 percent) find jobs in the service economy, and many of the newest jobs are in information technology. Employment in the private sector is growing as Europe's economies become better integrated and new jobs are created. Employment in the public sector is also common, because the governments of European countries provide many social services to their citizens. The European Union has generated hundreds of thousands of jobs, and international organizations associated with the United Nations and headquartered in Europe also employ thousands of permanent and contract workers. In the private sector, manufacturing and service employment in telecommunications is increasing.

Telecommunications and information technology are well advanced in Europe generally, with North Europe (Finland, Sweden,

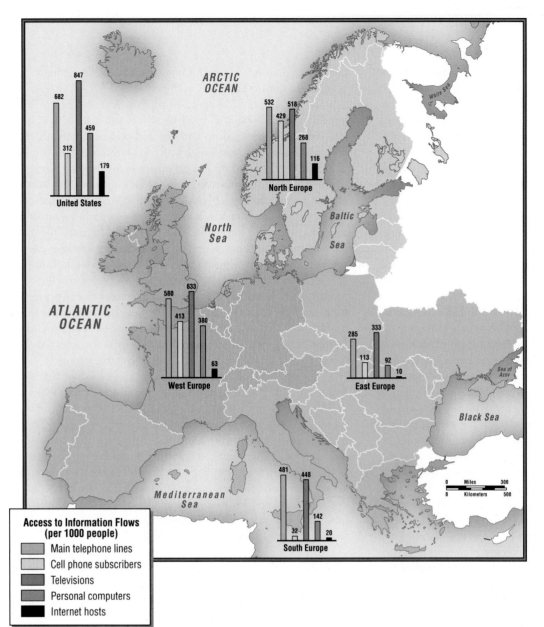

Figure 4.9 **Access to information flows: Europe.** Numbers are rounded averages for the subregions. Although Europe is behind the United States in most information access categories, cell phone subscribers (per 1000 people) in North and West Europe equal or outnumber their U.S. counterparts. Europe is also gaining in the number of Internet hosts: each subregion now has more than doubled its hosts (per 1000 people) in the last year. [*Sources: United Nations Human Development Report 2000*, Table 12; and *United Nations Human Development Report 2001*, Table A2.1.]

Access to Information Flows (per 1000 people)
- Main telephone lines
- Cell phone subscribers
- Televisions
- Personal computers
- Internet hosts

Norway, and Iceland) leading the way (Figure 4.9). Europeans provide an eager market for companies (European or not) able to supply such services. Cell phone ownership in Europe is considerably higher than in the United States. Personal computer use is also growing throughout Europe. As of 2000, however, there were fewer homes and businesses with computers in Europe than in the United States, where there were about 460 computers for every 1000 people (see Figure 4.9). Public computer facilities in cafés and libraries are common in parts of Europe where personal computer ownership is low, and the Internet is popular despite the relatively few servers per capita when compared to the United States. Overall, access to information is at a high level in Europe, with North and West Europeans having the best access and South and East Europeans having notably less access. A few countries in East and Central Europe are particularly well connected. Slovenia, for example, has five times as many computers per capita as Greece and considerably more than Spain or Italy.

Another important component of Europe's service economy is tourism. Europe is the most popular tourist destination on earth, and Europeans are themselves enthusiastic travelers, visiting one another's countries frequently. The long vacation is a European institution: most workers get at least four weeks off from the job every year, and many receive five or six weeks. In the last several decades, taking several weeklong trips has become popular, and this custom has increased the demand for urban hotels and guest houses, car rentals, and airline service. Not surprisingly, tourism is a substantial part of the European economy. One job in eight in the European Union is related to tourism, and the industry generates 13.5 percent of the EU's gross domestic product and 15 percent of its taxes.

The seaside is a favorite destination; for decades, Europeans flocked to Portugal, Spain, the south of France, Italy, the Adriatic, and Greece. These were once the poor parts of Europe where, because of low wages and cheap quarters along the sea, tourists could afford a long stay. The south of France is today one of the most posh vacation spots on earth. Spain and Italy are also becoming too expensive for many lower-middle-class Europeans. The Adriatic, once a popular destination for middle-class vacationers, has been inaccessible since the wars in Bosnia and Croatia during the 1990s. As a result, many less affluent Europeans are beginning to visit Central America and the Caribbean, especially Cuba, which by the late 1990s was attracting several hundred thousand working-class Europeans yearly.

Energy Resources. At the same time as Europe's economy has become oriented to services, its energy use has steadily moved away from coal to oil, natural gas, and nuclear power. Europe imports much of its oil and gas from outside the region, buying natural gas from Algeria and Russia and much of its oil from the Middle East. Within Europe itself, there are large oil and gas deposits in the North Sea and under the Netherlands. The North Sea deposit, controlled by Britain and Norway, now supplies much of the energy used in Germany and Britain, and the Netherlands ships gas to most of West Europe. The use of nuclear power to generate electricity has been more common in Europe than in North America—especially in France, where it accounts for 78 percent of the electricity generated compared to only 20 percent in the United States. Many European countries relied on nuclear power to achieve energy self-sufficiency, and nuclear energy remains an important resource in East and Central Europe. Nonetheless, most Western European countries are phasing out their use of nuclear energy. This movement is partly a response to public concerns about safety. Safety risks were dramatically illustrated by the Chernobyl accident in Ukraine in 1986, when an explosion sent a radiation cloud drifting across Western Europe. In addition, the desire for national self-sufficiency in energy is waning as the European Union fosters interdependence.

Building a Transportation Infrastructure. One goal of the EU has been to extend transportation networks to draw all of Europe together. Europeans have typically favored fast rail networks for both passengers and cargo rather than multilane highways for cars and trucks (see Figure 4.8). In 1994, the Eurotunnel under the English Channel, linking England and France, continued the reliance on rail transport. What was once a ferry crossing involving long waits is now a 20-minute rail trip for passengers, autos, and cargo. But even though fast rail transport remains important and new lines are planned (see Figure 4.8), there is a noticeable trend toward less energy-efficient but more flexible motorized vehicles. In the mid-1990s, leaders of the EU announced their intention to build **Corridor Five,** a major transportation roadway across southern Europe, stretching east 2000 miles (3500 kilometers) from Barcelona, Spain, to Kiev, Ukraine. It will match a similar route to the north that now extends across Germany and carries much of the east-west European trade. The completion of Corridor Five promises to change economic relationships among European countries, drawing more trade to the south and east.

Agriculture in Europe

Although only about 2 percent of Europeans are engaged in full-time farming, modern methods employing mechanization and chemical fertilizers, pesticides, and herbicides have greatly in-

creased production. As a result, most Europeans are eating as well as or better than ever before, and food is absorbing a smaller percentage of household budgets. Europeans like the idea of being self-sufficient in food, and most countries aid farmers with tariffs and **subsidies** (payments to farmers) and price supports (methods of maintaining high prices such as paying farmers not to farm land or setting legal minimum prices) on some agricultural commodities. These measures raise food costs for the consumer, but they also ensure a steady food supply. Although many countries, including the United States, Canada, and Japan, use agricultural tariffs and subsidies, they are unpopular internationally because the tariffs lock out producers from poorer countries and the subsidies encourage overproduction, causing a glut of farm products. European countries sometimes sell this overproduction cheaply on the world market. This practice, called **dumping,** lowers global prices and hurts producers of these same commodities elsewhere in the world.

Farms in Europe have tended to be smaller than those in the United States. The average European farm is 45 acres (18 hectares), less than one-tenth the average U.S. farm size of 487 acres (197 hectares). Geographer Ingolf Vogeler points out that small European family farms are disappearing, just as they did several decades ago in the United States, and the trend is toward larger, more profitable farms. In part this movement is driven by the use of expensive agricultural machines, a practice that to be profitable requires more acres in production. However, mechanization also raises living standards for the farm families who remain after consolidation. Hard labor is reduced, the workday is shortened, and greater profits can be made. Moreover, farmers have time to learn scientific techniques that improve all aspects of farming, from soil fertility to marketing.

In East and Central Europe, agriculture is developing somewhat differently. When communist governments gained power, they consolidated small, privately owned farms into large collectives. These farms were worked and managed by a group of laborers under the supervision of the state. Since the breakup of the USSR, the farms have been reclaimed by their original owners, who rent them out to large corporations. In the more open markets of present-day Europe, these larger, more profitable corporate farms may give the smaller family farms of Western Europe stiff competition, especially as Eastern European countries join the EU.

Because climates and cultural tastes differ across Europe, subregions specialize in raising particular crops and livestock. For example, the United Kingdom focuses on producing meat and dairy products partially because its wet, cloudy climate is not ideal for growing field crops such as corn or wheat but does promote the luxurious growth of grass for grazing. France is Europe's leading agricultural producer. With its diverse climates, France can turn out a wide variety of products, including grains, dairy products, fruits, vegetables, and its famous finished luxury products such as wine and fine cheese. The Netherlands, with its mild, cool climate, has some dairy farming and specializes in fine vegetables as well as flowers, seeds, and bulbs. Denmark produces fine dairy products. Farther east, colder winters and hotter summers are more conducive to wheat and cold-resistant vegetables such as potatoes, beets, and cabbages. In South Europe, the warm and dry growing season of the Mediterranean climate allows cultivation of such crops as olives and grapes as well as tomatoes, peppers, and oranges.

SOCIOCULTURAL ISSUES

Although the European Union was conceived primarily to promote economic cooperation and free trade, the economic integration that has resulted from its programs has social implications. Next we examine how the European Union is affecting attitudes toward immigrants, gender roles, and the evolution of social welfare programs.

Attitudes Toward Immigrants

Many Europeans fear that they will lose their national identity if their country joins the EU. Europe's countries have traditionally prided themselves on their cultural and social distinctiveness, and participation in the EU may threaten their identity.

Few Europeans mourn the demise of strict regulations at the borders between countries. Not long ago, border crossings meant tension-filled moments for travelers as armed border guards closely inspected passports, vehicles, and luggage, often with a rude edginess. Then, in the 1990s, the EU and many of its neighbors

GEOGRAPHER IN THE FIELD

Interview with Regional Consultant for Chapter 4, Stanley Brunn, University of Kentucky

What led you to become interested in Europe?
I have always had an interest in Europe because of its importance to an understanding of world geography and because of the contributions that Europeans have made to advances in the discipline. However, I became personally involved in European studies in the mid-1980s as a participant in geography conferences in Britain, Spain, and Belgium and as a member of the emerging IGU Study Group on the World Political Map. A two-week visit to the USSR in late 1989 triggered an interest in the rapid changes taking place in Eastern Europe and the former Soviet Union, which I followed up on in later trips. While I consider myself a latecomer to European studies, I find Europe to be a fascinating region because of the numerous changes occurring there.

Where have you done fieldwork in Europe, and what kind of work have you done?
The research I have done in the region has focused on the images, identities, and worldviews of new states, the voting patterns of nascent political parties, landscape and community changes associated with open borders, diaspora networks and communities, environmental quality and health issues, and the personal stories of those affected by war.

How has that experience helped you to understand the region's place in the world?
Historically, Western Europe has been one of the regions most connected to other parts of the world. However, what is fascinating to modern geographers is that former "closed" countries, especially in Eastern Europe, are now opening up to the rest of Europe and to other parts of the world. Having traveled throughout Europe many times during the past decade, I believe that new immigrant populations from outside the region, Europe's relationship with traditional trading partners, the growing significance of the European Community, and new digital technologies will continue to keep Europe well connected to the rest of the world.

What are some topics that you think are important to study in your region?
First, changing images and identities. Guest workers and new immigrants from southern and southeast Europe, the former Soviet Union, North Africa, the eastern Mediterranean, and Central Asia are changing the faces of countries in central and northern Europe that once had strong national majorities. These immigrants' growing presence in the workplace and marketplace, in schools, and on the playing fields is shaping political discussions and raising such issues as who may have access to health care and media. Many small towns and rural areas, and even large cities, are experiencing an influx of new religious, ethnic, and cultural faces, not to mention new language profiles. How will these continuing changes be felt by longtime residents and existing political parties? How will new residents of differing linguistic, religious, and cultural backgrounds interact with each other? One might expect that recent demographic, political, and boundary changes will result in a rewriting of the histories and geographies of both old and new states.

Second, the consequences of war. Southeast Europe has been racked by war during the past decade, and political and cultural flash points are likely to remain for some time. Contemporary geographers should not avoid examining war and conflict, even if they are unpleasant, horrific, and contentious. Specific topics to be addressed include forced and voluntary migration, resettlement of refugees, the psychosocial impacts of perpetual violence, distrust, and vengeance, the environmental consequences of war, economic reconstruction, the assistance provided by humanitarian groups, and efforts to promote community healing and reconciliation. To explore these issues, we must use not only archival data and governmental and intergovernmental reports but also the personal stories of those whose lives were directly affected by war.

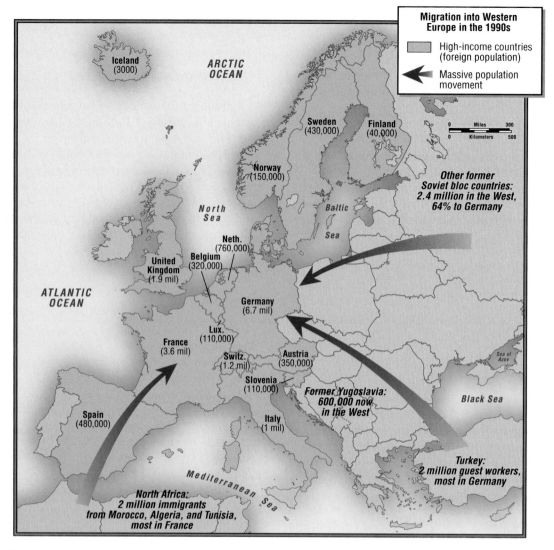

High-income countries (foreign population)

Massive population movement

ARCTIC OCEAN

Iceland (3000)

Sweden (430,000)

Finland (40,000)

Norway (150,000)

North Sea

Baltic Sea

Neth. (760,000)

Belgium (320,000)

United Kingdom (1.9 mil)

ATLANTIC OCEAN

Germany (6.7 mil)

Lux. (110,000)

France (3.6 mil)

Switz. (1.2 mil)

Austria (350,000)

Slovenia (110,000)

Spain (480,000)

Italy (1 mil)

Mediterranean Sea

Sea of Azov

Black Sea

Other former Soviet bloc countries: 2.4 million in the West, 64% to Germany

Former Yugoslavia: 600,000 now in the West

Turkey: 2 million guest workers, most in Germany

North Africa: 2 million immigrants from Morocco, Algeria, and Tunisia, most in France

Figure 4.10 **Interregional linkages: migration into Western Europe.** People flocked to Western Europe in the early 1990s. Some were seeking better lives or looking for jobs. Others were refugees fleeing oppression and ethnic cleansing. [Adapted from *National Geographic* (May 1993): 102.]

approved the **Schengen Accord,** an agreement for free movement across common borders. Now, even on crossings into Western Europe from Eastern European countries such as Slovenia, Poland, and Hungary, guards are usually scrupulously polite, uttering mild pleasantries while barely examining passports.

Relaxed borders have eased life throughout the region, facilitating trade, tourism, and social relationships. But open borders also mean that newcomers are entering European cities and towns more freely. Some immigrants are arriving from neighboring countries of Eastern Europe, others from Russia, Central Asia, North and sub-Saharan Africa, South and Southeast Asia, and the Caribbean (Figure 4.10). Some come legally as invited workers or refugees and are supported by community groups; others enter illegally. As noted in the population section previously, birth rates are declining in Europe, and these immigrants are needed to fill a wide range of jobs for which there are fewer and fewer young European workers. The taxes they pay are also welcome. Because so many come from the eastern fringes of Europe and from non-European places, however, Western Europeans are concerned that the immigrants will change long-cherished aspects of Western European culture.

The new Europeans are evident in schools, the workplace, sports arenas, and religious landscapes. Schools in Switzerland may have children from Sri Lanka, Bosnia, Algeria, Russia, China, and Iran. Workplaces in Austria and Germany may have migrants from Turkey and Iran, Ukraine, and Kazakhstan. Soccer teams may be made up of players originally from South Africa, Argentina, Zimbabwe, and Trinidad (see the box "Soccer: The Most Popular Sport in the World"). The presence of these new culture groups raises questions about identity in most countries. Is Germany no longer a German place? What history and geography curriculum should be taught in French schools, and in which languages? Questions of public welfare also arise as countries try to accommodate new value systems and customs in such social services as housing and health care.

Although many communities welcome newcomers and attempt to integrate them into their social and economic life, there are also rising fears of culture loss and dilution. Austrians worry about Italian and Slovene influence, the French worry about German influence; and the whole of Europe worries about the creeping homogenization (and also the Americaniza-

In Europe, soccer teams are a major component of national identity. They bring together citizens of vastly different cultural, racial, and class backgrounds to enjoy a common experience. Countries go to great lengths to recruit the best players for their teams, often importing them from South America, South Asia, Africa, and eastern Europe and granting them instant citizenship. Cities in Europe cultivate winning soccer teams in much the same way as cities in the United States support their professional basketball or football teams. Feelings can run so high in soccer matches that violence among the fans erupts easily.

Soccer, known as football in most countries except the United States, is a global game, played regularly by more than 240 million people, as reported in 2000 by the world soccer federation,

Fédération Internationale de Football Association (FIFA). During the 1999 women's World Cup matches played in the United States, it is possible that the entire population of the world could have seen some portion of the matches on worldwide television.

The oldest known precursors to soccer may have been kickball games played in China about 2500 years ago, but similar ball games were known in ancient times in other parts of the world (Figure 4.11). The Romans carried a version to England, where the game evolved. It emerged as the modern game of soccer in 1863, when the rules of the game were formalized. The first international match was played between teams from England and Scotland in 1872. British sailors then carried the game to mainland Europe, India, South America, and the South Pacific.

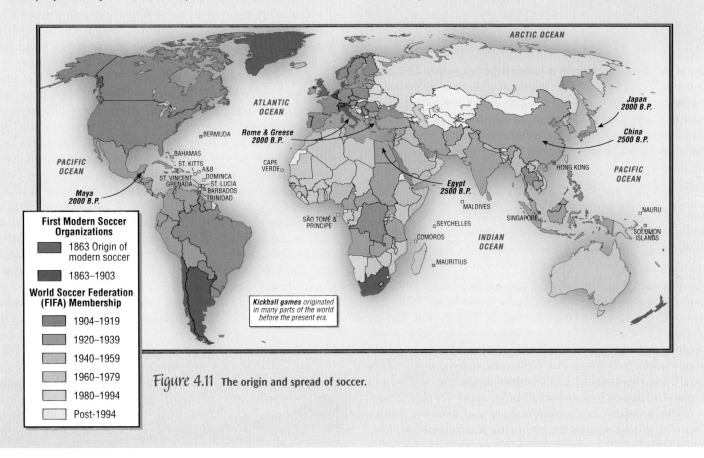

Figure 4.11 The origin and spread of soccer.

tion) of European popular culture via television, radio, and the Internet.

Guest Workers. Many immigrants were invited into Europe as **guest workers** to fill labor shortages. The hosts and guests often have different expectations of what this migration will mean in the long run. Some immigrants find that they wish to stay, while the Europeans usually expect the immigrants to earn a nest egg and then go home, as was the case with Mr. Barzey in the following vignette.

VIGNETTE In 1965, in the mountain village of Baker Hill on the island of Montserrat in the West Indies, Mr.

Barzey left his mate and young children in a tiny two-room house he had just built. With money saved from working as a laborer in cotton fields and from selling vegetables he grew in the mountains, he paid for passage on a steamer across the Atlantic. After arriving in England, he worked in the Kodak plant for 12 years without a break, sending home enough money to support his family and to feed a growing savings account in the Montserrat bank. In 1977, Mr. Barzey returned to his home island, where he settled back into a life of subsistence agriculture, a dream that had sustained him for more than a decade of long hours in a cold European industrial city. Using part of his nest

One of the continuing problems in Europe is the treatment of the Roma, or Gypsies. As in this Roma ghetto settlement in Hermanovce, Slovakia, the Roma are usually viewed with suspicion, disdain, and outright hostility. The issue is particularly acute for countries seeking admission to the European Union, for they must demonstrate improved treatment of minorities. [Tomasz Tomaszewski/National Geographic Image Collection.]

egg, he built several additions to his home, including a kitchen, bath, and walled flower beds. He christened the additions "Kodak" in honor of the company that had made possible his relative prosperity.

Many guest workers left much more difficult home situations than Mr. Barzey's; and some of the most recent fled from civil wars and no longer have homes to return to. Like the vast majority of European immigrants who permanently immigrated to North America from the seventeenth century on, many of these people hope to make their homes in a Western European country, raise a family, and eventually become accepted as citizens. The map in Figure 4.10 depicts migration into Europe in the 1990s.

The European host countries have very different policies regarding immigrant workers. Some, like the Netherlands, have lenient laws that allow immigrants from former colonial territories to stay indefinitely and to receive subsidized housing and other social welfare services equivalent to those available for citizens. France provides less adequate housing blocks in informally segregated working-class neighborhoods, and the atmosphere is often hostile. In Germany, Austria, and France, minority right-wing parties now openly speak of forcing immigrants to leave, but a wider movement favors accommodation and more acceptance.

Citizenship. Until recently, achieving legal citizenship in a European country has been very difficult for outsiders, especially those of non-European heritage. Germany, a country with a particularly large group of immigrants (in the late 1990s, 8 percent of its 82 million people were foreign born), has had especially stringent rules. Even if they were born in Germany, the children and grandchildren of Turkish or North African immigrant workers were

not considered citizens, and most were rejected if they applied for citizenship. In 1994, of the more than 2 million people of Turkish origin who lived in Germany and who ran more than 30,000 businesses, fewer than 1 percent had acquired citizenship. But attitudes and policies about German citizenship are changing. As of January 2000, all children born in Germany are citizens, and people born elsewhere who have lived in Germany for eight years can apply to become citizens. The law is retroactive to 1975.

France's citizenship laws are somewhat less generous than Germany's, although it has long been a culturally diverse country. The United Kingdom has absorbed many hundreds of thousands of West Indians, Indians, Pakistanis, and Africans from former colonies. Today, the United Kingdom is debating how to define the several hundred thousand citizens of its few remaining colonies. It is likely that they will be given British citizenship eventually. Many Europeans can now trace their ancestry to China, Nigeria, Thailand, Brazil, or Turkey, and these demographic changes are causing countries to redefine what it means to be a European.

Rules for Assimilation. In Europe, race and skin color play less of a role in defining the differences between people than does culture. An immigrant from Asia or Africa may be fully accepted into the community if he or she has adopted European ways. This is especially true for immigrants skilled in the host country's language and for those who have mastered the finer points of its manners and decorum. Yet certain European minorities that have been in Europe for thousands of years, such as the Basques in Spain or the Roma (Gypsies), who reside in many countries, will find it nearly impossible to blend into society if they retain their traditional ways. In Europe, **assimilation** usually means a comprehensive change of life. Immigrants give up the culture of home—language, dress, family relationships, food, customs, mores, and even religion —and take up instead the ways of their adopted country.

The geographer Eva Humbeck studied how Thai women who come to Germany explicitly to marry German men grapple with

Turkish migrants relax in Berlin's Tiergarten, the city's central park. A soccer game is under way. By the mid-1990s, one of every eight residents of Berlin was a foreigner. [Gerd Ludwig/National Geographic Image Collection.]

the demands of assimilation. Since 1990, about 1000 Thai brides have arrived in Germany every year. Although they willingly take on the role of housewife as opposed to career woman, they usually live isolated in apartments and sorely miss their female relatives—who, in Thai culture, are a woman's best friends. Most feel obliged to drop all Thai ways of doing things because they sense that Germans view Thai culture as inferior. Few Thai wives know one another; and many confess to daydreaming about the Thai hospitality and family conviviality they once enjoyed.

Note that the degree of adjustment expected of immigrants in Europe differs from that expected of immigrants in North America. In North America, the norm is **acculturation** (enough adaptation to the host culture for members of the minority culture to function effectively and be self-supporting), not total assimilation. In the United States and Canada, immigrants are expected to learn the language and to abide by the laws of the land. Canada encourages immigrants to retain their own languages, provided they learn English. Overall, immigrants are free to keep their native cultures and cuisine. Those who retain native dress and particularly exotic customs may encounter a certain amount of informal prejudice or even discrimination, but they will also find that the laws protect their right to be different and that some American customs—international street festivals featuring ethnic food and music, for example—actually encourage and reward a measure of cultural retention.

The changes in citizenship laws in Europe may herald a growing acceptance of non-Europeans. Cross-cultural marriage is increasingly common, and the multiethnic offspring of these marriages will undoubtedly influence their generation's attitudes. Furthermore, official EU agencies are encouraging public dialogue through frequent forums on cultural integration held at the local and international levels. The European Union is drawing up guidelines for the legal acceptance of cultural diversity by prospective EU members from East and Central Europe. This project has forced western Europeans to reflect on their own record as well.

European Ideas About Gender

Gender roles in Europe have changed significantly from the days when most women married young and worked in the home raising large families and tending to various agricultural duties. Table 4.1 lists several criteria for evaluating the changing role of women in European society. A large percentage of Europeans live in cities (73 percent), marry late, and have only one child. Increasing numbers of European women are working outside the home (Table 4.1, column 3). In all but a few European countries, women are seeking advanced training more often than men (Table 4.1, column 2)—an indication that European women may eventually exceed men in job qualifications.

Despite these changes, there are still distinct differences between men and women both in the public perception and in people's day-to-day lives. As is the case nearly everywhere on earth, Europeans generally accept the notion that men and women have innate abilities that determine the functions they should perform. European public opinion among both women and men holds that women are less able than men to perform the types of work typically done by men and that men are less skilled at domestic duties. These views continue to influence European social institutions. In most cases, men have greater social status, hold more managerial positions, earn higher pay (Table 4.1, column 4), and have greater autonomy in daily life (more freedom of movement across space, for example) than do women. Despite these gender differences, women increasingly work outside the home as factory laborers, service workers, teachers, textile producers, architects, physicians, and professors.

Often, women who work outside the home are also expected to do most of the domestic work, a situation called the **double day.** United Nations research shows that throughout Europe, women's workdays (including time spent in housework and child care) are five to nine hours longer than men's. Women burdened by the double day will generally operate with somewhat less efficiency on a paying job than do men, and they tend to choose employment close to home that offers more flexibility in the hours and skills required. These more flexible jobs almost always offer lower pay and less chance for advancement, but not necessarily fewer work hours, than typical male jobs.

Political Representation of Women. Despite strong EU emphasis on policies encouraging gender equality, the political influence and economic well-being of European women lags far behind that of European men, as Table 4.1 and Figure 4.12 illustrate. In

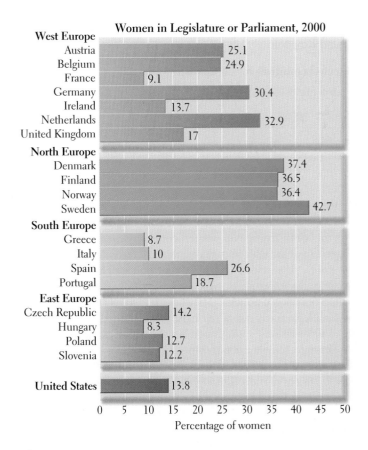

Figure 4.12 Women legislators. Percentage of women legislators in selected European nations (with U.S. comparison). [Data from *United Nations Human Development Report 2000,* United Nations Development Programme.]

TABLE 4.1 *Gender statistics from selected countries*

Country (1)	Education: advanced training, female students as % of male students, 1994–1997 (2)	Work: female income-earning economic activity rate as % of male activity rate, 1998[a] (3)	GDP per capita by gender, 1998 (PPP[b] $U.S.) (4)		Female administrators as % of total (most recent year available) (5)
			Female	Male	
Belgium	96	64.5	15,951	30,801	30.2
Denmark	114	83.6	19,965	28,569	23.1
France	116	75.2	16,437	26,156	N/A[c]
Germany	80	68.7	15,189	29,476	26.6
Greece	89	56.8	8,963	19,079	22.0
Hungary	104	71.5	7,452	13,267	35.3
Iceland	142	85.4	22,062	28,127	25.4
Italy	130	57.3	12,665	28,982	53.8
Netherlands	91	65.2	14,902	29,600	22.8
Norway	126	83.1	22,400	30,356	30.6
Poland	123	79.3	5,821	9,519	33.6
Romania	109	76.1	4,169	7,178	26.4
Spain	108	54.9	9,636	23,078	32.4
Sweden	124	88.8	18,605	22,751	27.4
United Kingdom	103	73.2	15,290	25,575	33.0
Selected countries for comparison					
United States	121	79.9	22,565	36,849	44.4
Japan	76	66.4	14,091	32,794	9.5

[a]Refers to paid work in the formal economy only.
[b]PPP = purchasing power parity.

[c]N/A = data not available.
Source: United Nations Human Development Report 2000, Tables 2, 3, 28, 29.

France, only 9 percent of elected representatives in parliament are women, and in the Czech Republic, only 15 percent are women. Only in North Europe do women come anywhere close to filling 50 percent of the seats in parliament or the legislature. Women are also poorly represented in government bureaucracies. They serve mostly in the lower ranks, which gives them little voice in the formation of national policies. They have had to rely on men to push for such legal measures as the right to work, equal job opportunities, and equal wages and fringe benefits. Progress has been slow. For example, female unemployment in the EU is 152 percent that of men, and 23 percent of women work part-time, as opposed to only 4 percent of men. Throughout Europe, women are paid less on average than men.

One might question why, given Europe's overall prosperity, women have not achieved more equality with men. The explanation is that although EU policies have officially ended legal discrimination against women, policy implementation lags. Also, some analysts point out that European women face certain obstacles to meeting together locally or internationally: geographic separation and differences in language, ethnicity, nationality, education level, and social background. Hence they have found it difficult to develop solidarity. This situation is changing as the EU opens up entrepreneurial options for women and encourages international conferences on women's interests. It is noteworthy that despite political underrepresentation, women in Europe hold a greater proportion of elected positions than do women in any other world region.

As Table 4.1 and Figure 4.12 illustrate, the role and status of European women vary greatly from country to country and from subregion to subregion. In the following discussion, this variability emerges in the differing approaches to social welfare in Europe.

Social Welfare Systems

In the United States, the term *welfare* refers to organized tax-supported or private services to the poor or disabled. In Europe, the term **social welfare** describes elaborate, tax-supported systems

Welfare Systems
- Social Democratic
- Conservative
- Former Communist
- Modest
- Rudimentary

Figure 4.13 European welfare systems. Several scholars have sought to categorize European countries according to their approaches to welfare. This map and the accompanying discussion are derived from several such studies and should be taken only as an informed approximation of the patterns.

that serve all citizens in one way or another. Social welfare is more widely accepted in Europe than in the United States, and such programs include many services to middle-class citizens. (Although some of these middle-class benefits exist in the United States and Canada—for example, social security, pension plans, tax-deferred savings programs, and mortgage tax deductions—there they are not considered to be welfare per se.) Europeans generally pay much higher taxes than Americans, and they expect a wide range of services and safety nets in exchange. Still, Europeans do not agree about the goals of these welfare systems, or about just how generous they should be. As Figure 4.13 illustrates, the various European countries differ in their approaches to welfare, and these differences have become a source of conflict in the European Union. The problem arises because unequal benefits may encourage those in need to flock to a country with a generous welfare system and overburden the taxpayers there. Conversely, a country with a less generous welfare system could find its best and brightest workers emigrating to more generous countries.

Using the work of the British sociologist Crescy Cannan and others, we have classified European welfare systems into five basic categories (see Figure 4.13). Each category makes certain assumptions about gender roles. All assume that the typical family has two parents and that the male is the breadwinner, even though in most parts of Europe the number of such families is decreasing. Whether married with children or single, most women take employment outside the home to support themselves and their families.

1. **Social democratic welfare systems** are strong in all of Scandinavia, but especially so in Sweden and Iceland. These systems attempt to achieve equality across gender and class lines by providing generous health, education, housing, and care benefits to all citizens from cradle to grave. Citizens pay high taxes to provide an environment, from prenatal care on, in which every child's physical and social potential is realized. Child care is widely available, not just to help women enter the labor market but also to provide the

state with a mechanism for instilling such values as polite behavior, good nutrition, the work ethic, and good study habits. Ideally, at adulthood, every citizen will be able to contribute to his or her highest capabilities, and social problems will not develop. Although gender equality is a stated goal, traditional gender roles are officially emphasized throughout this lifelong support system, and in the workplace and public forums women are not yet equal to men (see the wage gap indicated in Table 4.1 and the discussion on page 224).

2. **Conservative welfare systems,** such as those found in France, Germany, the Netherlands, Austria, and Switzerland, have a stated goal of ensuring just a minimum standard of living for all citizens. Hence, the state intervenes to assist those in need, but it does not see its mission as one of assisting upward mobility. For example, the state would not normally make a college education accessible to the children of poor families; although a university education is often free, there are strict entrance requirements that the poor find hard to meet. On the other hand, many state-supported services, such as health care and pension plans, are available to all economic classes. Like other public institutions in these countries, the welfare system reinforces the so-called housewife contract by assuming that women will stay home and take care of children. For example, the half-day school schedules and the lack of school lunches practically compel mothers to stay at home. Those mothers who do work outside the home choose part-time work or jobs with flexible hours. In France, women are likely to work full time, and there is support for the equality of men and women in the workplace. This view of equality does not extend to domestic duties, and the gender pay gap persists (see Table 4.1).

3. **Modest welfare systems,** presently found only in the United Kingdom, seek to encourage individual responsibility and the work ethic. This type of welfare system has been evolving in the United Kingdom since conservative reforms in the 1980s and 1990s, and it is roughly similar to the welfare system in Canada (the U.S. system is less generous). Benefits are modest, just enough for those who qualify to maintain a minimally adequate standard of living. Recipients are often stigmatized in jokes and conversation as being lazy and "on the dole." Welfare is thought to encourage dependency, whereas the ideal citizen is completely self-reliant. The state claims no direct interest in the quality of daily life of its citizens; how people live is a matter of individual choice in a free market. Women are free to work outside the home or not; but the state supports neither option. Critics point out that in reality, women cannot enter the labor market on the same terms as men. For example, women in the U.K. earn only about 70 percent of what men earn and occupy the lowest paying factory jobs and the lower echelons of the business and professional sectors—yet women are considered the proper, unpaid caretakers of children and the elderly. Only minimal state-supported child and elder care is provided, and its availability is unpredictable.

4. **Rudimentary welfare systems** are found primarily in South Europe—Portugal, Spain, Italy, and Greece—and in Ireland. These countries accept the idea that citizens do not have inherent rights to government-sponsored welfare support. Local governments provide some services or income for those in need, but the availability of such services varies widely even within one country. The traditional extended family and community are still common in these countries; and the state assumes that when people are in need, their relatives and friends will intervene to provide needed services, financial support, and care for the young, old, and disabled. The state also assumes that women kin are available to provide daily child care and other social services for free, despite the fact that in South Europe and Ireland women actually form the bulk of the flexible, low-wage workforce. Finally, there is the official assumption that the large, informal economy will provide some sort of employment for anyone who needs it. Some of these attitudes are now changing, especially in Italy, which experienced an economic boom in the 1990s. Young urban Italian women are very likely to work full time outside the home, and they are more likely to have managerial positions than men; nevertheless, as yet their average income is less than half that of Italian men (see Table 4.1).

5. **Communist welfare systems** prevailed in the countries of the old Eastern bloc allied with the USSR (Poland, the Czech Republic, Slovakia, Hungary, Romania, Bulgaria), and countries formerly part of Yugoslavia (Slovenia, Croatia, Bosnia-Herzegovina, Serbia, Macedonia, Montenegro). These systems were, in theory, comprehensive. Although the bureaucracy administering benefits was inefficient and the programs unevenly funded, the ideal resembled the cradle-to-grave social democratic system in Scandinavia, except that women were pressured to work outside the home. Benefits often extended to nearly free apartments and to meals provided on the job at nominal cost to the worker. In the 2000s, in all the former Eastern bloc countries, most women continue to work outside the home and accept nearly full responsibility for the home and all domestic duties. The double day is especially taxing for women in this part of Europe because labor-saving devices are not yet widely available: laundry is often done by hand, kitchens are rudimentary, food must be purchased daily because the small refrigerators cannot hold more than a day's food. The big change is that in the postcommunist era, state funding of these welfare systems has collapsed and people must cope with the loss of apartments, free meals, and other benefits of the welfare safety net once provided by the state at the same time as jobs are being lost. In the transition to the market economy, many people are going without such basic necessities as food, heating fuel, and health care; there is an increasing temptation to migrate to places where the basics are still provided.

As the European Union evolves and expands, Europe's social welfare systems will probably become more similar to one another, but it is still unclear just which models will prevail. Some observers suggest that the dominance of the German economy in Europe could mean that the conservative German model for welfare will spread. Others think that the entry of Scandinavian countries into the EU will inspire a shift to more active welfare policies with more equal treatment of women. There is also the possibility that state-provided benefits will shrink to the level of that in the modest welfare system in the U.K. as the EU adjusts to compete in global markets.

ENVIRONMENTAL ISSUES

Europe's environment has changed dramatically over the past 10,000 or more years, a result of human impact. Nearly all the original forests are gone; some have been gone for more than a thousand years (Figure 4.14). Many of Europe's seemingly natural landscapes are largely the creation of humans who have changed the landforms, the drainage systems, and the vegetation cover repeatedly over time. An extreme case is the Netherlands, where almost no natural landscapes are left. Environmental issues in Europe often focus less on setting aside pristine natural areas than on establishing livable environments for future generations.

Public awareness about the environment has increased over the past 30 years, and in all European countries there are **Green**

(environmentally conscious) political parties that influence policies at the national level as well as within the EU. In many ways, European lifeways result in lower resource consumption than elsewhere in the developed world, especially North America. Europeans live in smaller spaces, their yards often contain vegetable gardens rather than great expanses of mowed lawn, their cars are more fuel efficient, public transportation is widely used, and people walk or bike to many of their appointments. Nevertheless, these practices are related more to high population density and social customs in the region than to widespread explicit support for Green principles. Despite the successes of Green politicians, their concerns consistently take a backseat to issues of economic growth and social welfare. Europe's air, seas, and rivers remain some of the most polluted in the world.

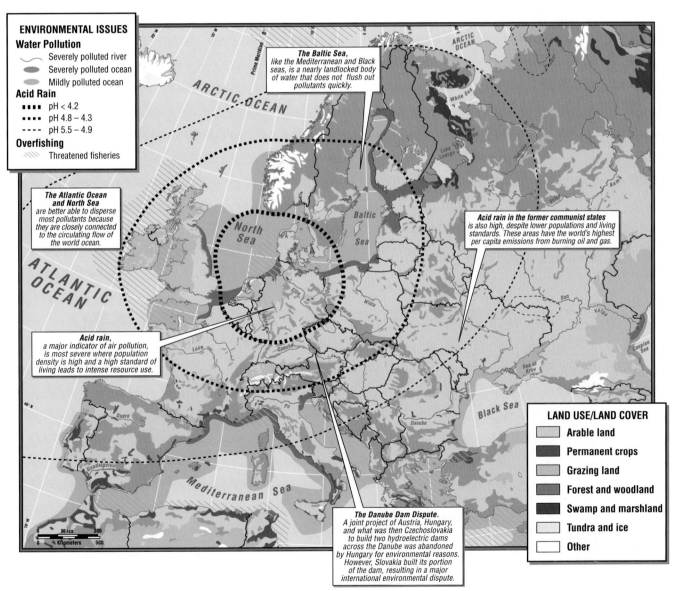

Figure 4.14 Environmental issues: Europe. Environmental issues in Europe encompass air and water pollution, disappearing and deteriorating natural habitats, and the unsustainable use of natural resources. The Mediterranean and Baltic seas exhibit particularly serious pollution. Acid rain is severe in northern Germany and Denmark. (pH 1 = high acidity; pH 7 = neutral.)

Pollution of the Seas

Europe is surrounded by seas—the Baltic Sea, the North Sea, the Atlantic Ocean, the Mediterranean Sea, and the Black Sea—and any pollutants that enter Europe's interior rivers, streams, and canals eventually reach these surrounding seas (see Figure 4.14). The Atlantic Ocean and the North Sea are able to disperse most pollutants dumped into them along European shores, because they are closely connected to the circulating flow of the world ocean. On the other hand, the Baltic, Mediterranean, and Black seas are nearly landlocked bodies of water that do not have the power to flush themselves out quickly: all three are prone to accumulate pollution. The surrounding countries are at different stages of development, which makes it difficult for them to cooperate on solutions.

Municipal and rural sewage (75 percent of it untreated), eroded sediment, agricultural chemicals, industrial waste, nuclear contamination, and oil spills are pouring into the Mediterranean Sea from adjacent lands. Many of the chemical pollutants add nitrogen to the water, which causes vast tracts of algae to bloom. The algal blooms render the sea environment inhospitable to natural life-forms and make some much-beloved seaside resorts unsafe for swimmers. Figure 4.15 shows the geographic pattern of Mediterranean algal pollution. Pollution in the Mediterranean is exacerbated by the fact that it takes about 80 years for a complete exchange of water with the Atlantic Ocean that could flush out pollutants. Atlantic seawater flows in through the narrow Straits of Gibraltar and moves eastward, evaporating as it goes. As a result, the water at the far eastern end of the Mediterranean is more saline and heavier than the water entering from the west. This heavier water sinks, flows back westward at a lower depth, and finally exits the Mediterranean at Gibraltar, many decades later. The ecology of the Mediterranean is attuned to this lengthy cycle,

but the balance has been upset by the 320 million people now living in countries surrounding the sea.

Relatively rich and industrialized Europe shares the Mediterranean Basin with relatively poor North Africa and Southwest Asia. Although most of the water pollution is generated by Europe, populations to the south are increasing rapidly, and soon their sheer numbers will also pose an environmental threat. In 1995, in an effort to address the water pollution problems, the EU inaugurated a cooperative movement with its southern neighbors to decide what kind of development is feasible for the welfare of the Mediterranean region. The developing countries of North Africa and the eastern Mediterranean, however, argue that they are being asked to make sacrifices for the environment that Europe was unwilling to make during its development era. They say that Europe has sounded the alarm only after its people have substantially polluted the water.

The Damming of Europe's Rivers

Europe's rivers are home to the region's oldest and largest cities, which today use the rivers both as a depository for municipal waste and as a source of energy. Many of Europe's rivers have hydroelectric dams built across them. It has been difficult to manage the use and cleanup of these rivers rationally because Europe is divided into small countries and many rivers run through or form the border of several different countries. The Danube River is a particularly interesting environmental case study. It rises in the Black Forest of southern Germany and flows east and south past Vienna, Budapest, and Belgrade, passing through or forming the border of seven countries before emptying into the Black Sea. Before the fall of the USSR, Austria, Hungary, and what was then Czechoslovakia had agreed jointly to build two hydroelectric dams across the river; these dams would have worked in tandem

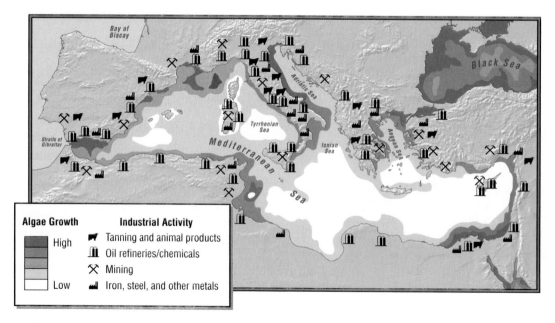

Figure 4.15 Sources and types of pollution in the Mediterranean Sea. [Adapted from *The Economist Atlas of the New Europe* (New York: Henry Holt, 1992), p. 210.]

to flush huge volumes of water through electricity-generating turbines twice daily. The project was pushed ahead without an accurate calculation of the environmental impacts along what had been a relatively unspoiled part of the river.

Then the political climate in Eastern Europe shifted, and in Hungary, remarkably, a new reformist government withdrew from the project, dismantled the newly built Hungarian portion of the dams, and even partially restored the landscape. Meanwhile, Czechoslovakia had split into two countries, the Czech Republic and Slovakia. Slovakia persevered in the project by diverting the Danube and sending it through a canal and turbines. The original riverbed has dried up, and species that once occupied the river's wetlands are dying. Agricultural chemicals are polluting a reservoir created upstream of the Slovak dam and are destroying its aquatic life. Although Hungary sued Slovakia in the World Court, the court eventually refused to make a judgment, ruling that the two countries would have to settle the case between themselves. Hungary argues that shared resources, such as rivers, can be exploited only by mutual consent, and that pollution of a common resource constitutes a breach of human rights. Slovakia cites customary European law that gives upstream users water rights over downstream users, a legal concept that has led to many a Wild West movie plot in the United States and to tensions in several parts of the world where this European concept of water rights applies.

Air Pollution in Eastern and Central Europe

Although there is significant air pollution across much of Europe, it is particularly heavy in eastern Europe. Figure 4.14 shows that acid rain, a major indicator of air pollution, is most severe over parts of Europe where population density and a high standard of living lead to intense resource use. However, the map also shows acid rain as a threat in areas of eastern and central Europe where population is not as dense and living standards are much lower. According to geographer Brent Yarnal, this high level of pollution exists in former communist states of East and Central Europe because they have the world's highest per capita emissions from burning oil and gas. Until they started to make the transition to market economies, these countries used energy at per capita rates averaging 50 to 150 percent higher than those in the United States (one of the world's largest per capita users of energy). In Poland in the late 1980s, air quality in the industrial centers was so poor that young children were relocated to rural environments to preserve their health. Reformers hoped that market economies would improve energy efficiency and reduce emissions, and in Bulgaria, the least efficient and most polluting of the lot, pollution did decrease briefly after 1990. But this happened only because of economic decline; by the mid 1990s, pollution levels were up again.

Yarnal found that in Bulgaria and elsewhere there is likely to be a long lag between the reduction of energy use and a reduction in air pollution. Reformist groups that might push for environmental improvements are not well organized, and antipollution activists can mobilize little support for regulations that might slow economic growth. Nevertheless, heavy pollution causes declines in human health, lower crop yields, and loss of water quality and

quantity. In the early 1990s, the World Bank had some success in addressing these problems when it encouraged five countries in eastern Europe—Bulgaria, the Czech Republic, Slovakia, Hungary, Poland, and Romania—to reduce fossil-fuel use by cutting government subsidies that lower the costs of fossil fuels to consumers and encourage wasteful use.

Why are environmental problems so serious in the countries of East and Central Europe? Part of the reason may be the Marxist theories and policies promoted by the USSR, which portrayed nature as existing only to serve human needs. The Soviet leader Joseph Stalin once said, "We cannot expect charity from Nature. We must tear it from her." Another reason may be that, with only superficial democratic institutions in place during the Soviet era, there was little opportunity for public outrage at pollution to be channeled into constructive political activism and change.

MEASURES OF HUMAN WELL-BEING

VIGNETTE The Parvanov family lives in a small apartment outside Sofia, Bulgaria. The family appears to have the material wealth of the middle class: adequate food, a nice home, a car in the garage. But looks are deceiving: these are largely leftover benefits of the communist economy. As the country adjusts to the free-market economy, inflation is high and Mr. Parvanov's salary is now worth only U.S. $30 a month, so he moonlights on a construction job. Mrs. Parvanov holds three part-time jobs: she works as a clerk and as an assistant in a building company, and at night she machine-knits clothes she sells in the local market. Theodora, the Parvanov's 19-year-old daughter, attends university by day and helps her mother knit at night. Yet with all this effort (six jobs in all), the family earns only about U.S. $100 a month. With this tiny amount, they must also support two grandparents whose pensions will no longer cover food, let alone medicine. The car hasn't been driven in months because gasoline is too expensive. They are able to survive only because the Bulgarian government still heavily subsidizes the costs of food, rent, and other basic necessities, a situation that will not last much longer. [Adapted from Robert Frank, "Price of isolation: impoverished Bulgaria is ready for reform," *Wall Street Journal* (February 28, 1997): 1.]

The governor of the central bank in Bulgaria, Lubomire Filipov, says that although Bulgaria likes to think of itself as European, its rate of inflation, mounting debt, and poverty mean that, except for Albania and possibly Yugoslavia, Bulgaria is now Europe's poorest country. We have already observed that there is considerable disparity in wealth across Europe. Table 4.2 compares the well-being of people like the Parvanovs with Europeans in 19 other countries and with people outside the region, in the United States, Japan, and Kuwait. The three indexes are those introduced in Chapter 1 and used throughout this book.

In the category of gross domestic product per capita (Table 4.2, column 2), you can see that most countries in North and West Europe compare favorably with the United States, Japan, and

TABLE 4.2 *Human well-being rankings of countries in Europe*

Country (1)	GDP per capita, adjusted for PPP[a] in 1998 $U.S. (2)	Human Development Index (HDI) global rankings, 2000[b] (3)	Gender Empowerment Measure (GEM) global rankings, 2000 (4)
Selected countries for comparison			
Japan	23,257	9 (high)	41
Kuwait	25,314	36 (high)	N/A[c] in 2000 (75 in 1996)
United States	29,605	3 (high)	13
North Europe			
Denmark	26,342	15 (high)	4
Estonia	7,682	46 (high)	27
Finland	20,847	11 (high)	5
Iceland	25,110	5 (high)	2
Latvia	5,728	63 (medium)	25
Lithuania	5,102	52 (medium)	29
Norway	26,342	2 (high)	1
Sweden	20,659	6 (high)	3
West Europe			
Austria	23,166	16 (high)	12
Belgium	23,223	7 (high)	10
France	21,175	12 (high)	N/A in 2000 (31 in 1998)
Germany	22,169	14 (high)	6
Luxembourg	33,505	17 (high)	N/A in 2000 (14 in 1998)
Netherlands	22,176	8 (high)	8
Switzerland	25,512	13 (high)	14
United Kingdom	20,336	10 (high)	15
South Europe			
Greece	13,948	25 (high)	49
Italy	20,585	19 (high)	31
Portugal	14,701	28 (high)	18
Spain	16,212	21 (high)	19
East and Central Europe			
Albania	2,804	94 (medium)	N/A
Bosnia-Herzegovina	1,700[d]	N/A	N/A
Bulgaria	4,809	60 (medium)	N/A in 2000 (43 in 1998)
Czech Republic	12,362	34 (high)	26
Hungary	10,232	43 (high)	42
Macedonia,Teyr	4,254	69 (medium)	N/A
Poland	7,619	44 (high)	36
Romania	5,648	64 (medium)	58
Slovakia	9,699	40 (high)	28
Slovenia	14,293	29 (high)	33
Yugoslavia (Serbia, Montenegro)	2,300[d]	N/A	N/A
World	6,526	NA[e]	NA

[a]PPP = purchasing power parity.
[b]The high and medium designations indicate where the country ranks among the 174 countries classified into three categories (high, medium, low) by the United Nations.
[c]N/A = data not available.

[d]2000 estimate (PPP), *CIA Factbook 2001.*
[e]NA = data not applicable.

Source: United Nations Human Development Report 2000.

Kuwait. The only exceptions are Estonia, Latvia, and Lithuania, separated from the Russian sphere in the early 1990s. But South Europe lies in the middle range of GDP per capita, and countries in East and Central Europe are way down the scale, falling well below those of several of the Middle and South American countries, for example (see Table 3.3, page 154). Yet despite these low GDP per capita figures, until recently in East and Central Europe there has not been great disparity in wealth between classes, because the communist system specifically sought to even out the distribution of wealth. As a result, GDP figures for this region are a somewhat more meaningful measure of real well-being than they are in Middle and South America and elsewhere, where class disparities are huge.

The United Nations Human Development Index (HDI) (Table 4.2, column 3) combines three components—life expectancy at birth, educational attainment, and adjusted real income—to arrive at a ranking of 174 countries that is more sensitive to more factors than just income. In many parts of Europe, state subsidies keep prices of necessities low and make salaries go further, as had been the case for the Parvanovs in Bulgaria. All countries in North, West, and South Europe (with the exception of Estonia, Latvia, and Lithuania) rank fairly high on the global HDI. Iceland, Norway, and Sweden stand out, undoubtedly because these countries have comprehensive social welfare systems. In contrast, East and Central Europe ranks relatively low because, as the communist system failed, most state-run firms cut jobs, social welfare programs went bankrupt, and environmental quality declined. People lost not only their jobs, but even health care, housing, and food subsidies. Those retaining their jobs saw their salaries rendered meaningless pittances as a result of high inflation rates—as did Mr. Parvanov in Bulgaria. Air and water pollution increased, daily stress mounted, overall quality of life declined drastically, and life expectancies plummeted.

The Gender Empowerment Measure (GEM) (Table 4.2, column 4) ranks countries by the extent to which females have opportunities to participate in economic and political life. It evaluates how women are employed in the economy and the extent to which women participate in society, especially as elected officials. The GEM figures show that North and West Europe, with the notable exceptions of Estonia, Latvia, and Lithuania, do better than most countries in empowering women. Estonia, Latvia, and Lithuania rank low in comparison because they were long part of the Soviet bloc and retain many customs and institutions that restrict women. In South Europe, the GEM ranks are lower than those of North and West Europe. Spain, Portugal, and Italy are changing rapidly as they enjoy an economic boom. Women who had long been overeducated for the lower echelon jobs they held have been encouraged to take new, more responsible positions; hence, the GEM rankings for these countries have risen and are likely to rise further. Nevertheless, Greece has one of the lowest ranks in Europe. In much of East and Central Europe, useful statistics have not been collected, in part because of underfunded statistics departments and in part because the problem of gender inequity is not recognized. Specific studies report that now, compared to men, a disproportionate number of women in East and Central Europe are unemployed and sinking into poverty (see pages 228–229).

SUBREGIONS OF EUROPE

In this section we look at the subregions of Europe as separate units, examining in greater detail some of the issues that affect them. The subregion in which a particular country is included was decided based on convention and on our best judgment of the situation in Europe at this time. Good arguments for other choices could be made. One could, for example, place the United Kingdom and the Republic of Ireland in North Europe; or Estonia, Latvia, and Lithuania, placed here in East and Central Europe, could be included in Chapter 5 with the Russian Federation.

Every subregion of Europe is undergoing changes; these changes result from the end of cold war politics, the struggles of East and Central European countries to establish national identities, moves toward greater economic and political cooperation throughout the region, and global shifts in economic power.

WEST EUROPE

Despite their economic success, the countries of West Europe (Figure 4.16)—the United Kingdom, the Republic of Ireland, France, Germany, Belgium, Luxembourg, the Netherlands, Austria, and Switzerland—confront several issues that will affect their development and overall well-being. These include

1. The continuing trend toward economic and political union

2. The resolution of long-standing cultural and religious conflicts, such as the one in Northern Ireland, and of recent strife between host countries and immigrant groups

3. The persistence of high unemployment despite general economic growth

4. The need to increase the global competitiveness of Europe's agricultural, industrial, and service sectors

Benelux

Often called the Low Countries or Benelux, Belgium, the Netherlands, and Luxembourg have achieved very high standards of living despite their dense populations. The three countries are well located for trade: they lie close to North Europe and the British Isles and are adjacent to the very active Rhine Delta and the industrial heart of Europe. The coastal location of Benelux and its great port cities of Antwerp, Rotterdam, and Amsterdam give these countries easy access to the global marketplace. Benelux countries have long played a central role in the European Union; Brussels, Belgium, is

Figure 4.16 The West Europe subregion.

considered the EU capital, and most EU headquarters are located there. The EU activity has drawn other business to Brussels.

International trade has long been at the heart of the economies of Belgium and the Netherlands. Both were active colonizers of tropical zones: Belgium in Africa; the Netherlands in the Caribbean and Southeast Asia. Their economies benefited from the wealth extracted from their colonies and from global trade in tropical products such as spices, chocolate, fruit, wood, and minerals. Private companies based in Benelux still maintain advantageous relationships with the tropical areas, which supply raw materials for European industries. The Benelux countries occasionally find themselves embroiled in conflicts related to both the colonial past

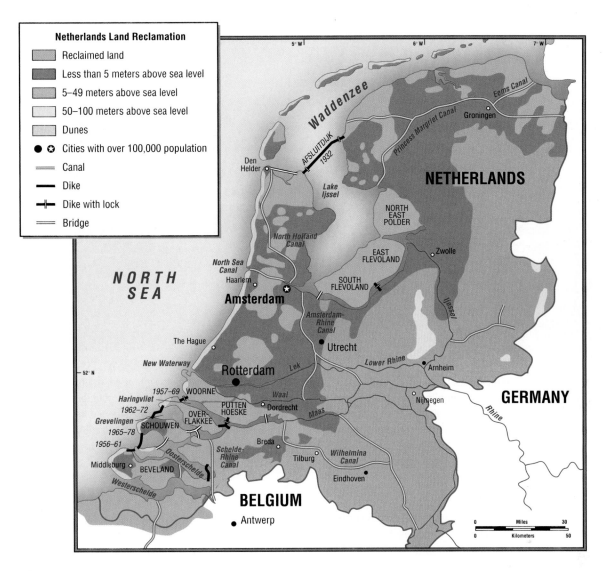

Netherlands Land Reclamation

- Reclaimed land
- Less than 5 meters above sea level
- 5–49 meters above sea level
- 50–100 meters above sea level
- Dunes
- ● ✪ Cities with over 100,000 population
- ═══ Canal
- ▬▬ Dike
- ▬┼▬ Dike with lock
- ═══ Bridge

Figure 4.17 Land reclamation areas in the Netherlands. [Adapted from William H. Berentsen, *Contemporary Europe: A Geographic Analysis* (New York: Wiley, 1997), p. 317.]

and pressures to integrate with their neighbors in the new EU era; the European "chocolate standards war" is a case in point.

The Chocolate Standards War. In Europe, chocolate is an important luxury product that sells at very high prices. Godiva, for example, a brand loved by American chocoholics, is a Belgian company. Chocolate first enters Europe in the form of dried cacao beans, which are grown mostly in Africa. Belgium, France, and six other EU nations agree that real chocolate (so labeled) must be made only of cocoa butter derived from cacao beans, with no oil or fat additives. But the United Kingdom, Denmark, and several other countries routinely add cheaper oils to their mass-produced candy. The EU Commission, charged with defining common manufacturing standards for the entire EU, is trying to reconcile the different national chocolate traditions. Some contend that with open borders, the cheaper chocolate made with oil additives will have an unfair advantage over pure chocolate both in the EU and abroad. Globally, six multinational firms, controlling well over half of all chocolate sales, side with the United Kingdom in arguing that chocolate recipes should not be standardized. The purists, on the other hand, argue that European chocolate will be more competi-

tive in global markets if it is a gourmet product that maintains the highest standards of purity.

This European chocolate debate has global implications: West African countries that produce cacao beans favor high standards of purity and the prohibition of other oils. Ghana, Cameroon, and Côte d'Ivoire are former European colonies that depend on cacao for at least a quarter of their export earnings. They estimate that the use of even as little as 5 percent of non-cocoa oils in the EU will mean a 10 percent drop in world cacao consumption and will devastate their economies. These three African countries have petitioned their former colonizers (Belgium, France, and Britain) to maintain the pure chocolate standards as part of agreements they made to help African economies rebound from colonial exploitation.

The Netherlands, a Human-Made Place. The largest Benelux nation, the Netherlands, is particularly noted for having reclaimed land that was previously under the sea (Figure 4.17). One result is that today its landscape is almost entirely a human construct. As populations grew during and after the Middle Ages, people created more living space by filling in a large natural

215

These two photographs of the Netherlands, taken from the air, show intensive land use and human manipulation of natural land and water interactions in coastal zones. The photograph at left shows part of the port of Rotterdam, looking northwest. Cranes transfer products between river vessels and oceangoing ships. The oil "tank farms" figure prominently in Dutch chemical industries. The photograph at right shows fields of flower bulbs on raised, drained lands close to the coast. Economically, the flowers, which are sold locally, are merely a by-product of the bulbs, which are marketed internationally. Dunes along the coast, planted with trees for stabilization, are the last defense against the sea. [Aeroview, Rotterdam.]

wetland. To protect themselves from devastating North Sea surges, they built dikes, dug drainage canals, pumped water with windmills, and constructed artificial dunes along the ocean. Today, a train trip through the Netherlands between Amsterdam and Rotterdam takes one past raised rectangular fields crisply edged with narrow drainage ditches and wider transport canals. The fields are filled with commercial flower beds of crocuses, tulips, daffodils, and hyacinths (see the photograph at the right on this page); there are also grazing cattle and vegetable gardens that feed the primarily urban population. Despite the almost complete absence of pockets of wilderness, the landscape has an open, bucolic ambience. In one vista, a train traveler can see huge cranes in the distance at the port of Amsterdam. On one side are high-tech office buildings and a new satellite town. In the foreground, one might see an elderly woman pushing her bicycle beside a sparkling blue, reed-lined canal, the rear basket stuffed with leeks and flowers.

The Netherlands has 15 million people, who enjoy a high standard of living. Yet it is the most densely settled country in Europe, and there is no land left for suburban expansion without intruding on agricultural space. People now travel great distances to reach their jobs, and long commutes add to the air pollution and miserable traffic jams. There is not nearly enough space for recreation; bucolic as they appear, transport canals and carefully controlled raised fields are not a venue for picnics and soccer games.

France

France has the shape of a rough hexagon bounded by water on the north and west, mountains on the southwest and southeast,

Mediterranean beaches on the south, and lowlands adjacent to Belgium on the northeast (see the map at the beginning of this chapter). The capital city of Paris lies in the north of this hexagon. It is the heart of the cultural and economic life of France, and it attracts worldwide admiration.

Paris is often called the City of Light because of the elegance of its architecture and urban design and its reputation as a center of culinary and cultural eminence. France's population is concentrated in Paris: together with its suburbs, the city has more than 11 million inhabitants, roughly a fifth of the country's total. It is also the hub of a well-integrated water, rail, and road transport system. Its central location and transport links attract a disproportionate share of trades and businesses. There is a steady stream of workers to firms that manufacture luxury and high-fashion items, to high-tech and automotive industries, and to distributors and processors of France's fine farm produce. In the 1970s, however, French planners decided that Paris was large enough, and they began diverting development to other parts of the country, especially Toulouse and farther south to the Mediterranean.

To the northeast of Paris, close to Belgium, is an older industrial area around Lille that has been in decline for some time. Some of its steel, electronics, and aluminum companies have gone out of business, but others have survived and become more efficient.

To the west and southwest of Paris, toward the Atlantic, are less densely settled lowland basins that are used primarily for agriculture. Here, though farmers grow wheat and other crops, grapes are an important crop. The Bordeaux region is a leading exporter of French wines. France has the largest agricultural output in the

EU and is second only to the United States in agricultural exports. Its climate is mild and humid, though drier to the south. Throughout the country, farmers have found profitable specialties: wheat (ranked fifth in world production), grapes (used to make wine), cheese (second in world production), fruit, olives, sugar beets and other vegetables, and sunflower seed oil (third in world production). Despite its leadership in agricultural production, agriculture accounts for only 8 percent of the national economy and employs only 5 percent of the population. Industry accounts for more than 30 percent and the service sector for 60 percent.

The Mediterranean coast is the site of France's leading port, Marseille, and of the famous French Riviera. Here, development is booming. Tourism, in particular, has inspired the building of marinas, condominiums, and parks that threaten to occupy every inch of waterfront. This densely populated region produces large amounts of the pollutants that contribute to Mediterranean environmental problems (see the discussion on page 210).

France derived considerable benefit and wealth from a large overseas empire, which it ruled until the mid-twentieth century. There were French colonies in the Caribbean, North America, North Africa, sub-Saharan Africa, Southeast Asia, and the Pacific. Many of the citizens of these former colonies live in France and bring to it their skills and a multicultural flavor. The French take pride in being cosmopolitan, but they are also unusually conscious of their distinctive European culture and wish to protect what they consider to be its purity and uniqueness.

Few would quarrel with the idea that French culture has a certain cachet and that France is an arbiter of taste. People of many different cultural backgrounds accept the idea that the French language is particularly refined. French style as exemplified in clothing, jewelry, perfume, and interior design is highly pursued, and French culinary arts remain among the most cultivated of cooking traditions. Its many cultural attributes have made France the leading tourist destination on earth, with 11 percent of global tourism arrivals. In 1999, France attracted 73 million international tourists, more than the population of the entire country (59.4 million). Other Europeans, Americans, Turks, and Japanese form the majority of tourists in France.

France's strong cultural identity is both a blessing and a curse. Many people in France are deeply concerned about what they perceive as a decrease in France's world prestige, about rising unemployment and falling social welfare benefits, about cultural adulteration from France's former colonies, and about a decline in French industrial competitiveness. Such feelings have resulted in the increasing popular support of the National Front, France's xenophobic right-wing party. But others in France are searching for a new way to think about themselves in a now much wider world—a potentially more egalitarian global community they helped to create.

Germany

The most famous image of Germany of the last several years is that of a happy crowd dismantling the Berlin Wall in 1989. This important symbolic end of the Soviet sphere of influence in East and Central Europe had particular significance for Germany, which for 40 years had been divided into two unequal parts. Russian troops occupied the smaller, eastern third of Germany after World War II; in 1949, the Soviets and their German counterparts created East Germany. Over the next decade or so, perhaps as many as 3 million East Germans migrated to West Germany. To retain the remaining population, in 1961 the Soviets literally walled off the border. East German skilled labor, mineral resources, and industrial capacities were used to buttress the socialist economies and support the military aims of the USSR and its allies.

Because Germany was regarded as the perpetrator of two world wars, West Germany had to tread a careful path after 1945: seeing to its own economic and social reconstruction and rebuilding a prosperous industrial base (see Figure 4.8), yet not seeming to become too powerful economically or politically within the European community. For the most part, West Germany played its complex role successfully. Since the early 1980s, it has been a leader in building the European Union, and it has borne the greatest financial burden of the European unification process.

After the fall of the Berlin Wall in 1989, the two Germanys were reunited. But East Germany came home with a suitcase of troubles. Its industries were outdated; inefficient; polluting; and, in large part, not redeemable. Its infrastructure of roads, bridges, dams, hydroelectric plants, nuclear energy plants, waterways, and buildings did not meet the standards of West Europe. They had to be remodeled or dismantled and replaced. Its industrial products were not competitive in world markets, and East German workers, though considered highly competent in the Soviet sphere, were undereducated by West European standards.

Reunified Germany is Europe's most populous country (82 million people) and is a leader globally and in Europe in industry and trade; but the costs of absorbing the poor eastern zone have dragged Germany down in many rankings within Europe. In the 1990s, unemployment rose sharply (especially among women) and then declined to 9.3 percent in 2000. Not only have there been layoffs in the east (where unemployment is nearly double the national average), but German workers, accustomed to high wages and lavish benefits, are losing jobs as firms move overseas, some to the United States. The German multinational corporation Daimler-Chrysler, headquartered in Stuttgart, is an example of a German firm that has moved operations out of Germany. In the late 1990s, Daimler-Benz bought Chrysler, the American auto company based in Michigan, and built a Mercedes-Benz plant in Tuscaloosa, Alabama. In 2000, it announced plans to expand the Alabama plant and add 2000 jobs. Globally, DaimlerChrysler has 467,000 employees (241,000 in Germany, 124,000 in the United States, 12,200 in Mexico City) and manufacturing plants in 37 countries.

The British Isles: The United Kingdom and the Republic of Ireland

The British Isles, located off the northwest coast of the main European peninsula, are occupied by two countries: (1) the United Kingdom of Great Britain (England, Scotland and Wales, and Northern Ireland), often called simply Britain, or the United Kingdom; and (2) the Republic of Ireland. The Republic of Ireland (not to be confused with Northern Ireland) was once a colony of Britain, and the two countries have long held contrasting positions in the world. Britain is a powerful industrialized country that has experienced a decline in its global influence and wealth. Ireland was a once poor agricultural country that now has one of Europe's fastest growing economies.

Ireland. In the seventeenth century, Protestant England conquered Catholic Ireland; England removed or killed many of the indigenous people and settled Lowland Scots and English farmers on the vacated land. For nearly 300 years, the remaining Irish resisted with guerrilla warfare until the Republic of Ireland gained independence from the United Kingdom in 1921. However, six counties in the northeast corner of Ireland with Protestant majorities remained part of the United Kingdom and became known as Northern Ireland. Here Protestants held political control and the minority Catholics experienced economic and social discrimination. Catholic nationalists unsuccessfully lobbied for a united Ireland by constitutional means. Other Catholic groups, the most radical being the Irish Republican Army (IRA), resorted to violence, often against British peacekeeping forces, who were seen as supporting the Protestants. These tactics were reciprocated by Protestants. More than 3000 people had been killed in the Northern Ireland dispute by 1995.

A peace accord was finally reached in 1998. The opposing groups were tired of seeing the development of the entire island—especially of Northern Ireland—blighted by the persistent violence. In May 1998, voters in Northern Ireland and the Republic of Ireland voted overwhelmingly to approve the peace accord (71 percent in Northern Ireland, more than 90 percent in the Republic of Ireland). Voter approval is a major step forward, even though problems of implementing the accord remain.

The Republic of Ireland has not participated in the violence of Northern Ireland; but poverty has plagued the country for centuries. The only physical resources in the Republic of Ireland are its soil, abundant rain, and beautiful landscapes. This combination has contributed to a continued dependence on agriculture and tourism, while industrialization has lagged. As a result, Irish people were for a long time the poorest in West Europe; and over the last two centuries some 70 million Irish migrated to find a better life. In the 1990s, however, a remarkable turnaround began. Ireland attracted foreign manufacturing companies by offering cheap, well-educated labor; accessible air transport; access to EU markets; low taxes; and other financial incentives. These foreign firms (such as Glaxo, Merck, Norsk Hydro, and Phillips Petroleum) have brought in small industries specializing in, for example, training software; food and drink processing; and the manufacture of pharmaceuticals, chemicals, and high-end giftware and textiles. The economy has grown so quickly that Irish labor is now in short supply. Some 200,000 workers are required, and the Irish diaspora is being invited back, while foreign workers are recruited in the Americas and South Africa. Already, Czech workers are packing meat, and there are Filipino nurses in most Irish hospitals. The cost of living has risen sharply in Ireland, however, and this could affect the country's ability to attract yet more foreign investment. Ireland does not favor the admission of Central and East Europe countries into the EU because these countries could attract investment away from Ireland, by offering even cheaper labor pools.

The United Kingdom. The United Kingdom, in contrast to Ireland, has been operating from a position of power for many hundreds of years. Like Ireland, the United Kingdom has a mild, wet climate and a robust agricultural sector—based, in its case, on grazing animals. The usable land is extensive, and the mountains contain mineral resources, particularly coal and iron. Beginning in the seventeenth century, however, Britain no longer depended on just its own resources. After colonizing Ireland, Britain extended its empire to the Americas, Africa, and Asia. These colonies gave Britain access to enormous resources of labor, agricultural products, minerals, and timber. These foreign resources plus its own were sufficient to make Britain the leader of the industrial revolution in the early eighteenth century. By the nineteenth century, the British Empire covered nearly one-quarter of the earth's surface. As a result, British culture was diffused far and wide, and English became the lingua franca of the world (displacing French).

Britain's widespread international affiliations, set up during the colonial era, positioned it to become a center of international finance as the global economy evolved. Today, London is Europe's leading financial center, and it handles 31 percent of the global foreign currency exchange, more than twice that of New York City, its closest rival. But although the United Kingdom remains Europe's financial center, it is no longer Europe's industrial leader. At least since World War II, and some say before, the United Kingdom has been sliding down from its high rank in the world economy toward the position of an average European nation. British manufactured goods no longer compete as well with goods from other world regions where labor and production costs are cheaper. The discovery of oil and gas reserves in the North Sea gave the United Kingdom a cheaper and cleaner source of energy for industry than the coal on which it had long depended, but it did not stop economic decline. Cities such as Liverpool and Manchester in the old industrial heartland, and Belfast in Northern Ireland, experienced long depressions.

Beginning in the 1970s, a series of conservative governments tried to make the United Kingdom more competitive by instituting two decades of budget cutbacks, reducing social spending, selling off unprofitable government-run firms, and tightening education budgets. Unemployment rose. Eventually, foreign investment began to pour into the United Kingdom because the skills of the labor force were high compared to the wages people would accept after years of cutbacks.

Like Ireland, Britain has now successfully established technology industries in parts of the country that previously were not industrial, for example, near Cambridge and west of London in a region called Silicon Vale. But service industries now dominate the economy. There are new jobs in health care, food services, sales, financial management, insurance, communications, tourism, and entertainment. Even the movie industry has recently discovered that England is a cheap and pleasant place for film editing and production. Britain's past experience with a worldwide colonial empire has left it well positioned to provide superior financial and business advisory services in a globalizing economy; but many of these new jobs are not highly paid. The jobs tend to go to young, educated, multilingual city dwellers, many of them from outside the United Kingdom, and not to middle-aged, unemployed ironworkers and miners left in the old industrial heartland. Because the EU allows the free movement of EU citizens, there are now more than 400,000 foreign nationals working in Greater London alone. Thirty-five percent of the J. P. Morgan brokerage house and bank staff are non-British.

Politically, the United Kingdom consists of the countries of Scotland, Wales, England, and Northern Ireland. Until recently, it was governed by a single parliament with delegates from the four countries, but England, with the largest population by far, held 85 percent of the votes. In national debates, the will of the English majority usually prevailed. In 1997, Scotland and Wales voted to have their own elected governing bodies—a parliament for Scotland and a national assembly for Wales. This decision for home rule is seen as an example of **devolution,** the weakening of

"Mixed relationships are more accepted here [in London] than in the U.S.," says Joanne Evans, as she spends Sunday afternoon with her friend, Neil Williams. Overall, Britain is a far more integrated society than America. Marriage often occurs across ethnic lines, and 74 percent of white people in England say they would not mind a close relative marrying someone of another color or ethnicity. [Jodi Cobb/National Geographic Image Collection.]

a formerly tightly unified state. It springs both from dislike of the retrenchment policies of recent U.K. conservative governments —which brought high unemployment and loss of social programs to both Scotland and Wales—and from rising feelings that the wishes of the voters in both countries were consistently overwhelmed by the English majority in the British parliament. It remains to be seen whether this change will result in the eventual independence of Scotland and Wales from the U.K. or merely the delegating of some powers to what amount to subassemblies within the U.K. Developments within the EU regarding allocation of political power and development funds will affect this situation in the U.K.

The Jamaican poet and humorist Louise Bennett used to joke that Britain was being "colonized in reverse." She referred to the many immigrants from the former colonies who now live in the United Kingdom. Indeed, it would be hard to overemphasize the international spirit of the U.K. London, for example, has become an intellectual center for debate in the Muslim world. Salman Rushdie, the controversial Indian Muslim writer, lives there; many bookstores carry Muslim literature and political treatises in a variety of languages; Israelis and Palestinians privately talk about peace there; and the Saudi Arabian government has a strong presence. On a leisurely stroll through London's Kensington Gardens, you will discover thousands of people from all over the Islamic world who now live, work, and raise their families in London. They are joined by people from the British Commonwealth: Indian, Malaysian, African, and West Indian families pushing baby carriages and playing games with older children. Impromptu soccer games may have team members from a dozen or more countries (see the box "Soccer: The Most Popular Sport in the World," page 203).

SOUTH EUROPE

The Mediterranean region of South Europe was once unrivaled in wealth and power. First Athens, in Greece, and then Rome, in Italy, were imperial seats whose influence extended widely. In the medieval period, from the fifth to the fifteenth centuries, the Italian cities of Venice, Florence, and Genoa were major trading centers with contacts stretching across the Indian Ocean through Central Asia to eastern China. During the sixteenth century, Spain and Portugal developed large colonial empires, primarily in the Americas but also in Africa and Southeast Asia. But times changed. As Europe's own economy became more global with the development of colonies in distant world regions, the Mediterranean was no longer a center of trade. South Europe declined. On the whole, the industrial revolution reached South Europe only recently, and even today, much of the region remains poor in comparison to the rest of the continent. In Europe, only the countries of East and Central Europe are poorer (see Table 4.2, page 212). In this section, we focus on Spain and Italy, contrasting the parts of each country that now are prospering with the parts that have remained poor (Figure 4.18).

In all four countries of the region—Portugal, Spain, Italy, and Greece—agriculture has long been the predominant occupation. Farmers often lived in poverty, using simple tools to produce crops for local consumption. Since the 1970s, however, there has been an effort to modernize and mechanize agriculture across the

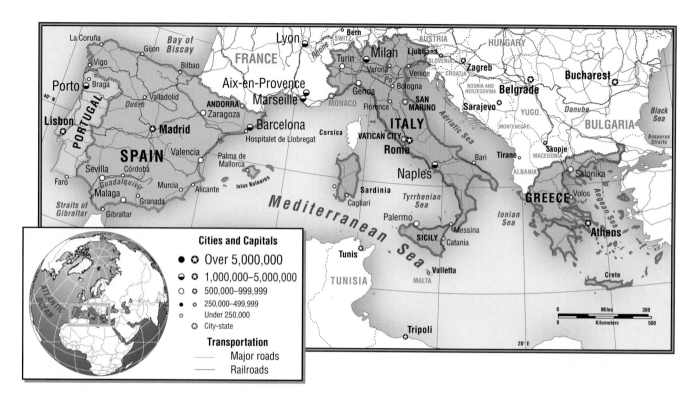

Figure 4.18 The South Europe subregion.

subregion. The focus has been on irrigating large holdings to grow commercial crops. Irrigation has increased output and opened up new semiarid locations for cultivation, but not everyone welcomes the abundance of produce flowing from South Europe. Today, both Spain and Italy mass-produce vegetables and fruits in quantities that exceed demand in the European Union. The surplus has driven down prices, which has hurt smaller family farmers across Europe and created ill feelings toward the European Union (see the vignette about Mrs. Kusmic on page 228).

As in the western United States, irrigation schemes result in water shortages, reservoirs and natural aquifers become depleted, and soil salinity increases over time. Farmers in northern Spain and Portugal complain because water originating there is transported by canal to irrigate more arid regions in the south. And despite extensive improvement projects, agriculture as a whole (measured as a percentage of GDP and as a percentage of labor force) has declined in importance as manufacturing and tourism have increased.

For several decades, middle-class tourist money has brought new life to the economy of South Europe. Masses of western and northern Europeans flock to see the monuments of past civilizations and to enjoy the south's warm climate and beautiful beaches. But tourist money has done little to elevate the lives of poor rural people. Although some localities have grown prosperous as resort sites, many have been bypassed. In Greece, for example, income from tourism has been confined to scenic coastal and island locations, and many interior rural areas remain poor and isolated. As in much of South Europe, the very qualities that attract tourists to Greece are being threatened by the sheer numbers of visitors and the impact they have on coastal environments and on rural culture.

Spain

In the last quarter of the twentieth century, Spain emerged from centuries of underdevelopment. Today, it is poised to take advantage of opportunities in a more integrated Europe, and its future is bright. Just 40 years ago, however, Spain was known as the Old Man of Europe. By the beginning of the twentieth century, Spain's once vast empire in the Americas and Asia was all but gone. This loss was followed by years of civil war and then a military dictatorship led by Francisco Franco. Even Spain's physical location was no longer an advantage, as commercial activity had shifted away from the Mediterranean to north central Europe. Mountain ranges slicing through the Iberian Peninsula kept Spain isolated from the more prosperous parts of Europe, making travel and commerce difficult.

In 1975, Spain made a cautious transition to democracy, and in 1986 it joined the EU. Since then, growth has accelerated with the help of EU funds aimed at bringing Spain into economic and social harmony with the rest of the EU through investment in infrastructure and in human resources. As a result of these improvements, foreign businesses are investing in industry and factories have been modernized. Barcelona (Catalonia) and the city of Madrid are now centers of population, wealth, and industry.

Nonetheless, some segments of Spanish society have not participated in the new prosperity. Many small farmers remain poor,

AT THE LOCAL SCALE   *The Moorish Legacy in Southern Spain (Andalucía)*

Muslim traders and conquerors (known as Moors) came to Spain in the 700s and remained until the 1400s. They deeply influenced all aspects of Spanish life: language, music, settlement patterns, architecture, principles of governance, and gender practices.

Tourists from all over the world are drawn to the many remains of Moorish culture in the landscape of southern Spain. Muslim tourists from Southwest Asia often comment about how "at home" they feel in the dry hills and walled courtyards of Andalucía. The palaces and mosques of Córdoba and Granada's Alhambra rank among the world's greatest architectural marvels. Strong Moorish influences can be heard in flamenco, the world-famous music of Andalucía's Roma (Gypsies), and Spanish contains countless Arabic words and sayings. The elaborate multigalleried home built to face inward around a central courtyard garden is a Moorish contribution. Its purpose was to house the extended family and to shield the women of the family from the public spaces (and gazes) of men, an important practice in Islamic culture.

Courtyard of the Parador de San Francisco in the Alhambra, Granada, Spain. [Mark D. Phillips/Photo Researchers.]

particularly in the northwestern and southwestern provinces, which lack water and have few resources. Even in the relatively prosperous zones of Catalonia and Madrid, workers have been laid off as a result of factory modernizations. In 2000, unemployment hovered around 15 percent, and the unemployed workers were willing to work for significantly lower wages than were paid in North and West Europe.

The low wage scales are one reason that Spain is attracting foreign investment from around the world. Other attractions include the now-stable democratic government, an educated and skilled workforce (partly the result of EU funds for training), and the absence of stringent environmental and workplace regulations. Additionally, Spain has good ports on both the Atlantic and the Mediterranean. Spain now has an industrial base that includes food and beverage processing and the production of automobiles, chemicals, metals, machine tools, and textiles. An interesting indicator of growth is Spain's increasing investment in its former colonies in Latin America. By 1997, Spanish banks had lent more than U.S. $7 billion to Cuba, Mexico, and various countries in South America. Between 1996 and 2000, Spanish corporations invested $14 billion in South America. Spain's telecommunications giant, Telefonica, holds more than U.S. $5 billion in Latin American telecommunications companies.

Regional disparities of wealth contribute to high levels of ethnic rivalry. Two of Spain's most industrialized and wealthy regions—the Basque country in the central north (adjacent to France) and Catalonia in the east (also bordering France)—have especially strong ethnic identities, and their inhabitants often speak of secession (Figure 4.19). They would like to avoid supporting Spain's poorer districts. A Basque separatist movement periodically resorts to violence. Meanwhile, Galicia in the far northwest corner of Spain also has a strong cultural identity, and the preservation of the Galician language is a popular issue. All three of these ethnic enclaves emphasize their distinctive languages as markers of identity and chafe against the use of Castilian Spanish as Spain's official language.

Thinking Globally: ON THE WEB

Figure 4.19 **The contentious cultural regions of the Iberian Peninsula. Both Basque and Catalan settlements extend into southern France, as indicated by the striped pattern.** [Adapted from Terry Jordan, *The European Culture Area* (New York: Harper & Row, 1988), p. 188.]

public buildings and empty factory hulks (called "cathedrals of the desert") dotting the landscape attest to the difficulties it has encountered. The money for development has been dissipated by corrupt officials, misguided policies, and a poorly skilled labor force.

Perhaps the most pervasive aspect of daily life in Italy is the informality of social and economic arrangements that sometimes extend into national politics. A parent contacts a teacher, an old friend, to make sure a child passes the upcoming exams; a factory worker punches the time card of a perpetually late friend in return for the free use of a large hall for his daughter's wedding reception. These kinds of arrangements permeate Italian society, making the country both inefficient and apparently corrupt in the larger sense and yet somehow warm, workable, and secure on a personal scale. In Italy, possibly as much as 20 to 25 percent of the economic activity takes place in the informal economy, much of that on the black market.

Italians seem to have limited faith in the effectiveness of government; they have elected several dozen different administrations since 1948. Italy has been a democracy since World War II, and it is a charter member of the EU. Yet it has been mired in bureaucracy for so many decades that a body based on informal relationships, the *sottogoverno*, or "undergovernment," is said to take over whenever a crisis develops. The infamous Mafia has permeated Italian society most deeply in southern Italy, where the absence of

Italy

Italy is by far the wealthiest country in the subregion and is, in fact, the fifth largest economy in the world. Nevertheless, huge disparities exist between the rich and dynamic north and the poor and stagnated south. In Roman times, the situation was reversed: the south—and especially the island of Sicily—was the most progressive and productive area in the Roman Empire, whereas the Po Valley in the north was an uninhabited, mosquito-infested swamp. But the people of northern Italy slowly and steadily converted the Po Valley into a productive agricultural region. By the thirteenth century, the northern trading center of Venice on the Adriatic had commercial links reaching around the Mediterranean and into Africa, Arabia, the Malabar Coast of India, and all the way to China. In the early twenty-first century, northern Italy has one of the most vibrant economies in the world, producing products renowned for their quality and beauty of design: Ferrari automobiles, Giorgio Armani clothing, Olivetti office equipment, and high-quality musical instruments. In Turin, Fiat automobiles produced in a fully robotic plant are shipped throughout Europe; in Milan, one of the largest cities in Italy, fashion designers have surpassed even their rivals in Paris.

South of the city of Naples, Italy, is a land of rural poverty. Archaic agricultural practices, steep slopes, deforestation, and an arid climate have made the south less and less productive over time. The government has made episodic efforts to promote infrastructure development and industry, but the unfinished roads and

Nestled within 10 miles of Mount Vesuvius, a volcano overdue for an eruption, is Naples. Many people consider this city of more than a million inhabitants to be the gateway to southern Italy. The beautiful, vibrant city, founded by Greeks in 600 B.C. as Neapolis, has many of the problems of the south: violence, drugs, and corruption. Yet it also houses what is perhaps Italy's finest opera house, luxurious palaces and gardens, and the Piazza del Plebiscito. This central square overlooking the Bay of Naples is the background for thousands of wedding portraits for southern Italy's brides. [David A. Harvey/National Geographic Image Collection.]

effective government is conspicuous. Only in the last few decades has the Mafia been recognized as something that people could combat collectively.

Many of these peculiarities of Italy seem to arise from an inherent distrust of any institutions not based on the family or on personal relationships. Most businesses are family-run, including some of the large corporations of northern Italy; as discussed earlier in this chapter (see page 208), the rudimentary welfare system found in Italy and other countries of South Europe assumes that strong families and social networks will provide what the state cannot.

NORTH EUROPE

North Europe (Figure 4.20) traditionally has been defined as the countries of Scandinavia—Iceland, Denmark, Norway, Sweden, Finland—and their various dependencies. Greenland and the Faroe Islands are territories of Denmark; the island of Svalbard in the Arctic Ocean is part of Norway. In this book; we expand North Europe to include the three Baltic states of Estonia, Latvia, and Lithuania, all of which were part of the USSR until September 1991. These small Baltic states have many remaining links to Russia, Belarus, and other parts of the former Soviet Union, but they are attempting to reorient their economies and societies in varying degrees to the West.

The countries of North Europe are linked by their location in the North Atlantic, the North Sea, or the Baltic Sea and by their intimate connections with the sea. Most citizens can drive to a seacoast within a few hours. The main cities of the region—Copenhagen, Stockholm, and Helsinki—are vibrant ports that have long been centers of shipbuilding, fishing, and the warehousing of goods. They are also home to legal and financial institutions related to maritime trade.

Scandinavia

The southern parts of Scandinavia are where most people live and where economic activity is concentrated. The warmer flatlands of southern Scandinavia are dedicated to agriculture. Denmark,

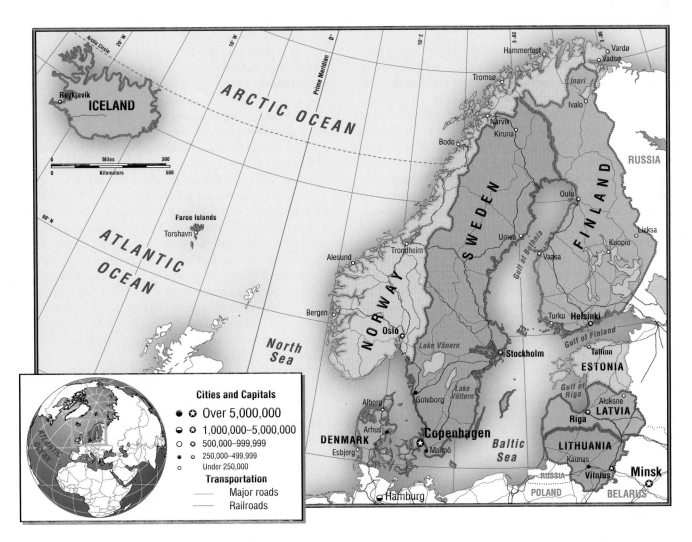

Figure 4.20 **The North Europe subregion.**

Even apartment complexes, such as the Tokarp development in the south central town of Jönköping, reflect the Swedish sense of modernist elegant simplicity. [Tomasz Tomaszewski/National Geographic Image Collection.]

although a small country, produces most of North Europe's poultry, pork, dairy products, wheat, sugar beets, barley, and rye. Despite Denmark's agricultural successes, two-thirds of the Danish economy is nonagricultural, consisting of the service sector (financial services and education, for example), high-tech manufacturing, construction and building trades, and fisheries. Sweden is the most industrialized nation in North Europe. The Swedes produce most of North Europe's transport equipment and two highly esteemed automobiles, the Saab and the Volvo, as well as furniture and housewares famed for their spare, elegant design. Helsinki, Finland, is home to some of the world's most successful information technology companies.

The northern part of Sweden and almost all of Norway are covered by mountainous terrain, while Finland is a low-lying land dotted with glacial lakes, much like northern Canada. Sweden and Finland contain most of Europe's remaining original forests. These well-managed forests produce timber and wood pulp. Norway, which has less forestland, had been considered poorly endowed with natural resources in comparison with the rest of Scandinavia, but the discovery of gas and oil under the North Sea in the 1960s and 1970s has been a windfall to that country. The exploitation of these resources dwarfs all other aspects of Norway's economy. Norway is able to supply its own energy needs through hydropower, and it exports oil and gas to the EU as well. It is now one of Europe's richest countries, with a high level of human well-being (see Table 4.2, page 212).

The fishing grounds of the North Sea, the North Atlantic, and the southern Arctic Ocean are also an important resource for North Europe countries (as well as for several other countries on the Atlantic). In recent decades, however, overexploitation has severely reduced fishing stock. Right of access to the fishing grounds remains a cause of dispute; but in 1994, the EU created a joint 200-mile (320-kilometer) coastal exclusive economic zone (EEZ)

that ensures equal access and sets fishing quotas for member states. In January 2001, EU members plus Norway and Iceland agreed to an annual three-month hiatus in cod fishing and in the taking of juvenile fish of all species to feed salmon in commercial farms. The agreement applies to a 40,000-square-mile (103,600-square-kilometer) swath of the North Sea lying between the British Isles, Denmark, and Norway. This conservation effort is aimed at giving North Sea fish populations a chance to rebound.

Two important characteristics distinguish the Scandinavian countries from other European nations. The first is the strong social welfare systems. The second is the extent to which Scandinavian countries have achieved equality of participation and well-being for men and women. Norway, Iceland, Sweden, Denmark, and Finland hold the top five positions in the United Nations Gender Empowerment Measure (GEM) rankings.

Social Welfare in Sweden. Sweden's cradle-to-grave social welfare system (see page 207) is founded on three concepts. First is the idea of security: all people are entitled to a safe, secure, and predictable way of life, free from discomfort and unpleasantness. Second, the appropriate life is ordered, self-sufficient, and quiet, not marred by efforts to stand out above others. Third, when the first two concepts are practiced as they should be, the ideal society, *folkhem* ("people's home"), is achieved. The three concepts help to explain why Swedes are willing to pay for a social welfare system that consumes two-thirds of their government's annual budget (as of 1993) and provides stability and a safety net for every one of the country's 8.8 million people.

Sweden's comprehensive welfare system is possible because it is a prosperous industrialized country. Generous social welfare systems work only in economies that produce substantial surpluses and where a democratic government ensures equitable distribution. Although the combination of a strong economy and a strong welfare system has generally worked well in Sweden, there have been concerns. In the 1990s, Swedes elected governments that cut taxes and benefits for a while, as many people, leery of how membership in the European Union might affect them, began to worry about an influx of outsiders interested in taking advantage of the generous welfare system. An anxious mood was reflected in an uncharacteristic rise in Swedish nationalism and some strong feelings against foreigners—both taboo political stances since World War II. But by 2000, with the economy flourishing, these feelings had dissipated noticeably, and some welfare benefits were restored.

Equal Rights in Norway. Unlike most societies, Norway has directed much of its most innovative work on gender equality toward advancing male rights in traditional women's arenas. Men, for example, now have the right to one month of paternity leave to stay home with a newborn. This policy recognizes a father's responsibility in child rearing and related family obligations; in addition, it provides a chance for father and child to bond during the early days of life. Even when addressing male violence against women, Norway recognizes in its public discussions that this global problem stems in part from cultural customs that suffocate sensitivity in individual men. For example, male children's games that include derision of boys who are not violent are now discouraged in Norway.

At the same time, equal rights for women are closer to being a reality in Norway than in any other country. In 1986, Gro Harlem Brundtland, a physician, became the country's first woman prime minister, and she appointed women to 44 percent of all cabinet posts. Norway became the first country in modern times to have such a high proportion of women in important government policy-making positions. She was reelected and served in the post for 10 years. In 1998, women made up 36 percent of Norway's parliament (second only to Sweden's 43 percent); they held nearly 33 percent of elected positions in municipal government and more than 40 percent of posts in county government. The growing power of women is perhaps best seen in the fact that in the 1990s, women were leaders of Norway's three major political parties.

It is not just in politics that gender equality is sought. Norway was the first country in the world to have an equal-status ombudsperson whose job it is to enforce the 1979 Equal Status Act. This act has a "60–40 rule" requiring a minimum of 40 percent representation by each sex on all public boards and committees. Even though the 40 percent rule has not yet been achieved in all cases, female representation averages over 35 percent for state and municipal agencies. In addition, employment ads are required to be gender-neutral, and all advertising must be nondiscriminatory. For example, an advertisement for a car cannot depict a bikini-clad male or female as an eye-catcher. Nude bodies can appear only in ads for products that are directly related to the body in question.

The Baltic States

The three small Baltic states of Latvia, Estonia, and Lithuania are situated together on the east coast of the Baltic Sea. Culturally, the countries are distinct, with different languages, myths, histories, and music. Recently they shared the experience of being part of the USSR. After World War II, all three became Soviet republics by force, then gained their independence when the Soviet Union collapsed in 1991. Despite the recent connection to Russia, their cultural ties are traditionally with West and North Europe. Ethnic Estonians, who make up 62 percent of the people in Estonia, are nominally Protestant, with strong links to Finland; they view themselves as Scandinavians. Ethnic Latvians, who are just 50 percent of the population of Latvia, are mostly Lutheran; they see themselves as part of the old German maritime trade tradition. Lithuanians are primarily Roman Catholic and also see themselves as a part of old Europe.

To secure its dominance in the Baltic states, the USSR settled many ethnic Russians in the main cities of all three countries. Today, Russians are about a third of the population in both Estonia and Latvia, and they wish to keep strong ties with Russia. In both countries, the indigenous populations are aging because they reproduce at very low rates, whereas the Russian minorities are having many children and could become the dominant culture group within a few decades. Poles, not Russians, are the dominant minority in Lithuania (somewhat under 20 percent), and ethnic Lithuanians will remain the dominant majority for some time to come. Ironically, although the Russian and Polish minorities were established in these three countries to secure the influence of the USSR, under EU rules their rights as minorities now

must be guaranteed before the Baltic states will be allowed to join the European Union.

Of all the countries of North Europe, the economic outlook for the Baltic states is the most precarious. Under Soviet rule, the Russians expropriated their agricultural and industrial facilities and directed them primarily for Russia's benefit. Until 1991, 90 percent of the Baltic states' trade was with other Soviet republics. Since then, in reorganizing their economies, the three countries have moved in different directions.

Estonia has a light industrial base that is relatively easy to reorient. It has undertaken the most radical economic change, moving toward a market economy and increasing its trade with the West, especially Finland and Germany. Of the three, Estonia is the only country to experience real economic growth, and its accomplishments have attracted foreign aid and private investors. Heavily industrialized Latvia and Lithuania still depend on Russia for the bulk of their trade. Their products do not meet Western standards of quality, and their factories are out of date and heavily polluting.

The ecological consequences of industrialization are widespread. One prominent example in Estonia is the oil shale production facility near Kohtla-Järve in the northeastern corner of the country. Smoldering hills of shale residue add to the noxious smoke from the plants. Oily acids seep into the groundwater and into the Baltic, causing skin ulcers on coastal fish. It is possible to ignite the chemicals floating on well water on nearby farms.

Riga, Latvia's capital, is spread along the Daugava River. Once known as the Paris of the Baltic, the city strives to regain its former economic prosperity. The return to private enterprise has helped many small entrepreneurs, but privatization of industrial production is proceeding more slowly. [Torleif Svensson/Corbis Stock Market.]

Gray snow floats down from the stacks of the Kunda cement factory in Estonia. As in most other former Soviet nations, environmental degradation in Estonia resulted from the Russian emphasis on industrial output as the key to economic growth and prosperity. [Larry C. Price.]

The Baltic states see national security as their major problem. Their strategic position along the Baltic Sea has long been coveted by Russia, which has no easy outlet to the sea except through St. Petersburg. In this regard, the status of the Russian **exclave** of Kaliningrad (old East Prussia) is crucial. Kaliningrad is called an exclave because, although it is an actual part of Russia, it lies far from that country along the Baltic between Lithuania and Poland. A relatively ice-free port, Kaliningrad is headquarters for the Russian Baltic fleet, and it is now also a center where Western business executives can travel to negotiate with Russians without having to visit Russia proper. Russia's strategic interest in Lithuania is unlikely to diminish, and Lithuania fears that changing power relationships in Russia could result in a reinvasion of its territory by Russia. The countries of Western Europe have been unwilling to commit to military support against any future invasion. They hope to resolve differences with the Russians, not antagonize them.

EAST AND CENTRAL EUROPE

No other part of Europe is in a more profound period of change than East and Central Europe (Figure 4.21). In the euphoria at the end of the cold war, many people believed that democracy and market forces would quickly turn around the region's centrally planned economies, which were sluggish, highly polluting, and inefficient. Instead, some parts of the region have experienced violent political turmoil, and virtually all experienced at least temporary drops in standards of living. It appears that two tiers of countries are emerging. Those countries closest to Western Europe are adjusting to a market economy fastest and are favored for early entry into the EU. The others, already somewhat behind their neighbors even during the communist era, are not advancing rapidly.

Those Eastern bloc countries with the most resilience and perhaps the brightest future are proving to be the Czech Republic, Hungary, Slovenia, and Poland, all bordering Western Europe. In the 1990s, these countries democratically elected new governments, made rapid progress toward market economies, and began to attract foreign investment. Access to consumer goods has improved markedly and, as discussed earlier in this chapter (see page 197), all are on the fast track to become part of the European Union. But conforming to the wide range of economic, environmental, and social requirements set forth by the EU is not easy.

The countries of the former Eastern bloc that are experiencing the greatest difficulties are Albania, Bosnia-Herzegovina, Bulgaria, Croatia, Macedonia, Romania, Slovakia, and Yugoslavia (Montenegro and Serbia, including Kosovo). For these countries, adjusting to democracy and to the privatization of formerly centrally planned economies has proved complicated. Plagued by corruption and economic chaos, many governments have relaxed the pace of reforms as they struggle merely to maintain civil peace. Others (all the countries of the old Yugoslavia except Slovenia) have been mired in a series of armed ethnic conflicts that have resulted in European and U.S. military intervention (see the upcoming section on conflict in southeastern Europe, pages 230–232). Not surprisingly, many people in these countries are looking back nostalgically to the communist era when jobs were stable; health care and education were free; and there was a strong, authoritative government. Still, many observers are optimistic about East and Central Europe. The countries in this region have useful natural resources, both for agriculture and industry, and all have large, skilled, and inexpensive workforces that will probably attract foreign investment once the political turmoil calms down.

The Difficulties of Economic Transition

The shift away from central planning toward a free-market economy has been rocky. One problem has been that democratic institutions are still immature, so old political bosses and high-level bureaucrats have been able to claim benefits for themselves at the expense of other citizens, as is demonstrated by the following case of agricultural reform in the Czech Republic. In the case of Slovenia, small farmers have also found it difficult to adjust to the requirements of a market economy.

Agricultural Reform in the Czech Republic and Slovenia. In the Czech Republic in the 1990s, land that had been expropriated after World War II and turned into state farms and cooperatives was either reorganized into private owner-operated cooperatives or restored to the original owners and their descendants. This latter

Figure 4.21 **The East and Central Europe subregion.**

group, rather than farming the land themselves, usually leased their regained land, either in small parcels to those who wished to become family farmers or in large parcels to the new private cooperatives. These private cooperatives are owned by a group of workers who labor on the farm and who share its profits. Eventually, 1.25 million acres (500,000 hectares), representing 30 percent of the agricultural land, went to individual private farmers, large and small. Private owner-operated cooperatives now account for the majority of the agricultural land in the Czech Republic.

German economist Achim Schlüter discovered that land redistribution was only the beginning of the reform process. Successful privatized farms also require assets other than land—such as animals, machinery, and buildings—and they need managers with marketing connections and agricultural expertise. Former agricultural bureaucrats, who had agricultural expertise and understood the old system, were able to make quick, informed decisions that allowed them to farm profitably. Several

bureaucrats became executives of large privatized farm cooperatives, while small family farmers languished for lack of capital, information, and market skills. Schlüter discovered that many of the new family farmers have given up because they had little understanding of market economies and were ill-equipped to negotiate the best terms for themselves. Consequently, throughout the 1990s there was a trend toward large quasi-corporate farms.

In Slovenia, unlike the rest of East and Central Europe, farms were not collectivized, so it has not been necessary to redistribute land. The problem instead is that farms are too small for efficient production. The average farm size is just 8.75 acres (3.5 hectares), and agriculture accounts for less than 5 percent of GDP. Although it has much rich farmland, Slovenia is a net importer of food, mostly from EU countries such as Italy, Spain, and Austria. Nonetheless, the new emphasis on private entrepreneurship has encouraged some Slovene farmers to seek a niche in the domestic market. The case of Vera Kusmic is illustrative.

VIGNETTE "Vera Kusmic" (a pseudonym) lives a two-hour drive south of Ljubljana, the country's capital. Her family has farmed on 12.5 acres (5 hectares) near the Croatian border for generations. During the communist era, the men usually worked in urban industrial jobs, commuting daily, while the women did much of the agricultural work, using both small tractors and hand labor. After Slovenia became independent in 1991, Mr. Kusmic lost his job when market reforms bankrupted many Slovene factories. The Kusmic family decided to try earning its living in market gardening. The adult children and Mr. Kusmic work on the land, and Mrs. Kusmic is in charge of marketing their produce.

Eventually, Mrs. Kusmic secured market space in a suburban shopping center in Ljubljana, where she and one employee maintain a small but orderly vegetable and fruit stall. Every day but Sunday, Mrs. Kusmic loads her station wagon with freshly picked produce and drives to the shopping center. She must compete with much less expensive Italian-grown produce sold elsewhere in the same shopping center—all of it produced on large corporate farms in northern Italy and trucked in daily. But Mrs. Kusmic counts on shoppers who value her guarantee that no pesticides or herbicides are used on her farm and that animal manure is the only fertilizer. For the time being, the organically grown produce has kept her in business. But she worries that if and when Slovenia joins the EU, many more corporate producers from western Europe will enter the Slovene market; at the same time, she won't be a large enough producer to sell her fine organic produce at competitive prices in wealthy western European markets.

Mrs. Kusmic in her market stall in Ljubljana, Slovenia. [Lydia Pulsipher.]

Challenges to Industrial Reform. As in agriculture, East and Central Europe's considerable industrial base is still burdened by many legacies of the communist era: a reliance on heavy industries that use energy and other resources inefficiently and hence are highly polluting; a dependence on imported raw materials because only a few parts of East and Central Europe are rich in minerals; and an emphasis on coal and steel production rather than consumer-oriented manufacturing and service industries. Industrial pollution has been a particular problem. For example, in Upper Silesia, Poland's leading coal-producing area, forests have succumbed to acid rain, soils yield contaminated crops, and pollution is at deadly levels. Residents have experienced birth defects, high rates of cancer, and lowered life expectancies. The pollution their industries create is the greatest obstacle to entry into the European Union. In addition, the EU requires evidence that prospective members can compete successfully in the EU marketplace, something that few East and Central European countries can do at present.

Corruption worsened during the last years of communism and then mushroomed in the post–cold war era. Involvement by organized crime has plagued the sale of formerly state-owned industries, especially in military sectors. One mobster based in Budapest has reportedly taken over virtually the entire Hungarian armaments industry and is selling guns globally. Organized crime as a whole has facilitated the spread of technologically advanced weapons, possibly including nuclear warheads smuggled from the military bases of the former Soviet Union, to terrorists and countries interested in bolstering their arsenals. Increasingly, politicians have connections to organized crime, and foreign businesses eager to invest in East and Central Europe reluctantly pay bribes to these groups.

Continuing Hardship. Throughout East and Central Europe, people have suffered directly from the privatization or collapse of state-owned industries. For example, because the new private owners usually have little experience in meeting regular payrolls or have actually gone bankrupt, some industrial workers in private firms have gone without pay for months. Because tax collection is far behind schedule, public funds are depleted and the remaining state industries have also been unable to meet payrolls. Other people have suffered layoffs or job losses as newly privatized industries have tried to cut costs. Many inefficient factories are being closed rather than updated because the owners lack the capital to remodel. The shutdowns bring the hardships of unemployment to countries that no longer have social safety nets. The situation in the once well-funded state hospitals is epitomized by First General Hospital in Sofia, Bulgaria, which in the late 1900s was running out of equipment and medicines, even intravenous fluids and painkillers. New patients are accepted only in extreme emergencies. "I can't look them in the eyes anymore," one doctor told a reporter, shuffling into an elevator. "I know we won't be able to treat them." Similar conditions still prevail in the region's underfunded schools and research facilities. In reaction, scientists, technicians, artists, writers, and highly skilled personnel are moving to wealthier parts of the world.

The effect of the economic transition on women has been somewhat ambiguous. The sociologist Alena Heitlinger, who studies gender in both the Czech Republic and Slovakia, writes that under state socialism, working women gained status mainly through a job in a state-run firm, but the traditional patriarchal village beliefs regarding gender roles were retained. Hence, most women were given lower paid "easy work" thought suitable for their sex. This easy work was not actually easy, it just had low status. For example, because agriculture was perceived as the most undesirable labor, farming became the task of older women, who often had only the use of manual tools. Urban cleaning jobs, too, were overwhelmingly performed by elderly women armed with rags and push brooms. Since the transition, the women have become more vulnerable than men to job loss. As state-run firms are sold off and made more efficient, women—who occupy the lowest ranks—are the first to be laid off. It is thought that in times of economic crisis and high unemployment, men need the jobs more than women, despite the fact that in many cases women are a family's sole support. On the other hand, a number of governments now have official agencies that look after the interests of women, especially encouraging entrepreneurship.

Steps Toward Economic Transition

For all the signs of trouble, however, many governments are taking "free-market friendly" steps to help their economies grow more quickly. One strategy is to encourage foreign investment. Virtually all the countries of East and Central Europe actively recruit foreign investors and offer a variety of incentives. Although there have been a number of false starts by investors unaware of the difficulties they would face, there have also been some successes. An example is Poland's joint venture with the French multinational glass company Saint-Gobain to construct one of the world's most modern glass factories. The factory is being constructed in the south, near the Czech border, in what has been Poland's most polluted region, Silesia. Significantly, it will be one of the most environmentally safe plants of its kind in Europe, with emissions well below the standards set by Germany, the industry leader. All waste glass will be completely recycled.

The efforts of the Coca-Cola Company provide an example of foreign investment in the subregion. Pepsi-Cola had long been the primary cola available in the Soviet bloc, having set up joint ventures with the communist regimes back in 1959. PepsiCo agreed to work through the state-run bottling and marketing companies and did not have control over the operations. When the Soviet Union began to disintegrate in the 1980s, Coca-Cola was invited to enter into a joint venture with Romania's largest soft drink bottler; but the Coca-Cola Company insisted on control of the operation. The company invested heavily in Romania, spending many millions on new plants and equipment, new uniforms, red-painted Coca-Cola trucks, and marketing. By 1995, the Coca-Cola Company had six bottling plants, 2500 employees, and annual sales of U.S. $120 million in Romania. Coke now outsells Pepsi by more than two to one. Coca-Cola's success has opened the way for thousands of small entrepreneurs who sell soft drinks in small shops. A University of South Carolina study in 1994 indicated that at least 23,000 jobs were supported by the sale of Coca-Cola products in Romania.

Another way to increase growth is to improve infrastructure, a strategy that Poland is implementing with the investment over the next 15 years of U.S. $9 billion on highways that will link it with the EU network. As is the case in many East and Central European countries, Poland's inadequate roads have been an obstacle to industrialization and commerce. Another U.S. $15 billion will be spent on upgrading energy production, which is now insufficient and highly polluting. And to meet EU pollution standards, another U.S. $9 billion will be earmarked to improve air and water quality. Other infrastructure improvement projects are under way. Because there is not enough tax money to cover these necessities, Poland is seeking direct foreign investment to upgrade energy and communications facilities. Among countries that border the northern Adriatic, there are plans to improve all the major ports and railroads and to link them with the Corridor Five roadway from Barcelona to Kiev, which is currently under construction.

Tourism is increasingly recognized as an important part of a well-rounded market-based economy. The remnants of medieval forts, castles, and other buildings are being refurbished and promoted as tourist attractions. The cities of East and Central Europe are some of Europe's most distinguished municipalities, and they are becoming magnets for tourists from America, Japan, and

The Fishermen's Bastion and its outdoor restaurant is one of Budapest's many tourist destinations. The structure was completed in 1905 on the site of the old fish market in memory of the fishermen who helped to defend the city during the Middle Ages. [Leo de Wys.]

western Europe. One example is Budapest, the capital of Hungary. Situated on the Danube River and built on the remnants of the second- to third-century Roman city of Aquincum, Budapest was one of the seats of the Austro-Hungarian Empire, which ruled central Europe until World War I. In the 1990s, Budapest reemerged as one of Europe's finest metropolitan centers, filled with architectural treasures, museums, concert halls, art galleries, discos, cabarets, theaters, and several universities of distinction. An added attraction is that Hungarians have revived their love affair with jazz, called "the music of the imperialists" under the communist regime. After the change to a market economy in 1989, dozens of jazz clubs opened. An interesting example of links with tourism in western Europe is the two-day round-trip Danube cruise that is now possible between Vienna and Budapest.

New Experiences with Democracy. The advent of democracy has reduced some of the tensions associated with the economic transition, because people have been able to vote against governments whose reforms create high levels of unemployment and inflation. Most countries have now had several rounds of parliamentary and local elections, and voters are becoming accustomed to Western-style political campaigns. However, the tendency to remove reformers before they can accomplish anything significant has dampened the enthusiasm of potential foreign investors who are waiting for real change before they offer their capital. On the other hand, in many cases, even crudely practiced democracy is leading to long-term economic growth and stability by encouraging governments to take a greater interest in the financial security of the population as a whole. In Hungary and the Czech Republic, largely because of pressure from voters and from the EU, there are plans for new national pension systems to replace defunct communist pension plans. New pension systems would give financial stability to people who have worked all their lives and political stability to states that cannot afford to alienate the older generation that labored under communism.

Southeastern Europe: Understanding an Armed Conflict

In 1991, a civil war erupted in the Balkans (now preferably called southeastern Europe) that has taken the lives of more than 250,000 people. This mountainous region north of Greece is home to many ethnic groups of differing religious faiths. The news media have asserted that violence and ethnic hatred is an age-old way of life in the Balkans. A deeper look shows that although ethnic tensions have existed for generations, they have greatly intensified in recent decades, suggesting that more immediate political and economic factors have contributed to the conflict.

After World War II, the country of Yugoslavia was formed out of six territories (Figure 4.22). In five of them—Serbia, Croatia, Slovenia, Macedonia, and Montenegro—a single ethnic group was

Dates of Secession from Yugoslavia	
Slovenia	1991
Croatia	1991
Bosnia/Herzegovina	1992
Macedonia	1991

Figure 4.22 **The Balkans.** [Adapted from "A survey of the Balkans," *The Economist* (January 24, 1998): 4.]

dominant, although other ethnic minorities were present in most of the five. The sixth territory, Bosnia-Herzegovina, did not have a majority ethnic group, although there were substantial populations of Muslims and Christians—some Serbian, some Croatian. In 1991, the first free elections ever held in Yugoslavia resulted in declarations of independence first by Slovenia, then by Croatia and Macedonia, and finally by Bosnia-Herzegovina. Although Slovenia and Macedonia were allowed to separate relatively peacefully, the now much smaller Yugoslavia, composed of Serbia and Montenegro, fought protracted wars against Croatia and Bosnia-Herzegovina. Both countries had Serb-populated territories that Serbia argued should be in one Serbian-led country. To rid the coveted territory of non-Serbs, some Serbs instigated genocidal "ethnic cleansing" campaigns against civilians, especially against Muslims in Bosnia-Herzegovina, in Croatia, and later in Kosovo (a province of Serbia).

The conflict in the former Yugoslavia caught most of its citizens off-guard. The experiences of the Serbian Djukanovic family illustrate how they once got along with people of different ethnicities and then found that, for reasons unknown to them, the situation changed. Now they struggle to reconstruct life in their old multicultural community.

VIGNETTE Mitra Djukanovic and her husband, Dusan, are Christian Serbian farmers who defied Serb leaders in 1996 by returning to their home village of Krtova, Bosnia-Herzegovina, in a territory now administered by Muslims. As they prepare for the coming hard winter, they tell their story amid bags of potatoes, walnuts, pickled peppers, and flour, as drying meat smokes over a smoldering tub. Dusan speaks: "I thought it was time to come back. I stopped worrying about living with Muslims. I lived with them most of my life. . . . My grandfather told me about the old times, and there was never trouble between us and the Muslims here. We all just worked a lot and had nothing. It was equal. Until this war, I never had trouble either. If there is a Muslim government, so what?" [Adapted from Mike O'Connor, "Serbs go home to Bosnia village, defying leaders," *New York Times* (International) (December 8, 1996): 1.]

Was Ethnic Violence Inevitable? If sentiments like those of the Djukanovics are common, as many observers say they are, why has there been so much violence between ethnic groups? To understand the recent ethnic politics in the former Yugoslavia, it is useful to look at how the political and economic changes of recent decades created ethnic hostilities where previously there were few.

Many of the so-called indigenous ethnic groups of southeastern Europe were themselves invaders over the last 1200 years or so, coming from Central Europe, Russia, Central Asia, and beyond. These disparate groups jointly occupied the Balkan Peninsula with groups who had been there even longer, and none of them were particularly concerned with precise land rights. For the most part, the various ethnic groups managed to live together peacefully, intermarrying, blending, and realigning into new groups. Historian

Charles Ingrao argues that coexistence was even easier in areas such as Bosnia, where three or more groups living side by side forestalled any one group from dominating the others. More recently, external wars exacerbated local ethnic tensions, as when the Nazis made an ally of Croatia, pitting it against Slovenia, Serbia, and others. Nevertheless, overall, over the last several hundred years ethnic rivalries were no worse than those in the rest of Europe.

After the formation of Yugoslavia, Marshal Tito, Yugoslavia's founder and leader until his death in 1980, tried through his personal leadership to gloss over ethnic differences by encouraging pan-Yugoslav nationalism. Although there was ethnic peace during Tito's rule, there was no national dialogue on the benefits of ethnic diversity. Other ethnic groups were made uneasy by the domination of Serbs, the most populous ethnic group in the military and in much of the federal government bureaucracy. Throughout the 1980s, resentment of the Serb-dominated central government increased as the centrally planned economy of the country deteriorated. By the late 1980s, in order to divert public attention away from economic deterioration, the Serbian leader, Slobodan Milosevič, claimed that ethnic (Christian) Serbs had long been threatened by Serbian and Albanian Muslims. When Slovenia and Croatia, two of the richest provinces, voted to declare independence in 1991, local populations were responding less to age-old hatreds and fierce nationalism than to a deteriorating Yugoslav economy, a government dominated by the single Serb ethnic group, and growing apprehension that the Serbian leadership was prepared to use repression to retain control.

Given the scale of interethnic violence that occurred in the following years, it is not surprising that many, especially outsiders, have seen ethnic hatred as the overriding cause of the civil war, but earlier lack of guidance on multicultural affairs by Marshal Tito and Serbian geopolitics were the main forces. The Serb-dominated government and military of Yugoslavia only reluctantly allowed the province of Slovenia to separate after just a short skirmish, in part because of its strategic location close to the heart of Europe and far from Serbia and in part because there were no significant enclaves of Serbian supporters in Slovenia. Slovenia was a major loss to Yugoslavia, because it had been by far the most modernized and productive of the provinces. The Yugoslav government feared that unless it made a strong show of force in dealing with Croats and Bosnians, the rest of the non-Serb populations of Yugoslavia would also move to secede and take valuable territory, resources, and skilled people with them. Eventually, this fear became a reality when Croatia and Bosnia-Herzegovina attempted to secede. Despite the undoubted culpability of the Serbs in starting the conflict and in perpetrating much of the brutality, they cannot be blamed exclusively for all the interethnic violence that has taken place. While suffering ethnic cleansing, Croats and Bosnians have, themselves, used brutality and ethnic cleansing in retaliation, as did Albanians and Kosovars later.

After agonizing delays, the war in the Balkans gained the attention of EU countries and the United States, both of which sent military peacekeepers. Peace accords were signed in Dayton, Ohio, in 1995. By this time the results of the war were devastating, and

Serbia, itself, was in shambles. It had lost significant territory, and the economy was ruined. Unemployment reached 60 percent during the war, and inflation was at 20 percent. Personal income was cut in half between 1990 and 1996. In Bosnia-Herzegovina, 200,000 people, 5 percent of the population, died in the war, many the victims of Serbian war crimes. The cosmopolitan city of Sarajevo, the site of the Olympics in 1984, was destroyed, as were countless villages.

Conflict sprung up again in the late 1990s in the province of Kosovo, the southern area of Serbia dominated by ethnic Albanian Muslims. Throughout the 1990s Milosevič's government in Serbia had been removing Albanians (Kosovars) from government positions and barring them from using public services. Increasingly, the Serbian government used arrests and violence to quell Kosovar activists. In response, the guerrilla Kosovo Liberation Army was formed, supported by expatriates of Albanian ethnicity. They also attempted to secede from Yugoslavia—perhaps to join the country of Albania, which lies just to the south. In spring 1999, Serbia forced most of the majority Albanian population out of the province of Kosovo. Several weeks thereafter, in their first-ever formal military action, NATO forces bombed Serbia in an effort to stop massacres and prolonged ethnic cleansing in Kosovo. The violence de-escalated, but hostilities between Kosovar Albanians and Serbs continued in 2002, fueled by covert aid to both sides from agents outside the region, some inspired by nationalism, some by Islamic fundamentalism. In November 2001, elections were held in Kosovo under UN supervision. Delegates from three parties were elected to the parliament, but no party had a majority.

The Balkan Stability Pact. The wars in and around Serbia have devastated the economies of southeastern Europe. The conflict destroyed the infrastructure of commerce, interrupting trade for all countries in the region. Dozens of bridges were bombed, and the Danube River, a major transport conduit through Europe to the Black Sea, is blocked with wreckage. In 2000, an election in Serbia resoundingly removed Slobodan Milosevič from power, and in 2002 he was standing trial for war crimes in the International Court of Justice (World Court) in The Hague, Netherlands. The European Union and the countries of southeastern Europe had earlier agreed to enact a Stability Pact once Milosevič was gone. The Balkan Stability Pact, funded by the European Union, is promoting recovery of the region through public and private investment and through social training to enhance the acceptance of multicultural societies and to strengthen informal democratic institutions. In the fall of 2000, Serbia and Montenegro were admitted to the United Nations under the name Yugoslavia.

REFLECTIONS ON EUROPE

Europe's position as a center of world economic and political power seems less formidable when the region is viewed close up. Europe is not a monolith; its geographic patterns are highly variable. The distribution of wealth, social welfare policies, approaches to environmental issues, treatment of outsiders, and participation of women in public and private affairs are just some of the aspects of life in Europe that reflect this variation.

Many of the issues confronting people in Europe are similar to those encountered elsewhere around the world. For example, the effort at economic integration is exposing European countries to the same problems of restructuring now faced by so many, much poorer, developing countries. The vicissitudes of dealing with an integrated world economy are apparent on every level in Europe. Europe's problems, although generally less severe, are the same as those faced by Latin America, Asia, and Africa. Europeans must answer the same questions: How can a country retain jobs for its people when there are qualified workers willing to work for less within relatively easy reach? How can more good jobs be created? How can a society help the poor and unemployed to a better life while at the same time lessening the drain on public funds by welfare programs? How can poor parts of the region be stimulated to develop? And how can development everywhere be made to conform to the requirements of an increasingly stressed environment?

No less complex are the cultural questions of how countries with distinctly different traditions and mores can find sufficient common ground to cooperate economically and thus achieve greater prosperity for them all. For all its current troubles, Europe's head start on development, its established world leadership, and the heightened awareness of global economic issues it has gained by its own efforts at integration—all position Europe to retain its global influence.

From the perspective of people in the Americas, Europe's current preoccupation with economic integration should prove instructive. Many of the same issues will be confronted if the economic integration of the Americas proceeds: the free movement of people, the loss of jobs to low-wage areas, and the need to strengthen and coordinate environmental policies, to name just a few.

The future of the relationship between Europe and the world region we will cover in Chapter 5—the Russian Federation, Belarus, Caucasia, and Central Asia—is not clear. Will these two regions integrate their economies and societies, or will Russia's very serious economic and social problems make it too unattractive a partner? Will new alignments and contentions in Central, South, and East Asia draw Russia away from Europe? The next chapter will discuss these and other geographic issues.

Thinking Critically About Selected Concepts

1. An influential notion in Europe is the idea of the nation, a large group of people bound together by common cultural traits and by their allegiance to the idea that they form a united group. The concept has influenced attitudes toward outsiders and the establishment of national boundaries. Some observers have suggested that in a globalizing world, the concept of the nation-state is defunct and that soon other allegiances will take precedence. *What evidence do you see that this might happen in Europe? Imagine such a world. Where would people's allegiances potentially lie? How would they be reinforced? Would such a world be more peaceful or more warlike than the one we know?*

2. Europe is one of the most densely occupied regions of the world. Its prosperity is related to its access to a resource base that is global in extent. *What other world regions resemble Europe in population density and access to global resources? Do you think regions with these features are likely to grow in number?*

3. Europe now has a negative rate of population growth (–0.1 percent), the lowest on earth. Many people are choosing to remain childless or to have only one child. Because Europe needs more workers, immigration is encouraged, but it is also feared. *What are some of the ramifications of low population growth in Europe? How is Europe likely to change as a result of people choosing to have no children, or only one?*

4. Definitions of what it means to be European are changing as various countries confront the ethnic diversity of guest workers in their midst and try to define new policies toward outsiders. In Europe, as in North America, the ethnic makeup of the population is changing. *What do you see as the most important factors bringing ethnic change in Europe? Is ethnicity more of a class or a race issue in Europe? Imagine yourself living in a European country; what would be the most important things for you to do to gain acceptance?*

5. Efforts to integrate Europe economically, the demise of the Soviet Union, and the move toward more open and competitive market systems in East and Central Europe have put Europe into a state of flux. European economic integration has brought many changes, some of which are expressed in the landscape. *Imagine or describe from your own travel experience what these changes are like. Be sure to include topics such as borders, store shelves, shopping opportunities, travel patterns, material culture, popular culture, the role of smuggling, and illegal immigration.*

6. The philosophies of social welfare systems and the abundance of benefits they provide vary greatly from region to region in Europe. This variability is an issue as economic integration increases in the EU. With more open borders, countries with many welfare benefits fear an influx of people from countries with fewer benefits and of immigrants from outside the region. *Discuss the role of social welfare systems in European economic integration. Why might these systems pose some problems as integration proceeds?*

7. European women confront many of the same issues that women all over the world confront: the double day, low status, low representation in government and policy-making bodies, and fewer employment opportunities and lower wages than are available to men. *Compare and contrast the issues facing European 20-year-old women or men with those facing American twenty-year-old women or men.*

8. Europe's long lead in industrialization is threatened by competition from abroad, but European countries are taking many steps to compete more efficiently. They are diversifying, seeking outside investors, investing abroad, and lowering some restrictions that have discouraged the creation of new jobs. *If you were an investor, would you be attracted to Europe? Why?*

9. Europe's agricultural sector is productive and highly diversified, and it is protected by price guarantees and subsidies. This sector has particular difficulties competing in world markets. *If you were planning to be a farmer in one of the countries of East and Central Europe, which one would you choose? Why? To whom would you be likely to sell your farm products?*

10. Europe's environment has been drastically modified by human activity. The Green movement is a region-wide effort to educate the public and influence public policy toward measures that will control pollution and the wasting of resources, which is particularly a problem in East and Central Europe. *What has attracted the attention of Europe's Green movement to East and Central Europe? To what extent do you think a citizens' environmental movement will succeed in this region? What handicaps might it face?*

11. Human well-being in Europe varies greatly, but it is generally acceptable, except in parts of East and Central Europe, where economic and social collapse have produced marked declines in general health and longevity. *How would you explain Europe's relatively good record? If you were to move to Europe, which human well-being indicators would be most important to you in choosing a country or subregion?*

Key Terms

acculturation (p. 205) adaptation of a minority culture to the dominant culture to the extent necessary for members of the minority culture to function effectively and be self-supporting; cultural borrowing

Age of Exploration (p. 187) a period of accelerated global commerce and cultural exchange facilitated by improvements in European navigation, shipbuilding, and commerce in the fifteenth and sixteenth centuries

assimilation (p. 204) the loss of old ways of life and adoption of the lifestyle of another culture

cold war (p. 190) the contest that pitted the United States and Western Europe, espousing free-market capitalism and democracy, against the former USSR and its allies, promoting a centrally planned economy and a socialist state

Corridor Five (p. 200) a planned and partially constructed major transportation roadway across southern Europe, stretching from Barcelona, Spain, to Kiev, Ukraine

democratic institutions (p. 189) institutions—such as constitutions, elected parliaments, and impartial courts—through which the common people gained a formal role in the political life of the nation

devolution (p. 219) the weakening of a formerly tightly unified state

double day (p. 205) the longer workday of women with jobs outside the home who also work as family caretakers, housekeepers, and cooks at home

dumping (p. 200) the cheap sale on the world market of overproduced commodities, which lowers global prices and hurts producers of these same commodities elsewhere in the world

economic integration (p. 191) the free movement of people, goods, money, and ideas among countries

economies of scale (p. 194) reductions in the unit costs of production that occur when goods or services are produced in large amounts, resulting in a rise in profits per unit

Euro, the (p. 195) the official currency of 12 of the 15 European Union countries, as of January 1, 1999

European Economic Community (EEC) (p. 194) an economic community created in 1958, when Belgium, Luxembourg, the Netherlands, France, Italy, and West Germany agreed to eliminate certain tariffs and promote mutual trade and cooperation in the interest of achieving stronger economic ties within Europe

European Union (EU) (pp. 182–183) a supranational institution including most of West, South, and North Europe, established to bring economic integration to member countries

exclave (p. 226) a portion of a country separated from the main part and constituting an enclave in respect to the surrounding territory

feudalism (p. 187) a social system established by A.D. 900 in Europe, which created a class of professional fighting men, or knights, ready to defend the farmers, or serfs, who cultivated their protectors' land

genetically modified organisms (GMOs) (p. 196) plants or animals whose genetic code has been modified to make them more attractive as commodities

Green (p. 209) name for environmentally conscious political parties

guest workers (p. 203) immigrants from outside Europe, often from former colonies, who came to Europe (often temporarily) to fill labor shortages; Europeans expected them to return home when no longer needed

infrastructure (p. 194) basic facilities that a society needs to operate (transportation and communication systems, power and water, etc.)

iron curtain (p. 190) a fortified border zone that separated Western Europe from Eastern Europe during the cold war

mercantilism (p. 188) the idea that countries increase their power and wealth by acquiring colonies with human and natural resources and by managing all aspects of production, transport, and trade to their own benefit

nation-state (p. 189) political unit, or country, formed by people who share a language, culture, and political philosophy

nationalism (p. 189) devotion to the interests or culture of a particular country or nation (cultural group); the idea that a group of people living in a specific territory and sharing cultural traits should be united in a single country to which they are loyal and obedient

Protestant Reformation (p. 187) a European reform (or "protest") movement that challenged Catholic practices in the sixteenth century, resulting in the establishment of Protestant churches

Renaissance (p. 187) a broad European cultural movement in the fourteenth through sixteenth centuries that drew inspiration from the Greek, Roman, and Islamic civilizations, marking the transition from medieval to modern times

resource base (p. 192) the selection of raw materials and human skills available in a region for domestic use and industrial development

Schengen Accord (pp. 201–202) an agreement signed in the 1990s by the EU and many of its neighbors that called for free movement across common borders

social welfare (pp. 206–207) in Europe, elaborate tax-supported systems that serve all citizens in one way or another

subsidies (p. 200) monetary assistance granted by a government to an individual or group in support of an activity, such as farming or housing construction, that is viewed as being in the public interest

welfare state (p. 189) a social system in which the state accepts responsibility for the well-being of its citizens

Pronunciation Guide

Adriatic Sea (ay-dree-AT-ihk)

Albania (ahl-BAY-nee-ah)

Alhambra (ahl-HAHM-brah)

Amsterdam (AM-sturr-dam)

Andalucía (ahn-dah-loo-THEE-ah)

Antwerp (AHN-twurrp)

Athens (ATH-ihnz)

Austria (AW-stree-uh)

Balkans (BAHL-kuhnz)

Baltic Sea (BAHL-tihk)

Basque (BASK)

Belfast (BEHL-fast)

Belgium (BEHL-juhm)

Belgrade (BEHL-grahd)

Benelux (BEHN-uh-luhks)

Bordeaux (bohr-DOH)

Bosnia-Herzegovina (BAWZ-nee-uh haint-suh-goh-VEE-nuh)

Brundtland, Gro Harlem (BROONT-lahnd, GROH HAHR-lehm ["oo" as in "book"])

Brussels (BRUH-suhlz)

Bucharest (BOO-kuh-rehst)

Budapest (BOO-duh-pehsht)

Bulgaria (buhl-GAIR-ee-uh)

Catalonia (kaht-uh-LOH-nee-uh)

Chernobyl (chyair-NOH-bihl)

Copenhagen (KOH-puhn-hay-guhn)

Córdoba (KOHR-doh-vah)

Croatia (kroh-AY-shuh)

Czech Republic (CHEHK)

Czechoslovakia (chehk-oh-sloh-VAH-kee-ah)

Danube River (DAN-yoob)

Daugava River (DAU-gah-vah)

Essen (EHSS-uhn)

Estonia (ehss-TOH-nee-uh)

Faroe Islands (FAHR-oh)

Finland (FIHN-luhnd)

Florence (FLOHR-uhntss)

folkhem (FOHLK-hehm)

Galicia (gah-LEESH-ee-ah)

Genoa (JEHN-oh-uh)

Gibraltar (jih-BRAHL-turr)

Granada (grah-NAH-dah)

Hague, The (HAYG)

Helsinki (hehl-SIHNG-kee)

Herzegovina (hairt-suh-goh-VEE-nuh)

Hungary (HUHNG-guh-ree)

Iberia (eye-BEER-ee-uh)

Jönköping (YEHN-keh-pihng)

Kaliningrad (kah-LYIH-nihn-graht)

Kiev (KEE-ehf)

Kohtla-Järve (KOHT-luh YAIR-vuh)

Kosovo (KOHSS-uh-voh)

Krtova (krah-TOH-vuh)

Latvia (LAT-vee-uh)

Lille (LEEL)

Lithuania (lihth-oo-AY-nee-uh)

Ljubljana (lyoo-BLYAH-nuh)

Luxembourg (LOOK-suhm-boork [first "oo" as in "book"])

Macedonia (mass-ih-DOH-nee-uh)

Madrid (mah-DRIHD)

Mafia (MAH-fee-ah)

Marseille (mahr-SAY)

Mediterranean Sea (meh-dih-tuh-RAY-nee-uhn)

Milan (mih-LAHN)

Montenegro (mawn-tih-NEH-groh)

Mt. Vesuvius (veh-SOO-vee-uhss)

Muslim (MOOZ-lihm ["oo" as in "book"])

Naples (NAY-puhlz)

Nazi (NAHT-see)

Neapolis (nay-AH-poh-leess)

Ottoman (AW-tuh-muhn)

Pannonian Plain (puh-NOH-nee-uhn)

Piazza del Plebiscito (PYAHT-suh dehl pleh-bih-SHEE-toh)

Portugal (POHR-chuh-guhl)

Renaissance (REHN-uh-sahnss)

Rhine (RINE) River

Riga (REE-guh)

Riviera (rih-vee-AIR-uh)

Romania (roh-MAY-nee-uh)

Rotterdam (RAWT-urr-dam)

Sarajevo (sah-rah-YAY-voh)

Scandinavia (skan-dih-NAY-vee-uh)

Scotland (SKAWT-luhnd)

Serbia (SURR-bee-uh)

Sicily (SIHSS-uh-lee)

Silesia (sih-LEE-zhuh)

Slovakia (sloh-VAH-kee-uh)

Slovenia (sloh-VEE-nee-uh)

Sofia (soh-FEE-uh)

sottogoverno (soh-toh-goh-VAIR-noh)

Svalbard (SVAHL-bahr)

Tito (TEE-toh), Josip Broz

Tokarp (TOH-kahrp)

Toulouse (too-LOOZ)

Turin (TOOR-ihn)

Ukraine (yoo-KRAYN)

Ural Mountains (YOOR-uhl)

Venice (VEH-nihss)

Vienna (vee-EHN-uh)

Warsaw (WOHR-saw)

Yugoslavia (yoo-goh-SLAH-vee-uh)

Selected Readings

A set of Selected Readings for Chapter 4, providing ideas for student research, appears on the *World Regional Geography* Web site at www.whfreeman.com/pulsipher.

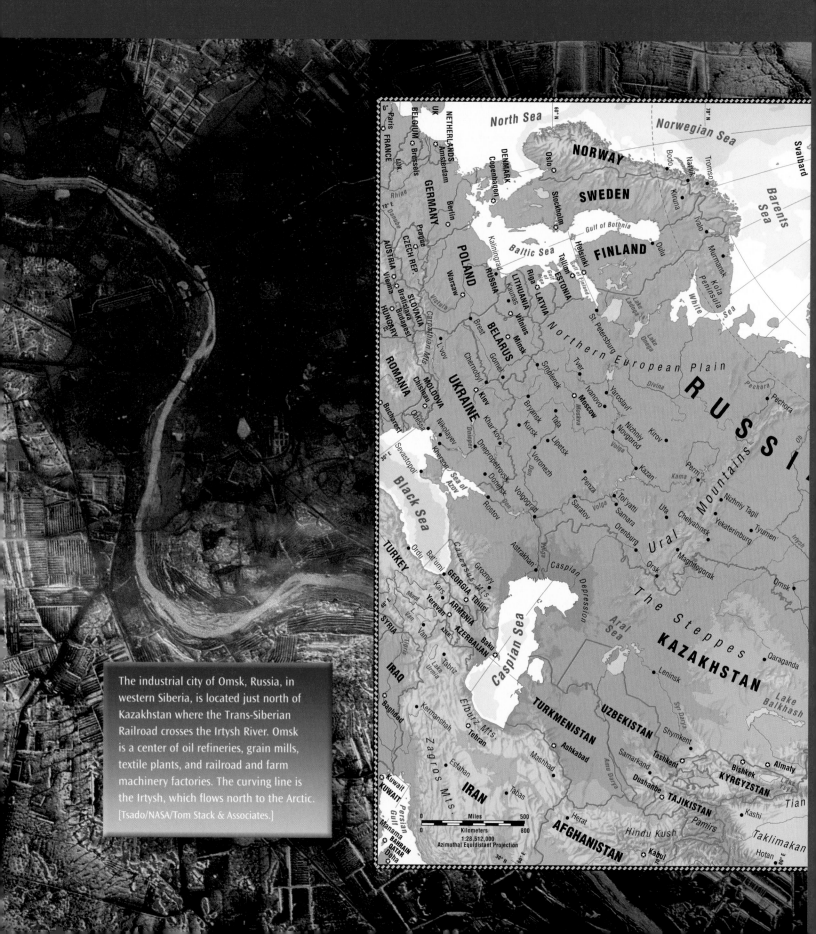

The industrial city of Omsk, Russia, in western Siberia, is located just north of Kazakhstan where the Trans-Siberian Railroad crosses the Irtysh River. Omsk is a center of oil refineries, grain mills, textile plants, and railroad and farm machinery factories. The curving line is the Irtysh, which flows north to the Arctic.

[Tsado/NASA/Tom Stack & Associates.]

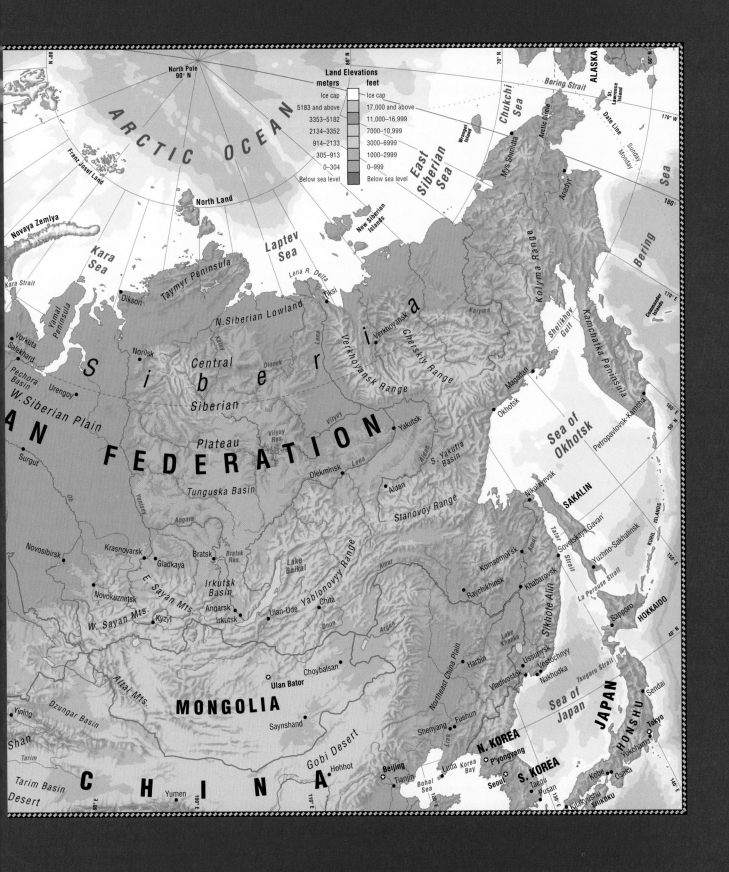

Chapter 5: Russia and the Newly Independent States

VIGNETTE When he was a teenager, Boris K. used to race go-carts, and he even produced films for a while. Still a young man, Boris owns several kiosks—which makes him, by most Russian standards, a gangster. Kiosks are mobile, modular stores, usually consisting of a large steel box about twice the size of a dumpster, with windows and a display case. They are usually crammed with an array of goods ranging from vodka and candy to toilet seats. Capable of being locked up as securely as a safe, kiosks can be loaded onto trucks and transported anywhere overnight. This portability allows the operators to take advantage of nearly any public space with commercial potential and to evade the authorities. The places where kiosks congregate are called *tolkuchkas,* or "places of pushing."

Why the gangster connection? Nearly all kiosks sell some sort of bootlegged product: counterfeit CDs, fake name brands, stolen property. Few pay taxes. Many are run by toughs who strong-arm other vendors or demand protection money. Some are in dirty, rubble-strewn squares, others next to pleasant parks. Nearly all are tempo-rary, disappearing and reappearing depending on the needs of those who own and supply them and on who is buying what. Many kiosks have names painted on the outside, usually in bright colors and in English. The names reflect their somewhat rebellious and capitalistic spirit: "Freedom," "My Valentine," "What You Want?" As Boris puts it, kiosks and their *tolkuchkas* are "spontaneous rackets," very responsive to the market trends of the moment. [Adapted from Mark Kramer, *Travels with a Hungry Bear: A Journey into the Russian Heartland* (Boston: Houghton Mifflin, 1996), pp. 262–268.]

Boris lives in a region that attempted to change its political and economic systems entirely in a few short years. The change offers opportunities to many people, like Boris, but brings hardships and uncertainty as well. Until a few years ago, the Union of Soviet Socialist Republics (USSR) was the largest country on earth, stretching from eastern Europe all the way across Central Asia and Asia to the Pacific Ocean. In December 1991, the USSR ceased to exist. In its place stood 15 independent political units joined in a loose economic alliance. The largest and most powerful was Russia. But soon, 3 of the 15—Latvia, Lithuania, and Estonia—having chafed under Russian domination for many years, abandoned their close association with Russia. In this book, they are discussed in Chapter 4.

Thus, the region presently consists of Russia and 11 loosely allied countries: Belarus, Moldova, and Ukraine to the west of Russia; Georgia, Armenia, and Azerbaijan, referred to jointly as the republics of Caucasia; and the 5 republics of Central Asia: Kazakhstan, Kyrgyzstan, Tajikistan, Uzbekistan, and Turkmenistan.

The breakup of the Soviet Union triggered a transformation in the region's political and economic systems. Authoritarian government gave way to experiments with democracy; an economy controlled by government bureaucrats was replaced by one that encourages its citizens to establish their own businesses. Boris K.'s roving kiosk is an example of the free-wheeling experimentation going on in this region. The questionable legality of his operation illustrates the pitfalls and temptations of an economic system that has yet to develop controls. His fondness for English phrases suggests the allure of the West's wealth and mass culture.

The demise of the USSR in the early 1990s has required a period of adjustment and reorganization around the globe because the cold war affected so many economic and political relationships worldwide. During the cold war, the West (western Europe and North America and their allies) and the East (the USSR and its allies) were locked in a struggle for power and influence. Although the confrontation between East and West has diminished drastically, it is not yet clear just what role the old components of the Soviet Union will play in world affairs.

Imported (probably smuggled) VCRs and TVs are sold in informal, non-tax-paying open markets; this one is in Moscow. They are tended by men affecting, if not actually living, a "mafioso" style. Police are paid to look the other way. [Gerd Ludwig/National Geographic Image Collection.]

During the 1990s, the countries that once made up the former Soviet Union began forging new relationships with one another and with the world outside. Russia's sphere of influence has diminished because it no longer has the economic means to court alliances through loans and assistance, yet Russia remains dominant within the region itself. The newly independent states are maintaining strong economic ties to Russia, but they are also developing new trading relationships with the outside world. Russia itself has had to make major adjustments. Once its economic system separated it from the West, and its satellite countries supplied food for its populace and raw materials for its industries. Now Russia must compete for resources and trading partners in the worldwide marketplace.

To understand the trauma and confusion of the many changes of the last decade, imagine that in the space of five years the geography of the United States changed in the following ways. First, old alliances with Canada and with Europe dissolved; then one-third of the states declared independence and formed associations with countries that had been viewed as the enemy. Resources on which the United States had depended, such as prime agricultural land and crucial minerals, now unexpectedly belonged to someone else. Whole sectors of the economy were abruptly declared obsolete, and in the restructuring, hundreds of thousands of people (especially women) lost jobs and with them a comprehensive support system of housing, daily meals, and medical care. Imagine further that in short order the disruptions began to take their toll in the form of earlier deaths for adults (especially men) and sharply rising infant mortality. Recurring food shortages and rising lawlessness, especially organized crime, added further dangers to daily life. Then imagine that an old enemy (say, the Soviet Union) began to send emissaries offering to restructure the United States along new lines—their lines.

All this happened in the former Soviet Union in the 1990s. Its own internal economic, political, and social systems are in flux, and important regions once under its sphere of influence have broken away. The people living in the former USSR are experiencing the exhilaration of new opportunities. The changes offer them greater freedom of expression and of movement, the satisfaction and challenge of entrepreneurship, the right to practice religion openly, and the availability of consumer products in far greater variety and quality than ever before. Nonetheless, the changes are so rapid and dramatic that many people feel anxious, particularly because the ultimate impact of these changes is unknown. Since the early 1990s, Russia, Belarus, Caucasia, and Central Asia have been overrun with Western businesspeople and lawyers, rock stars, franchise hawkers, international drug dealers and gun merchants, American academics and environmentalists, computer and cell phone vendors, U.S. Chamber of Commerce delegations—all missionaries in their own way for capitalism and the market economy. Even Christian missionaries have been holding revival meetings and seeking converts in many Russian cities. Occasionally, the word *transition* will be used to refer to the period from the collapse of the Soviet Union in 1991 to the present (the early 2000s), during which many economic, political, and social changes have been taking place.

Themes to Look For

Several themes appear throughout this chapter. Many of them are associated with the changing internal and external relationships just mentioned.

1. The move to a market economy. The most remarkable change in this region is the effort to move from a centrally planned economy to a market economy based on private enterprise. This change, in turn, has affected jobs, incomes, the delivery of social services, and the overall quality of life.

2. New alignments of people and trade. The breakup of the Soviet Union into smaller, more autonomous political units has resulted in population movements, new allocations of resources, and new trading patterns. These changes have left some parts of the old union isolated and disconnected from former trading partners, from needed resources, and from the transportation infrastructure.

3. The slow emergence of democracy. Market liberalization did not lead automatically to more popular participation in government, as Westerners had assumed. The newly independent states have all had to forge governing institutions in unstable circumstances, but regular and orderly elections are increasingly the norm at all levels from local to federal.

4. The reconfiguration of the institutions of a civil society. The USSR provided many social and civil safeguards for its citizens, but it did so in an authoritarian manner. During the disruptions of the transition to a market economy, crime and corruption flourished. It is taking time to build trained bureaucracies and civic-minded volunteer and service organizations.

5. Changing identities. Regional unity on political, economic, and social issues, once imposed by the centralized Soviet system, no longer exists. In place of this unity, new identifications—with new, smaller countries and with ethnic, social, and religious groups—are being forged. Hence, current linkages within Russia and among Russia, Caucasia, and the Eurasian republics are in the process of being broken or renegotiated.

6. Increased emphasis on environmental issues. The environment, never a priority during Soviet rule, is under considerable stress. It is getting somewhat more public attention in the transition period, because of public health effects that are now known and because some pollution has international implications.

Terms to Be Aware Of

At this time, there is no entirely satisfactory new name for the former Soviet Union. The formal name used in this chapter is *Russia and the newly independent states*. Russia is still closely associated eco- nomically with all of these states, but they are independent countries and their governments are legally separate from Russia's. Russia in- cludes more than 30 (mostly) ethnic *internal republics*—such places as Chechnya, Tatarstan, and Tuva (also spelled Tyva)—which con- stitute about a tenth of its territory and a sixth of its population.

THE GEOGRAPHIC SETTING

PHYSICAL PATTERNS

The physical features of Russia and the newly independent states vary greatly over the huge territory covered. The region bears some resemblance to North America in size, topography, climate, and vegetation.

Landforms

The landforms of this region follow a relatively simple pattern, as shown on the map at the beginning of the chapter. Moving west to east, there is first the eastern extension of the North European Plain, then the Ural Mountains, then the West Siberian Plain, fol- lowed by an upland zone called the Central Siberian Plateau, and finally a mountainous zone bordering the Pacific. To the south of these territories, from west to east, there is an irregular border of semiarid grasslands (**steppes**), barren uplands, and high moun- tains that separates the region from the southern reaches of the Eurasian continent.

The eastern extension of the North European Plain rolls low and flat from the Carpathian Mountains in Ukraine and Romania 1200 miles (about 2000 kilometers) to the Ural Mountains. This part of the region is called European Russia because the Ural Mountains are traditionally considered the border between Europe and Asia. It is the most densely settled part of the entire region and is its agricultural and industrial core. Its most impor- tant river is the Volga, which flows into the Caspian Sea. The Volga River is a major transport route; it is connected—via canals, lakes, and natural tributaries—to many parts of the North Euro- pean Plain, to the Baltic and White seas in the north, and to the Black Sea in the southwest.

The Ural Mountains extend in a fairly straight line south from the Arctic Ocean into Kazakhstan. The Urals are not much of a barrier to humans or to nature: there are several easy passes across the mountains, and winds carry moisture all the way from the Atlantic and Baltic across the Urals and into Siberia. Much of the once-dense forest that covered the Urals has been felled to build and fuel the new industrial cities established in the mountains during the twentieth century.

The West Siberian Plain, lying east of the Urals, is the largest plain in the world. A vast, mostly marshy lowland, it is drained by the Ob River and its tributaries, which flow north into the Arctic Ocean. Long, bitter winters mean there is a layer of permanently frozen soil (**permafrost**) just a few feet beneath the surface. Be- cause water doesn't sink through this layer, swamps and wetlands form, providing habitats for many birds in the summer months. Many species breed in the **tundra**—a treeless permafrost zone— and fly south for the winter. This plain has some of the world's largest reserves of oil and natural gas, although their extraction is made difficult by the harsh terrain and climate. The Central Siberian Plateau and the Pacific Mountain Zone are farther to the east; together they equal the size of the United States. Still farther east, on the Pacific coast, especially on the Kamchatka Peninsula, there are many active volcanoes. These are created by the sinking

Workers at an oil-extracting plant in the town of Raduzhny struggle with equipment in the cold environment of western Siberia. Raduzhny is in the northeastern district of the Khanty-Mansiysk Autonomous Okrug (territory). Khanty-Mansiysk is the richest oil region in Russia, producing about 58 percent of all oil in the country and about 5.4 percent of the world's oil supply. Some 40 joint stock companies (foreign and domestic combinations) operate there. The climate is severe: snow covers the ground from November to June, and the average temperature in January is between 0 and −10 degrees Fahrenheit. Temperatures may reach 65°F in July, the warmest month. [V. Kolpakov/TRIP.]

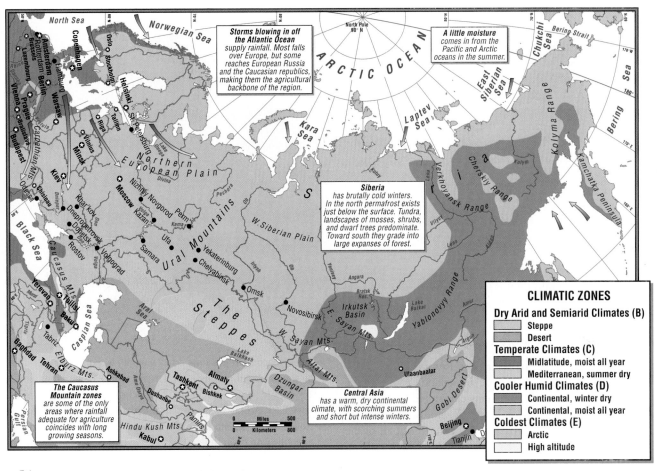

Figure 5.1 **Climates of Russia and the newly independent states.**

of the Pacific Plate under the Eurasian Plate, which causes gas and molten rock to rise to the surface through fissures and holes.

To the south of the arid grasslands that run in an irregular band from near the Caspian Sea to the Pacific is a wide, curving band of mountains, some lying within Russia and the newly independent states, others farther south and east. Those lying at least partly within the region include the Caucasus, Elburz, Hindu Kush, Tien Shan, and Altay mountains. Their rugged terrain has not deterred cultural influences that have penetrated north and south for literally tens of thousands of years. People have persistently crisscrossed the mountains, exchanging plants (apples, onions, citrus, wheat), animals (horses, sheep, cattle), technologies (agriculture, animal tending, portable shelter construction, rug and tapestry weaving), and religious belief systems (principally Islam and Buddhism, but also Christianity and Judaism).

Climate and Vegetation

Other than Antarctica, no place on earth has as harsh a climate as the northern part of the Eurasian landmass occupied by Russia (Figure 5.1). Here is a prime example of continental climate, in which the winters are long and cold, the summers short and cool

to hot. The seasonal continental "breathing" (described in Chapter 1, page 20) is dominated in winter by a long, dry, frigid exhaling of air that flows out of the arctic zones. A short inhaling in summer brings in only modest amounts of moist, warm oceanic air from the west and lesser amounts from the east. The mountain ranges along the region's southern edge block access to warm, wet air from the Indian Ocean far to the south.

Most rainfall in the region comes from storms that blow in from the Atlantic Ocean far to the west. By the time these initially rain-bearing air masses arrive, most of their moisture has been squeezed out over Europe. Nevertheless, sufficient rain reaches the fertile lands of European Russia and the Caucasian republics to make this area the agricultural backbone of the region. The Caucasian mountain zones are some of the only areas in the region where rainfall adequate for agriculture coincides with long growing seasons. Before the breakup of the Soviet Union, these areas were important sources of fruits and vegetables for the whole region. The natural vegetation of the northern part of the zone east of the Urals (European Russia) is mixed and boreal coniferous forest. The northern soils are generally infertile and fields are small. The best soils are in a region stretching from Moscow south toward the Black and Caspian seas, including much of Ukraine. The soils, of a type called **chernozem,** are black and as fertile as those in the American corn and wheat belt. Here the natural vegetation is open woodland and grassland (steppe).

East of the Urals, the lands of Siberia receive moderate precipitation (primarily from the east) but experience long, cold winters. Huge expanses of Siberia are covered with **taiga** (northern coniferous forest), which stretches to the Pacific. Agriculture is generally not possible because of the cold climate and poorly drained soils, though reindeer are tended in the far north beyond the taiga. Here in the tundra, a landscape of mosses, lichens, shrubs, and dwarf trees grows above permafrost that is only a foot or so below the surface. Plants are small and do not withstand much grazing or trampling.

East of the Caucasus Mountains, the lands of Central Asia have an arid and semiarid climate influenced by their location in the middle of a very large continent. Summers are scorching and short, the winters intense. These lands support grassland (steppe) vegetation. Large-scale cultivation is rarely possible without irrigation, and herding is practiced widely.

HUMAN PATTERNS OVER TIME

The core of this entire region has long been European Russia, the most densely populated area and the homeland of the ethnic Russians. From this center, the Russians conquered a large territory inhabited by a variety of other ethnic groups. These conquered territories remained under Russian control as part of the Soviet Union, which attempted to create an integrated social and economic unit out of the disparate territories. The breakup of the Soviet Union has reversed this gradual process of Russian expansion for the first time in centuries.

The Rise of the Russian Empire

For thousands of years, the politically dominant people in the region were nomadic herders who lived on the meat and milk of their herds of sheep, horses, and other grazing animals. Their movements followed the changing seasons in the wide grasslands stretching from the Black Sea to Lake Baikal. Nomads would often take advantage of their superior horsemanship and hunting skills to plunder settled communities. To defend themselves, permanently settled peoples gathered in fortified towns.

The towns arose in two main areas: the dry lands of Central Asia and the moister forests of Ukraine and Russia. As early as 5000 years ago, Central Asia was supporting settled communities, sometimes even large empires, that were enriched by irrigated croplands and by trade along the famed Silk Road, the ancient trading route between China and the Mediterranean (see Figure 9.19). The settled communities in the forests of European Russia eventually gave rise to the Russian Empire, which would one day dominate both the steppes and Central Asia. Powerful kingdoms arose in these forests after the arrival, about a thousand years ago, of the **Slavs,** a group of farmers who originated between the Dnieper and Vistula rivers, in what is now Poland, Ukraine, and Belarus. The Slavs expanded east, founding the towns of Kyiv (Kiev) and eventually Moscow. The threat of nomadic invasion remained, however, and the Mongol armies of Genghis Khan conquered these forested lands in the twelfth century. The **Mongols** were a loose confederation of nomadic pastoral people centered in eastern Central Asia; by the thirteenth century, they had conquered an empire stretching from Europe to the Pacific. Moscow's rulers became tax gatherers for the Mongols, dominating neighboring kingdoms and eventually growing powerful enough to end local Mongol rule in the late fifteenth century.

By the beginning of the seventeenth century, Russians centered in Moscow had conquered many former Mongol territories, integrating them into the Russian Empire, which by then stretched eastward from the Baltic over the thinly populated northern Eurasian landmass toward the Pacific (Figure 5.2). The outlying parts of the empire, south and east of the Ural Mountains, contained huge deposits of mineral resources and were inhabited by many non-Russian peoples. The first major non-Slavic area to be annexed was western Siberia (1598–1689). Although the Russians used many methods of conquest and administration developed by the Mongols, their expansion into Siberia in some ways resembled the spread of European colonial powers throughout Asia and the Americas. For example, much like Britain, France, and Spain, the Russian imperial colonists appropriated Siberian resources for

The archaeologist Jeannine Davis-Kimball, working in the Ural steppes on the border of northern Kazakhstan, found the burial ground of a 14-year-old young woman with bowed legs, possibly caused by a life on horseback. Artifacts recovered included arrowheads (1), a sword (2), a stone amulet (3), and shells (4). The finds are part of the evidence that nomadic women in the past participated fully in the economic life of their culture (herding, hunting, fighting, and playing games—often on horseback), as is the case in nomadic cultures of Tibet and Mongolia today. [Courtesy of Dr. Jeannine Davis-Kimball.]

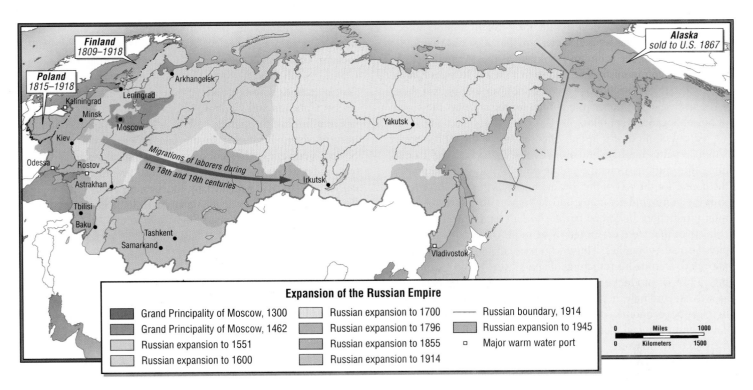

Figure 5.2 **Russian imperial expansion, 1300–1945.** [Adapted from Robin Milner-Gulland with Nikolai Dejevsky, *Cultural Atlas of Russia and the Former Soviet Union,* rev. ed. (New York: Checkmark Books, 1998), pp. 56, 74, 128–129, 177.]

their own use and invested little in local development. They upheld European notions of private property at the expense of indigenous patterns of communal land use. Perhaps most significant was that expansion into Siberia in the eighteenth and nineteenth centuries made way for massive migrations of laborers from Russia, who soon far outnumbered indigenous Siberians. By the mid-nineteenth century, Russia had also conquered Central Asia to gain control of cotton, its major export crop. Cotton was in high demand in global markets because the American Civil War was disrupting cotton exports from plantations in the United States.

The Russian Empire had great extremes of wealth and poverty. It was ruled by a **czar** (derived from "Caesar," the title of the Roman emperors), who, along with a tiny aristocracy, lived in splendor while the vast majority of the people lived in poverty. Many Russians were **serfs,** who were legally bound to live and farm on land owned by the lord. If the land was sold, they were transferred with it. Even though serfdom ended legally in the mid-nineteenth century, the brutal inequalities of Russian society persisted into the twentieth century, fueling opposition to the czar. By the early twentieth century, a number of violent uprisings had occurred.

A Communist Revolution

In 1917, at the height of Russian suffering during World War I, Czar Nicholas II was overthrown. Within a few months, a faction of revolutionaries called the **Bolsheviks** came to power. Though not very numerous, they were a highly disciplined organization inspired by a common ideology, **communism.** Based largely on the writings of the German revolutionary Karl Marx, communism criticized the societies of Europe as inherently flawed in that the **capitalists**—a wealthy minority who owned the factories, farms, businesses, and other means of production—dominated the impoverished, propertyless majority of people. With no other options, people had to work for low wages that undervalued their labor. Communism called on workers to unite to overthrow the capitalists and establish a completely egalitarian society with no government and no money, where people would work out of a commitment to the common good, sharing whatever they produced.

The Bolshevik leader, Vladimir Lenin, declared that the people of the former Russian Empire needed a transition period of centrally orchestrated drastic change in order to realize the ideals of communism. Accordingly, Lenin's Bolsheviks formed the **Communist party,** which set up a powerful government that ruled the former Russian Empire for the next 70 years. This highly authoritarian government was centered in Moscow, which had also been the center of power during most of the Russian Empire. Ethnic Russians held most key positions even in non-Russian areas. Lenin drastically increased the government's authority by creating a centrally planned **command economy** that gave government bureaucrats in Moscow authority to direct all economic activity. Private property was abolished, and the bureaucrats planned and managed all production and distribution. They decided where factories would be located, what and how much they would produce, who would manage them, where the products would be distributed, and what they should cost. Such a government-controlled system was termed **socialism.** The idea was that under a socialist system, the economy would grow quickly and hasten the transition to communism. The notion that the old Russian Empire was in a state of transition toward communism, but not yet there, was reflected in the new name chosen for the country: the Union of Soviet Socialist Republics.

After Lenin's death, Joseph Stalin's 26-year rule of the Soviet Union as party chairman and premier (1922–1953) brought a mixture of revolutionary change and despotic brutality that largely set the course for the rest of the Soviet Union's history. Stalin tried to shift the command economy into high gear. To increase agricultural production, he forced farmers to join large government-run collectives that were presumed to be more efficient than smaller plots farmed by families. Those who resisted collectivization were relocated. Despite his reshaping of agriculture, Stalin saw increasing industrial production as the true key to achieving economic growth. Accordingly, he ordered massive government investments in gigantic economic development projects, such as factories, dams, and chemical plants, some of which are still the largest of their kind in the world. The labor was largely supplied by former farmers. Eventually, government-controlled companies monopolized every sector of the economy, from agriculture to mining to clothing design and production.

This strategy of government control resulted in both massive successes and massive failures. During the Great Depression of the 1930s, the Soviet Union's industrial productivity grew steadily even while the economies of other countries stagnated. As was the case under the czar, however, much production was geared toward heavy industry (the manufacture of machines and transport equipment) and supplying the military with armaments, while little attention was paid to the demand for consumer goods and services that would have improved living standards for the Russian people. Conversely, Europe and North America were discovering at this time that everyday consumers, supplied with even modest wages, are a major source of economic growth. Meanwhile, Moscow gave scant consideration to the effects of heavy industrial development on either the natural environment or the health of Soviet citizens. During and after the Stalin regime, the Soviets created some of the world's worst industrial disasters (especially the unsafe disposal of hazardous materials) and most polluted and unlivable cities.

The most destructive aspect of Stalin's rule, however, was his relentless and ruthless amassing of personal power through fear. Sometimes called the Marxist czar, Stalin used the secret police, starvation, and mass executions to silence almost all opposition to his rule. The lucky survivors were those merely sentenced to labor in the frigid cold of Siberia. Stalin's atrocities, which resulted in the deaths of at least 20 million people, were a major reason that the Soviet Union had no real political revolution. The USSR continued nearly to the end to be governed by authoritarian power structures similar to those used by the czars.

The cold war, which pitted the USSR against the United States in an intense global geopolitical rivalry, further sapped the Soviet Union's resources. In an attempt to match the global military capacity of the United States, the Soviets diverted more and more resources to their military at the expense of much-needed economic

and social development. Internationally, the USSR promoted revolutionary communist ideology and its model of centrally planned socialist economic development in such far-flung countries as China, North Korea, Cuba, Vietnam, Nicaragua, and various African nations. Closer to home, especially in eastern Europe but also in Caucasia and Central Asia, the USSR maintained its pervasive influence through the use of political, economic, and military coercion. Beginning in the late 1960s, despite widespread Soviet assertiveness, the economies and political systems of countries under Soviet influence steadily drifted toward the free-market democratic model advocated by the United States and its allies. Beginning in 1979, a war in Afghanistan, launched to prop up a regime favorable to continued Soviet prominence in Central Asia, severely drained morale in the Soviet Union as a whole.

The Post-Soviet Years

In 1985, a reform-minded leader of the USSR, Mikhail Gorbachev, responded to pressures for change by opening up public discussions of social and economic problems. This innovation, labeled **glasnost,** ultimately brought about the end of the Soviet Union. Gorbachev's efforts to revitalize the Soviet economy through **perestroika,** or restructuring, resulted in little real change. But when he also began to democratize decision making throughout the Soviet Union in the late 1980s, long-silenced resentment of the government in Moscow boiled over and long-suppressed political and interethnic tensions emerged. Independence movements surfaced in the Baltic states. Gorbachev's liberalizing policies were strongly opposed by others who sought to unseat him. In 1991, the Soviet Union dissolved in an atmosphere of economic and political chaos.

In the post-Soviet years since 1991, one massive state has been replaced by 12 independent countries, each with its own agenda for reforming the failed systems inherited from the USSR. Russia is the chief successor of the Soviet Union and, although much smaller now, is still the largest country in the world (see Table 5.5, page 271). Russia has tried to maintain its strong influence on the rest of the former Soviet Union, but these efforts have foundered because the Russian leadership has been consistently unable to solve major problems. The transition to a free-market economy has proceeded rapidly and without much planning, leaving economies once centrally controlled drifting dangerously and democratic processes poorly developed. The region's already sad record of environmental abuse has worsened in response to intense financial strains. Nevertheless, there are some reasons for hope: public participation in democratic processes is increasing, there is now more open debate on a wide range of issues, elections are being held on schedule, and the relationships between Russia and the newly independent states are at least being discussed more openly than they were in the USSR.

POPULATION PATTERNS

European Russia is the center of population in Russia and the newly independent states (Figure 5.3). The settlement pattern in European Russia is the eastward extension of the band

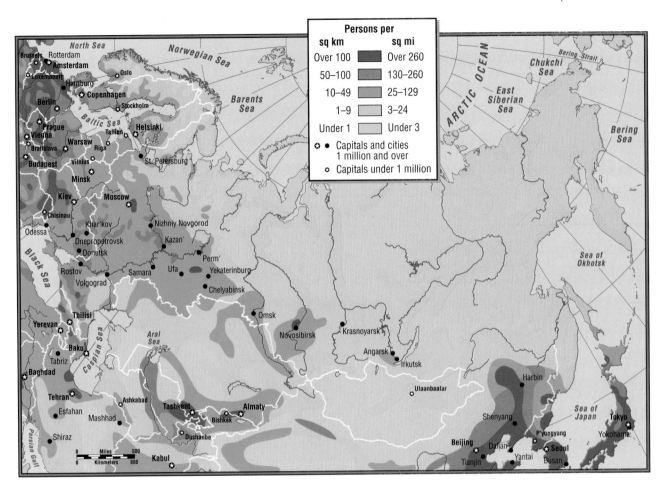

Figure 5.3 Population density in Russia and the newly independent states. [Adapted from *Hammond Citation World Atlas*, 1996.]

of heavy settlement (discussed in Chapter 4) that runs through the North European Plain beginning in Belgium and the Netherlands. A broad area of fairly dense population forms a wedge shape, with the base of the wedge stretching between Odessa in southern Ukraine on the Black Sea north to St. Petersburg on the Baltic Sea (see Figure 5.3). The irregular point of the wedge (medium rust color on the map) stretches across the eastern industrial cities of Yekaterinburg and Chelyabinsk to Novosibirsk, the largest city in Siberia. The area of relatively high population density corresponds roughly to the distribution of good agricultural soils.

Beyond Novosibirsk, light settlement (deep yellow on the map) follows roughly the irregular pattern of industrial development (including the processing of such minerals as coal, uranium, natural gas, and petroleum) across Siberia. These activities, in turn, are linked to the course of the Trans-Siberian Railroad to the Pacific. Such eastern industrial cities as Krasnoyarsk, Angarsk, and Irkutsk, all lying on the Trans-Siberian Railroad, each have several hundred thousand people. Beyond the cities, eastern Russia is very lightly populated territory—25 or fewer people per square mile (10 per square kilometer)—all the way to the Pacific Far East of Russia.

In the west, a secondary spur of dense settlement extends south from the main wedge into the Caucasus region between the Black Sea and the Caspian Sea. Another patch of relatively dense settlement is centered on Tashkent and Alma-Ata (Almaty) and along major rivers in the Central Asian republics. Here the development of irrigated agriculture (especially the cultivation of cotton) and mineral extraction resulted in high rural densities, fueled partly by ethnic Russian immigration. The correlation of good soil and population density does not hold everywhere: some industrial cities east of the Ural Mountains are on particularly infertile land.

Recent Population Changes

Before its breakup, the USSR was considered a developed region with fairly high standards of living and well-being, but in the early 1990s the well-being of the citizens of Russia and the newly independent states deteriorated significantly (Table 5.1). Not only did infants and children die in large numbers (column 7), but adult life spans also fell for a few years. In Russia, male life expectancy declined from 63.9 years in 1990 to 58.9 in 1993 (by 2000, it had rebounded to 61 (Table 5.1, column 5). Female life expectancy dropped from 74.4 years to 71.9 in the same period

TABLE 5.1 *National population statistics*

Country (1)	Population (millions), 1998 (2)	Births per 1000 (3)		Fertility rate (number of children per woman aged 15–45) (4)		Life expectancy at birth, male, 2000 (5)	Life expectancy at birth, female, 2000 (6)	Infant mortality rate (deaths per 1000 live births) (7)	
		1990	2000	1990	2000			1990	2000
Belarus	10.0	13.9	9	1.9	1.3	63	74	11.9	11
Caucasia									
Armenia	3.8	24	10	2.6	1.3	71	78	18.6	15
Azerbaijan	7.7	26	15	2.7	1.9	68	75	23.0	17
Georgia	5.5	17.0	9	2.2	1.2	69	76	15.9	15
Central Asia									
Kazakhstan	14.9	21.7	14	2.7	1.7	59	70	26.4	21
Kyrgyzstan	4.9	29.3	22	3.7	2.8	63	71	30.0	26
Tajikistan	6.4	38.8	21	5.1	2.7	66	71	40.7	28
Turkmenistan	5.2	34.2	21	4.2	2.5	62	69	45.2	33
Uzbekistan	24.8	33.7	23	4.1	2.8	66	72	34.6	22
Moldova	4.3	N/A[a]	11	N/A	1.5	63	70	N/A	18
Russia	145.2	13.4	8	1.9	1.2	61	73	17.4	17
Ukraine	49.5	N/A	8	N/A	1.3	63	74	N/A	13
Total	228								
For comparison									
United States	275.6	16	15	2.0	2.1	74	79	9.7	7
Europe	533.3	13	10	1.7	1.4	73	79	12	9

[a]N/A = data not available.

Sources: 2001 World Population Data Sheet (Washington, D.C.: Population Reference Bureau); and *Population Bulletin* 49 (December 1996): 48–49.

(in 2000, it was back up to 73). No other country has experienced a gap of 12 years between average male and female life expectancies during peacetime, and this gap indicates significant stress in the lives of the region's males. Male life expectancy in Russia (61 years) is now the shortest of all industrialized countries. On the other hand, in Caucasia—despite armed conflict, severe economic reversals, and drastic reductions in living standards—both males and females have a significantly longer life span than they do in Russia or Central Asia (Table 5.1, columns 5 and 6). This pattern was noticed well before the post-Soviet transition, and it may be partly a result of longer agricultural growing seasons, a more nutritious diet, little consumption of alcohol (many are Muslim and hence do not drink), and a genetic predisposition to long life.

The apparent cause of declining life expectancies is the physical and mental distress caused by lost jobs and social disruption,

TABLE 5.2 *Indicators of personal stress in selected countries*

Country (1)	Daily per capita supply of calories 1997 (2)	Food production index (1989–1991 = 100), 1998 (3)	Suicide rate per 100,000, 1998 (4)		Divorces as percentage of marriages, 1996 (5)
			Male	Female	
Armenia	2371	75	3.6	1.0	18
Belarus	3225	67	448.7	9.6	68
Russia	2904	59	772.9	13.7	65
Ukraine	2795	47	338.2	9.2	63
Uzbekistan	2433	114	9.3	3.2	12
For comparison					
United States	3699	121	119.3	4.4	49
France	3518	106	330.4	110.8	43

Source: United Nations Human Development Report 2000 (New York: United Nations Development Programme), Tables 22 and 27.

most notably reflected in high rates of alcoholism and suicide. By 1996, divorce rates—already high before the USSR dissolved—reached more than 6 out of every 10 marriages in Belarus, Russia, and Ukraine (Table 5.2, column 5). Alcohol abuse shortens and saddens many lives in this region, especially those of men. Male suicide rates in Russia, Belarus, and Ukraine are now the highest on earth (Table 5.2 column 4). After 1991, many people began to suffer nutritional deficiencies caused by sharply falling incomes and by scarcities, some of which were brought on by conflict in a few ethnic enclaves in Russia and in Caucasia. (The internal republics are more than 30 ethnic enclaves in Russia, whose political status and relationship to the Russian state are continuously being renegotiated republic by republic.) Daily per capita intake of calories was markedly below that of the United States and Europe (Table 5.2, column 2). The Canadian geographer James Bater has shown that per capita consumption of basic foodstuffs declined throughout Russia and the newly independent states in the mid-1990s. Food production declined overall after 1990 (Table 5.2, column 3), increasing only in those areas that had previously grown only nonfood cash crops. In the Soviet era, for example, Uzbekistan grew cotton and purchased wheat and other food from Russia; by the late 1990s, Uzbekistan was producing vegetables, fruit, grain, and livestock and exporting small amounts of food. Overall, during the 1990s people were eating fewer vegetables; less meat, eggs, and dairy products; and even less bread, long a staple of the diet in this region. The cost of food takes up as much as half of family budgets. In the mid-1990s, Bater wrote, "A diet of bread,

pasta products, tea, and the occasional egg or piece of sausage is now the daily reality for literally millions of people, in both cities and rural areas." Though entrepreneurial farmers (especially from Ukraine and the Caucasus region) supply attractive fresh food to the best city markets, ordinary people can rarely afford to buy these luxuries. Those who have access to land plant gardens; but there is a limit to what they can produce in the small spaces available and during the short growing seasons.

Environmental pollution also plays a large, if poorly understood, role in untimely illness and death (see page 267 in the Environmental Issues section). A third or more of the population may be affected by noxious emissions; heavy metal pollution of the ground, air, and water; and nuclear contamination. In Central Asia, the buildup of chemicals used in cotton cultivation has threatened the health of agricultural workers and helped to pollute river waters and the Caspian and Aral seas (see page 268). In the industrial cities of European Russia and Siberia, careless handling of industrial waste and inadequate water and sewage treatment endanger health through the pollution of air, water, soil, and food.

Declining birth rates are another significant change in population patterns. In the Soviet era, birth rates and fertility rates in Russia and Belarus were already low compared to birth rates in the United States and Europe; since the early 1990s, the rates have dropped still further (see Table 5.1, columns 3 and 4). Now the traditionally higher birth rates in the Caucasian and Central Asian republics are also dropping, ranging from just 14/1000 in Kazakhstan

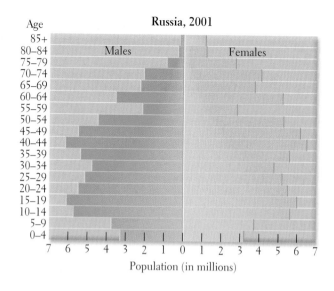

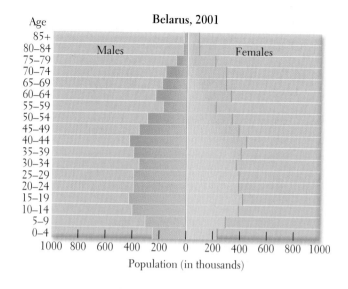

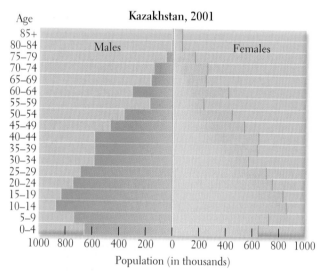

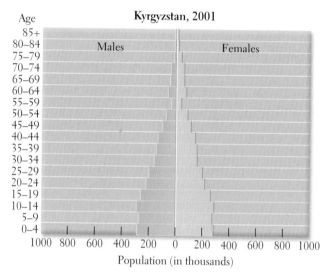

Figure 5.4 **Population pyramids for Russia, Belarus, Kazakhstan, and Kyrgyzstan.** The pyramid for Russia is at a different scale (millions) from the other three (thousands), but this does not unduly affect the pyramid shape. [Adapted from U.S. Bureau of the Census, International Data Base, at http://www.census.gov/cgi-bin/ipc/idbpyry.pl.]

to 23/1000 in Uzbekistan. Surveys show that people are choosing to have fewer children mostly out of concern for the gloomy economic prospects of the near future. In 1992, 75 percent of Russian women who had decided against childbearing cited insufficient income as the reason. Because contraceptives are in short supply and of poor quality, abortion is the method of choice for birth control.

By 2001, Russia, Belarus, and Ukraine all had death rates that were close to twice their birth rates, and their populations were actually declining (although very slowly). Population pyramids for several republics (Figure 5.4) show the overall population trends and reflect differences in patterns of family structure and fertility. The pyramids for Belarus and Russia resemble those of European

countries (for example, Germany, as shown in Figure 4.6). They are narrower at the bottom, indicating that birth rates in the last several decades have declined. The narrower point at the top for males depicts their much shorter average life span (see Table 5.1). The bases of the pyramids for Kazakhstan and Kyrgyzstan, in Central Asia, are narrowing as a drop in birth rates accompanies urbanization. But both pyramids still have substantial bases, reflecting the higher fertility rates common in a more rural, primarily Muslim society with a patriarchal family structure. All pyramids have a noticeable "waist" at the 55–59 age group, the result of high death rates and low birth rates during and just after World War II.

CURRENT GEOGRAPHIC ISSUES

The goals of the Soviet experiment begun in 1917 were unique in human history: to reform quickly and totally both a human society and its social, political, and economic systems. Now, in the aftermath of the Soviet Union's collapse, the people are undertaking a second unique experiment: the transformation of a communist society and a centrally planned economy into its antithesis—a society based on democracy and a free-market economy. At this point, the process has barely begun, and life is proceeding amid considerable chaos.

ECONOMIC AND POLITICAL ISSUES

VIGNETTE The kindly, gray-haired teacher tapped the blackboard under the words *profit* and *inventory* as the first-graders struggled to pronounce the Russian words. They were reviewing the story of Misha, a bear who opens a honey, berry, and nut store in the forest. He soon outsmarts the overpriced government-run Golden Beehive Cooperative, becoming first a prosperous bear and eventually the finance minister of the forest. [Adapted from Sarah Koenig, "In Russia, teaching tiny capitalists to compete," *New York Times* (February 9, 1997).]

Teaching economics in the first grade may seem a bit surprising, and the practice may not have been widespread, but this story

In an effort to supplement their government pensions (many of which have not been paid for months), some older citizens here at Petrovskiy Zavod sell food to Trans-Siberian Railroad passengers, using old baby carriages as carts. [Gerd Ludwig/National Geographic Image Collection.]

illustrates the marked shift in focus that took place in the 1990s throughout the region of Russia and the newly independent states. One real change is that schools now emphasize the teaching of free-market economics instead of the ideas of Marx and Lenin. This shift in education is but one of many ambitious market reforms pursued in some former Soviet states, especially Russia. To fully appreciate what free-market reforms have meant to this region, however, we must understand the Soviet institutions that were previously in place.

The Former Command Economy

As we have seen, although the long-term goal of the command (or centrally controlled) economy was to achieve communism, the shorter term goals were to end the severe deprivation suffered by so many under the czars and to distribute the benefits of economic development to all regions. To some extent, these goals were met, and until the 1960s the Soviet economy grew rapidly and abject poverty was largely conquered. Such basic necessities as housing, food, health care, and transportation were provided to all for free or at low cost. Nonetheless, the command economy allocated goods and services inefficiently. Scarcities of some goods and gluts of others made it clear that the Soviet economy was less efficient than market economies in allocating resources.

In a free-market economy, no single person or group decides what (and how much) will be produced or what prices will be. Production and pricing decisions are made independently by the managers of many privately owned companies, based on their judgments of what the market will bear. Hence, if a manager errs in predicting demand for a product, the mistake primarily affects his or her own company, not the entire economy. In the Soviet command economy, however, bureaucrats set production goals for the whole country, which meant that even small miscalculations of production types and amounts created country-wide shortages or gluts. Shortages of food or raw materials occurred regularly throughout the Soviet era. Clothing was boring and unimaginative, and most consumer goods—for example, washing machines, automobiles, and small appliances such as vacuum cleaners—were of poor quality and were available only to a privileged few. (Televisions and small refrigerators were an exception; they became widely available in the 1960s, and by the 1970s most urban households had these two appliances.)

Inefficient production methods were also typical in the command economy. In large part, this problem resulted from a lack of competition among producers. In free-market economies, competition among producers of goods usually encourages companies to supply cheaper, higher quality goods, because these are the goods that consumers tend to purchase. In contrast, Soviet bureaucrats faced little or no competition and often concentrated production of a given good in a single factory to minimize the labor, equipment, and raw materials needed for production. Because each Soviet factory tended to have a monopoly on the production of a particular product, Soviet goods became infamous for their high cost and low quality compared to similar goods produced in free-market countries.

Another major problem of the USSR was its failure to keep up with the rest of the industrialized world in technological and managerial innovation. Soviet engineers did develop advanced military and space exploration technologies, such as various nuclear weapons, world-renowned fighter jets, the first satellite (launched in 1957), and the first manned spacecraft (piloted by Yuri Gagarin in 1961). The Soviets also made technological breakthroughs in ferrous (iron) metallurgy and pioneered in mainframe computer technology (later abandoned when personal computers became available). But in the civilian economy, Moscow's bureaucrats often settled for outdated technologies to produce consumer goods. The equipment typically used more resources, produced more pollution, and was less productive than equipment used in capitalist, free-market economies. At the same time, Soviet leaders paid little attention to the need for rewarding hard work and innovation in the workforce. Instead, promotions and privileges were usually granted to individuals who had loyalty to and personal connections with the Communist party.

Soviet Regional Development Schemes. A central plan dictated the location of industrial and agricultural enterprises in the Soviet system. For several reasons, the USSR encouraged economic development in the farthest reaches of its territory. First, there was a desire to bring to all parts of the region the higher standards of living enjoyed in the industrialized western part of the country bordering Europe. The central planners and their military advisors also thought that dispersing industrial centers throughout the country would make them easier to defend militarily because it would be harder for an enemy to attack widely separated sites. In addition, there was some effort to locate new developments close to necessary resources. Thus the metal-processing firm Norilsk Nickel was situated near mineral beds in far northern Siberia (see the box "Norilsk: A City in Transition to a Global, Free-Market Economy" on page 262), and commercial cotton cultivation was linked to irrigable rivers in Central Asia (see Figure 5.15). Another factor in industrial location was the need to buttress Russia's claims to distant territory. Soviet leaders thought that economic development projects would placate ethnic minorities whose resources were being diverted to national uses and that the migration of ethnic Russians into minority areas would promote conformity with the dominant socialist ethic and with modernized ways of life.

In a market economy, many of the locations the Soviets selected for production centers would be seen as highly inefficient, because it costs more to ship a product from a remote factory to a market center than it does to ship it from a nearby location. Moreover, the far-flung industries were never sufficiently linked with ground transport to encourage the voluntary spread of settlement and the growth of local markets of the sort that fostered the spread of development in western North America.

Transport Problems. In locating industries in far-off corners of the region, the command economy purposely ignored the tremendous transport problems posed by the region's huge size and challenging physical geography. Throughout the region's history, long winters, permafrost, swampy forestlands, and complex upland landscapes, especially in Siberia, had limited the development of an efficient transport system. Although the former Soviet Union is more than two and a half times the size of the United States, it has less than one-sixth the number of hard-surface roads and virtually no multilane highways. Existing roads are of such poor quality that cities in eastern Russia and the Central Asian republics are linked only by rail and air. The Trans-Siberian Railroad remains the chief connection between Moscow and Vladivostok, the main port city on the Russian Pacific coast. The rail trip takes several days, and to drive the distance on roads would be practically impossible. In European Russia, feeder rails and roads north and south of the main line connect to industrial cities and resource supplies; by the late twentieth century, oil and gas pipelines supplemented the railroad. But, to this day, despite the Soviet plans to promote industry across the region, the transport system—roads, railroads, air service, and pipelines—remains highly concentrated in the more populous west and underdeveloped east and south of the Urals (Figure 5.5).

The methods used in the centrally planned economy to determine the cost of products did not always include full transport costs. Now factories operating in remote locations such as Siberia and Kazakhstan must include these costs in determining the prices of their goods, if they expect to earn profits. They must cut manufacturing costs to accommodate the high transportation costs if they want the prices of their products to be competitive with other suppliers. Another solution is to relocate factories closer to markets.

Reform in the Post-Soviet Era

Russia's economic reforms have been ambitious but haphazard. Many key institutions of the command economy have been dismantled, but so far this change has mostly benefited a small group of wealthy and politically connected individuals, many of whom are corrupt. In some respects, the lives of the majority have become more difficult.

Two key economic reforms have been privatization and the lifting of price controls. **Privatization** is the selling of industries formerly owned and operated by the government to private companies or individuals. Estimates vary, but it appears that in 2000 approximately 70 percent of Russia's economy was in private hands, a significant change from the 100 percent state-owned economy of 1991. Even the remaining state-owned industries have changed the way they operate to follow free-market rules. In the old command economy, the government set the prices at which goods could be bought and sold. Often, prices on certain basic necessities such as bread, milk, and specific vegetables were set lower than production costs, simply to make these goods affordable to all. Now prices are determined by supply and demand.

Lifting price controls has had unpredicted consequences. When price controls were removed and goods sold at their market prices, the prices skyrocketed and the managers or new private owners gained wealth. But with drastically higher prices, the

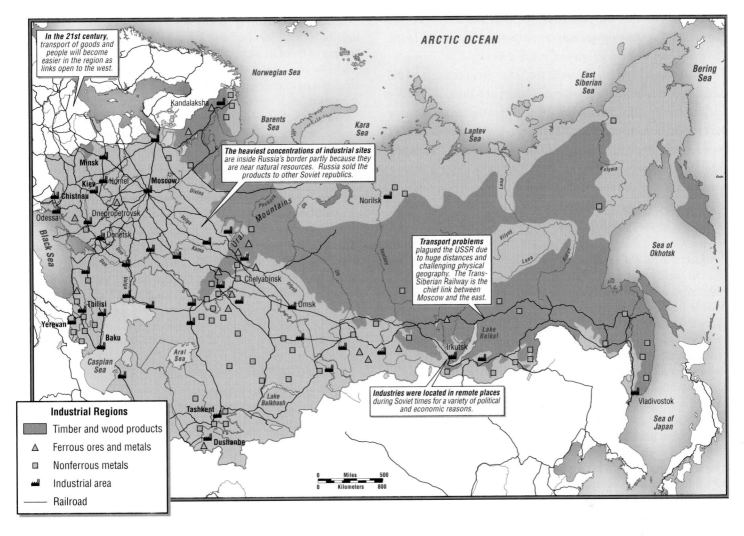

In the 21st century, transport of goods and people will become easier in the region as links open to the west.

The heaviest concentrations of industrial sites are inside Russia's border partly because they are near natural resources. Russia sold the products to other Soviet republics.

Transport problems plagued the USSR due to huge distances and challenging physical geography. The Trans-Siberian Railway is the chief link between Moscow and the east.

Industries were located in remote places during Soviet times for a variety of political and economic reasons.

Industrial Regions
- ▨ Timber and wood products
- △ Ferrous ores and metals
- □ Nonferrous metals
- ⛏ Industrial area
- — Railroad

Figure 5.5 Economic issues: principal industrial areas of Russia and the newly independent states. [Adapted from Robin Milner-Gulland with Nikolai Dejevsky, *Cultural Atlas of Russia and the Former Soviet Union,* rev. ed. (New York: Checkmark Books, 1998), pp. 186–187, 198–199, 204–205, 216–217.]

majority of people had to use their savings to pay for simple necessities. One consequence was that individuals who would have liked to become small business owners lacked the capital to launch these enterprises.

Nor has privatization had the intended outcome. Among the first enterprises privatized were the most lucrative ones: those producing raw materials for export, such as oil, timber, or metals. The profits from these exports are no longer used to fund government social service or pension programs as they once were. Moreover, many workers have lost their jobs as private companies reduced their workforces to save money and operate more efficiently. Meanwhile, many enterprises that produce consumer goods for Russian citizens have not been privatized because the changes needed to make them more efficient are so costly that investors find them unattractive. Hence state-made products remain expensive and of poor quality. Finally, the process of privatization has been plagued

with corruption: officials in the administrative hierarchy allowed buildings, factories, and access to resources to be sold for a fraction of their real value, in return for kickbacks or bribes.

VIGNETTE   It was the summer of 1995, and Yuri Ivanovich was ecstatic. In his driveway sat a brand new Lada automobile. With a few minor adjustments, it would soon be humming down the street. First he tackled an obvious little detail—the technicians at the government-run Lada assembly plant hadn't bothered to attach the gear lever to the gear box. By the end of the day, Yuri had fixed a host of similar problems. But he didn't mind the trouble, even if his brand-new car broke down a lot and looked a bit dated compared to the cars he saw in imported European magazines. (The 1995 Lada was a virtual carbon copy of a 1972 Fiat 124.)

During and just after the Soviet era, the Lada was known as "the people's car" because people like Yuri, a midlevel bureaucrat in the national department of education, could afford it.

By the mid-1990s, the demand for better cars produced in the competitive markets of Japan, Korea, Europe, and the United States was growing. By 2001, much-improved models of the Lada were being built in the Russian city of Tol'yatti by the Russian-owned AvtoVAZ Incorporated—some of them made to European Union standards and exported to Europe, the Americas, and North Africa. The elite who have prospered in the new economy, however, favor luxury cars such as Mercedes-Benz, Jaguar, and BMW. [Adapted from "Russia's car industry: modernizing the mastodon," *The Economist* (June 10, 1995).]

Corruption and organized crime have become major obstacles to economic advancement and effective government. Although crime and corruption existed during Soviet times, the decline in the government's power and authority since 1991 has allowed many crimes to go unchecked. Throughout the region, unsupervised upper-level provincial bureaucrats trade government vehicles, machinery, or raw materials for personal gain. The main buyers are highly organized criminals, often referred to as the Russian Mafia, who make large profits selling these items on the black market. In 2000, the Russian government estimated that through such corrupt arrangements, organized crime controlled 40 percent of private business and 60 percent of state-owned companies. Mafia-style violence deters legitimate foreign investors, who could fund needed economic development. Moreover, as reported repeatedly by the British Broadcasting Corporation (BBC), the activities of the Russian Mafia are global. Russian criminal organizations, some of the most sophisticated on earth, have used their access to Russian resources to expand their operations abroad to Israel, Hungary, London, Switzerland, parts of South America, the Caribbean, and the United States.

The Growing Informal Economy. The informal economy—that part of the economy that is unregulated, illegal, and untaxed—is a major problem throughout the former Soviet Union because governments are unable to collect revenue on informal economic activity. To some extent, the new informal economy is an extension of the old one that flourished under communism. A black market based on currency exchange and the selling of hard-to-find luxuries has long existed. For example, in the 1970s, savvy Western tourists could enjoy a vacation on the Black Sea paid for by a pair or two of smuggled Levi blue jeans and some Swiss chocolate bars. Today, however, the range of informal ventures is more extensive, as many people who have lost stable jobs due to privatization now depend on the informal economy for their livelihoods. Often they operate out of their homes, selling cooked foods, making vodka in their bathtubs, assembling telephones, or pirating and vending computer software, just to name a few informal activities. Indeed, so much of the economy is in the informal sector that in many countries of this region, people may be better off than official gross domestic product (GDP) figures suggest. For example, the true agricultural output is underreported because so many farmers, still employed on declining collective farms, now earn a considerable portion of their income from the private gardens they grow on the fringes of the state farms.

The loss of tax revenue means that the Russian government does not have enough money to pay the salaries of workers in the remaining state agencies, nor can it pay the pensions of retired workers or provide many essential public services such as electricity and garbage collection. To demonstrate the scope of the lost tax revenue, consider the effect of the underground economy on vodka production. Homemade and covertly produced vodka now accounts for 70 percent of the Russian market, drastically reducing sales of the legitimate vodka that was once a large source of tax revenue for the government. In the late 1990s, the average Russian was estimated to drink between 25 and 45 liters of vodka a year, and even with 45 percent of sales then untaxed, vodka produced nearly the tax revenue of oil. In June 2000, the government set up a state watchdog agency to oversee and perhaps return state control to the alcohol industry so taxes could again be levied and collected. Tax dodging, which does not carry the stigma it does in other free-market countries, has reduced tax revenues even from businesses in the formal economy, which often do not pay the taxes they owe (Figure 5.6).

Debt Crisis. Russia's large foreign debt dominates the country's financial relationships with other major countries. As of early 2001, Russia's debt was a staggering U.S. $148.7 billion. Of this sum, two thirds was incurred by the Soviet Union prior to 1991,

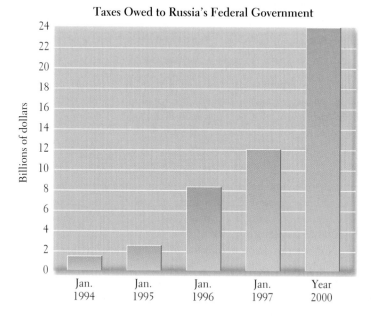

Figure 5.6 Tax debt in Russia. Even some large companies in Russia do not pay their lawful taxes, and enforcement of tax laws is hard to establish. The unpredictable state of the Russian economy has motivated some successful entrepreneurs to bank their profits outside the region, so the actual sum lost to taxation may exceed the figures indicated here. [Adapted from *New York Times* (February 19, 1997): A8; and *The Russia Journal* (June 17–23, 2000): 12.]

and theoretically the newly independent states should help Russia repay this money. But they have no surpluses to dedicate to this cause and little incentive to do so. In addition, the war in Chechnya (page 257) has added to Russia's debt at the rate of $200 million a year. The debt mounted after 1991 when the government had to borrow because tax collections waned at the same time that exports lagged, industries closed, and unemployment rose. The money is owed to private commercial lenders, foreign countries (Germany is Russia's biggest creditor by far), and the IMF and the World Bank. In early 2000, Russia defaulted on its debt payments and then requested debt rescheduling with all its creditors. Each request has led to intense negotiations. In particular, Western lenders want to make sure that any rescheduling or forgiveness of debt is tied to Russia's commitment to achieve a free-market economy, reduce theft and corruption, and encourage Russian entrepreneurs to reinvest their profits in Russia, not simply keep it in foreign banks. For its part, Russia worries that its annual obligatory payments, if not rescheduled, will undermine its economic growth. In 2000, private lenders agreed to cut 36 percent off the $32 billion Russia owes them and to extend payments over a 30-year period. It is interesting that foreign countries owe Russia, as the successor state of the Soviet Union, more than $100 billion. This money was loaned by the Soviets to their allies, such as Vietnam, Cuba, Mongolia, Mozambique, Angola, Algeria, and Yemen, but was never repaid.

Market Reforms and International Trade. For decades, Russia had a colonial-like relationship with its associates in eastern Europe, the Caucasus, and Central Asia, appropriating their resources in uneven exchanges that left some republics poorer and more dependent on Russia than they would have been otherwise. Today, despite Russia's struggle to maintain what is left of these arrangements, several former Soviet republics have already drifted far out of the sphere of Russian influence. The Baltic republics of Latvia, Lithuania, and Estonia are already so much more closely tied to Europe that they are not discussed in this chapter. Belarus retains close economic ties with Russia—the two countries trade freely without tariffs or customs duties. Events in the early 2000s in Ukraine and Moldova indicate that they may soon join this union. Other countries remain economically connected with one another and with Russia by arrangements that may not last. For example, the cotton growers of the Central Asian republics currently sell their product to textile mills located within Russia, which export cotton cloth back to Central Asia. In time, Central Asia will most likely build its own cotton mills and reap the profits of exporting finished textiles to the global marketplace. Similarly, Central Asian countries, whose oil and gas industries are unevenly distributed across their region, currently sell energy resources to Russia at prices below global market price, but in the future they are likely to sell to the global market as well. Multinational energy companies are now pressuring Central Asian countries to bypass Russia and send their oil instead to the world market through new delivery systems that would run through countries in Caucasia, Turkey, Afghanistan and Pakistan, or China.

Obstacles to International Trade. One goal of the economic reforms in Russia and the newly independent states is to increase

Russia has created the Vostochny-International Container Services (VICS) at Vostochny Port, in the Russian Far East, as one way of promoting closer ties with Europe and Asia. The facility connects to the Trans-Siberian Railroad, a high-capacity, double-track electric line that reaches Central Asia and Europe. The combination offers the shortest route from Far East Asia to central China, Central Asia, Russia, and Europe at competitive rates and transit times. A shipment from Shanghai to Finland, for example, takes 23 days, roughly half the time as sea routes. [Courtesy of Vostochny International Container Services.]

their exports, especially to the world market. But these countries face many problems as they seek to establish new international trade relationships. First, they are accustomed to trading with one another, under terms set by central planners. Each part of the region specialized in—and often held monopolies in—the production of specific products such as oil, cotton, electronics, and food crops. With new national borders now separating these regions, each country must establish new trade agreements to ensure sufficient food for its populace and adequate resources for its industries. Second, all must improve their products and manage production costs to compete globally with producers in Europe, America, and Asia. Most former Soviet countries lack marketing expertise, and they must acquire these skills rapidly. Many of these countries now maintain Web sites on the Internet to familiarize the public with their products and to seek investors.

Because of the difficulty of implementing trade changes rapidly, Russia and most of the newly independent states currently export raw resources—oil, gas, timber, minerals, chemicals, furs, wool, cotton—to generate earnings that they can use as they reorganize their production of manufactured goods and food. Some countries, such as Russia, Armenia, and Ukraine, have increased their trade with free-market countries, but the earnings from these exchanges have varied from year to year due to varying levels of production and varying competitive circumstances in the world market. These uncertainties have made it difficult to manage national budgets and to fund social programs and infrastructure projects.

The underdeveloped transport system has created further obstacles for the new, independent nations as they try to trade with partners in the global economy. Transportation routes are not always in ideal locations for present patterns of foreign trade. For example, Kazakhstan and Uzbekistan, independent since 1991, are increasing their direct trade with the outside world, but they are landlocked and still have only feeder rail connections with what are now Russian railroads (see Figure 5.5). Their oil and gas pipelines flow to Russia (Figure 5.7), not to other customers, and their air links have had to be refocused away from Moscow.

Supplying Oil and Gas to the World

Crude oil and gas are Russia's most lucrative export. The Soviet Union was the world's number one producer of crude oil from 1983 until its demise in 1991 and in the early 2000s is second in

the world only to Saudi Arabia. Despite large oil reserves recently discovered in Central Asia, Russia on its own remains the prime oil exporter in the region (see Figure 5.7). According to the U.S. Chamber of Commerce, by 1999 Russia was exporting oil at the rate of 6 million barrels a day and was consuming at least another 4 billion barrels itself. Gas exports were at the rate of 21.5 trillion cubic feet (608 billion cubic meters) per year.

In the past, the major customers for Soviet oil and gas were the countries of eastern Europe. After the breakup in 1991, however, these countries found it difficult to keep up payments for fuel from Russia, so Russian oil and gas exporters began cultivating markets in western Europe. By 1998, 89 percent of net exports were to countries outside the former Soviet Union, such as the United Kingdom, France, Italy, Germany, and Spain. The range of European customers is likely to expand as conduits for oil and gas transfers expand. Pipelines through central Europe are being improved and tanker depots on the Baltic expanded. India, China, and Japan are among other customers for Russian oil and gas and for oil and gas produced in the Central Asian countries.

In Russia, oil companies and Gazprom (Russia's joint government and private gas-extracting and exporting monopoly) account

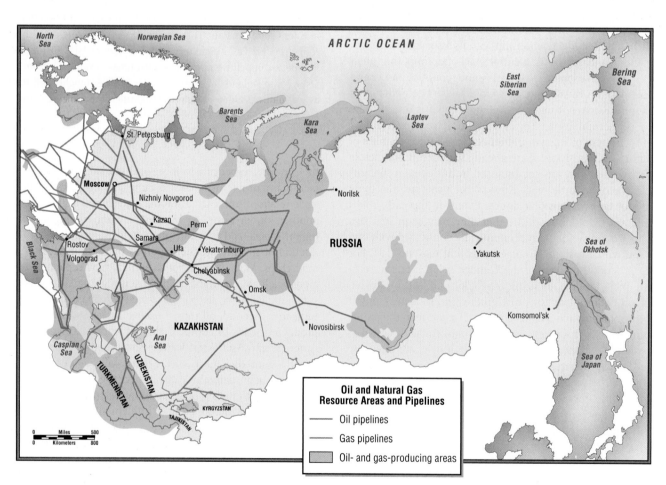

Figure 5.7 **Oil and natural gas: resources and pipelines.** [Adapted from William H. Berentsen, *Contemporary Europe: A Geographic Analysis* (New York: Wiley, 1997), pp. 625–627.]

for more than half of federal tax receipts. Because the oil companies provide so much revenue, the Russian government keeps a tight rein on the oil and gas sector, strengthening rules to ensure that the industry pays all its taxes regularly. As world oil and gas prices rose in the late 1990s and into the 2000s, tax proceeds increased, helping the budgets of Russia and of the newly independent states.

One measure of the increasingly global influence wielded by Russia's oil and gas industries is their expansion into retail marketing in the United States. In December 2000, Lukoil, a Russian oil company, announced the purchase of 72 percent of Getty Petroleum Marketing, a chain of 1300 gas stations in 13 eastern states. This is the first public acquisition by a Russian corporation of a U.S. firm, and although small ($71 million was paid), the deal is important psychologically and could lead to other Russian investments in the United States.

Greater Integration with Europe and the United States

The oil and gas trade with Europe, the United States, and Asia does much to improve Russia's economy and to connect Russia with the inner circle of industrialized nations. Yet the United States and Europe, motivated by the unspoken worry that Russia might return to foe status if its economic and political burdens are not eased, are seeking ways to bring Russia (as well as some of the newly independent states) into closer formal association with established trading institutions. Two such efforts have been the invitations to Russia to join the World Trade Organization and to participate more fully in meetings of the **Group of Seven,** an organization of the most highly industrialized nations. In part, these invitations are intended to reassure Russia as Poland, the Czech Republic, and Hungary— once Russia's close allies—become ever more closely linked to the European Union. These three have already joined the North Atlantic Treaty Organization (NATO), the old cold war military alliance against the USSR. Closer association with the European Union is also appealing to Russia, because the EU has been so successful in reorganizing the economic and political face of Europe.

Holding out the hope of trade and guest worker agreements, the EU has many projects that are aimed at bringing the countries of the former USSR into compliance with European trade-related standards. These standards include regulations on personnel training, the personal safety of workers, and product liability; requirements for carrying various kinds of insurance; rules governing finance, bankruptcy, and contract law; environmental codes on pollution and endangered species; border control standards; and licensing standards of various kinds. Although the first Eastern countries allowed to join the European Union will be such nations as the Czech Republic, Poland, Hungary, and Slovenia, those hoping to join eventually include Ukraine, Moldova, various Caucasian republics, and Russia itself. Their desire to join the European Union makes these countries willing to work toward conformity with European standards. For its part, Europe is interested in obtaining new trading partners, but it also sees upgrading conditions in such places as Ukraine, Russia, and the countries of Caucasia as a way of developing a buffer zone between itself

and unbridled illegal immigration from Central and East Asia. It is thought that hundreds of thousands of illegal immigrants enter the region each year from such trouble zones as Afghanistan, Iraq, Ethiopia, Congo, and Somalia, and many of these immigrants make their way to Europe.

Political Reforms in the Post-Soviet Era

Attempts to foster democracy in Russia and the newly independent states have been slower and more difficult than economic reform. Many observers in the West expected that the introduction of a market economy would go hand in hand with democratization, but this has not happened. Although several countries have held elections in which there were competing candidates for office, forms of authoritarian control remain. For example, Belarus and the Central Asian republics have very strong presidencies. In these countries, elected representatives act as rubber stamps and exercise only limited influence on policy. In Ukraine, Georgia, and Russia, even though high officials are elected, they sometimes seem overly influenced by **oligarchies**—powerful elites—that wield influence because of their enormous wealth. To be fair, it is not clear whether a majority of the people are prepared to participate more fully in the decision-making process: the last 80 years have not encouraged civic responsibility. Even today, ordinary citizens tend to think of political leaders as patriarchs who can dispense favors to individual citizens and produce simple solutions to complex problems. Furthermore, many citizens are now so concerned with daily survival that they have little energy for political participation.

The Media and Political Reform. In the Soviet era, all communications media were under government control. There was no free press, and public criticism of the government was punishable. Journalists who risked punishment by engaging in increasingly vocal critiques of public officials and policies were instrumental in bringing an end to the Soviet Union. Many observers thought an independent press would bring about democratic practices that would in turn ensure the formal establishment of free speech and public debate. Since 1991, the communications industry has been a center of privatization, and several media tycoons have emerged. Privately owned newspapers and television stations regularly criticize various leaders of Russia and the other republics and their policies. For the most part, it appears that a free press is developing, and such leaders as President Vladimir Putin openly state that a free press is vital to the growth of civil society. Nonetheless, Russian citizens are ambivalent about just how free the press is or should be. When media tycoon Vladimir Gusinsky, whose newspapers and broadcast facilities constantly criticized the government, was arrested and detained in June 2000, a survey on Moscow's streets showed little outrage among ordinary citizens. Most did not see the arrest as a government effort to squelch free expression, because Gusinsky himself was an oligarch. Since his detention, his media facilities have been acquired by both government and private interests, and local and international reports suggest that critical analysis of the government has diminished.

Closely related to the development of a free communications industry is the availability of communications technology to the

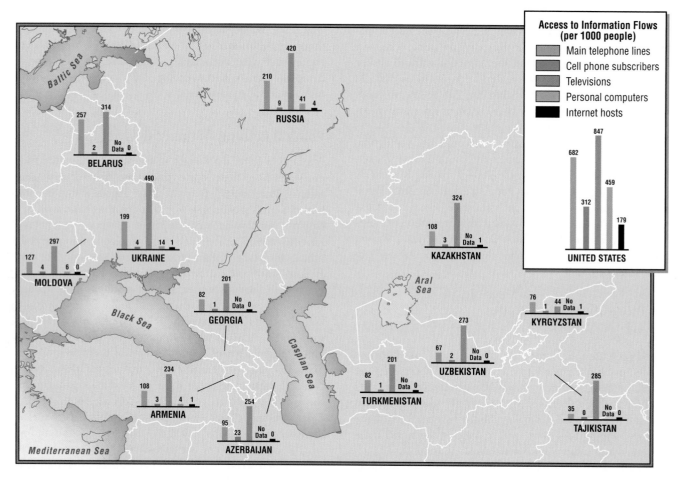

Figure 5.8 Access to information flows: Russia and the newly independent states. Access to information throughout the region is low when compared to more developed countries. In 2001, the United Nations designed a new measure of development, the Technology Achievement Index (TAI), based on information access, education level, and technology creation and exports. None of the countries in this region has a TAI rating (see Figure 6.1) as yet because they do not meet rating criteria at this time. [*United Nations Human Development Report 2000*, Table 12; and *United Nations Human Development Report 2001*, Table A2.1.]

general public (Figure 5.8). Television sets were widely available in the USSR, but programming was tightly controlled by the government and was free of advertisements. Now there are advertisements and programming is less tightly controlled. Access to telephone lines is still limited, but mobile phones have made instant personal communication possible in even the most remote cities of Siberia and Central Asia, where they are essential to many new enterprises. Mobile service, however, remains expensive and unreliable, and it is not available to most people.

Mobile phone companies are hoping to help the newly independent states leapfrog into the global economy by facilitating Internet commerce. Personal computers are still rare, however, although information technology expertise is available and demand is growing. Essential to the overall success of the Internet in the newly independent states is legislation that will provide for easy access and full disclosure in all Internet transactions. The Russian Academy of Science is participating in the framing of such legislation, and already Russia's leading newspapers have free access to Internet lines.

The Military in the Post-Soviet Era. The role of the military is one of the most important political issues facing the region today. The military was once the most privileged sector of society, and its officers had access to goods and services that ordinary citizens could only dream about. Now, after huge funding cuts, the Russian military has been reduced by hundreds of thousands. In 2000, there were just 1.5 million troops, and further cuts were proposed by President Putin. Military personnel are returning home from far-flung posts all across Central Asia and Siberia. With few skills, few job prospects, little decent housing for their families, and sometimes not even adequate clothing, these former soldiers form a huge reservoir of discontent. However, worries about military rebellions have not come to fruition, although it was a botched coup by top commanders against Gorbachev in August 1991 that led to the formal dissolution of the USSR. A more immediate concern is that the former Soviet nuclear arsenal could fall into the wrong hands if military cuts are poorly managed. Already there is some evidence that nuclear expertise and even some warheads may have been smuggled out of Russia to such countries as North Korea, Iran,

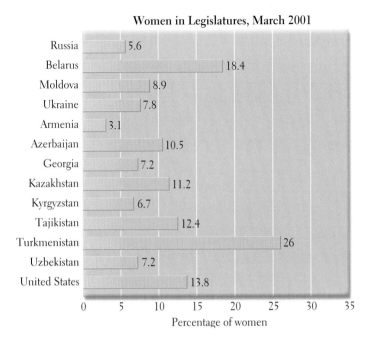

Women in Legislatures, March 2001

Country	Percentage of women
Russia	5.6
Belarus	18.4
Moldova	8.9
Ukraine	7.8
Armenia	3.1
Azerbaijan	10.5
Georgia	7.2
Kazakhstan	11.2
Kyrgyzstan	6.7
Tajikistan	12.4
Turkmenistan	26
Uzbekistan	7.2
United States	13.8

Percentage of women

Figure 5.9 **Women legislators. Percentage of women legislators in Russia and the newly independent states (with U.S. comparison).** [Adapted from Gender Empowerment Measure, *United Nations Human Development Report 2001,* at http://www.undp.org/hdr2001/.]

and Iraq, whose governments are widely distrusted by the international community.

The Political Status of Women. Although granted equal rights under the USSR's constitution, women never held much power in government and hence had little say in policy. In 1990, women accounted for 30 percent of Communist party membership, but they made up just 6 percent of the governing Central Committee. Since the fall of the USSR, women have lost political representation in the parliaments of Russia and the Caucasian republics, from 20 to 40 percent in Soviet times to just 3 to 10.5 percent in 2001 (Figure 5.9). Women in Belarus and Central Asia fared slightly better. But by the late 1990s, especially in Caucasia and the Central Asian republics, women were finding ways to influence policy by working through nongovernmental organizations (NGOs) and such agencies as the United Nations Development Program (UNDP). Hence, formal representation in parliament is no longer the only, or even the best, measure of women's political activity. Still, support for women's political movements is not widespread, despite strong evidence of sex discrimination. Many women fear that working for women's rights and well-being would be seen as anti-male or as a rejection of traditional feminine roles.

Russia's Internal Republics

Russia's centuries-long history of expansion into neighboring lands has left it with an exceptionally complex political geography. More than 75 percent of the country is divided fairly systematically into provinces (called oblasts), but Russia also contains more than 30 **internal republics** and more than 10 so-called autonomous regions. Russia and these internal republics and autonomous regions are often referred to as the **Russian Federation.** The internal republics and autonomous regions have been set aside as the homelands of non-Slavic peoples. They are only somewhat analogous to counties, states, or reservations in the United States.

The Russian Federation consists of much of the territory gradually claimed over the last 500 years by the European Russian state. As it expanded eastward, eventually reaching the Pacific Ocean, Russia absorbed a number of small non-Russian areas conquered by the czars or taken over by the Soviets, many of which have significant ethnic minority populations. The republics trace their origins to peoples who spoke such languages as German, Turkish, Finnish, Ugric, or Persian. The inhabitants of some republics in the far northern reaches of Siberia are analogous to Native Americans in that they are the descendants of ancient indigenous people. The peoples of some of these republics, particularly the more southerly ones, are Muslim—followers of the religion Islam, which is dominant in Southwest and Central Asia (see Chapter 6). The peoples of other republics are Christian, those of Kalmykia, between the Volga and Don rivers, are Buddhist, and some groups have animist beliefs that predate any of the organized religions. The lands of the Yakutiya (now Sakha) in Siberia cover large, resource-rich areas that are of special interest to Russia.

Even under the czars, assimilation of all minorities to Russian ways was the central policy, and during the Soviet era, many ethnic Russians were settled in the internal republics. Their stated mission was to modernize and civilize what were viewed as culturally marginal people. In the process, Russians tended to receive the best jobs and housing, and they often outnumbered native people. Despite the efforts of these Russian settlers, however, ethnic identities have not disappeared.

The internal republics have a bewildering array of political relationships to the Russian state; and their status is continuously being renegotiated republic by republic (see the box "An Ethnic Republic Reasserts Itself," page 278). The internal republics control some of their own affairs, and some republics, such as Tatarstan and Bashkortostan, are able to retain substantial federal taxes to subsidize the development of particular sectors, such as agriculture. As Figure 5.10 shows, some of the internal republics are tiny, and many lie entirely within the borders of Russia.

Russia has tried to avoid formally addressing the political status of the internal republics, because if these ethnic enclaves were to demand greater autonomy, or even independence, the territorial integrity of the vast Russian Federation would be threatened. Shortly after the breakup of the Soviet Union, several internal republics did demand greater autonomy, and two of them, Tatarstan and Chechnya, declared outright independence. Whereas Tatarstan has since been placated by significant efforts to increase its economic and political viability and autonomy, Chechnya has launched a rebellion against Russia that has led to the worst bloodshed of the post-Soviet era. For the time being, most of the other ethnic enclaves are content to stay closely associated with Russia.

The Conflict in Chechnya. Chechnya, a small republic on the northern flanks of the Caucasus Mountains, has a green and fertile, hilly landscape reminiscent of the Cumberland Mountains

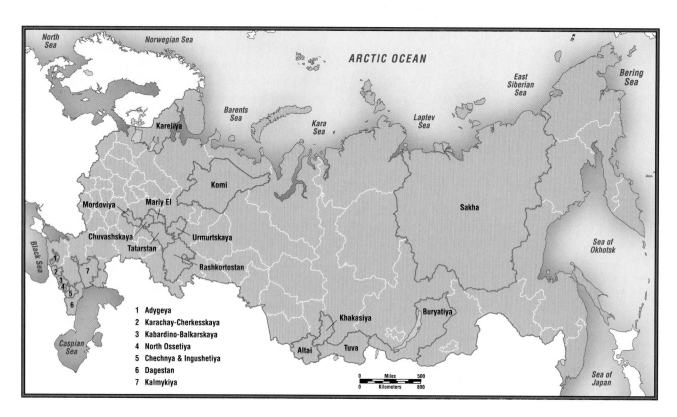

Figure 5.10 Russia's political subdivisions. The 20 internal republics in the Russian Federation subregion are outlined in green, and the oblasts in white. [Adapted from James H. Bater, *Russia and the Post-Soviet Scene* (London: Arnold, 1996), pp. 280–281.]

in eastern Tennessee. It is home to 800,000 people, about 130,000 of whom have been displaced by war since 1991. The Chechens converted to the Sunni branch of Islam in the 1700s, and ever since Islam has served as an important symbol of Chechen identity and of resistance against the Orthodox Christian Russians, who annexed Chechnya in the nineteenth century. After the Russian Revolution in 1917, the Chechens and other groups in the Caucasus formed the Republic of Mountain Peoples, hoping to separate themselves from Russia, but the Soviets abolished the republic in 1924. Chechens were forced onto collective farms in the 1930s, and during World War II they were accused of collaborating with the Germans.

In 1991, as the USSR was dissolving, Chechnya declared itself an independent state. To the Russians, this represented a dangerous precedent, because they feared that Chechen independence could spark similar demands by other cultural enclaves throughout Russia. In addition, Russia wished to retain the region's agricultural, oil, and gas wealth. Since 1991, Chechen guerrillas seeking independence have repeatedly challenged the Russian army, which has responded by bombing the capital of Groznyy and carrying out other reprisals that have left thousands of civilians homeless or dead. Although Chechens claim to be fighting simply for independence and the right to their own resources, Russia claims that the guerrillas have links to extremist Islamist groups in Central Asia, perhaps even global terrorists, and hence represent a threat to Russia's long-term security. Russia's brutal and ineffective attempts

to suppress Chechen independence have raised concerns about Russia's commitment to human rights and its overall ability to address internal political dissent.

SOCIOCULTURAL ISSUES

When the winds of change began to blow through the Soviet Union in the 1980s, political and economic repercussions were expected. But few anticipated that the transition to a market economy and away from central control of most aspects of life would happen so speedily and be so unstructured and so disruptive of social stability. On the one hand, the new freedoms have encouraged self-expression and individual initiative for some people, as well as the possibility of political participation and cultural and religious revival. On the other hand, the loss of structure once provided by the socialist state has resulted in the loss not only of livelihoods but also of housing, food, health care, and civil order. In the wake of uncontrolled change, widespread corruption has taken root.

Cultural Dominance of Russia

The Russians always regarded themselves as the heart of the Soviet Union, a point of view that originated with the czars when they extended the Russian Empire over adjacent regions. Throughout the twentieth century, Russians felt closest to the European repub-

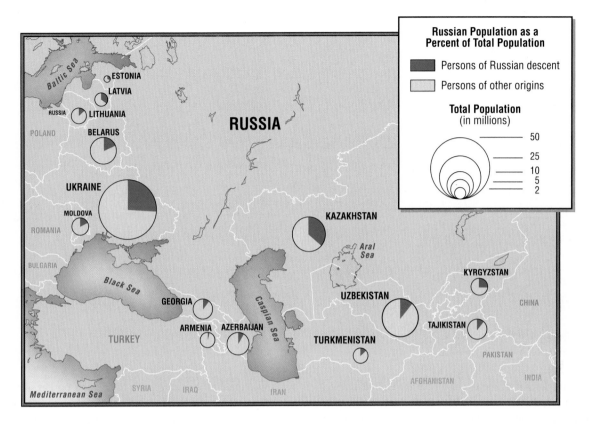

Russian Population as a Percent of Total Population

■ Persons of Russian descent

□ Persons of other origins

Total Population (in millions)

50
25
10
5
2

Figure 5.11 **Russian population in the newly independent states, early in the post-Soviet population shift.** More than 25 million Russians migrated to the USSR borderland states. After the breakup of the Soviet Union in 1991, most Russians living in Caucasia and Central Asia returned to Russia. In contrast, most Russians living in the newly created European countries remained in their new host countries, where they have become a significant minority. The pie-shaped wedges within the circles represent the percentage of Russians as a proportion of total population in the Eurasian republics, early in the post-breakup period. Why do you think the Russians in the European states chose not to return to their homeland? [Adapted from Graham Smith, *The Post Soviet States* (London: Arnold, 2000), p. 75.]

lics (Ukraine, Moldova, Belarus, and the Baltic states), whereas the Caucasian and Central Asian republics were of interest primarily for the resources they possessed and their strategic location. Their distinctive cultural features set them apart, and the Russians considered this distinctiveness to be a liability to national union. To ensure compliance with regional economic plans, to provide a skilled labor force, and to acculturate minorities to Russian customs and attitudes, the Soviets resettled Russian people in all the peripheral republics (now the newly independent states), just as they did in the non-Russian ethnic enclaves within Russia. Russian technicians, teachers, and professionals occupied the choice positions throughout the region, and they entered all parts of the USSR in such large numbers that in certain locations, Russians were significant and powerful minorities (Figure 5.11).

Across the region, Russian culture dominated all aspects of public life. For example, although more than 40 legally recognized languages were spoken throughout the USSR, Russian was the language of official business and was taught in the schools. By the 1980s, the use of minority languages was in decline everywhere in the Soviet Union. Ancient local customs of non-Russian peoples—including religion, family organization, domestic architecture, manner of dress, farming, and diet—were considered outmoded and were suppressed. In the school curriculum, Russian culture was presented as the norm. The Russians were especially intolerant of the Muslim religion in Caucasia and Central Asia.

Until the mid-1970s, migration within the USSR consisted primarily of ethnic Russians going to the various neighboring countries and republics to work and to help acculturate indigenous minorities; a comparatively few people from eastern Europe, Caucasia, Central Asia, and the internal republics moved to Russia temporarily for education and training. In the 1980s, migration between Russia and its neighbors apparently slowed. Then, in the 1990s, there was a shift in the migration pattern. Most Russians in the newly independent European states (Belarus, Ukraine, and Moldova) did not return to Russia. Instead, they remained in their new host counties, where they have become a significant minority. In the newly independent Caucasian and Central Asian states, on the other hand, a large majority of the Russian emigrants returned home to Russia, and ethnic minorities in Russia went home to the Caucasian and Central Asian states or to internal ethnic enclaves.

Cultural Revival in the Post-Soviet Era

Since the achievement of independence and the departure of many Russians, most of the newly independent countries have reasserted their pre-Soviet cultural identities. Cultural practices

Geographer Dmitri Sidorov traced the history of the Cathedral of Christ the Savior in Moscow from the 1810s to the present. The photo above shows the cathedral as it is being destroyed by the Bolsheviks in 1931; the photo at the right shows the restored cathedral in 1997. The restoration of the cathedral is often publicized as signaling the dismantling of the anti-religious Soviet state and the beginning of a new era in Russian history. [*left:* Dmitri Sidorov/E. I. Kirichenko/V. Mikosha. *right:* Dmitri Sidorov.]

that predate the Soviet era have been revived throughout the region in response to the relaxation of controls by the central government. The degree of change has varied from country to country, however, and Belarus remains the most Russian.

One example of cultural revival is the resurgence of interest in religion. Under the Soviets, religious practice was discouraged because religious beliefs were thought to inhibit the commitment of the people to revolutionary change. In European Russia (as well as in Georgia and Armenia in Caucasia), most people have some ancestral connection to Orthodox Christianity, and those with a Jewish heritage form a sizable minority. In the case of both Orthodox Christians and Jews, religious observance increased markedly in the 1990s. In Russia's internal ethnic republics—such as Tatarstan, Chechnya, Dagestan, and Ingushetiya—and in Azerbaijan in Caucasia, many people are Muslims. Among these and other Muslim minorities, observance of Islam has become more prevalent and Muslim identity more politically important.

As interest in religion revives and groups and individuals rediscover their religious roots, attendance at religious services has increased, and many sanctuaries are being built or restored. A spectacular example of restoration is Christ the Savior Cathedral in Moscow (see the photographs above). The original cathedral was completed in the 1880s and then completely destroyed in the 1930s to make way for the planned, but never built, Palace of the Soviets. During Gorbachev's policy of openness in the late 1980s, the idea of rebuilding the original cathedral took shape, and the new building was completed in 1997.

A countertrend to the revival of Orthodox Christianity is the modest spread of the new Christian fundamentalist sects from the United States. The magnitude of this spread may be exaggerated by the missionaries themselves, who count their success by numbers of converts. The movement also may be ephemeral, lasting only as long as the money the missionaries bring keeps flowing. However, in European Russia as in Latin America and Africa (see Chapters 3 and 7), these sects emphasize the role of the individual in the salvation of the soul and also in facing the rigors of everyday life. Some new capitalists see this emphasis on individualism as a particularly useful message. After years of relying on the state to meet their basic needs, people must now rely on their own resources, often for the first time. Many find the teachings particularly comforting not only during times of hardship but also as they adjust to new prosperity.

VIGNETTE   Valerii, age 35, once a government research scientist, is now a new capitalist. He makes a comfortable living importing and exporting goods in the informal economy. Though he has to bribe officials and pay protection money to the *reketiry* (racketeers or mobsters), his income places his family of three—himself; his wife, Nina, age 30; and their son, Mikhail, age

10—at a much higher economic level than their longtime friends. Nina is the only woman among them who does not work outside the home. Their new wealth, the precariousness of their position in the present economy, and the fact that their friends do not share their prosperity cause Nina and Valerii to be uneasy. In search of values that will guide them in these new circumstances, both have recently been baptized in a Christian fundamentalist sect. They say they chose this particular religious group because it promotes modesty, honesty, and commitment to hard work. [Adapted from Timo Piirainen, *Towards a New Social Order in Russia: Transforming Structures and Everyday Life* (Aldershot, U.K., and Brookfield, Vt.: Dartmouth Press, 1997), pp. 171–179.]

In the Central Asian republics, the return to religious practices is often a subject of contention. Here most people have a Muslim heritage; but some local leaders, schooled in Soviet theory, view traditional Muslim religious practices as obstacles to social and economic reform. Other devout Central Asian leaders worry that the extremists among modern Islamic revivalist movements in the region may be agents of religious states like Iran, Afghanistan, and Saudi Arabia and hence may endanger the security of Central Asian countries. Between 1992 and 1997, Tajikistan fought a civil war against Islamic insurgents with links to the Taliban in Afghanistan (see page 284). In 2000, Uzbekistan and Kyrgyzstan joined to eliminate an extremist Islamic movement. But many religious leaders and human rights groups say that the fervor to eliminate radical insurgents has resulted in persecution of ordinary devout Muslims, especially men.

VIGNETTE Tashkent, Uzbekistan; October 2000: Munira Nazarov, who wears the silk head scarf favored by devout Muslim women in many parts of the world, tells a visitor tearfully that her husband, an *imam* (religious leader) at one of Tashkent's biggest mosques, has been missing since 1998, when he was followed for several days by security police. She hopes he is in hiding. His family, including Munira, have been jailed for short periods, as the police try to obtain information on his whereabouts. As she pats the head of her youngest of seven children, Mrs. Nazarov says, "He used to talk to people about studying the Koran and other good books. He didn't talk about politics."

In Uzbekistan, men convicted of no more than possessing a religious pamphlet have died in prison, and young men who met to play soccer and eat together have been arrested and accused of planning a jihad (holy war). At their trials, defendants have shown bruises and torn-out fingernails. Human Rights Watch, which monitors rights abuses worldwide, reports that at least 4000 Muslim men have been arrested and detained in Uzbekistan alone. [Adapted from Douglas Frantz, "Persecution charged in ex-Soviet republic," *The New York Times* (International) (October 29, 2000): p. 6.]

Unemployment and Loss of Safety Net

Widespread unemployment and underemployment is now common. Nearly all families are affected by job loss in some way. In fact, the number of those who no longer have paid employment is probably much higher than the official unemployment figures. Numerous state firms simply no longer operate and have no money to pay employees who are still listed as workers. Estimates are that in the mid-1990s, three-fifths of the Russian labor force was not being paid in full and on schedule. For years, many families lived on their savings, but by now, such savings have been depleted. Economists have estimated that actual unemployment rates in Russia during 1996 were greater than 13 percent for all workers and 22 percent for young adults aged 18 to 25. The rate of underemployment was much higher. Figures for 1999 were virtually the same. In the United States, by comparison, an unemployment rate of 7 percent is considered a major economic and social problem.

The loss of job and status is difficult for many men, and, as discussed on page 247, some have turned to drinking. Cheap homemade liquor has become the drug of choice for those experiencing the boredom, anxiety, and loss of self-esteem of not working. It is estimated that half the men and perhaps one-third of the women in the region are alcoholics. Death by alcohol poisoning rose 25 percent in the 1990s.

Work once provided many benefits beyond a salary. It is especially devastating to lose a job in the former Soviet Union, because for many decades the social welfare system was organized around the workplace. The job was an individual's link not just to income but to necessities of life and general social services. The state-owned companies provided housing, two main meals a day in the company cafeteria, health care, and day care for children. They were also the center of community and social life. For many workers, when the job ended, there was no alternative safety net. With so many industries now retrenching, in some cases whole cities are facing not just widespread unemployment but the loss of many social benefits for their people. Such is the case in the city of Norilsk in the far north of the Central Siberian Plateau (see the box "Norilsk: A City in Transition to a Global, Free-Market Economy.").

Gender and Opportunity in Free-Market Russia

Soviet policy encouraged all women to work for wages outside the home. Leninist theory had always argued that women should contribute fully to national development, although the traditional attitude that women are the keepers of the home persisted. So women did both: they worked outside the home and did the housework. Most women put in especially long days, working outside the home for eight hours and then doing laundry and housework without the aid of household appliances. And because of shortages, they often had to stand in long lines to procure food and clothing for their families. By the 1970s, 90 percent of able women in Russia were working full-time, the highest rate of female paid employment in the world.

By the 1990s, the female labor force in Russia was, on average, better educated than the male force (an exception was Muslim

AT THE LOCAL SCALE   *Norilsk: A City in Transition to a Global, Free-Market Economy*

Norilsk lies in Siberia, 200 miles (320 kilometers) north of the Arctic Circle (see chapter opening map), where year-round average temperatures are below freezing. The city sits atop a rich deposit of minerals: 35 percent of the world's nickel supply, 10 percent of its copper, 40 percent of its platinum. The city's only industry is Norilsk Nickel, Russia's largest metal company. In 1998, Norilsk Nickel employed 110,000 workers out of a total population of about 270,000, and in 1999 it accounted for 4 percent of Russia's total exports. Because of its mineral production, Norilsk is important both to Russia and to the global economy; platinum, for example, is essential in the computer industry. Despite its prominence, Norilsk is a troubled company town in transition to a market economy, and it is a site of major industrial pollution and natural habitat destruction.

Before Norilsk Nickel was restructured to make its products competitive in a market economy, more than 18,000 of the 110,000 workers were employed in the social services division. They managed all the benefits provided to plant workers: housing, health care, saunas, sports clubs, holiday homes in warmer regions, daycare centers, cafeterias, and even farms that supplied some of the workers' food. In the old centrally planned economy, Norilsk Nickel merely had to produce the needed metals for the country's industries; any profits were used to provide for the workers, who had been attracted to the bleak, remote area only by the well-paying jobs and good social services. But because the environment is heavily polluted and dreary to live in, male life expectancy sank to just 50 years—despite sports clubs and health care.

In 1994, a private firm acquired 38 percent of the company and instituted cost-cutting measures in all branches, especially the social services division. By 1997, social services still absorbed all the plant's profits, an estimated U.S. $260 million, or about U.S. $2000 per worker. Few of the new, privately owned companies in Russia and the newly independent states offer any social services; the joint owners of Norilsk Nickel are trying to persuade the city of Norilsk to provide them instead. Whatever happens, the services are shrinking or disappearing just as plant workers are being laid off permanently. Those who lose jobs have no benefits and no pensions, and their few savings have been destroyed by recent high levels of inflation (rapidly rising prices for basic necessities).

By 2000, the fate of Norilsk Nickel was unclear. The new owners were promising to address international environmental concerns and to upgrade the facilities with equipment bought from Japanese, Swedish, and American suppliers to improve production. Improved efficiency was accompanied by the layoff of 20,000 workers. The average monthly salary at Norilsk Nickel was once again enough for a modest living (about U.S. $500 per month, plus a housing subsidy), and the company was repaying back wages. But the future of Norilsk Nickel is still uncertain; global stock investors may not put money in the company because they are wary of its history of environmental destruction and murky financial management—such as loans that may have come from underworld sources, yearly accounts that are filed late, and profits routinely overstated by half.

Source: The Economist (January 10, 1998): 59; "Russia's cooked books," *The Economist* (September 9, 2000); Norilsk Nickel Web page, http://www.nornik.ru/index.html/english/default.htm; and Marina Kamayeva, "Russia's Norilsk Nickel upgrades production facilities," *Business Information Service for the Newly Independent States* (Washington, D.C.: U.S. Department of Commerce, May 19, 2000), http://www.bisnis.doc.gov/bisnis/isa/000601nickel.htm.

The mountains that range over Norilsk (pop. about 200,000) were constructed from wastes produced by the mining and smelting operations around the city. [Randy Olson/National Geographic Image Collection.]

TABLE 5.3 *Women and wages in selected economic sectors in Russia, 1991*

Employment sector (1)	Percentage of workers who are women (2)	Average women's wage in sector as a percentage of the national average for all sectors (or particular sectors) (3)
Professional occupations		
Medicine	86	69 (all workers in health and social security)
Education and science	73	78 (workers in education) 113 (workers in science)
Planning and accounts	88	93–94 (all workers in administration, credit, and insurance)
Manual occupations		
Trades and catering	89	75
Services	89	76
Communications	84	89
Textiles	83	94
Clothing	93	81

Source: Sue Bridger, Rebecca Kay, and Kathryn Pinnick, *No More Heroines: Russia, Women, and the Market* (London and New York: Routledge, 1996), p. 42.

Central Asia). Women commonly held jobs as economists, accountants, scientists, and technicians. Three out of four physicians and one in three engineers were women. Many more, however, were simply laborers in factories and in the fields. Despite their generally higher qualifications, women were seldom found in senior supervisory positions, and as recently as 1996, the wages of women workers averaged about 20 percent less than those of men workers. For comparison, in the United States in 2000, women's earnings overall were about 25 percent less than men's, with the biggest gap in earnings experienced by those women with the most education: professors (21 percent lower), lawyers and judges (27 percent lower), physicians (42 percent lower), accountants (28 percent lower), and economists (32 percent lower). Table 5.3 shows the percentage of Russian women workers in selected occupations and the wages for those occupations as a percent of the national average for all economic sectors. In 1991, women were concentrated in sectors that consistently fell below the national wage average, female workers in science being the only exception (Table 5.3, column 3).

Market reforms reduced the number of jobs available to all citizens, and in the reshuffling, women were laid off in larger numbers than men; women with children were laid off first. Mikhail Gorbachev, general secretary of the Communist party until the end of 1991, publicly stated that the first role of Soviet women was domestic; hence, they could best serve their country by returning to their homes and leaving the increasingly scarce jobs to men. By the late 1990s, 70 percent of the registered unemployed were women. But many, if not most, of the women left jobless were—due to illness, death, or divorce—the sole support of their families, which often included three generations: themselves, their parents, and their children. In the Russian Federation, two out of three marriages end in divorce (see Table 5.2).

Corruption and Social Instability

Rampant corruption poses a dangerous threat to social stability in Russia and the newly independent states. Since independence in 1991, gangsters have taken over portions of the economy in every republic, and literally every citizen has to deal daily with the effects of corruption. In the mid-1990s, an American Fulbright scholar, John G. Stewart, was posted to teach public administration ethics in the Caucasian country of Georgia. He found himself learning from his students about the roots of corruption in daily life. First, they reminded him that even before rampant inflation, all government salaries were ridiculously low; by the mid-1990s, civil servants routinely supplemented salaries too low to live on by

AT THE REGIONAL SCALE   *The Trade in Women*

Among the less savory entrepreneurial activities in the new Russian market economy are those connected with the "marketing" of women. As observed in other chapters of this book (see especially Chapter 10), there is a growing demand for females as a commodity in global markets. Women from European Russia and Ukraine are deemed especially desirable. A relatively benign part of this market is the phenomenon of Internet-based mail-order bride services. A woman in her late teens or early twenties pays about $20 to be included in an agency's catalog of pictures and descriptions (one Internet agency advertises 30,000 women listed). She is then interviewed by the prospective groom, who travels—usually to Russia or Ukraine—to choose from the women he has selected from the catalog. Men over the age of 40 from the United States and Europe are the main customers for these services, for which they pay an entry fee of about $1500 (more may be due later) and all travel and legal costs. The United States issued 1700 visas to such brides in 1997.

A more unsavory activity is the hijacking of unsuspecting females for sex work outside Russia and the newly independent states. In 2000, *The Economist* magazine estimated that 300,000 such women are smuggled into the European Union yearly, many being held first at an intermediary stop in central Europe. The sex business in Europe is thought to generate $9 billion a year, and the supply side seems to be dominated by Russian-speaking gangs. Usually, the women in this market, desperate for a job, sign up to work as domestic servants or waitresses in Europe, only to find upon arrival that they are being held by force and are expected to work as strippers, dancers, and prostitutes. Most often, the women kidnapped into sex work have no well-connected relatives or friends who will report them missing. Once ensnared, the women may be sold for a few thousand dollars to brothels in parts of the world where their European looks will be considered exotic. They are usually unprotected from sexually transmitted diseases.

There are also women who, faced with unemployment and a contracting economy, choose sex work as a way to earn a living. A study by the Swedish Women's Forum in 1998 reported that many Russian, Belarus, and Ukraine sex workers said they chose this work because it was more dignified and remunerative than the routine sexual intimacy forced on them by bosses in their former jobs back home.

Studies of the men in this trade are rare, though their role as customers provides the demand. The Swedish report emphasizes, however, that many men are active in the organizations working to stop the illegal trade in women and to aid the victims. Furthermore, it is also clear that many men who marry mail-order brides treat their wives with respect, although there is little data on the ultimate success of these marriages.

Source: "Trafficking in women: In the shadows," *The Economist* (August 26, 2000): 38–39; and "Trafficking in Women for the Purpose of Sexual Exploitation: Mapping the Situation and Existing Organisations," and "Working in Belarus, Russia, the Baltic and Nordic States" (the last two studies by The Foundation of Women's Forum/Stiftelsen Kvinnoforum, Stockholm, August 1998).

accepting bribes. The government bureaucracy was and is nearly nonfunctional until a bribe is paid. Someone who wants to construct a building or open a shop can spend months waiting in lines to obtain the necessary permits—or shorten the process with a few well-placed bribes. Additionally, because laws are poorly written and contradictory, the most capable and ambitious people simply ignore restrictions and regulations. Many have adopted an attitude of noncompliance with all regulations—even those that might clearly serve the common good, such as speed limits, building codes, sanitary regulations, and limits on the transport and disposal of nuclear waste.

It was a short step from such ingrained systemic corruption to protection rackets, outright robbery, and violence, especially after cheap weapons became readily available in 1991. A market economy and a democratic society that rely on trust, ability, hard work, and the rule of law cannot flourish in such a setting. Also, it is dangerous for an individual to buck the system because the officials and the police have often been bought off by the gangsters.

Stewart's students wondered how they, as future public administrators, could practice ethical values and survive in the real world.

Since finishing the public administration course, his 28 Georgian students have set up a support group to help one another implement ethical procedures and to find legitimate supplementary employment, so that bribes need not become part of their income. Georgia's president, Edvard Shevardnadze, also mounted a government-wide campaign against corruption of all types.

Nostalgia for the Past

The years since 1991 have been fraught with the difficulties and hardships already discussed and with declining levels of well-being. Many people now fondly recall the times when life may have been a bit drab, but one could rely on having decent shelter, food, and even a few luxuries. There were some favorable aspects to everyday life in the Soviet era, such as a sense of economic security and of participation in a noble social experiment. There was the feeling that things were steadily improving: health care was good and getting better; life expectancies were getting longer, and infant mortality was falling. The government underwrote the costs of world-class research institutions, museums, and such cultural institutions as the Bolshoi Ballet and symphony orchestras. Even a moderately talented student could get a free university education, and all workers were entitled to several weeks of vacation time annually. The various parts of the Soviet Union saw radical improvements in their standards of living over the course of the twentieth century, and most areas moved from being truly poor places to ranking high among the middle-income people of the world.

Viewed against this backdrop, it is understandable that Russians would be impatient with the slow pace of reforms, with the hardships that the political and economic reforms have brought, and with the widening disparity of wealth. Observers in Europe and the United States tend to think that the difficulties are only temporary and will disappear as the transition proceeds, but those dealing with day-to-day hardships do not share this rosy perspective.

If nostalgia becomes an overpowering emotion, government policies could be affected. Public pressure to alleviate the immediate problems of lost jobs and income, obsolete skills, inadequate health care, deteriorating shelter, and declining nutrition could influence elected lawmakers and officials to abandon reforms. These leaders, perhaps experiencing many transition hardships themselves, may adopt policies that will thwart the ultimate development of democratic institutions and a market-oriented economy even before they have had a chance to succeed.

Although perhaps the majority of people experience this nostalgia for the past, many of these same people find the present era exciting. They enjoy the challenge of a more open society and are personally doing better, at least in some ways, than during Soviet times.

industrial facilities like the chemical plant near Katerina's home. Now Russia and the newly independent states have some of the worst environmental problems in the world.

The general decline in living standards that occurred after 1991 has placed environmental issues on the back burner throughout the region. As one Russian environmentalist recently put it: "When people become more involved with their stomachs, they forget about ecology." At present, attracting potential investors from Europe, America, and Asia is a higher priority for Russians. Such investment will create jobs, but it also will increase industrial production and resource exploitation. Investors may be attracted precisely because legal protections for the environment are lax, and some have bribed officials to ignore environmental degradation. These practices mean that in cases like Katerina's, leaders do not respond to complaints about pollution or other environmental problems because they are being paid by the polluters. At the beginning of the twenty-first century, in the former Soviet Union, more than 35 million people (15 percent of the population) live in areas where the air is dangerous to breathe, birth defects are rampant, and by some estimates only one-third of all schoolchildren enjoy good health.

Pollution controls are further complicated because instead of one government in Moscow coordinating economic development, there are now 12 independent countries with widely differing policies, and none of them has the money to correct even a few of the past environmental abuses. Notice in Figure 5.12 that environmental degradation crosses international borders, making the resolution of pollution issues especially difficult.

ENVIRONMENTAL ISSUES

VIGNETTE   Katerina, a mother of three, lives with her family in Azerbaijan, 20 miles (30 kilometers) from a massive chemical manufacturing complex on the Caspian Sea that was built during Soviet times. When the breezes off the Caspian are strong, a step outside in the morning brings only a momentary gasp as one's nose and throat are assaulted by the stench of sulfur and chlorine. The first sign of trouble for her family appears when her three-year-old son's nose starts to bleed. In a few minutes, the older kids get dizzy and they all detect the rotten egg smell of sulfur. The plant responsible for this pollution is one of dozens that line the Caspian. [Adapted from *The Economist*, supplement on the Caspian (February 7, 1998): 12.]

Soviet theories about nature are largely responsible for countless situations like Katerina's. The Soviets believed that nature had no value unless it could be exploited by humans for its resources. To Soviet leaders like Stalin, nature was the servant of industrial and agricultural progress, and the grander the evidence of human domination over nature, the better. Hence, the Soviets built huge

"Everything rots, everything dies" are the sentiments expressed by this Azerbaijani woman, referring to her garden near the Baku oil fields. [Reza Deghati/National Geographic Image Collection.]

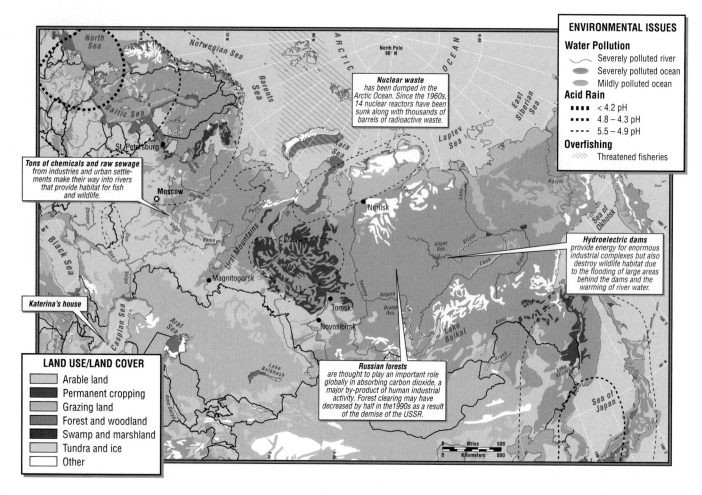

Figure 5.12 Environmental Issues: Russia and the newly independent states. Like most industrialized countries, the countries in this region have a wide range of environmental problems caused by the burning of fossil fuels, unwise disposal of hazardous wastes, and uncontrolled use of natural resources. These countries have inherited many problems from the Soviet era, when the negative consequences to the environment of Soviet development policies were downplayed. The pH of acid rain is explained on page 90 (see Figure 2.20 caption).

Resource Extraction and Environmental Degradation

Russia and the newly independent states have considerable deposits of natural resources (see Figures 5.5 and 5.7). As mentioned, Russia alone has the world's largest natural gas reserves, major oil deposits, and forests that stretch across the northern reaches of the continent. Russia also has major deposits of coal and industrial minerals such as iron ore, lead, mercury, nickel, platinum, and gold. It is the third largest producer of hydropower in the world. The Central Asian republics share a huge deposit of oil and gas that is apparently centered on the Caspian Sea and extends east toward China. The five independent states that border this sea, as well as Uzbekistan, are competing for rights to tap and transport these fossil fuels to the world market.

Resource extraction has brought environmental effects as impressive as the resources themselves. Pollution and environmental degradation are associated with the mining and industrial processing of minerals (including oil and gas) and with the gener-

ation of the energy necessary for mineral processing. In Siberia, some of the world's worst inland oil spills have contaminated lakes, rivers, and wildlife that provide sustenance for indigenous populations of hunters and fishers. Hydroelectric dams are the main source of energy for the enormous industrial complexes in Russia and the newly independent states, but the flooding of large areas behind the dams destroys wildlife habitat, and **thermal pollution** occurs when water warmed by the turbines is returned to the rivers below the dams.

The extraction of forest products causes pollution and soil erosion. Forest clearing, especially east of the Urals, is often accomplished by clear-cutting—all trees are cut over dozens of square miles. This technique exposes the soils to erosion; rivers become clogged with silt. Clear-cutting also destroys habitat for species of plants and animals. In addition, this practice may contribute to global warming. Russian forests are thought to play an important role in absorbing global carbon dioxide, a major by-product of human industrial activity.

For a while, after the USSR fell in 1991, forest clearing decreased by as much as a half for two reasons: during the ensuing economic crisis, domestic demand for wood declined; at the same time (but for unrelated reasons), international prices for wood declined. This period of low demand gave some respite to Russian forest resources. But environmentalists worry that as the Russian demand for wood rebounds and as global wood prices rise again, the economically strapped Russian government will issue contracts to private foreign concessionaires for very rapid and unsustainable clear-cutting, especially in Siberia, where environmental monitoring is difficult due to the size of the area.

The influx of people and industries, and the clearing of forests, have radically changed fragile natural habitats, endangering many species of plants and animals. Such is the case for both the Siberian tiger and the cranes of the Amur River basin. Siberian tigers once ranged throughout eastern Russia, China, and Korea. Now they are threatened not only by loss of habitat but also by the market for their fur and by Asian folk beliefs. Poachers receive $100 a pound for pulverized tiger bones, which are thought to increase the sexual prowess of men. By one estimate, in the late 1990s, only 430 tigers remained. Four species of rare cranes, birds that live and breed in wetlands, are threatened by timber harvesting on both the Chinese and Siberian sides of the Amur, the river that borders Russia and China in the Far East.

Urban and Industrial Pollution

Urban and industrial pollution was ignored during Soviet times. Cities expanded quickly to accommodate new industries and workers being relocated from the countryside. The dense concentration of workers alone was enough to generate lethal levels of many pollutants. Even today, there are few urban sewer systems and fewer still that actually process and purify the sewage. Many large apartment blocks have the equivalent of septic systems, meaning that they rely on the subsoil under and around the block to absorb sewage. Moreover, because cities were often built with residential areas located adjacent to industries producing harmful by-products, many people are exposed to industrial pollution. The city of Magnitogorsk in the Urals provides an example of the effect of extreme industrial pollution on residents. The largest steel mill in the world is the basis of the city's economy and simultaneously causes misery for tens of thousands of Magnitogorsk citizens. One in three has respiratory problems, such as asthma or bronchitis, from breathing smoke and airborne chemicals.

In some urban areas, it is less easy to link pollution directly to health problems because the sources of contamination are more diffuse. Often called **nonpoint sources of pollution,** these include untreated automobile exhaust, raw sewage, and agricultural chemicals that drain from fields into the urban water supplies. Moscow, for example, is located at the center of a large industrial area, where infant mortality and birth defects are unexpectedly high. Researchers are convinced that these effects result from a complex mixture of pollution that is difficult to trace but includes all the nonpoint sources just mentioned.

Rural pollution can often be just as severe as urban pollution. People in the countryside are subjected both to urban emissions that drift into rural areas and to fertilizer and pesticide pollution from agriculture. East of the Urals, industrial cities send tons of chemicals and raw sewage into the rivers that traverse Siberia and provide the habitat for fish and wildlife.

Nuclear Pollution

Nuclear pollution in Russia and the newly independent states is the worst in the world. In the thinly populated deserts of Kazakhstan, the Soviet military set off almost 500 nuclear explosions during the cold war era. Local populations, mostly poor Kazakh herders, suffered radiation sickness and birth defects but were not told the cause until 1989. The most famous nuclear disaster in the world occurred in north-central Ukraine in 1986, when one of four nuclear reactors at Chernobyl exploded. The explosion severely contaminated a vast area in northern Ukraine, southern Belarus, and Russia and spread a cloud of radiation over much of eastern Europe and Scandinavia. As a direct result of this incident, 5000 people died, 30,000 were disabled, and 100,000 were evacuated from their homes. Yet several Chernobyl-style reactors still operate in Ukraine, Russia, and Lithuania, and all are increasing safety risks.

Even the pollution released by Chernobyl pales in comparison with the radiation leaking from former Soviet military sites such as Tomsk-7 (a closed city east of the Urals, now renamed Seversk), where the soil alone holds 20 times the amount of radiation released by Chernobyl. These facilities have been linked to

Known locally as Atomic Lake, this shimmering place continues to exude radiation. It was created when a nuclear device was exploded in 1965 to make a reservoir northwest of Semey in Kazakhstan.
[Gerd Ludwig/National Geographic Image Collection.]

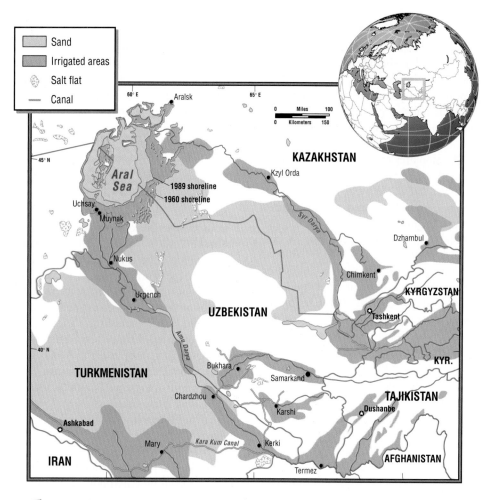

Figure 5.13 The decline and disappearance of the Aral Sea. [Adapted from *National Geographic* (February 1990): 72, 80–81.]

radiation pollution recorded thousands of miles away in the Arctic. The radiation was brought there by rivers and by migrating ducks, who carry it within their bodies. The Arctic Ocean and the Sea of Okhotsk in the northwestern Pacific are also polluted with large amounts of nuclear waste that have been dumped at sea. Although the Soviet government signed an international antidumping treaty, it sank 14 nuclear reactors and dumped thousands of barrels of radioactive waste in the world's oceans.

Irrigation and the Aral Sea

Once the fourth-largest lake in the world, the Aral Sea is disappearing as a result of large-scale irrigation projects in Central Asia. For millions of years, this landlocked inland sea was fed by the Syr Darya and Amu Darya rivers, which brought snowmelt from the lofty Hindu Kush and Tien Shan mountains to the southeast. In 1918, the Soviet leadership decreed that water diverted from the two rivers would irrigate millions of acres of land in Kazakhstan and Uzbekistan. This Soviet plan to grow cotton to meet Russian textile needs was so successful that by

1937, Russia had become an exporter of cotton, or "white gold." In 1956, millions more acres of cotton were planted in Turkmenistan, irrigated with Amu Darya water diverted into an open, 850-mile-long (1368-kilometer-long) desert canal. So much water was lost through evaporation that within four years, the Aral Sea had shrunk measurably. Yet planners, enticed by the income the crops generated, irrigated still more land. By the early 1980s, no water at all was reaching the Aral Sea. By early 1993, the sea had shrunk by more than 50 percent and had lost 75 percent of its volume (Figure 5.13). If irrigation continues, the sea will disappear entirely in a decade or two.

Shrinkage of the Aral Sea may have caused changes in climate and human health. Since the 1960s, the country around the sea has become drier. There is some evidence that the summers are 2°F to 3°F (1.1°C to 1.6°C) hotter and the growing season as much as three weeks shorter. Winters may also be cooler and longer. Winds sweeping across the landscape now pick up the newly exposed salt and seabed sediment, creating poisonous dust storms that deposit a deadly cargo of salty sediment in people's lungs, and drinking water is heavily polluted with agricultural

Thinking Critically: ON THE WEB

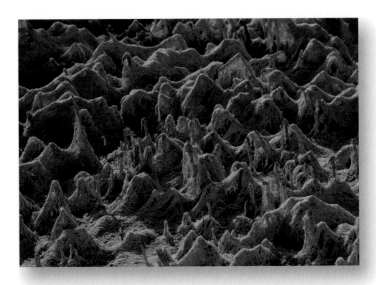

Near the Aral Sea, salts left from irrigation cover the stalks of dead cotton plants the water was meant to nourish. [David Turnley/Corbis.]

chemicals. In rural northern Turkmenistan near the Amu Darya, a majority of the population suffers from one or more chronic illnesses. At the south end of the Aral Sea, in Uzbekistan, 69 of every 100 persons are chronically ill, and in some villages, life expectancy is 38 years (the national average is 68 years).

What is being done to restore the Aral Sea? On the whole, very little. Because irrigated agriculture generates income, Central Asian governments are unreceptive to complaints from the public, and they suppress attempts to raise environmental issues. Drawing from both rivers, Uzbekistan irrigates 16 percent of its total cropland, most of which is in cotton. The country is the world's fifth-largest cotton grower; the crop is the country's leading hard currency earner (representing one-third of its exports) and employs about 40 percent of the national labor force. The situations in Turkmenistan and Kazakhstan differ in details but are comparable. Kazakhstan allows far more environmental activism and public input than its neighbors, but even here the people's desire for a better standard of living outweighs the desire to halt environmental degradation.

Green Movements

Citizen movements focused on environmental (Green) issues were organized across the region in the last half of the 1990s. There are more than 650 NGOs, at least three or four groups for each of the newly independent countries and hundreds for Russia. For example, in Belarus, several groups focus on environmental education, several on nuclear contamination, and one on cataloging the region's birds. The country of Georgia, in Caucasia, has groups promoting energy conservation, ecotourism, and environmental cleanups along transportation routes. These groups communicate with one another and with their members chiefly through the Internet, and several clearinghouses have sprung up

that list and catalog their activities. Despite the wide variety of concerns, a clear theme emerges from the statements and publications they offer: these NGOs see their activities as part of the larger effort to establish civil societies in their new countries. The following Web site, the home page of Civil Society International, offers links to environmental Web sites for the entire region: http://www.friends-partners.org/ccsi/.

MEASURES OF HUMAN WELL-BEING

As discussed in the section on population issues, average levels of well-being sank across the region after the breakup of the former Soviet Union. Disparities in well-being increased as the transition to a market economy created opportunities for a few and troubles for many. But by 1998, living standards began to improve. In Russia, the per capita GDP was up by about one-third over that in 1995, and GDP was up at least slightly in most of the newly independent states as well (Table 5.4, column 2). Only Armenia and Uzbekistan recorded losses between 1995 and 1998. Notice, however, that Russia and the newly independent countries all have GDP per capita figures well below the world average of $6526.

The GDP figures also show disparities among subregions, with Russia and Belarus each having a GDP per capita roughly three times (or more) that of some other countries in the region (Moldova, Armenia, Tajikistan, Uzbekistan). Russia retains most of the industrial capacity and much of the transport and trade infrastructure (banks, company headquarters, government agencies that supply statistics and customs services and that collect taxes), and these assets contribute to the disparities.

The GDP figures given here have been adjusted for differences in prices and other factors to obtain internationally comparable indications of purchasing power parity in U.S. dollars (PPP). What the GDP figures do not reveal is that there is also disparity within each republic. The former communist system's goal of equalizing the distribution of wealth is now being discarded as a few become very rich in the new market economy while the majority experience a decline in living standards, at least temporarily.

Well-being has not sunk as low as it might have, however, largely because citizens of this region help one another to find shelter, food, and personal assistance. Certainly purchasing power was severely limited throughout the region; but purchasing power is not the best measure of well-being because it ignores aspects of well-being other than income.

The Human Development Index (HDI) calculated by the United Nations tries to assess actual well-being more accurately by looking at GDP plus education and life expectancy (the lower the number the higher the rank). By 2000, the HDI ranking for nearly all of the republics was higher than it had been in 1998; two rankings in Caucasia jumped 20 points or more (Table 5.4, column 3). Only Uzbekistan had a lower rank in 2000 than it did in 1998. Part of these apparent improvements may be due to changes in statistical reporting; other factors are that production began to increase in the late 1990s and access to foreign markets improved, especially in Russia.

TABLE 5.4 *Human well-being rankings of Russia and the newly independent states*

Country (1)	GDP per capita, adjusted for PPP[a] in 1998 $U.S. (2)		Human Development Index (HDI) global rankings, 1998[b] (3)		Gender Empowerment Measure (GEM) and Gender Development Index (GDI) global rankings, 1998 (4)	
	1995	1998	1998	2000	GEM	GDI
Selected countries for comparison						
Japan	21,930	23,257	8 (high)	9 (high)	41	9
United States	26,966	29,605	4 (high)	3 (high)	13	4
Kuwait	23,848	25,314	54 (high)	36 (high)	N/A	34
World	5,990	6,526	N/A[c]	N/A	N/A	
Russia and the Eurasian Republics						
Russian Federation	4,531	6,460	72 (medium)	62 (medium)	53	54
Belarus	4,398	6,319	68 (medium)	57 (medium)	N/A	49
Moldova	1,547	1,947	113 (medium)	102 (medium)	N/A	81
Ukraine	2,361	3,194	102 (medium)	78 (medium)	55	63
Caucasia						
Georgia	1,389	3,353	108 (medium)	70 (medium)	N/A	N/A
Armenia	2,208	2,072	99 (medium)	93 (medium)	N/A	75
Azerbaijan	1,463	2,175	110 (medium)	90 (medium)	N/A	N/A
Central Asian Republics						
Kazakhstan	3,037	4,378	93 (medium)	73 (medium)	N/A	N/A
Turkmenistan	2,345	2,550	103 (medium)	100 (medium)	N/A	N/A
Uzbekistan	2,376	2,053	104 (medium)	106 (medium)	N/A	87
Tajikistan	943	1,041	118 (medium)	110 (medium)	N/A	92
Kyrgyzstan	1,927	2,317	109 (medium)	98 (medium)	N/A	N/A

[a]PPP = purchasing power parity.
[b]The high and medium designations indicate where the country ranks among the 174 countries classified into three categories (high, medium, low) by the United Nations.
[c]N/A = data not available.

Source: United Nations Human Development Report 1998 (New York: United Nations Development Programme), Tables 1–3; and *United Nations Human Development Report 2000* (New York: United Nations Development Programme), Tables 1–3.

The United Nations Gender Empowerment Measure (GEM) is available only for the Russian Federation and Ukraine. Gender Development Index (GDI) figures are listed instead (Table 5.4, column 4). They are not comparable to GEM, however, because they ignore the actual participation in government and society by females and measure only the extent to which males and females have access to the same basics, such as food, health care, and income. The GDI rankings show that women's access to health, education, and equal pay is closer to that of men in Russia and Belarus than in Caucasia and Central Asia; but (as is the case in all parts of the world) in no part of this region are women approaching equality with men in this respect.

SUBREGIONS OF RUSSIA AND THE NEWLY INDEPENDENT STATES

In this section, we discuss Russia first because it is the largest country and because it has dominated the entire region for many hundreds of years, in what most scholars now recognize as a quasi-colonial manner. The newly independent countries of Belarus, Moldova, and Ukraine are discussed next because of their location and their close social, cultural, and economic associations with European Russia. They are followed by the Caucasian countries of Georgia, Armenia, and Azerbaijan. We then discuss the countries of Central Asia: Kazakhstan, Kyrgyzstan, Tajikistan, Turkmenistan, and Uzbekistan. The themes covered for the various subregions echo those already discussed for the region at large. Here you will see what life is like in the various parts of this unusually large area.

RUSSIA

Within the former USSR, Russia was paramount in political and economic power, and it dominated culturally. In the post-Soviet era, Russia has remained influential in the region, even as the newly independent states have begun to construct their own economies and political identities. Russia is still the largest country on earth—nearly twice the size of Canada, the United States, or China (Table 5.5). It is also a leader in population; with 145 million people, it ranks sixth. It also ranks high in potential natural wealth.

European (or Western) Russia

European Russia is that area of Russia that shares the eastern part of the North European Plain with Latvia, Lithuania, Estonia,

TABLE 5.5	*Countries of the world by descending area size*

	Area	
Country	Square miles	Square kilometers
Russia	6,592,819	17,075,401
Canada	3,849,670	9,970,645
United States	3,717,796	9,629,092
China	3,696,100	9,572,899
Brazil	3,300,154	8,547,399
Australia	2,988,888	7,741,220
India	1,269,340	3,287,590

Source: 2000 World Population Data Sheet (Washington, D.C.: Population Reference Bureau).]

Belarus, Ukraine, and Moldova (Figure 5.14). It is usually considered the heart of Russia, because it is here that early Slavic peoples established what became the Russian Empire, with its center in Moscow. Although it occupies only about one-fifth the total territory of present-day Russia, European Russia has most of the industry, the best agricultural land, and about 70 percent of the population of the Russian Federation (105 million out of 147 million).

The vast majority of western Russians live in cities, their parents and grandparents having left rural areas when Stalin collectivized agriculture and established heavy industry. Most went to work in one of four major industrial regions chosen by central planners for ease of accessibility and location of crucial minerals. One region is centered on Moscow, another in the Ural Mountains and foothills (part of this zone lies in Siberian Russia), a third along the Volga River from Kazan' to Volgograd, and the fourth just north of the Black Sea extending into Ukraine around Donetsk.

The most heavily occupied part of European Russia—stretching south from St. Petersburg on the Baltic to the Russian Caucasus and east to the Urals—coincides with that part of the continental interior that has the most amenable climate (see Figure 5.3, page 245). Even so, because the region lies so far to the north (Moscow is at a latitude 100 miles (160 kilometers) north of Edmonton, Canada) and in a continental interior, the winters are rather long and harsh and the summers short and mild.

The best agricultural land is in the south of European Russia, the black soil (chernozem) region stretching from Moscow through Krasnodar and Stavropol to the northern parts of Russian Caucasia (Chechnya and neighboring ethnic enclaves between the Black and Caspian seas). Somewhat drier and less fertile agricultural land extends east on either side of the northern border of Kazakhstan (Figure 5.15, page 273). The really prime agricultural land of the former USSR now lies in the newly independent states of Ukraine and Moldova and in the Caucasian country of Georgia. Worries over loss of access to the agricultural resources of Ukraine, Moldova, and Georgia will play an important role in Russian policy decisions for years to come. Russians may be more open than they were in the past to striking trade and development agreements favorable to these countries in order to be certain of receiving agricultural exports from them.

Urban Life and Patterns. Moscow remains at the heart of Russian life. In the post-Soviet era, this city of about 9 million people has the best selection of goods and food and the most exciting nightlife. But prices are higher here than elsewhere (in 2001, prices in Moscow were rising at the rate of about 10 percent per year). The criminals are more concentrated, more innovative, and more violent. Entrepreneurs are more active, the market economy is more developed, and government jobs are particularly precarious. Moscow's economy is privatizing rapidly. By 1992, just one year after the end of the USSR, more than a quarter of Moscow's labor force of 4.4 million was in nonstate employment; by 1994, more than half the jobs were in the private sector. The retail

Figure 5.14 European Russia. Because of its great size, Russia itself is divided into three subregions: European Russia, Siberia, and the Russian Far East. The divisions are largely based on topographic features, so these features (primarily mountains) are included on the Russia subregion maps.

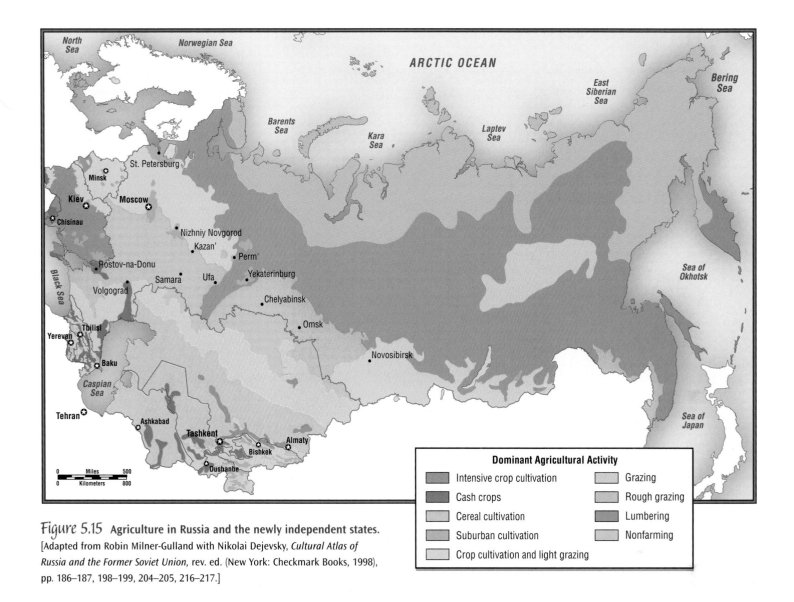

Figure 5.15 Agriculture in Russia and the newly independent states. [Adapted from Robin Milner-Gulland with Nikolai Dejevsky, *Cultural Atlas of Russia and the Former Soviet Union*, rev. ed. (New York: Checkmark Books, 1998), pp. 186–187, 198–199, 204–205, 216–217.]

Dominant Agricultural Activity

- Intensive crop cultivation
- Cash crops
- Cereal cultivation
- Suburban cultivation
- Crop cultivation and light grazing
- Grazing
- Rough grazing
- Lumbering
- Nonfarming

sector and small-scale services to the public—such as street food vending and sales of inexpensive clothes, home furnishings, and small appliances—created the most jobs (often in the informal economy). Today, for individuals adjusting to the loss of "cradle-to-grave" security, life in Moscow can be painful, frightening, exhilarating, or all three, as the following account of the experiences of Natasha and her customers illustrates.

VIGNETTE Natasha is an engineer in Moscow. She has managed to keep her job and the benefits it carries; but inflation has so diminished her buying power that in order to feed her family she sells used household items and secondhand clothes in a street bazaar on the weekends. "Everyone is learning the ropes of this capitalism business," she laughs. "But it can get to be a heavy load. I've never worked so hard before!" Asked about her customers, Natasha says, "Many are former officials and high-level bureaucrats who just can't afford the basics for their families any longer. Old people love the warm sweaters. Some of them who shop with me have to eat in soup kitchens.

"But, you know," she continues, brightening and changing the subject, "the bales of used clothes I buy now come from the United States. A guy drives a carload from a container ship in Amsterdam harbor every two weeks. And there are lots of sturdy clothes, especially for children, some quite new with the price tags still on; but [her voice registering disappointment] only occasionally is there a really fashionable item for a woman." [A composite story based on work by Alessandra Stanley, David Remnick, and David Lempert.]

The privatization of real estate in Moscow gives an interesting insight into pre- and post-Soviet circumstances. During the communist era, all housing was state-owned and there was a general shortage. Thousands of families were always waiting for housing,

Kaliningrad's busy port bristles with ship-loading cranes. It is the only year-round ice-free port on Russia's Baltic coast. [Dennis Chamberlin.]

some for more than 10 years. Most apartments were built after the mid-1950s, in large, shoddily constructed blocks; the units included one to three rooms and housed one to four (or more) people. But older buildings in the center city contained much more spacious, if bedraggled, dwellings. In 1991, the government allowed residents of Moscow to acquire, free, about 250 square feet (23 square meters) of usable living space per person—about the size of a typical American living room. Most people simply took ownership of the apartments in which they had been living; by 1995, a majority of Muscovites owned their apartments. A lively real estate market quickly emerged. Because of the housing shortage, apartments in the most desirable neighborhoods commanded very high prices. The demand was fueled by a rising wealthy elite that was willing to pay as much as U.S. $100,000 to $250,000 for large remodeled apartments in the center city. Rents also skyrocketed. Those who could find alternative housing for themselves made tidy sums simply by renting out their inner-city apartments The demand for commercial space for private businesses reduced the number of dwelling units, thus exacerbating the housing shortage.

Surrounding Moscow (and other cities) are many small parcels of land on which Muscovites maintain small garden plots and second dwellings (called dachas). Long important in urban middle-class Russian life, these modest holdings provide city dwellers with a place to relax on weekends and cultivate vegetables and fruits to supplement purchased food. The dacha, often only a tiny cottage (though some are quite large and well-appointed), may accommodate occasional overnight stays. Since privatization, the now wealthier and more numerous middle and upper classes have invested in substantial suburban dwellings that are beginning to ring the city, interspersed with the older pattern of dachas and gardens.

St. Petersburg, renamed Leningrad during most of the communist period (1924–1991), is European Russia's second-largest city, with about 5 million people. It is a shipbuilding and industrial city located at the eastern end of the Gulf of Finland, on the Baltic Sea. Czar Peter the Great ordered the city built in 1703 as part of his effort to Europeanize Russia; from 1713 to 1918, St. Petersburg served as the Russian capital. The city has a rich architectural heritage, including many palaces, and gardens. Since 1991, it has been undergoing a renaissance: the transport system is being refurbished, urban malls are being built, and historic sites are being renovated. The czars' Winter Palace, now called the Hermitage Museum, houses one of the world's most important art collections. Urban planners working in St. Petersburg maintain an elaborate Web site with numerous pictures that can be reached at http://www.kga.neva.ru/kga.html.

Kaliningrad. Kaliningrad is a 6000-square-mile (15,000-square-kilometer) exclave of Russia that lies beyond Belarus along the Baltic coast between Poland and Lithuania, about 200 miles (320 kilometers) from the closest Russian point (see Figure 5.14). Acquired from Germany at the end of World War II, Kaliningrad is now 90 percent Russian in population and is an important port on the Baltic for Russia, which has few year-round outlets to the world ocean. Kaliningrad's location on the Baltic also makes it ideally suited as a duty-free trade zone for Russia, and facilities there are being increased. Investors can import partially finished goods (such as appliances) into Kaliningrad, add up to 30 percent of the value within the zone (through such work as finishing and assembling, using Kaliningrad labor), and then ship the products duty-free into Russia.

Russia East of the Urals

To the east of European Russia lies a vast territory stretching from the Ural Mountains to the Pacific. It is usually divided into three physical zones: the western plain, the middle plateau, and the far

Figure 5.16 **Siberian Russia.**

eastern mountains (see the section on landforms earlier in this chapter, pages 240–241). Siberia has been the popular term for this area, but for our purposes here, Siberia refers to only about half the landmass east of the Urals: it encompasses the West Siberian Plain beyond the Urals and about half the Central Siberian Plateau (Figure 5.16) The rest of the landmass east to the Pacific is here called the Russian Far East.

Siberia. Siberia is so cold that 60 percent of the land is under permafrost. This vast central portion of Russia is home to just 33 million people, many of whom moved into the region from west of the Urals. Settlement is concentrated in the somewhat warmer southern quarter. For millennia, Siberia was a land of wetlands and quiet, majestic forests. But during the Soviet era, it became dotted with bleak urban landscapes and industrial squalor—

the legacy of the Soviet effort to establish industries to exploit Siberia's valuable mineral, fish, game, and timber resources (see the box on Norilsk, page 262).

Novosibirsk, Siberia's twentieth-century capital and financial center, is the largest Russian city east of the Urals (1.4 million people). Its location illustrates the Soviet policy of claiming space (and resources in that space) by distributing industrial cities across its vast landmass. Novosibirsk lies more than 1600 miles (2500 kilometers) east, as the crow flies, from Moscow. Although it has few natural resources in its immediate environs, this commercial center has the highest concentration of industry between the Urals and the Pacific. During the cold war, it was a center for strategic industries such as optics and weaponry and contained more than 200 heavy-industry plants. In the post-Soviet era, these

GEOGRAPHER IN THE FIELD

Interview with Regional Consultant for Chapter 5, Gregory Ioffe, Radford University

What led you to become interested in Russia and the newly independent states?
I was born and raised in Moscow. However, my true hometown is Kratovo, a dacha settlement 40 kilometers (about 25 miles) southeast of Moscow, where my grandmother owned a country house. As an economic geography major at Moscow State University, I decided to specialize in the Russian countryside because of the opportunity to study with the preeminent Russian rural geographer Sergei Alexandrovich Kovaliov. This somewhat arbitrary decision influenced my research interests for many years to come. Indeed, although I now live far away from my native country, I am still involved in the study of rural Russia.

Where have you done fieldwork in Russia, and what kind of work have you done?
I have traveled extensively in the former USSR. I have visited 13 of the 15 former republics and all of the economic regions of Russia, from Kaliningrad in the west to Chukotka in the east, and from Chechnya and Ossetia in the south to Murmansk in the north. I have made several field trips to the so-called *glubinka*, the outlying rural areas of Central Russia, where the sense of remoteness induced by general neglect, poor roads, and outdated transportation belies their relative proximity of 100 to 200 kilometers (about 60 to 120 miles) to major urban clusters.

What are some topics that you think are important to study about your region?

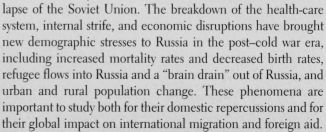

- The tensions between center and periphery, both within the federation and in the federation's relationship to the Eurasian republics.

- The demographic processes affected by the collapse of the Soviet Union. The breakdown of the health-care system, internal strife, and economic disruptions have brought new demographic stresses to Russia in the post–cold war era, including increased mortality rates and decreased birth rates, refugee flows into Russia and a "brain drain" out of Russia, and urban and rural population change. These phenomena are important to study both for their domestic repercussions and for their global impact on international migration and foreign aid.

- Methods of economic development in rural Russia. Policy makers and geographers must carefully examine contemporary reform efforts and study them in the context of European and Russian agrarian history and rural spatial development since the late nineteenth century. Is private farming the most viable option for the future of Russian agriculture? Or is a combination of Soviet-style subsidiary farming with traditional and reorganized collective farms a more beneficial path for rural Russia?

---

factories have become renowned for obsolescence and high rates of pollution, but they also represent a potential for future profit if privatized and reorganized. With its banks and financial services, Novosibirsk has become a center for outside investors interested in acquiring factories in the nearby surrounding region. In the 1990s, 400 joint ventures were established between Russians and Chinese, German, or South Korean partners. Many of them failed, but the successful ones were those that placed an emphasis on investing in Novosibirsk's human resources (rather than just its factory facilities) by offering retraining and education. An example of a still mostly state owned industry that seems to be making a successful transition to a market economy is Novosibirsk Instruments. This enterprise once produced artillery for the Soviet army, then moved to telescopes for civilian consumers in the 1980s, and most recently has focused on providing advanced amateur astronomers around the world with good-quality telescopes.

The Russian Far East. The Russian Far East is an extensive territory of mountain plateaus with long coastlines on the Pacific and Arctic oceans (Figure 5.17). In area, it makes up slightly over one-third of the entire country of Russia and is nearly two-thirds the size of the United States. The population of 8.16 million is sparsely settled: on average, there are just 2.3 persons per square mile (1 per square kilometer), compared to 23 per square mile (9 per square kilometer) for Russia as a whole. Almost 90 percent of the land is covered by permafrost; the coastal volcanic mountains stop the warmer Pacific air from moderating the Arctic cold that spreads deep into the continent.

The Russian Far East has many resources but has not been developed because it is difficult to reach from European Russia. Relatively unexploited stands of timber cover 45 percent of the area; the tree roots grow in the sun-warmed sediment above the permafrost. Soviet central planners, and now private investors, have been attracted by the region's resources of coal, natural gas,

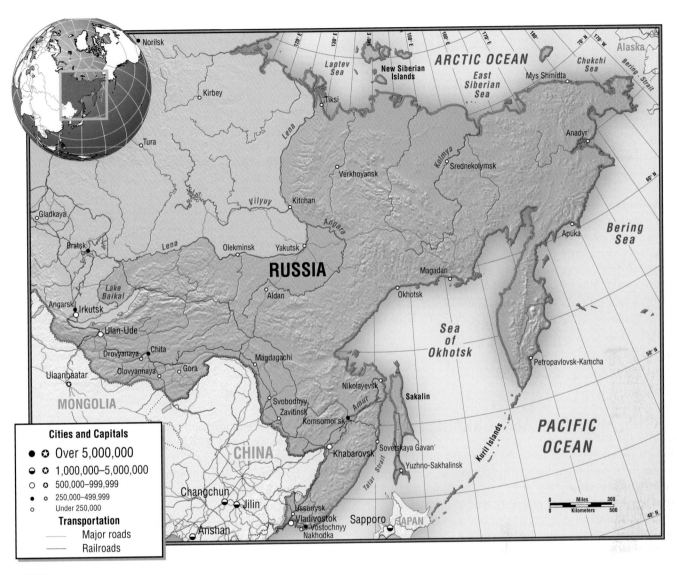

Figure 5.17 **The Russian Far East.**

oil, tin, antimony, gold, diamonds, iron, and other minerals. But, even more than in lowland Siberia to the west, development has been hampered by the distance from population centers and by the difficult environment. Only 1 percent of the territory, mostly along the Pacific coast and in the Amur River drainage basin, is suitable for agriculture (see Figure 5.15). Although forestry is now in a slump, both China and Siberia have plans to increase timber extraction and development of the Amur River basin drastically. Fruit and grain production is already growing in the basin. The Amur River is the world's largest remaining un-bridged and undammed river.

The Russian Far East is a land of immigrants and exiles—some of whom were sent there for a perceived misdeed in the Soviet era. Here, where serfdom never intruded, there is a spirit of independence that is rare elsewhere in Russia. The earlier migrants from western Russia worked in timber and mineral extraction enterprises and in isolated industrial enclaves; since the 1980s,

many immigrants have headed for cities along the Pacific coast, especially Nakhodka and Vladivostok. Today, 75 percent of the residents of the Russian Far East live in cities.

The people of the Russian Far East are those most likely to profit greatly in the transition to a market economy and freer international trade (primarily in minerals and other raw materials). In recent decades, port facilities on the Pacific coast have made the region accessible to such countries as Japan, Korea, and China (see photo on page 253). Japan, a rich country that is nonetheless poor in natural resources other than timber, accounts for 20 percent of the trade, and that portion is likely to increase. Japan has offered investment money to Russia in turn for several small islands it wishes to acquire near Sakalin Island. Some geographers and economists forecast that, because of its rich resource reserves and its location so far from the Russian core, the Russian Far East will eventually integrate economically with other Pacific Rim nations, perhaps detaching itself politically from European Russia.

AT THE LOCAL SCALE   *An Ethnic Republic Reasserts Itself*

Tuva is a narrow, mountainous republic just north of Mongolia (see Figure 5.10), with a population of 300,000 people. Traditionally, the Tuva were nomadic herdspeople who tended camels, reindeer, cattle, sheep, and horses. When the Soviets arrived in 1944, most of the inhabitants were forced to settle in wooden houses, enroll their children in Russian schools, and work on collective farms or in the coal, asbestos, cobalt, or gold mines. Russian bureaucrats settled in Tuva to run the various collectives, mines, and schools; today, Russians make up one-third of the population.

In summer, a few thousand Tuva nomads seek higher, cooler elevations where there is sufficient grass for their animals. Their homes are yurts—large, round, felt-covered tents stretched on circles of wooden lattice that contract into a tight bundle for moving. Many Tuva dream of returning to the nomadic life, but few retain the needed skills or the ability to face the rigor of the old ways.

Tuva has its own language and its own parliament, but the Tuvan economy generates only 18 percent of its national budget. The rest comes from Moscow, much of it to support Russian workers. Tuva's main export is its music, called "throat singing." The music has a distinctive harmonized sound created when one singer produces two notes at once, one a vibrating hum, the other a quaver. Touring throat singers have won acclaim in the United States and Europe, and they have recorded with such well-known musicians as Ry Cooder. The Tuva president hopes the singers' popularity will protect his republic from any Russian efforts to quash a mounting independence movement violently. "[If something bad happens], we want the world to know where Tuva is and who we are," he told interviewers.

Source: Richard Feynman, http://feynman.com/tuva.

BELARUS, MOLDOVA, AND UKRAINE

Sandwiched between Russia and the eastern European countries that were once part of the Soviet Union's sphere of influence are the newly independent states of Belarus, Moldova, and Ukraine (Figure 5.18). Each of these primarily agricultural countries is the home of a distinctive Slavic people. Because of their position between Russia and the rest of Europe, these countries have found themselves faced with the choice of maintaining their closest ties with Russia or turning toward Europe.

Belarus is a globe-shaped country surrounded by Latvia, Lithuania, Poland, Ukraine, and Russia. In size and terrain, Belarus resembles Minnesota: its flat, glaciated landscape is strewn with forests and dotted with thousands of small lakes, streams, and marshes that are replenished by abundant rainfall. During the twentieth century, much of the land was cleared and drained for agriculture, mostly on collective farms. The stony soils are not particularly rich; nor, other than a little oil, are there many known useful resources or minerals beneath the surface. Belarus absorbed a large amount of radiation contamination after the Chernobyl nuclear accident in Ukraine. Twenty percent of the agricultural land and 15 percent of the forestland were lost.

Belarus was forced into independence by the collapse of the Soviet Union in 1991 and now actively seeks to reintegrate with Russia. It shares cultural, economic, and political features with Russia. During the twentieth century, Belarus was rather thoroughly Russified—molded to accept Russian values and perspectives—by a relatively small but influential group of Russian workers and bureaucrats. Although 80 percent of the population of 10 million is Belarus (a Slavic group) and only 13 percent Russian, the Russian language predominates and Belarus culture survives

primarily in museums and historical festivals. The urban concrete landscape looks and feels Russian. The Belarus economy remains dominated by state firms that sell to Russia; only a few retail shops are now private. Its important petrochemical industry depends largely on Russian oil and gas for raw materials. In Russia, the Belarussian desire to reunite with Russia is welcomed by the remaining communists as well as by the military—who see it as a useful buffer state against the extension of NATO membership to Poland—and by all those who wish to see the former Soviet empire reconstituted. By 2001, Belarus and Russia had agreed to remove all customs and trading fees for each other—and the countries appeared to be united for all practical purposes.

Moldova and Ukraine are two closely related countries that share the low-lying territory between Belarus and the Black Sea. They enjoy a warmer climate than Belarus and Russia, and, with

TABLE 5.6   *Similarities between Ukraine and France*

	Ukraine	France
Size	223,690 sq miles	212,390 sq miles
Population	50.7 million	58.6 million
Value added in agriculture[a]	$37,873 million	$37,337 million

[a]1993 figures.

Source: World Development Report 1995, p. 169.

Figure 5.18 The Belarus, Moldova, and Ukraine subregion.

good agricultural resources, they have the potential to increase agricultural production significantly. Ukraine, which has rich soils, produces sugar (from beets), grapes for wine, sunflower oil and seeds, flax, nuts and fruits, meat, and dairy products. It is comparable to France in size, population, and agricultural output (Table 5.6). Moldova, lying to the west and bordering Romania, is much smaller (4.3 million people). It produces many of the same products as Ukraine and is now pursuing the possibility of emphasizing wine, nuts, and fruits to be marketed not only to Russia but also to the Baltic states that lie too far north to produce such items and to Europe via Romania.

In the 1930s, when the secretary of the Communist party, Joseph Stalin, was implementing socialist agricultural restructuring, Ukrainian farmers were forced to surrender title to their lands to large collective farms. Although there were some gains in effi-

ciency under the collectives, the Ukrainian farmers were required to give nearly their entire harvest to the central planners for redistribution. Those who refused, or who retained small amounts to feed their families or to reseed fields, were severely punished and often executed. The resulting famine in 1932 and 1933 killed 7 million to 8 million Ukrainians. As of 2000, much of the region is still collectivized, because government officials are wary of making changes that could jeopardize food supplies. But private individuals are changing Ukrainian agriculture on their own. Many families grow up to half their own food supply in small family gardens that have proliferated in the years since independence, especially in urban areas. Many analysts believe that Ukraine—which supplied food to the USSR, especially after World War II, and which continues to trade heavily with Russia—could enjoy a niche in the agricultural economy of Europe, giving France some serious

Farmlands like these in southwestern Ukraine (the Ivano Frankivsk region) evoke images of French Impressionist paintings and display the richness of soil and climate that made the Ukraine a prime agricultural producer for Soviet Russia. [Robert Semeniuk/Corbis Stock Market.]

competition in the production of vegetables, fruits, and wines. To do so, Ukraine would have to make many adjustments to meet EU standards. It would also need to refurbish its collective farms—which, for the most part, have outdated equipment (see the following vignette) and produce less efficiently than farms in western Europe. For the time being, therefore, Ukraine will probably continue to trade its agricultural products mostly with Russia.

VIGNETTE   Those Ukrainian farmers who spent their lives on collective farms have had difficulty making the transition to being private farmers. Vasyl Speiko, age 35, is a farmer in western Ukraine. After spending years on the Bukovyna collective farm, he recently obtained 12 acres (5 hectares) of his own. So far, the only equipment he can afford is a horse and wagon, but he is optimistic: "I want to know what it's like to be a free man." In working his land, he benefits from the counsel of his aged father, who lost his 61 acres (25 hectares) of farmland to the state collective in 1948. [Mike Edwards, "After the Soviet Union's collapse—A broken empire," *National Geographic* (March 1993): 49–52.]

Since the breakup of the Soviet Union in 1991, Moldova and Ukraine have experienced economic ups and downs that have left them ambivalent about whether they should aspire to join the European Union trading network or retain Russia as their chief trading partner. In the early 1990s, industrial and agricultural production rose in anticipation of trading outside the USSR and then fell as it became clear that Moldova and Ukraine would be unable

to compete with the better products of western Europe. Wage and pension payments fell into arrears, and inflation chopped the value of wages by half. Efforts to lower excessive public spending and make all parts of the economy more efficient (structural adjustment policies) resulted in job losses and declines in well-being. Ukraine still sells most of its wheat to Russia and relies on Russia for most of its oil and gas. Moldova, which appeared to be leaning toward Europe in the 1990s, had turned its attention back toward Russia by the time of the 2000 elections, as evidenced by the election of an ethnic Russian communist as head of state.

Moldova's choices are complicated by conflict over who will control its richest and most heavily industrialized western section, known now as Transnistria and previously as Bessarabia. Although Transnistria lies next to Romania in western Moldova, it has had a close relationship with Russia for decades, primarily because of its industrial capacity. Ethnic Russians held supervisory positions in Transnistrian industries and would like to retain their influence because of the potential for future industrial expansion, close to European markets. Neighboring Romania is equally interested in the productive resources of Transnistria and is more conveniently located than Russia. Some factions in Moldova would like to see closer, perhaps even political, connections with Romania; in 1992, these factions attempted an armed insurrection, which was put down with Russian troops. An apparent majority of Transnistrians and other Moldovans favor continuing relationships with Russia and also Ukraine, at least for the time being. Russia maintains several hundred troops in Transnistria to discourage future activities by pro-Romanian separatists.

The EU's interest in Ukraine and Moldova was rekindled in 2000 when it was revealed that much illegal migration into Europe from Sri Lanka, Bangladesh, Afghanistan, Chechnya, and West Africa passed through these two countries. The EU is now providing financial aid, especially to Ukraine, to tighten border security.

CAUCASIA: GEORGIA, ARMENIA, AND AZERBAIJAN

Caucasia is located around the rugged spine of the Caucasus Mountains, which stretch from the Black Sea to the Caspian Sea. It includes a piece of the Russian Federation on the northern flank of the mountains as well as the three independent countries of Georgia, Armenia, and Azerbaijan, which occupy the southern flank of the mountains in what is known as Transcaucasia.

Caucasia is surely one of the most ethnically complex places on earth. Between the Black Sea and the Caspian Sea, in a mountainous space the size of California, live more than 50 ethnic groups: Armenians, Chechens, Ossetians, Karachays, Abkhazians, Georgians, and Tatars, to name but a few (Figure 5.19). The groups vary widely in size: the Ginukh, for example, number just 200, whereas the Turkic Azerbaijanis number over 6 million. Some are Orthodox Christians, many are Muslims; Gypsies pass through; there are Jews; and, as in Russia, some groups retain ancient elements of local animistic religions. As in the part of northern Caucasia that is in Russia, many of the ethnic groups in the newly independent countries of Caucasia are remnants of

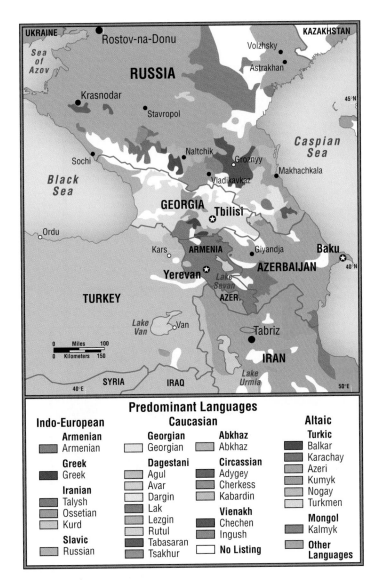

Predominant Languages

Indo-European	Caucasian		Altaic
Armenian	**Georgian**	**Abkhaz**	**Turkic**
Armenian	Georgian	Abkhaz	Balkar
Greek	**Dagestani**	**Circassian**	Karachay
Greek	Agul	Adygey	Azeri
Iranian	Avar	Cherkess	Kumyk
Talysh	Dargin	Kabardin	Nogay
Ossetian	Lak	**Vienakh**	Turkmen
Kurd	Lezgin	Chechen	**Mongol**
Slavic	Rutul	Ingush	Kalmyk
Russian	Tabasaran	No Listing	**Other Languages**
	Tsakhur		

Figure 5.19 The Caucasian subregion. The diversity of ethnolinguistic groups in the Caucasian subregion illustrates the cultural complexity of the area. [Adapted from http://www.geocities.com/southbeach/marina/6150/ethno.jpg.]

near the Black and Caspian seas. This is the landscape occupied by Georgia, Armenia, and Azerbaijan. Most of the inhabitants live in the valleys. Whereas much of the land in the rest of Russia and the newly independent states is arid or cold, these treasured pieces of rugged mountain slopes and narrow plains are blessed with warm temperatures and abundant moisture from the Black and Caspian seas. In these favorable conditions, farmers grow crops, such as citrus fruits and even bananas, that can be produced in no other part of the vast Russian and Central Asian expanse. Before 1991, most Soviet citrus and tea came from Georgia, as did much of its grapes and wine.

In Caucasia, a single ethnic group is often split up, and members of the group live in several different republics. After the Soviet collapse in 1991, the three culturally distinct republics of Georgia, Azerbaijan, and Armenia obtained independence as nation-states. Then, very quickly, several other ethnic groups took up arms to obtain their own independent territories: the Abkhazians and the South Ossetians against Georgia; the Christian Armenians in Nagorno Karabakh (an exclave of Armenia located inside Azerbaijan) against the Muslim majority. In the conflict over Nagorno Karabakh, 15,000 people died before a truce was signed in 1994. Although ethnic strife has been common, it is often fomented by larger powers—Turkey, Russia, Iran—seeking access to strategic military installations and agricultural and mineral resources.

In addition to cultural differences, access to oil and gas reserves has been a source of conflict in Caucasia. Estimates of Caspian oil reserves vary widely, from 28 billion to 200 billion barrels. A major issue is how to get the oil and gas out of Caucasia safely and into the world market. Pipelines? Trucks? Ocean

Jazz in Azerbaijan. Aziza Mustafa Zadeh is an Azerbaijani singer and jazz pianist, popular in Europe and the eastern Mediterranean. Her version of jazz is inspired by *mugam,* an ancient, mesmerizing modal system of traditional music from Caucasia. [Courtesy of Aziza Mustafa Zadeh.]

ancient migrations. For thousands of years, Caucasia was a stopping point for nomadic peoples moving between the Central Asian steppes, the Mediterranean, and Europe. Other ethnic enclaves were created more recently by Soviet-instigated relocation and then return of troublesome minorities. Today, many Caucasians maintain ties to Europe and North America, where hundreds of thousands of emigrants from the region live. Like Native Americans in North America, some of the ethnic groups in Caucasia are using the Internet as a way to gain publicity for their unique geographic location, disputes with neighboring republics, and complex historical heritage. A particularly elaborate example is the Web site maintained by the Abkhazians: http://www.abkhazia.com/.

Transcaucasia consists of a band of subtropical intermountain valleys and high volcanic plateaus that drop to low coastal plains

Figure 5.20 **The Central Asian subregion.**

Cities and Capitals

● ✪	Over 5,000,000	
◒ ✪	1,000,000–5,000,000	
○ ✪	500,000–999,999	
● ◦	250,000–499,999	
○	Under 250,000	

Transportation

— Major roads
— Railroads

tankers? All require passage through hostile territory or across difficult terrain. A consortium of Western oil companies has considered several pipeline routes to the Black Sea. Transit through Chechnya is problematic because of its continuing guerrilla war with Russia. Transit through Georgia and Russia exists now, and new lines are under construction. The United States and Europe favor a pipeline across Turkey. Turkey would use some of the energy supplied, but most would be sold to Europe.

THE CENTRAL ASIAN REPUBLICS

Since the fall of 1991, following nearly 12 decades of Russian colonialism, five independent nations have emerged in Central Asia. Kazakhstan, Kyrgyzstan, Tajikistan, Turkmenistan, and Uzbekistan stretch from the arid plains and steppes west of the Caspian Sea to the mountains of western China (Figure 5.20). Consistent with the mosaic of cultures that is found here, each of these new nations has traditions that are recognizably different from those of the others. Yet all of them draw on the deep, common traditions of this ancient region. For example, most people speak different Turkic languages that are often mutually intelligible; the Tajiks, however, speak a Persian language. After decades of efforts to align

their cultures and economies with those of the USSR, the countries of Central Asia are poised for change brought on by political independence; by the rise of ethnic and religious (Islamic) sensitivities; and, especially, by the chance to profit from the exploitation of recently discovered oil and gas reserves.

Central Asia is situated deep in the center of the Eurasian continent, in the rain shadow of the lofty southern mountains of Iran, Afghanistan, and Pakistan. Its dry continental climate is a reflection of this location. What rain there is falls mainly in the north, on the Kazakh plain, where wide grasslands support limited agriculture and large herds of sheep, goats, and horses. In the south are deserts crossed by rivers carrying glacial meltwater from the high peaks still farther to the south. These rivers are tapped to irrigate huge fields of cotton and smaller fields of wheat and other grains (see the discussion in the Environmental Issues section, page 268). In many areas of the south, gardens are attached to houses and are walled to protect against drying winds. They nurture apricots, peaches, apples (apricots and apples were first domesticated in Central Asia), and many types of melon.

Kazakhstan, Uzbekistan, and Turkmenistan are largely low-lying plains, but Kyrgyzstan and Tajikistan lie high in the Hindu Kush, Pamir, and Tien Shan mountains to the southeast of these plains. These two countries have exceedingly rugged landscapes. Tajikistan elevations go from near sea level to more than 22,000 feet (6705 meters), and those in Kyrgyzstan are similar. Both offer little in economic potential; Tajikistan is the poorest country in the entire region and has undergone five changes in government and a civil war since 1991. (We will discuss the civil war shortly.)

Civilization flourished in Central Asia long before it did in lands to the north. The ancient Silk Road, a continuously shifting ribbon of traders connecting China with the fringes of Europe, operated for thousands of years, diffusing ideas and technology from place to place. Modern versions of ancient trading cities still dot the land: two examples are Bukhara, renowned as a center of Islamic learning and culture, and Samarkand, which has been in existence for 5000 years. Commerce along the Silk Road diminished by the fifteenth century, as traders changed to shipping goods by sea. Central Asia then entered a long period of stagnation until, in the mid-nineteenth century, czarist Russia developed an interest in the region's major export crop, cotton.

The Russian imperial agents built modern mechanized textile mills, which quickly replaced small-scale textile firms that employed people to do traditional cotton and silk weaving. The new mills were often located in entirely modern Russian towns built alongside railroad lines. Consequently, they benefited few Central Asians. In rural areas, cotton fields replaced the wheat fields that had fed local people. Occasional famines struck whenever people could not afford imported food or when supplies were interrupted.

Russian domination intensified during the Soviet era. Although only a few Central Asians eventually joined the Communist party, they were thoroughly Russified. The practice of Islam, though officially tolerated, was undermined by the government's promotion of atheism and by restrictions on Islamic worship, pilgrimages to Mecca, and access to mosques (see page 261).

Nonetheless, the Soviet era did benefit Central Asians. The ancient patriarchal landholding system was partially dismantled, and some poor sharecroppers were given land. Health care improved, and literacy expanded dramatically, from less than 10 percent in 1915 to more than 50 percent by 1959 and more than 90 percent in 1998. The Soviet state encouraged Central Asian women to work outside the home. However, they have had little access to schooling beyond learning to read, and their main opportunities have been low-paying jobs in cotton fields and factories.

In the years since independence, many conditions have not changed, and others have worsened. Russian influence remains strong, although it is balanced somewhat by growing international interest in the region's oil and gas wealth. Transportation and services remain underdeveloped, as they were in Soviet times. Although elections have been held and free markets are opening opportunities for entrepreneurship, government bureaucracies remain authoritarian and patriarchal. Poverty is now more widespread, and health standards are the lowest of any in the former USSR. Although

Thinking Globally: ON THE WEB

One of the most remarkable cultural traditions of Central Asia is carpet weaving, an art practiced primarily by women and children. The carpets probably originated as the highly portable and useful furnishings of nomadic yurts and *gers* (tents); there are many different types and styles. Each region and culture group has distinctive patterns, with particular combinations of colors. The practice of weaving rugs stretches from interior China to Caucasia, Turkey, and the Balkans. The rug pictured here is an Ersari from Turkmenistan. [Courtesy of Charles W. Jacobsen, Inc., Syracuse, N.Y.]

Islamic fundamentalism is feared by governments and even moderate Muslims endure repression, the revival of Islam remains influential, if subdued (see the vignette on page 261).

A particularly troubling legacy of changes in the modern era survives in the complex mountainous borderlands where Central Asia touches China, Pakistan, and Afghanistan. There, Soviet efforts to develop commercial agriculture and industry and supply a labor force for this development resulted in the rearrangement of ethnic groups that previously had not been antagonistic. There are now, for example, new pockets of Tajiks in Uzbekistan and Afghanistan, Afghans in Tajikistan, and Kazakhs in Turkmenistan. Much of the cultural patchwork of this region is ancient and can be accounted for partially by the complexity of the mountainous terrain, which kept groups cut off from one another. But modernization created new patterns of settlement. The movement of large numbers of members of particular ethnic groups broke up families and clans and changed the groups' economic arrangements and trading patterns. The Soviet era did not prepare the people of this

Once Kazakhstan gained independence in 1991, ethnic Kazakhs returned home from self-imposed exile in Mongolia. Kazakhs and Mongolians share a cultural history of nomadic herding. The portable yurt homes are visible in the background. [Gerd Ludwig/National Geographic Image Collection.]

region to govern themselves, and when Soviet control abruptly diminished in the 1990s, armed conflict erupted.

Civil War in Tajikistan

High mountains and arid uplands account for nearly 90 percent of Tajikistan, which is about the size of Greece. There is one lowland zone in the southwest that is physically connected to the lowlands of northern Afghanistan and a second narrow lowland zone in the far north through which flow the headwaters of the Syr Darya River. Ancient trade routes have kept the people in this difficult terrain in social contact—the northernmost zone was part of the Silk Road—but for the most part, the complex physical landscape has fostered the development of many culturally distinct groups. A main division is that between mountain people and lowland people.

Shortly after independence in 1991, elections were held in Tajikistan. Nine main candidates representing a variety of ethnic, religious, and political factions ran for office. A former communist won, but many citizens did not accept his election. Fighting broke out between the government and a loose coalition of groups of varying interests. The opposition groups identified themselves as favoring democracy, but they had strong and differing loyalties: to the revival of Islamic values, to Tajik nationalism, or to autonomy for a particular ethnic group. Arms were supplied to the various protagonists by outside interests in Iran, Afghanistan, Russia, China, and other Central Asian countries.

Between 20,000 and 60,000 people were killed in the first year of fighting (1992). About 600,000 became internal refugees, and 80,000 fled to Afghanistan. Schooling was interrupted for years,

resulting in a decline in literacy and other skills. Unemployment reached 60 percent for young workers, and many women lost jobs, just as they were widowed by the conflict. The brutality of the civil war, the fear of invasion by the Taliban from Afghanistan (see discussion in Chapter 8, page 434), and the fear of losing their status as an independent country shocked the Tajik factions into negotiation. A peace accord was agreed to and signed in 1997. Since then, public debate has centered on the proper role of religion in government. A majority of Tajik people regard themselves as Muslim and consider their Islamic heritage important; but most apparently do not favor an Islamic state. The continued rise of the radical Islamist Taliban movement in Afghanistan during the 1990s gave the debate an air of high tension by increasing the prominence both of Tajik Islamists and of those who opposed the Taliban. The Tajik-led anti-Taliban fighting force, called the Northern Alliance, took up positions along the borders with Afghanistan The majority of Tajiks, however, appeared to prefer peaceful economic development to religious conflict.

The Free Market in Central Asia

Despite the overall slow pace of change and the difficulties of transition, the Central Asian republics appear to be finding their "capitalist legs" fairly quickly. As the Russians leave, Central Asians are taking up old market-based skills that have been part of their heritage since the ancient Silk Road days, when trade and long-

In Kyrgyzstan, Minavar Salijanova took advantage of a United Nations microcredit loan of 5000 SOM (about U.S. $100.00) to buy 60 chickens. Each day the chickens produced 60 eggs, which she sold in the market for about 6 cents each. Every three months she bought more chickens and now has 147. She has also bought some goats and is expanding the business. "The program has improved our vision for life," she says, and neighbors have begun to follow her example. [Staton R. Winter/New York Times Pictures.]

distance travel were the heart of the economy. For example, a largely contraband trade in consumer goods and illegal substances from drugs to guns is blossoming with Iran, Afghanistan, and China. (See references to the revival of the Silk Road and the ancient market at Kashgar in Chapter 9.) There are also opportunities to render such services as food, transport, and accommodations to tourists and traveling businesspeople who hope to set up legitimate enterprises in the region (one such effort is described in the vignette about Fahreeden, below). Meanwhile, Turkmenistan is already finding a market for its oil and gas in Iran. And Iran, certain that shoppers will once again materialize from across the steppes, as they have for thousands of years, is establishing a free-trade zone along its northern border with Turkmenistan.

VIGNETTE Fahreeden is a taxi driver who operates out of Samarkand, Uzbekistan, in a dusty and battered old Fiat that he starts by hot-wiring the ignition. The tires are worn down to the steel fibers. and the brakes no longer work. His client today is an American lawyer pursuing an interest in religious history. The destination is the train station in Bukhara, Fahreeden's hometown. Bukhara is four hours away, a distance that will bring three flat tires and five halts by highway police, who expect bribes (baksheesh) at each stop. Fahreeden and the lawyer have agreed upon a fee that includes the anticipated police bribes, flat tires, and a generous luncheon worthy of a Silk Road merchant.

The dusty ribbon of road through the semiarid landscape is conveniently punctuated with shops dedicated to tire repair. There are no signs advertising this service, just a bald tire sitting upright along the roadside. As you wait for your tire to be repaired, you may be treated to tea and conversation with the family of the repair person. Eventually, due to missed trains and changed plans, the lawyer will spend the night in the taxi driver's home where, to the guest's surprise, the walls and floors are covered with beautiful and expensive handmade carpets. Dinner, prepared by Fahreeden's wife, Miriam, is shish kebab, fresh fruit, tomato salad, round loaves of bread, rice pilaf, tea, and lassi (a yogurt-based drink, taken from a common bowl). The guest will sleep in splendor on the same soft rugs on which his meal was graciously served. [From the travels of Ron Leadbetter, personal communication, Autumn 1996.]

Fahreeden's efforts to adapt to the post-Soviet era have led him to reinvent ancient ways of life in Central Asia. As was the case for many families in the days of the ancient Silk Road trade, his income is based on accommodating travelers and merchants and steering them safely through a large, arid, and potentially dangerous territory.

REFLECTIONS ON RUSSIA AND THE NEWLY INDEPENDENT STATES

The region comprising the Russian Federation, Belarus, Ukraine, Moldova, Caucasia, and Central Asia has gone through dramatic changes over the last decade. As time passes, it may no longer be appropriate to combine these particular political units into one region. Just what the future holds is unclear. No other set of countries has ever attempted such a rapid and peaceful transformation to a free-market economy. As the command economy disappears and localities direct their own economic affairs, regional differences and disparities will increase. Although Russia retains its leadership position, its power and influence have weakened. Some of its closest allies in eastern Europe are no longer in its sphere of influence and are now closely connected with the increasingly powerful European Union.

It is likely that central governments will remain especially strong in most republics and that participatory democracies will evolve only slowly. Although some individuals are eager to learn from the West, there is widespread and understandable resistance to rapid culture change and to excessive influence from abroad. Most republics in the region will probably have an ambivalent relationship with the West for some years to come: embracing many aspects of market economies and some aspects of democracy with enthusiasm, yet mourning the loss of certain civilities and welfare guarantees of the old system. Rising crime, violence, and fraud, associated either with the breakdown of the old system or with the introduction of a market economy, have alienated many, especially in Russia, from Western models of development. In the Central Asian republics, the nuances of the market economy may be sorted out more rapidly because of the long heritage of entrepreneurial trade in the region. On the other hand, parochial ethnic rivalries and conflict between the secular influences of capitalism and the religious values of a reviving Islamic tradition in Central Asia may become the nexus of anticapitalist and anti-West feeling.

All parts of the region will be hampered for years to come by the aging and inefficient industrial infrastructure and by the severe environmental pollution that has accompanied industrialization. With the loss of central control, the rapid private development of the rich resource base will probably lead to increased pollution. It is possible, though, that some pollution may be ameliorated as the industrial infrastructure is first shut down and then modernized.

The region's people have a particularly challenging future as they deal with all that the move toward a market economy entails. On the one hand, jobs and social safety nets formerly associated with the workplace have been lost, skills have become obsolete, and changing gender roles are disorienting to both men and women. On the other hand, exhilarating opportunities exist for greater self-expression, the possibility of entrepreneurial activities, and the promise of the material well-being that market economies can provide.

Thinking Critically About Selected Concepts

1. Russia and the newly independent states form a huge territory that stretches across the northern reaches of the Eurasian continent from the Baltic Sea to the Pacific Ocean. The climate of this massive region is harsh, either especially cold or, in the south, especially dry. *How does the physical geography of this region compare and contrast with that of North America?*

2. Slavic Europeans and Central Asian nomads have interacted for thousands of years to create a complex pattern of cultures. Over the past several centuries, Russians have dominated; during the twentieth century, they imposed communism on the entire region. *Do you see similarities between Russian relations with non-Russian ethnic groups in surrounding territory and European colonization in the Americas, Africa, and Asia? What are some obvious differences?*

3. The population center of this region is in European Russia. The rest of the region is thinly settled, except for large industrial cities widely spaced in the Ural Mountains and on the plains of western Siberia. *What are the main reasons for this settlement and industrial pattern? How does the pattern express central planning ideologies?*

4. The communist command economy is in the process of being replaced with a free-market or capitalist economy. With the abrupt shift to capitalism, many state-owned enterprises could no longer compete and were closed or sold to private investors. As the new owners improved efficiency and cut costs, many jobs were lost. Unemployment and underemployment are now widespread. *Discuss the reasons that the industries and enterprises developed under the command economy are not competitive in the free-market economy. In this situation, to what extent is job loss a result of personal failings?*

5. Since the breakup of the Soviet Union, average living standards have declined as the socialist support system providing the basics of employment, housing, food, and health care has disintegrated. Nutrition and birth rates have declined, and death rates have risen. *Faced with living as a university student in a city in European Russia, what would your life be like? What would be your most pressing daily concerns?*

6. Economic and political relationships between Russia and the Caucasian and Central Asian republics are being renegotiated. The trend is toward greater political autonomy, control over resources, and self-sufficient production in the republics. The overall result is a lessening of the power of Russia. *How is this change in Russian influence affecting politics in Caucasia and Central Asia?*

7. The non-Russian internal republics and the ethnic groups in Caucasia and Central Asia are showing a renewed interest in reasserting their pre-Soviet cultural identities. *To what extent do their cultural heritages help them participate in the emerging capitalist economy?*

8. Russia and the newly independent states have many natural resources, especially oil, gas, minerals, and forests. But agricultural land and water are in short supply. Easily cultivable land is found primarily in the European and Caucasian parts of the region. Irrigation has made Central Asian lands temporarily cultivable, but the environmental effects of irrigation and chemical fertilizer use have been detrimental. *How is the uneven pattern of resource distribution and economic activity likely to change as the newly independent countries adjust to being autonomous? Do you think state control or popular pressure by ordinary people is more likely to result in resolving environmental problems in the region? Compare the likely effects of environmental protests in such places as Norilsk in Siberia with their likely effects in the area around the Caspian Sea, where many new countries are now involved.*

Key Terms

Bolsheviks (p. 243) a faction of communists who came to power during the Russian Revolution

capitalists (p. 243) a wealthy minority that owns the majority of factories, farms, businesses, and other means of production

chernozem (p. 242) black, fertile soils of the region stretching south from Moscow toward the Black and Caspian seas

command economy (p. 243) an economy in which government bureaucrats plan, locate, and manage all production and distribution

communism (p. 243) an ideology based largely on the writings of the German revolutionary Karl Marx, which calls on workers to unite to overthrow capitalists and establish an egalitarian society where workers share whatever they produce

Communist party (p. 243) the political party that ruled the former Russian Empire from 1917 to 1991

czar (p. 243) a ruler of the Russian Empire; derived from "Caesar," the title of the Roman emperors

glasnost ("openness") (p. 244) an opening up of public discussion of social and economic problems, which occurred under Mikhail Gorbachev in the late 1980s

Group of Seven (G7) (p. 255) an organization of the most highly industrialized nations, now sometimes called Group of Eight (G8) when Russia is invited to meetings

internal republics (p. 257) more than 30 ethnic enclaves in Russia, whose political status and relationship to the Russian state are continuously being renegotiated republic by republic

Mongols (p. 242) a loose confederation of nomadic pastoral people centered in eastern Central Asia, who by the thirteenth century conquered an empire stretching from Europe to the Pacific

nonpoint sources of pollution (p. 267) diffuse sources of environmental contamination, such as untreated automobile exhaust, raw sewage, and agricultural chemicals that drain from fields into the urban water supplies

oligarchies (p. 255) powerful elites that wield influence because of their enormous wealth

perestroika (p. 244) a restructuring of the Soviet economic system in the late 1980s in an attempt to revitalize the economy

permafrost (p. 240) permanently frozen soil a few feet beneath the surface

privatization (p. 250) the sale of industries formerly owned and operated by the government to private companies or individuals

Russian Federation (p. 257) Russia and its political subunits, which include 30 internal republics and more than 10 so-called autonomous regions

serfs (p. 243)   persons legally bound to live and farm on land owned by the lord

Slavs (p. 242) a group of farmers who originated between the Dnieper and Vistula rivers, in modern-day Poland, Ukraine, and Belarus

socialism (p. 244) a social system in which the production and distribution of goods are owned collectively, and political power is exercised by the whole community

steppes (p. 240)   semiarid, grass-covered plains

taiga (p. 242) northern coniferous forest

thermal pollution (p. 266) the return of unnaturally hot water to the environment, as by industrial processes or hydro and nuclear power plants

tundra (p. 240) a treeless permafrost zone

Pronunciation Guide

Amu Darya (ah-moo DAHR-yah)

Amur River (ah-MOOR)

Aral Sea (AHR-uhl)

Armenia (ahr-MEE-nee-uh)

Azerbaijan (ah-zair-bye-JAHN)

Baikal, Lake (bie-KAHL)

baksheesh (bak-SHEESH)

Baltic Sea (BAWL-tihk)

Belarus (byeh-lah-ROOS)

Bolshevik (BOHL-shuh-vihk)

Bukhara (boo-KAH-rah)

Caspian Sea (KASS-pee-uhn)

Caucasia (kaw-KAY-zhuh)

Caucasus Mountains (kaw-KUH-suhss)

Chechnya (chyehch-NYAH)

Chelyabinsk (chyehl-yah-BEENSK)

czar (ZAHR)

Dnieper River (DNYEH-purr)

Donetsk (duh-NYEHTSK)

Eurasia (yoo-RAY-shuh)

Genghis Khan (JEHNG-gihss kahn/ GEHNG-gihss kahn)

Georgia (JOHR-jah)

ger (GURR)

Gorbachev, Mikhail (gohr-buh-CHAWF, mee-KILE)

Kaliningrad (kah-LYIH-nihn-graht)

Kalmykia (kuhl-MIHK)

Karachay (kahr-ah-CHYE)

Karakalpakstan (kahr-ah-kahl-pahk-STAN)

Kashgar (kahsh-GAHR)

Kazakhstan (kah-zahk-STAHN)

Kazan' (kuh-ZAHN)

Kiev (kyee-EHF)

Kuznetsk Basin (kooz-NYEHTSK ["oo" as in "book"])

Kyrgyzstan (keer-gihz-STAHN)

Lenin, Vladimir (LYEH-nyihn, vlah-DEE-meer)

Magnitogorsk (mahg-nyee-tuh-GOHRSK)

Marx, Karl (MAHRKS, KAHRL)

Moscow (MAWSS-kau)

mugam (MOO-gahm)

Muslim (MOOZ-lihm ["oo" as in "book"])

Nagorno Karabakh (nah-GOHR-noh kah-rah-BAHK)

Nakhodka (nah-KAWD-kuh)

Nicholas II (NEE-koh-lahss)

Norilsk (nuh-REELSK)

Novgorod (NAWV-guh-ruht)

Novosibirsk (noh-voh-sih-BEERSK)

Ob River (AWB)

oblast (OH-blahsst)

Okhotsk (oh-KAWTSK)

okrug (OH-kroog ["oo" as in "book"])

perestroika (pyeh-ryih-STROY-kah)

pilaf (PEE-lahf)

reketiry (ryeh-kyeh-TEE-ree)

Samarkand (sah-mahr-KAHNT)

Semey (seh-MAY)

Shevardnadze, Edvard (shyeh-vahrd-NAHD-zyeh, EHD-vahrd)

shish kebab (SHEESH kuh-bahb)

Siberia (sye-BEER-ee-uh)

Soviet Union (SOH-vyeht)

Stalin, Joseph (STAH-lihn, YOH-sehf)

Syr Darya (SEER DAHR-yah)

taiga (TYE-gah)

Tajikistan (tah-jee-kyih-STAHN)

Tashkent (tahsh-KEHNT)

Tatar (TAH-turr)

tolkuchka (tohl-KOOCH-kah)

Transcaucasia (tranz-kaw-KAY-zhuh)

Turkmenistan (turrk-mehn-ih-STAHN)

Tuva (TOO-vah)

Ural Mountains (YOOR-uhl)

Uzbekistan (ooz-behk-ih-STAHN)

Vistula River (VIHSS-choo-luh ["oo" as in "book"])

Vladivostok (vluh-dyuh-vuh-STAWK)

Volga River (VOHL-guh)

Volgograd (vuhl-guh-GRAHT)

yurt (YURRT)

Selected Readings

A set of Selected Readings for Chapter 5, providing ideas for student research, appears on the *World Regional Geography* Web site at www.whfreeman.com/pulsipher.

The circles in this U.S. space shuttle image are irrigated grain and vegetable fields in the desert southwest of Riyadh, Saudi Arabia. The water is pumped from a deep aquifer and sprayed from a long pipe system that pivots on wheels from a center point. The Saudis hope to achieve food self-sufficiency with irrigation, but they are depleting a finite water source to do so. [NASA/Mark Marten/ Science Source/Photo Researchers, Inc.]

Land Elevations

meters	feet
Ice cap	Ice cap
5183 and above	17,000 and above
3353–5182	11,000–16,999
2134–3352	7000–10,999
914–2133	3000–6999
305–913	1000–2999
0–304	0–999
Below sea level	Below sea level

1:27,800,000
Lambert Equal Area Projection

Chapter 6: North Africa and Southwest Asia

 MAKING GLOBAL CONNECTIONS

In the spring of 2001, Ibrahim Betil, a former bank executive, was standing outside one of his mobile classrooms in rural southeastern Turkey when a teenage shepherd approached, trailed by his flock of several dozen sheep. The young shepherd asked Mr. Betil if he would take care of the sheep so that the shepherd could see the computers that his friends had described with awe.

It was not the first time Mr. Betil had filled in as a substitute shepherd. Betil is pioneering a nonprofit educational effort in Turkey, where 30 percent of the population is under the age of 15 and only one child in five attends school beyond the age of 14. International agencies report that lack of education is the chief cause of poverty in Turkey. Yet, because of economic troubles, the Turkish department of education has recently had to cut back its financial support of schools.

Betil's private Education Volunteers Foundation, funded by individuals and corporations and staffed mostly by volunteers, is focusing on what has been called leapfrogging: an effort to improve the prospects for poor urban and rural youth quickly and dramatically by sharply upgrading their training in math, science, and language skills in classrooms furnished with abundant computers. The free classes, which are meant to supplement public schools, emphasize stimulating students' curiosity. There are now 62 of the foundation's schools and 3 mobile classrooms in Turkey (Figure 6.1, insert). The program hopes to reach a million students in the next five years.

As Ibrahim Betil hands the care of the sheep back to the young shepherd, he reflects that, with the computers, attracting a million students will be no problem. Once they see their names on the computer screen, they are hooked, and the natural inclination of children to explore takes over from there. [Adapted from Douglas Frantz, "Poverty forces new methods for educating Turkish youth," *New York Times* (July 1, 2001): 7.]

Ibrahim Betil's efforts at educational reform, and those of the 2000 volunteers in the foundation he originated, are among the many types of social change that are taking place in the region of North Africa and Southwest Asia. Part of what unifies this region is the religion of Islam, which guides the daily lives of the vast majority of the region's inhabitants and has traditionally guided its educational and legal systems and other aspects of public life as well. In Turkey and elsewhere in the region, a tension exists between those who campaign for the teaching of Islamic values in public schools and those, like Mr. Betil, who strictly exclude religious teaching from the classroom. The debates sparked by educational reform in Turkey are part of wider debates over how these societies should respond to many issues. What should be the public role of religion-inspired leadership? To what extent should women be brought into full participation in society and government? How important is broad-based democratic consensus? What should be the response to cultural influences from outside the region, especially from Europe and America, that some people fear could undermine traditional religious beliefs and ways of life?

The 21 countries in the region achieved political independence only in the twentieth century, many of them not until after World War II. Although none were outright colonies of Europe, most submitted to some type of formal control by a European country that resembled colonialism. Many have found it necessary to fend off foreign efforts to appropriate the region's resources (especially oil and gas) and to influence their economies and cultural institutions: schools, universities, government systems, military establishments, and even religious practices. As a result, many people in the region mistrust Europe and the United States, and they see a need to protect their cultures from outside influences.

The region of North Africa and Southwest Asia stretches across an arid territory approximately twice as wide as the United States and bridges two continents, Africa and Eurasia. In this book the region includes the countries that line the Mediterranean coast of North Africa: Morocco, Algeria, Tunisia, Libya, and Egypt; and, to the south of Egypt, Sudan. In Southwest Asia, the region includes the countries that line the eastern Mediterranean Sea (Turkey, Lebanon, and Israel), as well as those that extend south into the Arabian Peninsula (Saudi Arabia and small countries along the Persian Gulf) and east into Central Asia (Syria, Jordan, Iraq, and Iran).

This region has long been subject to misrepresentation and unwarranted generalizations, both typical by-products of colonialism. Too often, outsiders see single issues as defining the entire region: the Arab-Israeli conflict, for example, or the isolated revival of particularly extreme versions of Islam. Or their mental images of this region may be sketchy and blurred, so they erroneously equate life in Morocco with that in Saudi Arabia, Sudan, or Iran. In fact, despite a common arid climate and a general religious preference for Islam, there is much geographic and cultural variety throughout this region. Oil is present in abundance in some countries but totally absent in others. In some parts of the region, nearly everyone lives in a city; in other parts, rural life is dominant. In some countries, civil laws are heavily influenced by religious law; others have secular legal systems. Women may be active in public life and prominent in government, or they may lead secluded

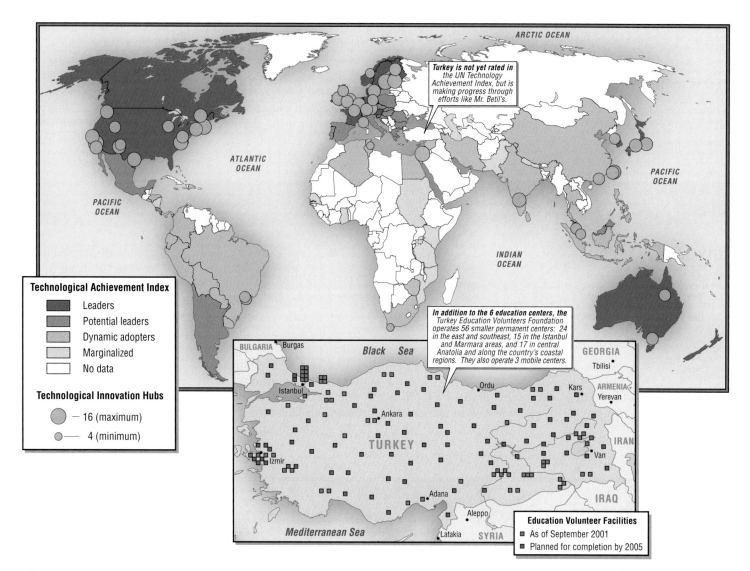

Figure 6.1 The geography of technological innovation and achievement. To become an innovative global technology hub, as defined by the United Nations (shown as a range of circle sizes on the map), a location must have each of the following elements: (1) a university or research center that can train skilled workers or develop new technologies; (2) an established company that provides expertise and stability; (3) local entrepreneurial drive to start new ventures; and (4) available venture capital to help take an idea to market. (Tunisia and Israel qualify.) Turkey does not yet have all these elements in place, but the Education Volunteers (EV) project developed by Mr. Betil helps the country make the transformation into the information age. By 2005, EV plans to have learning centers in every part of the country. [Adapted from *United Nations Human Development Report 2001*, Map 2.1, p. 45, @http://www.undp.org/hdr2001/chaptertwo.pdf. Insert from a September 2001 facilities map provided by the Education Volunteers (EV) Foundation of Turkey.]

Themes to Look For   Several themes appear throughout this chapter:

1. The importance of Islamic culture. A major source of debate in the region is the degree of influence Islamic beliefs should have on government, law, gender roles, and social behavior.

2. The impact of oil. The oil-rich countries of the Persian Gulf are major suppliers of fossil fuels to the rest of the world. Oil has provided wealth for a minority of citizens, but the economies of most countries in the region, especially the oil countries, remain undiversified and dependent on oil or a few other resources.

3. Water scarcity. A key challenge is to make this desert region inhabitable for an increasing population. Two pressing interrelated issues are how to obtain water and how to expand agriculture.

4. Political conflict. Rivalry for resources and territory has engaged a number of countries in armed conflict that is often too simplistically attributed to just religious contentions.

domestic lives and have few educational opportunities. And ethnicity varies greatly across the region.

Terms to Be Aware Of

The common term *Middle East*, often used for the eastern Mediterranean countries of the region, is not used in this chapter because it reflects the Eurocentric tendency to lump all of Asia together, differentiating it only by its distance (near, middle, far) from Europe. Also, the term *Middle East* does not usually include the western sections of North Africa or the eastern portions of Southwest Asia—Iran, for example—all of which we include in this region. On the other hand, the reader should know that some people who live in the region do use the term *Middle East* themselves.

THE GEOGRAPHIC SETTING

PHYSICAL PATTERNS

Landforms and climate are particularly tightly bound in this region. The climate is dry and hot in the vast stretches of relatively low, flat land and somewhat moister where mountains are able to capture rainfall.

Climate

No other region in the world is as pervasively dry as North Africa and Southwest Asia (Figure 6.2). A global belt of dry air that circles the planet between roughly 20° and 30° north and south latitudes cuts through this region, creating desert climates in the Sahara of North Africa, the Arabian Peninsula, Iraq, and Iran. The Sahara's size and southerly location make it a particularly hot desert; in some places, temperatures can reach 130°F (54°C) in the shade at midday. In the nearly total absence of heat-retaining water or vegetation, nighttime temperatures can drop quickly to below freezing. Nevertheless, in even the driest zones, humans survive at scattered oases where they maintain groves of drought-resistant date palms and some irrigated field crops. Traditionally, desert inhabitants have worn light-colored, loose, flowing robes

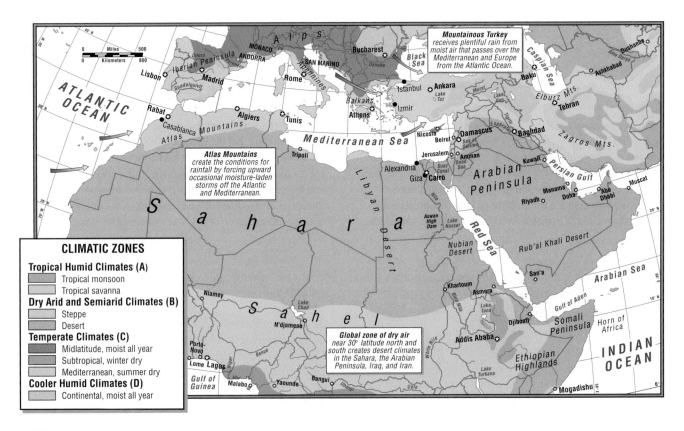

Figure 6.2 **Climates of North Africa and Southwest Asia.**

Climate change has repeatedly affected human settlement over the millennia. Grazing animals depicted in caves like this one at Tassili 'n Agger, Algeria, in the heart of the Sahara, would not survive there under present conditions. [Pierre Boulat/Woodfin Camp & Associates.]

Red Sea. The Arabian Peninsula lies to the east of this rift. There, mountains bordering the rift in the southwest corner rise to 12,000 feet (3658 meters), and the precipitation they capture from moisture-laden monsoon winds crossing Central Africa is the primary source of rain for the entire peninsula.

Behind these mountains, to the east, lies the great desert Rub'al Khali. Like the Sahara, it is virtually devoid of vegetation. Strong winds drive sand into every crevice and move huge dunes across the landscape. The dunes of the Rub'al Khali are the world's largest, some reaching more than 2000 feet (610 meters) in height. To protect themselves from the persistent winds, blowing sands, and temperatures that vary radically from day to night, the few remaining desert dwellers still line their tents with animal skins, rugs, and tapestries.

The landforms of Southwest Asia are more complex than those of North Africa and Arabia. The Arabian Plate, which carries the great desert peninsula of Arabia, is slowly rotating to the northeast, away from the African Plate. As it rotates, it is colliding with the Eurasian Plate and pushing up the widely spaced mountains and plateaus of Turkey and the Zagros Mountains in Iran. Mountainous

that reflect the sunlight during the day and provide insulation against the cold at night. Many people continue this practice.

On the uplands and margins of the desert to the north and south, enough rain falls to nurture grass and some trees. Generally too dry for cultivation, these areas have long been the prime herding lands of the region's nomads, such as the Berbers in North Africa and the Bedouins of the steppes and deserts on the Arabian Peninsula.

Landforms and Vegetation

The undulating surfaces of desert and steppe lands cover most of North Africa and Saudi Arabia. Fringes of mountains, located here and there across the region, capture moisture and make possible the survival and even flourishing of humans, plants, and animals. In northwestern Africa, the Atlas Mountains stretch from Morocco on the Atlantic to Tunis on the Mediterranean. They block and lift moisture-laden winds from the Atlantic, creating the conditions for rainfall of more than 50 inches (127 centimeters) a year in some places. Snowfall is sufficient in places to support a skiing industry. Behind these mountains, to the south and east, spreads the great Sahara Desert, where rain falls rarely and unpredictably.

Africa and Southwest Asia are separated by a rift formed between two tectonic plates (the African Plate and the Arabian Plate) that are moving away from each other. The rift, which began to form about 12 million years ago, is now occupied by the

A wadi is a dry riverbed in the desert. This is Wadi Rum in Jordan. [Adam Woolfitt/Woodfin Camp & Associates.]

Turkey receives considerable rainfall from the moisture masses that pass over Europe from the Atlantic Ocean. But by the time this air reaches the mountains of Iran, to the southeast of Turkey, only a little moisture remains. The interior of Iran is thus very dry.

There are only three major rivers in this entire region, and all have attracted human settlement for thousands of years. The Nile flows north from the central East African highlands across mostly arid Sudan and desert Egypt to a large delta on the Mediterranean. The Euphrates and Tigris rivers both begin in the mountains of Turkey and flow southeast to the swamps of the Shatt al Arab estuary terminating in the Persian Gulf. A fourth and much smaller river, the Jordan, starts as snow melt in the uplands of southern Lebanon and flows 223 miles (359 kilometers) through the Sea of Galilee to the Dead Sea. Most other streams (wadis) in the region flow only after the infrequent and usually light rains that fall between November and April. (See the photo on page 293.)

Although this region was home to some of the very earliest agricultural societies, agriculture today is possible only in a few places. In areas along the Mediterranean coast, where farmers can count on rain during the winter, they grow citrus fruits, grapes, olives, and many vegetables, often using supplemental irrigation. In the valleys of the major rivers, seasonal flooding and modern irrigation methods provide water for such crops as cotton, wheat, and barley. In occasional oases, and in a few places where groundwater is pumped to the surface with the help of technologically advanced equipment, such crops as dates, melons, apricots, and vegetables are grown.

HUMAN PATTERNS OVER TIME

Important advances in agriculture and in the organization of human society took place long ago in this part of the world: it was the location of early centers of crop agriculture, urban settlement, and civilization. Later, three of the world's great religions were born here. One of these, Islam, became the region's dominant religion and the center of its private and public life. The region's influence on Europe and the rest of the world continued as its learning and culture spread in the Middle Ages. Recently, however, it is emerging from a period of domination by western European countries and the United States. Western influences and modernization are challenging traditional ideas of Islam's role in society.

Agricultural Beginnings

Between 10,000 and 8000 years ago, nomadic peoples of this region began the earliest known agricultural communities in the world. These communities were located in an arc formed by the uplands of the Tigris and Euphrates river systems (in modern Turkey and Iraq) and the Zagros Mountains of modern Iran. This zone is often called the **Fertile Crescent** (Figure 6.3). There, people found bountiful environments: plentiful fresh water; open forests and grasslands; abundant wild grains; and goats, sheep, wild cattle, and other large animals, as well as rich fishing. As these early people adjusted to the environment, their emerging skills in domes-

ticating plants and animals allowed them to build more elaborate settlements and societies that were eventually based on widespread irrigated agriculture in lowland locations along the base of the mountains and along the Tigris and Euphrates.

Some researchers think that this period in human history may also mark the institutionalization of markedly different roles for men and women. As societies became more complicated, people with specialized skills were needed as food producers, engineers, builders, administrators, and scholars. Although there were undoubtedly many exceptions, it seems likely that men filled many of the new positions in public life, while women took over the domestic duties that increased in complexity and importance as settlements grew and family size expanded. According to this theory, men took over the more public roles because they could better endure the heavy labor required in some occupations. Women, on the other hand, were tethered to the home by the need to bear, nurse, and care for several small children, particularly as human fertility increased along with improved food yields.

Agriculture was also an early practice in the Nile Valley 6500 years ago, in the Maghreb (northwestern Africa) at about the same time, and in the eastern mountains in Persia (modern Iran). In adjacent steppe zones, societies of nomadic animal herders traded meat, milk, hides, and other animal products for the grain and manufactured goods of the settled areas, which eventually took on urban qualities: dense settlement, specialized occupations, centralized government, and concentrations of wealth. The city of Sumer (in modern southern Iraq), for example, existed 5000 years ago and gradually extended its influence over the surrounding territory. At times, nomadic tribes would band together, sweep over settlements with devastating cavalry raids, and set themselves up as a ruling class. These ruling nomads tended to adopt the culture of the sedentary peoples they conquered, and after a few generations they became almost indistinguishable from them. Usually rulers tried to expand their domains through a combination of military aggression, administrative consolidation, and trading acumen. These strategies occasionally led to the creation of vast but unwieldy empires that did not last long. In time, the conquerors themselves became vulnerable to defeat by new waves of nomadic peoples. Thus, the civilization of Sumer was succeeded by the Babylonian and Assyrian empires, the latter reaching its zenith about 3000 years ago.

The Coming of Judaism, Christianity, and Islam

The early religions of this region were based on belief in many gods linked to natural phenomena, but several thousand years ago **monotheistic** belief systems—those based on one god—emerged. Judaism, Christianity, and Islam have their origins in the eastern Mediterranean and are monotheisms.

Judaism was founded approximately 2000 years before Christianity. According to tradition, it was begun by the patriarch Abraham, who led his followers from Mesopotamia (modern Iraq) to the shores of the eastern Mediterranean (modern Israel and Palestine). Jewish religious history is recorded in the Torah (or Old

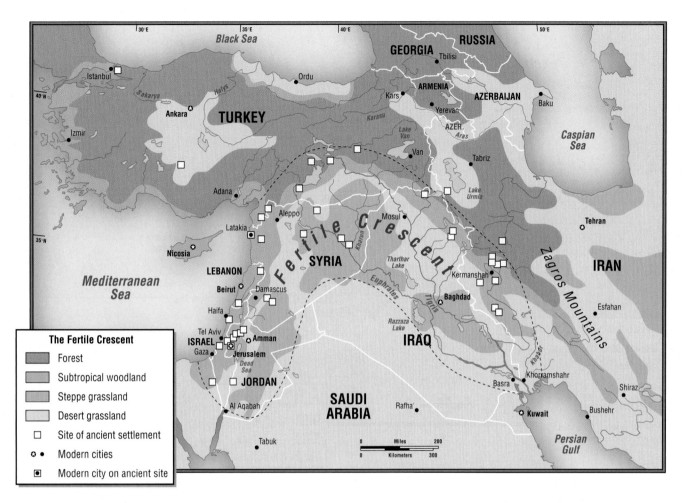

Figure 6.3 **The Fertile Crescent.** The Fertile Crescent is the name given to the area in this region where people first domesticated plants. It is an arc extending from the eastern Mediterranean coast into the mountains of southern Turkey and curving to the lands along the Tigris and Euphrates rivers and the Zagros Mountains. [Adapted from Bruce Smith, *The Emergence of Agriculture* (New York: Scientific American Library, 1995), p. 50.]

Testament of the Bible) and is characterized by the belief in one God, Yaweh; a strong ethical code summarized in the Ten Commandments; and an enduring ethnic identity. After the Jews rebelled against the Roman Empire at the Masada fortress on the Dead Sea in A.D. 73, they were expelled from the eastern Mediterranean and migrated to other lands in the **Diaspora.** Many Jews dispersed across North Africa, and some crossed to the Iberian Peninsula in Europe. Other groups of Jews made their way into eastern and central Europe. In the nineteenth century, Jews began to return to their ancestral land, and the stream of Jewish settlers greatly increased after 1948 with the formation of the Jewish state of Israel in the eastern Mediterranean.

Christianity is based on the teachings of Jesus of Nazareth, a Jew, also known as Christ, who gathered followers in the area of Palestine in the eastern Mediterranean. He taught that there is one God, whose relationship to humans is primarily one of love and support but who will judge those who do evil. The Jewish (religious) and Roman (governmental) authorities of the time saw Jesus as a dangerous challenge to their power. After his execution in about A.D. 32, his teachings spread and became known as Christianity. By A.D. 400, Christianity was the official religion of the Roman Empire. Christianity gained its strongest foothold in Europe, but some remnants of Christianity remained in the eastern Mediterranean (Maronites), in North Africa (Copts), and in enduring pockets elsewhere in Asia (such as the Malabar Coast of India). Christianity was spread through European colonization and missionary activity to the Americas, Africa, and Asia.

Islam is now the overwhelmingly dominant religion in the region of North Africa and Southwest Asia. According to tradition, Islam emerged in the seventh century A.D. when the archangel Gabriel revealed the principles of the religion to the Prophet Muhammad. Muhammad was a merchant and caravan manager in the small trading town of Mecca on the Arabian Peninsula near the Red Sea. Islam was, and is, considered an outgrowth of Judaism and Christianity. Followers of Islam, called Muslims, believe that Muhammad was the last in a long series of revered prophets,

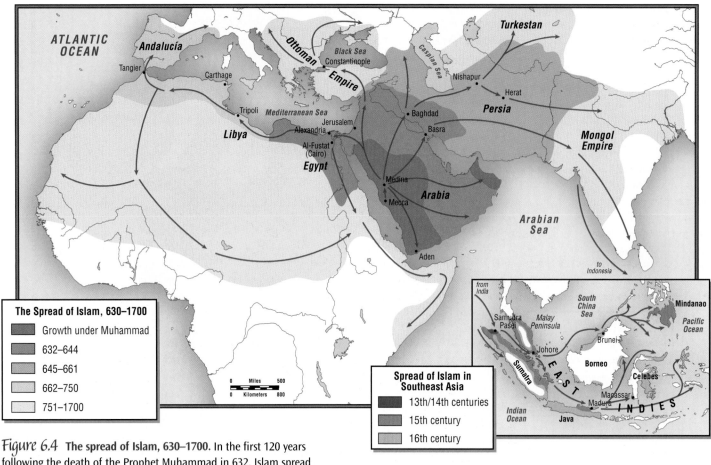

Figure 6.4 **The spread of Islam, 630–1700.** In the first 120 years following the death of the Prophet Muhammad in 632, Islam spread primarily by conquest. Over the next several centuries, Islam was carried to distant lands by both traders and armies. [Adapted from Richard Overy, ed., *The Times History of the World* (London: Times Books, 1999), pp. 98–99.]

which included Abraham, Moses, and Jesus. They also believe that the **Qur'an** (or **Koran**), the holy book of Islam, contains the words God (**Allah**) revealed to Muhammad. (*Allah* is the Arabic word for "God." Thus Arabic-speaking Christians as well as Muslims use the word *Allah* in their prayers.) In his early years as a messenger of God, Muhammad instructed his followers to pray toward Jerusalem (now they pray toward Mecca), the central holy place for Jews and Christians. This, and the fact that Jerusalem is the place from which the Prophet Muhammad is believed to have ascended into heaven, make the city particularly holy to Muslims around the world. Throughout the history of Islam, the Arabian Peninsula has maintained its central place in the Islamic world through the tradition of the **hajj,** the pilgrimage to the city of Mecca that all Muslims are encouraged to undertake at least once in a lifetime.

The Spread of Islam

The nomads of the Arabian Peninsula, the Bedouins, were among the first converts to Islam. They were already spreading the faith

and creating a vast Islamic sphere of influence by the time of Muhammad's death in A.D. 632. Over the next century, Muslim armies extended an Arab-Islamic empire (here defined as control of politics and trade) over most of Southwest Asia through Iran and then over North Africa, as well as into Spain (Figure 6.4).

The new empire established bonds between government and Islam that continued after the empire's demise. Islamic rules of conduct guided such matters as taxation policy and particularly the system of law, which was administered by religious specialists. The leaders of the empire often claimed a blood relationship with Muhammad. Arabs remained the dominant elite as Islam spread, but they placed local non-Arabs and non-Muslims as administrators of newly taken territory. Thus, they allowed local rule but collected taxes that supported the empire. They were particularly tolerant of Jews, Christians, and other monotheists.

While Europe languished in the Middle Ages (A.D. 500–1500), Muslim scholars traveled extensively throughout Asia and Africa, and they made important contributions in the fields of history, mathematics, geography, medicine, and other academic disciplines that flourished in centers of world learning in Alexandria,

Baghdad, Damascus, and Toledo (in Spain). These centers arose in vibrant economies that benefited from the early Islamic development of financial institutions and practices, such as banks, trusts, checks, and receipts. From India, Muslims imported the useful mathematical concept of the zero, further advancing trade, record keeping, and scientific knowledge in the region. Islamic medicine drew on a wide variety of traditions that had their origins from China to West Africa; Islamic medical practices were superior to those in Europe until the nineteenth century.

By the tenth century, the Arab-Islamic empire had begun to break apart into smaller units, as outlying provinces, often ruled by a local Islamic dynasty, escaped the control of central authorities. By then, subjects throughout the empire's former territories had gradually converted to Islam and many adopted the Arabic language. From the eleventh to the fifteenth centuries, Mongols from eastern Central Asia and other peoples conquered parts of Muslim territory. Meanwhile, beginning in the 1200s, nomadic herders of western Anatolia (Turkey) had begun to forge the Ottoman Empire, the greatest Islamic empire the world has ever known. By the fifteenth century, the Ottoman Muslims had defeated the Christian Byzantine Empire, which had succeeded the Roman Empire in about A.D. 300, and controlled the Balkan Peninsula and Asia Minor (modern Turkey) until 1453. The Ottomans renamed the Byzantine capital city of Constantinople "Istanbul." Very soon the Ottomans controlled most of the eastern Mediterranean, Egypt, and Mesopotamia, and they extended their control to much of central Europe in the sixteenth century.

The Ottoman Empire inherited the political and religious institutions of Islam from the late Arab-Islamic empire, including religious tolerance within its borders. Jews and Christians were allowed to practice their religion, although there were economic and social advantages to converting to Islam. The Ottomans prospered from the productivity of the empire's many different peoples. Drawing on trading networks that stretched from central Europe to Algeria to the Indian Ocean, Istanbul became a cosmopolitan capital, outshining most European cities until the nineteenth century.

Eventually the Islamic faith was extended to Turkey, the Balkans, the Indus Valley, and what is now northern India. Muslim traders carried the Islamic faith still farther, by land across Central Asia to western China and by sea to Indonesia and Malaysia, the southern Philippines, and southeastern China (see Figure 6.4). These areas, and those of the former Arab-Islamic empire, are still predominantly Islamic today, with the exception of Spain, some parts of the Balkans, and southeastern China.

In this area of ancient human occupation, the origins of cultural practices are often obscure and subject to misconceptions. For instance, a common belief is that restrictions on women, such as **seclusion** (the requirement that a woman stay out of public view) and veiling (the custom of covering the body with a loose dress and the head—and in some places, the face—with a scarf), originated with Islam. Yet scholars have found evidence that these practices predate Islam. Nor are the antiforeign attitudes associated with some movements in Islam today inherent in that religion. On the contrary, some past Islamic empires, such as the Ottoman, were known for their cosmopolitan culture, tolerance, and even appreciation of foreign ways. Antiforeign sentiment is a more recent phenomenon, coming after years of European and U.S. interference in the affairs of the region.

Western Domination

The Ottoman Empire ultimately withered in the face of a Europe made powerful by the industrial revolution. Throughout the nineteenth century, North African lands changed from Ottoman to European control. At first, European influence was exercised primarily through control of trade and finance. North Africa became a source of raw materials for Europe in a trading relationship dominated by European merchants, who sought the agricultural products of the coast, the cotton of Egypt, and the phosphates, manganese, hides, and wool of the western Mediterranean countries. In 1830, France became the first European country to exercise direct control of a North African territory when its military gained control of Algeria in the western Mediterranean. France eventually administered that land almost as though it were a part of France—although Muslims were not allowed the benefits of full citizenship. In the last two decades of the nineteenth century, France, Britain, and Italy sent their militaries to occupy several other North African countries, motivated in part by European power politics—the fear that a rival would step in first. Britain gained control of Egypt (1882) and then Sudan; France took over Tunisia (1881) and Morocco; and Italy moved into Libya. In these countries, Europeans set up a form of dependence that they termed a **protectorate:** local rulers remained in place, but European officials made the important decisions.

Europe also influenced by example. Reformist governments in some countries not under the direct control of Europe, including countries of the Ottoman Empire, introduced European systems of military organization, administration, taxation, free trade and travel, and law. Rulers hoped that by copying the Europeans their countries could become strong enough to resist European encroachment. The Europeans continued and widened these reforms in countries under their control. Thus, the old links between rulers and Islam began to be severed and new institutions introduced. One result was the extension of control by authoritarian governments strengthened by expanded armies and more efficient taxation.

Another result of European influence was that traditional systems of landownership were undercut to the disadvantage of small cultivators. European officials privatized tribal collective lands and gave title to tribal leaders in return for political support. Alternatively, newly wealthy merchants from the cities or members of ruling families bought up large tracts of land. Small farmers, often forced by tax-induced debt to sell, became sharecroppers or landless laborers. Thus European domination encouraged the concentration of wealth and power in the hands of a relative few.

World War I (1914–1918) brought the death of the Ottoman Empire. The Ottomans had allied with Germany, and after the war ended in defeat for that country, the League of Nations (a precursor to the United Nations) allotted almost all former Ottoman

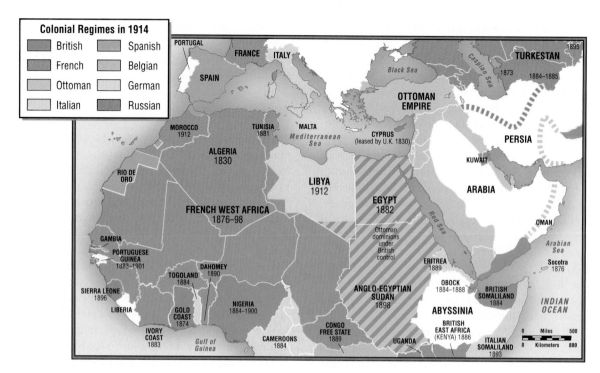

Figure 6.5 European control, 1914. European powers controlled most of this region at the onset of World War I (1914). Fifty years later, virtually all of the nations were independent, and the map of Southwest Asia had been redrawn to include Jordan, Israel, Lebanon, Syria, Iraq, and the new states of the Arabian Peninsula. [Adapted from *Hammond Times Concise Atlas of World History* (Maplewood, N.J.: Hammond, 1994), pp. 100–101.]

territories at the eastern end of the Mediterranean to France and Britain for supervision (Figure 6.5). Only Turkey gained true independence at that time. Syria and what is now Lebanon became mandated territories of France (dependencies thought incapable of self-rule). Palestine and Iraq, out of which the kingdom of Jordan was carved, became mandated territories of Britain. The Arabian Peninsula gained limited independence with the formation of Saudi Arabia (1932) and the smaller states bordering the Persian Gulf, but Britain retained influence over foreign relations.

The region achieved independence from overt foreign domination during the mid-twentieth century. Nevertheless, foreign influence remained strong in many countries. As the United States gained status as a world power, it backed local leaders sympathetic to U.S. policies. In Iran and Saudi Arabia, where vast oil deposits became especially lucrative by mid-century, European and U.S. oil companies played a key role in deciding who ruled. In both countries, an extremely wealthy elite retained oil revenues. Throughout the region, oil money was little taxed and seldom used to industrialize or improve general well-being. Rather, it was typically spent on opulent living for the elite or invested abroad in such assets as shopping malls, resorts, banks, and department stores in places such as the United States, Australia, and the United Kingdom. The region remained dependent on Europe and North America for the technology needed to exploit oil and to begin the mechanization of

manufacturing, transportation, and agriculture. Meanwhile, the wealthy bought their consumer goods and luxury items in London and Paris. During this period, disparity in wealth between the elite and the majority of the people increased dramatically.

An Era of Reform

In the first decade of the 2000s, the region is experiencing some attempts at economic, political, and environmental reform, although a number of problems are complicating these efforts. The region's regimes remain authoritarian and dominated by elites. Long-needed economic reform that would bring a more equitable distribution of wealth has just started in most countries. Most such reform efforts are aimed at reducing bloated and inefficient bureaucracies; redistributing agricultural land to poor farmers; privatizing state-run industries; and, to some extent, diffusing the power of wealthy elites. So far, the reforms have done little to reduce overdependence on oil and establish a broader range of industries, increase jobs for growing populations, or decentralize political power.

Because of these failures to reform in a timely manner, governments are vulnerable to criticism from revolutionary movements. The movement with the fastest growing influence across the region is **Islamic fundamentalism** (see the discussion below, on page 314). The **Islamist** movement builds on the discontent of the poor and the educated unemployed and on disenchantment with Western influences, while appealing to religious piety—devotion to and reverence of Islam. As problems of wealth disparity, inadequate infrastructure, and joblessness intensify in such countries as Algeria, Egypt, and Turkey, they bring increased levels of violence and social chaos. In this atmosphere, it is becoming more difficult to address the region's very severe social and environmental problems, which include inadequate education, especially for women; lack of access to technology (see Figure 6.1); various types of pollution; and water shortages. Increasing pressure on available resources by rapidly expanding populations is also of concern.

POPULATION PATTERNS

The great size of this region and its aridity have had important implications for human settlement and interaction. Although the region as a whole is nearly twice as large as the United States, the areas that are useful for agriculture and settlement are tiny by comparison. The land supports 100 million more people than the United States, and population densities in livable spaces can be quite high. Vast tracts of desert are virtually uninhabited. As you can see on the population map (Figure 6.6), most people live along coastal zones, in river valleys, and in mountain zones that capture orographic rainfall. Efforts are being made to extend livable space into the desert, but doing so may require extremely costly and environmentally questionable solutions to the problems of water scarcity and soil infertility.

Between 1980 and 2000, human fertility rates across the region fell dramatically. In the 1960s, when most women married shortly after reaching puberty, some countries had average fertility rates of close to nine children per woman. By 2000, the average woman married by age 21 and attended school for more than a year or two; the average fertility rate for the region was reduced to just over four children per woman. Nonetheless, this rate is still higher than the world average (only Africa's is higher), and in Iraq and the Arabian Peninsula, fertility rates are five children per woman and higher. One result is that well over 40 percent of the population is under the age of 15 in a number of countries in the region (Figure 6.7). Population will continue to grow rapidly as these children reach reproductive age, even if each couple has only two children.

At present rates of growth, the population of the region, now about 400 million, will double to 800 million by about the year 2030. It is difficult to imagine how the area will support the additional

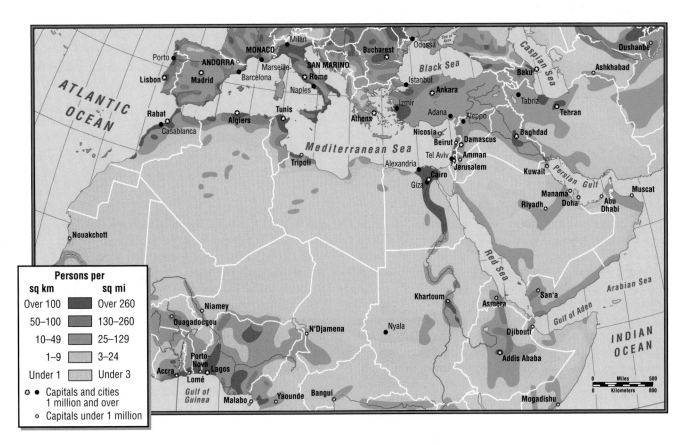

Figure 6.6 **Population density in North Africa and Southwest Asia.** [Adapted from *Hammond Citation World Atlas* (Maplewood, N.J.: Hammond, 1996).]

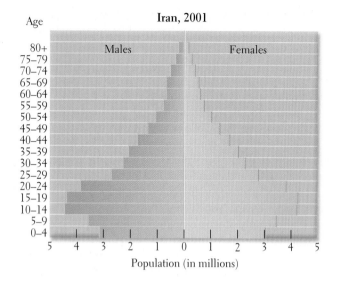

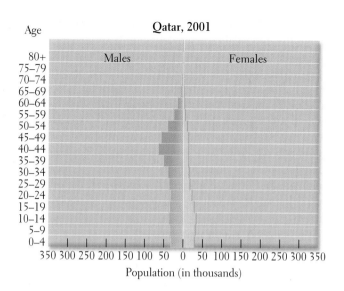

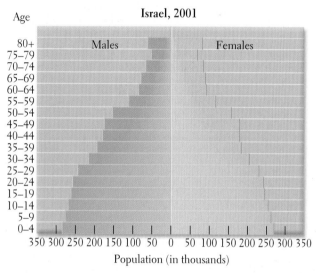

Figure 6.7 Population pyramids for Iran, Qatar, and Israel. The population pyramid of Iran is at a different scale (millions) from that of Israel and Qatar (thousands). Observe that Iran has had significantly fewer births over the last 10 years; that Qatar's pyramid is skewed by the presence of numerous male guest workers in the 25- to 54-year age groups; and note also that all three pyramids show missing females, especially in the age groups under 24. This is most easily observed by drawing lines from the ends of the male and female age bars to the scale at the bottom of the pyramid and comparing the numbers. [Adapted from "Population Pyramids for Iran," "Population Pyramids for Qatar," and "Population Pyramids for Israel" (Washington, D.C.: U.S. Bureau of the Census, International Data Base, May 2000.]

people. Fresh water is already in extremely short supply, and most countries must import food at great cost. Furthermore, 32 million more jobs would have to be created within the next decade to employ the added population—compared to just 4.5 million needed in Europe, where population is growing very slowly. If women start entering the workforce, even more new jobs will need to be created. As Figure 6.8 shows, by the year 2000, less than 25 percent of women across the region (except for Morocco and Turkey) worked outside the home (this figure excludes the large number of women who work on their family farms); in much of Southwest Asia, considerably less than 20 percent worked outside the home. If the status of women changes more quickly than anticipated, the demand for jobs will increase even beyond the 32 million figure.

Many inhabitants of the region believe that the industrialized world (Europe, Russia, North America, Japan, and parts of Oceania) is overly concerned with population growth in the developing world.

It is thought that these countries promote population control elsewhere so that they can continue to consume more than their share of the world's resources. On the other hand, some leaders in the region (especially in Algeria, Egypt, Iran, Jordan, Morocco, Oman, Sudan, Tunisia, Turkey, and Yemen) see the wisdom of population control and encourage family planning. Most specialists in Islamic law say that limiting family size is acceptable for a wide variety of reasons, so long as the motive is not to shirk parenthood altogether. As a method of birth control, abortion is very rare, but it is permitted in some circumstances.

Gender Roles and Population Growth

In societies where fertility rates remain high despite a decline in labor-intensive agriculture of the sort that depends on children as a labor supply, gender inequality may partially explain why couples are still choosing to have so many children. In urban contexts,

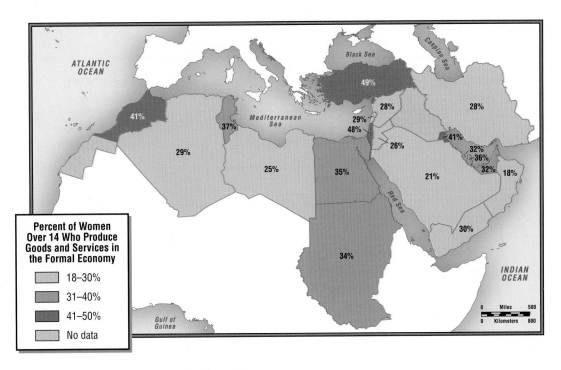

Figure 6.8 Percentage of women who are wage-earning workers. [Adapted from Joni Seager, *Women in the World: An International Atlas* (New York: Viking Penguin, 1997), pp. 66–67; and *United Nations Human Development Report 2001* (New York: United Nations Development Programme), pp. 92–95.]

especially where women have opportunities outside the home—to obtain knowledge, wages, or prestige on the job, or to participate in political activity (see Figure 6.8)—they almost always choose to have fewer children. Throughout the region, wherever men have a much higher education level than women (Table 6.1), greater access to employment, and greater power to make family and community decisions (see the discussion of the family on pages 305–306), children become an important source of power for women. Where women are secluded in the home and undereducated, they have little chance to accumulate wealth or to enhance the prospects and reputation of the family.

One result is a deeply entrenched cultural preference for sons. Because of their wider opportunities, sons are better positioned to increase the family wealth and social standing. Sons are expected to support their parents financially in their retirement (daughters are expected to render personal care). The strong preference for sons means that families sometimes continue having children until they have a sufficient number of sons, or young females simply may not survive, as is indicated by missing females in the 0- to 4-year age cohort of the population pyramids shown in Figure 6.7. (The issue of missing females is discussed at greater length in Chapter 8.)

Migration and Urbanization

Emigration is increasingly common across North Africa and Southwest Asia. Because jobs are hard to find, especially for work-ers with more than a basic education, many millions have emigrated to other parts of the world in search of work. Most of those leaving the region are young men, because women typically do not travel widely. Several million are guest workers in Europe; in 1999, Turkish guest workers alone numbered more than 1.2 million. Of the millions of emigrants, many intend to return home, in part because it is difficult to gain citizenship in European countries. In the meantime, remittances (earnings sent home) to their families in North Africa, Turkey, or Lebanon significantly increase local standards of living.

In some parts of the region, immigration rather than emigration is the trend. Over the past 30 years, the oil-rich states of the Arabian Peninsula have recruited large numbers of guest workers. When oil revenues poured in after price increases in the 1970s, the money was used to modernize some aspects of life rapidly. The labor force in these countries was too small and too undereducated to fill the jobs created. As the infrastructure for a modern economy was being completed in the 1980s, there was a marked shift in migration trends: manual workers were gradually replaced with skilled workers. By 1990, immigrants made up nearly 70 percent of the peninsula's labor force. They built roads, government buildings, housing, schools, universities and technical colleges, water desalinizing plants, oil and gas production facilities, and modern irrigation systems—and they ran the systems once completed.

These workers came from all over the world, but the employers preferred Muslims. Several hundred thousand workers—some technically skilled, many willing to do manual labor—arrived

Thinking Globally: ON THE WEB

TABLE 6.1 *Gender statistics from selected countries*

Country	Percent literate, female	Percent literate, male	Country	Percent literate, female	Percent literate, male
Selected countries for comparison			**The Arabian Peninsula**		
United States	99.0	99.0	Saudi Arabia	66	84
Brazil	85	85	Yemen	24	67
Portugal	89	94	Oman	60	79
World	73	85	United Arab Emirates	78	74
			Qatar	83	80
The Northeast			Bahrain	82	91
Turkey	76	93	Kuwait	79	84
Iraq[a]	43	64			
Iran	69	83	**Egypt and Sudan**		
			Egypt	43	66
Eastern Mediterranean			Sudan	45	69
Syria	59	88	**The Maghreb**		
Lebanon	80	92	Western Sahara	N/A[b]	N/A
Israel	94	98	Morocco	35	61
Jordan	83	95	Algeria	56	77
			Tunisia	59	80
			Libya	67	90

[a]Data for Iraq is from the *United Nations Human Development Report 2000*. Data for 2001 not available.
[b]N/A = data not available.
Source: United Nations Human Development Report 2001 (New York: United Nations Development Programme), pp. 210–213.

from Palestinian refugee camps in Lebanon and Syria, others from Egypt. In the 1980s, 1.5 million came from Pakistan alone. Many, especially female domestic workers, came from Muslim countries in Southeast Asia; they were attracted to Saudi Arabia by the relatively high wages and the chance to make the pilgrimage to the holy cities of Mecca and Medina. Texan Martha Kirk, the wife of an agricultural specialist employed by the Saudis, spent five years (in the late 1980s and early 1990s) living on an irrigated wheat farm in Saudi Arabia. She depicted the many Filipino and Sri Lankan female domestic servants as silent, shadowy figures, serving tea during lavish women-only parties in the harems of desert palaces and in urban villas in the city of Riyadh. Kirk also characterized the many male migrant workers she encountered: Turks drove bulldozers, Mexicans set up pumps and dug wells, Pakistanis and Americans specialized in grain cultivation, other Americans set up dairies, other Palestinians worked in construction and as farmhands.

Immigrant workers on the Arabian Peninsula will probably not become permanent residents. During and after the Gulf War in 1991, millions of foreign workers hurriedly left the Gulf states.

Although many have returned, some Arabian Peninsula governments are undertaking efforts to prepare their own citizens to take over jobs, especially at the managerial levels. As European- and American-trained nationals return from their studies abroad, they take jobs that immigrants once performed. The loss of income to the families of migrant workers forced to return to refugee camps in Syria or to rural villages in Indonesia, Turkey, Mexico, and Egypt has often meant a return to drastically lower standards of living.

Israel also has many immigrants, but those arriving are Jews settling permanently in Israel, many fleeing persecution. They began coming to Palestine before and after World War I; and huge numbers liberated from Nazi concentration camps arrived when Israel was created in 1948, just after World War II. Most recently, immigration surged in the 1990s, when more than 700,000 Jews were allowed to enter from the former Soviet Union. Another 25,000 fled the Ethiopian civil war. Despite already crowded conditions, Israel encourages Jewish immigrants, partly because its constitution grants sanctuary to Jews and partly because it believes that a large Jewish population is important for political weight in the region.

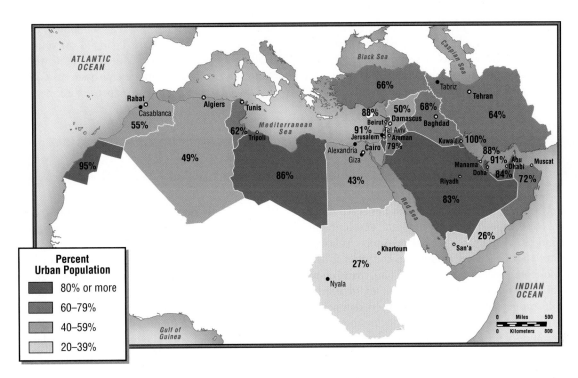

Figure 6.9 **Percentage of urban population for each country in North Africa and Southwest Asia.** [Data from *2001 World Population Data Sheet* (Washington, D.C.: Population Reference Bureau).]

Internal migration from rural villages to urban areas has also been an important population pattern across the region. Until recently, most people lived in small settlements; but by 2000, more than 60 percent of the region's people lived in urban areas, although the pattern varied greatly across the region (Figure 6.9). There are now more than 120 cities with populations of at least 100,000; 24 of them have more than 1 million people. Poor urban migrants often build their own dwellings hastily in peripheral shantytowns that lack such basic services as clean water and sewage disposal.

Refugees

There are hundreds of thousands of refugees in North Africa, the eastern Mediterranean, Turkey, Iraq, and Iran. Usually they seek to escape wars or environmental disasters such as earthquakes or long-term drought. In the last half of the twentieth century, many nomads became refugees when new national borders cut off seasonal migration routes, such as the one between the western Sahara and Mauritania, for example. When Israel was created, as many as 2 million Palestinians who were living there were displaced to refugee camps in Lebanon, Syria, Jordan, the West Bank, and the Gaza Strip. However, as of 2000, the country with the world's largest international refugee population was Iran (1.8 million in all), which sheltered 1.3 million Afghans and 510,000 Iraqis displaced by regional conflicts (see pages 331–332). Across the region, an even larger group of people are refugees within their home

countries, unable to occupy their homes, usually because of civil unrest. Sudan has 2 million internally displaced persons living in camps (see the photo below and discussion on page 323).

Al Salaam is a desert refugee camp near Khartoum, Sudan, for those who fled to the city from the south. Water is trucked in daily and distributed in plastic cans. [Robert Caputo/Aurora and Quanta Productions.]

In this region, refugee camps often have become semipermanent communities of stateless people, in which whole generations are born, mature, and die. The residents may show enormous ingenuity in creating a community and an informal economy under adverse conditions. Nevertheless, the cost in social disorder is high.

Children rarely receive enough schooling. Years of hopelessness and extreme hardship take their toll on youths and adults alike, leading some to violence. Moreover, even when international organizations contribute money for basic sustenance, large numbers of refugees represent a huge expense for their host countries.

CURRENT GEOGRAPHIC ISSUES

The region of North Africa and Southwest Asia has sharp differences in physical landscape, level of economic development, political and economic systems, and social relations. Although many social institutions are shared among the countries in this region, notably those based on the practice of Islam, many countries have different views on such issues as the role of the Islamic religion in daily life, proper roles for men and women, democracy as a political model, the pace and direction of economic development, the importance of oil, and rights to scarce water supplies.

SOCIOCULTURAL ISSUES

The countries of North Africa and Southwest Asia are experiencing broad social changes. Here we examine a few examples that illustrate these changes: religion, family, gender roles and female seclusion, the lives of children, and cultural diversity as reflected in language.

Religion in Daily Life

More than 93 percent of the people in this region subscribe to Islam. The Five Pillars of Islamic Practice (see the Cultural Insight box) embody the central teachings of Islam. All but the first pillar are things to do in daily life, rather than articles of faith. (Muslims who are poor, young, or sick, and pregnant women, are exempted from the physically demanding pillars.) Islamic practice in this region is thus a consistent part of daily life in ways rarely experienced in the West, where most people live by a personal ethical code but the formal practice of religion is often set aside for certain times (the Sabbath) and spaces (churches, synagogues, and mosques).

Saudi Arabia occupies a prestigious position in Islam. It is the site of two of Islam's three holy shrines, or sanctuaries: Mecca, the birthplace of the Prophet Muhammad and of Islam; and Medina, site of the Prophet's mosque and his burial place. (Jerusalem is the third holy site.) The fifth pillar of Islam, that all Muslims should make a pilgrimage (the hajj) to the two holy cities at least once in a lifetime, has placed Mecca and Medina at the heart of Muslim religious geography. Each year, a large private-sector service industry organizes and oversees the five- to seven-day hajj for more than 2.5 million foreign visitors. Most pilgrims visit during the month of Dhu al-Hijja on the Muslim lunar calendar. Although oil now overshadows the hajj as a source of national income, the event is economically, as well as spiritually, important to Saudi Arabia.

Beyond the Five Pillars, Islamic religious law, called **shari'a** ("the correct path"), guides daily life according to the principles of the Qur'an. Some Muslims believe that no other legal code is necessary in an Islamic society. There are, however, numerous interpretations of just what behavior meets the requirements of shari'a and what does not. The Islamic community is split into two major groups, with differing interpretations of shari'a: **Sunni**

CULTURAL INSIGHT *The Five Pillars of Islamic Practice*

1. A testimony of belief in Allah as the only God and in Muhammad as his Messenger (Prophet).

2. Daily prayer at one or more of five times during the day (daybreak, noon, midafternoon, sunset, and evening). Prayer, although an individual activity, is encouraged to be done in a group and in a mosque.

3. Obligatory fasting (no food, drink, or smoking) during the daylight hours of the month-long Ramadan, followed by a light celebratory family meal.

4. Obligatory almsgiving (*zakat*) in the form of a progressive "tax" of at least 2.5 percent that increases as wealth increases. The alms

are given to Muslims in need. *Zakat* is based on the recognition of the injustice of economic inequity; although it is usually an individual act, the practice of government-enforced *zakat* is returning in certain Islamic republics.

5. Pilgrimage (hajj) to the Islamic holy places, especially Mecca, during the twelfth month of the Islamic calendar. Rituals shared with the devout of all backgrounds, from around the world, reinforce the concept of *ummah*, the transcultural community of believers.

Source: Carolyn Fluehr-Lobban, *Islamic Society in Practice* (Gainesville: University of Florida Press, 1994).

A rare glimpse of the Grand Mosque in Mecca, packed with devout worshipers, at the height of the pilgrimage season. [Mohamed Lounes, Gamma-Liaison.]

Muslims, who today account for 90 percent of the world community of Islam; and **Shi'ite** (or **Shi'a**) Muslims, who are found primarily in Iran. Even within each of these groups, though, there are many different interpretations of shari'a.

Islam does not recognize a separation of religion and the state as this concept is understood in the West. In several of the countries in this region—Saudi Arabia, Yemen, the United Arab Emirates (UAE), Oman, and Iran, for example—the state is the defender and even the enforcer of the religious principles of Islam. In these **theocratic states,** Islam is the officially accepted religion, the leaders must be Muslim and are divinely guided, and the legal system is based on shari'a. Other countries—Algeria, Egypt, Morocco, Iraq, Turkey, and Tunisia—have declared themselves to be **secular states.** In these countries, theoretically, there is no state religion and no direct influence of religion on affairs of state. In practice, however, religion plays a public role even in the secular states. Most political leaders are at least nominally Muslim, and Islam plays a role in governmental affairs. (The exceptions are Israel, a Jewish theocratic state, and Lebanon, a multireligious state.) Still, here as elsewhere on earth, there is great variation in the stringency with which people practice their religion. Many people who consider themselves Muslim do not strictly observe the Five Pillars or shari'a, and they believe that a secular state is best.

Ancient settlements of Christians and Jews are also to be found in this region, many of them predating the spread of Islam. After the establishment of Israel in 1948, most North African Jews (more than 400,000) settled in Israel. Christian communities, often dating from the Roman era, exist in Egypt, Sudan, Jordan, Lebanon, Iraq, and Turkey. It is estimated that there are 10,000 Berber and Arab Christians in North Africa.

Family and Group Values

As in the United States and Canada, many people in North Africa and Southwest Asia worry that modernization is diminishing the role of the family and that the morals of the next generation are deteriorating as a result. Here, though, the traditional multigeneration patriarchal family is still very much the norm. Nonetheless, patterns are changing: families are becoming smaller, and no longer do several generations of one family always share the same household.

In Islamic culture, the family is both a physical space and a functional grouping. Physically, the family space is usually a walled compound that focuses inward, where food, shelter, and companionship are provided (see the photo below of Kalaa Sghrira, Tunisia, and photo of a house in Jeddah on page 307). The size and details of these compounds vary according to social class and geographic location. The family group consists of kin that span several generations and may include elderly parents as well as adult siblings and cousins and their children. A system of interlocking duties, obligations, and benefits, often assigned by gender and age, provides a role for each individual and solidarity for the whole family. All accomplishments or misdeeds become part of a family's heritage. The reciprocal responsibilities of family membership are enforced through informal social pressures that ensure no one becomes an obvious shirker. Whether a member receives just a meal or something as grand as a university education from pooled resources, the recipient knows that some measure of repayment will come due eventually. This informal contract between generations

In Kalaa Sghrira, Tunisia, founded in 670, streets are narrow and traditional courtyard houses focus family life inward. As in many crowded cities throughout the region, houses seldom rise more than one or two stories, and the skyline is broken only by the minaret towers of nearby mosques. [Roger Wood/Corbis.]

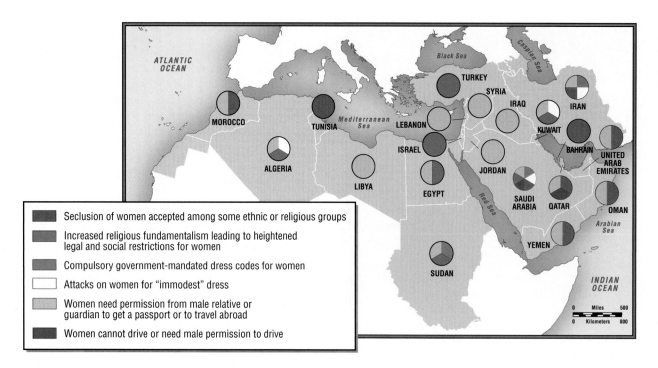

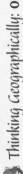

Figure 6.10 Restrictions on women. The map shows the variation of restrictions placed on women in the countries of North Africa and Southwest Asia. Some countries, such as Turkey, constitutionally guarantee equality for women, but religious fundamentalists are working for changes in these guarantees. [Adapted from Joni Seager, *Women in the World: An International Atlas* (New York: Viking Penguin, 1997), pp. 28–29.]

ensures the flow of financial remittances to elderly kinfolk; and although these remittances are formally expected only from sons, daughters who are able send them, too.

Gender Roles, Gender Spaces

Carefully specified gender roles are common in many cultures. Particularly strict versions of these roles are an integral part of Islamic culture; the differences between male and female roles are reflected in the organization of space within the home and within the larger society. In both rural and urban settings, men and boys go forth into **public spaces**—the town square, shops, the market. Here, men not only make a living, they also continuously transact alliances with other men to negotiate arrangements that will advance the interests of their particular families. It is through such networks, which span a range of social strata, that people get the best price for an appliance or a car, find a mate for a son or daughter, obtain a job or admittance to a professional school, find some scarce item, or unsnarl a particularly nasty bureaucratic problem. For any favors they receive, they incur future obligations. If men are successful at making arrangements, they also garner considerable respect and prestige within their own families.

By contrast, women usually inhabit secluded **domestic spaces.** In some parts of the region, particularly in Saudi Arabia, people believe that women should not be in public places except when on important business. In many countries, women must be accompanied by a male relative in public. Similarly, in some parts of

the region (Egypt, for example), seclusion is more strictly enforced in relatively affluent, urban neighborhoods than in rural villages. Rural women are less secluded because they have many tasks that they perform outside the home: agricultural work, carrying water, gathering firewood, and marketing. Although women are regarded as the rightful keepers of the home and the guardians and teachers of young children, the rules governing secluded women seem to make some negative assumptions about females in general. Women are thought to be more emotional and less capable and reliable than men.

The standards of seclusion vary greatly. In the more secular Islamic societies—such as Morocco, Tunis, Libya, Egypt, Turkey, Lebanon, and Iraq—women regularly engage in activities that place them in public spaces. And, increasingly, female doctors, lawyers, teachers, and businesswomen are found in even the most conservative societies. Figure 6.10 compares the various levels of restrictions on women across the region. Some restrictions are officially sanctioned, and some are the result of informal social pressures.

Architecture, window coverings, and clothing are examples of material culture designed to seclude women from the public gaze. In all parts of this region, homes are very private places, centered on an interior courtyard where household work is performed and family life takes place. The private female space within the home is usually this courtyard. Only male relatives may enter. For the upper classes in urban places, female space may be an upstairs set of rooms with latticework or shutters at the windows. Here it is possible to look out at street life and to enjoy

This house in Jeddah, Saudi Arabia, exhibits numerous versions of louvered and latticed bays that allow women to observe life on the streets from seclusion. Such architectural details are found wherever Islam has been an influence. [Hubertus Kanus/Photo Researchers.]

the custom of transferring all the wealth a woman brought to a marriage to her husband's control. In some parts of the United States similar state laws were changed only after pressure for equal rights in the 1960s.

States such as Tunisia and Morocco are enacting legislation to improve the status of women. In the past, men could divorce their wives without cause or a court proceeding, whereas women wishing a divorce had to have specific grounds and court approval. Contrary to the teachings of the Prophet Muhammad, some women were forced to marry and to undergo virginity tests, and they were left unprotected in cases of domestic violence. In 1999, the new king of Morocco, Mohammed VI, began to implement a program that encourages participation by women in civil society and reforms the legal code to grant women legal rights. Tunisia recognizes the equal right of either spouse to seek divorce. It also requires mutual spousal obligation in household management and child care, and it no longer permits preteen girls to marry. The penalties for spousal abuse have been increased, and there is a fund guaranteeing alimony for divorced women and their children. More than 36 percent of Tunisian women work in the formal economy, and Tunisia requires, but does not enforce, equal pay for equal jobs.

In some countries in this region (the Arab republics, Morocco, Algeria, Syria, Jordan, and Iraq) and elsewhere, Muslim men are permitted to take more than one wife, a custom called **polygamy**. But polygamy, although allowed under certain conditions by the Qur'an, is not encouraged, nor is it a common practice. It is estimated, for example, that less than 4 percent of males in North Africa have more than one wife. The occurrence is low partly because the Qur'an, which limits the number of wives to four, requires that all must be treated with scrupulous equality. This obligation alone discourages polygamy, but urbanization and modernization are more important factors. When agriculture was the main economic activity, extended families with several wives for each male and multiple children may have been more productive economically, but urban life with its small living spaces and cash requirements favors smaller families. According to the Ibn Khaldun Center for Development Studies in Cairo, democratization in North Africa and Southwest Asia has given voice to women who have successfully lobbied for bans against polygamy in Tunisia, Lebanon, and Palestine.

The Lives of Children

Two sweeping statements can be made about the lives of children in the Islamic cultures of North Africa and Southwest Asia. First, their daily lives take place overwhelmingly within the family circle. Girls and boys spend their time within the family compound with adult female relatives and with siblings and cousins of both sexes. Even teenage boys identify more with family than with age peers. Only the poorest of children play in urban streets. In rural areas, before puberty, girls often have considerable **spatial freedom**—the ability to move about in public space—as they go about their chores in the village. The U.S. geographer Cindi Katz found that until puberty, rural Sudanese Muslim girls have considerably more spatial freedom than do girls of similar ages in the urban United

breezes passing through the lattice, without being seen by people on the street (see the photo above).

There is considerable controversy over the origin of female seclusion. Many scholars of Islam, male and female, say that these ideas do not derive from the teachings of the Prophet Muhammad, who seems to have advocated equal treatment of male and female children. His first wife, Khadija, was an independent businesswoman whose counsel he often sought. Although Muhammad did suggest that both sexes dress modestly, the Qur'an contains only the barest hint about veiling and nothing about seclusion. Some modern scholars of Islam now state that interpretations of the Prophet Muhammad's sayings were influenced by the severely restrictive customs of non-Islamic cultures that were conquered by Arab Muslims in the first few centuries after the Prophet's death. Recently, Muslim scholars have shown that in some parts of the region, interpretations of Islamic law regarding female rights have become more conservative just during the twentieth century.

For Western readers, it is important to remember that European women had to wait more than a thousand years to acquire property-holding rights that were guaranteed by the Qur'an as far back as A.D. 700. The Qur'an allows a woman, married or not, to manage her own property and keep her wealth or monthly salary for herself. By contrast, it was not until 1870 that British laws broke

CULTURAL INSIGHT *The Veil*

The veil allows a woman to remain secluded when she enters a public space, and thus it actually expands the space she may safely occupy. Some garments totally cover the body, including the face; some allow the eyes to be seen; and some are a mere head covering. Among women who have worn the all-encompassing chador (full body veil), there are those who find that it offers an intriguing anonymity and satisfying privacy from the gaze of strangers. Women who wear the *hijab*—a long, loose dress and a scarf that leaves the face exposed—or even just the head scarf with blue jeans and a T-shirt, speak of liking the signal it sends that here is a devout Muslim woman. Paradoxically, many women throughout the region (including the mother of Iran's President Khatami) wear high-fashion Western clothes under the chador or *hijab*, or at special women-only gatherings (see the accompanying photo).

Egyptian women were among the first to take off their veils: when two Arab feminists returned from an international meeting in 1923, they discarded their veils in the Cairo train station. They started a trend that lasted until the late twentieth century, when some Cairenes (inhabitants of Cairo) began veiling with an enthusiasm born of a desire to assert Muslim identity and make a statement against Westernization.

Stores selling veils are often found in shopping malls. But for women who think of the veil as providing a chance for a fashion statement, Wafeya Sadek runs an Islamic designer-veil studio out of her home in Cairo. Women who find the black covering unattractive come from across the city to buy a custom-made veil. A best-seller is a leopard-print veil, favored by young women for day wear.

Some of Sadek's designs for evening wear seem to call for the type of bold individualism that is frowned upon by conservative Muslims: for example, a close-fitting black bonnet topped with a wildly colorful cock's comb of velvet. Operating in the informal economy, Madame Sadek counts on word-of-mouth advertising.

Source: Amy Docker Marcus, *Wall Street Journal* (May 1, 1997).

Only female family and friends attended this wedding party in Tehran. Some of the younger women shed their veils and danced in Western clothes, while most of their more conservative elders retained traditional garments. [Alexandra Avakian/Contact Press Images.]

States, where they now are rarely allowed to range through their own neighborhoods.

The second observation is that the lives of children in this region, like those of so many around the world, are increasingly circumscribed by school and television. Most children go to school (girls for at least a few years, many boys for a decade or more), and in most urban areas even the poorest families have access to a television. Furthermore, the set is often on all day, in part because it provides a window on the world for secluded women. TV may serve either as a reinforcer of traditional culture or as a vehicle for secular values, depending on which channels are watched.

Language and Diversity

Arabic is now the official language in all countries of this region except Iran, Turkey, and Israel (where Farsi, Turkish, and Hebrew, respectively, are spoken), but this uniformity of language masks considerable cultural diversity. There are numerous minorities within the region who have their own non-Arabic languages: Berbers, Tuareg, Sudanese, Nilotics, and Nubians, to name a few in North Africa; Kurds and Turkomans in Southwest Asia. There are also many dialects of Arabic that indicate deep cultural varia-

tions across the region. Yet, in this era of modern communications, dialects are disappearing as standardized Arabic becomes the language used by broadcasters. More important, Arabic is replacing the languages of many minorities as they become assimilated into mass media culture through radio, television, and movies. Nonetheless, the widespread use of Arabic facilitates region-wide communication and pan-Muslim solidarity. Meanwhile, the use of French and English as second languages, also very common especially in urban areas, and the dominance of English on the Internet, is contributing to the decrease of language diversity and to the intrusion of outside cultural influences. Figure 6.11 shows the distribution of languages in North Africa and Southwest Asia.

Islam in a Globalizing World

Many Muslims see modern culture, much of it originating in Europe and America, as undermining important values, such as the duty of the individual to the family and community. Hence, some Muslims object to the liberalization of women's roles, especially to their having careers and to their being active in community life outside the home. Many Muslims also lament the globalization of Western culture with its open sexuality and consumerism and its

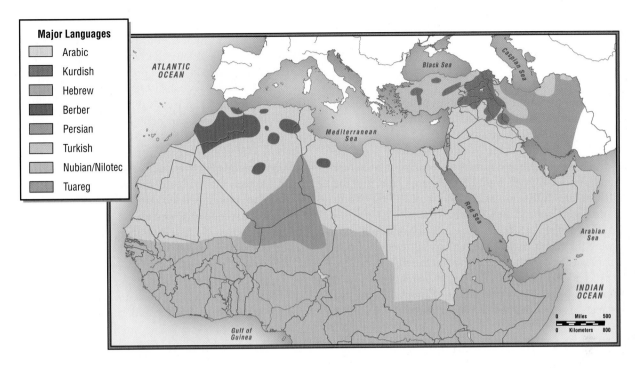

Figure 6.11 **Major languages of North Africa and Southwest Asia.** [Adapted from Charles Lindholm, *This Islamic Middle East: An Historical Anthropology* (Oxford: Blackwell, 1996), p. 9.]

tendency toward hedonism. They worry that such ways lead to family instability, alcoholism, drug addiction, and street crime by youth who have been poorly raised by inattentive parents. Consumerism is seen as leading to selfishness and to widening gaps between rich and poor. (A basic pillar of Islam is the duty to address the injustice of poverty.) Some social commentators argue that the basic concerns of Muslims about loss of responsibility to family and the obsession with materialism that can accompany modernization are similar to those of the more conservative believers in many of the world's religious and moral systems. All seek a reaffirmation of traditional values,

These beliefs are shared by many moderate Muslims who are not Islamists, but they are also strong among the Islamists. Islamists (also called Islamic fundamentalists) hark back to what are thought to be stricter interpretations of the Qur'an. They favor a simple, prayerful life focused on family and community, traditional gender roles, and respect for the elderly. But although all Islamists are reacting to some extent against what they regard as dangerous Western secular influences on Islamic society, the different Islamist factions vary greatly in the perspectives and fervor they bring to their causes. A few seek violent revolution against what they see as an all-powerful and insensitive West; most simply seek to insulate their communities from Western influence.

There are numerous moderate Muslims who believe that modernization can be combined with Muslim beliefs and values. The ideas of Abdullahi An-na'im, a Sudanese professor of religious studies at Emory University in Georgia, are echoed by many moderates. An-na'im calls for an open debate in Islam that includes a reevaluation of the ancient shari'a interpretations of the Qu'ran. He

notes that the shari'a were conditioned by the times in which they emerged, within several hundred years of Muhammad's life. They were progressive for the time—giving some limited rights to women and minorities that were not yet the norm in Europe or elsewhere. But, newly interpreted in light of modern experience, the Qur'an could become the basis of a modern society, where, for example, human rights are recognized as central to a civil society.

Ali Mazrui, a widely known Africanist and director of Global Cultural Studies at the State University of New York (SUNY) at Binghamton, sees the Internet as offering an opportunity for an Islamic reformation that would lead forward to modernization of Muslim culture and back to an original tenet of Islam: that all believers should be joined in a universal transnational community, or *ummah*. According to Mazrui, Internet technology is uniquely adapted to helping all people, even secluded women, join an open discourse within and beyond Islam. Muslims may use the Internet to bridge political borders and find common ground.

ECONOMIC AND POLITICAL ISSUES

There are major economic and political barriers to peace and prosperity within the region today. The oil wealth so prominent in Western minds is limited to a few elites; most people are relatively poor farmers, herders, or low-wage urban workers. The economic base of the region is unstable because oil and agricultural commodities are subject to wide price fluctuations on world markets. New bases of economic development are badly needed, but a more diverse range of industries is just beginning to emerge. Meanwhile, hard times loom for many poorer non-oil-producing countries as large national debts are forcing governments to restructure their economies.

The Oil Economy

This region contains the world's largest known petroleum (oil and gas) reserves. (Central Asia may prove to have yet larger reserves.) The reserves are located mainly around the Persian Gulf (Figure 6.12); thus, the petroleum wealth so often associated with the region as a whole is concentrated mostly in countries that border the Gulf: Saudi Arabia, Kuwait, Iran, Iraq, Oman, Qatar, and the United Arab Emirates. Oil and gas are also found in the North African countries of Algeria, Tunisia, and Libya.

Early in the twentieth century, European and North American companies were the first to exploit the region's oil reserves. These companies paid governments a small royalty for the right to explore and drill for oil. The oil was processed at on-site oil refineries owned by the companies and shipped primarily to Europe, the United States, and Japan. The governments of the region did not assume

control of their oil reserves until the 1970s, when they declared all oil resources and industries to be the property of the state. Even before this, in the 1960s, the oil-producing countries organized a **cartel,** a group able to control production and set prices for its products. The oil cartel is called **OPEC**—the **Organization of Petroleum Exporting Countries.** OPEC now includes all the oil states listed in the previous paragraph (except Tunisia), plus Venezuela, Indonesia, and Nigeria. (Gabon and Ecuador have left the organization.). OPEC members cooperated to restrict oil production, thereby significantly raising its price on world markets. Prices rose swiftly from less than U.S. $3/barrel before 1973 to a peak of U.S. $34/barrel in 1981; prices then fell during the 1980s and 1990s, sinking to $9/barrel. A second spike occurred in early 2001 ($32/barrel was the price on January 19, 2001) and then fell again. Note, however, that world petroleum prices result from a complicated set of factors. OPEC countries often fail to reach agreement on the management of prices (see the Gulf War discussion on page 334), and non-OPEC producers can influence petroleum supplies and, hence, prices. Consumers, by reducing demand, can also affect prices.

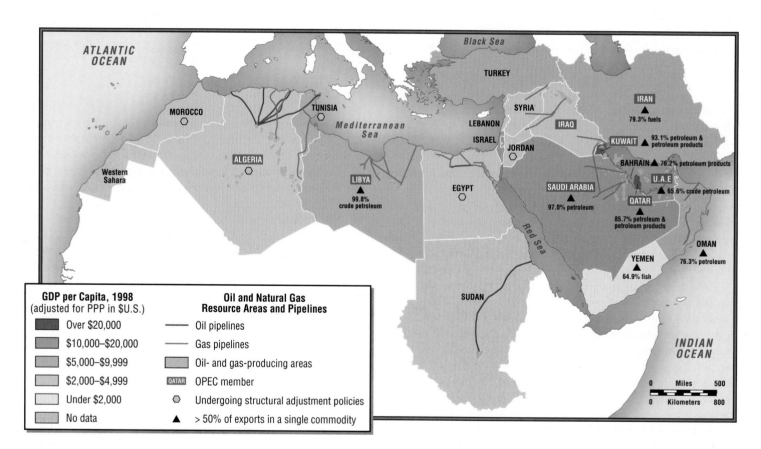

Figure 6.12   Economic issues: North Africa and Southwest Asia. Distribution of GDP per capita and oil and natural gas resources. Oil and gas resources are unequally distributed in North Africa and Southwest Asia. Much of the region remains poor and dependent on herding and agriculture, and many of these latter countries are now undergoing retrenchment due to structural adjustment policies imposed by international lending agencies. Notice that whereas Algeria has oil and gas resources, they account for less than 50 percent of exports. [Adapted from Rafic Boustani and Philippe Farques, *The Atlas of the Arab World—Geopolitics and Society* (New York: Facts on File, 1990), pp. 85, 88, 89; and Richard Overy, ed., *The Times History of the World* (London: Times Books, 1999), p. 304.]

Although petroleum price increases strain the economies of both industrialized and developing countries, for OPEC countries the huge profits could mean significant improvements in standards of living and well-being for all citizens. After the major Arab and non-Arab oil-producing countries raised oil prices dramatically in 1973, oil income in Saudi Arabia alone shot up from U.S. $2.7 billion in 1971 to U.S. $110 billion in 1981. Yet because the OPEC countries did not invest much of their oil wealth in basic human resources (by providing quality education and health care, for example) or in developing technical and natural resources, they have been unable to improve their economic bases. Like their poorer neighbors, they have remained dependent on the industrialized world for their technology and manufactured goods. Nevertheless, oil has brought significant benefits to the region as a whole. In this era of fabulous prosperity, those countries that had significant amounts of oil (Saudi Arabia, the UAE, Kuwait) sharply increased investment in some (but not all) parts of their long-neglected national infrastructures. They invested heavily in roads, airports, new cities, irrigated agriculture, and petrochemical industries, but much less in education, social services, housing, and health care. Even people from non-oil-producing countries were able to share in the wealth: many of the laborers on these projects came from Egypt, Jordan, Syria, and Turkey, none of which are oil producers.

During the 1980s and 1990s, however, much of this oil-driven growth slowed. OPEC was less able to control oil prices because the industrialized countries reduced their demand for OPEC oil by becoming more energy efficient and by developing their own oil resources. The very low oil prices in the 1990s triggered an "income-loss shock" in the oil-dependent economies of the Arab world. Governments had to scale back their development projects and reduce their demand for immigrant labor. The situation may shift again over the next several decades because demand for energy will surely rise as technology spreads; oil prices will rise if oil supplies dwindle worldwide, as expected. At some point in the future, the depletion of oil resources and the development of new sources of energy will force OPEC countries to find other ways of earning an income or risk economic ruin.

The Traditional Economy: Agriculture and Herding

Despite the abundance of oil and gas along the Persian Gulf and elsewhere, much of the region remains poor and highly dependent on agriculture and herding. It often appears from production (GDP) and export figures that economies are dominated by industries and services, but in several of the largest countries in this region, substantial portions of the population are still employed in agriculture and depend upon their own efforts for most of their daily nutrition. For example, in Morocco, industry accounts for 30 percent of GDP and agriculture for only 16 percent, yet 50 percent of the labor force works in agriculture. In Egypt, agriculture accounts for only 17 percent of GDP, yet 40 percent of the people work in agriculture. In Turkey, 18 percent of the GDP is in agriculture, yet agriculture employs 45 percent of the people. Herding remains an important aspect of agricultural life and land use in Morocco, Algeria, Egypt, Turkey, and Iran (in Morocco, for example, 47 percent of the land

is used for herding), but in none of these countries is it a significant income earner when compared to industry and services.

Agriculture. Crop agriculture in the region has traditionally been confined to moist river valleys and coastal areas and to upland areas where mountain ridges capture orographic rainfall. Grains, cotton, and sugarcane are raised in the Nile Valley of Sudan and Egypt, grains and fruit in the Tigris and Euphrates river valleys. Turkish farmers produce cotton, tobacco, sugar beets, and livestock; and Iraqi and Iranian farmers grow grains, nuts, tea, tobacco, and livestock. Citrus fruits, olives, dates, and wine grapes are grown on the northern coasts of Africa for export to Europe. (Alcohol consumption is not permitted by the Qur'an, but it does occur. Most wine is exported.) The farmers of Yemen and Oman grow coffee and grains. Israelis and Palestinians grow vegetables and fruits.

Ambitious irrigation schemes in Libya, Egypt, Saudi Arabia, Turkey, and Iraq are now expanding commercial, primarily export-oriented, agriculture to neighboring areas that had not been able to support large-scale cropping. These attempts at expansion have not always been well conceived. Many state-sponsored irrigation projects have damaged soil fertility through **salinization.** Salinization occurs when large amounts of water evaporate, leaving behind dissolved salts and other minerals that inhibit plant growth over time. It is particularly likely to occur in arid lands because there is little rain to wash away the salt. Israel has developed more efficient techniques of drip irrigation that curtail salinization, but poorer states have been unable to afford the technology, and other wealthy states have been wary of becoming dependent on a technology developed by a country they distrust.

The region's numerous political tensions have convinced many governments that they should try to be self-sufficient in food production no matter the expense. To that end, some oil-rich governments have wasted huge amounts of money on shortsighted and poorly engineered development schemes. Saudi Arabia and Libya

An irrigated wheat field in the Saudi Arabian desert. [Ray Ellis/Photo Researchers.]

are pumping massive volumes of precious groundwater to the surface at highly unsustainable rates to grow crops for home use and export. Libya is mining an ancient aquifer for irrigation and other purposes at eight times the rate of the natural replenishment. Saudi Arabia has even used the expensive process of desalinating seawater to provide irrigation for wheat fields. Wheat grown in this way costs 10 times the price of wheat on the world market.

Herding. The tending of grazing animals—particularly camels for transport and sheep for wool—was the economic mainstay of the region for thousands of years, but its economic importance has been in decline since the nineteenth century. Traditionally, camel breeders would take their herds from the sandy deserts of the Arabian Peninsula to cooler Syria or Iraq in the summer. Those in the Sahara would take their animals to summer in the southern fringes of the Atlas Mountains. Since the construction of railroads and highways, camels are no longer needed, but demand for sheep's wool continues. Sheep (and goats), however, cannot travel long distances, so the range of nomadic migrations of sheepherders was always smaller than that of camel herders. The settled herding of sheep is relatively common even in urban fringe areas, and many families engaged in nonagricultural employment maintain a small herd of sheep that the boys of the family tend (see the vignette at the beginning of this chapter).

Nomadic herders, such as the Kurds in Southwest Asia, have lost financial and spatial independence as the region has modernized. In establishing large irrigated agriculture projects, many governments take control of water sources previously available to the herders, or they may require the nomads to settle permanently in order to provide a labor pool for these projects. Herders are also forced to settle down because their tendency to cross national borders in search of grazing lands and water is perceived as a national security threat. Finally, nomads are hard to tax or to control in other ways; for example, it is difficult to enforce compulsory school attendance on their children.

Attempts at Economic Diversification

Greater economic diversification could bring broader prosperity to the region and limit the risk that a drop in the price of one or two commodities on the world market would bring economic disaster. **Economic diversification** is the expansion of an economy to include a wider array of economic activities in an effort to increase GDP and economic stability; but few countries in the region have achieved that goal. Perhaps the most successful is Israel. Compared to other countries in the region, Israel has a particularly solid manufacturing base, and its goods and those of its modern agricultural sector are exported worldwide. Turkey, Egypt, Morocco, and Tunisia are also starting to develop more diversified economies, although they are not as far along as Israel.

Economic diversification has not been achieved more widely for several reasons. One is that limited resources (other than oil)—such as water, arable land, minerals, and forests—inhibit diversification efforts. Another is that for decades the state, not the private sector, has driven economic development. Beginning in the 1950s, governments established state-owned enterprises for the production of such items as textiles, toilet paper, cement, and processed food, and then protected these enterprises from foreign competition with tariffs and other trade barriers. State-owned and protected industries were favored by ruling elites who often managed to gain profit from them. Also, in the postcolonial era, just as in Middle and South America, much of Africa, and India, the state has been viewed as the logical leader to spur development of local industries. European countries eager to sell their own industrial products to the region had often discouraged local private industries, and it was felt that government intervention was necessary to kick-start the previously impeded process of economic development. Finally, there was not enough local private capital to finance industrial development. Nonetheless, a number of problems prevented these countries from achieving widespread industrialization. Industries overprotected from foreign and domestic competition produced shoddy and expensive products with limited export potential. Furthermore, the extension of governmental control into so many parts of the economy nurtured corruption.

The desired economic diversification has not been achieved, but expensive blunders in state-led development have saddled the poorer governments of the region with crippling debt burdens. The need to make debt payments has forced governments to cut back their role in the economy. This trend has been under way since the 1970s, when international lending institutions, such as the International Monetary Fund and the World Bank, imposed structural adjustment programs (SAPs). As discussed elsewhere in this book, SAPs require that governments shift away from state-led development and toward a free-market economy, in return for guarantees of additional loans and for rescheduling the payments on existing debts. SAPs caused governments to privatize state-owned industries; streamline those that remained government-owned; remove import barriers; and reduce subsidies for food, housing, and agricultural rents.

Although there have been benefits from SAPs, they often have had a negative impact on poor people. In Egypt in the 1980s, for example, reductions in food and housing subsidies doubled poverty in rural areas and increased it by half in urban areas. To provide for their impoverished families, hundreds of thousands of men migrated to Jordan and Saudi Arabia to take up short-term construction labor contracts. Millions of others were pushed into Egypt's cities and into the informal economy. The streets of Cairo are now so clogged with vendors that traffic cannot pass. More than 6 million people live in illegal squatter communities on the outskirts of the city, without sanitation and other services. Throughout the region, governments in similar situations are depending more and more on international nongovernmental organizations (NGOs) to provide such services as education and health care. (One example is the NGO in Turkey mentioned in the vignette at the beginning of this chapter.) The resources of NGOs, however, are too small to provide a long-term solution.

Under these less than ideal circumstances, foreign investment in new industries that would lead to economic diversification has slowed, in part because investors can find other places where the return on their money will be greater. Nonetheless, in the 1990s a number of multinational corporations opened operations in Egypt (see page 324), and the countries of Lebanon, Egypt, Turkey, Jordan, Saudi Arabia, Israel, and Kuwait drew multinational invest-

GEOGRAPHER IN THE FIELD

Interview with Regional Consultant for Chapter 6,
Hussein A. Amery, Colorado School of Mines

What led you to become interested in North Africa and Southwest Asia?
North Africa and Southwest Asia are of great economic and geostrategic importance to the United States and to the rest of the world. However, as someone who was born in the Middle East, I have observed that the region's geography and cultures are not widely understood in Canada and the United States. Because my mother tongue is Arabic, I am able to access a large body of knowledge, both written and oral, that is available only in Arabic.

Where have you done fieldwork in North Africa and Southwest Asia, and what kind of work have you done?
I have done fieldwork in Lebanon, my place of birth, as well as Syria, Jordan, Egypt, Israel, and Palestine. I have studied the effect of immigrants' remittances on rural economic development and the relationship between water stress and political conflict. Some of these projects required door-to-door surveys to gather data; others involved interviews with ministers and other government decision makers. For the past three years, I have been researching and writing about Islamic perspectives on the natural environment.

How has that experience helped you to understand the region's place in the world?
My field and personal experiences in North Africa and Southwest Asia have convinced me that the region is on the verge of a violent

explosion, which, if understood properly, can be contained. This potential danger is the product of continuing Arab-Israeli violence, the huge income inequities that pervade the region, high levels of population growth and of unemployment, suppression of personal and political freedoms in many countries, and the tensions resulting from the spread of water stress.

What are some topics that you think are important to study about your region?
The following areas need closer examination:

• Can religion (and local cultural norms) be employed to influence public policy on environmental protection, water conservation, economic development, and conflict resolution?

• How do economic development and democracy contribute to political stability?

• How is environmental security (e.g., water stress, desertification, soil salinity) related to political stability in countries such as Syria?

• How will growing oil exports from the Caspian Sea states affect the West's interest and involvement in the Middle East?

---

ment in technology, food delivery, and auto manufacturing (see Figure 1.3 on page 8). Many people who live in the region do invest in business, including the informal economy, but the amounts invested are small, the businesses tiny, and the number of firms that become formalized and pay taxes is very few.

The Economic Legacy of Conflict

A complex tangle of hostilities between neighboring countries hinders needed economic changes in the region. Many of these hostilities are the legacy of a long history of outside interference in regional politics. In the early twentieth century, Britain and France carved up non-Turkish parts of the Ottoman Empire into a number of small countries dependent on Europe for defense and trade. Many countries today exist as a result of this era of strong European influence. One example is Israel; others are the oil-rich Persian Gulf monarchies of Kuwait, Saudi Arabia, Bahrain, Qatar, and the United Arab Emirates. Then, during the post–World War II cold war era, the region became the site of further geopolitical maneuvering by the West and the USSR. The attraction of the region's oil and gas resources and the conflict between Israel and its neighbors

over territory and resources led the superpowers to compete for influence. One result was a buildup of arms, provided by the West or the USSR, in each country. The fact that many countries possessed advanced weapons of war meant that localized disputes could escalate into regional warfare.

A case in point is the **Gulf War** (1990–1991), one of the most massive and economically destructive wars ever to occur in the region. (The Gulf War is described in more detail on page 334.) More than 1.2 million people died as a result of this war and its aftermath. Iraq invaded Kuwait in 1990, and this action prompted the formation of a military coalition of the United States and European and Arab countries. Iraq was quickly defeated, and shortly thereafter the United Nations imposed trade sanctions against Iraq in an effort to convince it to destroy its weapons, thought to consist of nuclear, chemical, and germ warfare devices. As a consequence of the war and the sanctions, Iraq has experienced severe economic depression, including loss of access to markets for its oil and the collapse of its industrial complex, which was focused primarily on armaments and chemicals manufacturing but which also included textiles, medicines, construction materials, and food processing.

The **Israeli-Palestinian conflict,** which has spawned several major wars and innumerable skirmishes, is a persistent obstacle to political and economic cooperation in the region. This conflict is described more fully on pages 328–330. Israel has the most modern and diversified economy in the entire region. Since the 1950s (after it took in hundreds of thousands of World War II refugees from Europe), its development has been facilitated by the immigration of relatively well educated middle-class Jewish settlers from the United States, Russia, and South America; by large aid contributions from the United States and private interests; and by the country's own excellent technical and educational infrastructure. Many Israelis would like to see their country become a major source of investment and technology for the region's poor but potentially growing economies. Although this possibility seemed to be gaining momentum, in recent years it has been derailed by mounting violent resistance to Israel's occupation of southern Lebanon and of territories long inhabited by Palestinians, such as the West Bank and Gaza Strip.

Another significant obstacle to improved economic relations within the region and with other parts of the world is a history of terrorism. Terrorism inhibits commerce by endangering those who trade across borders or in multicultural markets. It also increases suspicion, even hatred, between neighboring groups, causing them to eschew economic ties with one another. Turkey, Iran, Iraq, Libya, Egypt, Israel, Jordan, Lebanon, and Algeria have all experienced politically motivated bombings that have taken a high toll in lives, almost always those of innocent bystanders, and have chilled the climate for commerce. The governments of Iran, Iraq, and Libya are suspected of having instigated bombings outside the region, such as the 1988 bombing of a commercial jumbo jet over Lockerbie, Scotland. Private individuals (perhaps supported by one or more countries) are also suspected of planning and funding terrorist attacks, such as the 1993 bombing of the World Trade Center in New York and the September 2001 attack on the World Trade Center and the Pentagon carried out using hijacked commercial airplanes. Both of these attacks are thought to have been the work of Osama bin Laden, a Saudi millionaire with terrorist training camps in Afghanistan. All these activities have inhibited economic interchanges within the region and between the region and other parts of the world and have discouraged outside investment.

Islamism and Democracy

Many people fear that Islamic fundamentalism (Islamism) is the greatest threat to the region's political stability. Islamist movements are grassroots religious revivals that also seek political power. The leaders of these movements seek to take control of national governments currently dominated by secular parties. Islamism has been treated as a major threat since an Islamist revolution led by Ayatollah Ruhollah Khomeini overthrew the secular, though authoritarian, government of Iran in 1979, as described more fully on pages 332–333. Although such movements have not displaced any secular governments in the years since, they remain strong in certain countries (Algeria and Egypt are examples). Several governments in the region appear willing to sacrifice democracy in order to suppress Islamist movements. Yet the attempts to suppress these usually peaceful, popular movements may increase the like-lihood of violent revolutionary reaction, as has occurred for some years in Algeria (see page 322).

Understanding the popular base of Islamism can yield insights into its strength. Many recruits are young men from lower-class neighborhoods in the region's largest cities. Others are descendants of poor farmers and nomads who were forced off their land and into the city by state-sponsored development programs and imported labor-saving agricultural technologies. Refugee camps in Palestine and elsewhere are also sources of recruits. Given the crowded, polluted, and often chaotic living conditions in such cities as Cairo and Algiers, it is not surprising that many members of the younger generation have come to question the basic philosophy of the governments under which they live. Many Islamist activists, especially the leaders, come from the large pool of recent university graduates who are frustrated by their inability to find employment or to participate in the political process in their own countries and who are disenchanted with foreign interference in the region. Some Islamist leaders are respected religious men without much education who offer seemingly simple solutions to the region's problems. They say that all will be well if people return to fundamentalist versions of Islam and eschew secularism. Perhaps the greatest support for Islamism is gained by appealing to the widespread discontent of millions of increasingly poor people, whose plight has been further worsened by SAPs (see page 312). Although there are those who are drawn to the radical alternatives proposed by the Islamists, the movements are still small and based primarily in individual countries; only the most extreme are coordinated internationally (as was apparently the case with the organization that carried out the September 2001 air attack on New York City and Washington, D.C.; see the special box in Chapter 2, page 98).

Many governments are using the threat of Islamism as an excuse to keep political power in the hands of a small, wealthy elite. But non-Islamist forces dedicated to moderate reform may be able to remove political power from the elite and distribute it more widely. As more people become educated, fewer are content to submit passively to policies imposed by distant, inefficient, and often corrupt regimes. Ethnic minority movements, women's organizations, educated professionals, students, and workers returning from abroad have increasingly been voices for changes that would enhance democratic participation throughout the region. Many of these groups and movements are finding strong appeal among the people because they base their calls for political change on the Qur'anic concept of decision making through broad-based consultation with the larger Muslim community

ENVIRONMENTAL ISSUES

Salam, the Arabic root of the word *Islam,* means peace and harmony. Islam therefore calls on its followers to live in peace and harmony with both human and natural systems. The Islamic scholar and World Bank executive Ismail Serageldin writes that the Qur'an contains numerous references to the role of humans as stewards of the earth. In Muslim societies, these references are typically interpreted to mean that humans have the right to use the earth's resources, but only within the general limitations that Islam places on greed and personal ambition. The Qur'an requires Muslims to avoid spoiling or degrading human and natural environments, to share such resources as water equally with all forms of life, and to conserve natural resources even if they are abundant. Water purity is important in Islam. A few decades ago, a religious decree in Saudi Arabia approved the human use of wastewater if it had been completely and properly treated and impurities removed from it. This decree was important because it encouraged recycling of water and helped extend the life of existing freshwater aquifers.

In practice, Muslims are like people everywhere: they have not always followed their own religious teachings and cared for the earth. Urban crowding, mechanized agriculture, and the pursuit of material goods have resulted in pollution, species extinctions, and degraded environments. The region has a limited supply of arable land, water, forests, and minerals. Yet the people in the area expect to achieve higher living standards—which, along with population growth, will tax the region's resources. Careful management of the environment will undoubtedly become an important issue over the next several decades. Although environmental issues are not yet at the forefront of public debates, they are beginning to be discussed and are among the concerns of some Islamist critics of modern secular society. In this section, we look at two issues: water availability and use and desertification.

Water Issues

Water has always been in short supply in this arid region. Hence, cultural attitudes toward water differ greatly from those of North Americans. Here, people expect to use very little water, and they have devised many strategies to cope with the limited supply. Traditional ways to conserve water include designing buildings to maximize shade and conserve moisture; capturing mountain snowmelt and moving it to dry fields and villages in underground water conduits (*qanats*); bathing in public baths; and practicing countless recycling techniques, such as using bathwater for crop irrigation.

Despite a long tradition of careful water use, several factors have exacerbated the water shortage over the last few decades. Modernization has brought new ways to use water: in household plumbing; sewage treatment; industrial cooling and cleaning; and mechanized, irrigated agriculture. Moreover, the number of people who must share scarce water resources is growing rapidly. If natural supplies of water remain more or less constant, but water use rises, will supplies meet basic needs? Experts have established that for a country to maintain human health and to support development, it must have no less than 1000 cubic meters of water available per capita per year. Figure 6.13 shows that, in 1990, many

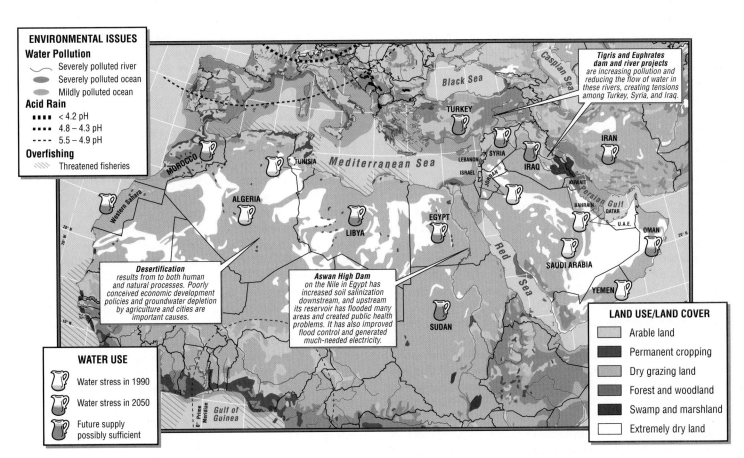

Figure 6.13 **Environmental issues: North Africa and Southwest Asia.** Lack of available potable water is the number one environmental issue facing most countries in the region. Over the next 50 years, only Sudan, Turkey, and Iraq (those with full jugs) are likely to have sufficient water to meet the annual minimum of 1000 cubic meters of water per capita considered essential to maintain public health and to support development. See Figure 6.14 for a map of dams on the Euphrates and Tigris rivers. Can you name three of the factors that contribute to growing water stress in the region? [Data from *World Resources, 1996–1997* (New York: Oxford University Press, 1996), pp. 302–303.]

countries in the region already fell well below this standard. By the year 2050, all but three countries will be below the standard.

Given its centrality to human and economic life, water is a strategic resource. Consequently, when demand exceeds supply, countries commonly follow one or a combination of the following strategies: storage of water in reservoirs behind dams, importation of water, water recycling, conserving the quality and quantity of water, or desalinating seawater.

The greatest use of water throughout the region is for irrigated agriculture. In a typical country, such as Tunisia, 89 percent of all the water withdrawn is used for agriculture, 9 percent for domestic purposes, and 3 percent for industry. In comparison, the United States uses 42 percent for agriculture, 13 percent for domestic purposes, and 45 percent for industry. In addition, the water in rivers and streams is being polluted by chemical fertilizers applied to nearby fields. Even strategies to increase the supply of water—damming rivers, desalinization, groundwater pumping—create their own problems. Damming wastes water through evaporation or leakage. The production and use of energy for desalinization creates air pollution. The pumping of groundwater lowers the water table and causes the land to subside, often resulting in the intrusion of seawater into the aquifer, which renders the groundwater useless.

Irrigated agriculture is at the heart of the water crisis, and yet it holds the seeds of a solution. Because the overwhelming majority of water is used for irrigation, even small increases in efficiency will yield large increases in the volume of fresh water available for domestic and other uses. Efficiency can be improved through such methods as applying drip irrigation more widely; reducing reliance on water-intensive crops such as rice, cotton and citrus fruits; gradually removing government subsidies to large water users; and upgrading water distribution networks to reduce leakage. Another approach, suggested by British geographer Tony Allan, calls for greater reliance on food imports. Allan argues that importing food for dry areas amounts to importing "virtual water," a reference to the water used in the production of all food and the water content found in foods. Allan suggests that water-poor regions give up water-intensive agriculture and concentrate instead on other economic activities, using the proceeds to import food from water-rich regions.

The most ambitious efforts to increase water supplies have involved the construction of dams and reservoirs on major river systems: the Euphrates, the Tigris, the Jordan, and the Nile (see the box "Turkey's Anatolian River Basin Project"). These projects raise questions about the rights of downstream and upstream water users. Damming a river removes water from huge tracts of land downstream and stops the deposition of silt during seasonal flooding. (The deposition of silt enhances soil fertility.) International laws governing water rights along river systems evolved in parts of the world, mainly Europe, where water was plentiful. Under existing laws, it is possible for a country to dam a river that flows through its territory, with little or no consideration of the people living downstream or upstream of the dam. Countries may have little recourse short of war when their lands are flooded or their source of water is cut off by a dam in a neighboring country.

Dams also create problems for the countries in which they are built. The huge artificial reservoir created upstream of a dam floods villages, fields, wildlife, and historic sites. The silt that used to serve as fertilizer downstream will fill up the reservoir behind a dam in just a few years. In a warm climate, the standing water behind a dam becomes a breeding environment for mosquitoes, which carry debilitating diseases such as malaria, and for dangerous parasites such as hookworm and schistosome species (see discussion on page 324). Also, huge volumes of water are lost from the reservoir through evaporation. Irrigated soil downstream, repeatedly saturated and dried out, will eventually become salty and infertile. Salinized soil and the accumulation of silt behind dams caused the demise of the great agricultural economies of the Tigris, Euphrates, and Nile river basins in ancient times. More recently, the Aswan High Dam project on the Nile in Egypt and numerous dam projects along the Tigris and Euphrates in Iraq, Syria, and Turkey have experienced most of these problems.

Sharing the water of rivers that flow across national boundaries has never been easy. In fact, the words *river* and *rivalry* come from the same root, and the original sense of "rivalry" was the conflict among users of a watercourse. Several questions bedevil these users: Who should receive how much water, and why? Should water allocations be based on a country's contribution to water flow (recharge) in the river? The distance the river travels in a country? The historically earliest use of the river's water (the "priority of use" or "first come, first served" principle that is widely applied in the United States)? The area of arable land within a certain country in the basin? Other variables that may be considered are per capita water use, current and projected population size, structure of the economy (agriculturally based economies might claim that they "require" more water), and availability of other sources of water. Naturally, each country emphasizes the variables that strengthen its case. For example, Turkey contributes the most to the flow in the Euphrates River and is upstream from the other users. However, the river travels the longest distance in Iraq; and the ancient inhabitants of Mesopotamia (in Arabic, as well as in Greek, the word means "lands between the two rivers"), the Sumerians, were clearly the earliest users of the Euphrates.

A number of factors complicate the chances of reaching water allocation agreements along the Euphrates and the Nile. Most countries in the region are experiencing rapid population growth and a rapid rise in living standards and urbanization. One consequence is the growing consumption of meat, yet raising cattle is more water-intensive than growing grains. Urbanites tend to consume more water and use more electricity (often hydroelectricity) than rural residents. Furthermore, the people in these basin states are from differing racial, ethnic, and religious backgrounds and have long had territorial and ideological disputes with one another. In the case of the Euphrates, the fact that Turkey has the upstream geographic location on the river basin and the strongest political and military might tilts the balance of power in the basin in Turkey's favor.

Desertification

The United Nations defines **desertification** as the ecological changes that convert nondesert lands into deserts. These changes include the loss of soil moisture, the loss of trees or the replacement

AT THE LOCAL SCALE   *Turkey's Anatolian River Basin Project*

The Euphrates River is very much on the mind of water engineer Olcay Unver, a native of Turkey. After years of study and work in Texas, he is back home managing the massive Anatolian River basin project (Figure 6.14), which is similar in many ways to the Tennessee Valley project in the United States, one of its models. The Euphrates River begins life as trickles of melting snow in the mountains of Turkey, then gathers volume from several tributaries as it flows southeast through the Anatolian Plateau and then through Syria and Iraq. In Iraq, the Euphrates is joined by a second major river, the Tigris, which also originates in Turkey. The two run roughly parallel for hundreds of miles, joining just before they reach the Persian Gulf.

The project that Mr. Unver manages was begun as a series of 22 dams to reserve Euphrates waters in Turkey for irrigation and hydropower generation; 19 power plants are planned. The enterprise has since broadened to include reforestation; instruction in farming; the establishment of fisheries; and the provision of such social services as literacy training, health care, and women's centers. Mr. Unver describes the project in this previously neglected part of southeastern Turkey as "an integrated socioeconomic development project which aims to enhance the standard of living in this area."

As important as the Anatolian project may be to Turkey, its effects on Syria and Iraq downstream could be disastrous. If completed as planned, the project will reduce the Euphrates flow to Syria by 30 to 50 percent and to Iraq by around 75 percent. Moreover, the water will be polluted with fertilizers, pesticides, and salts after having served to irrigate crops in Turkey. Syria, whose population will double in just 25 years, depends on the Euphrates for at least half of its water, and Syria also plans to use more of the Euphrates flow for irrigation. Iraq is the last to receive the river water and is the most vulnerable. Iraq and Syria demand a combined total of 700 cubic meters of water per second (entering Syria) as their rightful share of the Euphrates flow. Sixty percent of this flow would be released into Iraq. Turkey currently releases 500 cubic meters per second.

Source: Stephen Kinzer, "Restoring the Fertile Crescent to its former glory," *New York Times* (May 29, 1997); Christine Drake, "Water resource conflicts in the Middle East," *Journal of Geography* (January/February 1997): 4–12; and Hussein A. Amery, personal communication, August 2001.

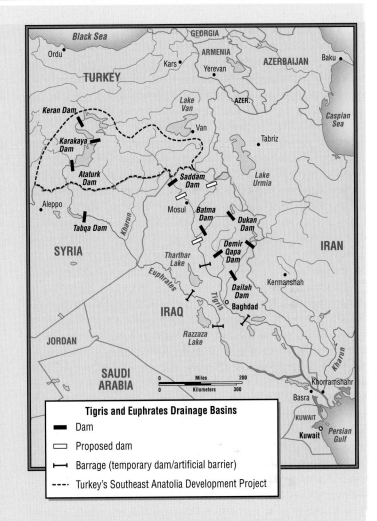

Figure 6.14 Dams of the Tigris and Euphrates drainage basins. [Adapted from Christine Drake, "Water resource conflicts in the Middle East," *Journal of Geography* (January/February 1997): 7.]

of plant species by other species that are more adapted to dry conditions, the erosion of soil by wind, and the formation of sand dunes on formerly vegetated land. These alterations can be brought on by long-term climatic change, by relatively short term climate variability, or by human activity that reduces moisture. Desertification is occurring in many places around the globe. Estimates of how much usable land is being lost to desert vary, but the ecologist and author G. Tyler Miller writes that globally, every year,

23,000 square miles (60,000 square kilometers) become new desert. It is not yet known just how much of this loss is due to natural climate change and how much to human activity through the greenhouse effect, human modification of forests and grasslands, and overuse of water resources.

In the region of North Africa and Southwest Asia, nomadic herders are often blamed for increasing desertification. They are said to overstock rangelands and allow their animals to overgraze.

AT THE LOCAL SCALE   *The Zabbaleen: An Example of Sustainable Development*

As is the case elsewhere on earth, environmentally sustainable development strategies—those that do not compromise the prospects of future generations—are just starting to be implemented in this region. Among the most effective are projects that are small in scale, place few demands on scarce resources, and are focused on the needs of relatively small communities. An example is the case of the Zabbaleen, a group in Cairo, Egypt, that has long earned a living by camping at urban dump sites and sorting through refuse looking for recyclable items. (*Zabbaleen* is Arabic for "garbage collector.") They perform a useful service, but their occupation puts them (especially children) at great risk from disease as well

as injuries from broken glass and metal. In the 1980s, several NGOs began helping the Zabbaleen to improve conditions in their settlements. With a small loan, they purchased machinery that converts rags and plastic into reusable materials, which bring higher income than reselling unprocessed materials. Also, improved compost systems allowed the Zabbaleen to recycle organic food and animal wastes into salable products, while clearing the streets of refuse they had previously ignored. The hope is that through such demonstration projects, people throughout the region can reclaim and sustain the environments that are important to them, without waiting for policy decisions from government bureaucracies.

With little grass remaining, the soil is unable to retain moisture. Studies are showing that the real story is more complicated. The geographer and veterinarian Diana Davis found that herders are usually careful to manage their herds so that the grasslands on which they depend are not destroyed. In her research among herders in Morocco, Davis observed that herdspeople were more likely to decrease rather than increase their herds when times were good. Their knowledge of range ecology has led to management practices that accord well with the arid environment.

More likely, a wide range of economic changes has contributed to a general drying of grasslands bordering the Sahara. As groundwater is lowered through urban use and irrigated agriculture, plant roots can no longer reach sources of moisture. The encouragement of settled cattle ranching by international development agencies leads to overgrazing of pastureland and to excessive demands on scarce water. Programs that encourage pastoral nomads to settle permanently and take up modern lifeways also result in greater per capita use of water.

MEASURES OF HUMAN WELL-BEING

As we have observed elsewhere in this book, the gross domestic product per capita is a less than adequate measure of the actual state of well-being in a region, but it nonetheless gives some indication of the distribution of wealth across the region. In Table 6.2, you can see that the pattern of wealth as measured by GDP per capita (column 2) is uneven. The primary oil-producing nations (Saudi Arabia, Iran, the UAE, Kuwait, Libya) and Israel have high GDP per capita figures. Many other countries in the region are well below the world average of about U.S. $6500. Even politically influential countries such as Turkey and Egypt have per capita incomes below the world average. GDP per capita does not indicate the internal distribution of a country's wealth but is only an average. Because the national income in most countries in this region is distributed unequally, many people live on an income much lower than that listed, while a very few people are extremely rich.

The United Nations Human Development Index (HDI) (Table 6.2, column 3) combines three indicators—life expectancy at birth, educational attainment, and income adjusted to purchasing power parity (PPP)—to arrive at a ranking of 174 countries that evaluates well-being. You may remember that in other chapters we cited Kuwait's HDI figures in our comparisons of well-being in countries in other regions. Despite its high GDP (highest in this region), Kuwait does not rank very high on the HDI scale in comparison to nearly all developed countries in the world. An important reason is its low literacy rates, especially among adult Kuwaiti women (just 78 percent have basic literacy). Until the past decade, education has been available to only a minority of females. Thus, in Kuwait and elsewhere in the Arab states, there is a large population of older illiterate women. Israel is the only country in this region with both a high GDP per capita and a high HDI rank (23). Notice that some of Israel's closest neighbors (Syria, Lebanon, Jordan, and Egypt) rank fairly low on both the GDP index and the HDI. Israel's ranking reflects its ability to provide the basics of a decent life—adequate income, shelter, food, education, and health care—to all its citizens, regardless of ethnicity, sex, or religion. Israeli culture emphasizes education, and Israel has attracted highly educated Jewish immigrants from around the world. Israel, however, has also received help in providing for its citizens through development aid, receiving $172.00 per capita in development assistance from the United States in 1998. Israel's much poorer neighbors, including Egypt, have received less aid per capita. Egypt, the largest recipient after Israel, received $29.00 per capita in U.S. development assistance in 1998.

The Gender Empowerment Measure (GEM) and the Gender Development Index (GDI) are combined in Table 6.2, column 4. GEM ranks countries by the extent to which women have opportunities to participate in economic and political life. For a variety of reasons, the statistics necessary to calculate GEM are not available for two-thirds of the countries in this region (only 70 countries are ranked worldwide). The Gender Development Index (GDI) measures the access of males and females to health care, education, and income. It does not measure the extent to

TABLE 6.2   *Human well-being rankings of countries in North Africa and Southwest Asia*

Country (1)	GDP per capita, adjusted for PPP[a] in 1998 $U.S. (2)	Human Development Index (HDI) global rankings, 2000[b] (3)	Gender Empowerment Measure (GEM)[c] and Gender Development Index (GDI) global rankings, 2000[d] (4)
Selected countries for comparison			
Japan	23,257	9 (high)	41 (GDI = 13)
United States	29,605	3 (high)	13 (GDI = 6)
Mexico	7,704	55 (medium)	35 (GDI = 49)
World	6,526	N/A[e]	N/A
The Northeast			
Turkey	6,422	85 (medium)	64 (GDI = 55)
Iraq	3,197	126 (medium)	N/A (GDI = 107)
Iran	5,121	97 (medium)	N/A (GDI = 84)
Eastern Mediterranean			
Syria	2,892	111 (medium)	65 (GDI = 95)
Lebanon	4,326	82 (medium)	N/A (GDI = 74)
Israel	17,301	23 (high)	23 (GDI = 22)
Jordan	3,347	92 (medium)	69 (GDI = N/A)
The Arabian Peninsula			
Saudi Arabia	10,158	75 (medium)	N/A (GDI = 76)
Yemen	719	148 (low)	N/A (GDI = 133)
Oman	9,960	86 (medium)	N/A (GDI = 82)
United Arab Emirates	17,719	45 (high)	N/A (GDI = 44)
Qatar	20,987	42 (high)	N/A (GDI = 41)
Bahrain	13,111	41 (high)	N/A (GDI = 42)
Kuwait	25,314	36 (high)	N/A (GDI = 34)
Egypt and Sudan			
Egypt	3,041	119 (medium)	68 (GDI = 99)
Sudan	N/A	143 (low)	N/A (GDI = 118)
The Maghreb			
Western Sahara	N/A	N/A	N/A (GDI = N/A)
Morocco	3,305	124 (medium)	N/A (GDI = 103)
Algeria	4,792	107 (medium)	N/A (GDI = 91)
Tunisia	5,404	101 (medium)	60 (GDI = 86)
Libya	6,697	72 (medium)	N/A (GDI = 65)
World	6,526	N/A	N/A

[a]PPP = purchasing power parity.
[b]Total ranked in world = 174.
[c]Total ranked in world = 70 (no data for many).
[d]Total ranked in world = 143 (no data for many).
[e]N/A = data not available.

Source: United Nations Human Development Report 2000 (New York: United Nations Development Programme), pp. 157–168.

which males and females actually participate in policy formation and decision-making within a country. As we have already observed, the region is characterized by rather stark differences between genders and the roles they play in society. Despite the importance of women in the home, only a very small percentage of women across the region participate in policy-making debates or as leaders in the larger society. Many countries have no women in parliament. Turkey has 4 percent, Egypt 2 percent, and Iraq 6 percent. In Tunisia, women hold 11.5 percent of the parliamentary seats, but gains are being made: women hold 23 percent of the country's judgeships and head 1500 businesses, managing 13 percent of those in greater Tunis. Israel's relatively high GEM rank of 23 is the result of its direct efforts to include females in all aspects of society, including the armed forces. Even in Israel, however, women hold only 12.5 percent of parliamentary seats.

SUBREGIONS OF NORTH AFRICA AND SOUTHWEST ASIA

The subregions of North Africa and Southwest Asia present a mosaic of the issues that have been discussed so far in this chapter. Although all countries in the region except Israel share a strong tradition of Islam, they vary in how Islam is interpreted in national life and the extent to which Westernization is accepted. They also vary in prosperity and in the extent to which wealth is evenly distributed in the general population. Conflict is not uncommon in this region, and the closer examination allowed by the subregional perspective may help you understand the factors that have led to the discord and, in some cases, violence.

THE MAGHREB (NORTHWEST AFRICA)

If you are interested in old movies, you have probably already formed intriguing images of North Africa: Berber or Tuareg camel caravans transporting exotic goods across the Sahara from Timbuktu to Tripoli or Tangier; the Barbary Coast pirates; Rommel, the World War II "Desert Fox"; the classic lovers played by Humphrey Bogart and Ingrid Bergman in *Casablanca*. These images—some real, some fantasy, some merely exaggerated—are of the western part of North Africa, what Arabs call the Maghreb ("the place of the sunset"). The reality of the Maghreb, however, is much more complex than popular Western images of it.

The countries of the Maghreb stretch along the North African coast from Morocco to Libya (Figure 6.15). A low-lying coastal zone is backed by the Atlas Mountains, except in Libya. These mountains trigger sufficient rainfall in the coastal zone to support export-oriented agriculture, despite the region's overall aridity. Except in Morocco, each country's interior reaches into the huge expanse of the Sahara Desert.

The cultural landscapes of the Maghreb reflect the long and changing relationships all the countries have had with Europe. European colonialism in this area lasted well into the twentieth century. During this time, the people of North Africa became acculturated to Europe, taking on European ways: consumerism; mechanized market agriculture; and manners of dress, language,

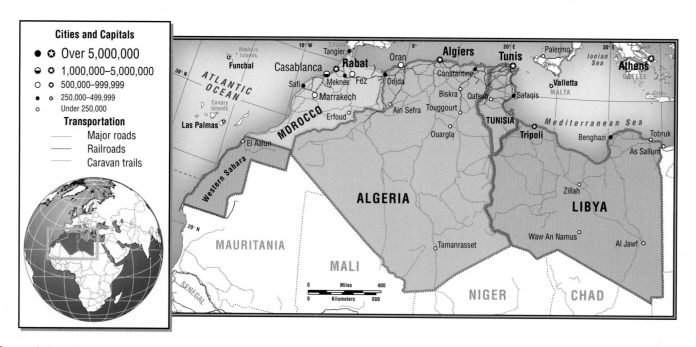

Figure 6.15 The Maghreb subregion (Northwest Africa).

Modern high-rise complex in Tripoli, Libya. [P. Robert/Corbis Sygma.]

Tunisia, and Libya have notable oil and gas reserves (see Figure 6.12). Libya and Algeria supply about one-quarter of the oil and gas used by the European Union. Oil makes up almost 99 percent of Libya's total exports, and the revenues have raised the standard of living for most of Libya's 5.6 million people.

Europe is also a source of jobs, both in local firms supported by European investment and tourism and for guest worker emigrants in Europe. *Migration News* (1998) estimates that there are half a million North African citizens in Spain and that France has 5 to 6 million North African migrants, many of them illegal.

Settlement and economic activities are concentrated along the narrow coastal zone (see Figure 6.6). Most people live in the cities that line the Atlantic and Mediterranean shores—Casablanca, Rabat, Tangier, Algiers, Tunis, Tripoli—and in the towns and villages that link them. The architecture, spatial organization, and lifestyles of these cities mark them as cultural transition zones between African and Arab lands and Europe. The city centers retain an ancient ambience, with narrow walkways fronted by family compounds (see the photo on page 305), while modern additions around the peripheries of these old cities sport highways, art galleries, shopping malls, restaurants serving international cuisine, and high-rise apartment buildings (see the photo of Tripoli, Libya, at the left).

Many North Africans are seeking an identity that is less European and retains their own distinctive heritage. In the Maghreb, the rejection of Europeanization by Islamists has been particularly hostile. As the historian John Ruedy puts it, for those people who were Westernized during the colonial era and who now make up the educated middle class, independence from Europe meant the right to establish their own secular nations with constitutions influenced by European models. But by the 1990s, the Islamist movement was seeking to challenge independent secular states by linking them in a negative way to European colonial domination. For these religious activists, Islamism is a positive cultural identity. Their interpretation of Islam, including the reassertion of Islamic law and its

and popular culture. The cities, beaches, and numerous historic sites of the Maghreb continue to attract millions of European tourists every year, who come to buy North African products for their homes—fine leather goods, textiles, handmade rugs, sheepskins, brass and wood furnishings, and paintings—and to enjoy a culture that, despite Europeanization, seems exotic. The agricultural lands of the Maghreb are strategically located close to Europe, where there is a strong demand for Mediterranean food crops: olives, olive oil, citrus fruits, melons, tomatoes, peppers, dates, grains, and fish.

North Africans depend on Europe as a market for exports—especially oil, gas, and agricultural produce. Agriculture employs about one-quarter of the population. With few exceptions, most farms are large, relatively modern operations that use irrigation, mechanization, and chemicals to enhance production. Algeria,

Although modern life in Algeria is centered on the coast, ancient strategies for living with the desert persist in inland areas. At Souf, an oasis in the Algerian Sahara, farmers plant date palms. The roots reach down for the water located near the surface in depressions between the dunes. The farmers also build fences against the prevailing winds, creating wavelike patterns in the sand. Sand collects against the fences, building higher dunes and further sheltering the trees from the desiccating wind. [Georg Gerster/Photo Researchers.]

CULTURAL INSIGHT *Algeria's Dangerous Political Music*

Algeria is the home of *rai*, a musical form born in the 1920s among poor urban youth. In the early years, *rai* was frankly Western, its lyrics often about pleasure, wine, women, and sex. But like calypso music in the Caribbean, *rai* songs began to celebrate political dissent and youthful rebellion. Many young French colonists and tourists liked the music, and it was imported into France. After the revolution, the early Algerian government restricted such free expression and tried to ban *rai*. Later, in the 1980s, controls were relaxed and the music revived. In the 1990s, however, both Islamists and conservative French disliked *rai*. Islamists objected to the secular qualities, and some extremists have killed *rai* musicians. Conservative French still associate it with the hundreds of thousands of North Africans living and working in France, whom they see as posing a threat to French ways of life.

Source: Joan Gross, David McMurray, and Ted Swedenburg, "Rai, rap, and Ramadan nights: Franco-Maghribi cultural identities," in Joel Beinin and Joe Stork, eds., *Political Islam* (Berkeley: University of California Press, 1997), pp. 257–268.

principle of public consultation, challenges the current quasi-democratic practices across the Maghreb. Secular leaders and some citizens who are religious but who prefer a secular government are fearful of the outcome.

Violence in Algeria

Algeria, with 28 million people, is arguably the Maghreb's most troubled country. By 1998, civil violence between Islamists and the military-controlled government had resulted in the deaths of at least 80,000 people. The violence has disrupted the society and ruined the already strained economy.

After a violent struggle with France for independence, won in 1962, Algeria was ruled by a single-party socialist dictatorship that repressed all viable opposition. At independence, Algeria's economy retained colonial features. Europeans remained in control of the most productive farmland. The new socialist military government expropriated (took over) the large European farms that produced dates, wine, olives, fruit, and vegetables for export, and soon these farms ended up in the control of the Algerian elite, either as private farms or as large cooperatives. Most of the large agricultural population remained on small, poor plots. By the early 1990s, nearly half the economy, including factories and utilities, was state owned and inefficiently operated. The economy declined to the point that by the late 1990s, unemployment among urban dwellers under the age of 30 was 80 percent or higher. The unequal division of wealth between those Algerians who replaced Europeans in positions requiring skills and education and those who remained poor, with only a few years of education, created social divisions. Those who are well-off and liberal in their attitudes toward economic reform, women's roles, and freedom of speech are often derided by the Islamists as secular; but a significant number of Islamist leaders were themselves technically and professionally educated people.

During the 1990s, the Algerian military government shifted its policy regarding ownership of industries and undertook major economic structural adjustment. In an attempt to lure foreign investment and create jobs, industries were privatized. During the lag that often occurs between structural adjustment reforms and economic resurgence, average personal income fell by a third. In 1991, under these dismal economic conditions, an Islamist party won the first free elections by an overwhelming margin. A peaceful transfer of power could have diffused the pent-up discontent of many Algerians. Instead, the military-influenced socialist government—motivated largely by fears that the newly elected government would pursue an extreme path—declared the parliamentary elections to be null and void. Leaders of the Islamist party were jailed, and the country was plunged into a devastating civil war.

Although some Islamist advocates joined religious social service agencies and worked to improve life in Algeria's shantytowns, extremist factions of Islamists began brutal terrorist attacks. Particularly targeted were journalists; musicians and artists; and educated, working women not observing seclusion or not wearing the veil. Several thousand such women were killed (see the box "Algeria's Dangerous Political Music"). The populations of whole villages were massacred, with apparently no attention paid to political or religious affiliation. Some studies and a few journalistic reports show that state-sponsored "militias" (troops in civilian clothing) also terrorized civilians to dissuade them from supporting the Islamists.

By mid-1997, public sentiment was turning against extremists in the Islamist movement. In a relatively fair election in June 1997, moderate Islamists, espousing nonviolence, won nearly one-third of the seats in the National Assembly. Since then the violence has slowly dissipated.

THE NILE: SUDAN AND EGYPT

The Nile River begins its trip north to the Mediterranean in the hills of Uganda and Ethiopia in central East Africa. The countries of Sudan and Egypt share the main part of the Nile system (Figure 6.16), and for these countries it is the chief source of water. Although Sudan and Egypt share the Nile River, an arid climate, and Islam (Egypt is 94 percent Muslim, Sudan 70 percent), they differ culturally and physically. Egypt, despite troubling social, economic, and environmental problems, is industrializing and plays an influential role in global affairs, whereas Sudan struggles with civil war and remains relatively untouched by development.

zones that become drier as the altitude decreases. The upland south, called the Sudd (the first zone), consists of vast swamps fed by rivers bringing water from the rainy regions of Central Africa. North of the Sudd, the much drier steppes and hills of central Sudan (the second zone) are home to animal herders. Yet farther north is the Sahara (the third zone), which stretches all the way through Egypt to the Mediterranean. Most Sudanese live in a narrow strip of rural villages along the main stream of the Nile and its two chief tributaries (the White Nile and the Blue Nile; see Figure 6.16). Between the two tributaries, irrigated fields produce cotton for export. Sudan's only cities are clustered around the famed capital of Khartoum, where the White and Blue Niles join.

There is animosity between southern and northern Sudan that has roots deep in the past and is fed by more recent events. Islam did not spread to the upland south until the end of the nineteenth century, when Egyptians, backed by the British, subdued that part of the country. Sudan then became a joint protectorate of Britain and Egypt until independence in 1955. The southern Sudanese are either Christian or hold various indigenous beliefs, and many are black Africans of Dinka and Nuer ethnicity. They have long feared the more numerous Muslim, Arabic-speaking Sudanese of the north, because for thousands of years northerners raided the upland south for slaves. In 1983, the Muslim-dominated government decided to make Sudan a completely Islamic state and imposed shari'a on the millions of non-Muslim southerners. Women, who were the most directly affected, were permitted to continue to work as nurses and primary school teachers but were barred from jobs in higher education and the civil service. Various militia groups that still operate in the country and in refugee camps will not allow women to travel without a male escort. In protest over Muslim domination, southern Sudan launched a war that has continued for nearly two decades. The hostilities have devastated Sudan. The situation worsened after a series of droughts and floods; to date, over one and a half million people have been killed and two million people are now refugees (see the photo on page 303).

Egypt

In Egypt, the Nile flows more than 800 miles (about 1300 kilometers) north in a somewhat meandering track to its massive delta along the Mediterranean. The valley and delta of the Nile are virtually the only habitable parts of the country. Ninety-six percent of Egypt is desert. Yet, for thousands of years, agriculture along the banks of the Nile has fed the country and provided high-quality cotton for textiles.

The Nile's flow is no longer unimpeded. At the border with Sudan, the river is captured by a 300-mile-long (483-kilometer-long) artificial reservoir, Lake Nasser, that stretches back from the Aswan High Dam, which was completed in 1970 (see Figure 6.16). The Aswan controls flooding along the lower Nile and produces hydroelectric power for Egypt's cities and industries. North of (below) the Aswan, the river environment is now greatly modified. No longer is the alluvial plain replenished every year by floodwaters carrying a fresh load of sediment from upstream. Irrigation canals keep some fields in nearly continuous production, and fertilizers are used to maintain productivity.

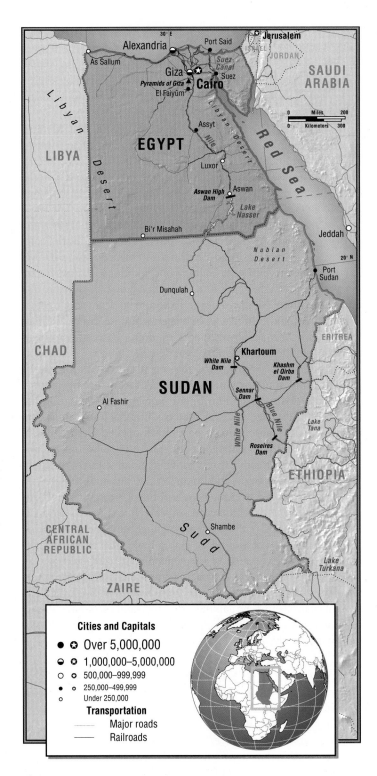

Figure 6.16 **The Egypt and Sudan subregion.**

Sudan

Sudan, slightly more than one-quarter the size of the United States, is the largest country in this region and, in fact, in all of Africa (see Figure 6.16). Yet, with 30 million people, it has less than half the population of Egypt. The country has three distinct environmental

The principal crops, now almost all irrigated, are cotton, grains, vegetables, and sugarcane, as well as animal feed. Egypt's population is increasing very quickly. Although it is the cities that are growing most rapidly, the number of people still living and working on the land has also increased. Despite the increased labor, the Nile fields are no longer sufficient to feed Egypt's 65 million people, and food must be imported. Meanwhile rural men often migrate to neighboring countries where they work to supplement the family income.

VIGNETTE Mohammed Abdel Wahaab Wad has just spent a week in jail for protesting the loss of the land his family has leased for several generations. In the 1950s, land was redistributed from rich to poor farmers at very low rents so that they would be able to support their families. In the late 1990s, a new law reversed the process—the result of structural adjustment policies (SAPs) imposed by the World Bank and the International Monetary Fund. The purpose of the new law is to increase the land's productivity by cultivating it in large tracts managed with machinery and controlled by the original owners. This reorganization of the agricultural system is expected to boost Egypt's ability to feed itself and grow water-intensive export crops such as cotton and rice. The change in control of the land is expected to affect 900,000 tenant farmers like Mr. Wad, who have continued to use labor-intensive techniques to cultivate crops of vegetables and wheat just sufficient to feed their families.

Mr. Wad, who has 11 children to help him cultivate, has never before even been asked for rent for the 8 acres (3.25 hectares) he occupies on the edge of the Sahara, northwest of the Aswan Dam. Now the new law says that the owner can tell the family to leave. The landowner is Gamal Azzam, a member of the urban aristocracy, who is adamant that he will recover his land and that the Wad family must go. There are growing worries that in the short run armed conflict will break out and in the long run the result will be another rush to the already crowded cities. Migration to the cities would result in a period of malnutrition for many people while the agricultural system is reorganized around mechanized production, which will undoubtedly produce more food per acre than traditional farmers have been able to produce. [Adapted from Douglas Jehl, "Egypt's farmers resist end of freeze on rents," *New York Times* (December 27, 1997): A5.]

Egypt's two main cities, Cairo and Alexandria, are both in the Nile Delta region. Cairo, at the head of the delta, has 14 million people in its environs and is one of the most densely populated cities on earth. Egypt's crowded delta faces a crisis of water availability and the threat of disease and pollution. In 1995, Egypt became one of the countries with less than 1000 cubic meters of water per capita per year (see Figure 6.13). The delta waters that irrigate some of Egypt's best land harbor a freshwater parasite that produces a disease called schistosomiasis (also known as bilharziasis). The parasite infests humans, causing internal bleeding, debilitating exhaustion, and often eventual death. One in 12 Egyptians is infected. For those who must work in the water, prevention is difficult and drug treatment is expensive.

In Chapter 4, we discussed pollution around the Mediterranean. To address its share of this problem, the Egyptian Ministry of the Environment, created in 1997, is seeking international contractors to treat industrial, agricultural, and urban solid and liquid waste that for years has been dumped untreated into the Nile and the Mediterranean.

An example from Cairo illustrates a different sort of pollution linkage between Europe and North Africa. A major industrial complex in Cairo recycles used car batteries collected from all over Europe. The complex releases lead concentrations 30 times higher than world health standards into the air that blows across Cairo and into the eastern Mediterranean. Exposure to lead affects brain development and, ultimately, intelligence. Some children's playgrounds in Cairo are so polluted they would be considered hazardous waste sites in the United States.

In the 1990s, Egypt's economy, long plagued with stagnation, inflation, and unemployment, appeared to be turning around. After years of structural adjustment policies (see the earlier discussion of SAPs, page 312), inflation was brought under some control, and many multinational companies opened subsidiaries in Egypt. Among these companies are Microsoft, Owens-Corning, McDonald's, American Express banking and investment divisions, Löwenbräu of Germany, and three German automakers.

For the ordinary working people of Egypt, however, this flashy development does not translate quickly into prosperity. Local export industries have difficulty competing internationally; unemployment remains at around 20 percent and may rise as population grows. Partly because of the general social and political malaise, but also because of personal values, many of Egypt's young people are attracted to Islamism. Overall, conservative interpretations of Islam seem to be gaining popularity; and as we have already discussed, many young women are choosing to wear the veil as a sign of their piety.

THE ARABIAN PENINSULA

The desert peninsula of Arabia has few natural attributes to encourage human settlement, and today large areas remain virtually uninhabited (Figure 6.17). The largest country, Saudi Arabia, has a population density of only 24 people per square mile (15 per square kilometer). The land is persistently dry and barren of vegetation over large areas. Streams flow only after sporadic rainstorms that may not come again for years. The peninsula's one significant resource, oil, amounts to more than 65 percent of the world's proven reserves. Saudi Arabia has by far the largest quantity, with approximately 25 percent of the world's reserves.

Traditionally, control of the land was divided among several tribal groups led by patriarchal leaders called **sheiks.** The sheiks were based in desert oasis towns and in the uplands and mountains

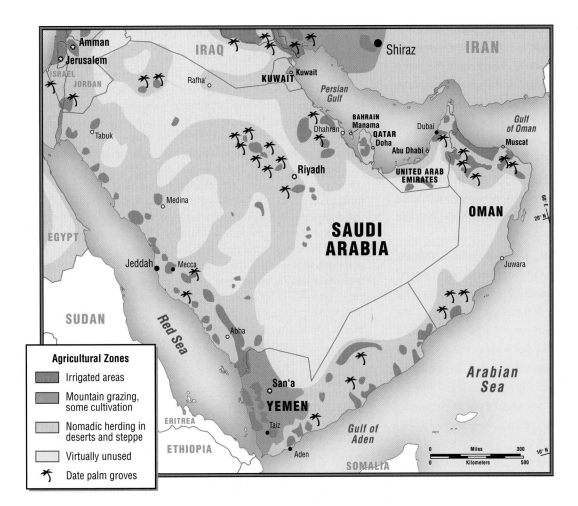

Figure 6.17 The Arabian Peninsula subregion.

Agricultural Zones
- Irrigated areas
- Mountain grazing, some cultivation
- Nomadic herding in deserts and steppe
- Virtually unused
- Date palm groves

bordering the Red Sea. The tribespeople they ruled were either nomadic herders, who traveled over wide areas in search of pasture and water for their flocks, or poor farmers settled where rainfall or groundwater supported crops. Until the twentieth century, the sheiks and those they governed earned additional income by trading with camel caravans that crossed the desert (once the main mode of transport) and by providing lodging and sustenance to those on holy pilgrimages to Mecca.

In the twentieth century, the sheiks of the Saud family, in cooperation with conservative religious leaders, consolidated the tribal groups to form an absolutist monarchy called Saudi Arabia. It now occupies the large midsection of the peninsula. Six much smaller Arab states are ruled by tribes that resisted the Saudi expansion; these occupy the peninsula's eastern and southern margins. The Saud family consolidated its power just as reserves of oil and gas were becoming exportable resources for Arabia. The resulting wealth has added greatly to their power and prestige.

Despite the overall conservatism of Arab society, oil money has changed landscapes, populations, material culture, and social relationships. Where once there were mud-brick towns and camel herds, there now are large, modern cities served by airports and taxis. The population of the peninsula burgeoned to 44 million people, half of whom live in Saudi Arabia. Pickup trucks not only have replaced camels but are now used to transport them. Irrigated agriculture has made the peninsula nearly self-sufficient in food; education is now promoted for young girls as well as boys; and increasing numbers of women work outside their homes, usually in work spaces secluded from men.

The other, smaller nations on the Arabian Peninsula have varying profiles. Kuwait and the United Arab Emirates, each with close to 10 percent of the world's oil reserves, are rapidly modernizing and are very affluent; their societies remain conservative. The modest oil and gas reserves of Oman and Qatar are sufficient to generate relatively high standards of living. Bahrain is virtually a city-state. It generates income not from its tiny oil reserves but from services related to oil production and transport and from providing entertainment and manufactured goods for neighboring wealthy Saudis. Yemen occupies about one-quarter as much space as Saudi Arabia but has a population nearly as large (16 million); it has only small and as yet undeveloped oil reserves. Yemen's standard of living remains by far the lowest on the peninsula and in the entire region (see Table 6.2).

VIGNETTE Raufa Hassan is 38 years old and holds a Ph.D. in social communications from the University of Paris. As Yemen's most outspoken feminist, her goal is to help women learn how to vote independently; Yemen's Islamist party, Islah, supports her efforts. Despite having the right to vote, Yemeni women do not participate in political processes. Nearly 70 percent of the population is still rural, and in rural areas only 1 in 10 women is literate. Typically, girls stay home and work rather than go to school. Women perform many essential tasks in the rural economy, such as herding cattle, grinding wheat, and carrying water. After marriage, the average woman bears seven children. Dr. Hassan has found that husbands generally keep their wives' and daughters' voter registration certificates because both men and women believe the women would be likely to lose them. As a result, men often control whether or not, and how, a woman votes.

Sheik Ahmed Abdulrahman Jahaf, whose daughter is running for parliament, worries that what he calls the "backward villagers" will think ill of her. He believes that an educated woman such as his daughter is more attractive as a bride. According to Sheik Jahaf, "The nature of women leads us to the conclusion that a woman's right place is home." In these waters, Dr. Hassan wades carefully, always getting the sheiks' permission before she talks to the village women; but lately the sheiks, too, seem to be changing their views. They see that they must begin to respond to the demands of their people, both men and women, if they wish to stay in power. They even recognize that if they support the women's right to vote and encourage them to do so, the sheiks' own sons may profit in future elections when female voters are more numerous. [Adapted from Daniel Pearl, "Yemen steers a path toward democracy with some surprises," *Wall Street Journal* (March 28, 1997): A1, A11.]

THE EASTERN MEDITERRANEAN

The part of Southwest Asia and North Africa now occupied by Jordan, Lebanon, Syria, and Israel (Figure 6.18) has been torn by conflict for much of the twentieth century. Yet if tensions were resolved, this area has the potential for economic leadership in the region. One state, Israel, is already technologically advanced and broadly prosperous. Another, Lebanon, was similarly developed and prosperous until disrupted by a civil war between Christians and Muslims in the 1970s. This war ended in 1990, and since then the country has been on an active path of rebuilding its economy and infrastructure. The countries of this subregion are strategically located adjacent to the potential markets of Central Asia and the Gulf states and the rich markets of Europe. Their climate makes them suitable for tourism and semitropical agriculture. Jordan already exports vegetables, citrus fruits, bananas, and olive products; if irrigation water can be found (not an easy matter; see the discussion on pages 315–317), Syria could do likewise. Syria lost an important trading partner after the breakup of the Soviet Union, but since the breakup, private investors, especially from the Gulf states, have been helping Syria expand its industrial base to include pharmaceuticals, food processing, and textiles, in addition to gas and oil production.

Jordan, Lebanon, Syria, and the Palestinian people (the Palestinians have no actual political state but live within these three countries and Israel) have all been preoccupied in recent decades with political and armed conflict centered on the establishment of the state of Israel, but encompassing wider issues of access to land and resources and religious rivalry. The Arab-Israeli conflict spilled over Lebanon's borders in the 1970s and exacerbated discord between Christian and Muslim Lebanese. Syria has conflicts with most of its neighbors, notably with Iraq and Turkey over water rights and with Israel over possession of the Golan

Sheik Rashid Terminal in Dubai's international airport is a futuristic, high-tech facility with clear glass exterior walls—an expensive building to air-condition in the hot, dry climate. On a stopover in June 2000, the author observed that Filipino young people staffed the dozens of upscale duty-free shops where every imaginable luxury is available—gold jewelry is sold by the ounce, not the piece. Other personnel were from India and Bangladesh. The tall palm trees and other plants that decorate the structure are artificial, thus requiring no water, a particularly scarce commodity in this Arabian Peninsula country.
[H. Rogers /TRIP.]

that have plagued the region since the end of World War II. More important factors were disputes deriving from competing claims for territory in the aftermath of World War I, when this part of the Ottoman Empire was dismantled.

After World War I, the League of Nations (a precursor to the United Nations) allotted the territories at the eastern end of the Mediterranean, once held by the Ottomans, to France and Britain for supervision. In 1946, Syria and Lebanon became independent republics and Jordan became an independent kingdom. A different fate awaited Palestine. Much of the land occupied by Palestinian Arabs would be used to create a new state, Israel, as a Jewish homeland.

The Jews of today base their claims to Palestine on the biblical kingdoms and on their expulsion from the region beginning in A.D. 73 until the fall of the Roman Empire in A.D. 476. Many Jews dispersed in the Diaspora (see pages 294–295) encountered persistent discrimination through the ages, especially in eastern Europe, where occasionally whole villages would be murdered in **pogroms** (episodes of ethnic cleansing). This deadly mistreatment culminated in the Nazi era before and during World War II, when more than six million Jews across Europe were rounded up, imprisoned, and made to work as slaves or killed in gas chambers. When the Jews left the Holy Land, more than 1500 years ago, they left behind several neighboring indigenous groups, many of whom converted to Islam shortly after the Prophet Muhammad founded the religion around A.D. 622. Those who now call themselves Palestinians came from these peoples and from Arab groups who joined them over the centuries.

In 1947, after World War II, the Western powers were searching for a place to settle the tens of thousands of Jewish refugees who were still in European displaced-persons camps after having narrowly escaped death in the Nazi gas chambers. Although some Jews were taken in by England, France, the United States, and countries in South America, no country stepped forward to offer a home to all Jews. Meanwhile, the idea of European Jews migrating to the ancestral homeland once known as Israel had been gaining popularity among Jews since it was proposed in nineteenth-century Austria. A small group of Jews who called themselves **Zionists** had begun to purchase land from wealthy Palestinian landholders and establish communal settlements called *kibbutzim*. Most of the people displaced by the Zionist land purchases were poor farmers and herders, who were accustomed to using bits of land held by village landlords for their fields, pastures, and houses. In 1917, the British government issued the Balfour Declaration, which committed it to support the establishment of "a national home for the Jewish people" in Palestine; and land purchases continued (see Figure 6.19A, page 329).

By 1946, strong sentiment had built among the world's Jews that a Jewish homeland should be created in Palestine, and many Jews already in Palestine took up arms to convince the British to stand by the Balfour Declaration. The Palestinian Arabs and their primarily Muslim supporters in the region fiercely objected to the formation of the state of Israel, out of fear that Palestinians would lose claim to the land they had long occupied and would be denied a voice in their own governance. Notice that the struggle between Zionists and Palestinian Arabs was not over religion, but over land, and this remains the case, with water now also an issue.

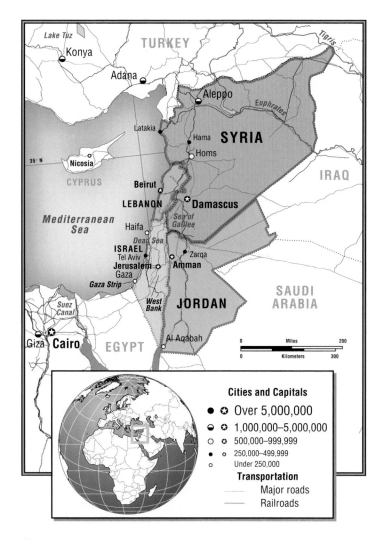

Figure 6.18 The Eastern Mediterranean subregion.

Heights. The core of the conflict, however, centers around Israel and the Palestinians.

Israel has the best-educated, most prosperous, and best-cared-for population in the region; a pool of unusually well-educated immigrants; and investment by the worldwide Jewish community. A range of sophisticated industries are located along the coast, between the large port cities of Tel Aviv and Haifa. Israeli engineering is world renowned, and Israeli innovations in cultivating arid land have spread to the Americas, Africa, and Asia. Some people argue that Israel could be a model of development for neighboring countries because it has managed to develop despite rather meager resources, but this possibility has been hindered by Israel's continual conflict with its neighbors.

The Emergence of Israel

The eastern Mediterranean is an area of ancient settlement that contains major holy places for Jews, Christians, and Muslims, but this fact is only marginally responsible for the wars and violence

AT THE LOCAL SCALE *Learning Palestinian History*

How do Palestinian children in the occupied territories learn about their own history? Until the 1993 peace agreements in Oslo, Israel had controlled the teaching of Palestinian history, geography, and culture in the occupied territories and had banned any content that criticized Israel. Since then, the Palestinian Authority, created to administer self-rule for the Palestinians in those territories, has been allowed to produce its own textbooks for primary and secondary school children. Palestinian textbook writers walk a thin line between the political reality of Israel and the dream of a Palestinian state of their own. The books contain poems and biographies of Palestinian writers and other artists and depict the Palestinian flag and anthem, once banned by the Israelis.

Historical accounts reveal some aspects of the Palestinian resistance movement against Israel but nothing about guerrilla leaders or violent activities. Yasser Arafat, who for decades led a guerrilla campaign in support of Palestinian self-determination, is depicted mildly as the current head of the Palestinian Authority. No mention is made of ancient Jewish ties to the land of Israel, nor are there accounts of the biblical kingdoms. The history jumps directly from pre-Jewish Canaan to the Muslim conquests in the seventh century A.D.

Source: Joel Greenberg, "Palestinian maps for a state that doesn't exist," *New York Times* (June 27, 1997).

The Palestinians were not unified, and they lacked connections in, and knowledge of, the West. By the late 1940s, British, European, and U.S. sentiment in the aftermath of World War II sided with the Jews, and in 1947 the United Nations created a Jewish state from part of Palestinian land (see Figure 6.19B). In May 1948, the British pulled out abruptly just as the state of Israel was proclaimed.

The warfare and violence that have since plagued the region began in 1948 on the very day that the last British soldier left. Forces from neighboring Arab countries moved into the mainly Arab parts of the country. In the ensuing conflict between Israeli and Arab forces, Palestinian land shrank yet further. By 1949, Israel had expanded and the remnants of Palestinian lands were incorporated into Jordan and Egypt (see Figure 6.19C). In the repeated conflicts that followed—such as the Six Day War (1967) and the Yom Kippur War (1973)—Israel not only defeated alliances of its much larger neighbors, including Egypt, Syria, and Jordan, but also expanded Israeli military control into the territories of these neighbors. The occupied lands comprised the West Bank (a part of the former Palestine that had become part of Jordan), the Gaza Strip and the Sinai Peninsula (which had become part of Egypt), and the Golan Heights (which had become part of Syria) (see Figure 6.19D).

As a result of the frequent warfare, hundreds of thousands of Palestinians fled the war zones or were removed to refugee camps in nearby countries. The Palestinians who stayed inside Israel in 1948 and became Israeli citizens have not been treated as equal to Jewish Israelis by the state (see the box "Palestinians in Israel"). In addition to those who remained in Israel in 1948, many thousands of Palestinians from the West Bank and Gaza enter Israel daily to work and then return to villages and refugee camps at night.

The Continuing Arab-Israeli Conflict

When Israel occupied Arab territories in 1967, the United Nations Security Council passed resolution 242. It required Israel to return these lands in exchange for peaceful relations between it and neighboring Arab states. This resolution, later dubbed the land-for-peace formula, formed the basis of peace talks between Egypt and Israel at Camp David (1979), which returned the Sinai Peninsula to Egypt. Land-for-peace was also the basis for the Madrid peace talks (1991) and the Oslo Accords (1993) between the Palestinians and Israel.

At the 1993 Oslo Accords, the Palestinian Liberation Organization (PLO), an important wing of the Palestinian Authority, acknowledged the right of Israel to exist in return for gaining control over the Gaza Strip and some portions of the West Bank. An important part of the 1993 peace accord was Israel's commitment to stop settling Jews in the occupied territories. These territories, consisting of the Gaza Strip and parts of the West Bank of the Jordan River, were Palestinian areas taken by Israel in the 1967 war. After 1967, Israel established hundreds of Jewish settlements, mainly on the West Bank, as part of its effort to absorb new Jewish immigrants and also to create a de facto situation that would secure Israel's continued control of the territory. They have continued to do so into 2001 by building housing units on the West Bank. The maps in Figure 6.19 show the progression of changes in territory since the 1920s. While Israel was consolidating its hold on disputed territories through settlement, the Palestinian people, many of whom had by then lived in refugee camps for 40 years or more, mounted a prolonged uprising against Israel, known as the **intifada.**

The 1993 Oslo Accords established the principles by which peace talks would be conducted. The crowning achievement of the peace talks, which have been ongoing since 1991, would be a final peace agreement between the Palestinians and Israel. Its major contours are more or less defined in the preliminary treaties agreed to by the Israelis and the Palestinians. The negotiators, however, decided early on to leave the thorniest issues until the "final status" negotiations. These remaining difficult issues are summarized here.

1. **Territory.** Palestinians want Israel to relinquish all of the West Bank, Gaza Strip, and East Jerusalem to the Palestinians. This would be comparable to Israel's return to Egypt and Jordan of

Thinking Geographically: ON THE WEB

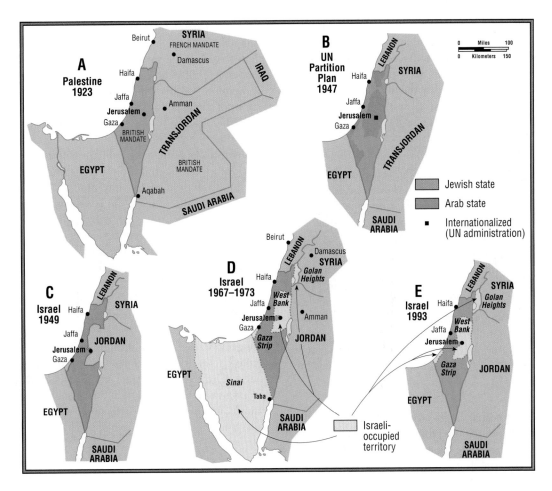

Figure 6.19 The changing map of Israel and Palestine. (A) In the 1920s, the British controlled what is now Israel and Jordan. (B) Following World War II, the United Nations developed plans for creating Jewish and Palestinian (Arab) states. (C) But the Jewish settlers did not agree; they fought and won a war, creating the country of Israel. Many of Israel's Arab neighbors were opposed to an Israeli state. (D) In 1967, Israel soundly defeated combined Arab forces and took control over the Sinai, the Gaza Strip, the Golan Heights, and the West Bank. (E) In subsequent peace accords, the Sinai was returned to Egypt, but Israel has maintained control over the Golan Heights and the West Bank, which the Israelis claim are essential to their own security. The Palestinians now have some autonomy in the Gaza Strip and the West Bank and would like more territory as living space. [Adapted from Colbert C. Held, *Middle East Patterns— Places, Peoples, and Politics* (Boulder, Colo.: Westview Press, 1994), p. 184.]

territories occupied during the 1973 war. Palestinians have accepted Israel's right to exist on portions of "historic Palestine" within the internationally accepted pre-1967 borders, and they agree to establish a Palestinian state only on the West Bank and Gaza Strip.

Israeli Prime Minister Ehud Barak (his term ended in 2001) was prepared to return more than 90 percent of the total territory of the West Bank and Gaza Strip. The next prime minister, Ariel Sharon, offered to return 40 percent of the territories. Returning to the pre-1967 borders is now unacceptable to Israel.

2. Jewish Settlements. Palestinians believe that all Jewish settlements in the West Bank or Gaza Strip are illegal because they violate UN resolutions and international law. Israel should not retain any settlements or allow Israelis to continue to live there unless the Jewish residents are willing to apply for and accept Palestinian citizenship.

Israelis say that the majority of some 200,000 Israelis who live in exclusive settlements in the West Bank and Gaza Strip must be incorporated into Israel. Israel is willing to discuss territorial exchanges with the Palestinians as a way of compensating them for this border adjustment.

3. Security Issues. Palestinians officially favor peace and disarmament and wish to rebuild their communities, but they have ready access to arms from neighboring Arab states, and extremists carry out frequent terrorist attacks.

Israelis receive military and political support from the United States and have an impressive armed force that is regularly used aggressively against Palestinians, often in response to terrorist attacks; but Israelis say they fear that when and if peace agreements are reached Israel will be vulnerable to attack from combined Arab forces.

4. Jerusalem. Palestinians say that Israel must return all of East Jerusalem, as well as the Islamic and Christian holy sites in the Old City. They also view all Israeli settlements in East Jerusalem as illegal. When they declare the creation of their independent state, its capital will be in a Palestinian neighborhood of East Jerusalem.

Israelis believe that a united Jerusalem must remain under Israeli control, but a high degree of autonomy may be extended to Palestinian neighborhoods in East Jerusalem.

5. Palestinian Refugees. Palestinians say that Israel must accept responsibility for Palestinian refugees who fled or were expelled from their homes as a result of the creation of Israel. The United Nations reported in 1995 that there were then 3.1 million Palestinian refugees in camps in the West Bank, the Gaza Strip, Jordan, Syria, and Lebanon. In accordance with UN resolution 194, Israel should grant them the right to return and compensate those who cannot.

Israelis state that the refugees have no "right of return" and should either be permanently settled in the countries where they reside or be resettled in another country. Refugees are due compensation, but most of it should be provided by Arab states, which Israel holds responsible for the refugee problem because most resulted from the 1967 and 1973 wars, for which Israel holds the Arab states responsible.

6. Water. Palestinians say that Israel must relinquish all of the West Bank and Gaza Strip, including the Jordan Valley. They are willing to share the aquifers of the West Bank but claim the lion's share of the resource because its recharge area—the area with the highest precipitation, which recharges the aquifers—is located on the hills of this Palestinian territory.

Israel believes that it must have access to the Jordan Valley and be allowed to maintain an ongoing security presence there. No foreign army will be allowed west of the Jordan River, and the Palestinian state must remain demilitarized. Israel says it should retain control over water resources in the West Bank. Desalinization projects should be undertaken to provide sufficient water for Israel, Jordan, and the Palestinians, instead of merely redistributing existing resources.

Although some of these positions seem irreconcilable, the two parties have in fact made significant progress on most of them. For example, Palestinians and Israelis are working to resolve water allotment issues. At present, Israel pumps the most water from the West Bank and the Jordan River, which it uses primarily in its agricultural sector. Israeli agriculture is technically advanced, but it is not a big employer (2.6 percent of the labor force) nor does it contribute much to Israel's GDP (2 percent). By contrast, the Palestinian economy remains agriculturally based (33 percent of GDP, 13 percent of labor force) and will be damaged if the Palestinians are not allocated sufficient amounts of water to sustain their production, maintain existing jobs, and create new ones.

Acceptable compromises on the issues of the right of Palestinian refugees to return to their homelands and on the uses of Jerusalem have been beyond reach so far. Allowing millions of Muslim and Christian Palestinian refugees back into Israel would threaten the very identity of the Jewish state. At the same time, Palestinians continue to wish to return, a notion that is found in Palestinians' prayers, literature, poetry, and song. Nonetheless, some observers think a compromise might include some form of symbolic, and very limited, return of a few thousand Palestinians.

The issue of Jerusalem is steeped in religious emotions on both sides. Jews, Christians, and Muslims all view the city as holy to them. In fact, the geographic proximity of the holy sites for the world's three monotheistic religions immensely complicates the territorial division of the city among its contestants. Although both the Palestinians and the Israelis want certain areas to come under their direct sovereign control, neither side wants to divide the city physically by erecting fences and border controls.

In the summer of 2000, President Bill Clinton brought the Israeli prime minister, Ehud Barak, and the leader of the Palestinian Authority, Yasser Arafat, to Camp David for intensive, marathon talks to reach a final peace agreement between the antagonists. His efforts failed primarily because the parties could not agree on how to share Jerusalem and on how resolve the Palestinian refugee question. By the summer of 2001, a new intifada, begun to protest new settlements on the West Bank, had degenerated into an outright war between Israel and the Palestinians, making peace seem a distant dream. The intifada, including numerous terrorist attacks, and Israeli military responses shattered most of the goodwill and trust that had been developing between the Palestinians and the Israelis, making both sides less inclined to compromise.

The current leaders of Israel and the Palestinian Authority are constrained by significant political baggage from violent struggles that date back five decades and more. When the ongoing intifada dies out, the precise framework of the Oslo Accords is likely to be superseded by a new framework for peace that takes into account the new realities of violence in the occupied territories.

Religious Fundamentalism in Israel

It is primarily conservative religious forces in Israel that have pushed to continue settlements in the West Bank and to take a hard-nosed approach to Palestinian demands. More secular groups favor rapprochement with the Palestinians. In a speech on May 14, 1997, marking Israel's independence day, Ammon Shahak, then chief of staff of the Israeli army, noted that the country is experiencing a division within its society between the religious and the secular that is reminiscent of the secular-Islamist division being experienced by many predominantly Muslim countries in the region. Shahak observed that the crucial struggles in the region are not between Arabs and Jews and not between countries, but within countries (Algeria, Turkey, Morocco, Egypt, Iran, Israel) along these religious lines. Since 1993, the issue is no longer whether Israel will exist, but rather how Israel and the Arab states can promote normal relations so that they can address common environmental, social, and economic issues nonviolently. Now the greatest roadblocks to real progress are the religiously based factions within the protagonist countries, not hostilities between the countries themselves.

THE NORTHEAST: TURKEY, IRAN, AND IRAQ

Turkey, Iraq, and Iran (Figure 6.20) are culturally and historically distinct. For example, a different language is spoken in each country: Farsi in Iran, Turkish in Turkey, and Arabic in Iraq. Yet they share some similarities. At various times in the past, each country was the seat of a great empire. Each was deeply affected by the spread of Islam, which is now the religion of more than 97 percent of the region's people. At present, each occupies the attention of Europe and the United States because of location, resources, or potential threats. All three countries share some common concerns, such as how to allocate scarce water and how to treat the large Kurdish minority. Moreover, each country has experienced a radical transformation at the hands of idealistic governments, although the paths pursued have varied dramatically.

Turkey

Turkey is more closely affiliated with Europe and the West than any other country in the region, with the possible exceptions of Israel and Lebanon. A strong faction in Turkey seeks to join the European Union, and many Turks have spent years in Europe as guest workers. Turkey was once the core of the Ottoman Empire, which, as discussed earlier, was dismantled after World War I. At the end of a civil war, in 1923, Turkey undertook a path of radical Europeanization, led by a military officer, Mustafa Kemal Atatürk. Kemal Atatürk is revered as the father of modern Turkey. He and his followers declared Turkey a secular state, modernized the bureaucracy, encouraged women to discard the veil and the custom of seclusion, promoted state-sponsored industrialization, and actively sought to establish connections with Europe. Dramatic transformations took place throughout the twentieth century, but backlashes and simple inertia have been obstacles to change. Although the state has sponsored industrialization in western Turkey, much of eastern Turkey remains agricultural and relatively poor. Islam, long deemphasized as a matter of state policy, is still an overriding influence on daily life, and fundamentalist versions of the religion are experiencing a resurgence.

Turkey straddles the Bosporus, a narrow passage from the Black Sea to the Mediterranean that is described (with exaggeration) as the separation between Europe and Asia. During the cold war, Turkey's location made it a strategic member of the North Atlantic Treaty Organization (NATO), an alliance of European countries and the United States whose mission was to contain Russia. In the post-Soviet era, Turkey's location gives it potential advantages if it can establish economic links with Europe,

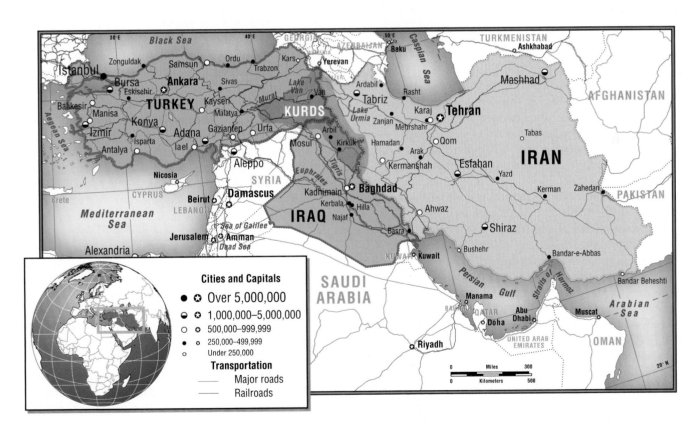

Figure 6.20 The Northeast subregion and the Kurds. The Kurdish people retain their own language and culture. Today, there may be 25 million Kurds, half of whom live in southeastern Turkey. Roughly 7 million live in Iran, more than 3.5 million in Iraq, close to a million in Syria, 300,000 in Armenia, 330,000 in Germany, and 60,000 in Lebanon; others are scattered elsewhere. [Adapted from Edgar O'Ballance, *The Kurdish Struggle 1920–1994* (New York: St. Martin's Press, 1996), p. 235.]

CULTURAL INSIGHT   *The Turkish Narghile*

One of life's simple pleasures for many retired Turks is relaxing with friends in one of Istanbul's salons dedicated to the narghile, a Turkish water pipe. The pipe base is a bell-shaped, water-filled glass bottle that sits on the floor next to a patron. An intricately engineered brass neck screws onto the bottle and connects with a pipe bowl and a long (5- to 6-foot; 2.5- to 3-meter), flexible smoking tube and mouthpiece. The aficionado puts a plug of strong Turkish tobacco in the bowl. The server then places one or two red-hot coals on the tobacco, and the smoker puffs away; the smoke reaches the mouth only after bubbling through the water, where it is cooled and some particles are removed. It takes one to two hours to smoke a pipeful; meanwhile, the male patrons (women may partake occasionally) sip tea or coffee (no alcohol), play dominoes or backgammon, and enjoy the companionship of old friends.

Men meet at a social club in Istanbul, Turkey, to enjoy conversation and the narghile water pipe. [Robert Frerck/Woodfin Camp & Associates.]

the Caucasus, and Central Asia. Istanbul, already a booming city of more than 10 million, is now the regional headquarters for hundreds of international companies. There are more than 250 U.S.-based companies alone, virtually all of them joint ventures of one kind or another: Coca-Cola, IBM, Eastman Kodak, Kraft Foods, Levi Strauss, and Citibank, to name a few. Turkey offers a large market of potential consumers: 66 percent of Turkey's 65 million people are city dwellers. Many of them once lived in rural areas of Turkey, then traveled to Europe to work in factories and on farms; they have returned eager to pursue a standard of living similar to what they observed in Europe. Also, remittances from Turkish guest workers in Europe add substantially to the country's GDP.

Turkey's future depends on its ability to manage relatively abundant water resources. With its mountainous topography, Turkey receives the most rainfall of any country in the region, and the headwaters of the economically and politically significant Tigris and Euphrates rivers are in the mountains of southeastern Turkey. As new irrigation and power generation systems on the Euphrates come on line (see the discussion of water issues on pages 316–317), production of cotton, grains, fruits and vegetables, soybeans, and seed oils will increase dramatically. Agriculture employs half of Turkey's workforce and accounts for 20 percent of its exports (mostly cotton and luxury food items).

The cheap hydropower will also expand industry, which already accounts for 80 percent of exports. Until now, a lack of electrical power has hampered industrialization in Turkey, yet the country's diversified economy still manages to produce a wide range of goods, including steel, textiles, leather goods, cement, automobiles, and tires, all of which are exported to Europe and Central Asia. Hydropower could spur rapid growth in Turkey, which has a moderately strong industrial base complemented by a

well-trained workforce. Turkey's plans to appropriate a lion's share of the flow of the Euphrates River, however, are a source of conflict with its neighbors to the southeast.

Perhaps the greatest obstacle to Turkey's stability is its ongoing civil war with the Kurdish minority. The Kurds are tribal peoples (some are nomadic) who have lived in the mountain borderlands of Iran, Iraq, and Turkey (see Figure 6.20) for at least 3000 years. The division of the Ottoman Empire after World War I by France and Britain dispersed Kurdish lands among Turkey, Iran, Iraq, and Syria, leaving the Kurds without a state of their own. All four countries have had hostile relations with their Kurdish minorities. In Turkey, the conflict has escalated to the point of continuous armed strife and repression of noncombatant Kurds; Turkey's harsh treatment of the Kurds are among the issues that may block its entrance into the European Union.

Iran

Iran occupies a transitional geographic position in this region: Iran's western parts are occupied by people of Arab, Turkish, and Caucasian heritage, its eastern parts by people with roots in South Asia, especially Afghanistan and Pakistan. Because its northern flank borders the Caspian Sea, Iran shares in the debate over the future of this inland sea and its resources. Iran is close to Central Asia, and its border regions have common language and ethnic features with Turkmenistan. During the cold war, Iran was a zone of contention between the United States and the Soviet Union.

Like Turkey, Iran has abundant natural resources, a strategic location, and a large population (66 million) that could make it a regional economic power. Iran's large petroleum reserves give it influence in the global debate on energy use and pricing. Yet social and economic turmoil have long held the country back.

Iran's present situation in the region and in global politics results from its geographic location, its recent history, and its status as a theocratic state based largely on Islamism. The theocracy was formed under the leadership of the Islamic fundamentalist Ayatollah Ruhollah Khomeini, a Shi'ite spiritual leader who led a revolution against the previous ruler, Shah Reza Pahlavi, in 1979. Pahlavi's father had seized control of the country in the 1920s and introduced secular and economic reforms patterned after those instituted by Kemal Atatürk in Turkey; he abdicated in favor of his son in 1941. After a reformist coup attempt in 1953, Shah Reza Pahlavi retained power through U.S. support and continued his father's efforts at Europeanization. But his emphasis on military might and on royal grandeur paid for with oil money overshadowed any genuine efforts at agricultural reform and industrialization. The disparity of wealth and well-being in Iran grew ever wider. Finally, with the Khomeini-led Iranian revolution of 1979, political conditions in Iran changed rapidly and radically. Resistance to the new theocratic state was crushed by massive imprisonments and executions. Among those most affected were women. Many women had lived emancipated lives under the shah's reforms, studying abroad and returning to serve in important government posts. Now all women past puberty had to wear long black chadors, and they could not travel in public alone, drive, or work at most jobs. Yet even some highly educated women supported the return to seclusion, seeing it as a way to counter the unwelcome effects of Western influence.

The turmoil of revolution was characterized by resentment against the West, which led to a decade of isolation. During that period (the 1980s), Iran engaged in a devastating war with Iraq (see the discussion on page 334). As of 2001, Iran has begun to change both internally and in its relationships with the outside world. Punitive economic sanctions imposed by the United States and Europe in response to Iran's suspected funding of terrorism in the West may soon be lifted. Although many essential products must be imported, the government has begun to work toward economic diversification and industrialization. The country's out-of-date transport system is being updated (see the box "The Bandar Express") so that Iran can begin again to participate in trade with its Arab and Central Asian neighbors. Agriculture remains Iran's weakest economic sector: even though it accounts for 25 percent of the GDP and employs one-third of the workforce, Iran must import a large part of its food.

Elections in May 1997 brought the landslide victory of a moderate leader, Mohammad Khatami. His election success in 1997 and again in 2001 is credited to women and young voters who were attracted by his subtle hints that he favored liberalizing reforms; because of strong conservative resistance, however, they have been slow to materialize.

Partly because of Iran's location at the borders of several regions undergoing social and economic transitions (the former Soviet Union, Central Asia, Southwest Asia, and South Asia), it has the unhappy distinction of being the country with the largest number of international refugees—more than 1.8 million. More than half a million are Iraqis who have fled the regime of Saddam Hussein; some are Kurds who have fled discriminatory treatment in Iraq and Turkey; and 1.3 million are Afghans who left during the Soviet invasion of Afghanistan and its aftermath. For the most part, Iran has at least temporarily absorbed all these people, and very few live in refugee camps. There have been benefits to the Iranian economy: Afghans often do the undesirable jobs of curing leather, cleaning wool, and digging ditches; many refugees work long hours at lower pay than Iranians receive. Still, as of 2001, Iran was seeking to repatriate as many refugees as possible, hoping to lower the 25 percent unemployment rate in Iran.

Iraq

Iraq, more than any other country in the region, is in a state of crisis. Once an oil-rich socialist state, the country has suffered from the devastation of two wars: the Iran-Iraq War (1980–1988) and the Gulf War (1990–1991). Ten years of severe economic sanctions imposed by Europe and the United States have crippled the country. By 2000, European countries were lifting sanctions and Iraq was tacitly allowed to sell some oil to buy necessities, but U.S. sanctions remained in place. Iraq is now significantly poorer than it was in 1990, with a lower level of human well-being and technological development.

Most of Iraq's people (27 million in all) live in the area of productive farmland in the country's eastern half, on the alluvial plain of the Tigris and Euphrates rivers. Sunni Muslims (about 7 million) reside in the northern two-thirds of the plain, around the capital of Baghdad. About 4 million Kurds, who are also Sunnis, live in the northern mountains in the border regions with Turkey and Iran. The southern third of the country is occupied by at least 11 million Shi'ite Muslims, concentrated around Basra at the head of the Persian Gulf. Iraq's main oil fields are also in the south.

Iraq, home to one of the earliest farming societies on earth and to the Babylonian Empire of biblical times, was carved out of the dying Ottoman Empire after World War I by Britain and the Arab sheiks. Iraq remained a British mandate until 1932, when it became an independent monarchy. Although close to half of the Iraqi people are Shi'ite Muslims, since 1932 the government and most of the wealth have been controlled by Sunni Muslims. Like many governments in the region, the Iraqi monarchy maintained strong alliances with Britain and the United States and gradually lost touch with its people. A tiny minority monopolized the wealth generated by increasing oil production but invested little in development.

In 1958, a group of young officers, including Saddam Hussein, overthrew the government and created a secular socialist republic that prospered over the next 20 years despite occasional political disruptions. Hussein founded Iraq's secret police, assumed leadership of the Ba'ath Party, and eventually became president of Iraq. Large proven oil reserves (second in size only to those of Saudi Arabia) became the basis of a very profitable state-owned oil industry, and the profits from oil financed a growing industrial base. Agriculture on the ancient farmlands of the Tigris and Euphrates alluvial plain also prospered. The proceeds from oil and agriculture produced a decent standard of living for most people, who benefited from excellent government-sponsored education and health-care systems.

The Iran-Iraq War (1980–1988). In 1980, the Iraqi president, Saddam Hussein, invaded Iran, apparently thinking that the country was in such chaos that he could acquire territory along the Shatt al Arab estuary and the fertile lowlands between the Zagros Mountains and the Tigris River. Hussein calculated correctly that other countries would not come to Iran's aid. The historian Bernard Lewis explains the many issues. There was an ethnic dimension: Arabs against Persians; a sectarian side: Shi'ite Muslims versus Sunni Muslims; and an economic side: control over oil extraction and sale. It was also a war between the Islamism of Iran and the secularism of Iraq. And it was a political war: a battle over territory and regional domination. Finally, there was a geopolitical dimension, in that the United States and Arab countries, fearing the spread of Islamism from Iran, had financed Saddam Hussein's regime and equipped Iraq with sophisticated weapons. But Hussein, expecting a relatively quick victory, had not realized the extent to which Iranian leaders were willing to use the bodies of their religiously inspired young men against the war machine of Iraq. Iran lost hundreds of thousands of people, many of them young boys barely past puberty. Eventually, with heavy financing from wealthy Arab neighbors and with U.S. military and intelligence support, Hussein forced Iran to the peace table. Hussein himself emerged in a marginally better position, but the infrastructures and oil industries of both countries were left in serious disrepair.

The Gulf War (1990–1991). In 1990, Iraq, using the weapons and knowledge garnered from the West, invaded Kuwait and initiated the Gulf War. Kuwait had been unwilling to restrain its oil production in accordance with OPEC guidelines. Kuwait's behavior was driving down the price of oil on global markets at a time when Iraq was badly in need of oil revenues to fund its recovery from the war with Iran. It is believed that Iraq's need to boost oil prices was a major factor in the invasion; but Iraq was also interested in acquiring more territory along the Persian Gulf. Iraq's decision to invade Kuwait ultimately proved disastrous, because it brought retaliation from a coalition of European and Southwest Asian forces led by the United States in an operation known as Desert Storm. The result was resounding defeat of Iraq and its withdrawal from Kuwait. More than 100,000 Iraqis died in combat, and 300,000 were wounded. Iraq lost 4000 of its 4230 tanks, whereas the coalition lost just 4 tanks. The United States suffered 148 combat deaths, 121 a result of accidents, and 458 wounded.

Despite its victory, the United States is often criticized throughout the region for having rejected several opportunities to solve the conflict peacefully. In the years following the war, Iraq's refusal to give up its lethal chemical weapons, capable of killing on a massive scale, led the UN to impose the most crippling economic sanctions ever leveled against a country. The resulting shortages of food, medicine, and other necessities in Iraq caused massive deaths of civilians, especially children, one in five of whom are now malnourished.

The Aftermath of War. The long war with Iran and the subsequent Gulf War impoverished Iraq. According to European private aid agencies that began to enter the country in 2000, the economy has remained weak as a result of international sanctions imposed by the United States and Europe after the Gulf War. These measures severely curtailed the sale of the country's oil, and the ensuing economic depression has hit the poor, the ill, the elderly, and children the hardest. But even doctors and lawyers, now unemployed, are working as street vendors. Meanwhile, smugglers formed a new economic elite; many of them are desert people with a knowledge of ancient transport routes into Jordan and Syria. Another group of the newly rich are buying up state-owned industries at bargain prices. Money from such sales has enabled the regime of Saddam Hussein to survive.

The situation in Iraq raises important ethical questions. How should the global community handle cases of abuse of power within countries when normal avenues of censure prove ineffective? Is it best to allow conditions to become so bad that the people themselves rise up against a despot? Or does such a strategy lead to unconscionable losses of life and resources? At what point can the world community pass judgment on one country alleged to have developed chemical and biological weapons when those standing in judgment have themselves breached the principles they are now enforcing? (For example, the United States and European nations have researched and developed similar weapons for years.) Given that it is a violation of international law to seek the removal of a head of state surreptitiously or to destabilize a country purposely, is it ethical for the United States and Europe to seek the same goal through economic sanctions, which hurt the poor more than the offending despot?

AT THE LOCAL SCALE   *The Bandar Express*

The Bandar Express is an Iranian train that runs from Tehran to Iran's main port city on the Persian Gulf, Bandar-e-Abbas. As part of its economic recovery, Iran is building an extensive rail network connecting all its major towns and cities. The most recently completed link to the Turkmenistan border now connects, indirectly, the countries of the old Silk Road to the sea at Bandar-e-Abbas. Already, trainloads of Uzbeki and Kazakhi cotton are reaching the port; and autos, refrigerators, and other consumer goods make the return trip from duty-free Dubai, just across the Gulf. A newer and shorter direct route is being planned across the arid central Iranian Plateau.

Source: The Economist (June 21, 1997): 49–50.

REFLECTIONS ON NORTH AFRICA AND SOUTHWEST ASIA

The region of North Africa and Southwest Asia extends across portions of two continents yet demonstrates considerable spatial cohesion. People in this part of the world face common environmental problems defined primarily by water scarcity; settlement patterns reflect the necessity to live close to the few water sources available. The people also share a common religion: Islam. Although interpreted differently from place to place, Islam nonetheless unites by virtue of its Five Pillars of Islamic Practice. In all parts of the region, there is tension between strict adherence to Islam and more secular ways of life. And gender roles are at the heart of social debate everywhere, because these roles raise such basic questions about how individuals will relate to family and society.

Additionally, most of the region has endured some form of foreign domination over the last several hundred years; and it is presently in transition from outside control to regional or local systems of governance and economic development. Although not united since the Ottoman Empire, the 21 countries described in this chapter share long histories that have intersected repeatedly and often violently. The success of OPEC over the last 30 years has led to other efforts to set up associations that work for common goals. Although the nations of North Africa and Southwest Asia are far from achieving economic or political union, the possibilities are occasionally discussed within the region.

Just how long the countries of this region will remain in their present relationships is debatable. To some extent, Iran is now reaffirming ancient connections with Central Asia, Afghanistan, and Pakistan. Turkey may become more closely associated with Europe if secular forces dominate; or it may be aligned with Central Asia and Iran if Islamist forces gain power. Though it seems unlikely just now, North Africa might begin to see its future as more closely linked with European countries rimming the Mediterranean, a scenario envisioned by those in southern Europe who see possibilities in organizing countries bordering the Mediterranean as an economic and environmental unit (see Chapter 4).

The trends to look for in the news about this region are a growing environmental movement, political maneuverings over water rights, efforts to develop participatory democracy interwoven with further efforts to establish Islamist governments, a continuation of the decline in birth rates, and slow but relentless redefinitions of gender roles. As always, the geographic patterns of these trends will be uneven but related to earlier social and physical patterns.

Thinking Critically About Selected Concepts

1. More than 93 percent of the people in the region subscribe to Islam, and it is the dominant religion in all countries but Israel. *Discuss some of the common misperceptions of Islam in Europe, the United States, and other places outside the region.*

2. Some countries have become theocratic states that require all leaders to be Muslim and all citizens to follow Islamic law (in the case of Israel, the leaders are Jewish and citizens must follow Jewish law). Other countries are secular states in which there is theoretically no state religion and no direct influence of religion on affairs of state. Islamism, a fundamentalist movement, appears to be growing in influence. *How does daily life in the theocratic states of this region differ from daily life in the secular states? Is there variation among the states within each category? Do the secular and theocratic states differ in their ability to provide for the well-being of their people?*

3. Throughout the region, men tend to occupy public spaces (the workplace, the mosque, and the marketplace), whereas women occupy domestic spaces (the home). In some parts of the region, women's activities, behavior, and ability to work outside the home are restricted by social custom. *How is modernization of economic life being affected by gender roles and customs? How are the roles and customs changing? How might varying social gender customs across the region affect both men's and women's choices of how they spend leisure time?*

4. Both immigration and emigration are important in this region. The oil-rich countries have imported laborers and specialists from within the region and from Southeast Asia and the Americas, while young men from poorer countries often move to European countries or wealthy regional states as guest workers. *Discuss how these various kinds of migration affect life and family economies within the region.*

5. Rich petroleum deposits (oil and gas) are located along the Persian Gulf and in some places along the northern coast of Africa. Petroleum wealth has been used to fund the luxurious lifestyles of a small elite and to build communications and transportation infrastructure. But the economies of most countries remain undiversified. *What are the major factors that limit economic diversification?*

6. The region has recently endured instances of hostility between states—Iran versus Iraq, Iraq versus the Gulf states and the United States, Israel versus the Palestinians—and within countries—in Egypt, Sudan, Algeria, Lebanon, Turkey, and Iraq. *To what extent are the roots of hostility in various parts of the region related? How do they differ? Why do so many analysts say religion is not the root cause of strife?*

7. The shortage of water is a looming environmental issue. Attempts to obtain water through aquifer pumping and dam construction create their own environmental difficulties, while the use of water in irrigation may lead to soil salinization and loss of fertility. *What is the pattern of water availability across the region? In those places that use the most water per capita, what are the potentials for limiting this use? What factors seem to govern who gets access to water?*

Key Terms

Allah (p. 296) the Arabic word for "God," used by Arabic-speaking Christians and Muslims in their prayers

cartel (p. 310) a group that is able to control production and set prices for its products

Christianity (p. 295) a monotheistic religion based on the teachings of Jesus of Nazareth, a Jew, who described God's relationship to humans as primarily one of love and support, as exemplified by the Ten Commandments

desertification (p. 316) a set of ecological changes that convert non-desert lands into deserts

Diaspora (p. 295) the dispersion of Jews around the globe after they were expelled from the eastern Mediterranean by the Roman Empire beginning in A.D. 73

domestic spaces (p. 336) spaces in the home relegated to women

economic diversification (p. 312) the expansion of an economy to include a wider array of economic activities

Fertile Crescent (p. 294) an arc of lush, fertile land formed by the uplands of the Tigris and Euphrates river systems and the Zagros Mountains, where nomadic peoples began the earliest known agricultural communities in the world

Gulf War (1990–1991) (p. 313) one of the most massive and economically destructive wars that has ever taken place in the region, in which a military coalition of the United States and European and Arab countries quickly defeated Iraq after it invaded oil-rich Kuwait in 1990

hajj (p. 296) the pilgrimage to the city of Mecca that all Muslims are encouraged to undertake at least once in a lifetime

intifada (p. 328) a prolonged Palestinian uprising against Israel

Islam (p. 295) a monotheistic religion considered an outgrowth of Judaism and Christianity; it emerged in the seventh century A.D. when, according to tradition, the archangel Gabriel revealed the tenets of the religion to the Prophet Muhammad

Islamic fundamentalism (p. 298) a grassroots movement to replace secular governments and civil laws with governments and laws guided by Islamic principles

Islamists (p. 299) fundamentalist Muslims who favor a religiously based state that incorporates conservative interpretations of the Qur'an and strictly enforced Islamic principles into the legal system

Israeli-Palestinian conflict (p. 314) a long-running conflict over the control of territory between Israel and Palestinians who live on lands now occupied by Israel

Judaism (p. 294) a monotheistic religion characterized by the belief in one God, Yaweh, a strong ethical code summarized in the Ten Commandments, and an enduring ethnic identity

monotheistic (p. 294) pertaining to the belief that there is only one god

OPEC (Organization of Petroleum Exporting Countries) (p. 310) a cartel of oil-producing countries—including Algeria, Indonesia, Iran, Iraq, Kuwait, Libya, Nigeria, Oman, Qatar, Saudi Arabia, the United Arab Emirates, and Venezuela—that was established to regulate the production, and hence the price, of oil and natural gas

pogroms (p. 327) episodes of persecution, ethnic cleansing, and sometimes massacre, especially conducted against European Jews

polygamy (p. 307) the practice of having more than one wife at a time

protectorate (p. 297) a relationship of partial control assumed by a European nation over a dependent country, while maintaining the trappings of local government

public spaces (p. 336) social spaces such as the town square, shops, and markets, which are defined in contrast to the domestic (home and private) sphere

Qur'an (or **Koran**) (p. 296) the holy book of Islam, believed by Muslims to contain the words Allah revealed to Muhammad through the archangel Gabriel

salinization (p. 311) damage to soil caused by the evaporation of water, which leaves behind salts and other minerals

seclusion (p. 297) the requirement that a woman stay out of public view, a regional cultural practice that possibly predates Islam

secular states (p. 305) countries that have no state religion and in which religion has no direct influence on affairs of state or civil law

shari'a (p. 304) literally "the correct path"; Islamic religious law that guides daily life according to the principles of the Qur'an

sheiks (p. 324) patriarchal leaders of tribal groups on the Arabian Peninsula

Shi'ite (or **Shi'a**) (p. 305) the smaller of two major groups of Muslims with differing interpretations of shari'a; Shi'ites are found primarily in Iran

spatial freedom (p. 307) the ability to move about without restrictions

Sunni (p. 304) the larger of two major groups of Muslims with differing interpretations of shari'a

theocratic states (p. 305) countries that require all government leaders to subscribe to a state religion and all citizens to follow certain rules decreed by that religion

Zionists (p. 327) European Jews who worked to create a Jewish homeland (Zion) on lands once occupied by their ancestors in Palestine

Pronunciation Guide

Al Salaam (ahl sah-LAHM)

Alexandria (al-ihk-SAN-dree-uh)

Algeria (ahl-JEER-ee-uh)

Algiers (ahl-JEERZ)

Allah (AH-luh)

Anatolia (an-uh-TOH-lee-uh)

Arabia (uh-RAY-bee-uh)

Arafat, Yasser (AHR-uh-faht, YAH-seer)

Aswan High Dam (ahss-WAHN)

Ayatollah (eye-yah-TOH-lah)

Baghdad (bahg-DAHD/BAG-dad)

Bahrain (bah-RAYN)

Bandare Abbas (BAHN-dahr AH-buhss)

Barbary Coast (BAHR-buh-ree)

Basra (BAHZ-rah)

Bedouin (BEHD-oo-ihn)

Berber (BURR-burr)

Bosporus (BAWSS-puh-ruhss)

Byzantine Empire (BIHZ-uhn-teen)

Cairo (KYE-roh)

Canaan (KAY-nuhn)

Casablanca (kah-suh-BLAHNG-kuh)

Caspian Sea (KASS-pee-uhn)

chador (chah-DOHR)

Constantinople (kawn-stan-tih-NOH-puhl)

Coptic Christians (KAWP-tihk)

Damascus (duh-MASS-kuhs)

Dinka (DIHNG-kuh)

Euphrates River (yoo-FRAY-teez)

Farsi (FAHR-see)

Gaza Strip (GAH-zah)

Golan Heights (GOH-lahn)

hajj (HAHJ)

hijab (hee-JAHB)

Hussein, Saddam (hoo-SAYN, sah-DAHM)

intifada (ihn-tih-FAH-duh)

Iran (ih-RAHN)

Iraq (ih-RAHK)

Islam (ihz-LAHM)

Israel (IHZ-ree-uhl)

Istanbul (ihss-STAHN-bool / ihss-tahn-BOOL ["oo" as in "book"])

Jeddah (JEHD-uh)

Jerusalem (juh-ROO-suh-lehm)

Jordan (JOHR-duhn)

Khartoum (kahr-TOOM)

Khatami, Mohammad (kah-TAH-mee)

Khomeini, Ruholla (koh-MAY-nee, roo-HAWL-uh)

Koran (koo-RAHN ["oo" as in "book"])

Kurd (KURRD)

Kuwait (koo-WAYT)

Lake Nasser (NAH-surr)

Lebanon (LEH-buh-nuhn)

Libya (LIH-bee-uh)

Maghreb (MAH-gruhb)

Mecca (MEH-kuh)

Medina (meh-DEE-nuh)

Mediterranean Sea (meh-dih-tuh-RAY-nee-uhn)

Mesopotamia (meh-suh-puh-TAY-mee-uh)

Mongol (MAWNG-gohl)

Morocco (muh-RAW-koh)

Muhammad (moo-HAHM-ihd ["oo" as in "book"])

Muslim (MOOZ-lihm ["oo" as in "book"])

narghile (NAHR-guh-leh)

Nilotic (nye-LAW-tihk)

Nubian (NOO-bee-uhn)

Oman (oh-MAHN)

Ottoman Empire (AW-tuh-muhn)

Pahlavi, Reza (PAH-luh-vee, RAY-zah)

Palestine (PAL-uh-stien)

Persia (PURR-zhuh)

pogrom (puh-GRAWM)

qanat (KAH-naht)

Qatar (KAH-tahr)

Qur'an (koo-RAHN ["oo" as in "book"])

Rabat (rah-BAHT)

rai (RYE)

Ramadan (rahm-uh-DAHN)

Riyadh (ree-YAHD)

Rub'al Khali (roob ahl KAH-lee)

Sahara (suh-HAHR-uh)

Saud (sah-OOD)

Saudi Arabia (sah-OO-dee uh-RAY-bee-uh)

shah (SHAH)

shari'a (shah-REE-ah)

Shatt al Arab (SHAHT ahl AHR-uhb)

sheik (SHAYK)

Shi'a (SHEE-uh)

Shi'ite (SHEE-eyt)

Sinai Peninsula (SYE-nye)

Sudan (soo-DAN)

Sudd (SOOD)

Suez Canal (SOO-ehz)

Sunni (SOO-nee)

Syria (SEER-ee-uh)

Tangier (tahn-JEER)

Tassili 'n Agger (tah-see-LEE nah-JAIR)

Tehran (teh-RAHN)

Tel Aviv (tehl ah-VEEV)

Tigris River (TYE-grihss)

Tripoli (TRIH-puh-lee)

Tuareg (TWAH-rehg)

Tunis (TOO-nihss)

Tunisia (too-NEE-zhuh)

Turkoman (TURR-kuh-mahn)

ummah (OO-muh)

United Arab Emirates (EHM-uh-ruhts)

wadi (WAH-dee)

Yemen (YEH-muhn)

Zabbaleen (zah-bah-LEEN)

Zagros Mountains (ZAH-grohss)

zakat (zah-KAHT)

Selected Readings

A set of Selected Readings for Chapter 6, providing ideas for student research, appears on the *World Regional Geography* Web site at www.whfreeman.com/pulsipher.

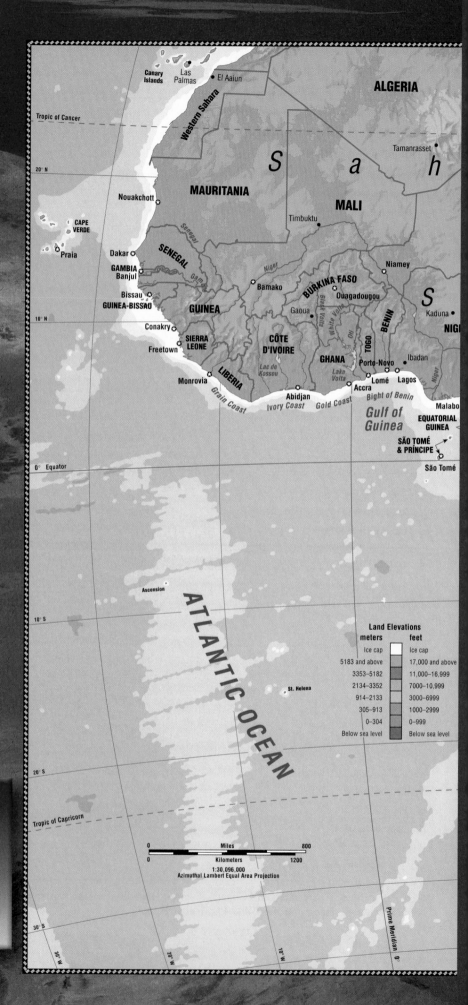

ALGERIA

Canary Islands
Las Palmas
El Aaiun

Western Sahara

Tropic of Cancer

S a h

Tamanrasset

20° N

MAURITANIA

Nouakchott

MALI

Timbuktu

CAPE VERDE

Praia

Dakar

Senegal

Niamey

SENEGAL

S

GAMBIA
Banjul

Bamako

BURKINA FASO

Kaduna

Bissau

Ouagadougou

NIG

GUINEA-BISSAO

GUINEA

Gaoua

Niger

10° N

Conakry

CÔTE D'IVOIRE

BENIN

SIERRA LEONE

TOGO

Ibadan

Freetown

GHANA

Porto-Novo

Lac de Kossou

Lake Volta

Lomé   Lagos

LIBERIA

Accra

Monrovia

Abidjan

Malabo

Ivory Coast   Gold Coast

Bight of Benin

EQUATORIAL GUINEA

Grain Coast

Gulf of Guinea

SÃO TOMÉ & PRÍNCIPE

0° Equator

São Tomé

10° S

Land Elevations

meters	feet
Ice cap	Ice cap
5183 and above	17,000 and above
3353–5182	11,000–16,999
2134–3352	7000–10,999
914–2133	3000–6999
305–913	1000–2999
0–304	0–999
Below sea level	Below sea level

Ascension

ATLANTIC OCEAN

St. Helena

20° S

Tropic of Capricorn

0   Miles   800

0   Kilometers   1200

1:30,096,000
Azimuthal Lambert Equal Area Projection

30° S

30° W   20° W   10° W   Prime Meridian 0°

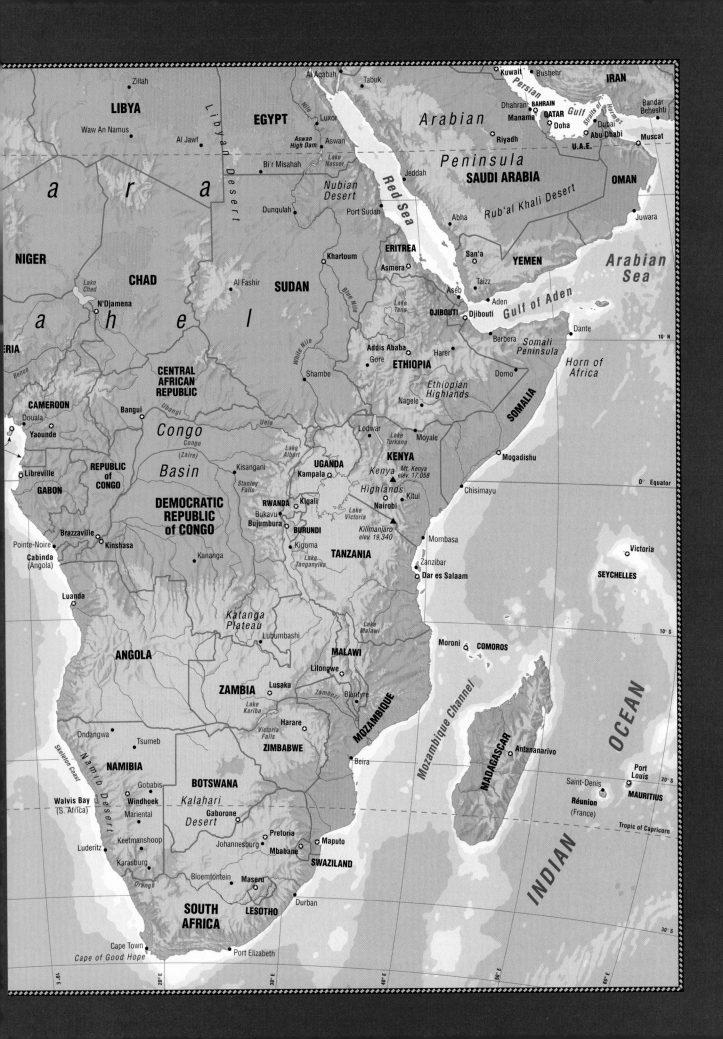

MAKING GLOBAL CONNECTIONS

VIGNETTE "I never thought the sun could do all this!" exclaimed Melusi Zwane, the principal of Myeka High School, located two hours outside Durban, South Africa.

Principal Zwane was referring to the school's new photovoltaic solar panels, which can generate 2.4 kilowatts of power. They were installed in the late 1990s by Solar Electric Light Fund (SELF), a non-profit group from the United States. SELF, in cooperation with Dell Computer and Infosat Telecommunications, set up the school with computers and a satellite uplink so that the students could have Internet access.

The school lies deep in the Valley of a Thousand Hills, where there is no power grid to supply electricity service. The South African government cannot afford to string power lines to the valley from Durban. Students used to read their lessons by candlelight and had only a few textbooks among them. Less than a third of the students graduated, and those who did had little chance of employment other than agricultural labor. But that has changed. By 2001, 70 percent of the students were graduating. The arrival of solar power, computers, and connections to the Internet gave them access to a world of information and ideas that helped them pass the national exams. Being connected to the Internet also generated ambition among the students to set up small enterprises and seek employment that makes use of their education, either within the valley or in Durban.

Myeka High School's experience is a vivid illustration of the impact that solar power can have on a society, deeply changing its economic and social dynamics. Two billion people, or roughly one-third of the world's population, live in rural areas that are not connected to the energy grid. [Adapted from David Lipshultz, "Solar power is reaching where wires can't," *The New York Times* (Business Section, September 9, 2001): 4.]

The story of Myeka High School illustrates some of sub-Saharan Africa's challenges: poverty, lack of education, and missing infrastructure needed for economic progress. But it also illustrates the promise of solutions.

Students outside Myeka High School eagerly wait for school to start so they can begin using the computers. The solar panels supplied by SELF provide electricity to light the classrooms and power the computers. [Courtesy of Solar Electric Light Fund.]

Africa is a diverse continent, with great natural wealth and a long, rich history of peoples and their civilizations. It is the birthplace of humankind and one of the world's cultural hearths, where agriculture and metalworking developed and where people first put down their thoughts in writing. The world's longest continuous religious and philosophical traditions are also found in Africa.

Sub-Saharan Africa, the region discussed in this chapter, is home to about 700 million people. Several of the fastest growing economies in the world and some of the world's richest deposits of oil, gold, platinum, copper, and other strategic minerals are found here. Yet Africa is not often described in positive terms. The news that comes out of Africa is usually about disease, environmental devastation, war and genocide, political corruption, and poverty. And, all too often, these reports do describe conditions in large areas of the continent.

It has been barely a full generation—less than 50 years—since the first sub-Saharan African country, Ghana, achieved independence from British colonial rule in 1957. During the era of European colonialism (1850s–1950s), wealth flowed out of Africa; and although colonialism officially ended with the granting of political independence in the 1950s and 1960s, the net flow of wealth is still out of Africa. Investors from the rich countries of the world continue to reap Africa's wealth by extracting minerals and other natural resources under unfair terms of trade—a practice called **neocolonialism**. An example is provided by the fishing industry of Guinea-Bissau, a country that ranks near the bottom on the United Nations Human Development Index (HDI) (169 out of 174). It lies on the west coast of Africa, adjacent to the rich Canary Current. In 1996, 97 percent of the fish caught in the waters of Guinea-Bissau were taken and sold abroad by high-tech foreign fishing fleets from Japan, Korea, Russia, Spain, and Portugal. The owners of these fleets paid just U.S. $11 million a year to Guinea-Bissau for licenses to take a catch worth U.S. $130 million annually.

Similar deals have been the rule throughout Africa for 500 years and more. It is not surprising, then, that Africa is now impoverished and often at war with itself. The average per capita income in sub-Saharan Africa is the second lowest in the world. (The region with the dubious distinction of having the lowest per capita income is South Asia, which includes India.) From time to time, ethnic conflict, often arising over access to resources, has turned the generally low standard of living into outright destitution.

But today, there are many hopeful signs that conflicts can be resolved and the well-being of the majority enhanced. In recent years, a number of Africans have been recognized as among the best leaders the world has to offer: Mary Balikungeri of Rwanda; UN Secretary-General Kofi Annan from Ghana; South Africa's former president, Nelson Mandela; former president Julius Nyerere of Tanzania; and the late Nigerian writer Ken Saro-Wiwa, to name but a few. Since independence, African countries and societies have been engaged in determining the appropriate pathways to economic development and progress. Some of them are forging ahead, others are struggling to find a sustainable path. It is too soon to dismiss the ability of the people in sub-Saharan Africa to achieve sustainable economic, political, and social progress.

Themes to Look For

Several themes are developed in this chapter. Among the most important are

1. The roots of African poverty. Sub-Saharan African countries rank among the poorest on earth. The roots of this poverty lie in past exploitative colonial relationships with European countries and in modern global trade patterns that leave Africa as a provider of cheap raw materials and low-cost labor.

2. Persisting circumstances of colonialism. Although colonialism officially ended in the 1950s and 1960s, modern economic patterns perpetuate the circumstances of colonialism, especially the unsustainable use of resources and an increasing disparity of wealth.

3. Disadvantageous position in the global economy. Africa interacts with the global economy only as a supplier of resources; because of its pervasive poverty, it has little to invest in upgrading the skills and earning power of its citizens. Large debt payments to governments and banks in the developed world take money from needed social services.

4. Disadvantageous population dynamics and disease patterns. Generally, Africa is not densely populated, but population growth rates are very high and in some places densities are too high to be supported by local resources. Disease has always been a factor limiting African population growth, but now the human immunodeficiency virus (HIV)–acquired immunodeficiency syndrome (AIDS) epidemic is severely reducing growth, affecting social conditions, and slowing improvements in standards of living.

5. Distinctive gender roles. Across Africa's wide range of ethnic groups, it is common for men and women to have distinctly different roles. At least since colonial times, women have performed most of the labor connected with growing and preparing food, most local transport duties, and most marketing of surpluses; men have prepared fields for cultivation, taken responsibility for cash crop production, and often migrated to work in mines or urban factories.

6. Encouraging developments in sub-Saharan Africa. Increasingly, African countries are joining together to define the region's problems and design solutions for them. The private sector, both informal and formal, is emerging as a vital force in the economy, as direct government involvement in agriculture and industry decreases. Political reform movements have resulted in democratically elected governments in a majority of sub-Saharan African countries.

Terms to Be Aware Of

The language used to describe Africa has been particularly prone to **ethnocentrism,** the often subconscious belief in the superiority of one's own culture or ethnic group. There is now a movement to purge colonial terminology from descriptions of Africa because, too often, these terms are based on European misperceptions of cultures very different from their own. When words such as *tribe, tribalism, primitive,* and *savage* are used to refer to Africa, they present a biased view of African culture. In this book, we use the terms **ethnic group** and **ethnicity** instead to describe groups of people and the cultures they share. *Civilization* is another term often misused in relation to Africa. Europeans considered only certain people to be civilized: the people of ancient China, Greece, Mesopotamia, the Indus Valley, the Maya and Inca empires in the Americas, and most of Europe since A.D. 1000. Africa was regarded as uncivilized. Today, most scholars recognize the limitations of this European concept of civilization. Africa has had several ancient societies that fit common definitions of civilization. Moreover, Europeans often grossly misunderstood African cultural practices that they labeled "uncivilized," and they seldom evaluated cultural practices in Europe and elsewhere closely enough to determine how "civilized" they were.

Because there are two neighboring countries called Congo—the Democratic Republic of the Congo and the Republic of the Congo—and because these designations are both lengthy and easily confused, in this text the names will be abbreviated. The Democratic Republic of the Congo (formerly Zaire) will carry the name of its capital in parentheses: Congo (Kinshasa); and the Republic of the Congo will carry the name of its capital, Brazzaville: Congo (Brazzaville).

THE GEOGRAPHIC SETTING

PHYSICAL PATTERNS

The African continent is big—in fact, it is the second largest after Asia. At its widest point, it stretches for 4000 miles (6400 kilometers) from east to west, and its length from the Mediterranean to its southern tip is nearly 5000 miles (8000 kilometers). But Africa's great size is not matched by its surface complexity: it has no major mountain ranges—the highest point in Africa is Mount Kilimanjaro, at 19,324 feet (5890 meters)—and more than one-quarter of the continent is covered by the Sahara Desert, which comprises many thousands of square miles of homogeneous landscape.

Landforms

Geologists usually place Africa at the center of the ancient supercontinent of Pangaea. Over the past 200 million years, several continents broke off from Africa and moved away: North America to the northwest, South America to the west, India to the northeast. As Africa shed these continent-sized pieces, it readjusted its position only slightly, drifting gently into Southwest Asia. Although Africa remains in approximately the same place on the globe's surface that it has occupied for more than 200 million years, it is still breaking up, especially along its eastern flank. The Arabian Plate has rifted away from Africa and drifted to the northeast, leaving the Red Sea, which separates Africa and Asia. The Great Rift Valley, another series of developing rifts, curves inland from the Red Sea and extends more than 2000 miles (3200 kilometers) south to Mozambique, near the east coast. At some future time, Africa is expected to break apart along this rift, which is a formation marked by chains of lakes, including two long and narrow ones, Lake Tanganyika and Lake Malawi.

The continent of Africa can be envisioned as a raised platform, or plateau, bordered by fairly narrow and uniform coastal lowlands and covered by an ancient mantle of rock in various stages of weathering. The platform slopes downward to the north. Thus, the southeastern third of the continent is an upland region with several substantial mountain ranges, whereas the northwestern two-thirds of the continent is a lower lying landscape, interrupted only here and there by uplands and mountains. Despite their lack of complexity, the landforms of Africa have obstructed transport on the continent to this day and have hindered connections to the outside world. Routes from the plateau to the coast must negotiate steep escarpments (long cliffs) around the rim of the continent, and the long, uniform coastlines have few natural harbors.

Climate

Most of sub-Saharan Africa has a tropical climate (Figure 7.1), because 70 percent of the continent lies between the Tropic of Cancer and the Tropic of Capricorn. Average temperatures are generally high, staying above 64°F (18°C) year-round everywhere except at the more temperate southern tip of the continent and in upland zones. Seasonal climates in Africa differ more by the amount of rainfall than by temperature.

Most rainfall comes to Africa by way of the **intertropical convergence zone (ITCZ),** a band of atmospheric currents that circles the globe roughly around the equator (see Figure 7.1). At the ITCZ, warm winds converge from both north and south. These winds push against each other, causing the air to rise, cool, and release moisture in the form of rain. The rainfall produced by the ITCZ is most abundant in central and western Africa near the equator. There, in places such as the Congo Basin, the frequent rainfall nurtures the dense vegetation of tropical rain forests.

The ITCZ shifts north and south seasonally, generally following the area of the earth's surface that has the highest average temperature at any given time. The tilt of the earth's axis causes the most direct rays from the sun to fall in a band that sweeps into the Northern Hemisphere during its summer (June–September) and

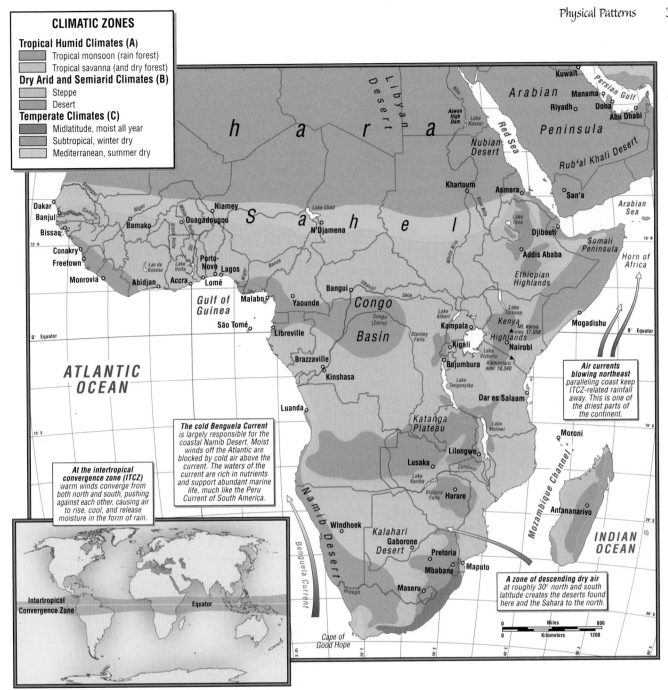

Figure 7.1 **Climates of sub-Saharan Africa.** The inset at lower left shows the position of the intertropical convergence zone. [Inset adapted from Tom L. McKnight, *Physical Geography* (Upper Saddle River, N.J.: Prentice Hall, 1996), p.125.]

into the Southern Hemisphere during its summer (December–March). Hence, during the height of the Southern Hemisphere summer in January, the ITCZ might bring rain far enough south to water the dry grasslands, or steppes, of Botswana. During the height of the Northern Hemisphere summer in August, the ITCZ brings rain as far north as the fringes of the Sahara, an area called the Sahel where steppe and savanna grasses grow. Poleward of both of these extremes, a belt of descending dry air blocks the effects of the ITCZ and creates the deserts found in Africa and other conti-

nents at roughly 30°N latitude (the Sahara) and 30°S latitude (the Namib). Across central Africa, between the wet equatorial rain forests and the dry scrub of the deserts, there are almost mirror image bands of ecosystems that reflect the bands of differing rainfall. The two major examples of these ecosystems are the seasonally dry tropical forest that borders the true rain forest and the moist savanna beyond, where tall grasses and trees intermingle. Both environments have provided suitable land for agriculture for thousands of years.

The banded pattern of African ecosystems is modified in many areas. Winds blowing north along the east coast of Africa keep ITCZ-related rainfall away from the **Horn of Africa,** the triangular peninsula that juts out from northeastern Africa below the Red Sea. As a consequence, the Horn of Africa is one of the driest parts of the continent. Along the very dry coast of the Namib Desert of southwestern Africa, moist winds off the Atlantic are blocked by cold air above the Benguela Current. This northward-flowing current, much like the cold Peru Current off South America, is chilled by its passage past Antarctica; it is rich in nutrients and supports a major fishery.

The climate of Africa presents a number of challenges to human beings. Insects and parasites that breed prolifically in warm, wet climates cause debilitating diseases, such as river blindness, schistosomiasis, and malaria. In drier tropical climates, water for drinking, farming, and raising animals is often in short supply; the soils, lacking organic matter, are not particularly fertile, even if irrigated. In humid tropical climates, cultivated soil loses its fertility easily. Wherever both temperature and moisture are high, **organic matter** (the remains of any living things) in the soil decays rapidly, and the nutrients it releases are quickly absorbed by surrounding plants or **leached** (washed) out into groundwater and runoff. In a standing forest, the organic matter shed every day decays and provides a continual source of nutrients for plant growth. Removing the forest also removes the source of nutrients, and the soil quickly deteriorates. The minerals leach out and soil particles wash downslope. The direct rays of the sun bake the soil into a permanently hard surface called **laterite.**

Over the ages, cultivators in the wet tropics developed the method of farming called **shifting cultivation** to maintain soil quality. They clear only small patches, an acre or two at a time. They use the cleared vegetation as "fertilizer," sometimes burning it to release nutrients. And they plant their gardens as a sort of miniature forest with many species—often 20 or more per garden—of various sizes, shapes, and requirements that cover the soil quickly to prevent it baking hard in the hot sun. Because the soil loses fertility quickly, the small plots produce well only for two or three years and are then allowed to revert to forest for several decades. Typically, because of soil infertility, tropical diseases, and pests, commercial agriculture that depends on large cleared fields and long-term production has not succeeded for extended periods in tropical zones.

There is evidence that Africa has experienced repeated changes in climate over the last 20,000 years and that a new, drier cycle is starting. Droughts have been frequent since the 1970s. They may simply be a consequence of the natural cycle or, as many scientists think, at least partly the result of global warming caused by the rising levels of greenhouse gases produced in the industrialized parts of the world. Scientists expect present climatic patterns in Africa to intensify: hot, wet places may become hotter and wetter; hot, dry places may become hotter and drier. Grasslands on the desert margins may become deserts; low-lying coastal areas may be inundated by seawater as melting glaciers at the poles raise sea level. Because so many Africans live in arid or low-lying zones, as many as one-quarter of the people could find their homelands becoming uninhabitable.

HUMAN PATTERNS OVER TIME

Almost 2000 years ago, the Roman Empire spread its influence into northeast Africa as far south as Nubia (present-day Sudan) and Abyssinia (Ethiopia). In precolonial times, other Europeans had some awareness of African societies in the areas that border the southern Sahara Desert, because Arab traders brought products from north central Africa across North Africa to Europe. In East Africa, the coastal zone was part of the Indian Ocean trading network by the eighth century. There were trade routes up the coast and along the Red Sea into the Mediterranean or into the Persian Gulf and Persia, eventually connecting with the Central Asian Silk Road trade (see Chapter 5, page 283, and Chapter 9, page 495). As early as the tenth century, the Manda Kingdom (in present-day Kenya) was receiving Chinese porcelain through this trade route. At about the same time, gold from Zimbabwe in southeastern Africa began to attract foreign traders to the interior of southern Africa. According to historian Ali Mazrui, "The interaction between Egyptians, Mesopotamians, Assyrians, Nubians, ancient Greeks and Romans resulted in the explosion of one of the most dazzling galaxies of cultures in human history."

Despite this rich past, African culture has often been misunderstood and dismissed over the last several hundred years. Euro-

This drawing, made in 1686 by Olfert Dapper, a seventeenth-century Dutch explorer, shows Loango, a sizable coastal city just north of present-day Pointe-Noire, Congo (Brazzaville). Little of it remains today. [From Graham Connah, *African Civilizations* (Cambridge: Cambridge University Press, 1987). Original source O. Dapper, *Description de l'Afrique . . . Traduite du Flamand* (Amsterdam: Chez Wolfgang, Waesberge, Boom & van Someren, 1686), p. 321. John Carter Brown Library, Brown University.]

pean slave traders and colonizers resisted evidence of the historical achievements of Africa's peoples. They justified their invasion and colonizing of sub-Saharan Africa by calling Africa the Dark Continent, a place where little of significance in human history had occurred. For example, the substantial and elegantly planned city of Benin in West Africa, encountered by European explorers in the 1500s, never became part of Europe's image of Africa, nor did the city of Loango in the Congo Basin (see the photograph on page 344). Even today, most people in North America are unaware of Africa's history or contributions to world civilization.

The Peopling of Africa

Africa is the original home of the human race. It was in eastern Africa that the first humans evolved more than 2 million years ago, although they differed anatomically from human beings today. Anatomically modern *Homo sapiens sapiens* evolved in Africa between 200,000 and 100,000 years ago. The earliest *Homo sapiens* remains yet recovered in Africa come from southern Ethiopia and may be as old as 100,000 years. By 90,000 years ago, modern humans had reached North Africa and the eastern Mediterranean. From there, they spread into Asia and eventually into Europe.

People in sub-Saharan Africa began to cultivate plants about 7000 years ago in the southern belt of the Sahara Desert, which was less arid at that time. The highlands of Ethiopia in East Africa were another early center of plant cultivation and food production. The Bantu peoples seem to have brought plant cultivation to equatorial Africa when they migrated from West Africa to the Congo Basin within the last 2500 years. Within a thousand years, descendants of people from both coastal West Africa and the highlands of East Africa were converging and migrating into southern Africa, displacing local hunter-gatherers.

In western Central Africa, several influential centers made up of dozens of linked communities developed in the forest and the savanna. One of several powerful kingdoms was the empire of Ghana (A.D. 700 to 1000), which was located not in the modern country of Ghana but in present-day Mali north of the headwaters of the Niger River. The Ghana Empire was succeeded by the kingdoms of Mali (1200 to 1400) and Songhay (1400 to 1500), with each succession or takeover enlarging the empire's territorial jurisdiction. The kingdoms' wealthy and prominent converts to Islam periodically sent large entourages on pilgrimages to Mecca, where their opulence was a source of wonder.

In East Africa, the kingdom of Aksum in present-day Ethiopia began its conversion to Christianity around A.D. 350. At that time Aksum ranked with the empires of Persia, China, and Rome. For more than 400 years, Aksum thrived in its location near the Red Sea and grew rich from trade with Rome, southern Arabia, and Gaul (modern-day France). Farther south, cities along the shore of the Indian Ocean were linked in a huge annual cycle of cross-ocean trade, stretching north to the Persian Gulf and east to India, Southeast Asia, and even China. In the mid-1800s, British explorers entered the capital of the Buganda Kingdom in that region. They found a town of 40,000 people with a central government that included officials who collected taxes throughout the kingdom to support the construction of roads, bridges, and viaducts.

The archaeological ruins of The Enclosure at Great Zimbabwe are surrounded by the ruins of numerous other structures. [Don L. Boroughs/ The Image Works.]

Iron smelting, including the making of steel, has long been practiced in the highlands of eastern Central Africa and at locations in West Africa. By the fifteenth century, a remarkable civilization marked by substantial stone buildings and supported by agriculture and mining had developed in the highlands of Zimbabwe in southeastern Africa. Nineteenth-century British explorers were mystified by their discovery of the ruins of stone structures, tens of thousands of mine shafts, and even fragments of Chinese pottery. The European archaeologists of the time did not believe that indigenous African civilizations could have existed, let alone have trading links to China, so they mistakenly credited the Zimbabwe culture to outsiders, not Africans.

In both East and West Africa, Africans traded ivory, iron, gold, and slaves for animals, dyes, food, and manufactured goods such as fine leather products, ceramics, and textiles. It is estimated that between 600 and 1900, close to 9 million African slaves were exported just to parts of Asia and to Islamic areas around the Mediterranean. There was a long-standing custom in Africa of enslaving certain classes of people as a result of hostilities between two or more ethnic groups. Usually the treatment of slaves within Africa, governed by local custom, was reasonably humane. Although such protections were lacking when slaves were traded to non-African strangers, many African slaves who worked in the homes and fields of wealthy Mediterranean Muslims were treated as family members, though with notably lesser status.

The Coming of the Europeans

The course of African history shifted dramatically in the mid-1400s, when Portuguese sailing ships began to appear off the West African coast. The names given this coast by the Portuguese and other early European maritime powers reflected their interest in trading Africa's resources: the Gold Coast, the Ivory Coast, the Pepper Coast, the Slave Coast. By the 1530s, the Portuguese had organized the slave trade with the Americas. The trading of slaves by the Portuguese, British, Dutch, and French was far more widespread and brutal than either the small-scale taking of slaves that had occurred within Africa or the limited slave trade conducted with Muslims and Asians. The most distinctive feature was the commercial motivation of the Europeans, who needed cheap labor for their American plantations, which in turn supplied raw materials and money for the industrial revolution in Europe. Slavery as managed by Europeans was more impersonal than slavery as practiced in Africa, and slaves were often treated strictly as a commodity.

To acquire slaves, the Europeans established forts on Africa's west coast and paid nearby African kingdoms with weapons, trade goods, and money to make slave raids into the interior. As in the pre-European slave trade, some of the captives were taken from enemy kingdoms in battle. Many more, however, were kidnapped from their homes and villages in the forests and savannas. Most slaves traded in the international market were male, because the raiding kingdoms preferred to keep captured women for their reproductive capacities. From 1600 to 1865, some 11 million to 12 million captives were packed aboard cramped and filthy ships and sent to the Americas. Up to a quarter of them died at sea. Those who arrived in the Americas went primarily to plantations in South America and the Caribbean; about one-quarter were sent to the southeastern United States (Figure 7.2).

The European slave trade severely drained the African interior of human resources and set in motion a host of damaging social responses that are not well understood even today. Slavery also made Africans dependent on European trade goods and technologies, especially guns used by raiding kingdoms. Much African culture survived the cruelties of slavery, however, and enriched the language, cuisine, religion, literature, music, agricultural technology, and art of other world regions, especially the Americas.

The Scramble to Colonize Africa

European involvement in Africa gradually increased, culminating in the establishment of formal colonies in the late nineteenth century. The British transatlantic slave trade officially ended in 1809, but other European nations continued trading in slaves until 1865. By that time, Europeans had found it more profitable to use African labor *in Africa* to extract raw materials for Europe's growing industries. European interests extended inland to include the exploitation of fertile agricultural zones, areas of mineral wealth, and places with large populations that could serve as sources of labor.

Colonial powers competed avidly for territory and resources. To prevent the eruption of armed conflicts, it became necessary to establish official boundaries between the claimed African territories. The result was the virtually complete seizure and partition of the continent by the time of World War I (Figure 7.3). Africans were not consulted about the partitioning of their continent, and only two African countries retained independence: Liberia on the west coast and Ethiopia in East Africa. Otto von Bismarck, the

Thinking Critically: ON THE WEB

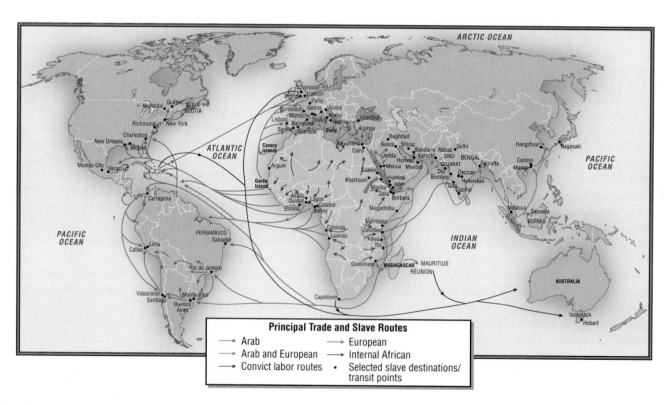

Principal Trade and Slave Routes

→ Arab   → European
→ Arab and European   → Internal African
→ Convict labor routes   • Selected slave destinations/ transit points

Figure 7.2 African diaspora resulting from the slave trade. The arrows show the general direction of the slave trade within Africa and of the external American, Arab, and European slave trade until 1873. [Adapted from the work of Joseph E. Harris, in Monica Blackmun Visona et al., eds., *A History of Art in Africa* (New York: Harry N. Abrams, 2001), pp. 502–503.]

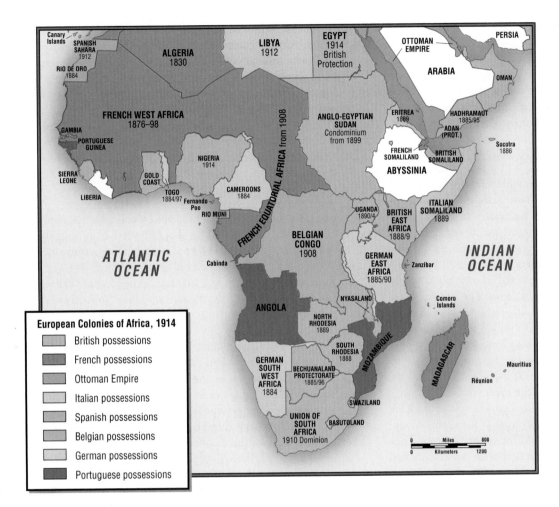

Figure 7.3 The European colonies of Africa in 1914. Some countries (those without dates) had been occupied by colonial powers for a few centuries. In the latter part of the nineteenth century, as European countries began to compete on a large scale for natural resources, cheap labor, and new markets, they expanded into Africa. The dates on the map indicate the years of officially recognized control by the colonizing power. [Adapted from Alan Thomas, *Third World Atlas* (Washington, D.C.: Taylor and Francis, 1994), p. 43.]

German chancellor who convened the 1884 Berlin Conference, at which the competing powers first gerrymandered Africa, declared: "My map of Africa lies in Europe." With some notable exceptions, the boundaries of most African countries today derive from the colonial boundaries set up between 1884 and 1916 by European treaties. These territorial divisions lie at the root of many of Africa's current problems.

There were a few basic patterns in the European domination of Africa during the colonial period. The geographer Robert Stock, a specialist in sub-Saharan Africa, identifies three:

1. European settlers occupied land on a large scale in only a few places. These were mainly areas with especially attractive resources or places where Europeans considered the climate comfortable—such as the relatively cool highlands of Kenya in East Africa and the upland plateau of South Africa. In these places, Europeans forced indigenous farmers and herders onto marginal land in reservations or made them labor on European-owned farms and plantations. Areas taken over by Europeans were privileged: taxes were kept low and roads were constructed to provide ready access to local and foreign markets.

2. Africans continued to farm small plots in many densely populated agricultural regions and in places considered disagreeable by Europeans. They were often encouraged or directed to switch from their traditional food crops to export crops. Some African farmers complied eagerly, in the hope of becoming wealthier; others changed grudgingly, only because they now needed cash to pay taxes to the colonial regimes and to buy food. Many Africans suffered malnutrition as more and more land was shifted from food to cash crops or livestock production. Despite the disruption of their economies, these areas of indigenous farming received little assistance from Europeans. Transportation was only modestly upgraded, and Africans were often required to pay higher taxes than the Europeans who had taken their land.

3. Remote areas that were difficult to exploit economically were treated as labor reserves from which young men were often removed and put to work on government projects such as railroad construction. They were separated from their families, subjected to dangerous working conditions, and paid very little. Because they were not given enough food or allowed the time to grow or hunt their own, many died of malnutrition. In the Belgian Congo, as many as 10 million people starved during the reign of King Leopold of Belgium (1865–1909). (In 1960, the Belgian Congo became an independent country that was called Zaire until the mid-1990s, when the name was changed to the Democratic Republic of the Congo.) In all areas colonized by Europe, the main objectives of colonial administrations were to extract as many raw materials as possible, create markets for European manufactures, and keep the costs and the commitments of European-administered governments to an absolute minimum. It is useful to examine the case of South Africa to appreciate how the European incursion into Africa led to the expropriation of land; the subjugation of African peoples; and, in this case, the infamous system of **apartheid** (racial segregation).

The Colonization of South Africa

In the 1650s, the Dutch took possession of the Cape of Good Hope at the southern tip of Africa from the Portuguese, and shortly thereafter the first significant migrations of Europeans to Africa began. The Dutch immigrant farmers, called Boers, practiced herding and farming techniques that used large tracts of land. As they spread into the interior, the Boers pushed indigenous people off their land, despite strong resistance. The Boers enslaved laborers elsewhere in Africa and brought them to work on their farms. The British gained control of South Africa in 1795 and outlawed slavery in 1834. At that time, to elude British control, large numbers of Boers migrated to the northeast in what was called the Great Trek. There were often intense and violent conflicts with the Africans in these interior areas. The discovery of extremely rich diamond deposits in the 1860s and gold in the 1880s secured the economic future of this land, which became known as the Orange Free State and the Transvaal. Africans were often forced to work in the diamond and gold mines under unsafe conditions and for minimal wages. They lived in unsanitary compounds that travelers of the time compared to large cages. Britain, eager to claim the wealth of these mines, invaded the Orange Free State and the Transvaal in 1899, waging the bloody Boer War. The war gave them control of the mines briefly, until resistance by Boer nationalists forced the granting of independence to all of South Africa in 1910.

In 1948, apartheid laws were enacted to reinforce the longstanding segregation of Boer society. These laws required everyone except whites to carry passbooks and live in racially segregated rural townships, specific urban sections, or workers' dormitories attached to mines and industries. Eighty percent of the land was reserved for the use of Europeans, who then made up 15 percent of the population. Blacks were assigned to ethnic-based "homelands" that were considered independent enclaves within the borders of, but not part of, South Africa. African people were treated similarly in other areas of the continent, such as Northern and Southern Rhodesia (modern Zambia and Zimbabwe), Tanzania, and Kenya.

The fight to end racial discrimination in South Africa began even before the apartheid laws were formally introduced in 1948. The African National Congress (ANC), one of the major organizations participating in this struggle, was formed in 1912 to work nonviolently for civil rights for black Africans. After the apartheid laws were passed, the ANC began to recruit more outspoken leaders. Overt resistance to the apartheid system intensified in the 1960s and grew steadily despite heavy-handed repression. Among the key leaders who fought for racial justice in South Africa were Steven Biko, an attorney who died while in police custody; Anglican archbishop Desmond Tutu; ANC leader Winnie Madikizela-

Mandela; and Nelson Mandela, an ANC leader who was jailed by the South African government for 27 years. The resistance was supported by millions of ordinary people: professionals, migrant workers, schoolchildren. Finally, in the early 1990s, the white-dominated government realized that it could no longer resist majority rule. It released Nelson Mandela and began to negotiate with the ANC and other political organizations for an end to apartheid. Majority rule arrived in 1994 with the election of Nelson Mandela as South Africa's president.

The Aftermath of Independence

In Africa, the era of formal European colonialism was relatively short. In most places it lasted for about 80 years, from roughly the 1880s to the 1960s. In 1957, Ghana in West Africa became the first African colonial state to achieve its independence. The last sub-Saharan African country to gain independence was Eritrea, in 1993, although Eritrea won its independence from its neighbor Ethiopia, rather than from a European power, after waging a three-year civil war.

Colonial rule has continued to influence the postindependence history of Africa. Many African countries have struggled to find the most appropriate economic systems and forms of government. Like their colonial predecessors, most independent African governments became authoritarian, antidemocratic, and dominated by privileged and Europeanized elites. Nation building has been a rocky road for these countries, but most of them have been on this road for less than 40 years. Many of the countries remain economically dependent upon their former European colonizers. In the last several years, however, pro-democracy movements have begun to spread across Africa, and by 2001 some 26 countries had democratically elected governments, up from 11 in 1970.

Africa enters the twenty-first century with a complex mixture of enduring legacies from the past and looming challenges in the future. Although Africa has been liberated from colonial domination, it is still strongly influenced by the world's wealthy countries. Because of its reliance on exports and imports, it remains inextricably linked to the global economy, in which it has long operated at a disadvantage. Poverty is expanding rapidly as solutions to Africa's problems are slow to come. At present, Africa faces declining economic productivity and rising debt; severe periodic drought and famine; major health problems, including the world's worst AIDS epidemic; and the challenge of having the fastest growing population on earth. The brightest spot in Africa's future is that the recent decades of rapid and often wrenching change have made many people more willing to consider innovative alternatives to conventional solutions.

POPULATION PATTERNS

A look at the population density map of sub-Saharan Africa (Figure 7.4) will surprise many readers, who may have the erroneous impression that Africa is densely populated. In fact, the population is distributed very unevenly, but generally sparsely, over the continent. Only a few places exhibit the densities that are wide-

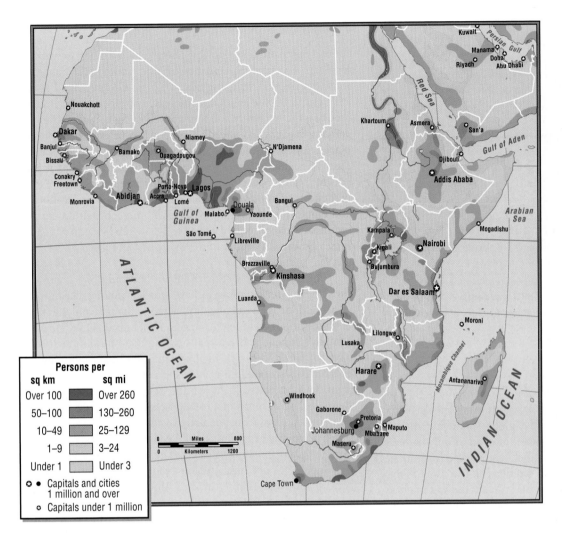

Figure 7.4 **Population density in sub-Saharan Africa.** [Adapted from *Hammond Citation World Atlas* (Maplewood, N.J.: Hammond), 1996.]

spread in Europe, India, and China. Nonetheless, there are serious population problems in Africa. Some countries—for example, Rwanda, Burundi, and Nigeria—have pockets of very high density but have not developed existing resources to support their people. A number of countries (Eritrea, Chad, Liberia, and Niger, for example) have annual population growth rates above 3 percent; the populations of these countries could double by 2025. High population growth rates make it difficult for already poor countries to supply adequate educational and health services to the increasing numbers of people. The result is that living standards decline.

Defining Density

The density of a population depends on a region's **carrying capacity**: the maximum number of people a given territory can support sustainably with food, water, and other essential resources. Several environmental factors affect the carrying capacities of African lands. In some persistently dry places, such as areas bordering the Sahara Desert, the lack of water limits cultivation and grazing. In some persistently wet places, such as the Congo Basin, the leached soils cannot sustain long-term cultivation. Countries with alternat-

ing wet and dry seasons—such as South Africa, Kenya, and other countries in eastern Africa—can support fairly dense populations as long as people use the land and water sustainably. Some otherwise habitable African ecological zones, such as the high plains of the Serengeti southeast of Lake Victoria, are sparsely populated because they are breeding grounds for such diseases as sleeping sickness (trypanosomiasis), borne by tsetse flies, and malaria, borne by mosquitoes (see page 352).

Ultimately, the carrying capacity of any place is affected by cultural, social, economic, and political factors as well as by physical features. For example, in most of Africa, people must depend largely on the local agricultural carrying capacity of the land for a subsistence living. There is insufficient wealth to import such items as food, building materials, and industrial raw materials. Such wealth would allow many more people to be supported than can be supported by relying solely on the carrying capacity of the land. To understand why this is so, compare an American suburban family of four on a half-acre lot, supporting themselves on the adults' annual salaries of $50,000 to $100,000, with a rural African family of seven or eight subsisting almost entirely on 4 acres (1.6 hectares) of land, often with no reliable outside source of cash.

Exhausted and dehydrated Rwandan refugees receive emergency rehydration therapy at a refugee camp in Congo (Kinshasa) run by the United Nations High Commission for Refugees. [Courtesy of UNHCR.]

The Africans must live off their land, whereas the American suburbanites do not subsist on their lawns and flower gardens. The inhabitants of wealthy societies (Europe, North America, Australia, Japan) tap into resources around the globe (including Africa), often at favorable prices. Their standard of living is only partially linked to the carrying capacity of their country's environment. Thus, people in regions like North America can live in cities that are both dense and affluent. On the other hand, the vast majority of people in Africa, like many people in Asia and in Central and South America, have much less access to resources in distant lands because they do not have the political power or cash to acquire them.

Political unrest places additional burdens on a given carrying capacity. War or oppression often creates refugees. Over the last decade of the twentieth century, for example, refugees in such places as Somalia, Ethiopia, Uganda, Liberia, Sierra Leone, Rwanda, and Mozambique have poured back and forth across borders to escape **genocide**—the deliberate destruction of an ethnic, racial, or political group—instigated by rival political factions. According to the United Nations High Commission for Refugees, in the 1990s Africa hosted about one-third of the world's refugees (Asia hosted 44.6 percent, Europe 19.3 percent, North America 5.2 percent, Oceania 0.6 percent, and Latin America and the Caribbean 0.3 percent).

Demographers estimate that if people who are displaced within their home countries are counted, Africa has half the world's refugee population, and about three-quarters of Africa's refugees are women and children. As difficult as the burden is on the refugees, the burden on the countries that host them is also severe. Even with help from international agencies, the hosts find their own development plans derailed by the arrival of so many distressed people, who must be fed, sheltered, and given health care. Large portions of economic aid to Africa have been diverted to deal with the emergency needs of refugees.

Population Growth

African populations are growing faster than any others on earth. In less than 50 years, the population of Africa south of the Sahara has more than tripled, reaching nearly 657 million in 2000. Between 1997 and 2000, the population increased by 50 million. This rapid growth is the main threat to human well-being in places where the carrying capacity has already been reached or exceeded. In some places, Africa's already low standards of living are declining further. The addition of increasing numbers of individuals to feed, educate, house, vaccinate, and employ is outstripping even the best efforts to improve nutrition, education, housing, health care, and employment possibilities. The UN projects that the number of primary school students (aged 6–11) in sub-Saharan Africa will increase from 76.9 million in 1985 to 144.2 million in 2005 and to 195.3 million in 2025.

The geographer Ezekiel Kalipeni has found that many Africans are not yet choosing to have smaller families because they view children as an essential link between the past and the future. Childlessness is considered a tragedy. Not only do children ensure a family's genetic and spiritual survival, they still do much of the work on family farms. In regions with high infant mortality, parents have many children in hopes of raising a few to maturity. In short, in most places in Africa, the demographic transition—the sharp decline in births that usually accompanies economic development—has not yet happened.

In a handful of countries scattered throughout the region, however, the population growth rate is starting to decline. These countries include South Africa, Gabon (on the west coast of Central Africa), and Côte d'Ivoire (in West Africa). In these countries, circumstances have changed sufficiently to make smaller families desirable. First, in places such as South Africa, the education of women has improved and gender role restrictions have been relaxed (as measured by the United Nations Gender Empowerment Measure). Hence, as in other world regions (Middle and South America, East Asia, Southeast Asia, Oceania, and parts of South Asia), women are choosing to use contraception because they have options beyond motherhood. Second, because infant mortality is now relatively low, parents can expect their two or three children to live to adulthood.

VIGNETTE Mary, a Kenyan farmer, has just had her third son. The father of Mary's children works in a distant city and visits the family only several weeks a year. He supports himself with his earnings and buys occasional nonessentials for the family. Mary tells an interviewer that three children are enough for happiness and so today, at age 29, she is having surgery that will prevent conception. She owns only one cow and a small piece of land that can't be further divided, so all she can provide for her children is an education. Mary says that she can afford to educate only three children.

Such attitudes are spreading in Kenya, where food, health care, and jobs are in short supply. Mary plans to augment her farm income by starting a sanitary pit toilet business. She has applied for a small loan (U.S. $150) for this purpose. The success of her business could mean that her children will become well educated and that she herself will gain prestige. If Mary accomplishes her goals, she could become a role model for other women seeking to limit their families so that they can become self-sufficient owners of small businesses. [Adapted from Jeffrey Goldberg, *The New York Times Magazine* (March 2, 1997): 39; and *World Resources, 1996–1997* (New York: Oxford University Press, 1996), p. 5.]

Some factors contribute to reduced family size in any culture. Education helps people understand that under modern conditions, additional children are unlikely to add to a family's wealth as they did in the past. Communication between husband and wife also influences family size. Surveys reveal that men often wish they had fewer children and that they would suggest birth control to their wives if they were encouraged to do so. Breast-feeding after childbirth retards fertility, and in some African ethnic groups it is the custom to abstain from sex for several years after a birth. On the other hand, polygamous relationships, which are rather common in parts of Africa (see page 372), produce more children than monogamous relationships do. In addition, men who choose polygamy tend to be less well educated and to be especially interested in having many children as a status symbol, because children have traditionally been a source of wealth. However, men who choose polygamy also tend to be wealthier, because a husband is expected to provide at least part of the financial support for his wives, children, and multiple households.

The population pyramids of Nigeria, Africa's most populous country, and South Africa, the most developed country on the continent, demonstrate the classic wide-bottomed shape of countries experiencing rapid growth (Figure 7.5). Notice, however, that South Africa's pyramid has contracted at the bottom in the last several years. The birth rate in South Africa has slowed markedly, dropping from 35 per 1000 births to 25 per 1000 just since 1990. This effect undoubtedly results partly from economic and educational improvements and the social changes that have come about since the end of apartheid in the early 1990s. But this pattern in the South African pyramid may also be a consequence of the spread of AIDS among young adults.

The AIDS epidemic is now the main cause of the slowing population growth rates in Africa, especially in the Southern Africa subregion. In 2001, 20.7 percent of Southern Africa's adult population between the ages of 15 and 49 was infected with HIV-AIDS. In other words, one-fifth of those adults who should be in their most productive years are sick and dying. (Africa's HIV-AIDS epidemic is discussed on pages 352–354.)

Population and Public Health

Sub-Saharan Africa has long been troubled by infectious diseases that have been particularly harmful and difficult to control, including schistosomiasis, malaria, river blindness, and cholera. Observers report that infectious diseases (including HIV-AIDS) are by far the largest killers in Africa and are responsible for about 50 percent of all deaths (Figure 7.6). Some of these diseases are linked to particular ecological zones. For example, people living between the 15th parallels north and south of the equator are routinely exposed

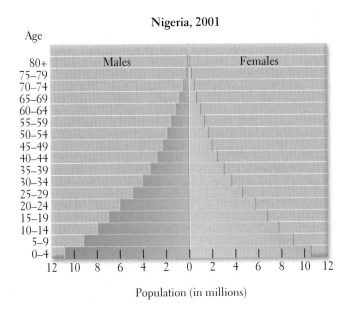

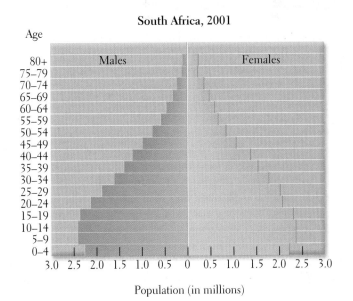

Figure 7.5 **Population pyramids for Nigeria (123 million) and South Africa (43 million).** The fact that the two pyramids are at different scales does not affect their overall patterns. [Adapted from "Population Pyramids for Nigeria" and "Population Pyramids for South Africa" (Washington, D.C.: U.S. Bureau of the Census, International Data Base, May 2000.]

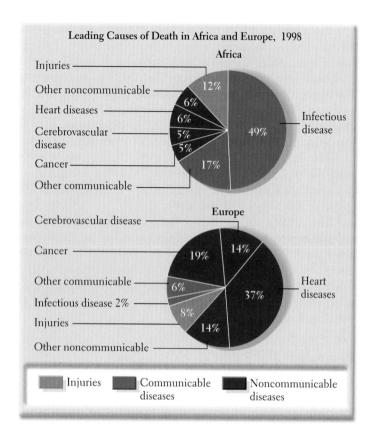

Figure 7.6 Leading causes of death in Africa and Europe, 1998.
[Adapted from *Attaining Global Health* (Washington, D.C.: Population Reference Bureau, April 2000), p. 8.]

to sleeping sickness (trypanosomiasis). It is within this range that the tsetse fly lives, inhabiting the vegetation near rivers and lakes, in open woodlands, and in grasslands. The disease, spread by the bite of this fly, leaves the victim exhausted and, if left untreated, attacks the central nervous system, resulting in death. Sleeping sickness affects both humans and cattle, and the presence of the tsetse fly has limited animal husbandry. Several hundred thousand people, mostly rural residents, are thought to suffer from sleeping sickness, and most of them are not treated because they cannot afford the expensive drug therapy developed in Germany.

Schistosomiasis is Africa's most common chronic tropical disease and malaria its second most common. Both diseases are linked to standing freshwater, and their incidence has increased with the construction of dams and the growing of rice in paddies. Schistosomiasis develops when a particular snail larva enters the skin of a person standing in freshwater. The victim first experiences a rash, then a cough and diarrhea, headaches, fever, and general malaise. If the disease is left untreated, the liver and spleen are impaired and the belly swells and is painful. The condition can lead to bladder cancer. Malaria is spread by the anopheles mosquito, and its victims experience chills and fever alternately. Left untreated, malaria will lead to anemia and jaundice and eventually to enlargement of the liver and spleen. In some cases, damage

to the brain leads to death. Until recently, little research in Africa was focused on how best to control the most common chronic tropical diseases. Now, several groups based in Kenya are working on control measures for both schistosomiasis and malaria.

The bite of the blackfly introduces a parasite into the human body that causes river blindness (onchocerciasis). This disease afflicts millions of Africans between the 20th parallels north and south, especially those who live along fast-moving rivers because the blackfly requires fast-moving water to breed. The parasite first causes itching, then disfigurement, and finally eye lesions that lead to blindness. Nineteen countries in the affected zone participate in the Onchocerciasis Control Programme funded by the World Health Organization (WHO), the United Nations Development Program (UNDP), and the World Bank. French scientists and scientists from the member countries have devised ways to control blackfly breeding with insecticides; and a new drug, Ivermectin, which kills the parasites, has been administered to more than 20 million people since 1997. It appears to act both as a preventative and as a cure.

HIV-AIDS in Africa

HIV-AIDS is the most severe public health problem in sub-Saharan Africa, but it is also a serious social issue in a large number of countries. According to a joint report by the United Nations and the World Health Organization, in the year 2000, sub-Saharan Africa had an estimated 70 percent of the worldwide total of 36.1 million HIV-positive cases (Figure 7.7). For the world as a whole, only 1.1 percent of the adult population is infected with HIV-AIDS. In North America, the figure is 0.6 percent and in East Asia and the Pacific just 0.07 percent; but in sub-Saharan Africa, 8.8 percent of adults are infected, and in certain countries the rate is much higher. By 2001, 36 percent of adults aged 15 to 49 were infected in Botswana; 25 percent in Zimbabwe and Swaziland; about 20 percent in Zambia, Namibia, and South Africa; and 16 percent in Malawi.

The disease threatens to change sub-Saharan Africa's population growth and life expectancy patterns drastically. According to a Population Reference Bureau report issued in 1997, AIDS will probably reduce life expectancy in Zimbabwe from the 1997 level of 51 years to 33 years by 2010; by 2000, life expectancy had already fallen to 40 years. Had HIV-AIDS not occurred, Zimbabwe could have achieved a life expectancy of about 70 years by 2010; instead, life expectancy may fall to what it was 200 years ago. Zimbabwe's population growth rate was 3 percent in 1993, but by 1997 it had dropped to 2.5 percent, primarily in response to rising standards of living. Then the AIDS epidemic reached Zimbabwe, and by 2000 the country's population growth rate was only 1.0 percent. If the epidemic is not stopped, Zimbabwe could lose 20.5 percent of its population by 2010. The demographic effects of HIV-AIDS are expected to be comparably dramatic in such countries as Uganda, Nigeria, Kenya, Burkina Faso, Rwanda, Zambia, Namibia, Tanzania, Ethiopia, and South Africa.

In Africa, AIDS affects both men and women: 55 percent of infected adults in sub-Saharan Africa are women, and four-fifths of

Thinking Globally: ON THE WEB

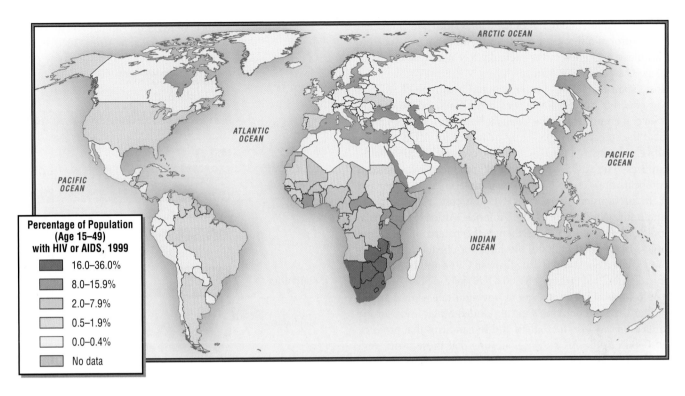

Figure 7.7 **Global prevalence of HIV and AIDS.** The map shows that the countries with the highest percentage of people aged 15 to 49 infected with HIV or suffering from AIDS are in Africa.
[Adapted from *2001 World Population Data Sheet* (Washington, D.C.: Population Reference Bureau.]

the world's women with AIDS are in Africa. Many of them are young mothers infected by their husbands, who may visit sex workers when they travel for work or business. In some cities, virtually all sex workers are infected. In Uganda, one-quarter of married urban women, who are normally considered at low risk, are HIV-positive. Because AIDS develops slowly in its victims, controlling the epidemic is particularly difficult. UN officials suspect that 90 percent of African carriers of HIV do not learn they have been infected until several years after they acquired the virus. During this time, they may produce HIV-positive children and pass the disease to other adults. The overwhelming majority of Africans with HIV-AIDS cannot afford the costly combination of drugs that keeps victims alive in such places as North America, Europe, and Japan.

The rapid urbanization of Africa (see the discussion on page 368) has contributed to the rapid spread of HIV-AIDS. Transportation between cities and the countryside has improved, and bus and truck drivers are thought to be major carriers of the disease, as are urban migrants who come home for visits. Young men and women encounter each other more easily in urban settings, often without the community pressure that would have precluded intimacy in the countryside. Women may be coerced to have sex, and they have great difficulty insisting that their mates use condoms. According to custom, men often have multiple partners. A popular belief is that only sex with a mature woman can cause AIDS, so very young girls are increasingly sought as sex partners (this is sometimes referred to as "the virgin cure"). For many poor

urban women, removed from village support systems, occasional sex work is part of what they do to survive economically.

Education has played a changing role in the spread and control of HIV-AIDS in Africa. Early in the epidemic, educated Africans were more susceptible to infection, partly because they were likely to live in urban areas and to be free of community-enforced sanctions against multiple sex partners. Now that the epidemic has grown, those who can read and understand explanations of how HIV is spread have an advantage, and rates of infection are now lower among educated people than among people who lack education. Nonetheless, perhaps as a result of the lag between exposure and the development of obvious symptoms, university campuses in some countries are being decimated by the disease. In 2001, Nelson Mandela reported in speeches that in a given year some countries in southern Africa have lost more than a third of their teacher trainees to AIDS.

Despite the devastating effects of AIDS in some regions, demographers predict that in the end, the total population of Africa will be only about 5 percent smaller than it would have been without AIDS (Figure 7.8). Nonetheless, the social consequences of the AIDS epidemic are enormous. By 2000, more than 15 million Africans had died. Young adults, parents, teachers, skilled craftspeople, and trained professionals are lost to a slow, agonizing death in the prime of their lives. Elderly grandparents must provide for orphaned dependent children (who may also have been infected since birth). Time, money, and energy are being channeled to

HIV-AIDS prevention and treatment and away from education and from the control of other diseases such as schistosomiasis, malaria, river blindness, and cholera that actually threaten a much larger percentage of sub-Saharan Africans.

Most experts agree that prevention and treatment of AIDS is the most effective strategy for reducing AIDS-related illness and death. Several countries, led by Uganda, have launched programs to combat the spread of AIDS. Drug treatment for AIDS remains prohibitively expensive for many patients, even within the wealthiest countries. The AIDS "drug cocktail" produced by American and European drug companies costs around $10,000 a year per patient. In 2000, in an attempt to lower this cost, a global movement including countries in Latin America, Africa, South Asia, and Southeast Asia began to challenge the patents of the drug companies. By the end of 2000, the hope for cheaper drug treatments was given a boost when drug firms in Cuba and India began to offer a regimen that costs just $1.00 a day ($365 a year)—a price that people in most countries where infection rates are high still cannot afford. Drug companies in the United States and Europe brought lawsuits against the Cuban and Indian firms that produce cheaper copies of the AIDS drugs, arguing that those firms were illegally using drug formulas that had been developed and patented at great cost. Those suits were dropped, however, when the plight of untreated, poverty-stricken AIDS patients in Africa and elsewhere was broadcast worldwide and the drug companies found themselves accused of putting profits ahead of human lives. By mid-2001, several private foundations were seeking ways to provide affordable drug treatment to the more than 36 million AIDS patients worldwide.

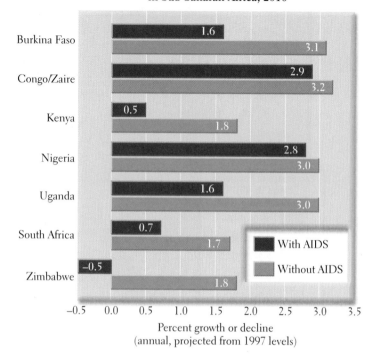

Effect of AIDS on Population Growth in Sub-Saharan Africa, 2010

Percent growth or decline (annual, projected from 1997 levels)

Figure 7.8 **AIDS and population growth.** By 2010, AIDS is expected to reduce population growth in many countries of sub-Saharan Africa. [Adapted from S. Jay Olshansky et al., "Infectious diseases—New and ancient threats to world health," *Population Bulletin* 52, no. 2 (July 1997): 20.]

CURRENT GEOGRAPHIC ISSUES

Most countries of sub-Saharan Africa have been independent of colonial rule for less than 40 years. Many are only beginning to build their nations and must come to terms with such issues as arbitrary national boundaries and rapid urbanization. Africans are still developing appropriate systems of government organization and economic policies that will lead to sustainable growth and development. Because the effects of the colonial era in Africa run deep, it is too soon to tell just how some of the African countries will find ways to become politically, economically, and socially viable. Still, some countries are achieving higher standards of living for their citizens, and many researchers are optimistic about the future. A broad consensus is developing that Africa's prospects will improve if development policies give Africans the major power in seeking and implementing solutions to the problems they face.

ECONOMIC AND POLITICAL ISSUES

Africa cannot continue to produce what it does not consume and consume what it does not produce.

(THE AFRICANS, CO-PRODUCED BY WETA-TV, WASHINGTON, D.C., AND THE BRITISH BROADCASTING CORPORATION.)

Geographer Robert Stock writes that the 1960s were a time of hope and optimism for Africa south of the Sahara. In that decade, 31 countries became independent; the prevailing view among economists and political leaders was that modernization and development would at last give Africa a chance to benefit from its own resources and labor. In the 1970s, the optimism faded as development proved much harder to achieve than expected. By the 1980s, the news was bleak: Africa was not just failing to thrive, it was clearly in decline. Videos of starving refugees from civil wars appeared on the nightly news in Europe and America, and countries that had shown promise began to incur huge debt. Roads and water systems deteriorated for lack of maintenance, and agricultural production leveled off and even declined. The cost of food and basic necessities rose beyond what ordinary people could afford. In the 1990s, civil wars spread and then the AIDS epidemic hit. Efforts by outside agencies such as the United Nations and the International Monetary Fund to assist with aid packages, peacekeepers, and economic restructuring helped to stabilize governments and economies, but only after major losses and at great cost to African well-being. Often the prescribed cures seemed worse than the sickness, as governments were advised to tighten budgets and cut not just jobs but basic health and education services.

After decades of dictatorship, ethnic strife, and elite dominance, all of which brought economic instability, some countries (Botswana, Namibia, Swaziland, and South Africa, for example) are beginning to move toward more democratic forms of government. Although democratic gains in these countries remain tentative and fragile, many people hope that democracy will provide the stability needed for economic development. With that hope, Africans are now pursuing alternative strategies to strengthen their local and regional economies. So far, though, these efforts provide only partial solutions. The long slide that began in the 1960s demands deep changes in Africa and in its economic relations with the rest of the world.

Stagnating Economies

After years of declining output and shrinking economies, many Africans are poorer today than they were in the 1960s. Figure 7.9 compares GDP per capita for the decades from 1960 to 1990 for selected African and Asian countries. These economies began at roughly similar GDPs per capita in 1960. Between 1960 and 1990, GDPs rose modestly but very steadily in India (65 percent) and dramatically in South Korea (638 percent). In Africa, GDP per capita rose very modestly in Kenya, Tanzania, and Zimbabwe, shown in the figure, and sank in Uganda and Ghana. In nearly all countries, GDP had risen by 2000.

British development specialist Elizabeth Francis agrees with Stock that there are several reasons for Africa's difficulties in the postindependence era. First, the long-term impact of colonialism placed Africa in the global market as a supplier of cheap resources and low-cost labor. That role persists today, leaving Africa in competition with other suppliers of raw materials and with little influence on market prices. Second, political chaos followed independence. The political leaders and citizens of the new states had little or no experience in allocating resources, power, and opportunity fairly. Consequently, corruption became widespread and people were unable to assert their will through democratic channels. Third, decades of war in many countries retarded both social and economic progress. During the cold war, interventions by the United States, Europe, and the Soviet Union intensified the wars in Africa as these world powers attempted to draw new African nations to one side or the other of the cold war (see pages 364–365). Fourth, steep oil price increases in the 1970s forced governments into debt to pay for fuel. Finally, ill-advised economic reorganization and structural adjustment programs were promoted by outside agencies. These belt-tightening strategies were widely prescribed in the 1980s and 1990s, often by international financial institutions, to remedy indebtedness and dysfunctional development (see SAP discussion, page 358).

Subsistence Agriculture: Source of Livelihood for Most Africans. Most Africans, 70 percent of whom live in rural areas, produce their own food by farming small plots, raising livestock, or a combination of both. Many also fish, hunt, and gather some of their food. Often, small-scale cash cropping, adopted during the colonial era to make money to pay taxes, is part of the system of production. Today people may grow cash crops of peanuts, cocoa beans, rice, or coffee to obtain money to pay school fees or purchase a bicycle or other useful item.

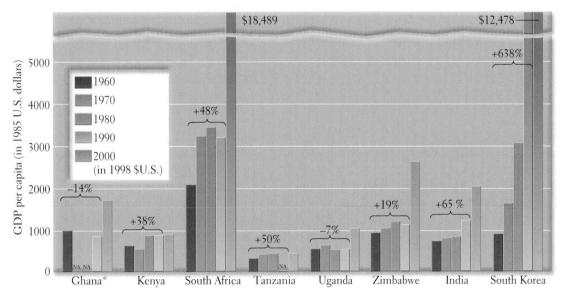

Real GDP per Capita, 1960–1990

Figure 7.9 **Real GDP per capita (in 1985 U.S. dollars; 2000 GDP in 1998 U.S. dollars) for selected sub-Saharan African countries, plus India and South Korea.** Notice that some African countries experienced very modest growth or even decline in 1960–1990. Some grew significantly in 2000 while others did not. [Adapted from Elizabeth Francis, *Making a Living: Changing Livelihoods in Rural Africa* (London and New York: Routledge, 2000), p. 2.]

Farms are usually 2 to 10 acres (1 to 4 hectares), the size being limited more by lack of available labor than by scarcity of land. Fields given over to food production are often planted with many different species (intercropped), in a planned arrangement that is rotated over the course of the field's use. After a few years of cultivation, the field is allowed to lie fallow (rest) so that the soil can be replenished by the growth of natural vegetation. In some places, population density has increased to the point that sufficient fallow times are no longer possible and soil fertility is decreasing. Farmers in many environments—the tropical forest fringes of Nigeria and the high volcanic plateaus of southern Uganda, for example—cultivate their fields permanently. Permanent cultivation requires a high degree of specialized knowledge: fields are intercropped, and the needs of particular species are carefully assessed and met. Great care is taken to fertilize the soils continuously and intensively. Sometimes commercial fertilizer is used.

Most African farmers also raise livestock; but they usually limit their holdings to a cow or two, some poultry, and several goats and sheep. Herding is practiced primarily on savanna grasslands and desert margins and also in the mixture of grass and shrubs called open bush. Ideally, herders move their animals seasonally to the freshest pasture to minimize damage to the grazable grasses. Increasingly, however, drought or the encroachment of field agriculture has forced pastoralists to reduce or divide their herds. Some herders have even taken up farming, though to do so is considered a step down in status.

Agriculture is an economically significant part of the urbanization process in Africa. Tiny vegetable gardens (with maize, beans, potatoes, yams, greens, and bananas) and fowl pens can be seen even in the hearts of central business districts. A study has shown that in Mombasa, Kenya, 15 percent of households grow food on urban land. The produce provides nutrition to city dwellers; generates employment, especially for women; and puts derelict urban land to sustainable productive use.

Exporter of Raw Materials to the Global Economy. After independence, African economies were still centered around the export of one or two raw materials, and this pattern has continued (Figure 7.10). Southern Africa exports a wide range of minerals and produces 27 percent of the world's gold and 50 percent of its platinum. West Africa exports oil (from Nigeria) and produces about half of the world's cocoa beans, nearly all of which supply the European market. Peanuts and fish are other West African exports. East Africa exports small percentages of the world production of copra, palm oil, and soybeans; Central Africa exports some tropical hardwoods.

Large-scale commercial cultivation for export of tree crops such as palm oil, copra, and cacao; rice; and various dessert fruits is increasingly common as African countries search for crops that will sell well in the global marketplace. Commercial agriculture has expanded in sub-Saharan Africa despite the infertile soils, thanks to improved crop varieties, more effective environmental management techniques, and greater investment in roads and the water supplies needed for agriculture. Large commercial farms produce cotton, coffee, tea, maize, groundnuts, poultry, and dairy products. These farms are usually several hundred acres in size and use mechanized equipment and chemical fertilizers and pesticides.

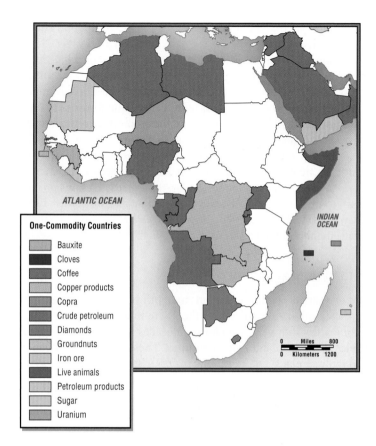

One-Commodity Countries

- Bauxite
- Cloves
- Coffee
- Copper products
- Copra
- Crude petroleum
- Diamonds
- Groundnuts
- Iron ore
- Live animals
- Petroleum products
- Sugar
- Uranium

Figure 7.10 One-commodity countries. Most exports from African countries are shipped as raw materials (metallic ores and timber, for example) that require further processing and manufacturing before they become consumer products. Because most African countries have only one to four of these primary products, they are dependent on just a few finite resources. This map depicts those countries that depend on just one commodity for more than 50 percent of their export earnings. [Adapted from George Kurian, ed., *Atlas of the Third World* (New York: Facts on File, 1992), p. 76; and data from *United Nations Human Development Indicators 2000* (New York: United Nations Development Programme).]

Recently, some countries (for example, Kenya and Zimbabwe) have become producers and exporters of horticultural products such as vegetables and cut flowers. Commercial agriculture is sorely needed to meet the demands of growing populations and to supply products for export.

Africa is currently ranked at the bottom of world regions in terms of global trade. Its share of world trade is only 2 percent, which presents Africans with the challenge of playing "catch-up." Sub-Saharan Africa has had a weak position in the world economy partly because global prices for its raw materials fluctuate according to supply and demand, and for several decades prices for many of these materials have been down. The production of commodities such as palm oil, cacao, coffee, and bauxite has increased both in Africa and in relatively poor countries elsewhere, while the consumption of raw materials has not expanded to meet supply. Meanwhile, prices have risen for the products that Africa imports, including the machinery required by mines and the fertilizers, special seeds, equipment, and pesticides required by modernizing farms and plantations.

Botswana: A Case Study. A look at the country of Botswana in southern Africa demonstrates that a rich resource base may not translate into prosperity for the majority of a country's people. Since independence in 1966, Botswana has been hailed as a rising economic star in Africa. The GDP per capita had been U.S. $535 in 1960; by 1998, it was U.S. $6103 (adjusted for purchasing power), second only in the southern part of Africa to South Africa (at U.S. $8488). How is an arid, landlocked country able to prosper so well? More important, to what extent does the general population share in the country's wealth?

Botswana derives its wealth from mineral resources. Diamond deposits were found shortly after independence, and by 1995 Botswana was the world's third largest diamond producer and ranked first in diamond output value. The country also generates income from deposits of gold, soda ash and potash, copper, nickel, and coal. Diamonds are the mainstay, however, and account for over 63 percent of government revenues and 80 percent of export dollars. When such wealth is statistically apportioned among just 1.5 million people, the per capita GDP is high by African standards.

The actual allocation of wealth in Botswana, however, is uneven. Fewer than 1 percent of the people have become very wealthy, and perhaps as many as 20 percent have reasonably well paid positions in government or the mining industry. The remaining 80 percent work in agriculture, and most suffer dire poverty.

The Jwaneng mine in Botswana is one of the world's richest. The diamond-laden earth is loaded into trucks and taken to a treatment plant for crushing and sifting. The diamonds are then separated into industrial and gem-quality stones. [Peter Essick/Aurora and Quanta Productions.]

The lot of the poor is made worse because some of the newly wealthy people have purchased productive land for large-scale cattle ranching. Owners of small farms who sell their land to cattle ranchers have more cash temporarily, but they are pushed onto less productive lands and soon cannot support themselves adequately. Although meat production has increased, much of it is exported, and overall food production has decreased. By the late 1990s, the country produced only 50 percent of the food needed to feed its population; the rest had to be imported, and one-third came in as food aid. So even though there is an appearance of increasing wealth in the statistics, only a small percentage of the population does well, a situation that is reflected in Botswana's relatively modest global ranking of 122 on the Human Development Index—which, nonetheless, is relatively high for Africa.

South Africa's Success. South Africa has become an exception to the general pattern because it managed to create a large and diversified economy. South Africa's total output in 1998 ($137 billion) was only slightly less valuable than the output of all other sub-Saharan African countries combined. A major reason for South Africa's success is that for centuries it had a large European population with the skills and external connections to foster economic development. In 1910, the Dutch (Boer) and British settlers of South Africa became the rulers of an independent South Africa. During the twentieth century, they established many lucrative industries, such as the refining of minerals, gemstones, and metals from the country's rich mines. These enterprises kept profits in the country, allowing the construction of a large industrial economy.

Nonetheless, only a small minority reaped the benefits of this effort. Although the labor of black workers helped build South

Although not yet one of Mozambique's principal exports, tea is becoming an important cash crop in Gurue, Zambezia Province. For example, Chazeira de Mozambique, the largest tea grower in the region, exports tea to Pakistan and Yemen and plans to expand to other countries. In this photo, workers who have picked the tea leaves by hand to preserve quality and flavor wait in line to have the tea weighed. [Brian Seed.]

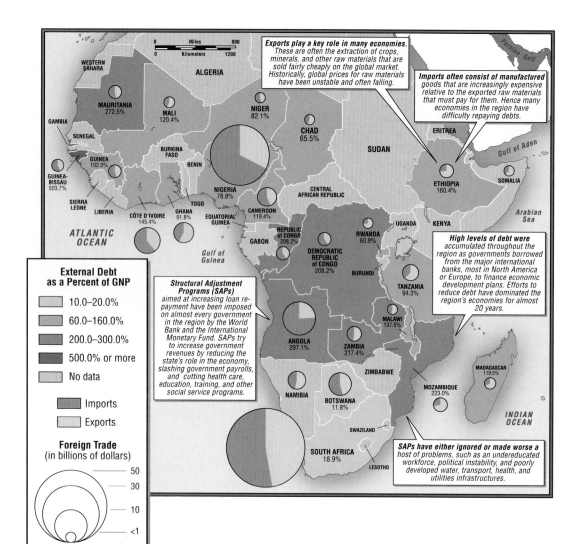

Figure 7.11 Economic issues: sub-Saharan Africa. Increasingly, imports are exceeding exports even in countries, like Nigeria and South Africa, with major exportable resources. When imports exceed exports and money is borrowed for large projects, debt rises. [Debt data from *United Nations Human Development Report 2000.*]

Africa's exceptional economy, only a few black South Africans benefited from relatively high wages. Very low wages and the apartheid system kept the vast majority poor. Essentially, 100 percent of the population worked to provide a high standard of living for 16 percent. With the end of apartheid, this pattern has begun to change somewhat (see pages 348, 391).

The Current Economic Crisis

More than half of sub-Saharan Africa's countries averaged economic growth rates of at least 4 percent a year in the late 1990s. However, this growth has not been enough to erase the effects of the downward spirals of the last two decades. As a result of the low prices paid for their products on the global market, and the high prices of imported industrial products, most countries have found it increasingly difficult to repay loans extended for mechanized agriculture, small industries, hydroelectric plants, roads, and so forth (Figure 7.11)

In the early 1980s, the major international banks and financial institutions that had made loans to African governments developed a plan to obtain repayment. Working through the World

Bank and the International Monetary Fund, the lending institutions threatened to stop all further lending to the debtor countries unless they enacted specific packages of economic reform called **structural adjustment programs (SAPs).** SAPs try to increase government revenues by reducing state interference in the economy. According to supporters of SAPs, economies guided solely by free-market mechanisms will allocate investment money and resources better, will be more vibrant and entrepreneurial, and hence will produce more tax revenues with which to pay off the loans. To raise money, governments are also required to slash their own payrolls by cutting health care, education, training, and other social service programs.

SAPs have accomplished some good. They tightened bookkeeping procedures and thereby curtailed corruption and waste in bureaucracies. They reduced the power of the elites to commandeer resources for their own profit. They eliminated some inefficient and corrupt state agencies that were cheating farmers in the process of getting their crops to market. (In Ghana, for example, cacao producers can now negotiate directly with buyers from Europe.) And they stopped the practice of capping food prices to appease urban dwellers; as a result, local food producers received fair prices for their crops and were encouraged to produce more. SAPs closed state-owned monopolies in industries and services, and they opened some sectors of the economy to medium- and small-scale business entrepreneurs. They also made tax collection more efficient.

SAPs have also exacerbated some problems. Researchers in Africa, including Robert Stock and Elizabeth Francis, cited previously, agree that SAPs have made it harder for the poor majority to make a decent living and stay healthy, and they have failed to reduce the debt burden. On the contrary, the debt has continued to grow, despite the fact that the region as a whole is now spending more on debt payments than on all health care and education—more than U.S. $13 billion a year. One reason the debt persists is that SAPs have generally eliminated more jobs than they have created. To maximize revenues and achieve greater efficiency, governments have sold off state-owned businesses to the private sector and removed protections for local industries. Both steps usually bring firings and layoffs. Additional jobs have been lost as large state bureaucracies have been significantly reduced. Moreover, contrary to SAP theory, Africa did not attract foreign investment once the cuts had been made. Investors have been discouraged by problems that SAPs have either ignored or made worse: an undereducated workforce; political instability; and poorly developed water, transport, health, and utility infrastructures. The result has been rising unemployment and economic decline that has left more than two-thirds of the population in poverty and prone to social unrest. By 1998, foreign investment in sub-Saharan Africa was less than one-tenth of that in East Asia or Latin America and only slightly higher than that in South Asia (Figure 7.12). South Africa

and Nigeria together accounted for half the direct foreign investment in all of sub-Saharan Africa.

Agriculture and Economic Restructuring. Economic restructuring has especially affected the agriculture sector, which employs 70 percent or more of Africans in one way or another. Modern agricultural development programs provide incentives to encourage investment in growing cash crops—such as cocoa, sugar, bananas, and palm oil—for export. These incentives may include lower taxes on profits, decreased government regulation, and low-cost loans for equipment and land. At the same time that export cropping has been encouraged, however, structural adjustment policies have inadvertently reduced the profitability of food production for Africa's own use.

For example, the often recommended policy of **currency devaluation** lowers the value of African currencies relative to the currencies issued by other countries. Devaluation promotes the sale of African export crops by making them cheaper on the world market, but it also makes all imports more expensive. Therefore, farmers who grow food for the African market must spend much more on imported seeds, fertilizers, pesticides, and farm equipment. Also, with the shift to export production, many farmers no longer have the time or space to grow their own food or food for local markets—so, increasingly, Africans must pay for expensive imported food. Meanwhile, because SAPs have also mandated production increases in many competing countries, world market prices for agricultural export crops have declined, eroding farm incomes.

In some countries, land has been taken away from local people outright and made available to foreign-owned export-agriculture ventures. This is one reason that critics equate SAPs with colonial exploitation. In northern Mozambique, for example, white Afrikaner (of Dutch descent) plantation owners who emigrated from South Africa have bought some of the best lands for large, chemical-intensive farms that produce grain, beef, and fruit. Local farmers, who had been accustomed to cultivating land they did not own, have lost access to lands they had farmed for generations. They now face a future as laborers or sharecroppers on the new plantations, where they are paid little and endure abusive treatment. The government of Mozambique supports the program because it increases the value of Mozambique's exports and allows it to pay off its debts. Paradoxically, the government of South Africa also supports the Mozambique program. The South African government intends to redistribute the lands left by emigrating white farmers among the much poorer South African black majority.

Agricultural scientists, with their European and U.S. backgrounds, typically neglected to consider that in Africa women grow most of the food for family consumption. Hence, they did not consult women or include them in development projects, except possibly as field laborers. The displacement of women from the land and food production and the focus on cash crops contributed to the economic decline in sub-Saharan African countries after 1960. Between 1961 and 1995, per capita food production in Africa decreased by 12 percent, making it the only region on earth where people are eating less well than they used to.

The overwhelming emphasis that SAPs place on the energy- and resource-intensive farming techniques common in the United States and Europe disregards growing evidence that indigenous

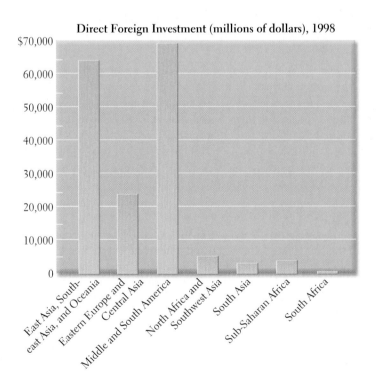

Direct Foreign Investment (millions of dollars), 1998

Figure 7.12 **Direct foreign investment in several world regions, plus South Africa, 1998.** [Data from *World Bank Development Report, 2000/2001: Attacking Poverty* (New York: Oxford University Press, 2000), pp. 314–315.]

GEOGRAPHER IN THE FIELD

Interview with Regional Consultant for Chapter 7, Barbara E. McDade, University of Florida

What led you to become interested in sub-Saharan Africa?
During the 1960s and 1970s, many African-Americans of my generation became increasingly curious about Africa and our relationship to the region. Although Africa is the historical origin of African-Americans, few of us knew much about it. I first received a nuanced picture of Africa in college, when I enrolled in a course taught by the renowned artist Dr. John Biggers, whose visit to Ghana in 1957 produced a beautiful book of illustrations, *Ananse*. Rather than presenting Africans as strange specimens of some bygone era, his drawings of the people of Ghana showed them as "regular folks" working, educating their children, playing, and going about their daily lives. Dr. Biggers's book inspired me to see Africa for myself. I first visited Ghana in 1985 and have returned many times. Recently, I have expanded my research interests in visits to southern Africa (Botswana and South Africa) and East Africa (Kenya and Uganda).

Where have you done fieldwork in sub-Saharan Africa, and what kind of work have you done?
I began my fieldwork studying a traditional business sector in Ghana. The subject of my dissertation was the entrepreneurship of cane and rattan weavers, an artisanal industry, and the factors that lead to their success. I have also studied networking among automobile repair shop workers in an industrial district in Ghana. My recent fieldwork looks at regional networking among entrepreneurs in the modern business sectors of information technology, telecommunications, horticulture exports, retail stores, and electronics manufacturing and assembly.

How has that experience helped you to understand the region's place in the world?
Visiting Africa, and meeting and interviewing Africans as they go about their daily lives, have given me a better understanding of contemporary Africa. In the United States, much of what one hears about Africa is sensational and crisis-based. Although there are certainly many alarming or "ugly" situations there, there is so much more to Africa than suffering, poverty, ethnic conflict, and malaise.

What are some topics that you think are important to study about your region?

• Precolonial history. The long period of cultural development that occurred before contact with Europeans continues to influence contemporary events in Africa more than has been generally understood.

• History written from African perspectives, which provides fresh ways to conceptualize events and their impact.

• Educational reform. Many African governments are reworking their educational systems to make them more relevant to their societies' needs for development.

• Research on indigenous crops and agricultural practices. During the past several decades, the introduction of "modern" or imported agricultural techniques often discounted or ignored existing knowledge that farmers in Africa had developed over centuries to adapt to their particular environments. Programs now link indigenous technology with "Western" or imported technology to form a best-practices approach.

• Private sector development. My recent field research indicates a growing understanding among African policy makers that the state and the private sector must become cooperative rather than competing actors if there is going to be sustainable economic development and improvement in the quality of people's lives. Geographers can explore how this emerging awareness will affect development in Africa.

African farming techniques are more sustainable and better adapted to African conditions. Commercial farming techniques often are not sustainable, because soils do not remain fertile under continuous use in many moist, humid African environments. In Nigeria, Dr. Bede Okigbo, a botanist at the International Institute of Tropical Agriculture (IITA) in Ibadan, was eventually able to convince aid officials of what he had long known to be true: traditional farmers (many of them women), growing complex tropical gardens, are highly skilled systems analysts. They can successfully cultivate 50 or more species of plants at one time, sustainably producing more than enough food for their families on a continual basis. "At IITA we were working at a much simpler level than the local farmer," says Okigbo. "We [could only manage] one or two systems, while they had to deal with something far more complex." In the late 1990s, Dr. Okigbo and other scientists were able to persuade international aid agencies to design farm aid projects along lines suggested by experienced local farmers.

Industry and Economic Restructuring. Africa's debt crisis was triggered partly by attempts, prompted by the World Bank, to develop industries for export rather than for domestic markets. For example, in the 1980s, the World Bank financed a shoe factory in Tanzania that was to use that country's large supply of animal hides to manufacture high-fashion shoes for the European market. Machines for this type of footwear had to be imported at high cost. Due to miscalculations, by both Tanzanians and the World Bank, the factory never managed to produce shoes suitable for Europe.

Unfortunately, the machines could not produce the type of shoes Africans would care to wear, even though there is a large market in Africa for good shoes. The unused factory deteriorated, but Tanzania was still expected to repay the cost to the World Bank.

The trend toward export-oriented industry was furthered in other ways. For example, SAP policies encouraged the massive Maputo Corridor highway project between South Africa and Mozambique. This road mainly connects ports, plantations, and mines, bypassing local communities whose farm-to-market roads remain in disrepair. In Ghana, foreign-owned mining companies are taking land away from local farmers and pushing out the small-scale mining sector, diverting money away from local economies.

Whether export-oriented manufacturing industries can succeed is uncertain. The rich industrialized countries place tariffs on African manufactured goods such as footwear and clothing, so Africa lacks access to prosperous markets. But, if Africa sells its products only to other poor regions, the demand for Africa's products may be highly unpredictable because poor people's access to cash fluctuates. Some aspects of SAPs have actually worked against the African export-oriented manufacturing industries they were intended to help. For example, SAPs and belief in the free market have inspired governments to reduce tariffs that formerly kept out textiles from Asia. The flood of imported cloth has led to factory closings and massive job losses in the textile industries of Tanzania, Zimbabwe, and Uganda. Meanwhile, throughout Africa, currency devaluations have made imported equipment and spare parts for locally focused industries prohibitively expensive.

Economic and industrial recovery began in Africa around 1995, after two decades of stagnation, but the recovery will have to gain momentum if Africa is to raise its standards of living and improve the well-being of the majority of its people. Investment from both domestic and foreign sources will be required if countries are to broaden and diversify their industrial bases. In its 2001 report on business opportunities in Africa, the Mbendi Group, which provides services for those wishing to invest in Africa, noted that although investment risks are high, the return on successful investments is also higher than in most other world regions, including eastern Europe and Asia. Some observers are now describing African countries such as Ghana, Mozambique, Nigeria, and South Africa as "the last emerging markets." Financial stock markets are opening all over Africa, a sign of increased confidence in the potential for business growth.

The Informal Economy and Economic Restructuring. All in all, the various strategies dictated by SAPs have resulted in hardship for Africa's working people: lower wages, lost jobs, fewer social services, deteriorating working conditions. For many, the escape hatch has been the informal economy. Most people who work in the informal economy perform useful and productive tasks, such as selling craft items, prepared food, or vegetables grown in one of Africa's prolific urban gardens. Others make a living by smuggling scarce or illegal items, such as drugs, weapons, endangered animals, or ivory.

In most African cities, informal trade once supplied perhaps a third to a half of all employment. Now it often provides more than two-thirds. Informal employment has offered some relief from abject poverty, but it cannot resolve the overarching problems of African economies. Because the informal economy is very difficult to tax, it does not help pay for government services or repay debt. Its profits are unreliable, even declining. More and more people compete to sell goods and services to people with less and less disposable income.

One result is a decline in living standards that is disproportionately severe for women and children. As large numbers of men have lost their jobs in factories or the civil service, they have crowded into the streets and bazaars as vendors, displacing the women and young people who formerly dominated there. Women and children have turned to contract work, manufacturing clothing, shoes, or other items in their homes. Often they work long hours for little pay, using chemicals or techniques that are hazardous to their health. With families disintegrating under the pressure of economic hardship, some women turn to sex work, putting themselves at high risk of contracting AIDS. More and more children from these disintegrating families must fend for themselves on the streets. Cities such as Nairobi, Kenya, which had very few street children even as recently as 1985, now have thousands.

Alternative Pathways to Economic Development

For many reasons, Africans seek alternatives to past development strategies. Regional economic integration along the lines of the European Union, South America's Mercosur, or Southeast Asia's ASEAN is one such strategy. Another is locally designed and based grassroots development.

Regional Economic Integration. Many African countries, especially in West Africa, are simply too small to function efficiently in the world economy. Their national markets and their resources cannot nurture a significant industrial base. These countries have remained heavily dependent on Europe, especially the former colonial powers, for most of their trade. A measure of this dependency is that only 4 percent of the total trade of sub-Saharan Africa is conducted between African countries. It is difficult to establish regional trade links because the necessary transport and communication networks are not in place. Air travel and even long-distance telephone calls from one African country to another often must go by way of Europe, and roads are poor even between major cities. All these impediments make doing business in Africa about 50 percent more expensive than doing business in Asia.

Governments in Africa have seen the bad results of surrendering their authority to form economic policy to distant international lending and finance institutions, such as the IMF and World Bank. Therefore, governments are reluctant to surrender what little power they still have to a new regional bureaucracy that would implement economic integration. Nonetheless, there are a number of subregional organizations working toward economic integration. Three of the best known are the **Economic Community of West African States (ECOWAS)**, the **Southern African Development Community (SADC)**, and the **East African Community (EAC)** (Figure 7.13).

ECOWAS, which is dominated by Nigeria, has worked toward removing or reducing tariffs, forming a common currency, and forging cooperation among former British and French colonies. SADC, which has grown to include 14 countries in southern and East Africa and includes more than 200 million

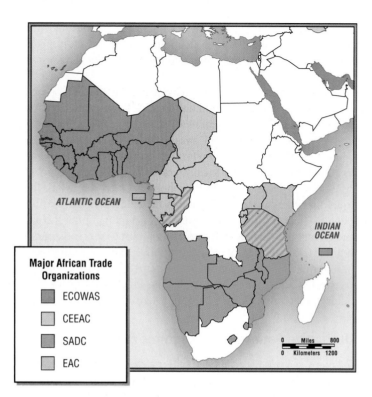

Figure 7.13 Principal trade organizations in sub-Saharan Africa. Congo (Brazzaville) and Tanzania belong to two trade organizations.

Major African Trade Organizations

- ECOWAS
- CEEAC
- SADC
- EAC

ATLANTIC OCEAN

INDIAN OCEAN

people, has major projects under way to link transport and communications infrastructures and to work toward building regional industrial capacity and freer trade. The EAC unites Kenya, Tanzania, and Uganda in an effort to form a common economy, with the possibility of political federation in the future. Participants in this effort, which is supported by the European Union, hope that a vibrant coastal economy in this region will ultimately help the landlocked and still unstable countries of Rwanda, Burundi, Malawi, Zambia, and Congo (Kinshasa) to find investors for more sustainable development. Another subregional organization, the Economic Community of Central African States (CEEAC), is presently not active because of widespread civil unrest since the early 1990s. All of these efforts at subregional integration owe a debt to the Organization of African Unity (OAU), a pan-African movement founded by President Kwame Nkrumah of Ghana in the late 1950s. The OAU, which includes all the countries on the African continent, plus nearby islands, was broadly conceived as an antidote to European dominance. Currently, it is a loose union that promotes economic cooperation and general social welfare. Its role continues to evolve and may become greater.

Grassroots Development. Another alternative development strategy for Africa is **grassroots rural economic development,** which provides sustainable livelihoods in the countryside as an alternative to urban migration. One of the most promising ideas is **self-reliant development,** which consists of small-scale projects in rural areas. These projects use local skills, create local jobs, produce products or services for local consumption, and maintain local control so that participants retain a sense of ownership. One

district in Kenya has more than 500 such self-help groups. Most members are women who terrace land, build water tanks, and plant trees. They form farm cooperatives for small-scale producers of poultry, honey, dairy products, and crafts or for those engaged in fishing. They also build houses, schools, and barns; run bookshops, nurseries, and restaurants; and form credit societies. Nonetheless, there are limitations to such efforts. For one thing, they require numerous facilitators with extraordinary skills in managing people. Moreover, the success of locally funded self-reliant development strategies has tempted governments to ignore the needs of rural communities for their fair share of financial support.

Examining the issue of rural transport illustrates how an African perspective and a focus on local needs might generate improvements that differ from those that would be advocated by large external development agencies. When non-Africans learn that transport facilities in Africa are in need of development, they usually imagine building and repairing roads, railroads, airports, buses, and the like. But a recent study that analyzed transport on a local level—the time spent, the distances traveled, the loads carried—discovered that women provide a major part of village transport and that most of the goods moved are carried on their heads. Women "head up" firewood from the forests and water from wells to their homes, as well as goods going to market. Eighty-seven percent of this domestic load bearing is on foot; on average, an adult woman moves the equivalent of 44 pounds (20 kilograms) more than 1.25 miles (2 kilometers) a day and spends about 1.5 hours a day doing so. Men tend to carry much fewer of these local burdens. In Kasama, Zambia, for example, an adult woman transports the equivalent of 35.7 tons (32.4 metric tons) a year, compared

Women returning from market at Kisangani, Congo (Kinshasa), pass by Tshopo Falls on a tributary of the Congo River. [Tom Friedmann/Photo Researchers.]

AT THE REGIONAL SCALE   *The Internet Comes to Africa*

The Internet has grown rapidly in Africa over the past five years. At the end of 1996, only 11 African countries had Internet access. By the end of 2000, all 54 countries in Africa had achieved some permanent connectivity and had installed some local full-service dial-up Internet service providers (ISPs).

Although Internet access is largely confined to capital cities and other large urban areas, Internet cafés that provide e-mail service are becoming as common as the letter writers and typists who used to set up shop near downtown post offices years ago. Of the more than 1 million ISP subscribers, 85 percent are in South Africa and in North Africa along the Mediterranean.

The largest ISP is Africa Online. It was started in 1994 by three young Kenyans who were working in the United States after having graduated a few years earlier from MIT, Princeton, and Harvard. Inspired by America Online, they originally set out to create a service by which African students abroad could exchange information and get news from home. They raised $170,000 from their own funds to start Africa Online, then took it to Kenya. Africa Online has become the leading pan-African Internet service, with operations in Kenya, Uganda, Ghana, Namibia, Zimbabwe, and several other African countries. African Lakes PLC purchased the company for $12 million in 2001. One of the original founders remains as the company's business strategy and development director.

---

with 7.1 tons (6.4 metric tons) transported by the typical man, no matter what his conveyance. Yet even when development agencies have examined transport at the household and village level, they have focused exclusively upon transport by men, suggesting bicycles, wheelbarrows, donkey carts, and pickup trucks to help men out. As forests are depleted and water becomes scarcer, the time spent, loads carried, and distances traveled by women increase. Mothers usually recruit their daughters to help them carry their loads, thus cutting into the girls' time in school.

Analysis of grassroots development shows that tiny changes can make big differences in people's lives. Giving women access to low-tech transport such as donkeys and bicycles would free some of their time for education, short training courses, and perhaps small-scale businesses.

Technological Development. A generation of modern technological entrepreneurs is emerging in Africa. Well educated and familiar with global communication, they are buoyed by Africa's potential in human and natural resources rather than intimidated by past problems. These entrepreneurs, who are in their 20s, 30s, and 40s, were educated in Europe or North America and first experienced success as young professionals in these wealthy economies. They developed a desire to find similar opportunities in their own countries and have established businesses in telecommunications, radio production, graphics and publishing, Internet services, furniture manufacturing, glass manufacturing, bookstore franchises, horticulture, and cut flowers exports, among others. Many of these entrepreneurs are public-spirited and want to provide good services and produce good products as well as to make profits. Some have organized regional networks to share information and financial support and to develop their entrepreneurial skills. There are currently three such networks: the West African Enterprise Network (WAEN), which began in 1993 and by 2000 had members who owned businesses in 11 of the 16 countries in West Africa; the Southern African Enterprise Network (SAEN), which includes Botswana, South Africa, Zambia, Zimbabwe, Mozambique, Malawi, and Mauritius; and the East African Enterprise Network (EAEN), made up of Kenya, Uganda, Tanzania, Rwanda, and Ethiopia.

Political Issues: Colonial Legacies and African Adaptations

Few places in the world today seem as politically turbulent as Africa. The news media bring us reports of civil wars, genocide in Rwanda, and military despots who execute members of the opposition. We also see the lingering effects of despotic presidents such as Mobutu Sese Seko, who plundered the national treasury of the Congo (Kinshasa) for decades while millions of Congolese went without food, shelter, health care, and education (see page 385). Africa's civil wars are partly a legacy of European colonialism and partly the result of Africa's own ethnic and regional discrimination. Fortunately, there are some reasons for optimism on the horizon. Political and economic changes are reducing the power of entrenched elites and forcing them to allow democratic freedoms that, in the future, may bring more responsible government.

Origin of Conflict. In 1990, 45 percent of the population of sub-Saharan Africa was considered at risk of suffering harm resulting from armed conflicts of some sort. In many ways, the prevalence of conflict is related to Africa's national borders, which were established by Europeans during the colonial era. Some borders split ethnic groups, others resulted in very different and sometimes hostile groups sharing the same country (Figure 7.14). In many cases, years of carnage followed independence, as African governments, though less racist than their colonial predecessors, nonetheless often practiced ethnic discrimination. Civil wars arose from attempts to crush ethnic or regional separatist movements in Nigeria, Congo (Kinshasa), and Somalia, and from general repression of disadvantaged minorities. The state even encouraged interethnic violence bordering on genocide in Congo (Kinshasa), Uganda, Rwanda, and Burundi. Often, conflicts were worsened by

Thinking Geographically: **ON THE WEB**

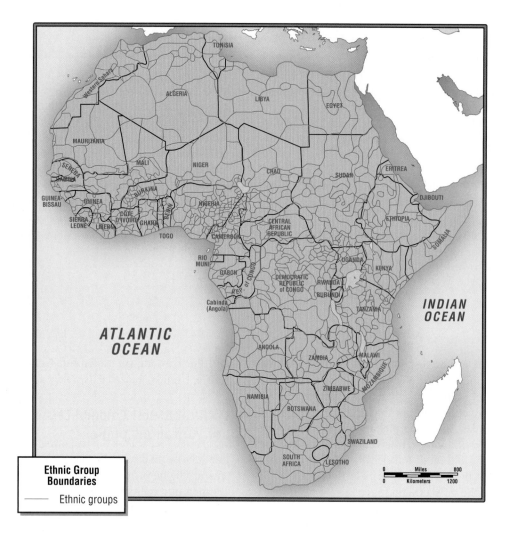

Ethnic Group Boundaries

—— Ethnic groups

Figure 7.14 **Ethnic groups in sub-Saharan Africa.** The map shows the large number of distinct ethnic groups spread across the continent of Africa. Superimposed on the pattern of ethnic groups are the present country boundaries. Very rarely do ethnic group and political boundaries match. Often a group is split by a country boundary, or groups with little in common share a political unit. [Adapted from James M. Rubenstein, *An Introduction to Human Geography* (Upper Saddle River, N.J.: Prentice Hall, 1999), p. 246.]

military establishments that were only loosely under the control of civilian governments or that seized the state through coups d'état—as in Nigeria, Liberia, and Sierra Leone.

The European colonial strategy of using local elites to control local populations often created or exacerbated ethnic tensions within a colony. The British in Uganda, the Germans in Tanzania, and the Belgians in Rwanda and Burundi gave local African elites the responsibility for collecting taxes and drafting laborers for various colonial construction projects. In Rwanda and Burundi, for example, the elite Tutsi were empowered to take advantage of the more numerous Hutu. This practice destabilized long-standing relations between these two groups and was partly responsible for ethnic bloodshed in the 1990s.

The conflicts in Nigeria and Congo (Kinshasa) are examined in more detail in the subregional sections (see pages 382 and 383ff).

The Cold War in Africa. The cold war between the United States and the former Soviet Union deepened and prolonged many conflicts in sub-Saharan Africa. After independence, many African governments sought to redress the problems of economic underdevelopment and exploitation during the colonial period by turning to socialism, often receiving economic and military aid from the Soviet Union. The United States and its allies tried to undermine these regimes by arming and financing rebel groups—as it did in Angola, Congo, and Namibia—or by pressuring other governments in the region to intervene. In the 1970s and 1980s, the United States encouraged South Africa's apartheid government, which itself was facing increasing political pressure from the Soviet-backed African National Congress of Nelson Mandela, to carry out military interventions against socialist governments in Namibia, Angola, and Mozambique. Another major area of cold

During the Mozambique civil war, the world's arms makers supplied enough weapons to arm nearly every person in the country with a rifle. More than 6 million AK-47s alone entered Mozambique (a country with only 18 million people). In a recent move to reduce guns among the population, the Anglican church began buying back the weapons in a Guns for Plowshares program. Guns are exchanged for tools and agricultural implements, sewing machines, bicycles, schoolbooks, and other useful items. The weapons are immediately destroyed. On any given day, one might encounter the unlikely sight of a church employee sawing guns in half in the churchyard. [Joao Silva/Corbis Sygma.]

war tension was the Horn of Africa, where the governments of Somalia and Ethiopia alternately received aid from the United States and the Soviet Union, much of which went to fund fighting between the two African nations.

Elite Rule. Many independent African countries are controlled by African elites that continue colonial traditions of authoritarian and corrupt rule. Often a single political party has a monopoly on power, and its leaders repress any viable political opposition. The elites who control these parties argue that the open political

debates of Western democracies are incompatible with African traditions of communal identity and collective decision-making. Similarly, elites have warped African traditions of patronage (wherein the leader is expected to look after the needs of the people) to create huge, inefficient government bureaucracies, encouraging rampant corruption that has penetrated to the highest levels. For example, in Nigeria, which has sub-Saharan Africa's second largest economy (after South Africa), political and economic elites have appropriated immense wealth from the country's large oil industry. Hence, despite Nigeria's oil wealth, most Nigerians are no better off than any other Africans. When poor Nigerians take actions that could reduce oil profits—for example, by asking for compensation for damages created by oil spills—they often suffer brutal government repression.

Shifts in African Geopolitics

Recently, some positive developments have shown that sub-Saharan Africa could become more stable and democratic. Signs of progress are especially visible in the southern part of the region. For years, the white-dominated government of South Africa tried to prolong white rule throughout Southern Africa. Now that white political control has ended in South Africa, the country's leaders are pursuing friendlier and more cooperative relations with neighboring countries, bringing greater stability to Southern Africa as a whole. The end of the cold war is also aiding political reform, although the situation is still complicated. Civil wars in several countries have subsided because the opponents can no longer secure military aid by promising to side with either the United States or the Soviet Union in exchange for arms. On the other hand, Africa now receives less foreign aid because more resources from developed countries are being directed toward promoting free-market reforms in eastern Europe. The reduction in aid, along with Africa's deepening economic crisis, has contributed to urban protest movements that have threatened governments throughout Africa.

Nevertheless, legal opposition parties have been able to express dissent more openly and freely. There is hope for a "second independence" as autocratic, corrupt ruling elites are replaced or moderated by political parties that enjoy the electoral support of the majority of the people. South Africa is an example of this trend. At least one encouraging statistic supports hopes for democracy elsewhere in Africa: in 1990, only four countries in sub-Saharan Africa were pluralist democracies; during the 1990s, 47 states promised or allowed multiparty elections (Figure 7.15). By the end of the century, even Nigeria, long under a military dictatorship, had held elections. Nonetheless, there remained several states—including Ethiopia, Rwanda, and Congo (Kinshasa)—where violence or open warfare made fair elections impossible.

Democracy and Economic Development

In view of the daily suffering of Africa's people, it is not unusual for observers to wonder if African governments shouldn't meet their citizens' basic needs for food, shelter, and health care before they

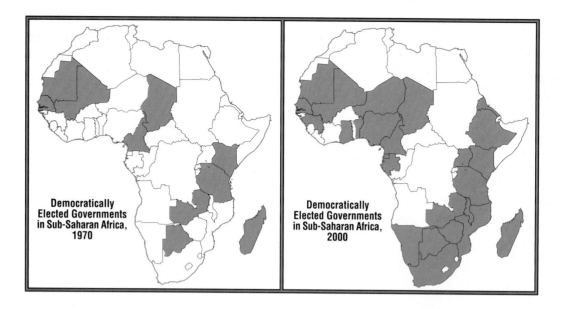

Figure 7.15 **Democratically elected governments in sub-Saharan Africa, 1970, 2001.** Between 1970 and 2001, the number of democratically elected governments (color) in sub-Saharan Africa increased nearly two and one-half times, from 11 to 26. [From Barbara E. McDade.]

open up the electoral process. The question implies that undemocratic governments could satisfy these immediate needs better than elected governments. The Nobel Prize–winning economist Amartya Sen, of Harvard University, disagrees. He points out that famine is rarely a problem in independent democratic countries where there is a free press. If there are no forums for public criticism of leaders, no chance for an opposition to speak, and no opportunity to replace leaders through elections, those who rule don't have to be accountable for the consequences of bad or corrupt policies. Famines never kill the rulers, so inept rulers have little incentive to solve the problems. Some observers think it is no accident that Botswana, where democracy is beginning to flourish, has prevented famine and begun to raise standards of living. By comparison, Congo (Kinshasa), Rwanda, and Ethiopia, all governed by repressive dictatorships, have not only fallen far short of providing basic food and shelter but have descended into horrendous civil violence. The truth is that the processes of development and democratization probably work best when they go hand in hand.

AT THE REGIONAL SCALE *African Women in Politics*

Charity Kaluki Ngilu (pronounced GEE-loo) will never forget the day she became a professional politician. She was washing dishes in her kitchen in Kitui, about 75 miles (120 kilometers) east of Nairobi, Kenya, when she saw a group of women approaching the back door with leafy branches in their hands. The women had worked with her to build better waterworks and health clinics in the town. Mrs. Ngilu answered the knock on the door, drying her hands on an apron. The women said they wanted her to run for parliament in Kenya's first multiparty elections. She assumed they were joking.

That was in 1992. Mrs. Ngilu not only beat the governing party's incumbent, she then became a thorn in the side of President Daniel arap Moi, criticizing his government for doing little or nothing for the poor, especially women.

On July 6, 1997, she announced that she would run for the presidency. She was the first woman to run for president in Kenya, a country not known for women's equality. She faced an uphill fight: as in most of Africa, Kenyan women are not likely to vote at all, and the majority of men are not likely to support female candidates. Nevertheless, many men supported her because they felt she was capable of making bigger changes than a man could, her very candidacy being a blow to convention. Although she didn't win (she came in fifth), she made an important statement with her campaign, which focused on repealing colonial-era laws that enable the president to cripple opposition by jailing dissidents and breaking up political demonstrations. Because so many problems are beyond the capacity of any one country, or even any one region, to solve, in 1999 she became a member of the board of Parliamentarians for Global Action—a group of legislators focused on environmental planning efforts at the global scale. In 2000, Ngilu was in her second term in parliament.

Source: Adapted from James C. McKinley, Jr., "A woman to run in Kenya? One says, 'Why not?'" *The New York Times* (August 3, 1997): 3; and Kennedy Graham, ed., *The Planetary Interest: A New Concept for the Global Age* (London: Rutgers University Press, 1999).

SOCIOCULTURAL ISSUES

A majority of sub-Saharan Africa's people appear to live traditional lives in rural villages. A closer look, however, reveals the influences of the Western colonizers and the modern world on such aspects of traditional culture as religion and gender roles. The growing numbers of people who are moving to the cities find their lives transformed still further as they exchange the isolated village for crowded shantytowns. Yet even in the cities, traditional African culture is visible in the plentiful vegetable gardens

Settlement Patterns

Human settlements can take many forms, from single isolated agricultural homesteads to high-rise modern cities. The various types of living arrangements reflect how people relate to one another economically, politically, and socially. This section discusses the two main settlement patterns found in sub-Saharan Africa: small rural villages and dense low-rise cities. There is a trend toward more urban settlement and the high-rise model of European, Asian, and American cities.

Rural Villages. More than 70 percent of sub-Saharan Africa's people still live in villages, making this region by far the most rural area in the world (along with South Asia). There are thousands of versions of the African village (see the photographs on this page), with many different types of houses and village arrangements, all depending on the cultural heritage of the inhabitants. But there are some constants in African villages, too. People usually live in extended family compounds consisting of several houses arranged around a common space in which most activities take place. These compounds may stand alone, dispersed across an agricultural landscape, or they may be grouped with other compounds and surrounded by fields designated for each family. Villages, and the compounds that make them up, are economic units. Subsistence agriculture is almost always an important component of village life, with labor carefully apportioned by custom. Increasingly, rural people also work seasonally on commercial farms for pay or have several other ways of earning small amounts of cash.

VIGNETTE   African villages usually have a leader or chief. In some African cultures, the leader is autocratic; in others, decisions are reached through consensus. Traditionally, the village head has looked after both the civic and the spiritual well-being of the community.

Twenty-two-year-old Chief Sinqobile Mabhena succeeded her father as head of the Ndebele village of Nswazi in Zimbabwe in December 1996. Her father suggested her as his successor before he died, and she was the villagers' choice. Nonetheless, there was strong opposition from some villagers, who objected to a woman sitting in judgment of men accused of wrongdoing or becoming privy to the secret knowledge normally presided over by a chief. It took a year to resolve the difficulties. Among the items on the agenda at her first village meeting were discussions of why rain had been sparse and how to solve the problem, who should receive more agricultural land, and

Here are three of the many types of villages found in sub-Saharan Africa. *Top:* One of the very colorful Ndebele villages in Zimbabwe. *Middle:* Cattle are returned to a Masai village compound in eastern Africa after a day's grazing. *Bottom:* Granaries with removable thatch roofs sit alongside houses in this Dogon village in Mali. [*Top:* Walter Bibikow/Index Stock Imagery/PictureQuest; *middle:* Frans Lanting/Minden Pictures; *bottom:* Henning Christof/Das Fotoarchiv.]

who should be sent to a shrine to inform the ancestors that the old chief had died and she was taking his place.

Chief Mabhena, a teacher in a nearby college, has several goals for her village, including a new high school and a poultry cooperative. She does not see herself as a fighter for women's rights so much as a practitioner of them. When asked how she reconciles traditional beliefs (such as that charms will bring rain or cure illness) with her knowledge as a trained teacher, she says: "In every subject I teach, there must be an aspect of culture. The children need that. I am for [the charms] if [they] will improve things." [Adapted from Donald G. McNeil, Jr., "Zimbabwe tribal elders air a chief complaint," *The New York Times* (January 12, 1997): 1.]

Urbanization. Although the majority of Africans live in rural areas, cities are not a recent settlement pattern in Africa. Many cities are long established, with histories extending back to the age of the great empires of West Africa and the kingdoms of East Africa. In the past, the cultural and social institutions of these cities focused on maintaining social and economic continuity with the surrounding rural areas. Although the rural village remains the predominant settlement type, African cities have grown very quickly indeed in recent decades, and their links with rural areas are more tenuous. The average annual growth rate of African cities—more than 5 percent—exceeds even the rates of cities in Asia (3 percent) and Latin America (2.5 percent). In the 1960s, only 15 percent of Africans south of the Sahara lived in cities; now about 30 percent do, but the percentage of urban population in specific countries varies considerably (Figure 7.16).

African cities have attracted massive migration for at least two reasons. First, life in rural villages and towns is perceived as offer-

ing few jobs and little opportunity for upward mobility. Second, there is a widely held misconception that life in the cities offers quick access to money and prestige. As discussed on page 361, a majority of Africa's city dwellers work at very low wages in the informal economy much of the time. On the other hand, urbanization can add a new dynamism to economic development in Africa by attracting talented young people to educational institutions and then to jobs in industry and services.

As is often the case in developing countries, most African countries have one very large primate city, usually the capital, that attracts virtually all migration. (The primate city phenomenon is discussed in Chapter 3, pages 131–132.) Kampala, Uganda, for example, is almost 10 times the size of Uganda's next largest city. The population of another typical primate city, Kinshasa in the Democratic Republic of the Congo, has soared from 450,000 in 1960 to an estimated 4.35 million in 2000. The population of Lagos, Nigeria, was 48 times larger in 2000 (11 million) than it was in 1950 (230,000). Such rapid and concentrated urbanization usually overwhelms a country's urban infrastructure: housing, sewage treatment, schools, utilities, and transport.

Typically, governments have paid little attention to the housing needs of the throngs of poor urban migrants. Most migrants have had to construct their own shanties on illegally occupied land, using materials that would be discarded as trash in North America, Europe, or Japan. They live in vast squatter settlements surrounding the older urban centers. Transport in these huge and shapeless settlements is a jumble of government buses and private vehicles. In most cities, people often have to travel long hours to reach distant jobs. They spend a good deal of their travel time moving at a snail's pace through extremely congested traffic. But for some members of the small but growing middle class, housing construction is providing both new homes and jobs. Some industries are locating close to such housing developments, resolving the transport problem for their workers.

The harshness of urban squatter communities is softened by the ubiquitous African urban agriculture—tiny gardens grown to supply food for the household and spiritual enrichment.

Religion

Religion is particularly important in African daily life. The region's rich and complex religious traditions derive from three main sources: the indigenous belief systems, Islam, and Christianity. Hinduism, like Islam and Christianity, was introduced from outside Africa, but its influence is found primarily in specific locations along Africa's east coast and on the eastern coastal islands, where there are significant East Indian or mixed African-Indian populations.

Indigenous Belief Systems. Traditional African religions, sometimes called **animism,** probably have the most ancient heritage of any religion on earth and are to be found in every part of Africa. Figure 7.17 shows the countries in which traditional beliefs remain particularly strong. According to these beliefs, all beings, including plants and animals, are part of a whole that is situated in, and part of, the physical environment. African beliefs and rituals seek to bring the vast host of departed ancestors into contact with

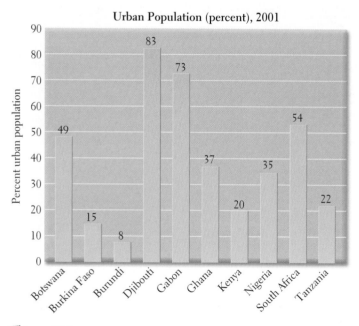

Figure 7.16 **Urbanization in selected countries, 2001.** [Data from *2001 World Population Data Sheet* (Washington, D.C.: Population Reference Bureau).]

African urban landscapes vary a great deal, ranging from contrasting patterns of shantytowns and modern mid-rise office buildings to single-family homes and apartment towers to elegant villas set in landscaped grounds. Markets, both formal and impromptu, are common features and are the sites of important social interactions. Above, closely packed houses in a shantytown near downtown Lagos, Nigeria, leave little room for ventilation in the tropical heat. The dwellings surround a parking lot for the cars of office workers. To the right is an urban view of Nairobi, Kenya, a major economic center in East Africa. [*Above:* L. Gilbert/Corbis Sygma; *right:* B. E. McDade/University of Florida.]

people now alive—who, in turn, are the connecting links in a time-less spiritual community that stretches into the future. The future is reached only if present family members procreate and perpetuate the family heritage in other ways, such as by storytelling,. The people in a living community are usually led by a powerful man, or occasionally a powerful woman, who combines the roles of politician, patriarch or matriarch, and spiritual leader.

According to traditional African beliefs, the spirits of the deceased are all around—in trees, streams, art objects, for example. In return for respect (expressed through ritual), these spirits offer

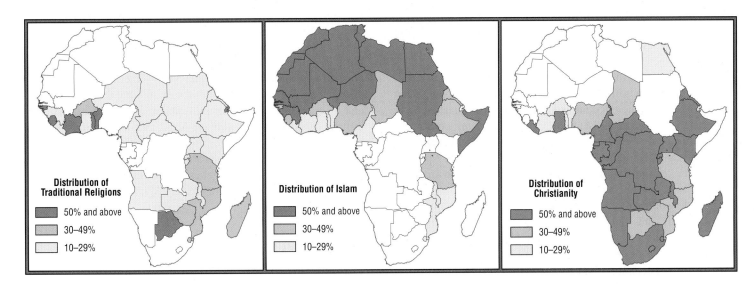

Figure 7.17 **Religions in Africa.** Traditional belief systems persist to varying degrees across Africa, even in those areas that are heavily Muslim or Christian. As elsewhere in the world, many African peoples have integrated traditional beliefs and practices with Islam and Christianity. [Adapted from Mathew White, "Religions in Africa," in *Historical Atlas of the Twentieth Century* (October 1998), at http://users.erols.com/mwhite28/afrorelg.htm.]

Coffin art in Ghana. Ghana's coffin carvers accomplish an apparently seamless interweaving of religion, art, and modern life. They are experts at interpreting the life of the deceased artistically: a Mercedes-Benz for a chauffeur, a plane for a pilot, a boat for a fishing fleet captain. Both rich and poor prize the coffin carver's art; people may commission a coffin long before death comes, often spending their life savings to buy a niche in the cultural memory of the community. [Carol Beckwith and Angela Fisher/Robert Estall Photo Library, U.K.]

protection from life's vicissitudes and from the ill will of others. Rosalind Hackett, a scholar of African religions and art, writes that "African religions are far more pragmatically oriented than Western religions, being concerned with explaining, predicting and controlling misfortune, sickness and accidents. [They give believers a way] to make sense of and act upon adverse forces in their public and private lives." African religions remain fluid and adaptable to changing circumstances. Osun, the god of water, traditionally credited with healing powers, is now invoked also for those suffering economic woes (job loss, cuts in funding for social services) caused by recent structural adjustment measures imposed by the International Monetary Fund.

Religious beliefs in Africa are not static but evolve continually as new influences are encountered. If Africans who practice traditional beliefs convert to Islam or Christianity, they commonly retain parts of their indigenous religious heritage and blend them with aspects of the new faith, creating a new entity—a process sometimes referred to as **syncretism** (or fusion). The maps in Figure 7.17 show an overlap of beliefs, but they do not convey the blending of two or more belief systems, which is widespread. In the Americas, new belief systems developed from the fusion of Roman Catholicism and African beliefs: Voodoo in Haiti; Obeah in Jamaica; Condomble in Brazil; and, more recently, Santeria in Cuba, Puerto Rico, and North America. In Africa, Christian missionaries of the established churches—Roman Catholic, Episcopal, Baptist, and so forth—often refused to allow Africans to incorporate their folk beliefs. This rejection helped to spawn independent, charismatic Christian sects that suited the need of Africans for reli-

gions that specifically represented their culture and their experience. These independent churches have given rise to "gospel of success" churches that have grown quickly in recent decades (see the box "The Gospel of Success").

Islam and Christianity in Africa. Islam began to extend into Africa south of the Sahara, especially East Africa, soon after Muhammad's death in 632 (see the discussion of Islam and Christianity on pages 295–296, 304). As Figure 7.17 shows, Islam is now the predominant religion throughout North Africa and is important in much of West Africa and East Africa, including the central coastal zone on the Indian Ocean where Islamic traders were especially active. When the British began to colonize West Africa formally in the late 1800s, they obtained the help of Islamic leaders in governing the countryside and towns, especially in the drier northern areas. The descendants of these Islamic administrators are still politically powerful in Nigeria and some other parts of West Africa.

Christianity came to northeastern Africa before it spread to Europe. In the fourth century, Christianity was adopted by the people of the kingdom of Aksum, the antecedent of modern-day Ethiopia. Today, the Ethiopian Coptic church maintains a tradition of Christian practices derived from this long heritage. Christianity began to spread into sub-Saharan Africa in the nineteenth century, when Methodist missionaries from Europe and America became active along the coast of West Africa. Roman Catholic missions followed, and today Christianity is prevalent along the coast of West Africa and in Central and Southern Africa. Christian missionaries in Central and Southern Africa found a niche as providers of the education and health services that colonial administrators had neglected. Christians are either the majority or an important minority in much of East Africa. There, colonial administrators in the British-held territories discouraged missionaries from going north into Muslim territory because Islamic leaders were already aiding the colonization process.

In the 1980s, old-line established churches began to gain adherents in East and West Africa. For example, the Church of England (Anglican church) has been growing so rapidly in Kenya, Uganda, and Nigeria that by 2000 there were more Anglicans in these countries than in the United Kingdom. African Anglicans appear to be conservative on social issues—they are against female priests and the acceptance of homosexuals—but they are often liberal on economic issues. For example, in 1998, African Anglican bishops persuaded the worldwide Anglican communion to oppose holding the world's poor countries to crippling debt payments. The Anglican church in Africa attracts the educated urban middle class, whereas modern evangelical versions of Christianity appeal to less educated, more recent urban migrants (see the box "The Gospel of Success").

Gender Relationships

There are long-standing African traditions of a fairly strict division of labor and responsibilities between men and women. The exact allocation of work, power, and family wealth by gender varies from subregion to subregion. In general, however, women are responsible for domestic activities, including rearing the children, tending

AT THE REGIONAL SCALE *The Gospel of Success*

We first encountered the "gospel of success" in Chapter 3, "Middle and South America." In Africa, this adaptable version of evangelical Christianity combines certain interpretations of Christianity and capitalism with Central and West African beliefs in the importance of sacrificial gifts to ancestors and in miracles. The message is simple. According to preachers such as those at the Miracle Center in Kinshasa, Congo: "The Bible says that God will materially aid those who give to Him.... We are not only a church, we are an enterprise. In our traditional culture you have to make a sacrifice to powerful forces if you want to get results. It is the same here." The supplicants accept the message that generous gifts to such churches will bring divine intervention to alleviate their miseries, whether physical or spiritual. They donate food, television sets, clothing, and money—one woman gave three months' salary in the hope that God would find her a new husband. The collection plates are large plastic bags. By combining spiritual ministration with a touch of hucksterism, the leaders of this largely urban movement receive the respect and adulation Africans used to reserve for rural village leaders.

Like all religious belief systems, the gospel of success is best understood within its cultural context. Many of the believers, new to the city, feel isolated and are seeking a supportive community to replace the one they left behind. People view the material contributions to the church as similar to the labor and goods they previously donated to maintain their standing in a village compound. In return for these dues, members receive social acceptance and community assistance in times of need.

Worshipers at the Miracle Center in Kinshasa, Congo. [Robert Grossman/Corbis Sygma.]

the sick and elderly, and maintaining the house. Women carry about 90 percent of the water, collect 80 percent of the firewood, and produce and prepare nearly all the food. In those countries where rice is popular (and it is increasingly the staple of choice for urban working women with families to feed), rural women are the primary cultivators of rice. Labor studies show that women also handle about 50 percent of the care of livestock and that when there are small agricultural surpluses or handcrafted items to trade, it is women who transport and sell them in the market. Throughout Africa, married couples tend to keep separate accounts and manage their earnings as individuals, so when a wife sells her husband's produce at market, she will usually give the proceeds to him.

Men are usually responsible for preparing all land for cultivation, a job that often entails clearing large trees and heavy brush with hand tools. Men then stack the refuse, let it dry and burn it, and turn the soil. In the fields intended to produce food for family use, women sow, weed, and tend the crops as well as process them. In the fields where cash crops are grown, men perform most of the work. Women may help to weed and harvest these fields, but it is usually understood that any earnings belong to the men. When husbands in search of cash income migrate to work in the mines or in urban jobs, women take over nearly all agricultural work. Studies have shown that the majority of agricultural laborers in Africa are women, and they contribute about 70 percent of the total time spent on African agriculture. They tend to work with hand tools in the fields and in the home, and during their reproductive years they often do field labor with a child strapped on their backs.

In rural areas, African men, on average, do not have as many responsibilities as women, nor do they work as hard or for as many hours. The reasons for this situation are complicated. Over the last century, many men have migrated to the cities or the mines to work, at least seasonally. There they may endure long hours of hard labor, difficult commutes, and crude living conditions. But men's labor has been mitigated more often than women's by labor-saving machines. As the geographer Robert Stock observes, rural African women do not have the "double day" that women in industrial societies have, but rather a "double double day," in that so much of their work entails backbreaking physical labor and time-consuming treks for water and fuel, usually without the aid of even a wheelbarrow or bicycle.

The retreat of men from virtually all tasks directly related to supporting domestic life seems to have started with European colonialism. At that time, men became increasingly engaged in activities to earn cash, as laborers or as cash-crop cultivators. At first, this change took place to meet the colonial administration requirement that taxes be paid in cash; later, men worked for cash to pay for school fees, basic electricity, certain consumer goods, and even food, when families lost access to land due to agricultural modernization. European Christians colonized Africa during the

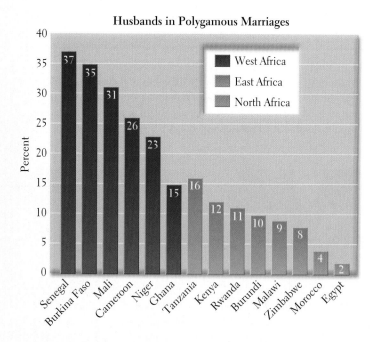

Figure 7.18 Polygyny in Africa. Polygyny is less common in East Africa than in West Africa, and it is uncommon in Southern African families. [Adapted from Farzaneh Roudi and Lori Ashford, *Men and Family Planning in Africa* (Washington, D.C.: Population Reference Bureau, 1996), p. 7.]

Female Genital Excision

A practice known as female circumcision is widespread in at least 27 countries throughout the central portion of the African continent (Figure 7.19). In the procedure, usually performed without anesthesia, the labia and the clitoris are removed and the vulva is often stitched nearly shut (called infibulation). Female circumcision eliminates any possibility of sexual stimulation. Because the practice results in the permanent removal of a healthy organ; many physicians refer to it as **female genital mutilation**. The custom, which was reported by Herodotus in 500 B.C., predates Islam by more than 1000 years and is practiced only by some Muslims, Christians, and followers of traditional religions. It is probably intended to ensure that a female is a virgin at marriage (and hence not pregnant by another male), but there are many complex symbolic meanings.

A controversial movement to eradicate the practice has arisen. In some culture groups, such as the Kikuyu of Kenya, nearly all females would have been "circumcised" 40 years ago; today only about 40 percent of Kikuyu schoolgirls have been. But in Africa as a whole, as many as 130 million girls and women presently living have undergone the procedure (2 million a year, 6000 per day). The practice also takes place surreptitiously among African émigrés in Europe, North America, and elsewhere. For those who have undergone female genital excision, urination and menstrua-

Victorian era, when the prevailing attitude in Europe was that women were lesser creatures who should remain in the home and let their husbands deal with the outside world. This idea influenced the colonial policy in Africa of recruiting men for cultivating cash crops and doing work for wages, while leaving women to shoulder all the domestic work as well as what had formerly been the shared work of subsistence agriculture.

In the precolonial past, there were social controls that tempered the effects of male domination over women's lives. Most marriages were social alliances between families; therefore, husbands and wives spent most of their time doing their tasks with family members of their own sex rather than with each other. Traditional gender relationships, however, were modified wherever Muslim or Christian religious and cultural practices were introduced. In most cases, it seems that men gained power and women lost freedom. Muslim women, restricted to domestic spaces, could no longer move about at will, seize opportunities to trade in the markets, or engage in public activities. In Islamic Africa, the strong patriarchy and the practice of **polygyny**—the taking by a man of more than one wife at a time—left many junior wives in servitude, with ambiguous legal rights. Christian women operate in the public sphere more than Muslim women, but they too are socialized to restrict their activities to the home. Although having multiple wives seems to be more common in sub-Saharan Africa than in Muslim northern Africa, as Figure 7.18 shows, Africans in polygnous marriages are a minority. Those countries with the highest rates of polygyny—Senegal, Burkina Faso, Mali, and Cameroon—have relatively small populations.

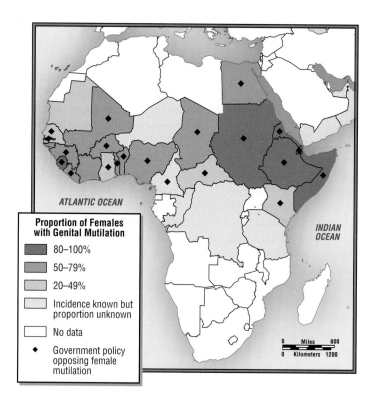

Figure 7.19 Female genital mutilation. The practice occurs all across the center of the African continent in spite of government policies against it. [Adapted from Joni Seager, *The State of Women in the World: An International Atlas* (London: Penguin, 1997), p. 53.]

tion are difficult, intercourse is painful, and childbirth is particularly devastating because the scarred flesh is inelastic. Recent research shows that the practice also leaves women exceptionally susceptible to HIV infection. Many African and world leaders, men and women, have concluded that the custom constitutes an extreme human rights abuse, but because it is deeply ingrained in some value systems, the most successful eradication campaigns have emphasized the threat posed to a woman's health.

Defenders of the custom cite its symbolic importance as ritual purification; often girls who have not undergone excision are considered unclean and unmarriageable. The *New York Times* reporter Celia W. Dugger interviewed a 12-year-old in Côte d'Ivoire, who said that she wants more than anything to be "cut down there," gesturing to her lap. All her friends have had it done, and afterward they were showered with gifts and money and there was a huge celebration for relatives and friends. The father of this girl said that if he did not have it done, he would not be allowed to speak in village meetings and no man would marry her. Nonetheless, the girl's mother, who has undergone excision, says she hates the custom because it deprives a woman of sexual sensitivity.

Eliminating female genital mutilation will require enforcement of existing laws against it in many African countries. There have also been successful efforts to change people's attitudes toward the practice and to create acceptable alternatives. A growing number of Kenyan families, for example, are turning to a new rite

known as "circumcision through words." This new ritual involves a weeklong program of counseling, ending with a community celebration and affirmation of a girl's passage to adulthood. In Ghana, one of the most successful programs encourages community leaders to perform alternative rite-of-passage ceremonies. In Uganda, a health education program seeks to show the public that some cultural practices can change without compromising the society's values. It is reported that from 1994 to 1996, female genital mutilation in Uganda declined by more 30 percent, largely as a result of this program.

Ethnicity and Language

Ethnicity refers to the shared language, cultural traditions, and political and economic institutions of a group. The term does not refer to race or color, only to culture.

The map of Africa presented in Figure 7.20 shows a rich and complex mosaic of ethnic languages. Yet despite its complexity, this map does not adequately depict Africa's cultural diversity. Most ethnic groups have a core territory in which they have traditionally lived; but very rarely do groups occupy discrete and exclusive spaces. Often, several groups will share a space, perhaps practicing different but complementary ways of life and using different resources. For example, one ethnic group might be subsistence cultivators, another might herd animals on adjacent grasslands, and a third might be

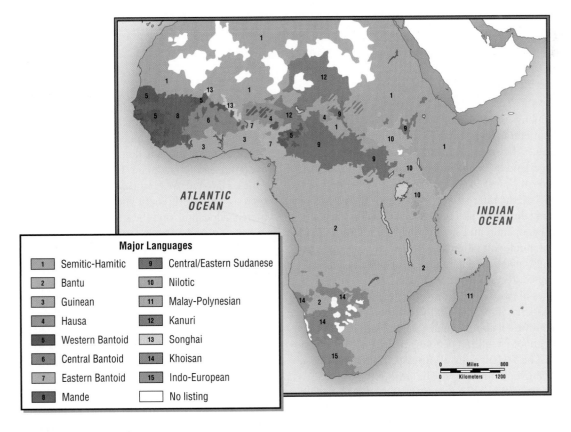

Figure 7.20 **Major languages of sub-Saharan Africa.** [Adapted from Edward F. Bergman and William H. Renwick, *Introduction to Geography—People, Places, and Environment* (Englewood Cliffs, N.J.: Prentice Hall, 1999), p. 256.]

CULTURAL INSIGHT *Language Heritage in a Trunk*

It was the kind of day every scholar dreams of. Some years ago, Professor John Hunwick of Northwestern University was in Timbuktu, Mali, once a major trading town where Arab Africans and black Africans from Central and Southern Africa met to trade. A young man who knew of Hunwick's interest in African history invited him to see the family library. The man lifted the lid on an old trunk filled with manuscripts. The contents of these manuscripts are profoundly altering long-accepted views of African history and civilization.

The manuscripts, dating from the mid-fifteenth century, show that the knowledge of writing passed from Arabs of the continent's northern regions through Timbuktu and other places in West Africa and then farther south. People in Central Africa adapted the Arabic alphabet to create written forms of local languages, such as Fulani. This process is similar to the way in which people in western Europe adopted writing. The Romans adapted their alphabet from the Greeks, and then French, Spanish, German, and English speakers used the Romans' Latin alphabet to give written form to their spoken languages.

Several cultures in Africa produced indigenous written languages. Ethiopia has been literate for at least 2000 years. The Ethiopian language, Amharic, has its own distinctive alphabet (see the photo at the right). The ancestors of the Nubian people established the Kushite kingdoms thousands of years ago and developed a system of writing similar to Egyptian hieroglyphs. Nubians today are spread across southern Egypt and Sudan. In the 1850s, Euro-

pean missionaries came across the Vai people (in present-day Liberia and Sierra Leone) and found them using their own alphabet system as they have done for many generations.

Source: Adapted from Ron Grossman, *The Gainesville Sun* (April 11, 2001).

An Ethiopian teacher (with her baby on her back) points to Amharic script on her classroom blackboard. [Sarah Errington/The Hutchison Library.]

craft specialists working as weavers or smiths. On the other hand, people can share an ethnic identity yet have little in common culturally or in the spaces they occupy. In Kenya, for example, Kikuyu villagers in the hinterland live very different daily lives from Kikuyu urbanites who reside in Nairobi. People may also be very similar culturally and occupy overlapping spaces but identify themselves as being from different ethnic groups. Hutu and Tutsi cattle farmers in Rwanda share an occupation and similar ways of life but think of themselves as having different ethnicities. In recent years, external pressures have exaggerated their differences to the point that the two groups have attempted to eradicate each other.

Some countries in Africa have only one or two ethnic groups; others—such as Nigeria, Tanzania, and Cameroon—have many. Nigeria has three groups of about equal size: the Hausa, Igbo, and Yoruba, each with about 25 million people; but there are hundreds of smaller groups. The vast majority of African ethnic groups have peaceful and supportive relationships with one another.

To a large extent, language correlates with ethnicity. More than a thousand languages, falling into more than a hundred language groups, are spoken in Africa. Some are spoken by only a few dozen people; others, such as Hausa, by several million. Language use is

always in flux, so some African languages are now dying out. They are being replaced by languages that better suit people's needs or have become politically dominant. Increasingly, a few **lingua francas** (languages of trade) are taking over: Swahili and Arabic in East Africa, Hausa in West Africa. In addition, former colonial languages such as English, French, and Portuguese are widely used in commerce, education, and the exchange of scientific knowledge.

ENVIRONMENTAL ISSUES

Africa is home to many unique animal and plant species that warrant protection and preservation. But attempts to ensure the well-being of Africa's human citizens often threaten the habitats that support endangered animals and plants. For example, Africans who cannot afford kerosene or gas to cook food must use wood—but in some densely populated areas, so many people are collecting wood that the forests have disappeared and even shrubs are constantly pruned for firewood. The wisest African planners design strategies that both provide for basic human needs and address Africa's environmental issues. This section presents a sample of the environmental issues facing Africa today (Figure 7.21).

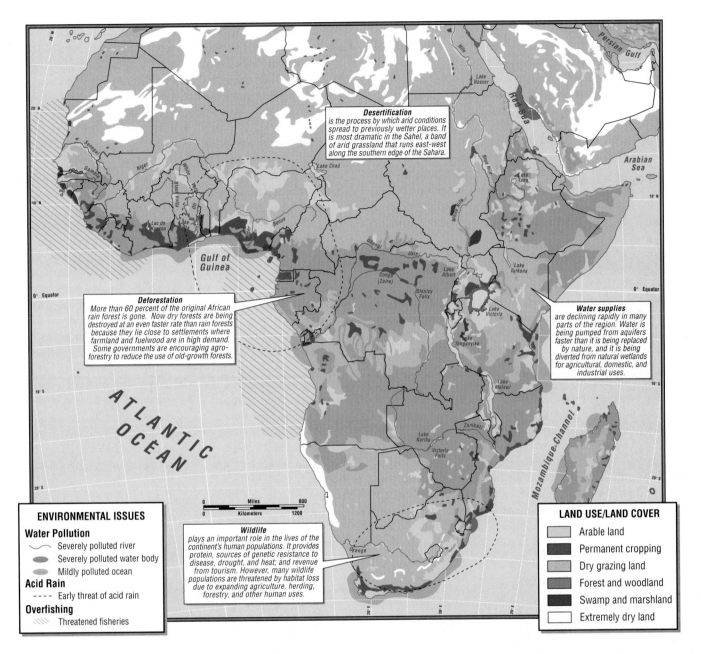

Desertification
is the process by which arid conditions spread to previously wetter places. It is most dramatic in the Sahel, a band of arid grassland that runs east-west along the southern edge of the Sahara.

Deforestation
More than 60 percent of the original African rain forest is gone. Now dry forests are being destroyed at an even faster rate than rain forests because they lie close to settlements where farmland and fuelwood are in high demand. Some governments are encouraging agro-forestry to reduce the use of old-growth forests.

Water supplies
are declining rapidly in many parts of the region. Water is being pumped from aquifers faster than it is being replaced by nature, and it is being diverted from natural wetlands for agricultural, domestic, and industrial uses.

Wildlife
plays an important role in the lives of the continent's human populations. It provides protein, sources of genetic resistance to disease, drought, and heat; and revenue from tourism. However, many wildlife populations are threatened by habitat loss due to expanding agriculture, herding, forestry, and other human uses.

ENVIRONMENTAL ISSUES

Water Pollution
~ Severely polluted river
⬤ Severely polluted water body
◯ Mildly polluted ocean

Acid Rain
- - - - Early threat of acid rain

Overfishing
/// Threatened fisheries

LAND USE/LAND COVER
☐ Arable land
■ Permanent cropping
☐ Dry grazing land
▦ Forest and woodland
■ Swamp and marshland
☐ Extremely dry land

Figure 7.21 **Environmental issues: sub-Saharan Africa.** Africa's environmental issues are varied and are less connected to urbanization and industrialization than those of other regions, but they are no less worrisome. The drying out of environments (desertification) is probably the result of both natural and human-induced processes, some stemming from bad development advice. Deforestation and loss of plant and wildlife species are also related both to global market demands and more local circumstances that result in forest clearing for fuelwood. In some places, water resources are being depleted or polluted by growing urban demands and irrigated export agriculture.

Desertification

Climatologists think that parts of Africa are now in a natural cycle of increased aridity. The effects are most dramatic in the region called the **Sahel** (Arabic for "shore" of the desert), a band of arid grassland 200 to 400 miles (320 to 640 kilometers) wide that runs east-west along the southern edge of the Sahara. Over the last century, the Sahel has shifted to the south in a process known as desertification, by which arid conditions spread to areas that previously were quite moist. For example, the *World Geographic Atlas* in 1953 showed Lake Chad situated in a tropical rain forest well south of the southern edge of the Sahel. By 1998, the Sahara itself was encroaching on Lake Chad. This shift of aridity to the south—and its opposite, the shift of moister conditions to the north—is part of a long-term natural cycle, but human activity may be speeding up the cycle and broadening the area affected.

The Sahel and other dry grasslands in Africa are what geographers call **fragile environments** because they exist in zones that are barely sufficient for the needs of native plants and animals. Rainfall is already low in these grasslands, at 10–20 inches (25–50 centimeters) per year, and there are only low levels of organic matter in the soil to provide nutrients. Consequently, any further stress such as fire, plowing, or intensive grazing of animals may cause dry-adapted grasses to give way to introduced species such as tough bunchgrass and thorny shrubs that are more tolerant of stress. But these alien species do not cover the soil as well as the native plants, so they allow rain to evaporate more quickly, and they are not useful for grazing. Soon, the dry, denuded soil is blown away by the wind, and the remaining sand piles up into dunes as the grassland becomes more like a desert.

Indigenous animal herders are often blamed for desertification. Although overgrazing can trigger this process, natural cycles are often the underlying cause. Moreover, economic development activities can advance desertification. For example, agencies such as the World Bank have sometimes advised animal herders to raise cattle instead of traditional animals, such as camels and goats, because cattle can be sold as meat in distant urban markets. But raising cattle places greater stress on native grasslands than more traditional herding. The conditions for desertification can also be set up by the use of high-speed drilling equipment to bore holes for wells, around which nomads and their herds settle permanently. When herders stay in a place permanently, their herds may overgraze their pastures. Desertification can also occur where marginal lands are cleared and plowed for agriculture. A few years of irrigated agriculture in a dry environment can leave the soil poisoned by waterborne minerals that are left behind as crystals when the irrigation water evaporates. The overharvesting of trees and shrubs for fuel also contributes to a general drying out of the soil and the advance of desertlike conditions.

Forest Vegetation as a Resource

Africans have long made extensive use of forest products for construction, furniture, tools, fencing, medicines, religious carvings, fuel, and food. Like forests all over the world, the forests of Africa are disappearing. It is estimated that by 2000, 1.15 million square miles (3 million square kilometers)—more than 60 percent of the original African rain forest—were gone. Most of the loss of African forests is attributed to the demand for farmland and fuel by growing populations, especially in West Africa (see Figure 7.21). In recognition of the rapid loss of forests, governments in some countries (such as Ghana, Uganda, and Tanzania) are encouraging **agroforestry**—the raising of economically useful trees—to take the pressure off old-growth forests.

So far, logging for export is only a minor part of African economies, but it is responsible for the severe depletion of forests in easily accessible countries such as Côte d'Ivoire in West Africa. The harvested logs are sold as raw wood, even though the financial return would be much greater if the trees were harvested sustainably and processed in Africa into refined wood products. The most extensive remaining tracts of African rain forest are in the Congo River basin, most of which lies in the country of Congo (Kinshasa), where sustainable forest management is not practiced at all. Commercial logging is expected to expand and the loss of forested land to increase when the armed conflict that has engulfed the country and its neighbors since the mid-1990s ends.

Dry forests—those that lose their leaves during the dry season—once covered nearly twice as much territory as the rain forests. Now dry forests are being destroyed even faster than rain forests because they lie close to settlements and their products are in great demand by growing populations. Their greatest use is as fuel. Africans still use wood or charcoal—a prepared wood product that burns hotter than wood—to supply nearly all domestic and industrial energy. Making and selling charcoal is an important industry in the informal economy of most African countries. Much of the wood is harvested free because Africans traditionally have considered the forests to be a resource held in common. In urban areas, wood and charcoal remain the cheapest fuels available. Even in Nigeria, which has large reserves of oil, most people use fuelwood because they cannot afford petroleum products as fuel. In fact, Nigeria is the leading producer of fuel wood in Africa.

For a decade or more, many Africans have recognized the need to use fuelwood more sustainably. Fast-growing trees such as Leucana are farmed for use in making charcoal. Economic development planners have tried to promote fuel-efficient stoves, but users find that the stoves lack the convenience and nostalgic associations of old-fashioned cooking fires. Some Africans are turning to alternative energy sources—hydropower, bottled gas, solar power, and wind power—to meet their growing energy needs; but the trend is not yet strong.

Wildlife

Africa's wildlife has always played an important role in the lives of the continent's human populations. Many Africans depend on wild game and fish as their main sources of protein, and there is growing interest in game farming because wild animals are resistant to African diseases, are adjusted to the climate, and can withstand occasional droughts or heat waves better than domesticated animals. African wildlife protected in national parks in Kenya, Tanzania, South Africa, Zimbabwe, Botswana, and Zambia attracts tourists. African animal products such as skins, taxidermy specimens, and ivory have been important exports to Europe, China, and the Americas for several hundred years, and live animals have supplied the world's zoos.

Sometimes, this trade in animal parts and in live animals threatens the survival of species. Reports documenting the diminishing numbers of gorillas, elephants, lions, zebras, and giraffes and the decline of their rain-forest and grassland habitats have raised public concern about Africa's diminishing wildlife heritage. Although increasing international interest helps, it may not be enough to save many animals. The cheetah, for example, is very near extinction, with only 12,000 individuals left. About 7000 have been killed in the last 10 years, mostly by farmers on the fringes of nature reserves who kill cheetahs to protect their livestock. The use of dogs to protect livestock, an old African strategy, has peacefully resolved the conflict between cheetahs and farmers with some success. Nevertheless, as more and more areas are turned over to agriculture, the number of cheetahs inevitably will continue to decline.

Water

In a growing number of African countries, the amount of water per capita is declining rapidly. In many parts of arid Africa, water is being pumped from aquifers faster than it is being replaced by nature; it also is being diverted from natural wetlands. Large economic development projects may redirect water for industrial, agricultural, or urban uses, leaving rural inhabitants of arid lands with insufficient water and causing the loss of complex ecosystems. Kenya, for example, has only 830 cubic yards (635 cubic meters) of water per capita, compared to the accepted global standard of 1300 cubic yards (1000 cubic meters); its water supply is expected to drop to just 250 cubic yards (190 cubic meters) per capita by the year 2050.

Safe water is scarce even in the moist areas of the continent. Water is being polluted by salinization, human waste, and chemical poisons (see Figure 7.21). In rural areas, most households must carry all their water from springs, wells, pools, or streams; in urban areas, water is drawn from standpipes. It is a long-standing tradition that women are the procurers of water. As sources become depleted or polluted, women must walk farther and farther to collect safe water. The difficulty of carrying water limits the amounts available for people's needs, and illness is spread when insufficient water is used to clean dishes, diapers, clothing, and other materials. Because plumbing and sewage treatment are usually not available, even in cities, human wastes often find their way into water sources, another cause of illness.

Saving what is now being wasted is perhaps the greatest source of "new" water for most places on earth. Agriculture accounts for about 70 percent of global water use. In Africa, many large agricultural projects are irrigated, and most modern irrigation systems lose a great deal of water through leakage and evaporation because water often runs through unlined open ditches and is sent into fields via sluices or sprinklers. In addition, standing pools of water are sources of disease. Culturally aware development thinkers say that the solutions to Africa's water problems lie less in high-tech applications than in rediscovering folk ideas about water conservation that fit with local customs. A decade or two ago, West African women who delivered irrigation water directly to the roots of their plants by using human water brigades were seen as inefficient by foreign development planners. Today, the water these practices save (and the camaraderie they foster) are better appreciated. The installation of "old-fashioned" sanitary roof catchments and cisterns in place of complex piped delivery systems can save construction costs and water, as can the use of sanitary hand-flushed pit toilets in place of public sewer systems.

At present, the goals of economic development and structural adjustment are often in conflict with environmental conservation. When people are as poor as they are in Africa, structural adjustment often drives them and their governments to spend down environmental capital—renewable and nonrenewable resources such as forests, oil, gas, and water—to stave off imminent disaster. In other words, just to survive they use their resources at grossly unsustainable rates, often exporting them for prices that are much too low.

MEASURES OF HUMAN WELL-BEING

The very low gross domestic product (GDP) per capita figures in Africa (Table 7.1, column 2) indicate a desperately poor continent. However, GDP per capita does not include economic activities that take place outside the formal market system: sharing, bartering, participation in the informal economy, raising food in family plots, and conserving resources. Within the low range of GDP per capita figures, there is still considerable variation. Botswana, Gabon, Mauritius, Cape Verde, Namibia, South Africa, Swaziland, and Seychelles all have per capita GDPs above U.S. $3000 per year, but together they account for only 51 million people, less than 10 percent of Africa's population. Moreover, there is a wide spread in income within these countries. Many people live on much less.

The Human Development Index (HDI) combines three measures—life expectancy at birth, educational attainment, and adjusted real income—to arrive at a ranking of 174 countries that is sensitive to more factors than just income (Table 7.1, column 3). Virtually all African countries rank in the lowest third of world nations on the HDI. Because GDP figures are so low, there is little tax money to invest in basic health care, sanitation, and education. The literacy figures (columns 5 and 6) reflect both the generally low level of education and the discrimination against women. During the 1980s and 1990s, per capita government spending fell, often in response to cutbacks mandated by SAPs, and the introduction of fees limited the number of people who could afford services, including health care and education. One result has been that more mothers and infants have died during and after childbirth. Spending on education fell from U.S. $11 billion in 1994 to U.S. $7 billion in 1998. When school fees were introduced, enrollment in elementary schools fell from 77.1 to 66.7 percent.

For various reasons, such as the low status of women in African countries and official hesitancy to reveal the data, the statistics necessary to calculate the Gender Empowerment Measure (GEM) are available for only five countries, so this measure is not included in Table 7.1. Nonetheless, there has been progress for women. In Botswana, Lesotho, Mozambique, South Africa, and Zimbabwe, women occupy 25 percent or more of professional and technical positions. Moreover, in these same countries, women have some seats in parliament—17 percent of the seats in Botswana, 11 percent in Lesotho, 30 percent in Mozambique, 28 percent in South Africa, and 14 percent in Zimbabwe—so they may be able to influence future policies that affect women.

The Gender Development Index (GDI) ranks 36 sub-Saharan African countries on the world scale of 174 countries (Table 7.1, column 4). This index measures the access of men and women to health care, education, and income, but not the degree to which they participate in public policy formulation and decision making.

TABLE 7.1 *Human well-being rankings of countries in sub-Saharan Africa*

Country (1)	GDP per capita, adjusted for PPP[a] in 1998 $U.S. (2)	Human Development Index (HDI) global rankings, 2000[b] (3)	Gender Development Index (GDI) global rankings, 2000[c] (4)	Female literacy (percentage), 1998 (5)	Male literacy (percentage), 1998 (6)
Selected countries for comparison					
Japan	23,257	9 (high)	13	99	99
United States	29,605	3 (high)	6	99	99
Mexico	7,704	55 (medium)	49	88.7	92.9
West Africa					
Benin	867	157 (low)	132	23	54
Burkina Faso	870	172 (low)	142	13	32
Cape Verde Islands	3,233	105 (medium)	88	65	84
Côte d'Ivoire	1,598	154 (low)	129	36	53
Gambia	1,453	161 (low)	134	35	48
Ghana	2,077	129 (medium)	105	60	79
Guinea	1,817	162 (low)	N/A[d]	19	36
Guinea-Bissau	616	169 (low)	140	17	57
Liberia	1,100	N/A	N/A	22	54
Mali	681	165 (low)	137	23	39
Mauritania	1,563	147 (low)	122	31	52
Niger	739	173 (low)	143	7	22
Nigeria	795	151 (low)	124	53	70
Senegal	1,307	155 (low)	128	26	45
Sierra Leone	458	174 (low)	N/A	18	45
Togo	300	145 (low)	120	38	73
Central Africa					
Cameroon	1,474	134 (medium)	111	67	80
Central African Republic	1,118	166 (low)	138	32	58
Chad	856	167 (low)	N/A	20	41
Congo (Brazzaville)	995	139 (medium)	114	72	86
Congo (Kinshasa)	822	152 (low)	125	47	71
Equatorial Guinea	1,817	131 (medium)	109	72	91
Gabon	6,353	123 (medium)	N/A	53	74
São Tomé and Príncipe	1,469	132 (medium)	N/A	62	85

Country (1)	GDP per capita, adjusted for PPP[a] in 1998 $U.S. (2)	Human Development Index (HDI) global rankings, 2000[b] (3)	Gender Development Index (GDI) global rankings, 2000[c] (4)	Female literacy (percentage), 1998 (5)	Male literacy (percentage), 1998 (6)
Horn of Africa					
Djibouti	1,266	149 (low)	N/A	51	74
Eritrea	833	159 (low)	131	24	30
Ethiopia	574	171 (low)	141	31	42
Somalia	600	N/A	N/A	14	36
East Africa					
Burundi	570	170 (low)	N/A	37	55
Kenya	980	138 (medium)	112	74	88
Rwanda	660	164 (low)	135	57	72
Tanzania	480	156 (low)	127	64	83
Uganda	867	158 (low)	130	23	54
East African Islands					
Comoros	1,398	137 (medium)	113	57	66
Madagascar	756	141 (low)	116	58	72
Mauritius	8,312	71 (medium)	61	80	87
Seychelles	10,600	53 (medium)	N/A	60	56
Southern Africa					
Angola	1,821	160 (low)	N/A	28	56
Botswana	6,103	122 (medium)	101	78	73
Lesotho	1,626	127 (medium)	104	93	71
Malawi	523	163 (low)	136	44	73
Mozambique	782	168 (low)	139	27	58
Namibia	5,176	115 (medium)	98	80	82
South Africa	8,488	103 (medium)	85	84	85
Swaziland	3,816	112 (medium)	93	77	80
Zambia	719	153 (low)	126	69	84
Zimbabwe	2,669	130 (medium)	106	83	92

[a]PPP = purchasing power parity.
[b]The high, medium, and low designations indicate where the counry ranks among the 174 countries classified into three categories (high, medium, low) by the United Nations.
[c]Only 143 out of a possible 174 ranked in the world; no data for many.
[d]N/A = data not available.

Source: United Nations Human Development Report 2000 (New York: United Nations Development Programme), pp. 157–168.

SUBREGIONS OF SUB-SAHARAN AFRICA

We divide sub-Saharan Africa into four subregions that predominantly reflect geographic location: West Africa, Central Africa, East Africa, and Southern Africa. Such factors as the current geopolitical situation and tradition have also guided the placement of a particular country in a particular subregion. Chad, for example, could have been discussed with the arid countries of northern West Africa, along with Niger and Mali. We have placed it in Central Africa, chiefly because most of Chad's people live in the far southern reaches of the country, are closer in culture to Cameroon and the Central African Republic, and are affected primarily by political and economic circumstances in Central Africa. For each subregion, we provide a short discussion of its physical setting and then concentrate on specific circumstances that have led to current conditions. (See the chapter opening map for landform features.)

WEST AFRICA

West Africa, which occupies the bulge on the west side of the African continent, includes 15 countries (Figure 7.22). It is bordered on the north by the Sahara Desert, on the west and south by the Atlantic Ocean, and on the east by Lake Chad and the mountains of Cameroon. The region can be thought of as a series of hor-

izontal physical zones grading from dry in the north to moist in the south. The north-south division also applies to economic activities—herding in the north, farming in the south—and, to some extent, religions and cultures—Muslim in the north, Christian in the south. Most of these cultural and physical features do not have distinct boundaries but rather zones of transition and exchange.

Stretching across West Africa from west to east is a series of horizontal vegetation zones that reflect the climate zones shown in Figure 7.1. Remnants of tropical rain forests still exist along the Atlantic coasts of Côte d'Ivoire, Ghana, and Nigeria, although much forest has been lost as a result of intensifying settlement and agriculture. To the north, this moist environment grades into drier woodland mixed with savanna (grassland). Farther north still, as the environment becomes drier, the trees thin out and the savanna dominates. The savanna grades into the yet drier Sahel region of arid grasslands. In the northern parts of Mali, Niger, and Mauritania, the arid land becomes actual desert, part of the Sahara.

People of many different cultures live in these environments. Along the coast, and inland where rainfall is abundant, crops such as coffee, cacao, yams, palm oil, corn, bananas, sugarcane, and cassava are common. Farther north and inland, the crop complex

Figure 7.22 The West Africa subregion.

is adapted to the drier climate: millet, peanuts, cotton, and sesame grow in the savanna environments. Cattle are also tended in the savanna and north into the Sahel; farther south, however, the threat of tsetse flies prevents cattle raising (see page 352). People in the Sahel and Sahara are generally not able to cultivate the land, except along the Niger River during the summer rains. Food, other than that provided by animals and the fields along the riverbanks, must be traded for with coastal zones.

West African environments have dried out as more and more people have cleared the forests and woodlands for cultivation and to obtain cooking fuel. Forest cover no longer blocks sunlight and wind. As a result, sunlight evaporates moisture, and desiccating desert winds blow south all the way to the coast. The drying out of the land is exacerbated by the recurrence of a natural cycle of drought. The ribbons of differing vegetation running west to east are not as distinct as they once were, and even coastal zones occasionally experience dry, dusty conditions reminiscent of the Sahel.

The boundaries set by the European colonizers interrupted natural economic flows. Moreover, because colonial economies were based on the extraction of resources and their export primarily to European markets, each colony became linked more to its European colonizer than to neighboring territories. Today, laborers who migrate from the poorer states in the interior to the wealthier states on the coast and send money home to their families are breaking down the formal economic and political colonial divisions. In this way, the people are reestablishing the informal economic links within West Africa that predated colonization. Against the historical backdrop of colonial division, ECOWAS (see page 361) was founded to promote freer trade among West African countries. It also seeks the eventual formal economic integration of West Africa into a trade bloc with a common tariff on goods from outside the region. Although intraregional (sub-Saharan) trade in 2001 was still only 11 percent of the amount of trade with countries outside the region, ECOWAS reaffirmed its economic goals and began also to consider a regional court of justice, a regional parliament, and a social council.

The geographies of religion and culture in West Africa also reflect the north-south physical patterns to some extent. Generally, the north is Muslim, and people are light-skinned and tend to have angular Caucasian features. The south is populated by darker skinned people who practice a mix of Christian and traditional religions. Some large ethnic groups once occupied distinct zones, but as a result of migration, members of specific groups live today in many different areas and in a number of West African countries. There are hundreds of smaller ethnic groups, each with its own language; the map depicting ethnic languages in Nigeria (Figure 7.23) is

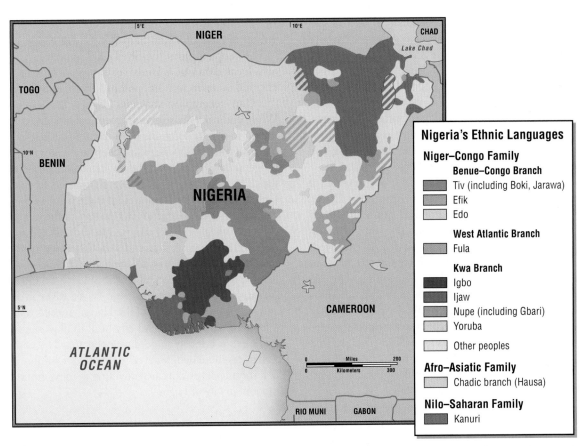

Nigeria's Ethnic Languages

Niger–Congo Family

Benue–Congo Branch
- Tiv (including Boki, Jarawa)
- Efik
- Edo

West Atlantic Branch
- Fula

Kwa Branch
- Igbo
- Ijaw
- Nupe (including Gbari)
- Yoruba
- Other peoples

Afro–Asiatic Family
- Chadic branch (Hausa)

Nilo–Saharan Family
- Kanuri

Figure 7.23 **Nigeria's ethnic languages.** Nigeria's languages reflect the variety of ethnic groups in the country. The 11 main languages are indicated in color, but many smaller groups speak distinct languages or Creole dialects not shown on this map. Striped pattern indicates zones of two or more main languages. [Adapted from James M. Rubenstein, *An Introduction to Human Geography* (Upper Saddle River, N.J.: Prentice Hall, 1999), p. 167.]

representative of the cultural complexity of the entire subregion. Nigeria illustrates the potential difficulties faced by a nation attempting to unite a multitude of ethnic groups.

Nigeria

Nigeria is the dominant country in West Africa economically because of its large population, its oil reserves, and its relatively diversified economy. It is the most populous country in Africa, and its 123 million people outnumber the total of those who live in the rest of West Africa (see Figure 7.4). Like the other countries of West Africa, Nigeria is a culturally complex nation. It presents an interesting case study of how colonial and postcolonial rule has led to increasing tensions among the various ethnic groups that have long occupied the area that is now Nigeria.

As geographer Robert Stock observes, "Nigeria was, and still remains, a creation of British imperialism." When Europeans drew the borders of African countries, they did not place them along natural physical boundaries, or along traditional borders between different ethnic groups. Instead, they deliberately plotted borders to split up some ethnic groups (see Figure 7.14). Conversely, the Europeans also often forced disparate groups into one unusually diverse country. This is what the British did with Nigeria, which has 395 indigenous languages in 11 language groups (see Figure 7.23).

In present-day Nigeria, there are four main ethnic groups. The Muslim Hausa (21 percent of the population) and Fulani (9 percent) have, until recently, lived mostly in the north, where the moist grassland grades into dry savanna and then semidesert. The Hausa are traders, dryland farmers, and animal herders. Since independence in 1960, they have dominated both the government of Nigeria and its military. The Yoruba (20 percent) live in southwest Nigeria in the moister woodland and grassland environments. The Igbo (17 percent) are centered in the southeast in and near the tropical rain-forest zone. Hundreds of other ethnic groups live among and between the larger groups.

Colonial administration of the four main groups created a north-south dichotomy, especially in education. Among the Hausa, the British let local Muslim leaders administer community life. These leaders did not encourage public education. In the south, among the Yoruba and Igbo, Christian missionary schools were common. At independence, the south had more than 10 times as many primary and secondary school students as the north. This disparity in education led to a disparity in income, with southerners holding most government civil service positions, yet the poorer northern Hausa continued to dominate politically. Bitter disputes have erupted over the allocation of profits from Nigeria's oil-related industries and over the distribution of general development funds and jobs. Accusations of rigged census data and election fraud usually fall out along ethnic and north-south divisions.

Oil is Nigeria's primary resource: 90 percent of its foreign trade earnings come from the sale of oil. Nigeria's oil reserves have the potential to make it the richest country in Africa, but this resource has been mismanaged to enrich military leaders and foreign oil companies. A Nigerian commission reported that the mil-

itary regime in power at the time stole $12 billion in oil money between 1990 and 1994, and a subsequent regime is thought to have taken another $12 billion. Officially, Nigeria earns about $10 billion a year from oil.

High oil prices in the 1970s and 1980s encouraged the government to overspend on expanded social services and construction of a new federal capital. The government borrowed heavily against future oil revenues but invested little in productive activities such as agriculture or industry. By 2000, Nigeria was 36 percent urban, and the growing cities became increasingly dependent on expensive imported food. When oil prices fall, as they do periodically, the government cannot make debt payments to international lending agencies out of current earnings. Its response is often to cut back or abandon basic social services and new development projects so that it can redirect money into debt payments.

There has been little pretense of democracy in Nigeria for several decades, and virtually every form of protest or opposition has been suppressed through beatings, imprisonment, and killings. Universities have been shut down for years at a time after campus protests. In 1966, after an Igbo-led military coup and a Hausa-led countercoup, the slaughter of 30,000 Igbos ignited the three-year Biafran War. The Igbo portion of southeastern Nigeria tried to secede (unsuccessfully), and more than 200,000 people died as a result of the war and ensuing food shortages. During the 1990s, ethnic and regional differences continued to mold Nigerian politics. In 1995, the world community was outraged by the summary execution of the writer Ken Saro-Wiwa, who had led a protest of the Ogoni people against the Shell Oil Company. Much of Nigeria's oil is located on Ogoni lands, yet the Ogoni have received little benefit, and oil extraction has polluted their lands.

Can geographic strategies resolve the tensions in Nigeria? While enduring divisive and violent civil disputes, Nigeria has continued to pursue solutions to its unwieldy ethnic and political spatial patterns. One strategy has been to create more political states (Nigeria now has 30) and thereby reallocate power to smaller units. Large, wealthy states have been subdivided to ensure more equitable distribution of development money and oil profits and to prevent any one state from seizing control of the central government. Although dividing the country into more states has increased administration costs, it also seems to have eased regional ethnic and religious hostilities. Another strategy to defuse ethnic conflict has been to relocate the capital city from Lagos on the Atlantic coast to Abuja, an interior village on the Niger River in neutral territory away from the heartlands of the Hausa, Yoruba, and Igbo. Abuja is a **forward capital:** a capital city built to draw migrants and investment to a previously underdeveloped area The cost of establishing Abuja has increased Nigeria's debt by $4.5 billion, and it turns out that most civil servants are unwilling to move from the vibrant cosmopolitan city of Lagos.

CENTRAL AFRICA

If you were to look at a map of Africa without any prior knowledge, you might reasonably expect the countries of Central Africa—

AT THE SUBREGIONAL SCALE *Water Management in Mali and Niger*

The Niger is the most important river of West Africa. It carries summer floodwaters northeast into the normally arid, clay-lined lowlands of Mali and Niger. There the waters spread out into lakes and streams that nourish wetlands (see Figure 7.21). For a few months of the year (June through September), these wetlands ensure the livelihoods of millions of fishers, farmers, and pastoralists who share the territory in carefully synchronized patterns of land use that have survived for millennia. These desert wetlands along the Niger produce eight times more plant matter per acre than the average wheat field and provide seasonal pasture for millions of cattle, goats, and sheep—the highest density of herds in all of Africa.

Because of mounting population pressure, international experts are advising the governments of Mali and Niger to dam the river and channel it into irrigated agriculture projects that will help to feed the more than 20 million overwhelmingly poor people of Mali and Niger. Even though 80 percent of the people are farmers, Mali and Niger already must rely on imported food. But the dams will forever change the seasonal rise and fall of the river that has supported an intricate mix of wildlife and human uses for thousands of years. The impact will be felt in Europe as well, because the Niger is a seasonal migration stopover for birds that spend summer in Europe, north of the Mediterranean.

Every rainy season, the Bani River, a major tributary of the Niger River, floods and turns the city of Djenne, Mali, into a series of islands. Djenne, with a population of 20,000, is the oldest known urban settlement in sub-Saharan Africa (about 2200 years old). The annual flood revitalizes surrounding soil nutrients and the river's marine life and permits the population to feed itself. A proposed upstream dam threatens the city's self-sufficiency. [Sarah Leen/Matrix.]

Cameroon, the Central African Republic, Chad, Congo (Brazzaville), Gabon, São Tomé and Príncipe, and Congo (Kinshasa)—to be the hub of continental activity, the true center around which life on the continent revolves (Figure 7.24). This is not the case. Paradoxically, Central Africa plays a peripheral role on the continent. Its dense tropical forests, tropical diseases, and difficult terrain make it the least accessible part of the continent, and recent armed conflict over resources and power have left it in a deteriorating state that discourages visitors and potential investors.

Central Africa consists of a core of wet tropical forestlands surrounded on the north, south, and east by bands of drier forest, and then grasslands (savanna) (see Figure 7.1). The core is the drainage basin of the Congo River, which flows through this area and is fed by a number of lesser tributaries. The forests of Central Africa, although rich in resources, form a barrier that has isolated Central Africa from the rest of the continent for centuries and has impeded the development of transportation, commercial agriculture, and even mining. Moreover, the soils of the forest are not suitable for long-term cash-crop agriculture (see page 344). Most Central Africans are subsistence farmers who live in rural settings along the river courses and the occasional railroad line. Cities are only beginning to grow; too often, the newcomers are people escaping civil violence in the countryside.

Violence and its attendant corruption are another reason for the lack of development in Central Africa. Although this subregion possesses significant mineral and forest wealth, a long string of corrupt governments has pocketed the profits, and smugglers of minerals and other resources have leaked away the potential wealth to foreign interests. Persistent warfare between rival groups has discouraged most legitimate investors, Africans or outsiders.

The region's colonial heritage has contributed to these difficulties. In what is now known as the Democratic Republic of the Congo (Kinshasa), for example, the Belgian colonists left a terrible legacy of violence. In the 1880s, Belgium was a small but wealthy and industrialized European country, and its king, Leopold II, had a personal fortune to invest. After failing to establish a colony in Latin America or Asia, King Leopold obtained international approval in 1885 for his own personal colony, the Congo Free State, with the promise that he would end slavery, protect the native people, and guarantee free trade. He did none of these things. The Congo Free State consisted of 1 million square miles (2.5 million square kilometers) of tropical forestland rich in ivory, rubber, timber, and copper, with a population of 10 million people, whose labor King Leopold would soon appropriate to extract Congo's resources for his own profit. The Belgians used physical violence and terror to control the people of the Congo and gain access to their resources. The cruel behavior of Belgian colonial officials was witnessed by the young writer Joseph Conrad, when he obtained a job on a steamer headed up the river in 1890. Conrad's novella, "Heart of Darkness" (published in 1902), and a pamphlet by

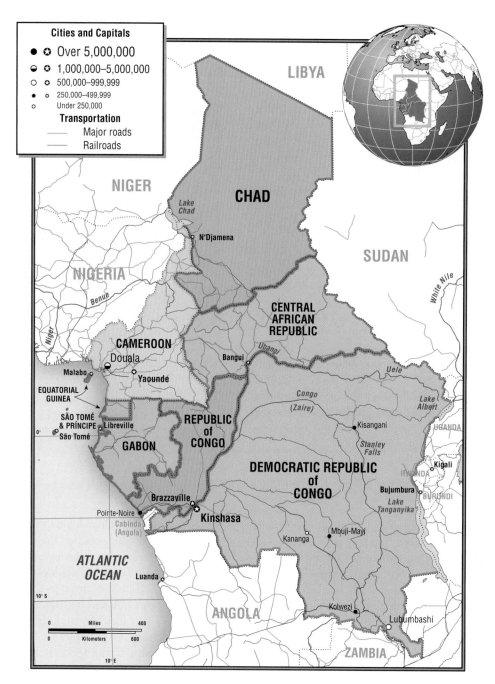

Cities and Capitals
- ● ✪ Over 5,000,000
- ◖ ✪ 1,000,000–5,000,000
- ○ ✪ 500,000–999,999
- • ◦ 250,000–499,999
- ∘ Under 250,000

Transportation
- —— Major roads
- —— Railroads

Figure 7.24 **The Central Africa subregion.**

Mark Twain, *King Leopold's Soliloquy* (published in 1905), awakened Europeans and Americans to Leopold's perfidy, and efforts began to end his regime. French colonists in what is now Congo (Brazzaville), Gabon, and the Central African Republic; the Germans in Cameroon; and the Portuguese in neighboring Angola used similar tactics. For example, a chronicler of the 1890s recorded that after a group of Africans rebelled against the French, their heads were used as a decoration around a flower bed in front of a French official's home.

The indigenous people of Central Africa form an overlapping ethnic mix, and they speak some 700 languages. The artificial polit-

ical borders imposed by the Europeans and their policy of deliberately pitting Africans against one another exacerbated tensions between ethnic groups. Indigenous governance systems were disrupted when Europeans removed chiefs and installed new leaders who would do their bidding. Long-standing traditional restraints on unwise leadership were lost, thus opening the door for the type of power abuses that developed in the twentieth century.

Most of the countries of Central Africa received their independence in the 1960s, but the Europeans stayed on to continue their economic domination. Tax-free profits were sent home to France, Belgium, or Portugal; rarely were profits reinvested in Cen-

tral Africa. With no money to develop an infrastructure, no experience with multiparty democracy, and no educated constituency, many Central African countries, like Congo (Kinshasa), fell into the hands of corrupt leaders.

Political upheavals that resulted when some groups were left out of representation in government were accompanied by downward-spiraling economies. As in other parts of sub-Saharan Africa, the average person's purchasing power has decreased over the last few decades. These countries have little tax revenue because of tax concessions to foreigners as well as government corruption. So maintenance of roads, railroads, waterways, telephone networks, electricity grids, and buildings is negligible. Industries, both government and private, cannot survive for lack of electricity and raw materials. People are so poor that there is little market for any type of consumer product.

Congo (Kinshasa): A Case Study

VIGNETTE For women in Kisangani, the saying, *"Se débrouiller"*—"Make do with what you have"—is the way of life. Mauwa Funidi, 45, the first college graduate in her once prominent family, now sells small bags of charcoal on the street in Kisangani, Congo (Kinshasa), to earn a few pennies to help support her extended family. She still holds her real job as a university librarian, which paid U.S. $300/month in 1976, but after two decades of inflation paid her just U.S. $11/month in 1997. The university library is empty, the windows are broken, books and other publications are rotting in the humidity. Thirty years ago, education was celebrated in Kisangani. Now there are no funds for the university, and the government has stopped funding kindergarten through twelfth grade. The majority of children do not finish primary school because their families cannot afford the tuition of U.S. $7 a month.

Many of the 47 million inhabitants of Congo (Kinshasa) are in circumstances similar to those of Mauwa Funidi's family. The average GDP per capita is U.S. $130 a year. Yet the country has an enormous wealth of untouched natural resources: copper, gold, diamonds, and oil. Congo (Kinshasa) is sub-Saharan Africa's fourth largest oil producer, and oil is its major foreign trade commodity. It also has enough undeveloped hydroelectric resources to supply power to every household in the country and to those of some neighboring countries as well.

The problems in Congo (Kinshasa) have resulted from European colonization, the cold war, and the desire of local leaders to stay in power and enrich themselves. When independence came in 1960, severe political and economic collapse was imminent as Belgian colonial officials withdrew, leaving untrained Congolese in charge. The new government of Patrice Lumumba tried to correct the inequalities of the colonial era by nationalizing foreign-owned companies so that the profits could be kept in the country. Lumumba's strategies, reminiscent of communism and adopted

Mauwa in her defunct library, Kisangani, Congo (Kinshasa).
[Stephen Crowley/NYT Pictures.]

during the height of the cold war, led to foreign intervention. The Soviet Union supported and armed Lumumba's pro-socialist movement, and the Western allies supported and armed opposing politicians, such as the extraordinarily corrupt Mobutu Sese Seko.

After five years of violent political struggle, Mobutu Sese Seko seized power in 1965. Once firmly established, Mobutu nationalized the country's institutions, businesses, and industries—but he did not intend to use the previously foreign-owned profits for the good of the people. Instead, he and his supporters began to extract personal wealth from the newly reorganized ventures. Foreign experts, trained technicians, and capital left the country. The country's infrastructure, economy, educational system, food supply, and social structure declined rapidly. As conditions worsened, Mobutu and his supporters used their money and their paid army to hold onto power. By the 1990s, the vast majority of Congolese people were impoverished. Mobutu was finally overthrown in 1997 by Laurent Kabila, a new dictator supported by Uganda and Rwanda, neighbors of Congo (Kinshasa) to the east. Kabila's inept regime quickly led to widening strife in Central Africa. By 1999, Angola, Namibia, Zimbabwe, Chad, Rwanda, and Uganda all had troops in Congo (Kinshasa) and were aggressively competing for parts of its territory and resources.

To some extent, the conflict is ethnically based. For example, Kabila's Rwandan allies are primarily Tutsi. They have attacked Congolese Hutus, whom they hold responsible for genocidal attacks against Tutsis in Rwanda in 1997. But the prospect of economic and political power is also a strong incentive for conflict. In January 2001, Kabila was assassinated, reportedly by previous supporters from neighboring countries. His son, Joseph Kabila, was installed as head of state and appeared to be amenable to efforts to establish peace in the subregion.

EAST AFRICA

East Africa (Figure 7.25) is a land of relatively dry coastal zones that contrast with moister interior uplands. A vital and productive farming economy has long existed in the uplands. On the coast, there is a lively trading economy with links across the Arabian Sea and Indian Ocean. This commerce has especially influenced the culture and economy of the southern half of the subregion.

In the northeast, the Horn of Africa wraps around the southern tip of the Arabian Peninsula. This more northerly section of East Africa includes the countries of Djibouti, Eritrea, Ethiopia, and Somalia. South of the Horn of Africa are the countries of Kenya, Uganda, and Tanzania; the tiny interior highland countries of Rwanda and Burundi in the Great Rift Valley; and the islands of Madagascar, Comoros, and Mauritius in the Indian Ocean across the Mozambique Channel. Africa's ability to support its growing population is being tested in East Africa. The entire Horn of Africa has suffered periodic famine since the mid-1980s as the result of naturally occurring drought, ineffective government, and civil conflict. Even in the moister southern portion of East Africa, growing populations are straining the region's carrying capacity.

The Interior Highlands

The landforms of East Africa's interior are dominated by the Great Rift Valley. This irregular series of developing tectonic rifts curves inland from the Red Sea and extends southwest through Ethiopia and into highland "Great Lakes" country that curves around the western edges of the subregion and extends into Mozambique to the south. The rift valley is covered primarily with temperate and tropical savanna, although patches of forest remain on the high slopes and in southwestern Tanzania. Most rural people make a living by cattle herding, although herding and cultivation are combined in the wetter central uplands of Ethiopia and Kenya and in the woodlands of southern Tanzania. A mixture of shifting subsistence and plantation cultivation sustains populations in the south.

In the north of East Africa, near the rift, is an area of mountains and plateaus called the Ethiopian Highlands, which range in elevation from 4000 feet (1200 meters) to over 10,000 feet (3000 meters). The highlands are drained by deep river valleys, including the valley where the headwaters of the Blue Nile lie. Rain falls during the summer in the semiarid highlands, and the area includes a mixture of environments that vary with elevation. The Amhara and Gurage live here; these ethnic groups are known for their complex and productive systems of **mixed agriculture,** meaning that they raise cattle and use the manure to fertilize a wide variety of crops that are closely adapted to particularly fragile environments. Pastures and fields are rotated systematically. Agronomists have recently begun to appreciate the precision of these ancient and specialized mixed-agriculture systems, which are now threatened by development schemes and population pressure.

| VIGNETTE | Kadija and Hassain Saide Adem have a 10-acre (4-hectare) farm in the Ethiopian Highlands |

150 miles (400 kilometers) north of Addis Ababa, the capital. They grow mostly sorghum and some corn. The Adems' sorghum plants are 15 feet (4.5 meters) tall; the heads of the plants contain seeds that are *landraces,* the term for the genetic descendants of plants carefully nurtured in a particular place for thousands of years.

Hassain says he prefers to grow the landraces rather than purchase the seeds because he is impressed with the results. His seeds are highly productive: he once had 29 stalks of sorghum sprout from a single seed. His wife, Kadija, usually sorts the seeds, determining which are best for planting, for storing, and for cooking. Successful food farmers like this couple, who produce enough for their families and surpluses for sale, are especially important to Ethiopia as it attempts to mitigate the effects of a prolonged drought and the economic and political disruption that dominated the 1980s and 1990s. During this time, agricultural productivity was drastically reduced; the resulting famine left many thousands dead and others seriously malnourished. [Adapted from Pattie Lacroix, "Ethiopia's living laboratory of biodiversity" (Ottawa, Ontario: International Development Research Centre of Canada, 1995).]

The Coastal Lowlands

To the east and south of the Ethiopian Highlands is a broad, arid apron that descends to the Gulf of Aden and the Indian Ocean. Along the coasts of the Horn of Africa, there are large stretches of sand and salt deserts and few rivers. The country of Somalia, which is shaped like a **V** with its point jutting into the Indian Ocean, occupies the majority of the low, arid coastal territory in the Horn. The people who live in this area of oasis cultivation and dry-grass herd-

In this Italian-built factory in Asmera, Eritrea's capital, women make shoes for foreign markets. Eritrea has new laws guaranteeing gender equality: women now have the right to divorce, to vote, to acquire and own land, and to work outside the home. [Robert Caputo/Aurora & Quanta Productions.]

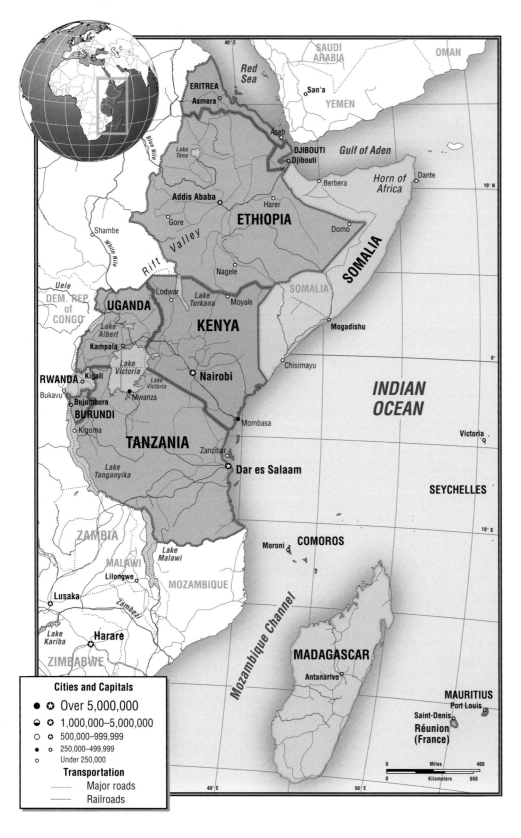

Figure 7.25 **The East Africa subregion, including the Horn of Africa.**

ing can eke out only a meager living. Somalia's Muslim population of 10.2 million, virtually all ethnically Somali, are aligned in six principal clans that have engaged in a violent rivalry for power over the last decade. Political turmoil has created refugees who cannot be productive, and it has blocked access to some lands that are cru-

cial to the food production system. Thus, cultivators must overuse the land that is available. Population has increased significantly in all countries of the Horn of Africa, but Somalia has seen the largest increase. Growing populations are especially difficult to accommodate in these arid countries because of the scarcity of water and other resources and because the complex agricultural system requires careful management to remain productive.

For thousands of years, people who occupied the lands around the Indian Ocean—China, Indonesia, Malaysia, India, Persia (Iran), Arabia, and Africa—have met and traded along East Africa's shores, from southern Somalia southward. The result has been a grand blending of human genetics, human cultures, plants, animals, and languages. In the sixth century, Arab traders brought Islam to East Africa, where it remains an important influence in the north and along the coast. Traders speaking Swahili, a lingua franca that is a pragmatic blend of Arabic and African languages, established networks that linked the coast with the interior. These networks penetrated deeply into the farming and herding areas of the savanna and into the semifeudalistic cultures of the highland lake country. Trade across the Indian Ocean continues today: India and Japan are East Africa's most important Asian trade partners, and trade with Indonesia and Malaysia is increasing. East Africa exports mostly agricultural products (coffee, cashew nuts, tobacco, and cotton) and imports machinery, transportation equipment, industrial raw materials, and consumer goods.

The Islands

Madagascar, which lies off the coast of East Africa, is the fourth largest island in the world; only the islands of Greenland, New Guinea, and Borneo are larger. Its unique plant and animal life—the result of *diffusion* (transmission by natural or cultural processes) from mainland Africa and Asia and subsequent evolution in isolation from the mainland—is highly prized by biologists and greatly threatened by environmental degradation. The other East African island nations—Comoros, Seychelles, and Mauritius—are much smaller and less physically complex than Madagascar. All the islands have a cosmopolitan ethnic makeup, the result of thousands of years of trade across the Indian Ocean. The people and languages of Madagascar, for example, have ancient origins in Southeast Asia. Further cultural mixing in the islands took place during European (primarily French and British) colonization. During the colonial era, these islands supported large European-owned plantations, worked by labor brought in from Asia and the African mainland. More recently, islanders have added tourism to the economic mix. Most visitors come from the African mainland, from Asian countries surrounding the Indian Ocean, and from Europe.

Kenya: A Modern Case Study in Competition for Land

Kenya is the wealthiest and most industrialized country in East Africa, producing cement, petroleum products, automobiles, beer, and processed food. Yet it remains one of the poorest countries in the world (see Table 7.1, pages 378–379).

As far back as ancient times, traders in this part of Africa exchanged commodities—slaves, ivory, rare animals—that can no longer be legally traded. By the early nineteenth century, the depletion of elephant herds, hunted for ivory, and changing attitudes toward slavery had pushed the region into a period of stagnation that made it easier for the British to gain control. In the reorganization of East African economies under colonialism, the British relocated native highland farmers in Kenya, such as the Kikuyu, to make way for European coffee plantations, and they displaced herders, such as the Masai in Kenya and Tanzania, from the savanna. The lands of the Masai became game reserves designed to protect the savanna wildlife for the use of European hunters and tourists.

Today the former colonial game reserves of Kenya and Tanzania are among the finest national parks in Africa, offering protection to wild elephants, giraffes, zebras, and lions, as well as many less spectacular but no less significant species. Tourism, mostly visits to wildlife parks, accounts for 20 percent of foreign exchange earnings. The future of the parks is precarious, however, because of competing demands for the land by herders and cultivators.

After independence, government policies resettled black African farmers on highland plantations formerly owned by Europeans, giving the country a strong base in export agriculture through the production of coffee, tea, sugarcane, corn, and sisal. Nomadic peoples have fared less well than farmers. Policies for managing the national wildlife parks have inadvertently resulted in the spread of tsetse fly infestations into the local pastoralists' herds. Disputes over scarce water supplies have broken out between cultivators who wish to irrigate their crops and herders who need regular sources of water for their animals. So far, the government has sided with the farmers, even to the point of policing the nomadic herders from armed helicopters.

Agriculture, especially export agriculture, is important to Kenya's economy, but the country's dependence on this sector has also put people's livelihoods and the famous national parks at risk. One problem is related to population growth. In the mid-twentieth century, populations throughout East Africa began to grow extremely rapidly. In 1946, Kenya had just 5 million people; by the late 1990s, it had 28 million trying to eke out an existence from the same land and resource base. In the past, the primarily arid land could adequately support relatively small populations when managed according to traditional customs practiced by herders and subsistence cultivators. Today the land's carrying capacity is severely stressed. Competition for cultivable land has pushed thousands of small farmers out of the more fertile uplands into the drier and more fragile savannas that had been communal rangelands and habitats for wild animals. The small plots plowed in the savanna are only marginally productive and have displaced nomads, elephants, impala, and giraffes alike. While Kenyans try desperately to increase agricultural production, volatile world market prices for some export crops have sent the national economy plunging; world coffee prices fell sharply in 2001, for example. These problems are threatening Kenyan and Tanzanian game reserves and parks. Population pressure has forced farmers to plow under and overgraze wildland buffer zones surrounding the parks. Hungry people have begun to poach park animals for food, and there is now a large underground trade in bush meat.

SOUTHERN AFRICA

At the beginning of the new millennium, optimistic voices are being heard in Southern Africa. Just a few years ago, racial conflict in South Africa, the largest and wealthiest country in this region, dominated the news. Today, racial reconciliation is happening in South Africa, and the competent work of the democratically elected governments of Nelson Mandela and his successor, Thabo Mbeki—the first governments led by indigenous Africans since before the colonial era—is bringing an air of cautious confidence to the entire region.

The Southern Africa subregion consists of Angola, Zambia, Malawi, Mozambique, Zimbabwe, Botswana, Namibia, South Africa, Swaziland, and Lesotho (Figure 7.26). Southern Africa is a plateau ringed on three sides by mountains and a narrow lowland coastal strip. At its center is the Kalahari Desert, but for the most part, Southern Africa is a land of savannas and open woodland. Population density is low; the highest densities are found in the southeastern part of South Africa (see Figure 7.4). Africa's mineral wealth is concentrated in this part of the continent: there are rich deposits of diamonds, gold, chrome, copper, uranium, and coal.

Angola and Mozambique are both former Portuguese colonies. Their wars of independence evolved into civil wars, with Marxist governments on one side and rebels supported by South Africa and the United States on the other. Mozambique ended its civil war in 1994. By 1996, the once devastated economy was experiencing one of the highest rates of growth on the continent. But in Angola, the fighting continues. The officially recognized Marxist-influenced government holds the coastal regions where most Angolans live and where the country's oil and fishing resources are located. The opposition, known as Union for the Total Independence of Angola (UNITA), draws its strength from among the Ovimbundu-speaking people in the south-central plateau, where gold, diamonds, uranium, and manganese are located. UNITA held the upper hand for years, controlling 70 percent of the country. It previously garnered support from the now defunct Mobutu regime in Congo (Kinshasa)

Figure 7.26 The Southern Africa subregion.

and from the sales of diamonds and other resources. In early 1998, UNITA and the Angolan government agreed to a peace treaty, but episodic violence continues in the countryside.

The three interior countries of Malawi, Zimbabwe, and Zambia are still overwhelmingly rural. Agriculture and related food-processing industries form the basis of the economy, and their products are a major portion of exports. Over the last two decades, however, agricultural productivity has fallen, and the case of Malawi is illustrative. Malawi had enough food for its people and surpluses for export until the 1980s, when land productivity began to decline as a result of overuse. After independence from Britain in 1964, the government chose to focus on large-estate agriculture to produce cash crops (tobacco, tea, sugar) for export. Fragile tropical soils that ordinarily need years of rest between plantings were expected to produce continually. After a few years, the soil was depleted and production on plantations declined. Small holders, who had lost land to commercial producers, found themselves with insufficient land to let some of it lie fallow for the necessary periods. Their productivity, too, began to fall. The historian F. Jeffress Ramsay writes that, by the 1990s, 86 percent of Malawi rural households had less than 5 acres (2 hectares) of land on which to grow food for themselves. Yet these small holders produce nearly 70 percent of the food for rural and urban consumption in Malawi. In addition, the country must now feed more than 600,000 refugees from Mozambique's civil war.

Although Zambia and Zimbabwe are struggling to reconstitute their cultural heritage after generations of British colonialism (see the box "The Thumb Piano"), their economies remain focused on long-established extractive industries. In Zimbabwe, chromium, gold, coal, nickel, and silver mining accounts for 40 percent of the country's exports; a platinum mine opened in 1994 is expected to make Zimbabwe the world's second largest producer, after South

Africa. Zambia has rich copper resources and has been a major world producer, but the industry lost money in the 1990s. The loss resulted from a combination of high debt, falling world prices in the 1980s, inefficient management, and the low productivity of miners hampered by poor working conditions. Botswana's situation was discussed earlier in the chapter (see page 357).

Majority Rule Comes to South Africa

South Africa's land area is about twice the size of Texas. It has a population of 39 million people: 75 percent are of African descent, 14 percent of European descent, and 2 percent of Indian descent; the rest have mixed backgrounds. After more than 300 years of colonization by Europeans and rule by the white minority, free elections were held in 1994 and black Africans gained control of the government. The end of white minority rule came after nearly 50 years of strict segregation by color enforced by the apartheid system.

South Africa is a country of beautiful and striking vistas. Just inland of its long coastline lie uplands and mountains that afford dramatic views of the coast and the ocean. The interior is a high rolling plateau reminiscent of the Great Plains in North America. South Africa is entirely within the midlatitudes and has consistently cool temperatures that range between 60°F and 70°F (16°C and 20°C). The extreme southern coastal area around Cape Town has a Mediterranean climate: cool and wet in the winter, dry in the summer.

The country has long been deeply divided by race. As discussed earlier in the chapter (see pages 348, 357), white supremacy lasted from 1652 until 1994. President Mandela was unyielding in his efforts to heal the racial divide. He insisted that the country first face the reality of what had happened under apartheid, through Truth

---

CULTURAL INSIGHT *The Thumb Piano*

In Zimbabwe, the Shona people have considered the mbira (thumb piano) a core element of their culture for several hundred years. The mbira has both a sacred and a secular role in society. Practitioners use it in *bira* ceremonies to contact the spiritual realm. In the 1970s, the mbira was one of the principal instruments used in the subtle songs and traditional parables of protest against white rule. These songs infuriated members of the white government because they could not understand them. Missionaries viewed the music as pagan and tried to ban it. But now it is once again used extensively both for entertainment and to lift the spirit.

Musicians play the mbira by depressing and releasing the ends of the finely tuned metal keys with their thumbnails. The keys vibrate against the carefully crafted wooden resonator and produce warm, metallic tones that add mysterious, melodious sounds to any composition. There are many different styles of mbiras, and the number of keys can range from 8 to 52. In contrast to many modern instruments, which are box-shaped, traditional mbiras are made of resonant calabash bowls.

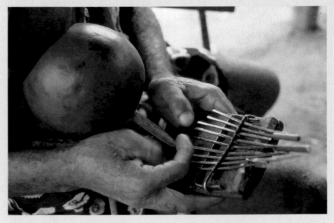

Thumb piano. [Jose Azel/Contact/Woodfin Camp & Associates.]

and Reconciliation Commissions, and then forgive those who admitted their misdeeds.

South Africa is now considered to be one of the world's 10 emerging markets. The country has a balanced economy that contributes 45 percent of the entire continent's production. It has built competitive industries in communications, energy, transport, and finance (it has a world-class stock market), and supplies goods and services to neighboring countries. For example, 80 percent of Namibia's and Botswana's imports come from South Africa. In return, products from all over Africa are sold in South Africa's markets, and many people throughout the region of Southern Africa are dependent on jobs in South Africa.

The Angolan peace treaty, economic successes in Zimbabwe and Botswana, reconciliation, and political and economic progress in South Africa have prompted forecasters to speculate that Southern Africa, more than any other region, holds the potential to lead sub-Saharan Africa into a more prosperous age. Certainly, poverty is still widespread, resources are overstressed, and the HIV/AIDS epidemic, which is so severe in this region (see pages 352–354), is bringing terrible losses. But with creative leadership and outside help, South Africa and a few of its neighbors may be able to move directly from agrarian economies into the information age of the twenty-first century, bypassing the industrial phase experienced by Europe, North America, and Japan.

REFLECTIONS ON SUB-SAHARAN AFRICA

Late one evening in a restaurant in central Europe, after a lengthy conversation that touched on some of the earth's perplexing problems, my colleague leaned across the elegant white linen tablecloth and asked, "But don't you think that Africa is, after all, better off for having been colonized by Europeans?"

How does one reply to such a question? Africa, the ancient home of the human race, is of all regions on earth the one in greatest need of attention. Africa is poor and by some measures getting poorer. Yet the reasons for Africa's poverty are not immediately apparent to the casual observer. Africa is blessed with many kinds of resources: agricultural, mineral, and forest. Nonetheless, it does not get much income from these resources. And although most of Africa is not densely occupied, rapid population growth is thwarting efforts to improve standards of living. Africa's people work hard, but their productivity, at least as measured by the standards of the developed world, is low.

An understanding of the reasons for Africa's present poverty and its social and political instability begins to emerge only through an exploration of its history over the last several centuries of European colonialism and from a careful analysis of how that history is still affecting the organization of African economies and societies. Colonialism methodically removed Africans from control of their own societies and turned Africa's peoples and resources to the service of distant countries. Even today, with colonialism officially dead for more than three decades, outsiders in the global marketplace continue to view African resources as available for the taking at less than a fair price. In view of all this, it is hard to believe that anyone would think Africa is better off for having been colonized.

Africans are only beginning to devise new economic development strategies and political institutions to replace the ones imposed by outsiders over the last 500 years. Africa remains a continent of countries created out of foreign perceptions of how Africans should be organized politically, but now African leaders are articulating wider visions. Pan-African movements for economic integration and regional cooperation are emerging.

It is tempting to suggest that the rest of the world should leave Africa to the Africans for the next era. But that view may be too simplistic. Although Africa had plenty of help getting into its current predicament, and outsiders have prospered from its wealth, it has received very little foreign aid. Europe, America, and Japan together give about U.S. $16.2 billion a year to sub-Saharan Africa, of which the United States gives only about 4 percent, or U.S. $665 million, annually. Egypt alone receives four times more U.S. foreign aid ($2.4 billion). U.S. citizens give less in aid to sub-Saharan Africa than Americans reap in profits from private investment in Africa. For example, in 2000, the Coca-Cola Company alone took profits of roughly $148 million from its African operations.

Many free-market economists, such as Jeffrey Sachs, director of the Center for International Development at Harvard University, think wealthy nations should provide financial support for Africa's comeback so that Africans can themselves become a market for imported products. Experts suggest several strategies. First, African debt to foreign governments, international lending agencies, and some private lenders should be cancelled so that tax money can once again support schools, health care, and social services. As of January 2001, American Baptists, Presbyterians, and Methodists, and the international Anglican and Roman Catholic leadership, were publicly on record as supporting this idea. Soon thereafter, the leaders of the G8 countries (the world's most developed economies) agreed in principle to forgive debt owed their governments by the poorest countries in Africa. What remained to be decided was how to ensure that the money would go instead to badly needed health, education, and social programs. A second strategy suggests that the developed world should lower tariffs against African manufactured products to foster indigenous industries. This proposal is opposed by people who worry that jobs in the developed world will be lost if cheaper African consumer goods can enter the developed countries free of tariffs. A third idea is that future aid to Africa should be designed to take advantage of indigenous skills and knowledge and that development planning should address local needs as defined and managed by local experts. This fundamentally conservative perspective puts its faith in the ultimate ability of people to look out for themselves.

Thinking Critically About Selected Concepts

1. Most of sub-Saharan Africa has a tropical climate that varies primarily by amount of rainfall, distance from the equator, and the annual north-south movements of the intertropical convergence zone. *Discuss the validity of the idea that Africa's present state of development is the result of its climate.*

2. Africa's arid ecosystems are particularly fragile environments. *Why are these environments fragile? Describe some of the human practices that contribute to the loss of moisture.*

3. The era of formal European colonialism lasted less than 100 years, but it resulted in widespread reorganization of sub-Saharan African economies and settlement patterns and the dislocation and separation of families. *Describe and discuss what happened in one or two specific places. What do you think may have been going through the minds of the Europeans when they made these changes? How might they have justified them?*

4. Gender relationships in Africa today are the result of various sets of influences over time: long-standing gender-role traditions, the more recent effects of European colonial rule on family organization, and modern circumstances that affect work and residence patterns. *Discuss the differing expectations for men and women under each of these three sets of factors. Using some of the vignettes in the chapter, show how individuals reflect these differing influences.*

5. Africa has been particularly adversely affected by European colonialism and continues to endure neocolonialism. *Try to define some circumstances in which it would be reasonable for non-Africans to have relatively inexpensive access to African resources. Would there be responsibilities and restrictions connected to this access? What might those responsibilities and restrictions be, and who would rightfully define them?*

6. During the cold war, newly independent African countries were often courted and armed by either the United States or the Soviet Union as these two powers competed for global dominance. *How are present conflicts in Africa related to the cold war? What other circumstances have led to African warfare in recent years?*

7. Africa is not particularly densely populated; but there are population issues. *Discuss how the concept of carrying capacity figures into sub-Saharan Africa's economic and environmental issues.*

8. Most countries in Africa rank low in global standards of living. However, although poverty remains widespread, there are definite patterns, with some countries providing a decent life for most citizens. *Using Table 7.1 and information in the subregional sections, compare the records of two or more of the following countries: Nigeria, Congo (Kinshasa), Kenya, Zimbabwe, Botswana, and South Africa.*

9. Several debilitating diseases afflict a large number of sub-Saharan Africans. AIDS is only one and so far not the most deadly of them. *Characterize the HIV-AIDS epidemic in Africa and its likely effects, using the most recent news reports on the situation. Devise an appropriate response on the part of the world community.*

10. International development agencies such as the World Bank have often encouraged African countries to go into debt to implement development schemes, only to find that, due to bad planning, the schemes are unworkable. *Why do you think that structural adjustment programs were seen as a remedy for the debt incurred? Recognizing the problems that structural adjustment programs have represented for Africans, discuss what might be a better approach.*

Key Terms

agroforestry (p. 376) the raising of economically useful trees

animism (pp. 368–369) a belief system in which natural features carry spiritual meaning

arable land (p. 375, map) land suitable for cultivation

apartheid (p. 348) a system of laws mandating racial segregation in South Africa, in effect from 1948 until 1994

carrying capacity (p. 349) the maximum number of people that a given territory can support sustainably with food, water, and other essential resources

currency devaluation (p. 359) the lowering of a currency's value relative to the U.S. dollar, the Japanese yen, the European Euro, or other currency of global trade

dry forests (p. 376) forests that lose their leaves during the dry season

East African Community (EAC) (pp. 361–362) an organization formed by Kenya, Tanzania, and Uganda to promote economic links among the countries of East Africa

Economic Community of West African States (ECOWAS) (p. 361) an organization of West African states working toward forming an economic union

ethnic group (p. 342) a group of people who share a set of beliefs, a way of life, a technology, and usually a geographic location

ethnicity (p. 342) the quality of belonging to a particular culture group

ethnocentrism (p. 342) the belief in the superiority of one's own cultural perspective

female genital mutilation (p. 372) the removal of the labia and the clitoris and often the stitching nearly shut of the vulva

forward capital (p. 382) a capital city built to draw migrants and investment to a previously underdeveloped area

fragile environment (p. 376) an area that contains barely enough water, soil nutrients, or other resources essential to meet the needs of plants and animals; human pressure in such environments may result in long-term or irreversible damage to plant and animal life

genocide (p. 350) the deliberate destruction of an ethnic, racial, or political group

grassroots rural economic development (p. 362) programs established to provide sustainable livelihoods in the countryside

Horn of Africa (p. 344) the part of Africa that juts out from East Africa and wraps around the southern tip of the Arabian Peninsula.

intertropical convergence zone (ITZC) (p. 342) a band of atmospheric currents circling the globe roughly around the equator; warm winds from both north and south converge at the ITCZ, pushing air upward and causing copious rainfall

laterite (p. 344) a permanently hard surface left when minerals in tropical soils are leached away

leaching (p. 344) the washing out into groundwater of soil minerals and nutrients released into soil by decaying organic matter

lingua franca (p. 374) a language of trade used to communicate by people who don't speak one another's native language

mixed agriculture (p. 386) the raising of a variety of crops and animals on a single farm, often to take advantage of several environmental niches

neocolonialism (p. 341) the practice whereby wealthy countries of the world reap the wealth of poorer countries through contracts and circumstances that favor the wealthy countries

organic matter (p. 344) the remains of any living thing

polygyny (p. 372) the taking by a man of more than one wife at a time

Sahel (p. 375) a band of arid grassland that runs east-west along the southern edge of the Sahara Desert

self-reliant development (p. 362) small-scale development schemes in rural areas that focus on developing local skills, creating local jobs, producing products or services for local consumption, and maintaining local control so that participants retain a sense of ownership

shifting cultivation (p. 344) multicrop gardening using land prepared by cutting and burning small plots of rain forest; the cultivator moves to a new plot of land after soil fertility is exhausted in a few years

Southern African Development Community (SADC) (pp. 361–362) an organization of 14 countries in Southern and East Africa working together for regional development and freer trade

structural adjustment programs (SAPs) (p. 358) belt-tightening measures implemented in debtor nations that are meant to release money for loan repayment

syncretism (or **fusion**) (p. 370) the blending of elements of a new faith with elements of an indigenous religious heritage

Pronunciation Guide

Abuja (ah-BOO-jah)

Addis Ababa (AH-dihss AH-buh-bah)

Aden, Gulf of (AH-duhn)

Afrikaner (af-rih-KAH-nurr)

Amhara (ahm-HAHR-uh)

Angola (ang-GOH-luh)

apartheid (uh-PAHRT-hyt)

Asmara (ahz-MAHR-ah)

Bantu (BAN-too)

Benin (beh-NEEN)

Biafra (bee-AH-fruh)

Boer (BOHR)

Botswana (bawt-SWAH-nah)

Burkina Faso (burr-KEEN-uh FAH-soh)

Burundi (boo-ROON-dee [both "oo" as in "book"])

Cameroon (kahm-uh-ROON)

Cape Verde Islands (VAIR-day)

Comoros (KOHM-uh-rohz)

Congo (KAWNG-goh)

Côte d'Ivoire (KOHT deev-WAHR)

Djibouti (jih-BOO-tee)

Eritrea (air-ih-TREE-uh)

Ethiopia (ee-thee-OH-pee-uh)

Fulani (foo-LAH-nee)

Gabon (gah-BOHN)

Gambia (GAHM-bee-uh)

Ghana (GAH-nah)

Guinea (GIH-nee)

Guinea-Bissau (GIH-nee bih-SAU)

Hausa (HAU-zuh)

Hausaland (HAU-zuh-land)

Hutu (HOO-too)

Igbo (IHG-boh)

Jwaneng (JWAY-nehng)

Kalahari Desert (kah-lah-HAH-ree)

Kampala (kahm-PAH-luh)

Kasama (kah-SAH-muh)

Kikuyu (kih-KOO-yoo)

Kinshasa (kihn-SHAH-suh)

Kisangani (kee-sahn-GAH-nee)

Kitui (kee-TOO-ee)

Lagos (LAH-gohss)

Lesotho (leh-SOO-too)

Liberia (lye-BEER-ee-uh)

Loango (loh-AHNG-goh)

Lumumba, Patrice (loo-MOOM-buh, pah-TREESS [both "oo" as in "book"])

Madagascar (mad-uh-GASS-kurr)

Malawi (muh-LAH-wee)

Mali (MAH-lee)

Maputo (muh-POO-toh)

Masai (mah-SYE)

Mauritania (mawr-ih-TAY-nee-uh)

Mauritius (maw-RIHSH-uhss)

mbira (uhm-BEER-uh)

Mobutu Sese Seko (moh-BOO-too SAY-see SAY-koh)

Mogadishu (moh-guh-DEE-shoo)

Mombasa (mawm-BAH-suh)

Mozambique (moh-zuhm-BEEK)

Nairobi (nye-ROH-bee)

Namibia (nuh-MIHB-ee-uh)

Niger (NYE-jurr)

Nswazi (uhn-SWAH-zee)

Obeah (OH-bee-uh)

Ogoni (oh-GOH-nee)

Ovimbundu (oh-vihm-BOON-doo [first "oo" as in "book"])

Pointe-Noire [PWA(NT) NWAHR]

polygyny (puh-LIHJ-uh-nee)

Rwanda (roo-AHN-duh)

Sahel (suh-HAYL)

Santeria (sahn-tuh-REE-uh)

Saro-Wiwa, Ken (SAHR-oh WEE-wah)

Senegal (SEHN-ih-gahl)

Seychelles (say-SHEHLZ)

Shona (SHOH-nuh)

Sierra Leone (see-AIR-uh lee-OHN)

Somalia (soh-MAHL-ee-uh)

Swahili (swah-HEE-lee)

Swaziland (SWAH-zee-land)

Tanzania (tan-zuh-NEE-uh)

Timbuktu (tihm-book-TOO [first "oo" as in "book"])

Togo (TOH-goh)

Transvaal (tranz-VAHL)

tsetse (TSEH-tsee)

Tshopo Falls (TCHOH-poh)

Tutsi (TOOT-see)

Tutu, Desmond (TOO-too)

Yoruba (YOH-roo-bah ["oo" as in "book"])

Zaire (zah-EER)

Zambia (ZAM-bee-uh)

Zimbabwe (zihm-BAHB-way)

Selected Readings

A set of Selected Readings for Chapter 7, providing ideas for student research, appears on the *World Regional Geography* Web site at www.whfreeman.com/pulsipher.

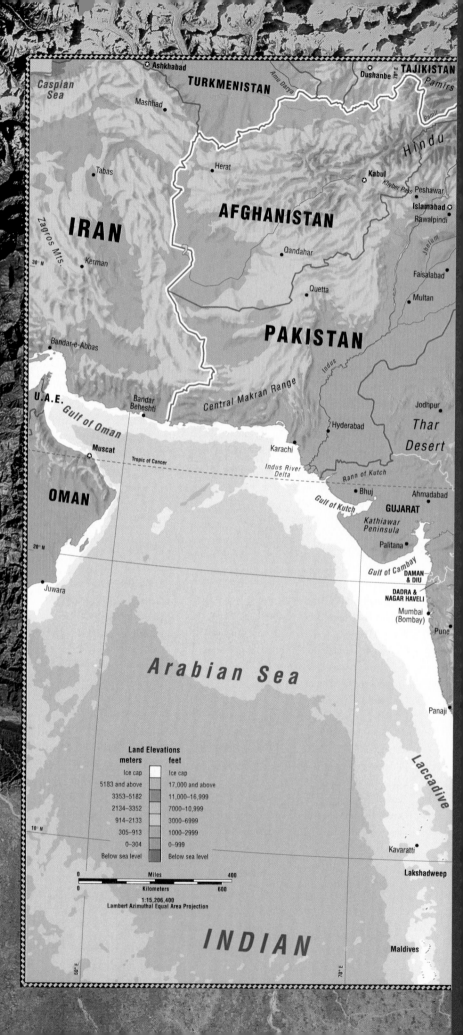

The image shows part of Nepal. The high peaks of the Himalayas are at the top, and the foothills appear in the lower middle. The bottom of the photo shows the tributaries of the Ganga River joining and flowing south across the floodplain of the Ganga. [Geospace/Science Photo Library/Photo Researchers, Inc.]

Land Elevations

meters	feet
Ice cap	Ice cap
5183 and above	17,000 and above
3353–5182	11,000–16,999
2134–3352	7000–10,999
914–2133	3000–6999
305–913	1000–2999
0–304	0–999
Below sea level	Below sea level

Miles 400

Kilometers 600

1:15,206,400
Lambert Azimuthal Equal Area Projection

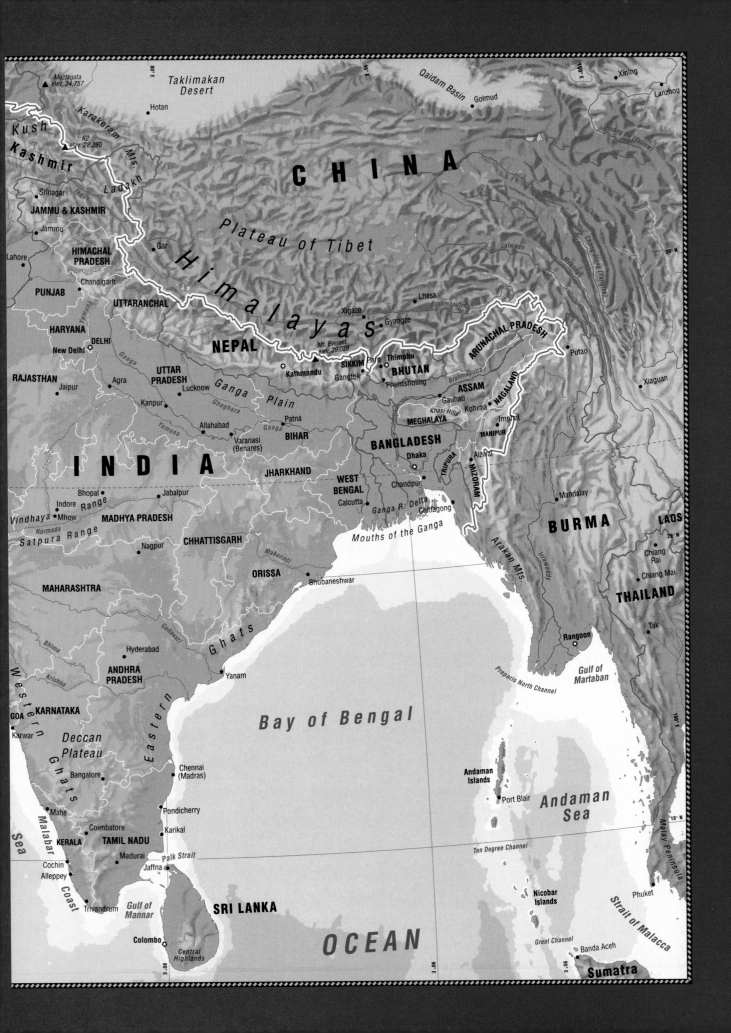

Chapter 8: South Asia

MAKING GLOBAL CONNECTIONS

VIGNETTE Ramesh Seshadri leans back from his computer at the Indian Institute of Technology, a prestigious institution of higher learning in Chennai (Madras), South India. He stares out his window, which faces the Bay of Bengal. The beach is strewn with litter and dotted with the makeshift tents of poor people who call the beach home. This reminder of his country's increasing population and decreasing environmental quality makes Ramesh think about a job opportunity in the United States, as a network technician, that a friend has just written him about. Many of Ramesh's classmates dream of a job in the United States, where salaries and living standards far surpass those in India. Alternatively, Ramesh could take a job in India's own booming information technology sector, which would allow him to remain close to his family and to Hindu religious traditions.

Then there is an opportunity offered by Lal Krishnan, a wealthy businessman of Indian descent who is now minister of environment for a Caribbean island nation. On a recent trip to Chennai, Krishnan visited a student group that Ramesh leads. The group organizes demonstrations urging India's government to base its rule on Hindu principles and not, as they feel it has in the past, on the secular ideals of Western-style democracy. Krishnan took note of a campaign that Ramesh organized to ban the thin plastic shopping bags that are commonly discarded in the streets. The cows that roam the streets of Chennai and most urban areas in India often eat the bags and develop lethal digestive problems. Devout Hindus like Ramesh, who hold cows to be sacred, are concerned about this problem. After the campaign, Krishnan offered Ramesh a position as an organizer and fundraiser for the Caribbean branch of the World Council of Hindus. This global organization works with Indians living abroad to raise awareness about issues of concern to Hindus all over the world, including newly emerging religion-based concerns about environmental deterioration. The organization also seeks financial support for Hindu political parties in India. Even though the salary is much lower than what he could earn in a computer job, Ramesh is excited about this opportunity because it would enable him to work against what he sees as the degradation of Hindu culture and the Indian environment by foreign ways and ideas. Nevertheless, a guaranteed job in the United States and a well-paid technical career are very tempting. [Adapted from the field notes of Alex Pulsipher, India, 1999.]

Ramesh's concerns and opportunities typify the global forces influencing South Asia today as well as the more local problems South Asians face. This region is the second poorest in the world, after sub-Saharan Africa, and the most densely populated. These pressures have motivated many South Asians to emigrate to other nations. Figure 8.1 shows the distribution of South Asians throughout the world (also see the box "Indian Diaspora" on page 402). Should Ramesh choose to go abroad, he will be joining a migration that began more than 2000 years ago as traders from the Indian subcontinent ventured to Arabia, Africa, and Southeast Asia. The migration continued under British colonization. In some places, people of South Asian origin have worked their way into positions of power and prosperity and can aid contemporary South Asians who wish to migrate.

Themes to Look For Here are some themes to follow as you read this chapter.

1. The ancient and layered pattern of cultural influences. This region has experienced multiple waves of cultural and religious influences since prehistoric times. The most recent is globalization.

2. The importance of village life. South Asia has many large cities, but 70 percent of the region's people live in hundreds of thousands of villages, and rural modes of spatial organization and interaction persist even in the cities.

3. The lingering influence of British colonization. Although British colonization ended 50 years ago, it has left its mark on the landscape and culture. South Asians are struggling to define national identities that can accommodate some British and other outside influences while retaining strong South Asian traditions and moral values.

4. Extremes between rich and poor. Disparities of wealth and startling contrasts between traditional and technically advanced ways of life are commonplace across the region and pose dangers to the continuation of democracy in these plural societies.

5. The challenge of rapid population growth. South Asia now has a population comparable to that of China. It lags in providing the basics of life for its people, a lag that will only increase with continued growth.

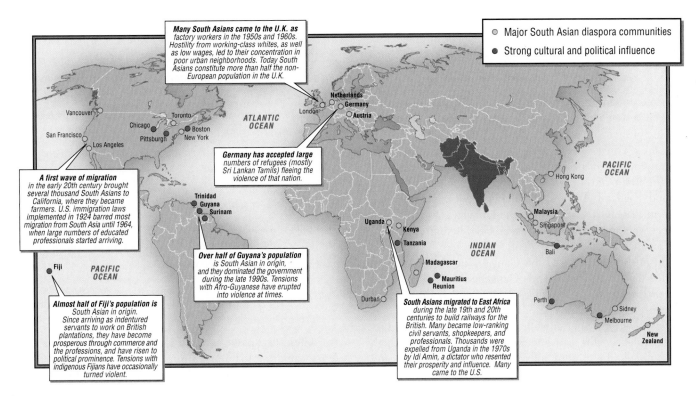

Major South Asian diaspora communities

Strong cultural and political influence

Many South Asians came to the U.K. as factory workers in the 1950s and 1960s. Hostility from working-class whites, as well as low wages, led to their concentration in poor urban neighborhoods. Today South Asians constitute more than half the non-European population in the U.K.

Germany has accepted large numbers of refugees (mostly Sri Lankan Tamils) fleeing the violence of that nation.

A first wave of migration in the early 20th century brought several thousand South Asians to California, where they became farmers. U.S. immigration laws implemented in 1924 barred most migration from South Asia until 1964, when large numbers of educated professionals started arriving.

Over half of Guyana's population is South Asian in origin, and they dominated the government during the late 1990s. Tensions with Afro-Guyanese have erupted into violence at times.

Almost half of Fiji's population is South Asian in origin. Since arriving as indentured servants to work on British plantations, they have become prosperous through commerce and the professions, and have risen to political prominence. Tensions with indigenous Fijians have occasionally turned violent.

South Asians migrated to East Africa during the late 19th and 20th centuries to build railways for the British. Many became low-ranking civil servants, shopkeepers, and professionals. Thousands were expelled from Uganda in the 1970s by Idi Amin, a dictator who resented their prosperity and influence. Many came to the U.S.

 Figure 8.1 Interregional linkages: South Asian diaspora communities (see the box "Indian Diaspora," page 402).

South Asia is a relatively easy region to define. It is bordered by the Indian Ocean on the south and by extraordinarily high mountains on the north. Although it is complex culturally and politically, there are a number of unifying features: the village is one; the common experience of British colonialism is another.

Physically, South Asia is far smaller than Africa: it would fit into the African continent five times. Yet, with nearly 1.35 billion people, it has twice Africa's population. The countries that make up the region are Afghanistan and Pakistan in the northwest; the Himalayan states of Nepal and Bhutan; Bangladesh in the northeast; India; the island country of Sri Lanka; and several sets of islands in the Arabian Sea and the Bay of Bengal. Because its clear physical boundaries set it apart from the rest of the Asian continent,

the term **subcontinent** is often used to refer to the entire Indian peninsula, including southeastern Pakistan and Bangladesh.

Terms to Be Aware Of

South Asians are reclaiming the original names of many of their important places. Under British colonial rule, which lasted until 1947, many of the ancient place names were changed or modified. In the last decade, Indians, especially, have begun to use the old names. The city of Bombay, for example, is now officially *Mumbai*, the city of Madras is *Chennai*, and the Ganges River is the *Ganga River*. In the spirit of returning to indigenous names, the country neighboring far eastern India is referred to as *Burma* rather than Myanmar.

THE GEOGRAPHIC SETTING

PHYSICAL PATTERNS

The physical geography of South Asia was created by ancient and continuing shifts in the earth's crust. Many of the landforms and even the climatic features of this region are the result of forces that positioned the Indian subcontinent along the southern edge of the Eurasian continent, where the warm Indian Ocean surrounds it and the Himalayan Mountains shield it from cold airflows from the north.

Landforms

The Indian subcontinent and its surrounding territory comprise some of the most spectacular landforms on earth and illustrate dramatically what can happen when two tectonic plates collide. About 180 million years ago, the Indian-Australian Plate, which carries India, broke free from the eastern edge of the African continent and drifted to the northeast. It began to collide with the Eurasian Plate about 60 million years ago, and India became a giant peninsula jutting into the Indian Ocean. As the relentless pushing from the south continued, both the leading (northern) edge of India and the southern edge of Eurasia crumpled and buckled. The massive Himalayan mountain range, rising more than 28,000 feet (8500 meters), was formed out of this collision, but the effects extended far beyond it. To the west and east of the Indian landmass, the pressure pushed curved crinkles into the "fabric" of Eurasia. In the west, these crinkles became the Hindu Kush, the Pamir, and the other mountains of Pakistan and Afghanistan; in the east, they became the mountains of far eastern India and adjacent Burma, China, and Thailand. As a result of the continuous compression, the Tibetan Plateau rose up behind the Himalayas to more than 15,000 feet (4500 meters) in some places. Elsewhere in Asia, as far north as Siberia, the land arched, bent, and cracked.

South of the Himalayas is the Ganga (Ganges) River basin (also called the Ganga Plain or the Gangetic Plain). Still farther south, near the tip of the Indian subcontinent, is the Deccan Plateau, an area of modest uplands, 1000 to 2000 feet (300 to 600 meters) high, interspersed with river valleys. This upland region is bounded on the east and west by two moderately high mountain ranges, the Eastern and Western Ghats. They descend to a long but narrow coastline interrupted by extensive river deltas and alluvial plains. The river valleys and coastal zones are densely occupied; the uplands only slightly less so. Because of the high degree of tectonic activity and deep crustal fractures, South Asia is prone to devastating earthquakes, such as the magnitude 7.7 quake that shook the state of Gujarat in western India in January 2001. In August 2001, seismologists announced that measured pressures in the earth's crust indicate that a very large earthquake is overdue in the Himalayan region.

Climate

The end of the dry [winter] season [April and May] is cruel in South Asia. It marks the beginning of a brief lull that is soon overtaken by the annual monsoon rains. In the lowlands of eastern India and Bangladesh, temperatures in the shade are routinely above a hundred degrees; the heat causes dirt roads to become so parched that they are soon covered in several inches of loose dirt and sand. Tornadoes wreak havoc, killing hundreds and flattening entire villages. Even the wind provides little relief, as it whips up sandstorms, making it impossible to see farther than six feet in any direction. Inhaling the sand and dust leads to widespread respiratory problems that cause many to spend long stretches of the summer ill.

This is also a time of hunger, as with each passing day thousands of rural families consume the last of their household stock of grain from the previous harvest and join the millions of others who must buy their food. Each new entrant into the market nudges the price of grain up a little more, pushing millions from two meals a day to one, from 90 percent of the minimal caloric intake needed to sustain life, to 70 percent.

(ALEX COUNTS, *GIVE US CREDIT*, 1996, P. 69.)

From mid June to the end of October [summer] is the time of the river. Not only are the rivers full to bursting, but the rains pour down so relentlessly and the clouds are so close to village roofs that all the earth smells damp and mildewed, and green and yellow moss creeps up every wall and tree. . . . Cattle and goats become aquatic, chickens are placed in baskets on roofs, and boats are loaded with valuables and tied to houses. Cooking fires are impossible . . . so the staples [are] precooked rice, a dry lentil called dal, and jackfruit, a large smelly melon that ripens on trees during this season. . . . Because most villages are built on artificial mounds raised above the fields, as the floods rise villages become tiny islands, . . . self-sustaining outpost[s] cut off from civilization . . . for most of three months of the year.

(JAMES NOVAK, *BANGLADESH: REFLECTIONS ON THE WATER*, 1993, PP. 24–25.)

These two passages highlight the contrasts between South Asia's winter and summer **monsoons** (dominant wind patterns) (Figure 8.2). In winter, cool, dry air flows from the land to the ocean; in summer, warm, moisture-laden air flows from the Indian Ocean over the land of the Asian subcontinent. Although this process happens around the globe, the huge size of the Eurasian landmass greatly intensifies it in South Asia. The summer monsoon brings heavy rains, and the abundance of rainfall is amplified by yet another process taking place at the equator.

Air masses moving south from the Northern Hemisphere and north from the Southern Hemisphere converge at the equator. This warm, moisture-laden air rises, forming clouds that produce copious precipitation as they rise and cool. This belt of warm, rising air that circles the earth around the equator is called the intertropical convergence zone (ITCZ) (see Figure 7.1). In the southern winter (northern summer), the ITCZ shifts north of the equator; in the southern summer (northern winter), it shifts south of the equator. Climatologists now think that the powerful "inhalation" of air over the Eurasian landmass in summer actually bends the ITCZ belt northward near the Horn of Africa. This bending allows the ITCZ to move north rapidly over the Indian Ocean, picking up enormous amounts of moisture that are then deposited throughout Asia.

The monsoons and the ITCZ are major influences on South Asia's climate (Figure 8.3, page 400). In early June, the ITCZ-generated warm, moist air first reaches the mountainous Western Ghats and Sri Lanka's central highlands. The rising air mass cools as it moves over the mountains, releasing rainfall that nurtures dense tropical rain forests. After dumping huge amounts of rain, the moisture in the air mass is reduced, and somewhat less rain (but still significant amounts) falls to the east of the Western Ghats. Once on the other side of India, the monsoon gathers additional moisture and power in its northward sweep up the Bay of Bengal, sometimes turning into a tropical cyclone. As the monsoon system reaches the hot plains of West Bengal and Bangladesh in late June, rising air causes strong updrafts in the moist atmosphere. These updrafts create massive, thunderous cumulonimbus rain clouds that drench and then flood the parched countryside. Precipitation

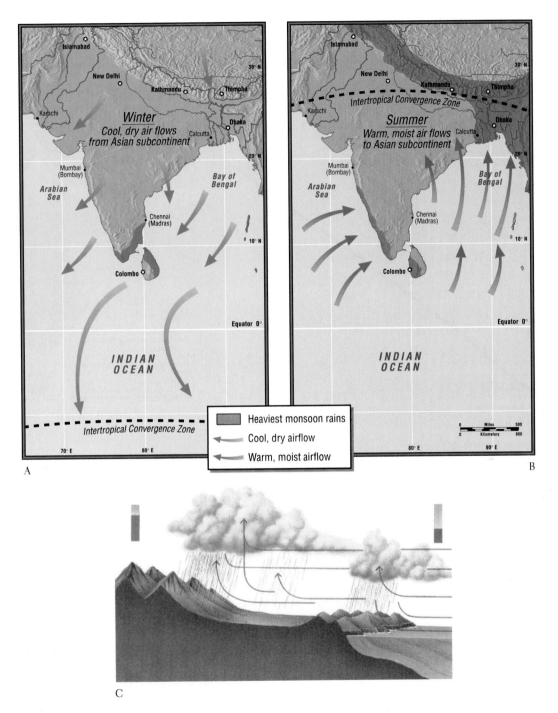

Figure 8.2 Winter and summer monsoons in South Asia. (A) In the winter, cold, dry air blows from the Eurasian continent south across India toward the ITCZ, which is much farther south. (B) In the summer, the ITCZ moves north across India, picking up huge amounts of moisture from the ocean, which it then deposits over India. (C) Landforms affect rainfall: in summer, moisture-laden air from the ocean rises and cools when it reaches the coastal mountains, releasing copious rain. The rising and cooling process is repeated when the wet air reaches the Himalayas.

is especially intense in the foothills of the Himalayas. The northeastern Indian state of Meghalaya (see the chapter opening map) has the highest average annual rainfall in the world: about 35 feet (10.6 meters). Rainfall diminishes to the west of Nepal, but seasonal rain is still significant in a band parallel to the Himalayas that reaches across northern India all the way to northern Pakistan by July. The differences in rainfall are reflected in the varying climate zones (see Figure 8.3) and agricultural zones (see Figure 8.13). Rice is generally grown in the wettest areas, and cereals and especially wheat are the dominant crops in the drier areas. Pasture and

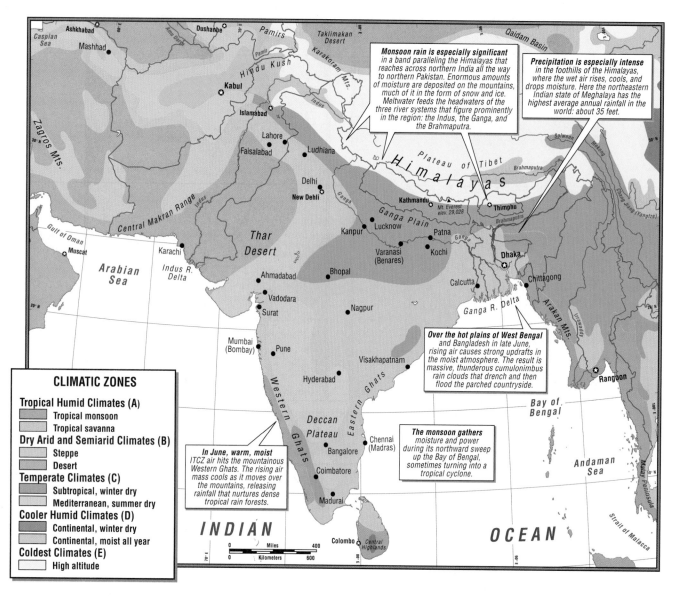

Monsoon rain is especially significant in a band paralleling the Himalayas that reaches across northern India all the way to northern Pakistan. Enormous amounts of moisture are deposited on the mountains, much of it in the form of snow and ice. Meltwater feeds the headwaters of the three river systems that figure prominently in the region: the Indus, the Ganga, and the Brahmaputra.

Precipitation is especially intense in the foothills of the Himalayas, where the wet air rises, cools, and drops moisture. Here the northeastern Indian state of Meghalaya has the highest average annual rainfall in the world: about 35 feet.

Over the hot plains of West Bengal and Bangladesh in late June, rising air causes strong updrafts in the moist atmosphere. The result is massive, thunderous cumulonimbus rain clouds that drench and then flood the parched countryside.

The monsoon gathers moisture and power during its northward sweep up the Bay of Bengal, sometimes turning into a tropical cyclone.

In June, warm, moist ITCZ air hits the mountainous Western Ghats. The rising air mass cools as it moves over the mountains, releasing rainfall that nurtures dense tropical rain forests.

CLIMATIC ZONES

Tropical Humid Climates (A)
- Tropical monsoon
- Tropical savanna

Dry Arid and Semiarid Climates (B)
- Steppe
- Desert

Temperate Climates (C)
- Subtropical, winter dry
- Mediterranean, summer dry

Cooler Humid Climates (D)
- Continental, winter dry
- Continental, moist all year

Coldest Climates (E)
- High altitude

Figure 8.3 **Climates of South Asia.**

nonagricultural land dominate in the driest areas and in the areas where altitude inhibits plant growth.

By November, the cooling Eurasian landmass sends cooler, drier air to South Asia. This heavier air from the north pushes the warm, wet air back south to the Indian Ocean. Although very little rain falls during this winter monsoon, parts of South India and Sri Lanka receive winter rains as the ITCZ drops moisture picked up on its now southward pass over the Bay of Bengal.

Monsoon rains deposit large amounts of moisture over the Himalayas, much of it in the form of snow and ice. Meltwater feeds the headwaters of the three river systems that figure prominently in the region: the Indus, the Ganga, and the Brahmaputra. All three rivers begin within 100 miles (160 kilometers) of one another in the Himalayan highlands near the Tibet-Nepal-India borders. The Ganga flows generally south through the mountains

and then east across the Ganga Plain to the Bay of Bengal. The Brahmaputra flows first east, then south, and then west to join the Ganga in forming the giant delta of Bengal. The Indus, in the far west, takes a southwesterly course across arid Pakistan and empties into the Arabian Sea. These rivers, and many of the tributaries that feed them, are actively wearing down the surface of the Himalayas; they carry an enormous load of silt, especially during the rainy season when their volume increases. Their velocity slows when they reach the lowlands, and much of the sediment settles out as silt. It is then repeatedly picked up and redeposited by successive floods. As illustrated in the diagram of the Brahmaputra River (Figure 8.4), the seasonally replenished silt nourishes much of the agricultural production in the densely occupied plains of Bangladesh. The same is true on the Ganga and Indus plains.

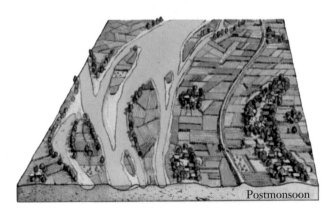

Figure 8.4 **The Brahmaputra River in Bangladesh at various seasonal stages.** In the premonsoon stage (top), the river flows in multiple channels across the flat plain. During peak flood stage (middle), the great volume of water overflows the banks and spreads across fields, towns, and roads. It carves new channels, leaving some places cut off from the mainland. In the postmonsoon stage (bottom), the river returns to its banks, but some of the new channels persist, changing the lay of the land. As the river recedes, it leaves silt and algae that nourish the soil. New ponds and lakes form and fill with fish. The people learn to adapt their farms to a changing landscape. Throughout much of the country, farmers are able to produce rice and vegetables nearly year-round. [Adapted from *National Geographic* (June 1993): 125.]

HUMAN PATTERNS OVER TIME

A variety of groups have migrated into South Asia over the millennia, many of them invaders who conquered peoples already there. The merging and blending of different cultural, religious, social, and political elements have given South Asia a unique and rich heritage. One result has been a level of cultural diversity that few other world regions can equal. Another is that South Asia is one of the most politically dynamic and contentious places in the world.

The Indus Valley Civilization

There are indications of early humans in South Asia as long as 200,000 years ago, but evidence of modern humans is dated at about 38,000 years B.P. (before the present). The first substantial settled agricultural communities, known as the **Indus Valley civilization** (or Harappa culture), appeared around 4500 years B.P. along the Indus River in modern-day Pakistan and along the Saraswati River in modern-day India. The architecture and urban design of this very early civilization were quite advanced for the time: there were multistory homes with piped water and sewage disposal; planned towns with wide, tree-lined boulevards laid out in a grid; and a high degree of consistency in building materials over a region that covered more than 1000 square miles (2500 square kilometers). Fine beadwork and jewelry, evidence of a trade network that extended to Mesopotamia and eastern Africa, continue to fascinate students of this region, as does a writing system that has yet to be deciphered. Moreover, much of the Indus Valley civilization's agricultural system survives to this day, including techniques for storing monsoon rainfall to be used for irrigation in dry times; methods for cultivating wheat, barley, oilseeds, cotton, and other crops adapted for arid conditions; and the use of wooden plows drawn by oxen.

The reasons for the decline of the Indus Valley civilization after about 800 years (3700 years B.P.) are debated. Some scholars believe that complex ecological changes brought about a gradual demise; others argue that foreign invaders brought a swift collapse. Although the civilization disappeared, many aspects of its culture blended with subsequent foreign influences. Thus, the Indus Valley civilization is the foundation of modern South Asian religious beliefs, social organization, linguistic diversity, and cultural traditions.

A Series of Invasions

The richness of the Ganga and Indus river valleys and the agricultural societies that lived there attracted people from the drier surrounding areas of Central and Southwest Asia. The first recorded invaders into South Asia called themselves **Arya** (Aryan is the contemporary term), and they moved into the Indus Valley and Punjab from Central and Southwest Asia probably as early as 3500 years ago. Many scholars believe that the Aryas, in conjunction with the indigenous Harappa culture and other indigenous cultures, instituted the remarkably influential caste system and

AT THE GLOBAL SCALE   *Indian Diaspora*

From the tropical island of Bali in Indonesia to Queens in New York City, one can hear South Asian temple bells ringing and catch the pungent fragrance of spices being ground for curry and kabob dishes. Both the Balinese and the South Asian Americans, though separated by thousands of miles, belong to what is called the **Indian diaspora**, the set of all people of South Asian origin living (and often born) abroad. Some are the descendants of groups who have been living outside India for a thousand years or more. The term *Indian diaspora* is used because for many centuries, India was the popular name for the whole region of South Asia.

The earliest migrations of South Asians were to eastern Africa and the islands of southeastern Asia. Archaeological evidence shows that people from the Indus Valley in what is today Pakistan traded with people in eastern Africa and Mesopotamia as early as 4000 years ago. The Balinese trace their cultural roots to trade networks between southern India and Southeast Asia established over 2000 years ago. Their primarily Hindu culture and religious practices are a remnant of that time.

The next major wave of South Asian emigration occurred from about 1850 to 1920 under British colonial rule. Hundreds of thousands of people were recruited, and sometimes even kidnapped, from various parts of the Indian subcontinent to labor as indentured servants on British plantations in such places as Trinidad and Guyana in the Caribbean, Fiji in the Pacific Ocean, and Malaysia in Southeast Asia. Some were recruited to build railroads in East Africa (especially in Kenya and Uganda) or to serve with the British army in other parts of the world. Many of these migrants remained in their new homes permanently.

After World War I, another wave of migrants from the west coast of India (mainly merchants and civil servants) arrived along the coast of East Africa, where they became successful merchants and exporters. They were pushed out of countries such as Uganda during African nationalistic movements in the 1970s. Their descendants live in the United Kingdom, the United States, Canada, and Australia. These communities are called "twice-migrants." A substantial number of twice-migrants in the United States and elsewhere opened shops and restaurants or purchased small mom-and-pop motels. The films *Mississippi Masala* (1992) and *Miss India Georgia* (1998) illustrate South Asian struggles to adapt to life in the United States.

Over the past 30 years, another great wave of migrants has left South Asia. Since the oil price rises of the early 1970s, tens of thousands of Muslims from India, Pakistan, Bangladesh, and Sri Lanka have found work in the oil-rich nations of the Arabian Peninsula and Southeast Asia. Others have found jobs in England's factory towns (Oldham, Leeds, Manchester) or as taxi drivers and construction workers in Canadian and U.S. cities. Probably the most socially visible members of South Asian diasporic communities are those professionals in medicine and information technology who have settled in North America, Australia and New Zealand, and Europe. In the United States, citizens and residents of South Asian origin now form the wealthiest ethnic group, as defined by the U.S. census. Recent migrants, who maintain particularly close contact with their homelands, remit millions of dollars back to their families and make regular return visits.

Although migration allows people to escape from the relative poverty and environmental degradation of South Asia, it also contributes to the **brain drain**—the flight of the best and brightest South Asians to wealthier regions. Trends that may counter the brain drain are increasing foreign investment in India and greater access to foreign markets by Indian companies, both of which create desirable jobs in India but also intensify the foreign influence that alarms many Indians. Ironically, some of the strongest reactions against foreign influence are funded by members of the South Asian diaspora in the United States and the Caribbean. Surrounded on a daily basis by what they see as degrading Western influences, some diasporic South Asians try to strengthen the cultural and religious traditions they cherish back home by financially supporting religiously based political parties and activism of the type that Ramesh Seshadri practices (see the vignette that opens this chapter).

---

brought some of the early elements of classical Hinduism, the major religion of India (see pages 412–414). Thus, contemporary Hinduism and the caste system represent more than 3000 years of the blending of ritual concepts and practices from many communities. The caste system divides members of society into a social status hierarchy (see the discussion of caste on pages 414–416). Over time, the caste system may have weakened South Asia's ability to resist invaders, because certain castes were easily assimilated into the invaders' ruling structures. In particular, members of the highest priestly caste, known as the Brahmins, served as advisors and administrators for foreign conquerors. Under Brahmin influence, however, invaders often adopted many of the very cultural practices and institutions (including the caste system itself) that had made the South Asians vulnerable to conquest in the first place. Thus, over generations, as invaders settled into their role as a caste of ruling elites, their realms became more vulnerable to waves of new invaders.

In addition to the Aryas, other invaders who arrived from Central Asia included the Persians and Alexander the Great, the Greek warrior-explorer who penetrated as far as Punjab in 326 B.C. but was stopped by a mutiny of his own soldiers. After Alexander, waves of Turkic and Mongolian peoples continued to enter the region via the Indus Valley. Starting around 1000 years ago, they introduced the religion of Islam from Central and Southwest Asia.

The great fortress at Agra, built 1565–1571, had walls 72 feet (22 meters) high enclosing an area about 1.5 miles (2.4 kilometers) in circumference. [J. H. C. Wilson/Robert Harding Picture Library.]

The invasion in 1526 by the **Moguls,** a later group of related Turkic people from Persia, intensified the spread of Islam by giving it imperial stature and artistic grandeur. The Moguls reached the height of their power and influence in the seventeenth century and controlled the north central plains, but farther south their influence was much weaker. During the last century of Mogul rule, the British began to establish their colonial influence; by 1857, the British were able to supplant the Moguls.

The primary legacy of the Moguls is the more than 300 million Muslims now living in South Asia. The Moguls also left a unique heritage of architecture, art, literature, and linguistics that includes the Taj Mahal, the fortress at Agra, miniature painting, and the tradition of lyric poetry. The Moguls helped to produce the **Hindustani** language, which became the lingua franca (language of trade) of the northern Indian subcontinent. Hindustani is still used by more than 400 million people (see pages 411–412). In the northern reaches of South Asia, the Islamic ideals of the Moguls influenced and in some cases supplanted the largely Hindu cultural complex, making a lasting impact on aesthetics, social interactions, gender roles, and religious architecture. In all these ways, the Moguls contributed important elements to modern definitions of what it means to be Indian, Pakistani, or Bangladeshi.

Today, few people can claim to be descendants of the Moguls themselves; most Muslims descend from Hindus who converted to Islam for a variety of reasons, including a wish to escape their low-status castes. And although the Moguls introduced many of their changes hundreds of years ago, there is a contemporary back-lash against Mogul influence in India that is connected to the rise in Hindu nationalism (see page 425). Nevertheless, it is because of the Moguls that Islam is central to the national character of three countries in the region: Afghanistan, Pakistan, and Bangladesh. In India, Muslims are the largest religious minority, making up about one-eighth of the Indian population (more than 120 million people).

The Legacies of Colonial Rule

Great Britain was the most recent influential invader of South Asia. Britain was already a strong trading presence in the 1750s, when the British East India Company established its control over a number of coastal trading centers. The British controlled most of South Asia from the 1830s to 1947, profoundly influencing South Asia politically, socially, and economically.

Economic Influence. British colonial policy in South Asia, as elsewhere, was to use the region's resources primarily for the benefit of Britain. This policy often resulted in detrimental effects on South Asia. A typical example was the textile industry in Bengal (modern-day Bangladesh and the Indian state of West Bengal), where Britain made one of its first inroads into South Asian economies. Bengali weavers, long known for their high-quality muslin cotton cloth, initially benefited from the greater access British traders gave them to overseas markets in Asia, in the Americas, and on the European continent, as well as in Britain itself. The high quality of Bengali cloth and the great demand for it reflected the advanced manufacturing economy of South Asia, which in 1750 produced 12 to14 times more cotton cloth than did Britain alone and more than all of Europe combined. However, as Britain's own highly mechanized textile industry developed during the second half of the eighteenth century, cheaper British cloth replaced Bengali muslin, first in world markets and eventually throughout South Asia. The British East India Company encouraged the supplanting of indigenous textile manufacturing by instituting a rule against looming. Those who continued to run their own looms rather than work in British-owned plants were severely punished. Thus, as one British colonial official put it, while the mills of Yorkshire prospered, "the bones of Bengali weavers bleached the plains of India."

The British-induced reversal of India's fortunes in the eighteenth century profoundly affected its people's lives in the nineteenth century. Many people who were pushed out of their traditional livelihood in textile manufacturing were compelled to find work as landless laborers in an economy that already had an abundance of agricultural labor. Others migrated to emerging urban centers. Increasingly, bandits (called *dacoits* in India) roamed the countryside, while small landowners lost their land to large landowners as a consequence of a highly corrupt and ill-conceived tax system instituted by the British East India Company. In the 1830s, a drought exacerbated the privations of colonial rule, and more than 10 million people starved to death in the heart of the Gangetic Plain. It was during these trying times that South Asian laborers were pressed into joining the stream of indentured laborers immigrating to other British colonies in the Americas, Africa, Asia, and the Pacific (see pages 396 and 402). Ironically, at the same time as Britain's own labor-exploitative feudal system was dying out (see Chapter 4, page 189), India's caste-based feudal system was expanding and becoming more entrenched.

What little economic development did take place in South Asia was allowed only if it benefited Britain. The use of tariffs is an example. Whereas Europe and the United States, and later Japan, used protective tariffs to provide a shield against cheaper imported manufactured goods, the British prohibited protective tariffs in South Asia. Hence, imported British products ruined many established South Asian industries and channeled their workers into agricultural pursuits. For the remainder of the colonial period, Britain encouraged the production mainly of tropical agricultural raw materials, such as cotton, jute (a fiber used in making gunnysacks and rope), tea, sugar, and indigo (a blue dye). These products were intended to supply Britain's own growing industries and to fit in with British consumption patterns.

The building of railroads provides another example of how colonialism distorted development in South Asia. Thousands of miles of track were installed, but almost all the locomotives, cars, and rails were manufactured in Britain, which thus received a boost for its coal, steel, and manufacturing industries. The railroads were not primarily intended to be a transport network linking South Asian places. Rather, they were meant to carry raw materials to ports and to transport British manufactured goods into the hinterland markets of South Asia. Nonetheless, internal linkages were eventually made, and by independence, India, especially, had a viable rail system.

The economic historian Dietmar Rothermund argues that this British tendency to inhibit and reverse Indian industrial development also encouraged population growth, now such a prominent issue in all of South Asia. Poor farmers produced as many children as possible, both for farm labor and as insurance against destitution in their old age. Had they moved from being farmers to being industrial workers and business owners, they might have favored smaller families, like their industrial counterparts elsewhere; and they might simply have saved their earnings to provide for their old age.

Nonetheless, the British Empire did bring some benefits. Trade with the rest of the empire brought prosperity to a few areas, especially the large British-built cities on the coast, such as Bombay (Mumbai), Calcutta, and Madras (Chennai). The railroad boosted trade within South Asia, and it greatly eased the burden of personal transport. In addition, English became a common language for South Asians of widely differing backgrounds, assisting both trade and cross-cultural understanding. Moreover, most of today's South Asian governments retain institutions put in place by the British to administer their vast empire. South Asian governments have also inherited many of the shortcomings of their colonial forebears, such as highly bureaucratic procedures, a resistance to change, and a tendency to remain distant and aloof from the people they govern. Nonetheless, these governments have proved functional. In particular, democratic government, though it was not instituted on a large scale until the final years of the empire, has

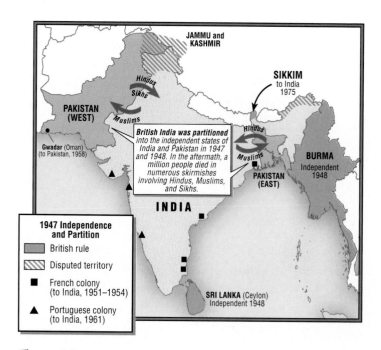

Figure 8.5 South Asian independence and partition. The European colonies in South Asia were broken up in 1947 with the partition of India and Pakistan. [Adapted from *National Geographic* (May 1997): 18.]

given people an outlet for voicing their concerns and has enabled many peaceful transitions of elected governments since 1947.

Partition and Independence. Perhaps the most enduring and damaging outcome of colonial rule was the partition of British India into the independent countries of India and Pakistan in 1947 (Figure 8.5). As part of the independence agreement, the British decided that northwestern and northeastern India, where the population was predominantly Muslim, would become a single country, Pakistan. This new country thus consisted of two parts, known as West and East Pakistan, separated by all of north central India. Although both India and Pakistan maintained secular constitutions, with no official religious affiliation, the general understanding was that Pakistan would have a Muslim majority and India a Hindu majority. Fearing that they would be persecuted if they did not move, millions of Hindus and Sikhs migrated from their ancestral homes in Pakistan to India; similarly, many Muslims left their homes in India for Pakistan. In the process, families and communities were divided, looting and rape were widespread, and more than a million people were killed in innumerable skirmishes between the religious groups.

Many historians argue that partition could have been avoided had it not been for deliberate British efforts throughout the colonial era to heighten tensions between South Asian Muslims and Hindus, thus creating a role for themselves as indispensable and benevolent mediators. There is much evidence for these so-called divide-and-rule tactics. British administrators at the local level commonly favored the interests of minority communities in order to weaken the power of the majority community, which could have threatened British authority. At the national level, the British worked out special political agreements with Muslim leaders as a way of weakening Muslim support for the Indian Congress party. The Congress party led the independence movement against the British and claimed to represent the interests of both Muslims and Hindus; thus, it posed a threat to Britain. Partition in 1947 spawned repeated wars and skirmishes, and relations between India and Pakistan remain strained in the best of times. Currently there is sporadic armed conflict over the territories of Kashmir and Punjab in northeastern India.

Since Independence

In the more than 50 years since the departure of the British, South Asians have experienced both progress and setbacks. Democracy has expanded steadily, albeit somewhat slowly. India has maintained its status as the world's most populous democracy and is gradually dismantling age-old traditions that hold back poor, low-caste Hindus and other disadvantaged groups. Agricultural advances have brought relative prosperity to a few rural areas; and a small but increasing number of educated Indians are reaping the economic benefits of the global growth in information technology. Nonetheless, all South Asian countries have continued to suffer from low levels of economic development; they have a collective annual gross domestic product per capita of about U.S. $2112 (as of 1998, adjusted for PPP). Only sub-Saharan Africa has a lower average regional per capita GDP ($1607).

Although threats from outside the region have diminished, the danger of violence originating from within has steadily increased. After 1947, East and West Pakistan, despite their common religion, struggled to overcome their stark cultural differences, economic disparity, and geographic separation. East Pakistan was the more populous and was the chief earner of the country's foreign exchange through the export of primary products. Yet East Pakistan saw much of its earnings spent by officials in West Pakistan. West Pakistan, on the other hand, dominated government and military positions. In 1970, the first general election in 12 years gave a leader from East Pakistan a landslide victory. The leaders of West Pakistan refused to turn over the government. After protracted negotiations, West Pakistan attacked East Pakistan, which resulted in a cry for independence. Thus, as the result of a bloody civil war and intervention by India in 1971, Pakistan was divided and East Pakistan became the independent country of Bangladesh.

In the years since 1971, civil wars have plagued Sri Lanka, Afghanistan, and parts of India; and the possibility of nuclear confrontation between Pakistan and India has loomed ominously whenever the dispute over Kashmir has heated up (see page 427). Notably, however, disputing parties have been known to lay aside conflict in emergencies. Such was the case after the earthquake in Gujarat in 2001, when Pakistan sent much-appreciated aid to India. Future threats to the region's stability may result from stresses that a huge and still growing population place on the region's already precarious natural environment. Hence, despite the considerable gains of the past 50 years, South Asia is one of the world's two poorest regions and periodically one of the most unstable.

POPULATION PATTERNS

Although Africa has the fastest-growing population on earth, South Asia is already so densely populated (Figure 8.6, compare with Figure 7.4) and continues to grow so quickly that the consequences can be seen everywhere. At night, especially during the hot season, thousands of sleepers occupy virtually all urban public spaces; commuters cling to the outside of packed buses as they travel lopsidedly through the streets of Peshawar, New Delhi, and Mumbai; rural hills are denuded of trees and bushes because people have cut wood for cooking and grass for animal fodder; and crowded urban shantytowns, where village life is reconstituted, have but a few water spigots for thousands of residents. Nearly everywhere, there are people—throngs of people—the vast majority of whom, despite the crush, are congenial and good-natured with one another.

South Asia has several of the world's largest cities—Mumbai with 13 million, Calcutta with 11 million, Delhi with 14 million, Dhaka with 7 million—and the stream of migrants from the countryside seems never-ending. The stream is likely to increase because, so far, less than 30 percent of the population is urban.

People come to South Asian cities for all the reasons that cities attract people everywhere. Some hope for business opportunities or better jobs with higher pay, others come for an education or training. Some seek the anonymity and individualism—perhaps even a rise in caste (see page 414)—that are not possible in close-knit rural communities. Many are pushed into urban migration by agricultural modernization that reduces the numbers of rural jobs and by the simple relentless rise of rural populations that makes access to land and resources ever more difficult. Some are refugees who have left drought-stricken or flooded countryside.

VIGNETTE One consequence of South Asia's rapid urban growth is that the cities have become so crowded that housing has long since been exhausted and many people simply

A rickshaw bicycle and driver in Delhi. [Lindsay Hebberd/Woodfin Camp & Associates.]

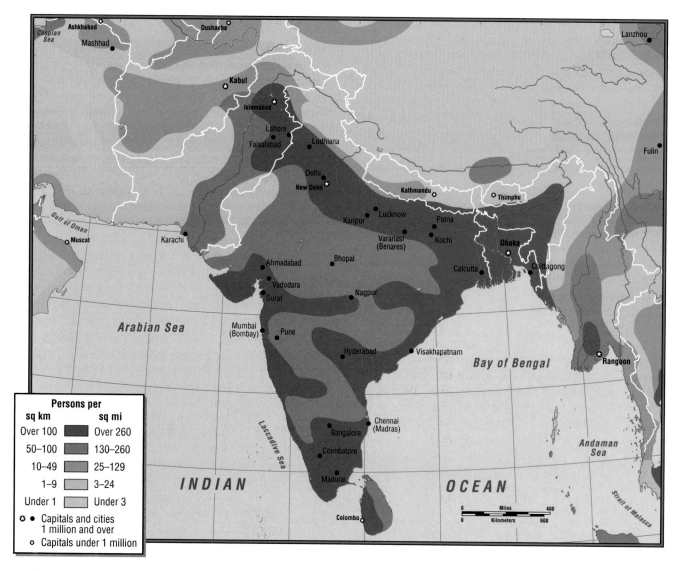

Figure 8.6 Population density in South Asia. [Adapted from *Hammond Citation World Atlas* (Maplewood, N.J.: Hammond, 1996.]

live on the streets. In a National Public Radio interview in August 1997, an Indian journalist recalled once impulsively asking a bicycle rickshaw driver about himself as he peddled her through Delhi. He replied that his belongings—a second set of clothes, a bowl, and a sleeping mat—were under the seat where she was sitting. He had come to Delhi from the countryside 14 years before, and he had never found a home. He knew virtually no one; he had few friends and no family; and no one had ever inquired about him before. He worked virtually around the clock and slept here and there for two hours at a time. [Gagan Gill, "Weekend Edition," National Public Radio (August 16, 1997).]

Because South Asia's population is so young—more than a third of the region's people are under age 15—rapid population growth will continue to strain efforts to improve life for South

Asians. As a region, South Asia already has more people (1.35 billion) than China (1.26 billion). By 2020, India alone is expected to overtake China, where the rate of natural increase (0.9 percent) is half that of India's (1.8). Although the Indian economy is not particularly efficient, it has grown at a steady annual rate of close to 5 percent for several decades; but each year India adds more than 18 million people, approximately the population of the state of Texas, to its ranks. To accommodate these new Indians adequately, every single year the country would need to build 127,000 new village schools, hire 373,000 new schoolteachers (50 students per teacher), build 2.5 million new homes (7 people per home), create 4 million jobs, and produce 180 million new bushels of grain and vegetables. Only in food production has India come close to keeping pace with population growth, although its ability to continue doing so is not secure. The technological advances in agriculture responsible for increased yields may not be sustainable

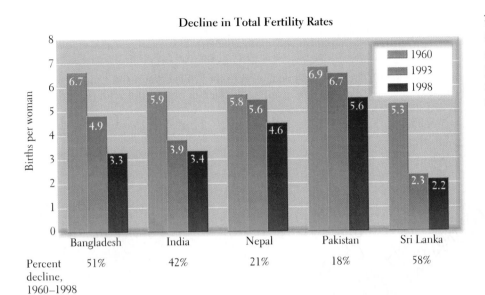

Decline in Total Fertility Rates

Figure 8.7 **Decline in total fertility rates.**
Some South Asian countries have experienced large declines in fertility rates in the last 30 years.
[Adapted from *A Demographic Portrait of South and Southeast Asia* (Washington, D.C.: Population Reference Bureau, 1994), p. 8.]

because of high costs and negative impacts on the environment and soil. In all other categories, India falls further and further behind, despite a respectable economic growth rate.

South Asia has been trying to reduce births since 1952. India spends over a billion dollars a year on population control programs. Unlike many other countries, which rely on financial aid from developed countries to fund such programs, India pays for nearly all of this effort on its own. Fertility rates have indeed declined significantly in India and Bangladesh, and especially in Sri Lanka, although much less in Pakistan (5.6) and Nepal (4.8) (Figure 8.7). Why does population continue to boom despite such efforts? The answers are similar to those we have given in earlier chapters.

In Pakistan, for example, a look at the population pyramid (Figure 8.8) shows that a significant portion of the population is in the early reproductive years, so even a one-child-per-couple policy would result in population growth for years to come. Of equal importance, poor, rural, uneducated people see children as their only source of wealth. Babies not only bring joy, they quickly become productive family members: 3-year-olds are already able to perform some useful tasks, and 10- or 12-year-olds pull their weight, quite literally, in the fields or even accomplish such highly skilled tasks as carpet weaving (see the box "Should Children Work?"). Furthermore, grown children are the only retirement plan that most South Asians will ever have. Because access to health care is limited, infant mortality rates hover around 75 per 1000 live births

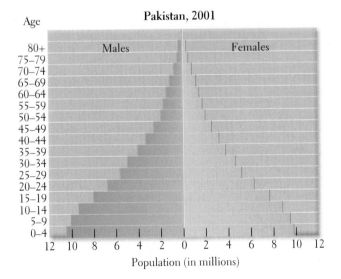

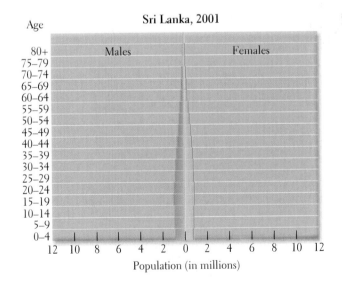

Figure 8.8 **Population pyramids for Pakistan and Sri Lanka, 2001.** Both Pakistan and Sri Lanka have experienced decreasing fertility rates since 1960 (see Figure 8.7), but the great size of Pakistan's population and its youth mean that Pakistan's population will continue to grow for years. Sri Lanka

has a much smaller population, and growth rates have slowed markedly. Hence Sri Lanka's pyramid no longer has a pyramidal shape. [Adapted from "Population Pyramids for Pakistan" and "Population Pyramids for Sri Lanka" (Washington, D.C.: U.S. Bureau of the Census, International Data Base, May 2000).]

AT THE REGIONAL SCALE   *Should Children Work?*

Beginning in the mid-1990s, a number of reports questioned the exploitation of child labor in Asia, especially in the handwoven carpet industry, where there have been reports of outright enslavement of kidnapped children ("Weekend Edition," National Public Radio, March 17, 2001). Carpet weaving is an ancient artistic and economic enterprise in Central and East Asia, as well as in South Asia. Traditionally, it has been a family-run enterprise, with women and children the weavers and men the merchants. International trade has increased the demand for fine handwoven carpets, but it is South Asian middlemen and foreign traders from Europe and America who profit, not the weavers, and some unscrupulous carpet merchants have apparently resorted to forced labor to produce carpets.

Children have always participated in the home-based carpet industry. For thousands of years, young children have learned weaving skills from their parents and have become proud members of the family's production unit. Yet, in today's society, children must balance their role as family workers with the need to attend school, where they learn the skills that will enable them to survive in a modern economy.

Should children be allowed to work? It is clear that kidnapping and enslavement must be stopped, but is there room for different cultural views of what childhood should be like? When consumers buy goods made by children, are they supporting family values or greedy factory owners and middlemen? The United Nations and South Asian governments are now addressing these questions. There is an active program to curb child labor abuses, while remaining open to the positive experience that learning a skill and being part of a family production unit can be for a child. In India, for example, there is now a national system to certify that exported carpets are made in shops where the children go to school, have an adequate midday meal, and receive basic health care. Such carpets will bear the label "Kaleen."

If abuses can be eliminated, the custom of child labor is likely to continue because of significant psychological and economic benefits to children and their families. Until the modern era, such work was considered an important part of a child's training for adulthood, even in the United States and Europe. Some experts argue that it is the lack of meaningful roles for children that leads to juvenile crime.

Two young boys tie knots at a carpet loom in India. [Cary Wolinsky/Stock, Boston.]

(only sub-Saharan Africa's are higher); hence, couples often choose to have more than two or three children to ensure that some reach maturity. The pyramid for Sri Lanka, on the other hand, shows a significant decline in fertility. Sri Lanka has a much lower infant mortality rate (17 per 1000 live births) than India (72 per 1000). This low rate in Sri Lanka is related to education for women, discussed below.

Another reason families favor a large number of offspring in regions like South Asia is that they view sons as more beneficial than daughters. Sons, it is believed, contribute more to a family's wealth than daughters, and many couples want at least two sons to provide for them in old age. Daughters are thought to have little economic value, in part because the domestic work they do rarely earns cash. In patriarchal societies, women are rarely educated, and most women have few ways to achieve fulfillment except as prolific mothers. A popular toast to a new bride is "May you be the mother of a hundred sons."

Even the middle class becomes caught up in the desire for sons. Many couples who wish sons patronize high-tech labs that specialize in identifying the sex of an unborn fetus. Their intention is to abort the fetus if it is female. Several Indian states have

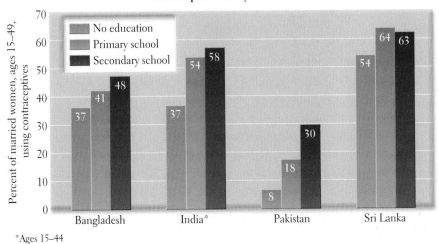

Contraceptive Use by Level of Education

Ages 15–44

Figure 8.9 **Contraceptive use by level of education.** The positive relationship between education and the use of contraception is virtually universal. In South Asia, women with at least some education use contraception at higher rates than do women with no education. In countries such as Pakistan, where the overall use of contraception is low, education drastically changes a woman's attitude toward birth control. In countries where education and contraceptive use are already fairly high, as in Sri Lanka, some education still increases the rates of use significantly (from 54 to 64 percent in Sri Lanka), but the effects of even more education—through secondary school, for example—are negligible. [Adapted from *A Demographic Portrait of South and Southeast Asia* (Washington, D.C.: Population Reference Bureau, 1994), p. 18.]

banned this use of technology, but enforcement is difficult. The practice of selective abortion and the even less savory practices of neglecting girl children and of female infanticide have resulted in the odd circumstance that throughout India, among adults, men outnumber women. (As a result of their longer natural life span, adult women usually outnumber men.) The 1991 Indian census showed 929 females for every 1000 males. In India, women outnumber men only in the state of Kerala—where there are 1048 females for every 1000 males. (In the United States there are about 1035 females per 1000 males.)

Indian social scientists explain the exception of Kerala by noting that its entire population is more educated. The literacy rate in Kerala is 90 percent, versus 44 percent for women and 67 percent for men in India as a whole. In Kerala, the elected communist state government has for many years provided education and broad-based health care for all. Education is credited with the fact that 63 percent of women in Kerala use contraception; in India as a whole, the rate is only 48 percent. Women who limit the number of children they bear have healthier children and are less likely to die as a result of childbirth. The country of

Sri Lanka, like the Indian state of Kerala, has also succeeded in bringing down population growth rates while enhancing health for women by focusing on basic literacy. In Sri Lanka, literacy rates in 1998 were 88 percent for women and 94 percent for men (Figure 8.9).

Education, especially of women but also of men, reduces the incentives for large families. Educated women are able to add significantly to family incomes and to the well-being of elderly parents, thus countering the preference for sons. Yet according to the *United Nations Human Development Report 2000*, during the 1990s India spent an average of just U.S. $17 per capita annually on education, and very little of that on girls (only 44 percent of eligible girls were in school). By comparison, South Korea—where prosperity has leapt upward while family size has shrunk—spends about $320 per capita annually, and the state of Tennessee (ranked 48th in the United States in education spending per capita) spent about $3000 per capita annually in the late 1990s. Paradoxically, although education has the power to reduce fertility rates quickly, the increasing costs of rapid population growth make adequate education harder and harder to provide.

CURRENT GEOGRAPHIC ISSUES

For the Westerner first learning about South Asia, the size of the region and the immense diversity of its people and landscapes seem overwhelming. The writer Gordon Johnson notes that once "confronted by the massive physical scale of the subcontinent, the size of the population, and the long reach of its traditions," the outsider is wise to try to "catch a glimpse of the whole by pursuing the specific and studying the particular." That is the approach we take here. In the section on sociocultural issues, we start with the feature that most characterizes the region and makes it distinctive: vil-

lage life. We then examine language, religion, caste, and gender relationships—for the most part, within the context of South Asian village life. These cultural characteristics touch all lives, yet vary greatly in practice across the region. The sections on economic and on political issues touch on those matters that are receiving coverage in the local South Asian and international press: economic change and income disparities, relative degrees of democratization, agricultural change, technologically advanced industrialization, and innovative economic development strategies.

SOCIOCULTURAL ISSUES

Many travelers to South Asia say that life there is best understood if it is observed in the intimate setting of villages, where relationships among individuals, and among and within groups, are easier to discern than they would be in the cities. We begin our coverage of sociocultural issues by visiting a few villages. We then explore how religion, caste, and gender influence life in the village and how leaving the village is one way people escape from traditional religious, caste, and gender restrictions.

Village Life

Joypur. The writer Richard Critchfield, who has studied village life in more than a dozen countries, writes that the village of Joypur in the Ganga Delta is set in "an unexpectedly beautiful land, with a soft languor and gentle rhythm of its own," where in the early evening, mist rises above the rice paddies and hangs there, "like steam over a vat." In the heat of the day, the village is sleepy: naked children play in the dust, women meet to talk softly in the seclusion of courtyards, and chickens peck for seeds. Here and there, under the trees, people ply their various trades: one may encounter the village tailor sewing school uniforms on his hand-cranked sewing machine, for example. But it is at dusk that the village comes to life. The men return from the fields, and after a meal in their home courtyards, they come "to settle in groups before one of the open pavilions in the village center and talk—rich, warm Bengali talk, argumentative and humorous, fervent and excited in gossip, protest and indignation," as the men discuss their crops, an upcoming marriage, or national politics. The men come to these gatherings barefoot, wearing turbans and "clad only in sarong-like longis—from habit and the high price of cloth."

Ahraura. The anthropologist Faith D'Aluisio and her colleague Peter Menzel give us another peek into village life as night falls. This village is in the state of Uttar Pradesh in north central India. In the enclosed women's quarters of the village, lamps flicker and the chirping of cicadas rises from the surrounding fields. In one walled compound, Mishri is finishing up her day by the dying cooking fire, as her year-old son tunnels his way into her sari to nurse himself to sleep. Mishri, who is 27, lives in a tiny world bounded by the walls of the courtyard she shares with her husband and five children and several of her husband's kin. Almost all her activity takes place in seclusion. Her village, though Hindu, long ago adopted the Muslim practice of **purdah,** in which women keep themselves apart from male gazes and community life. Mishri feels fortunate that her husband does not need her help in the fields, making possible her seclusion, a mark of status. Within the compound, she works from sunup to sundown, only taking an hour off in the early afternoon to chat with two women friends who cover their faces and move quickly from their own courtyards to hers for the short visit. Mishri is devoted to her husband, who was chosen for her by her family when she was 10; out of respect, she never says his name aloud.

The vast majority—about 70 percent—of South Asians live in hundreds of thousands of villages like Joypur and Ahraura. The vil-

Several generations of a North Indian family, from the village of Dharamkot in Himachal Pradesh province, enjoy one another's company. [David Morgan.]

lage way of life unites the entire region. Even many of those now living in South Asia's giant cities were born in a village or visit one regularly. In fact, were one to observe city life in places as widely separated and culturally different as Chennai, Mumbai, Kathmandu, and Peshawar, one would discover that they are hardly cities at all. Rather, each comprises thousands of tightly compacted, reconstituted villages, where daily life is intimate and familiar, not anonymous as in Western cities.

Language and Ethnicity

Within the life of one village, there is often room for considerable cultural variety. Differences based on caste, economic class, religion, and even language are usually accommodated peacefully by long-standing customs that guide cross-cultural interaction. One equalizing factor is that everyone in South Asia is, in one way or another, a minority. As the Indian writer and diplomat Shashi Tharoor writes:

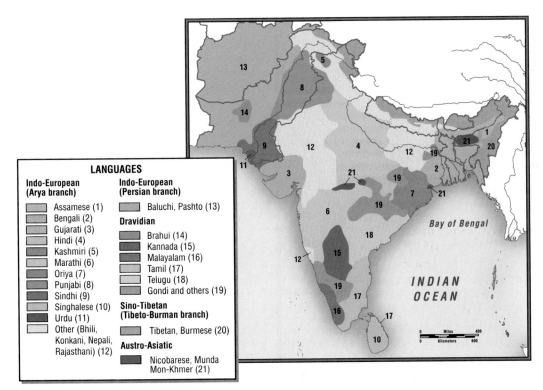

Figure 8.10 **Languages of South Asia.** [Adapted from Alisdair Rogers, ed., *Peoples and Cultures* (New York: Oxford University Press, 1992), p. 204.]

LANGUAGES

Indo-European (Arya branch)

- Assamese (1)
- Bengali (2)
- Gujarati (3)
- Hindi (4)
- Kashmiri (5)
- Marathi (6)
- Oriya (7)
- Punjabi (8)
- Sindhi (9)
- Singhalese (10)
- Urdu (11)
- Other (Bhili, Konkani, Nepali, Rajasthani) (12)

Indo-European (Persian branch)

- Baluchi, Pashto (13)

Dravidian

- Brahui (14)
- Kannada (15)
- Malayalam (16)
- Tamil (17)
- Telugu (18)
- Gondi and others (19)

Sino-Tibetan (Tibeto-Burman branch)

- Tibetan, Burmese (20)

Austro-Asiatic

- Nicobarese, Munda Mon-Khmer (21)

A Hindi-speaking male from the Gangetic plain state of Uttar Pradesh might cherish the illusion that he represents the "majority community," . . . but he does not. As a Hindu he belongs to the faith adhered to by some 82 percent of the population, but a majority of the country does not speak Hindi; a majority does not hail from Uttar Pradesh; and if he were visiting, say, Kerala, he would discover that a majority is not even male. . . . Our archetypal Hindu has only to step off a train and mingle with the polyglot polychrome crowds thronging any of India's five major metropolises to realize how much of a minority he really is. Even his Hinduism is no guarantee of majorityhood, because his caste automatically places him in a minority as well: if he is a Brahmin, 90 percent of his fellow Indians are not; if he is [of the] Yadav [caste], 85 percent of Indians are not, and so on."

(SHASHI THAROOR, *INDIA: FROM MIDNIGHT TO MILLENNIUM*, 1997, P. 112.)

There are many distinct ethnic groups in South Asia, and some of them have their own languages, dialects, and subdialects. In India alone, 18 languages are officially recognized, but there are actually hundreds of separate languages and literally thousands of dialects. Figure 8.10 shows the complexity of the distribution of languages in South Asia. That complexity results from the history of multiple invasions from outside, the relentless movement and rearrangement of people, and the long periods of isolation experienced by particular groups. In Figure 8.10, the patches of

lavender (20) and the patches of dark red (21) indicate some of the most ancient culture groups in the region. The dark red patches are remnants of Austro-Asiatic languages that were once more widely distributed and were left as isolated pockets despite the sweeping cultural changes brought by invaders. These aboriginal languages are distantly akin to others found farther east in Southeast Asia. The languages represented by numbers 1–12 are linked to various groups of Arya people who entered South Asia from Central Asia during prehistory and brought with them distinctive languages and other cultural features.

The Dravidian language-culture group (identified by numbers 14–19) is another ancient group that predates the Arya invasions by a thousand years or more. Today, Dravidian is found mostly in southern India, but the language was widely used across the subcontinent more than 4000 years ago. A small remnant of the Dravidian past can still be found in the Indus Valley in south central Pakistan (number 14 in Figure 8.10), where a language derived from Dravidian (Brahui) is still spoken. Dravidians appear to have had a matrilineal social organization, meaning that descent was figured through the female line and important family decisions were made by the mother and mother's brother. This female-centered family model was modified but not completely erased by Arya patriarchal models coming from the north. Remnants of these matrilineal ideas are still to be found among isolated groups in South India.

The Moguls used Persian (an Indo-European language) as their court language, but it eventually blended with Sanskrit-based northern Indian languages to form Hindustani. By the time of British colonialism, Hindustani was the lingua franca (language of

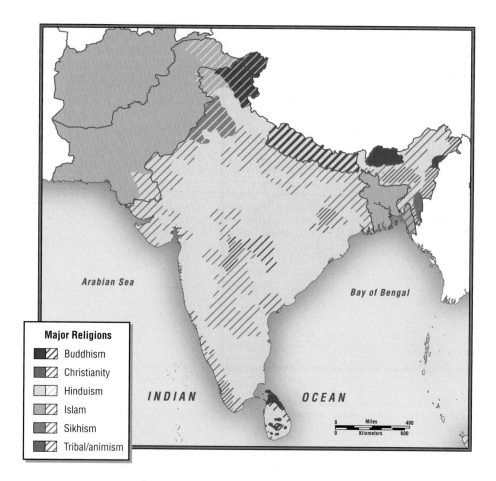

Figure 8.11 **Religions in South Asia.** The map attempts to convey the overlapping patterns of religion in this region, showing brownish gray where Islam is dominant, light tan where Hinduism is dominant, and so forth. The color of the hatching indicates which two or more religious traditions are practiced in a particular part of the region. [Adapted from Gordon Johnson, *Cultural Atlas of India* (New York: Facts on File, 1996), p. 56.]

trade) of all of northern India and what is today Pakistan. Its distribution is not shown in Figure 8.10 because of its status as a relatively new lingua franca, not a discrete language. The Muslims wrote Hindustani in a form of Arabic script and called it Urdu, whereas the Hindus and other groups wrote it in a script derived from Sanskrit and called it Hindi. Today, Urdu is the national language of Pakistan and Hindi the national language of India, but the people who speak the common forms of these languages can understand one another in the same way as speakers of American and British English do.

Hindi, because of its origins in Hindustani and the popularity of romantic Hindi-language films, is understood by most Pakistanis and by about 50 percent of India's population. Only Chinese and English are spoken by more people worldwide. But Hindustani is not the first language of the majority: as Tharoor noted, no Indian language can make that claim.

English is a common second language throughout the region. For years, it was the language of the colonial bureaucracy, and it remains a language used at work by professional people. From 10 to 15 percent of South Asians speak, read, and write English.

Religion

The main religious traditions of South Asia are Hinduism, Buddhism, Sikhism, Jainism (see discussion on page 413), Islam, and Christianity. Buddhism and Jainism originated as reformist movements in Hinduism. The geographic distribution of the adherents of these various faiths is uneven, as Figure 8.11 shows.

Hinduism. Hinduism is a major world religion practiced by approximately 900 million people, 800 million of whom live in India. It is a complex belief system, full of seeming contradictions that often make it difficult for outsiders to understand. These contradictions result largely from the fact that Hinduism includes a broad range of beliefs and practices that have their roots in highly localized folk traditions (known as the Little Tradition), as well as a classical system (known as the Great Tradition). These traditions are based on the amalgam of Harappa and Arya ritual beliefs recorded in ancient Sanskrit texts. For example, most Hindus worship a number of gods and goddesses, and some differences exist between the worshipers of the god Shiva, the creator and destroyer of the universe, and the god Vishnu, the preserver of the universe. However, many Hindus primarily worship gods found only in one region, in one village, or even in one family. Scholars explain this diversity as a product of Hindu notions of divinity, which are quite flexible. Hence, although local areas have retained their own deities, over time some local gods have become incorporated into the classical Hindu pantheon.

A major tenet of classical Hindu philosophy, as described in the 4000-year-old Hindu scriptures called the Vedas, is that all gods are merely illusory manifestations of the ultimate divinity, which is formless and infinite. Indeed, many devout Hindus worship no gods at all. These Hindus may engage in meditation, yoga, and other spiritual practices designed to liberate them from illusions and bring them closer to the ultimate reality, described as infinite consciousness. The average person, however, is seen as needing to personify divinity in the form of a god.

Nevertheless, there are some things that almost all Hindus have in common, such as the belief in reincarnation—the idea that any living thing that desires the illusory pleasures (and pains) of life will be reborn after it dies. Hindus usually participate in the caste system, which is deeply interwoven with classical Hindu ritual. Dietary rules commonly distinguish caste groups, and a reverence for cows, which are seen as only slightly less spiritually advanced than humans, binds all groups together.

Geographic Patterns in Religious Beliefs. There is an overlapping distribution pattern of religions in this region. Hindus, as we have observed, are the most numerous, and they are found mostly in India. The Ganga River plain is considered the **hearth** (place of origin) of Hinduism, and every 12 years, during India's largest religious festival, millions of Hindus converge on the city of Allahabad to bathe at the confluence of the Ganga and Yamuna rivers as an act of devotion. Other religions are important in various parts of the region.

Muslims are followers of **Islam,** which is described in Chapter 6 on pages 304–309. The 300 million Muslims in the region form the majority in Afghanistan, Pakistan, Bangladesh, and the Maldives. Muslims are also a large and important minority in India (numbering about 120 million there); they live mostly in the northwestern and central Gangetic Plain but are also scattered throughout the country.

Buddhism began as a reform and reinterpretation of Hinduism. Its origins are in northern India, and it flourished there early in its history before spreading eastward to eastern and southeastern Asia. Only 1 percent of the region's population, or about 10 million people, are Buddhists; they are a majority only in Bhutan and Sri Lanka.

The 18 million **Sikhs** in the region combine the beliefs of Islam and Hinduism. **Sikhism** was founded in the fifteenth century by Guru Nanak as a challenge to contemporary socioreligious systems, including Hinduism and Islam. The new religious philosophy was inspired by both Hindu and Islamic ideals. Sikhs espouse belief in one God, high ethical standards, and meditation. They reject the caste system but accept the Hindu idea of reincarnation. Sikhs live mainly in Punjab in the northwest of India but are found elsewhere. Their influence in India is greater than their numbers because many Sikhs hold positions in the military and police. More Sikhs live in diaspora communities than live in India itself, and some have financed a Sikh separatist movement (see the section on regional conflict on pages 426–427).

The faith tradition of **Jainism** is more than 2000 years old. Jains (about 10 million people, or 1 percent of the region's population) are found mainly in western India and in large urban centers. They are known for their strict vegetarianism.

Parsis, though few in number, are a highly visible minority in India's western cities, where they have distinguished themselves in business, politics, and the arts. Zubin Mehta, a famous symphony conductor, is a Parsi. Parsis are descended from Persian migrants who did not give up their traditional religion of Zoroastrianism when Iran became Muslim.

The first Christians in the region are thought to have arrived in the far southern Indian state of Kerala with St. Thomas, the Apostle of Christ, in the first century. (Christianity is described in Chapter 6, on page 295). Today, Christians are an important minority along the west coast of India, dating back to the time of St. Thomas and to Portuguese influence in Goa. In some places in northeastern India, more than half the descendants of the ancient aboriginal inhabitants are Christian, the result of British colonial mission efforts in the late nineteenth and twentieth centuries. At that time, ethnic minorities adopted Christianity, partly as a hedge against the encroachment of Hindu power. Christians are also a small minority elsewhere in India, as well as in Sri Lanka, Pakistan, and Bangladesh.

Small communities of Jews are found along the Malabar Coast and in such major cities as Mumbai, Calcutta, and Ahmadabad. Some are thought to be the descendants of ancient migrants who arrived perhaps 2000 years ago or even earlier. Others came via Europe in the modern period.

Animism is practiced throughout South Asia, especially in central and northeastern India, where there are indigenous people whose occupation of the area is so ancient that they are considered aboriginal inhabitants. The animist beliefs of these peoples often incorporate aspects of Hinduism, Islam, Buddhism, or Christianity.

The Hindu-Muslim Relationship. The different religious traditions of South Asia have influenced one another, and this is especially true of the two largest faiths. Where Hindus have lived in close association with Muslims—often within the same villages—they have absorbed Muslim customs, such as the seclusion of women. For their part, Muslims have adopted some of the Hindu ideas of caste. Although Muslims make up only 11 percent of the Indian population, the political and social relationships between Hindus and Muslims are enormously complex. The great independence leaders Mohandas Gandhi and Jawaharlal Nehru both emphasized the common cause that once united Muslim and Hindu Indians: throwing off British colonial rule. Since independence, members of the Muslim upper class have been prominent in Indian national government and the military; Muslim generals served India willingly, even in its wars with Pakistan after partition. In the upper echelons of society, Hindus and Muslims often socialize amicably, live in the same neighborhoods, share recreational facilities, and join the same clubs. In urban areas, middle-class Hindus and Muslims go to school together and occasionally marry one another.

But there is an opposite side to the Hindu-Muslim relationship. Especially in Indian villages, some upper- and middle-caste Hindus regard Muslims as members of low-status castes. Fueling this perception is the occasional conversion of entire low-caste Hindu villages to Islam. Those who convert seek to escape the hardships of being members of a low caste. Often, religious rules about food are the source of discord, because dietary habits (such as differing degrees of vegetarianism) are a primary means of distinguishing caste. Hindus regard the cow as sacred and use its products with reverence: milk for food and dried excrement for fertilizer and fuel. Although Hindus do not kill cows for food or hides, people at the lowest rungs of Hindu society process and consume individual animals that die by other means. Muslims, on the other hand, run slaughterhouses and tanneries (though discreetly), eat beef, and use cowhide to make shoes and other items. Indigent cows, picked up off the streets of India's cities, are sometimes exported to nearby Muslim countries.

The Hindu-Muslim relationship is no less complex in Bangladesh. After the separation of Bangladesh from Pakistan, most upper-class Muslims moved to Pakistan, leaving Bangladesh with a preponderance of poor Muslim farmers. Meanwhile, many of the Hindu landowners remained, and some lower-caste Hindus converted to Islam. In Bangladeshi villages, the Muslims are usually a majority, but the Hindus are often somewhat wealthier and disdainful of them. Although the two groups may coexist amicably for many years, they view themselves differently, and **communal conflict**—that resulting from religious differences—can erupt.

VIGNETTE The sociologist Beth Roy, a specialist in conflict resolution who studies communal conflict in South Asia, recounts an incident in the village of Panipur, Bangladesh, in her book *Some Trouble with Cows*. The incident started when a Muslim farmer carelessly allowed one of his cows to graze in the lentil field of a Hindu. The Hindu complained, and when the Muslim reacted complacently, the Hindu seized the offending cow. By nightfall, Hindus had allied themselves with the lentil farmer and Muslims with the owner of the cow. More Muslims and Hindus converged from the surrounding area, and soon there were thousands of potential combatants lined up facing each other. Fights broke out. The police were called. In the end, a few died in the ensuing riot, and relationships in the village were forever molded by the incident. In the words of Roy, the dispute "delineat[ed] distinctions of caste, class, and [religious] culture so complex they intertwine[d] like columbines climb-

ing on an ancient wall." [Adapted from Beth Roy, *Some Trouble with Cows—Making Sense of Social Conflict* (Berkeley: University of California Press, 1994), pp. 18–19.]

Caste

Caste, the ancient system for dividing society into hereditary hierarchical classes, seems alien to many people outside South Asia, and yet social division and inequality are common in all societies, including Europe and America. In fact, all human groups have deeply ingrained concepts of relative social status. The old often have more authority than the young; or conversely, as in U.S. pop culture, the young sometimes have greater influence than the old. In many situations in Europe and America, men still have more power and prestige than women. Nearly everywhere on earth, social difference is indicated by clothes, hairstyle, body decorations, manner of speaking, material possessions, residential location, and religion. In America, something so simple as the use of "ain't" or "youse" or a certain kind of headgear (baseball cap, motorcycle helmet, straw hat) gives immediate information about one's origins, role, or point of view. In some quarters, race still carries overtones closely akin to those of caste in South Asia. Thus, South Asia's customs associated with caste should not be beyond comprehension to outsiders.

Caste is a custom associated primarily with Hindu India, but some Muslims and Buddhists have incorporated the system into their cultures. One is born into a given subcaste (called a *jati*), and that happenstance largely defines one's experience for a lifetime—where one will live, where and what one can eat and drink, with whom one will associate, one's marriage partner, and one's livelihood (see the box "Food Customs in South Asia"). The caste system has four main divisions and many hundreds of subcategories (*jatis*) that vary from place to place; there are also important groups that fall outside the caste system. Thus, caste represents groups of *jatis* that are tied together by a set of associated social

CULTURAL INSIGHT *Food Customs in South Asia*

Customs governing food and its preparation demonstrate how religion and caste influence the everyday life of South Asia. The best known of India's food rules is that Hindus do not eat beef. This prohibition may have grown from the fact that, historically, cattle have been more valuable as living resources because they were the primary source of labor, fuel, and fertilizer and are a symbol of life. In Afghanistan, Pakistan, and Bangladesh, it is pork that is proscribed, a prohibition that predates Islam. But there are countless other restrictions that vary from place to place.

The long tradition of highly seasoned, meatless cuisine in South Asia is particularly strong in southern India. Many notions of purity and pollution surround food preparation throughout the

region of South Asia. According to tradition, Hindus should not accept food prepared by anyone from a lower caste, and only the right hand is to carry food to the mouth because the left hand is reserved for cleaning the body after defecating and hence is ritually impure. Menstruating women are also not allowed to prepare food. These taboos are fading in urban areas where caste and gender rules are now less important.

In rural areas, where the vast majority of South Asians still reside, the serving of food is also culturally prescribed. Women and girls eat after men and boys, a practice that often leaves them undernourished in poorer families and partly accounts for the higher death rate of female children than of male children.

CULTURAL INSIGHT *Mohandas Gandhi: The Founder of Nonviolent Protest*

Mohandas Gandhi's emphasis on nonviolent resistance against social ills has influenced and transformed civil society everywhere on earth. Gandhi (1869–1948) came from a wealthy merchant family in Gujarat, in west central India, where relations among Hindus, Muslims, and Jains were tolerant and where social reform was often discussed in community life. British colonialism had failed to penetrate Gujarat significantly; the region was prosperous and little affected by Western culture. After studying law in England, Gandhi had difficulty finding his niche back home in India and found work in South Africa. There, discrimination based on race was intense, and for 20 years he worked to promote social reform and developed a personal philosophy of self-control and nonviolence. In protesting the treatment of both Africans and Indians by the white South African government, Gandhi developed the strategy of gathering a large group of sympathizers to publicly, but peacefully, break a

discriminatory law—first notifying the authorities of the coming civil disobedience. If the authorities ignored the act, the demonstrators would have made their point, and the law would have been rendered moot. If the authorities used force to break up the peaceful demonstration, the government's moral sway was undermined, and it lost respect.

Gandhi later used his nonviolent techniques in India against the British, where he made hand-loomed cloth the central symbol of resistance. Considered the father of Indian independence, he did not support the partition of India into Muslim Pakistan and Hindu India and greatly lamented the bloodshed that followed. Nonviolence as a social reform technique has since diffused around the world. It has been used successfully in the civil rights movement in the United States; in labor movements in Asia; in the women's movement worldwide; and, especially, in South Africa to bring an end to apartheid.

---

characteristics; a caste category, in and of itself, is not a discrete group of people.

Brahmins, the priestly caste, are the most privileged in ritual status. Because Hinduism and caste are interlinked, groups of higher status must conform to behaviors that are considered ritually purer (for example, strict vegetarianism, abstention from alcohol, and abstention from certain types of work). Then, in descending rank, are **Kshatriyas,** who are warriors and rulers; **Vaishyas,** who are landowning farmers and merchants; and **Sudras,** who are low-status laborers and artisans. A fifth group, the **Harijans** (also called Dalits or untouchables), are actually considered to be so lowly as to have no caste. The Harijans perform tasks that caste Hindus consider the most despicable: killing animals, tanning hides, sweeping, and cleaning. A sixth group outside the caste system is composed of descendants of very ancient pre-Hindu ethnic groups (for example, see numbers 20 and 21 on the language map, Figure 8.10).

The *jatis* (subcastes) are often associated with a certain place, such as a village or a neighborhood in a city, and with a particular language or dialect. Thus, the *jatis* are segregated into distinct spaces. The spatial segregation of *jatis* arises out of the higher-caste communities' fears of ritual pollution through sharing water or food. When one stays in the familiar space of one's own *jati*, one is enclosed in a comfortable circle of families and friends that becomes a mutual aid society in times of trouble. This cohesion within *jatis* and the attachment to place help to explain the persistence of a system that seems to put such a burden of shame and poverty on the lower ranks.

It is important to note that caste and class are not the same thing. There are class differences within caste groups because of differences in wealth, and being lower in the caste system does not always mean one is lower in socioeconomic class. Many

Kshatriyas are wealthy landowners, and some Vaisyas are extraordinarily wealthy businesspeople. Brahmins, as a group, might be considered upper middle class by American standards, though some Brahmin families live in poverty or struggle to achieve a middle-class standard of living. Still, by and large, Harijans are very poor.

Caste is slowly declining as a force in everyday life wherever people are able to separate themselves from their ancestral villages, occupations, and *jati*-related dialects. Historically, entire *jatis* could improve their status by changing such customs as their diet or their occupation. Today, even individuals can alter their status by migrating far from home and disappearing into the anonymity of city life. More often, though, new migrants to India's cities maintain *jati* ties by settling in slum neighborhoods called *bustees* that often have a strong regional and *jati* identity. In this way, they reconstitute village customs and find support among people with similar social, cultural, and linguistic backgrounds. Campaigning politicians often take advantage of this clustering of people who share similar backgrounds.

In the twentieth century, Mohandas Gandhi, India's leader for independence (see the box "Mohandas Gandhi: The Founder of Nonviolent Protest"), began an official effort to eliminate the caste system as the organizing principle of daily life and of social relationships. As a result, India's constitution bans caste. In the late 1940s, India began an affirmative action program that reserves a portion of government jobs, places in higher education, and parliamentary seats for the lowest *jatis* and Harijans. The Harijans now constitute approximately 16 percent of the Indian population. Today, many reject the name Harijan as patronizing and prefer the term Dalit ("the oppressed"). The descendants of pre-Hindu groups are also protected by affirmative action laws. They account for about 8 percent of the population and are guaranteed 22.5 percent of the government jobs and 15 percent of the parliamentary seats. The Dalits and pre-Hindu groups are sometimes referred to as "scheduled castes."

Among educated people in urban areas, the campaign to eradicate caste has been remarkably successful. Some Dalits are now powerful officials throughout the country. Members of high and low castes now ride the city buses side by side, eat together in restaurants, use the same rest rooms, drink from the same water fountains, and attend the same schools and universities. More remarkably, for some urban Indians, caste is disappearing as the crucial factor in finding a marriage partner. Newspaper matrimonial ads, which are commonly used in cities to make connections for an arranged marriage, often bear the proviso that "caste is no bar." Nonetheless, it would be incorrect to conclude that caste is now irrelevant in India. Less than 5 percent of registered marriages cross *jati* lines. Nearly everyone notices social clues that reveal an individual's caste; and in rural areas, where the majority of Indians still reside, the divisions of caste remain prevalent.

Geographic Patterns in the Status of Women

The overall status of women in South Asia is notably lower than the status of men. There are highly successful women in such occupations as business, the media, academia, medicine, politics, law, and law enforcement—some in very high positions of power. Nevertheless, the vast majority of women in the region have little status within their families or in society and very little free choice. Throughout the region, most women are partners in marriages arranged for them, often without their wishes being consulted. Especially in the villages, a girl may be married as a very young child to a husband who is considerably older. The girl continues to live with her parents until puberty. At age 12 or so, and without much warning or discussion, she must abruptly leave her parents' home and join her husband in his family compound, where she becomes a source of labor for her mother-in-law. Most junior brides work at domestic tasks for many years until they have produced enough children to have their own crew of small helpers, at which point they gain some prestige and a measure of autonomy. Their power increases when they become mothers-in-law themselves. But the death of a husband is a disgrace to a woman and can completely deprive her of all support and even of her home, children, and reputation. Widows are often ritually scorned and blamed for their husbands' death. Widows of higher caste rarely remarry and, in some areas, become bound to their in-laws as household labor or are asked to leave the family home.

Purdah is the practice of concealing women from the eyes of nonfamily men, especially during their reproductive years. It is a custom in rural villages in areas of South Asia where Islam is the main religion or where Islam has influenced Hindu practices (as was case with Mishri in the village of Ahraura, discussed on page 410). Purdah is weaker where there is less Islamic influence—that is, among Hindus in central and southern India and in the larger cities across the region, especially among highly educated upper-caste women. The custom is not observed by the aboriginal people of the region, many of whom live in central and northeastern India. Low-caste Hindus throughout the region also do not observe

purdah. Nonetheless, purdah is sometimes a mark of prestige for those subcastes that are making an effort to upgrade their overall class and caste status, because the ability to seclude women signals the possession of surplus wealth and increased ritual purity. Purdah accounts for the relatively limited participation of women in agriculture in the north and northwest of India, because work in the fields would put them in view of nonfamily males.

There is geographic variability in the relative well-being of women in South Asia. Women in Afghanistan arguably have had the most difficult lives since an archconservative Islamist movement, the **Taliban,** gained control of the government there from the mid-1990s to November of 2001 (see page 435). The Taliban (meaning "God's students") supported strict interpretations of Islamic law (see the discussion of the Taliban on page 434). Both men and women had to follow strict dress codes (men had to wear beards and wear robes), and females had to live in seclusion: girls and women were not to work outside the home or attend school, and they had to wear a completely concealing heavy black veil, called a *burqa*, whenever they were out of the house. (*Burqa* is the name used throughout South and Southeast Asia, *chador* is the Persian name, and *abiya* is the Arabic name.) The Taliban even decreed that women whisper and not make noise as they walked, because the sound of their footsteps was distracting and potentially erotic to men.

Bangladesh is another part of South Asia where women are under particular limits on their freedoms. Islamist village councils, called *salish*, operate outside the Bangladesh constitution with apparent impunity, as community court and punishment agencies. According to investigations conducted in the late 1990s by Amnesty International, an agency that investigates human rights abuses, very young women have been stoned to death for alleged relationships with men. In several cases, the victims, as young as 13, became pregnant after being raped by village men, who remained unpunished. Village courts have reportedly carried out similar patterns of violence against girls and women in Afghanistan and

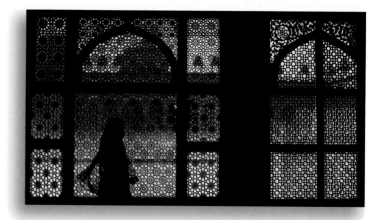

Lattice screens known as *jalee* are an architectural feature in South Asian areas where women are in purdah. Like the louvers and latticed bay windows of Saudi Arabia (see the photo on page 307 in Chapter 6), *jalee* allow ventilation and let in light but shield women from the view of strangers. [Lindsay Hebberd/Woodfin Camp & Associates.]

Pakistan. In addition, the offices of nongovernmental agencies (NGOs) that provide schooling, health care, family-planning services, or economic aid to women have been sacked and burned after Islamist clerks issued *fatwas*—legal opinions based on interpretations of Muslim law—against them. Islamist leaders have said publicly that NGOs like the famous Grameen Bank in Bangladesh, known for its small loans to poor rural women, "alienate women from their proper social roles and Islamic lifestyle," and hence should be banned.

Bride Burning and Female Infanticide. For some years now, a rare practice in India called bride burning or dowry killing has been reported, in which a husband and his relatives stage an "accidental" kitchen fire that kills his wife. Her death enables the widower to marry again and collect the dowry that, by custom, comes from a wife's family. The government of India released figures in late 1987 that affirmed 1786 such deaths in that year alone. Although not widespread—India has hundreds of millions of married women—this practice is a disturbing reminder of the ancient practice of *sati*, the burning of a high-caste widow on her husband's funeral pyre. In some cases, the threat of bride burning is used to extort a wife's family into providing further dowry in the form of cash or durable goods. More commonly, a young bride might be made to feel undeserving of her status in the groom's family if the family feels cheated in the marriage negotiations. A related practice is that of female infanticide, in which girl babies, deemed unaffordable because of the dowry investment they will require, are killed.

Changing customs regarding the marriage dowry appear to be a cause of the growing incidence of both bride burning and female infanticide in India. Until the last several decades, it was the custom among the lower castes to pay a **bride price,** not a dowry: a groom paid the family of the bride a relatively small sum that symbolized the loss of their daughter's work to the family's economy and the gain of her labor to his. **Dowry,** on the other hand, is wealth provided by the bride's family at time of marriage; it originated as an exchange of wealth between landowning, high-caste families. With her ability to work reduced by seclusion, an upper-caste female was considered a liability. The dowry that went with the wife to her new husband gave her dignity as a bride and leverage in her husband's household, because the dowry had to be returned if the marriage dissolved. As the practice of purdah spread to poorer, lower-caste families wanting to upgrade their status, the dowry spread with it. Moreover, the payment of a substantial dowry could further increase the status of a bride's family by attracting a groom of higher status. A good dowry settlement for the first child in a family (male or female) set the tone for future marriage settlements of siblings. Thus, caste and class became conflated as dowry was redefined in practice to be a nonrefundable endowment paid by the bride's family to a higher status groom and his family.

Oddly, increasing education for males and family affluence reinforced the custom of dowry. Young men, even those of low caste, came to feel that their diplomas increased their worth as husbands and put them in the category of those who deserved a bride with a substantial dowry. This upgrade in status through education gave them the power to demand larger and larger dowries. Soon the practice spread through the lower castes, and now the poorest of families are crippled by the dowries they must pay to get their daughters married, which in turn influences the matches they can make for their sons.

Despite the fact that the Indian government bans the practice of dowry, the social duty to provide a dowry for a daughter is taken extremely seriously, because the stigma of having an unmarried daughter is huge. Some ambitious families are giving their daughters a graduate school education with its promise of earning power in lieu of dowry. But for an increasing minority of families, the birth of more than one daughter threatens the family with hopeless impoverishment, whereas the birth of sons promises future daughters-in-law who will bring dowry wealth with marriage. A village proverb captures this inequitable relationship: "When you raise a daughter you are watering another man's plant." Some families view the birth of a daughter as a calamity, to such an extent that second and third daughters may be fed poison soon after birth. The circumstances that lead parents to take such a horrifying step are illustrated in the following example.

VIGNETTE Muya and Mohan were married and had one daughter. They worked as agricultural laborers and together earned $1730 a year. Muya was paid half as much as Mohan for the same work. The couple lived on an acre of land they cultivated for themselves, and they owned a cow. Mohan's mother lived with the family, and the four shared a collection of neat mud huts surrounded by coconut trees. They were just well off enough to plan for a brighter future. They had managed to save enough money so that when their daughter was old enough they could pay the going dowry rate, an amount equal to what the two parents earned in a year. This outlay of money would be balanced by the dowry any future son would receive upon marriage. When Muya became pregnant again (she knew nothing about birth control methods), they decided that they could not afford the dowry for a second girl. Thus, if a girl were born, they would "put her to sleep." Eventually, Muya delivered a girl after four hours of labor. The next day, the family gathered and fed the baby cow's milk laced with five sleeping pills. Then they buried the child along the road in view of the surrounding hills. [Adapted from Elisabeth Bumiller, *May You Be the Mother of a Hundred Sons* (New York: Fawcett Columbine, 1990), pp. 110–112.]

Education and the Status of Women. Islamist leaders are correct in their belief that there are agencies (such as Oxfam, India Literacy Project, and South Asian Women's Network) seeking to improve the status of women and their earning opportunities and to diminish the influence of purdah, which keeps women illiterate and confined to their houses. These organizations believe that freeing women from purdah and other ancient strictures will increase the educational attainments of all of South Asia's children and improve the health and nutrition of families. Research has shown that even with as little as two years of education,

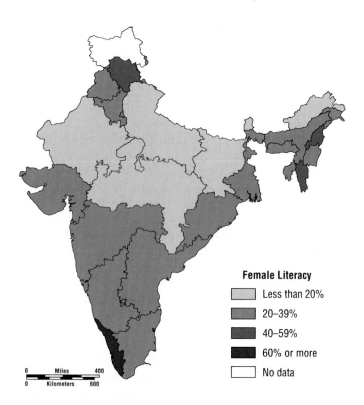

Figure 8.12 **Female literacy in India.** Female literacy is below 40 percent in most of India. [Adapted from Bina Agarwal, "Gender, environment, and poverty interlinks: regional variations and temporal shifts in rural India, 1971–91," *World Development* 25 (1997): 32.]

Female Literacy

- Less than 20%
- 20–39%
- 40–59%
- 60% or more
- No data

women choose to have fewer children (see Figure 8.9) and begin to make decisions that enhance the well-being of their children. United Nations researcher Martha Nussbaum found that women who can read often seek a way to earn some income and generally invest their earnings in food, medicine, and schooling for their children. Over time, freeing women from purdah will also encourage lower fertility and greater economic growth, because women will channel less of their energies into reproduction and more into such activities as business, innovative agriculture, recycling of resources, teaching, and other service occupations. Figure 8.12 shows the extent to which women are now literate in India.

The Internet has become a forum for discussion within South Asian communities about issues related to the low status of women: purdah, the preference for sons, bride burning, and female infanticide. For example, descriptions of daughters performing skilled mechanized agricultural work and conducting Hindu death ceremonies have been posted on the India Family Net Talk Web site at http://indiafamily.net/ (home page) and http://indiafamily.net/talk/messages/54/578.html (specific article).

Both tasks are typically done only by sons. Also posted on this site are testimonies about the value of educated daughters and per-

sonal accounts of young women who have experienced son preference and the low status of being a daughter. Web sites like this one, and access to global Internet connections in general, lend strength to social change efforts.

Gender Equality at the Village Level and Beyond. In India, sanctions against women comparable to those described above for Bangladesh and Afghanistan sometimes occur, and bride burning and female infanticide have received wide publicity. But a strong activist movement in India has led to more enthusiastic enforcement of constitutional protections, at least after the fact. In the 1980s, Prime Minister Rajiv Gandhi introduced *panchayati raj* (village government) to encourage gender equality in village life. These local councils differ from the impromptu Bangladeshi *salish* not only because they have official standing but because 30 percent of the councils must reserve seats for women during a given election cycle. Furthermore, since 1993 there has been training for women *panchayati raj* members to improve their effectiveness on the councils. Support is growing for legislation that would reserve one-third of the seats in the lower house of the Indian parliament and in state assemblies for women, for a 15-year trial period. After that time, it is hoped, women will have achieved the political experience to win elections without the aid of quotas.

A minority of upper- and middle-class urban professional women in India, Pakistan, Sri Lanka, and Bangladesh may have

In this Mumbai neighborhood Internet café, people of all ages and both sexes are welcome and receive instruction and assistance as they need it. Virtually every city of more than 10,000 people now has such cafés. The Internet technology itself may empower women: many of the workers in India's Internet cafés are young women, who meet many international travelers in the course of a day and learn to use the technology themselves because they must help customers. [Dinodia.]

more in common with their counterparts in Europe and America than they do with village women in their own countries. In such cities as Karachi, Delhi, Mumbai, Bangalore, Chennai, Dhaka, and Colombo, there are growing numbers of highly successful businesswomen, female directors of companies, highly qualified female technicians, high-ranking female academics, and women who serve prominently in government. India, Bangladesh, Sri Lanka, and Pakistan have all had women heads of state.

ECONOMIC ISSUES

South Asia is a region of startling economic contrasts, where a country (India) that is home to hundreds of millions of poor people can also foster a growing computer software industry and a space program. These extremes reflect the propensity of South Asian economies to favor the interests of a privileged minority over those of the poor majority. Although the British colonial system deepened the extent of South Asia's poverty and widened the gap between rich and poor, the current wealth disparities in the region result mostly from economic policies favored by postindependence leaders. Despite India's celebrated democratic traditions, for example, the poor have often been left out of the political process and bypassed or hurt by economic reforms.

Agriculture remains the basis of South Asian economies, but rapid industrialization and self-sufficiency have been the dream since independence. Recent strategies adopted to encourage economic development include an emphasis on information technology in India and innovative finance strategies pioneered in Bangladesh.

Agriculture and the Green Revolution

Well over 60 percent of the region's population is still rural and engaged in agricultural labor. In tiny Nepal, 90 percent of the population is so occupied. Nonetheless, the contribution of agriculture to national economies (GDP) hovers below 30 percent. Although production per unit of land has increased dramatically over the past 50 years, agriculture is the least efficient sector of the regional economy. The service sector occupies just 18 percent of the workforce in India and 39 percent in Pakistan, yet it accounts for close to 50 percent of GDP in both countries. Similar figures apply elsewhere in South Asia.

Figure 8.13 shows the distribution of agricultural zones in South Asia. Except in the far northwest and highland areas, double-cropping is common throughout the region: crops adapted to dry conditions are planted in the winter, and those adapted to wet conditions are planted in the summer when monsoon rains

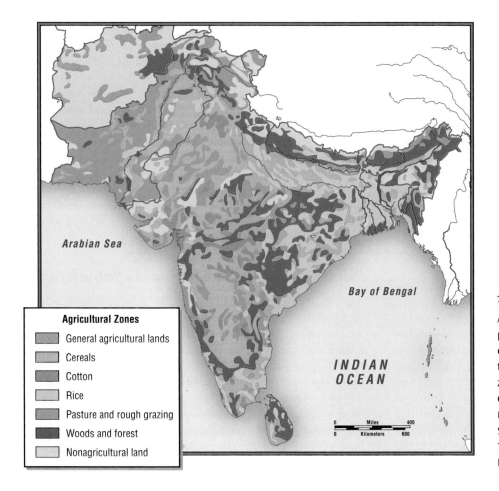

Agricultural Zones

- General agricultural lands
- Cereals
- Cotton
- Rice
- Pasture and rough grazing
- Woods and forest
- Nonagricultural land

Arabian Sea

Bay of Bengal

INDIAN OCEAN

Figure 8.13 **Agricultural zones in South Asia.** Agricultural zones in this region form a complex pattern influenced by landforms, climate, cultural customs, and recent development theories. Note especially the distribution of rice zones as opposed to cereal zones. [Adapted from Gordon Johnson, *Cultural Atlas of India* (New York: Facts on File, 1996), p. 34; and "Afghan Economy (map)," SESRTCIC (Statistical, Economic and Social Research and Training Centre for Islamic Countries) InfoBase at http://www.sesrtcic.org/members/afg/afgmapec.shtml.]

Thinking Geographically: **ON THE WEB**

are plentiful. Farmers in some areas of Bangladesh can harvest three crops a year. Rice is the main crop wherever rainfall is plentiful, especially in the flat Ganga Basin and around the eastern and southern rim of India. Consumption of rice is increasing because it is viewed as a high-status food and because it is favored by urban women who work outside the home and appreciate the ease with which it can be prepared. Rice occupies about one-third of the total land planted in grain. Wheat, grown in the west central and northwest parts of the region, is the second most important crop, and its cultivation is spreading with the use of irrigation systems. Other grains include millet and grain sorghum. Cotton remains an important cash crop and is often grown with the aid of irrigation. Animal grazing is predominant in the dry areas of Pakistan, Afghanistan, and northwestern India.

Until the 1960s, agriculture in this region was based largely on traditional small-scale systems that managed to feed families in good years but often left them hungry in years of drought or flooding. These systems did not produce sufficient surpluses for the region's growing cities, which even now rely on imported food. Large-scale mechanized agriculture, where it existed, was aimed at producing export crops such as cotton, flax, jute, tea, and rice. Much of South Asia's agricultural land is still cultivated by hand; and overall agricultural development has been neglected in favor of industrial development, especially in India. Nonetheless, by the 1970s, important gains in agricultural production had begun.

Beginning in the late 1960s, a so-called **green revolution** boosted grain harvests dramatically through the use of seeds selected for high yield and for resistance to disease and wind damage. Other components of the revolution were fertilizers; mechanized equipment; irrigation; pesticides; herbicides; double-cropping; and, to a lesser extent, an increase in the amount of land under cultivation. One result is that Pakistan is now self-sufficient in wheat, rice, and sugarcane and produces surpluses for export. India has become one of the world's leading producers of grain and now exports rice to the Americas, Asia, and Africa. Where the new techniques were used, yield per unit of farmland improved by more than 30 percent between 1947 and 1979. International loans financed the building of dams to store monsoon water for irrigation and to create hydroelectric power. These projects, in turn, boosted industrial growth and created jobs. India was able to repay its green revolution–related loans, which boosted its reputation as creditworthy.

The green revolution, however, has not made India reliably self-sufficient in food. There are still years (2000 was one) when output falls short of demand, and faminelike conditions can occur in local areas, such the eastern state of Orissa. In addition, the main beneficiaries of the green revolution innovations have not been South Asia's millions of very small producers. Poor farmers are unable to afford the special seeds, fertilizers, pesticides, and new equipment. Many have been forced off rented or borrowed land and into low-wage farm labor or sharecropping; the land they once cultivated is now under new green revolution managers. Other farmers who have lost access to land have migrated to the cities, where they have difficulty finding jobs of any sort.

Although the new technologies result in dramatically higher production rates, the increased food supplies tend to bypass the poorest and hungriest. Between 1970 and 1990, there was a 9 percent increase in the amount of food produced per capita in South Asia and also a 9 percent increase in the numbers of hungry people. The hungry could not afford food, while the increased supplies of food were sold to those with money or were exported. Some South Asian rice, for example, finds its way to specialty stores in the United States. Another factor in food distribution is that rural areas are bypassed during food scarcities. South Asian governments place a high priority on ensuring a sufficient supply of food to the cities because urban unrest poses a greater threat to the interests of government and to the middle and upper classes than does rural unrest.

Meanwhile, increasing soil salinity and other kinds of environmental damage created by chemical fertilizers, pesticides, and high levels of irrigation are reducing yields in many areas. One such area is the Pakistani Punjab, that country's most productive, but highly irrigated, agricultural zone.

Green revolution technologies also inadvertently reduced the utility of many crops for the rural poor, especially women. For example, the new varieties of rice and wheat yield more grain but less of the other components of the plant previously used by women, such as the wheat straw used to thatch roofs, make brooms and mats, and feed livestock. Moreover, women's already low status in agricultural communities often erodes further because their contribution to household production is supplanted by new technology, such as small tractors and mechanized grain threshers, which are usually controlled by male members of the family.

A potential remedy for some of the failings of the green revolution style of agriculture is **agroecology**: the use of traditional methods to fertilize crops and of natural predators to control pests. Unlike green revolution techniques, the methods of agroecology are not disadvantageous to poor farmers because the necessary resources are readily available in most rural areas. To participate, the farmers do not need access to cash but only to knowledge, which can be taught orally to small groups and over the radio. Studies in South India that compared agroecology techniques with those of the green revolution found their productivity and profitability to be equal—but agroecology techniques reduced soil erosion and loss of soil fertility.

Industry over Agriculture: A Vision of Self-Sufficiency

After independence from Britain in 1947, South Asia's new leaders favored industrial development over agriculture. Influenced by socialist ideas, especially the model of the Soviet Union, they concluded that agriculture was incapable of supplying the growth and technological innovation that poor countries needed. Government involvement in industrialization was considered necessary to ensure the levels of job creation that would cure poverty. Another motive, especially in India, was to create a self-sufficient economy that did not need to import manufactured goods from the industrialized world. The new South Asian leaders engineered government takeovers of the industries they believed to be the linchpins

GEOGRAPHER IN THE FIELD

Interview with Regional Consultant for Chapter 8, Carolyn V. Prorok, Slippery Rock University

What led you to become interested in South Asia?

I did not know that I would become a geographer until I became one. Likewise, my interest in studying the peoples and places of the Indian subcontinent blossomed surreptitiously as I searched for a topic for my master's thesis 20 years ago and found it at a Hare Krishna commune in West Virginia. My advisor encouraged me to study New Vrindaban as a sacred place, and voila!—I began an investigation that continues to stimulate my intellectual curiosity.

You see, I met some Trinidadians there. Much to my surprise I found that the Caribbean island of Trinidad was home to thousands of Hindus. Determined to learn more about the lives and faith traditions of Hindus in the Caribbean, I decided to pursue my Ph.D. research in this area. In order to understand their circumstances more fully a trip to India was essential. With a three-month rail pass in my backpack, and 6000 miles (10,000 kilometers) under my belt, I circumambulated the heart of the subcontinent from Mumbai (Bombay) in the west, to Darjeeling in the north, Calcutta in the east, and Kannyakumari in the south, with many stops in between. Since this journey of discovery, I have been drawn back again and again to a land whose people inherited the splendor of ancient civilizations and transcendental meditations yet hold the keys to a future South Asia of high technology and emergent political clout.

Where have you done fieldwork in South Asia, and what kind of fieldwork have you done?

Much of my fieldwork has centered on temple architecture in diasporic communities (Trinidad, Malaysia, Suriname, Zimbabwe). Thus I have visited rural areas in western Bihar and central Tamil Nadu in India, where many of the nineteenth-century migrant laborers once lived.

What has that experience taught you about the region's place in the world?

South Asia will eventually play a central role in global politics and trade. Over a billion people live there, and they are poised to make a powerful place for themselves in the twenty-first century. Powerful nations to the west would do well to take notice. A better understanding of South Asia's dynamic political life, its potential to expand the technology sector of its economy, and its extraordinary cultural history will not only expand our horizons but will likely contribute to a future of amity and cooperation with South Asian nations on the cusp of renewed greatness.

---

of a strong economy: steel, coal, transport, communications, and a wide range of manufacturing and processing industries.

For the most part, South Asian industrial policies failed to meet their goals. There was enough economic growth in the early years of independence to make India the eighth most industrialized country in the world in terms of the relative output of the industrial sector compared to that of the agricultural and service sectors. But the emphasis on self-sufficiency in industry was ill-suited to countries that had been primarily agricultural for years. In India, for example, governments invested huge amounts of money in a relatively small industrial sector that even today employs only 15 percent of the population (agriculture employs 67 percent). Only a small portion of the population directly benefited, so industrialization failed to increase South Asia's overall prosperity significantly.

Another problem was that the measures governments took to boost employment often contributed to inefficiency and ignored market incentives. One policy encouraged industries to employ as many people as possible, even if they were not needed. So, for example, it still takes 250,000 Indian workers to produce the same amount of steel as 8000 Japanese workers; consequently, Indian steel costs much more to produce than Japanese steel. In addition,

decisions about which products particular manufacturing industries should produce have been made by ill-informed government bureaucrats rather than being driven by consumer demand. Until the 1980s, items that would improve daily life for the poor majority were produced only in small quantities. Such items include cheap cooking pots, buckets, cheap yet sturdy bicycles, and simple tools. At the same time, there is a relative abundance of such items as vacuum cleaners, watches, TVs, kitchen appliances, and cars—but only a very few can afford to purchase them (although their numbers are increasing). Figure 8.14 shows the distribution of manufacturing and service industries in South Asia as of the mid-1990s.

Economic Reform: Achieving Global Competitiveness

During the 1990s, much of South Asia began to undergo a wave of economic reforms intended to make national economies more efficient and productive. It is still unclear whether these reforms will benefit the vast majority of the poor. India's structural adjustment programs (SAPs) began in 1991. They are aimed at privatizing industries, removing government regulation, and opening up the economy to foreign goods and foreign investment. In this more

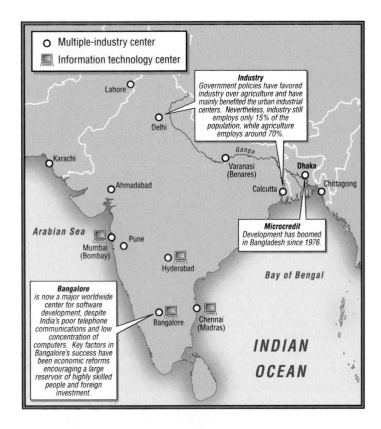

Figure 8.14 **Economic issues: industrial and information technology centers in South Asia.** The symbols mark locations where three or more major industrial activities (heavy industry, extractive industry, light manufacturing, or service industry) take place. [Adapted from Gordon Johnson, *Cultural Atlas of India* (New York: Facts on File, 1996), pp. 184–185, 190–191, 198–199, 204–205, 210–211, 217, 219–220, 222.]

ers in services, but this sector's contribution to GDP in India, Pakistan, and Bangladesh is nearly 50 percent. Hence, the service sector is seen as having the best chance of competing successfully in the global economy. Within the service economy, facilities that engage in trade, transport, storage, and communication (including information technology) show the most growth; finance, insurance, real estate, business services, and tourism have also grown quickly. All these activities are connected in some way with international commerce and benefit from India's success at developing information technology. Tourism is also growing and has considerable potential as South Asia's own middle class expands and outsiders are attracted. Tourism sites abound in both coastal and interior mountain locations.

Differing Views of Globalization. The growth of the service sector and information technology seems to foretell increasing connections to the global marketplace. The development of such connections is both desired and feared across South Asia. The ambivalence is particularly strong in India, the country with the strongest links to the global economy. Globalization is desired because it is expected to increase the number of high-paying jobs, raise standards of living, and fuel local production of consumer products and services. Cities throughout the region have experienced these benefits; the economic liberalization begun in 1991 has stimulated investment by international companies such as Coca-Cola, Microsoft, Cisco, Sony, Union Carbide, and Lever Brothers, Ltd.

Some small rural places see information technology as the answer to their geographic isolation. For example, the small community of Dhab (40,000 people), located on an island in the Ganga River downstream of Varanasi (Benares), has no electricity and few jobs. The shifting course of the Ganga between the rainy

competitive situation, productivity has increased in the export and industrial sectors of the economy. In particular, India has achieved spectacular successes in high technology (see the box "Bangalore: India's Silicon Valley"). As a result, rates of economic growth are now 5 to 8 percent per year, compared to 3 to 4 percent before the reforms.

Most of the new growth has occurred in the Indian Ocean states of Maharashtra, Gujarat, and Karnataka, where the cities of Mumbai, Ahmadabad, and Bangalore, respectively, are growth centers (see Figure 8.14). The rest of the country lags behind but will gain ground if a strong global economy persists. Like SAPs in Middle and South America and Africa, however, the new policies are producing wider disparities in income. The middle class, now estimated to be nearly 25 percent of the population, has received most of the gains. Another result is that the new private owners of formerly state-owned industries, who seek efficiency rather than maximum employment for India's multitudes, have fired hundreds of thousands of employees.

South Asia's service economies have been expanding as the agricultural and industrial sectors have gone through ups and downs. As a whole, South Asia has close to 20 percent of its work-

As the number of Indians with a bit of disposable income grows (estimates are that this number may be as high as 250 million people, or 25 percent of the population), opportunities for all kinds of entrepreneurial capitalism increase. Indian pharmaceutical companies are eyeing this group as potential consumers of basic modern medicines such as aspirin, cough syrup, and ointments. [Pablo Bartholomew/Gamma-Liaison.]

AT THE LOCAL SCALE *Bangalore: India's Silicon Valley*

As of 2001, an air of excitement surrounds the Indian state of Karnataka in southwestern India, particularly the city of Bangalore. That city is a center of information technology research, development, and practice. Bangalore holds what is probably India's highest concentration of scientists and technicians, universities, colleges, and technical schools. In and around the city, India's information technology industry is thriving.

India's institutes of technology are highly regarded, and many Indians have earned graduate computer engineering degrees at home, while others have done so abroad at such universities as Stanford in California and MIT and Harvard in Boston. Some of the most successful have returned home to Bangalore to take part in establishing India's fast-growing computer-based industries. At the same time, American and European software companies have recruited large numbers of information technology specialists from Bangalore schools and businesses to work in the United States and Europe. The success of Bangalore illustrates that it is possible for an agricultural state such as Karnataka to leap directly from an agricultural economy to the information age. Some observers believe that India as a whole is poised to become a technologically advanced nation without following the standard economic practice of having a strong local market for its own products.

Bangalore is now ranked eleventh in the world as a global hub of technological innovation (see Figure 6.1), even though India as a whole has very poor telephone communications and one of the world's lowest concentrations of computers: 3 per 1000 people. The world average is 25 computers per 1000 people; Singapore has 458 per 1000 and the United States 459 per 1000. Lacking a sufficient home market, Bangalore sells its products on the other side of the world. Although annual per capita income in Bangalore is only a few dollars higher than India's average of $2077, it seems poised to increase rapidly. There are now more than a dozen industrial parks outside Bangalore, housing such firms as Hewlett-Packard, Compaq, American Express, Citibank, Reebok, and IBM and employing thousands of highly educated Indian computer experts and engineers. Besides developing new software for the international market, these people maintain many of the everyday systems that govern financial life in the United States. For example, while you are sleeping, a software specialist in Bangalore may be fixing the glitch in the automatic teller machine that ate your bank card yesterday. Or, if you need to call a company's toll-free number for assistance late in the evening, the person who answers your queries may be in Bangalore, half a world away. Some companies have gone so far as to train their employees to speak with an American accent. Just one company, GE Capital Services, employs more than 6000 people in its international call center.

Some of Bangalore's success has resulted from the economic reforms that opened up the Indian economy to foreign investment and foreign goods, especially computers. However, much of Bangalore's remarkable progress can be attributed to the highly skilled people (primarily Brahmins) who have benefited from years of relatively high levels of government funding for secondary and higher education.

Unfortunately, although the high-tech sector has dramatically increased the number of millionaires in and around Bangalore, it has done little to bring down levels of overall poverty.

Source: Authors' visit to Bangalore, June 2000; and Tom Field, "On the road to Bangalore," *CIO.com, The Magazine* (December 1, 2000): 1.

---

and dry seasons has isolated the island community and inhibited development. Dhab's citizens are lobbying for dikes that would control the river's course and, more important, would support electricity and Internet cables and thus bring Dhab from the era of candles and oil lamps to the information age in just a few months.

Those who worry about globalization include environmental activists and conservative community leaders who occasionally find common cause in opposing projects like the one proposed for Dhab. Environmentalists warn against unduly modifying the river's natural seasonal oscillations with dikes, and community leaders fear that access to information on the Internet will cause a breakdown of traditional social and economic relationships and bring a flood of Westernization.

Economic Development and Poverty Rates. Poverty continues to decline in South Asia, but poverty rates have decreased much more slowly since 1991 than they had since about 1968. In India, the poor are defined as that portion of the population that cannot afford to eat a basic minimum diet. As we have noted, population growth over the last 50 years has outstripped economic growth. Despite all the policy efforts that have reduced poverty from nearly 45 percent in 1952 to less than 35 percent by the late 1990s, there are now nearly twice as many people who qualify as poor—simply because there are now so many more people overall. In 1998, there were 310 million poor, whereas 50 years ago there were only 180 million.

Innovative Help for the Poor

In recent years, South Asians have developed some promising strategies for helping poor people. One of these is **microcredit,** a program that makes very small loans available to poor would-be business owners in both rural and urban areas. Throughout South Asia, and indeed in much of the world, poor people have a difficult time obtaining loans. They do not have much collateral, so lending to them seems risky; nor do they need large sums, so lending to them is unprofitable. Hence, the poor must rely on small-scale moneylenders who often charge extremely high interest rates of 30 percent or more *per month.*

In the late 1970s, Mohammed Yunnis, an economics professor in Bangladesh, responded to this problem by starting the Grameen Bank, or "People's Bank," which makes small loans mostly to people in rural villages who wish to start businesses. Loans often pay for the start-up costs of small enterprises, such as chicken raising, small-scale egg production, or construction of pit toilets. The problem of collateral is resolved by having potential borrowers arrange themselves in small groups that are responsible as a whole for paying back the loans. If one member fails to repay a loan, then the group will be denied loans in the future. This system, reinforced with weekly meetings, creates incentives for mutual support among group members as well as peer pressure to repay loans. The weekly meetings are often the only time that women in purdah leave the confines of their homes and may be their first contacts with women (and sometimes even men) who are not kin. The repayment rate on the loans is extremely high, averaging around 98 percent—much higher than most banks obtain. Hence, the Grameen Bank can afford to charge interest rates of around 13 to 14 percent per year, much lower than those of village moneylenders.

So far, the Grameen Bank has been an enormous success in Bangladesh, where it has loaned over U.S. $1.8 billion to more than 2 million borrowers. Similar microcredit projects have been established in India and Pakistan and throughout Africa, Middle and South America, North America, and Europe.

In this village in Bangladesh, women borrowers gather weekly at the local Grameen Bank, usually in someone's house, to pay their loan installments. [Alex Pulsipher.]

VIGNETTE   In a small hamlet in Bangladesh not too far from the Indian border is the house of Mosamad Shonabhan, a 32-year-old married woman whose life has been changed by her 11-year participation in the Grameen Bank. Everyone agreed that she had been the smartest of her brothers and sisters, but because her father earned only 50 cents a day as a farm laborer, she could not go to school and was instead married at the age of 14 to a young barber. For a year she lived in her father-in-law's house, but financial problems soon forced her to move back into her father's house. There she faced increasingly dire circumstances as his health deteriorated. After a few years, a local political leader suggested that she join the Grameen Bank's lending program. She was afraid to go, as she had never handled money and had heard a local rumor that the bank's real purpose was to convert people to Christianity. Nevertheless, she went and eventually took out a loan for $40 that would allow her to set up a small rice-husking operation in her father's backyard. Eleven years and 11 loans later, she earns about $1.50 every day, three times what her father had made, and is a pillar of the local community. Her main source of income is a small shop inside her father's old house, which she bought from her siblings after his death. She also leases an acre of land, which produces enough rice to feed her family and the numerous guests and friends who now come by to see her. She plans to open another shop that her husband will run. She is being taught to read by her 15-year-old daughter, whom she plans to send to university. [Adapted from the field notes of Alex Pulsipher, 2000.]

POLITICAL ISSUES

Since independence in 1947, South Asia has peacefully resolved many conflicts, smoothed numerous potentially bloody transitions of power, and nurtured vibrant public debate over the issues of the day. India is often cited as a bastion of democracy that serves as an example of political enlightenment to the rest of the developing world. For all these successes, much tension has resulted from the incorporation into a modern society of older, non-democratic traditions, such as caste. In recent years, there have been increasing signs that corruption, demagogic leadership, and accompanying violence are eroding democracy in the region and leading many people to embrace religious nationalisms that threaten the generally peaceful relations within India and between India and Pakistan.

Caste and Democracy

Since the beginning of broad-based democracy in India in the 1930s, caste has been a defining yet contradictory factor in both local and national politics. At the local level, most political parties design their vote-getting strategies to appeal to subcaste (*jati*) loyalties. Often, they secure the votes of entire *jatis* with such political favors as bringing in new roads, schools, or development projects to communities dominated by these groups. These arrangements fly in the face both of official political party ideologies that deny any caste loyalties and of Indian government policies that actively work to undermine discrimination on the basis of caste. Currently, the role of caste in politics seems to be increasing, with the emergence of several new political parties that explicitly support the interests of low castes who have formed group alliances. Hence, caste has been absorbed into the political system in ways that create and maintain tension. We will see shortly that the same is true for religious loyalty.

Corruption

In South Asia, virtually any transaction involving the government, such as having a phone installed, requires a bribe. Traffic police, judges, government bureaucrats, and (in recent years) even prime ministers have been implicated in bribery scandals. One reason bribery exists is that civil servants are poorly paid. However, bribery has very old and deep roots in South Asian culture. Even before the British arrived, state officials accepted payment or favors in return for preferential treatment. British participation in this system reinforced the practice. Indeed, the British tendency to favor British-owned businesses introduced the practice into the modern free-market economy of the region. Nevertheless, corruption has become so rampant in recent decades that most South Asians now see it not as an ancient practice that must be tolerated but as a problem. Corruption favors the rich because it nurtures demands for larger and larger bribes of officials at higher and higher levels. The wealthy are the only ones able to pay such huge amounts, which sometimes reach into the millions of dollars. In return, they receive exemption from many regulations and access to immense state-controlled resources, such as national forests. In the process, the great majority of people, who cannot bribe on such a scale, are shut out of the benefits of economic development. Corruption has also nurtured violence in politics because it allows organized criminals,

who often have access to large amounts of cash from illegal activities, to achieve a level of influence not possible by other means. In some cases, politicians themselves are virtual gangsters; for example, a political cartoonist, Bal Thackeray, founded a political party that, since 1995, has dominated the politics of Mumbai and the Indian state of Maharashtra with violent attacks on opponents.

Religious Nationalism

Increasingly, people disgusted with corruption are joining religious nationalist movements and regional separatist movements because they see such efforts as purifying sources of morality. Although most people in South Asia live under legally secular governments, religious nationalism has long been a reality, shaping relations between people and their governments. **Religious nationalism** is the belief that a particular religion is strongly connected to a particular territory, perhaps even to the exclusion of other religions, and that those who share a belief system should have control over their own political unit—be it a neighborhood, part of a country, or a separate country. Although both India and Pakistan were formally created as secular states, India is increasingly thought of as a Hindu state and Pakistan and Bangladesh as Muslim states. Many people in the dominant group strongly associate their religion with their national identity. Since partition, religious nationalism has periodically fueled wars and conflicts between India and Pakistan over disputed territories. Through manipulation of the poor and dispossessed, political parties based on religious nationalism have gained popularity throughout South Asia, often fueling conflicts between religious majorities and minorities within a particular state. Although their members think of these parties as forces that will purge corruption and violence, they are, in fact, usually only slightly less corrupt and certainly no less violent than other parties (see the box "Babar's Mosque"). The

AT THE REGIONAL SCALE   *Babar's Mosque: The Geography of Religious Nationalism*

Proponents of religious nationalism often try to gain mass support through political campaigns that interweave South Asian history, mythology, and landscape. In late 1992 and early 1993, a series of riots occurred throughout South Asia that were the culmination of a long campaign waged by India's leading Hindu nationalist party, the Bharatiya Janata Parishad (BJP). The riots were triggered by the destruction of a Muslim mosque in the town of Ayodhya, in Uttar Pradesh on the Ganga Plain (see Figure 8.15). The Mogul emperor Babar had built the mosque during his early invasions of India in the sixteenth century. It supposedly stood on the ruins of a Hindu temple believed by local residents to mark the birthplace of the Hindu god Ram. After years of campaigning for the mosque's destruction, the BJP finally succeeded in late 1992, when a highly organized Hindu mob of 300,000 razed the structure. The destruction led to riots that ripped through most major northern Indian cities. Mobs,

commanded by urban Hindu nationalist political parties and social organizations (many of them funded by Indian Hindus living abroad), burned and looted selected Muslim businesses and homes, often with the complicity of the police. Nearly 5000 people died.

The BJP's selection of the Ayodhya mosque, an important Muslim heritage site, as a locus of protest was calculated to elicit violent reactions from Muslims. The location of the riots also reflected the urban base of Hindu nationalist parties, and their highly organized criminal actions reflected the violent gangster-style tactics that were common on the South Asian political scene in the 1990s. The violence soon spread across international borders: in retaliation, Muslim mobs in Bangladesh and Pakistan harassed Hindu communities and destroyed their temples.

Source: Alex Pulsipher, field notes, India, 1993.

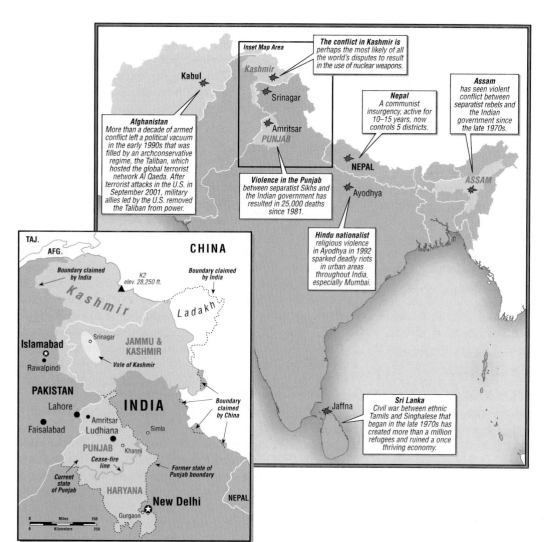

Figure 8.15 **Political issues: conflicts in South Asia.** Well-known conflicts in the region include the dispute between India and Pakistan over Kashmir and Punjab (see inset); the civil war and the 2001 "war on terrorism" in Afghanistan; the internationally funded Hindu nationalist movement, for which Ayodhya is a symbol; the Assam secession rebellion in far northeastern India; and the conflict between Tamils and Singhalese in Sri Lanka. [Insert adapted from *National Geographic* (May 1997): 19.]

Ganga River basin has emerged as the center of Hindu nationalist activism because of the sacred status of the Ganga River, the plethora of Hindu sacred places in the basin, and the long history of Muslim rule. Periodically, there is violence against Muslims, who make up about 15 percent of the population in this part of India.

Regional Political Conflicts

The most intense armed conflicts in South Asia today are **regional conflicts** in which an ethnic or religious minority actively resists the authority of a national or state government or between nations over territorial boundaries (Figure 8.15). The resisting group sees the government officials as depriving it of a voice in its own governance. Two regional conflicts in the neighboring Indian states of Punjab and Kashmir can serve as examples. Both states are in far northwestern India along the Pakistan border. Both conflicts have religious and political components, and both are sources of unrest that could destabilize the region. These two conflicts are at the heart of intermittent hostilities between India and Pakistan.

Conflict in Punjab. Punjab is the ancestral home of the Sikh community. When India and Pakistan were partitioned in 1947, Punjab was divided between the two countries. Caught between majority populations of Muslims and Hindus during the violence that followed partition, large numbers of Sikhs chose to live in the Indian part of Punjab or elsewhere in India because they thought the secular Indian constitution would better allow them to preserve their unique identity. The Sikh community has since pressed India for greater recognition of its distinct religious and ethnic identity and for greater regional political autonomy. Despite a long history of peaceful coexistence and intermarriage with Hindus in Punjab, Sikh conflicts with Hindus over access to land and water, over development policies and funds, and over the control of religious sites have created communal tensions.

After supporting India in its 1947 war with Pakistan, Sikhs successfully agitated for the creation of two states out of Punjab: a Sikh majority state called Punjab and a Hindu majority state called Haryana (see the inset in Figure 8.15). The borders between the two states had to be drawn on the basis of language (Punjabi and Hindi) because division based on religious difference is unconstitutional. Since then, however, Sikhs have felt alienated from the rest of India. From the early 1980s through the early 1990s they were governed by leaders federally appointed rather than by democratically elected officials.

Tensions escalated in 1983 and 1984, when the members of a militant Sikh group barricaded themselves in the holiest Sikh shrine, the Golden Temple in Amritsar, which they then used as a base of operations to agitate for Sikh political autonomy. In 1984, government forces (headed by a Sikh general) attacked the shrine, killed the militants, and damaged the temple, thereby deeply alienating and further radicalizing the Sikh community. Shortly thereafter, Prime Minister Indira Gandhi, who had called for the attack, was assassinated by two of her Sikh bodyguards. Riots and organized mob violence spread throughout India over the next few days, resulting in the deaths of more than 2700 Sikhs, who are a wealthy and influential minority in many cities outside of Punjab. Since this low point, conditions have improved somewhat as the government has begun an investigation into the organized nature of the mob attacks, but government forces and Sikh militants continue to battle periodically in parts of Punjab. Since 1981, 25,000 people have died from political violence there.

Conflict in Kashmir. In Kashmir, Hindu and Muslim differences date back to British rule before partition in 1947. At that time, Kashmir was one of hundreds of independent states that had surrendered much of their autonomy to the British, although their leaders retained nominal control. All these states were eventually absorbed into either Pakistan or India. Kashmir has long been a Muslim-dominated area, but its ruler in 1947 was a Hindu king. Pakistan's leaders believed that Kashmir should be turned over to them based on the rules for partition established by the British. But the kings and princes of the autonomous states also wanted a say in where their territories would be allotted. Kashmir's Hindu king wanted Kashmir to remain independent, but in 1948, when Pakistan invaded and there was an uprising of villagers in western Kashmir, he quickly agreed to join India. A brief war between Pakistan and India resulted in a cease-fire line that became a tenuous boundary (see Figure 8.15 inset).

A vote that was supposed to determine Kashmir's fate was never held. India argued that because Kashmiris had already voted in Indian elections, they had, in effect, voted to be part of India. Pakistan attempted another invasion of Kashmir in 1965 but was defeated. The two countries are technically still waiting for a UN decision on where the final border will be, but Pakistan effectively controls thinly populated mountain areas in the northern and western third of the densely populated Vale of Kashmir. India holds nearly all the rest and maintains an ominous presence of more than 500,000 troops. China claims the Ladakh part of Kashmir (see Figure 8.15 inset).

Civil war has erupted repeatedly over the years because many Kashmiris support independence from both India and Pakistan, and as many as 20,000 people have been killed. As in Punjab, much of the conflict has centered around the right of Kashmiris to elect their own state leaders; the national government in New Delhi has often appointed its own favorites. A number of anti-Indian Kashmiri guerrilla groups, equipped with weapons and

training from Pakistan, have carried out many bombings and assassinations. Blunt counterattacks launched by the Indian government have killed large numbers of civilians. The deaths have alienated more Kashmiris, including the local police force, which is now seen as sympathetic to the militants. Sporadic fighting between India and Pakistan continues along the boundary line, where at one location the two countries intermittently clash in the world's highest battle zone, at an altitude of 20,000 feet (6000 meters). Another development in the Kashmir dispute is the increasing involvement throughout the 1990s of Islamic militant mujahedeen, soldiers who took part in Afghanistan's war of resistance against the USSR in the 1980s. The mujahedeen are a powerful destabilizing force because, although they receive weapons and training from Pakistan (as do the Kashmiri militants), it is difficult to tell if their actions are always controlled directly by Pakistan or perhaps by more global terrorist networks.

A further complication in the Kashmir conflict is that both India and Pakistan have nuclear weapons. Both countries tested their nuclear weapons as recently as 1998, and both have threatened to use them should the other invade. Many international security analysts believe that, of all the world's current disputes, the conflict in Kashmir is the one most likely to result in the use of nuclear weapons, because of the nationalistic religious fervor of the protagonists.

The Future of Democracy

Although there are many political hot spots in South Asia as a whole, countries such as India and Sri Lanka have elected democratic governments regularly since their independence. Moreover, there are a number of reasons to expect greater tranquility in the future. Signs that democracy is expanding, thus making better government a possibility, include the recent peaceful creation of three new states in India and the fact that a more competitive multiparty arena has been taking shape in that country. There, voters are increasingly intolerant of corruption and violence. Although expanding democracy has resulted in greater influence for Hindu nationalists in the short run, it has also led to fairer and more peaceful elections and a clearer focus on providing opportunities for women and other disadvantaged groups. In Bangladesh, after years of military dictatorship, democratic elections have occurred with some regularity. In Nepal, a more freely elected legislature and a reduction of the king's power are giving ordinary citizens a greater political voice. In 2000, the last monarchy, the tiny Himalayan kingdom of Bhutan, was planning to give its people the vote.

ENVIRONMENTAL ISSUES

In 1973, in the Chamoli district of Uttar Pradesh, India, a sporting-goods manufacturer planned to cut down a grove of ash trees so that his factory, in the distant city of Allahabad, could use the wood to make tennis racquets. The trees were sacred to nearby villagers, however, and when their protests were ignored, a group of local women took a dramatic action that became a symbol of the struggle to protect South Asia's environmental quality. When the loggers came, they found the women hugging the trees and refusing to let go until the threat to their grove ended. Soon the manufacturer located another grove. The women's action grew into the **Chipko** (or social forestries) **movement,** which has spread to other forest areas, slowing deforestation and increasing ecological awareness.

Deforestation

Deforestation is not new to the Indian subcontinent. The spread of agriculture at the time of the Indus Valley civilization, as well as the expansion of newly developing Hindu kingdoms after the Arya migrations, caused a major loss of forested lands more than 3000 years ago. Ecological historians Madhav Gadgil and Ramachandra Guha show how the western regions of the subcontinent (from India to Afghanistan) became drier and drier as the forests vanished. They suggest that the early success of Buddhism and Jainism in these areas resulted, in part, from the focus of these two religions on vegetarianism, which possibly served as an adaptive response to a decline in game supplies as habitat was lost to deforestation. Under British colonialism, deforestation intensified once again. Starting in the mid-nineteenth century, perhaps a million trees a year were felled for use in building the railroad alone. Laxman Satya, another ecological historian, has shown that the massive deforestation of the nineteenth century contributed to the increasing aridity and heat of twentieth-century Deccan environments.

In the twenty-first century, the subcontinent's forests are still under stress, the result of village populations expanding into forestlands for living and cultivation space. Commercial uses of forest products and tourism are also factors. Forests are cleared to make trails for tourist trekkers in mountainous forest zones, such as those in Nepal, Kashmir, and the Nilgiri Hills of southern India, and trekkers consume wood for cooking and heat. Among the results of forest clearing are massive landslides that often close railroad lines and roads during the rainy season. Landslides increase erosion, and the eroded mountain environments are less able to absorb rainfall during future summer monsoons. Another result of forest clearing in mountainous areas is that flooding and silt deposition increase in lowland places such as Bangladesh.

Unlike China and many African nations facing similar problems, South Asia has a healthy and vibrant climate of environmental activism that brings the consequences of deforestation to the attention of the public. Activism focused on saving forests is a reaction to a pattern found throughout South Asia, in which the resources of rural areas are channeled to urban industries without consideration of the needs of local rural people. The proponents of the social forestries movement argue that management of forest resources should be turned over to local communities. They say that people living at the edges of forests possess complex local knowledge of these ecosystems, gained over generations—knowledge about which plants are useful for building materials, for food, for medicinal use, and for fuel. These people have the incentive to manage carefully, because they want their progeny to benefit from forests for generations to come. In contrast, the government forestry departments typically shut local people out of their traditional forestlands, forcing them to depend on smaller and more marginal common-access lands that may now contain only a very few of the useful species.

The fact is, though, that burgeoning local populations themselves carry out deforestation as they try to obtain ever more firewood for themselves and fodder for their animals. It is difficult to convince impoverished people to conserve resources. Their need for income predisposes them to collaborate with poachers of rare forest products. Moreover, the powerful industrial and government interests that monopolize forest reserves are not likely to yield control to local people easily. Despite these problems, several natural reserves have been established in India, and local residents are gaining increasing influence in decisions about control of these areas (see the box "A Visit to the Nilgiri Hills," page 430).

Water Issues

One of the most controversial environmental issues in South Asia today is the battle over water. South Asia has more than 20 percent of the world's population but only 4 percent of its fresh water. It is not surprising, then, that conflict exists between India and Bangladesh over access to the waters of the Ganga River (Figure 8.16).

Conflict over Ganga River Water. In recent years, during the dry season, India has diverted 60 percent of the Ganga's flow to Calcutta to flush out channels where silt is accumulating and hampering river traffic. India's policies, however, deprive Bangladesh of normal flow. Less water is available for irrigation, causing crop yields to fall, and salt water from the Bay of Bengal penetrates inland, ruining fields. The diversion has also caused major alterations in Bangladesh's coastline, damaging the small-scale fishing industry. Thus, to serve the needs of Calcutta's 11 million people, the livelihoods of 40 million have been put at risk, triggering protests in Bangladesh.

In the late 1990s, India signed a treaty promising to reduce the scale of the diversions, but as of 2001, they had not been reduced. As with other major environmental problems, a solution has been hard to achieve because the population adversely affected is not only poor and rural but is located in a different region—in this case, in a different country—from the politicians and bureaucrats who are in a position to respond to protests.

Similar water use conflicts occur between states within India and between the wealthier and poorer sectors of the population. Just 17 five-star hotels in Delhi use about 210,000 gallons (800,000 liters) of water daily, enough to serve the needs of 1.3 million slum dwellers. At the state level, Harayana diverts water from the Ganga River, depriving farmers downstream in Uttar Pradesh of the means to irrigate their crops.

Conflict over Dams. Hydroelectric dams are planned for many of India's rivers to supply the electricity needed to support a modern infrastructure and India's new information technology industries. An example is the Sardar Sarovar hydroelectric dam, one of 30 planned for the Narmada River in Madhya Pradesh (see Figure 8.16). The environmental problems posed by these dams have drawn national and international attention. As many as 320,000 people will have to be relocated to make room for new reservoirs. The law requires government agents to provide land of

Thinking Geographically: ON THE WEB

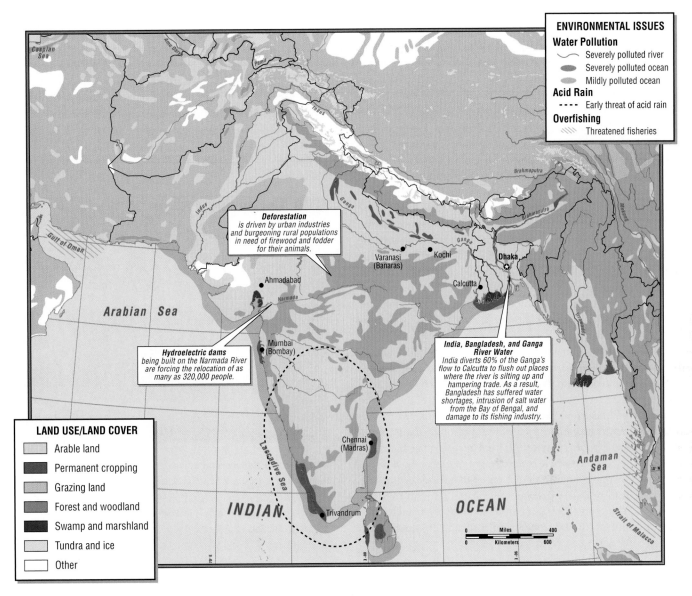

ENVIRONMENTAL ISSUES
Water Pollution
- Severely polluted river
- Severely polluted ocean
- Mildly polluted ocean

Acid Rain
- - - - Early threat of acid rain

Overfishing
- Threatened fisheries

Deforestation
is driven by urban industries and burgeoning rural populations in need of firewood and fodder for their animals.

Hydroelectric dams
being built on the Narmada River are forcing the relocation of as many as 320,000 people.

India, Bangladesh, and Ganga River Water
India diverts 60% of the Ganga's flow to Calcutta to flush out places where the river is silting up and hampering trade. As a result, Bangladesh has suffered water shortages, intrusion of salt water from the Bay of Bengal, and damage to its fishing industry.

LAND USE/LAND COVER
- Arable land
- Permanent cropping
- Grazing land
- Forest and woodland
- Swamp and marshland
- Tundra and ice
- Other

Figure 8.16 **Environmental issues: South Asia.** Environmental issues in South Asia are closely related to the ways in which land and sea resources are used by an ever more densely settled population. Forests are cleared for settlements and agriculture, increasing the likelihood of erosion. Modern agricultural methods and industrialization increase pollution in the water, soil, and air. This pollution, in turn, affects coastal zones, including fisheries. Government water management efforts to clear streams for navigation, harness waterpower, and impound water for irrigation all affect downstream users, who are often poor and politically powerless.

equal or better value to displaced people, but popular resistance hardened in the late 1980s when the land given to farmers displaced by the Sardar Sarovar Dam turned out to be barely arable. Facing starvation on these new lands, the farmers and their families returned to their old villages and fields in 1989 and eventually marched, 80,000 strong, on the capital in New Delhi; there they demanded that the project be halted. This protest and others culminated in a 1993 decision by India's supreme court to turn down World Bank loans of U.S. $450 million for the Narmada project. Nonetheless, after a brief hiatus, construction resumed, although legal challenges intermittently interrupted work on the dam as late as 2001. In resuming construction, the Indian government was bowing to pressure from the state government of Gujarat. Although the disputed dam is in Madhya Pradesh, most of the more prosperous technology industries that would use the electric power and the farmers who stand to benefit from the irrigation waters provided by the project live in Gujarat (see Figure 8.21, page 443). The "Save the Narmada" movement has persuaded the international funders of such projects to demand environmental impact studies more frequently. No such study was provided for the Narmada hydroproject.

AT THE LOCAL SCALE *A Visit to the Nilgiri Hills*

As part of their research trip through southern India in the summer of 2000, Lydia and Alex Pulsipher visited several natural reserves in the Nilgiri Hills. These hills, part of the Western Ghats south of Mysore in Karnataka, harbor some of the last scraps of forest in southern India. One of the most successful of the Nilgiri reserves is Longwood Shola, a tiny remnant of ancient tropical evergreen forest. Here, two local naturalists, members of ethnic minorities native to the Nilgiri Hills, manage the 287-acre (116-hectare) park, which harbors 13 mammal species, 52 bird species, and 118 plant species, many of them found only in the Nilgiri. One bird species, the Grey Jungle Fowl, is ancestor to the domestic chicken. Tiny as it is, the forest of Longwood Shola protects three perennial streams that supply fresh water to 16 downstream villages. Among the projects the naturalists have initiated is a reforestation effort for which they themselves are growing the seedlings from local native species.

Phillip Mulley, a naturalist, Christian minister, and leader of the Badaga ethnic group, helps visitors to understand the issues that face the people native to the Nilgiri Hills. He noted that in addition to being a region of wildlife reserves, this part of India is the homeland of indigenous peoples (Badaga, Toda, Kota, Kurumba) who must now compete for space with a burgeoning tourist industry (1.2 million visitors a year in 1991) and huge tea plantations that have been established where forests stood until recently. Some of the newest tea plantations were cut out of forestlands by the state government to provide employment for Tamil refugees from the Tamil-Singhalese conflict in Sri Lanka. So while one branch of the state government (the forestry department) and citizen naturalists try to preserve forestlands, another branch (social welfare), faced with a huge refugee population, is cutting down the forests to accommodate the employment needs of the refugees.

Source: Lydia and Alex Pulsipher, field notes, Nilgiri, June 2000.

Increasingly, geographers are using digital photography and video in their research. This rare photo of a Bengal tiger in the Mudumalai Wildlife Sanctuary in the Nilgiri Hills of southwestern India is from a mini–digital video filmed by Alex Pulsipher in June 2000. The Bengal tiger is an endangered species. [Alex Pulsipher.]

Water Purity of the Ganga River. Water purity is an issue in historic pilgrimage towns such as Varanasi, where each year millions of Hindus come to die, be cremated, and have their ashes scattered over the adjacent Ganga River. The Ganga is believed to be a goddess that literally pours life and purifies the world. As the number of such final pilgrimages has increased, wood for the cremation fires has become scarce, and incompletely cremated bodies are being dumped into the river, where they pollute water used for drinking, cooking, and bathing. In addition, dead cows are often consigned to the river.

Recently, the government installed an electric crematorium with its own generator on the banks of the river. The cost is only 70 rupees (Indian currency) per funeral compared to the 2000 rupees for a traditional funeral pyre. The line at the electric facility is much longer than the one at the pyres; thus, there is hope for reduced levels of unburned human remains in the river.

Of greater concern, however, is the amount of industrial waste and sewage dumped into the river. Most sewage enters the river in raw form, because Varanasi's sewage system (built by the British early in the twentieth century) long ago exceeded its capacity. The danger posed by sewage is measured in terms of the river's biochemical oxygen demand, a figure that indicates the degree of oxygen consumption caused by decomposing matter such as fecal material or animal remains. The Ganga's biochemical oxygen demand is 375,000 times the level of safety (see Figure 8.16). Pumps have been installed to move the sewage up to a new and expensive sewage processing plant built on the Varana, a small stream that enters the Ganga on the city's southern boundary. There the sewage is processed and released back into the Ganga. Unfortunately, the plant is so overwhelmed during the rainy season that it can process only a small fraction of the city's sewage. During the dry season, the municipal electrical supply is haphazard at best. Thus, it seems that traditional Western technologies for sewage treatment are unsuited to the extreme conditions of India.

Veer Bhadra Mishra, a Brahmin priest and professor of hydraulic engineering at Banaras Hindu University in Varanasi, is on a mission to clean up the Ganga using unconventional methods. He is working with engineers from the United States to build a series of processing ponds that will use India's heat and monsoon rains to clean the river at half the cost of more technologically

industries up and down the Yamuna and Ganga rivers is destroying good farmland and such great monuments as the Taj Mahal—which, scientists say, has "marble cancer": its soft, intricately carved surfaces are becoming increasingly pockmarked. M. C. Mehta, a Delhi-based lawyer, became an environmental activist partly in response to the condition of the Taj Mahal. For more than 20 years, he has successfully promoted environmental legislation that has removed hundreds of the worst polluting factories from the river valleys. His efforts are also a response to a horrible event that took place in central India in 1984, when an explosion in a pesticide plant in Bhopal, India, produced a gas that killed at least 3000 people and severely damaged the lungs of 50,000 more. The explosion was largely the result of negligence on the part of the U.S.–based Union Carbide Corporation, which owned the plant, and the local Indian employees who ran it. In response to the tragedy, the Indian government launched an ambitious campaign to clean up poorly regulated factories.

MEASURES OF HUMAN WELL-BEING

We have emphasized throughout this book that gross domestic product (GDP) per capita (Table 8.1, column 2) is at best a crude indicator of well-being and is best used in conjunction with other measures. GDP per capita in South Asia is nearly as low as it is in Africa (and in some cases lower), and a large majority of people in South Asia are extremely poor. Most people in this region are frugal and resourceful. Because they recycle nearly everything, their villages tend to be extremely clean. They are entrepreneurs in the informal economy; they grow their own food whenever possible; and they strictly limit cash expenditures through reciprocal exchange agreements with one another. Through such efforts, the people of South Asia manage to give themselves a somewhat higher standard of living than the GDP figures would indicate.

South Asia does not exhibit the wide variations of GDP seen in some world regions (East Asia, Southeast Asia, and Oceania, for example). In fact, the one country with even a marginally higher GDP, the Maldives, has only a tiny population, whose income is increased by tourism. Sri Lanka's slightly higher GDP per capita results from a physical environment that is favorable to agriculture (tea is a chief product) and from the presence of exportable minerals. More important, a history of investing in the education and health of its citizens has helped Sri Lanka to equalize wealth distribution among its 19 million people. The country would undoubtedly be even more prosperous if unrest between the Tamil and Singhalese ethnic groups had not hindered development for several decades (see pages 447–448).

The United Nations Human Development Index (HDI) (Table 8.1, column 3) combines life expectancy at birth, educational attainment, and adjusted real income to arrive at a ranking of 174 countries. As of 2000, the largest South Asian countries (India with 1 billion and Pakistan with 150 million) had advanced to the lowest ranks of countries in the medium range (those ranking between 47 and 139). This is a notable accomplishment because of the huge populations involved (more than one-sixth of the earth's population). Again, the rankings are based on averages, so there are still staggering numbers of people who remain very poor.

Women who must wash clothes in the watercourse flowing through Mumbai's Dharavi section, Asia's largest slum, have a difficult time because garbage and raw sewage pollute the water daily. [Steve McCurry/National Geographic Image Collection.]

sophisticated methods. In addition, he preaches a contemporary religious message to the thousands who visit his temple on a bank of the sacred Ganga. The belief that the Ganga purifies all it touches leads many Hindus to think that it is impossible to damage this magnificent river. Mishra reminds them that because the Ganga is their symbolic mother, it would be a travesty to smear her with sewage and industrial waste.

Industrial Pollution

Elsewhere in the region, the air as well as the water may be endangered by industrial activity. Emissions from vehicles and coal-burning industries are so bad that breathing Delhi's air is equivalent to smoking 20 cigarettes a day. The acid rain caused by

TABLE 8.1   *Human well-being rankings of countries in South Asia*

Country (1)	GDP per capita, adjusted for PPP[a] in 1998 $U.S. (2)	Human Development Index (HDI) global rankings, 2000[b] (3)	Gender Empowerment Measure (GEM) global rankings, 2000[c] (4)	Female literacy (percentage), 1998 (5)	Male literacy (percentage), 1998 (6)	Life expectancy, 1998 (7)
Selected countries for comparison						
Japan	23,257	9 (high)	41	99	99	80
United States	29,605	3 (high)	13	99	99	76.8
Mexico	7,704	55 (medium)	55	88.7	92.9	72.3
South Asia						
Afghanistan[d]	N/A[e]	N/A	N/A	15	47	46
Bangladesh	1,361	146 (low)	67	28.6	51.1	58.6
Bhutan	1,536	142 (low)	N/A	N/A (28 in 1995)	N/A (56 in 1995)	61.2
India	2,077	128 (medium)	N/A	43.5	67.1	62.9
Maldives	4,083	89 (medium)	N/A	96	96	65
Nepal	1,157	144 (low)	N/A	21.7	56.9	57.8
Pakistan	1,715	135 (medium)	N/A	28.9	50	58
Sri Lanka	2,979	84 (medium)	66	88.3	94.1	73.3

[a]PPP = purchasing power parity.
[b]The high, medium, and low designations indicate where the counry ranks among the 174 countries classified by the United Nations.
[c]Total ranked in world = 70; no data for many.
[d]Statistics for Afghanistan from *CIA Country Reports, 2000.*
[e]N/A = data not available.

Source: United Nations Development Report 2000.

With the exceptions of the Maldives and Sri Lanka, both literacy (columns 5 and 6) and life expectancy (column 7) are very low in this region, especially for women. And, as in Africa, governments have not provided even the most basic services. According to the United Nations, only 60 percent of the eligible children in India are in secondary school; in Pakistan, the figure is only one-quarter (1995); and in Afghanistan, just one-eighth (1995). In India, Bangladesh, and the Himalayan states of Nepal and Bhutan, clean water is available only to about 80 percent of the population. In the entire region, however, less than a quarter of the people have access to sanitary toilets, including outhouses and pit toilets.

Even as economic development proceeds in South Asia, the likelihood is that the gap between rich and poor will continue to grow. Most wealth is being created in the middle and upper classes, and the spending and reinvestment of these groups tend neither to create jobs for the poorest nor to enhance their well-being in other ways. Often, development means fewer jobs for the poorest and increased reliance on costly imported goods.

The United Nations Gender Empowerment Measure (GEM) (Table 8.1, column 4) ranks countries by the extent to which women have opportunities to participate in economic and political life. For all but Bangladesh (ranked 67th out of 70) and Sri Lanka (ranked 66th), these figures were not available in 2000 (worldwide, only 70 of 174 countries reported data). It is reasonable to assume that the missing figures for the unranked countries would also be low.

SUBREGIONS OF SOUTH ASIA

Here, we subdivide the region of South Asia for closer examination. The subregions are grouped roughly according to their physical and cultural similarities. In several cases, parts of India are grouped with adjacent countries. This is true in the Himalayan region, in northeastern South Asia, and in the southernmost South Asian region, where parts of India and the country of Sri Lanka are treated as a subregion.

Among the themes that emerge in the following sections are

1. The physical and cultural diversity of the region

2. The pervasive rural quality of life for most people

3. The very large and distinctive cities

4. The still small, but growing, industrial economy of the region

5. The lingering religious and ethnic conflict that is exacerbated by lack of economic opportunities

AFGHANISTAN AND PAKISTAN

Afghanistan and Pakistan (Figure 8.17) share location, landforms, and a history, and they have been involved in recent political disputes over global terrorism. Many cultural influences have passed through these mountainous countries into the rest of South Asia: Arya migrations, Alexander the Great and his soldiers, the

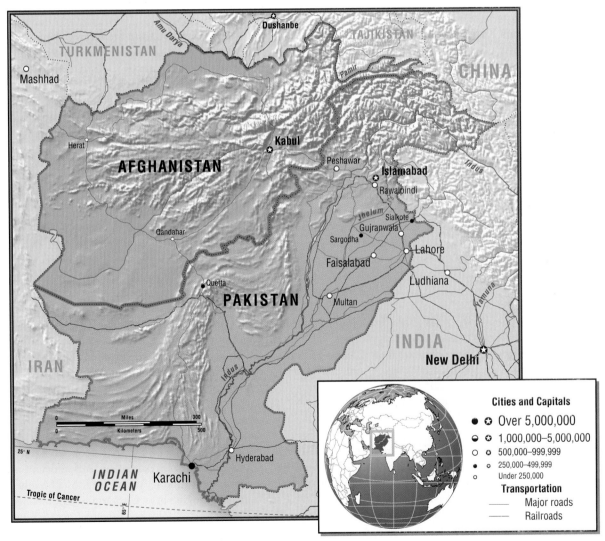

Figure 8.17 The Afghanistan and Pakistan subregion.

continual infusions of Turkish and Persian peoples, and the Turkish-Mongol influences that culminated in the Mogul invasion of the subcontinent from the beginning of the sixteenth century. Today, both countries are primarily Muslim and rural; 80 percent of Afghanistan's people and 67 percent of Pakistan's live in villages and hamlets. Both countries must cope with arid environments, scarce resources, and the need to find ways to provide rapidly growing populations with higher standards of living. Both countries are home to conservative Muslim movements. Afghanistan was ruled until late 2001 by a fundamentalist religious group, the Taliban; Pakistan by a military dictatorship that needs the support of fundamentalists to stay in power. Afghanistan's more extreme poverty is made worse by the armed strife from which it has suffered for more than 20 years.

The landscapes of Afghanistan and Pakistan are best imagined in the context of the ongoing tectonic collision between India and Eurasia. At either end of the Himalayan mountain mass, the collision uplifted curved crinkles into the Eurasian landmass. The lofty Hindu Kush, Pamir, and Karakoram mountains of Afghanistan and Pakistan are the western manifestation of this buckling (see the map that opens this chapter). The system of high mountains and intervening valleys swoops away from the Himalayas and bends down to the southeast toward the Arabian Sea. Landlocked Afghanistan, bounded by Pakistan on the east and south, Iran on the west, and Central Asia on the north, is entirely within this mountain system. Pakistan has two contrasting landscapes: the north, west, and southwest are in the mountain and upland zone just described; the central and southeastern sections are arid lowlands watered by the Indus River and its tributaries.

Afghanistan

Afghanistan's Hindu Kush mountains in the north fan out into lower mountains and hills and then into plains to the north, west, and south. In these gentler but arid landscapes, characterized by steep, sparsely vegetated meadows and pasturelands, most of the country's people struggle to earn a subsistence living from grazing and some cultivation. The main food crops are wheat, fruit, and nuts. Historically, opium poppies, native to this region, have been an important cash crop.

As of mid-2001, there were more than 26 million Afghans, 43 percent of whom were 14 years of age or younger. Life expectancy is only 46 years, yet the population is growing by 2.5 percent per year. Literacy in Afghanistan is among the lowest on earth. Only 20 percent of women and 50 percent of men over age 15 can read. Women bear six children on average. Generally, low life expectancy, low literacy, and a high birth rate are the markers of an extremely poor country.

The less mountainous regions in the north, west, and south are associated with the three main ethnic divisions. In the northwest are the ethnic groups who share culture and language traditions with the Turkmen and Uzbeks of Central Asia. In the northeast are the Tajiks and in the west the Hazara, both of whom are closely aligned with Iranian culture and languages. In the south, the Pashto-speaking Pathans are culturally akin to groups farther south across the Pakistan border. The various ethnic groups have

remained separate and competitive, but there is significant variation within ethnic (often called tribal) groups, especially regarding views toward religion, education, and gender roles.

In the 1970s, political debate in Afghanistan became polarized between urban elites, who favored industrialization and democratic reforms, and rural conservative religious leaders, whose positions as landholders and ethnic leaders were threatened by the proposed reforms. Some urban elites allied themselves with the Soviets, who, fearing that a civil war in Afghanistan would destabilize the Central Asian states of the Soviet Union, invaded Afghanistan in 1979. This was still the cold war era; the United States, Pakistan, and Iran supported the anti-Soviet movement (mujahedeen) formed by rural conservative men—a collection of ethnic leaders and their followers who were strongly influenced by militant Islamist thought. The mujahedeen were tenacious fighters, and in 1989, after heavy losses, the Soviets gave up and left the country. Anarchy prevailed for a time as the Afghan factions fought each other, but the rural conservatives eventually defeated the reformist urban elites.

In the early 1990s, a radical religious-political and military movement, called the **Taliban,** emerged from among the mujahedeen. For the most part, the Taliban are young men from remote villages, many in southern and eastern Afghanistan. They are led by students from the *talibs* (Islamist schools of philosophy and law); but most followers have had little chance even to learn to read. The Taliban saw their role as controlling corruption and bringing stability and peace by strictly enforcing the **shari'a**, the Islamic social and penal code (see the explanation of shari'a in Chapter 6, pages 304–305). They maintain their public expression of political control through a series of proclamations (*fatwas*). The Taliban view themselves as guardians of the social order and see Western influence as so fundamentally corrupting that it must be rooted out. In the 1990s, the Taliban took particular aim at urban professional women, whom they saw as symbolic of Westernization. All women were forced into domestic seclusion; whenever they appeared briefly, in public, they had to wear the completely concealing *burqa* (see page 416 in this chapter and the discussion of veiling in Islam in Chapter 6, page 308). Other efforts by the Taliban to purge their society of non-Muslim influences included restricting education for everyone, especially girls and young women; destroying the 1500-year-old Buddhist sculptures in the Bamiyan Valley; and banning the production of opium, to which many Afghan men are addicted. By 2001, the Taliban controlled 95 percent of the country, including the capital, Kabul, and were moving north, where an alliance of non-Pashto ethnic groups was mounting a counteroffensive. Meanwhile, an active resistance movement of Afghan exiles, many of them educated Pashto, labored against the Taliban in Pakistan, Europe, and America.

The traditional rural subsistence economy of Afghanistan has proved remarkably resilient and self-sufficient in the face of continuing hostility among Afghan factions. Nonetheless, the sufficiency of that system has been compromised by an ongoing drought beginning in the late 1990s that threatens over a million people with starvation. During the intermittent armed conflicts, many male farmers and herders have left home to fight; some have not returned, which has left women increasingly in charge of family subsistence.

The Antiterrorism War in Afghanistan, Fall 2001

Shortly after the terrorist attacks on the United States on September 11, 2001, U.S. officials announced that Osama bin Laden and his Al Qaeda network of radical Islamic fundamentalists were suspected of organizing the assault. Bin Laden, known to be hosted by the Taliban regime in a secret location near Kandahar in southern Afghanistan, soon confirmed in videotapes and interviews with journalists that he supported the concept of the attacks as a symbolic gesture against the growing global cultural and economic domination of the West.

Officials in the United States moved quickly to organize an alliance for a military attack on bin Laden and the Taliban in Afghanistan. The unusually broad alliance included nearly all countries in the Americas, Europe, and Oceania, as well as Russia, China, and Japan. Also supporting the alliance were the predominantly Muslim countries of Turkey, Egypt, Jordan, Saudi Arabia, Indonesia, Pakistan, and the newly independent Central Asian republics, as well as India, a country with a large Muslim minority. These latter countries faced considerable domestic opposition from conservative Muslim citizens, who sympathized with Al Qaeda's attempt to curb the spread of Western secular influence and economic power.

The war strategy was to render bin Laden's position in the country untenable by toppling the Taliban regime with modern air strikes followed by ground support from the Northern Alliance, a fighting force of anti-Taliban Tajiks, Uzbeks, Turkmen, and Hazara from the north of Afghanistan. In the process of defeating the Taliban, bin Laden and his followers were to be killed or captured. In late 2001, the Taliban were overpowered. The United Nations helped establish an interim coalition government under the leadership of Hamid Karzai. It is the first step toward a constitutional democracy.

The ethnic diversity of Afghanistan and its neighbors (see the map on this page) has thwarted many previous efforts to unite the country under one government. Contrary to media reports, Afghanistan's ethnic groups have not been at continuous war with one another, but they are distinct entities with different languages, interpretations of Islam (most are Sunni), values, and traditional social roles. The events of the last several decades have disturbed long-standing roles and feelings of trust. As the map illustrates, most of the ethnic groups located in Afghanistan extend into surrounding countries. At times in the past, ethnic loyalties across national borders have been stronger than allegiances to home countries, and neighboring states have sought to take parts of Afghanistan based on ethnic affiliations. These states also fear that civil war in Afghanistan could spread to them along ethnic corridors. Therefore, Afghanistan's ethnic geography must be considered in the postwar economic and political reconstruction of the country.

The dominant ethnic groups in Afghanistan are the Pashtuns, Tajiks, Hazara, Uzbeks, Baluch, Kyrghiz, and Turkmen. Although no census has been taken for many years, the Pashtuns are the largest group, making up from 40 to 60 percent of the Afghan population (about 11 million). Pashtuns have been the kings, lesser political leaders, and large landowners of Afghanistan, and they were among the most educated and modernized. But, during the Soviet occupation in the 1980s, most of the Afghan elite went into exile, leaving a

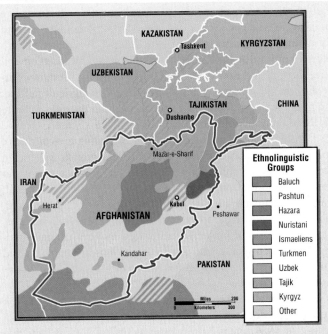

Ethnolinguistic groups in Afghanistan and surrounding countries. [Phillippe Rekacewicz and Cecile Marin, "A tangle of nations," *Le Monde diplomatique,* January 2000, www.monde.diplomatique.fr.]

void of leadership within the country. It was in this void that a small group of uneducated southern Pashtuns, numbering at their strongest about 50,000, formed the extremist Taliban group (see page 434) with the aid and encouragement of Pakistan. The Tajiks are the second largest group (about 3.5 million) and are culturally and linguistically akin to Iran. Many Tajiks in Afghanistan are urban dwellers and are noted for their skills as traders, merchants, and civil servants. Hazara, living in central Afghanistan, are Shi'ites, speak a Persian-related language (Farsi), and have close ties with Iran. Of Mongolian origins, the Hazara were traditionally servants and providers of water and fuel, but recently they have turned to selling kerosene and petroleum products, some becoming wealthy. Uzbeks, descended from Central Asian nomads, are associated in Afghan minds with communism because Uzbekistan was a Soviet republic and Uzbeks supported the Soviet invasion of Afghanistan.

As Afghans, their neighbors, and the U.S. alliance turn their attention to the reconstruction of Afghanistan, geography issues will be central to discussions: the need for broad participation in government and in the benefits of development across ethnic, class, and gender lines; the need to decentralize development so no one group, geographic area, or city dominates; and the need to avoid ethnically based autonomous economic regions because they would inhibit trade and threaten the unity of Afghanistan. Although aid from many sources will undoubtedly flow into Afghanistan over the next decade, the country could generate some funds for development from its location. It is strategically situated to host income-producing oil and gas pipelines running from Central Asia through Afghanistan and Pakistan to the coast of the Arabian Sea.

Pakistan

Although not much larger in area than Afghanistan, Pakistan has nearly six times as many people (151 million). Pakistanis also live primarily in villages. Some are sprinkled throughout the arid mountain districts associated with herding and subsistence agriculture, but it is the lowlands that have attracted most settlement. Here, the ebb and flow of the Indus River and its tributaries during the wet and dry seasons form the rhythm of agricultural life. The river brings fertilizing silt during floods and provides water to irrigate millions of cultivated acres during the dry season.

Pakistan enjoyed an overall annual economic growth rate of 6 percent in the 1980s and early 1990s. This growth was fueled substantially by agriculture in the Indus Valley basin, especially in the Pakistani Punjab area, where such cash crops as cotton, wheat, rice, and sugarcane are grown on large tracts of irrigated land. Pakistan has had irrigation systems for many thousands of years, but recent irrigation projects, encouraged by international funding, overstressed the system. By the mid-1990s, waterlogged and salinized soil had drastically reduced production, and some fields were entirely barren.

Other problems are related to financial issues. In 1996, the International Monetary Fund withheld U.S. $70 million of a loan because of governmental mismanagement and apparent fudging on budgetary figures. Currently, over 70 percent of the national budget goes to pay off debt and to finance Pakistan's standing military of over half a million and its nuclear weapons program, thought necessary primarily because of Pakistan's disputes with India over Punjab and Kashmir (see pages 426–427). Until the Afghan war of 2001, Pakistan had to pay for an increasing proportion of its military budget because the United States, which funded Pakistan's military expansion for decades, reduced its support in the early 1990s. This reduction was largely motivated by the collapse of the Soviet Union, which increased U.S. friendliness with India, and Pakistan's insistence on developing its own nuclear bomb. Following September 11, 2001, however, the United States sought to repair relations with Pakistan. Significant military and social aid was offered in return for support against the Taliban in Afghanistan. Partly because of Pakistan's huge investment in its military rather than in education, health care, and economic modernization, annual per capita GDP is only U.S. $1715—lower than both India's (U.S. $2077) and China's (U.S. $3105).

Most Pakistanis do not reap the benefits of the country's overall economic growth. A third of the population lives on less than U.S. $1.00 a day. The productive farmland of Punjab is owned by a small elite. Although textile and yarn-making industries are growing around the cities of Lahore and Karachi, providing jobs for rural workers, the wealth generated is not passed on to the workers in the form of higher wages; and the reinvestment of profits in Pakistan is low. In addition, international drug dealing is a major activity, and corruption among officials is widespread. Arif Nizami, editor of the Lahore newspaper *The Nation*, is quoted by *New York Times* correspondent John Burns (August 1997) as saying: "Pakistan is a country where millions cannot get two square meals a day, yet the prime minister has a fleet of planes, flies to his place in the country in a personal helicopter, and lives in a palace [that would] shame the White House."

HIMALAYAN COUNTRY

The northern border of South Asia is the Himalayan mountain chain (Figure 8.18). The map at the beginning of this chapter

Figure 8.18 **The Himalayan country subregion.**

AT THE LOCAL SCALE *Salt for Grain and Beans*

The Dolpo-pa people are yak herders and caravan traders who live in the high, arid part of Nepal. In that difficult environment, they can produce only enough barley and corn to feed themselves for half a year. Through trade, the Dolpo-pa parlay a half-year's supply of grain into enough food to feed them for a whole year. At the end of the summer harvest, they load a portion of their grain onto yaks and head for Tibet. There, they trade grain to Tibetan nomads in return for salt, a commodity in short supply in Nepal. They leave some of the salt in their village for their own use; the rest they carry on to Rong-pa villages in the central Nepal foothills. Then they trade this salt for enough grain to last them through the winter. Bargaining is fierce, and it can take days before a price is agreed upon; prices average 3 to 5 measures of corn for 1 measure of salt. The trading season ends in November, so the Dolpo-pa usually winter over in a convenient field, paying for the privilege.

The Rong-pa people are sheep and goat herders, and salt is a necessary nutrient for both them and their animals. In recent years, because of salt shortages in Tibet, the Dolpo-pa have not brought enough salt to meet all the Rong-pa's needs. The Rong-pa, therefore, load goats and sheep with bags of red beans and set out for Bhotechaur, where they meet Indian traders at a large bazaar and trade their beans for iodized Indian salt; a good price is 1 measure of beans for 3 measures of salt. Often they will also sell a sheep or two to buy cloth, or perhaps a copper pan, to take home.

Source: Adapted from Eric Valli, "Himalayan caravans," *National Geographic* (December 1993): 5–35.

The Dolpo-pa people trading with Tibetan nomads. [Eric Valli.]

shows the mountainous zone running from west to east through the northern borderlands of India and continuing through Nepal and Bhutan. Notice that the country of India actually extends east in a narrow corridor running between Bhutan to the north and Bangladesh to the south and then balloons out in a far eastern lobe. Physically, this mostly mountainous region grades from relatively wet in the east to dry in the west, because the main monsoons move up the Bay of Bengal and strike the east first and hardest (see Figure 8.2, page 399).

Life in this region is dominated by the spectacular mountain landscape. This strip of Himalayan territory can be viewed as having three zones: (1) the high Himalayas, (2) the foothills and lower mountains to the south, and (3) a narrow strip of southern lowlands running along the base of the mountains. This band of lowlands is actually the northern fringe of the Indus and Ganga river plains in the west and the narrow Assam Valley of the Brahmaputra River in the east. Although some people manage to live in the high Himalayas, most of the population lives in the foothills, where they cultivate subsistence strips and terraces in the valleys or herd sheep and cattle on the hills. This area is so rural in character that the Bhutan capital city of Thimphu, with just 27,000 inhabitants, is the largest town in that country; the capital city of Nepal, Kathmandu, has just 52,000 people.

Culturally, the region is Muslim in the west, Hindu and Buddhist in the middle; indigenous beliefs are important throughout but are especially strong in the far eastern part. Throughout the region, but especially in valleys in the high mountains and foothills, indigenous cultures continue to live in traditional ways, isolated from daily contact with the broader culture. There are many of these groups in such places as the Indian state of Arunachal Pradesh, at the eastern end of the region; there, a population of less than 1 million speaks more than 50 languages. Despite language differences, the Himalayan people have learned to survive in the difficult mountain habitat by relying on one another. An example of this reciprocity between cultures comes from central Nepal, where two indigenous groups engage in a complicated cycle of trade that links them with both Tibet and India (see the box "Salt for Grain and Beans").

Most people in the Himalayan mountain region are very poor. Statistics for the various Indian states in this region hover near those of Nepal, which has an annual per capita GDP of just U.S. $1157. In Nepal, the average life expectancy is 57, literacy is about 40 percent, and some 65 percent of the children under three years of age are malnourished. At present, only 18 percent of the land is cultivated, and the country's high altitude and convoluted topography make agricultural expansion unlikely. The government's strategy, therefore, is to improve crop productivity and lower the population growth rate.

The Indian state of Arunachal Pradesh is one of the region's more prosperous areas. Moist air flowing north from the Bay of Bengal brings plentiful rain as it lifts over the mountains. This is one of the most pristine regions in India. Forest cover is abundant, and a dazzling array of flora and fauna occupies habitats in descending elevations: glacial terrain, alpine meadows, subtropical mountain forests, and fertile floodplains. Conditions at different altitudes are so good for cultivation that lemons, oranges, cherries, peaches, and a variety of plants native to South America—pineapple, papaya, guava, beans, maize, and potatoes—are now grown commercially for shipment to upscale specialty stores in Indian cities.

NORTHWEST INDIA

Northwest India stretches almost a thousand miles from the Punjab-Rajasthan border with Pakistan to somewhat east of the famous Hindu holy city of Varanasi on the Ganga (Figure 8.19). It is dry country, yet contains some of the wealthiest and most fertile areas in India.

In the western part of this subregion, there is so little rainfall that houses can safely be made of mud with flat roofs. Widely spaced cedars and oaks are the only trees, and the landscape has a dusty khaki color. Yet filling the landscape between the trees are fields of barley and wheat, potatoes, and sugarcane, plowed by villagers using oxen and humpbacked cattle. Along the northern reaches of this region (Punjab and Uttar Pradesh), the rivers descending from the Himalayas compensate for the deficient rainfall. The most important is the Ganga; its many tributaries flow east and water the whole of Uttar Pradesh, bringing not only moisture but fresh soil from the mountains.

The western half of the region contains one of India's poorest states—Rajasthan—as well as its wealthiest—the agriculturally productive Indian part of Punjab. Rajasthan, with only a few fertile valleys, is dominated by the Thar (Great Indian) Desert in the west, which covers more than a third of the state. Historically, small kingdoms were established wherever water could be contained (Rajasthan means "land of kings"). Seminomadic herders of goats and camels still cross the desert with their animals. Perhaps the best known are the Rabari. Originally a caste of camel herders and dung gatherers, today the Rabari, about 250,000 strong, continue their annual migrations during the dry season in search of green pastures. As they travel, they sell or trade animal dung for grazing rights, thus keeping farmers' fields fertile and their fires burning. By the 1990s, however, there were more and more small farmers occupying former pasturelands. Although these farmers want the dung, they cannot afford to lose one bit of greenery to the passing herds, so increasing population pressure means the benefits of reciprocity between herders and farmers are being lost.

In this arid state, less than 1 percent of the land is arable, yet agriculture, poor as it is, produces 50 percent of the state's domestic product. It is not surprising that the annual per capita GDP is just U.S. $500. Crops include rice, barley, wheat, oilseeds, peas

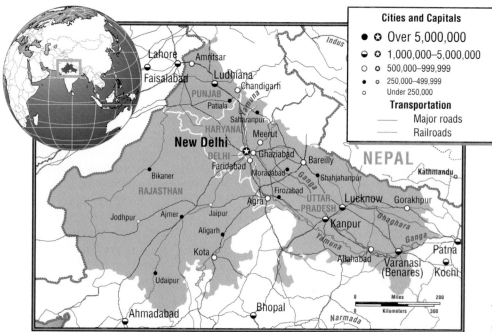

Figure 8.19 The Northwest India subregion.

In the Rabari village of Bhopavand, near the border of Rajasthan and Gujarat, camels are taken out to graze nearby. Many Rabari still lead a partially nomadic life, traveling northward through Rajasthan and across the Thar Desert in the dry season. [Dilip Mehta/Contact Press Images.]

and beans, cotton, and tobacco. A thriving tourist industry, focused on the exotic palaces and fortresses constructed by warrior princes of the past who fought off invaders from Central Asia, accounts for much of the other half of the GDP.

Punjab, one of India's most prolific agricultural states, has an annual per capita GDP of U.S. $1105, double that of Rajasthan. The differences in wealth result partly from differences in physical geography and the introduction of green revolution technology. Rajasthan is a desert; Punjab receives water and fresh soil carried down by rivers from the mountains. Although Rajasthan is about three times the size of Punjab, only 1 percent of its land is cultivated. Nearly 85 percent of Punjab's land is cultivated, and it is particularly productive. In 1994 and 1995, Punjab alone provided 62 percent of India's food reserves. The typical crops are maize, potatoes, sugarcane, peas and beans, onions, and mustard.

Although agriculture is the most important economic activity throughout the region and employs about 75 percent of the people, industry is also important in the city-state of Delhi and on the Ganga River plain. Here, and elsewhere in the region, typical industries are sugar refining and the manufacture of cotton cloth and yarn, cement, and glass. People also make craft items and hand-knotted wool carpets. Urban areas have more modern forms of industry. For example, in the city of Chandigarh, there are 15 medium-sized and large-scale industrial facilities that produce electronic and biomedical equipment, household appliances, tractor parts, and cement tiles and pipes.

The city-state of Delhi, with about eight times the population of Washington, D.C., is located approximately in the center of the region and is the site of New Delhi, India's capital. Built by the British in 1931 on the remains of eight ancient cities, it has all the monumental hallmarks of an imperial capital—and all the problems one might expect of a big city in a country as poor as India. This city of more than 14 million people (in 1999) attracts a continuing stream of migrants. Most of the new arrivals have left nearby states to escape conflict or poverty, or both. The city is also home to refugees who fled Tibet to escape Chinese oppression and to those who fled the Afghan-Soviet war of the 1980s. Annual per capita GDP in Delhi (U.S. $1335) is below average for the country, but actual incomes for most people are lower still; Delhi's tiny minority of the very wealthy pulls up the average. The low literacy rate for an urban area, just 75 percent, is partly a result of the continual arrival of migrants from poor rural areas, but Delhi also has far fewer schools than it needs for its population. In fact, the city has difficulty providing even the most basic services: there are not sufficient water, power, or sewer facilities, and 75 percent of the city structures violate local building standards. Many people have no buildings to inhabit at all; like many of the poor worldwide, they live in shanties constructed from refuse.

Pollution is a particular problem: Delhi has been designated the world's fourth most polluted metropolitan area. The annual death rate from pollution-related illness has reached 7500. One estimate is that on any given day, each citizen inhales the toxic equivalent of 20 cigarettes. Most of the pollution comes from the more than 3 million unregulated motor vehicles. Taxis, trucks, buses, motorized rickshaws, and scooters, most without pollution control devices, all compete for space, for cargo, and for passengers.

Despite this grim picture, life in Delhi is vibrant and upbeat, in part because the middle class is growing and many of the poor sense there is a chance for upward mobility. At 250 million people, the Indian middle class is nearly as large as the total U.S. population; many live in Delhi. The global market economy, evident across the nation, is especially visible in cities like Delhi, where clubs, department stores, boutiques, cinemas, and video parlors share space with traditional businesses; and such brand names as McDonald's, Louis Vuitton, Nike, Baskin-Robbins, and Compaq are common.

NORTHEASTERN SOUTH ASIA

Northeastern South Asia, in strong contrast to Northwest India, has a wet tropical climate. This subregion bridges national and state boundaries, encompassing the states of Bihar, Jharkhand, and West Bengal in India; the country of Bangladesh; and the far eastern provinces of India—all clustered at the north end of the Bay of Bengal (Figure 8.20). Their dominant shared features are two rivers, the Ganga and the Brahmaputra, and the giant drainage basin and delta region created by these rivers. The wet climate and fertile land have nourished a population that is now among the densest on earth, often struggling to support itself in the overcrowded conditions.

The Ganga-Brahmaputra Delta is the largest on earth. Every year, the two rivers deposit enormous quantities of silt, building up

Figure 8.20 The Northeastern South Asia subregion.

the fluid landmass so that the delta extends farther and farther out into the bay. The rivulets of the bay repeatedly change course, until the bay is periodically flushed out by a huge tropical storm (a cyclone in 1991 left 130,000 delta dwellers dead). The people of the delta have learned never to regard their land as permanent (see Figure 8.4). Their dwellings, means of transport, and livelihoods are adapted to drastic seasonal changes in water level and shifting deposits of silt. Villages on river terraces are clusters of houses on mounds, surrounded by lush green forests and fields. The houses are intricately woven from bamboo and palm material, with sloping thatched roofs able to withstand heavy rainfall. In the lowlands, the houses are usually raised on posts above the high-water line. Transport is by small boats; people fish during the wet season and farm when the land emerges from floods. Even beyond the delta, most people have to cope with flooding at some time of the year, either when the summer monsoon rains come or, in the foothills of the Himalayas, when the spring snowmelt in the mountains swells rivers beyond their banks.

The Case of West Bengal

With a population of over 68 million packed into an area slightly larger than Maine, West Bengal is India's most densely occupied state: 1982 people per square mile (765 people per square kilometer). Its population has been swelled by refugees from eastern Bengal (now in Bangladesh) since partition in 1947, and even more immigrants have come in since the Chinese takeover of Tibet and the Pakistani civil war that gave Bangladesh its independence. Today, better employment opportunities in West Bengal and the possibility of farming in the eastern Indian states draw a continual flow of Bangladeshi migrants (most of them illegal). The result is a mix of cultures that is frequently at odds: there are often Hindu and Muslim religious demonstrations and disputes with indigenous people over land occupied by new migrants.

Nearly 75 percent of the people in this crowded state earn a living in agriculture, yet agriculture is not a major contributor to West Bengal's GDP, accounting for only 35 percent of domestic production. In addition to growing food for their own consumption, many people work as rice and jute cultivators or tea pickers, all labor-intensive but low-paying jobs. Twenty-five percent of India's tea comes from West Bengal—the plant is grown in the far north of the state around Darjeeling, a name well known to tea drinkers. Each leaf must be selected and picked by hand, and it is often women and children who do this work. One result of a dense population dependent on agriculture is that continuous intensive cultivation often overstresses the land severely, and soil fertility declines over time. Also, surrounding woodlands are depleted by impoverished farm laborers who must gather firewood because they cannot afford kerosene.

Calcutta, one of the most famous cities on earth, is in the delta region of West Bengal. This giant city of 12 million people is known for its legendary beggars, Mother Teresa's ministrations to the poor at Nirmal Hriday (Home for Dying Destitutes), and opulent but crumbling marble palaces. Calcutta was built on a swampy riverbank in 1690 and served as the first capital of British India. But its sumptuous colonial-built environment, substantially upgraded in the nineteenth century, has become lost in squatters' settlements that surround the city and have invaded its parks and boulevards. It is often characterized as a city in such a state of decline that nothing can be done to improve conditions. The reasons for Calcutta's decline are economic and social, brought on by massive immigration since 1947. Also, outmoded regulations limit incentives to start businesses and to improve private property. For example, apartment owners have little reason to make improvements when, under rent controls presently in place, a Calcuttan can pay as little as U.S. $1.40 a month for a four-room apartment and pass the lease down to the next generation. On the other hand, for those only precariously in the middle class, rent hikes would be disastrous. Such conundrums produce an inertia in governance that the city's educated young people find stifling, and many are leaving Calcutta—a situation that makes it difficult to revive the city.

Somraj Kundu, a 20-year-old who has been accepted for graduate studies at Oxford University in England, identifies four reasons he is leaving Calcutta: the breakdown of traditional extended families; the much too rapid increase in population as immigrants pour in; corruption at many levels (one can bribe an official to get a passport or a professor to avoid receiving a low grade in a course); and, perhaps most important, the lack of opportunity for economic advancement for him and others like him.

Eastern India

Eastern India has two physically distinct regions: the river valley of the Brahmaputra as it descends from the Himalayas, and the mountainous uplands stretching south of the river between Burma on the east and Bangladesh on the west. Although migrants have recently arrived from across South Asia, traditionally this region has been occupied by ancient indigenous groups that are related to the hill people of Burma, Tibet, and China.

The Indian state of Assam encompasses the river valley of the Brahmaputra, eastern India's most populated and most productive area. Hindu Assamese are two-thirds of the population of 29 million, and indigenous Tibeto-Burmese ethnic groups make up another 16 percent. The rest are recent migrants. To reduce the proportion and influence of the Assamese people, who have continually objected to being under central Indian control, the Indian government has made large tracts of land available to outsiders: for example, Bengali Muslim refugees fleeing the civil war in East Pakistan (now Bangladesh) in 1971, Nepali dairy herders, and Sikh merchants. There were violent disputes in the late 1970s and early 1980s between Assamese and the new ethnic settlers and between Assam and India; the disputes continue, even though the violence has lessened.

More than half the people in Assam work in agriculture by growing and producing food; another 10 percent are employed on tea plantations or in forestry (forests cover about 25 percent of the

This street scene in Calcutta shows a small marketplace and several modes of transportation. Relics of European influence are visible in the architecture. Note the mosque in the background. [Steve McCurry/ National Geographic Image Collection.]

land area). Two-thirds of the cultivated land is in rice; but tea is the main cash crop—Assam produces half of India's tea. By the 1990s, Assam's oil and natural gas accounted for more than half that produced in all of India. Given India's shortage of energy, this alone could explain India's efforts to dominate the Assamese politically with a flood of settlers from outside.

Colorful names such as "Land of Jewels" and "Abode of the Clouds" convey the exotic beauty of the emerald valleys, blue lakes, dense forests, carpets of flowers, and undulating azure hills in the mountainous sections of eastern India that surround Assam. In these uplands, occupied largely by indigenous ethnic groups, people produce primarily for their own consumption, with 80 percent or more of the inhabitants making a living from the land. Many practice a particularly complex version of traditional tropical horticulture specifically adapted to the region; others cultivate rice on permanently terraced fields.

Although literacy rates are not high in most of the region, the state of Mizoram ranks second in India, at 82 percent, because of the influence of Christian missionary schools.

Bangladesh

The American James Novak, who has lived and worked in Bangladesh, writes:

Bangladesh is not so much a land upon water as water upon a land. One-third of Bangladesh's physical space of fifty-five thousand square miles is comprised of water in the dry season, while in the rainy season up to 70 percent is submerged. Water is the central reality of Bangladesh, just as its shortage is the central reality of Saudi Arabia. At least 10 percent of the people live in boats, up to 40 percent depend on the sea and rivers for a livelihood, and 100 percent depend on rain and floods for food. Water is the main source of protein [fish], the major provider of crop fertilizer and transport, and unquestionably the greatest source of wealth. Bangladesh's main crops—rice, jute, and tea—cannot exist without huge amounts of water.

(James J. Novak, Bangladesh: Reflections on the Water, 1993, pp. 22–23.)

Today, Bangladesh is one of South Asia's poorest countries; only Bhutan and Nepal rank lower on most indexes. It is also one of the region's most populous democracies and the world's most densely populated agricultural nation. More than 122 million people live in an area slightly smaller than Alabama. Population density is 2432 people per square mile (939 per square kilometer)—compared to 75 people per square mile (29 per square kilometer) in the United States—yet all but 18 percent of the people live in rural areas, trying to manage as farmers on a severely overcrowded land. Some 50 million people live below the poverty line, meaning that their daily caloric intake is below 2122, the minimum standard for adults.

Although desperately poor, the country is better off today than it was just a few years ago. The percentage of rural people living in

Called *nodi bhanga lok* ("people of the broken river"), the people who occupy the constantly shifting silt of the Ganga Delta region are looked down upon by more permanent settlers on slightly higher ground. Because they must often flee rising floodwaters, they are known to be less secure financially and are thought to lack the qualities of thrift and good citizenship that come from living in one place for a lifetime. The delta floods come from the Brahmaputra and from storms that sweep up the Bay of Bengal. [James Blair/National Geographic Image Collection.]

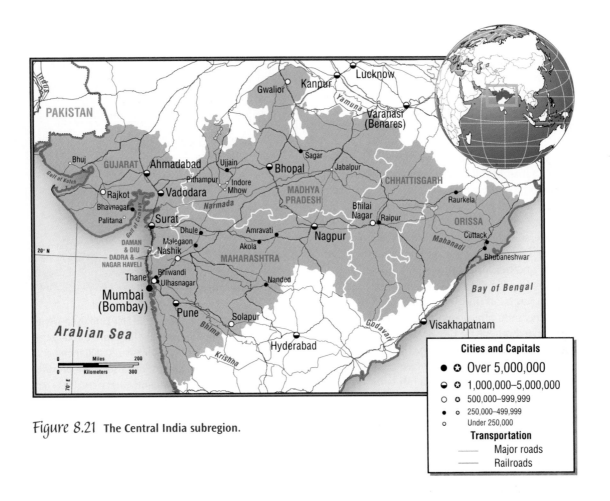

Figure 8.21 **The Central India subregion.**

poverty dropped 9 percent in the years from 1989 to 1994, according to the U.S. Agency for International Development. Higher literacy and the availability of contraceptives (50 percent of women use some form of birth control), both funded primarily by foreign aid, have brought a significant reduction in fertility rates, from 7 children per woman in 1974 to 3.3 in 2000. Infant mortality has also dropped: from 128 per 1000 in 1986 to 82 per 1000 in 2000. Economically, there are signs that the textile industry, once the source of considerable wealth in Bangladesh, is reviving. Bangladesh now ships more than U.S. $2 billion worth of garments to the United States and Europe annually. Although Bangladesh remains in distress, there is no reason to conclude that further progress is not possible. In fact, Bangladeshi innovation in microcredit is contributing to progress among the poor in many countries through the model provided by its Grameen Bank (discussed on page 424).

CENTRAL INDIA

The Central India subregion stretches across the widest part of India, from Gujarat in the west to Orissa in the east (Figure 8.21). It contains both India's last untouched natural areas and much of its industrialized area. The Narmada River, site of many hydroelectric dams (see pages 428–429), flows across the region and empties into the Gulf of Cambay.

Some of India's most significant environmental battles are being fought in the central highlands of this subregion. Central India has most of India's remaining forest cover and a concentration of national parks and sanctuaries (most notably the tiger reserve). Yet already the forest cover is merely patchy, not continuous; areas between patches are densely occupied by people in dispersed rural villages. The isolation of small populations of wild plants or animals in forest patches weakens the gene pools, and extinction becomes inevitable, even if somewhat delayed by protection. In recognition of this problem, there are now plans to reconstitute corridors between parks. Estimates of future human population growth, however, do not bode well for the future of wildlife anywhere in this subregion.

Central India is notable for its several industrial areas (see Figure 8.14). The state of Gujarat in the west, on the Arabian Sea, has light industries in all of its major cities. Gujarat's service and industrial sectors account for 71 percent of its domestic production; the rate is even higher, at 77 percent, in the neighboring state of Maharashtra. Even the central plateau and eastern areas of Central India, although more rural in character, have pockets of industrial activity. An example is the city of Indore, a commercial and industrial hub whose residents think of it as a mini-Mumbai. Although currently in decline, cotton textiles remain the city's main product. Nearby, the town of Pithampur, known as India's Detroit, houses several automobile plants, a steel plant, a container plant, and small appliance factories. Industry and urbanization are clearly connected, and in fact the western part of the region is exceptionally urbanized for India. In Gujarat, more than 35 percent of the people live in urban areas; in Maharashtra, the location of the megacity Mumbai, the figure is about 40 percent.

Even though this region might be considered newly industrializing, it too can lose industries if it fails to compete effectively in the mobile global economy. For example, the city of Ahmadabad in Gujarat, with 4.8 million people, used to be known as the Manchester of India because of its large textile mills. Although the Arvind denim mills are still in operation, most of the old mills are closed today, having lost out to newer mills and cheaper labor elsewhere in Asia. But another aspect of globalization may help Gujarat keep its economic edge. Many Gujarati live abroad in a worldwide diaspora of merchants and businesspeople who maintain ties with their homeland. The diaspora could prove to be a major source of foreign investment for Gujarat.

Bombay is the name by which most people know Maharashtra's capital, but since 1995 its official name has been Mumbai, after the Hindu goddess Mumba. In the sixteenth century, when a local sultan gave the bay, with its seven small islands, to the Portuguese, it became known as Bom Bahia, or "Beautiful Bay." Eventually the British joined the islands with bridges and landfill and built the largest deepwater harbor on India's west coast. Mumbai, with 13 million people, is now India's most prosperous city. It hosts India's largest stock exchange and the nation's central bank. It pays about a third of the taxes collected in the entire country and brings in nearly 40 percent of India's trade revenue;

In Ahmadabad, a member of the Self-Employed Women's Association signs a withdrawal slip with her fingerprint because she is not able to read or write. This trade union has 220,000 members. In addition to helping set up bank accounts, the association provides child care, job training, legal assistance, and other services. [Steve McCurry/National Geographic Image Collection.]

its annual per capita GDP is three times that of Delhi. Yet the wealth is not readily visible. By some estimates, lack of housing has reduced half the population to living on the streets. Real estate developers have a booming trade in high-rise condominiums

One day during a recent visit to India, Lydia and Alex Pulsipher visited Koli, a fishing village located on the bay in the heart of Mumbai. At sunset, the fishing crews often retire to their boats to celebrate the day's events. The Koli were the fisherfolk living on the seven islands in 1600 when the city was founded. The village is now a crowded low-rise labyrinth between a busy urban thoroughfare and the bay. Many people in the village work at bureaucratic jobs in the city. Although the village had the look of a shantytown and was one of the more densely occupied places we visited, its 400 years on the site and the relative affluence of the inhabitants gave the village an air of permanence. More than a few living quarters boasted TVs and white marble floors, and we saw an occasional computer. [M. M. Navalkar/Dinodia.]

built for the city's rapidly growing middle class on land reclaimed from the sea.

Mumbai is also home to Asia's largest slum. The community of Dharavi houses more than 600,000 people on less than 1 square mile, and most people in Dharavi have no plumbing. People work at three or four jobs, and the community is known for its inventive entrepreneurs. One young man, for example, collects and sells aluminum cans that once held ghee, India's form of butterfat. He says he makes about 15,000 rupees a month (U.S. $480), nearly twice that of the average college professor in India and much more than he made as a truck driver. Hence, despite widespread poverty and ethnic and religious tensions, Mumbai has more than a few success stories.

Mumbai is known popularly in India as "Bollywood" because it produces popular Hindi movies portraying love, betrayal, and family conflicts. The stories, played out on lavish sets and accompanied by popular music and dance, serve to distract the huge audiences from the physical difficulties of daily life, at least temporarily. Mumbai produces many more films than Hollywood, on much smaller budgets. The stars make six or more films a year and are so popular that movie posters are everywhere, in public and private spaces alike.

SOUTHERN SOUTH ASIA

Southern South Asia (Figure 8.22) resembles the rest of South Asia in that the majority of people work in agriculture—ranging from about 70 percent in the west to somewhat more than 50 percent in the east. But the region is set apart by its relatively high proportion of well-educated people, its advanced technology sectors, its strong tradition of elected communist state governments (primarily in Kerala), its focus on environmental rehabilitation and preservation, and the higher status of women. The cultural mix here also sets it apart. It is the center of ancient Dravidian cultures and languages that predate Arya influences.

This part of India receives consistent rainfall and is well suited for growing tobacco and rice; also grown are peanuts, chilies, limes, cotton, spices (cinnamon and cloves), and castor-oil plants (the source of an intestinal medicine and a skin lubricant). The southwestern coast of India (the Malabar Coast) is a narrow coastal

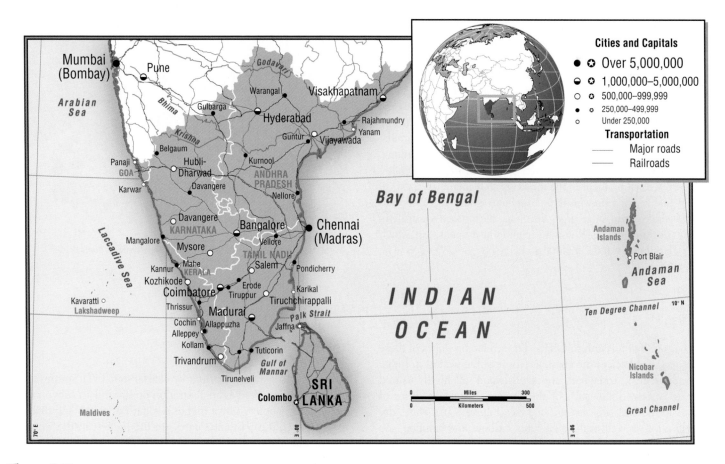

Figure 8.22 The Southern South Asia subregion.

AT THE LOCAL SCALE   *Kerala's Fishing Industry*

The geographer Holly Hapke has studied how recent structural changes in the fishing industry of Kerala have altered family economics among Christian fisherfolk in and around Trivandrum, in the far south. Before independence in 1947, fishing in Kerala was carried out almost entirely on a small scale. Thousands of fishers worked from boats that held crews of 1 to 40 men. The catches were small and consisted of multiple species of fish. Once the boats had returned, the wives of the fishermen sorted and cleaned the fish, then sold them on the beach to wholesalers or in town at the market. Unlike most women elsewhere in South Asia, Kerala fishwives regularly dealt with the public as businesswomen; at the same time, they performed their multiple household and child-rearing tasks.

In the 1950s, the Indian government hoped to improve the incomes of Kerala's fishers by increasing the catch for the local market. It introduced large mechanized vessels to villages north of Trivandrum. Very soon, the project shifted to harvesting prawns for the world market; but mechanization also spread to the fishing economy elsewhere along the coast. By the mid-1970s, the mechanized fleet was encroaching on traditional fishers, damaging their gear, and competing for their catch. The intensification of fishing has led to declining fish stocks and pronounced decreases in income for Trivandrum's fishing villages, which continue to use traditional technology. Men no longer harvest sufficient fish, so the women no longer have fish to clean and sell. Although the men occasionally find wage labor in the mechanized fishing industry, the role of marketing the catch has shifted away from women and into the hands of male fish brokers. At the same time, because the fishermen are no longer bringing in sufficient income, the families have become increasingly dependent on the women's earnings. The women, therefore, have started buying fish from wholesalers and then reselling them in the market or to regular customers, all the while keeping up with their household duties. To compete as fish vendors against the larger operators, the women must themselves deal in larger volumes; this requires investment cash, which they must borrow at the high rates charged by moneylenders. The exorbitant interest depletes the women's earnings, leaving them operating on the margins, with their family economies at risk.

Hapke concludes that state planners ignored women's roles in processing and marketing the fish catch; when the system was reorganized, women were left out entirely. They must now bear more of the responsibility for their family incomes, but they have little access to institutional and societal supports. Despite their considerable entrepreneurial spirit, their situation is likely to worsen as the traditional sector disappears.

Source: Adapted from Holly M. Hapke, *Fish Mongers, Markets, and Mechanization: Gender and the Economic Transformation of an Indian Fishery* (Syracuse, N.Y.: Syracuse University, 1996).

Women sell fish in a Trivandrum neighborhood market. [Holly Hapke.]

---

plain backed by the Western Ghats. The sea-facing slopes of these mountains, just back of the coast, are some of the wettest in India; they support teak, rosewood, and sandalwood, all highly valued furniture woods. Small parts of the Deccan Plateau, a series of uplands to the east, are also forested (see Figure 8.16). Here, dry deciduous forests yield teak, eucalyptus, cashew, and bamboo (see the box "A Visit to the Nilgiri Hills," page 430). Several large rivers and numerous tributaries flow toward the east across this plateau and form rich deltas along the lengthy and fertile coastal plain facing east to the Bay of Bengal.

The Special Case of Kerala

Kerala is a well-watered, primarily coastal state in far southwestern India. It is often cited as an aberration on the Indian subcontinent because its people enjoy a higher standard of well-being than the rest of the country. Literacy rates are the highest in the whole of South Asia, at close to 100 percent (see the discussion of Kerala in the section on population issues, page 409). Women enjoy relatively high status and outnumber men in the population—there are no "missing females" as elsewhere in India. Moreover, women actively participate in the economy and are far less secluded than elsewhere. Standards of health are also better in Kerala.

Just why Kerala stands out in these ways is not entirely understood, but the state has a relatively unique history. Since the 1950s, it has had a series of elected communist governments that have strongly supported broad-based social services, especially education for both males and females. Also, it has a long history of a particular traditional family structure that gives considerable power to women; for example, a husband often resides with his wife's family (a pattern also found in Southeast Asia), instead of the reverse. The concept of female seclusion, while present, is less stringently practiced; women are often on the street, and it is not unusual to see a lone woman going about her shopping duties. Agriculture employs less than half the population; instead, fishing has long figured prominently in the state's economy. This livelihood strategy leaves women in charge of families while men are away at sea, and it gives women even more power because tradition assigns them the jobs of processing and selling the catch (see the box "Kerala's Fishing Industry").

Another feature of life in Kerala that may have added to the openness of society and to greater freedom for women is significant contact with the outside world. Beginning thousands of years ago, traders from around the Indian Ocean, and especially Southeast Asia, made calls along the Malabar Coast, possibly bringing other ideas about gender roles. Then, in the first century A.D., the Apostle Thomas is said to have come to spread Christianity. Eventually Muslims brought their faith by sea, as well. The state's 29.1 million people have a history of cultural and religious pluralism that seems to have led to somewhat more tolerance than exists in neighboring states. Still, some groups, especially fisherfolk, experience a pervasive sort of discrimination. Fishers are one of the very lowest Hindu castes, and there is a broad unwritten assumption that they will always be an underclass.

Sri Lanka

Sri Lanka, known as Ceylon until 1972, is a large island country off India's southeast coast known for its beauty. From the coastal plains, covered with rice paddies and with nut and spice trees, the land rises to hills where tea and coconut plantations are common. At the center is a mountain massif that reaches nearly 8200 feet (2500 meters) at its highest elevation. The southwest monsoons bring abundant moisture to the mountains, giving rise to forests and several unnavigable rivers that produce hydroelectric power. Close to 30 percent of the land is cultivated, and just over 25 percent remains forested. Despite its natural beauty, Sri Lanka has suffered particularly violent civil unrest that grew out of ethnic rivalry, postcolonial politics, and high unemployment.

The original hunter-gatherers and rice cultivators of Sri Lanka, today known as Veddas, number less than 5000. They were joined several thousand years ago by settlers from northern India who built numerous city-kingdoms. Today known as Singhalese, these descendants of northern Indians make up about 74 percent of the population of 18.7 million. The Singhalese brought Buddhism to Sri Lanka, and today 70 percent of Sri Lankans (most of them Singhalese) are Buddhist.

Jewelry stores line a portion of Sea Street in Colombo, the capital of Sri Lanka. Gemstones have long been one of Sri Lanka's resources, and polishing and setting them is a major industry. Other businesses include textiles and clothing. Tourism would be a substantial industry but for the conflict between Singhalese and Tamils. Britain, Germany, and the United States are Sri Lanka's primary export trading partners. [Steve McCurry/National Geographic Image Collection.]

About 1000 years ago, Dravidian people from southern India, known as Tamils, began migrating to Sri Lanka; by the thirteenth century they had established a Hindu kingdom in the northern part of the island. Later, in the nineteenth century, the British imported large numbers of poor Tamils from Tamil Nadu in southeastern India to work on British-managed tea, coffee, and rubber plantations. Known as "Indian" Tamils, the plantation-based population shares linguistic and religious traditions with the "Sri Lankan" Tamils of the north and east, but each community considers itself distinct because of its divergent historical, political, economic, and social experiences. Together, the two Tamil populations make up about 18 percent of the total population of Sri Lanka. The Sri Lankan Tamils have done well, dominating the commercial sectors of the economy, whereas the Indian Tamils have remained isolated and largely poverty-stricken laborers on interior plantations.

Sri Lanka once had a thriving economy, led by a vibrant agricultural sector, and a government that made significant investments in health care and education. It was thought to be poised to become one of Asia's most developed economies. But conditions in rural areas took a drastic turn for the worse in the 1960s. Declining prices for the chief agricultural exports of tea, rubber, and coconuts prompted the government to shift investment away from rural development and toward urban manufacturing and textile industries, dominated by Singhalese. This shift exacerbated a political conflict that already existed because the legal status of the plantation laborers imported by the British had not been guaranteed at the time that the British withdrew as a colonial government. The Singhalese alienated all of Sri Lanka's Tamils by calling into question the status of the Indian Tamils as citizens of Sri Lanka and threatening to bar them from participation in elections. The Singhalese also claimed that only Singhala was an official language, not Tamil or English. Moreover, Buddhist Singhalese were privileged in obtaining university admissions, job opportunities, and other social services. In news stories protests against this favoritism by Tamil plantation workers have been confused with brutal acts of terrorism by guerrilla forces, known as the Tamil Tigers, operating mostly in Sri Lanka's far northeast. In 1987, Indian troops intervened at the request of the Sri Lankan government, but that action failed to end the violence.

Meanwhile, resentment grew in southern Sri Lanka in response to the government's lack of attention to growing poverty and corruption. In 1989, Sri Lanka appeared to be near collapse as the result of an insurrection by southern Singhalese guerrillas and the withdrawal of the Indian forces in the north. Armed conflict flared repeatedly throughout the 1990s, and the civil war has produced more than a million refugees and severely impeded Sri Lanka's economic development. Urban industrial development has lagged despite efforts to create a free-trade zone to attract industries. Repeated episodes of terrorist bombings have frightened off international investors. As the conflict continues in the first decade of the new millennium, security remains tight on major highways and at the airport.

REFLECTIONS ON SOUTH ASIA

Conflicting images of life in South Asia abound. The writer Elisabeth Bumiller, who has written about women in India, made a comment that could be applied to the entire region. She suggested that any one statement made about a region as complex as South Asia can be matched by an opposite statement that is equally true. For example, South Asia has some of the largest and most crowded cities on earth, yet more than 70 percent of the population lives in rural villages. Poverty is endemic and the gaps between rich and poor are widening, but South Asia is also the home of one of the most potentially empowering and far-reaching development strategies ever conceived—microcredit. Although telephones are missing from most homes, information technology is flourishing. And while religious conflict between Muslims and Hindus threatens to precipitate nuclear war between Pakistan and India, the region is a mecca for those seeking spiritual enlightenment. It is also the home of the highly effective strategy of nonviolent social resistance, begun by Mohandas Gandhi and adopted by advocates for human rights all over the world, including the American South, South Africa, and Beijing's Tiananmen Square. One more contrast is that India is the world's largest democracy, yet it is also a place where religious intolerance has led to thousands of deaths in the last few years. In South Asian democracies, a person's sex significantly determines whether that person has access to sufficient food, education, income, opportunity, and indeed to life itself. Yet in no other region have more women been elected head of state, and South Asian women are among the most articulate supporters of women's rights in the global forum.

Perhaps one of the most provocative characteristics of South Asia is that it has spawned eloquent and prolific writers who frequently publish their work in English. Many Indian and Bengali writers make the best-seller lists in Europe and North America (see the bibliography on this textbook's Web site: www.whfreeman.com/pulsipher). Most have something to add to the global conversation about poverty, human rights, and development. This literary tradition is exemplified by Bengali Nobel laureate Rabindranath Tagore, who wrote prophetically before his death in 1941 about the need for South Asia to assess carefully its Western-led development paths:

We have for over a century been dragged by the prosperous West behind its chariot, choked by the dust, deafened by the noise, humbled by our own helplessness and overwhelmed by the speed. We agreed to acknowledge that this chariot-drive was progress, and the progress was civilization. If we ever ventured to ask, "progress towards what, and progress for whom," it was considered to be peculiarly and ridiculously Oriental to entertain such [reservations] about the absoluteness of progress. Of late, a voice has come to us to take count not only of the scientific perfection of the chariot but of the depth of the ditches lying in its path.

(RABINDRANATH TAGORE, "CRISIS OF CIVILIZATION,"
IN COLLECTED WORKS OF RABINDRANATH TAGORE, VOL. 18.)

It took more than three decades for Tagore's critique of development policy to gain wider acceptance. The "progress for whom" question is only now being asked in the highest halls of policy formation at the World Bank and the United Nations. Strategies for aiming development at the poorest, rather than at those who are already reasonably well off, are finally being invented; and as we have seen, South Asians are some of the most innovative creators of these strategies. Readers have only to check the newspapers in their towns and cities to discover small groups of cooperative borrowers meeting regularly to support and encourage one another in entrepreneurial ventures, emulating their million-plus counterparts in the Grameen Bank in Bangladesh, where microcredit got its start.

At the close of a chapter about a region that is arguably one of the poorest on earth, it is illuminating to note that this region is also a leader in inventive ideas for development as well as in provocative thought about the present trajectory of human society.

Thinking Critically About Selected Concepts

1.   About 75 percent of the people of South Asia live in one of the hundreds of thousands of villages in this region. *What are the strengths of these rural communities? Why might South Asian governments be looking for ways to interest their people in staying in rural areas? What changes will be necessary to achieve this goal?*

2.   South Asia has experienced many cultural influences from outside the region: some, such as Islam, were brought by invaders; others, such as Hinduism, Buddhism, Sikhism, and Jainism, were the result of indigenous development. *How does this cultural complex affect landscapes, daily life, and political issues in modern South Asia?*

3.   British colonial rule had a pervasive influence on South Asia, and the effects can still be found in many aspects of South Asian life. *What do you see as the most important consequences of British colonization in India? What is some of the evidence that colonization negatively impacted South Asian indigenous industries and agriculture?*

4.   Caste is a custom associated primarily with Hindu India, but ideas connected with caste have been absorbed into Muslim and Buddhist culture, too. Caste is declining as a force in everyday life wherever people are able to separate themselves from their ancestral villages, occupations, and caste-related dialects. *What arguments could you give against the idea that caste is likely to disappear entirely?*

5.   The overall status of women in South Asia is extraordinarily low, as measured by education, life expectancy, and access to jobs. *What facts in* each of these categories would you use to illustrate the low status of women? What facts would you use to argue that women's status is not so bad, or at least to support the idea that change is on the way?*

6.   Agriculture in South Asia has an ancient heritage and is the occupation of the majority of citizens to this day. Despite its importance to the economy and some recent scientific advances, South Asian agriculture is not as productive as it might be or as it needs to be to feed the region's hungry. *If you were to argue for agricultural change in India, how would you characterize the present system? What features of that system would you work on first? What position would you take on the green revolution and the advantages and disadvantages of making it the centerpiece of agricultural reform?*

7.   Microcredit, one of the most promising strategies for helping poor people, was developed in South Asia and has spread around the world. *Make a list of the needs that microcredit can address and discuss its strengths as a development strategy. What are some of the limitations you can see on the potential of microcredit to solve South Asia's problems? Can you think of situations in which it might be considered socially disruptive?*

8.   South Asian countries have influenced and continue to influence European and American ways of life. *Make a list of all the ways you can think of that this influence is felt by the ordinary college student. Allow your mind to range freely, reflecting on both material and nonmaterial culture.*

Key Terms

agroecology (p. 420) the practice of traditional methods of crop fertilization and the use of natural predators to control pests

Arya (p. 401) an ancient people of Southwest Asia; the first recorded invaders of South Asia

Brahmins (p. 415) members of the priestly caste, which is the most privileged caste in ritual status

brain drain (p. 402) the flight of the best and brightest South Asians to wealthier regions

bride price (p. 417) a price paid by a groom to the family of the bride; the opposite of dowry

Buddhism (p. 413) a religion of Asia that originated in northern India in the sixth century B.C. and emphasizes modern living and peaceful self-reflection leading to enlightenment

caste (p. 414) an ancient system for dividing society into hereditary hierarchical classes

Chipko (social forestries) **movement** (p. 428) an Indian environmental movement that attempts to slow down deforestation and increase ecological awareness

communal conflict (p. 414) a euphemism for religion-based violence in South Asia

dowry (p. 417) a price paid by the family of a bride to the groom; the opposite of bride price

fatwa (p. 417) a legal opinion based on an interpretation of Muslim law

green revolution (p. 420) increases in food production brought about through the use of new seeds, fertilizers, mechanized equipment, irrigation, pesticides, and herbicides

Harijans (or Dalits or untouchables) (p. 415) members of a social group considered so lowly as to have no caste

hearth (p. 413) place of origin (of Hinduism)

Hindustani (p. 403) during Mogul rule, the lingua franca (language of trade) of all of northern India and what is today Pakistan; the precursor of modern Urdu and Hindi

Indian diaspora (p. 402) the set of all people of South Asian origin living (and often born) abroad

Indus Valley civilization (or Harappa culture) (p. 401) the first substantial settled agricultural communities in South Asia, which appeared around 4500 years ago along the Indus River in modern-day Pakistan and along the Saraswati River in modern-day India

Islam (p. 413) a monotheistic religion considered an outgrowth of Judaism and Christianity; it emerged in the seventh century A.D. when, according to tradition, the archangel Gabriel revealed the tenets of the religion to the Prophet Muhammad

Jainism (p. 413) a faith tradition that is more than 2000 years old; Jains are found mainly in western India and large urban centers and are known for their strict vegetarianism

jati (p. 414) the subcaste into which a person is born, which largely defines the individual's experience for a lifetime

Kshatriyas (p. 415) members of the warrior and ruler caste

microcredit (p. 423) a program that makes very small loans available to poor entrepreneurs

Moguls (p. 403) Muslim invaders of South Asia

monsoons (p. 398) opposing winter and summer patterns of air and moisture circulation in South Asia, Southeast Asia, and parts of East Asia; warm, wet air coming in from the Indian Ocean brings copious rainfall during the summer, and cool, dry air moves south and east out of Eurasia in winter

Muslims (p. 413) followers of Islam

Parsis (p. 413) a highly visible religious minority in India's western cities; they are descendants of Persian migrants who did not give up their traditional religion of Zoroastrianism when Iran became Muslim

purdah (pp. 410, 416) the practice of concealing women, especially during their reproductive years, from the eyes of nonfamily men

regional conflict (p. 426) a conflict created by the resistance of a regional ethnic or religious minority to the authority of a national or state government

religious nationalism (p. 425) the belief that a certain religion is strongly connected to a particular territory and that adherents should have political power in that territory

shari'a (p. 434) the Islamic social and penal code

Sikhism (p. 413) a religion of South Asia that combines beliefs of Islam and Hinduism

Sikhs (p. 413) adherents of Sikhism

subcontinent (p. 397) term used to refer to the entire Indian peninsula, including southeastern Pakistan and Bangladesh

Sudras (p. 415) members of the caste of low-status laborers and artisans

Taliban (pp. 416, 434) an archconservative Islamist movement that gained control of the government of Afghanistan in the mid-1990s

Vaishyas (p. 415) members of the landowning farmer and merchant caste

Pronunciation Guide

Agra (AH-gruh)

Ahmadabad (AH-muh-duh-bahd)

Allahabad (ah-luh-huh-BAHD)

Amritsar (ahm-RIHTZ-surr)

Arabian Sea (uh-RAY-bee-uhn)

Arunachal Pradesh (ah-ROON-ah-chahl prah-DAYSH)

Aryan (AIR-ee-uhn)

Assam (ah-SAHM)

Ayodhya (ah-YOHD-yah)

Bangalore (bang-guh-LOHR)

Bangladesh (bahng-gluh-DEHSH)

Benares (beh-NAHR-ehss)

Bengal (behn-GAHL)

Bhopal (boh-PAHL)

Bhopavand (boh-pah-VAHND)

Bhotechaur (boh-teh-CHAWR)

Bhutan (boo-TAHN)

Bihar (bee-HAHR)

Bodhgaya (boh-duh-GAH-yah)

Bom Bahia (bohm bah-EE-ah)

Bombay (bawm-BAY)

Brahmaputra River (brah-mah-POO-truh)

Brahmin (BRAH-mihn)

Ceylon (say-LAWN)

Chamoli (chuhm-OH-lee)

Chandigarh (CHUHN-dih-gurr)

Chennai (chehn-IE)

Chipko Movement (CHIHP-koh)

Colombo (koh-LOHM-boh)

Dalit (DAH-liht)

Darjeeling (dahr-JEE-lihng)

Deccan Plateau (DEHK-uhn)

Delhi (DEHL-ee)

Dhaka (DAHK-uh)

Dharamkot (duh-RUHM-koht)

Dharavi (duh-RAH-vee)

Dolpo-pa (DOHL-poh-pah)

Dravidian (drah-VIHD-ee-uhn)

fatwa (FAHT-wah)

Gandhi, Indira (GAHN-dee, ihn-DEER-ah)

Gandhi, Mohandas (GAHN-dee, moh-HAHN-dahss)

Ganga River (GAHNG-gah)

Ganges River (GAN-jeez)

Ghats Mountains (GAHTSS)

ghee (GEE)

Gujarat (GOO-juh-raht)

Harijan (HAH-ree-jahn)

Haryana (hahr-YAH-nah)

Himachal Pradesh (hih-MAH-chahl prah-DAYSH)

Himalayan Mountains (hih-MAHL-ee-uhn/ hih-muh-LAY-uhn)

Hindi (HIHN-dee)

Hindu (HIHN-doo)

Hindu Kush (HIHN-doo KOOSH [second "oo" as in "book"])

Hindustan (hihn-doo-STAHN)

Indore (ihn-DOHR)

Indus River (IHN-duhss)

Jainism (JYE-nihz-uhm)

Joypur (joy-POOR)

Karachi (kuh-RAH-chee)

Karakoram Mountains (kahr-uh-KOHR-uhm)

Karnataka (kahr-nah-TAH-kah)

Kashmir (kash-MEER)

Kerala (KAIR-uh-luh)

Lahore (luh-HOHR)

longi (lohn-ZEE)

Madras (muh-DRAHSS)

Mahabodhi Temple (mah-hah-BOH-dee)

Maharashtra (mah-hah-RAHSH-trah)

Malabar (MAL-uh-bahr)

Maldives (mawl-DEEVZ)

Meghalaya (mehg-AH-lah-yah)

Mizoram (mee-ZOH-rahm)

Mogul (MOO-guhl)

mujahedeen (moo-jah-hu-DEEN)

Mumbai (moom-BYE)

nankeen (nan-KEEN)

Narmada River (nahr-MAH-duh)

Nehru, Jawaharlal (NAY-roo, jah-wah-HAHR-lahl)

Nepal (neh-PAHL)

New Delhi (DEHL-ee)

Nirmal Hriday (NEER-mahl HREE-day)

nodi bhanga lok (NOH-dee BAHNG-gah LAWK)

Orissa (oh-RIH-sah)

Pakistan (pah-kih-STAHN)

Pamir Mountains (pah-MEER)

panchayati raj (pahn-chah-YAH-teh RAHJ)

Panipur (pahn-ee-POOR)

Pashto (PAHSH-toh)

Pathan (puh-TAHN)

Peshawar (peh-SHAH-wurr)

Pithampur (pee-tahm-POOR)

Punjab (poon-JAHB)

purdah (PURR-duh)

Rabari (rah-BAHR-ee)

Rajasthan (rah-jah-STAHN)

Rong-pa (RAWNG-pah)

salish (SAH-lihsh)

Sanskrit (SAN-skriht)

Shari'a (shah-REE-ah)

Sikh (SEEKeek)

Sikkim (SIHK-ihm)

Singhalese (sihn-ghuh-LEEZ)

Sri Lanka (shree LAHNG-kah)

Sudra (SOO-druh)

Tajik (tah-JEEK)

Taliban (TAHL-ee-bahn)

Tamil (TAH-mihl)

Tamil Nadu (TAH-mihl NAH-doo)

Thar Desert (TAHR)

Thimphu (THIHM-boo)

Trivandrum (trih-VAHN-druhm)

Udaipur (oo-dye-POOR)

Uttar Pradesh (OO-turr prah-DAYSH)

Vaishya (VYE-shyuh)

Varanasi (vuh-RAH-nuh-see)

Vedda (VEHD-uh)

Yamuna River (yah-MOO-nah)

Selected Readings

A set of Selected Readings for Chapter 8, providing ideas for student research, appears on the *World Regional Geography* Web site at www.whfreeman.com/pulsipher.

This satellite image depicts only a small part of Tokyo, Japan. Home to over 28 million people, Tokyo is the world's largest metropolitan settlement. Runways of Tokyo's Haneda Airport, the nation's busiest, are visible in the lower center. [RESTEC, Japan/Science Photo Library/Photo Researchers, Inc.]

East Asia

KAZAKHSTAN

W. Sayan Mts.

Gorno-
Altaysk

Irtysh
Semey
Lanimogor

Kyzyl
Yenisey
Tannu Ola

Balqash

Lake
Balkhash
Saryesik
Atyrau Desert

Uys
Nuur
Uliastay
Har Us
Nuur

Zavhan

Ulungar

Altay Mts.

Junggar Basin

Tashkent

Bishkek
Alma-Ata
Yining
Urumqi

KYRGYZSTAN

UZB.
Kokand
Naryn

Lake
Issyk
Victory Peak
elev. 24,700

Tian Shan
Kuqa

Bosten
Hu

Turfan
Depression

Kongi
Lop
Nor

Yumen

TAJIKISTAN

Kashi
Tarim

Tarim Basin

AFG.

Muztagata
elev. 24,757

Taklimakan
Desert

XINJIANG UYGUR

Qaidam Basin

Gilgit

PAKISTAN
Islamabad

(Godwin-Austen)
elev. 28,250

Yutian

QINGHAI

Srinagar
Leh

Tuotuo

C

XIZANG
(TIBET)

Nu (Salween)

Delhi
Brahmaputra

New Delhi

Xigaze
Lhasa

H i m a l a y a s

NEPAL
Kathmandu

Mt. Everest
elev. 29,028

Gyangze
Thimphu
BHUTAN

Kanpur

Ganga Plain
Ganga

Darjeeling

Brahmaputra

BANGLADESH

Nagpur

INDIA

Dhaka

Calcutta

Mandalay

Eastern Ghats

Arakan Yoma

BURMA

Bay
of
Bengal

90° E

THAIL

Chapter 9: East Asia

VIGNETTE Eighteen-year-old Hong Xiaohui (Hong is her family name) and two friends eagerly signed up when the labor officials came to her remote farming village in Sichuan Province. They had come to hire workers for Mattel's Barbie doll plant in Changan, a factory town between Guangzhou (Canton) and Hong Kong in the southern province of Guangdong.

When the bus left a few days later, it held 100 young women. Most of them had never left their home villages before, but now they were on a 1200-mile (1900-kilometer) trip that would take five days, most of it along bumpy, winding roads. Exhaust fumes and a lack of ventilation would bring constant nausea. Although nearly all the women were short of cash, most had a supply of food provided by their anxious mothers: sugarcane, bags of oranges, salted duck eggs, peanuts, and bottles of orange soda. While chatting with her friends, Xiaohui said, "I've always itched to get away from home. Life there is empty. I want to be independent. I want to learn something so I can come back and change things." These sentiments echo those of migrants everywhere.

On the fifth day, the bus reached Changan. When Xiaohui arrived in 1996, the city's population had grown from 30,000 to 180,000 in less than a decade. The Mattel Barbie doll factory, in the middle of town, is in a modern white building known locally as "Paradise of Girls" because nearly all 3500 employees are female. Within a day, Xiaohui had completed her training, signed a three-year contract, and learned how to put brown or blonde hair into Barbie doll heads with varying skin colors. The pay, 200 yuan a month (about U.S. $24),

was considerably less than the 350 yuan the recruiters had promised. Xiaohui learned by asking around that no one earns more than 400 yuan.

When a reporter inquired about the women's pay, a Mattel official replied, "We're confident that we provide these workers with a better standard of living than they would have without Mattel." [Adapted from Kathy Chen, "Boom-town bound," *The Wall Street Journal* (October 29, 1996): 1, A6.]

Hong Xiaohui's first trip to China's booming southeastern coast illustrates some of the major transitions under way in the vast region of East Asia. She and the other rural villagers from one of China's interior agricultural provinces are among the millions who are making East Asia's coastal cities some of the biggest and most rapidly growing cities on earth. Not only will these young women work for wages that are among the lowest in the global economy, but many will make sacrifices to send remittances home to poor relatives. Although the remittances help, poor rural interior provinces are deprived of the energy, initiative, and leadership of their young adults. Thus, the migration of rural people from the interior to the coast contributes to the growing disparity in wealth between China's poor interior and its booming coastal provinces (Figure 9.1). Hong Xiaohui's story also hints at how gender roles are changing throughout East Asia. Although many of the old restrictions on women persist, increasingly it is young women who are the workers in the factories of multinational firms such as Mattel.

East Asia, home to nearly one-quarter of humanity, is a vast territory that stretches from the Taklimakan Desert in the far west

Themes to Look For Current trends in the economy, migration, population, and environment of East Asia are reflected in the following themes, which appear repeatedly throughout this chapter.

1. Rapid economic growth. Free-market economies, pioneered by Japan, are growing rapidly throughout the region.

2. Influence of ideas from ancient China. Ancient Chinese forms of government and philosophy have influenced the entire region. One can see these influences in the development of economies, in the dominance of bureaucracies, and in gender roles. Yet countries such as Japan and China have incorporated these influences in very different ways.

3. Population concentration on coasts and in lowlands. In all parts of East Asia, the population is unevenly distributed, with the greatest concentrations in coastal areas and other lowlands.

4. Environmental stress. Large, densely settled populations are straining the environment, causing floods, dangerous levels of pollution, and overcrowding.

5. Urban-rural disparities. Throughout the region, there are disparities in wealth and well-being between rural and urban areas. This pattern is especially true in China. The disparities are growing as cities gain special economic concessions and attract millions of rural migrants.

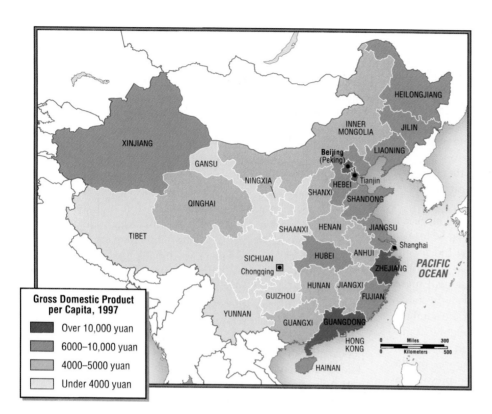

Figure 9.1 **Regional disparity in wealth in China.** U.S. $1.00 = 8.28 yuan (May 18, 1999). [Adapted from *The Economist* (January 16, 1999): 40.]

Gross Domestic Product per Capita, 1997

- Over 10,000 yuan
- 6000–10,000 yuan
- 4000–5000 yuan
- Under 4000 yuan

to Japan's rainy Pacific coastline and from the frigid mountains of Mongolia in the north to the steaming subtropical forests of China's southeastern coastal provinces. The East Asia world region consists of the countries of China, Mongolia, North Korea, South Korea, Japan, and Taiwan. East Asia also includes Hong Kong and its smaller neighbor, Macao; both are now parts of China but previously had long histories as European colonies. Several of the world's most rapidly growing economies of the recent past are in East Asia. China, especially, will provide a huge market for products from other world regions. East Asia is also the most populous world region and is among the most precarious environmentally. Japan has achieved economic success and a high standard of living as a supplier of consumer products to the United States and much of the developed world. Burgeoning rates of consumption in this region will add to local environmental stress and to such problems as the worldwide accumulation of greenhouse gases. For all these reasons, East Asia will figure prominently in global relationships for years to come.

Terms to Be Aware Of

East Asian place names can be unusually complicated, especially those taken from Chinese. Whenever possible, we will give place names in English transliterations of the appropriate Asian language. We also avoid redundancies. For example, *he* and *jiang* are both Chinese words for *river*. Hence the Yellow River (so called because it is yellowed by the heavy sediment load it carries) is the Huang He, and the Yangtze River is the Chang Jiang. It is incorrect to add the term *river* to either name, because it is already there. Pinyin forms of Chinese place names are now commonplace. For example, the city once called Peking in English is Beijing, Canton is Guangzhou, and the region known as Manchuria is here referred to simply as China's Northeast. Also, although China refers to Tibet as Xizang, people around the world who support the idea of Tibetan self-government avoid using the word. This text uses Tibet for the region (with Xizang in parentheses) and Tibetan for the people who live there.

THE GEOGRAPHIC SETTING

PHYSICAL PATTERNS

A quick look at the map of East Asia reveals that its topography is perhaps the most rugged in the world. Powerful tectonic forces have produced a wide range of complex landforms. East Asia's var-

ied climates result from a dynamic interaction of huge warm and cool air masses with the land and sea. Rapidly expanding human populations have affected the variety of ecosystems that evolved over the millennia and that still contain many important and unique habitats.

Landforms

There are few flat surfaces to be found in the rugged landscapes of East Asia. Many of those that do exist are too dry or too cold to be very useful to humans. The large numbers of people who occupy the region have had to be particularly inventive in creating spaces for agriculture. They have cleared and terraced entire mountain ranges using only simple hand tools, irrigated drylands with water from melted snow, drained wetlands using elaborate levees and dams, and applied their complex knowledge of horticulture and animal husbandry to help plants and animals flourish in difficult conditions.

A simple way to visualize the varied landforms of East Asia is to think of them as analogous to the shapes that would be formed in a huge carpet if a grand piano (representing the Indian subcontinent) were shoved deeply into one side of it. The mountain ranges, plateaus, depressions, fissures, and bulges across Eurasia resulted from the slow-motion collision of the Indian-Australian Plate, which carries the Indian subcontinent, with the southern edge of Eurasia (the Eurasian Plate) that began roughly 50 million years ago and continues today. The Himalayan Mountains are the most dramatic result of the collision. The area on the north side of the Himalayas absorbed some of the pressure by bulging upward and spreading outward to the northeast, forming the Tibetan Plateau (also known as the Xizang-Qinghai Plateau or Northern Plateau). It is depicted in rust and lavender on the map at the beginning of the chapter. To the north of the plateau, in Xinjiang, the land responded to the tectonic forces by sinking

In the midst of the Taklimakan Desert, international petroleum company trucks are driving to a remote oil field that has potential reserves far greater than those in the United States. [Reza/National Geographic Image Collection.]

to form basins, including the Qaidam Basin and the sea-level Tarim Basin in the Taklimakan Desert. Farther north, it buckled again to form the mountains on either side of the Dzungar Basin and those of northern Mongolia and southern Siberia. To the east and west of the Himalayan impact zone, wrinkled mountain and valley formations curve away to the southwest and southeast from the Tibetan Plateau.

The landforms of East Asia form four steps. The top step is the Tibetan Plateau. Many of the rivers of China and the Southeast Asian mainland have their headwaters along the eastern rim of this plateau.

The second step down is a broad expanse of basins, plateaus, and low mountain ranges—depicted in deep yellow and light greens on the map at the beginning of this chapter. These include the deep, dry basins and deserts of Xinjiang Province, to the north of the Tibetan Plateau, and the broad, rolling highland grasslands and deserts of the Mongolian Plateau northeast of Xinjiang. East of Xinjiang, this step also includes the upper portions of China's two great river basins, through which flow the Huang He and Chang Jiang. To the south is the rugged Yunnan Plateau. The topography of this plateau is dominated by a system of deeply folded mountains and valleys that bends south through the Southeast Asian peninsula. The middle portions of the Salween (Nu), Mekong, and Red rivers are found here.

The third step, directly east of this upland zone, consists mainly of broad coastal plains and the deltas of China's great rivers (shown in green on the map), with intervening low mountains and hills (shown as yellow patches on the green areas on the map). Starting from the south is a series of three large lowland river basins: the Xi Jiang (Pearl) basin, the massive Chang Jiang basin, and the Huang He lowland basin on the North China Plain. The Far Northeast of China, the Korean Peninsula, and the westernmost parts of southern Siberia are also part of this third step.

The fourth step consists of the continental shelf, covered by the waters of the Yellow Sea, the East China Sea, and the South China Sea. Numerous islands—including Hong Kong, Hainan, and Taiwan—are anchored on this continental shelf; all are part of the Asian landmass. The islands of Japan have a different geological origin: they are volcanic, rather than being part of the continental shelf. These islands lie along a portion of the Pacific Ring of Fire. They rise out of the waters of the northwestern Pacific where the Pacific Plate grinds into the Eurasian Plate. The dramatic volcano Mount Fuji is perhaps Japan's most recognizable symbol. The entire Japanese island chain is particularly vulnerable to disastrous eruptions and earthquakes. Mount Fuji last erupted in 1707. However, in 2001, there were deep internal rumblings, suggesting that the period of dormancy may be ending.

Climate

East Asia has two contrasting climate zones (Figure 9.2): the dry continental west and the monsoon east. Recall from Chapter 1 that *monsoon* refers to the seasonal reversal of surface winds that flow from the continent outward during winter and from the surrounding oceans inward during summer.

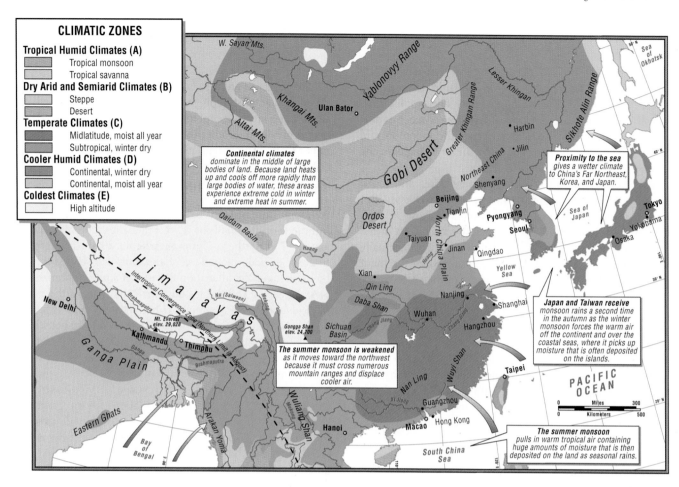

CLIMATIC ZONES

Tropical Humid Climates (A)
- Tropical monsoon
- Tropical savanna

Dry Arid and Semiarid Climates (B)
- Steppe
- Desert

Temperate Climates (C)
- Midlatitude, moist all year
- Subtropical, winter dry

Cooler Humid Climates (D)
- Continental, winter dry
- Continental, moist all year

Coldest Climates (E)
- High altitude

Continental climates dominate in the middle of large bodies of land. Because land heats up and cools off more rapidly than large bodies of water, these areas experience extreme cold in winter and extreme heat in summer.

Proximity to the sea gives a wetter climate to China's Far Northeast, Korea, and Japan.

Japan and Taiwan receive monsoon rains a second time in the autumn as the winter monsoon forces the warm air off the continent and over the coastal seas, where it picks up moisture that is often deposited on the islands.

The summer monsoon is weakened as it moves toward the northwest because it must cross numerous mountain ranges and displace cooler air.

The summer monsoon pulls in warm tropical air containing huge amounts of moisture that is then deposited on the land as seasonal rains.

Figure 9.2 **Climates of East Asia.**

The Dry Interior. The dry western zone lies in the interior of the East Asian continental landmass. Because land heats up and cools off more rapidly than water, locations in the middle of large bodies of land in the midlatitudes tend to experience intense cold in winter and intense heat in summer. The western part of China is an extreme example of a midlatitude continental climate. This part of China includes the Mongolian Plateau, the basins of Xinjiang, and the Tibetan Plateau—all of which are very dry and cold in winter and very hot during the day in summer. Summer nights can be cold because there is too little vegetation or cloud cover to retain the warmth of the sun after nightfall. Heat is lost so rapidly that the difference between summer daytime and nighttime temperatures may be as much as 100°F (55°C).

Grasslands and deserts of several varieties cover most of the land in this dry region. Only scattered forests grow on the few relatively well watered mountain slopes and in protected valleys supplied with water by snowmelt. In all of East Asia, humans and their impact are least conspicuous in the large, uninhabited portions of the deserts of Tibet (Xizang), the Tarim Basin in Xinjiang, and the Mongolian Plateau. The grasslands of Mongolia and the basins of Far Northwest China traditionally supported only scattered tribes of nomadic herders. However, they are increasingly

being put to more intensive uses, such as irrigated agriculture in the Taklimakan and Dzungar basins. In Mongolia, many nomads now live on large-scale, stationary livestock cooperatives, also supported with irrigation.

The Monsoon East. The monsoon climates of the east are influenced by the extremely cold conditions of the huge Eurasian landmass in the winter and the warm maritime temperatures of the surrounding seas and oceans in the summer. In what is commonly called the winter monsoon, dry, frigid arctic air sweeps south and east through East Asia, producing long, bitter winters on the Mongolian Plateau, on the North China Plain, and in China's Northeast. The winter monsoon also causes occasional freezes in southern China. Central and southern China have shorter, less severe winters because they are protected from the advancing arctic air by the east-west ranges of the Qin Ling Mountains and because they lie close to the warm waters of the South China Sea.

In the summer, continental warming pulls in wet tropical air from the Pacific Ocean and its adjacent seas (see Figure 8.1). The warm wet air from the ocean deposits moisture on the land as seasonal rains. As the summer monsoon moves northwest, it must cross numerous mountain ranges and displace cooler air. Consequently, its effect is weakened toward the northwest. Thus, the

Xi Jiang basin in the far southeast is drenched with rain and enjoys warm weather for most of the year, whereas the Chang Jiang basin, which lies in central China to the north of the Nan Ling Mountains, receives only about five months of summer monsoon weather (see the map that opens this chapter). The North China Plain, north of the Qin Ling and Dabie Shan ranges, receives only about three months of monsoon rains. Very little monsoon rain reaches the southern Mongolian Plateau or Inner Mongolia. The Far Northwest gets very little rain.

China's Far Northeast is wet in summer; Korea and Japan have wet climates all year because of their proximity to the sea. But they still have hot summers and cold winters because of their northerly location and exposure to the continental effects of the Eurasian landmass. Japan and the more southerly Taiwan actually receive monsoon rains twice: once in the spring when the main rains move toward the land, and again in autumn as the winter monsoon forces the warm air off the continent and over the coastal seas. The warm air, in retreat, again picks up moisture that is then deposited on the islands. Japan, however, has a much longer and more severe winter than semitropical Taiwan. Much of Japan's autumn precipitation falls as snow.

The eastern areas of the Asian landmass, watered by the seasonal rains, once supported rich forests with diverse plant and animal life, much of it unique to East Asia. The character of the forests changed from north to south: coniferous forests covered the Far Northeast and graded to deciduous broad-leafed forests in Korea, Japan, and the North China Plain. Broad-leafed evergreen forests and even tropical rain forests grew in the south. Over the past two or three millennia, however, agriculture has transformed the landscape. Many lowland ecosystems were wiped out as farmers expanded into any well-watered or irrigable land, be it flat or hilly. Hills and low mountainsides were rarely left covered with undisturbed forest; they were logged continuously, planted in orchards, or completely cleared and terraced for agriculture. Today, the few undisturbed natural areas in the humid zone are remote, deeply convoluted, and increasingly threatened by development.

HUMAN PATTERNS OVER TIME

East Asia is home to some of the most ancient civilizations on earth. Archaeological evidence indicates that settled agricultural societies have flourished in China for over 4000 years. A little more than 2000 years ago, the basic institutions of government that still exist across the region today were established in eastern China. Until the twentieth century, China was the main source of wealth, technology, and culture for the people of East Asia.

The hearth of East Asian culture is thought to be the North China Plain, a rich agricultural area that was unified under successive empires and has come to be regarded as **China proper.** Bordering China proper to the north and west were the lands of inner Asia, inhabited by Mongolian peoples considered alien and uncivilized by the Chinese. The huge, dry Mongolian territories were unsuitable for crop farming but ideal for nomadic animal herding. The nomads, for their part, pitied the Chinese farmers who were tied to a particular place for a lifetime. On East Asia's

eastern fringe, the Korean Peninsula and the islands of Japan and Taiwan were profoundly influenced by the culture of China; nonetheless, they were isolated enough that each developed distinctive cultures and usually maintained political independence. These characteristics proved crucial during the twentieth century, when these areas (except for North Korea) leaped ahead of China economically and militarily, partly by integrating European influences that China disdained.

The Beginnings of Chinese Civilization

Although humans and their ancestors have lived in East Asia for hundreds of thousands of years, the earliest complex civilizations appeared in northern China only about 2000 B.C. Here, agricultural societies developed a feudal system that shaped the political hierarchy and the allocation of land. A small militarized aristocracy controlled vast estates on which the mass of the population worked and lived as impoverished farmers and laborers. The landowners usually owed allegiance to one of the petty kingdoms that dotted northern China. These kingdoms were relatively self-sufficient and well-defended with private armies, and they often proved insubordinate to central authority.

Between 400 and 221 B.C., a new order emerged that eradicated feudalism and laid the foundations for the great Chinese empires that dominated East Asia for more than 2000 years. Following a long period of war between the petty kingdoms of northern China, a single dominant kingdom emerged. What gave this kingdom, known as the Qin dynasty, its advantage was the use of a trained and salaried bureaucracy to extend the monarch's authority into the countryside. The old feudal alliance system was scrapped, and the estates of the aristocracy were divided into small units and sold to the previously semienslaved farmers. The kingdom's agricultural output increased because the people worked harder to farm land they now owned. In addition, the salaried bureaucrats who replaced their former masters were more responsible about building and maintaining levees, reservoirs, and other public works that reduced the threat of floods and other natural disasters. Although the Qin dynasty was short-lived, subsequent empires maintained its bureaucratic ruling methods, which have proved essential to the governing of China as a whole right up to the present.

Confucianism

Closely related to China's bureaucratic ruling tradition is the philosophy of **Confucianism.** Confucius, who lived about 2500 years ago (prior to the Qin dynasty), taught that the best organization for the state and society was a hierarchy modeled after the patriarchal family. In the Chinese extended family, the oldest male held the seat of authority. All other family members were aligned under him according to age and sex. Extending these conventions, Confucianism held that commoners must obey imperial officials, and everyone must obey the emperor. As the supreme human being, the emperor was seen as the source of all order and civilization — in a sense, the grand patriarch of all China.

After unifying China in 211 B.C., Qin Shi Huang became its emperor. One of his many public works, which included China's Great Wall, was the creation of a life-size terra-cotta army of thousands of soldiers in a vast underground chamber that became his mausoleum. Since 1974, archaeologists have been excavating, restoring, and preserving this site, near modern-day Xian, which attracts more than 2 million visitors a year. [Wally McNamee/Woodfin Camp & Associates.]

Over the centuries, this elitist and patriarchal Confucian ideology (not really a religion but certainly a value system) penetrated all aspects of Chinese society, altering its social, economic, and political geography and circumscribing the lives of the masses at the base of the hierarchical pyramid, who were marshaled as a labor force. An early literary interpretation of the ideal woman demonstrates how women were confined to the domestic spaces of the home and almost always placed under the authority of others, where they performed many economic tasks. A student of Confucius wrote: "A woman's duties are to cook the five grains, heat the wine, look after her parents-in-law, make clothes, and that is all! When she is young, she must submit to her parents. After her marriage, she must submit to her husband. When she is widowed, she must submit to her son." The system also rested on the labor of large numbers of low-status men and children, who worked in agriculture and as servants.

Confucian ideology had a lasting impact on China's human geography because the elite had strong, self-interested reasons for maintaining the status quo prescribed by Confucianism, even at the expense of others in society. Elites, for example, claimed the produce raised by impoverished farmers and the wealth produced by artisans and craft workers. To block the emergence of competing powerful segments of society, elites saw to it that merchant wealth was limited wherever possible. In parable and folklore, merchants were characterized as a necessary evil, greedy and disruptive of the social order. For 2000 years, the expropriation of farmers' and artisans' wealth through high taxes and the curtailment of merchants' prosperity left these groups little incentive or investment capital for agricultural improvements, industrialization, or entrepreneurialism.

Although the Confucian bureaucracy allowed empires to expand, its weaknesses resulted in eventual imperial decline (Figure 9.3). The heavy tax burden on farmers led to periodic revolts that weakened imperial control. In addition, invasions by nomadic peoples from what is today Mongolia and western China often dealt the final blow to successive Chinese empires—despite the defensive walls, such as the Great Wall of China, built along China's northern border. So weakened were the Chinese in the 1200s that the Mongolian military leader Genghis Khan and his descendants were able to conquer all of China and then push west as far as Hungary and Poland (see page 510). It was during this empire (Yuan) that the first direct contacts between China and Europe took place.

China's Preeminence

In the tenth century, China proper was the world's most developed region: it had the world's wealthiest economy, its largest cities, and the highest living standards. Improved strains of rice allowed dense farming populations to expand throughout southern China and supported larger urban industrial populations. Metallurgy flourished. As early as 1078, people in northern China were producing twice as much iron as people in England did 700 years later. Nor was innovation lacking: Chinese inventions included papermaking, printing, paper currency, gunpowder, and improved shipbuilding techniques. All these advances contributed to the growth of remarkable urban areas. When the Venetian trader Marco Polo wrote about his travels in China between 1271 and 1295 (during the Yuan, or Mongolian, dynasty), Europeans were stunned to learn that the imperial city of Hangzhou (near modern Shanghai) was 100 miles in circumference; had a population of 900,000; was dotted with parks and served by a sewer system; had a large canal system; and supported several principal markets, each frequented daily by 40,000 to 50,000 people. Wares included rhinoceros horns from Bengal, ivory from India and Africa, coral, agate, pearls, crystal, rare woods, and spices.

Nevertheless, the overall pace of change in China remained slow compared to the economic growth that occurred later during Europe's industrial revolution. In part, the slow pace of change was the result of a continued focus on agriculture and the failure to encourage the growth of a merchant class. With commerce not

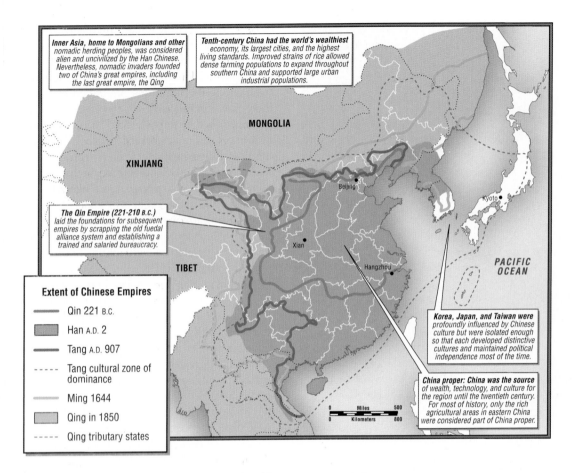

Inner Asia, home to Mongolians and other nomadic herding peoples, was considered alien and uncivilized by the Han Chinese. Nevertheless, nomadic invaders founded two of China's great empires, including the last great empire, the Qing

Tenth-century China had the world's wealthiest economy, its largest cities, and the highest living standards. Improved strains of rice allowed dense farming populations to expand throughout southern China and supported large urban industrial populations.

The Qin Empire (221–210 B.C.) laid the foundations for subsequent empires by scrapping the old fuedal alliance system and establishing a trained and salaried bureaucracy.

Korea, Japan, and Taiwan were profoundly influenced by Chinese culture but were isolated enough so that each developed distinctive cultures and maintained political independence most of the time.

China proper: China was the source of wealth, technology, and culture for the region until the twentieth century. For most of history, only the rich agricultural areas in eastern China were considered part of China proper.

Extent of Chinese Empires

Qin 221 B.C.
Han A.D. 2
Tang A.D. 907
Tang cultural zone of dominance
Ming 1644
Qing in 1850
Qing tributary states

Figure 9.3 The extent of Chinese empires, 221 B.C.–A.D. 1850. The Chinese state expanded and contracted throughout its history. The colors on the map indicate the extent of its dominion at various times. [Adapted from *Hammond Times Concise Atlas of World History*, 1994.]

a respected occupation, trade did not develop rapidly or nurture entrepreneurial activity. The overwhelmingly conservative nature of the Chinese empires, as well as their periodic collapses, meant that technologies were improved and diffused only slowly. Thus, despite the high level of technological know-how that China attained early on, European technology eventually surpassed that of China in the eighteenth century.

Rising European Influence

By the mid-1500s, Spanish and Portuguese traders had found their way to East Asian ports. They were interested in acquiring the silks, spices, and ceramics of East Asia, and they brought a number of new food crops from the Americas, such as corn, peppers, and potatoes, to offer in exchange. The earliest visible European impact on East Asia was a rapid rise in the Chinese population. Although the Confucian bias toward agriculture had always nurtured large populations (estimated at 120 million by A.D. 1000) to provide a labor pool, after 1500 the new American food crops fueled further population growth. By the mid-1800s, China's population stood at over 400 million, and many Chinese were migrating to the frontiers of the empire: to Yunnan in central southwestern China and to

Manchuria in the northeast. By the nineteenth century, European influence had increased markedly as foreign merchants lobbied for access to the huge Chinese market. Between 1839 and 1860, Qing dynasty leaders fought the Opium Wars in an unsuccessful attempt to resist the European incursion. The Chinese were trying to prevent British merchants from exchanging opium imported from India for Chinese wares such as silks and fine ceramic dinnerware, which were much prized in Britain.

In the 1800s, the Qing dynasty's authority declined rapidly. After China was defeated in a series of conflicts with Britain and France, most of the country was carved up into spheres of influence in which Britain, France, Germany, and Russia established exclusive trading interests. Two important British enclaves were the port cities of Shanghai and Hong Kong. The final blow to China's preeminence in East Asia came in 1895, when a rapidly modernizing Japan won a spectacular naval victory over China in the Sino-Japanese War. After its defeat by the Japanese in 1895, the Qing dynasty made only halfhearted attempts at modernization, and it collapsed in 1912 after a coup d'état. Until China's communists took control in 1949, much of the country was governed by provincial warlords in rural areas and by a mixture of Chinese and Western administrative agencies in the major cities. During this era, radical ideologies gained popularity in new Western-style universities, as intellectuals searched for a new basis of political authority to replace Confucianism. Of particular interest were various forms of socialism and communism.

China's Turbulent Twentieth Century

Two rival reformist groups arose in the early twentieth century. One was the Nationalist party, known as the Kuomintang (KMT), led by Chiang Kai-shek. The KMT united the country in 1924 and at first reorganized it politically and economically according to the socialist views of Lenin in the Soviet Union. After 1927, however, the KMT increasingly became the party of the urban upper and middle classes. The rival group, the Chinese Communist party (CCP), appealed most to the far more numerous rural laborers. Japan took advantage of these internal struggles to seize control of northeastern China (Manchuria) in 1931. Then, in 1937, while the Western powers were distracted by European fascism, Japan invaded China, conquering most of its major cities and coastal provinces. For a while, the KMT and CCP tried to unite against this common enemy, but once Japan surrendered to the Allied forces at the end of World War II (in 1945), the two Chinese reformist parties no longer cooperated. Although the KMT was backed by the United States (which was then particularly preoccupied with opposing communist movements), it was quickly pushed out of the country by the now very popular CCP, led by Mao Zedong. The KMT and many of its urban supporters fled to Taiwan. There they formed a government-in-exile opposed to the mainland communists. Even today Taiwan is set apart from the rest of China.

On the Chinese mainland, Mao Zedong's tactics of revolutionary mobilization among the rural majority resulted in the historic proclamation of the People's Republic of China on October 1, 1949, in Beijing's Tiananmen Square. Before long, Mao's government became the most powerful and expansionist government China had ever had. The Communist party assumed total control over the economy. It established a dominant presence in the outlying areas of the northeast, Inner Mongolia, and Xinjiang, and it launched a brutal occupation of Tibet (Xizang). The People's Republic of China was in many ways similar to a traditional Chinese dynasty. The Confucian bureaucracy was replaced by the Chinese Communist party, and Mao Zedong became a sort of emperor with unquestioned authority over the party, the military, and virtually all other activities of government.

The communist revolution substantially changed Chinese society. The chief early beneficiaries were the masses of Chinese farmers and landless laborers. It would be hard to exaggerate the desperate plight of the Chinese people on the eve of the revolution in 1949. Huge numbers lived in abject poverty, there were frequent famines, massive floods regularly devastated villages and farmland, and infant mortality rates were high. For the millions outside the small middle and upper classes, life was short and miserable. The vast majority of women and girls held low social status and spent their lives in unrelenting servitude.

The revolution drastically changed this picture. All aspects of economic and social life became subject to central planning. Land and wealth were reallocated, often—although certainly not always—resulting in improved standards of living for those who needed it most. Heroic efforts were made to improve agricultural production and to ameliorate the age-old twin afflictions of floods and droughts. Huge numbers of Chinese, regardless of age, class, or sex, were mobilized to construct massive public works projects almost entirely by hand. Chinese laborers terraced and reterraced mountainsides with hand tools. Bucket brigades unclogged silted waterways. People built roads across difficult terrain with virtually no machinery. Industries were founded in many regions with an eye to providing jobs to a broad segment of society. The famed "barefoot doctors"—people with rudimentary medical training—dispensed basic medical care, midwife services, and nutritional advice to the remotest locations. Schools were built in the smallest of villages. Opportunities for women opened up, and some of the worst abuses against them—such as the crippling binding of women's feet to make them small and childlike—stopped. Until recent decades, famine had occurred somewhere in the country every few years. Since the mid-1970s, however, China has managed to feed its own people, only occasionally having to buy grain on the world market or accept food aid. The vast majority of Chinese who are old enough to have witnessed the changes say that, materially, life is now better by far than before the revolution.

Nonetheless, the progress made under Communist party rule has come at enormous human and environmental costs. Millions died needlessly, persecuted for dissenting from party policies or killed by famine or other avoidable disasters that resulted from the rush to fulfill poorly planned development objectives. (The worst famine, from 1958 to 1961, is thought to have killed 30 million people and was entirely the result of ill-advised agricultural policies.) Similarly, many of China's considerable natural resources have been sacrificed to its vision of progress.

In the aftermath of famine and other disasters, some Communist party leaders tried to correct the inefficiencies of the centrally planned economy, only to be purged themselves as Mao Zedong attempted to stay in power. One early ploy to keep political control was the initiation in 1966 of the **Cultural Revolution,** a series of highly politicized and destructive mass campaigns to force the entire population to support the continuing revolution. Educated people and intellectuals were a main target because they were thought to instigate dangerously critical evaluations of Communist party central planning. Many Chinese scientists and scholars were summarily sent to labor in mines and industries or to jail, where they often died. The public was encouraged to purge suspected dissidents. People who had engaged in any type of petty capitalism were severely punished, as were those who adhered to any type of organized religion. The Cultural Revolution so disrupted Chinese society that by Mao's death in 1976, the communists had been thoroughly discredited.

Two years later, a new leadership formed around Deng Xiaoping. In the early 1980s, he initiated a series of reforms to give China's economy some characteristics of open-market economies, while at the same time maintaining Communist party political control. The pace of reform, however, did not keep up with popular expectations. In April and May of 1989, more than 100,000 students and workers agitating for democratic freedoms marched on Tiananmen Square in Beijing. On June 3 and 4, the government suppressed the demonstrations by firing into the crowd, killing hundreds and injuring thousands. Arrests, hasty trials, and executions followed. By 2000, many observers were concerned that the strains of rapid movement to a market economy might be creating further political instability in China that could lead the government to abridge civil and human rights once again.

East Asia's Communist-Capitalist Split

The history of the entire region is to some extent grounded in what transpired on the mainland. All East Asian countries, for example, have been affected by ancient Confucian philosophy, the relics of which can be found in present-day folk beliefs and social policy. More recently, the 1949 communist revolution in China has had ramifications for the rest of the region. North Korea and Mongolia have joined China in the communist experiment. Japan, Taiwan, and South Korea have chosen to develop free-market economies but must contend with a very large hostile neighbor.

The Korean Peninsula. Korea was a unified country until a cold war clash led to its division into North and South Korea in the years after World War II. From 1910 to 1945, Japan occupied Korea and treated it as a colony. Korea became a source of cheap minerals and agricultural raw materials for Japan's own growing manufacturing centers. Resentment toward the Japanese grew in Korea, and nationalist sentiment increased. Western ideas, including communism, flooded into Korea, competing with the more traditional Confucian and Buddhist beliefs.

As World War II was drawing to a close in August 1945, the Soviet Union declared war against Japan and invaded Manchuria and the northern part of Korea. Rather than allow the entire country to come under Soviet control, the United States proposed dividing Korea at the 38th parallel. The United States took control of the southern half of this peninsula and the Soviet Union the northern half. Over the next few years, the United States held elections in the south, recognized South Korea's independence, and withdrew its troops. Meanwhile, in the north, the Korean and Soviet communists established a military government. In June 1950, North Korean forces attacked South Korea. The United States soon came to the south's defense, leading a United Nations force that fought a three-year war against North Korea, the Soviet Union, and China. The Korean War, which ended in a truce in 1953, caused huge losses of life on both sides and devastated the peninsula's infrastructure. Since then, North Korea has closed itself off from the rest of the world and remains impoverished. South Korea has evolved a prosperous free-market economy, with U.S. assistance in its early years and the model of Japan as a guide afterward.

Taiwan. In the late nineteenth century, Taiwan (once known as Formosa) was a poor agricultural island on the periphery of the Chinese Qing Empire. In 1895, as the Qing Empire weakened, Japan annexed the island as a colony and exploited its resources for more than half a century. Then, in 1949, the Chinese Nationalists (the Kuomintang) lost to the Chinese communists and had to flee the mainland. They chose to settle in Taiwan, despite opposition from the indigenous people, and renamed the island the Republic of China (ROC). Taiwan became home to the exiled government of Chiang Kai-shek and more than a million migrants from the mainland. Over the next 50 years, it became a modern, crowded, and highly industrialized society and played a leading role, through investment and technology diffusion, in the rapid transformation of East Asian economies.

Mongolia. Mongolia is traditionally a land of nomadic herders. It long posed a threat to China, and China colonized the region in 1691 and controlled it until the early twentieth century. A communist revolution followed soon after independence from China in the 1920s, and Mongolia continued to follow a communist system under Soviet guidance until the breakup of the Soviet Union in 1989 launched it on a difficult road to a free-market economy.

The Transformation of Japan

Japan has transformed itself from an isolationist and militaristic society into a democratic nation that has achieved impressive economic success.

The Japanese islands were probably first inhabited about 20,000 years ago, and the principal components of the modern Japanese population are migrants from the Asian mainland, the Korean Peninsula, and the Pacific islands. By A.D. 300, Japanese society was divided into military clans that had established their rule over most of the islands that are now part of Japan. These islands included the central and largest island of Honshu, the island of Hokkaido to the north, and several smaller islands to the south. Settlement was concentrated in central Honshu and the southern islands. Leaders of the Yamato clan, who claimed divine approval, ruled until about 700. During the Yamato period, ideas and material culture imported from China and the Korean Peninsula transformed the everyday lives of Japan's people and their use of the land. These imports included Buddhism, Confucian bureaucratic organization, architecture, Chinese writing, the arts, and agricultural technology.

Between 800 and 1300, Japan turned inward, and influences from the Asian mainland waned. The Japanese successfully fended off invasion by Mongols, whose empire in the 1300s stretched from coastal China to Europe (see page 510). During the ensuing period of isolation, a feudal system with a rigid class structure evolved, and Japan was divided into mini-kingdoms. This was the situation when Portuguese sailors landed in Japan in 1543, followed by Christian missionaries. Active trade with Portugal brought new ideas and technology that strengthened the wealthier feudal lords, who then seized control and unified Japan under a military bureaucracy. Between 1600 and 1868, the military elite (known as **shoguns**) imposed a strict four-tier social class system, expelled foreigners, and imposed isolation once again, especially rejecting all European influence.

In 1853, a small fleet of naval vessels commanded by U.S. Commodore Matthew C. Perry arrived unexpectedly near the mouth of Tokyo Bay. Japan was quickly forced to recognize that more than two centuries of self-imposed isolation from the rest of the world would have to end. The foreigners, who were stronger in military might and much more advanced in technology, were in a position to dictate terms of trade and foreign relations. The threat

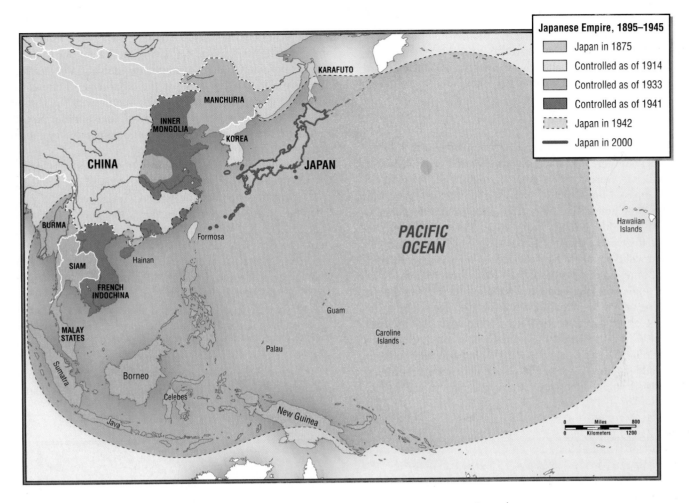

Figure 9.4 Japan expansions, 1875–1942. [Adapted from *Hammond Times Concise Atlas of World History*, 1994.]

was compounded by domestic unrest over various political and economic issues that had been building for decades. Consequently, by 1868, the family line of shoguns who had ruled Japan since 1600 lost power to a group of reformers, the Meiji, who restored the authority of the emperor and moved his seat from the ancient capital Kyoto to Tokyo (see the box "A New Capital for Japan?" on page 502). In the emperor's name, they set Japan on a crash program of modernization and industrial development. They sent Japanese students abroad and imported experts from around the world, especially from Western nations, to teach everything from foreign languages to modern technology. An important geographic dimension of the new era was the expansion of Japanese settlement from central Honshu and the southern islands to northern Honshu and Hokkaido, where new farmlands and cities were developed and strategic resources were exploited for heavy industry, such as coal mining.

By 1905, Japan was strong enough to defeat a European power, Russia, in war. To expand its resource base further and to gain a labor force for its mines and factories, Japan colonized Korea, Taiwan (then known as Formosa), and Manchuria, and then pushed farther into China and into Southeast Asia (Figure 9.4). Japan's

imperial expansion eventually led to its defeat in World War II (1941–1945) and to the loss of all its foreign territories.

Japan's imperial ambitions ended in 1945, and the country was occupied by U.S. military forces until 1952. The U.S. government imposed many social and economic reforms and required Japan to create a democratic constitution. Japan's postwar constitution allows only a very limited military, and for more than 50 years the country has relied primarily on U.S. forces based in Japan, South Korea, and the western Pacific to protect it from attack. Some observers believe that China could one day pose a military threat to Japan and motivate the country to reassess both its strategic alliance with the United States and the post–World War II constitutional provisions that prohibit an active military. The alliance with the United States is strained by Japanese rightists, who, for reasons of national pride, want to close American bases and rearm Japan. An additional stress on the alliance has been caused by a series of crimes, including rape and murder, committed by U.S. military personnel against Japanese citizens.

Japan rebuilt rapidly after World War II, aided in the early years by the United States. It eventually became a giant in global business, exporting automobiles, electronic goods, and many other

Thinking Geographically: ON THE WEB

products to the United States, Western Europe, and other parts of the world. By 2001, Japan was the world's second most technologically advanced economy, after the United States, and the third largest economy, after the United States and China.

POPULATION PATTERNS

East Asia is the most populous world region. China, with 1.26 billion people, has more than one-fifth of the world's population. Yet, throughout East Asia, people are having markedly fewer children than in the past. Most Chinese couples are limited by law to one child, and the Chinese are now reproducing at a rate considerably lower than the world average: 15 births per 1000 people yearly, as opposed to the world average of 22 per 1000. Nevertheless, China's population will continue to grow for years to come, simply because so many Chinese are just entering their reproductive years.

If the one-child family pattern continues, there is some hope of curtailing growth significantly. With relatively fewer births, the average age of the population will rise fairly quickly. As the population ages, China's growth will eventually slow and possibly even stop—but not until perhaps the year 2050, and only after millions more are added. However, a halt in growth depends on continued and improved compliance with control measures. Couples who

are more affluent often choose to have two or even three children and pay the fines imposed for having more than one child. If this trend continues, China may double its population in less than the presently estimated 79 years. On the other hand, if China follows Japan's lead and very small families—some with no children at all—become common, then China will face another problem: a very large elderly population that must be supported by a relatively small group of people of working age.

Elsewhere in East Asia, population growth rates hover around China's rate of 0.9 percent. Japan has the lowest rate in the region, at 0.2 percent (one-third the rate in the United States). In fact, Japan's population is growing so slowly that it will not double for 462 years. Thus, Japan's population, like China's, is rapidly aging (see page 503). By 2025, Japan will have a ratio of one pensioner for every two workers—comparable to several European countries with aging populations. Only in Mongolia (with 2 million people) and North Korea (with 22 million) are women still averaging more than two children each. But even in these two countries, increasing opportunities for women outside the home mean that family size is shrinking, family life and gender roles are changing, and the numbers of dependent elderly are growing.

As you can see in the population map of East Asia (Figure 9.5), people are not evenly distributed on the land. In fact, many parts

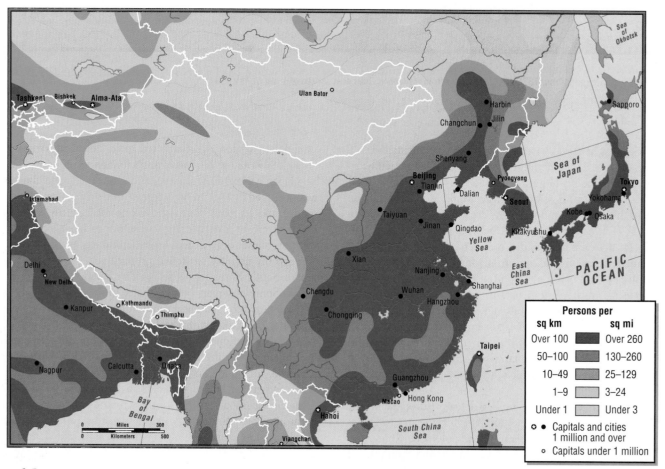

Figure 9.5 **Population density in East Asia.** [Adapted from *Hammond Citation World Atlas,* 1996.]

of Japan, the Koreas, Taiwan, and China are very lightly settled—either because the terrain is very rugged or (in the case of western China and Mongolia) because the land is extremely dry or cold. Ninety percent of China's people are clustered on only one-sixth of the total land area. They extract a very high level of agricultural production from this land, though at considerable environmental cost. People are concentrated especially densely in the eastern third of China: in the North China Plain, the middle and lower Chang Jiang basin, the delta of the Xi Jiang in South China, and the Sichuan Basin. South Korea is also densely settled, as are northern and western Taiwan. In Japan, settlement is concentrated in a band stretching from the cities of Tokyo and Yokohama on Honshu Island through the coast of the Inland Sea to the islands of Shikoku and Kyushu. The rest of Japan is mountainous and more lightly settled. Because the overall population of East Asia is so large and population issues are especially important to this part of the world, a number of demographic topics are discussed in the sociocultural issues section below.

CURRENT GEOGRAPHIC ISSUES

East Asia, like other world regions, faces the difficult task of providing a decent standard of living for its growing population—and doing so equitably and without causing environmental damage. Although East Asia's countries adopted different economic systems after World War II, most of them are making progress in creating a better life for their citizens. In fact, Japan has one of the highest standards of living in the world. The great challenge in this crowded part of the world is to restrain population growth and the environmental damage that results from it.

POLITICAL AND ECONOMIC ISSUES

After World War II, the countries of East Asia adopted different economic systems. The communist regimes of China, Mongolia, and North Korea relied on central planning to set production quotas and to allocate goods among their citizens. In contrast, Japan, Taiwan, and South Korea established government-aided market economies, with the aid and support of the United States and Europe. More recently, the differences among East Asian countries have diminished as China and Mongolia have adopted reforms that allow market forces—such as competition for customers, supply and demand—rather than government planning quotas to set production quantities and prices. This transition has caused hardships for many people, because reforms have not been completely accomplished and a free market has brought income inequality and the loss of social services. The older market economies of East Asia are in a recession that has lasted since the mid-1990s.

The Free-Market Countries

Throughout the nineteenth century, the economies of Japan, Korea, and Taiwan were minuscule compared to China's. Then, during the twentieth century, all three grew tremendously. Despite economic setbacks in the 1990s, Japan's economy remains the second largest in the world (as measured by the value of goods and services produced). It is four and a half times larger than China's economy, even though Japan's population is only one-tenth as large as China's. South Korea, with one-twenty-sixth the population of China, has the world's twelfth largest economy. Tiny Taiwan's economy is one-third as large as mainland China's. The credit for these countries' economic success belongs mostly to Japan, which—as far back as the late nineteenth century—began to develop the model of a government-guided free-market economy that all three countries used to achieve prosperity.

Japan's Economic Rise. Japan is renowned for its successful export industries, which produce automobiles, cameras, computers, video and audio equipment, and many other products. Yet, less than a century and a half ago, the country was one of the most isolated in the world, its doors closed tightly to all trade and international travel. Until 1868, it was essentially a feudal society ruled by hereditary shoguns. At a time when the industrial revolution and urbanization were transforming Europe and North America, Japan was primarily a nation of poor farmers, essentially unchanged from the Middle Ages. Industrial technology was primitive and far behind that of other leading countries. Thus, Japan's rapid rise to world prominence in trade and technology since the middle of the nineteenth century is one of the most remarkable tales in modern history, all the more so because of its limited base of natural resources and its crippling defeat at the end of World War II.

Japan first rose to economic prominence in the late nineteenth century (see pages 462–463) and steadily built its economy and acquired new colonial territories until its bombing of U.S. warships in Honolulu in 1941 brought it fully into World War II. It would be hard to overstate the extent of devastation in Japan at the end of the war. Except for Kyoto, which was spared because of its historical and architectural significance, all major cities had been mercilessly bombed and destroyed—most notably, Hiroshima and Nagasaki with nuclear weapons and Tokyo with incendiary bombs. Factories, roads, bridges, rail lines, power stations, and other infrastructure were in ruins. The people were hungry because food production and distribution were disrupted, and they were homeless and jobless because of the destruction wreaked by the bombing. At the same time, several million Japanese from lost colonies returned home as refugees. According to a description in the Japanese press, the condition of the country immediately after the war was "one hundred million people in a state of trauma."

Thus, it was literally from ashes and despair that Japan rebuilt itself in the second half of the twentieth century into one of the world's wealthiest and most influential nations. The key ingredients of its success were the incredibly hard work and sacrifice of the

GEOGRAPHER IN THE FIELD

Interview with Regional Consultant for Chapter 9, Roman Cybriwsky, Temple University

What led you to become interested in East Asia?
My interest in East Asia was accidental. In 1984, my employer, Temple University, in Philadelphia, opened a campus in Tokyo and asked me to be one of the first faculty. I have since spent a number of years at the Tokyo branch and have traveled widely throughout East and Southeast Asia. As a geographer, I am always interested in place. My new place fascinated me and I made a point of learning as much as I could about it. It turned out to be a good mid-career transformation from earlier interests in U.S. urban geography.

Where have you done fieldwork in East Asia, and what type of fieldwork have you done?
I specialize in urban geography, and I focused on the neighborhoods of Tokyo. I am particularly interested in studying how places and the lives of people change. I do a lot of my work by walking, observing, starting conversations and asking questions, taking notes, and shooting photographs.

I have also done fieldwork in Batam, a small island in Indonesia. In the last decade or so the island has grown tremendously, as new factories and a budding tourism economy have attracted migrants looking for work. There are enormous social and environmental impacts. For me, Batam is a window on the third world. Research there

complements work in the United States and Japan.

How has that experience helped you to understand East Asia's place in the world?
Both sides of the Pacific are closely linked in a global economy. There are economic linkages, too, between Japan and developing countries in Pacific Asia like Indonesia. It has been interesting to see these linkages in person: to see the home base of a big corporation in New York or Tokyo and to see its factories in Batam, Indonesia; to see and compare the life rhythms of workers on both sides of the divide; and, as a result, to understand better my own place in the world.

What are some topics that you think are important to study about your region?
There are lots of critical issues: the spread of global commercial culture, preservation of local cultures, inequality between and within countries, the true meaning of "quality of life," and environmental impacts.

Japanese citizenry and a benign period of occupation (1945–1952) during which the United States imposed needed democratic and economic reforms on the country. Procurements for the Korean War (1950–1953) were a particular stimulus for Japanese industry. Also contributing to rapid recovery were the opening of markets abroad for Japanese products and cozy relations between Japanese business and political institutions. A strong state bureaucracy helped wealthy private capitalists to start industrial enterprises that became the foundation of early economic development in Japan and, later, in Taiwan and South Korea. The government provided advice, financial assistance, and protection from internal and external competition.

Japanese innovation in manufacturing was another factor in its success. Japanese manufacturers organize firms into highly efficient spatial arrangements, and their practices have been imitated globally. Firms that supply components such as engine parts for automobiles are often clustered around a single internationally renowned company where final assembly takes place. The proximity allows suppliers to deliver their products "just in time," literally minutes before they are needed. Defects in production runs can be spotted more quickly, so fewer defective parts are produced and less space is taken up by warehouses. This **kanban system** has influenced economic geography globally. Rather than outsourcing

parts to cheap labor pools in developing countries, many firms now seek to locate the assembly of both components and final products in a single area, either abroad or at home.

For much of the post–World War II period, Japan's economy grew roughly 10 percent a year, with heavy industry and manufacturing the leading sectors. By 1964, when it hosted the Summer Olympics in Tokyo, Japan was eager to engage in trade. The Olympics gave Japan a chance to show off its event-planning abilities and its new international hotels, light rail system, high-speed bullet train linking Tokyo with major cities, and ultramodern Olympic sports facilities. Japan's economic growth continued steadily through the next three decades.

Recession. In the early 1990s, the Japanese economy showed signs of strain, and by the end of the decade all three of East Asia's free-market countries had experienced significant economic downturns. Japan's economy was the weakest, shrinking by several percentage points a year. Its economic problems are linked to the economic crisis in Southeast Asia that began in the mid-1990s and became apparent by 1997 (discussed in Chapter 10). Japan ships 45 percent of its exports to Southeast Asia and is the single largest investor in Southeast Asian ventures, including agriculture, forestry, manufacturing, mining, and tourism. When companies in Southeast Asia began to close because of bad debts, Japanese in-

In Kyushu, Japan, a highly mechanized Toyota plant turns out cars for the domestic market. Here precision robot arms weld the car bodies. [Mike Yamashita/Woodfin Camp & Associates.]

vestors lost money. Similarly, when millions of Southeast Asians lost their jobs, they could no longer afford to buy Japanese products. As a result, many Japanese companies began to experience significant business losses.

Corruption also contributed to the Japanese economic crisis. The long-standing close relationship between government and industry had nurtured favoritism. Politicians often encouraged economic development projects that were not needed, because the projects brought money and jobs to voters in their home districts. For example, many rural areas in Japan now have elaborate transportation systems, far beyond the needs of the small rural populations. Another problem was that Japan overexpanded its productive capacity. Between 1988 and 1992, Japan increased its production capability by an amount equal to the entire economy of France—just when the demand for its products began to contract at home and abroad. Bureaucrats who had nurtured enterprises with advice and financial support were often unwilling to let them go bankrupt, even when they were being poorly managed or no longer had the potential to make profits. The Japanese government forced banks to make large loans to troubled firms. Eventually, both the firms and the banks went bankrupt. Korea and Taiwan have experienced financial crises similar to Japan's, though Taiwan has suffered less, largely because it has been more willing to let its weak enterprises go bankrupt.

Some experts on Japan link the Japanese financial crisis to overprotected Japanese workers who have little incentive to perform at their full potential. Good workers are rewarded with job security rather than higher pay, a policy that has encouraged conformity at the expense of innovation and creativity. As a result, Japanese industry depends largely on technologies developed elsewhere, usually in North America and Europe. Another consequence of the relatively low wages is that middle- and lower-class Japanese consumers cannot buy many of the high-quality items they themselves produce, whereas their counterparts in other industrialized economies form a large internal market for locally made wares. Japanese high-end goods are sold instead to consumers in North America and Europe. Many experts argue that these characteristics of the Japanese economy make it inherently unstable, because they render the country dependent on foreign technology and on just a few large external markets. Should the economies of North America and Europe falter, so would the economies of Japan and all the countries following its economic policies. It is therefore argued that workers in Taiwan, South Korea, and Japan should be paid more, with raises contingent on performance, so that there will be incentives for workers in these countries to generate more of their own technological innovations. Also, more prosperous workers would create an internal market for domestic products, thus strengthening the economy.

The Communist System

The communist economic systems of China, Mongolia, and North Korea emerged in the mid-twentieth century. After World War II, all three countries abolished private property, and the state took control of virtually every aspect of the economy: agricultural and industrial production; construction; service industries such as transportation and distribution, utilities, social services and education; and sales of food and consumer goods. Many leaders saw this course as the only way to salvage societies that had been badly damaged by a grossly self-centered elite, by Japanese and European colonial domination, and by civil war. The sweeping changes ultimately proved less successful than was hoped.

The Commune System. When the Communist party first came to power in China, in 1949, its top priority was to make large improvements in both agricultural and industrial production. The first strategy was land reform, which to the communists meant taking large, unproductive tracts out of the hands of landlords and putting them into the hands of the millions of landless farmers. By the early 1950s, much of China's agricultural land was divided into tiny plots. But it soon became clear that this system was inefficient. The farmers needed to produce enough to feed people who were being drawn out of agriculture to work in the many expanding industries. Communist leaders decided to band small landholders into cooperatives so that they could share their labor and pool their resources to acquire tractors and other equipment to increase agricultural production. In time, these cooperatives were expanded into full-scale communes with an average of 1600 households each. The communes, at least in theory, took over all aspects of life. They were the basis of political organization (all within the Communist party), and they provided health care and education. In addition, the communes were responsible for building rural industries to supply themselves with such items as fertilizers, gunnysacks, pottery, and small machinery. They also had to fulfill the ambitious expectations of the leaders in Beijing for better flood control, expanded irrigation systems, and the generation of surplus funds for investment elsewhere.

The commune system had several difficulties. Farmers were required to spend so much time building roads, levees, and drainage ditches, or working in the new rural industries, that they had much less time to farm. Local Communist party administrators often compounded the problem by reporting that harvests were larger than they actually were, to impress their superiors in Beijing. The leaders in Beijing responded by requiring larger grain shipments to the cities, which created food shortages in the countryside. These measures pushed the system to its limits in the late 1950s, during the "Great Leap Forward," a government-sponsored national promotion to implement communist reforms. The result was a massive famine in which 30 million people starved to death.

Focus on Heavy Industry. Industrial production beyond that of small farm-linked industries was focused on heavy industry rather than on consumer goods. The leadership felt that China needed first to improve its infrastructure—to build roads, railways, dams—all of which required heavy equipment. So the government emphasized mining for coal and other minerals, producing iron and steel, and building heavy machinery. The state obtained funds for industrial development from the already inefficient agricultural sector in several ways. Farmers were required to sell their crops to the state at artificially low prices, and the state kept any profits gained in reselling the food. In addition, the state diverted funds that had been earmarked for agricultural improvements to industrialization. Other funds for industry came from profits in mining and forestry and from savings on workers' wages, which were kept very low.

By design, most people in communist China were not allowed to consume more than the bare necessities. On the other hand, nearly everyone was guaranteed a job for life and the means to obtain the basic essentials. As was the case in Japan, the emphasis on job security rather than on rewarding creativity and hard work led to overwhelming conformity and a lack of innovation. Hence, the level of overall productivity remained low. Central planners in Mongolia and North Korea also favored giving workers job security rather than higher wages.

Regional Self-Sufficiency in China. One objective of China's leaders was to reduce regional variations in prosperity. For centuries, agricultural and herding areas in the interior, especially those in more remote areas to the west, had been much poorer than the coastal areas to the east, where industries and trade were concentrated. After the revolution, the communist leadership instituted a policy of **regional self-sufficiency.** It encouraged each region to develop independently, building agricultural and industrial sectors of equal strength, in the hope of creating jobs and evening out the national distribution of income. Government funds were used to set up industries in nearly every province, regardless of practicality. For example, a steel industry was established in Inner Mongolia at great expense, when other provinces already had steel industries that could have been improved with far greater payoff. Similarly, agriculture was extended into marginal physical environments far from markets, often in areas where continuous irrigation was necessary or where growing seasons were very short. Huge mechanized grain farms were developed on cleared forestlands in far northern Heilongjiang, when the same effort elsewhere would have yielded better results and the forests could have been saved for other purposes. This policy of regional self-sufficiency also encouraged individual communities to develop backyard smelters, kilns, tool and die factories, or their own tiny tractor factories.

Certainly these efforts were a tribute to community ingenuity, but ultimately they wasted time and resources while producing relatively little.

Cold war tensions also encouraged the dispersion of industry across China and regional self-sufficiency. Communist leaders thought that a dispersed pattern of industrial development and the ability of each region to feed itself would foil an enemy's effort to destroy China's productive capacity quickly.

Globalization and Market Reforms in China

Dramatic changes have occurred in China in recent decades, and their effects have reverberated throughout the world. In the late 1970s, the failure of China's communist economic policies to make dramatic improvements in living standards motivated China's leaders to enact market reforms. Initially, these reforms were made domestically at the local level, and they changed China's economy in three ways. First, economic decision making was decentralized. Second, farmers and small businesses were permitted to exchange their produce and goods in markets. The markets encouraged competition and improved the efficiency with which goods were manufactured and distributed to consumers. Finally, regional specialization rather than regional self-sufficiency was encouraged.

Eventually, the new markets expanded to allow foreign investment in Chinese enterprises and the sale of foreign products in China. This shift to a market-based economy has transformed East Asia and the surrounding areas as China has become a participant in the global economy. In the 1990s, it emerged as a significant producer of manufactured goods, and it represents a market of more than 1 billion potential consumers. Nearly every major company in the world is eager to sell its goods to customers in China.

The Reforms in Overview. The key change in China's economic policies was the decentralization of decision making, because it enabled the market system to function and encouraged regional specialization—and it subsequently led to the globalization of Chinese markets. To decentralize decision making, the government introduced **responsibility systems,** which gave managers of many state-owned enterprises the right and responsibility to improve the efficiency of their operations. The responsibility systems were linked to the introduction of markets, in that managers set production quotas to meet market demands. In addition, they also could set the prices for goods and services according to supply and demand. To some extent, anyone could create a new enterprise. Those who earned money could either spend it or reinvest it. The reforms increased regional specialization, as managers and entrepreneurs (sometimes with assistance from the state) took advantage of the different resources and opportunities offered by different areas of the country. For example, most new manufacturing has been located in coastal areas, such as the city of Shenzhen (see the box "Pearl River Megalopolis," page 498), that are more accessible to ocean trade and the global economy. In contrast, interior areas have refocused on agriculture and raw materials industries.

Since the reforms, the Chinese economy has become more productive overall, and many citizens have achieved greater pros-

Thinking Globally: ON THE WEB

perity. The per capita gross domestic product has doubled several times, from U.S. $300 in the mid-1980s to U.S. $3104 (PPP) by 1998. But the reform of state-owned enterprises has proceeded slowly, largely because new private enterprises have not emerged fast enough to employ those who would lose their jobs if state-owned factories reduced their workforces. Hence, many state-owned enterprises still flounder, using labor and resources inefficiently, making products with limited market appeal, and adding to pollution. And rural workers' wages are falling far behind the wages of workers in urban areas, thus increasing the disparity in standards of living between urban and rural regions.

Regional Trends in Agriculture. The market reforms have brought new opportunities to some agricultural sectors. Throughout China, farmers have been encouraged to organize family-sized or larger operations that meet the local food demand. Because agricultural potential varies greatly across China, regional specialization in agriculture has increased. Those farmers who live close to cities are producing food for urban markets. Instead of selling the grain they grow, increasing numbers of farmers are feeding animals with it to satisfy a rising demand for meat among the affluent. In places where the growing season is long enough, such as the far

southern provinces of Yunnan and Guangdong, farmers are now producing fresh produce for distant markets in northern cities.

There is a danger that these new developments could compromise China's **food stability**: its ability to supply sufficient basic food to all the people consistently. Several factors are putting pressure on the food production system. Only a portion of China's vast territory can support agriculture, because much of the land is too cold, too dry, or too steep to cultivate (Figure 9.6). China's huge population has already stretched the productive capacity of many fertile zones well beyond sustainability. And demands are increasing—in some parts of China, there are six times more people living on productive farmland today than there were just 50 years ago. As urban populations grow more affluent, their taste for meat and for fresh fruits and vegetables will place an added burden on China's agricultural land. For example, in the North China Plain and along the middle and lower reaches of the Chang Jiang, 500 to 600 people may subsist on and produce surpluses for sale on a plot of 500 acres (202 hectares), which in America would provide income for just one farm family of five or six people. The raising of animals for meat requires more land and resources than required to supply the plant-based national diet of the past. As

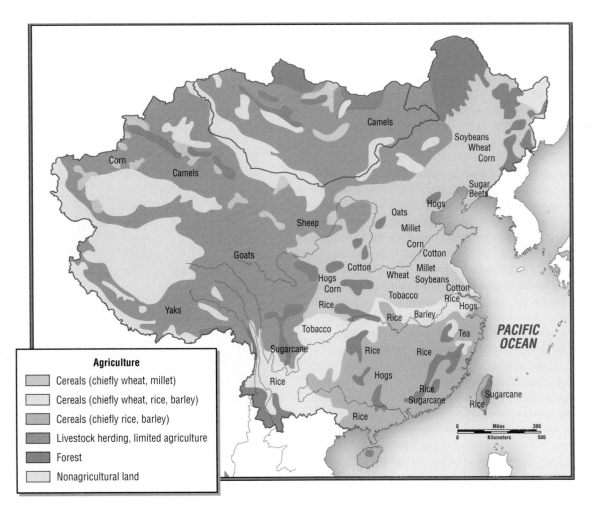

Figure 9.6 China's agricultural zones. [Adapted from *Hammond Citation World Atlas,* 1996.]

agriculture becomes mechanized and agricultural collectives close, farm laborers will lose their jobs, and many of them will not have access to land to grow their own food. They are likely to flee to the already overburdened cities to look for work. The result may be a markedly larger urban population that must depend on a food production system that meets the desires of the more affluent for meat and fine produce but does not provide nutritious basic food at low enough prices for the masses.

A Market Focus for Rural Enterprises. One of the most remarkable recent economic developments in China is the growth of rural enterprises that provide a wide variety of goods and services—from operating coal mines to making plastic flowers to assembling electronic equipment. These enterprises have become the mainstay of many rural economies, especially in the eastern and southern coastal provinces of Jiangsu, Fujian, and Guangdong. Such enterprises may still be township or village collectives, although some are privately owned. All of them leave major decisions to managers (rather than to the collectives) and price their products according to market demand. Rural enterprises have bloomed so quickly and successfully that they now constitute a quarter of the economy, and they may actually produce over half of China's industrial output and 40 percent of its exports. Statistics are not easy to come by, but it is known that these enterprises now employ more than 125 million people, more than the Chinese government itself, and account for more than 30 percent of farm household income.

Nonetheless, environmental pollution and corruption have accompanied the growth of these rural enterprises. In the mountains west of Beijing, for example, where many such industries are located, pollution exceeds that in the city. Clouds of exhaust from trucks that link the rural industries with their markets contaminate the air. Paper mills pollute the waterways, farmers' fields, and aquifers. Some managers (perhaps even a majority) steal funds from their own enterprises, try to evade taxes, and pay officials to look the other way when they breach environmental regulations. The government, although moving to enact legislation that could lessen these problems, is wary of driving rural enterprises out of business because the whole country is now dependent on their success.

The Persistence of Regional Disparities. Since China's market reforms, long-term regional disparities in wealth are once again increasing (see Figure 9.1, page 455). Remote provinces such as Tibet (Xizang) and Yunnan, and those with little mineral or agricultural wealth (such as Inner Mongolia and Qinghai), have lost special support from the central government yet have had difficulty generating rural enterprises. Increasingly, so many young people and skilled laborers are leaving the interior provinces (often without official permission) that the only people left in some rural communities are children, some women, and the very old. Meanwhile, the eastern and southern coastal provinces—with their larger cities, more developed infrastructure, skilled workforces, and ready access to the sea—are enriching themselves through manufacturing industries and foreign trade.

International Trade and Special Economic Zones. Central to China's new market reforms are the **special economic zones (SEZs)** and related **economic and technology development zones (ETDZs)** that now stretch into the interior of China (Figure 9.7).

In the early 1980s, China's government, wary of the disruption that could result from a rapid opening of the economy to international trade, selected five coastal cities to function as free-trade zones: Zhuhai, Shantou, and Shenzhen in Guangdong Province (near Hong Kong); Xiamen in Fujian; and Hainan Island. Industries in these cities were allowed to practice management methods that were not permitted in the rest of the country until recently. For example, foreign investors and their companies were given income tax breaks, reductions in import duties, and the freedom to take their profits out of China. In the years since they were established, these SEZs have succeeded in attracting international investors and industry. Industry, in turn, has spawned related activities in these trade zones, such as insurance, general commerce, information-gathering and processing, and consulting services.

Since 1990, many more cities along the coast and inland have become SEZs. Shanghai, for example, long a window on the outside world for China, was chosen to spearhead the expansion of free-trade zones along the Chang Jiang all the way to Chongqing. By the late 1990s, the Chinese authorities had designated 32 provincial locations in the interior of China as ETDZs, a distribution that stretches all the way to Urumqi in Xinjiang Province. These ETDZs provide footholds for international investors and multi-

Trucks from Hong Kong wait to pass through the portal into Shenzhen SEZ. [Mike Yamashita/Woodfin Camp & Associates.]

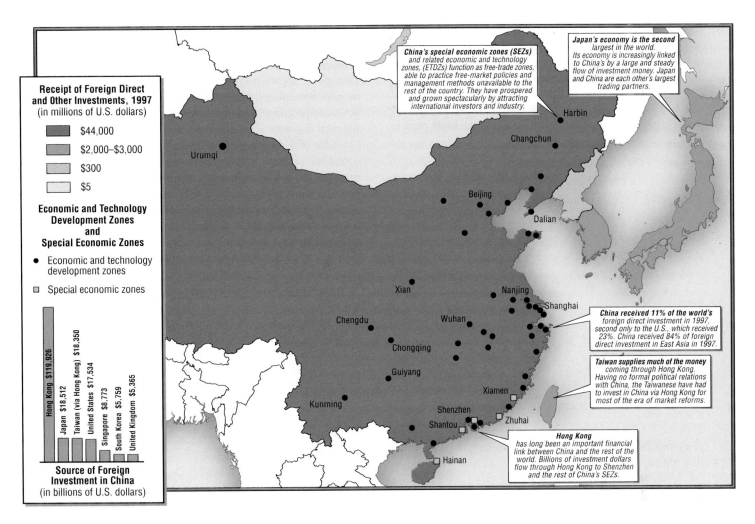

Figure 9.7 Economic issues: East Asia. The map shows China's special economic zones (SEZs) and economic and technology development zones (ETDZs). The colors on the map reflect levels of foreign investments (direct and others) in each of the countries of the region. Also shown are the principal countries from which the investments came. [Adapted from http://www.harper.cc.il.us/~mhealy/geogres/maps/eagif/chsez.gif and http://www.sezo.gov.cn/.]

national companies eager to establish operations in a country with such a huge pool of cheap labor and such a large number of potential consumers. Today the original SEZs and related development zones (there are now at least 48 altogether) are major **growth poles** in the country, meaning that their success is drawing yet more investment and migration. Coastal cities such as Shenzhen, Qingdao, Wenzhou, and Changan are attracting skilled people from the rest of the country. In just 20 years, many coastal cities have grown from midsize towns or even villages into some of the largest urban areas in the country (see the box "Pearl River Megalopolis," page 498).

Life in the Growing Cities. In the vignette that opens this chapter, we met Hong Xiaohui, who left her home in Sichuan Province to work in Mattel's factory in Changan. She is typical of millions of young migrants who are leaving sheltered rural village life to work in the highly competitive industrial zones of the SEZs. Many arrive intending not only to succeed as individuals but to send money home to their impoverished families and possibly even

return one day to improve their home communities. We return now to Xiaohui's story because her triumphs and travails over the first year after she left home demonstrate what life is like for the migrants from the hinterlands to China's booming coastal cities.

VIGNETTE When Xiaohui first arrived in Changan, she learned that, in addition to her pay being lower than she had been promised, the shifts at Mattel were usually 12 hours long, with no overtime pay and just one day off a month. More important, when Xiaohui rapidly mastered her task of putting hair on Barbie dolls, she began to entertain dreams of becoming a supervisor—only to discover that such a position might require sexual favors to the plant manager. With an income equaling just U.S. $24 a month, Xiaohui and her fellow workers could afford only a crowded dormitory room and canteen food. There were 11 women in one small concrete room with a tiny bath.

In less than two months, Xiaohui broke her contract with Mattel and returned home to her village in Sichuan, where—using ideas and the assertiveness she had gained from her time in the south—she opened a kebab stand. In just one month, she made 10 times her investment of U.S. $12. Soon she also opened a Hot Pot soup shop in her family home. But Xiaohui yearned to return to the south coast to seek employment on her own. So, leaving the kebab shop to her sister and brother-in-law, she eventually returned to Changan.

In one year, the city had grown by 20 percent and now had 1400 foreign companies trying to hire workers. Xiaohui asked one of her old friends to help her find work. In a matter of days, she found a job in a Japanese-owned factory—and also one for her sister, who rushed to Changan, leaving her husband and baby in Sichuan. The sisters quickly made friends with several young women from various parts of China, and together they enjoyed an active urban social life hampered only by a tight budget and their generally conservative village ways. Just a year after Xiaohui's first trip to Changan, she was making enough to live relatively comfortably, though still in a crowded dormitory. She was sending money home to her parents and once again saving to start a business. This time she dreamed of opening a karaoke bar back in her village. [Adapted from Kathy Chen, *The Wall Street Journal* (July 9, 1997).]

Hong Xiaohui's story illustrates why the central government in Beijing thinks of southeastern China as "the golden coast" and hopes it will become a model for the rest of the country. Yet the golden coast is not without problems. Policing the activities of thousands of foreign firms is already an impossible task for the Beijing government. Finding ways to tax businesses and yet not discourage small capitalists like Xiaohui Hong is difficult. Just

housing the flood of migrants and providing them with sanitation and needed transportation is expensive and poses environmental problems. Developing the infrastructure of the coastal zone fast enough to accommodate all these new residents will divert resources from places where life is still unbearably hard. As a result, yet more people will move to the cities from the interior, and the regional disparity of wealth will increase.

One of the most visible and potentially dangerous results of the economic reforms is the growing number of impoverished and unemployed people now crowding into China's cities. Chinese citizens are not allowed to migrate at will and must always obtain a permit to go to a particular place. This regulation is increasingly ignored, however. A growing problem for the government is this so-called **floating population,** made up of jobless or underemployed people who have, without permission, left economically depressed rural areas for the cities and are now unaccounted for. By some estimates they may number 100 million people, many virtually homeless. Although some find work and contribute to the support of rural relatives, many are suspected of engaging in petty crime. Moreover, there is the worry that this floating population may become a source of social instability. Some large industrial towns already have 20 percent unemployment, a figure that is likely to rise. Mongolia is facing similar dilemmas as it, too, implements market reforms (see page 510). Even in the much wealthier economies of Japan and South Korea, the economic downturn of the 1990s and early 2000s has created large numbers of unemployed.

China and the World Trade Organization

China's dramatic economic changes over the last 20 years, its strong growth rate, and its present ability to attract investment while maintaining a low ratio of external debt to GNP have led to the proposal that China be admitted to the **World Trade**

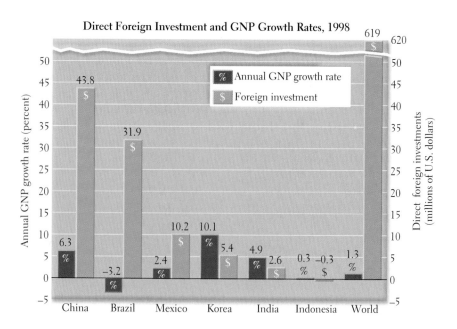

Figure 9.8 Direct foreign investment and economic growth rates (GNP) in China and other countries, 1998. China received more foreign investment capital than any other country in 1998, and its GNP growth rate was also relatively high. [*World Bank World Development Report 2000–2001,* p. 314, Table 21.]

Organization (WTO). The WTO, the world's preeminent trade body, seeks to remove barriers to global trade. A poor and relatively isolated country in 1980, China is increasingly prosperous and linked to the rest of the world. As of 1998, it had one of the fastest growing economies in the world, partly because it received more foreign investment than any other developing country (Figure 9.8). Many governments and large multinational corporations see China's entry into the WTO as essential to the goal of global economic integration. In 2001 both China and Taiwan were admitted.

The inclusion of China in the WTO became highly controversial as awareness spread that much of China's growth is based on environmentally destructive activities and on abuses of workers. Moreover, China has brutally suppressed separatist movements in Tibet (Xizang) and Xinjiang and has committed human rights abuses against citizen groups agitating for democracy and greater political freedom. In 2001, most governments of the world and some business leaders were willing to ignore these problems to avoid angering the Chinese government and jeopardizing trade. They argued that once China was in the WTO, increased trade would result in more openness and prosperity, which would ultimately improve environmental and labor standards. In fact, there may already be signs of a new openness (see the box below, "The Information Highway in China"). Consumers, environmentalists, and workers outside China counter that increased trade may only result in more widespread abuses. They suggest that China's entry into the WTO should have been conditioned on improvements in human rights, worker safety, and environmental protection. Workers in the United States and Europe, some of whom are losing their jobs as a result of competition from China, occasionally led demonstrations against China's admission into the WTO, and its inclusion remains controversial.

AT THE REGIONAL SCALE
The Information Highway in China: Bringing Openness to Government?

Before the economic reforms of the past two decades, the central government in China controlled the news media. By the late 1990s, however, the expanding use of electronic communication devices was loosening central control over information. Many more people now have personal computers, which provide access to the Internet. Just a few years ago, only a very few people had telephones; but now anyone can buy temporary cellular phone access without showing identification. As a result, many more people have anonymous access to an international network of information. Greater public access to information—which many people believe is essential in a market economy—has had a number of important social effects.

One such effect is that it is difficult for the government to promulgate inaccurate explanations for disasters or problems caused by inefficiency and corruption. In March 2001, for example, the government first attributed an explosion in a rural primary school, which killed at least 38 schoolchildren, to the act of a suicide bomber. But soon reporters in Beijing randomly dialed numbers in the village area code to reach some of the children's parents directly. The parents explained that their dead children had been forced to work in a fireworks factory during and after school hours. Proceeds from the factory went to the school budget and to school officials. The Chinese prime minister apologized publicly on March 15. He acknowledged that there had been an "industrial accident" that was causing him to reflect on his own work, perhaps meaning that he felt implicated.

Analysts inside and outside China think the apology was a watershed event that demonstrates the power of technology to make information more accessible. In the future, government will have to be more open because at least 23 million people are now connected to the Internet, and many more have access to telephones (Table 9.1).

Source: Craig S. Smith, "China arrives at a moment of truth," *The New York Times* (April 1, 2001): 5.

TABLE 9.1 *Changes in access to telecommunications in China, 1990 to 1998–2000*

Telecommunications	1990	1998–2000
Rural telephone lines	1.4 million	34 million
Internet hosts	0	1264
Internet users	0	23 million
Mobile phone subscribers	0	50 million
Regular phone lines	7.5 million	100 million
Personal computers	0	11.3 million

Sources: United Nations Human Development Report 2000 (New York: United Nations Development Programme), Table 12; R. Benewick and S. Donald, "Access to information flows," in *The State of China Atlas* (New York: Penguin, 1999), pp. 198–200; and Craig S. Smith, "China arrives at a moment of truth," *The New York Times* (April 1, 2001): 5.

SOCIOCULTURAL ISSUES

East Asia's economic progress has led to social and cultural change throughout the region. Here we discuss how societies that are homogeneous and inward-looking manage relationships with strangers, both foreign and domestic. Also, successful attempts to control population growth, made possible by prosperity, are creating new problems—such as an aging society and an imbalance in the numbers of males and females. Modernized economies are changing work patterns and family structures. Here we look at how countries in the region are responding to these changes.

Cultural Adaptation to Capitalism: Doing Business with Strangers

China's embrace of the free market is an episode of radical social engineering analogous to the communist revolution of 1949. Although engaging in business for profit was part of Chinese culture long before the 1949 revolution, more than 50 years have elapsed since then. Now, in addition to adjusting their ways of life away from the communist system, China's industrial laborers and new business owners must manage the risks of doing business with strangers—both their fellow Chinese and investors and firms from free-market economies in Europe, the Americas, and elsewhere in Asia. People who have spent a lifetime in rural villages must travel to distant cities for the first time to find a job or negotiate deals, they may have to hire competent and efficient workers from among masses of unfamiliar applicants, or they may deal with impersonal financial institutions. Surprisingly, there are ancient cultural institutions that are useful in the modern business era. One such venerable institution is the third-party **guarantor**, a person who vouches for the honesty and reliability of one party in his or her dealings with another.

The traditional preference for dealing with family members or fellow villagers grew out of the desire to minimize risk. Guarantors were used for all kinds of negotiations when the parties to a deal did not know each other. For instance, before the communist revolution, a person new to Shanghai who sought employment in a bank, department store, or factory needed a guarantor—someone who would convince the employer that the prospective employee was honest and had needed skills. The guarantor also gave the employee some protection from exploitation. These days, ideally, one still first seeks out business associates who are known in some way—people from the same region, if nothing else. But when two total strangers must do business, the *xinyong*, or trustworthiness and financial solvency, of each party is vouched for by the *xinyong* of a professional guarantor. The guarantor is paid for this service and, if things go wrong, is obliged at least to mediate and sometimes to cover losses. Once we understand the role of the guarantor as someone who assumes (for a price) the risk in business deals—which in simpler times was taken on by the family—it is easier to appreciate Asian business practices. In China, the Koreas, Japan, and Taiwan, guarantors or similar functionaries play a role in nearly every aspect of business dealings, from finding a factory job to building a skyscraper. Some Western businesspeople, perhaps feeling shut out by this system, tend to view guarantors as evidence of favoritism and cronyism, even corruption. Those with more insight simply find themselves a guarantor.

At the same time, there is ample evidence that East Asian business practices are becoming more **transparent,** a Western term meaning open to public scrutiny. This is especially true for firms that are partially owned by European or American stockholders. These stockholders have insisted that managers now be paid partly with stock options, so that they will share an interest in making the firm efficient and profitable. Managers whose decisions affect their own self-interest can more easily be held to wise and transparent management practices.

Population Policies and the East Asian Family

Population patterns in East Asia began to change after World War II, and by the 1970s government policy aided by rapid urbanization had reduced fertility in Japan and Taiwan to just over 2 children per woman aged 15 to 45. South Korea's rate was still high, at 4.3 children per woman. By 2000, women in Japan, Taiwan, and South Korea were bearing fewer than 2 children on average. In China, Mongolia, and North Korea, fertility rates remained very high into the 1970s: in China and North Korea, nearly 5 children per woman and in Mongolia, 7 children per woman. In China, especially, leaders realized that the rapidly rising population was sapping the country's ability to progress. Strict population control measures were deemed necessary, and by 1980 China had adopted a policy of one child per family.

The One-Child Policy in China. Chinese demographers and the people at large soon realized that the one-child policy would dramatically affect family life and the social fabric. For example, within two generations, the kinship categories of sibling, cousin, aunt and uncle, and sister- and brother-in-law disappeared from all families that complied with the policy—a major loss in a society that has long placed great value on the extended family. But it is the prospect of the one child being a daughter, with no possibility of having a son in the future, that has caused the most despair. Despite strong conditioning by the communist government that the sexes should be considered equal, the preference for sons remains strong. Sons are preferred partly because of the belief that only a son can inherit property, pass on the family heritage, and provide sufficient income for aging parents (even though daughters or daughters-in-law are expected to provide daily personal care of the aged). For years, the makers of Chinese social policy have sought to eliminate the preference for males by empowering women economically and socially. They believe that female children will be just as desirable as males when it is clear that well-trained, powerful daughters can bring honor to the family name and earn sufficient income for their families.

The offspring in one-child families, whether male or female, have few kinfolk with whom to share tasks when they reach adulthood. They have increased responsibilities for both child care and elder care. And in China, the elderly are especially dependent on

This army officer enjoys some moments with his only child, a son.

[Alain le Garsmeur/Panos Pictures.]

the younger generation. There was never a state-wide pension system. Pensions are even less common now, since the closing of many collectives during the economic reforms of the 1980s and 1990s. Hence, as the single-child family spreads through the generations, elderly parents will remain dependent on their one child.

The one-child family policy has reduced population growth rates: between 1950 and 1990, the Chinese birth rate dropped dramatically, and the one-child family is now the most common family form. Nevertheless, the policy has never been very popular—for the reasons just discussed and because the policy is enforced unevenly. From time to time, the government has offered various incentives to young couples in different parts of the country. In some urban areas, couples who have only one child receive a monthly subsidy and a housing allowance. In other areas, complying couples receive special chances for promotion. In some cases, those who have additional children may lose these benefits, receive demotions, and have to pay a fine; or a pregnant mother may be forced to abort her second child. In other cases, the infractions may be ignored. Some Chinese—those newly affluent in the market economy and some private enterprise farmers—may have additional children and simply pay the substantial fines. Some ethnic minorities are officially exempted from family-size restrictions, presumably to answer political charges that the majority Han ethnic group is too dominant. Still, there is evidence that family-size limits have been imposed on ethnic Tibetan women, some of whom have been forced to have abortions.

There is no doubt that population control has been successful in reducing births, but much of the drop in fertility is simply the result of modernization of life in China and would have happened without the one-child policy. In almost all situations, people who learn to read, move to the city, and take up a lifestyle based on cash income will choose to have fewer children because children are expensive in the new urban circumstances. They do

not help to produce income as they did on the farm; in fact, they require a major investment of time and money to prepare for urban adulthood.

The population pyramid on the left in Figure 9.9 reflects two sharp past declines in birth rates. The first took place during the Great Leap Forward (1959–1961), when policy errors resulted in famine and malnutrition led to lower fertility. After that era, China's birth rate grew rapidly. The second decline began 30 years ago and continued as the result of strict population control efforts in the early 1980s. In the mid-1980s, birth rates increased when market reforms were instituted. Increasing prosperity prompted some Chinese to have two or more children despite official sanctions. By the mid-1990s another significant decline in birth rates was under way, perhaps inspired more by the realities of raising children in urban situations than by the one-child rule.

Gender Attitudes and Population Control. The left-hand pyramid in Figure 9.9 shows another interesting phenomenon. If you look carefully, it appears that for 50 years or more the number of male infants born (or reported) has significantly exceeded the number of female infants. The pyramid shows that in 2001, among those 0–4 years of age, there were 40+ million girls and 50+ million boys, indicating a deficit of about 8 million girls, or a ratio of 119 boys to every 100 girls. (Without intervention, the normal sex ratio at birth is 105 boys to 100 girls.) What happened to the girls? There are several possible answers. Given the preference for male children, the births of these girls may simply have gone unreported as families hoped to conceal the girl and try again for a boy. There are many anecdotes of girls being raised secretly or even disguised as boys. Adoption records indicate that girls are given up for adoption much more often than boys. Or the girl babies may have died in early infancy, either through benign neglect or actual infanticide. Finally, some parents have access to medical tests that can identify the sex of a fetus, and there is evidence that in China, as elsewhere around the world, some of these parents will chose to abort a female fetus. A side effect of the preference for sons is that there is now a growing shortage of women of marriageable age throughout East Asia. In China alone, between 1991 and 1996, 88,000 young women and girls were freed by the police from kidnappers who, for a fee, provide mates for men who have had difficulty finding one.

Elsewhere in East Asia, the cultural preference for sons persists, with the deficit of girls showing up on the 2001 population pyramids for Japan, the Koreas, Mongolia, and Taiwan. Nonetheless, evidence shows that attitudes are changing. Increasingly, girls are being educated as well as or even better than boys: in Japan, South Korea, and Mongolia, the percentage of women receiving higher education equals or exceeds that of men. In China, only half as many women as men receive higher education. Data is not available for North Korea and Taiwan.

Family and Work in Industrialized East Asia

In Japan, Taiwan, and South Korea, the institution of marriage must meet the needs of urban life because close to 80 percent of

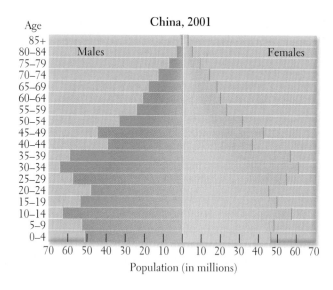

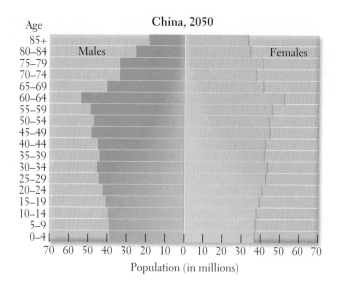

Figure 9.9 **Population pyramids for China, 2001 and 2050.** The population pyramid on the left shows China's sex and age distributions in the year 2001; the one on the right shows them projected to the year 2050 (based on a low estimate of China's growth rate). China's population growth rate does appear to be slowing. Note on the right pyramid, however, that as the population ages, there will be fewer and fewer to care for the large numbers who are aged. [Adapted from U.S. Bureau of the Census, International Data Base, at http://www.census,gov/cgi-bin/ipc/idbpyry.pl.]

the population now lives in cities. In China, more than 65 percent of the people still live in small towns and rural areas, but urbanization is increasing rapidly. Life in a small city apartment is different from life among a host of relatives in a farming village. Nonetheless, in urban situations throughout East Asia, the wife still performs most domestic duties and dedicates herself primarily to looking after her husband and children. Often she also has responsibility for the care of elderly parents or parents-in-law. Surveys show that East Asian women who do work outside the home do not yet desire a job with full responsibilities. Although there is a small but growing group of trained young women who seek careers and are as ambitious as men, even many university-educated married women say they wish to earn only supplementary income for the household.

One reason that the primary duties of family life have been assigned to women is that jobs in East Asian industrial economies are particularly demanding. Workdays are long, and commuting often adds more than three hours to the time away from home. Even more important is the "culture of work," which in Japan, Taiwan, and South Korea (and increasingly in China) is based on male camaraderie and demands an especially high level of loyalty to the firm. In Japan, for example, a firm will have a corporate song, a corporate exercise program, and a strict dress code. After work, leisure time often must be spent with business colleagues.

AT THE LOCAL SCALE   *Chinese Marriage Customs in Cities and Villages*

In most Chinese cities today, a new couple must apply for housing long before they plan to marry. Because women are not allowed to apply for housing until they have worked for a firm for 15 years, the couple will usually apply through the prospective husband's job. While they wait to obtain housing, the couple tries to acquire the necessary household goods. By custom, the groom must provide "48 legs and the whole chicken." This saying refers to the fact that the full furnishings for an apartment have a total of 48 legs (the bed has 4 legs, two bedside stands 8 legs, and so forth). The "whole chicken" is a play on words: *ji* is the Chinese word for *chicken* and is a homonym (same sound, different meaning) for *machine*. Hence, the "whole chicken" refers to all the household appliances, from the sewing machine and television set to the stove and refrigerator. Meanwhile, the bride must provide the "soft goods," such as bedclothes, kitchen and laundry equipment, and a new set of clothes for herself. The groom's contribution is usually far greater than the bride's, a holdover from the prerevolutionary custom of the bride price, a sum paid by a groom to the parents of the bride to compensate them for the costs of raising her and for the loss of her labor.

Management is a vertical hierarchy that is accessible only to those who are perceived as diligent and socially interactive after hours. Rivalry between firms is often so strong that it elicits a siege mentality among workers. It is considered disloyal to refuse overtime, and a man who takes off early to spend time with his wife or to help her with care of the children or elderly is not a "real man." Women, when hired, are there to support men as secretaries and assistants, not to participate as full members of corporate teams.

By 2000, the East Asian urban work ethic and family structure were being publicly challenged, not least by men themselves. In Tokyo, for example, a group calling itself Men Concerned about Child Care meets regularly to discuss ways to participate more fully in home and community life. Another group has formed to lobby for four-hour workdays for both men and women, so that both can spend time with the family. This group has filed lawsuits against employers who routinely require employees to accept job transfers to distant locations where they cannot take their families. These movements may eventually result in some important changes in gender and work roles throughout the region. It is particularly significant that urban men are the ones challenging the extremes of the lockstep East Asian work ethic. They are the group that has been most regimented and most deprived of personal time and family life, and they are also the group that has the power to change the system. If work rules are liberalized and workdays shortened, these policies will add to the cost of East Asian products and lessen their competitiveness in the marketplace. But if employers don't make work situations more attractive to middle managers, efficiency and innovation may suffer, which will also reduce competitiveness.

Indigenous Minorities

Cultural diversity exists throughout East Asia even though most countries have one dominant ethnic group. In China, for example, 93 percent of the population call themselves "people of the Han." The name harks back about 2000 years to the Han dynasty, but it gained currency only in the early twentieth century, when nationalist leaders were trying to create a mass Chinese identity. The term *Han* does not denote an ethnic group but rather connotes a general and unspecified pride in Chinese culture and a sense of superiority over ethnic minorities and outsiders. The main language of the Han is Mandarin, although there are many dialects. (All versions of Chinese use the same writing system.) The Han are overwhelmingly rural agriculturists. About 75 percent of them still live in the countryside, primarily in the eastern half of the country.

Because China is such a populous country (1.2 billion people in 2000), the non-Han minorities, though only 9 percent of the population, still number more than 108 million people. There are more than 55 different minority groups scattered across the vast expanse of China; as Figure 9.10 shows, most live outside the Han heartland of eastern China. They are located in the hinterlands: in the borderlands with Mongolia and Russia in the north and with Central Asia and Pakistan in the west; in Xinjiang and Tibet (Xizang); and in the southern mountains that lead to Southeast

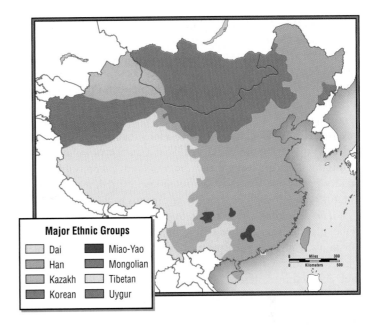

Figure 9.10 Major ethnic groups of China. [Adapted from Chiao-min Hsieh and Jean Kan Hsieh, *China: A Provincial Atlas* (New York: Macmillan, 1995), p. 12.]

Asia. Officially, the largest minority groups have been granted autonomous status and theoretically can manage their own affairs. In practice, however, the Han-dominated Communist party has not allowed self-government. The central government in Beijing has controlled the fate of the minorities, especially those who live in zones that are believed to pose security risks or that have resources of economic value.

Turkic-Speaking Peoples. One such zone is Far Northwest China, where Turkic-speaking peoples live, such as the Uygurs and Kazakhs. Many of these peoples remain nomadic. The Beijing government has sent troops and hundreds of thousands of Han settlers to this area of China. The Han settlers manage mineral extraction and run military bases, facilities for nuclear testing, and power generators. Communist party officials are frank in their assertion that an important role of these Han settlers is to dilute the power of minorities within their own lands and to assist these minorities in ridding themselves of "unacceptable" cultural practices and distinctive national identities. In the case of the Turkic-speaking peoples, cultural assimilation to Han ways would include giving up Islam. Although assimilation may be the long-term outcome, for now there has been a rebirth of Islamic culture, inspired in part by the resurgence of Central Asia in the aftermath of the Soviet decline. Turkic people have revived trade across the borders. The old Silk Road market of Kashgar (now Kashi) is back in business. (The Silk Road was the ancient trading route between eastern Asia and regions to the west, including Europe.) Islamic prayers are once again heard five times a day, Muslim women are again wearing Islamic dress, and Islamic architectural traditions are being revived.

The Hui. The Hui are the descendants of ancient Muslim traders who plied the Silk Road across Central Asia from Europe to Kashgar (Kashi) to Xian. They still live as distinct groups, each numbering a million or more, on the Loess Plateau, in Inner Mongolia, on the North China Plain, in Sichuan Province, and in southeastern China. Some are the descendants of Muslim traders who made their way to China from Malaysia beginning in the twelfth century. The Hui, who have a long tradition of commercial activity, are particularly successful in China's new free-market economy. They are using their money not only for consumption of luxury goods but also to revive religious instruction and to fund their mosques, which are now more obvious in the landscape.

The Tibetans. In contrast to the prosperous Hui are the Tibetans, an impoverished ethnic minority of nearly 5 million individuals scattered thinly over a huge and high mountainous region in western China. Their homeland is widely known as Tibet, but since Chinese troops invaded in 1950, the territory has been divided and assigned to what the Chinese government designates as Xizang and Qinghai provinces. The Chinese government also suppressed the Tibetan Buddhist religion, long the mainstay of most Tibetans' daily life, by destroying thousands of temples and monasteries and massacring many thousands of monks and nuns. The spiritual leader of Tibet, the Dalai Lama, was forced into exile in India in the 1950s, along with thousands of his followers. The Potala, his palace in Tibet's ancient capital Lhasa, is now a museum run by the Chinese government. To dismantle traditional Tibetan society further, hundreds of thousands of Han Chinese were resettled in Tibet, where they control the economy and major cities, exploit mineral and forest resources, and force native Tibetans to adopt Chinese ways. We discuss Tibet in greater detail on pages 494–495.

Ancient Native Peoples of China. In China's far southeast, in Yunnan Province, there are more than 20 groups of ancient native peoples, including the Bai, Yi, Li, Miao, Yao, Dai, and Zhuang. These peoples speak many different languages and live in remote areas tucked away in the deeply folded mountains. Many of these groups have ethnolinguistic connections to the Tibeto-Burmese people, others to Thai and Cambodian people. Interestingly, women are treated more equally here than in Han culture areas. A crucial difference may be that among several groups, most notably the Dai, the husband moves in with the wife's family at marriage and provides her family with labor or income. A husband inherits from his wife's family rather than from his birth family. In contrast, in most of China, a woman traditionally leaves her birth family and moves to her husband's family compound, where she works under the direction of her mother-in-law. Anthropologists think it is not a coincidence that the minorities in Yunnan and neighboring provinces, unlike the majority of Chinese, value female children just as highly as males. Similar patterns of family structure and gender values are noted among indigenous peoples in Southeast Asia, as discussed in Chapter 10.

Aboriginal Peoples in Taiwan. In Taiwan and adjacent islands, the Han account for 95 percent of the population, but this region is also home to 60 minority groups. Nine of these are aboriginal groups that have lived in Taiwan for thousands of years. They have some cultural characteristics—languages, certain types of weaving, the making of bark cloth, iron smithing, and agricultural and hunting customs—that indicate a strong connection to ancient cultures in far Southeast Asia and the Pacific. Mountain dwellers have resisted assimilation better than plains dwellers; both are now protected and may live in mountain reserves if they choose. But increasingly, these Taiwanese aboriginal peoples, who numbered 365,000 in 1994, are attending schools and participating in mainstream urbanized and modernized Taiwan life. Hence native cultures—embodied in languages, beliefs, skills, and modes of family support—are dying out. Like native groups in other societies worldwide, the average incomes of the Taiwanese aboriginals lag far behind those of the majority. In addition, they experience the acute social problems typical of the underclass everywhere (alcoholism, low self-esteem, unemployment, undereducation, poverty).

The Ainu in Japan. Japan has a particularly strong sense of cultural solidarity and a tendency to suppress difference. Nonetheless, there are several indigenous cultural minorities in Japan.

At the market in Kashgar (often called Kashi), an ancient Silk Road city, you can still buy a camel, a fine Oriental rug, the original bagel, or the skin of an endangered animal. You can also get a haircut or get your teeth fixed, as this person is doing. [Reza/National Geographic Image Collection.]

Not surprisingly in a country that prizes sameness, they have suffered considerable discrimination, although the topic is sensitive and rarely discussed. A small and distinctive minority group is the **Ainu.** Now numbering only about 16,000, the Ainu are a Caucasoid group physically distinguished from Japanese by their light skin, heavy beards, and thick, wavy hair. They are thought to have migrated many thousands of years ago from the northern Asian steppes. The Ainu, who lived by hunting, fishing, and some cultivation, once occupied the northern parts of Honshu and Hokkaido islands. They are being displaced by forestry and other development activities. Today there are very few full-blooded Ainu because, despite prejudice, they have been steadily assimilated into the mainstream population. Some now make a living by demonstrating traditional crafts to tourists visiting "Ainu villages."

ENVIRONMENTAL ISSUES

The countries of East Asia share a number of environmental concerns. Some result from high population density and rapid economic development, others from long-term patterns of resource management or ineffective planning in more recent times. China has the widest array of environmental problems, some of which also affect Mongolia and North Korea. Japan, Taiwan, and South Korea are affected by issues that result primarily from their high degree of industrialization, their efforts to overcome limited resource bases, and their high rates of urbanization.

China's remarkable record of improved well-being is now slipping, as a result of the very activities that produced its dramatic economic turnabout. It is now apparent that China has the most severe environmental problems on the planet—so severe that they could prevent future progress (Figure 9.11). Accomplishments such as industrialization, improved transport, increased agricultural production, better housing, widespread home heating, and urbanization are producing air and water pollution, water scarcity, and other negative environmental effects. Pollution alone causes degenerative health conditions such as asthma, chronic bronchitis, intestinal diseases, and various cancers. Adding to the problem is population growth that is still significant, despite vigorous efforts at

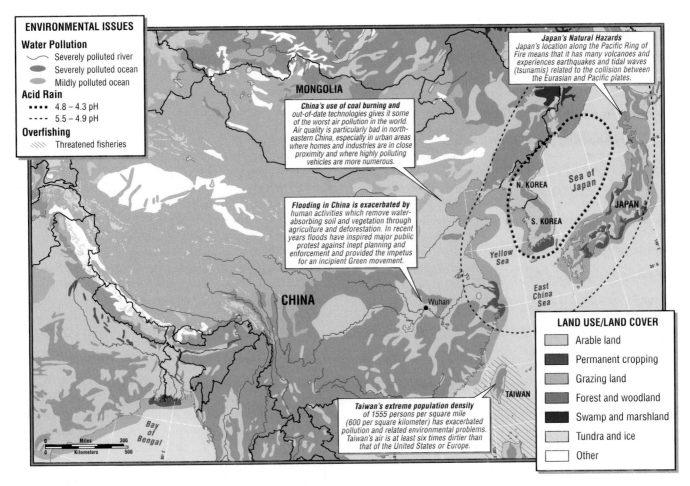

Figure 9.11 **Environmental issues: East Asia.** Environmental issues in East Asia are extensive and include air and water pollution, deforestation (25 percent of China's forests have been clear-cut in just three decades), and flooding. Air pollution in this region often results in acid rain (see Figure 2.20, page 90).

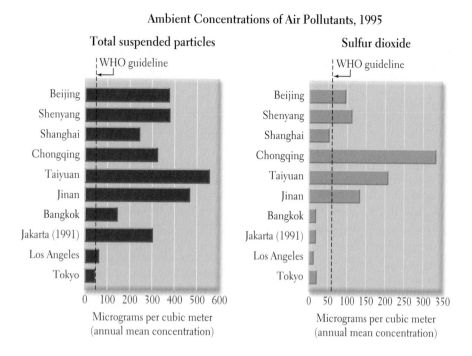

Ambient Concentrations of Air Pollutants, 1995

Figure 9.12 Air pollution concentrations in selected Chinese and world cities, 1995. The vertical dotted lines indicate World Health Organization safe guidelines. [Adapted from *World Resources 1989–99* (New York: Oxford University Press, 1998), p. 117.]

birth control. Of the several environmental problems that exist in China, including resource depletion and erosion, we focus here on just two: air pollution and water availability and quality.

Air Pollution in China

Since the late 1970s, industrial output in China has been growing at an annual rate of 18 percent. Such rapidly developing industry requires large amounts of energy. China's consumption of fossil fuel energy has risen sharply over the last 20 years: between 1977 and 1999, coal consumption doubled and coal now fulfills 75 percent of China's energy needs. The use of coal is likely to increase despite more use of oil, gas, and other sources of energy such as hydroelectricity generated by the new Three Gorges Dam (see the box on page 483). China has considerable reserves of coal.

Coal burning is the primary cause of China's poor air quality and is blamed for a critical toll in respiratory ailments. The combustion of coal releases high levels of two pollutants: suspended particulates and sulfur dioxide (SO_2). In Chinese cities, these emissions can be 10 times higher than World Health Organization guidelines. Figure 9.12 shows emission levels of particulates and sulfur dioxide for a number of Chinese cities and compares them with those of other large world cities.

The geographic pattern of air pollution in China correlates with population density and urbanization. In the lightly populated plateaus and mountains of the north, northwest, and west, air pollution occurs primarily in urban areas where industries are concentrated and where there are large numbers of homes to be heated. In the Sichuan Basin and in most of eastern China, air pollution is prevalent in both cities and rural areas. Air pollution is the worst in northeastern China because homes and industries are

in close proximity and both depend heavily on coal for fuel. Less heating is necessary in the warmer climate of southeastern China. But the air is still laden with sulfur and suspended particles because there are many industries in this region and they burn a softer, highly polluting coal. Because precipitation rates are higher in the south, air pollution also becomes water pollution. The rain absorbs sulfur in the air and becomes acid rain, which is known to harm many species of plants and animals.

Throughout eastern China, large numbers of people are exposed to high levels of gaseous emissions from vehicles. And yet the use of personal cars has hardly begun: Beijing has just one-tenth the vehicles that Los Angeles has. Yet pollution attributable to vehicles in Beijing already equals that of Los Angeles. The reason Beijing has such high auto emissions is that most of China's vehicles are fueled with leaded gasoline, and lead emissions can reach 14 to 25 times the suggested maximum per vehicle. Chinese cars also emit 6 to 12 times more carbon monoxide than foreign vehicles.

Water in China: Too Much or Too Little, and Polluted

Flooding. Like many places in the world, China suffers from having too much water at certain times and locations and too little at others. During the summer monsoon, huge amounts of rain are deposited on the land, often causing catastrophic floods along the rivers and their tributaries. Flooding is a common and particularly severe problem along the Chang Jiang and Huang He (see pages 488 and 491). Engineers have constructed elaborate systems of dikes, dams, reservoirs, and artificial lakes to help control flooding and to provide water for irrigation and power for industry and urban populations. Since 1949, for example, the government has built 120,000 miles (193,000 kilometers) of dikes.

Despite these efforts, the flood control system failed repeatedly in the late 1990s. Along the Chang Jiang, heavy rains in June and July of 1996 caused some of the worst flooding in two centuries. Several thousand people drowned, and millions were still homeless months later. Crops that were ready for harvest rotted in the fields. Much of the farmland and many essential fish-farming facilities remained unusable for months. This catastrophe was repeated with even more severity in 1998.

Farther north, along the Huang He, the rains are less heavy, but levels in water channels are still reaching the highest marks ever recorded because silt is building up in the riverbed. The situation is so precarious that the silt must be removed by hand to prevent levees from bursting—a disaster that could kill more than a million people.

Flooding is a natural phenomenon, but in China, as in many other places, it has been exacerbated by human activities. Many environmental scientists attribute the sharp rise in worldwide deaths and damage from floods in the last three decades to deforestation, land cultivation, urbanization, mining, and overgrazing—all human activities that remove water-absorbing soil and vegetation. Another suggested factor is a change in rainfall patterns caused by global warming, the result of burning fossil fuels.

Drought. At the other end of the water availability scale, droughts occur somewhere in China every year and often cause more suffering and damage than any other natural hazard. Droughts are triggered by periods of abnormally low rainfall and abnormally high temperatures, with attendant higher evaporation rates. They are made worse by many of the same human factors that contribute to flooding—overgrazing, irrigation, deforestation, global warming, overpopulation, and urbanization. When people begin to live or farm in dry environments, as many millions have done in China in the twentieth century, they clear natural vegetation and tap underground aquifers for water. Removing vegetation increases evaporation rates, and the more intense demand for water depletes the aquifers. Ecological forecasters predict that northern China, parts of the western United States, and other naturally dry areas of the world that are subject to increased human occupation and use of water for irrigation are mining groundwater from aquifers at grossly unsustainable rates. These areas are likely to confront severe water shortages in the next several decades.

Research by China's Geological Environmental Monitoring Institute (GEMI), first reported in August 2001, sounds an urgent alarm about the rapidly falling deep water table under the North China Plain. In 2000 alone, the table dropped 10 feet (3 meters), apparently as a result of greatly increased use of water for irrigation, industry, and public consumption. Nearly 100,000 wells ran dry and had to be abandoned. Of the 220,000 new wells dug, those near Beijing had to go down half a mile (800 meters) to reach water.

The decrease in available groundwater is coming at a crucial time in China. Water shortages threaten China's food security and economic growth. The North China Plain is China's breadbasket, producing half of the country's wheat and one-third of its corn. Production already fell short of need in 2000, so coming shortages in irrigation water raise questions of food security. China's population is projected to grow by another 126 million over the next 10 years, and because of urbanization and rising water consumption, the country's urban water demand will increase from 65 billion cubic yards (50 billion cubic meters) in 2001 to 105 billion cubic yards (80 billion cubic meters) by 2010. Industrial water needs are expected to expand by 62 percent. In China, it takes 20 to 40 tons (18 to 36 metric tons) of water to produce a single ton of steel, and 1200 to 1700 tons (1100 to 1500 metric tons) to produce a single ton of textile fiber. The United Nations reports that in China's cities, water shortages are curtailing industrial production by more than U.S. $10 billion a year. Thus, water shortages retard development.

With the main aquifer under a key region being depleted, China has limited options: find new sources of water or find ways to conserve. Conservation is the less expensive option by far and is much likelier to reap results. The question remains: can conservation save enough water to make a difference? If China's grain production fell and it had to import even 10 percent of its current needs, it would quickly deplete present exportable world grain supplies, and the price of grain products would rise worldwide.

Water Shortages and Pollution. Water use is commonly divided into three basic categories: industrial, public (domestic and municipal), and agricultural. In China, industries use 7 percent; public use accounts for 6 percent; and agriculture uses the

This is the cityscape of Beijing as it adapts to the market economy. Advertisements swathe old buildings from the communist era, cranes mark the locations of future skyscrapers of glass and steel, and bicycles compete with pedestrians, who must pick their way around hazardous potholes and building materials. Meanwhile, dust and exhaust mix to create a constant polluted haze. [Stuart Franklin/Magnum Photos.]

remaining 87 percent. By comparison, in the United States, industry and power plant cooling uses 49 percent of the water; domestic and municipal uses account for 10 percent; and agriculture uses 41 percent. China's allocations, so heavily tilted toward agriculture, resemble those of arid Tunisia in North Africa, which expends 3 percent of its water for industry, 9 percent for domestic uses, and 88 percent for agriculture. The use of so much water for agriculture indicates that China is overly dependent on irrigation for agricultural production, a technology that results in the waste of water through evaporation, the buildup of harmful salts in soil, and the pollution of water with dissolved agricultural chemicals.

The United Nations reports that pollution of China's water, as well as the overall scarcity of water, causes health problems costing U.S. $4 billion a year. Fully one-third of the population does not have safe drinking water. Raw sewage and chemical pollution have contaminated lakes, reservoirs, and 29,000 miles (47,000 kilometers) of China's waterways. Much of the pollution is from synthetic nitrogen fertilizers, but some of the worst water polluters are the village enterprises that have been so successful in creating jobs and alleviating rural poverty since 1978. Furthermore, only 24 percent of the Chinese population has access to adequate sanitation. In Sichuan Province, only 37 percent of the 9 million people who live in the city of Chengdu are connected to any sort of sewer system. In rural areas, 90 percent of the people have only latrines or more primitive facilities. In the east, coastal cities have depleted underground reserves to the extent that new construction is collapsing on subsiding land, and salt water is intruding into water supplies. Some Chinese scientists estimate that one-half of the groundwater supplying Chinese cities is contaminated.

Efforts to Improve Environmental Health. The desire for improved environmental health is fueling efforts at both the national and local levels to alleviate the problems. The 1998 floods along the Chang Jiang, coming on the heels of the 1996 floods, prompted a major public protest against inept planning and inadequate enforcement, both acknowledged by the government. One result was the beginning of a Green movement. High literacy rates and a freer press are creating an informed public that is beginning to press for environmental cleanup. Although only 30 percent of urban water is now recycled, many of China's cities are taking steps to improve recycling procedures. And a system of permits, incentives, and penalties aimed at removing pollutants from industrial wastewater discharge is being imposed on Chinese industry and farming at all levels. Ultimately, this processed water will be safe to use elsewhere.

But even the water management solutions have their own negative environmental effects. One example is a plan to supply water to cities, which includes conservation incentives. Some of China's rivers are being diverted (at least partially) to provide water for growing metropolitan areas. River diversion has enabled stricter monitoring of groundwater usage: those who use more than their planned allotment are fined, those who save water are rewarded. In the Loess Plateau, water from the Huang He is being diverted to several growing cities. River diversion can lower the quality of water downstream and deprive downstream users, such as farmers and small villages, of sufficient water. Also, river diversion destroys natural biological habitats. Damming rivers to generate electricity and control flooding also solves some problems while creating others. China has built 22,000 to 24,000 large dams—half of all the large dams in the world—most of them since 1949. The most spectacular and potentially damaging project is the Three Gorges Dam on the Chang Jiang (see the box "The Three Gorges Dam"). The problems it is creating include most of the typical environmental concerns that arise with dam projects.

Environmental Problems Elsewhere in East Asia

Many environmental problems found in China also exist elsewhere in East Asia. Normally, though, Japan, the Koreas, and Taiwan do not suffer water deficits because they are located in the wet northwestern Pacific. In the late 1990s, North Korea suffered flooding as an immediate consequence of deforestation and soil depletion as a long-term consequence. That country has also recently suffered a series of devastating crop failures related to environmental mismanagement, and it is thought that many thousands of people have died. Mongolia, like northern and northwestern China, has always had to cope with arid conditions; much of the country is desert and grassland. Mongolia's forests are made up of species especially adapted to cold, dry environments; in the late 1990s, hundreds of thousands of acres of these forests were lost to fires.

All these countries have experienced the ills of water and air pollution connected with modern agriculture, industrialization, and urbanized living, but to varying degrees. Mongolia and North Korea have lower overall levels of pollution only because they are not yet heavily industrialized. Air pollution in the largest cities of Japan (Tokyo and Osaka), Taiwan (Taipei), and South Korea (Seoul) and in adjacent industrial zones is severe enough to pose public health risks. Antipollution legislation is being passed and enforcement is increasing, but in these cities as elsewhere around the globe, high population density and rising expectations for better standards of living make it difficult to improve environmental quality. Taiwan's case is an example.

Taiwan's extreme population density of 1555 people per square mile (600 per square kilometer) and high rate of industrialization have exacerbated pollution and related environmental problems. In the far north, around the metropolitan area of Taipei, seven cities house a total of 3.26 million people, and densities can rise to 5000 people or more per square mile (1930 per square kilometer). As standards of living have increased, so has per capita consumption of water, sewage facilities, energy, and material possessions of all sorts.

Automobile ownership has been rising rapidly in Taiwan, to the point where there are now four motor vehicles (cars or motorcycles) for every five residents. This adds up to more than 16.5 million exhaust-producing vehicles on this small island. In addition, there are nearly eight registered factories for every square mile (three per square kilometer), all emitting waste gases. As a result, Taiwan

AT THE LOCAL SCALE   *The Three Gorges Dam*

The Chang Jiang carries about two-thirds of all goods shipped on China's inland waterways. Before the first locks and dams were built at Gezhouba above Yichang in the 1980s, barges, junks, and tugs traveling upstream to Sichuan could not handle the grade and the current at certain points, so they had to be winched up with human power. Crews of as many as 200 men were harnessed to long towlines of braided bamboo extending from the bow of the boat to the shore. "For the whole month's work we were paid three silver dollars in addition to our food," reported one tow crewman when interviewed in the 1980s, "but we had to furnish our own towing harness and sandals."

A workforce of more than 100,000 laborers, most working only with their hands, constructed the Gezhouba lock and dam. The lock improved navigation, and the 21 dam turbines generate three times more electricity than the entire Chinese system supplied in 1949. Despite these improvements, the amount of electricity supplied is now woefully insufficient.

All this will change after mid-2009, when the Three Gorges Dam is scheduled for completion (see the map at beginning of the chapter and Figure 9.17). The Three Gorges Dam is the largest engineering project in history: it will be 600 feet (183 meters) high and 1.4 miles (2.3 kilometers) wide. It is designed to improve navigation on the Chang Jiang, generate electricity equivalent to at least one-tenth of the national total, and control flooding. The Three Gorges Dam will transform Sichuan and other provinces in the heart of China by opening the way for industrialization and global trade. The large city of Chongqing is expected to benefit considerably from economic development brought by the dam and to emerge as the leading metropolis of China's interior.

The Three Gorges Dam will also have enormous costs. Not only is construction expensive, but there are incalculable costs associated with relocating the 1.3 million people who lived where the dam will form a reservoir some 370 miles (600 kilometers) long. Thirteen major cities, 140 large towns, hundreds of small villages, as many as 1600 factories, and at least 62,000 acres (25,000 hectares) of farmland will be submerged. The new lake will also destroy important archaeological sites, as well as some of China's most spectacular natural scenery.

Other concerns include the possibility of disaster brought on by a major earthquake, the problem of siltation of the reservoir, and disruptions of the ecology of the river itself. One example of eco-

logical loss is the giant sturgeon, a fish that can weigh as much as three-quarters of a ton. Huge numbers of sturgeon used to swim more than 1000 miles (1600 kilometers) up the Chang Jiang past the Three Gorges location to spawn, but now the reproductive process of the sturgeon is being irretrievably interrupted. Hundreds of other species of plants and animals will also be affected, but there is little money allotted to fund research or salvage efforts.

Consequently, the dam has many critics both in China and abroad. In 1992, many Chinese and foreign scientists and engineers tried to dissuade the Chinese government from beginning the project. Moreover, international funding sources such as the World Bank withdrew support because of the risks. However, construction of *Da Ba* (the Big Dam) is a top government priority and is proceeding quickly, even ahead of schedule.

Known as the Pearl of the Three Gorges (and considered to be one of the world's eight unique structures), Shibaozhai Temple is a 12-story wooden pagoda built about 1545 during the Ming dynasty (1568–1644). The temple stands atop a huge rectangular rock, hugging the side of a rock face some 720 feet tall. Shibaozhai Temple will just barely survive the construction of the Three Gorges Dam; the water in the dam reservoir will lap at the temple's base, and the town of Shibaozhai, 300 feet below, will be completely submerged. [Wolfgang Kaehler.]

has some of the dirtiest air on earth—publicly acknowledged by the government as six times dirtier than that of the United States or Europe (see the box "Leisure Time in Taiwan," on page 505).

In the category of natural hazards, Japan is a special case. Its location along the northwestern edge of the Pacific Ring of Fire means that it has many volcanoes (Figure 9.13), and it experiences earthquakes and seismic sea waves (**tsunamis**) resulting from the collision between the Eurasian Plate and the Pacific Plate. These natural hazards are a constant threat in Japan, and the heavily populated zone from Tokyo southwest through the Inland Sea is particularly endangered by earthquakes, volcanoes, and tsunamis.

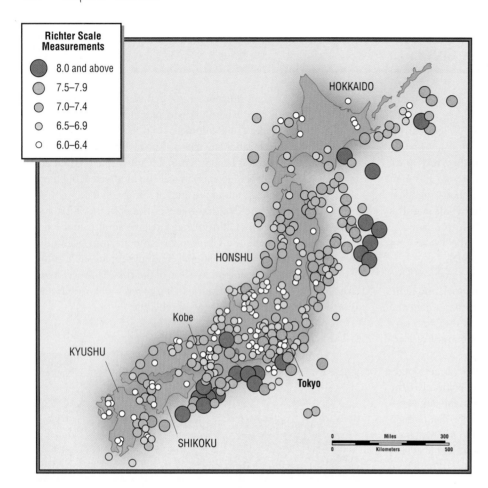

Richter Scale Measurements

- 8.0 and above
- 7.5–7.9
- 7.0–7.4
- 6.5–6.9
- 6.0–6.4

HOKKAIDO

HONSHU

Kobe

KYUSHU

Tokyo

SHIKOKU

Miles 300
Kilometers 500

Figure 9.13 **Earthquakes in Japan.** This map shows the location of dozens of severe earthquakes (above 6.0 on the Richter scale) that have devastated Japan since the year 599. For example, the 1995 Kobe earthquake, which measured 6.8, leveled much of the city and killed over 6000 people. On the map the larger and darker the circle, the higher the scale and the greater the intensity. [Adapted from *New York Times* (January 13, 1998): B14.]

MEASURES OF HUMAN WELL-BEING

More than in most other regions on earth, there is extreme variation from country to country in East Asia on all indexes of human well-being. Table 9.2 combines several indicators that are widely used by international agencies to compare how different countries are doing in providing basic health care and welfare for their citizens.

Human well-being is a complicated assessment that includes social, cultural, and political values as much as economic measures. The most general indicator of well-being is gross domestic product (GDP) per capita calculated at purchasing power parity (PPP) (Table 9.2, column 2). This value ranges from U.S. $1000 per capita per year in North Korea to U.S. $23,257 per capita per year in Japan. This is the widest spread between countries (23 to 1) in any world region. Yet the apparently enormous regional disparity of wealth is somewhat misleading. Socialist governments in North Korea, Mongolia, and China have attempted for half a century to keep abject poverty at bay by providing basic necessities for their citizens. In fact, mass poverty such as that found in India or Bangladesh, or during famine in Africa, has not existed in China or Mongolia for several decades. Mongolia and China now fall in the medium category of human development. Figures for North Korea are not available. Although this country is known to be extraordinarily poor, the life expectancy value indicates that basic needs are being met. North Korean infant mortality rates are also lower than

those for China or Mongolia. Complete figures for Taiwan are not available because of its unresolved political status vis-à-vis China. Nonetheless, Taiwan has a relatively high GDP per capita and provides many social services for its citizens. Were Taiwan ranked on the HDI, it would doubtless be in the high category.

United Nations Gender Empowerment Measure (GEM) figures are missing for all East Asian countries but South Korea and Japan, and Gender Development Index (GDI) figures are not available for all countries. As a result, it is difficult to assess progress toward gender equity; the patterns appear to be irregular, at best. In China, nearly 100 percent of young girls and boys attend primary school, but only 65 percent of the eligible girls attend secondary school, compared to 75 percent of boys. Only 0.003 percent of women go to college, compared to 0.006 percent of men. Women workers earn about two-thirds of what men workers earn. In Mongolia, women are 25 percent more likely than men to receive a high-school education, and more than twice as many women as men go beyond high school. Yet, as in China, women earn about two-thirds of what men earn. In the industrialized societies of Japan and South Korea, nearly all women attend high school, and more than two-thirds attend college or receive training beyond the high school level., Yet, in both countries, women earn less than half of what men earn.

China has made the most spectacular improvements in human well-being. Since the 1970s, China has reduced infant mortality by

TABLE 9.2 *Human well-being rankings of East Asian countries*

Country (1)	GDP per capita, adjusted for PPP[a] in 1998 $U.S. (2)	Human Development Index (HDI) global rankings, 2000[b] (3)	Gender Development Index (GDI) global rankings, 2000[c] (4)	Female literacy (percentage), 1998 (5)	Male literacy (percentage), 1998 (6)	Life expectancy (years), 2000 (7)
For comparison						
United States	29,605	3	11	99	99	76
Mexico	7,704	55	50	89	93	72
East Asia						
China	3,104	99 (medium)	79	75	91	71
Hong Kong	20,763	26 (high)	26	89	96	80
Japan	23,257	9 (high)	9 (GEM = 41)	99	99	80
North Korea[d]	1,000	N/A[e]	N/A	95	95	72
South Korea	13,478	31 (high)	30 (GEM = 63)	96	99	71
Macao[d]	17,500	N/A	N/A	86	93	82
Mongolia	1,541	117 (medium)	N/A	77	88	63
Taiwan[d]	16,100	N/A	N/A	94	94	75

[a]PPP = purchasing power parity.
[b]The high, medium, and low designations indicate where the country ranks among the 174 countries classified into three categories (high, medium, low) by the United Nations.
[c]Only 143 out of a possible 174 ranked in the world.
[d]Data constructed from several sources.
[e]N/A = data not available.

Sources: Human Development Report 2000 (New York: United Nations Development Programme); and *2000 World Population Data Sheet* (Washington, D.C.: Population Reference Bureau).

50 points, and life expectancy has risen from 63 to 71 years. These advances are the result of efforts to improve basic health care delivery, including massive immunization and family planning programs, programs to control infectious and parasitic disease, and improved nutrition. Also contributing to well-being are better housing, improved water quality, and literacy training. These advances were mirrored for a time in Mongolia and North Korea, but both countries have lost ground in the 1990s. In Japan, Taiwan, and South Korea, basic health indicators such as life expectancy and infant mortality are comparable to those of Europe and the United States.

SUBREGIONS OF EAST ASIA

This section discusses five of the six countries of East Asia—Japan, Taiwan, North and South Korea, and Mongolia—as distinctive subregions. Because the sixth country, China, is so large in area and population and contains so many physically and culturally distinctive parts, it is divided into four subregions (Figure 9.14). The focus here is mainly on the modern period in each of these countries (from the twentieth century on); brief historical summaries were given early in the chapter (see pages 458–464). Although China dominates East Asia in size and population and wields great influence, all the countries have their own unique physical environments and social circumstances. Taiwan, South Korea, and Japan are now far more developed than China, and their experiences with

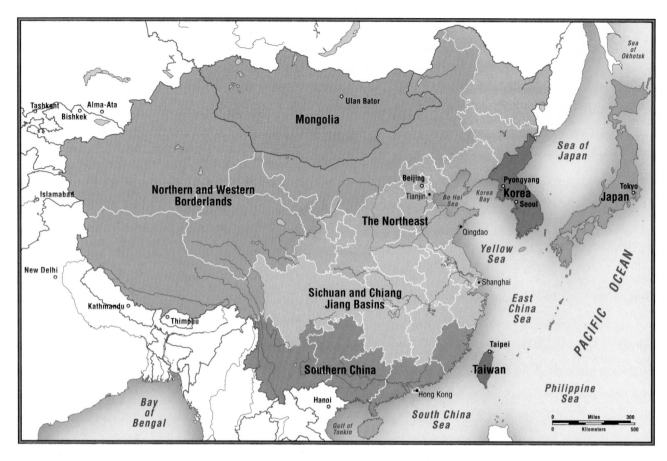

Figure 9.14 **Subregions of East Asia.** This map shows the four subregions of China, as well as Taiwan, Japan, North and South Korea, and Mongolia.

rapid economic growth have influenced China and many other developing countries. Mongolia and North Korea are both markedly less developed and less prosperous than all other countries in the region.

CHINA'S NORTHEAST

China's Northeast consists of the Loess Plateau, the North China Plain, and the Far Northeast (Figure 9.15). The Loess Plateau and the North China Plain are the ancient heartland of China. By the eighth century, the city of Chang'an (not to be confused with Changan in South China)—near modern Xian in the southeastern Loess Plateau—was an imperial capital and cultural center for peoples of both the Loess Plateau and the North China Plain. At that time, with 2 million inhabitants, Chang'an may have been the largest city on earth. After A.D. 900, the center of Chinese civilization shifted from the plateau down and east to the North China Plain, but the Loess Plateau remained a crucial part of China. Chang'an and Xian served as the eastern terminus of the lucrative Silk Road trade that connected China with Central Asia and Europe at least by the time of the Roman Empire, and perhaps earlier. In addition, the Loess Plateau acted as a **buffer zone**—

a neutral area that serves to prevent conflict—between the more populous plain in the east and the arid areas to the west and north occupied by often hostile nomadic herders.

The Loess Plateau and the North China Plain are linked physically as well as culturally. Both are covered by fine yellowish **loess,** or windblown soil. For millennia, dust storms have picked up loess from the surface of the Gobi and other deserts to the north and west, then carried it east. The loess has drifted into what used to be deep mountain valleys in Shanxi and Shaanxi provinces, creating an undulating plateau. The Huang He, in an earth-moving process that might be even more massive, has transported enormous amounts of loess sediment from the Loess Plateau to the coastal plain. Over the millennia, this river has created the North China Plain by depositing its heavy load of loess sediment in what was once a much larger Gulf of Bo Hai.

The Loess Plateau

This dry upland region is sheltered from the moisture-bearing monsoon rains by higher surrounding mountains. An unusually diverse mixture of peoples from all over Central Asia and East Asia has found it a fertile, though challenging, place to farm and

Figure 9.15 China's Northeast subregion.

herd. Among the many culture groups that share this populous and productive plateau are the Hui and the Han of China and, particularly in the western reaches, Mongols, Tibetans, and Kazakhs. Many of the inhabitants farm cotton and millet in irrigated valley bottoms or raise sheep on the drier grassy uplands. China's largest coal reserves, also found here, supply energy for the nation's industries.

Soil erosion is a particularly severe problem in the Loess Plateau. Thousands of years of human occupation have stripped the land of ancient forests, leaving only grasses to anchor the loess. Because the loess is so thick—hundreds of feet deep in many areas—deep gullies form after torrential rains. Such gullies now cover much of the landscape. Although hillsides have been terraced for agriculture, they are prone to landslides, and so many of them are no longer farmed. For decades, the government has maintained a reforestation campaign to stabilize slopes, and the new forests are spreading. But cropland is too badly needed to restore all the land to forests.

In some places in Shanxi Province, the loess is so thick and firm that people build energy-efficient, cavelike houses in it. [Wolfgang Kaehler.]

The North China Plain

The North China Plain is the largest and most populous expanse of flat, cultivable land in China. Since the sixth century, it has been home to most of the imperial dynasties. From this seat, the Han Chinese have dominated for the longest period, spreading through the various river basins of China.

Today the North China Plain is one of the most densely populated spaces in the country: it has an average of more than 1400 people per square mile (540 per square kilometer). More than 370 million people—more than live in the whole of North America—live in a space about the size of France. The majority are farmers who work the relatively small fields that cover the plain. Most produce wheat, which grows well in the relatively dry climate. The food historian E. N. Anderson says of the North China Plain that there is not

one square inch of natural cover left. The forests that once blanketed the plain were cut down millennia ago, and now the only trees that survive are those around temples or planted as windbreaks.

The Huang He (Figure 9.16) is the most important physical feature of the North China Plain. The river is a major transport artery; it provides some irrigation water but is particularly famous for its disastrous floods. After the river descends from the Loess Plateau, its speed slows drastically on the flat surface of the plain, and its load of silt begins to settle out. The silt steadily raises the level of the riverbed year after year. To contain the river, people dredge the channel and bank the silt into levees along both sides of the streambed. Over time, deposited silt has built up the riverbed until, in some places, it lies higher than the plain. During the spring surge, the river occasionally breaks out of the levees and rushes across the surrounding plain in a destructive flood. The spring surge has helped the Huang He cut many new channels over time, as can be seen on the map in Figure 9.16. In one way, the spring floods are a gift that endows the plain with a new layer of fertile soil as much as a yard (a meter) thick. But over 3000 years of recorded history, some 1500 floods have wiped out crops, brought famine to millions of people, and destroyed whole cities. For these reasons, the river is referred to as both the Mother of China and China's Sorrow.

Far Northeast China (Manchuria)

The Far Northeast corner of China, also known as Manchuria, has long been considered a peripheral region, partly because of its location and partly because of its harsh climate. The winters are long and bitterly cold, the summers short and hot. The frost-free season is often less than 120 days, so the types of crops that can be grown are limited. Nonetheless, Manchuria has become increasingly important for its agriculture and natural resources. The region is endowed with fertile plains; rain that falls primarily in the summer; thickly forested uplands; and mineral resources of oil, coal, gas, gold, copper, lead, and zinc. Heilongjian, the most northerly province, is now covered with huge state farms that grow wheat, corn, soybeans, sunflowers (for oil), and beets. With the discovery of oil at Daqing, the province has become the country's leading oil and gas producer. The rich mineral base is the foundation for a growing industrial sector. The huge iron and steel complex in Anshan produces more than 20 percent of the national output.

Development in Far Northeast China has been helped by its extensive rail and road network and is increasingly linked to river traffic on the Amur. In the nineteenth century, the Russians built a part of the Siberia-to-Vladivostok railway through the fishing village of Harbin in Heilongjiang. Harbin, located on the Amur, is now an industrial city of more than 4 million. The Amur River, which runs along more than 1000 miles (1600 kilometers) of the border between Russia and China, is the conduit for increasing trade among Russia, China, Japan, and the United States, and ports on the Amur already handle more than 20 percent of the trade between Russia and Japan. U.S. trade passing through Amur ports doubled several times over between 1995 and 2000. Although China has ample resources (fish, petroleum, and wheat) to sell along the Amur, trade

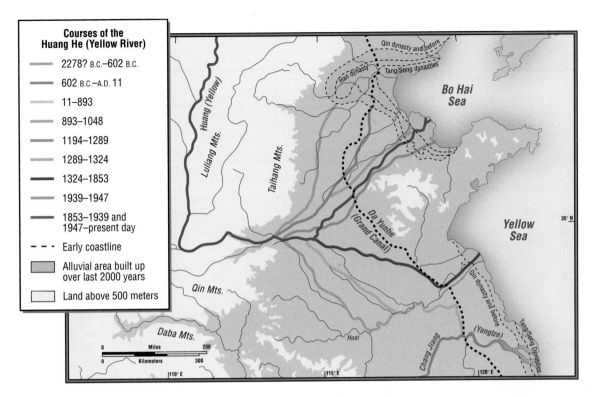

Figure 9.16 The Huang He's changing course. The lower course of the Huang He has changed its direction of flow many times over the last several thousand years. In 2000 B.C., it flowed north and entered the Gulf of Bo Hai, south of Beijing. Then it repeatedly shifted to the east, then southeast, like the hand of a clock, until it joined the Huai, and finally cut south all the way to the delta at Shanghai. Now it once again flows into the Gulf of Bo Hai. [Adapted from Caroline Blunden and Mark Elvin, *Cultural Atlas of China,* rev. ed. (New York: Checkmark Books, 1998), p. 16.]

alliances along the river have been fragile so far, and disputes with Russia over fishing rights have been frequent. Nonetheless, through the efforts of the World Wildlife Federation, Russia and China have agreed to species preservation along the Amur, where huge and diverse wetlands provide habitat for endangered storks, leopards, and the Siberian tiger (see Chapter 5, page 275).

Beijing and Tianjin

Beijing, China's capital city, lies between the North China Plain and the Far Northeast. It is the administrative headquarters of the People's Republic of China; has several of the nation's most prestigious universities; and, with the nearby port city of Tianjin, is becoming an industrial and transport center. Some of the grandest Chinese architectural masterpieces have been preserved there, notably the Forbidden City, the former imperial headquarters. Nonetheless, Beijing has lost some of its character under the communists, who have torn down many older monuments. Supposedly, the demolition was intended to improve the flow of traffic, but it also served to blot out some of China's politically out-of-fashion history. Nowadays, in the context of China's acceptance of capitalism, much of Beijing is being reconstructed, this time as a center for international commerce, with new neighborhoods of high-rise office towers and hotels replacing older neigh-

borhoods of tile-roofed single homes and communist-era apartment blocks.

CENTRAL CHINA: THE BASINS OF THE CHANG JIANG

Central China consists of the upper, middle, and lower portions of the Chang Jiang (Figure 9.17). Like many of Eurasia's rivers, the Chang Jiang starts in the Tibetan (Xizang-Qinghai) Plateau. It then skirts the southeastern rim of the Red Basin in Sichuan Province (this is the upper basin, also known as the Sichuan Basin). From there, the Chang Jiang flows rapidly through the Witch Mountains (Wu Shan) to China's densely occupied central plain (the middle basin). Finally, it winds through the coastal plain (the lower basin) to the Pacific Ocean, exiting in a huge delta on which sits the famous trading city of Shanghai.

Sichuan Province

Sichuan is China's most populous province, with 107.2 million people, and one of its richest in resources. It has fertile soil, a hospitable climate, inventive cultivators, and sufficient natural resources to support diversified industry. Of all the provinces of

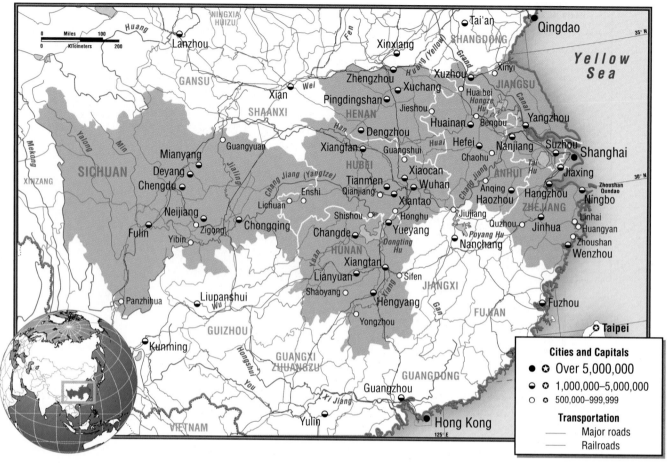

Figure 9.17 The Central China subregion: the Chang Jiang Basin.

China, Sichuan perhaps best embodies the revolutionary ideal of complementary agricultural and industrial sectors.

The heart of the province is the Sichuan (or Red) Basin, so named because of the underlying red sandstone. This is a fertile region of hills and plains crossed by many rivers that drain south toward the Chang Jiang. The shape and angle of the Sichuan Basin result in a milder and wetter climate than is usual at such a midlatitude, continental location. Because the basin is tilted toward the south and therefore receives comparatively direct sunlight during much of the year, the overall temperatures are higher than would be expected. Furthermore, the basin is surrounded by mountains that are highest in the north and west (the latter rising to 24,000 feet, or 7300 meters), which form a barrier against the arctic blasts of winter. The mountains also trap and hold the moist, warm air moving up from the southeast all year round. For these reasons, the Sichuan climate is generally mild and humid, and the basin is so often cloaked in fog or low cloud cover that it is said: "A Sichuan dog will bark at the sun."

For the last several thousand years, the basin has been home to a dense population of relatively affluent farmers. They have cleared the native forests, manicured the landscape, and chan-neled the rivers into an intricate system of irrigation streams. Only a few patches of old-growth forest are left in the uplands and mountains. The geographer Chiao-Min Hsieh vividly describes how the wet "rice fields in [the lowlands] are shaped like squares on a chessboard. Everywhere one can hear water gurgling like music as it brings life and growth to the farms."

The use of irrigation in such a wet climate may seem surprising, but it is an effort to maximize production. By diverting the river water into many small sluiceways, the farmers can control flooding and make water available during the winter dry season, when temperatures are still high enough for some crops to be cultivated. Moreover, this is a region where **wet rice cultivation** is practiced. This type of rice production, which results in particularly prolific yields, requires that the roots of the plant be submerged in water for part of the growing season. It would be risky to depend solely on rainfall to keep the rice paddies filled.

Most of the people of Sichuan live in rural areas in the basin, where the population density is more than 800 people per square mile (300 per square kilometer). Their crops include rice, wheat, and maize. Most farmers raise such animals as silkworms, pigs, and fowl as a sideline; and in the new market economy, it is often

AT THE LOCAL SCALE *Silk and Sericulture*

Silk has long been an important crop in central and southern China. Silk cloth has figured prominently in China's interactions with the outside world—indeed, it gave its name to the Silk Road trade with Central Asia and Europe. Silk production spread to Japan in the seventh century and is now a major industry in many parts of South and Southeast Asia. Although the overall economic importance of silk has been reduced because China has turned to other industries and because fabrics made of artificial fibers and cotton have largely replaced silk fabrics, silk is still an important commodity in Chinese commerce.

The silk fiber is made by caterpillars (of the *Bombyx* moth family) that produce a fine, strong, continuous fiber from which they spin a cocoon. The caterpillars are raised from eggs and are fed on the leaves of the mulberry tree. The caterpillars that hatch from one ounce of eggs will consume the leaves of 25 to 30 trees! To make silk, the completed cocoons, with the caterpillar inside, are plunged into hot water to kill the insect. The cocoons are then carefully dried and unspun so the fiber can be used to weave cloth.

The process of raising the caterpillars and harvesting the cocoons is called sericulture. People who tend the cocoons must take great care to control temperature, humidity, and disease. Losses of 70 percent of the product are not uncommon, although technical assistance has improved production. Sericulture is primarily a sideline cottage industry run by women—one of the ways that Chinese women, and now women throughout East Asia, have gained access to their own cash.

these sidelines that earn them the most cash income (see the box "Silk and Sericulture"). In the mountain pastures to the west of the basin, farmers raise cattle, yaks, sheep, and horses, which produce meat, hides, wool, and animal power. Many of the mountain herders are Tibetan (Sichuan has the largest group of Tibetans outside Tibet—close to 1 million).

The two main cities of the Sichuan Basin are Chengdu, the capital, and Chongqing. Chengdu, with a current population approaching 9 million, was an ancient trading center. It remains a main hub for national rail and highway service. Chengdu is also home to light industry, especially food processing and the manufacturing of textiles and precision instruments. Some of the finer quality products are sold in the global market, as a check of labels in gourmet food stores or discount clothing stores will attest. Chongqing, with a population of well over 3 million, has a high concentration of heavy industry. Its thousands of iron and steel manufacturers and machine-building industries take advantage of Sichuan's rich deposits of iron, coal, copper, and lead.

Sichuan's historic ability to support its dense population has made it a productive region, but with increasing population and growing expectations for yet more affluence, degradation of the environment may cause intolerable pollution levels. This side effect is a result of the basin's tendency to retain warm, moist air, which allows industrial and vehicle emissions to build up. For example, sulfur dioxide emissions in Chongqing in 1995, measured in micrograms per cubic meter, were five times greater than the guidelines recommended by the World Health Organization (see Figure 9.12, page 480).

Sichuan's age-old connection with the outside world has been the Chang Jiang. The river used to leave the basin in the southeast in a turbulent rush through the famous Three Gorges of the Wu Shan. The changes to the river caused by the Three Gorges Dam project are discussed in the box "The Three Gorges Dam" (see page 483).

The Central Plain (the Middle and Lower Basins and Shanghai)

After the Chang Jiang emerges from its concrete shackles in the Three Gorges Dam, it traverses an ancient, undulating lake bed that is interrupted in many places by low hills. This is the middle basin of the Chang Jiang. It and the river's lower basin, leading to the coast, are rich agricultural regions dotted with industrial cities.

The lake bed of this middle basin and the bed of the lower basin (a coastal plain) have been silted in with **alluvium** (river-borne sediment) carried down from the Sichuan Basin. Other rivers entering the Chang Jiang Basin from the north and south also bring in loads of alluvium along with significant volumes of water, which are added to the main river channel. The Chang Jiang system carries a huge amount of sediment—as much as 186 million cubic yards (142 million cubic meters) per year—past the large industrial city of Wuhan. This sediment, which under natural conditions was deposited on the floor of the basins during annual floods, formerly enriched agricultural production. But because the floods often destroyed people, animals, and crops, the Chang Jiang, like the Huang He, now flows between levees. Still, it occasionally breaches the levees and floods both rural and urban areas, as it did twice in the late 1990s.

The climate of the middle and lower basins, though not as pleasant as Sichuan's, is milder than that of the North China Plain. The Qin Ling Mountains and other, lower hills that extend eastward across the northern limits of the basin block some of the cold northern winter winds. They also trap warm, wet southern breezes, so the basins retain significant moisture during most of the year. The growing seasons are long: nine months in the north and ten in the south. The natural forest cover has long since been removed to make agricultural land for the ever-increasing rural population. Summer crops are rice, cotton, corn, and soybeans;

AT THE LOCAL SCALE *Shanghai's Urban Environment*

Shanghai has a long history as a trendsetter. Here we describe three examples: a penchant for stock trading that is not typical of communist regions; the recent, but ephemeral, popularity of pajamas as street fashions; and some tentative experiments with alternative sexual identity.

The Secrets of Wall Street was one of the hottest titles in Shanghai bookstores in the late 1990s. Housewives, barmaids, and bureaucrats were buying the book, taking investment courses, and placing orders at automated brokerage machines or trading via the telephone. Shanghai investors tend to be highly speculative, most holding a stock for only a short time.

Shanghai has long been a leading arbiter of fashion. Until recently, it was only the wealthy who indulged in the effort to keep abreast of trends. In the late 1990s, however, ordinary people began to participate in an extraordinary fashion statement by wearing pajamas on the street. Pastel printed, light cotton pajamas became the leisure-time garment of choice for males and females, young and old. The reason? Comfort, convenience, and good looks. In Shanghai's steamy climate, the light cotton fabric is undeniably comfortable. And for many migrants newly employed in urban firms, the ability to afford pajamas was a statement about well-being and upward mobility. But, like most modern fashions, the popularity of street pajamas did not last long. By 2001, pajamas had vanished and synthetic microfiber clothing, by then current in Sidney, Singapore, and Seattle, was favored.

Shanghai is also introducing China to a more broadminded multicultural approach to modern living. Although homosexual orientation has been taboo in China in the past, many gay and lesbian young people are now finding that Shanghai offers them their first opportunity to live their preference openly. Official tolerance of gay and lesbian sexual identity is quietly growing, to the point where there are now public gathering spots, restaurants, and bars where people can meet openly—one being a park across from a Shanghai police precinct station. In response to the apparently more open attitudes, some gays and lesbians are choosing to tell their friends and coworkers about their lifestyles. Some report acceptance by heterosexual friends; most, however, are hesitant to tell their families back in the villages, where sexual discussions of any sort are considered to be in poor taste.

Source: Adapted from *The Wall Street Journal* (August 17, 1997); *The New York Times* (August 6 and September 2, 1997); "Lights, culture, action, Shanghai is back," *Asiaweek* (March 30, 2001): 1.

Busy shopping street scene, Shanghai. [Keren Su/China Span.]

winter crops include barley, wheat, rapeseed, sesame seed, and broad beans. As the river approaches the Pacific, it deposits the last of its sediment load in a giant delta. The old and rapidly reviving trading city of Shanghai sits at the delta's outer limits (see the box "Shanghai's Urban Environment").

There are more than 400 million people in the hinterland drained by the lower Chang Jiang and its tributaries. In the past, many of them were farmers, growing wheat and rice, but every year more people migrate to the many industrialized cities of the region. Shanghai overshadows them all in importance. For many centuries, Shanghai's location facing the East China Sea, with the whole of central China at its back, positioned the city well to participate in whatever international trade was allowed. When the British forced trade on China in the 1800s, Shanghai became their base of operations. Before the communist revolution, Shanghai was among the most refined and elegant cities on earth, home to well-educated literati, wealthy traders, connoisseurs of fine antiquities—and a goodly portion of underworld figures as well. After the revolution, the communists designated the city as the nexus of capitalist corruption and it fell into disgrace.

Shanghai was well situated to take advantage of the changes that began in the 1980s. In the early twenty-first century, the Chinese government has given developers incentives (such as reduced taxes, help in preparing sites, and permission to take profits out of the country) to build factories, workers' apartment blocks, international banks, luxury apartment houses, shopping malls, and elegant postmodern-style villas. Today, some of the Overseas Chinese—those who fled the repression of the Communist party to places such as Taiwan, Singapore, and Kuala Lumpur—are returning to invest in everything from television stations, production companies, and discotheques to factories, shopping centers, and entertainment parks. Europeans and both North and South Americans have been attracted to Shanghai, too, hoping to find joint ventures with Chinese partners so they can tap into the huge pool of consumers that is emerging in China.

CHINA'S FAR NORTH AND WEST

The northern and western interior provinces have long been considered the periphery of China (Figure 9.18). Here the land is especially dry and often cold, and today's cultures retain influences from a long history of nomadic pastoralism. Settlements are widely dispersed, and crop agriculture usually requires irrigation. Much of this large interior zone is designated as autonomous regions rather than as provinces because of the high percentage of ethnic populations. As mentioned in the earlier discussion of minorities in China, the designation of an autonomous region has not meant that the region's people have ultimate power over their affairs, nor can they even count on being the majority population in the future. The central government in Beijing retains control over economic and political policy and over the region's industrial assets. Furthermore, millions of Han from eastern China are being encouraged to immigrate to the Far North and West, where they are offered the best jobs. In comparison to eastern (coastal) China, the native citizens of this region are very poor—for example, farmers here earn one-third the income of farmers in the eastern regions.

Xinjiang

The Xinjiang Uygur Autonomous Region, in the Far Northwest of China, is the largest of all the provinces and autonomous regions. It accounts for one-sixth of China's territory. Once considered extraordinarily remote from eastern China, it is now, paradoxically, becoming a region of lively trade and enterprise.

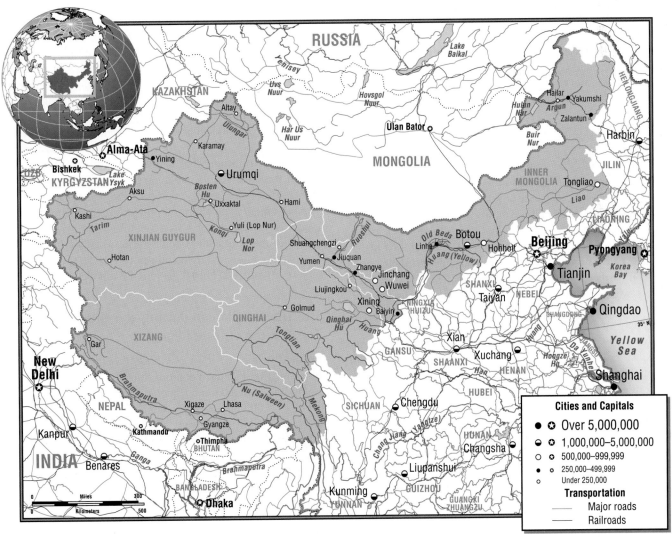

Figure 9.18 **China's Northwest subregion.**

In Yining, northwest Xinjiang, a Kazakh girl plays "kiss the maiden." In the game, a suitor gallops after the girl and tries to steal a kiss. If he succeeds, she chases him and tries to beat him with a whip. The contest tests the riding skills of both future spouses. [Jay Dickman.]

Xinjiang has only 17.5 million inhabitants, and their roots lie mainly in Central Asia. Of this number, the most numerous, at 8 million, are the Turkic-speaking Uygurs. There are also Mongols, Persian-speaking Tajiks, Kazakhs, Kirghiz, Manchu-speaking Xibe, and Hui. New migrants of Han Chinese origin number about 6 million. The peoples native to the region once made their living as nomadic herders and animal traders, moving with their herds and living in **yurts** (or *gers*)—round, heavy felt tents stretched over collapsible willow frames. These cozy houses can be folded and carried on horseback or in horse-drawn carts. In the last few decades, many nomadic people have taken jobs in the emerging oil industry and now live in apartments provided for workers. The Han migrants live primarily in the cities. They work as bureaucrats; in the oil, gas, or nuclear power industries; in state-owned agricultural colonies; or on highway and railroad building projects.

Xinjiang consists of two dry basins: the Tarim Basin, occupied by the Taklimakan Desert, and the smaller Dzungar (Junggar) Basin to the northeast. Both are virtually surrounded by 13,000-foot-high (4000-meter-high) mountains topped with snow and glaciers. In this distant corner of China, far from the world's oceans, rainfall is exceedingly sparse. Snow and glacier melt from the high mountain peaks are important sources of moisture. Much of the meltwater makes its way to underground rivers, where it is protected from the high rates of evaporation on the surface. Long ago, people built conduits called *qanats* deep below the surface to carry groundwater dozens of miles to areas where it was needed.

Qanats have made some of the hottest and driest places on earth productive. The Turfan Depression, situated between the Dzungar and the Tarim basins, is a case in point. Temperatures often reach 104°F (40°C), and evaporation rates are extremely high. But because the *qanats* bring irrigation water, this region produces some of China's best luxury foods: melons, grapes, apples, and pears. The produce is destined for urban populations in eastern China.

Xinjiang once occupied a central place in the global economic system. Traders carried Chinese and Central Asian products (silk, rugs, spices and herbs, ceramics) over the Silk Road (Figure 9.19) to Europe, where they were exchanged for gold and silver. Today, the province is recovering some of its ancient economic vitality, but just who will profit is not yet clear.

Xinjiang is now one of a growing number of places on earth where the local and the global, the very ancient and the very modern, confront each other daily. Herdspeople still living in yurts and *gers* dwell under high-tension electric wires that supply new oil rigs. Tajik women weave traditional rugs that are sold to merchants who fly from as far as New Jersey in the United States to the ancient trading city of Kashgar (now Kashi) for the Sunday market, which can draw more than 100,000 shoppers (see the photo on page 478). With the breakup of the Soviet Union, citizens of the new republics of Central Asia are eager to return to their trading heritage, and China is welcoming them and attracting outside investors from Europe and the Americas by establishing SEZs in cities such as Kashgar and Urumqi.

The Uygur and other ethnic leaders of Xinjiang are wary of Beijing's intentions, especially regarding Xinjiang's oil and gas resources. Uygur and other minority leaders have posted Web pages on the Internet in which they join Uygurs in Kazakhstan and Kyrgyzstan in warning that Beijing's true intent is not to improve life for the local people of Xinjiang, but rather to exploit their oil and gas resources and to appropriate settlement land for eastern China's excess population. A young Uygur man in Urumqi, Xinjiang's capital, spoke of his discontent: "I am a strong man, and well-educated. But [Han] Chinese firms won't give me a job. Yet go down to the railroad station and you can see all the [Han] Chinese who've just arrived. They'll get jobs. It's a policy to swamp us."

The Tibetan Plateau (Xizang and Qinghai): The Tibetan Culture Region

Situated in the Far West of China, this region is the traditional home of the Tibetan people. Officially, it includes the Xizang Autonomous Region and Qinghai Province. Xizang and Qinghai average 13,000 feet (4000 meters) and 10,000 feet (3000 meters) above sea level, respectively. They are surrounded by mountains that soar thousands of feet higher. Tibet (Xizang) and Qinghai have rather cold, dry climates (late June can feel like March does on the American Great Plains) because of their high elevation and because the Himalayas to the south block warm, wet air from moving in. Across the region, but especially along the northern foot of the Himalayas, snowmelt and rainfall are sufficient to support a short growing season for barley and such vegetables as peas and broad beans. Snowmelt also forms the headwaters of some major rivers: the Indus, Ganga (Ganges), and Brahmaputra begin in the

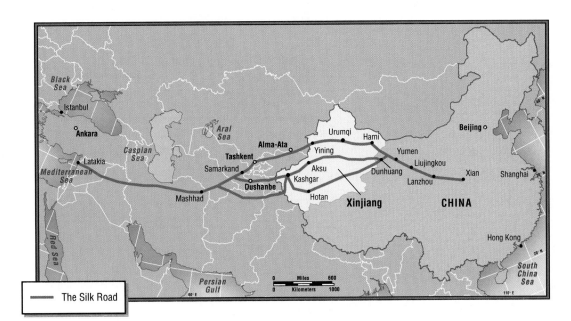

Figure 9.19 The ancient Silk Road. Merchants who plied the Silk Road rarely traversed the entire distance. Rather, they moved back and forth along only part of the road, interacting with other merchants to the east or west. [Adapted from *National Geographic* (March 1996): 14–15.]

western Himalayas; the Nu (Salween), Irrawaddy, Mekong, Chang Jiang, and Huang He all begin along the eastern reaches of the Tibetan Plateau.

Traditionally, the economy of Tibet (Xizang) and Qinghai has been based on the raising of grazing animals. The yak is the main draft animal, and it also provides meat, milk, butter, cheese, hide, fur, and hair, as well as dung and butterfat for fuel and light. Other animals of economic importance are sheep, horses, donkeys, cattle, and dogs. Animal husbandry on the sparse grasses of the plateau has required a mobile way of life so that the animals can be taken to the best available grasses at different times of the year. Yet, for several decades, the government has pressured Tibetan herdspeople to give up their portable houses (yurts and *gers*) and settle in permanent locations so that their wealth can be taxed, their children schooled, their sick cared for, and dissidents curtailed. Still, throughout the region, many native (non-Han) peoples continue to live mobile yet solitary lifestyles as they have for centuries, occasionally adopting some of the accoutrements of modern life and adapting to its restrictions.

By the 1990s, the Beijing government's strategy was to overwhelm Tibetans with secular social and economic modernization instead of with military force. China is attempting to bring its economic boom to Tibet by spending hundreds of millions of dollars on housing and on roads and other infrastructure, by trying to attract trade and tourism to this remote region, and by capitalizing on European and American interest in Tibetan culture. China sees its actions in Tibet as part of its overall strategy to integrate the entire country economically and socially. Schools are being built and the job markets opened up to young Tibetans. Increasingly, Tibetans are accepting the continued Chinese presence and are channeling their Tibetan cultural pride into efforts to preserve the Tibetan language and religion.

Women have always had a relatively high position in traditional Tibetan society. Among the nomadic herders, they were free to have more than one husband, just as men were free to have more than one wife. Also, as with some other minority groups in China, ancient marriage customs call for the husband to join the wife's family. This custom alone allowed women to attain a higher status than they had among the Han. Buddhism introduced patriarchal attitudes from outside Tibet, yet encouraged female independence. At any given time, up to one-third of the male population was living a short-term monastic life, so Tibetan Buddhist women have been particularly self-sufficient, often spending days herding on horseback as well as performing most other daily duties that support community life.

Chinese culture has typically regarded the women of the western minorities as barbarian, precisely because gender roles were not clearly defined: they rode horses and worked alongside the men in herding and agricultural activities. An effect of residual Confucian ideas in modern policy is that the highest paying technical or supervisory jobs are often awarded to men, whereas the low-paying, labor-intensive jobs—such as those in agriculture and factories—are offered to women.

SOUTHERN CHINA

Southern China has two distinct sections. The first is made up of the mountainous and mostly rural provinces of Yunnan and Guizhou to the southwest. The second includes the coastal provinces of Guangxi, Guangdong, and Fujian to the southeast, where the booming cities of China's evolving economic revolution are located.

The Yunnan-Guizhou Plateau

Yunnan and Guizhou provinces share a plateau noted for its natural beauty and mild climate and for being the home of numerous indigenous groups that are culturally distinct from the Han

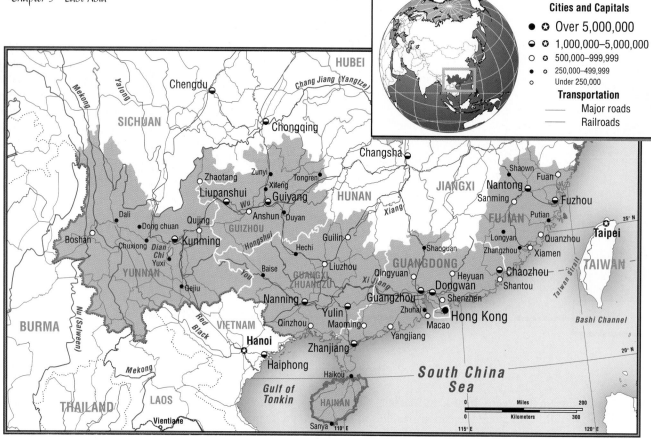

Figure 9.20 **The Southern China subregion.**

Chinese (see Figure 9.10, page 477). The plateau is a rough, mountainous land of deeply folded mountains that trend north-south and carry the headwaters of the Nu (Salween) and Mekong rivers (Figure 9.20). The heavily forested valleys may be as deep as 5000 feet (1500 meters), yet only 1300 feet (400 meters) across. Although people can call to one another across the valleys, it may take more than a day's difficult trip to reach the other side. In some places, rope and bamboo bridges have been slung across the chasms. The landforms here are unstable, and earthquakes cause heavy damage to the carefully constructed rice paddy terraces, which must be absolutely level to receive gravity-fed irrigation water.

The valuable natural resources of the Yunnan-Guizhou Plateau are primarily biotic. Yunnan Province is one of the most heavily forested regions in the country. Although Yunnan is required to supply wood for China's industrialization (25 percent of China's forests have been clear-cut in just three decades), it still has many untouched zones. Yunnan is called "The National Botanical Garden" because it is home to many of China's native plant species and one-third of its bird species (400 or more). Both flora and fauna are now threatened by recent human disturbance of the environment. A British team of biologists traveling in Yunnan in the mid-1980s reported that by then "birds were absent even in the reserves." Until the 1970s, the tropical forests of far southern Yunnan (close to Burma, Thailand, and Laos) harbored elephants, bears, porcupines, gibbons, and boa constrictors.

The lower lying, eastern parts of the Yunnan-Guizhou Plateau are composed of karst (limestone) deposits that are eroded by water to form fantastic landscapes of jagged peaks jutting out from surrounding flat valleys. This phenomenon is found in its most extreme case in Guilin, Guizhou Province, where there is a karst "forest" of 90-foot-high (27-meter-high) jagged limestone columns interspersed with lakes. [Keren Su/China Span.]

Kunming, capital of Yunnan Province, lies at the heart of a booming heroin trade that flows from major producers in the mountains of Burma and Laos to China's northern and east-coast cities, where the heroin is shipped to the global market. Kunming police say that heroin can be bought for 100 yuan (U.S. $12.50) per gram in Yunnan, a price that has left many local youth addicted. Kunming is now also a center of drug treatment facilities, where herbal medicine is the chief therapy.

The Southeast Coast

The southeast coastal zone of China has long been a window to the outside world. Very likely it was the coastal fishing culture that first encouraged contact with strangers on the high seas and provided the opportunity for trade. Arab traders began to visit this part of China more than 1200 years ago. By the fifteenth century, some of the first Europeans in the region described a string of flourishing trading towns all along the coast. People from this region have long ventured out to seek their fortunes throughout Southeast Asia and beyond. The overwhelming majority of Overseas Chinese—who live in places as widely scattered as Singapore; West Africa; London; São Paulo, Brazil; and Atlanta, Georgia—have their roots along China's southeast coast. Even today, fisher folk sail their legendary junks far out to sea to meet officially forbidden trading partners from Taiwan, Vietnam, and elsewhere. In the 1980s, the central government of China decided to take advantage of this long tradition of outside contact by designating several of the old coastal fishing towns as SEZs, anticipating that they would quickly grow into major trading cities. This designation gives them special rights to conduct business with the outside world and allows them to create conditions that will attract foreign investors (see Figure 9.7, page 471).

The principal river of southern China, the Xi Jiang (Pearl River), is joined by several tributaries to form one large delta below the city of Guangzhou (Canton) called the Pearl Delta. The lowlands along the rivers and delta have a subtropical climate and a perpetual growing season. The rich delta sediment and the interior hinterland are used to cultivate sugarcane, tea, fruit, vegetables, herbs, mulberry trees for sericulture (the raising of silkworms), and timber—products that are sold in the various SEZs and exported to global markets.

Guangzhou has long been the most important trading center in South China; but since the nineteenth century, Hong Kong (Xianggang)—which until 1997 was a British crown colony—has posed stiff competition. Now there are several other trading and manufacturing centers along the southern coast that have SEZ status: Xiamen in Fujian Province; Shantou in far eastern Guangdong Province; Shenzhen, adjacent to Hong Kong; Zhuhai, close to the old Portuguese trading colony of Macao, across the Pearl River estuary from Hong Kong; and Haikou, on Hainan Island (see Figure 9.7).

Hong Kong

Hong Kong, one of the most densely populated cities on earth, has packed its 6.4 million people into just 23 square miles (60 square

Rainy and warm Yunnan Province is increasingly important nationally for its market gardening and raising of small animals. Here a family transports its wares in the produce baskets typical of the subregion. [Eastcott/Momatiuk/Woodfin Camp & Associates.]

kilometers) of the territory's total of 380 square miles (985 square kilometers). This very small place has the world's eighth largest trading economy and the world's largest container port. It is also the world's largest producer of timepieces. Hong Kong's populace has achieved an annual gross domestic product per capita of more than U.S. $20,000, with an annual economic growth rate of 10 percent at the end of 2000.

In July 1997, Hong Kong's niche as a British trading enclave ended when the 150-year lease agreement ran out and Hong Kong became a special administrative region (SAR) in China. There had been considerable worry that China would absorb Hong Kong and no longer allow it the economic freedoms it had enjoyed. However, Hong Kong has long played an important role as China's unofficial link to the outside world of global trade. Some 60 percent of foreign investment in China was funneled through Hong Kong before 1997. It seems that although Hong Kong's situation may shift a bit, dramatic change in its fortunes is unlikely. Given the very obvious success of many cities along China's Pacific coast, from Macao to Shanghai, Hong Kong will probably continue its role as a world financial hub in the development of this very fast growing region. Its new state-of-the-art airport, called

AT THE REGIONAL SCALE *Pearl River Megalopolis*

The Megalopolis along the eastern seaboard of the United States, described in Chapter 2 (see page 77), is but one of the world's great concentrations of urban settlement. Another has formed along the lower Pearl River (Xi Jiang) around Guangzhou, Hong Kong, Shenzhen, Macao, and the SEZ cities, and it is now one of the world's largest and fastest growing metropolitan areas. The total population is probably greater than 25 million. This region is often called the South China Megalopolis or the Pearl River Megalopolis.

The SEZ city of Shenzhen has experienced particularly dramatic growth. Located just north of Hong Kong at the neck of the Hong Kong Peninsula, Shenzhen has received spillover investment from Hong Kong and has emerged as a leading center of labor-intensive manufacturing plants that make an array of products for the global export market. As recently as the late 1980s, it was no more than a local farmers' market center with a population in the range of tens of thousands. Now there are 3 million or more residents, as migrants have poured in from a wide area of China to find work and seek their fortunes. Rice and vegetable fields have been paved over and now sprout some of Asia's tallest high-rise buildings. Densities run as high as 23,500 people per square mile (9000 per square kilometer). Shenzhen has grown so quickly that the government has not been able to keep track of the population and can give only crude census estimates.

Many problems accompany such rapid growth. As in North America, in the Pearl River Megalopolis, government jurisdictions overlap, and it is difficult to address problems effectively. Although Shenzhen and other SEZ cities represent vast improvements in living conditions for many migrants, many others find poor working conditions; exploitative pay; and poor, overcrowded, and overpriced housing. Moreover, there are social problems such as crime, prostitution, and alcoholism, as well as a general decline of traditional values and culture. Some people see what they consider an unhealthy emphasis on money and material possessions. Environmental problems include traffic congestion, air and water pollution, and poor building construction, which has led to some disastrous fires and building collapses.

Hong Kong International or Chep Lap Kok, opened in 1999 on a nearby artificial island and is evidence of expectations that the city will continue to play a major role in global trade.

Macao

Macao (also spelled Macau) is the oldest permanent European settlement in East Asia. Portuguese traders first arrived in about 1516, and by 1557 they had established a colonial trading center. Macao remained a Portuguese colony until the Chinese regained control of it at the end of 1999. The city is built on a series of three islands that lie adjacent to the southern China coast across the Pearl Delta from Hong Kong. Macao grew rapidly after 1949, with the influx of refugees from the mainland communist revolution; today, most of its 1.7 million people are Chinese from nearby southern provinces. Many earn a living from the manufacture of clothing, textiles, toys, and plastic products and from the tourism industry. Macao is known as a center of casino and dog-track gambling, and every weekend thousands of visitors from the mainland and Hong Kong cross the water to Macao by high-speed ferry. Although not quite as prosperous as Hong Kong (see Table 9.2) the standard of living in Macao vastly exceeds that of the mainland.

JAPAN

When Japan began its rapid economic and social transformation after its disastrous and humiliating defeat in World War II, few expected it to become a world economic leader in less than 25 years. But by 1975, Japan was turning out some of the most popular consumer products in the world, Japan's workers were adequately paid and had access to health and education services, and the entrepreneurs reaping profits from its industries were investing their money in emerging Asian economies, such as Taiwan, Malaysia, and Thailand. Japan had become an admired model for how to leap from poverty into prosperity.

Nonetheless, Japan faltered during the 1990s. As neighboring Southeast Asia entered a difficult era of economic and social reform, Japan lost markets there. Many Japanese companies that had invested heavily in the expectation of continued rapid growth were faced with staggering losses. Furthermore, the exposure of corruption and cronyism at the top levels of Japanese politics and business, the failure of a number of banks and industries, widespread environmental problems, and the persistence of long-entrenched patterns of social prejudice and discrimination have all tarnished Japan's image, both at home and abroad, as an economic miracle.

Landforms, Climate, and Vegetation

The Japanese Archipelago is a chain of four main islands and hundreds of small ones (Figure 9.21). Most are volcanic and mountainous, rising steeply out of the highly unstable zone where the Pacific, Philippine, and Eurasian tectonic plates grind against one another. Prone to severe earthquakes, seismic sea waves (tsunamis), and disastrous tropical storms, and so mountainous that only

Figure 9.21 The Japan subregion.

18 percent of its land can be cultivated, Japan might seem an unlikely place to find 126 million people living in affluent comfort. Yet its citizens have learned to cope with these limitations and have made the most of crowded conditions in cities and countryside alike.

Counting its southernmost tiny islands, Japan stretches over a range of latitudes roughly comparable to that between Canada's province of Nova Scotia and the Florida Keys (46°N–24°N). Although half a world away, it has a climate similar to the east coast of North America. The climate is cool on the far northern island of Hokkaido; temperate on the islands of Honshu, Kyushu, and Shikoku; and tropical in the southern Ryukyu Islands. Despite the moderating effect of surrounding oceans, the great seasonal climatic shifts of nearby continental East Asia give Japan a more extreme seasonal variation in temperature than would be the case if it lay farther out to sea.

The large central island of Honshu, the most densely populated of Japan's islands, also has the most mountains and forests. Because it has gained access to forest products in Southeast Asia and other parts of the world, Japan has been able to keep much of its own forestland in reserves. Today the interior mountains of Honshu (and Hokkaido) remain largely forested, although many of the slopes are planted with a monoculture of *sugi*, Japanese cedar, to be harvested commercially. The few flat lowlands and coastal areas of northern and southern Honshu were once intensively cultivated and supported a dense rural population. Industrial cities now fill many of these same lowlands, especially in the south. Tokyo-Yokohama is the largest urban agglomeration in the world, with 26.2 million people. It lies on the east-central coast of Honshu on the Kanto Plain, Japan's largest flatland. Kobe-Osaka, Japan's second largest metropolitan area with 10.6 million people, is one of several major urban areas located on the flatlands that ring the Inland Sea.

Japan's other three main islands are Hokkaido, the most northerly and least populated island, referred to as Japan's northern frontier, and Kyushu and Shikoku, two smaller southern islands. The northern slopes of Kyushu and Shikoku islands, facing the Inland Sea, have narrow strips of land put to agricultural, industrial, and urban uses. The southern slopes are more rural and agricultural. Off Kyushu's southern tip, the very small, mountainous Ryukyu Islands stretch out in a 650-mile (1000-kilometer) chain, reaching almost to Taiwan. Although many areas are densely settled and are important for tourism and specialty agriculture, such as the growing of pineapples, the Ryukyu Islands are considered a poor rural backwater of the four big islands. Okinawa, the site of a strategic U.S. military base that has recently been the source of U.S.-Japanese tensions, lies at the southern end of this island chain.

The Japanese Economy

Japan's "economic miracle" has had, and continues to have, an immense worldwide impact. Resources from all parts of the world are shipped to Japan: minerals and ore from Africa; coal and steel from Australia; timber from the Philippines, Indonesia, Malaysia (see Figure 10.12), Canada, and the United States; and oil from Southwest Asia. Japan's purchases of resources are mainstays of local economies in the producing countries and make Japan something of a global employer. Japanese know-how and technology turn the imported raw materials into inexpensive yet high-quality manufactured goods that are in demand throughout the world. Profits from these industries have often been invested abroad in factories, hotels, and resorts, thus securing multiple markets for the country. At the same time, Japan has restricted foreign investment and foreign trade within its borders.

Economic Challenges. Experts on Japan now debate whether the economic and political model that served the country so well for most of the second half of the twentieth century can continue to be successful. Indeed, the Japan of the early twenty-first century is an increasingly different place from the Japan of the economic miracle. The country has endured a decade-long economic recession that began in the early 1990s. Purchasing power has declined, and urban populations—though still affluent by the standards of most other countries—are growing uneasy. The Japanese people, long forced to endure extended workdays, overcrowding, pollution, high prices, and only modest buying power, in the interest of rapid economic growth, are agitating for changes that will improve their living standards. The greater participation of women in the workforce and the need for employees to be more creative and innovative are leading to changes in the workplace (see the discussion below, on page 503). Moreover, increased life expectancy is producing an aging population. The ever larger proportion of the elderly will soon require more social spending, which will further burden the working-age population (see page 503).

Internationally, China's rising economic power promises opportunity for Japanese investors, but China could also pose a competitive threat in the marketplace. Japan is also experiencing increasing economic competition from neighboring South Korea and Taiwan and the newly industrializing countries of Southeast Asia—all countries with strong export economies. Meanwhile, the United States has been pushing Japan to remove its barriers against foreign imports. Since 1950, Japan's access to North American markets has been a major factor in its economic success, but it has consistently blocked incoming trade goods.

Food Production. With 126 million people and only 18 percent of its land suitable for agriculture, Japan has nearly the largest ratio of people to farmland in the world. More than 7000 people depend on each square mile (more than 2700 on each square kilometer) of cultivated land. Thus, Japanese agriculture depends heavily on high-yield varieties of rice and other crops as well as on irrigation, fertilization, and mechanization. Because flat land is scarce, Japanese farmers have had to make efficient use of hill slopes by planting tea bushes and orchards and by terracing for rice paddies.

Although the agricultural way of life plays a large part in Japanese national identity, today only 4 percent of the population farms for a living. Japanese farms are increasingly mechanized, and they are highly productive in terms of output per acre. But the produce is some of the most expensive in the world. A single cantaloupe can cost as much as U.S. $10, a head of lettuce U.S. $5, a pound of beef more than U.S. $50. These high prices are partly a reflection of the very superior quality of the produce, but they are

CULTURAL INSIGHT *Japanese College Life*

Being a college student in Japan is a very different experience from being one in the United States. In Japan, once a student is admitted to a university, he or she is almost certain to graduate on schedule after four years, regardless of grades or class attendance. Students simply don't flunk out as they sometimes do in America. What they don't learn in college, they learn on the job after graduation. Employers, who normally hire young workers to stay with them until retirement, prefer in-house training and often shuffle new employees from position to position to expose them to all aspects of their operations. What they look for during job interviews is personality: the ability to get along with others, be part of a team, and have a sense of responsibility. Students hone these traits in college by being active in various groups: clubs for sports such as tennis or golf and interest clubs ranging from photography to foreign affairs to environmental issues. Thus, the first weeks at a Japanese university are spent joining various groups and building new friendships that will carry students through their four years.

Employers believe that the score a student receives on the university entrance exam is evidence of that student's potential to learn. Therefore, the biggest academic pressure on students is not in college but near the end of high school, when they must take a test that will determine the kind of university they can enter and what employers will think of them after graduation. The most prestigious universities in Japan are Tokyo University, Keio University, and Waseda University, all in Tokyo. Studying at one of these universities provides entry to important business networks and employment in big companies. Many Japanese students who don't like their country's approach to higher education elect to study abroad in the United States, Europe, Australia, or elsewhere. Alternatively, they can enroll in a foreign university operating in Japan. The largest of these is Temple University, an American institution based in Philadelphia with a campus in central Tokyo.

also the result of government efforts to keep Japan self-sufficient in food production. The government awards huge subsidies to farmers to persuade them to stay on the land. These payments discourage competition, as do trade barriers that keep imported food out of Japan. The system is increasingly unpopular with city dwellers, who are aware that imported food would be much cheaper.

The seas surrounding Japan, where warm and cold ocean currents mix, are an especially important source of food. There are some 4000 coastal fishing villages and tens of thousands of small craft that work these waters and bring home great quantities of tuna, halibut, mackerel, salmon, and other fish. Japan also sends out larger fishing ships, complete with canneries and other processing equipment, to harvest oceans around the world. Tokyo's early morning fish market is the largest fish market in the world. In addition, a large aquaculture industry produces oysters, seaweed, and other foods in shallow bays and freshwater fish in artificial ponds.

Living in Japan

Perhaps more than any other country on earth, Japan is renowned for its ability to confound foreigners. In an attempt to penetrate some of Japan's cultural complexity, we will look at several aspects of life in Japan: its unique cultural patterns, its urban life, the home, and its elderly population.

Distinctive Culture. Japan is a land of fascinating cultural combinations. The cultural base is a substantial foundation of long-standing Japanese customs and traditions, but atop that is a plentiful mix of borrowings from around the world, creating a unique blend of cultural contrasts and apparent contradictions. We see the mix in many ways: the kinds of clothes people wear, the foods they eat, the music they listen to, the sports they enjoy, and

even how they practice religion. Thus, among other contrasts, Japan is a land of kimonos *and* blue jeans, sushi *and* hamburgers, the koto (a traditional stringed instrument) *and* rap music, sumo wrestling *and* baseball. When they marry, Japanese couples often change their clothing after the ceremony from traditional wedding kimonos to a Western-style tuxedo and a white bridal dress for the reception, and they will have posed during the day for formal photographs in both outfits.

There are many attempts to explain Japan's unique cultural patterns. Some experts trace the contrasts to the forced opening of Japan by U.S. Commodore Matthew Perry in 1853. At that time, many Japanese, particularly in Tokyo, took on exaggerated international trappings to accommodate foreign visitors and protect what was truly Japanese. There is often a distinction in Japanese life between what is displayed on the outside, *omote*, and what is private on the inside, *ura*. The former often protects the latter, as in the contrasts between *omote-ji*, the outer layer of cloth in a kimono, and *ura-ji*, the layer closest to the skin, and between *Omote-Nippon*, the urban-industrial, eastern (Pacific Ocean) side of Japan that trades with the world, and *Ura-Nippon*, the more traditional and secluded, western (Sea of Japan) side of the country. Still other observers of Japan emphasize that the country's penchant for copying foreign fashions, trends, and technologies, often slavishly, is a response to cultural forces—such as a stodgy educational system (see the box "Japanese College Life")—that stifle creativity and reward conformity. There are no easy ways to explain Japan's culture; the only guarantee is that a visit to Japan will be both fascinating and mystifying.

Urban Japan. Nearly all of Japan's major cities are located along the coastal perimeter. The cities are where Japan's rapid industrialization has taken place, and their coastal location facilitates

AT THE LOCAL SCALE   *A New Capital for Japan?*

Tokyo has been the capital of Japan since 1868, the year when Japan's emperor moved his seat of power to the city from the ancient imperial capital of Kyoto. He took over control of the county from the hereditary line of shoguns, members of a military elite who had ruled from their castle in central Tokyo (then called Edo) since about 1600. The government buildings of modern Japan, including the National Diet (equivalent to the Capitol Building), are in the Kasumigaseki district of central Tokyo near the ruins of the old castle and the present-day Imperial Palace that occupies part of the site.

There has been much discussion in Japan about the possibility of moving the capital to another location. The move could be to another city, such as Sendai on the Pacific coast north of Tokyo, or back to Kyoto. Another option would be to construct an entirely new city to be a "tailor-made" capital. In the past, other countries, such as Belize, Nigeria, Pakistan, Brazil, Australia, and the United States (with the move in 1800 from Philadelphia to the newly constructed city of Washington, D.C.), have built new capitals with the idea of improving the social and economic life of the country.

One reason to move Japan's capital would be that Tokyo is simply too crowded. If the government of Japan moves away, other employers whose business is tied to government would have to move too, and Tokyo would become smaller and less dense.

Housing vacancies would increase, rents would fall, and land in the abandoned government district would become available for new uses. In addition, economic development would be stimulated in whichever part of the country was chosen for the new capital. Japan has many comparatively poor regions that are begging for investment. Some of their politicians have been lobbying for years to attract the national government.

Tokyo's vulnerability to earthquakes is another reason that the Japanese talk about moving the capital. Although all of Japan shares this hazard, Tokyo's site is in particular danger because of its proximity to three offshore tectonic plates and because much of the city center is built on relatively weak landfill. The city has a long history of disastrous quakes; the most recent was the Great Kanto Earthquake of 1923, which leveled the city and killed more than 100,000 people. Major earthquakes have struck Tokyo about every 70 years on average, so another might come at any time. The small tremors that one feels routinely in Tokyo are constant reminders that the city is in peril.

Moving the capital would be extremely expensive. Therefore, discussion of this topic has quieted somewhat since the economic slowdown in Japan became a serious problem in the late 1990s. The first priority is to get the national economy on track again, and then perhaps the debate can continue about if, when, and where the capital should be moved.

---

the import of raw materials and the export of finished products. Ideas from the outside world can penetrate easily, aiding technological advancement and cultural synthesis. Tokyo, for example, has the world's largest stock exchange, numerous centers for research and development, and some of the world's most beautiful modern architecture. It is also a major international cultural center: museum exhibitions, concert tours, and leading intellectuals and other dignitaries from around the world all visit the city regularly.

Nonetheless, Japanese cities also suffer from overcrowding and pollution. Overcrowding is especially bad. In Tokyo, it is not uncommon for a middle-class family of four, plus a grandparent or two, to live in a one-bedroom apartment. Japanese cities cannot expand to relieve crowding because they are limited by surrounding mountains and the ocean and by regulations protecting Japan's scarce agricultural lands. But perhaps the greatest causes of overcrowding are corruption and a lack of competition in the construction industry, which keep housing in short supply. Although there is growing pressure for change, most Japanese stoically endure minuscule, expensive apartments and long commutes to work. In the Tokyo metropolitan area, it is common to travel two or even three hours one way in trains that are so overcrowded that stations employ "shovers," who physically push as many passengers as possible into a single car.

The greatest source of discomfort for urban dwellers is pollution. Several cities have endured major episodes of poisoning from mercury and polychlorinated biphenyls (PCBs), and all large cities have chronic air pollution from automobiles and factories, as well as noise pollution. In the 1970s, local grassroots campaigns led to the first serious government efforts to limit pollution and to establish a moderately effective environmental agency. During Japan's boom years in the 1980s and early 1990s, many cities, including badly congested Tokyo, were able to expand public park space and open other recreational amenities such as artificial beaches, sports fields, and new aquariums.

The Home. Japanese domestic space is known for its simple, elegant aesthetics and its highly functional use of space. Traditional homes are made of wood and are usually roofed with heavy tiles. Even in larger rural homes, interior space is much smaller than most Westerners are accustomed to. The three or four rooms of a typical middle-class home are used for many purposes during the course of the day: they are transformed as needed by sliding doors of paper and wood or by folding, decorative room dividers. Furniture is simple, with much of family life centered on a low table and floor cushions. Sleeping is done on a firm mattress, called a futon, which is often rolled out on whatever floor space is available. However, such patterns are changing in many newer houses and apartments in favor of Western-style arrangements of rooms and furniture. Most Japanese kitchens are generously equipped with a full range of electric appliances, including the all-important automatic rice cooker, and almost every Japanese family has a color TV, video or DVD player, and sound system. Many Japanese also have automobiles, personal computers, cell phones, and other consumer items.

An Aging Society. Because of a healthy, low-fat, high-protein diet; good medical care; and perhaps even lucky genes, among other factors, Japanese life expectancy is the longest in the world. For men, life expectancy is now 77.62 years; for women, 84.15 years. Longevity combined with a low birth rate (in 2001, about 1.41 children were born per woman) has given Japan's population a large proportion of elderly people. According to estimates for 2001, the percentage of the Japanese population aged 65 years and over is 17.53 percent. This figure compares to 15.7 percent for the United Kingdom, 12.61 percent for the United States, and 7.11 percent for China. Thus, Japan has a disproportionately large social security obligation that is becoming larger and harder to fund over time. The country also faces a challenge in providing a full range of services and facilities for the aged: affordable medical care, recreational and cultural opportunities, and easier physical access to mass transportation and public buildings.

As has long been the practice in Japan, most older people who cannot fully take care of themselves are cared for by their adult children. However, this pattern is being eroded, as younger generations find that they lack space in their dwellings for their parents or grandparents, or prefer privacy to traditional family obligations. Thus, a relatively new industry of nursing homes and other care facilities is growing in Japan.

The suburbs of Japanese cities are densely packed with family homes. During the day, housewives go about their many duties: shopping, laundry, child care, and care of the elderly. [Michal Heron/Woodfin Camp & Associates.]

Working in Japan

Japan's distinctive culture is reflected in the working lives of its people: their attitudes toward work and toward outsiders who perform many of the menial jobs.

Traditional Work Patterns. The Japanese are renowned for their hard work and high-quality output, and most are devoted to their jobs. The lives of many people are shaped by the companies that employ them. The employer provides lifetime employment; regular paid vacations; a pension upon retirement; and, often, subsidized housing. Some companies even have a graveyard for employees. Job-hopping is considered reprehensible, and loyalty is reinforced by personnel management that instills camaraderie and conformity. Individualism is regarded as selfish, and innovation is discouraged. Often, employees who have good ideas will not present them for several years or will give credit to someone else. The Japanese language does not even have a word for entrepreneur; only recently has the English term become a buzzword in Japanese.

Traditional work patterns are changing in the face of the current economic crisis, as many companies have had to reduce their workforces and cut back employee benefits. Moreover, some younger college graduates have turned away from the old pattern of lifetime employment in favor of temporary employment. These workers, called "freeters," move from job to job, earning what they need, then taking a break to pursue other interests. They settle for a lower standard of living and give up job-related benefits, but they are unencumbered by contracts and do not feel compelled to acquiesce to their superiors or work overtime. This freelance approach to the world of work allows them time to start their own businesses. Whether by choice or necessity, accepting a temporary or part-time job is risky. The old loyalty system makes it difficult to find a new job. Firms worried about industrial security are often unwilling to hire someone who has worked for a competing company.

Japan's Dual Economy. Japan's kanban system has given the country a sort of dual economy. In this system, firms that supply components are clustered around a single final assembly plant. The high-status final assembly firms offer better wages and benefits and more secure employment than do the supplier firms. Supplier firms tend to hire more women, who are almost always paid at lower rates. Often, even though these women actually work more than full time, they are formally classed as part-time workers and hence receive fewer or no benefits and can be laid off and rehired relatively easily.

Foreign Workers. Foreign workers from poor countries are also on the wrong side of the dual economy. They come from China, the Philippines, Bangladesh, Iran, Brazil, and other places as guest workers to take some of the hard, dirty, dangerous, and low-paying jobs that today's Japanese workers avoid. Many of the migrants from Brazil and other Latin American countries are descendants of Japanese immigrants who went abroad in search of work some generations ago, when Japan was a poor country. Foreign men often work in construction, in factories, and as dishwashers in restaurants; women are hotel maids, cleaners, and other low-status workers. There are also many foreign women employed as hostesses in bars and as sex workers. The recruitment of women for sex work has become a controversial issue, because unscrupulous "agents," who are actually gangsters, force many of these women into sex-industry jobs even though they had initially promised them other kinds of work.

Because Japan's population is aging, there is a shortage of Japanese workers of prime employment age. Foreign workers have become necessary to keep Japan's economy productive. Consequently, the country is debating whether to begin allowing permanent immigration from poor countries (as opposed to the temporary guest workers). Permanent immigrants would eventually change Japanese society, making it less homogeneous and less distinctive.

TAIWAN

Taiwan is located a little over 100 miles (160 kilometers) off China's southeastern coast (Figure 9.22). With an area of 14,000 square miles (36,000 square kilometers) and more than 21 million people, Taiwan is a crowded space; it has a population density of over 1500 people per square mile (580 per square kilometer). The island, which is shaped somewhat like a leaf, has a mountainous spine running from the northeastern corner to the southern tip. A rather steep escarpment faces east, and a long, gentler slope faces west. Most of the population lives on the western side, especially in lower elevations along the coastal plain. Increasingly, as farming declines, people are concentrated in a few urban centers. The greatest concentration is in the far north, in the area surrounding Taipei, the capital.

Taiwan today is a geopolitical hot spot. Its status is ambiguous: the United Nations does not recognize it as a country, because of

Technology and electronics industries are a core of Taiwan's capitalist economy. Stan Shih, chairman and CEO of Acer, displays products ready for shipping in the company plant in Taipei. [Lincoln Potter/Gamma-Liaison.]

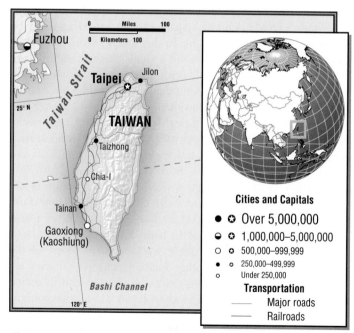

Figure 9.22 **The Taiwan subregion.**

China's opposition, yet Taiwan operates as a country in nearly every way, and its economic significance in the region cannot be ignored. Taiwan's existence is a source of political tension between it and China because it is a reminder of old cold war hostilities. China wishes to regain control of Taiwan, much as it has done in Hong Kong and Macao (see pages 497–498), but the Taiwanese people, through the electoral process, have recently expressed the desire to become an officially recognized independent country. The United States tacitly supports this position and is willing to sell arms to Taiwan, but its strong interest in trade with mainland China and its desire to avoid military entanglements keep the United States from taking an overt stand on behalf of Taiwan. The degree to which the United States would support Taiwan should armed conflict erupt appears to vary with U.S. administrations.

Taiwan, despite its small size and small population, is one of the most prosperous countries of the **Asia-Pacific region,** a huge trading area that includes all of Asia and the countries around the Pacific Rim. It is fourteenth globally in the size of its foreign trade, and its GDP per capita in 2000 was U.S. $16,100. After the communist revolution in China, Taiwan took advantage of its ardent anticommunist stance, its burgeoning refugee population, and its geographic location so close to the mainland to draw aid and

AT THE LOCAL SCALE *Leisure Time in Taiwan*

A social issue in Taiwan to which many urban dwellers around the world can relate is the lack of leisure time and of space in which to enjoy it. Almost all Taiwanese employers expect people to work half a day on Saturday. In addition to the long work week, gridlock traffic severely limits the enjoyment of leisure time in Taiwan. And places to go are scarce, too. Taipei and the surrounding densely packed cities have very few public spaces where people can relax and spend time together or be alone. There are a few national parks, but dense traffic makes them inaccessible. It is not unusual for a family to set out early to visit such a park, only to give up and return home after spending half a day stalled in traffic and breathing exhaust fumes. As a result, Taiwanese urban dwellers (the vast majority of the population) increasingly spend their leisure time inside, either watching videos at home or visiting high-tech entertainment centers, karaoke bars, and restaurants.

Source: Arthur Zich, "The other China changes course—Taiwan," *National Geographic* (November 1993): 2–33.

A family masked against air pollution goes for a drive in a park; Kaohsiung, southwestern Taiwan. [Jodi Cobb/National Geographic Image Collection.]

investment from Europe and America. The island's economy quickly changed from overwhelmingly rural and agricultural to mostly urban and industrial. Many industries made products for Taiwan's impressive export markets. Then, slowly, the economic emphasis changed again, from labor-intensive industries to high-technology and service industries requiring education and skill. By 1999, only 8 percent of the people were still working in agriculture.

By the mid-1990s, the domestic economy was expanding rapidly and local buyers were absorbing a major proportion of Taiwan's own production: home appliances, electronics (including the latest in computers and related communications devices), automobiles, motorcycles, and synthetic textiles. The Taiwanese performed a neat pirouette to make this high rate of local consumption possible. As Taiwan lost its labor-intensive industries to cheaper labor markets throughout Asia—including mainland China—Taiwanese entrepreneurs built computer chip and electronics factories in the very places that were giving them such stiff competition: Thailand, Indonesia, the Philippines, Malaysia, Vietnam, and the

new SEZs in South China. It is estimated that in the mid-1990s, Taiwanese businesses invested about U.S. $25 billion in China alone. Hence, Taiwan remained competitive as a rapidly growing economy with strong export markets and also profited from dealing with the less advanced, but emerging, economies in the region.

Life for Taiwanese women has changed radically since the 1950s. At that time, women made up less than one-third of the high school students, and their low social status reflected Confucian values. After World War II, many Taiwanese women left their homes in towns and villages and took low-paying urban factory jobs in the textile and garment industries that fueled Taiwan's economy. By the 1990s, Taiwanese women had equal access to education, at least through high school. In addition, 40 percent worked outside the home, and most were choosing to have only one or two children. Despite these gains, women are still greatly underrepresented in supervisory and management positions throughout the government bureaucracy and in private firms, and very few are politicians.

Today, the republic's likely future role in the wider world is a hotly debated question in Taiwan. As mainland China emerges as a world power with a huge population and enormous market potential, tiny Taiwan can no longer promote the idea that it speaks for (or is) China. On the other hand, Taiwan is perfectly located to participate in the development of mainland markets and in related efforts to integrate the Asia-Pacific economy. Young Taiwanese voters have recently elected officials who support an independent status for Taiwan rather than the long-held hope of eventual reunification with a noncommunist China.

KOREA, NORTH AND SOUTH

Some of the most enduring international tensions in East Asia have focused on the Korean Peninsula. As discussed in the "Human Patterns over Time" section earlier in this chapter, after centuries of unity under one government, the peninsula was divided in 1953 between communist and inward-looking North Korea, which does not participate in the global economy, and more cosmopolitan and newly democratic South Korea, which is often cited as a model of free-market capitalism.

Physically, Korea juts out from the Asian continent like a down-turned thumb (Figure 9.23). Two rivers, the Yalu (Amnok in Korean) and the Tumen, separate the peninsula from the Chinese mainland and a small area of Russian Siberia. Low-lying mountains cover much of North Korea and stretch along the east into South Korea, covering nearly 70 percent of the peninsula. There is little level land for settlement in this mountainous zone. The rugged terrain disrupts ground communications from valley to valley. Along the western side of Korea, alluvial plains slope toward the Yellow Sea, and most people live on these western slopes and plains. Although the peninsula is surrounded by water on three sides, its climate is essentially continental because it lies so close to the huge Asian landmass. The same cyclic monsoons that "inhale" and "exhale" over the Asian continent (see pages 456–458) bring hot, wet summers and cold, dry winters.

The Korean Peninsula was a unified country as early as A.D. 668. Scholars think that present-day Koreans are descended primarily from people who migrated from the Altai Mountains in western Mongolia. The Korean language appears to be most closely related to languages from this region. Other groups—Chinese, Manchurians, Japanese, and Mongols—have invaded

Figure 9.23 The Korea subregion.

the peninsula, sometimes as settlers, other times as conquerors. The Chinese Confucian values and Buddhism brought by some of these groups have influenced Korea's educational, political, and legal systems. Korea is noted for early advances in mathematics and medicine and for a system of printing by movable type. This form of printing, developed by 1234, predated the Gutenberg press in Europe by 200 years.

Contrasting Political Systems

Although an armistice brought the Korean War to an end in 1953, North Korea and South Korea each assumed a strong nationalist stance that has resulted in 50 years of hostile competition between the two. Both governments adopted the Korean concept of *juche*, which means self-reliance or "the right to govern yourself in your own way." In North Korea, *juche* was interpreted as unquestioned loyalty to the Great Leader, Kim Il Sung, and later to his son, Kim Jong Il, who succeeded him in 1997. In North Korea, there is no pretense of democracy, and the government remains a military dictatorship. North Korea limits its involvement with its continental neighbors, China and Russia, and declines to trade with them. Its restrictive internal and external economic policies make it one of the poorest nations in the world. Despite its apparent desire for isolation, North Korea has occasionally threatened South Korea and its allies, especially Japan.

In South Korea, *juche* has come to mean vigorous individualism, coupled with pride and loyalty to one's own people and nation. Social criticism and even violent protest from labor unions and other social movements are allowed, with the understanding that one's ultimate loyalty is still to South Korea. Politically, South Korea allied itself with the United States, Japan, and Europe shortly after World War II and sought economic revival through foreign aid and capitalist development. But despite its economic success, South Korea was governed by a series of military dictatorships until 1997, when democratic elections were held. Kim Daejung, an outspoken critic of the military who endured years of harassment and jail for his beliefs, was elected prime minister.

Gender Roles

The cultural history of Korea bears a certain resemblance to that of China, primarily because both countries have been shaped by Confucian thought. Society is organized hierarchically. Elder males have the greatest authority, and women are expected to be subservient to men in all public situations. Parents prefer sons, partly because, like the Chinese, they believe that sons will provide better income for them in their old age and that sons' wives will provide the necessary personal care. Within the household, women wield financial power, even to the point of controlling men's spending. Yet this responsibility often means that women must make insufficient funds stretch so that the earning power of the husband will not be questioned and the family lose face. Women are responsible for most of the household work, even though most women in both North and South Korea now work outside the home. Although data is not available for North Korea,

On this April morning in Pyongsung City, about 20 kilometers (12.5 miles) from North Korea's capital, Pyongyang, one person is doing tae kwon do exercises while others sweep the sidewalk along the stream. [Hyungwon Kang.]

women who work in the formal economy of South Korea earn less than half what their male counterparts earn. Few managers and administrators are women, and women hold just 4 percent of the seats in parliament. Men are expected to maintain the family's public social and economic connections. After work, they meet along the streets and in shops to drink, eat, tell stories, play games, and make personal connections that can facilitate everyday life for the entire family.

Contrasting Economies

Table 9.3 compares several statistics for North and South Korea (see also Table 9.2, page 485). Economically, the countries differ dramatically. North Korea has better physical resources for industrial development, including forests and deposits of coal and iron ore. Its many rivers, which descend from the mountains, have considerable potential to generate hydroelectric power. South Korea has better resources for agriculture, because its flatter terrain and slightly warmer climate make it possible to double-crop rice and millet. Nevertheless, in recent decades, South Korea has surpassed North Korea not only in agricultural production, as expected, but also in industrial development. Its manufacturing industries now export goods throughout the world.

A major reason for South Korea's economic success has been the formation of huge corporate conglomerates known as **chaebol.** These conglomerates include such companies as Samsung, Hyundai, and Daewoo. South Korea's governments have assisted the *chaebol* by making credit easily available to them and by allowing

TABLE 9.3	Comparative statistics for North and South Korea	
	North Korea	South Korea
Land area (square miles)	46,541	38,324
Population	21,700,000	47,300,000
Population per square mile	466	1,234
Human Development Index (HDI) rank	75 (in 1998)[a]	30 (in 1998)[b]
GDP per capita (adjusted for PPP in 1998 $US)	1,000	13,478
Percent natural increase	1.5 to 9	0.9
Infant mortality (deaths per 1000 live births)	26	11
Life expectancy (years)	70	74

[a]2000 data not available for North Korea.
[b]1998 data used to match that for North Korea.

Sources: 2000 World Population Data Sheet (Washington, D.C.: Population Reference Bureau); and United Nations Human Development Report 1998 (New York: United Nations Development Programme).

them to purchase foreign patents so that Koreans could focus on product quality and marketing, rather than on inventing. Today there are numerous facilities in South Korea that produce consumer electronic equipment (televisions, stereos, computers, videocassette recorders, and microwave ovens), as well as cars, ships, vehicles, clothing, shoes, and iron and steel. Nevertheless, the chaebol have come under increasing criticism in recent years, because their close connections to the government have led to corruption. Unwise loans and investments have disrupted the South Korean economy, and laws restricting chaebol finance deals were enacted in 2000. One of the largest chaebol, Daewoo, was dismantled because it was deeply in debt and had too many layers of management.

Despite these problems, the South Korean economic system has worked well enough that today most South Koreans, who were extremely poor at the end of the Korean War, can now afford the products they formerly only exported. For example, even in rural areas where incomes are comparatively low, 90 percent of households have a refrigerator, an electric rice cooker, and a propane gas range, and many have telephones and color television sets.

North Koreans have benefited from some communist economic policies, particularly access to basic health care and education. In general, however, the economy has deteriorated. North Korea's industries are inefficient and its workers poorly motivated; and since the demise of the Soviet Union, the country has suffered a loss of markets, raw materials, fuel, and technical training. Poor harvests and recurrent cycles of floods and droughts have brought extensive famines, beginning in 1995 and continuing through 2001. Although the government has been unwilling to release much information, it is possible that up to 2 million people have died of hunger over this six-year period. The government has responded ineffectively to this dire situation. In general, North Korean farms are organized as centrally controlled collectives, and these collectives have not been able to meet the demand for food. Only very small plots are allowed for the private production of vegetables and fruit and for raising small animals. In August 2001, a United Nations World Food Program visitor to North Koreas reported "no significant improvement in the country's ability to feed itself" in the past few years and warned that present UN programs, which feed 7.6 million North Koreans, would have to continue for years to come.

The future of the Koreas remains uncertain. Closer cooperation and even reunification are increasingly discussed, not only because so many families have been painfully split by the long dispute but also because closer cooperation would bring obvious economic benefits to both sides. For example, South Korea needs more energy to grow its industrialized economy so that it can better compete with Japan, Taiwan, and eventually China. North Korea also needs energy to begin its development, and it has numerous sites at which hydroelectric dams could be built, especially if South Korea were to provide the investment capital. But reunification would require much sacrifice from South Koreans, just as it has from West Germans as they attempt to merge their economy with that of the much poorer, formerly communist East Germany. Because of the stark difference in wealth between the two Koreas, reunification would mean that capital that might have been used to develop South Korea further would have to be used to pay for basic improvements in North Korea. Poor refugees might flood into South Korea, lowering standards of living there even more.

MONGOLIA

VIGNETTE   In July 1991, as their truck descended onto a broad plain within the high Altai Mountains of Mongolia, the visitors saw two horsemen herding hundreds of sheep. One of the horsemen, a suntanned man in his 60s, rode over, dismounted, and greeted the two Americans (and their interpreter) with a smile and a query about their state of health. He wore a traditional knee-length coat, fastened at the side and the shoulder with buttons and belted with a silk sash, and Western-style black leather boots. Anthropologists Cynthia Beall and Melvyn Goldstein were to discover

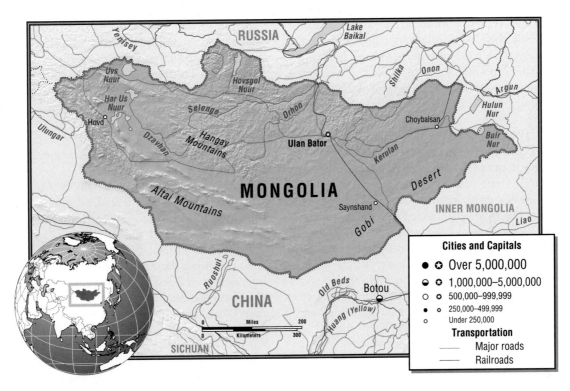

Figure 9.24 **The Mongolia subregion.**

that his easy, confident demeanor was typical of Mongolians. He seemed amused to have found these strangers popping up among the sheep on this normally lonely sweep of grazing land.

They had spoken for only a few moments when the Mongolian herder said cheerfully, "You know, I heard on the radio that your Foreign Minister Baker has visited our capital and that our two countries are now friends. That is good. Please come to visit my camp later. It's not far from [where you are headed]. We can talk more then. I have many questions to ask you about America, and I have a lot to say about [how things are changing in Mongolia]."

In their book, *The Changing World of Mongolia's Nomads* (1994), Goldstein and Beall admit that this encounter shattered their assumptions that Mongolians would be wary of outsiders, hesitant to express themselves, and unaware of the outside world.

With just 2.5 million people, Mongolians occupy a territory so large (Figure 9.24) that the average density is nearly the lowest on earth, at 4 people per square mile (1.5 per square kilometer). For thousands of years, the economy has been based on the nomadic herding of sheep, goats, camels, horses, and cattle (yaks). Today, nearly half the population is still engaged in rural activities in some way related to this traditional lifestyle. The other half now live in cities, primarily the capital of Ulan Bator. Urban workers are employed in a wide range of services and in industries related to the processing of mined minerals and such animal products as

hides, fur, and wool. In the 1990s, Mongolia's attempted transition from communist central planning to a market economy was interrupted by an economic recession that forced some people working in urban jobs to return to the countryside.

The Mongolian Plateau lies in the heart of Central Asia, directly north of China and south of Siberia. It is high, cold, and dry, with an extreme continental climate. The physical geography of Mongolia can be broken down into four major zones. In the far south and extending to the border with China is the Gobi Desert—actually a very dry grassland that grades into true desert in especially dry years or where it is overgrazed. To the east and northeast of the Gobi is a huge, rolling, somewhat moister grassland. The remaining two zones are Mongolia's two primary mountain ranges: the forested Hangai (Khangai) in north-central Mongolia and the grass- and shrub-covered Altai, which sweep around west and south and into the Gobi Desert.

History

Scientists are not sure where the ancestors of the Mongolians came from, but it is reasonable to assume that they descended from groups of people who have occupied the massive mountains and plains of Central Asia for more than 40,000 years. **Nomadic pastoralism**—a way of life and economy centered on the tending of grazing animals—is one of the most complex agricultural traditions and dates at least as far back as the cultivation of plants (8000 to 10,000 years). Nomadic herders understand the complex biological requirements of the animals they breed, and they must also know the ecological and seasonal vegetation changes in the landscapes they traverse with their herds. Mongolian pastoralists have perfected management practices and equipment, such as the portable *ger*, that make it possible to move whole communities and hundreds of animals frequently over the course of the year.

The present country of Mongolia is the northern part of what was once a much larger culture area in eastern Central Asia known by the same name. The height of Mongolian influence was between 1235 and 1366. During this period, the Mongols created a land-based empire that stretched from the eastern coast of China to central Europe, including parts of modern Russia and Iran. While in control of China, the Mongols improved the status of farmers, merchants, scientists, and engineers; fostered international trade; and broke the control of traditional elites by abolishing their automatic access to privilege. They refined the manufacture of textiles, jewelry, and blue and white porcelain, and trade of these wares along the Silk Road flourished. Although the Mongols brought many important changes to China, they alienated many people with their authoritarian rule and discrimination against ethnic Chinese. The Mongols were deposed in 1366. They retreated to the north and, over the next 300 years, lost control of territory in Asia and Europe. Eventually, the southern part of Mongolia became part of China.

By 1900, after long years under Manchurian colonization, Mongolians were among some of the poorest people in Asia, despite their herding skills. Profits from herding went to the feudal elite, and trade was monopolized by Chinese merchants who sent their profits home rather than investing in Mongolian development. In 1921, after considerable turmoil, Mongolian independence from China was declared, and three years later a communist revolution took place.

The Communist Era in Mongolia

From 1924 to 1989, Mongolia was a communist country that sought guidance from the Soviet Union. Russian advisors and technicians wielded considerable influence in setting up a system of central planning that organized the economy and allocated resources. In time, the nomadic herder economy was reorganized according to socialist policies, without drastic disturbance of traditional lifeways. The old rural households made up of extended families became economic collectives. Some Mongolians continued to herd animals, others engaged in sedentary livestock production, and others farmed crops. During this time, Mongolia was partially industrialized and urbanized. By 1985, nomadic herding and forestry accounted for less than 18 percent of the Mongolian national income and somewhat less than 30 percent of the labor force, whereas employment in mining and industry, communications, transport, and construction accounted for about 48 percent of the national income and more than a third of the labor force.

The Soviets subsidized an elaborate system of social services. By 1989, there was nearly universal education through middle school, and adult literacy reached 93 percent. Health care became available to all, and life expectancy rose dramatically. The previously high infant mortality rates had dropped and family size decreased. That nomadic herder who impressed the American anthropologists Melvyn Goldstein and Cynthia Beall in the early 1990s was undoubtedly a beneficiary of these educational and social reforms.

Recent Economic Issues

At the end of the 1980s, the collapse of the Soviet Union brought drastic social and economic downturns. Several natural disasters, and lower world prices for the minerals that Mongolia produces, led to increased national debt and declines in economic growth, human well-being, and social order. Few Mongolians were trained to replace the Soviet managers who had run the economy. Because of layoffs and factory closings, everyone experienced a decline in living standards. Between 1990 and 1992, total public expenditures dropped by 58 percent, and education was cut by 69 percent. Kindergartens and rural boarding schools closed, and many older students left school to help their families economically. Throughout the 1990s, young adults remained unemployed and rootless, some turning to substance abuse for solace. Street children, the outcasts of disintegrating families, proliferated in Ulan Bator. By 2000, more than one-quarter of the population lived in poverty. Nonetheless, there were signs of a better future. Mongolia's new membership in the World Trade Organization was bringing increased international attention and development

The Batsuur family lives in a *ger*, the traditional tentlike Mongolian house that can easily be dismantled, packed up, and moved to another place. Located on the outskirts of Ulan Bator, the house has electricity (for the hot plate and television), a coal-burning stove, and a variety of furniture and decorative and functional items for the six people who live here. [Leong Ka Tai and Peter Menzel/Material World.]

assistance, and private foreign investors were beginning to take an interest in Mongolia.

Gender Roles

Traditionally, Mongolian women have enjoyed a status approximately equal to that of men, but outside forces such as Lamaist Buddhism in the sixteenth century and Manchurian rule in the seventeenth century eroded their status. Nevertheless, in today's pastoral economy, men and women share many tasks. Women usually breed animals and tend the mothers and their young, preserve meat and milk, and produce many essentials from animal hides and hair, including felt, yarn, and furs. Men also perform many of these tasks. Women provide the materials for house (*ger*) construction and also dye and weave beautiful furnishings for the interiors. Both boys and girls are taught to ride horses at an early age, and they help with the herding. As they mature, women become more responsible than men for the routines of daily life in the home: child care, food preparation and serving, home maintenance, and care of the elderly and infirm. Men engage primarily in pasturing the herds and arranging the marketing of the animals and their by-products. Men also perform other tasks—such as the occasional planting of barley and wheat—that are carried out beyond the home compound.

Under communism, much of this system remained intact or was enhanced. The daily work of women herders was valued and, like their husbands, they were eligible for old-age pensions. A woman could leave an unsuccessful marriage because there was support for her and her children. A few women became teachers, judges, and party officials, and soon women were better represented in institutions of higher education.

As a result of the jolting changes of the past decade, women have lost some of the status they enjoyed under communism. Nonetheless, women are regaining legal protections and access to education, jobs, health care, housing, and credit. Women entrepreneurs are emerging. For example, an out-of-work truck mechanic received a small loan and put herself and her brothers to work repairing trucks. A woman who used to make sausages in a state-run meat plant now makes them in her kitchen, employing 14 of her former fellow workers. The sausages are peddled on the streets. As in so many other countries that are moving from communism to a market economy, the informal economy of street peddlers and small-scale manufacturing and service people has been an essential component of economic transition.

REFLECTIONS ON EAST ASIA

The opening of China to market forces and to outside influences presents huge economic opportunities to other East Asian countries, but it also means stiff competition for their economies. Most of China's East Asian competitors have already begun to invest in the rapidly growing SEZs in Chinese coastal areas. But the expectation of benefits from China's debut into capitalism could be premature. It is not at all clear yet how the rapid social changes occurring in China will be worked out politically, either within the country or internationally. Will economic liberalization lead to more political freedom for ordinary people and more public input into such issues as environmental policy? Will China's emerging policies toward Hong Kong and Taiwan bring continued rapid economic growth, social change, and outside contact? Or will they result in a reversion to greater control of people and economies and a break with the international community?

Leadership within the region is in question. Residual animosities toward Japan linger for the excesses it committed during its imperial period before and during World War II. Whether Japan can earn the confidence of fellow Asians and play a leadership role in East and Southeast Asia may depend on how it solves its own internal financial crises and on whether it opens its own markets to its neighbors. China may bolster its position as a regional leader if it decides to woo the Overseas Chinese, many of whom were forced to leave China during the revolution but have since prospered in Taiwan, Singapore, and Malaysia. Politically, there is no clear leader in the effort to bring more participatory democracy to East Asia. Although variations on parliamentary democracy appear to be gaining ground in East Asia, actual power overwhelmingly tends to remain with a male elite.

A combination of factors related to population control will profoundly affect the future. These factors include shrinking family size, the increase in the proportion of retired and elderly people, and the fact that women are no longer available to stay home to care for children and elders. To cope with these changes, governments may have to spend more on social services and be content with slower economic growth. The population issue most emphasized in East Asia today is the need to curtail population growth. Yet the issue that may most influence the future is finding a way to care for large numbers of aging people.

There is also the question of how East Asia will deal with the world's cultural diversity. Barring a major reversal of policy by China, both personal and electronic contact between all of East Asia and the rest of the world will increase significantly. Not only will outsiders from all backgrounds be more prominent, but women and minority groups within each country are bound to become more politically assertive. A possible outcome is that East Asian society will become more welcoming to other cultures and to the idea of women acting in expanded roles.

Finally, East Asia, along with the rest of the industrialized world, will be considering how to balance the desire for a clean, safe environment with the desire for consumer lifestyles. In this regard, the tremendous talents of innovation and synthesis found throughout East Asia may be called upon even more in the future than they have been in the past.

Thinking Critically About Selected Concepts

1.   East Asia is a huge, physically diverse region. *Explain the techniques you use to remember the main physical and human features of this region.*

2.   East Asia has two contrasting climate zones: the dry continental west and the monsoon east. These main regional divisions have considerable relevance to the social and economic features of East Asia. *Which regional features do you think have some useful correlations with the climate zones? What are some of the limitations of these correlations?*

3.   The philosophy of Confucianism teaches that the best organizational model for the state and society is a hierarchy based on the patriarchal family. *Describe some of the ways in which Confucianism has influenced both family and government structures throughout the region. To what extent do you think these ideas will persist into the future? What are the factors now counteracting Confucianism?*

4.   In the past, the region has been split between two economic models. Japan, South Korea, and Taiwan practice a model of free-market capitalism developed by Japan. China, Mongolia, and North Korea practice a form of communism influenced by the Soviet Union. *What do you think are the most important and interesting changes now under way? How do you see influence and power shifting in the region?*

5.   The Chinese communists implemented a policy of regional self-sufficiency: each region was expected to build agricultural and industrial sectors of equal strength to even out the national distribution of income. *How is China handling regional inequality now? To what extent do you think the Chinese people will have to handle regional inequalities in the future? What patterns and problems do you foresee related to regional disparities?*

6.   China is the most populous country in the world, and Japan, South Korea, and Taiwan are also densely populated. Most East Asian countries have dramatically lowered their fertility rates as a result of urbanization, industrialization, and public policies encouraging one child per family. *What do you think are the most interesting and consequential results of these policies? What unforeseen effects can you imagine in the future?*

7.   China, Japan, and Taiwan each has a single dominant ethnic group that shares the country with a diversity of other ethnic groups in an often uneasy relationship. *What do you think are the spatial and statistical indicators that best illustrate the marginality of these minority ethnic groups?*

8.   The Chinese people have battled floods and droughts for centuries. *Explain what you see as the major environmental issues in China and the rest of East Asia today. If you were in charge of gathering community support for democratic solutions to environmental problems, where would you start?*

Key Terms

Ainu (p. 479) an indigenous cultural minority group in Japan characterized by their light skin, heavy beards, and thick, wavy hair, who are thought to have migrated thousands of years ago from the northern Asian steppes

alluvium (p. 491) river-borne sediment

Asia-Pacific region (p. 504) a huge trading area that includes all of Asia and the countries around the Pacific Rim

buffer zone (p. 486) a neutral area that serves to prevent conflict

chaebol (p. 507) huge corporate conglomerates in South Korea that receive government support and protection

China proper (p. 458) the North China Plain, a rich agricultural area that was unified under successive empires and that is considered the hearth of East Asian culture

Confucianism (p. 458) a Chinese philosophy that teaches that the best organizational model for the state and society is a hierarchy based on the patriarchal family

Cultural Revolution (p. 461) a series of highly politicized and destructive mass campaigns launched in 1966 to force the entire population to support the continuing revolution in China

economic and technology development zones (ETDZs) (p. 470) zones in China with fewer restrictions on foreign business, established to encourage foreign investment and economic growth

floating population (p. 472) jobless or underemployed people who have left economically depressed rural areas for the cities and move around looking for work

food stability (p. 469) the ability of a state to supply a sufficient amount of basic food to the entire population consistently

growth poles (p. 471) zones of development whose success draws more investment and migration to a region

guarantor (p. 474) in China, a third party who vouches for the honesty and reliability of one party in his or her dealings with another party

kanban system (p. 466) an efficient Japanese manufacturing system in which suppliers are located in close proximity to the main factory where final assembly takes place

loess (p. 486) windblown soil

nomadic pastoralism (p. 509) a way of life and economy centered on the tending of grazing animals who are moved seasonally to gain access to the best grasses

qanats (p. 494) ancient underground conduits that carry groundwater for irrigation

regional self-sufficiency (p. 468) an economic policy that encouraged each region to develop independently in the hope of evening out the national distribution of production and income in communist China

responsibility systems (p. 468) economic reforms that gave the managers of Chinese state-owned enterprises the right and the responsibility to make their operations work efficiently

shogun (p. 462) a member of the military elite of feudal Japan

special economic zones (SEZs) (p. 470) free-trade zones within China

transparent (policies or practices) (p. 474) open to public scrutiny

tsunami (p. 483) a large sea wave caused by an earthquake

wet rice cultivation (p. 490) a prolific type of rice production that requires the submersion of the plant roots in water for part of the growing season

World Trade Organization (WTO) (pp. 472–473) the world's preeminent global trade body, which seeks to remove barriers to global trade

yurt (or *ger*) (p. 494) a round, heavy felt tent stretched over collapsible willow frames used by nomadic herders in northwestern China and Mongolia

Pronunciation Guide

Ainu (IE-noo)

Altai Mountains (AHL-tye)

Amnok River (AHM-NAWK)

Amur River (ah-MOOR)

Bai (BYE)

Beijing (BAY-JYIHNG)

Bo Hai, Gulf of (OH-HYE)

Canton (KAHN-tawn)

chaebol (CHYE-BOHL)

Chang Jiang (CHAHNG JYAHNG)

Changan (CHAHNG-AHN)

Chengdu (CHUHNG-DOO)

Chiang Kai-shek (JYAHNG KYE-SHEHK)

Chongqing (CHOHNG-CHIHNG)

Confucianism (kuhn-FYOO-shuhn-ihz-uhm)

Dai (DIE)

Dalai Lama (DAH-lye LAH-mah)

Daqing (DAH-CHIHNG)

Deng Xiaoping (DUHNG SHYAU-PIHNG)

Dzungar Basin (DZOONG-GAHR ["oo" as in "book"])

Edo (AYD-oh)

Fujian (FOO-JYAHN)

Fukuoka (foo-koo-OH-kah)

ger (GURR)

Gezhouba (GUH-JOH-BAH)

Gobi Desert (GOH-bee)

Guangdong (GWAHNG-DAWNG)

Guangxi (GWAHNG-SHEE)

Guangzhou (GWAHNG-JOH)

Guizhou (GWEE-JOH)

Haikou (HYE-KOH)

Hainan (HYE-NAHN)

Han (HAHN)

Hangai Mountains (HAHNG-GIE)

Hangzhou (HAHNG-JOH)

Harbin (HAHR-BIHN)

he (HUH)

Hebei (HUH-BAY)

Heilongjiang (HYE-LAWNG-JYAHNG)

Himalayan Mountains (hih-MAHL-yuhn/ hih-muh-LAY-uhn)

Hiroshima (hih-ROH-shih-muh)

Hokkaido (haw-KYE-doh)

Honshu (HAWN-shoo)

Huang He (HWAHNG HUH)

Hui (HWEE)

ji (JEE)

jiang (JYAHNG)

Jiangsu (JYAHNG-SOO)

Jilin (JEE-LIHN)

Kanto Plain (KAHN-TOH)

Kaohsiung (GAU-SHYOONG ["oo" as in "book"])

Kashgar (kahsh-GAHR)

Kobe (KOH-bay)

Kunming (KOON-MIHNG ["oo" as in "book"])

Kuomintang (KWOH-mihn-TAHNG)

Kyushu (KYOO-shoo)

Lijiang (LEE-JYAHNG)

Loess Plateau (LUHS)

Macao (muh-KAU)

Manchu (MAN-CHOO)

Manchuria (man-CHOOR-ee-uh)

Mandarin (MAN-duh-rihn)

Mao Zedong (MAU DZUH-DAWNG)

Meiji (MAY-jee)

Miao (MYAU)

Mongolia (mawng-GOH-lee-uh)

Mount Fuji (FOO-jee)

Nagasaki (nah-guh-SAH-kee)

Nan Ling Mountains (NAHN LIHNG)

Nanjing (NAHN-JYIHNG)

Ordos Desert (OHR-dohss)

Osaka (oh-SAH-kah)

Peking (PEE-KIHNG)

Pudong (POO-DAWNG)

Qaidam Basin (KYE-DAHM)

qanat (KAH-naht)

Qin (CHIHN)

Qin Ling Mountains (CHIHN LIHNG)

Qingdao (CHIHNG-DAU)

Qinghai (CHIHNG-HYE)

Ryukyu Islands (RYOO-kyoo)

Seoul (SOHL)

Shaanxi (SHAH-AHN-ZHEE)

Shanghai (SHAHNG-HYE)

Shantou (SHAHN-TOH)

Shanxi (SHAHN-SHEE)

Shenyang (SHUHN-YAHNG)

Shenzhen (SHUHN-JUHN)

Shikoku (SHE-kaw-koo)

Sichuan (ZIH-CHWAHN)

Taipei (TYE-PAY)

Taiwan (TYE-WAHN)

Taiyuan (TYE-YUAHN)

Takla Mountains (TAHK-luh)

Taklimakan Desert (TAHK-luh-muh-KAHN)

Tarim Basin (TAH-REEM)

Tianjin (TYAHN-JIHN)

Tibet (tih-BEHT)

Tian Shan (TYEHN SHAHN)

Tiananmen Square (TYEHN-AHN-MUHN)

Tokyo (TOH-kee-oh)

Tumen River (OO-MUHN)

Turfan Depression (TOOR-FAHN)

Ulan Bator (OO-lahn BAH-tawr)

Urumqi (OO-ROOM-CHEE [both "oo" as in "book"])

Uygur (WEE-gurr)

Wenzhou (WUHN-JOH)

Wu Shan (WOO SHAHN)

Wuhan (WOO-HAHN)

Xi Jiang (SHEE JYAHNG)

Xiamen (SHYAH-MUHN)

Xian (SHEE-AHN)

Xianggang (SHYAHNG-GAHNG)

Xibe (SHEE-BUH)

Xinjiang (SHEEN-JYAHNG)

Xinjiang Uygur (SHEEN-JYAHNG WEE-gurr)

Xizang (SHEE-DZAHNG)

Yalu River (YAH-LOO)

Yamato (YAH-mah-toh)

Yangtze River (YAHNG-TSUH)

Yi (YEE)

Yichang (YEE-CHAHNG)

Yokohama (yoh-kaw-HAH-mah)

Yunnan-Guizhou (YOO-NAHN GWEE-JOH)

Yunnan Plateau (YOO-NAHN)

yurt (YURRT)

Zhuang (JWAHNG)

Zhuhai (JOO-HYE)

Selected Readings

A set of Selected Readings for Chapter 9, providing ideas for student research, appears on the *World Regional Geography* Web site at www.whfreeman.com/pulsipher.

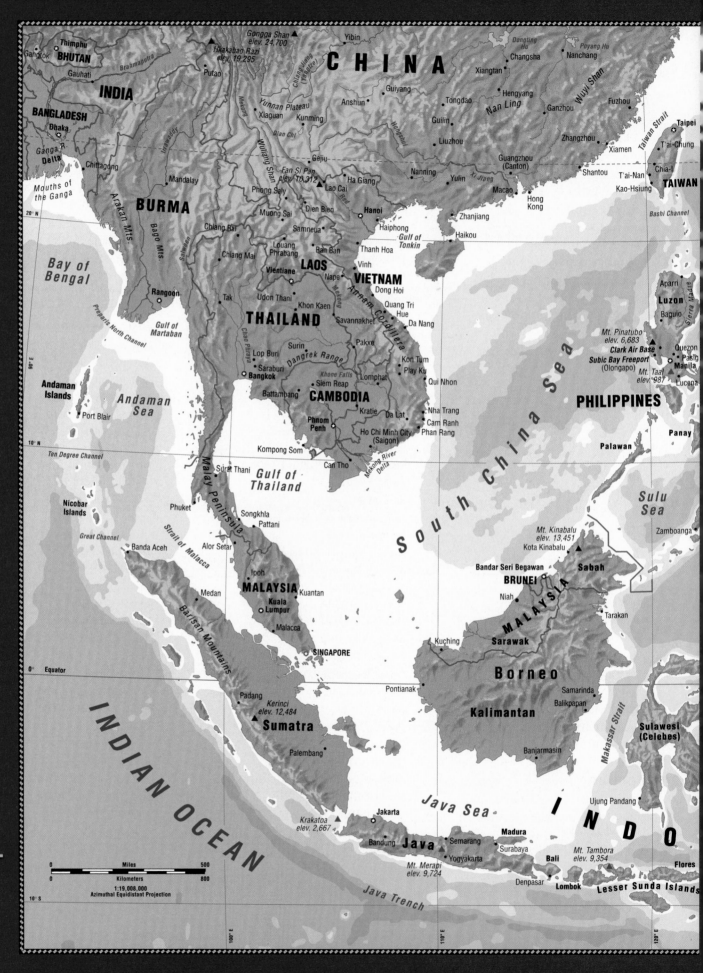

BHUTAN
Thimphu
Gangtok
Gauhati
Brahmaputra
INDIA
BANGLADESH
Dhaka
Ganga R.
Ganga
Delta
Chittagong
Mouths of
the Ganga
20° N
Arakan Mts.
Mandalay
Bago Mts.
BURMA
Irrawaddy
Salween
90° E
Bay of
Bengal
Preparis North Channel
Rangoon
Gulf of
Martaban
Andaman
Islands
Port Blair
Andaman
Sea
10° N
Ten Degree Channel
Nicobar
Islands
Great Channel

CHINA
Gongga Shan
elev. 24,700
Hkakabao Razi
elev. 19,295
Putao
Yibin
Chiang Jiang (Yangtze)
Dongting
Ho
Changsha
Poyang Hu
Nanchang
Yunnan Plateau
Xiaguan
Kunming
Guiyang
Anshun
Xiangtan
Hengyang
Tongdao
Nan Ling
Ganzhou
Wuyi Shan
Fuzhou
Dian Chi
Gejiu
Guilin
Liuzhou
Zhangzhou
Xiamen
Taiwan Strait
Taipei
Wuliang Shan
Fan Si Pan
elev. 10,312
Lao Cai
Red R.
Ha Giang
Nanning
Yulin
Xi Jiang
Guangzhou
(Canton)
Macao
Shantou
T'ai-Chung
Chia-I
Kao-Hsiung
TAIWAN
Mekong
Phong Saly
Chiang Rai
Dien Bien
Muong Sai
Samneua
Hanoi
Zhanjiang
Haiphong
Hong
Kong
Haikou
Bashi Channel
Chiang Mai
Louang
Phrabang
Ban Ban
Vinh
Thanh Hoa
Gulf of
Tonkin
LAOS
Nape
Vientiane
VIETNAM
Dong Hoi
Tak
Udon Thani
Khon Kaen
Annam Cordillera
Quang Tri
Hue
THAILAND
Chao Phraya
Lop Buri
Surin
Dangrek Range
Savannakhet
Da Nang
Pakxe
Mekong
Kon Tum
Play Ku
Mt. Pinatubo
elev. 6,683
Clark Air Base
Subic Bay Freeport
(Olongapo)
Aparri
Luzon
Baguio
Sierra Madre
Quezon
Pasig
Manila
Lucena
Saraburi
Bangkok
Khone Falls
Siem Reap
Lomphat
Qui Nhon
Mt. Taal
elev. 987
Battambang
CAMBODIA
Kratie
Da Lat
Nha Trang
Cam Ranh
Phan Rang
PHILIPPINES
Panay
Phnom
Penh
Ho Chi Minh City
(Saigon)
Palawan
Kompong Som
Gulf of
Thailand
Can Tho
Mekong River
Delta
South China Sea
Sulu
Sea
Zamboanga
Surat Thani
Phuket
Malay Peninsula
Songkhla
Pattani
Banda Aceh
Strait of Malacca
Alor Setar
Mt. Kinabalu
elev. 13,451
Kota Kinabalu
Sabah
Bandar Seri Begawan
BRUNEI
Ipoh
MALAYSIA
Kuantan
Niah
MALAYSIA
Medan
Kuala
Lumpur
Tarakan
Barisan Mountains
Malacca
Kuching
Sarawak
SINGAPORE
Borneo
0°
Equator
Pontianak
Samarinda
Padang
Kerinci
elev. 12,484
Balikpapan
Kalimantan
Sumatra
Palembang
Banjarmasin
Makassar Strait
Sulawesi
(Celebes)
INDIAN OCEAN
Java Sea
INDO
Ujung Pandang
Krakatoa
elev. 2,667
Jakarta
Bandung
Java
Semarang
Surabaya
Madura
Bali
Mt. Tambora
elev. 9,354
Flores
Mt. Merapi
elev. 9,724
Yogyakarta
Denpasar
Lombok
Lesser Sunda Islands
Java Trench

10° S

Miles
0   500
Kilometers
0   800
1:19,008,000
Azimuthal Equidistant Projection

100° E
110° E
120° E

East
China
Sea

Ryukyu Islands

Amimi
Islands

Okinawa
Islands

Naha

Sakishima Islands

Tropic of Cancer

20° N

PACIFIC
OCEAN

Philippine Sea

Mt. Mayon
elev. 9,810

Cebu

10° N

MICRONESIA

Negros

Koror

Mindanao
Davao

PALAU

Land Elevations

meters	feet
Ice cap	Ice cap
5183 and above	17,000 and above
3353–5182	11,000–16,999
2134–3352	7000–10,999
914–2133	3000–6999
305–913	1000–2999
0–304	0–999
Below sea level	Below sea level

*Celebes
Sea*

*Molucca
Sea*

Manado

Halmahera

Equator   0°

Sorong

Jayapura

Molucca Islands

West Papua

Buru

Ceram

N e w   G u i n e a

Puntiak Jaya
elev. 16,499

Maoke Mountains

PAPUA
NEW
GUINEA

Banda Sea

Aru
Islands

N E S I A

*Arafura
Sea*

Merauke

10° S

E. TIMOR

*Timor
Sea*

AUSTRALIA

The Irrawaddy River delta in
southern Burma, one of the
world's great rice-producing
regions, consists of fertile river
mud and sand deposited over the
past 2 million years. The light
pink shades in this infrared image
are fields; the deep red zones are
tropical forests and mangrove
swamps. The light blue shows the
extent of river sediment deposition
offshore. [Tsado/NASA/ Tom
Stack and Associates.]

Chapter 10: Southeast Asia

 MAKING GLOBAL CONNECTIONS

VIGNETTE It is breakfast time, and the food stand in the village in northeastern Thailand is crowded. The middle-aged woman cooking the food is friendly, her portions large, and the price right. For the equivalent of 5 cents, she serves up a huge mango leaf filled with rice, fish paste, and fried beetles.

One of the half-dozen people sitting on a bench eating his breakfast is a sinewy, bare-chested laborer in his late 30s. It is a hot, lazy day, so the two reporters from the *New York Times* take the opportunity to chat idly with the laborer about the food and their families. Soon he mentions his daughter, who is 15. His voice softens as he speaks of her. She is beautiful and smart, and he has invested his hopes in her.

"Is she in school?" the reporters ask.

"Oh, no," the father replies, mildly amused. "She is working in a factory in Bangkok . . . making clothing for export to America." He explains that she makes U.S. $2.00 a day for a nine-hour shift, six days a week. It's dangerous work: twice, the needles of her sewing machine went right through her hands. But the managers bandaged her hands and she soon went back to work. He is pleased that she has kept her job despite widespread factory closings, and he worries about what she would do if her factory shut down.

The reporters noted that the man was not indifferent to his daughter's injuries; he appeared to care for her deeply. Her persistence in the face of injury made him proud. From his point of view, the long hours were an advantage because they meant more pay for her, and her income promised a better future for their entire family. The reporters' subsequent conversations with factory managers revealed that although some enforce particularly long work days, others find workers' enthusiasm for long hours exasperating because it is hard to provide supervisors and security guards around the clock. [Adapted from the writings of Nicholas D. Kristof and Sheryl Wu Dunn, "Two cheers for sweatshops," *New York Times Magazine* (September 24, 2000): 70–71.]

The country of Thailand, where the laborer and his daughter live, and several other Southeast Asian countries—Burma, Laos, Cambodia, Vietnam, and Malaysia—occupy a substantial peninsula that extends from the southeast corner of Asia. The rest of the region consists of thousands of islands, most of which belong to one

Themes to Look For This chapter discusses the characteristics of Southeast Asia and the successes and problems it has faced in recent years. Our discussion will include the following themes.

1. Distinct culture groups. Many distinct culture groups in Southeast Asia have lived side by side for centuries, absorbing a spectrum of cultural influences from outside, yet retaining their uniqueness.

2. Disparities in well-being. Cultural and physical diversity have contributed to stark disparities in wealth and well-being, often within a given country.

3. Management of the tropical environment. Social and economic factors have influenced human strategies for managing physical resources, but so has the tropical environment. The rich soils of river valleys and deltas, and the much poorer soils of tropical rain forests in the uplands and islands, are being farmed intensively. Farms and plantations support large numbers of people, but there have been environmental consequences, including soil erosion, flooding, and loss of biodiversity.

4. Availability of labor and resources for consumer product manufacturing. The countries of the region range from agri-cultural societies that are just modernizing (Laos and Cambodia) to fiercely competitive industrialized societies that are active in the global marketplace (Indonesia, Malaysia, Thailand, and Singapore). Throughout the region, workers receive low wages and work under unhealthy conditions while producing products for consumers around the globe who are usually much wealthier and healthier. Resources are extracted at unsustainable rates.

5. Challenges to national unity. Authoritarian governments throughout the region—some fairly elected, some not—are encountering increasing difficulty in maintaining control over countries that have conflicting interest groups. Separatist movements are challenging several governments, which are accused of violating human rights in responding to the threats.

6. Persistence and change in family structure. Although traditional family structures are patriarchal, they have some unique features that give women more power and freedom than they have in many other world regions. Furthermore, as a result of recent economic changes and urbanization, the extended family is giving way to the nuclear family. These changes help to account for the generally higher levels of education among women in the region and for reductions in fertility.

of two island countries: the Philippines and Indonesia. Southeast Asia is easily connected to neighboring regions by sea and has long provided its spices and other agricultural products to the rest of the world. In the past half-century, it has become a key link in the global economy and supplies many of the consumer and electronic goods sold at inexpensive prices in the malls and galleries of the world.

Beginning in the 1960s, Southeast Asia became a center of rapid industrialization when governments and local investors began to build factories that employed thousands of willing workers. The workers' wages were low and the raw materials they used inexpensive, so the prices of the finished products made them highly competitive in the world market. By the 1970s, businesses and governments were courting investors from Japan, Europe, Australia, and North America, and by the 1990s, industrial facilities had greatly expanded. Millions of rural workers were attracted to urban areas throughout the region. Although the wages of a few dollars a day were very low by Western standards, they were enough to make a substantial difference to people's standard of living. Parents could afford simple medicines, better shelter, and better nutrition for their families. With increasing government revenues and private profits, a building boom ensued. Magnificent banks, hotels, and office buildings, such as Petronas Towers in Kuala Lumpur, began to rise in capital cities across the region.

By the early 1990s, Southeast Asia was regarded as a model of rapid development that other regions might emulate. But in the late 1990s, the region began to appear less adequate as a model of success. First, Western consumers became aware of the unhealthy and exploitative conditions under which the products they were buying from Southeast Asia were made. The factories became known as sweatshops. Soon protesters appeared on American college campuses and at meetings of the World Bank and World Trade Organization, agencies that were seen as encouraging the growth of such factories (see the photographs on this page). Second, the environmental costs of rapid industrialization were revealed: deforestation to make room for rubber trees and oil-producing palms, soil erosion caused by loss of forest cover, and heavy air pollution from the burning of fossil fuels. Finally, in 1997, a widespread economic recession hit. Thousands of factories closed at least temporarily, and families that depended on factory wages had to turn to the informal economy to pay for necessities.

By 2001, demand for Southeast Asian products was reviving and factories were increasing production. It was clear by then, however, that world events would continue to have an impact on the job security of Southeast Asia's factory workers.

Terms to Be Aware Of

Many of the region's inhabitants choose to dispense with place names that connote their colonial past. As a result, many names that are familiar to Westerners are not the names now preferred by the people of the region. In this chapter, we generally use the place name officially preferred by each locality. The older, more familiar Western name is given in parentheses the first time that a place name appears. In the case of Burma, however, we made an exception to this practice: Burma, the old name for Myanmar, is now being revived popularly; hence, *Burma* is used in this text.

Rapid industrialization in Southeast Asia has created contrasting scenes in urban landscapes. *Left:* Money from profitable industries and eager outside investors has been invested in sleek buildings, such as the upscale shopping mall at the base of the Petronas Towers in Kuala Lumpur, Malaysia. *Center:* Widespread layoffs force many former urban factory workers to fall back on other skills to earn a living. Here, a couple in Vientiane, Laos, makes simple iron tools in a street-side blacksmith shop. *Right:* After years of working under difficult conditions, thousands of factory workers began to demonstrate at trade meetings, like this one in Bangkok, Thailand, in spring 2000. [Mac Goodwin (left), Alex Pulsipher (center and right).]

THE GEOGRAPHIC SETTING

PHYSICAL PATTERNS

The physical patterns of Southeast Asia have a continuity that is not immediately apparent when looking at a map of the region. On a map, one sees a unified mainland region that is part of the Eurasian continent and a vast and complex series of islands arranged in chains and groups. The apparent contrast between mainland and islands obscures the fact that these landforms are related in origin. Moreover, despite covering a large territory, most of the region has a tropical or subtropical climate.

Landforms

Southeast Asia is a region of peninsulas and islands (see the map that opens this chapter, pages 514–515). Although the region stretches over an area nearly as extensive as Europe, most of that space is ocean. In fact, the area of all the land in Southeast Asia amounts to less than half that of the contiguous United States. There is a large mainland peninsula, sometimes called Indochina, that extends to the south of China. It is occupied by Burma, Thailand, Laos, Cambodia, and Vietnam. This peninsula itself sprouts a very long, thin peninsular appendage that is shared by outlying parts of Burma and Thailand; a part of Malaysia; and the city-state of Singapore, which is actually built on a series of islands at the southern tip. This long, thin peninsula is usually considered part of the archipelago that lies to the south and east of the mainland. The Southeast Asia **archipelago** is a series of large and small islands, fanning out over an area larger than the continental United States. These islands are grouped into the countries of Malaysia, Indonesia, and the Philippines. Indonesia alone has some 17,000 islands, the Philippines 7000. The independent country of Brunei shares the large island of Borneo with Malaysia and Indonesia. The entire region, except for the northernmost part of Burma, lies in the tropics, with the islands located on and near the equator.

The irregular shapes and landforms of the Southeast Asian mainland and archipelago are the result of the same tectonic forces that were unleashed when India split off from the African Plate and crashed into Eurasia (discussed first in Chapter 8). The collision, which is still under way, has pushed up the Himalayan Mountains and has forced up folds in the land to the east and west of the Himalayas. The eastern folds are a series of high mountainous ridges and intervening gorges that bend out of the Tibetan Plateau and turn south, then fan out to become the peninsula of Indochina. The gorges widen into valleys that stretch toward the sea, each valley containing a river or two that begins in the mountains of China to the north. The major rivers are the Irrawaddy and the Salween in Burma; the Chao Phraya in Thailand; the Mekong, which flows through Laos, Cambodia, and Vietnam and forms the border of Thailand and Laos for more than 620 miles (1000 kilo-

meters); and the Black and Red rivers of northern Vietnam. On the mainland, the mountain ranges descend from the high peaks of the north, which reach heights of almost 20,000 feet (6100 meters), to lower ranges of 2000 to 3000 feet (600 to 900 meters) toward the south, and then yet lower hills.

The landforms of the archipelago are related to those of the mainland. The islands of Southeast Asia sit on a submerged lobe of the Eurasian Plate that extends south from the mainland bulk of the plate. In fact, the curve formed by the islands of Sumatra, Java, Bali, Timor and the other islands of the Lesser Sunda group, and New Guinea conforms approximately to the shape of the lobe's leading southern edge. As the Indian-Australian Plate plunges beneath the Eurasian Plate along this curve, hundreds of volcanoes have been created, especially on the islands of Sumatra and Java. Volcanoes are also being created in the Philippines, where the Philippine Plate is pushing against the eastern edge of the Eurasian Plate's southern (island) lobe. The volcanoes of the Philippines are part of the Pacific Ring of Fire (see Chapter 1, page 19). Volcanic eruptions, and the mudflows and landslides that occur in their aftermath, endanger and complicate the lives of many Southeast Asians. In the long run, though, the volcanic material creates new land and provides minerals that can enrich the soil for farmers.

The now-submerged shelf of the Eurasian continent that extends under the Southeast Asian peninsulas and islands was above sea level during recurring ice ages, when much of the world's water was frozen as glaciers (Figure 10.1). The exposed shelf, known as Sundaland, allowed ancient people and Asian land animals (elephants, tigers, rhinoceros, proboscis monkeys, and orangutans) to travel south into the islands of Southeast Asia. Sundaland remained exposed until about 16,000 years ago. A second exposed continental shelf, known as Sahulland, was attached to Australia and New Guinea; its western boundary stopped near the islands of Timor in the south and the Moluccas in the north. Sundaland and Sahulland never met because they were, and are, separated by a deep ocean trench. This trench extends north from Bali through the Makassar Strait and along the western side of the Philippines. The trench (and the nearby islands) is known as **Wallacea.** Wallacea is a **biogeographical transition zone** between Asian and Australian flora and fauna: within this narrow zone there is some mixing of the two groups; but beyond it they remain distinct. Humans using watercraft were probably the only major land creatures to move across to New Guinea and Australia; dogs apparently made the trip with them. But Balinese tigers, for example, were not able to spread east, and Australian kangaroos could not move west.

Climate

The tropical climate of this region is distinguished by continuous warm temperatures in the lowlands—consistently above 65°F

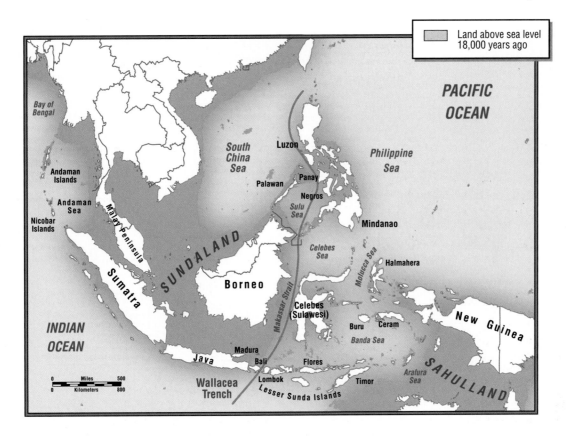

Figure 10.1 Sundaland, Sahulland, and Wallacea 18,000 years ago. [Adapted from Victor T. King, *The Peoples of Borneo* (Oxford: Blackwell, 1993), p. 63.]

(18°C)—and heavy rain (Figure 10.2). The rainfall is the result of two major processes: the continental "breathing" (see Chapter 1) that brings the monsoons, and the intertropical convergence zone (ITCZ), a band of rising warm air that circles the earth roughly around the equator (see Figure 7.2). The wet summer season extends from May to October, when the warming of the Eurasian landmass sucks in moist air from the surrounding seas. Between November and April, there is a long dry season on the mainland, when the seasonal cooling of Eurasia causes dry air from the inner continent to blow out toward the sea. On the many islands of Southeast Asia, however, the winter can also be wet because the air that blows out from the continent picks up moisture as it passes south and east over the seas. The air releases its moisture as rain after ascending high enough to cool down. With rains coming from both the monsoon and the ITCZ, the island part of Southeast Asia is one of the wettest regions of the world. Its copious year-round rainfall supports dense tropical vegetation over much of the landscape.

Periodically, the normal patterns of rainfall are interrupted, especially in the islands, and severe droughts result. These periodic droughts, which occur on an 8- to 13-year-long cycle, are part of the El Niño phenomenon (see Figure 11.3). In an El Niño event, the usual patterns of air and water circulation in the Pacific are reversed. Ocean temperatures are cooler than usual in the

Dayak farmer Abdur Rani plants cassava in a section of Kalimantan rain forest ravaged earlier by loggers. The haze—like the one that covered most of Indonesia, Malaysia, and parts of other Southeast Asian countries in 1997—is smoke from forest fires. [Michael Yamashita/Woodfin Camp & Associates.]

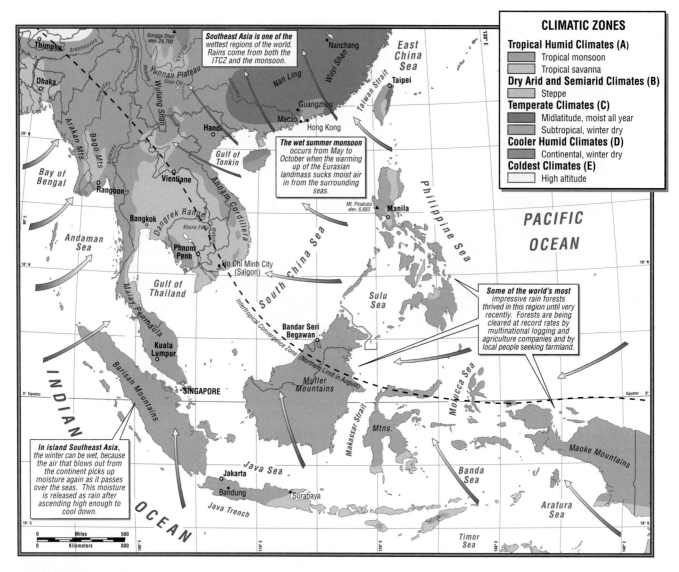

Figure 10.2 Climates of Southeast Asia. Southeast Asia experiences few periods of prolonged dryness. The islands are washed by rains that result both from monsoon patterns originating on the Eurasian continent and from rising warm, wet air along the intertropical convergence zone (ITCZ). The arrows on the map show airflows during the northern summer.

western Pacific near Southeast Asia. Instead of warm, wet air rising and condensing as rainfall, cool, dry air sits at the ocean surface. In 1997, an El Niño event brought drought and cool temperatures to the countries of Malaysia and Indonesia. Crops failed; springs and streams dried up; and the heavy, cool air prevented contaminants from venting into the upper atmosphere—especially the pollution caused by automobile and industry exhaust and forest fires. The result was the worst case of widespread air pollution ever experienced in the region.

The soils in Southeast Asia are typical of the tropics. Although not particularly fertile, they will support dense and prolific vegetation when left undisturbed for long periods. The high temperatures and damp conditions promote the rapid decay of **detritus**—dead plant material and insects—and the quick release of useful minerals. These minerals are taken up directly by the living forest rather than enriching the soil. Some of the world's most impres-

sive rain forests thrived in this region until very recently. Today, the forests are being cleared at record rates, especially by multinational logging companies but also by local people seeking farmland (see the section on deforestation later in this chapter, pages 545–546). Often the refuse is burned, creating smoke that contributes to severe air pollution under certain conditions. The fires that played a role in the severe pollution episode of 1997 had been set in Indonesia to clear land for commercial purposes, most notably for oil palm plantations. Winds carried the copious smoke to Malaysia, Singapore, and the Philippines (see the photograph on page 519).

HUMAN PATTERNS OVER TIME

Southeast Asia's position as a group of islands and peninsulas surrounded by seas has made it easily accessible to ocean trade and to outside cultural influences. First settled by migrants from the Eurasian continent, it later was influenced by Arab, Indian, and Chinese traders. Later still, it was colonized by Europe (sixteenth to twentieth centuries) and Japan (World War II). In the late twentieth century, the region emerged from a long period of colonial domination ready to profit from selling manufactured goods to its former colonizers. Its economic success has not yet been sufficient to catapult the majority of its people out of poverty, but signs of the region's development include rapid industrialization, declining birth rates (the demographic transition), growing urbanization, and rising rates of literacy.

The Peopling of Southeast Asia

The modern indigenous populations of Southeast Asia arose from two migrations widely separated in time. In the first, a group of hunters and gatherers called **Australo-Melanesians** moved from the present northern Indian and Burman parts of southern Eurasia into the exposed landmass of Sundaland about 40,000 to 60,000 years ago. They were the ancestors of the indigenous peoples of New Guinea, Australia, and Indonesia's easternmost islands. Very small numbers of their descendants still live in small pockets in upland areas elsewhere, notably in the Philippines, Borneo, Java, Sumatra, the Malay Peninsula, and the Andaman Islands.

In the second migration, Mongoloids from southern China began moving into Southeast Asia about 10,000 years ago, at the end of the Ice Age that closed the Pleistocene epoch. Their migration gained momentum about 5000 years ago, when a culture of skilled farmers and seafarers from South China, named **Austronesians,** migrated first to Taiwan, then to the Philippines, and then into island Southeast Asia and the Malay Peninsula. Other groups of Austronesians appear to have traveled southward along the coast of China and Vietnam. Some of these sea travelers eventually moved westward to southern India and as far as Madagascar off the east coast of Africa. They also moved eastward to the far reaches of the Pacific islands (see Chapter 11). Today there are no clear biological or geographic boundaries between the descendants of the early Australo-Melanesians and the more recent Austronesians and others who drifted south from China. Rather, there is a gradual grading from one to the other, the result of the continual intermixing of these people and the many others who have since migrated to Southeast Asia.

The migrations of many different peoples to this highly accessible region, together with its extremely fragmented physical geography, have led some geographers to compare Southeast Asia with another region that also has a highly complex cultural geography: the Balkans of southeastern Europe.

Other Cultural Influences

Southeast Asia has been and continues to be shaped by a steady stream of cultural influences, both internal and external. These influences have been so numerous partly because the sea that touches or surrounds all the countries except Laos brought ships and travelers. The shifting winds of the monsoon also played a role. In spring and summer, winds blowing from the west brought seaborne traders, religious teachers, and occasionally even invading armies from the coasts of India and Arabia. These newcomers brought religions—Hinduism, Buddhism, and Islam—trade goods such as cotton textiles, and food plants such as mangoes and tamarinds deep into the Indonesian archipelago. The traders and teachers penetrated as far as Cambodia, Laos, Vietnam, and China on the mainland as well as reaching the islands farther south. The merchant ships sailed home on the monsoon winds of autumn and winter that blow from the northeast. They carried Southeast Asia's people—as well as spices, bananas, sugarcane, silks, domesticated pigs and chickens, and other goods—to the wider world.

Islam came to the region mainly through South Asia, after India fell to Muslim conquests in the thirteenth century. Today Islam is a major religion in Southeast Asia, and Indonesia is the country with the world's largest Muslim population. China was also influential, especially in northern and central Vietnam, which was part of the Chinese empire from 111 B.C. until A.D. 939. After the tenth century, the Vietnamese expelled the Chinese, becoming a major bulwark that kept China's empire from expanding into the rest of the region. The blending of these and other cultural influences is obvious today in such places as Singapore and parts of Malaysia, where Buddhist, Taoist, Confucian, and Hindu temples exist side by side with Islamic mosques. Other areas are more homogenous, such as largely Buddhist Burma, Thailand, and Laos and mostly Muslim Indonesia.

As important as external influences have been, indigenous cultural characteristics also have shaped the character of the region. Several powerful urban empires emerged in Southeast Asia, including Angkor (A.D. 800–1400), Pagan (A.D. 800–1100), and Srivijaya (A.D. 600–1000). The cities of these kingdoms were impressive for their size and architecture. At its zenith in the 1100s, the city of Angkor was among the largest in the world, and its ruins are still an important heritage site in Cambodia.

As we shall see, ideas about family structure and gender roles that appear to date from the earliest inhabitants of the region can still be observed. These ideas may account for the relatively high status of women in the region (compared to parts of South Asia and China, for example)—a status that declined during the period of Western colonialism.

Colonialism

Over the last five centuries, Europe, Japan, and the United States have also influenced Southeast Asia. All three established colonies or quasi colonies in the region (Figure 10.3). The Portuguese were drawn to Southeast Asia's fabled spice trade in the fifteenth century. By sailing around Africa, they reached India in 1498 and the Strait of Malacca by 1509, and in 1511 they established the first permanent European settlement at the port of Malacca. Although their better ships and weapons gave them an advantage, their anti-Islamic and pro-Catholic policies provoked strong resistance in Southeast Asia. Only in their colony of East Timor did the Portuguese establish Catholicism, and today the people of East Timor, which is scheduled to become an independent country in 2002, are predominantly Catholic.

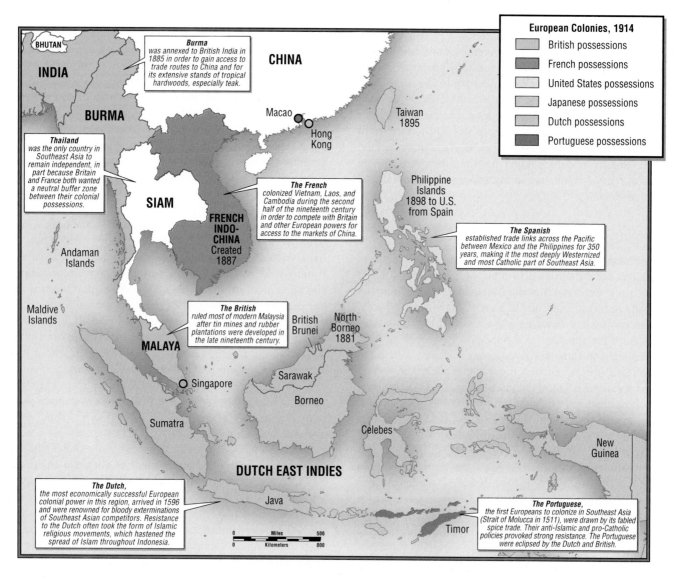

Figure 10.3 **European colonies in Southeast Asia, 1914.** [Adapted from *Hammond Times Concise Atlas of World History*, 1994, p. 101.]

By 1540, the Spanish had established trade links across the Pacific between the Philippines and their colonies in the Americas, especially Mexico. Like the Portuguese, they practiced a style of colonial domination grounded in Catholicism; the Spanish met less resistance, however, because their attitude toward non-Christians was somewhat more tolerant. The Spanish ruled the Philippines for 350 years, and as a result the Philippines are the most deeply Westernized and certainly the most Catholic part of Southeast Asia.

The Dutch were the most economically successful of the European colonial powers in Southeast Asia. From the sixteenth to the nineteenth centuries, under the auspices of the Dutch East India Company, they extended their control of trade over most of what is today called Indonesia. Their privately chartered trading company was headquartered in Batavia (now Jakarta) on the island of Java. Dutch colonists were far less interested in territory than in

profits, and they avoided direct administrative rule by placing local leaders in charge. In exchange for granting local control, the Dutch obtained exclusive trading rights, and they were renowned for their often bloody extermination of Southeast Asian trade competitors. Eventually, the Dutch became interested in export cash crops, primarily coffee, sugar, and indigo. From 1830 to 1870, the Dutch diverted farmers from producing their own food to working without pay in Dutch enterprises, especially plantations. Periodic severe famines were the result of this disruption of local food-producing systems. Resistance to the Dutch often took the form of Islamic religious movements, and the rise of these movements hastened the spread of Islam throughout Indonesia. As in South Africa, the Dutch settlers made little effort to spread Christianity.

Like their Dutch rivals, the British were commercially motivated. In the early nineteenth century, the British East India Company, operating with government and military powers sanctioned by the British Crown, controlled a few key ports on the Malay Peninsula. The British held these ports both for their trade value and to protect the Strait of Malacca, through which passed trade between China and Britain's empire in India. In the nineteenth century, Britain extended its rule over the rest of modern Malaysia to benefit from the tin mines and plantations in the interior. Once Malaysia came under their rule, the British argued that Malays, and in fact all "Orientals," were incapable of governing themselves effectively and needed a "benevolent" European power to rule over them. Similar justifications were offered for the violent annexation of Burma to British India in 1885. The British takeover of Burma gained Britain access to trade routes to China and to extensive stands of tropical hardwoods.

In the early seventeenth century, French Catholic missionaries began actively seeking converts in the eastern mainland area of Southeast Asia—comprising the modern states of Vietnam, Cambodia, and Laos. The French eventually colonized the region in the late nineteenth century, spurred by rivalry with Britain and other European powers for greater access to the markets of nearby China.

In all of Southeast Asia, the only country to remain independent was Thailand (formerly Siam). Like Japan, it protected its sovereignty by undergoing a massive drive toward European-style modernization. In addition, Thailand was adept at using diplomacy to prevent European colonization. Another factor in Thailand's ability to maintain its sovereignty was that Britain and France both wanted Thailand to serve as a neutral **buffer zone,** a territory separating adversaries. The existence of a neutral zone between their colonial possessions on mainland Southeast Asia lessened the possibility of direct confrontation between the two colonial powers.

Struggles for Independence

Agitation against colonial rule began in the late nineteenth century, when Filipinos fought first against Spain and then, after the United States took control of the Philippines in 1899, against the Americans. However, independence for the Philippines and the rest of Southeast Asia had to wait until the end of World War II. By then, Europe's ability to administer its colonies was weakened, partly because its attention was diverted by the devastation of the war. Also, during the war, Japan had exploded the myth of European superiority by conquering most of European-held Southeast Asia. At first, the Japanese were welcomed as liberators from European domination. But the Japanese soon imposed harsh administrative policies, and Southeast Asians turned their attention to ridding themselves of all outside control. By the mid-1950s, the colonial powers had granted self-government to most of the region. Only Singapore had to wait until the mid-1960s to become independent.

The most bitter battle for independence took place in the French-controlled territories of Vietnam, Laos, and Cambodia. Although all three became nominally independent in 1949, France retained political and economic power. Various nationalist leaders led resistance movements against continued French domination. Although these leaders did not begin as communists, and in fact shared ancient antipathies toward China for its previous efforts to dominate the region, the resistance leaders accepted military assis-

tance from communist China and the Soviet Union when the French refused to release control. Thus, the cold war was brought to mainland Southeast Asia.

In 1954, with their defeat by Ho Chi Minh at Dien Bien Phu, the French lost what had become a guerrilla war centered in Vietnam. In 1954, the United States, increasingly worried about the spread of international communism, stepped in and eventually replaced the French in the effort against the guerrillas. What was known as the Vietnam War in the United States became known as the American War in Vietnam. The resistance leaders controlled the northern half of Vietnam, while contesting control of the southern half with the United States and a U.S.-supported South Vietnamese government. The pace of the war accelerated in the mid-1960s. Many people in the United States were turned against the war by nightly broadcasts showing gruesome pictures of burning children and village massacres, by the loss of their own sons, and by revelations of the ineptness of U.S. officials. After several years of brutal conflict, public opinion in the United States forced the country to withdraw from the conflict in 1973. The war finally ended in 1975. During the Vietnam War, more than 4.5 million people died in Indochina, including more than 58,000 Americans. Another 4.5 million on both sides were wounded, and bombs, napalm, and defoliants ruined much of the Vietnamese environment. Land mines continue to be a hazard.

Although the withdrawal of its forces in 1973 ranks as one of the most profound defeats in U.S. history, the suffering of Vietnam and its neighbors, Cambodia and Laos, continued. The United States imposed economic sanctions against Vietnam that lasted until 1993 and crippled its recovery from the war. Moreover, the war spilled over the border into Cambodia, where a particularly violent revolutionary faction called the Khmer Rouge seized control of the Cambodian government in the mid-1970s. Inspired by a vision of rural communist society, they attempted to destroy virtually all traces of the European colonial era, including over 2 million people, or one-quarter of the Cambodian population. Many people were targeted simply for being educated urbanites. In 1978, Vietnam invaded Cambodia in an effort to drive out the Khmer Rouge. Laos, Cambodia's neighbor to the north, was also drawn into the conflict, and eventually the Vietnamese established an interim government that replaced the Khmer Rouge in the capital. Despite United Nations–monitored elections in 1993, however, hostilities continue to occur from time to time between rival factions from Vietnam, Laos, and Cambodia.

Although the independence era has brought violence to some areas, it has also brought relative peace and economic development to much of Southeast Asia. Since the 1960s, the economies of Thailand, Malaysia, Singapore, Indonesia, and the Philippines have grown considerably, aided largely by industries that export manufactured products to the rich countries of the world. Some critics regard this situation as a form of neocolonialism because the countries of the region remain dependent on markets in distant wealthy nations and often compete with one another to sell their products to their former colonizers. This stiff competition often results in wages that are too low to provide a decent standard of living and in weak pollution controls and worker safety regulations.

POPULATION PATTERNS

Today more than half a billion people occupy the peninsulas and islands of Southeast Asia. Twice as many people as live in the United States are packed into a land area half its size. The population map in Figure 10.4 reveals, however, that several parts of the region are only lightly populated. Relatively few people live in the upland reaches of Burma, Thailand, and northern Laos, where the land is particularly rugged and difficult to traverse. Much of Cambodia is also lightly settled. In some islands of Southeast Asia, wetlands, dense forests, mountains, and geographic remoteness have inhibited settlement for centuries. Small groups of indigenous people have lived here for thousands of years, supported by shifting subsistence cultivation and by hunting, gathering, and small-plot permanent agriculture. Over the last several decades, however, in places such as Kalimantan in Borneo, Sulawesi (the

Celebes), the Moluccas, and West Papua (until recently called Irian Jaya), new settlers have been lured by promises of farmland and jobs in commercial agriculture and in extraction of forest products and minerals.

Most of the people of Southeast Asia (about 60 percent) live in patches of particularly dense rural settlement along coastlines, on the floodplains of major rivers, and in the river deltas of the mainland. In the islands, settlement is concentrated in the northern Philippines and on Java. These places are attractive because people can pursue several ways of making a living: the rich and well-watered soils allow intensive agriculture on small plots, and there are also opportunities for small business, petty trade, wage labor, and enjoyable social interaction. About 36 percent of the region's people live in cities, and the cities of Southeast Asia are among the most crowded on earth. The most populated metropolitan areas are Jakarta, Manila, and Bangkok (Table 10.1). These and other cities

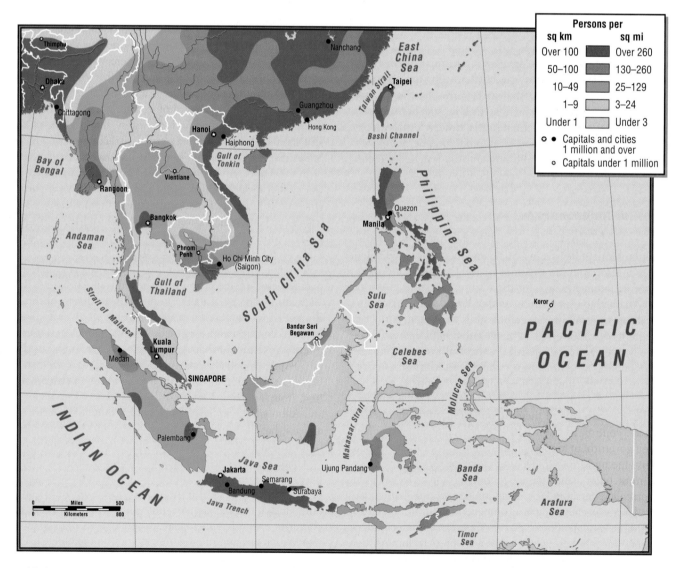

Figure 10.4 **Population density in Southeast Asia.** [Adapted from *Hammond Citation World Atlas,* 1996.]

TABLE 10.1 *Southeast Asia's largest metropolitan regions*

Rank (2000)	Metropolitan region	Population, in thousands (1980)	Population, in thousands (2000)	Average annual growth rate, 1980–2000, percent[a]
1	Jakarta, Indonesia	6,503	13,000	3.46
2	Manila, Philippines	5,926	11,200	3.18
3	Bangkok, Thailand	4,960	7,700	2.20
4	Ho Chi Minh City (Saigon), Vietnam	2,701	4,600	2.66
5	Rangoon, Burma	2,200	4,500	3.58
6	Singapore	2,410	4,000	2.53
7	Hanoi, Vietnam	820	3,100	6.65
8	Bandung, Indonesia	1,463	2,900	3.42
9	Kuala Lumpur, Malaysia	938	2,200	4.26
10	Medan, Indonesia	1,379	2,100	2.10
11	Palembang, Indonesia	787	1,550	3.39
12	Ujung Pandang, Indonesia	709	1,250	2.84
13	Phnom Penh, Cambodia	530	1,200	4.09

[a]Growth rates calculated using natural logarithm method: $\dfrac{\ln\,(2000\ \text{population}/1980\ \text{population})}{20}$

of the region are growing very rapidly (see page 541 for a discussion of the reasons for this growth). Rural migrants stream into the increasingly dense slum and squatter areas that exist in all of the region's cities except Singapore. Indeed, in some cities, as much as one-third of the population resides in these temporary, usually self-built, settlements.

Population Dynamics

Overall, Southeast Asians have smaller families today than they did in the past, but the patterns vary geographically. In some places, fertility rates have declined sharply since the 1960s (Figure 10.5). Singapore, an unusually wealthy city-state, had a fertility rate in

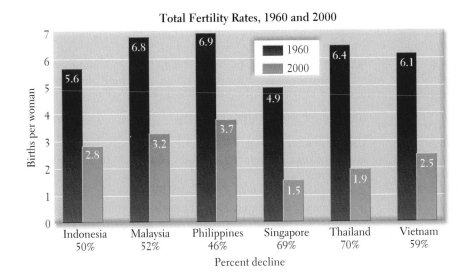

Figure 10.5 Total fertility rates, 1960 and 2000. Total fertility rates declined from 1960 to 2000 for several Southeast Asian countries. The percentage decline for each country is indicated below the name of the country. [Adapted from *A Demographic Portrait of South and Southeast Asia* (Washington, D.C.: Population Reference Bureau, 1994), p. 9; and *World Population Data Sheet,* 2000 (Washington, D.C.: Population Reference Bureau).]

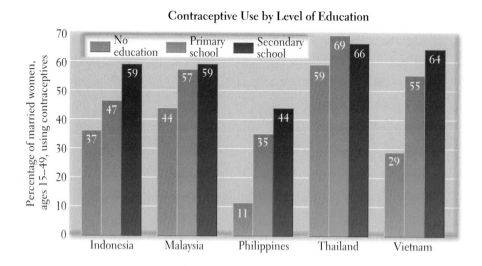

Contraceptive Use by Level of Education

Figure 10.6 **Contraceptive use by level of education.** The correlation between contraceptive use and level of education was also noted in discussing South Asia. In Southeast Asia, patterns are similar: as education rises, use of contraception increases, but the strength of the correlation varies from country to country. [Adapted from *A Demographic Portrait of South and Southeast Asia* (Washington, D.C.: Population Reference Bureau, 1994), p. 19.]

2000 below replacement level: 1.5 children per adult woman. The Singapore government is so concerned about the low fertility rate that it offers young couples various incentives, such as priority for the best public housing, to marry and procreate. Singapore's population continues to grow despite low birth rates because Singapore attracts a steady stream of highly skilled immigrants and less skilled illegal immigrants. Thailand, with one of the region's most successful family planning programs, has the next lowest fertility rate in the region: 1.9 children per adult woman. Compared to Singapore, Thailand is very poor, with just one-fifteenth the income per capita. Nevertheless, literacy is high, even for women. Although wages are low and working conditions often exploitative, Thai citizens have many employment opportunities and are allowed to move freely throughout the country to find work.

Two extremely poor countries in the region show the more usual correlation between poverty and high fertility. In Cambodia and Laos, women average more than five children each, and infant mortality rates are 80 per 1000 births for Cambodia and 104 per 1000 for Laos. Vietnam, where per capita incomes are as low as in Cambodia and Laos, has a relatively low fertility rate (2.5) and infant mortality rate (37). Vietnam's lower rates are explained by the fact that this socialist state provides basic education and health care to all its people, regardless of income. In Vietnam, literacy rates are more than 90 percent for both men and women, whereas in Cambodia only 45 percent, and in Laos only 30 percent, of women can read. Also, Vietnam's rapidly growing economy is attracting foreign investment; hence employment for women is expanding, replacing child rearing as the central role in their lives.

The use of modern birth control by women in Southeast Asia (56 percent) is significantly more prevalent than it is in South Asia (46 percent). Generally, couples tend to use birth control when access to health services, education, and jobs—especially for women—improves. Figure 10.6 shows the correlation between education and contraceptive use for five countries in the region. It is very likely that fertility rates would drop further if birth control technology were more readily available. According to the Population

Reference Bureau, in several of the most populous countries—including Indonesia, the Philippines, and Vietnam—over half the women say they do not want to become pregnant again, yet many do not yet have access to effective contraception.

Southeast Asia's population is young and hence is likely to keep growing even as family sizes shrink. Those 15 and younger make up 34 percent of the population and have not yet begun to reproduce. Figure 10.7 shows the population pyramids for Malaysia, the most developed country in the region (except for the city-state of Singapore), and for Indonesia, the country with the largest population (212.2 million). Malaysia's pyramid remains wide at the bottom, indicating that the population is young and thus will continue to grow even though family size is shrinking. Indonesia's birth rate slowed sharply 25 years ago but then increased again about 5 years ago, giving the lower slopes of its pyramid a blocky, irregular shape. Analysts anticipate that population growth will slow markedly in the next 25 years, but because so much of the population is now young, the number of people in the region will still double in about 40 years or less. Another trend is that more people will live long lives, possibly creating problems for younger workers who must give care and support to the elderly. Given Southeast Asia's relatively small land area, the rising level of consumer demand, and the fragility of its tropical environments, it seems likely that population pressure on land, soil, water, and forest resources will remain a major public issue for many years.

Southeast Asia's AIDS Tragedy

Human immunodeficiency virus (HIV) infection is growing across Southeast Asia: at present, Cambodia, Thailand, and Burma have the highest rates. In Thailand, acquired immunodeficiency syndrome (AIDS) is the leading cause of death, overtaking accidents, heart disease, and cancer, and more than a million people are thought to be infected. Men between the ages of 20 and 40 are the most common victims, but the number of women victims is rising. Many observers think that the disease may soon increase rapidly in places throughout the region that are only lightly affected so far—

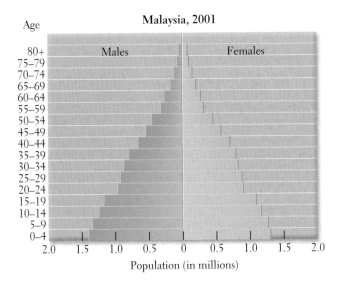

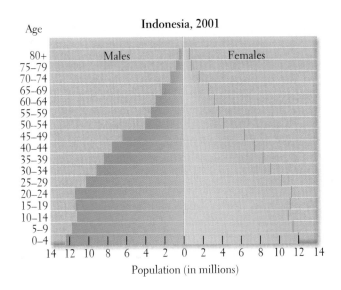

Figure 10.7 **Population pyramids for Malaysia and Indonesia.** Note that the scale for the population axis is roughly eight times greater for Indonesia than for Malaysia. In 2001, Indonesia had 206.1 million people while Malaysia had just 22.7 million people.

rural areas and secondary cities just being brought into closer economic and social contact with the region's big cities. The growth of the sex industry and especially of sex tourism is also a factor in the spread of AIDS (see the discussion on pages 544–545).

There are several reasons for the gloomy forecast of increasing AIDS and related infections. Southeast Asian societies have a number of religious traditions, especially Islam and Catholicism, that strictly limit both expressions and discussions of sexuality. At the same time, there are ancient, popularly accepted traditions that allow a fair amount of latitude in sexual practices that facilitate the spread of the disease. Conservative religious leaders restrict public sex education and AIDS prevention programs because these are viewed as unseemly topics, but popular customs support sexual experimentation, at least among some segments of the population. In Thailand, for example, a visit to a brothel has long been a rite of passage for young Thai men, and brothels are found in most neighborhoods. Studies in 1991 showed that more than 60 percent of female sex workers in urban locations were infected with AIDS and that 80 percent of young male soldiers visited brothels in that year. Also, just as in South Asia and Africa, truck drivers in Southeast Asia commonly spread AIDS because they travel far from home and may have sexual partners in several different locales. The UN AIDS Epidemic Update for 2000 noted that casual sex (between those who do not know each other well) is the most significant way in which AIDS is spread. A newly infected carrier is particularly contagious, yet is unaware of his or her condition and is relatively unconcerned about spreading the disease to a stranger. In Thailand, a rigorous government education campaign to use condoms is reducing infection rates. Neighboring Malaysia, in contrast, has a rapidly growing AIDS problem that is made worse by religious prohibitions on public discussion of sexuality, condom use, and drug abuse.

CULTURAL INSIGHT *Caring for the Ill*

In Thailand, Buddhist monks often provide intermittent care for the gravely ill in temples, but the high incidence of AIDS in recent years has strained this custom. In the crematorium in Lopburi, Thailand, dozens of fist-sized white cotton bags sit unclaimed at the foot of the gilded statue of Buddha. The Buddhist monk in charge of the hospice, himself infected with HIV, explains that because of concern about the contagion of AIDS, families leave those infected to be cared for entirely by the monks. When the patients die, the families leave the ashes there because they fear getting sick themselves.

In 1998, the Thai government announced a plan to have the Ministry of Agriculture and Cooperatives and Buddhist agencies cooperate to provide care for AIDS patients, because the traditional temple care system is overwhelmed. An AIDS community rehabilitation center, as it is referred to by the abbot of Lopburi's Bat Namphu temple, is to be built 75 miles north of Bangkok; it will shelter approximately 10,000 patients. But AIDS activists have denounced the plan, saying that the center will ostracize AIDS patients and project the image that they are difficult to care for and should be kept quarantined.

CURRENT GEOGRAPHIC ISSUES

Not surprisingly, some themes identified elsewhere around the world are also found in Southeast Asia. Like the Americas, Africa, and South Asia, Southeast Asia has a history of colonial rule, has experienced uneven development resulting in wide disparities of wealth, is expanding its links to the global economy through manufacturing and migration, and is home to a wide array of culture groups and religions, with social conflict an increasing reality. Environmental concerns—especially those related to urbanization and to unsustainable extraction of forest and soil resources—are also a theme. Nonetheless, there are some surprises. Southeast Asia has an economic, political, cultural, and social repertoire with a distinctly different flavor from all other regions; and, of course, the space it occupies is unique.

Because Southeast Asia has developed rapidly since World War II, the region is often cited as a proving ground for development strategies that may be usable elsewhere. But recent economic recession and political instability have cast some doubt on the stability of Southeast Asian systems and on their usefulness as a model for other regions.

ECONOMIC AND POLITICAL ISSUES

The economic and political situation in the region has changed dramatically in recent years. From the mid-1980s to the late 1990s, Southeast Asian countries had some of the highest economic growth rates in the world. Several countries earned a reputation as economic "tigers," aggressively embracing capitalism and poised to leap into a new era of prosperity. In Malaysia, Thailand, Indonesia, the Philippines, and Vietnam, gross national product figures soared and poverty was reduced significantly. Many countries that once struggled to feed themselves exported rice as well as a range of increasingly valuable manufactured goods. But beginning in 1997, economic growth stagnated and political instability increased. Among the reasons for the economic setback were decisions by country leaders to respond harshly when some minorities mounted protests against government policies, and corruption by these same leaders and members of their families and administrations. The perception that political instability was likely to increase deterred foreign investors, who had financed much of the earlier growth. Another important reason for the crisis was the huge debt burden accumulated by governments and private businesses. Much of the borrowed money went into real estate (especially in Thailand) and inadequately planned businesses or was used by elites in corrupt ways.

The leadership failures, corruption, and loss of investment have made life difficult for the millions of people whose hard work over the last several decades fueled these dynamic economies. For them, even during boom times, poverty was never a distant memory because the fierce competitiveness of the "tiger" economies often created low wages and unhealthy working conditions. By the late 1990s, poverty was again an increasingly painful reality for many people; and thwarted hopes for prosperity led to rising political violence.

Although popular images of Southeast Asia link the region to industrialization, in fact the economies of all the countries in this region are grounded in agriculture and services as well as industry. It is not uncommon for individuals to earn a living by plying skills in all three sectors at once.

Agriculture

Agriculture has a long history as a mainstay of life in Southeast Asia. As we have noted, about 60 percent of Southeast Asians still live in rural settlements. These people depend on agriculture for part of their support (Figure 10.8), even though many of them also fish, make and sell crafts, perform services, or work part time for wages. Agriculture also yields important products for export (especially rice, rubber, sugar, coconut products, and palm oil) and provides important components of urban diets (rice, fruits and vegetables, fish, meat, and dairy products).

Several forms of agriculture are practiced in Southeast Asia. So-called **slash-and-burn** (**shifting** or **swidden**) **cultivation** is an ancient form of multicrop gardening using small plots, often on rain-forest lands and in uplands that do not support more intensive types of agriculture. This system requires intimate knowledge of plant and soil characteristics. The farmer intercrops multiple species with careful attention to local soil and climate and attends to the needs of individual species and individual plants with hand tools. This type of cultivation is practiced today by subsistence farmers in the hills and uplands of mainland Southeast Asia and in coastal Sumatra, Borneo, Sulawesi, the southern Philippines, and West Papua—that is, in areas where population densities are relatively low (see Figure 10.4). Shifting agriculture is possible only in low-density areas because farmers move their fields every three years or so, as soil fertility is depleted. Where populations are sparse, plots can be abandoned for decades before they are used again, allowing the forest to regrow.

Another venerable and highly productive form of agriculture in the region is **wet rice** (or **paddy**) **cultivation.** Wet rice cultivation is permanent, as opposed to shifting. It entails the hand planting (usually by women) of rice seedlings in flooded fields that are first cultivated (usually by men) with hand-guided plows pulled by water buffalo. First laborers terrace and bank the soil by hand to create multiple adjacent ponds on a hillside or plain. Then they apply animal manure to the soil to maintain fertility and deliver water to the paddies by directing natural runoff or by pumping water from streams. Wet rice cultivation has transformed landscapes throughout Southeast Asia (see the photograph on page 537). It is practiced in river valleys and deltas, where rivers bring a yearly supply of silt, and on rich volcanic soils on such islands as Java and Sumatra in Indonesia and Luzon in the Philippines. Although wet rice farming has extensive environmental

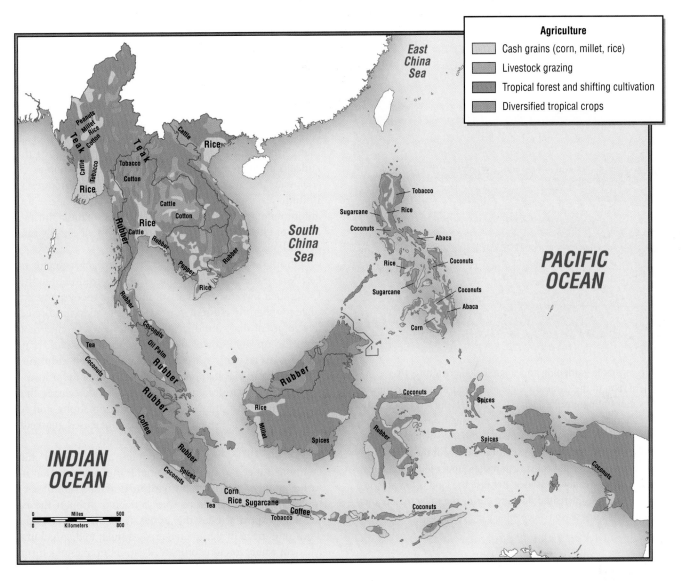

Figure 10.8 **Agricultural patterns in Southeast Asia.** [Adapted from *Hammond Citation World Atlas,* 1996, pp. 74, 83, 84.]

effects—in the form of forest clearing, landscape manipulation, and waterlogging of soils—this type of agriculture is sustainable over long periods and yields are fairly high. It is often cited as one of the better examples of nature being harnessed effectively in the service of humans.

Many farmers are abandoning paddy cultivation for work in the cities. The rice fields are now worked mostly on the weekends. Increasingly, they are prepared quickly with a type of tractor ("iron buffalo"), and rice is broadcast into the paddy, not hand-planted as seedlings. Weeding and the maintenance of water systems and terraces are increasingly neglected, and in some areas yields are declining as old methods are abandoned.

During the recent decades of prosperity, some of Southeast Asia's farming families left the countryside after selling their small farms to corporate plantations. The result has transformed rural ecology. Small farms that were once operated by families have been combined into large commercial farms owned by local or multinational corporations and producing such crops as rubber, African oil palm trees, bananas, pineapples, and rice. This style of agriculture entails combining huge tracts of land into one system; clear-cutting patches of forest that once occupied the fringes of farms; deep plowing the soil; planting usually one species of crop plant over many square miles; bolstering soil fertility with chemicals; using mechanized equipment; and, in the case of rice, using large quantities of water. Commercial farming reduces the need for labor, and its goal is large-scale production for cities and for export. Quick profit, not long-term sustainability, is the objective. Many commercial farmers have achieved dramatic boosts in harvests (especially of rice) by using high-yield crop varieties, the result of green revolution research that has been applied in many parts of the world. Although *all* types of farming, except some forms of small-scale shifting cultivation, have significant negative effects,

Large-scale vegetable gardening and single-crop agriculture, like this tea plantation in the Cameron Highlands, are transforming the higher elevations of peninsular Malaysia. Although the terraces supply food to urban markets both within the country and overseas and provide a profitable livelihood for farm and corporate owners, there are signs of environmental damage amid the beauty of the pastoral settings. Erosion in fields and construction sites (upper left) and along roadsides is causing deadly landslides. Because the land is less able to hold moisture, there is increased flooding during the rainy season and drought during the dry season. [Alex Pulsipher.]

the impact of large-scale commercial farming appears to be wider and longer lasting. Commercial farming can lead to displacement of traditional farmers and forest dwellers, loss of wildlife habitat and plant and animal species, soil erosion and loss of soil fertility, flooding, and the chemical pollution and depletion of land and groundwater resources.

Poor farmers usually cannot afford to become green revolution farmers because the new technologies are too expensive for them. Often they cannot compete with green revolution farms and are forced to migrate to the cities to look for work. In contrast, the farmers and corporations who can afford the new technologies can earn high profits, at least until soil resources and water are depleted, by selling their agricultural products to the millions in Southeast Asian cities. Commercial agriculture, therefore, gains importance as economies industrialize.

Patterns of Industrialization

Despite the widespread economic growth throughout Southeast Asia, there are significant differences in the types of industrial enterprises that dominate the economies in various countries and related differences in levels of wealth and well-being (see Table 10.4). Generally, the poorer countries—such as Vietnam, Laos, Burma, and Cambodia—and the rural parts of the other countries depend primarily on agriculture, fishing, and forestry. Even in some rural areas, however, construction and maintenance jobs on such projects as new hydroelectric dams are raising incomes. In cities and towns, garment and shoe making, food processing, and many types of light manufacturing are expanding. In the urban and suburban areas of the wealthier countries—such as Thailand, Malaysia, and Singapore—and in a few areas in Indonesia and the Philippines, more elaborate types of manufacturing are important: automobile assembly, textile and garment manufacturing, the refining of chemicals and petroleum, and the assembly of computers and other electronic equipment.

From the 1960s to the 1990s, national governments in Southeast Asia restricted and controlled foreign investment in the region. Using their own tax revenues, along with a modest amount of regulated foreign investment, governments tried to nurture economic development that would steadily create new jobs and be shielded from the ups and downs of global economic cycles. Export agriculture was favored, as were import substitution industries. These are local industries that produce consumer goods for local use; the use of tariffs excludes competing goods made outside the region. For the most part, these policies resulted in strong and sustained economic growth. Standards of living increased markedly, especially in Malaysia, Indonesia, and Thailand.

In the 1970s, the region adopted an additional strategy for encouraging economic development, this one aimed at manufacturing products for export, primarily to developed countries. Foreign multinational corporations were attracted through the establishment of **export processing zones (EPZs)**, or free-trade zones. These are specially designated areas, similar to the maquiladoras of Mexico, where foreign companies can set up manufacturing plants using inexpensive labor and, sometimes, inexpensive local raw materials to produce items only for export. Subcontractors, often locally owned, also may set up factories to supply needed parts. Taxes are eliminated or greatly reduced. The first export processing zone was established in Malaysia in 1971 and attracted electronics industries. Since then, EPZs have been established in Indonesia and Vietnam for the production of garments, textiles, and shoes. EPZs generally employ women because they will work for lower wages than men and are thought to be more reliable workers.

Among the newest free-trade and industrial zones is Subic Bay Freeport, the former U.S. naval base at Olongapo City in the Philippines. The naval base had many features attractive to industry: a deepwater port, a large airport, broad expanses of cleared land, in-place utilities, and a large group of people eager to work. Some 42,000 Filipinos lost their jobs when the U.S. Navy left Subic Bay and other facilities in the Philippines in 1992, but by 1996 many of them were working again. The new industrial zone had attracted some 200 multinational manufacturing and service firms that invested $1.2 billion in factories, warehouses, and offices and employed nearly 20,000 workers. Companies with facilities at Subic Bay Freeport include Federal Express (Subic Bay has been Asia's FedEx air hub since 1996) and Acer (a Taiwanese computer company).

In many ways, the star of economic growth in the region is Singapore, which has been a center of trade since its days as a nineteenth-century free port created by the British. Today Singapore is considered a highly developed country. It has a high standard of living; a well-established financial sector; a burgeoning computer software industry; a world-class airline; and the world's busiest harbor, where goods are moved from one shipping line to another.

Singapore is part of a new trend that is taking shape in the region: the emergence of larger transnational economic regions called **growth triangles** or economic zones. The Singapore-Johor-Riau (or Southern) Growth Triangle was created in about 1989 when low-cost manufacturing began to be relocated from Singapore to Johor, just across the border in southern Malaysia, and eventually to the nearby Indonesian Riau Archipelago. The move was prompted by Singapore's relatively high labor costs, land shortages, and expensive freshwater. Multinational corporations locating in the Southern Growth Triangle strategically pick the advantages of Singapore's efficient business and port infrastructure and highly skilled labor force (workers earned about U.S. $600 per month in 1990) for some tasks, and for other tasks choose the lower costs of the cheaper land and water and semiskilled labor in nearby Johor and Riau. (Labor costs are U.S. $220 per month in Johor and U.S. $90 per month in Riau.) A single multinational firm can locate core high-technology telecommunications, managerial, and shipment activities in Singapore; less technical activities (for example, textiles and electronics manufacturing) in Johor; and assembly operations in Riau.

Not everyone has benefited equally from Southeast Asia's "economic miracle." In the region's new factories and other enterprises, it is not unusual for employees to work 10 to 12 hours a day, 7 days a week for less than the legal minimum wage and with no benefits. Labor unions that would address working conditions and wage grievances are rare, and international consumer pressure to improve working conditions has been only partially effective. For example, in the late 1990s, the U.S. shoe manufacturing firm Nike became the focus of worldwide outrage over the treatment of its workers in Vietnam, Indonesia, and elsewhere across Asia; the employees were frequently exposed to hazardous chemicals, as well as to physical abuse and psychological cruelty on the job. In 1997, Nike laborers in Vietnam were paid $10 per week for 65 hours of work. The production cost for a pair of Nike shoes was U.S. $5.60, while the retail price in America was $90 to $145. Nike's labor tactics became widely known to American university students when student activists revealed that many campus stores were selling equipment and garments with the Nike logo that were made under substandard conditions in Southeast Asian factories. The students also noted that Nike promoted its products on college campuses, in high schools, and in boys and girls clubs by sponsoring athletic events and providing equipment with prominent logos to athletic teams. Student activist groups succeeded in finding out where Nike's factories were located in Asia, and investigations of factory working conditions revealed the abuses. Although the students' exposure of abuses in 1997 resulted in some improvements in working conditions, a report issued by a consortium of U.S. university students and labor and human rights groups in April 2000 notes that Nike workers were still earning less than U.S. 20 cents an hour, far less than a living wage, and were often required to work 72-hour weeks. Managers still physically abused workers and called them degrading names, and workers still worked in environmentally hazardous conditions. When questioned, Nike placed the blame on subcontractors in Taiwan and Korea who set up and administer Nike factories throughout Southeast Asia.

Economic Crisis: The Perils of Globalization

Beginning in 1997, a financial crisis swept the region, forcing millions of people back into poverty and creating tensions that disrupted Southeast Asia's political order. A major cause of the crisis was the rapid shift of most Southeast Asian economies away from government regulation and toward the free market. Another factor was widespread corruption.

By the late 1980s, most governments were opening up their national economies. One change was a greater willingness to allow the marketing of foreign products in Southeast Asia. Foreign governments in Japan, North America, Europe, and elsewhere

This view from atop the Raffles Hotel captures the active and prosperous spirit of Singapore. In the foreground, you can see the old colonial town center and, across the river, the skyscraper banking district. The pleasure boat harbor is to the left, port facilities at upper left. Singapore is a sleek and modern planned environment, but here and there historic buildings have been adapted to modern uses, usually connected with tourism. Singapore is the world's largest port, and from the hotel's penthouse restaurant one can see dozens of container ships waiting far out in the open ocean for permission to come into the docks for loading or unloading. [D. Saunders/TRIP.]

AT THE REGIONAL SCALE   *Building Cars in Asia*

Before the Asian recession of the late 1990s, many of the world's largest automobile manufacturers built factories in Southeast Asia, in expectation of widespread prosperity that could make family cars affordable for a large portion of the population (Figure 10.9). Foreign automakers set up plants throughout the region, with the highest concentration in Thailand. But in Indonesia, which is a major market and a major site of new plants, GM, Ford, Chrysler, Toyota, and Honda lost their ability to compete when President Suharto announced in 1995 that his son Hutomo and KIA Motors of South Korea would build a "national car." Because the car company was to be exempted from duties on imported Korean parts, its auto could be sold for thousands of dollars less than those produced by competitors. Furthermore, the car would be required for government fleets, guaranteeing a market.

KIA Motors went bankrupt in the Korean recession of 1996, and the IMF's rules for helping Indonesia with its debt crisis stopped all special tax and customs privileges for the national car project in January 1998. However, other countries produce their own national cars, and some of them (Malaysia, for example) still maintain high tariffs on foreign-made cars, including those made by foreign-owned companies located in the region. Hence, the market for foreign-made cars remains modest, especially in view of the general economic uncertainties in the region. The small motorbike remains the vehicle of choice for most people, so many foreign carmakers are planning to export their products—built with Southeast Asian assembly-line wages of $1.00/hour—to other regions such as Europe, Australia, and even South America.

Foreign carmakers are among those pressing for the creation of export processing zones (see page 530). Such a zone would allow them to outsource the manufacture of parts to the places with the lowest costs, while retaining assembly plants in such centrally located places as Thailand, where labor is skilled yet cheap. Foreign car manufacturers hope that eventually the market for cars in the region will increase radically and that Southeast Asian countries will reduce tariffs on cars made in Southeast Asia.

Source: Adapted from Philip Shenon, "For Asian nation's first family financial empire is in peril," *New York Times* (January 16, 1998); Robyn Meredith, "Auto giants build a glut of Asian plants, just as demand falls," *New York Times* (November 5, 1997); "Southeast Asia's Motor City," *Business Week Online* (May 8, 2000); and "Car making in Asia: politics of scale," *The Economist* (June 22, 2000).

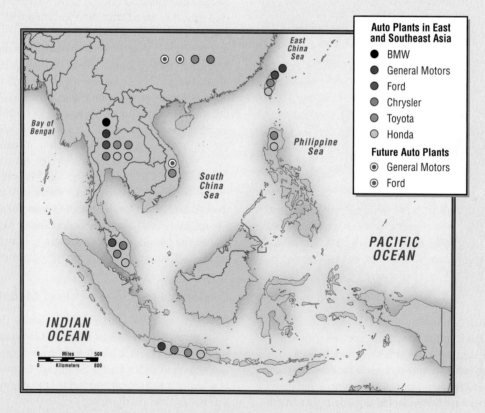

Figure 10.9 Auto plants in East and Southeast Asia. A number of automobile plants are already built or under construction by leading carmakers. [Adapted from *The New York Times* (November 5, 1997): 1.]

negotiated with Southeast Asian governments to gain local markets for their countries' products (such as cars and motorbikes; see page 532). Multinational corporations, such as automobile manufacturers, saw opportunities to save on labor, land, transportation, and resource costs by locating their facilities and selling their products within the region. Thus, factories manufacturing products for export in the EPZs were joined by foreign-owned factories manufacturing products they hoped to sell in Southeast Asia.

In opening national economies to the free market, governments focused especially on the crucial financial sector. Emboldened by their years of success in creating jobs and raising incomes and living standards, Southeast Asian firms were eager to expand; to do so, they needed investment capital. Southeast Asian banks were given greater freedom to determine the kinds of investments they would make and were allowed to accept more money from foreign investors. Soon, these banks were flooded with money from investors in the rich countries of the world, who hoped to profit as the banks invested in the region's growing economies. To make quick profits, the bankers made risky loans to real estate developers, often for high-rise office building construction. As a result, by the mid-1990s, many cities had a glut of office space—in Bangkok alone, there was U.S. $20 billion worth of unsold office space. A further consequence was that the stock prices of many companies were inflated far beyond their real value.

However, the lifting of controls on investments might not have been such a problem had the region not been plagued by a kind of corruption called **crony capitalism.** In most Southeast Asian countries, high-level politicians, bankers, and wealthy business owners often have close personal relationships or are part of the same extended family, and these personal connections can encourage corruption. For example, in Indonesia, the most lucrative government contracts and business opportunities were for decades reserved for the children of former President Suharto, who ruled Indonesia from 1967 to 1997 (see the box "Building Cars in Asia"). His children are now some of the wealthiest people in Southeast Asia. With the opening of Southeast Asian economies to foreign investors, this kind of corruption expanded considerably as foreign investment money was easily diverted to bribery or unnecessary projects that glorified political leaders.

The cumulative effect of crony capitalism and the pursuit of quick returns on risky investments was that many ventures failed to produce profits at all. In response, foreign investors withdrew their money on a massive scale. Before the crisis, in 1996, there was a net inflow of U.S. $96 billion to Southeast Asia's leading economies as well as to South Korea, which was also affected by the crisis. In 1997, there was a net outflow of U.S. $12 billion. The panicked withdrawal of investment money was the immediate cause of the economic crisis (Figure 10.10).

The International Monetary Fund (IMF) made a major effort to keep the region from sliding deeper into recession. In 1998, the IMF implemented a bailout package worth some U.S. $65 billion, much of it used to rescue banks. The IMF considers the banks essential to the region's economic stability, but critics argue that the bailout gives these banks little incentive to improve their disastrous record.

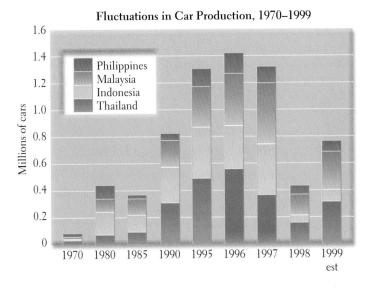

Fluctuations in Car Production, 1970–1999

Figure 10.10 **Car production in Southeast Asia, 1970–1999.** The sudden decline in car manufacturing in 1998 is an illustration of the depth of the sudden economic decline in the region. [From "Car making in Asia: politics of scale," *The Economist,* (June 22, 2000).]

By 2000, the economic recovery of Southeast Asia was apparently under way. Southeast Asian countries were successfully expanding older strategies, such as the establishment of export processing zones to attract additional multinational corporations to Southeast Asia.

Consequences of the Economic Crisis

The Southeast Asian economic crisis has had a number of disturbing consequences. Low-income people have suffered the most. Many have lost their jobs as a result of the rapid withdrawal of foreign investment money, which shut down many foreign-backed businesses. Moreover, the crisis sparked a decline in the value of most Southeast Asian currencies. As a result, prices have risen for the imported food and consumer goods on which many urban people depend and for the fertilizers and pesticides that agriculture requires. As prices rise, consumption is curtailed, firms can no longer operate profitably, and workers face layoffs and firings. The IMF bailout is to be repaid with tax dollars, so low-income people are likely to suffer cuts in health care, education, and food subsidies for years to come. Many urban workers no longer have families farming in the countryside to whom they can return during hard times. In Indonesia, for example, rates of poverty during the crisis climbed from roughly 11 to 60 percent. Frequent outbreaks of rioting, looting, and general violence troubled urban and rural areas.

The IMF's bailout package, which consisted of structural adjustment programs like those discussed in Chapters 3 (Middle and South America) and 7 (sub-Saharan Africa), required countries to reduce tariffs and abandon other policies intended to protect domestic industries. The goal was to lessen corruption and increase efficiency in order to win back the confidence of foreign investors. Critics point out that too great a dependence on foreign investment leaves the region open to a recurrence of a panicked withdrawal of funds. Also, if foreign investors take over the most profitable sectors of the economy, local enterprise will suffer, and profits made by foreign investors are likely to be invested outside the region.

The Association of Southeast Asian Nations

Southeast Asian countries trade on a much larger scale with the rich countries of the world than they do with one another. They export food, timber products, minerals, processed commodities, and manufactured finished goods and components to Japan, Australia, New Zealand, Europe, and the United States. They import manufactured products that they do not make themselves, including high-fashion consumer products, industrial materials, machinery, and parts. Trade among countries within the region has been inhibited by the fact that they all export similar goods and traditionally have imposed tariffs against one another. It is in this context that a region-wide free-trade zone is being established by the **Association of Southeast Asian Nations (ASEAN)**, an increasingly important bulwark of both economic growth and political cooperation in the region. Started in 1967 as an anticommunist, anti-China association, ASEAN now includes all the states in the region except East Timor and focuses on nonconfrontational

accords that strengthen regional cooperation. An example is the Southeast Asian Nuclear Weapons–Free Zone treaty signed in December 1995 by all 10 Southeast Asian nations.

ASEAN is assembling a free-trade association patterned after the North American Free Trade Agreement and the European Union. This trade alliance, known as the **ASEAN Free Trade Association (AFTA)**, was launched in 1992. According to current plans, most tariffs on items manufactured from at least 40 percent regional components will be reduced to just 5 percent by the mid-2000s. The main objective of AFTA is to help ASEAN members compete better in the world marketplace. Eliminating tariffs and other trade barriers between countries will lower production costs and make ASEAN's manufacturing industries more efficient. This will allow ASEAN products to be priced lower and hence to gain a competitive edge in the global market. At the same time, the region's consumers will be more likely to buy the more efficiently produced regional products, thus expanding trade among ASEAN countries. Member states agree that an integrated ASEAN economy would be a potent attraction for international investors, who prefer large, unified, and efficient markets to small, fragmented, and less efficient ones. As the cost competitiveness of manufacturing industries improves and the market grows larger, investors will be attracted by good returns on their money and by the dynamic business climate that continually encourages innovation.

Intraregional trade in ASEAN is growing steadily, if slowly. In 1990, 18 percent of the region's total trade was within the

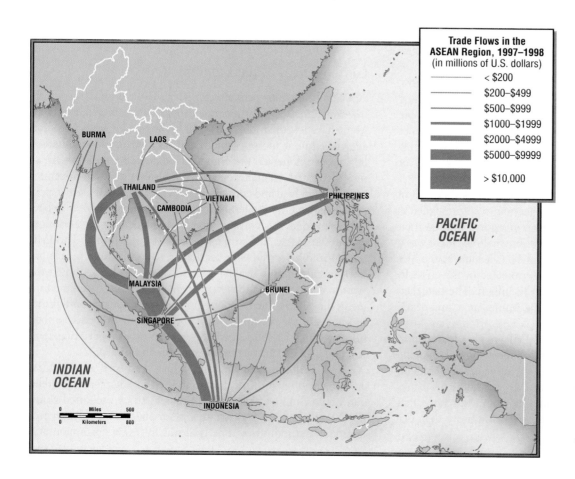

Figure 10.11 **Trade flows in the ASEAN region, 1997–1998.** Intraregional trade in Southeast Asia remains lopsided. In 1997–1998, 95 percent of such trade was between just five countries: Malaysia, Singapore, Indonesia, Thailand, and the Philippines. [Data from *The Far East and Australia, 2001* (London: Europe Publications, 2001).]

AT THE LOCAL SCALE *Pancasila: Indonesia's National Ideology*

Indonesia is the largest country in Southeast Asia and the most fragmented—it comprises 3000 inhabited islands stretching over 3000 miles (8000 kilometers) of ocean. It is also the most culturally diverse, with dozens of ethnic groups and multiple religions. Until the end of World War II, Indonesia was not a nation but rather a loose assemblage of distinct island cultures, which Dutch colonists managed to hold together as the Netherlands East Indies. When Indonesia became an independent country in 1945, its first president, Sukarno, hoped to forge a new nation out of disparate parts. To that end, he articulated a famous and controversial national philosophy known as **Pancasila,** based on tolerance, particularly in matters of religion.

The five precepts of *Pancasila* are belief in God and the observance of conformity, corporatism, consensus, and harmony. It is easy to see that all four of these words could be interpreted to discourage dissent or even loyal opposition; and they seem to require a perpetual stance of boosterism. For some people, the strength of *Pancasila* is that it ensures there will never be either an Islamic or a communist state. Others note the chilling effect that the philosophy has had on participatory democracy (the two political parties are state controlled) and on criticism of the president and the army.

And, in fact, there has been no orderly democratic change of government for many years. Sukarno was in power from 1945 until he was deposed by Suharto in 1965. Suharto declared himself president and was in power until forced to step down in 1998. A series of appointed successors followed him, including Megawati Sukarnoputri, Sukarno's daughter, in 2001.

An example of how *Pancasila* intrudes into the daily lives of the people and counters democratic ideals is provided by the state-sponsored women's organization Dharma Wanita (Women's Duty). The organization is modeled on an organization of American military wives, and membership is obligatory for all female government employees and the wives of male government employees. Its purpose is to teach women to serve as good wives and to support their husbands' careers, regardless of their actual marital or career status. There is a chapter in every government office, and a woman holds office in Dharma Wanita according to her husband's position in an agency or firm.

Source: Adapted from Saskia Wieringa, "The politicization of gender relations in Indonesia," *The Colonial Widow* (The Hague: The Digital City Project, 1995).

10 ASEAN countries; by 1999, it had increased to over 22 percent. But participation in intraregional trade remains skewed, with Singapore alone accounting for 43 percent of the total. The original ASEAN member states (Thailand, Indonesia, Singapore, the Philippines, and Malaysia) account for over 95 percent of the total intraregional trade. Burma, Laos, Cambodia and Vietnam hardly participate at all. Figure 10.11 shows the direction of trade within the region.

Pressures for Democracy and Self-Rule

Although there are strong demands for a greater public voice in the political process in Southeast Asia, significant barriers to democratic participation exist. Undemocratic socialist regimes control Laos and Vietnam, for example, and a military dictatorship runs Burma. There are few democratic institutions even in slightly wealthier countries, such as Indonesia, or in the rich city-state of Singapore. Some Southeast Asian leaders, such as Singapore's former prime minister Lee Kuan Yew, have argued that Asian values are not compatible with Western ideas of democracy. Asian values are supposedly grounded on the Confucian view that individuals should be submissive to authority. Hence, Yew and other leaders have argued that Asian countries should avoid the highly contentious public discourse of electoral politics. (For more on this view of Asian values, see the box "*Pancasila*: Indonesia's National Ideology.") But this position fails to account for the many people living under undemocratic, authoritarian regimes who have re-

belled. Many people in East Timor, Aceh, the Molucca Islands, Sulawesi, and elsewhere in Indonesia risked their lives to resist the authoritarian policies of President Suharto.

Is Indonesia Breaking Up? In recent years, rising violence and instability in Indonesia have many people wondering if this multi-island country of 206 million might be headed for disintegration. Independence movements have sprouted in four distinct areas (Figure 10.12), and one has already resulted in the formation, at the end of a long, bloody conflict, of a new independent nation, East Timor. The large far eastern province of West Papua has a growing separatist movement that formed in response to discrimination against the local Melanesian population by Javanese who have moved into the area. The movement also resists attempts to relocate people to West Papua from more densely populated parts of the country. Just to the west in the Moluccas, clashes between Christians and Muslims have killed thousands and encouraged the revival of a Christian-led separatist movement. In the far western province of Aceh, separatists have battled Indonesian security forces for decades, largely because they resent that most of the wealth yielded by their province's oil industry goes to the central government in Jakarta on the island of Java.

Jakarta is trying to appease these movements by giving people in separatist areas greater control over their own affairs, but it is also wary of strengthening separatism by allowing too much autonomy. One strategy is to bypass the provincial governments, which are the most prone to separatism, and instead give more power to local and municipal authorities. Many people fear that this strategy

Thinking Geographically: ON THE WEB

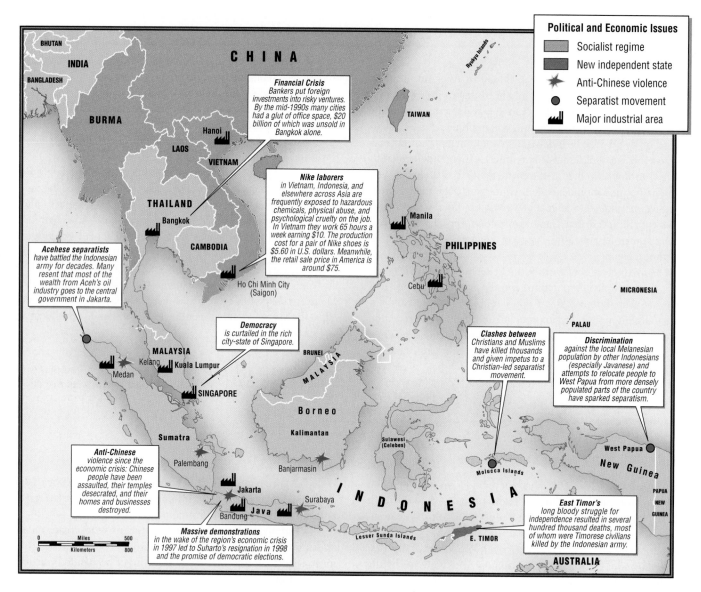

Figure 10.12 Political and economic issues: Southeast Asia. Economics, politics, and culture interact to create particularly complex circumstances in Southeast Asia. While rapid industrialization has improved standards of living, incomes vary widely. Some countries seek development with socialist policies; others pursue capitalist policies and democracy, with varying degrees of success. Indonesia has experienced the most violence and political unrest in the recent past, as the boxes in the map indicate. Thailand, Malaysia, and Singapore are, for now, the most stable politically and the most prosperous.

will increase corruption at the local level while doing little to improve the behavior of the Indonesian armed forces, whose often brutal tactics have alienated locals throughout Indonesia, not just in separatist areas.

The Indonesian army, which until recently enjoyed considerable financial backing from the United States, was responsible for killing 200,000 East Timorese (one-third of the population) during the country's long and bloody struggle for independence. Then, in 1999, the Indonesian army went on a rampage after 98 percent of the registered voters gave a 78 percent majority vote

for independence in a UN-sponsored referendum. The army killed an unknown number of people and displaced more than 250,000, over 100,000 of whom were still detained in Indonesia as of 2001. The United Nations has established a Transitional Administration of East Timor to exercise legislative and executive authority in the territory. During a transition period, UN administrators are working in close cooperation with the East Timorese people to deliver humanitarian aid, develop civil and social services, and build the capacity for self-government and sustainable development.

Progress Toward Democracy. Pressure for democracy has been rising in several Southeast Asian countries, in response to both economic development and economic crisis. In Thailand, years of economic growth have increased literacy levels and awareness of the democratic privileges that exist in many other countries. In the late 1990s, Thais successfully agitated for constitutional provisions that reduce corruption and provide for regular elections. In Malaysia, increased education and prosperity have similarly pushed the government to act in a slightly more democratic and less authoritarian manner. In Indonesia, after three decades of semidictatorial rule by President Suharto, massive demonstrations in the wake of the region's economic crisis led to Suharto's resignation and the promise of democratic elections. And in Burma, more than a decade of protest has followed the failure of the military regime to step aside when the people elected a civilian reformist government in 1990.

SOCIOCULTURAL ISSUES

The cultural complexity of Southeast Asia supplies people in the region with an array of strategies for responding to rapid and pervasive change. But, as we will see, culturally complex societies can often be in precarious balance if national mores fail to help citizens relate well across ethnic and religious lines. We will also see that attitudes about family size and gender roles are changing in response to new economic circumstances.

Cultural Pluralism

Anthropologists like to say that Southeast Asia is a place of **cultural pluralism** (or **cultural complexity**). They mean that it is inhabited by groups of people from many different backgrounds who have lived together for a long time yet have remained distinct, partly because they lived in isolated pockets separated by rugged topography and spans of ocean. Small groups of descendants of the Australo-Melanesians who came into the region some 40,000 to 60,000 years ago still live primarily by hunting, gathering, and swidden agriculture in the interior forests of several larger islands, primarily Sumatra, Java, the Malay Peninsula, Borneo, Sulawesi, New Guinea, and the Philippines. Far more numerous today are descendants of the Austronesians who migrated from the South China mainland into the islands about 5000 years ago. Past isolation has contributed to distinctive cultural differences among groups of these people as well. They tend to live in coastal zones, although some, such as the Igorots of Luzon in the Philippines, live in the mountainous interiors. Language is one of the clearest markers of ethnicity among groups of both Australo-Melanesian and Austronesian descent; each group speaks its own distinctive language. The descendants of these two major divisions of the earliest inhabitants of the region are usually referred to together as *indigenous peoples.* They are distributed in small patches from northern Burma and Thailand to southern Indonesia.

Over the last 2000 years, newer immigrants have arrived, first from Arabia, Central Asia, India, China, Japan, and Korea, and then (after 1500) from Europe. Today, Southeast Asia is one of the most ethnically diverse regions in the world.

Look very closely at the bag the young woman is carrying and the drink the young man is holding, as they pause amid manicured rice fields in Indonesia. Even in remote areas it is difficult to escape the symbols of globalization in Southeast Asia. [Ian Lloyd/Black Star.]

Southeast Asia entered an era of increasing interaction with the outside world when Europeans began colonizing the region after 1500. Most recently, Southeast Asia's cultural mix has been broadened by ideas from around the globe: free-market capitalism, communism, nationalism, consumerism, and environmentalism. These ideas and the material culture associated with them—from advertising to motorbikes to spandex to green politics—have also modified the life and landscapes of Southeast Asia.

Given the introduction of such new ideas, one might expect that by now Southeast Asia would be a melting pot. To some extent, notably in the cities, it is showing some signs of moving in that direction. In some of the region's countries, however, dozens of different languages are still spoken. (Indeed, it is estimated that 1000 of the world's 6000 or so languages are spoken here.) People practice many different religions, and neighboring families may trace their roots to remarkably different racial and ethnic origins. Today, the barriers separating groups are falling as modernization and industrialization draw large numbers of people into the cities and resettlement projects move several ethnic groups together onto newly opened lands. Migration and cross-cultural marriage are creating ethnically mixed groups who are unsure of their heritage and may never have seen their ancestral landscapes.

The Overseas Chinese. One group, prominent beyond its numbers, is the so-called Overseas (or Ethnic) Chinese. Small groups of traders from southern and coastal China have been active in Southeast Asia for thousands of years; some got their start as fisher folk who met their customers on the high seas. Over the

TABLE 10.2	Ethnic Chinese in Southeast Asia	
Country	Number of Ethnic Chinese in country (millions)	Percentage of Ethnic Chinese in country populations
Indonesia	7.2	3.8
Thailand	5.8	10
Malaysia	5.2	28
Singapore	2.0	70
Burma	1.5	3.4
Vietnam	0.96	1.4
Philippines	0.8	1.3
Laos	0.008	1.3
Cambodia	0.33	4.0
Brunei	0.075	25
Southeast Asia, total	23.873	5.0

Source: Jonathan Rigg, Southeast Asia (New York: Routledge, 1997), p. 123.

centuries, there has been a constant trickle of these Chinese immigrants. The ancestors of today's Overseas Chinese, however, began to arrive in large numbers during the nineteenth century, when the European colonizers needed labor for their plantations and mines. And many of those who fled China's communist revolution (1949) sought permanent homes in Southeast Asia's trading centers (Table 10.2). Today the Chinese are artisans, merchants, middlemen, bankers, transport owners, developers, and laborers.

The Chinese in Southeast Asia have the reputation of being rich and influential in government and commerce yet strongly loyal to their own group, often importing kin as employees rather than hiring local people. Some in the region think the Overseas Chinese should be forced to change their habits, assimilate, and—especially—hire non-Chinese workers. Occasionally, the proponents of these ideas have turned to violence.

Many low- and middle-income workers who have been hurt by the recent financial crisis (see the discussion beginning on page 531) blame their sorrows on the Overseas Chinese. The Chinese are industrious and more prosperous than the average; with their region-wide connections and access to start-up money, they are able to take advantage of the new growth sectors of modernizing economies. Sometimes new Chinese-owned enterprises have put out of business older, more traditional establishments that many local people depend on. For example, in the town of Klaten in central Java, an open-air bazaar where local Indonesians once sold an

array of goods has been shut down to make way for a government-sponsored air-conditioned shopping mall. Only the Chinese can afford the expensive store rents. Since the economic crisis, resentment against the Chinese has brought a wave of violence. Chinese people have been assaulted, their temples desecrated, their homes and businesses destroyed. Indeed, since the 1970s, conflicts involving the Chinese have occurred in Vietnam, Malaysia, and many places in Indonesia (Sumatra, Java, Kalimantan, Sulawesi).

Despite recent hostilities, Chinese commercial activity throughout the region has reinforced the perception that the Chinese are diligent, clever, and civic-minded businesspeople, often working especially long hours. Some Overseas Chinese, increasingly aware that the global economy presents dangers as well as opportunities to their group, are attempting to alleviate their situation through public education. There are Overseas Chinese study centers in Thailand, Vietnam, Malaysia, Singapore, Indonesia, and the Philippines. They maintain a higher profile as civic participants within their local communities or establish more formal international connections and services, such as investment banking companies, that will give them financial flexibility and spatial mobility should life in one particular place become too inhospitable.

Religious Pluralism

The major religious traditions of Southeast Asia include Hinduism, Buddhism, Confucianism and Taoism, Islam, Christianity, and animism (Figure 10.13). Of these, only animism is native to the region. Animism is the traditional belief system of the original settlers whose descendants are now rural subsistence cultivators and forest-dwelling peoples living in groups scattered across the region. For animists, such natural features as trees, rivers, crop plants, and the rains all carry spiritual meaning and are the focus of festivals and rituals to mark the passing of the seasons.

The other religions were brought by immigrants, traders, colonists, and missionaries, and their patterns of distribution reveal an island-mainland division (see Figure 10.13). Buddhism is dominant on the mainland, especially in Burma, Thailand, and Cambodia. In Vietnam, people practice a mix of Buddhist, Confucian, and Taoist beliefs originating in China. Islam is dominant on the southern Malay Peninsula and in the islands of Malaysia and Indonesia, and it is growing in the southern Philippines. In the islands, Christianity is a distant second in importance, dominant only in East Timor and in the Philippines north of Mindanao. Catholicism was imported to the Philippines by the Spanish colonists who arrived there to open transpacific trade with Mexico in the 1540s. Hinduism first arrived with Indian traders thousands of years ago and was once much more widespread; now it is found only in small patches, chiefly on the islands of Bali and Lombok, east of Java. Recent Indian immigrants who came as laborers in the twentieth century during the later part of the European colonial period have reintroduced Hinduism to Burma, Malaysia, and Singapore.

All of Southeast Asia's religions have undergone change as a result of exposure to one another and to the traditional animist beliefs. Filipino Catholics and other Christians in Burma and Thailand (the Karen and Wa hill people, for instance) believe in

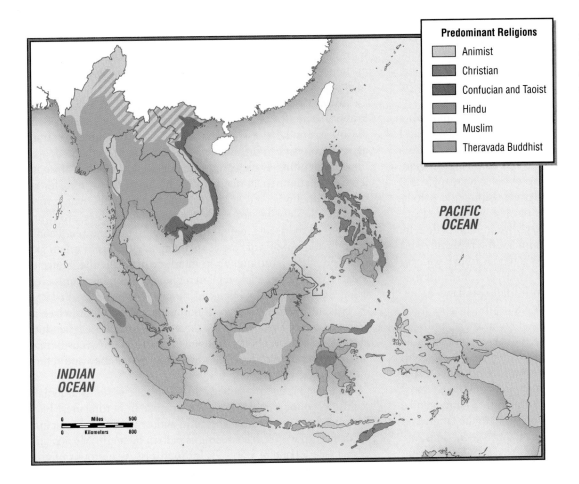

Figure 10.13 Religions of Southeast Asia. [Adapted from Alisdair Rogers, ed., *Peoples and Cultures* (New York: Oxford University Press, 1992), p. 222.]

Predominant Religions
- Animist
- Christian
- Confucian and Taoist
- Hindu
- Muslim
- Theravada Buddhist

PACIFIC OCEAN

INDIAN OCEAN

spirits, and they practice rituals that have their roots in animism. Often, missionaries for Christianity or Islam deliberately pointed out ways in which the new faith was similar to indigenous beliefs. Hindus and Christians in Indonesia, surrounded as they are by Muslims, have absorbed many ideas from Islam, such as the seclusion of women. On the other hand, Muslims have absorbed ideas and customs from indigenous belief systems, especially ideas about kinship and marriage, as illustrated in the following vignette.

VIGNETTE Marta is a *dukun,* a woman who prepares a bride for a traditional Javanese wedding. Nearly all Javanese are Muslims, but Islam does not have elaborate marriage ceremonies, so colorful animist rituals have survived. Usually, the bride and groom have never met, their families having worked out the match. Hence, the first meeting at the wedding ceremony is surrounded by a great deal of mystery. Marta's job is to reinforce this mystery and to prepare the couple for a life together. She bathes and perfumes the bride, puts on her makeup, and dresses her, all the while making offerings to the spirits of the bride's ancestors and counseling her about how to behave as a wife and how to avoid being dominated by her husband.

The groom, meanwhile, is undergoing ceremonies that prepare him for marriage. Each will sign the marriage certificate before they meet. In the traditional Javanese view, it is best that a young couple not be in love—that way, they will not fall out of love. Rather, from the start they will hold each other at arm's length, not investing all their emotional capital in a relationship that is bound to change over the course of the decades as they mature and their family grows up. Marta's role is to create a magically reinforced bond between these two strangers. After marriage, as before, a man's closest friends will be his male age-mates and kin, and a woman's will be her female friends and kin.

Divorce in Malaysia and Indonesia is fairly common among Muslims, who often go through one or two marriages early in life before they settle into a stable relationship. Although the prevalence of divorce is lamented by society, it is not considered outrageous. Apparently, ancient indigenous customs allowed for mating flexibility early in life, and this attitude is still tacitly accepted. [Adapted from Walter Williams, *Javanese Lives* (Piscataway, N.J.: Rutgers University Press, 1991), pp. 128–134.]

Family, Work, and Gender

Family organization in Southeast Asia is variable, as one might expect in a region of so many cultures. There are some patterns that are remarkably different from the patriarchal patterns found elsewhere around the world. Nonetheless, there is a patriarchal overlay that derives from many sources: Islam, Hinduism, and Christianity, to name a few. In this section, we describe some of these differences in family life in Southeast Asia.

Family Patterns. Throughout the region, it is common for a newly married couple to reside with the wife's parents, or at least to be most closely associated with them. This ancient family structure survives today even where Islam is the dominant religion. Along with this custom there is a range of behavioral rules that empower the woman in a marriage, despite some basic patriarchal attitudes. For example, a family is headed by the oldest surviving male, usually the wife's father. Only when he dies does he pass on his wealth and power to his son-in-law, the husband of his oldest daughter, not to his own son. (A son goes to live with his wife's parents and inherits from them.) Hence, a husband may live for many years as a subordinate in his father-in-law's home. Instead of the wife being the outsider under the hegemony of her mother-in-law, as in Southwest Asia, China, India, and elsewhere, it is the husband who must kowtow. There is an inevitable tension between the wife's father and the son-in-law, which is resolved by the custom that the two men must practice ritual avoidance. The wife manages communication between them by passing messages and even money back and forth. Consequently, she has access to a wealth of information crucial to the family and has the opportunity to influence each of the two men. This role gives her power that wives in Muslim families in India or Southwest Asia do not have.

Urban families tend to be nuclear in the early years of a marriage. In urban families influenced by modernization, the young couple may choose to live apart from the extended family. This arrangement takes the pressure off the husband in daily life. Because this nuclear family unit is often entirely dependent on itself for support, wives usually work for wages outside the home. The drawback of this compact family structure, as many young families are now finding out in Europe and America, is that there is no pool of extended kin (grandparents, siblings, aunts, and uncles) available to help working parents with child care. There are not even any kin or neighbors to be hired for child care and housework because most women work, and grandmothers often live back in the village. By the same token, no one is left to help elderly parents maintain the old rural family home. So, increasingly, tiny urban apartments must accommodate a young family and one or more aged grandparents in need of care. On the brighter side, working grandmothers in their vigorous middle years can often help their adult children financially.

"Women are good with money" is a common saying in Indonesia. The social researchers Hannah Papanek and Laurel Schwede have written about working women and money in Indonesia. Women in that country must take into account the high divorce rate: already, in the 1970s, 38 percent of urban women over the age of 35 and 43 percent of all rural women had been married more than once. Thus, Indonesian women often start planning at the beginning of their marriage for the eventuality of having to support themselves. They tend to keep their spend-

ing low. In 70 percent of the families studied by Papanek and Schwede, women were the family money managers. Women are customarily the managers of family budgets elsewhere around the region as well.

Work Patterns. As the economies of Southeast Asia are transformed, so are the work lives of its people. The changing pattern of work in Southeast Asia is more complex than a simple shift from agricultural work to factory work or from rural to urban occupations. Agriculture is changing from small-plot family farming to large-scale commercial operations. The larger farms need fewer, but more highly skilled, workers. Rural communities, which used to be inward-looking and self-sufficient, now produce crops primarily for export: rice, palm oil, or bananas, for example. People cycle in and out of rural areas, spending a few years in urban factory work, then returning for an interlude of village life. Statistics indicate that many people remain registered as rural residents even when they are actually living in the city. For example, in 1992, 72 percent of the people of Indonesia identified themselves as rural, yet only 56 percent of the labor force worked in agriculture in that year. Households counted as rural in the census may include several individuals who live and work in cities and return only on weekends, if then.

A related emerging pattern is that, despite customs of female seclusion, young single women who migrate from rural areas have become the employees of choice in industries that manufacture goods for export. We have seen this pattern before in Mexico and China (see Chapters 3 and 9). In Southeast Asia, groups of girls often migrate together and live several to a room, close to their jobs (see the vignette on page 561). The geographer Jonathan Rigg has analyzed several studies of the region and concluded that bosses prefer to hire young, single women because they are perceived as the least troublesome employees. Statistics do show that, at least for now, women will work for lower wages than men, will not complain about poor and unsafe working conditions, will accept being restricted to certain jobs on the basis of sex, and are not as likely as men to agitate for promotions. But the situation varies from country to country. For example, female Filipino workers are more assertive than Malaysians. Supervisors are somewhat more likely to be female in U.S.- and European-owned firms than in Japanese- or Korean-owned ones, though both male and female bosses in this region generally tend to be authoritarian, and factory workers usually have little say in workplace policies.

Men continue to work in agriculture, construction, and services as well as in manufacturing. They have a higher rate of employment than women throughout the region, but women are catching up in the number of paid jobs they hold. However, their wages (as reflected by GDP per capita) average less than half those of men (Table 10.3). In Brunei, Burma, the Philippines, and Thailand, more women than men are completing training beyond secondary school. Hence, if training qualifications were the sole consideration, women would appear to have an advantage for future employment in Southeast Asian information technology industries. It will be interesting to see if women eventually do become the majority in this industry and if their wages are commensurate with those of men with similar qualifications.

TABLE 10.3   *Gender comparisons for Southeast Asia: GDP and education level*

Country (1)	GDP per capita (adjusted for PPP in U.S. $) Female (2)	GDP per capita (adjusted for PPP in U.S. $) Male (3)	Female secondary school enrollment as percentage of male enrollment (4)	Female tertiary school enrollment as percentage of male enrollment (5)
Brunei	10,135	22,790	105	156
Burma	1,011	1,389	96	156
Cambodia	974	1,822	66	23
Indonesia	1,780	3,526	91	53
Laos	1,390	2,073	72	44
Malaysia	4,501	11,674	115	91
Philippines	2,512	4,580	102	133
Singapore	15,966	32,334	98	81
Thailand	4,159	6,755	97	111
Vietnam	1,395	1,991	97	Not available
United States	22,565	36,849	100	121

Source: United Nations Human Development Report 2000 (New York: United Nations Development Programme), Tables 2 and 28.

Migration

Migration has long been a common feature of life in Southeast Asia. Well before the era of European colonization, outsiders arriving from Arabia, India, and China contributed major components to the region's cultural mix. Today, many people leave the region in search of work or move to new areas within the region as part of vast resettlement schemes. Recently, rural-to-urban migration within the region has been particularly significant, as is the case elsewhere in the world.

Rural-to-Urban Migration. Southeast Asia as a whole is just 36 percent urban, but the balance between rural and urban employment is changing quickly in response to global market forces. Malaysia is already 57 percent urban, the Philippines 47 percent, and Singapore 100 percent. Often, the focus of migration is the capital of a country, which may quickly become a primate city— one that is two or three times the size of the second largest city and overwhelmingly dominates the economic and political life of the country. Indeed, Bangkok is nearly 30 times larger than Thailand's next largest metropolitan area (Chang Mai), and Manila is 10 times larger than the Philippines' second city (Cebu).

Individuals and families leave rural areas to escape poverty in the countryside where multinational firms now cultivate oil palms, rubber trees, or fruit for urban and export markets. Commercial agriculture often displaces subsistence cultivators who must then obtain cash to buy what they previously provided for themselves. These are the **push factors** in migration. On the other hand, people are attracted to the city by the higher living standards they expect to find there, perhaps working for multinational producers of shirts or sneakers such as Gap Inc. or Nike or for a semiconductor manufacturer such as Applied Materials in Singapore. These are the **pull factors.** Migrants often leave behind very young and very aged relatives whom they must then support with earnings sent home (remittances). Rarely, however, can primate cities such as Jakarta, Kuala Lumpur, Bangkok, or Manila provide sufficient jobs, housing, and services for the new arrivals. Of all the primate cities in Southeast Asia, only Singapore provides well for nearly all of its citizens, but even there uncounted illegal immigrants live on the margins.

VIGNETTE   Mak (age 33) and Lin (age 27), husband and wife, left their two sons in the care of her parents in a village north of Bangkok, Thailand. They could not afford the school fees for even one of their sons on their rural wages, yet they hoped to educate both boys, even though educating just the eldest is the custom. So Mak and Lin traveled by bus for 10 hours to reach Bangkok, where—after several anxious days—they both found grueling work unloading bags of flour from ships in the harbor.

GEOGRAPHER IN THE FIELD

Interview with Regional Consultant for Chapter 10, Richard Ulack, University of Kentucky

What led you to become interested in Southeast Asia?
I have been interested in exotic, foreign places since I was a young boy, but by the time I entered the geography graduate program at Penn State I had developed a strong interest in eastern and southern Asia. What has made Southeast Asia so appealing to me is its tremendous physical geographic and cultural geographic diversity. While still a student, I spent six months on the island of Mindanao in the Philippines, where I gathered the data and information necessary to finish my Ph.D.

Where have you done fieldwork in Southeast Asia, and what kind of work have you done?
Since leaving Penn State I have been back to the Philippines to do fieldwork, and to teach under the auspices of a Fulbright Award, on three more occasions. My fieldwork has focused on my interests in population geography (internal migration and fertility), urbanization (especially of middle-sized cities), and, most recently, tourism. In 1991 a Fulbright Award allowed me to conduct research on tourism in the island country of Fiji in the South Pacific.

How has that experience helped you to understand the region's place in the world?
Traveling and living abroad for extensive periods has allowed me to appreciate how interconnected the world is and how greatly the

Western world has impacted Southeast Asia and other world regions. One can see the interconnectedness of today's world in as simple a matter as food. Even Lexington, Kentucky, has several Thai restaurants, and Vietnamese, Indonesian, Malay, and Filipino restaurants are found in large cities throughout the United States. Conversely, American fast-food restaurants such as Kentucky Fried Chicken are common in Southeast Asia.

What are some topics that you think are important to study in your region?
There are numerous interesting topics that one could pursue in studying a region like Southeast Asia. One excellent topic is the role of women in economic development. Women play a major role in the region's national economies, and the status of women is generally higher in this region than in nearby East and South Asia. Other topics worthy of study include tourism and development, the impact of the rapid industrialization on population phenomena such as migration and urbanization, and the effects of the region's version of a "common market," ASEAN, on development.

Each day when they finished their work, they walked miles to their quarters in one part of a tiny houseboat anchored, with thousands of others, on the Chao Phraya River. That river is Bangkok's low-income housing site, its source of water, its primary transport artery, and its sewer. Mak and Lin had to step gingerly across dozens of boats to get to theirs, intruding repeatedly on the privacy of their fellow river dwellers. They bathed in the dangerously polluted river, washed their sweaty, flour-covered work clothes by hand, and cooked their dinner of rice over a Coleman stove. Their only entertainment was provided by the private lives of their too numerous and too near-at-hand neighbors, who were often drunk on cheap liquor. For two years, Mak and Lin sent remittances to their family and managed to save enough to buy a bicycle, to pay the school fees, and to tide them over for several months at home. Although they were glad to be out of Bangkok, both nevertheless agreed that they would eventually go back if they couldn't find work near their village.

Resettlement. To alleviate rural unemployment, to forestall migration to the crowded cities, and for many other reasons, governments in Southeast Asia have developed **resettlement schemes** to move large numbers of rural people from one part of a country to another. (**Transmigration** is the local term for resettlement in Indonesia.) Beginning in 1904, Dutch colonists in Indonesia began to move people from the islands of Java, Bali, and Madura to land only lightly occupied by indigenous people in the outer islands of Sumatra, Kalimantan in Borneo, Sulawesi, and West Papua in New Guinea. These immigrants supplied labor on plantations growing such export crops as rubber, coconut, and palm oil. Much larger resettlement efforts on these same islands began after 1950. More than 4 million people have been relocated, making the Indonesian effort one of the largest land resettlement schemes ever.

The reasons for resettlement in Indonesia have changed over time, from promoting food production, especially national self-sufficiency in rice, to more general goals of regional development, national integration, and population redistribution. Resettlement also dovetails with government policies to assimilate Indonesia's outlying indigenous culture groups into mainstream society. Such

Designated SP6, this new settlement carved out of the West Papua rain forest is part of the government's resettlement program. People who elect to move here receive a one-way air ticket, a house, 5 acres (2 hectares) of land, and a year's supply of rice. Estimate are that 5 million to 6.5 million people have participated in transmigration in Indonesia since 1960. [George Steinmetz.]

policies are carried out by relocating people, especially Javanese, from the densely occupied cultural heartland to areas occupied by indigenous people. It is also Indonesian policy to disperse people from the crowded core in western Java to avoid political unrest. Many thousands of people not part of the formal resettlement schemes have joined the stream of people to the outer islands. The

results have been unfortunate. In some cases, thousands of informal migrants moving onto the lands of indigenous peoples have built shantytowns and destroyed native habitats for plants and animals. Many thousands of acres of forest have been cleared for agricultural resettlement in areas where the tropical soils are too fragile to sustain cultivation for more than a year or two (see photo at left). Severe environmental damage, ethnic discord, and breaches of human rights have been common.

In 2000, the indigenous people began to fight back in the Molucca and Sulawesi islands and in Kalimantan. Violent clashes occurred on several occasions. For example, the Dayak people of Kalimantan, many of whom are now Christians but continue to live in traditional ways, launched an attack on settlers from Madura (near Java), who happened to be Muslim. The Dayaks complain that their lands were taken without their consent and for far too little payment. Dayaks have killed hundreds of Madurese settlers.

Migration out of the Region. **Extraregional migration,** or migration to countries outside the region, is also important in Southeast Asia because of its effects on families, income, and employment. Extraregional migration may be either short term or permanent. Recently, women have constituted a major proportion of these migrants. In the Philippines, women emigrants now outnumber men, and many are skilled nurses and technicians who work in European and North American cities and in Southwest Asia. Many Southeast Asian Muslim women (chiefly from Indonesia) have become part of the "maid trade" of 1 million or more who work in the homes of wealthy citizens in the Arab states of Saudi Arabia, Kuwait, and United Arab Emirates (Figure 10.14). Between 1983 and 1988, 193,000 Indonesian women migrated to Saudi Arabia alone to work as maids.

Skilled male workers from Southeast Asia are especially well known in the world merchant marine (see the vignette on page 28, Chapter 1). They typically serve in the lower echelons of shipboard occupations, working as seamen, cooks, or engine mechanics. In recent years, some have advanced to officer status.

Thinking Globally: **ON THE WEB**

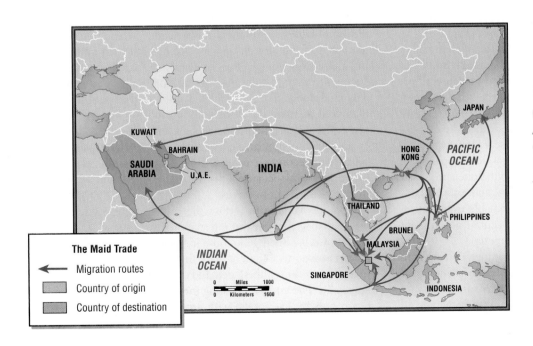

Figure 10.14 Interregional linkages: the "maid trade." In 1996, between 1 million and 1.5 million Southeast Asian women were working elsewhere in Asia (Japan, Malaysia, Brunei, Singapore, Kuwait, Bahrain, the United Arab Emirates, and Saudi Arabia) as foreign domestic workers. Educated women from Southeast Asia also migrate and send remittances home. Wealthy countries within Southeast Asia, such as Malaysia and Singapore, are, in turn, destinations for women from India, Sri Lanka, Indonesia, and the Philippines. [Adapted from Joni Seager, *Women in the World: An International Atlas* (London: Penguin, 1997), p. 65.]

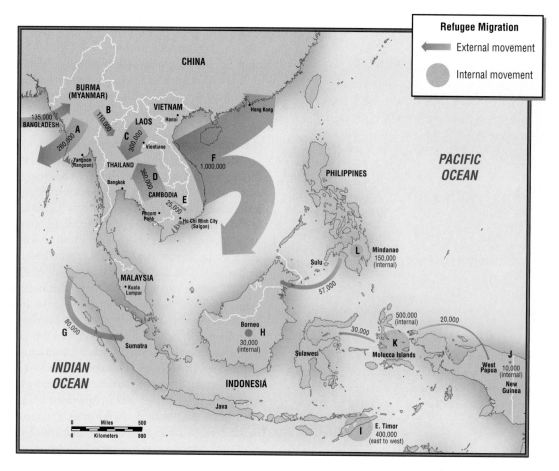

Figure 10.15 Flow of refugees in Southeast Asia. Refugees fled violence in Burma, Laos, Cambodia, and Vietnam in the decades following the end of the Vietnam War. (A) 1992–1995: Rohingya Muslims fled from Burma to Bangladesh, but about half have since returned. (B) 1995: Burman refugees in Thailand. (C) 1973–1975: Laotians fled communism. (D) 1979–1993: Cambodian refugees in Thailand. (E) 1995: Cambodian refugees in Vietnam. (F) 1975–1989: Boat people from Vietnam. (G) 2000–2001: Aceh refugees in Sumatra. (H) 2000: Resettled Madurans displaced by indigenous Dayak. (I) 1999–2000: Timor/East Timor, postcolonial conflict. (J) 2000: West Papua. (K) 2000: Molucca. (L) 1978–2002: Philippines internal conflict: some refugees in Malaysia, others displaced within the Philippines. [Adapted from Jonathan Rigg, *Southeast Asia* (London: Routledge, 1997), p. 130; and http://www.refugees.org/.]

Refugees from Conflict. In this region, as in others, **forced migration** still plays a role in the movement of people (Figure 10.15), and war is perhaps its most prevalent cause. During the last half of the twentieth century, many millions of mainland Southeast Asians fled into neighboring states to escape protracted conflict. Thailand, Laos, and Cambodia, in particular, received many refugees during and after the Vietnam War. More recently, refugees have fled discord on many of the islands, especially those of Indonesia. On the surface, it may appear that the discord is between ethnic or religious groups because the groups in conflict happen to be of different ethnicities or religions. In fact, disputes often arise over such issues as resettlement (Molucca, Sulawesi, and Kalimantan), political autonomy (East Timor), and unsustainable extraction of resources from peripheral provinces by the central government (West Papua and Aceh, both provinces of Indonesia).

The Sex Industry. Several forms of forced migration are connected to the sex industry, which has two aspects: the international and the local. The international aspect, **sex tourism,** is in some ways an extension of the sexual entertainment industry that served foreign military troops stationed in Asia after World War II and during and after the conflicts in Korea and Vietnam. Now civilian men arrive from around the globe to live out their fantasies during a few weeks of vacation in Thailand, Cambodia, Indonesia, Vietnam, or the Philippines. In 2000, 9 million tourists visited Thailand alone, up from 250,000 in 1965, and some observers estimate that as many as 70 percent are looking for sex. In recent years, Thai government officials have encouraged sex tourism to create jobs, and they praise the sector's role in helping the country weather the economic crisis of 1997. Corrupt public officials also support sex work because it provides them with a source of illegal, untaxed income from bribes.

NGOs working to further the human rights of sex workers report that women in the international sex tourism industry are paid somewhat better than those in the local market and may have greater autonomy to join or leave the industry. Yet it is in the local market that the greatest demand as well as the most grievous abuses lie. The custom of young men visiting a brothel when they reach puberty has spread with growing prosperity and with the possibility of anonymity in impersonal urban life. In Thailand, a quarter to a third of adult males visit a brothel at least once a month, and a recent survey of male Thai soldiers found that 81 percent had visited a sex worker within the past 6 months. One result of this high local demand is that organized crime has entered the field, and girls and women are being coerced into sex work. Some research indicates that girls may have been sold by their families to pay off family debt. Others have been kidnapped, often at a very young age. Demographers estimate that 20,000 to 30,000 Burmese girls taken against their will (some as young as 12) are working in Thai brothels; their wages are too low to make buying their own freedom possible. In the course of their work, they must service more than 10 clients a day and are routinely exposed to physical abuse and sexually transmitted diseases, especially HIV. The industry is found throughout the region, although perhaps not to the extent it is in Thailand.

VIGNETTE Twenty-five-year-old Watsanah K. (not her real name) awakens at 11 every morning, attends afternoon classes in English and secretarial skills, and then goes to work at 4 P.M. in a bar in Patpong, Bangkok's red light district. There she will meet men from Europe, North America, Japan, Taiwan, Australia, Saudi Arabia, and elsewhere who will pay to have sex with her. She leaves work at about 2 A.M., studies for a while, and then goes to sleep.

Watsanah was born in northern Thailand to an ethnic minority group whose members, like many others in the area, are poor subsistence farmers who have recently become involved in the global drug trade by growing opium poppies. Watsanah married at 15 and had two children shortly thereafter. Several years later, her husband developed an opium addiction. She divorced him and left for Bangkok with her children. There she found work at a factory that produced seat belts for a nearby automobile plant. In 1997, Watsanah lost her job when a financial crisis ripped through Thailand and the rest of Southeast Asia. She became a sex worker (prostitute) to feed her children.

Although the pay, between $400 and $800 a month, is much better than the $100 a month she earned in the factory, the work is dangerous and demeaning. Sex work is illegal in Thailand, so Watsanah must live in constant fear of going to jail and losing her children. Moreover, she can't always make her clients use condoms, which puts her at high risk of contracting AIDS or other sexually transmitted diseases (STDs). Sex work, though widely practiced and generally accepted in Thailand, is nonetheless looked down upon. "I don't want my children to grow up and learn that their mother is a prostitute," says Watsanah, "that's why I am studying." [Adapted from the field notes of Alex Pulsipher and Debbi Hempel; Donald Wilson and David Henley, "Prostitution in Thailand: facing hard facts," *Bangkok Post* (December 25, 1994); and "Sex industry assuming massive proportions in Southeast Asia," *International Labor Organization News* (August 19, 1998).]

ENVIRONMENTAL ISSUES

Virtually everywhere in Southeast Asia, environments are deeply affected by population pressure and by such commercial activities as mining, logging, and export agriculture. Most Southeast Asian countries are rich in resources: oil and natural gas, timber and other forest products, minerals (gold, gemstones, tin, and copper), and fertile soils in river valleys and volcanic areas. These resources are being rapidly depleted—often extracted by foreign-owned companies and sold abroad, with only a fraction of the profits being reinvested in the host countries. In this region, as elsewhere, it is useful to consider the market forces that influence the rate of resource use and to ask whether or not the rate is sustainable and whether the citizens of a region are the ones benefiting from the development of their resources.

People in Europe and North America may inadvertently be part of the problem. We, may, for example, have beautiful tropical hardwood paneling in our dens, purchased at an attractively low price at the local branch of a multinational home improvement warehouse, or we may use computer paper by the case. We are on the demand side of the Southeast Asian resource equation. Though we live far from Southeast Asia, we often support our lifestyles with their resources. In the United States, for example, we consume 690 pounds (313 kilograms) of paper per capita (the highest rate in the world). Japan's rate of paper use is 500 pounds (225 kilograms); and most of Europe consumes 220 to 440 pounds (100 to 200 kilograms) per person. Most Southeast Asian countries consume less than 67 pounds (30 kilograms) per capita. Singapore is an exception, consuming 450 pounds (218 kilograms) of paper per person. Much of the wood fiber in paper products comes from Brazil, the Philippines, Indonesia, or other tropical pulpwood producers. The map in Figure 10.16 shows the

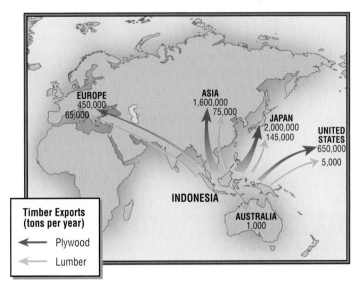

Figure 10.16 **Interregional linkages: annual timber exports from Indonesia.** [Adapted from map insert, *Indonesia*, Cartographic Division, National Geographic Society, February 1996.]

destinations for Indonesia's timber products. Although we have paid for our paneling and paper, we must ask if the price was fair, especially in light of the environmental damage that results from logging.

Deforestation

Environmentalists estimate that 13 to 19 square miles (34 to 50 square kilometers) of Southeast Asia's rain forests are destroyed each day, a rate 50 percent faster than in the Amazon. About 20 percent of the forests are destroyed by legal logging, road and pipeline construction, and utilities installation. Another 25 percent are destroyed for resettlement schemes. Fifty-five percent disappear to gain land for commercial agriculture and shifting cultivation. Companies that have legal rights to log the land over a period of 25 years sometimes choose to "cut and run" in just a few months and thus avoid fulfilling already woefully ineffective conservation agreements. Moreover, illegal logging often outpaces legal logging, and its impact on the environment is even worse because it is done secretly and in a matter of hours, using heavy equipment, and with no attention to preserving natural habitat.

Indirectly, much of the deforestation is linked to population growth and poverty. Traditional, nonintensive tropical swidden or shifting agriculture techniques on forested lands were especially effective and sustainable in Southeast Asia when the population was only about 10 million, as it probably was in about 1800. Until the early twentieth century, population density remained low enough to allow for the use of small cultivation plots and long fallow periods, during which the forest could fill in the clearings. But recently, population has grown rapidly. For example, in the 19 years between 1973 and 1992, the population of Southeast Asia increased by 48 percent, from 317 million to 469 million. By 2000, it stood at 528 million (more than 50 times that of 1800). To feed the increased numbers, cultivation is expanding to the steepest of slopes. Trees are cut down for sale and to make room for farmland. Erosion is drastically changing the shape of the denuded land. Without a forest cover, water moves across the landscape too rapidly to be absorbed, carrying away precious topsoil, which is very thin in tropical environments. Watercourses cut new gullies and ravines in the uplands and deposit tons of silt in the lowlands and surrounding oceans.

The Philippines serve as an example of widespread deforestation. In 1900, 75 percent of the Philippines was covered with **old-growth forests,** which are forests dominated by large trees in the mature stages of their life cycle. By 1990, only 2.3 percent of the country was so forested. In 1991, 3000 people on the island of Leyte were killed by floods blamed on illegal logging that had been allowed by corrupt officials. In the Philippines, original forests are now confined to small ribbonlike patches on higher mountainous terrain. **Secondary forest** will regrow after the first cutting if conditions are right, although it will not have the diversity of species of the original forest. On Sumatra, and in Sarawak and in parts of Kalimantan on Borneo, even secondary forest has disappeared, and in many places former forestlands are so degraded that they are useless for agriculture. The map in Figure 10.17 shows a range of environmental concerns in the region.

The Freeport-McMoRan Grasberg open-pit mine. [George Steinmetz.]

Mining

Mechanical strip-mining is probably the extractive activity that most disrupts the land. This technique is increasingly used in Southeast Asia to extract such minerals as copper, silver, and gold. The land is cleared of forest, and heavy equipment then peels away the layers of soil and rocks until the desired mineral is exposed. In the United States, companies must take measures to rehabilitate the landscapes they bulldoze, and they must dispose of mining and processing waste without polluting watercourses. Strip-mined land can never be truly restored, however, and despite regulations, many mining operations in the United States are still heavy polluters. One reason that international mining companies are interested in Southeast Asia is that environmental regulations do not exist or are not enforced.

The most publicized mine in the region is the 13,000-foot-high (4000-meter-high) "Mine in the Clouds," an open-pit multimineral mine on Grasberg Mountain in West Papua. The mountain contains the world's largest known gold reserve. Every day in the late 1990s, the Freeport-McMoRan international consortium (a group with strong ties to firms in the United States) extracted U.S. $7.2 million worth of copper, gold, and silver from this mine. Freeport, Indonesia's largest foreign taxpayer, now has contracts to mine over 9 million acres (3.6 million hectares) in the mountain range. Mining will continue here for at least another century, according to Freeport president George Mealey. Local villagers, however, complain that they have been driven off their land without compensation. And the Komoro, who live on the coast, have demonstrated that tailings from the mines have polluted the Aikwa River, causing flooding and ruining their stands of sago palm. Mining by Freeport has also threatened the nearby Lorenz National Park, in direct violation of Indonesian law.

In Indonesia (Kalimantan and West Papua) and the Philippines (Mindanao), national armed forces have been called out to quell protests against loss of native habitats to mining operations.

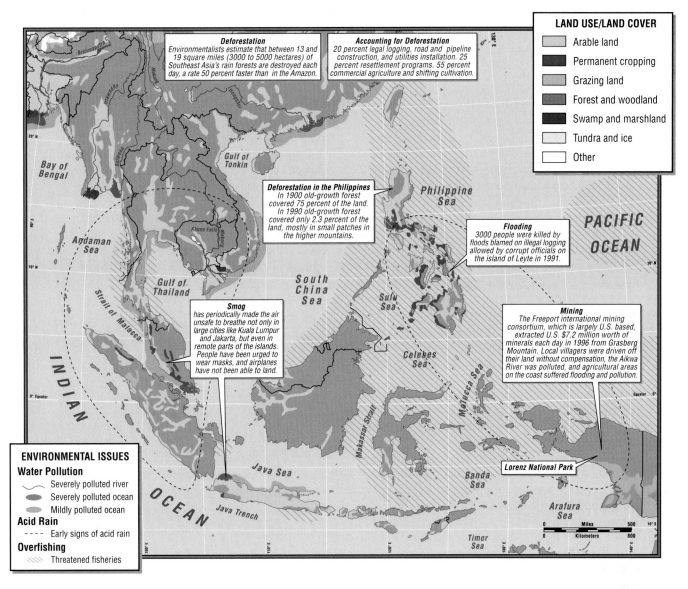

LAND USE/LAND COVER
- Arable land
- Permanent cropping
- Grazing land
- Forest and woodland
- Swamp and marshland
- Tundra and ice
- Other

Deforestation
Environmentalists estimate that between 13 and 19 square miles (3000 to 5000 hectares) of Southeast Asia's rain forests are destroyed each day, a rate 50 percent faster than in the Amazon.

Accounting for Deforestation
20 percent legal logging, road and pipeline construction, and utilities installation. 25 percent resettlement programs. 55 percent commercial agriculture and shifting cultivation.

Deforestation in the Philippines
In 1900 old-growth forest covered 75 percent of the land. In 1990 old-growth forest covered only 2.3 percent of the land, mostly in small patches in the higher mountains.

Flooding
3000 people were killed by floods blamed on illegal logging allowed by corrupt officials on the island of Leyte in 1991.

Smog
has periodically made the air unsafe to breathe not only in large cities like Kuala Lumpur and Jakarta, but even in remote parts of the islands. People have been urged to wear masks, and airplanes have not been able to land.

Mining
The Freeport international mining consortium, which is largely U.S. based, extracted U.S. $7.2 million worth of minerals each day in 1996 from Grasberg Mountain. Local villagers were driven off their land without compensation, the Aikwa River was polluted, and agricultural areas on the coast suffered flooding and pollution.

ENVIRONMENTAL ISSUES
Water Pollution
- Severely polluted river
- Severely polluted ocean
- Mildly polluted ocean
Acid Rain
- Early signs of acid rain
Overfishing
- Threatened fisheries

Figure 10.17 Environmental issues: Southeast Asia. Most Southeast Asians live frugally and use far fewer resources per capita than people in Europe or North America. Nonetheless, rapid industrialization and dependence on income from extractive activities such as export agriculture, forestry, and mining have led to environmental crises over the last two decades. The observant traveler, seeing evidence of environmental stress, may wonder to what extent consumers in the developed world are implicated.

Such clashes reflect the conflict between the interests of local, often remote, politically weak forest dwellers and traditional cultivators on the one hand and national governments focused on rapid economic development on the other.

Air Pollution

Usually, we think of places surrounded by the sea as pristine and their air as invigorating. But twice in the 1990s, the islands of Southeast Asia have been covered with a putrid and persistent cloud of pollution. The most recent instance occurred in the fall of 1997. Although large cities such as Kuala Lumpur and Jakarta experienced the worst pollution, air pollutants surged beyond safe levels even in remote parts of the islands (see Figure 10.17). People were urged to wear masks, and airplanes could not land. One plane crash is thought to have been caused by the smog.

The immediate cause of the poisonous smog was smoke from fires set on forestland that had been recently logged and was being prepared for the planting of palm oil trees and lucena, a fast-growing tree used for paper pulp production. Unfavorable winds, drought, and a delayed monsoon combined to spread the fires. A high-pressure air mass (possibly linked to El Niño) kept the smoky haze and the region's normal industrial air pollution from diffusing, as it usually does, into the upper atmosphere. The ASEAN environmental ministers meeting to deal with the crisis had to move their meeting to avoid the unhealthy air.

Energy Production

There is a huge unmet need for energy in Asia. With 60 percent of the world's population, Asia as a whole now uses only about one-quarter of the world's energy. Long brownouts are common in the largest Asian cities, and power shortages are cited as a leading obstacle to further industrialization and to further development of information technology industries. Likely sources of future energy are oil and gas, of which there are major deposits in the region.

Oil and Gas. Southeast Asia produces about 5 percent of the world's oil and natural gas, with Indonesia alone accounting for half the region's production. Overall, the region produces more oil and natural gas than it consumes. However, oil and gas are unevenly distributed: Brunei, Burma, Indonesia, and Malaysia have large surpluses; the Philippines, Thailand, and Singapore have large deficits. (See the discussion of the Burma-Thailand Yadana natural gas project on page 553.) The Philippines and Thailand must spend about 10 percent of their export earnings on imported oil and natural gas. Potentially, there are large offshore oil and natural gas deposits north of Sumatra and in the South China Sea. Indeed, because the sea may hold valuable oil and gas, sovereignty over the Spratly Islands in the South China Sea is being contested by all the countries that border it, including China.

Pollution from the use of oil and gas is already a severe problem. Few cars, trucks, or buses have pollution-control devices, and the respiratory effects of breathing exhaust are felt by everyone who walks along streets or rides in vehicles, even in small towns and rural areas. A brown haze of pollution obscures the skylines of all cities most of the time, and many pedestrians choose to wear masks. Yet, despite already critical levels of automotive pollution, all indications are that automobile use is likely to rise sharply as personal incomes grow.

Hydropower. Hydroelectric power plants are another source of future power. Many of the poorest countries of Southeast Asia are building dams, some to provide irrigation water and most to generate power that they plan to sell to more rapidly developing neighbors. For example, landlocked Laos, one of the poorest countries in Asia, hopes to develop hydropower on the Mekong River to sell to Thailand.

Swedish and Norwegian investors are privately financing a 210-megawatt hydroelectric dam being built as a profit-making operation in Laos on the Theun River, a tributary of the Mekong. Laos will receive a share of income from the power sales but will not use the electricity itself. Nonetheless, Laotian villagers living near the dam sites will bear the environmental costs. By blocking the natural flow of water, dams stop the seasonal deposition of silt downstream that refreshes crop soil. Also, the turbines transfer heat to water flowing downstream, degrading plant and animal habitats. The standing reservoir water may leak into geological structures, polluting aquifers or causing erosion at some distance from the dam itself, and flooding behind the dam may destroy human and natural habitats. The new availability of irrigation water may stimulate agricultural clearing nearby. Reservoirs may silt up rapidly, making water storage and power generation impossible. There is little incentive for the investors in faraway Sweden and Norway to bear the costs of the environmental damage they will never experience.

Urban Pollution

As we have seen, Southeast Asia is home to some remarkably densely populated and rapidly growing cities. Many of these urban environments lack the basics of housing, sanitation facilities, and safe drinking water. The factories that attract migrants dispense industrial wastes into rivers that bubble with stinking gases. Proliferating motor vehicles and a lack of emissions control make the air dangerous to breathe. Rain passing through polluted air becomes dangerously acidic. In the cities and in the countryside beyond, this acid rain threatens aquatic life, human health, and the garden plants meant to feed city people. The vignette below demonstrates one way in which Jakarta's effort to control automobile emissions has intersected with the city's informal economy. Like Washington, D.C., New York, and other U.S. cities, Jakarta tries to relieve pollution and traffic congestion by requiring carpooling.

> **VIGNETTE** Traffic regulations in Jakarta require that on some streets, all cars must carry at least three occupants. At the entrances to such streets wait the "jockeys," young boys who will provide a driver with the necessary number of passengers. The driver pays the boys 500 rupiah for the ride and then drops them on the other side of the restricted area, where they may make the return trip with another needy driver. From time to time the boys are rounded up, banned, fined, imprisoned, sometimes even beaten by the police. But they always return. In fact, the traffic generates a great deal of employment. Young men and women at the traffic lights, wearing masks against the pollution, sell bottled water, candy, and cigarettes. Most are financing their education. Other young men earn a living by stopping the traffic on main roads to enable vehicles to enter from small side roads. By placing himself bodily in front of the oncoming vehicles, a young man stops traffic long enough to allow the cars from side streets to join the mainstream of vehicles. The grateful driver gives the "traffic cop" 100 rupiah, sometimes as much as 500 rupiah (equivalent to twenty cents). [Adapted from Jeremy Seabrook, *In the Cities of the South—Scenes from a Developing World* (London: Verso, 1996), p. 296.]

Bangkok is considered one of the most polluted cities on earth. Thirty percent of its rubbish is deposited in rivers and streets, and many of the waterways of the Chao Phraya are considered biologically dead. Although there has been little locally based mass environmental activism, individuals throughout the region are developing an understanding of the systems that have brought about such severe environmental degradation. Jeremy Seabrook recounts standing by the polluted Chao Phraya River in Bangkok and being joined by an elderly gentleman, who explained:

In Thailand we used to use banana leaves as plates and to wrap food. Then we could discard them in the streets and fields because they were natural products that decayed, fertilized the fields, left no pollution. Now we buy buns in polystyrene containers, Coke in plastic bottles; but we still treat these things as if they were natural products and throw them down in the streets and streams. We haven't noticed yet that it is not the same, and that is why Bangkok is drowning in a sea of filth.

(JEREMY SEABROOK, IN THE CITIES OF THE SOUTH—
SCENES FROM A DEVELOPING WORLD, PP. 296–297.)

Some people suggest that the repeated occurrence of preventable disasters such as the air pollution crisis of 1997 highlights the weaknesses of Southeast Asian social institutions. Environmental regulations are not strongly enforced, and those who pay bribes may avoid them entirely. Meanwhile, the political systems do not encourage public debate. Throughout the region, great value is placed on gentle, consensual persuasion as embodied in the principles of *Pancasila* in Indonesia (see the box on page 535)

rather than on legal sanctions. This is especially true for environmental issues, which—as elsewhere in the world—consistently take a backseat to economic development. Ultimately, success in addressing Southeast Asia's environmental problems will depend both on consumers outside the region giving serious thought to the environmental consequences of their purchases and on the Southeast Asian general public becoming more aware of how they can change unsustainable development policies.

MEASURES OF HUMAN WELL-BEING

Human well-being is a complicated measure that is determined by sociocultural and political factors as much as by economic ones. As stated in earlier chapters, gross domestic product (GDP) per capita (Table 10.4, column 2), the most common index used to compare countries, is not an accurate indicator of well-being, so

TABLE 10.4 *Human well-being rankings of countries in Southeast Asia*

Country (1)	GDP per capita, adjusted for PPP[a] in 1998 $U.S. (2)	Human Development Index (HDI) global rankings, 2000[b] (3)	Gender Empowerment Measure (GEM) and Gender Development Index (GDI) global rankings, 2000[c] (4)	Female literacy (percentage), 1998 (5)	Male literacy (percentage), 1998 (6)
For comparison					
Japan	23,257	9 (high)	41 (GDI = 9)	99	99
United States	29,605	3 (high)	13 (GDI = 4)	99	99
Mexico	7,704	55 (medium)	55 (GDI = 50)	89	93
Southeast Asia					
Brunei	16,765	32 (high)	N/A[d] (GDI = 31)	87	94
Burma	1,199	125 (low)	N/A (GDI= 102)	79	89
Cambodia	1,257	136 (low)	N/A	53	80
Indonesia	2,651	109 (medium)	N/A (GDI = 90)	80	91
Laos	1,734	140 (low)	N/A (GDI = 117)	30	62
Malaysia	8,137	61 (medium)	47 (GDI = 57)	82	91
Philippines	3,555	77 (medium)	44 (GDI = 64)	95	95
Singapore	24,210	24 (high)	38 (GDI = 24)	88	96
Thailand	5,456	76 (medium)	N/A (GDI = 62)	93	97
Vietnam	1,689	108 (medium)	N/A (GDI= 89)	91	95

[a]PPP = purchasing power parity.

[b]The high, medium, and low designations indicate where the country ranks among the 174 countries classified into three categories (high, medium, low) by the United Nations.

[c]Total ranked in the world = 104 (no data for many).

[d]N/A = data not available.

Sources: United Nations Human Development Report 2000 (New York: United Nations Human Development Programme), Tables 1, 2, and 3; and *1998 World Population Data Sheet* (Washington, D.C.: Population Reference Bureau).

additional measures are included here. Southeast Asia's countries, unlike those of sub-Saharan Africa and South Asia, have greatly varying GDP per capita figures. The GDP figures for some countries (Singapore and Brunei) are very high, for others (Malaysia, Thailand, and the Philippines) are only moderate, and for still others (Burma, Cambodia, Laos, and Vietnam) are very low. Singapore and Brunei enjoy widespread prosperity, but elsewhere in the region, GDP figures mask wide variations in well-being. To survive, those with the lowest incomes depend on subsistence cultivation, reciprocal exchange of labor and services, and the informal economy. But often these very poor people live in close proximity to a vastly better-off minority who live in fine houses, travel widely, and consume at high levels. Protracted war in Vietnam, Laos, and Cambodia has hindered development in those countries. Their economies, especially Vietnam's, are expected to improve as low wages attract foreign investment. In Burma, a military dictatorship and widespread corruption have been major factors in the continuing poverty of this country, which is among the richest in the region in natural resources.

As with GDP, the countries of Southeast Asia vary in their rankings on the United Nations Human Development Index (HDI) scale (Table 10.4, column 3). Singapore's HDI rank is very high indeed (24), whereas Cambodia (136), Laos (140), and Burma (125) all rank in the bottom third, globally. In those countries with the lowest HDI rankings (Laos and Cambodia), literacy figures (columns 5 and 6) are exceptionally low, especially for women. Governments in these countries have not provided even the most basic education or health-care services. Laos spends only U.S. $6 per capita on education, compared to U.S. $179 per capita spent in Malaysia. Indonesia is also surprisingly low, at just U.S. $16 per capita.

United Nations Gender Empowerment Measure (GEM) figures (Table 10.4, column 4) are missing for all Southeast Asian countries except Malaysia (47), the Philippines (44), and Singapore (38). Because only 70 of the possible 174 countries are ranked on GEM, these scores place all three in the bottom half of the rankings. (But notice that even very rich countries can have relatively low GEM rankings—as does Japan, with a ranking of 41.) Singapore ranks the highest in the region. Women in Singapore have many opportunities for employment, though at lower wage scales than men (see Table 10.3). In Malaysia and the Philippines, a few well-connected women have access to some high-ranking jobs as administrators and as professional and technical workers. Many Filipino women obtained experience working as bureaucrats and in private companies that served the U.S. military until several large American military bases were closed. Now, many well-educated and experienced Filipino women, especially in health-related fields, are migrating to work in the United States or other wealthy nations. Elsewhere in the region, most women lag well behind men in education level, wages, and access to jobs. Still, compared to South Asia and Africa, women are gaining some ground toward equal treatment; Table 10.3 shows that young women are now staying in school, sometimes longer than young men (see page 541).

SUBREGIONS OF SOUTHEAST ASIA

The subregions of Southeast Asia all have tropical environments that can support dense rain forests as well as rice paddy and other cultivation wherever the soil is rich enough and water is sufficient. All the subregions are ethnically diverse, retain vestiges of European colonialism, and have dominant religions—Buddhism on the mainland, Islam and Catholicism in the islands. In all subregions, especially in cities and towns, Overseas Chinese minorities are especially active in commerce. One obvious difference among the subregions is the pace of modernization. Another interesting difference is the response of various countries to the ethnic diversity within their borders. In some countries, cultural differences are tolerated; in others, people are encouraged to conform to one standard. In some countries, the various ethnic groups are more or less equal in wealth and power; in others, just one or two groups dominate. Countries also differ in their approaches to achieving national unity, a major concern throughout Southeast Asia.

MAINLAND SOUTHEAST ASIA: THAILAND AND BURMA

Burma and Thailand occupy the major portion of the Southeast Asian mainland and share the long, slender peninsula that reaches south to Malaysia and Singapore (Figure 10.18). Although the two countries are adjacent and share similar physical environments, Burma is poor, depends on agriculture, and has a repressive military government, whereas Thailand is rapidly industrializing and has a more open, less repressive society. Both trade in the global economy, but in different ways: Burma supplies raw materials; Thailand provides low-wage labor and hosts multinational manufacturing firms.

The landforms of Burma and northeastern Thailand consist of a series of ridges and gorges that bend out of the Tibetan Plateau and descend to the southeast, spreading out across the Indochina Peninsula. Rivers such as the Irrawaddy and the Salween, which originate far to the north on the Tibetan Plateau, flow south through narrow valleys to the huge delta at the southern tip of Burma or to the large plain in central Thailand. Most agriculture takes place in the Burmese interior lowlands around Mandalay and in Thailand's large central plain.

Ancient migrants from southern China, Tibet, and eastern India settled in the mountains of the northern reaches of these two countries. The rugged topography has protected these indigenous peoples—Shans, Karens, Mons, Chins, and Kachins are among the largest such groups—from outside influences. Hence, they follow traditional ways of life and practice animism. The valleys and

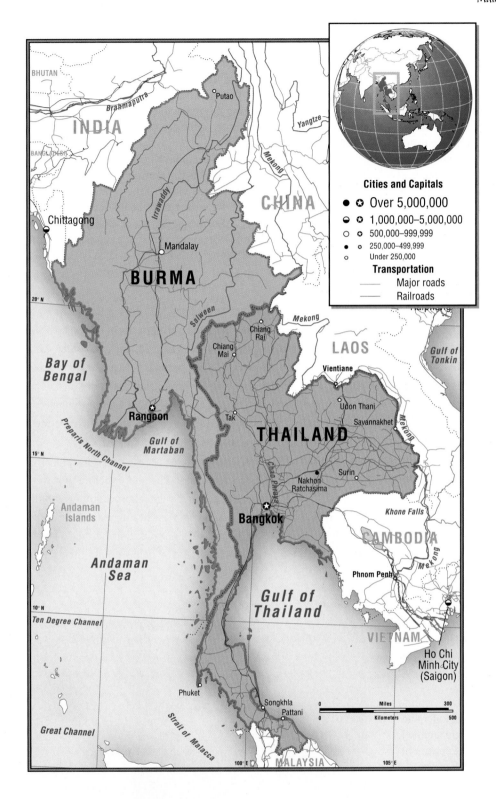

Figure 10.18 **Mainland Southeast Asia: Thailand and Burma (Myanmar) subregion.**

lowlands to the south in Burma, and the large central plain in Thailand, are places of urban development, Buddhism, and modernization. Burma is named for the Burmans, who constitute about 70 percent of the population and live primarily in the lowlands. Thailand is named for the Thais, a diverse group of indigenous people who originated in southern China.

From 70 to 80 percent of the world's remaining teakwood still grows in the interior uplands of Burma. A single tree can be worth

U.S. $200,000. Other resources include oil, tin, antimony, zinc, copper, tungsten, limestone and marble, precious stones, and natural gas. Although Burma is rich in natural resources, it ranks as one of the region's poorest countries, with an estimated per capita annual income of U.S. $660 and an HDI of 125. About 75 percent of the people still live in rural villages, and the economy is based primarily on rural activities: the cultivation of paddy rice, corn, oilseed, sugarcane, and legumes and the logging of teak and other tropical hardwoods. Burma is thought to be the world's largest producer of illicit opium, the source of heroin, and is estimated to supply over 60 percent of the heroin market in the United States. The U.S. Government Accounting Office reported in 1996 that an increase in the global supply of opium had lowered the cost of heroin in New York City to less than one-quarter of the cost in the 1980s; this lower cost may be fueling increased usage. Opium is the major source of income for a number of indigenous ethnic groups, especially the Wa people. This group also manufactures synthetic drugs, especially methamphetamine. The Wa, who number about 700,000, are said to protect their territory in northern Burma with surface-to-air missiles. Those who are persuaded to stop growing poppies experience a 90 percent drop in income.

The military government of Burma has manipulated the drug traffic and has even harvested poppies, especially in the lightly settled uplands and mountains near the Thai border. When indigenous inhabitants have protested, they have been silenced by assassination, plundering, resettlement, and the abduction of their young women into the sex industry (see the discussion on pages 544–545). There are more than 100,000 refugees from Burmese military violence in camps in Thailand and Bangladesh. In 1997, the international community applied economic sanctions against the military government, including a U.S. embargo on future investment in Burma.

Is Thailand an Economic Tiger?

During the boom years of the mid-1990s, Thailand was known as one of the Asian "tiger" economies, meaning that it was rapidly approaching widespread modernization and prosperity. Thailand takes pride in having avoided colonization by European nations and in having transformed itself from a traditional agricultural society into a modern nation. But this self-image of Thailand is not entirely accurate. Although Thailand's rapid industrialization and phenomenal urban growth are impressive, the majority of working people are still farmers who, according to official statistics, produce only 11 percent of the nation's wealth. (Remember, though, that much of what farmers, especially women, produce is not counted in the statistics.) Thirty percent of the population, or nearly 20 million people, live in cities, and most of them endure crowded, polluted, and often impoverished conditions.

According to the United Nations, Thailand had the world's fastest growing economy between 1975 and 1998, averaging an annual growth in GDP per capita of 4.9 percent (U.S. annual growth averaged 1.9 percent). At the same time, the country kept unemployment relatively low for a developing country—at 6 percent—and inflation in check at 5 percent or lower. The growth rate climbed as a result of the rapid industrialization that took place in Thailand because the government provided attractive conditions for large multinational corporations from Europe, Japan, Taiwan, and the United States. These firms located in Thailand to take advantage of its literate, yet low-wage, workforce and its lenient environmental laws and other laws affecting manufacturing and trade.

The Bangkok metropolitan area, which lies at the head of the Gulf of Thailand, contributes 50 percent of the country's wealth

and has 14 percent of its population. Still, cities elsewhere in Thailand are also growing and attracting investment: Chiang Mai in the north, Khon Kaen in the east, Surat Thani on the southern peninsula. And despite the significant downturn in the economy and rise in inflation during the regional economic recession commencing in 1997, the country still has a growing middle class with money to spend and invest.

The life circumstances of Buaphet Khuenkaew in the vignette following illustrate a standard of living common among Thai people who live on the urban fringe. Many of them retain some agricultural ways of life and at the same time take up opportunities for employment in the city. Also illustrated are several points made earlier about typical family organization, residence patterns, and the relative autonomy of women in Southeast Asia, as compared to other places in Asia.

VIGNETTE   Buaphet Khuenkaew, 35, lives in Ban Muang Wa, a village near the northern city of Chiang Mai. She married at 18 and has two children: a son, 10, and a daughter, 17. A Buddhist with a sixth-grade education, she is both a homemaker and a seamstress. Six days a week she drives the family motor scooter 30 minutes to her job in Chiang Mai, where she sews buttonholes in men's shirts for 2800 baht (U.S. $118) a month. The children perform weekday household chores when they return from school.

Buaphet's husband, Boontham, is a farmer who is about five years older than she. The couple knew each other before they were married. He would visit her at her parents' home, and eventually they fell in love. One day he and his parents came to the house with a bride price of 10,000 baht ($420) in gold, and he asked her to marry him. She accepted. They used the gold to build their house

Buaphet shares the morning meal with her family before setting off to work. [Joanna Pinneo/Material World.]

on land owned by her mother, across the street from where she was born. They have electricity and a small television set. Drinking water comes from a well, is filtered through stones, and is stored in ceramic jars. Many household activities, including bathing, washing clothes, and washing dinner dishes, take place in small shelters outside, but meals are eaten inside. Although her husband feels that men are rightly regarded as superior in Thai society, Buaphet reports that she and her husband have an egalitarian marriage in which all decisions are made jointly. As is common for married couples in the region, they do not spend much of their leisure time together. She regularly spends time with her female friends and relatives, he with his male friends and family. Buaphet says she is happy with her life but also regularly complains about not having appliances and more up-to-date furnishings like her friends have. [Adapted from Faith D'Alusio and Peter Menzel, *Women in the Material World* (San Francisco: Sierra Club Books, 1996), pp. 228–239.]

Industrialization has had unanticipated detrimental side effects in Thailand. Thousands of rural people have been drawn to the cities seeking jobs, status, and an improved quality of life. Many of them, like the Bangkok migrants Mak and Lin described earlier (see pages 541–542), arrive only to find a difficult existence and meager earnings. Catherine Shepherd and Julian Gearing, reporters for *Asiaweek*, a regional weekly newsmagazine, tell the story of Luan Srisongpong, a 33-year-old farmer from the southeast province of Ubon Ratchathani. He, along with 20,000 others, is a member of the Assembly of the Poor, a protest group that has convened three times since 1995 on the streets of Bangkok to protest living and working conditions. Many are protesting large development programs, such as the Yadana pipeline (see the box "The Yadana Project"), that have forced them off the land and into the city. Some protesters are farmers who suffer from food price controls meant to quiet hungry urban workers. Others are objecting to factory layoffs, which take place when industries produce more than they can sell or move to parts of Asia that have even cheaper labor. The Assembly of the Poor is effective: half the problems on its list have been resolved to its satisfaction. The fact that public protest is permitted in Thailand is noteworthy, because protest is forbidden in Burma and many other parts of the region.

The geographer Jim Glassman challenges the idea that Thailand is an Asian tiger. First, its infrastructure is far from

AT THE LOCAL SCALE   *The Yadana Project: A Natural Gas Joint Venture*

Bangkok, Thailand, needs power. Neighboring Burma has natural gas reserves. The Yadana natural gas field, located some 43 miles (70 kilometers) off Burma's coast in the Andaman Sea, contains reserves of some 5 trillion cubic feet (142 billion cubic meters). These reserves are being tapped in a joint venture led by Total of France (holding a 31 percent interest), Unocal of California (28 percent), the Burmese government oil company (15 percent), and the Petroleum Authority of Thailand (26 percent). The liquefied petroleum gas (LPG) will be piped to the port at Daminseik, Burma; from there it will pass through 155 miles (250 kilometers) of buried pipeline to a power plant outside Bangkok. Beginning in 1998 and continuing for the next 30 years, the gas from the Yadana field is expected to generate up to 2800 megawatts of electricity for Bangkok. (A megawatt is the amount of electricity required to light 10,000 100-watt lightbulbs.) By comparison, San Francisco, Berkeley, and Oakland, California, together require about 1300 megawatts for residential use.

Burma will gain a natural gas facility near Rangoon that will provide 125 million cubic feet (3.5 million cubic meters) a day for domestic consumption. It will also gain an electricity-generating plant that uses the gas and an accompanying fertilizer plant (fertilizer is a by-product of LPG production). Burma will also receive U.S. $200 million annually from sales of natural gas to Thailand. In addition, Thailand will pay another U.S. $200 million a year to Total and Unocal. The project has additional benefits for Burma. It will create 2000 high-wage jobs; provide much-improved medical care and renovated hospitals; pay for new schools; rebuild roads; improve village water systems; provide electricity; and fund several village-scale development projects, primarily the breeding of pigs, chickens, shrimp, and cattle for profit. Burmese villagers displaced by the pipeline are to be compensated.

As good as all this sounds for Burma, a major problem remains. Burma is ruled by a military government that has committed numerous human rights violations. In acquiring land for the pipeline, the Burmese government is thought to have moved whole villages forcibly and then used forced labor to prepare the land for the pipeline. Meanwhile, the environment is being altered by clearing and bulldozing for the pipeline construction. At shareholder meetings, critics in the United States protested Unocal's position. They filed court cases suggesting that the company indirectly participated in human rights abuses that the Burmese government may have committed against its own citizens in connection with the Yadana pipeline project. Unocal's position is that it is simply an oil company and is not involved in, nor should it be held accountable for, political problems. The company argues that its expenditures to bring jobs and improvements in quality of life to local villagers should outweigh other, negative effects.

Source: Adapted from "Country Report: Myanmar," *The Economist* (November 11, 1999): 16.

developed: the number of paved roads lags far behind the number found in other middle-income countries; it lacks a second international airport; and there are only six telephone lines per 100 people. (But, as in rapidly modernizing parts of the old Soviet Union, the cellular phone industry is leaping into the breach; there are now more than 3 million cellular phones in Thailand.) Glassman's main point is that too large a percentage of Thailand's population is not experiencing prosperity, and the levels of income disparity are increasing. Unlike Korea and Taiwan, Thailand has not devoted significant funds to alleviating inequalities in education, basic health care, and housing. Even though Thailand has literacy rates above 92 percent, true education levels are minimal. Only 33 percent of the children attend secondary school, and just a tiny minority go on to higher education. As a result, Thailand's workers are less able to compete in the global wage market, and there is a deficit of highly trained scientists and engineers, compared to such countries as the Philippines and Malaysia. Glassman adds that Thailand has received much less foreign aid than, for example, Korea and Taiwan, which were given significant U.S. aid during the cold war era to fund education, infrastructure improvements, and some land reform.

MAINLAND SOUTHEAST ASIA: VIETNAM, LAOS, CAMBODIA

On the map in Figure 10.19, the countries of Vietnam, Laos, and Cambodia appear to be ideally suited for peaceful cooperation, sharing as they do the Mekong River and its delta on the southeastern peninsula of Eurasia. Nonetheless, the three have suffered a disruptive half-century of war that pitted them against one another. Until the end of World War II, all three were colonies, known collectively as French Indochina. All three have fought a long struggle to transform themselves from colonies controlled by the French to independent nations. In the early 2000s, they remain essentially communist states; yet free-market capitalism has been allowed in measured doses. Visitors and investors began arriving in the 1990s, and by 2000 once-somber city streets were abuzz with people and enterprise.

Vietnam is a long, slender country with nearly 1000 miles of coastline on the South China Sea. Both the northern and southern extremities of the country are marked by deltas that are important rice-growing regions: the Red River delta in the north and the Mekong River delta in the south. A long, curved spine of mountains runs through Laos and into Vietnam (see the map that opens this chapter). In the south, these mountains are flanked on the west by the broad alluvial plain that is occupied by Cambodia and on the east by the long coastline and fertile river deltas of Vietnam. The rugged mountainous territory in the north is the least densely occupied area in mainland Southeast Asia. Laos, almost entirely mountainous, has only 59 people per square mile (23 per square kilometer). Most of the subregion's population is in the Mekong River delta and coastal zones of Vietnam. There, people take advantage of the wet tropical climate and flat terrain to cultivate rice. Throughout the subregion, at least 75 percent of the people support themselves as subsistence farmers; in Cambodia, only 16 percent of the people live in cities.

The Mekong River is the region's major transport artery. It originates in China, high on the Tibetan Plateau, then flows south along the border of Burma and Laos and then Thailand and Laos. It then crosses central Cambodia and far southern Vietnam. The southern part of the subregion is a large wetland formed by the

People who live along the Mekong River in Cambodia and southern Vietnam usually build their houses on stilts. During the rainy season, the river can rise 10 feet (3 meters) or more, broadening into a lazy flow 2 to 3 miles (3 to 5 kilometers) wide. [Michael Yamashita/Woodfin Camp & Associates.]

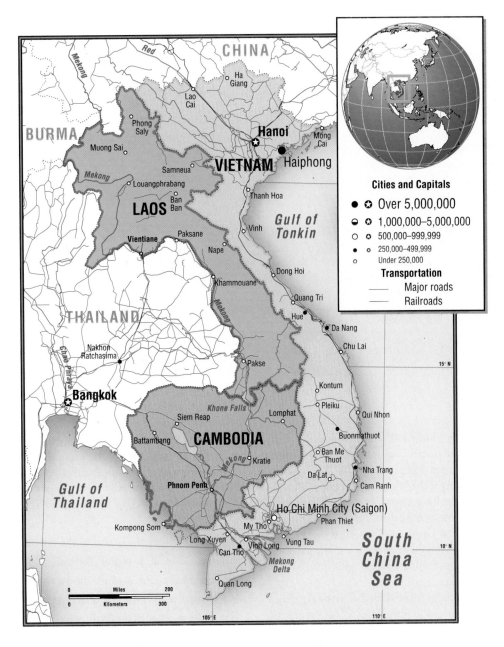

Figure 10.19 **Mainland Southeast Asia: Vietnam, Laos, and Cambodia subregion.**

Mekong Delta. Here, people live in houses on stilts to accommodate the seasonal floods, as in the Ganga Delta. Until recently, the Mekong River had but one bridge for automobile traffic and one dam, located in Yunnan, China, and completed in 1995. China now plans eight Mekong dams and several bridges on its side of the border.

The main trading partner of Laos and Cambodia is Thailand, immediately to their west. Thai investors wield considerable power because they have broad investment interests in such resources as timber, gypsum, tin, gold, precious gems, and hydropower and in the production of rice, vegetables, coffee, sugarcane, and cotton for export. Some investors are hoping to promote the Mekong Basin as a secondary destination for European tourists who visit Thailand. Despite the regional economic downturn in the late

1990s, all three countries, but especially Vietnam, appear to be entering a period of rapid economic growth.

Vietnam

Vietnam is by far the most populous of the three eastern countries of the Southeast Asian mainland and the most developed. It has a population of 79 million people, 24 percent of whom live in cities. In contrast, Laos has 5.2 million people and Cambodia 12 million.

After the United States withdrew its military forces in 1973 (see page 523), the communist government, assisted by the Soviet Union, began to invest aggressively in health care, basic nutrition, and basic education. In the 1990s, only about 214 people out of

Vietnam's capital city of Hanoi is home to more than 3 million people. The city's tree-lined streets and architecture retain a French colonial flavor. The 1911 opera house, in a central square at the end of the boulevard, serves today as the municipal theater. [David Allen Harvey/National Geographic Image Collection.]

every 100,000 attended college, but most completed elementary school and literacy was above 90 percent for both women and men. (Cambodia and Laos have much lower rates of well-being; see Table 10.4.) Because 76 percent of the Vietnamese people are cultivators living in rural areas, they have extensive knowledge of such practical matters as useful plants and animals, home building and maintenance, fishing, medicinal herbs, and related subjects. The Vietnamese have developed a national flair for excellent cuisine. When times are good, the dinner table is filled with artfully prepared fish, vegetables, herbs, rice, and fruits.

Vietnam has mineral resources (phosphates, coal, manganese, offshore oil) and at one time had a lush forest cover, much of which was destroyed by defoliants in the later years of the Vietnam War. Despite its resources, Vietnam's economy languished for nearly two decades after the war, partly because the United States imposed an economic blockade and partly because of the inefficiencies of communism. In the mid-1980s, the Hanoi leadership began to introduce elements of a market economy in a program called *doi moi*, meaning economic and bureaucratic restructuring (very similar to perestroika in Russia or even the SAPs imposed on so many countries by international lending agencies in recent decades). With the lifting of the U.S. embargo in 1994, firms from all over the world began to open branches in Vietnam. By the late 1990s, about 90 percent of the country's industrial labor force worked in the private sector producing such exportable products as textiles and clothing, cement, fertilizers, and processed food.

Wages remain extremely low, partly because of an oversupply of workers. One-third of Vietnam's population is under the age of 15, which means that more than a million new workers enter the labor force every year. They are joined by several hundred thousand displaced agricultural workers. The Vietnamese government hopes that local entrepreneurs and multinational firms investing in Vietnam will provide jobs for the expanding workforce. Multinationals that are exploring new or expanded investment in Vietnam include Boeing, Citigroup, Coca-Cola, General Electric, General Motors, Cisco Systems, J. Walter Thompson, Nike, and Proctor & Gamble.

VIGNETTE   Hong Ahn works for the multinational ad agency J. Walter Thompson in Ho Chi Minh City and is pleased to have a job in which she uses her business education. But she is also concerned about the effect her employer will have on the success of local advertising firms, which only began to spring up after Vietnam's opening-up process began in the 1990s. Under the communists, advertising didn't exist. "Local firms are young and [will] be overwhelmed as they just do not have the experience, techniques, and skills of their foreign counterpart," she observes. But multinational corporations operating in Vietnam want sophisticated advertising that the local firms are not producing. Hence J. Walter Thompson has gained a foothold, so long as it agrees to joint ventures with local firms.

Ms. Hong finds herself with ambivalent feelings. She recognizes that unless local firms have an incubation period protected from stiff outside competition from firms such as her employer, they may go out of business and her friends will lose jobs. But she also realizes that competition will force local firms to upgrade their products more quickly. Ultimately, the profits of local firms are more likely than those of multinational firms to stay within Vietnam to generate yet more jobs. [Adapted from *The Vietnam Investment Review* (October 16–20, 2000): 1; and *The Economist* (March 24, 2001): 48.]

Although economic growth in Vietnam was a rapid 8 percent per year in the early 1990s, the factors that brought growth caused considerable dislocation elsewhere in society. One focus of the government's market economy program has been the privatization of land and other publicly held assets. The farmers who own newly privatized land have a greater security of tenure and hence pay more attention to conservation and efficient production. But the process often leaves out the poorest, who do not get land. In Vietnam, there was an attempt to accommodate the landless by generating employment in nonfarm rural enterprises, but the downturn in Asian economies in the late 1990s put a halt to this effort. Those left without land or agricultural work ended up in the informal service economy, and their well-being has declined dramatically. Also, new enterprises that promise to bring in direct foreign investment have sometimes taken precedence over local interests. For example, in the late 1990s, small-scale Vietnamese farmers felt the squeeze when their taxes were raised and some of their lands were confiscated to be used in such foreign-funded enterprises as tourist hotels, golf courses, and oil refineries.

ISLAND AND PENINSULA SOUTHEAST ASIA: MALAYSIA AND SINGAPORE

Malaysia and neighboring Singapore and Brunei are the most economically successful countries in Southeast Asia. Malaysia was created in 1963 by combining the previously independent Federation of Malaya (the lower portion of the Malay Peninsula) with Singapore (at the peninsula's southernmost tip) and the territories of Sarawak and Sabah (on the northern coast of the large island of Borneo to the east) (Figure 10.20). All had been British colonies since the nineteenth century. The city-state of Singapore became independent from Malaysia in 1965. The tiny and wealthy sultanate of Brunei, also on the northern coast of the island of Borneo, refused to join Malaysia and remained a British colony until independence in 1984. Brunei does not release statistics, but the estimated per capita income is U.S. $16,765, and virtually all citizens have a high standard of living supported by wealth from oil and natural gas. A look at the human well-being table (Table 10.4) will confirm that Singapore is one of the richest countries on earth, with Brunei not far behind. And Malaysia, though much less wealthy than the other two, still ranks relatively high (at 61) in human well-being.

Malaysia

Malaysia is home to 22 million ethnically diverse people. Most live in West Malaysia, a portion of Malaysia at the end of the Malay Peninsula that has 40 percent of the country's land area and 86 percent of its people. Nearly 60 percent of Malaysia's people are Malay, and nearly all Malays are Muslims. Ethnic Chinese make up 28 percent of the population, and they are mostly Buddhist. Eight percent are Tamil- and English-speaking Indian Hindus, and 2 percent are forest-dwelling indigenous peoples who live primarily on Borneo. Until the 1970s, conflicts among these groups divided

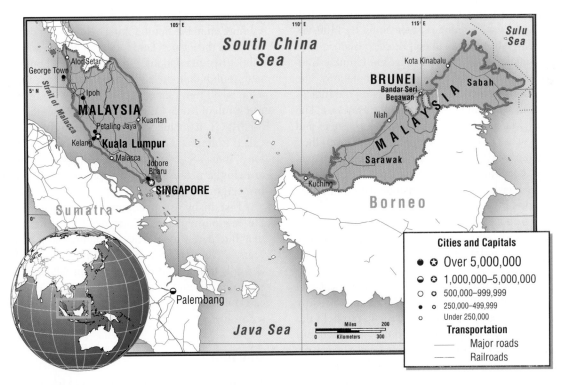

Figure 10.20 The Southeast Asia Island and Peninsula subregion.

Members of some of Malaysia's culture groups can be seen in the capital city, Kuala Lumpur, which has slightly more than 1 million inhabitants.
[Stuart Franklin/Magnum Photos.]

the country socially and economically. Malaysians have worked hard to improve relationships among these groups, and they have had noteworthy success.

Throughout most of its history, West Malaysia was inhabited by Malays and a small percentage of Tamil-speaking Indians. The Indians were traders who plied the Strait of Malacca, the ancient route from India to the South China Sea. Even before the eastward spread of Islam in the thirteenth century, Arab traders of the ninth and tenth centuries began using this route, stopping at small fishing villages along the way to replenish their ships. By 1400, virtually all ethnic Malay inhabitants had converted to Islam.

During the colonial era, the British brought in Chinese Buddhists to work as laborers on their West Malaysian plantations. The Chinese eventually became merchants and financiers; the Indians achieved success in the professions and in small businesses. The far more numerous Malays remained poor village farmers and plantation laborers. The British maintained control over their Southeast Asian colonies by promoting ethnic, religious, and economic rivalries. As economic disparities widened, antagonisms between the groups increased. After independence, animosities exploded with the onset of widespread rioting by the poor in 1969. Political rights were suspended, and it took two years for the situation to calm down. The violence so shocked and frightened Malaysians that they agreed to address some of its fundamental social and economic causes.

After a decade of discussions among all groups, Malaysia launched a long-term affirmative action program in the early 1980s. Its core was a new economic policy designed to help the Malays and indigenous peoples gain economic advancement. The policy required Chinese business owners to have Malay partners. It set quotas that increased Malay access to schools and universities and to public jobs. As the plan evolved in the 1980s, the goal became to bring Malaysia to the status of a fully developed nation by the year 2020, a plan known as Vision 2020. The program has succeeded in narrowing social and economic inequalities. A sense of nationhood and national pride has developed, and many Malaysians now make a point of noting their cross-cultural origins. One measure of success in building nationhood is that during the affirmative action reform period (1980s and 1990s), when the rights of indigenous Malays were being championed, the prime minister, Mahathir Mohamad, was of Indian heritage. *Asiaweek,* a regional newsmagazine, reported that in late 1997, the real standard of living in Malaysia—adjusted for per capita purchasing power—was more than double what it had been at the beginning of the 1990s.

Nonetheless, poverty is still prominent in Malaysia. The capital, Kuala Lumpur, boasts beautifully designed skyscrapers like those in Singapore, including the two tallest buildings in the world in 2000—Petronas Towers I and II. Yet at least one-fifth of the city's total population, amounting to several hundred thousand people, are squatters. In most of Malaysia, squatters reside on land, but in crowded urban areas that are close to water, many of them live in raft houses on rivers or bays. Rafts are an ingenious way of overcoming the acute housing shortage. In 1993, the geographer Asmah Ahmad studied urban raft dwellers in Malaysia and found that the typical domicile is a wood structure of 36 by 24 feet (11 by 7.3 meters), with a zinc roof. A rear platform provides space for domestic chores, and a wooden plank gives access to the riverbank. People often cultivate fast-growing crops such as sweet potatoes on the riverbank, so that if they must move, not too much will be lost.

The economy of Malaysia has traditionally relied on export products such as palm oil, rubber, tin, and iron ore. In contrast, much of the growth of the 1990s has come from Japanese and U.S. investment in manufacturing, especially of electronics, and from offshore oil in the South China Sea. Timber exports are also important, and the cleared land is developed into additional palm oil and rubber tree plantations. Now Malaysia is hoping to benefit from a new, multibillion-dollar, high-tech manufacturing corridor 10 miles wide by 30 miles long (16 kilometers wide by 48 kilometers long) just south of Kuala Lumpur. Called the Multimedia Supercorridor (or MSC), it includes two new cities: the new national capital, called Putrajaya, and another city named Cyberjaya. Like Bangalore, "India's Silicon Valley," (see Chapter 8), the MSC is expected to help the nation spring fully prepared into a prosperous twenty-first century.

Like many Asian economic tigers, Malaysia is subject to volatile economic patterns. Currency values and stock market prices may fluctuate widely. Lavish building projects may be halted for years, as they were in Kuala Lumpur from1998 through 2001. Overall, however, it does seem that Malaysia is more balanced economically and its wealth distributed more equitably than is the case for many of its neighbors. In spite of downturns, Malaysia seems likely to achieve increasing prosperity.

Singapore

Singapore occupies a hot, flat, humid island just off the southern tip of the Malay Peninsula. Its 4 million totally urban inhabitants live at a density of over 16,714 people per square mile (6500 per square kilometer), yet it is one of the wealthiest countries on earth. Singapore's wealth is derived from the manufacture of highly technical pharmaceutical, biomedical, and electronic products; financial services; oil refining and petrochemical manufacturing; and oceanic transshipment services. Unemployment is low and poverty unusual, except among temporary and often illegal immigrants, primarily from Indonesia. Singaporeans seem to share the notion that financial prosperity is a worthy goal. The free-market economy reigns, but there is strict government control over virtually all aspects of society. Nonetheless, daily life for most people is secure, relaxed, and affluent.

Ethnically, Singapore is overwhelmingly Chinese (77.5 percent). Malays account for 14.2 percent of the population, Indians for 7.1 percent, and mixed or "other" for 1.2 percent. Singapore's people are religiously diverse: 42 percent are Buddhist or Taoist, 18 percent Christian, 16 percent Muslim, and 5 percent Hindu; the rest include Sikhs, Jews, and Zoroastrians. The government encourages unity among this diversity. Singapore officially subscribes to a national ethic not unlike *Pancasila* in Indonesia. The nation commands ultimate allegiance. Loyalty is to be given next to the community and then to the family, which is recognized as the basic unit of society. Individual rights are respected but not championed. The emphasis is on shared values, community consensus rather than conflict, and racial and religious harmony.

Singapore has a meticulously planned cityscape with safe, clean streets and little congestion. Permits are required for virtually any activity: having a car radio, owning a copier, working as a journalist, having a satellite dish, being a sex worker, or performing as a street artisan or entertainer. Eighty percent of the people live in government-built housing, and workers are required to contribute up to 25 percent of their wages to a government-run pension fund. Home care of the elderly, however, is the responsibility of the family and is enforced by law. Each child, on average, receives 10 years of education and can continue further if grades and exam scores are high enough. Literacy averages over 90 percent. Law and order is strict. A few years ago, a U.S. teenager was sentenced to a caning for spray painting graffiti. Drug users are severely punished, and drug dealers are sentenced to life imprisonment or death. But virtually all citizens of Singapore seem to have accepted this control and strictness in return for the next highest per capita income in Asia after Japan.

INDONESIA

Indonesia is a recent amalgamation of island groups, inhabited by people who before the colonial era never thought of themselves as a national unit (Figure 10.21). Rivalry between island peoples remains strong, and many resent the dominance of the Javanese in government and business. This resentment and other burgeoning

Figure 10.21 **The Indonesia subregion.** Many of the Indonesian island names have changed in the past three decades. We are using the Indonesian names in this book, but here include in parentheses the name better known in the West—for example, Sulawesi (Celebes).

issues of identity and allegiance are serious threats to continued unity (see the section titled "Is Indonesia Breaking Up?" on pages 535–536).

The archipelago of Indonesia consists of the large islands of Sumatra and Java; the Lesser Sunda Islands east of Java; Kalimantan, which shares Borneo with Malaysia and Brunei; West Papua on the western half of New Guinea; Sulawesi; and the Moluccas. In all, there are some 17,000 islands, but many of them are small uninhabited bits of coral reef. The term *Indonesia* was coined by James Logan, an Englishman residing in Singapore, in 1850. He combined two Greek words, *indos* (Indian) and *nesoi* (islands), to create the name. Indonesians themselves now refer to the archipelago as *tanah air kita,* meaning "Our Land and Water." This name conveys a sense of the archipelago environment but falsely indicates a universal feeling of togetherness and environmental concern.

Java and Sumatra are the two most economically productive islands and are home to 80 percent of Indonesia's population. Java alone has 60 percent of the nation's 212 million people, but only 7 percent of the country's available land. Over thousands of years, the ash from Java's 17 volcanoes has made the soil rich and productive, capable of supporting large numbers of people. Another

21 percent of the people live on the adjacent volcanic island of Sumatra. Recently, nearly 3 million people were resettled from Java to Sumatra, most under government sponsorship (see the discussion on pages 542–543). The resettlement of Javanese to other islands in Indonesia is an increasingly contentious issue because indigenous people resent being inundated with Javanese culture and concepts of economic development. They see the resettlement as a Javanese effort to gain access to and profit from indigenously held natural resources, such as oil, precious metals, and forestlands.

Harvesting the Rain Forest

Indonesia contains close to 40 percent of the total Southeast Asian land area. In about 1800, it was occupied by only 10 million people. At that time, it boasted huge tropical rain forests with more than 4000 species of trees and at least 1500 bird and 500 mammal species—including such large mammals as Asian elephants, tigers, rhinoceros, proboscis monkeys, and orangutans. Now, 200 years later, Indonesia has 20 times the population (212 million) and much of its forest has disappeared.

Rain forests still cover more than 56 percent of the total Indonesian land area, but logging operations and commercial farming strip at least 4700 square miles (12,000 square kilometers) of rain forest a year—twice Brazil's rate of rain forest destruction. Only a few individuals of the larger mammal species are left in the shrinking habitat. Some 30 national parks and 150 nature preserves protect about 10 percent of the forest habitat. In theory, legislation protects much of the remainder, but lack of government personnel, weak enforcement of regulations, and absence of environmental awareness jeopardize conservation efforts outside the fully protected 10 percent.

Official government policy is to seek sustainable development of forest resources and to discourage the export of logs. According to policy, wood is not to be exported as raw timber; rather, local factories are supposed to turn harvested trees into products for export. The most commonly made product is plywood from the sought-after dipterocarp, the dominant hardwood in the forest canopy. This strategy creates jobs for Indonesians—who, the government argues, might otherwise be clear-cutting the forest for subsistence farming.

Timber harvesting in Kalimantan and West Papua is a representative example of timber operations throughout Southeast Asia. Kalimantan, the southern part of the island of Borneo, has about 10.5 million people and about one-third of Indonesia's forests. In the mid-1980s, East Kalimantan province supplied 30 percent of the country's timber. West Papua, with a population of about 2 million, is the western section of the island of New Guinea. Some 85 percent of West Papua is covered with rain forests and thick mangrove swamps. These two places are wet, warm, and mountainous. Such tropical environments are at their most productive and sustainable when supporting their natural vegetative cover. They do not easily sustain large human populations, nor can they support export-oriented agriculture, forestry, or mining without being damaged.

Two members of the Dani, an indigenous group in West Papua, try to adjust to urban life in the highland trading town of Wamena. The Dani have little in common with recent Muslim migrants from Java, who were promised acreage carved out of the Dani's forested lands. [George Steinmetz.]

Logging companies have obtained leases to huge tracts of rain forest for timber operations on Kalimantan and West Papua. These companies are usually owned by foreigners, sometimes in partnership with Javanese companies controlled by political leaders. In contradiction to official government policy, the favored method of operation is clear-cutting, which means that loggers cut every tree in an area instead of taking only the valuable species and preserving the forest in some form. Clear-cutting is fast and gives quick returns. It lets companies avoid the surveillance of human rights advocates and environmentalists that would come with long-term forestry efforts. The indigenous peoples who live in the forest are quickly driven off the land and are often resettled near state-owned factories affiliated with the timber interests or on newly developed state rubber plantations. In either case, they become a source of cheap, compliant labor. Sometimes settlers from overcrowded rural and urban areas are also brought in to subdue the indigenous peoples more quickly. These transmigrants are expected to cultivate the newly cleared land, work for which they may have no prior experience.

In this way, the now deforested land becomes contested space between displaced indigenous peoples and incoming settlers (see the discussion of violence against transmigrants by Dayak forest people in Kalimantan, page 543). The government's argument that resettlement provides jobs is true, strictly speaking. But the habitat is degraded beyond repair, and people are left in impoverished circumstances, in tiny shacks on muddy tracts of land, with little sense of community. The central government invokes the principles of *Pancasila*, arguing that resettlement unifies the people as Indonesians by spreading modernization and eliminating ways of life that differ from the government's vision of the norm.

Urban Factories

Levi Strauss is one of several prominent multinational companies in Jakarta, Indonesia. The company prides itself on requiring fair labor practices for its foreign employees and for employees of Levi Strauss subcontractors (small companies that make snaps, buttons, and other garment parts). In Indonesia, many of the garment subcontractors are owned by Koreans. Working conditions and wages at these companies are far below U.S. standards. Levi Strauss and other companies have a code of conduct that they issue to subcontractors, but the subcontracts almost always go to the lowest bidder, regardless of code enforcement.

The writer Jeremy Seabrook observes that Levi Strauss, like other companies, reacts only when the poor labor practices of its subcontractors are publicized. In one case, for example, rather than try to enforce its code of conduct and improve working conditions, Levi Strauss canceled the contract with the offender and went to another company that has not (yet) received bad publicity.

VIGNETTE Hira and Mirim are two single women in their 20s who work in a Singapore-owned Jakarta garment factory that was once a Levi Strauss subcontractor. Bad publicity

Shantytowns, like this one on the outskirts of Jayapura, the capital of West Papua, rise on the outskirts of most Indonesian cities as thousands of people flock to urban areas looking for jobs. Notice the mosque: native Papuans are rarely Muslim. Jayapura is a city of more than 310,000 people. On any given day, an interisland ferry may arrive with another 2000 people—peddlers, contract laborers, students, sex workers, and soldiers—looking for work. [George Steinmetz.]

caused Levi Strauss to void the contract, but the factory remained in operation by supplying garments to other companies. The two women earn the minimum Indonesian wage of 3800 rupiah (U.S. $1.80) a day. Their factory has an uninsulated metal roof and is crammed with 200 workers. There is only one toilet. The women are often forced to work an additional six hours of overtime to fill an order. For this they earn the equivalent of 10 cents an hour. There is no allowance for food, transportation, or health care. Independent observers say that people need to earn a minimum of U.S. $5 a day to meet basic needs in Jakarta.

Hira and Mirim share a single room of 60 square feet (5.6 square meters) in the upper story of a slum shack that costs them U.S. $8 a month each. The showers and toilets are down the street a bit, on the edge of a foul, black canal. Water is available only after dark. To avoid hunger pangs, the two women must each spend a minimum of U.S. $24 a month on food and bottled water; but to maintain basic nutrition, they need to spend more. Thus, each woman's minimum essential needs of food, water, and shelter cost her about U.S. $32 a month out of her U.S. $45 monthly take-home pay. Like the vast majority of their coworkers, Hira and Mirim come from villages in central Java. They migrated to further their education and make something of their lives. They do not return home because they are embarrassed that they have not accomplished what they set out to do and would bring shame on their families. [Adapted from Jeremy Seabrook, *In the Cities of the South—Scenes from a Developing World* (London: Verso, 1996), pp. 90–92.]

THE PHILIPPINES

The Philippines, lying at the northeastern reach of the Malay Archipelago, encompasses more than 7000 islands spread over about 500,000 square miles (1.3 million square kilometers) of ocean (Figure 10.22). The two largest islands are Luzon in the north and Mindanao in the south. Together, they make up about two-thirds of the total land area, which is about the size of Arizona. The Philippine Islands, part of the Pacific Ring of Fire, are volcanic. The violent eruption of Mount Pinatubo in June 1991 devastated 154 square miles (400 square kilometers) and blanketed most of Southeast Asia with ash. Volcanologists had predicted the eruption and precautions were taken, so although more than a million peo-ple were threatened, only 250 died. Over time, volcanic eruptions have given the Philippines relatively fertile soil and rich mineral deposits (gold, copper, iron, chromate, and several other elements).

The Philippines had 7 million people when it became a U.S. protectorate in 1898. Just over 100 years later, it has more than 80 million and is home to one of the largest and most densely set-tled urban agglomerations on earth. Close to 50 percent of Filipinos live in cities, a high proportion compared to Southeast Asia as a whole. The metropolitan area of Manila, the country's capital, has a population of at least 11 million. Urban densities exceed 50,000 peo-ple per square mile (19,000 per square kilometer). Many people in the cities are unemployed squatters living in shelters built out of scraps. Some of the newest urban dwellers are people displaced by

Figure 10.22 The Philippines subregion.

the eruption of Mount Pinatubo in 1991. Others were displaced by rapid deforestation, dam projects, and the mechanization of commercial agriculture. The masses of urban poor in the Philippines (close to 38 percent of the population is classed as poor by local standards) represent a particular political and civil threat, because the people's faith in the government has not yet been restored after the excesses of the dictator Ferdinand Marcos, who maintained a flamboyant and brutal regime from 1965 until he was deposed in 1986.

The radical increase in population density over the last century might well have caused social problems by itself, but Philippine society is also complex culturally. Within the Malay majority—about 95 percent of the population—there are at least 60 distinct ethnic groups. The Chinese make up about 2 percent of the population, and the remaining 3 percent include Europeans, Americans, other Asians, and non-Malay indigenous peoples. Most people in the Philippines are Roman Catholic (83 percent), a result of nearly 400 years as a Spanish colony (1520–1898). In contrast, the southern and central portions of the island of Mindanao and the Sulu Archipelago are predominantly Muslim. Beginning in the 1990s, Muslim extremists, possibly linked to international terrorist movements, were operating in the southern Philippines. In 2002 the United States sent Special Forces soldiers to help train Philippine military personnel in combating these terrorists.

In the Philippines, as elsewhere in Southeast Asia, wealth is apportioned by ethnicity and religion. The vast majority of the wealthy are descendants of Spanish and Spanish-Filipino plantation landowners or of Chinese business and financial people. It is estimated that 30 percent of the top 500 corporations in the islands are controlled today by ethnic Chinese-Filipino people (those of Chinese extraction but resident in the Philippines for generations). This group is less than 1 percent of the population. The poor are overwhelmingly Malay people, and, in the southern islands, they are often Muslim as well.

By 1972, family conglomerates controlled 78 percent of all corporate wealth in the Philippines. Unlike Malaysia, the Philippine government under Marcos did not embark on a new economic policy to give the more disadvantaged ethnic groups access to jobs and education, nor were there consistent efforts to improve roads, ports, railways, and other infrastructure. Instead, after violent protests erupted in the early 1970s, President Marcos declared martial law in 1972 and managed to hang on to power for another 14 years until he was overthrown. During that time and since, there has been little progress in infrastructure development, job creation, or wealth redistribution. The Philippines still spends comparatively little on education for its populace, just U.S. $25 per capita in the 1990s.

Violent social unrest has been very much a part of life in the Philippines. Filipinos are keenly aware that a better social and political system is possible. The long association of the Philippines with the United States (beginning in 1898 as a protectorate, and after World War II as the site of large U.S. military installations) gave the Philippine people experience with U.S. institutions and culture. But some of the worst aspects of U.S. culture were also imported (drug use, sex work, environmental abuse, wasteful consumerism), especially around the six U.S. military bases. Nonethe-

less, the Subic Bay Naval Base alone employed 32,000 local people and indirectly created 200,000 jobs. Many of the workers, especially women office workers, were exposed to information, education, and opportunities to travel and migrate that other Southeast Asians have not had. Furthermore, some of the urban protest strategies employed in the United States have been brought to the Philippines by liberation theology Catholics (see Chapter 3, page 148) and by community developers.

Geographers are often interested in the strategies used to gain and maintain control of contested space. Denis Murphy, coordinator of Urban Poor Associates, is an American who helps urban squatters resist government eviction plans in Manila. He notes that the urban poor in cities such as Manila form ad hoc organizations to take care of one another and to challenge those who would displace them—be it government, developers, the police, or the military. In an area of Manila called Bulaklak, meaning "flower," squatters foiled urban developers by building their own houses and naming all the streets after flowers. Then they embarked on a self-financed project to develop a drainage system and to mark individual lots, even though ultimate landownership remains questionable. Each household contributes 5000 pesos (U.S. $200) for the project, apportioned over a 15-year period. As of 1996, construction was under way on the drainage system. Nonetheless, Bulaklak risks violent demolition, a common fate of squatter settlements in Manila and elsewhere. Murphy believes that although the squatters' movement is not motivated by socialist ideology, they represent a threat because they demonstrate their ability to construct autonomous, egalitarian alternatives to the status quo.

In their efforts to settle social unrest and attract investment, Marcos's successors have been hampered by the triple economic disasters of the Mount Pinatubo eruption, the closing of the American military bases, and a drastic cut in U.S. foreign aid in 1993. One of the few bright spots in the Philippine economy is the success of microcredit programs that encourage the founding of small businesses by the poor. The successes of these small entrepreneurs illustrate the spirit that Denis Murphy talks about and that too often remains untapped in Philippine society.

VIGNETTE When Jesusa Ocampo made her first batch of *macapuno* candy for sale 14 years ago, she had no idea that the venture would one day grow into a full-fledged business. Neighbors snapped up all 20 packets of her coconut-based sweet. It wasn't long before Ocampo was making over 100 packets of candy daily. There was just one problem: she was too poor to pay for the expansion needed to meet rising demand. Then a friend told her about a microcredit program financed by the Philippine government. Beginning with a loan of just U.S. $145 obtained through a local cooperative, Ocampo gradually built up her tiny operation. She flourished, and so did her credit rating. In the spring of 1997, she obtained a loan of U.S. $3000, her nineteenth loan. She and her family are now building their own home on land they have purchased outside Manila.

The case of the Philippines shows how poor leadership and corrupt government can hold a country back. Despite the rampant sale of its once-rich timber resources (see page 546), the economy has been slow to grow. The agricultural sector—rice, coconuts, corn, sugarcane, bananas, pineapples, mangos, fish, and animals—has not flourished over the past 20 years, sinking from 25 percent of the total economy to just 20 percent, but employing at least 40 percent of the labor force. In 1998, industry accounted for 32 percent of the economy and employed 15.6 percent of the labor force; but perhaps the most meaningful figure is that although officially only 10 percent of the population is unemployed or underemployed, real unemployment figures may be as high as 30 percent.

Paradoxically, because the Philippines remained underdeveloped longer than its neighbors, the country has been less affected than others by the financial crisis and turmoil of the late 1990s across most of the region. Foreign investors are now building facilities, construction in Manila is growing, and an industrial park is rising on the former U.S. naval base at Subic Bay (see page 530). Some economists predict that even though the country is less developed than most of its neighbors, it may be the leader in terms of real economic growth into the twenty-first century.

REFLECTIONS ON SOUTHEAST ASIA

For years, Southeast Asia has been touted as a model of fast, efficient development, following the peerless example of Japan. Many countries in Southeast Asia—especially Malaysia, Thailand, Singapore, and Indonesia—have grown more quickly and consistently than countries in any other developing region. Table 10.5 shows that the distribution of wealth in Southeast Asia has been far more equitable than in Latin America.

Southeast Asia is often held up as a model to regions such as Latin America and sub-Saharan Africa, in terms of rate of growth, distribution of wealth, and rate of personal savings. It is important to note, though, that what is often called the miracle of Southeast Asian development is almost always calculated in economic terms. Those who do not think only in those terms (geographers, political scientists, anthropologists) have advocated a more thorough analysis of the miracle. They would also look at how rapid economic progress affects the environment, cross-cultural relations, human rights and political participation, investment in social programs and education, families, and the status of women.

For a time, it appeared that human and environmental issues were not likely to receive much attention in Southeast Asia. Then, in the late 1990s, the situation began to change. A major economic downturn in conjunction with an alarming environmental episode—a persistent poisonous haze throughout island Southeast Asia—turned the region's attention to factors that had escaped notice for some time. The threatening economic recession precipitated appalling revelations of bad loans based on widespread corruption in high places. Under scrutiny, the poisonous smog was shown to have resulted from rampant environmental exploitation, including the clear-cutting of timber and the subsequent burning of the resulting debris for plantation cultivation, as well as from industrial and vehicle emissions. Soon it seemed as if the region's competitive edge had been founded on wildly unsustainable practices. None of the information was really new; it was just that the supposed economic prowess of the region had always overridden indications that all was not well.

During the same period, ethnic and religious hostilities broke out in numerous places in Indonesia and also in the Philippines and northern Burma and Thailand. Most often, these problems were linked to resource allocation, with indigenous people complaining either about the loss of their ancestral lands to new settlers moved in by the government or about the appropriation of their resources (especially minerals and forestlands) by multinational corporations with the collusion of government officials.

The question now is, will the crises in Southeast Asia lead to a different definition of development, to a recognition that the means of development—democratic participation and widespread access to opportunity and information—are more im-

TABLE 10.5	Income spread ratio of selected countries in Middle and South America and Southeast Asia
Country	Income spread ratio[a]
Brazil	25.5
Peru	11.6
Chile	17.4
Mexico	16.2
Costa Rica	13
Malaysia	12.0
Singapore	9.6
Thailand	7.6
Philippines	9.7
Vietnam	5.6
Indonesia	5.6

[a]The higher the number, the greater the spread between the income of the wealthiest 20 percent and the income of the poorest 20 percent.

Source: *United Nations Human Development Report 2000* (New York: United Nations Human Development Programme), Table 4.

portant than such ends as quick profits and high mass consumption? Certainly the people of the region have amply demonstrated that they have the fortitude to work hard, but it seems possible that the national philosophies of conformity and harmony may thwart deep reforms and gloss over real issues between

ethnic and religious groups too easily. If the social, economic, and environmental issues are addressed, the region may very well pull off another miracle. So far, no other region in the world has achieved development that is both socially and physically sustainable.

Thinking Critically About Selected Concepts

1. Southeast Asia's geographic location on the southern edge of Eurasia, surrounded by seas, has made it easily accessible to trade and outside cultural influences for thousands of years. *Discuss the contrasting influences of those who have arrived just over the last 1000 years: Arab traders, the Overseas Chinese, and European colonizers.*

2. Since the 1960s, the economies of some countries in the region—Thailand, Malaysia, Singapore, Indonesia, and the Philippines—have grown considerably, aided largely by industries that export manufactured products to the rich countries of the world. Despite wide disparities of wealth and well-being, the region is often suggested as a model for other developing countries. *Discuss which aspects of Southeast Asian development might be transferable to other regions. What are some drawbacks of using this region's development as a model?*

3. Only about 35 percent of Southeast Asia's people live in urban areas, yet the region's cities are among the most densely occupied cities on earth. *In what ways do people in urban areas continue to interact with people in rural areas? How would you expect life in Singapore to compare with life in cities such as Bangkok, Kuala Lumpur, and Manila?*

4. Several types of migration have led to major changes in the region: rural-to-urban migration, government resettlement or transmigration schemes, forced migration, and extraregional migration. *Why is migration so important in this region? To what extent has each of these types of migration proved problematic for particular countries?*

5. Despite patriarchal customs, women play central roles in Southeast Asian families and have considerable autonomy. *Discuss the cultural and demographic factors that account for the greater autonomy of Southeast Asian women.*

6. Democratic processes are found nearly everywhere in the region, but democracy flourishes in some countries more than others. *What characteristics would you choose as indicators of the growth of democratic processes? Where do you think these features are found in Southeast Asia? What do you see as the greatest threats to the development of democratic participation in the region?*

7. Southeast Asia's cultural diversity poses two political threats to democratic government in the region: diversity may make consensus impossible, or one culture group may impose its will on the others. *Discuss the various ways in which governments have attempted to resolve the political problems of diversity. How have national philosophies that encourage conformity and harmony sometimes resulted in undemocratic repression?*

8. Southeast Asia is rich in natural resources, although its tropical environments are fragile and do not withstand disturbance well. Loss of forest and soil resources is evident in every country, and resettlement schemes often exacerbate environmental damage. *Discuss how the global market contributes to environmental degradation in the region.*

Key Terms

archipelago (p. 518) a group, often a chain, of islands

ASEAN Free Trade Association (AFTA) (p. 534) a free-trade association of Southeast Asian countries launched in 1992 by ASEAN and patterned after the North American Free Trade Agreement and the European Union

Association of Southeast Asian Nations (ASEAN) (p. 534) an organization of Southeast Asian governments established to further economic growth and political cooperation

Australo-Melanesians (p. 521) a group of hunters and gatherers who moved from the present northern Indian and Burman parts of southern Eurasia into the exposed landmass of Sundaland about 40,000 to 60,000 years ago

Austronesians (p. 521) a Mongoloid group of skilled farmers and seafarers from southern China who migrated south to various parts of Southeast Asia between 10,000 and 5000 years ago

biogeographical transition zone (p. 518) an area in which there is some mixing of two or more distinct types of flora and fauna

buffer zone (p. 523) a neutral territory separating adversaries

crony capitalism (p. 533) a type of corruption in which politicians, bankers, and entrepreneurs, sometimes members of the same family, have close personal as well as business relationships

cultural pluralism (or **cultural complexity**) (p. 537) the cultural identity characteristic of a region where groups of people from many different backgrounds have lived together for a long time, yet have remained distinct

detritus (p. 520) dead organic material (such as plants and insects)

doi moi (p. 556) economic and bureaucratic restructuring in Vietnam

export processing zones (EPZs) (p. 530) specially designated areas where foreign companies can set up manufacturing plants using inexpensive labor and (sometimes) low-cost local raw materials to produce items only for export

extraregional migration (p. 543) short-term or permanent migration to countries outside a region

forced migration (p. 544) the movement of people against their wishes

growth triangles (p. 531) large transnational economic regions where firms attempt to find the best skills and resources for the best prices

old-growth forests (p. 546) forests that have never been logged

Pancasila (p. 534) the Indonesian national philosophy based on tolerance, particularly in matters of religion; its precepts include belief in God and the observance of conformity, corporatism, consensus, and harmony

pull factors (p. 541) positive features of a place that attract people to move there

push factors (p. 541) negative features of the place where people are living that impel them to move elsewhere

resettlement schemes (p. 542) government plans to move large numbers of people from one part of a country to another to relieve urban congestion, dispense political dissidents, or accomplish other social purposes

secondary forests (p. 546) forests that grow back after the first cutting, often with decreased diversity of species

sex tourism (p. 544) the sexual entertainment industry that services civilian men who arrive in Southeast Asia from around the globe to live out their fantasies during a few weeks of vacation

slash-and-burn (**shifting** or **swidden**) **cultivation** (p. 528) an ancient form of multicrop gardening using small plots, often on lands that do not support more intensive types of agriculture

transmigration (p. 542) the local term used for resettlement in Indonesia

Wallacea (p. 518) a trench extending from Bali through the Makassar Strait and along the western side of the Philippines; within this zone there is some mixing of Asian and Australian flora and fauna

wet rice (or **paddy**) **cultivation** (p. 528) a highly productive type of rice cultivation in which the roots of the plants are submerged in water during the first part of the growing season

Pronunciation Guide

Aikwa River (IKE-wah)

Andaman Sea (AN-duh-muhn)

Angkor (AHNG-oor)

Austronesian (aw-stroh-NEE-zhuhn)

Bali (BAH-lee)

Ban Muang Wa (BAHN MWAHNG WAH)

Bandung (BAHN-doong ["oo" as in foot])

Bangkok (bang-KAWK)

Batavia (buh-TAY-vee-uh)

Borneo (BOHR-nee-oh)

Brunei (broo-NYE)

Burma (BURR-muh)

Cambodia (kam-BOH-dee-uh)

Cebu (say-BOO)

Celebes (seh-LAY-behss)

Chao Phraya River (chau prah-YAH)

Chiang Mai (CHYAHNG MYE)

Dayak (DYE-ak)

Dharma Wanita (DAHR-muh wah-NEE-tuh)

Dien Bien Phu (dyehn-byehn-FOO)

doi moi (DOY MOY)

East Timor (TEE-mohr)

El Niño (ehl NEEN-yoh)

Filipino (fihl-ih-PEE-noh)

Grasberg Mountain (GRASS-burg)

Hanoi (hah-NOY)

Ho Chi Minh City (HOH CHEE-MIHN)

Indonesia (ihn-doh-NEE-zhuh)

Irian Jaya (EER-ee-ahn JYE-yah)

Irrawaddy River (eer-uh-WAWD-ee)

Jakarta (juh-KAHR-tuh)

Java (JAH-vuh)

Jayapura (jye-yuh-POOR-uh)

Johor (juh-HOHR)

Kalimantan (KAH-LEE-MAHN-TAHN)

Karen (kuh-REHN)

Khmer Rouge (KMAIR ROOZH)

Khon Kaen (KAWN GAN)

Klaten (KLAH-tehn)

Kuala Lumpur (KWAH-luh loom-POOR [first "oo" as in "book"])

Laos (LAH-ohss)

Lee Kuan Yue (LEE KWAHN YWEH)

Leyte (LAY-tay)

Lombok (lawm-BAWK)

Lopburi (LAWP-BOO-REE)

Luzon (loo-ZAWN)

Madura (mah-DOOR-uh)

Makassar Strait (muh-KASS-urr)

Malay (muh-LAY)

Malaysia (mah-LAY-zhuh)

Manila (mah-NIHL-uh)

Marcos, Ferdinand (MAHR-kohss, FAIR-dee-nahnd)

Medan (may-DAHN)

Mekong River (MAY-KAWNG)

Mindanao (mihn-dah-NAU)

Molucca Sea and Islands (maw-LOOK-uh ["oo" as in "book"])

Mon (MOHN)

Myanmar (myahn-MAHR)

New Guinea (GIH-nee)

Palembang (pay-lehm-BAHNG)

Pancasila (pahng-kuh-SEE-luh)

Philippines (FIHL-uh-peenz)

Phnom Penh (NAHM pehn)

Pinatubo, Mount (pee-nah-TOO-boh)

Rangoon (rahn-GOON)

Riau (REE-au)

Sabah (SAH-bah)

Saigon (sye-GAWN)

Salween River (SAHL-ween)

Sarawak (suh-RAH-wahk)

Singapore (SIHNG-uh-pohr)

Strait of Malacca (mah-LAHK-uh)

Subic Bay (SOO-bihk)

Suharto (SOO-HAHR-toh)

Sulawesi (soo-lah-WAY-see)

Sulu Archipelago (SOO-loo)

Sumatra (soo-MAH-truh)

Sunda Islands (SOON-duh ["oo" as in "book"])

Sundaland (SOON-duh-luhnd ["oo" as in "book"])

Surat Thani (suh-RAHT TAH-nee)

Thailand (TYE-land)

Ubon Ratchathani (OO-buhn RAH-chayt-nee)

Ujung Pandang (OO-joong pahn-DAHNG [second "oo" as in foot])

Wa (WAH)

Wallacea (woh-LAH-see-yuh)

West Papua (PAP-uh-wah)

Selected Readings

A set of Selected Readings for Chapter 10, providing ideas for student research, appears on the *World Regional Geography* Web site at www.whfreeman.com/pulsipher.

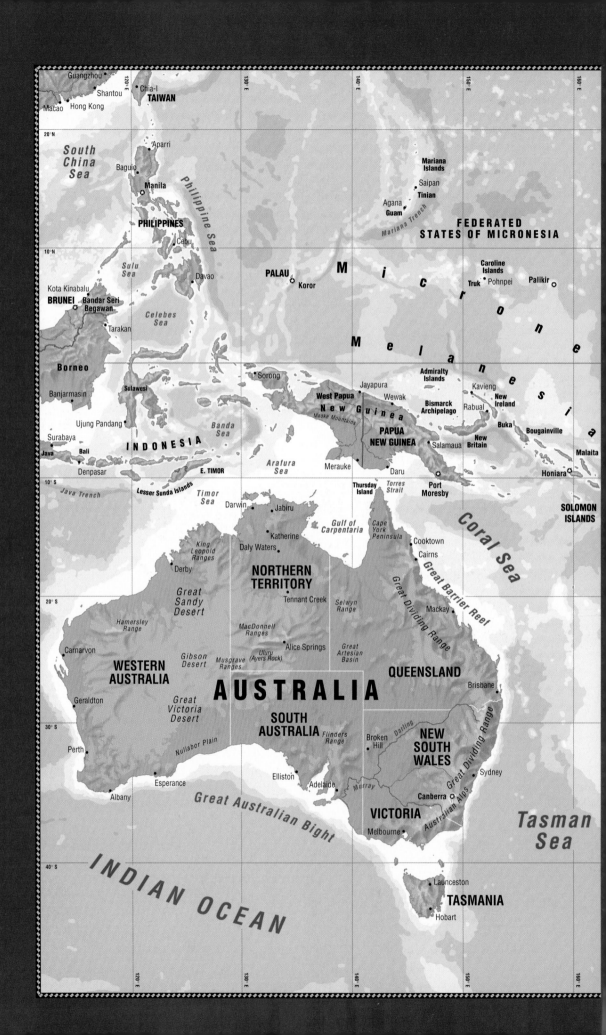

NORTH PACIFIC
OCEAN

The landforms at the geographic center of the Australian continent are low and subdued, but Uluru (Ayers Rock), to the left of center in this satellite image, is a rounded, deep red sandstone monolith that rises more than 1000 feet (300 meters) above the desert plateau. Uluru is surrounded by sites that are sacred to the Aborigines of Australia, and it is a popular tourist destination (see photo on page 572). Also visible at the top of the image is the dry bed of Lake Amadeus. [Tsado/NASA/Tom Stack & Associates.]

MARSHALL ISLANDS
Majuro

Kosrae

KIRIBATI
Tarawa

NAURU
Yaren

Phoenix
Islands

P o l y n e s i a

Line Islands
Kiritimati

Equator   0°

Funafuti
TUVALU

10° S

Mata Utu   **Tokelau**   Apia
Wallis   • Pago Pago
VANUATU   **FIJI**   Vanau   **WESTERN**   **AMERICAN**
                                Levu   **SAMOA**   **SAMOA**
Port-Vila   Viti   Suva   **Cook**   Papeete
                Levu   **Islands**   Tahiti
                        Alofi
                TONGA   Futuna   20° S
NEW   Noumea   Nuku'alofa   Avarua
CALEDONIA   Rarotonga

Kingston
Norfolk   SOUTH PACIFIC
                                                OCEAN   30° S

Kermadec Trench

Tonga Trench

Land Elevations
meters   feet
Ice cap   Ice cap
5183 and above   17,000 and above
3353–5182   11,000–16,999
2134–3352   7000–10,999
914–2133   3000–6999
305–913   1000–2999
0–304   0–999
Below sea level   Below sea level

Auckland
North   Gisborne
Island
NEW
ZEALAND

Westport   Wellington
Christchurch   Chatham
South   Islands
Island   Southern Alps

Dunedin
Invercargil   Bounty
                Islands

Miles   800
Kilometers   1200
1:36,748,800
Mercator Projection

Hawaii
(U.S.A.)
Kauai
Oahu
Honolulu   Molokai
                Maui
        Hawaii

20° N

Chapter 11: Oceania: Australia, New Zealand, and the Pacific

*It is not the koalas or the Olympics which make Australia such a
wonderful country. What makes Australia the best place in the world
is the diversity of the people who live here. . . . There are indigenous
Australians, Japanese migrants, transgender[ed people], middle aged
hippies, and many others, who all belong to subcultures.*

—MAYU KANAMORI

VIGNETTE Mayu Kanamori was born and raised in Tokyo
and educated in English and Japanese. She
moved to Australia in 1981 and has since become a much needed
cultural mediator between Japan and Australia. Today she lives in
Sydney and works as a freelance photographer and writer for
Australian and Japanese newspapers and other publications. Much
of her work focuses on helping Australians and Japanese better
understand and appreciate each other. The quote cited above and
the photograph below were presented at a photographic exhibition
in Tokyo titled "Australia Without Make-Up." The exhibition was
directed, in part, at the many Japanese people who visit Australia
but see little more than the friendly facade of blue skies and the
Sydney Opera House. Kanamori similarly tries to reveal an unfamiliar side of Japan to Australians through exhibitions such as "Unseen
Faces of Japan," which features images—poor Japanese laborers,
Japanese Christians, and a gay pride parade in Tokyo—that break
down Australian stereotypes about Japan. Another of Kanamori's
exhibitions illustrates the environmental and cultural consequences
of uranium mining for Australia's indigenous people. (Japan buys
Australian uranium for nuclear-generated electricity.) [Adapted from
http://www.mayu.com.au/info/mayu.html.]

This photo is of the Buchannon family, who live on Eveleigh Street,
Redfern, commonly known as "the Block," in Sydney, Australia. In her
exhibit in Tokyo, photographer Mayu Kanamori had this caption for her
photo: "The Block is where many of the aboriginal people in inner Sydney
live. It is considered a slum, and the media have called it a 'police no go'
area. There are many problems of poverty, alcoholism, and drugs on the
Block, and many white people will not dare to go near. Despite media
reports, however, there are aboriginal families who live on the Block with
hopes, dreams, and aspirations like the rest of Australia. This photo is
when grand dad said, 'Moggi, look at the camera.'" [Mayu Kanamori.]

Cultural mediators like Kanamori are increasingly important
in Oceania because dramatic shifts are reorienting the region as a
whole toward Asia and away from old colonial powers and allies in
Europe and North America. Oceania, which comprises Australia,
New Zealand, Papua New Guinea, and the myriad Pacific islands,
has been dominated politically and economically by people of
mainly European descent for more than 400 years. Beginning in
the early 1800s, thousands of Europeans made the long ocean voyage to Australia and New Zealand (far fewer to the Pacific islands)—most of them motivated by the simple desire to make a
new life for themselves. Almost everywhere they went, they dominated, often violently, the indigenous people of the region, some
of whom had been living there for perhaps as long as 50,000 years.

Indigenous people throughout Oceania, as well as minority
immigrants such as the Japanese, are challenging European domination today by taking steps to revive traditional cultures and by
agitating for greater recognition of their rights. The people of the
Pacific have developed an attitude or point of view they call the
Pacific Way. This concept has different meanings in different parts
of the region; but it includes the sense that traditional methods of
reaching consensus provide a way for Pacific people to work together to address the environmental, economic, and other issues
of their society.

Asia poses a much greater challenge to European influence in
the region than do the movements for indigenous or migrant
rights. Roughly three-quarters of Oceania's people are of European
descent, and migration rates show no signs of changing that fact in
the near future. Rather, Asia is becoming prominent primarily because the economies of the region are becoming more dependent
on Asian markets and Asian investment. For example, Australia
now exports more to Japan and Southeast Asia than it does to the
European Union and the United States combined (see Figure
11.8). Tourism is an important component of the economies of
most countries in Oceania, and Japanese are the largest group of

Themes to Look For Several themes characterize this region:

1. The growth of Pacific regional consciousness. People of the Pacific are forging a regional identity by reviving diverse traditional cultures while encouraging cooperation that brings people of diverse backgrounds into close contact.

2. Reorientation toward Asia. Trade and social interaction with Asia is increasing as the emphasis on European roots declines.

3. Growing recognition of environmental issues. Pacific Islanders are cooperating to address nuclear and other types of pollution. Nevertheless, some of the most dangerous environ-

mental risks to the region—future rising sea levels and the depletion of the ozone layer—are the result of global trends beyond the control of the people of the Pacific.

4. Increasing attention to the rights of indigenous peoples. Australia and New Zealand are beginning to recognize indigenous claims to lands and resources, as people of both indigenous and European descent learn to take pride in their countries' heritage.

5. Familiarity with mobility. Travel among the islands for education, sports, cultural events, and trade is facilitating unity and cooperation among peoples of formerly isolated places.

tourists. Asia may never dominate Oceania politically (as Japan attempted to do during World War II) or demographically (although Asians are now a majority on some Pacific islands, such as Hawaii). However, Oceania's 31 million people can no longer ignore the almost 3 billion people to the north in Asia, as they have tried to do in the past.

A look at the globe reveals that the Pacific Ocean covers more than a quarter of the earth's surface. Oceania is located in the central and southwestern Pacific. This part of the Pacific contains the continent of Australia and more than 20,000 islands, many of which are uninhabited atolls barely rising above the surface of the sea. Setting Australia aside for a moment, the single island of New Guinea and the two islands of New Zealand account for 90 percent of the land area in the Pacific. The other 10 percent is shared among thousands of islands lying mostly in the central and southern Pacific. Oceania's relatively small population (31 million) has prompted us to omit the usual treatment of the subregions—Australia, New Zealand, and the Pacific islands—at the end of the chapter. Instead, we have written about these subregions in the various sections on geographic issues.

THE GEOGRAPHIC SETTING

PHYSICAL PATTERNS

The huge expanse of the Pacific Ocean is of supreme importance in this region as both an avenue and a barrier. For living things, the Pacific serves as a link between widely separated lands. Plants and animals have found their way from island to island by floating on the water; and humans have long used the ocean as a conduit for communication: visiting, trading, and raiding. Sea life is a source of food; and today it is a source of income, too, as catches are sold in the global market. The movement of the water and its varying temperatures influence climates in all the region's landmasses. But the wide expanses of water have also acted as a barrier, profoundly limiting the natural diffusion of plant and animal species and keeping Pacific Islanders relatively isolated from one another. The vast ocean has imposed solitude and has fostered self-sufficiency and subsistence economies well into the modern era.

Continent Formation

The largest landmasses of the Pacific are the frozen continent of Antarctica (not considered here) and the ancient continent of Australia at the southwest perimeter of the region. The continent of Australia is partially composed of some of the oldest rock on

earth and has been a more or less stable landmass for more than 200 million years, with very little volcanic activity and only an occasional mild earthquake. Australia was once a part of the great landmass, called **Gondwana,** that formed the southern part of the ancient supercontinent Pangaea. Australia broke free from Gondwana and drifted until it eventually collided with the Eurasian Plate, on which Southeast Asia sits. The impact created the mountainous island of New Guinea to the north of Australia; downwarping created the lowland that is now filled with ocean between New Guinea and Australia.

Australia is shaped roughly like a dinner plate, with a lumpy, irregular rim and two portions missing: one in the north (the Gulf of Carpentaria) and one in the south (the Great Australian Bight). The center of the plate is the great lowland Australian desert. The lumpy rim is composed of uplands: the Eastern Highlands are the highest and most complex of these. Over millennia, the forces of erosion—both wind and water—have worn Australia's landforms into low, rounded formations, some quite spectacular.

Off the northeastern coast of the continent lies the **Great Barrier Reef,** the longest coral reef in the world. A coral reef is an intricate structure composed of the skeletons of tiny, calcium-rich living creatures called coral polyps. The tiny polyps that make up the reef are abundant beyond enumeration. These extremely

Uluru (Ayers Rock) and the nearby Olgas of central Australia are smooth remnants of ancient sedimentary rock mountains that have resisted erosion. These sites are held sacred by central Australian Aborigines. They are also among the most popular tourist destinations. [Susan Metros.]

fragile mollusks are unable to live in water that is too cold, too full of sediment, or polluted. The Great Barrier Reef stretches in an irregular arc for more than 1000 miles (1600 kilometers). The giant reef interrupts the westward-flowing ocean currents in the mid–South Pacific circulation pattern, shunting warm water to the south, where it warms the southeastern coast of Australia. In and around the reef, there is a great profusion of aquatic environments, some protected and warm, some cool and wildly active with pounding surf.

Island Formation

The islands of the Pacific were created (and are being created still) by a variety of processes related to the movement of tectonic plates. Some islands are part of the Ring of Fire, the belt of frequent volcanic eruptions and earthquakes circling the Pacific. (Central America and California are also a part of this belt.) Many Pacific islands are situated in boundary zones where tectonic plates are either colliding or pulling apart. For example, the Mariana Islands east of the Philippines are volcanoes formed when the Pacific Plate plunged beneath the Philippine Plate. The two much larger islands of New Zealand were created when the eastern edge of the Indian-Australian Plate was thrust upward by its convergence with the Pacific Plate.

There are two basic types of Pacific islands: continental and oceanic. Continental islands are remnants of the great southern landmass Gondwana. These are found in the western reaches of Oceania, west of the so-called Andesite Line, which generally follows the boundary between the Indian-Australian Plate and the Pacific Plate. These large, mountainous, geologically complex islands include New Guinea, New Caledonia, and the main islands of Fiji.

Oceanic islands are volcanic in origin. These islands exist in three forms: volcanic high islands, low coral atolls, and raised or uplifted coral platforms known as *makatea*. **High islands** are usually volcanoes that rise above the sea into mountainous, rocky formations that contain a rich variety of environments. New Zealand, the Hawaiian Islands, Tahiti, the Samoa Islands, and Easter Island are among the many examples of high islands. Those that lie in the mid-Pacific were created when magma pushed up through weakened spots in the earth's crust called hot spots (Figure 11.1). An **atoll** is a low-lying island formed of coral reefs that have built up on the circular or oval rims of submerged volcanoes. These reefs are arranged around a central lagoon that was once the volcanic crater. As a consequence of their low elevation, these islands tend to have only a small range of environments and very limited supplies of freshwater. Subsequent uplift of an atoll may result in the formation of a *makatea.* These raised coral platforms may have cliffs over 65 feet (20 meters) high.

Climate

Although the Pacific Ocean stretches nearly from pole to pole, Oceania is situated within a primarily tropical or subtropical region of that ocean. Oceania is widest through its tropical zones, and most of the islands of Oceania lie within, or close to, the tropics. The tepid water temperatures of the central Pacific bring year-round mild climates to nearly all the inhabited parts of the region

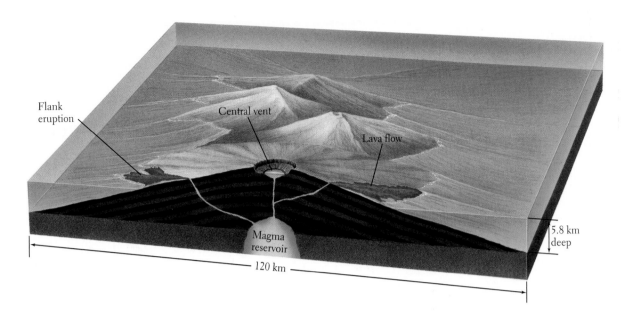

Figure 11.1 Formation of high islands. Some islands in the Pacific, such as the Hawaiian chain, are formed by shield volcanoes. As this model of Mauna Loa on the Big Island of Hawaii illustrates, magma bursts through hot spots, or weaknesses, in the surface of a tectonic plate. Wide, thin layers of lava spread from the point of eruption, slowly building undersea mountains that emerge as islands. [Adapted from Frank Press and Raymond Siever, *Understanding Earth,* 2nd ed. (New York: W. H. Freeman, 1998), p. 111.]

(Figure 11.2). The seasonal variability in temperature is greatest in the southernmost reaches of Australia and New Zealand.

Moisture and Rainfall. With the exception of the arid interior of Australia, much of Oceania is warm and humid nearly all the time. New Zealand and the high islands of the Pacific receive copious rainfall and once supported dense forest vegetation. Most travelers approaching New Zealand, either by air or by sea, notice a distinctive long, white cloud that stretches above the two islands. A thousand years ago, Polynesian settlers called the Maori also noticed this phenomenon, and so they named the place *Aotearoa,* "land of the long white cloud." The distinctive mass of moisture is brought in by the legendary **roaring forties,** powerful westerly air and ocean currents that speed around the far Southern Hemisphere virtually unimpeded by landmasses. These westerly winds deposit 130 inches (330 centimeters) of rain a year in the New Zealand highlands and more than 30 inches (76 centimeters) a year on the coastal lowlands. At the southern tip of North Island, the site of New Zealand's capital, Wellington, the wind averages more than 40 miles per hour (64 kilometers per hour) about 118 days a year. Cabbages have to be staked to the ground or they will blow away.

An atoll in the Tuamotu Archipelago of French Polynesia. As is the case here, usually the perimeter land area of an atoll is not continuous, so the parts form a sort of necklace of flat islets known as *motu* around a central lagoon. Often the necklace surrounds one or more islands that are the remnants of the old volcanic core. Some of these lagoons are home to *Pinctada margarifera,* the black-lipped pearl oyster, which produces famous and expensive black pearls. [David Doubilet.]

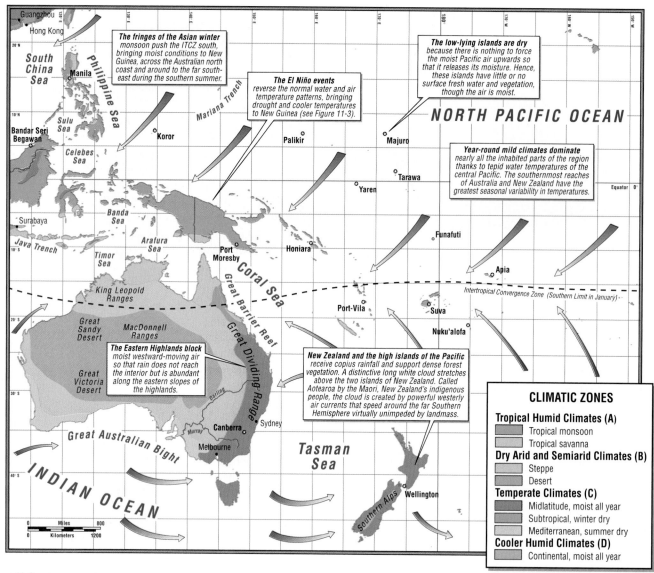

The fringes of the Asian winter monsoon push the ITCZ south, bringing moist conditions to New Guinea, across the Australian north coast and around to the far south-east during the southern summer.

The El Niño events reverse the normal water and air temperature patterns, bringing drought and cooler temperatures to New Guinea (see Figure 11-3).

The low-lying islands are dry because there is nothing to force the moist Pacific air upwards so that it releases its moisture. Hence, these islands have little or no surface fresh water and vegetation, though the air is moist.

Year-round mild climates dominate nearly all the inhabited parts of the region thanks to tepid water temperatures of the central Pacific. The southernmost reaches of Australia and New Zealand have the greatest seasonal variability in temperatures.

The Eastern Highlands block moist westward-moving air so that rain does not reach the interior but is abundant along the eastern slopes of the highlands.

New Zealand and the high islands of the Pacific receive copius rainfall and support dense forest vegetation. A distinctive long white cloud stretches above the two islands of New Zealand. Called Aotearoa by the Maori, New Zealand's indigenous people, the cloud is created by powerful westerly air currents that speed around the far Southern Hemisphere virtually unimpeded by landmass.

CLIMATIC ZONES

Tropical Humid Climates (A)
- Tropical monsoon
- Tropical savanna

Dry Arid and Semiarid Climates (B)
- Steppe
- Desert

Temperate Climates (C)
- Midlatitude, moist all year
- Subtropical, winter dry
- Mediterranean, summer dry

Cooler Humid Climates (D)
- Continental, moist all year

Figure 11.2 Climates of Oceania.

Australia's relatively low landforms and dry climate are remarkably different from New Zealand's mountainous topography and wet temperate climate. Two-thirds of the continent of Australia is overwhelmingly dry. The Eastern Highlands block the movement of moist westward-moving air and deflect it to the south, so rain does not reach the interior. As a result, the large, dry portion of Australia receives less than 20 inches (50 centimeters) of rain a year (see Figure 11.2), and humans have found rather limited uses for this territory. But the eastern slopes of the highlands receive abundant moisture. This moister eastern rim of Australia was favored as a habitat by the Aborigines and also by the Europeans who displaced them after 1800. During the southern summer, the fringes of the monsoon that operates over Southeast Asia and Eurasia bring moisture across the Australian north coast. There rainfall varies in a given year from 20 to 80 inches (50 to 200 centimeters).

Overall, Australia is so arid that there is only one major river system—in the temperate southeast, where most Australians live. There, the Darling and Murray rivers drain one-seventh of the continent and flow into the ocean at Adelaide. A measure of the overall dryness of Australia is that the entire average *annual* flow of the Murray-Darling river system is equal to just one day's average flow of the Amazon in Brazil.

Many of the mountainous high islands in the Pacific exhibit orographic rainfall patterns, with a wet windward side and a dry leeward side. Because low-lying islands across the region are not high enough to intercept regular orographic rainfall, they tend to have no surface fresh water and little vegetation, though the air is moist.

El Niño. Every 8 to 13 years, a number of changes take place in the circulation of air and water in the Pacific (Figure 11.3). These cyclical changes (also known as oscillations), which are not

A. Normal Equatorial Conditions

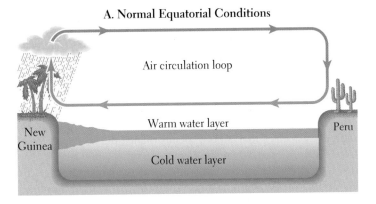

B. Developing El Niño Conditions

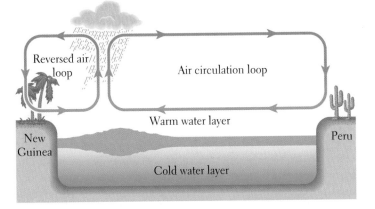

C. Fully Developed El Niño

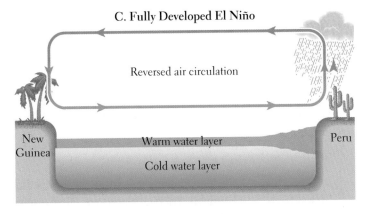

Figure 11.3 **The El Niño phenomenon.** [Adapted from Environmental Dynamics Research, Inc., 1998; and Ivan Cheung, George Washington University, Geography 137, Lecture 16, October 29, 2001: http://www.gwu.edu/~/geog137/download/lecture16.ppt.]

yet well understood, have been given the popular name **El Niño** by Peruvian fishers. Normally (Figure 11.3A), water in the western Pacific (the New Guinea–Australia side) is warmer than water in the eastern Pacific (the Peru side). Warm air rises in the west, and rain clouds form. In the east, cool air descends and there is little

rainfall. As an El Niño event develops (Figure 11.3B), ocean temperatures (and temperatures of the air above the water) are cooler than usual in the west and warmer than usual in the east. Instead of warm, wet air rising over the mountains of New Guinea and condensing as rainfall, the cool, dry air sits at the ocean surface. The result is less cloud cover and less rainfall in the west but more in the east (Figure 11.3C). The El Niño event of 1997–1998 illustrates the effects. By December 1997, New Guinea had received very little rainfall for almost a year. Crops failed, springs and streams dried up, and fires broke out in tinder-dry forests. The cloudless sky allowed heat to radiate up and away from elevations above 7200 feet (2200 meters), so temperatures at high elevations dipped below freezing at night for stretches of a week or more. Tropical plans died and people unaccustomed to chill weather sickened. Meanwhile, along the coasts of North, Central, and South America, the warmer than usual weather brought unusually strong storms, high ocean surges, and damaging wind and rainfall.

Flora and Fauna

The fact that Oceania comprises an isolated continent and numerous islands has had a special effect on its animal and plant life. Many species are **endemic**: they exist in a particular place and nowhere else on earth. This is especially true for Australia, but many Pacific islands also have endemic species.

Plant and Animal Life in Australia. The uniqueness of Australia's plant and animal life is the result of the continent's long physical isolation, its large size, its relatively homogeneous landforms, and the strikingly arid climate. Since Australia broke away from the southern supercontinent of Gondwana more than 65 million years ago, its plant and animal species have evolved in isolation (see Figure 1.7). Australian species evolved to fill not only the humid zones similar to those on Gondwana but all the various arid niches that eventually dominated the Australian landscape.

One spectacular result of the long isolation has been the development of more than 144 living species of marsupial animals. **Marsupials** are mammals that give birth to their young at a very immature stage and then nurture them in a pouch equipped with nipples. The most famous marsupials are the kangaroos; other species include wallabies, wombats, phalangers, the tiger cat, possums, the koala, numbats, and bandicoots. These various marsupials fill ecological niches that in other regions of the earth are occupied by rats, badgers, moles, cats, wolves, ungulates (grazers), and bears. One other type of mammal is exclusive to Australia and New Guinea. These are the **monotremes**, egg-laying mammals that include the duck-billed platypus and the spiny anteater.

Birds are also unusually varied, and parrot species are especially diverse. Some of the 750 species of birds known on the continent migrate in and out, but more than 325 species are endemic. Because birds had few natural predators in Australia, they rarely needed to make quick escapes. Several species, therefore, lost the ability to fly. The large emu and cassowary are examples of flightless birds.

Most of Australia's endemic plant species are adapted to dry conditions. Many of the plants have deep taproots to draw moisture from groundwater and small, hard, shiny (sclerophyll) leaves

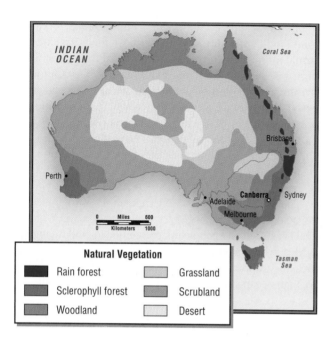

Figure 11.4 Australia's natural vegetation. [Adapted from Tom L. McKnight, *Oceania* (Englewood Cliffs, N.J.: Prentice Hall, 1995), p. 28.]

to reflect heat and to hold moisture. Much of the continent has only grass- and scrubland, with bits of open woodland; there are only a very few true forests—pockets along the Eastern Highlands and in Tasmania (Figure 11.4). Two plant genera account for nearly all the forest and woodland plants. They are the *Eucalyptus* (450 species, often called "gum trees") and the *Acacia* (900 species, often called "wattles").

Plant and Animal Life in New Zealand and the Pacific Islands. The prehuman biogeography of the Pacific islands has long interested those who study evolution and the diffusion of plants and animals, including the great scientist Charles Darwin, who formulated many of his ideas about evolution after visiting the Galápagos Islands of the eastern Pacific. Islands have to snare their plant and animal populations from the sea and air around them. Birds and large storms drop seeds and spores. Dead logs and other debris from storms carry small animals, seeds, and plant shoots from larger islands and continents to distant atolls and volcanic islands. Some seeds—such as large, buoyant coconuts—can float to a random destination on their own. Then, in a complicated process, these organisms "colonize" their new home. For example, coconut trees accomplish the long process of spreading inland from the beach by dropping the ripe fruit of each succeeding generation onto ever slightly higher ground. Over time, isolated plant and animal colonizers may evolve into new species that are endemic to (found only on) one island.

High, wet islands generally contain more and varied species because their more complex environmental mosaics provide niches for a wider range of wayfarers and thus a greater range of circumstances for evolutionary change. The flora and fauna are also modified by human inhabitants, once they arrive. In prehistoric times,

Asian explorers in ocean-going sailing canoes brought plants such as taro, bananas, and breadfruit and animals such as pigs, chickens, and dogs—as well as less desirable passengers such as rats, fleas, and diseases. Today, human activities from tourism to military exercises to city building continue to change the plants and animals of Oceania.

Generally, the distribution of land animals and plants is richest in the western Pacific, near the larger landmasses, which are the sources of species that diffuse to the islands. Biological diversity thins out to the east, where the islands are smaller and farther apart and hence less likely to be encountered by plant or animal wayfarers drifting on the wind or water. Although the natural rain-forest vegetation (*flora*) of New Zealand, New Guinea, and the high islands in the Pacific is abundant, the animals (*fauna*) in New Zealand and the Pacific islands are often described as simply "absent" because there are no indigenous land mammals, almost no indigenous reptiles, and only a few indigenous species of frogs. The reason for the absence of large animals is that New Zealand and the islands were never connected to the landmass of Eurasia as were Australia and the islands of Southeast Asia. There was never a land bridge that animals (and eventually humans) could cross. On the other hand, indigenous birds (especially waterfowl) in New Zealand are numerous and varied. In the absence of predators, several species of flightless birds have evolved. New Zealand is the home of the kiwi and (in the past) the huge moa. Some species of this bird grew 12 feet (3.7 meters) high. The moa was a major source of food until the Maori people hunted it to extinction before Europeans arrived. Today, New Zealand may well be the country on earth most characterized by introduced species of mammals, fish, and fowl. Nearly all were brought in by European settlers.

HUMAN PATTERNS OVER TIME

With courage, you can travel anywhere in the world and never be lost. Because I have faith in the words of my ancestors, I'm a navigator.

—MAU PIAILUG

VIGNETTE In 1976, Mau Piailug made history by navigating a traditional Pacific island voyaging canoe through the 2400 miles (3860 kilometers) of deep seas that separate Hawaii and Tahiti. He did so without a compass, charts, or other modern instruments, using methods passed down through his family. He relied mainly on observations of the stars, the sun, and the moon to find his way. When clouds covered the sky, he used the patterns of ocean waves and swells, as well as the presence of sea birds, to tell him of distant islands over the horizon.

Piailug reached Tahiti 33 days after leaving Hawaii and made the return trip in 22 days. His voyage settled a major scholarly debate. For centuries, scholars and explorers had debated how people had managed to settle the many remote islands of the Pacific without navigational instruments, thousands of years before the arrival of Euro-

In the 1990s the *Hokule'a* sailed from Hawaii to Easter Island, a 14,000-mile round trip, guided by traditional navigational techniques, such as the reading of wave patterns. [Cary Wolinsky.]

peans. Many scholars did not believe that navigation without instruments was possible and argued that would-be settlers might have simply drifted about on their canoes at the mercy of the winds, most of them starving to death on the seas, with but a few happening on new islands by chance. It was hard to refute this argument because local navigation methods had died out almost everywhere. However, in isolated Micronesia, where Piailug is from, indigenous navigation traditions had survived.

Piailug learned his methods from his grandfather in secret, because the Germans who first colonized Micronesia, and later the Japanese, banned long-distance navigation. The colonizers had wanted to prevent their Micronesian forced laborers from sailing away. After World War II, however, Piailug was free to sail and began making small voyages of several hundred miles. Eventually he attracted the attention of some Hawaiians who were building a traditional voyaging canoe, *Hokule'a*, with the aim of proving that traditional Pacific Islander sailing and navigational technologies were adequate for long-distance Pacific travel. Since the successful voyage of 1976, Piailug has trained several students in traditional navigation techniques. Throughout the Pacific, traditional sailing and navigational techniques have become a symbol of rebirth and a source of pride. [Adapted from Richard Nile and Christian Clerk, *Cultural Atlas of Australia, New Zealand, and the South Pacific* (New York: Facts on File, 1996), pp. 63–65.]

The story of humans in Oceania began in ancient times with a series of sea voyages across huge distances. Then, in historical times, the region experienced a process of European colonization that resembled colonialism in other world regions in its land use, settlement patterns, exploitation of resources, and treatment of indigenous people. Although a minority of the Pacific islands remain under the jurisdiction of a Western power, the majority of the independent nations of Oceania are interacting less with their former colonizers and more with Asia, their nearest neighbor.

The Peopling of Oceania

The longest surviving inhabitants of Oceania are Australia's **Aborigines.** As discussed in Chapter 10, the ancestors of the Aborigines were the Australoids, who migrated from Southeast Asia possibly as early as 50,000 years ago, over the Sundaland landmass that was exposed during the ice ages (see Figure 10.1). Amazingly, some memory of this ancient journey may be preserved in Aboriginal oral traditions, which recall mountains and other geographic features that are now submerged under water. At around the same time that the Aborigines were settling Australia, related groups of Australoids were settling nearby areas. These were the **Melanesians,** so named for their relatively dark skin tones, a result of high levels of the protective pigment melanin. The Melanesians spread throughout New Guinea and other nearby islands, giving this region its name, Melanesia. One indication of the great age of the Melanesian settlement of New Guinea is the existence of hundreds of distinct yet related languages on that island. Like the Aborigines, the Melanesians survived mostly by hunting, gathering, and fishing, although some groups—especially those inhabiting the New Guinea highlands—also practiced agriculture. There, agricultural features excavated in the terrain of a swampy basin have been dated at 9000 years, making this region one of the oldest places on earth where plant domestication has been firmly documented.

Much later, about 5000 to 6000 years ago, groups of linguistically related **Austronesians** migrated out of Southeast Asia and continued the settlement of the Pacific. By about a thousand years ago, the Austronesians had reached most of the remaining far-flung islands of the Pacific. They sometimes mixed with the Melanesian peoples they encountered (Figure 11.5). The Austronesians were renowned for their ability not only to survive at sea but also to navigate over vast, featureless distances by using the stars and reading the waves. They were fishers, hunter-gatherers, and cultivators who developed complex cultures and maintained trading relationships among their widely spaced islands.

In the millennia that have passed since first settlement, humans have continued to circulate throughout Oceania. Some apparently set out because their own space was too full of people and conflict, food reserves were declining, or they wanted a life of greater freedom. It is also likely that Pacific peoples were enticed to new locales by the same lures that later attracted some of the more romantic explorers from Europe and elsewhere: sparkling beaches, magnificent blue skies, beautiful people, scented breezes, and lovely landscapes.

The vast area settled by these people encompasses three distinct subregions. **Micronesia** refers to the small islands lying east

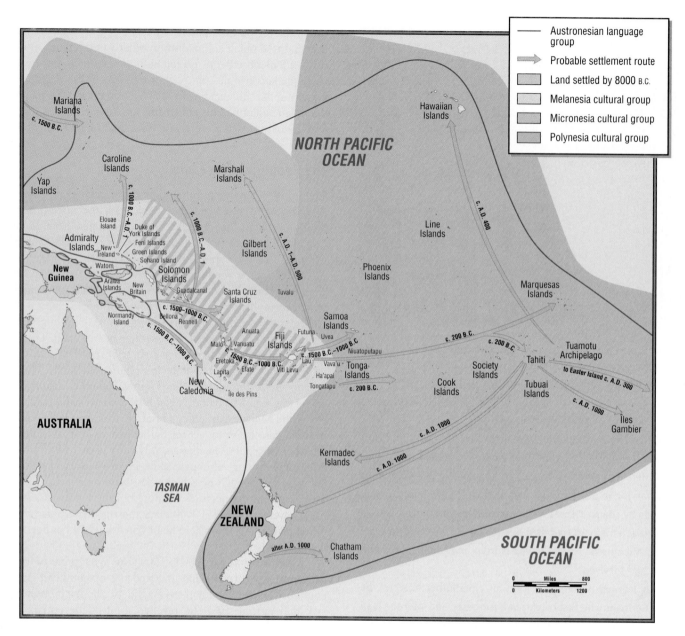

Figure 11.5 **Primary culture groups in island Pacific.** Modern *Homo sapiens,* scholars agree, reached coastal New Guinea around the same time they reached Australia, about 50,000 years ago. By about 25,000 years ago, people were spread across a large part of New Guinea and had even begun moving across the ocean to nearby Pacific islands. Archaeologists have found a site on Buka in the northern Solomon Islands dated to about 26,000 years ago. Movement into the more distant Pacific islands apparently began with the arrival of the Austronesians, who went on to inhabit the farthest reaches of Oceania. This map shows the division of Pacific islands into Melanesia, Micronesia, and Polynesia. [Adapted from Richard Nile and Christian Clerk, *Cultural Atlas of Australia, New Zealand, and the South Pacific* (New York: Facts on File, 1996), pp. 58–59.]

of the Philippines and north of the equator. **Melanesia** includes New Guinea and the islands south of the equator and west of Tonga (the Solomon Islands, New Caledonia, Fiji, and Vanuatu). **Polynesia** refers to the numerous islands situated inside an irregular triangle formed by New Zealand, Hawaii, and Easter Island (a tiny speck of land in the far eastern Pacific). Polynesia is the most recently settled part of the Pacific; some Polynesian influence remains in Melanesia in such places as Fiji and the Solomon Islands (see Figure 11.5).

Arrival of the Europeans

The earliest recorded contact between Pacific peoples and Europeans took place in 1521, when the first Europeans to cross the Pacific, led by the Portuguese explorer Ferdinand Magellan, landed on the island of Guam in Micronesia. That encounter ended badly. The islanders, intrigued by European vessels, tried to take a small skiff. For this crime, Magellan had his men kill the offenders and burn the village to the ground. By the 1560s, the Spanish had set up a lucrative trade route between Manila in the Philippines and Acapulco in Mexico. Explorers from other European states followed, at first taking an interest mainly in the region's valuable spices. The British and French explored extensively in the eighteenth century, including three separate scientific expeditions by the region's greatest voyager, James Cook.

The Pacific was not formally divided among the colonial powers until the nineteenth century, and by that time, the United States, Germany, and even Japan had joined France and Britain in taking control of various island groups. European colonization of Oceania proceeded according to the models developed in Latin America, Africa, South Asia, and Southeast Asia, with the major emphasis being on extractive agriculture and mining. Native people were often displaced from their lands or succumbed to exotic diseases to which they had developed no immunity, and their populations declined. This invasive early contact between Europeans and Pacific Islanders has been termed "the fatal impact."

Many enduring notions about the Pacific arose from the European explorations of the eighteenth and nineteenth centuries. This was a period of intense questioning in Europe about whether or not civilization actually improves the quality of life for human beings. Some Europeans argued that civilization corrupts and debases people. This point of view grew partly out of news filtering home to Europe about the negative effects of colonization and partly out of the experience with industrialization in Europe, where crowded, dirty, impersonal, often crime-ridden cities were proliferating. Romanticists glorified what they termed "primitive" people living in distant places supposedly untouched by corrupting influences; they coined the term **noble savage** to describe such people. These ideas influenced Pacific explorers, who often spoke of the people they encountered as being in a pristine state of nature that was more conducive to virtuous moral conduct. Europeans were caught off guard when, from time to time, the very same islanders rebelled, armed themselves, and attacked those who were taking their lands and resources. Usually the surprised Europeans quickly revised their opinions and relabeled the "noble savages" as brutish and debased.

The realities of life in Oceania were much more balanced than the positive and negative extremes that Europeans perceived. In Australia, and on New Guinea and the other larger islands of Melanesia, a relatively plentiful resource base made it possible for people to live in small, simple societies, less subject to the stratification and class tensions of so much of the world. On the smaller islands of Micronesia and Polynesia, in contrast, land and resources were scarcer. On these islands, many societies were **hierarchical,** with layers of ruling elites at the top and undifferentiated commoners at the bottom. Moreover, many of the peoples of Oceania coexisted in a state of moderate antagonism. Although warfare was not uncommon, hostilities were often settled ritualistically and by means of annual tribute-paying ceremonies, rather than by resorting to mortal violence. Individual rulers rarely amassed large territories or controlled them for long.

Perhaps the most enduring myth Europeans created was their characterization of the women of the Pacific islands as gentle, simple, compliant love objects. (Tourist brochures still promote this notion.) Although there is ample evidence to suggest that Pacific Islanders did have more sexual partners in a lifetime than Europeans did, the reports of unrestrained sexuality related by European sailors were no doubt influenced by the exaggerated fantasies one might expect from all-male crews living at sea for months at a time. The notes of Captain James Cook are typical: "No women I ever met were less reserved. Indeed, it appeared to me, that they visited us with no other view, than to make a surrender of their persons."

Over the years, such notions about Pacific island women have been encouraged by the paintings and prints of Paul Gauguin, the writings of the novelist Herman Melville (*Typee*), and the studies of the anthropologist Margaret Mead (*Coming of Age in Samoa*), as well as by movies and musicals such as *Mutiny on the Bounty* and *South Pacific*. In reality, gender roles in the Pacific varied considerably from those in Europe, but not in the ways European explorers imagined. Women often exercised a good bit of power in family and clan, and their power increased with motherhood and advancing age. In Polynesia, a woman could achieve the rank of ruling chief in her own right, not just as the consort of a male chief. In everyday Polynesian life, men were the primary cultivators of food as well as the usual cooks. Women were primarily craftspeople but also contributed to subsistence by gathering fruits and nuts and by fishing for shellfish in reefs and lagoons near the shore. And in some places—Micronesia, for example—lineage was established through women, not men.

Arearea ("Amusement") by Paul Gauguin, 1892. The painting shows Tahitian women rendered in a European Romantic pastoral style that emphasizes their gentle, compliant demeanor. [Musée d'Orsay, Paris. Photograph by Erich Lessing, Art Resource, New York.]

The Colonization of Australia and New Zealand

Though all of Oceania has experienced European or American rule at some point, the most Westernized parts of the region are Australia and New Zealand. The colonization of these two countries by the British has resulted in many parallels with North America. In fact, the American Revolution was the major impetus for "settling" Australia because it deprived Britain of a place to send its convicts. In early nineteenth-century Britain, a relatively minor theft—for example, of food or a piglet—might be punished with a term of seven years' hard labor in Australia. After their sentences were served, most former convicts chose to stay in the colony. They were joined by a much larger group of free immigrants from the British Isles who were attracted by the availability of inexpensive farmland. Waves of these immigrants arrived until World War II. Nevertheless, there was a steady flow of English and Irish convicts until 1868, and they are given credit for Australia's rustic self-image and egalitarian spirit. New Zealand also has a rustic self-image and an egalitarian spirit, although it was colonized somewhat later, in the mid-1800s. And, although its population too was overwhelmingly British, it was never a penal colony.

Another similarity among Australia, New Zealand, and North America was the treatment of indigenous peoples by European settlers. In both Australia and New Zealand, native peoples were killed outright, annihilated by infectious diseases to which they had little immunity, or shifted to the margins of society. The few who lived on territory deemed undesirable by Europeans were able to maintain their traditional way of life; but the vast majority who survived lived and worked in grinding poverty, either in urban slums or on cattle and sheep ranches. Today, native peoples still suffer from pervasive discrimination and maladies such as alcoholism and malnutrition that are common in an **underclass,** the lowest social stratum composed of the disadvantaged. Even so, some progress is being made toward improving their lives, as discussed on pages 583–584. New Zealand has made great strides in appreciating the culture and rights of its native Maori population. Australia is beginning to do the same with regard to the Aborigines, although in Australia there is vocal resistance to this project among some people of European background.

Oceania's Growing Ties with Asia

During the twentieth century, Oceania's relationship with the rest of the world changed at least three times: from a predominantly European focus, to identification with the United States and Canada, to its currently emerging linkage with Asia. Up until roughly World War II, the colonial system gave the region a European orientation. In most places, the economy depended largely on the export of raw materials to Europe. Thus, even when a colony gained independence from Britain, as Australia did in 1901 and New Zealand did in 1907, people remained strongly tied to their "mother countries." During World War II, however, the European powers could provide only token resistance to Japan's invasion of much of the Pacific and bombing of northern Australia. Hence, after the war, the United States became the domi-

nant power in the Pacific and throughout much of Asia. Although U.S. dominance did not sever the region's economic linkages with Europe, U.S. investment became increasingly important to the economies of Asia and Oceania. Australia and New Zealand joined the United States in a cold war military alliance known as ANZUS. Both fought alongside the United States in Korea and Vietnam, suffering considerable casualties and experiencing significant antiwar movements at home. U.S. cultural influence was strong, too, as its products, technologies, movies, and pop music penetrated much of Oceania.

By the 1970s, another shift was taking place as Oceania became steadily drawn into the growing economies of Asia. One could consider this development long overdue, given Asia's proximity to Oceania, its potential as a market for Oceania's raw materials, and some cultural connections. Many Pacific islands have significant Chinese, Japanese, Filipino, and Indian minorities; and even the small Asian minorities of Australia and New Zealand are increasing. Since the 1960s, Australia's thriving mineral export sector has become increasingly geared toward supplying Japan's burgeoning manufacturing industries. Similarly, since the 1970s, New Zealand's wool and dairy exports have gone mostly to Asian markets. As we shall see, these transformations are accompanied by considerable cultural and economic strain. Nonetheless, despite occasional backlashes against "Asianization," Australia, New Zealand, and the rest of Oceania are becoming more open to Asian influences.

POPULATION PATTERNS

Oceania occupies a huge portion of the planet yet has few people compared to other world regions. Whereas Southeast Asia has more than half a billion people and East Asia well over a billion, just over 31 million people are spread throughout Oceania —from New Guinea, Australia, and New Zealand across the Pacific to Hawaii and Easter Island. The Pacific islands, including Hawaii, have somewhat over 8 million people; Australia has 19.2 million and New Zealand 3.8 million.

As the population density map (Figure 11.6) indicates, the people of this region are unevenly distributed. Most Australians live in a string of cities along the country's well-watered and relatively fertile eastern and southeastern coasts—Brisbane, Newcastle, Sydney, Canberra, Melbourne, Adelaide—and on the southwest tip of the continent in and around the city of Perth. Australia and New Zealand have among the highest percentages of city dwellers outside Europe: 85 percent for both countries in 2000. The vast majority of people in these two countries live in modern, affluent cities and work in a range of occupations typical of highly industrialized societies. New Zealand has one very large city, Auckland, and several medium-sized cities on both North Island and South Island. Nonetheless, overall densities in both countries are low: Australia averages just 6 people per square mile (2.3 per square kilometer) and New Zealand 37 per square mile (14.3 per square kilometer).

The smaller islands of the Pacific are much less urbanized, and the population density of these islands is often linked to spe-

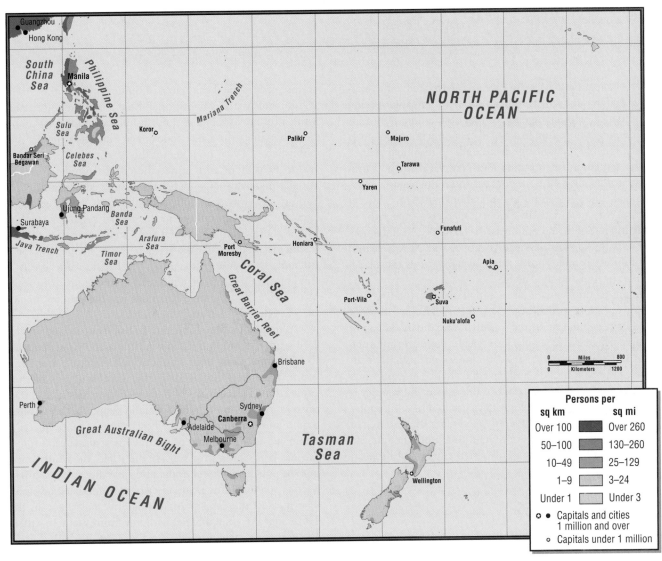

Figure 11.6 **Population density in Oceania.** [Adapted from *Hammond Citation World Atlas* (Maplewood, N.J.: Hammond, 1996).]

cific features and to the history of settlement and development. Some are sparsely settled or uninhabited; others—including some of the smallest Pacific islands, such as the Marshall Islands and Tuvalu—have 900 to 1000 people per square mile (347 to 386 per square kilometer). The highest density is found on the tiny island of Nauru, with 1360 people per square mile (525 per square kilometer). The people of Nauru live in relative prosperity on the proceeds from the mining of phosphate derived from eons of bird droppings. The phosphate is exported for use in the manufacture of fertilizer.

The urban poor of Oceania, whether they live in the developed economies of Australia and New Zealand—where they tend to have Aboriginal or Maori origins—or in cities across the Pacific—such as Port Moresby in Papua New Guinea, Suva in Fiji, Papeete in Tahiti, and Honolulu in Hawaii—may have higher cash incomes than rural Pacific Islanders but much lower standards of living. In the cities, there is little they can do to supplement their incomes with self-produced food or other necessities (see the discussion of subsistence affluence on page 589).

There are two different population profile patterns in Oceania. The Pacific islands tend to have high fertility rates, between 2.5 and 6.6 children per woman. Their populations are young, with 30 to 50 percent under the age of 15 and just 3 to 5 percent elderly. In contrast, the population profiles of Australia and New Zealand closely resemble those of North America, Europe, and Japan. There, fertility rates average less than 2 children per woman, and people tend to live into their mid- to late 70s. The overall trend throughout the region, however, is toward smaller families and aging populations—although this trend is just beginning in the Pacific islands and is much more obvious in Australia and New Zealand.

CURRENT GEOGRAPHIC ISSUES

Many current geographic issues in Oceania are related to the transition now under way from European to Asian and inter-Pacific cultural influence. This change in cultural orientation, in turn, is related to the larger transition to a global economy now under way everywhere. The old relationships were built on historical factors, such as the settlement of Australia and New Zealand by Europeans. The new relationships, in contrast, are influenced by practical financial and geographic considerations, such as physical proximity to Asia and recently opened markets—especially in China but also in South and Southeast Asia. Hence, Australia now trades more with Asia and less with Europe; and New Zealand, while continuing to trade mostly with Australia, finds increasing opportunities in Asia and the Pacific.

Emerging technologies such as e-mail and the Internet and rapid air travel have a special impact on the far-flung locations of Oceania. They encourage cultural sensitivity and the modification of old prejudices because people have greater access to information about unfamiliar ways of life, can correspond instantly with distant colleagues and friends, and have the chance to experience distant places personally. One result of this greater interaction is that New Zealand and the Pacific islands are finding philosophical grounds for a closer mutual identity, including the fostering of greater public awareness of environmental issues and the renewal and acceptance of their common Polynesian as well as European cultural heritage.

SOCIOCULTURAL ISSUES

The cultural sea change away from Europe and toward Asia and the Pacific has been accompanied by new respect for indigenous peoples: the Aborigines of Australia; the Maori of New Zealand; and the Melanesians, Micronesians, and Polynesians of the Pacific islands. In addition, the growing sense of common economic ground with Asia has heightened awareness of the attractions of Asian culture.

Ethnic Roots Reexamined

Weakening of the European Connection. Until very recently, most people of European descent in Australia and New Zealand thought of themselves as Europeans in exile. Many considered their lives incomplete until they had made a pilgrimage to the British Isles or the European continent. In her book, *An Australian Girl in London* (1902), Louise Mack wrote: "[We] Australians [are] packed away there at the other end of the world, shut off from all that is great in art and music, but born with a passionate craving to see, and hear and come close to these [European] great things and their home[land]s."

These longings for Europe encouraged Australians and New Zealanders to think of themselves as transients in their own countries, separate from the region in which they resided; such feelings

were accompanied by racist attitudes toward the Aborigines, Maori, and Asians. The historian Stephen H. Roberts managed to write a history of settlement in Australia without even mentioning the Aborigines. In a later book, *The Squatting Age in Australia*, published in 1935, he noted of the Aborigines: "It was quite useless to treat them fairly, since they were completely amoral and usually incapable of sincere and prolonged gratitude." The prevailing idea was that both Australia and New Zealand should preserve European culture in this nether region of the Southern Hemisphere, warding off not only indigenous peoples but also Pacific Islanders and Asians in general. Hence, in the 1920s, migrants from Asia, Africa, and the Pacific were legally barred. Trading patterns further reinforced connections to Europe: until World War II, the United Kingdom was the primary trading partner of both New Zealand and Australia.

When migration from the British Isles slowed in the mid-1940s, after World War II, both Australia and New Zealand began to lure immigrants from southern and eastern Europe, many of whom had been displaced by the war. Hundreds of thousands came from Greece, parts of the former Yugoslavia, and Italy. The arrival of these non-English-speaking people began a shift toward a more multicultural society. The election of more liberal governments led to a loosening of restrictions against Asian migrants in the late 1950s and early 1960s. In the 1960s, the whites-only immigration policies were set aside, and migration from China, India, Vietnam, South Africa, and elsewhere in the Pacific was allowed. The greatest increase in Asian migrants took place in the early 1970s after the United States withdrew from Vietnam, when many Vietnamese refugees sought a safe haven.

As a result of the new immigration policy, the ethnic makeup of the populations of Australia and New Zealand is changing. Australia is now among the most ethnically diverse countries on earth (Figure 11.7). In both Australia and New Zealand, immigrants from Asia, especially from Vietnam and China, are a significant portion of new arrivals. Nonetheless, Asians remain a small percentage of the total population in both countries (less than 7 percent in Australia). People of European descent are declining as a proportion of the population but are expected to remain the most numerous segment throughout the twenty-first century.

Strengthening of the Connections with Indigenous Peoples. The population makeup in Australia and New Zealand is also changing in another respect. In Australia, for the first time in recent memory, the number of people who claim indigenous origins is increasing. Before Europeans came at the end of the eighteenth century, the sole inhabitants of Australia were about 750,000 Aborigines. In the intervening two centuries, the overall population has grown to more than 18 million, but the Aboriginal population, just 2 percent of that total, is only half what it was in 1800. Therefore, when those claiming Aboriginal origins rose by 33 percent between 1991 and 1996, it was a surprise. In New Zealand, those claiming a Maori background rose by 20 percent.

Percentage of Foreign-born Australians from Various Countries

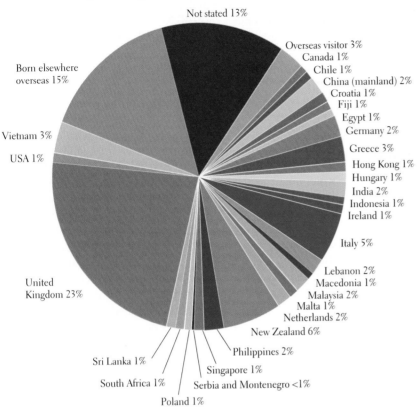

Not stated 13%
Overseas visitor 3%
Canada 1%
Chile 1%
China (mainland) 2%
Croatia 1%
Fiji 1%
Egypt 1%
Germany 2%
Greece 3%
Hong Kong 1%
Hungary 1%
India 2%
Indonesia 1%
Ireland 1%
Italy 5%
Lebanon 2%
Macedonia 1%
Malaysia 2%
Malta 1%
Netherlands 2%
New Zealand 6%
Philippines 2%
Singapore 1%
Serbia and Montenegro <1%
Poland 1%
South Africa 1%
Sri Lanka 1%
United Kingdom 23%
USA 1%
Vietnam 3%
Born elsewhere overseas 15%

Figure 11.7 Australia's cultural diversity. About 74 percent (13.2 million) of Australia's people were born in the country; the rest were born in other places, making Australia one of the world's most culturally diverse nations. The chart, based on the 1996 census, shows the birthplace of the 26 percent of Australia's residents born elsewhere. [Data from 1997 Commonwealth of Australia.]

These increases are thought to be linked in two ways to changing attitudes toward indigenous peoples in the region. First, more people now understand that colonial attitudes are largely responsible for the low social standing and impoverished state of indigenous peoples. Hence, those who have some Aborigine or Maori ancestors are more willing to claim them. Another explanation is that mating relationships between European and indigenous peoples are more common and more open, and so the number of people with mixed heritage is increasing.

In 1988, during a bicentennial celebration of the founding of white Australia, a contingent of some 15,000 Aborigines protested that they had little reason to celebrate. During the same 200 years, they had been excluded from their ancestral lands, from basic civil rights, and even from the national consciousness. Into the 1960s, Aborigines had only limited rights of citizenship, and it was even illegal for them to have a drink of alcohol. Until 1993, Aborigines were thought to have no prior claim to any land in Australia. During settlement in the late eighteenth century, the British had deemed all Australian lands to be available for British use because the Aborigines were thought to be too primitive to have concepts of land tenure, since their nomadic cultures had "no fixed abodes, fields or flocks, nor any internal hierarchical differentiation." After the Australian High Court declared this position void in 1993, Aboriginal groups began to win some land claims, mostly for land in the arid interior previously claimed only by the Australian government.

In New Zealand, relations between the majority European-derived population and the indigenous Maori have proceeded somewhat more amicably. Today, New Zealand proudly portrays itself as a harmonious multicultural society. Yet there is deep ambivalence about the foundations of this supposed harmony. The geographer Eric Pawson writes that when the Maori signed the Waitangi Treaty with the British in 1840 and accepted European "guns, goods and money," they thought they were granting rights of land usage in return for a trading relationship. The Maori did not regard land as a tradable commodity, but rather as an asset of the people as a whole, used by families and larger kin groups to fulfill their needs. Pawson writes: "To the Maori the land was sacred . . . [and] the features of land and water bodies were woven through with spiritual meaning and the Maori creation myth." The British, on the other hand, assumed that the treaty had transferred Maori lands to them, giving them *exclusive* rights to settle the land with British migrants and to extract wealth through farming, mining, and forestry. Therein lay the grounds for a conflict that has resulted in significant losses for the Maori.

In the 50 years after 1840, the Maori lost control of more than 84 percent of their former lands; by 1950, they had lost all but 6.6 percent. European settlers and the government owned and occupied the rest. Maori numbers shrank from a probable 120,000 in the early 1800s to 42,000 in 1900; and they came to occupy the lowest and most impoverished rung of New Zealand society. In the 1990s, however, the Maori began to reclaim their culture, and they established a tribunal that forcefully advances Maori interests through the courts. New Zealand formalized its efforts to right past wrongs and bring greater equality and social participation to the

There were many different groups of Aborigines in various parts of Australia when the Europeans arrived, but all seem to have based their moral laws and daily customs on the idea that the spiritual and physical worlds are intricately related. The dead are everywhere present in spirit, and they guide the living in how to relate to the physical environment. *Dreamtime* refers to the time of creation when the human spiritual connections to rocks, rivers, deserts, plants, and animals were made clear. This representation (in Kakadu National Park, Northern Territory) is of Djawok, a creator who left his image of a cuckoo on this rock. Aboriginal people who have remained close to their heritage still read the landscape as a complex sign system conveying spiritual meaning. Territory is recognized and made familiar by following the tracks of ancient "song lines." Particular tribal groups associate with specific animals or landscape features from which they gather solace and inspiration. When people die, they are said to "go into the country." [Belinda Wright/National Geographic Image Collection.]

viously hid their Maori origins but are now proud to claim them. Many New Zealanders now embrace the Maori component of their cultural heritage; overall, New Zealand may lead the world in addressing past mistreatment of indigenous peoples. Nonetheless, the Maori still have higher unemployment, lower educational attainments, and poorer health than the population as a whole.

Balancing Indigenous Rights. The Fiji island group in the southern Pacific exemplifies an issue that is typical in the Pacific islands: how to balance the rights of indigenous people with those of "outsiders" who may have lived on Pacific islands for several generations. Fiji, one of the first colonial domains in the Pacific to achieve independence, has a population that is about evenly divided between indigenous Fijians and the descendants of indentured sugar plantation workers brought in by the British from India more than a century ago. Fijians of Indian origins have flourished as the owners of tourism facilities and other businesses and as growers of sugarcane, one of the country's leading exports. They hold significant economic and political power, especially in western urban centers and areas of tourism and sugar cultivation. On the other hand, the indigenous Fijians who are governed by traditional chiefs tend to live on rural islands in the east of the island group and to be less prosperous. Complicating matters is a codified system established by the British whereby land rights are held by indigenous clans and land cannot be alienated or sold. Indian Fijians who grow sugarcane or run businesses and tourism facilities do so on leased land. In 1987 and again in May 2000, rivalry between these two main groups of Fijians led indigenous Fijians to attempt coups d'état against legally elected governments dominated by Indian Fijians. The conflict has ruined the tourism economy and stymied economic and social development. Many Indian Fijians have left the islands, fearing that indigenous rights will weaken egalitarian policies. Political leaders around the region, from Australia to Hawaii, have reminded indigenous Fijians that indigenous rights cannot be held superior to fundamental human rights.

Forging Unity in Oceania

In embracing its Maori roots, New Zealand has in effect also begun to embrace its connection to the wider Pacific, especially Polynesia. Although the majority of New Zealand immigrants continue to be from the United Kingdom, the second largest influx has come from the Pacific islands. Auckland now has the largest Polynesian population of any city in the world.

A sense of unity with Oceania as a whole is developing throughout the region, as people begin to appreciate the region's cultural complexity and cooperate in activities ranging from schooling to sports to environmental activism. One way in which this unity is manifest is through interisland travel. Today, people usually fly rather than travel by sea. They often travel in small planes from the outlying islands to hubs such as Fiji, where jumbo jets can be boarded for Auckland, Melbourne, or Honolulu. Cook Islanders call these little planes, which carry 5 to 30 passengers, "the canoes of the modern age." New Zealanders migrate to Australia to teach or train. A businessman from the Kiribati group in Micronesia flies to Fiji to take a short course at the University

Maori and other minority groups in the country in 1996, when the government agreed to settle several long-standing Maori claims to land and fishery rights.

During the twentieth century, Maori numbers rebounded, reaching 523,000 in 1996. This number includes those who preof the South Pacific. A Cook Islands teacher takes graduate training in Hawaii.

Who funds all this travel, especially in the Pacific islands? Many travelers do so themselves. A person may earn money by selling food or handmade crafts at public events. A popular way for a group to collect plane fares is to sell raffle tickets for baskets of food or a bicycle. (There may be plenty of customers in one's own extended family.) Governments help, too, with scholarships to the various universities in Australia, New Zealand, Hawaii, and Fiji. Because there is often a waiting list for these grants, a college education may come later in life, when one is 30 or 40.

The arts have become a powerful medium for forging a regional identity, through events such as the Festival of Pacific Arts. First held in 1972 in Fiji, this weeklong gathering takes place once every four years in a different location. The festival features indigenous music, dance, and other traditional forms of artistic expression such as wood carving and tattooing. Recently, these events have also included the skills of ocean-going navigation and canoe voyaging. Several other trends are encouraging a growing sense of unity throughout the region: the use of pidgin languages, the movement called the Pacific Way, and the popularity of sports.

Languages in Oceania. The Pacific islands, and most notably Melanesia, have a rich variety of languages. In some cases, islands in a single chain will have several different languages. A case in point is Vanuatu, a chain of 80 mostly high volcanic islands to the east of northern Australia. At least 108 languages are spoken by a population of just 180,000—averaging one language for every 1600 people! New Guinea is the largest, most populous, and most ethnically diverse island in the Pacific. No less than 800 languages are spoken on New Guinea by a populace of 4.1 million. Languages are both an important part of a community's cultural identity and a hindrance to cross-cultural understanding.

In Melanesia, and elsewhere in the Pacific, the need for communication with the wider world is served by several **pidgin** languages that are sufficiently similar to be mutually intelligible. Pidgins are made up of words borrowed from the several languages of people involved in trading relationships. Over time, they can grow into fairly complete languages, capable of fine nuances in expression. When a particular pidgin is in such common use that mothers talk to their children in it, then it can literally be called a "mother tongue." In Papua New Guinea, pidgin English is the official language.

The imposition of European languages in Oceania extinguished a number of native tongues. For example, there were 500 or more Aboriginal languages in Australia at the time of first contact; most are now gone. Attempts to revive and preserve indigenous languages have crystallized an effort at regional solidarity known as the Pacific Way.

The Pacific Way. The **Pacific Way** is a term used since 1970 to convey the idea that Pacific Islanders and their governments have a regional identity growing out of their own particular social experience. It includes the idea that Pacific Islanders have the ability to control their own development and solve their own problems. The concept of the Pacific Way first emerged when Pacific peoples were grappling with the problems presented by school curricula imposed by the colonial powers, which suppressed the use of native languages and often depicted the region as peripheral and backward. They developed the Pacific Way as a philosophy to guide them in writing their own school texts. It embodied the idea that Pacific island children should first learn about their own cultures and places before they studied Britain, France, or the United States.

Increasingly, the Pacific Way has grown into an integrated approach to economic development and environmental issues, with an emphasis on consensus as a traditional approach to problem solving. In some ways, the concept resembles *Pancasila* in Southeast Asia (see Chapter 10), and it has some of the same potential for abuse. But there have been constructive steps to implement the Pacific Way. The South Pacific Regional Environmental Program (SPREP), in existence since 1980, emphasizes pan-regional cooperation and grassroots environmental education. The geographer Carolyn Koroa, formerly of the University of the South Pacific in Fiji, reports that the organization is extraordinarily egalitarian and that SPREP programs routinely incorporate traditional environmental knowledge into efforts to promote sustainable livelihoods, often based on traditional crafts.

Sports as a Unifying Force. Sports and games are a major feature of daily life throughout Oceania, and the region has both shared with and borrowed from cultures around the world. Long-distance sailing, now a world-class sport, was an early skill in this region. Surfing evolved in Hawaii from ancient navigation customs that matched human wits against the power of the ocean.

The Australian Aboriginal Embassy is a collection of colorful wooden structures erected on the lawns in front of Old Parliament House in Canberra as a protest against the seizure of Aboriginal lands by European settlers at the end of the eighteenth century and the continuing refusal of the government to acknowledge the injustice. The protest began in 1972 and was contining in 2001, when the authors visited. Nearby is a permanent encampment of Aboriginal people who staff the facility and welcome the many visitors each year. [Mac Goodwin.]

Thinking Globally: ON THE WEB

In the Trobriand Islands near New Guinea, the British game of cricket has been reformulated to include local traditions. Village teams of as many as 60 men dress in traditional garments and decorate their bodies in ways that are reminiscent of British cricket uniforms, such as the painting-on of white kneesocks, yet also include elements of magical decoration. Chants and dances are part of Trobriand cricket matches. The matches can go on for weeks, and they attract crowds of young onlookers, as seen in this picture. [Robert Harding Picture Library, London.]

Sporting traditions introduced from outside have been embraced and modified to suit local needs. The Scottish game of golf is a favorite of Japanese and South Korean tourists in Hawaii, Tahiti, and Guam. On hundreds of Pacific islands and in Australia and New Zealand, the rugby field, the volleyball court, the soccer field, and the cricket pitch are important centers of community activity. Baseball is a favorite in the parts of Micronesia that were U.S. trust territories. Women compete in the popular sport of netball, which is played like basketball but without a backboard. Australia or New Zealand often fields the world champion netball team.

Sports competitions, including native dance, are the single most common and resilient link among the countries of Oceania. Such competitions encourage regional identity and provide the opportunity for ordinary citizens to travel extensively around the region and to sports venues in other parts of the world. The South Pacific Games—featuring soccer, boxing, tennis, golf, and netball, among other sports—are held every four years. The Micronesians hold periodic games that incorporate many traditional tests of skill, such as spearfishing, climbing coconut trees, and racing outrigger canoes.

In New Zealand, the Maori and Samoan cultures are central to that country's world dominance in rugby. At the beginning of a game, the players perform the Maori war dance, the Haka, to arouse feelings of aggression. Many of the New Zealand players are Samoan in origin; Samoans are Polynesians, noted for their size, strength, and acuity in games of skill. In Oceania, as in societies the world over—in North America, the Caribbean, Africa, Europe, and South Asia—sports have smoothed the way for multicultural acceptance, especially at the Olympic level. Nonetheless, here as elsewhere, there is a danger that a particular group can be considered good at sports yet still find full participation in society blocked.

Gender Roles

There is great variation in gender roles in Oceania, and they are changing significantly throughout the region. In Australia and New Zealand, women's access to jobs and policy-making positions has improved over the last few decades. Among the results is that women bureaucrats in Australia have been credited with materially improving the living conditions of the poorest segments of Australian society. Increasingly, young women are choosing careers and postponing marriage until their 30s. In New Zealand, 48 percent of women aged 24 to 29 are not married, nor are 26 percent of those aged 30 to 34. Often, these women have mates but choose not to marry them. Nonetheless, Australian and New Zealand societies continue to reinforce the housewife role for women in a variety of ways. For example, the expectation is that women, not men, will interrupt their careers to stay home to care for young or very old family members.

In the Pacific, recent research has shown that, historically (and prehistorically), gender roles and relationships varied greatly from island to island and were in a continual state of flux. Moreover, gender roles changed over the course of life, and this is still very much the case. Today, many young women fulfill traditional roles as wives and mothers and practice a wide range of domestic crafts such as weaving and basketry. Then, in middle age, they may return to school and take up careers. Some Pacific women, with the aid of government scholarships, pursue higher education or job training that takes them far from the villages where they raised their children, yet their incomes are often shared with children and grandchildren. Accumulating age and experience may boost Pacific women into community positions of considerable power.

The New Zealand geographer Camilla Cockerton reports that throughout their lives Pacific women contribute significantly to family assets through the formal and informal economies. Most traders in marketplaces are women, and the items they sell are usu-

CULTURAL INSIGHT *Waltzing Matilda*

The words of Australia's informal national anthem, "Waltzing Matilda," were composed by A. B. ("Banjo") Paterson in 1895. The tune is thought to be an old Scottish hymn. The term *swagman* encompasses the many types of itinerant laborers who wandered across the outback in search of work, walking, bicycling, or hitching rides. A swagman carried all his possessions on his person, often wrapped in a blanket and formed into a backpack called a *swag*. Swagmen were not denigrated as tramps or hobos—as were such itinerants in North America—but rather were immortalized in legends and songs.

The song requires a bit of deciphering. To "waltz Matilda" means to take to the road with a bedroll, which was often given a girl's name as an ironic reference to the scarcity of women. A billabong is a section of still water adjacent to a river (an oxbow lake) where water usually stays longer than in the watercourse itself. A billy is a tin can used for boiling water to make tea or soup. A coolibah is a type of eucalyptus tree that grows beside billabongs. A jumbuck is a sheep; and a tucker-bag is a sack in which a swagman carried his food supply. In Australia, "squatters" were settlers who usually got claim to the land they occupied, and the authorities backed them up against incursions by swagmen.

Once a jolly swagman camped by a billabong
Under the shade of a coolibah tree

And he sang as he watched and waited 'til his billy boiled
"You'll come a-waltzing Matilda with me"

Refrain:
Waltzing Matilda, Waltzing Matilda
You'll come a-waltzing Matilda with me
And he sang as he watched and waited 'til his billy boiled
"You'll come a-waltzing Matilda with me"

Down came a jumbuck to drink at that billabong
Up jumped the swagman and grabbed him with glee
And he sang as he stuffed that jumbuck in his tucker-bag
"You'll come a-waltzing Matilda with me"
(refrain)

Up rode the squatter, mounted on his thoroughbred
Up rode the troopers, one, two, three
"Where's that jolly jumbuck you've got in your tucker-bag?"
"You'll come a-waltzing Matilda with me"
(refrain)

Up jumped the swagman and sprang into that billabong
"You'll never take me alive!" said he
And his ghost may be heard as you pass by that billabong
"You'll come a-waltzing Matilda with me"
(refrain)

---

ally made and transported by women. Yet, like women everywhere, they have trouble obtaining credit for their businesses. For example, of the 2039 loans approved by the Agricultural Bank of Papua New Guinea in January 1991, only 4 percent (or 91) went to women. Cockerton also reports that Pacific women who migrate play a huge role in Pacific economies, sending larger remittances to their home countries than men do, despite their lower incomes.

Being a Man: Persistence and Change

Because of its cultural diversity, Oceania encompasses many acceptable roles for men, but images of supermasculinity are popular throughout the region. In the Pacific, men traditionally were cultivators, deepwater fishermen, and masters of seafaring. In Polynesia, they also were responsible for many aspects of food preparation, including cooking. Men fill many positions in the modern world, but idealized male images continue to be associated with vigorous activities such as rugby, cricket, and canoe building and navigating.

In Australia and New Zealand, the cult of the supermasculine white working-class settler has long had prominence in the national mythologies. Only recently have the stories of female immigrants,

ordinary men, Aborigine laborers, and the many early Chinese settlers come to public attention. In New Zealand, the classic male settler was a farmer and herdsman; in Australia he was more often an itinerant laborer—stockman, sheepshearer, cane cutter, or digger (miner). This many-skilled drifter, who possessed a laconic, laid-back sense of humor, went from station (large farm) to station or from mine to mine, working hard but sporadically, gambling, and then working again, until he had enough money or experience to make it in the city. There, he often felt ill at ease and chafed to return to the wilds. Now immortalized in songs, novels, films such as the *Crocodile Dundee* series, and American TV auto advertisements, these men are portrayed as a rough and nomadic tribe (see the box "Waltzing Matilda"). Although they may have had families somewhere, they traveled as loners, often outnumbering women in a given place by 100 to 1. Not surprisingly, male camaraderie, a demonstrated loyalty to their "mates" (male friends), and frequent brawls dominated their social life. No small part of this characterization derived from the fact that many of Australia's first immigrants were convicts who seized the opportunity for a new identity and measure of freedom "down under."

Today, as part of larger efforts to recognize the diversity of Australian society, new ways of life for men are emerging and are

On the Cape York Peninsula, Queensland, a ringer, or station hand, takes a break before heading home at the end of the workday.
[Sam Abell/National Geographic Image Collection.]

breaking down the national image of the tough male loner. Nonetheless, the old model persists and remains prominent in the public images of Australian businessmen, politicians, and movie stars.

ECONOMIC AND POLITICAL ISSUES

Although the reorientation away from Europe and toward Asia and the Pacific has many cultural ramifications in the region, the process is driven largely by global economic forces such as trade, tourism, and migration.

Export Economies

For decades, natural resources and agricultural products have been the backbone of the national economies of Australia and

New Zealand. Australia, for example, supplies about 25 percent of the world's wool; it is the world's largest supplier of coking coal used in steel manufacturing; it is first in bauxite production, third in gold, and fourth in nickel. New Zealand specializes in dairy products, meat, fish, wool, and lumber products. But neither country has been a major supplier to the world market of more profitable manufactured goods. There are several reasons for this pattern of exports. First, during the early and mid-twentieth century, tariffs were used to protect manufacturing industries from outside competition, so jobs could be created for new immigrants. At the same time, domestic markets in both countries were (and are) too small to support multiple competitors. Thus, with limited outside sources of manufactured goods and few domestic ones, the type of competition among manufacturers that would result in quality improvement and price reductions rarely occurred. As a result, imported products were more popular than those produced locally, despite their higher costs because of tariffs. Second, most of the major trading partners of Australia and New Zealand purchase only raw materials to supply their own manufacturing industries. Japan, for example, now buys 20 percent of Australia's exports (Europe buys just 14 percent), and almost all purchases are raw materials. A third factor that has discouraged the development of manufacturing in Australia and New Zealand is that Japan and other Asian markets have restrictive import policies, so Australia cannot sell many of its own manufactured goods in those markets.

New Zealand has attempted to create more profitable and reliable international markets for its agricultural products by lowering transportation costs and by producing luxury products. Milk is sold as whole milk powder, whey, and casein because these forms are cheap to ship. Fine dairy products such as butter and cheese and exotic gourmet foods such as venison and the bright green, lemon-banana-tasting kiwi fruit have found markets in Europe, Asia, and North and South America. New Zealand is now the world's leading exporter of dairy products, and more than 90 percent of its production is sold overseas.

The Pacific islands also depend on raw natural resources for their exports. But there is a major difference between their economies and those of Australia and New Zealand: on many of the islands, subsistence lifestyles are still common. On the island of Fiji, for example, part-time subsistence agriculture engages more than 60 percent of the population. But a desire for modern amenities has made it imperative to find methods of generating cash. In Fiji, sugar and other cash crops—copra, cassava, ginger, and rice, along with timber—make up about 22 percent of the economy; but the importance of these commodities is decreasing, partly because wages are low. Gold mining, however, is still a significant sector of the Fijian economy, and tourism has assumed additional importance. Today, tourism is the primary source of income in many areas of the Pacific, and fisheries and seafood products are second.

Security and Vulnerability in the Pacific

In the tiny islands of the Pacific, economic growth is hindered by small resource bases, small populations, and remote locations. Although many islands have economies based on tourism; mineral

AT THE REGIONAL SCALE   *The Role of Taro in Pacific Subsistence*

On Pacific islands, tubers form the basis of subsistence, and on many islands no food plant is more important than taro (*Colocasia esculenta*). Each taro plant produces a starchy corm or root that provides important nutrition, especially carbohydrates.

Taro is best cultivated within a wet "pond-field" environment, which is provided either by a naturally swampy site or through irrigation technology. On many islands across the Pacific, Austronesian settlers found it necessary to apply irrigation to produce adequate yields. Villagers would build dams to divert stream water into aqueducts, which would carry water to terraces constructed with logs, stone-faced walls, or earthen embankments. Unlike terraces in other parts of the world, which may be built to retard soil erosion or simply to provide a level planting area, the irrigated taro terraces of the Pacific are designed to control the flow of water. Water must never be allowed to stagnate but must always flow slowly through the gardens, typically at a depth of 4 to 6 inches (10 to 15 centimeters). This continuous run of water regulates the temperature of the garden, cooling the taro stalks and helping to prevent corm rot. The flowing water also delivers nutrients that nourish the rich saturated soil layer.

The terraces, which are a communal resource, are planted on a sophisticated rotation schedule that allows portions of the system to rest and recover. Villagers repeatedly stress that both the method and the design used in building and maintaining these terraces are *neimami qase mai liu mai liu,* or "passed down to us from the ancestors." They invest significant time and labor in the careful management and constant monitoring required by irrigated terraces to maintain long-term stability. The gardens themselves are beautiful to behold and are a source of pride among the villagers. Irrigated taro terrace technology is valuable knowledge that may help small island nations achieve some measure of self-sufficiency in the face of rising populations and a shrinking resource base.

Source: Robert Kuhlken, "Intensive agricultural landscapes of Oceania," *Journal of Cultural Geography* 19(2) (Spring/Summer, 2002).

resources; and sugar, coconut, and other cash crops, few have any significant manufacturing industries. Hence they must import manufactured goods from Asia. What few Pacific manufacturers there are have to mount aggressive "buy Pacific" appeals to maintain markets for their wares, which almost always carry higher prices than imports.

Many Pacific islands are cushioned to some extent from the stresses of reorientation to Asia and to the global economy by customs that contribute to self-sufficiency. Official income figures (see Table 11.1) do not reflect the fact that many households still rely on fishing and subsistence cultivation for much of their food supply, nor do they include income from thriving informal economies and from remittances sent regularly by family members overseas. These resources mean that Pacific Islanders can have a safe and healthy life with relatively little formal income. Some people have called this state of affairs **subsistence affluence,** meaning that diets supplied by home-grown food can be more than adequate and nutritious, and the limited cash can be saved for travel and occasional purchases of manufactured goods. If there is poverty, it is related to geographic isolation, which often means lack of access to information and opportunity. But computers and the revolution in global communication networks (satellite media, cell phones, and the Internet), although not yet widely available in the Pacific islands, have the potential to alleviate this isolation for some island groups.

Some places—such as Papua New Guinea, the Solomon Islands, and Tuvalu—find that they are not cushioned by local customs. Once they too had vibrant local cultures, but in comparison to other parts of Oceania they are now poor and undereducated (see Table 11.1). They have high birth rates (a fertility rate of 5.4 children per woman) and low literacy and life expectancy rates. Their economic potential is limited to tourism or extractive activities such as mining, export agriculture, fishing, or forestry. Conditions on many of the smaller Pacific islands typify what has been termed a **MIRAB** economy—one based on migration, remittance, aid, and bureaucracy. Many families depend on the remittances sent by family members living and working overseas, and foreign aid from former or present colonial powers supports bureaucracies that supply employment for the educated and semiskilled.

Tourism in Oceania: The Hawaiian Case

Tourism is a growing part of the economy throughout Oceania, but perhaps nowhere in the region are the issues raised by this industry clearer than in Hawaii. Travel and tourism is the largest industry in Hawaii, and in 2000 it produced 26.07 percent of the gross state product (GSP). By comparison, travel and tourism account for 10.8 percent of GDP worldwide. In 2000, this segment of Hawaii's economy provided 181,050 jobs, employed nearly one-third of Hawaii's workforce, and accounted for U.S. $9.4 billion in output. Tourism accounted for 26.8 percent of Hawaii's tax receipts.

In the 1990s, the Hawaiian economy began to falter, partly because of a decline in tourism. The downturn in the Japanese economy was the chief factor contributing to this decline. Although the total number of visitors to Hawaii increased in the early 1990s, the point of origin of these visitors had been shifting from North America to Asia, particularly to Japan, the Philippines, Indonesia,

AT THE LOCAL SCALE   *Native Hawaiian Rights versus Residential Tourism*

Although tourism is important to Pacific economies, it can threaten the local people's way of life. An example is provided by a recent effort to build a retirement home for mainlanders near Honolulu. In the early 1990s, the national officials of a major U.S. Protestant denomination voted to build a retirement home for their church members on the island of Oahu in Hawaii. The idea was to acquire land and build a multilevel care facility to which members from the mainland could retire, living independently until they needed nursing home care. This type of retirement relocation to sunny climes is often called "residential tourism." The church officials proceeded to look for affordable land close to Honolulu, yet with landscapes of rural tropical beauty. They found a suitable tract in Pauoa Valley, one of the last valleys near Honolulu where rural people of Native Hawaiian origins still live in extended family compounds and grow their traditional gardens.

In earlier years, the Pauoa land might have been acquired and the facility built before the public had time to reflect on the development's impact. But this time, Native Hawaiians and their supporters led a public discussion that linked the retirement home plan to past episodes of colonialism, especially to issues of indigenous land rights and environmental justice. They pointed out that the people of Pauoa Valley live as *ohana* —that is, as members of a traditional Hawaiian community. Although many of them work in Honolulu, they still grow much of their own food and hunt and gather in the adjacent forest. They are surrounded with the stone ruins of sacred sites and burial grounds from the pre-European days; and they believe that trees and other natural features are inhabited by spirits and that those who lived long ago still visit now and then. The residents have a general feeling of living a special way in a special place. Were the Native Hawaiians the only inhabitants, their solidarity and Hawaiian laws would protect the land against sale; but new immigrants from North America and Asia now share the valley, and some were eager to sell. Yet, if the retirement home was built, life would change irreversibly even for those residents who chose not to sell. The ecology and ambience of the val-

ley would be transformed by the introduction of a large complex with 141 apartments and a 106-bed nursing home, large concrete parking lots, and manicured grounds. Once the church officials understood the issues raised by the Native Hawaiians, they decided to build the retirement home in the mainland United States rather than in Pauoa Valley.

Archaeologists Tom Dye and Mac Goodwin assist Pauoa Valley residents to establish the archaeological features of the valley. The documentation of such sites established long-standing occupation of the valley by Native Hawaiian ancestors and helped residents fend off the acquisition of their land by a mainland-based church. [Conrad "Mac" Goodwin.]

and South Korea. By 1995, 40.3 percent of all visitors to Hawaii came from Asia, mostly from Japan. Thus, the dramatic slump in Asian economies in the late 1990s had a major impact on the economy of Hawaii. The decline in tourism hurt not only the tourist industry (travel companies, airlines, hotels, restaurants, tour companies, and so forth) but also the construction industry, which had been thriving on the building of office towers, condominiums, hotels, and resort and retirement facilities. By December 2000, arrivals of tourists had returned to record levels because of a surge in Japanese visitors (83,000), and arrivals from the U.S mainland were up 5.6 percent to 355,774. But by the summer of 2001, arrivals were down again, because of a continuing recession in Japan and a slowdown in the United States. The terrorist attacks of September 11,

2001, sharply affected Hawaii's economy (see map, page 99). During Fall 2001, hotel occupancy ran at 50 percent of expected rates, and the Hawaiian economy contracted by an estimated 1.5 percent for the year. These fluctuations illustrate the vulnerability of tourism to economic downturns and political events.

Many other islands in Oceania also depend on tourism for a significant part of their income, so global economic ups and downs, especially in Asia, cause declines and surges in their economies, too. Guam, for example, depends on middle- and lower-class Japanese and South Koreans—laborers, clerks, students, midlevel managers, and assembly line workers—for its tourism clientele. In 1995, more than 1.3 million tourists came to Guam, 77 percent from Japan and 14 percent from South Korea. During the recession in

Asia in the late 1990s, several hundred thousand Japanese and South Koreans cut the yearly Guam vacation from their budgets—a loss to Guam of several hundred million dollars per year.

New Asian Orientations

The increasing influence of Asia in Oceania is seen most clearly in the realm of economics. Throughout the Pacific, Asians dominate the tourist trade, both as tourists and as investors, making a vital contribution to Pacific economies and impacting land use significantly. For example, an important segment of the Honolulu tourist infrastructure—hotels, golf courses, specialty shopping centers, import shops, nightclubs—is geared to visitors from Japan,

and many such facilities are actually owned by Japanese investors (although Japan's economic downturn in the 1990s forced some to sell their holdings). In addition, increasing numbers of Asians are taking up residence in the region. In Hawaii, for example, the strongest force favoring reorientation toward Asia in the future may come from the simple fact that Asians now make up about 60 percent of the population, outnumbering both native Pacific Islanders and transplanted North Americans.

The rise in Asian influence is also seen in the increasing exports of raw materials to Asia (Figure 11.8). In the Pacific, coconut, forest resources, and fish products sell mostly to Asian markets; increasingly, Asian companies control these industries. Fishing fleets from Asia are able to purchase licenses, and they regularly

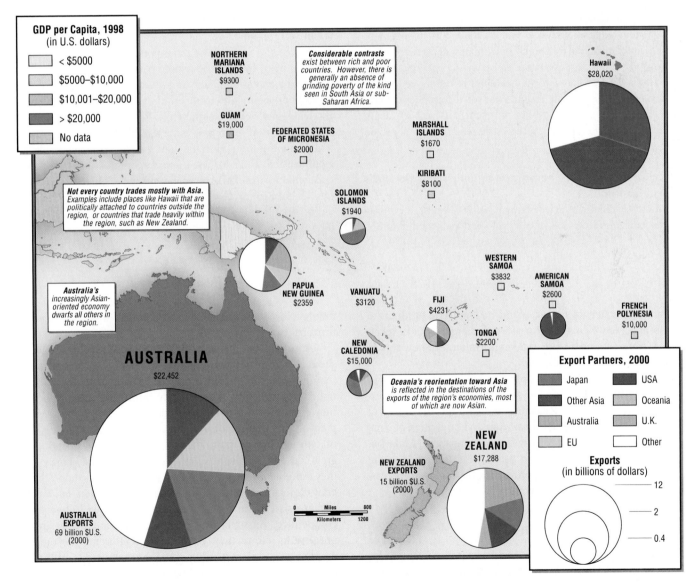

Figure 11.8 Economic issues: Oceania. Note that color of country indicates GDP per capita, and colors of circle wedges indicate export partners. Oceania's exports are diversifying. This is expecially true for Australia, New Zealand, and Hawaii. The white sections of the export circles include trade with Canada, Mexico, the Caribbean, sub-Saharan Africa, and other locales, many of them new trade partners.

GEOGRAPHER IN THE FIELD

Interview with Regional Consultant for Chapter 11, Robert Kuhlken, Central Washington University

What led you to become interested in Oceania?
I first became familiar with the people and landscapes of Oceania during a year-long hiatus from my work as a regional planner in Oregon, when I traveled throughout Australia, New Zealand, and many of the islands of the South Pacific. Whether one is visiting remote Polynesian *motu* by sailboat or tramping the rugged tracks of New Zealand's Southern Alps, these places take a firm hold on the imagination.

Where have you done fieldwork in Oceania, and what kind of work have you done?
I have conducted fieldwork on several islands in Fiji, to document the form and function of traditional agriculture systems required for the irrigation of taro, a staple in the diet of most Pacific island peoples. These sophisticated gardens made use of canals, aqueducts, and terraced fields to provide the saturated soil medium for wet taro production. Although most of the old gardens have been abandoned across the archipelago, several irrigated terrace systems are still being used.

How has that experience helped you to understand Oceania's place in the world?
The study of traditional agrosystems requires learning the much broader culture history of a people: who they believe themselves

to be, what traditions and cultural-ecological adaptations they have developed, and how they have come to interact with the world at large. Basic food production strategies must always be placed within the context of political economy and a society's connectivity and interdependence with other places.

What are some topics that you think are important to study about your region?
Today, because of increasing transportation costs to their remote location, many small island nations need to explore alternatives to imported food commodities, and thus a knowledge of indigenous agricultural techniques may contribute to greater self-sufficiency. Elsewhere across the region, while New Zealand is blessed with adequate natural resources, its remote location is a hindrance to trade opportunities. Australia has learned to adapt to its geographic extent and harsh environment, but must now find ways to cope with urban sprawl and accommodate an increasingly diverse mix of immigrants. And indigenous rights and traditional land tenure have become a signature concern for societies throughout the vast realm of Oceania.

---

ply the offshore waters of Pacific island nations. The products of Australian and New Zealand farms—wool, dairy products, meat, and hides— once found a ready market in Britain but now supply Asian consumers and factories.

Australian coal and iron supply the rapidly growing Asian manufacturing industries. Asia buys 60 percent of Australia's exports and 40 percent of New Zealand's. Increasingly, the islands of the Pacific sell to Japan and Southeast Asia and buy much of what they need from Japan, South Korea, Taiwan, Singapore, Malaysia, Thailand, and Indonesia. Although Asia's rising influence in the region is associated with growing prosperity, the region is increasingly vulnerable to economic downturns in Asia.

The Stresses of Reorientation to Asia

The shift toward greater trade with Asia has posed challenges for Oceania. Throughout the region, local industries that used to enjoy protected trade with Europe have lost that advantage because new European Union regulations prohibit preferential trade agree-

ments. Now these industries face competition from much larger firms in Asia.

Australia and New Zealand are experiencing considerable stress as a result of reorientation to Asia. Their export sectors—mainly wool, cattle, dairy products, coal, and metals—have maintained a competitive edge largely through mechanization; thus, they employ only a small proportion of the population. Most people work in other sectors of the economy, such as services or manufacturing. In the past, the manufacturing and service sectors grew, despite relatively small domestic demand, partly because they benefited from government subsidies, tariffs, and other barriers to foreign investment. Today, as trade barriers fall worldwide and as Asian manufacturing industries continue to become more competitive, Australia and New Zealand are under increasing pressure to reform their economies by reducing the numbers of employees and streamlining operations, while also cutting social services and eliminating subsidies and tariffs.

Reform measures have been especially traumatic for the Australian and New Zealand labor movements, which historically have been among the world's strongest. Australian coal miners' unions successfully agitated for the world's first 35-hour workweek. Other labor unions won a minimum wage, pensions, and aid to families with children long before such programs were enacted in many other industrialized countries. For decades, these arrangements were highly successful: both Australia and New Zealand enjoyed living standards comparable to those in North America but with a more egalitarian distribution of income. Since the 1970s, however, competition from Asia has meant that increasing numbers of workers have lost jobs and many hard-won benefits. In Australia, unemployment has risen from 2 percent in the 1960s to between 8.5 and 9.5 percent in the 1990s; it hovered around 7 percent by 2001. Meanwhile, previously high rates of social spending have been cut across the board to address mounting government deficits. The loss of social support, especially for those who have lost jobs in the more competitive economy, has contributed to rising income disparity in recent years. Nonetheless, Australia had about half the poverty rate of the United States throughout the 1990s.

Social tensions related to the reorientation toward Asia are especially pronounced in New Zealand. Because of its small size and limited resource base, New Zealand has always found it harder to compete in world markets than has Australia. For decades, it managed well by selling its major exports of wool, meat, and dairy products to the United Kingdom on preferential trading terms. When Britain joined the European Common Market in 1973, however, the other members insisted that the United Kingdom curtail its protected agricultural imports from New Zealand, which France felt competed unfairly with its own agricultural products. The loss of the U.K. market sent New Zealand's economy into decline. By the mid-1980s, the government was forced to cut agricultural subsidies, funding for the welfare state, and the government payroll. At the same time, New Zealand's whites-only immigration policies were dropped and the doors were opened to Asian immigrants, especially to wealthy professionals—who, it was hoped, would reinvigorate the economy with their savings and investments. The new immigration policy has resulted in considerable social tension; some long-time residents of New Zealand resent the presence of wealthy newcomers at a time when they themselves are jobless or strapped for cash. According to the United Nations, income disparity in New Zealand during the 1990s was roughly twice that in Australia.

The Future: A Mixed Asian and European Orientation?

Despite the powerful forces pushing Oceania toward Asia, important factors still favor a strong Western influence. The economic recession that swept through Asia in the late 1990s emphasized the need for Oceania to maintain broad contacts with economies outside Asia, especially in Europe and the United States, because they seem to be more stable. Another factor is the lingering fear of Chinese aggression, justified to some extent by expansionist moves that China has made in the South China Sea toward Taiwan and the potentially oil-rich Spratly Islands. In all likelihood, Australia

and New Zealand will retain a cultural affinity with the West for generations to come. In the Pacific, several islands, especially those in the northwestern part of the region (Micronesia), may be strongly drawn into the Asian sphere. But lingering vestiges of colonialism will prevent others from straying very far from a Western orientation. Both the United States and France maintain possessions in Polynesia and Micronesia, and any desire for independence in these possessions has not been sufficient to override the financial benefits of aid, subsidies, and investment money from the former colonial powers. Hence, as Oceania is being steadily drawn closer to Asia, strong ties remain with the United States and Europe.

ENVIRONMENTAL ISSUES

A case could be made that there is more public awareness of environmental issues in Oceania than in any other world region. Despite its relatively small population, Oceania faces many environmental problems (Figure 11.9), and they are often discussed in such public forums as the press, exhibits, and the broadcast media. The human introduction of many nonnative species has ravaged the region's unique ecosystems, as has the expansion of the human population itself. Even remote areas have been subjected to intense pollution from mining and the testing of nuclear weapons, especially during the cold war era. Finally, global environmental crises related to climate change and ozone depletion are particularly threatening to this region.

Despite the increasingly severe environmental problems facing the Pacific islands, there are a few bright spots on the horizon. In many areas, a revived pride in cultural heritage has encouraged Pacific Islanders and other indigenous peoples to reestablish more sustainable lifestyles and to become active in Green movements, especially as cultural allegiances shift away from distant Europe. Because environmental conditions vary around the region, we will address Oceania's environmental issues regionally.

Australia: Human Settlement in an Arid Land

Unique organisms found nowhere else occupy many ecological niches in Australia's unusually arid environments (see pages 575–576). Since settlement by Europeans, nonnative species have displaced or even driven to extinction many of Australia's native plants and animals. At least 41 bird and mammal species and more than 100 plant species have become extinct. In all likelihood, many more disappeared before they were biologically classified.

European rabbits are among the most destructive of the introduced species, and their story illustrates the complex problems these species can cause. Rabbits were brought to Australia by early British settlers who enjoyed hunting and eating them. With no natural predators in Australia, the rabbits multiplied quickly, browsing on many local plants and destroying the food supply for indigenous species. The holes they dug to live in created considerable soil erosion. Attempts to control the rabbit population by introducing European foxes and cats only increased the threat to native species. The vigorous European predators drove many native

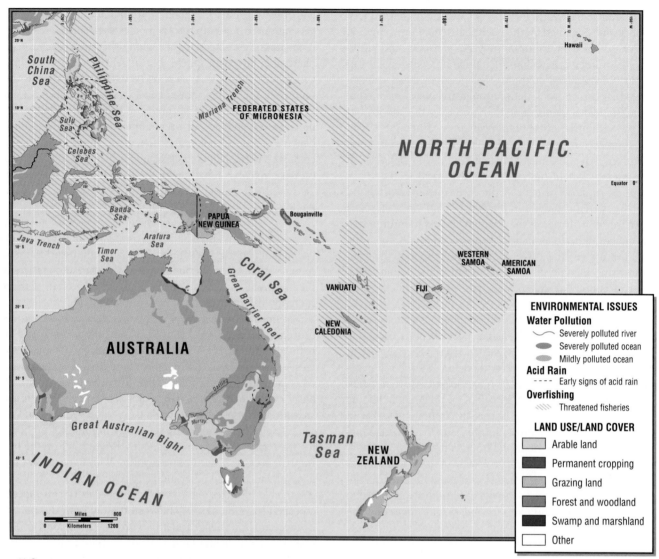

Figure 11.9 Environmental issues: Oceania. In Oceania, environmental issues are often the center of public debate. Deforestation and loss of species diversity are common issues. Overfishing is widely recognized as a growing problem, often because of abuse by fishing fleets that come from outside the region. Air pollution is locally a problem in Australia and New Zealand, but overall less problematic than in neighboring Southeast Asia.

predator species to extinction without having much impact on the rabbit population. In recent years, intentionally introduced diseases have destroyed as much as 90 percent of the target rabbit population. Yet, without rabbits to eat, the cats and foxes now prey more heavily on indigenous species. Hence, the eradication of feral cats and foxes has been stepped up as well. Moreover, some rabbits have proved resistant to disease, and the rabbit population may rebound.

The cane toad (*Bufo marinus*) is another example of an introduced species that endangers native Australian life. This large amphibian was released along the north Queensland coast in 1935 in an attempt to control beetle pests in the sugarcane fields. It

feeds on ground-dwelling small animals (including some endangered species) and insects—but not on the cane beetle it was intended to eradicate. The cane toad has no natural enemies, lives up to 20 years, and lays up to 40 times as many eggs as native toads. The cane toad carries highly toxic glands on its shoulders that kill most animals that come in contact with it.

Agriculture has also had a huge impact on Australian ecosystems. Because the climate is arid and soils are infertile, the dominant land use in Australia is the grazing of introduced species—primarily sheep, but also cattle. More than 15 percent of the land is given over to grazing; increasingly, ranchers are using irrigation, herbicides, and pesticides to extend the seasonal availability of pas-

The dingo fence. The world's longest fence snakes for 3307 miles (5322 kilometers) between South Australia and New South Wales. Made of tough wire attached to wood posts, it separates the indigenous dingo dog from sheep herds. A single dingo can kill as many as 50 sheep in one night of marauding. The fence requires constant patrolling and mending, but it has reduced the dingoes' killing of sheep dramatically. Kangaroos, the dingoes' natural prey, have learned to live on the sheep side of the fence, where their population has boomed beyond sustainable levels. Kangaroos compete with sheep for the same scarce grass and water. [Medford Taylor/National Geographic Image Collection.]

tureland. Crop agriculture is limited to just 6 percent of the land, most of it in the southeast. Crops, too, now rely on irrigation and chemicals. One result is that natural underground water reserves are being depleted. Another is that many rivers and lakes have become polluted by agricultural chemicals and the algae that feed on them.

In a place where the overwhelming majority of the people live in cities, urban activities are the most likely to create environmen-

tal problems. Australia's major cities form a series of nodes along the eastern and southeastern coasts from Brisbane to Adelaide. The nodes themselves are of fairly low density, and they are connected by dispersed towns and coastal developments. As a result, commuters typically travel long distances daily and depend primarily on private cars for transport (although urban streetcars and light rail systems and interurban commuter lines are more common than they are in the United States). The dependence on cars has a variety of environmental impacts. Streets and parking lots not only consume space but also seal the soil surface. Paving thus reduces the amount of rainfall that the ground can absorb, decreasing the retention of scarce rainfall and increasing flooding and erosion. Because of the great distances traveled, more automobile fuel is used in Australia than elsewhere, and per capita energy consumption is higher in Australia than in most western European countries. Consequently, per capita greenhouse emissions are higher than in any other industrialized countries except the United States, Luxembourg, and Singapore; and photochemical smog and chemical haze are common.

Water scarcity is another problem associated with large urban populations. Domestic per capita water consumption has risen markedly over the last 25 years, due mostly to lifestyle changes: more flushing toilets, more washing machines, more showers, more irrigation of ornamental plants and crops. Australians have had to build giant water catchment and storage facilities, which have displaced natural habitats and diverted the natural flow of streams.

New Zealand: Loss of Forest and Wildlife

New Zealand shares many of Australia's environmental problems: loss of endemic plants and animals, pollution, and water shortages. There are some differences, however, which result from New Zealand's different set of physical and human features. There were no human beings in New Zealand until about 1000 years ago, when Polynesian Maori people settled there. When they arrived, dense midlatitude rain forest covered 85 percent of the land. The Maori were cultivators who brought in yams, taro, and sweet potatoes, as well as other nonindigenous plants and birds. By the time of European contact, forest clearing and overhunting by the Maori had already degraded the environment and driven several species of the moa bird to extinction. There is also archaeological evidence that intense competition for increasingly scarce resources had led to warfare.

European settlement in New Zealand drastically intensified land use well beyond Maori levels (Figure 11.10). The settlers made concerted efforts to recreate the British landscape and farming system in this new land. Today, just 23 percent of the country remains forested. The Europeans used most of the land they cleared for export agriculture: pastures and cropland. Grazing has become so widespread that today there are 15 times as many sheep as people, and 3 times as many cattle. The Europeans also brought many new crops and inadvertently introduced damaging pests. Soils exposed by the clearing of forests proved thin and acidic, and ranchers buttressed them with chemicals to maintain pastures. As in Australia,

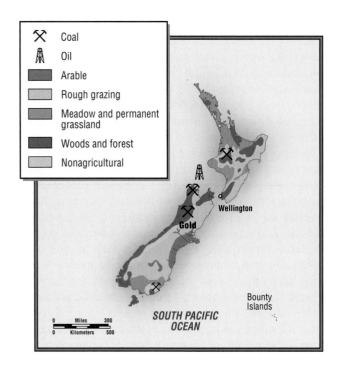

Figure 11.10 **Land use and natural resources of New Zealand.** [Adapted from Richard Nile and Christian Clerk, *Cultural Atlas of Australia, New Zealand, and the South Pacific* (New York: Facts on File, 1996), p. 194.]

chemicals have polluted waterways. New Zealand's environment has actually become hostile to many native species. Ranches, farms, roads, cities, and towns claim more than 90 percent of the lowland area, resulting in unusually high extinction rates for endemic plants and animals. Hydroelectric schemes have disrupted the natural flow of rivers, particularly on South Island.

The Pacific Islands: At Risk from Global Trends

The Pacific islands, though widely dispersed, share a number of environmental concerns and opportunities. Their small sizes, often tiny populations, and insignificant influence on the global economy give them little say about international forces that nonetheless affect their environments. Mining, nuclear pollution, ozone depletion, global warming, economic globalization, and increasing tourism all have environmental effects on the Pacific islands.

Resource Extraction and Pollution. Pollution from mining has disrupted traditional lifeways in Papua New Guinea. Two mines in particular have had devastating effects on the country's environment. Both were operated by Australian mining companies that took advantage of poorly enforced or nonexistent environmental codes to dump large amounts of mine waste in New Guinea's rivers. As a consequence, entire river systems have been devastated and indigenous subsistence cultivators displaced.

In the case of the Ok Tedi mine on mainland Papua New Guinea, 30,000 local villagers sued the Australian parent mining company, BHP, for U.S. $4 billion. Two villagers, Rex Dagi and

Alex Maun, traveled to Europe and America to explain their cause and meet with international environmental groups. They and their supporters convinced American and German partners in the Ok Tedi mine company to divest their shares. The international pressure also resulted in objections to BHP's plans to mine diamonds in Canada's Northwest Territories. The parties reached an out-of-court settlement that sets important precedents: (1) the plaintiffs won U.S. $125 million in trust funds for damage mitigation; (2) the villagers won a 10 percent interest in the mine; and (3) the accord provides that any further disputes will be heard in Australian rather than New Guinea courts, making it less likely that the mining company can manipulate the justice system.

The second case involved a copper mine on the island of Bougainville (administered as part of Papua New Guinea, but lying in the Solomon Islands). Soon after an Australian-owned company began mining in Bougainville in the 1960s, on land deforested and stripped for the purpose, the erosion of mine waste into rivers poisoned downstream cultivation plots, forestlands, and fisheries. People who lost the resources on which they had subsisted were forced into the cash economy, many of them fleeing to new mining market towns where their subsistence skills are of little use (see the photograph on page 597). By the 1990s, protests against the environmental and social damage done by the copper mine developed into a war for the independence of Bougainville Island from Papua New Guinea, in which 1 in 8 Bouganivilleans (some 20,000 people) lost their lives. Australia and New Zealand aided the Papua New Guinea government by supplying helicopters and munitions to use against the rebels, until public pressure in Australia put a stop to this support. Then the Papua New Guinea government hired Executive Systems, a mercenary group (soldiers for hire) of white South Africans to suppress the secessionist movement. This inflammatory action convinced the Australian government to support the closing of the mine and supply peacekeeping forces. By January 2001, peace negotiations were under way, with Bougainville seeking to become an autonomous unit within Papua New Guinea with its own judiciary and police force.

Another major environmental issue for the Pacific islands is radioactive pollution from nuclear weapons tests and reactor waste. During World War II, U.S. nuclear bomb experiments destroyed Bikini atoll in the Marshall Islands and caused cancer among the islanders. Similarly, Mururoa atoll in French Polynesia has been the site of 180 nuclear weapons tests and the recipient of numerous shipments of nuclear waste from France. The result has been widespread cancer, infertility, birth defects, and miscarriages among the native populations of nearby islands. One response to this pollution was the establishment of the South Pacific Nuclear Free Zone by the Treaty of Rarotonga in 1985. Most independent countries in Oceania have signed this agreement, which bans weapons testing and waste dumping on their lands. As a result of political pressure from France and the United States, French Polynesia and U.S. territories such as the Marshall Islands have not signed the agreement. Waste dumping and weapons testing continue. Japan, North Korea, South Korea, France, and the United States have all explored the possibility of depositing their radioactive waste in the Marshall Islands. Since the 1990s, Pacific leaders have coordinated their efforts against nuclear testing and waste dumping

Supermarkets, such as this one in Tari in the Papua New Guinea highlands, are rearranging the ways people have traditionally obtained food and other goods. Among the customers are forest dwellers displaced by mining operations. The stores are doing away with long-established personal trading networks and are rapidly bringing the impersonal global economy into one of the last frontiers on the planet. Pacific people are happy to have access to the goods, though many worry about the changes that result. [Sassoon/Robert Harding Picture Library, London.]

through the United Nations Working Group on Indigenous Populations (UNWGIP).

The United Nations Convention on the Law of the Sea. This 1994 UN convention (called the Law of the Sea Treaty, for short) establishes rules governing all uses of the world's oceans and seas (157 countries have signed this treaty; the United States is not one of them). There is no overarching enforcement agency, however. The treaty, based on the sound idea that all problems of ocean space are interrelated and need to be addressed as a whole, is an example of how the globalization of island economies has thwarted environmental protection for the islands. The treaty allows islands to claim rights to sea resources 200 miles (320 kilometers) out from the shore, and island countries can now make money by licensing privately owned foreign fishing fleets from Japan, South Korea, Russia, and the United States to fish within the offshore limits. However, protecting the fisheries from overfishing by these rich and powerful licensees has turned out to be an enforcement nightmare for tiny island governments with few resources. Similarly, it has proved difficult to monitor and control the exploitation by foreign companies of seafloor mineral deposits. Even mining operations conducted legally outside the 200-mile limit have the potential to pollute fisheries and other ocean resources.

Multiple Impacts of Tourism. Even tourism, which until recently was considered a "clean" industry, has now been revealed to create environmental problems. Foreign-owned tourism enterprises have often accelerated the loss of wetlands and worsened beach erosion. Tourism increases the use of scarce water resources; the production of sewage; and the consumption of environmentally polluting products, such as gasoline, kerosene, fertilizers, plastics, and paper. As discussed in other chapters, the widely promoted concepts of ecotourism and cultural tourism are now common elements of development in the Pacific, but they have quickly become commercialized by mass tourism. All tourist ventures, no matter how ecologically aware, have some impact on the local environment, and the invasive presence of visitors invariably makes demands on local cultures.

Global Warming and Ozone Depletion. Global warming, which may melt the polar ice caps and raise sea levels (see page 45), is of obvious concern to islands that already rise barely above the waves (Figure 11.11). Many of the lowest lying atolls will simply disappear under water if seas rise the 4 inches (10 centimeters) per decade predicted by the International Panel on Climate Change. Other islands, some already very crowded, will be severely reduced in area and will become more vulnerable to storm surges and cyclones. (The highest point on the densely occupied island of Nauru is 200 feet, or 61 meters.) The peoples of the Pacific have little direct control over the forces that appear to be producing these precarious circumstances, because they themselves produce only a very small percentage of the earth's greenhouse gases.

People in the whole of Oceania are particularly at risk from depletion of the ozone layer. Ozone is a type of oxygen molecule that is normally heavily concentrated in a layer of the upper atmosphere. This ozone layer filters out biologically harmful ultraviolet radiation emitted by the sun. Since the mid-1980s, the ozone layer has been thinning, and a hole in the layer has appeared periodically over Antarctica. The cause is thought to be a buildup of a class of manufactured chemicals called chlorofluorocarbons (CFCs), which float up to the ozone layer and destroy it. Increasing amounts of ultraviolet radiation reaching the surface of the earth through the damaged ozone layer are likely to increase the prevalence of skin cancer. Australia already has the highest incidence of skin cancer in the world, which is explained by the fact that its largely white, transplanted population has little protective skin pigment (melanin) yet lives under intense sunlight. Higher levels of ultraviolet radiation are also likely to increase the incidence of eye disease and weaken the immune systems of a variety of organisms, including humans. Increased ultraviolet radiation could also decrease agricultural and marine productivity, and it is one of Australia's most significant natural hazards.

MEASURES OF HUMAN WELL-BEING

The overall status of human well-being in Oceania is probably higher than the statistics indicate, especially for the Pacific islands, because subsistence agriculture and reciprocal exchange remain important in the everyday economy. Statistics for Australia, New Zealand, and several of the larger islands of the Pacific are easy to obtain; but data for many of the smaller island groups is missing or only partial data is available. Keep in mind that the sizes and populations of many of the island states are very small.

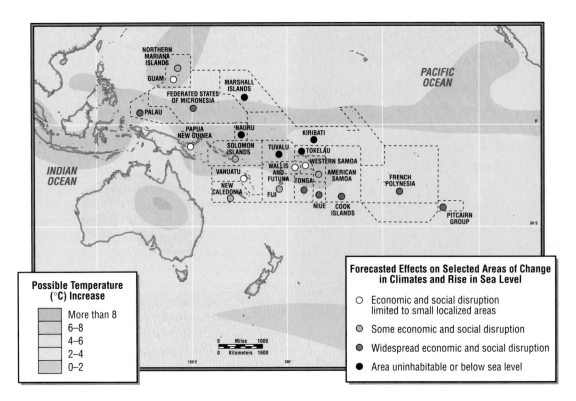

Figure 11.11 **Global warming as it may affect Oceania.** Scientists studying global warming have forecast its effects on land temperatures and sea level rise. Should these forecasts materialize, how will living conditions in various parts of Oceania change? [Adapted from Richard Nile and Christian Clerk, *Cultural Atlas of Australia, New Zealand, and the South Pacific* (New York: Facts on File, 1996), p. 223.]

As discussed throughout the text, gross domestic product per capita (Table 11.1, column 2) is at best a crude indicator of well-being. Oceania, like most other regions, has a wide variation in GDP per capita. Australia and New Zealand provide their citizens with a high standard of living that is further enhanced by publicly financed social programs such as health care and subsidized housing. The high standard of living in these societies is demonstrated by their long life expectancy and high literacy rates. Further evidence of their high living standards is the fact that 67 percent of the dwellings in Australia are owned by their occupants. Generally, the Pacific islands have relatively low GDP figures. Still, even the lowest figures of U.S. $800–$1000 probably support an adequate standard of living for most people. The GDP statistics do not reflect income from the informal economy, remittances from family members overseas, or the fact that many households practice subsistence fishing and agriculture.

The United Nations Human Development Index (HDI) (Table 11.1, column 3) is not particularly useful for Oceania. Notably, Australia and New Zealand rank high, fourth and twentieth, respectively. Hawaii is not ranked because it is part of the United States. If it were ranked, Hawaii would have a higher score than the United States as a whole (which is ranked third behind Canada and Norway), because it ranks among the very top of the U.S. states in life expectancy, educational attainment, and income. Fiji, in Melanesia, has a respectable ranking of 66; but all other places ranked by the United Nations are low on the HDI index. Their low rankings are partly the result of the difficulty of providing services such as medical care and advanced training to very small, remote populations. Changing diets (more sugar and fat) and lifestyles (more sedentary) have provoked an increase in diseases such as diabetes and heart disease. A result is significantly lower life expectancies. Specialized, high-tech health care facilities are generally not to be found outside New Zealand, Australia, and Hawaii.

The United Nations Gender Empowerment Measure (GEM) figures (Table 11.1, column 4) are also missing for most of the Pacific islands, but, as we have observed, women participate actively in island life and enjoy increasing prestige as they age. New Zealand's high rank on gender empowerment (9) results partly from the high percentage of women in parliament (29 percent of the seats) and from the fact that more than one-third of administrators and managers, and more than half the professional and technical workers, are women. In 1997, New Zealand acquired its first woman prime minister when the transportation minister, Jenny Shipley, became head of the ruling coalition. In 1999, Helen Clark was elected prime minister after 25 years of leadership in parliament as a member of the Labour (liberal) party.

Literacy rates are high throughout Oceania, with the exception of Papua New Guinea and the Solomon Islands in Melanesia. This statistic is particularly meaningful because it indicates that most of the citizenry is equipped to participate in democratic processes and in public debate on regional issues.

TABLE 11.1 *Human well-being rankings of countries in Oceania*

Country (1)	GDP per capita, adjusted for PPP[a] in 1998 $U.S. (2)	Human Development Index (HDI) global rankings, 2000[b] (3)	Gender Empowerment Measure (GEM) global rankings, 2000 (4)	Female literacy (percentage), 1998 (5)	Male literacy (percentage), 1998 (6)	Life expectancy (years), 2000 (7)
Selected countries for comparison						
Japan	23,257	9 (high)	41	99	99	80
United States	29,605	3 (high)	13	99	99	76.8
Kuwait	25,314[c]	36 (high)	N/A (75 in 1998)	78	83	76
Australia	22,452	4 (high)	11	99.0	99.0	78
Melanesia (selected units)						
Fiji	4,231	66 (high)	61	90	94	73
Papua New Guinea	2,359	133 (medium)	N/A (91 in 1998)	55	71	58.3
Solomon Islands	1,940	121 (medium)	N/A	62	62	72
Vanuatu	3,120	118 (medium)	N/A	N/A	N/A	68
New Caledonia (French Overseas Territory)	15,000	N/A[d]	N/A	90	90	73
Micronesia (selected units)						
Guam (U.S. dependency)	19,000	N/A	N/A	99	99	78
Federated States of Micronesia	2,000	N/A	N/A	88	91	69
Kiribati	8,100	N/A	N/A	N/A	N/A	60
Marshall Islands	1,670	N/A	N/A	93	93	66
Northern Marianas	9,300	N/A	N/A	96	97	76
Nauru	10,000	N/A	N/A	N/A	N/A	61
Palau	8,800	N/A	N/A	90	93	69
Polynesia (selected units)						
French Polynesia (French Overseas Territory)	10,000	Shares French rank of 12 (high)	N/A Shared French rank of 31 in 1998	98	98	75
Hawaii (U.S.)	28,020	Shares U.S. rank of 3 (high)	Shares U.S. rank of 13	99	99	77
Tonga	2,220	N/A	N/A	99	99	66
Tuvalu	800	N/A	N/A	N/A	N/A	66
Western Samoa	3,832	95 (medium)	N/A	97	97	72
American Samoa	2,600	N/A	N/A	97	98	75
New Zealand	17,288	20 (high)	9	99.0	99.0	77

[a]PPP = purchasing power parity.
[b]The high and medium designations indicate where the country ranks among the 174 countries classified into three categories (high, medium, low) by the United Nations.
[c]1999 data.

[d]N/A = data not available.

Sources: United Nations Human Development Report 2000 (New York: United Nations Development Programme); and 2000 *World Population Data Sheet* (Washington, D.C.: Population Reference Bureau).

REFLECTIONS ON OCEANIA

One way to assess Oceania and its future prospects is to think of it in relation to its nearest neighbors. In other chapters, we have noted the economic vigor of East and Southeast Asian economies, the energy and potential they seem to have for future expansion, and the possibility that they will either subvert or pull along other economies around the globe. We have also considered the many challenges these regions face, not the least of which are cronyism, corruption, inattention to environmental issues, and lack of commitment to enhancing democratic processes. Will Oceania's economic relationships with Asia leave it wealthier and more secure—or more precarious economically, environmentally, and socially?

Some people have suggested that Oceania might implement the Pacific Way effectively and, perhaps, teach it to others. These observers are encouraged by Oceania's somewhat superior environmental record, its leadership in the movement to ban nuclear weapons, and the strength of Pacific indigenous cultures. Although the concept of the Pacific Way is only just emerging; it seems to hold considerable promise in the post–cold war era, when many people are in a mood to demilitarize, to find less wasteful lifestyles, and to discover something grander than national chauvinism. The Pacific Way is based on ancient techniques for conflict resolution through cooperation and consensus.

It is, in fact, more of an ideal and a legend than a reality, because traditional Pacific cultures were not particularly democratic. Many of the region's multi-island nations are not accustomed to collective decision-making for the whole, and the disruptive forces of political factionalism and ethnic conflict have yet to be overcome. But it appears that most parts of Oceania are at a stage where cooperation and consensus may seem desirable. Most Pacific islands are redefining their relationships with former colonial governments and are choosing independence of some sort; at the same time, they recognize that they can hardly go it alone as tiny states. Meanwhile, Australia and New Zealand are newly aware of their "Pacificness." As New Zealand political scientist Steven Levine has noted: "If in time a genuinely consensual 'Pacific Way' does emerge, it will no doubt be based on the extension of the values of empathy and social harmony found in some island cultures to the ethnically distinct yet native-born 'newcomers' in their midst." Why not, then, become a region with a splendid mission—a place where value is placed on sustainable development, where affluence is defined in nonmaterial ways that focus on the quality of essentials rather than the quantity of nonessentials, where the primary goal is to enhance human and social capital, and where collaboration and common consent are honored?

Thinking Critically About Selected Concepts

1.   Oceania consists of Australia, New Zealand, and a huge number of high and low islands spread over a large portion of the Pacific Ocean. *Discuss how the various components of Oceania differ from one another in the types of physical environments they provide as a context for human life.*

2.   The vast Pacific Ocean plays a major role in the climate of Oceania. Since prehistoric times, it has been the route for human communication and migration. *Compare the role of the Pacific as a physical barrier that can impose solitude and foster self-sufficiency with its role as an avenue for communication and diffusion.*

3.   Oceania's composition as a region of islands has had a special effect on its animal and plant life. *Explain the physical features that account for the fact that many species are endemic, that is, they exist in a particular place and nowhere else on earth.*

4.   Over the course of the twentieth century, Oceania's relationship with other parts of the world shifted. *Discuss why the connections to Europe are loosening and why other links are being forged.*

5.   The Pacific Way conveys the idea that Pacific Islanders and their governments have a regional identity that grows out of their own particular social experience. Pacific peoples have overcome European judgments that the Pacific is peripheral and backward, and they now take pride in their ability to control their own development and solve their own problems. *Do you think this philosophy could inspire similar movements elsewhere around the world? What components do you see as most crucial to the success of the Pacific Way?*

6.   The overall state of human well-being in Oceania is probably higher than the statistics indicate, especially for the Pacific islands, because of subsistence affluence. *To what extent do you see subsistence affluence as having a link to the idea of the Pacific Way?*

7.   Australia and New Zealand remain producers and exporters of raw materials, both agricultural and mineral, rather than manufacturers of finished products. Yet they have also managed to achieve high levels of wealth and well-being. It has long been thought that a country's economy must move away from resource exports and toward industrial manufacturing if the country is to become affluent. *How have Australia and New Zealand been able to achieve high levels of wealth and well-being while still remaining primarily producers of raw materials? How are these patterns now shifting?*

8.   Since settlement by Europeans, many of Australia's native plants and animals have been displaced or even driven to extinction by nonnative species. *To what extent can the same can be said of New Zealand and the Pacific islands?*

9.   The Pacific islands are affected by a number of environmental problems that originate in the world's industrialized economies: pollution from mining and from nuclear weapons testing and nuclear waste; ozone depletion and global warming; the effects of overusing resources, which are linked to the global reach of rich economies; and increasing tourism. *How can the Pacific islands gain some control over the forces that affect their environments?*

Key Terms

Aborigines (p. 577) the longest surviving inhabitants of Oceania, whose ancestors, the Australoids, migrated from Southeast Asia possibly as early as 50,000 years ago, over the Sundaland landmass that was exposed during the ice ages

atoll (p. 572) a low-lying island, formed of coral reefs that have built up on the circular or oval rims of submerged volcanoes

Austronesians (p. 577) a group that migrated out of Southeast Asia and continued the settlement of the Pacific about 5000 to 6000 years ago

El Niño (p. 575) a pattern of periodic climate-altering changes, especially in the circulation of the Pacific Ocean, that are now understood to operate on a global scale

endemic (p. 575) belonging or restricted to a particular place

Gondwana (p. 571) the great landmass that formed the southern part of the ancient supercontinent Pangaea

Great Barrier Reef (p. 571) the longest coral reef in the world, located off the northeastern coast of Australia

hierarchical societies (p. 579) social structures in which groups are ranked according to power, prestige, and wealth, with layers of ruling elites at the top and undifferentiated commoners at the bottom

high islands (p. 572) volcanoes that rise above the sea into mountainous, rocky formations that contain a rich variety of environments

makatea (p. 572) raised coral platforms formed when atolls are uplifted

marsupials (p. 575) mammals that give birth to their young at a very immature stage and then nurture them in a pouch equipped with nipples

Melanesia (p. 578) New Guinea and the islands south of the equator and west of Tonga (the Solomon Islands, New Caledonia, Fiji, Vanuatu)

Melanesians (p. 577) a group of Australoids named for their relatively dark skin tones, a result of high levels of the protective pigment melanin; they settled throughout New Guinea and other nearby islands

Micronesia (pp. 577–578) the small islands lying east of the Philippines and north of the equator

MIRAB (p. 589) an economy based on migration, remittance, aid, and bureaucracy

monotremes (p. 575) egg-laying mammals such as the duck-billed platypus and the spiny anteater

noble savage (p. 579) a term coined by European Romanticists to describe what they termed the "primitive" peoples of the Pacific, who lived in distant places supposedly untouched by corrupting influences

Pacific Way (p. 585) the idea that Pacific Islanders have a regional identity and a way of handling conflicts peacefully, which grows out of their own particular social experience

pidgin (p. 585) a language used for trading made up of words borrowed from the several languages of people involved in trading relationships

Polynesia (p. 578) the numerous islands situated inside an irregular triangle formed by New Zealand, Hawaii, and Easter Island

roaring forties (p. 573) powerful westerly air and ocean currents that speed around the far Southern Hemisphere virtually unimpeded by landmasses

subsistence affluence (p. 589) the ability of Pacific Islanders who rely primarily on a home-grown diet, informal local economies, and remittances from overseas to maintain a safe and healthy lifestyle on relatively little formal income

underclass (p. 580) the lowest social stratum, composed of the disadvantaged

Pronunciation Guide

Aborigine (ab-uh-RIHJ-uh-nee)
Adelaide (AD-uh-layd)
Aoraki (ah-oh-RAH-kee)
Aotearoa (ah-oh-tay-ah-RŌH-ah)
Auckland (AWK-luhnd)
Australia (aw-STRAYL-yuh)
Austronesian (aw-stroh-NEE-zhuhn)
Ayers Rock (AIRS)
billabong (BIHL-uh-bawng)
Bougainville (boo-gahn-VEEL)
Brisbane (BRIHZ-buhn)
Buka (BOO-kah)
Canberra (KAN-burr-uh)
Carpentaria, Gulf of (kahr-puhn-TAIR-ee-uh)
coolibah (KOO-luh-bah)
Fiji (FEE-jee)
Gauguin, Paul [goh-GA(N)]

Haka (HAH-kah)
jumbuck (JUHM-buhk)
Kiribati (keer-ih-BAH-tee)
Maori (MAU-ree)
Marianas (mahr-ee-AH-nahss)
Marquesas (mahr-KAY-zahss)
Mauna Loa (MAU-nah LOH-ah)
Melanesia (mehl-uh-NEE-zhuh)
Micronesia (mie-kroh-NEE-zhuh)
Nauru (NAU-roo)
Oahu (oh-AH-hoo)
Oceania (oh-shee-AN-ee-uh)
ohana (oh-HAH-nah)
Olgas (OHL-guhss)
Palau (pah-LAU)
Papeete (pah-pee-AY-tay)
Papua New Guinea (PAH-poo-ah)

Pauoa Valley (PAU-oh-ah)
Polynesia (pawl-ih-NEE-zhuh)
Port Moresby (MOHRZ-bee)
Saipan (sie-PAHN)
Samoa (sah-MOH-ah)
Spratly Islands (SPRAT-lee)
Sundaland (SOON-duh-luhnd ["oo" as in "book"])
Suva (SOO-vah)
Tahiti (tah-HEE-tee)
Tonga (TAWNG-gah)
Trobriand Islands (TROH-bree-uhnd)
Tuamotu Archipelago (too-uh-MOH-too)
Tuvalu (too-VAH-loo)
Uluru (oo-LOO-roo)
Vanuatu (vah-noo-AH-too)
Waitangi (wie-TAHNG-gee)

Selected Readings

A set of Selected Readings for Chapter 11, providing ideas for student research, appears on the *World Regional Geography* Web site at www.whfreeman.com/pulsipher.

The following table of population data reproduces most of the data in the *2000 World Population Data Sheet*, published by the Population Reference Bureau. This table lists data for individual countries and also for world regions. Notice that many of the world regions are equivalent to those in this textbook, but that there are some notable differences. Notice, also, that while gross national product (GNP) figures are given in the *Data Sheet*, gross domestic product (GDP) figures, adjusted for purchasing power parity, are used in the textbook whenever possible. For some countries there is relatively little difference between GNP and GDP, but for others the differences are wide.

.

	Population, mid-2000 (millions)	Births per 1000 pop.	Deaths per 1000 pop.	Rate of natural increase (annual, %)	Doubling time at current rate[a] (years)	Projected population, 2025 (millions)	Infant mortality rate[b] (per 1000 births)	Total fertility rate[c]	Percent of population of age <15	Percent of population of age 65+	Life expectancy at birth (years) Total	Male	Female	Data available code[d]	Percent urban	Percent of adult pop. aged 15-49 with HIV/AIDS, 1997	Contraceptives All methods	Modern methods[e]	Government view of birth rate[f]	GNP per capita, 1998 (U.S.$)	Land area (sq mi)	Population per sq mi
WORLD	**6,067**	**22**	**9**	**1.4**	**51**	**7,810**	**57**	**2.9**	**31**	**7**	**66**	**64**	**68**		**45**	**1.0**	**60**	**52**		**4,890**	**51,789,516**	**117**
AFRICA	**800**	**38**	**14**	**2.4**	**29**	**1,258**	**88**	**5.3**	**43**	**3**	**52**	**51**	**53**		**33**	**5.8**	**25**	**18**		**670**	**11,698,111**	**68**
SUB-SAHARAN AFRICA	**657**	**41**	**16**	**2.5**	**27**	**1,053**	**94**	**5.8**	**44**	**3**	**49**	**48**	**50**		**29**	**7.1**	**18**	**13**		**520**	**9,379,573**	**70**
NORTHERN AFRICA	**173**	**27**	**7**	**2.0**	**34**	**251**	**51**	**3.6**	**38**	**4**	**64**	**63**	**66**		**46**	**0.2**	**46**	**40**		**1,200**	**3,286,031**	**53**
Algeria	31.5	29	6	2.4	29	46.6	44	3.8	39	4	69	68	70	B	49	0.1	47	43	H	1,550	919,591	34
Egypt	68.3	26	6	2.0	35	97.4	52	3.3	37	4	65	64	67	B	44	0.0	55	52	H	1,290	386,660	177
Libya	5.1	28	3	2.5	28	8.3	33	4.1	40	4	75	73	77	B	86	0.1	45	26	S	—	679,359	8
Morocco	28.8	23	6	1.7	41	39.3	37	3.1	34	5	69	67	71	B	54	0.0	59	—	H	1,240	172,413	167
Sudan	29.5	33	12	2.2	32	46.3	70	4.6	43	3	51	50	52	C	27	1.0	10	7	H	290	967,494	30
Tunisia	9.6	22	7	1.6	44	12.9	35	2.8	35	5	69	67	70	B	61	0.0	60	49	H	2,060	63,170	152
Western Sahara	0.3	46	18	2.9	24	0.4	150	6.8	44		47	46	48	D			—	—	—		97,344	3
WESTERN AFRICA	**234**	**42**	**14**	**2.8**	**25**	**390**	**89**	**5.9**	**45**	**3**	**51**	**50**	**52**		**35**	**4.1**	**14**	**7**		**340**	**2,370,015**	**99**
Benin	6.4	45	17	2.8	24	11.7	94	6.3	49	3	50	49	51	B	38	2.1	16	3	S	380	43,483	147
Burkina Faso	11.9	47	18	2.9	24	21.6	105	6.8	48	3	47	47	47	B	15	7.2	12	5	H	240	105,792	113
Cape Verde	0.4	37	9	2.8	25	0.5	77	5.3	46	6	68	65	72	C	44	—	—	—	H	1,200	1,556	258
Côte d'Ivoire	16.0	38	16	2.2	32	23.3	112	5.2	43	3	47	45	48	B	46	10.1	15	7	H	700	124,502	128
Gambia	1.3	43	19	2.4	29	2.2	130	5.6	41	3	45	43	47	C	37	2.2	15	7	H	340	4,363	299
Ghana	19.5	34	10	2.4	29	26.5	56	4.5	43	3	58	56	59	C	37	2.4	22	13	H	390	92,100	212
Guinea	7.5	42	18	2.4	29	12.6	98	5.5	44	3	45	43	47	C	26	2.1	6	4	H	530	94,927	79
Guinea-Bissau	1.2	42	20	2.2	31	1.9	130	5.8	43	4	45	43	44	C	22	2.3	—	—	H	160	13,946	87
Liberia	3.2	50	17	3.2	21	6.0	139	6.2	45	4	50	49	52	C	45	3.7	7	5	H	—	43,000	74
Mali	11.2	47	16	3.1	22	21.3	123	6.7	47	4	53	55	52	C	26	1.7	7	5	H	250	478,838	23
Mauritania	2.7	41	13	2.7	25	4.8	92	5.5	45	3	54	52	55	C	54	0.5	3	1	S	410	395,954	7
Niger	10.1	54	24	3.0	23	18.8	123	7.5	48	2	41	41	41	C	17	1.5	8	5	H	200	489,189	21
Nigeria	123.3	42	13	2.8	24	204.5	77	6.0	44	3	52	52	53	C	36	4.1	15	7	H	300	356,668	346
Senegal	9.5	41	13	2.8	25	16.7	68	5.7	45	3	52	51	54	C	41	1.8	13	8	H	520	75,954	125
Sierra Leone	5.2	47	21	2.6	26	9.9	157	6.3	45	3	45	42	47	C	37	3.2	—	—	H	140	27,699	189
Togo	5.0	42	11	3.1	23	7.6	80	6.1	46	3	49	48	50	C	31	8.5	24	7	S	330	21,927	229
EASTERN AFRICA	**246**	**42**	**18**	**2.4**	**29**	**390**	**103**	**6.0**	**45**	**3**	**46**	**45**	**47**		**20**	**10.4**	**19**	**14**		**260**	**2,456,184**	**100**
Burundi	6.1	42	17	2.5	28	10.5	75	6.5	48	3	47	46	47	C	8	8.3	—	—	H	140	10,745	563
Comoros	0.6	38	10	2.8	25	1.1	77	5.1	42	3	59	57	62	B	29	0.1	21	11	H	370	861	671
Djibouti	0.6	39	16	2.3	30	1.0	115	5.8	41	3	48	47	50	D	83	10.3	—	—	H	—	8,958	71
Eritrea	4.1	43	13	3.0	23	8.4	82	6.1	43	3	55	52	57	C	16	3.2	8	4	H	200	45,405	91
Ethiopia	64.1	45	21	2.4	29	115.0	116	6.7	45	3	46	45	47	C	15	9.3	4	3	H	100	426,371	150
Kenya	30.3	35	14	2.1	33	34.4	74	4.7	46	3	49	48	49	B	20	11.6	39	32	H	350	224,081	135

	Population, mid-2000 (millions)	Births per 1000 pop.	Deaths per 1000 pop.	Rate of natural increase (annual, %)	Doubling time at current rate[a] (years)	Projected population, 2025 (millions)	Infant mortality rate[b] (per 1000 births)	Total fertility rate[c]	Percent of population of age <15	Percent of population of age 65+	Life expectancy at birth (years) Total	Male	Female	Data available code[d]	Percent urban	Percent of adult pop. aged 15-49 with HIV/AIDS, 1997	Percent of married women using contraceptives All methods	Modern methods[e]	Government view of birth rate[f]	GNP per capita 1998 (U.S.$)	Land area (sq mi)	Population per sq mi
Madagascar	14.9	44	14	2.9	24	28.5	96	6.0	45	3	52	51	53	B	22	0.1	19	10	H	260	226,656	66
Malawi	10.4	41	22	1.9	36	12.6	127	5.9	46	3	39	38	40	C	20	14.9	22	14	H	210	45,745	227
Mauritius	1.2	17	7	1.1	66	1.4	19	2.0	26	6	70	67	74	A	43	0.1	75	60	S	3,730	788	1,510
Mozambique	19.1	41	19	2.2	32	20.6	134	5.6	46	2	40	40	39	B	28	14.2	6	5	H	210	309,494	62
Réunion	0.7	20	5	1.4	49	1.0	9	2.2	30	6	74	70	79	B	73	0.0	73	67	—	—	969	739
Rwanda	7.2	43	20	2.3	30	8.0	121	6.5	45	3	39	39	40	D	5	12.8	21	13	H	230	10,170	711
Seychelles	0.1	18	7	1.1	65	0.1	9	2.0	28	7	70	67	73	B	59	—	—	—	H	6,420	174	472
Somalia	7.3	47	18	2.9	24	14.9	126	7.0	44	3	46	45	48	D	24	0.3	—	—	S	—	246,201	29
Tanzania	35.3	42	13	2.9	24	59.8	99	5.6	45	3	53	52	54	C	20	9.4	24	16	H	220	364,900	97
Uganda	23.2	48	20	2.9	24	48.0	81	6.9	49	2	42	42	43	B	15	9.5	15	8	H	310	93,066	251
Zambia	9.6	42	23	2.0	35	14.3	109	6.1	45	3	37	37	38	B	38	19.1	26	14	H	330	290,583	33
Zimbabwe	11.3	30	20	1.0	69	9.5	80	4.0	44	3	40	41	39	B	32	25.8	54	50	H	620	150,873	75
MIDDLE AFRICA	**96**	**46**	**16**	**3.0**	**23**	**185**	**106**	**6.6**	**46**	**3**	**49**	**48**	**51**	**—**	**32**	**4.3**	**10**	**3**	**—**	**320**	**2,553,151**	**38**
Angola	12.9	48	19	3.0	23	25.1	125	6.8	48	3	47	45	48	D	32	2.1	—	—	S	380	481,351	27
Cameroon	15.4	37	12	2.6	27	24.7	77	5.2	43	3	55	55	56	C	44	4.9	19	7	H	610	183,568	84
Central African Republic	3.5	38	18	2.0	34	4.9	97	5.1	44	4	45	43	46	C	39	10.8	15	3	S	300	240,533	15
Chad	8.0	50	17	3.3	21	17.3	110	6.6	44	3	48	46	51	B	22	2.7	4	1	S	230	495,753	16
Congo	2.8	40	16	2.4	29	4.6	109	5.3	43	3	48	45	50	C	41	7.8	—	—	H	680	132,046	21
Congo, Dem. Rep. of	52.0	48	16	3.2	22	105.3	109	7.2	48	3	49	47	50	C	29	4.4	8	3	S	110	905,351	57
Equatorial Guinea	0.5	41	16	2.5	28	0.8	108	5.6	43	4	50	48	52	D	37	1.2	—	—	S	1,110	10,830	42
Gabon	1.2	38	16	2.2	32	2.0	87	5.4	39	6	52	51	54	C	73	4.3	—	—	L	4,170	103,347	12
São Tomé and Príncipe	0.2	43	9	3.4	20	0.3	51	6.2	47	4	64	63	66	B	44	—	—	—	H	270	371	432
SOUTHERN AFRICA	**50**	**26**	**13**	**1.3**	**52**	**43**	**51**	**3.1**	**35**	**5**	**54**	**53**	**55**	**—**	**42**	**13.5**	**53**	**52**	**—**	**3,100**	**1,032,730**	**48**
Botswana	1.6	32	17	1.6	45	1.2	57	4.1	41	4	44	43	45	C	49	25.1	—	—	H	3,070	224,606	7
Lesotho	2.1	33	13	2.1	33	2.4	85	4.4	41	5	53	52	55	C	16	8.4	23	19	H	570	11,718	183
Namibia	1.8	36	20	1.7	42	2.3	68	5.1	44	4	46	47	45	B	27	19.9	29	26	H	1,940	318,259	6
South Africa	43.4	25	12	1.3	55	35.1	45	2.9	34	5	55	54	57	B	45	12.9	56	55	H	3,310	471,444	92
Swaziland	1.0	41	22	1.9	37	1.6	108	5.9	47	3	38	36	39	C	22	18.5	21	19	H	1,400	6,703	150

(continued)

[a]The number of years it will take for the population to double assuming a constant rate of natural increase, based on the unrounded rate of natural increase.

[b]The annual number of deaths of infants under age 1 per 1000 live births.

[c]Average number of children born to a woman during her lifetime.

[d]A = completed data ... D = little or no data.

[e]"Modern" methods include clinic and supply methods, such as the pill, IUD, condom, and sterilization.

[f]H = too high; S = satisfactory; L = too low.

Source: Population Reference Bureau, 2000.

	Population, mid-2000 (millions)	Births per 1000 pop.	Deaths per 1000 pop.	Rate of natural increase (annual, %)	Doubling time at current rate[a] (years)	Projected population, 2025 (millions)	Infant mortality rate[b] (per 1000 births)	Total fertility rate[c]	Percent of population of age <15	65+	Life expectancy at birth (years) Total	Male	Female	Data available code[d]	Percent urban	Percent of adult pop. aged 15-49 with HIV/AIDS, 1997	Percent of married women using contraceptives All methods	Modern methods[e]	Government view of birth rate[f]	GNP per capita, 1998 (U.S.$)	Land area (sq mi)	Population per sq mi
NORTHERN AMERICA	306	14	9	0.6	124	374	7	2.0	21	13	77	74	80	—	75	0.7	77	70	—	28,230	7,699,508	40
Canada	30.8	11	7	0.4	178	36.0	6	1.5	19	12	79	76	81	A	78	0.3	80	66	S	19,170	3,849,670	8
United States	275.6	15	9	0.6	120	337.8	7	2.1	21	13	77	74	79	A	75	0.8	76	71	S	29,240	3,717,796	74
MIDDLE & SOUTH AMERICA	518	24	6	1.8	39	703	35	2.8	33	5	70	66	73	—	74	0.6	68	59	—	3,880	7,946,649	65
CENTRAL AMERICA	136	26	5	2.1	33	192	34	3.1	38	4	71	68	74	—	67	0.4	62	53	—	3,230	957,452	143
Belize	0.3	32	5	2.7	26	0.4	34	3.9	41	4	72	70	74	B	50	1.9	47	42	S	2,660	8,865	29
Costa Rica	3.6	22	4	1.8	39	5.8	13	3.2	33	5	77	75	79	B	45	0.6	75	65	S	2,770	19,730	182
El Salvador	6.3	30	7	2.4	29	9.8	35	3.6	36	5	70	67	73	B	58	0.6	60	54	H	1,850	8,124	773
Guatemala	12.7	37	7	2.9	24	22.3	45	5.0	44	3	64	61	67	B	39	0.5	38	31	H	1,640	42,042	301
Honduras	6.1	33	6	2.8	25	8.6	42	4.4	42	3	68	66	71	B	45	1.5	50	41	H	740	43,278	142
Mexico	99.6	24	4	2.0	36	132.5	32	2.7	37	5	72	69	75	B	74	0.4	65	56	H	3,840	756,062	132
Nicaragua	5.1	36	6	3.0	23	8.7	40	4.4	44	3	68	66	71	B	63	0.2	60	57	H	370	50,193	101
Panama	2.9	22	5	1.7	41	3.8	21	2.6	32	5	74	72	77	C	56	0.6	—	—	S	2,990	29,158	98
CARIBBEAN	36	22	8	1.3	52	46	47	2.6	30	7	69	66	71	—	61	1.8	—	—	—	—	90,618	401
Antigua and Barbuda	0.1	22	6	1.6	45	0.1	17	2.2	28	8	71	69	74	B	37	—	—	—	S	8,450	170	400
Bahamas	0.3	21	5	1.5	45	0.4	18	2.2	32	5	74	70	77	A	84	3.8	—	—	H	—	5,359	58
Barbados	0.3	14	9	0.5	130	0.3	14	1.8	24	10	75	72	77	A	38	2.9	—	—	S	—	166	1,560
Cuba	11.1	14	7	0.7	103	11.7	7	1.6	22	9	75	73	78	C	75	0.0	—	—	S	—	42,803	260
Dominica	0.1	16	8	0.8	83	0.1	15	1.9	38	7	78	75	80	A	—	—	—	—	S	3,150	290	262
Dominican Republic	8.4	28	6	2.2	32	12.1	47	3.1	36	4	69	67	71	B	62	1.9	64	59	H	1,770	18,815	449
Grenada	0.1	29	6	2.3	30	0.2	14	3.8	43	5	71	68	73	B	34	—	54	49	H	3,250	131	747
Guadeloupe	0.4	17	6	1.1	61	0.5	10	2.0	26	9	77	73	80	A	48	—	—	—	—	—	660	680
Haiti	6.4	33	16	1.7	40	9.6	103	4.7	40	4	49	47	51	C	34	5.2	18	14	H	410	10,714	599
Jamaica	2.6	22	7	1.6	45	3.3	24	2.6	31	7	71	70	73	B	50	1.0	66	63	H	1,740	4,243	615
Martinique	0.4	15	6	0.9	81	0.5	9	1.8	24	11	78	75	82	C	81	—	—	—	—	—	425	961
Netherlands Antilles	0.2	17	6	1.1	62	0.3	14	2.2	27	7	75	72	78	B	—	—	—	—	—	—	309	715
Puerto Rico	3.9	17	8	0.9	75	4.2	11	2.1	25	10	74	70	79	A	71	—	78	68	—	—	3,456	1,133
St. Kitts—Nevis	0.04	20	11	0.9	82	0.1	24	2.2	31	9	67	64	70	C	43	—	—	—	H	6,190	139	309
Saint Lucia	0.2	19	6	1.2	56	0.2	17	2.5	33	6	72	71	72	A	48	—	—	—	H	3,660	239	656
St. Vincent and the Grenadines	0.1	19	7	1.2	59	0.1	20	2.0	37	7	73	71	74	A	44	—	—	—	H	2,560	151	744
Trinidad and Tobago	1.3	14	7	0.7	103	1.5	16	1.7	28	6	71	68	73	A	72	0.9	—	—	H	4,520	1,981	654

	Population, mid-2000 (millions)	Births per 1000 pop.	Deaths per 1000 pop.	Rate of natural increase (annual, %)	Doubling time at current rate[a] (years)	Projected population, 2025 (millions)	Infant mortality rate[b] (per 1000 births)	Total fertility rate[c]	Percent of population of age <15	Percent of population of age 65+	Life expectancy at birth (years) Total	Male	Female	Data available code[d]	Percent urban	Percent of adult pop. aged 15–49 with HIV/AIDS, 1997	Percent of married women using contraceptives All methods	Modern methods[e]	Government view of birth rate[f]	GNP per capita, 1998 (U.S.$)	Land area (sq mi)	Population per sq mi
SOUTH AMERICA	345	23	6	1.7	42	465	34	2.7	32	5	69	66	73	—	78	0.6	72	63	—	4,270	6,898,579	50
Argentina	37.0	19	8	1.1	62	47.2	19	2.6	29	9	73	70	77	A	90	0.7	—	—	S	8,030	1,073,514	35
Bolivia	8.3	30	10	2.0	34	12.2	67	4.2	40	4	60	59	62	B	62	0.1	48	25	S	1,010	424,162	20
Brazil	170.1	21	6	1.5	45	221.2	38	2.4	30	5	68	64	71	B	78	0.6	77	70	S	4,630	3,300,154	52
Chile	15.2	18	5	1.3	54	19.5	11	2.4	29	7	75	72	78	A	85	0.2	—	—	S	4,990	292,135	52
Colombia	40.0	26	6	2.0	34	58.3	28	3.0	33	4	69	65	73	B	71	0.4	72	59	S	2,470	439,734	91
Ecuador	12.6	27	6	2.1	33	17.8	40	3.3	35	4	69	67	72	B	63	0.3	57	46	H	1,520	109,483	116
French Guiana	0.2	27	3	2.4	29	0.4	18	3.4	36	4	74	71	77	C	79	—	—	—	—	—	34,749	6
Guyana	0.7	24	7	1.7	40	0.8	63	2.7	35	4	66	63	69	C	36	2.1	—	—	S	780	83,000	8
Paraguay	5.5	32	6	2.7	26	9.4	27	4.3	41	4	70	68	72	B	52	0.1	57	48	H	1,760	157,046	35
Peru	27.1	27	6	2.1	32	39.2	43	3.4	34	5	68	66	71	B	72	0.6	64	41	H	2,440	496,224	55
Suriname	0.4	26	7	1.9	37	0.5	29	2.4	33	5	70	68	73	D	69	1.2	—	—	S	1,660	63,039	7
Uruguay	3.3	16	10	0.7	107	3.9	15	2.3	25	13	74	70	78	A	92	0.3	—	—	L	6,070	68,498	48
Venezuela	24.2	25	5	2.0	34	34.8	21	2.9	37	4	73	70	76	A	86	0.7	—	—	S	3,530	352,143	69
ASIA	3,684	22	8	1.4	48	4,723	56	2.8	32	6	66	65	68	—	35	0.3	62	57	—	2,130	12,262,691	300
ASIA (Excl. China)	2,420	26	8	1.7	40	3,292	64	3.3	35	5	64	63	65	—	38	0.5	50	43	—	2,910	8,566,591	282
WESTERN ASIA	189	28	7	2.1	33	300	55	4.0	37	4	68	66	70	—	65	0.0	—	—	—	3,620	1,823,873	103
Armenia	3.8	10	6	0.4	161	4.1	15	1.3	25	9	75	71	78	B	67	0.0	22	—	L	460	11,506	331
Azerbaijan	7.7	15	6	0.9	77	9.8	17	1.9	33	6	72	68	75	B	52	0.0	—	—	S	480	33,436	231
Bahrain	0.7	22	3	1.9	37	1.7	8	2.8	31	2	69	68	71	B	88	0.2	62	31	S	7,640	266	2,594
Cyprus	0.9	14	8	0.6	124	1.0	8	1.9	24	10	77	74	79	C	64	0.3	—	—	L	11,920	3,571	247
Georgia	5.5	9	8	0.2	462	4.8	15	1.2	24	11	73	69	76	B	56	0.0	41	20	L	970	26,911	203
Iraq	23.1	38	10	2.8	25	41.0	127	5.7	43	3	59	58	60	C	68	0.0	—	—	S	—	169,236	137
Israel	6.2	22	6	1.5	45	8.3	6	2.9	29	10	78	76	80	A	90	0.1	—	—	L	16,180	8,131	766
Jordan	5.1	33	5	2.9	24	8.8	34	4.4	42	3	69	68	70	B	78	0.0	53	38	H	1,150	34,444	148
Kuwait	2.2	24	2	2.2	32	3.8	13	3.2	29	1	72	72	73	B	100	0.1	—	—	S	—	6,880	318
Lebanon	4.2	23	7	1.6	43	5.6	35	2.4	30	6	70	68	73	D	88	0.1	—	—	S	3,560	4,015	1,046
Oman	2.4	43	5	3.9	18	5.2	25	7.1	46	3	71	69	73	B	72	0.1	24	18	H	—	82,031	29
Palestinian Territory	3.1	41	5	3.7	19	7.4	27	6.0	47	3	72	70	73	B	—	—	—	—	—	1,560	2,417	1,283
Qatar	0.6	20	2	1.8	38	0.8	20	4.2	27	1	72	70	75	C	91	0.1	—	—	S	—	4,247	139
Saudi Arabia	21.6	35	5	3.0	23	40.0	46	6.4	42	3	70	68	71	C	83	0.0	—	—	S	6,910	829,996	26
Syria	16.5	33	6	2.8	25	26.9	35	4.7	45	3	67	67	68	B	51	0.0	40	28	S	1,020	71,498	231
Turkey	65.3	22	7	1.5	46	88.0	38	2.5	30	5	69	67	71	B	66	0.0	64	38	H	3,160	299,158	218
United Arab Emirates	2.8	24	2	2.2	32	3.8	16	4.9	33	2	74	73	76	C	84	0.2	28	24	S	17,870	32,278	88
Yemen	17.0	39	11	2.8	25	38.6	75	6.5	49	3	59	58	61	B	26	0.0	21	10	H	280	203,849	84

(continued)

	Population, mid-2000 (millions)	Births per 1000 pop.	Deaths per 1000 pop.	Rate of natural increase (annual, %)	Doubling time at current rate[a] (years)	Projected population, 2025 (millions)	Infant mortality rate[b] (per 1000 births)	Total fertility rate[c]	Percent of population of age <15	Percent of population of age 65+	Life expectancy at birth (years) Total	Male	Female	Data available code[d]	Percent urban	Percent of adult pop. aged 15–49 with HIV/AIDS, 1997	Percent of married women using contraceptives All methods	Modern methods[e]	Government view of birth rate[f]	GNP per capita, 1998 (U.S.$)	Land area (sq mi)	Population per sq mi
SOUTH CENTRAL ASIA	**1,475**	**28**	**9**	**1.9**	**37**	**2,037**	**75**	**3.6**	**38**	**4**	**61**	**60**	**62**	**—**	**29**	**0.6**	**46**	**40**	**—**	**510**	**4,157,320**	**355**
Afghanistan	26.7	43	18	2.5	28	48.0	150	6.1	43	3	46	46	45	D	20	0.0	—	—	H	—	251,772	106
Bangladesh	128.1	27	8	1.8	38	177.3	82	3.3	43	3	59	59	58	B	20	0.0	49	42	H	350	55,598	2,305
Bhutan	0.9	40	9	3.1	22	1.4	71	5.6	43	2	66	—	—	D	15	0.0	8	—	H	470	18,147	48
India	1,002.1	27	9	1.8	39	1,363.0	72	3.3	36	4	61	60	61	B	28	0.8	48	43	H	440	1,269,340	789
Iran	67.4	21	6	1.4	48	90.8	31	2.9	39	5	69	68	71	B	63	0.0	73	56	H	1,650	630,575	107
Kazakhstan	14.9	14	10	0.4	161	14.6	21	1.7	29	7	65	59	70	B	56	0.0	66	54	L	1,340	1,049,151	14
Kyrgyzstan	4.9	22	7	1.5	47	5.8	26	2.8	37	6	67	63	71	B	34	0.0	60	49	S	380	76,641	64
Maldives	0.3	35	5	3.0	23	0.5	27	5.4	45	3	71	71	72	B	25	0.1	18	—	H	1,130	116	2,469
Nepal	23.9	36	11	2.5	28	38.0	79	4.6	41	3	57	58	57	B	11	0.2	29	26	H	210	56,826	421
Pakistan	150.6	39	11	2.8	25	227.0	91	5.6	43	3	58	58	59	B	33	0.1	18	13	H	470	307,375	490
Sri Lanka	19.2	18	6	1.2	60	23.9	17	2.1	35	4	72	70	74	C	22	0.1	66	44	S	810	25,332	757
Tajikistan	6.4	21	5	1.6	43	8.4	28	2.7	44	4	68	66	71	B	27	0.0	21	—	H	370	55,251	115
Turkmenistan	5.2	21	6	1.5	48	6.8	33	2.5	40	4	66	62	69	B	44	0.0	20	—	S	—	188,456	28
Uzbekistan	24.8	23	6	1.7	40	31.5	22	2.8	40	4	69	66	72	B	38	0.0	56	51	S	950	172,741	143
SOUTHEAST ASIA	**528**	**24**	**7**	**1.7**	**41**	**717**	**46**	**3.0**	**34**	**4**	**65**	**63**	**67**	**—**	**36**	**0.6**	**56**	**48**	**—**	**1,240**	**1,735,448**	**304**
Brunei	0.3	25	3	2.2	32	0.5	24	3.4	34	3	71	70	73	B	67	0.2	—	—	S	—	2,228	149
Cambodia	12.1	38	12	2.6	27	21.2	80	5.3	43	3	56	54	58	B	16	2.4	22	16	H	260	69,900	173
East Timor	0.8	34	16	1.8	39	1.2	143	4.6	42	2	46	45	47	B	—	—	—	—	—	—	5,741	137
Indonesia	212.2	24	8	1.6	44	273.4	46	2.8	34	4	64	62	66	B	39	0.1	57	55	H	640	735,355	289
Laos	5.2	41	15	2.6	26	8.4	104	5.6	44	4	51	50	52	C	17	0.0	25	21	H	320	91,429	57
Malaysia	23.3	25	5	2.1	34	37.0	8	3.2	34	4	72	70	75	C	57	0.6	—	—	H	3,670	127,317	183
Myanmar	48.9	30	10	2.0	35	68.1	83	3.8	37	4	54	53	56	C	26	1.8	17	14	S	—	261,228	187
Philippines	80.3	29	7	2.3	31	117.3	35	3.7	38	4	67	66	69	B	47	0.1	49	32	H	1,050	115,830	693
Singapore	4.0	13	5	0.8	84	8.0	3	1.5	22	7	78	76	80	A	100	0.2	65	—	L	30,170	239	16,714
Thailand	62.0	16	7	1.0	70	72.1	22	1.9	24	5	72	70	75	B	31	2.2	72	70	S	2,160	198,116	313
Vietnam	78.7	20	6	1.4	48	109.9	37	2.5	34	6	66	63	69	B	24	0.2	75	56	H	350	128,066	615
EAST ASIA	**1,493**	**15**	**7**	**0.8**	**85**	**1,669**	**29**	**1.8**	**24**	**8**	**72**	**70**	**74**	**—**	**38**	**0.1**	**81**	**78**	**—**	**3,880**	**4,546,050**	**328**
China	1,264.5	15	6	0.9	79	1,431.0	31	1.8	25	7	71	69	73	B	31	0.1	83	81	S	750	3,696,100	342
China, Hong Kong SAR	7.0	7	5	0.3	256	8.6	3	1.0	17	11	80	77	82	A	95	0.1	—	—	—	23,660	413	16,949
China, Macao SAR	0.4	10	3	0.7	96	0.6	6	1.2	25	8	77	75	80	B	99	—	—	—	—	—	8	57,628
Japan	126.9	9	8	0.2	462	120.9	4	1.3	15	17	81	77	84	A	78	0.0	64	57	L	32,350	145,869	870
Korea, North	21.7	21	7	1.5	48	25.7	26	2.3	28	6	70	67	73	C	59	0.0	—	—	S	—	46,541	466
Korea, South	47.3	14	5	0.9	82	53.3	11	1.5	22	7	74	71	78	B	79	0.0	77	66	S	8,600	38,324	1,234
Mongolia	2.5	20	7	1.4	50	3.4	34	2.7	35	4	63	60	66	C	52	0.0	57	41	S	380	604,826	4
Taiwan	22.3	13	6	0.7	97	25.3	7	1.5	21	8	75	72	78	A	77	—	—	—	—	—	13,969	1,593

	Population, mid-2000 (millions)	Births per 1000 pop.	Deaths per 1000 pop.	Rate of natural increase (annual, %)	Doubling time at current rate[a] (years)	Projected population, 2025 (millions)	Infant mortality rate[b] (per 1000 births)	Total fertility rate[c]	Percent of population of age <15	Percent of population of age 65+	Life expectancy at birth (years) Total	Male	Female	Data available code[d]	Percent urban	Percent of adult pop. aged 15-49 with HIV/AIDS, 1997	Percent of married women using contraceptives All methods	Modern methods[e]	Government view of birth rate[f]	GNP per capita, 1998 (U.S.$)	Land area (sq mi)	Population per sq mi
EUROPE:	**728**	**10**	**11**	**-0.1**	**—**	**714**	**9**	**1.4**	**18**	**14**	**74**	**70**	**78**	**—**	**73**	**0.2**	**74**	**55**	**—**	**13,420**	**8,875,866**	**82**
NORTHERN EUROPE	**96**	**12**	**11**	**0.1**	**653**	**101.4**	**6**	**1.7**	**19**	**15**	**77**	**74**	**80**	**—**	**83**	**0.1**	**73**	**64**	**—**	**21,640**	**675,794**	**141**
Denmark	5.3	12	11	0.1	472	5.8	5	1.7	18	15	76	74	79	A	85	0.1	—	—	S	33,040	16,637	320
Estonia	1.4	8	13	-0.5	—	1.3	9	1.2	19	14	70	64	75	B	69	0.0	70	56	L	3,360	17,413	82
Finland	5.2	11	10	0.2	433	5.3	4	1.7	18	15	77	74	81	A	60	0.0	84	57	S	24,280	130,560	40
Iceland	0.3	15	7	0.9	81	0.3	3	2.0	23	12	79	77	82	A	92	0.1	—	—	S	27,830	39,768	7
Ireland	3.8	15	9	0.6	116	4.5	6	1.9	22	11	76	73	79	A	58	0.1	—	—	S	18,710	27,135	140
Latvia	2.4	8	14	-0.6	—	2.1	11	1.2	19	14	70	64	76	B	69	0.0	85	51	L	2,420	24,942	97
Lithuania	3.7	10	11	-0.1	—	3.5	9	1.3	20	13	72	67	77	B	68	0.0	66	25	L	2,540	25,174	147
Norway	4.5	13	10	0.3	217	4.9	4	1.8	20	15	78	76	81	A	74	0.1	—	—	S	34,310	125,050	36
Sweden	8.9	10	11	-0.1	—	9.3	4	1.5	19	17	79	77	82	A	84	0.1	—	—	S	25,580	173,730	51
United Kingdom	59.8	12	11	0.1	546	64.1	6	1.7	19	16	77	74	80	A	89	0.1	72	68	S	21,410	94,548	632
WESTERN EUROPE	**183**	**11**	**10**	**0.1**	**612**	**188**	**5**	**1.5**	**17**	**16**	**78**	**74**	**81**	**—**	**79**	**0.2**	**80**	**73**	**—**	**26,160**	**427,702**	**429**
Austria	8.1	10	10	0.0	2,310	8.1	5	1.3	17	15	78	75	81	A	65	0.2	68	53	L	26,830	32,378	250
Belgium	10.2	11	10	0.1	770	10.3	6	1.6	18	17	78	75	81	A	97	0.1	84	74	S	25,380	11,787	869
France	59.4	13	9	0.3	204	64.2	5	1.8	19	16	78	75	82	A	74	0.4	75	68	L	24,210	212,934	279
Germany	82.1	9	10	-0.1	—	80.2	5	1.3	16	16	77	74	80	A	86	0.1	85	79	L	26,570	137,830	596
Liechtenstein	0.03	14	7	0.7	105	0.0	18	1.6	19	10	72	67	78	A	—	—	—	—	S	—	62	534
Luxembourg	0.4	13	9	0.4	198	0.6	5	1.7	19	14	77	74	80	A	88	0.1	—	—	L	45,100	999	438
Monaco	0.03	20	17	0.3	239	0.0	—	—	12	22	—	—	—	C	100	—	—	—	S	—	1	45,333
Netherlands	15.9	13	9	0.4	193	17.3	5	1.6	19	14	78	75	81	A	61	0.2	74	71	S	24,780	15,768	1,010
Switzerland	7.1	11	9	0.2	315	7.6	5	1.5	18	15	80	77	83	A	68	0.3	—	—	L	39,980	15,942	448
EASTERN EUROPE	**304**	**9**	**13**	**-0.5**	**—**	**287**	**14**	**1.2**	**19**	**13**	**69**	**64**	**74**	**—**	**68**	**0.1**	**68**	**41**	**—**	**2,340**	**7,264,035**	**42**
Belarus	10.0	9	14	-0.5	—	9.4	11	1.3	20	13	68	63	74	B	70	0.2	50	42	L	2,180	80,154	125
Bulgaria	8.2	8	14	-0.6	—	6.6	14	1.1	16	16	71	67	74	A	68	0.0	—	—	L	1,220	42,822	190
Czech Republic	10.3	9	11	-0.2	—	10.2	5	1.1	17	14	75	71	78	A	77	0.0	70	45	L	5,150	30,448	337
Hungary	10.0	9	14	-0.5	—	9.2	9	1.3	17	15	71	66	75	A	64	0.0	73	68	L	4,510	35,919	279
Moldova	4.3	11	11	0.0	1,733	4.5	18	1.5	26	9	67	63	70	B	46	0.1	74	50	S	380	13,012	329
Poland	38.6	10	10	0.0	—	38.6	9	1.4	20	12	73	69	78	B	62	0.1	76	12	L	3,910	124,807	310
Romania	22.4	11	12	-0.2	—	20.6	21	1.3	19	13	69	66	73	A	55	0.0	64	30	L	1,360	92,042	244
Russia	145.2	8	15	-0.6	—	136.9	17	1.2	19	13	67	61	73	B	73	0.1	67	49	L	2,260	6,592,819	22
Slovakia	5.4	11	10	0.1	866	5.4	9	1.4	20	11	73	69	77	A	57	0.0	74	41	L	3,700	18,923	285
Ukraine	49.5	8	14	-0.6	—	45.1	13	1.3	19	14	68	63	74	B	68	0.4	67	37	L	980	233,089	212

(continued)

	Population, mid-2000 (millions)	Births per 1000 pop.	Deaths per 1000 pop.	Rate of natural increase (annual, %)	Doubling time at current rate[a] (years)	Projected population, 2025 (millions)	Infant mortality rate[b] (per 1000 births)	Total fertility rate[c]	Percent of population of age		Life expectancy at birth (years)			Data available code[d]	Percent urban	Percent of adult pop. aged 15-49 with HIV/AIDS, 1997	Percent of married women using contraceptives		Government view of birth rate[f]	GNP per capita, 1998 (U.S.$)	Land area (sq mi)	Population per sq mi
									<15	65+	Total	Male	Female				All methods	Modern methods[e]				
SOUTHERN EUROPE	145	10	10	0.0	2121	137	7	1.3	16	16	77	74	80	–	70	0.3	–	–	–	15,340	508,336	285
Albania	3.4	18	5	1.3	55	4.5	22	2.2	33	6	71	69	74	B	46	0.0	–	–	S	810	11,100	309
Andorra	0.1	11	3	0.8	92	0.1	6	1.2	15	12	0	–	–	C	95	–	–	–	S	–	174	386
Bosnia–Herzegovina	3.8	13	8	0.5	141	4.2	12	1.6	22	8	73	71	76	D	40	0.0	–	–	L	–	19,741	193
Croatia	4.6	11	12	-0.1	–	4.4	8	1.5	20	12	72	69	76	A	54	0.0	–	–	L	4,620	21,830	211
Greece	10.6	10	10	0.0	–	10.4	7	1.3	16	17	78	75	81	A	59	0.1	–	56	L	11,740	50,950	208
Italy	57.8	9	10	-0.1	–	52.4	6	1.2	15	17	78	75	81	A	90	0.3	91	56	L	20,090	116,320	497
Macedonia	2.0	15	8	0.6	112	2.2	16	1.9	23	9	73	70	75	A	59	0.0	–	–	H	1,290	9,927	205
Malta	0.4	12	8	0.4	182	0.4	5	1.8	21	12	77	74	80	B	89	0.1	86	43	S	10,100	124	3,157
Portugal	10.0	11	11	0.1	990	9.3	5	1.5	17	15	75	72	79	A	48	0.7	–	–	L	10,670	35,514	282
San Marino	0.03	11	7	0.4	193	0.04	9	1.2	15	16	80	76	83	C	88	–	–	–	S	–	23	1,166
Slovenia	2.0	9	10	-0.1	–	1.9	5	1.2	17	14	75	71	79	A	50	0.0	84	54	S	9,780	7,819	252
Spain	39.5	9	9	0.0	6931	36.7	6	1.2	15	16	78	74	82	A	64	0.6	72	70	L	14,100	195,363	202
Yugoslavia	10.7	11	11	0.1	866	10.7	10	1.6	21	13	72	70	75	D	52	0.1	–	–	S	–	39,448	270
OCEANIA	31	18	7	1.1	65	39	29	2.4	26	10	74	72	77	–	70	0.1	61	56	–	15,400	3,306,692	9
Australia	19.2	13	7	0.6	110	22.8	5	1.7	21	12	79	76	82	A	85	0.1	67	63	S	20,640	2,988,888	6
Federated States of Micronesia	0.1	33	7	2.6	27	0.2	46	4.7	44	4	66	65	67	C	27	–	–	–	H	1,800	270	440
Fiji	0.8	22	7	1.5	46	1.1	13	3.3	35	3	67	65	69	C	46	0.1	–	–	S	2,210	7,054	115
French Polynesia	0.2	21	5	1.6	44	0.3	10	2.6	31	3	72	69	74	C	54	–	–	–	–	–	1,544	150
Guam	0.2	28	4	2.4	29	0.2	9	3.5	32	5	74	72	77	A	38	–	–	–	–	–	212	720
Kiribati	0.1	33	8	2.5	28	0.2	62	4.5	40	3	62	59	65	C	37	–	–	–	H	1,170	282	326
Marshall Islands	0.1	26	4	2.2	31	0.2	31	6.6	49	3	65	63	67	C	65	–	–	–	H	1,540	69	978
Nauru	0.01	19	5	1.4	48	0.0	25	3.7	43	1	61	57	65	B	100	–	–	–	S	–	9	1,360
New Caledonia	0.2	21	5	1.7	42	0.3	7	2.7	31	5	72	69	77	C	59	–	–	–	–	–	7,174	30
New Zealand	3.8	15	7	0.8	89	4.4	6	2.0	23	12	77	74	80	A	85	0.1	75	72	S	14,600	104,452	37
Palau	0.02	18	8	1.0	68	0.0	19	2.5	28	6	67	64	71	C	71	–	–	–	S	–	178	108
Papua–New Guinea	4.8	34	10	2.4	29	7.7	77	4.8	40	4	56	56	57	C	15	0.2	26	20	H	890	178,703	27
Solomon Islands	0.4	37	6	3.1	23	0.8	25	5.4	43	3	71	69	74	C	13	–	–	–	H	760	11,158	39
Tonga	0.1	27	6	2.1	33	0.2	19	4.2	43	4	71	70	72	C	32	–	–	–	S	1,750	290	372
Vanuatu	0.2	35	7	2.8	25	0.3	39	4.7	44	3	65	64	67	C	18	–	–	–	S	1,260	4,707	41
Western Samoa	0.2	31	6	2.5	28	0.2	25	4.2	39	4	68	65	72	C	21	–	–	20	H	1,070	1,097	161

Aborigines (p. 577) the longest surviving inhabitants of Oceania, whose ancestors, the Australoids, migrated from Southeast Asia possibly as early as 50,000 years ago, over the Sundaland landmass that was exposed during the ice ages

acid rain (p. 89) acidic precipitation that has formed through the interaction of rainwater or moisture in the air with sulfur dioxide and nitrogen oxides emitted during the burning of fossil fuels

acculturation (pp. 146, 205) adaptation of a minority culture to the host culture enough to function effectively and be self-supporting; cultural borrowing

age distribution or **age structure** (p. 36) the proportion of the total population in each age group

Age of Exploration (p. 187) a period of accelerated global commerce and cultural exchange facilitated by improvements in European navigation, shipbuilding, and commerce in the fifteenth and sixteenth centuries

agroecology (p. 420) the practice of traditional methods of crop fertilization and the use of natural predators to control pests

agroforestry (p. 376) the raising of economically useful trees

Ainu (p. 479) an indigenous cultural minority group in Japan characterized by their light skin, heavy beards, and thick, wavy hair, who are thought to have migrated thousands of years ago from the northern Asian steppes

air pressure (p. 20) the force exerted by a column of air on a square foot of surface

Allah (p. 296) the Arabic word for "God," used by Arabic-speaking Christians and Muslims in their prayers

alluvium (p. 491) river-borne sediment

altiplano (p. 171) an area of high plains in the central Andes of South America

animism (pp. 368–369) a belief system in which natural features carry spiritual meaning

apartheid (p. 348) a system of laws mandating racial segregation in South Africa, in effect from 1948 until 1994

aquifer (p. 91) a natural underground reservoir

arable land (p. 375, map) land suitable for cultivation

archipelago (p. 518) a group, often a chain, of islands

Arya (p. 401) an ancient people of Southwest Asia; the first recorded invaders of South Asia

ASEAN Free Trade Association (AFTA) (p. 534) a free-trade association of Southeast Asian countries launched in 1992 by ASEAN and patterned after the North American Free Trade Agreement and the European Union

Asia-Pacific region (p. 504) a huge trading area that includes all of Asia and the countries around the Pacific Rim

assimilation (pp. 146, 204) the loss of old ways of life and the adoption of the lifestyle of another culture

Association of Southeast Asian Nations (ASEAN) (p. 534) an organization of Southeast Asian governments established to further economic growth and political cooperation

atoll (p. 572) a low-lying island, formed of coral reefs that have built up on the circular or oval rims of submerged volcanoes

Australo-Melanesians (p. 521) a group of hunters and gatherers who moved from the present northern Indian and Burman parts of southern Eurasia into the exposed landmass of Sundaland about 40,000 to 60,000 years ago

Austronesians (pp. 521, 577) a Mongoloid group of skilled farmers and seafarers from southern China who migrated south to various parts of Southeast Asia between 10,000 and 5000 years ago

average population density (p. 35) the average number of people per unit area (for example, square mile or square kilometer)

Aztecs (p. 124) native people of central Mexico noted for advanced civilization before the Spanish conquest

baby boomer (p. 87) a member of the largest age group in North America, the generation born in the years after World War II, from 1947 to 1964, in which a marked jump in the birth rate occurred

barriadas (p. 132) see **favelas**

barrios (p. 132) see **favelas**

biogeographical transition zone (p. 518) an area in which there is some mixing of two or more distinct types of flora and fauna

birth rate (p. 35) the number of births in a given population per unit time, especially per year

Bolsheviks (p. 243) a faction of communists who came to power during the Russian Revolution

boreal forests (p. 104) northern coniferous forests

Brahmins (p. 415) members of the priestly caste, which is the most privileged caste in ritual status

brain drain (pp. 134, 402) the migration of educated and ambitious young adults to cities, depriving the sending communities of talented youth in whom they have invested years of nurturing and education

bride price (p. 417) a price paid by a groom to the family of the bride; the opposite of dowry

brown field (p. 77) an old industrial site whose degraded conditions pose obstacles to redevelopment

Buddhism (p. 413) a religion of Asia that originated in northern India in the sixth century B.C. and emphasizes modern living and peaceful self-reflection leading to enlightenment

buffer zone (pp. 486, 523) a neutral area that serves to prevent conflict

capital (p. 30) wealth in the form of money or property used to produce more wealth

capital controls (p. 31) restrictions on business investment by foreigners requiring that investment capital stay in the host country for a minimum length of time or that companies be partially controlled by local investors

capitalists (p. 243) a wealthy minority that owns the majority of factories, farms, businesses, and other means of production

carrying capacity (p. 349) the maximum number of people that a given territory can support sustainably with food, water, and other essential resources

cartel (p. 310) a group that is able to control production and set prices for its products

caste (p. 414) an ancient system for dividing society into hereditary hierarchical classes

chaebol (p. 507) huge corporate conglomerates in South Korea that receive government support and protection

chernozem (p. 242) black, fertile soils of the region stretching south from Moscow toward the Black and Caspian seas

China proper (p. 458) the North China Plain, a rich agricultural area that was unified under successive empires and that is considered the hearth of East Asian culture

Chipko (social forestries) **movement** (p. 428) an Indian environmental movement that attempts to slow down deforestation and increase ecological awareness

Christianity (p. 295) a monotheistic religion based on the teachings of Jesus of Nazareth, a Jew, who described God's relationship to humans as primarily one of love and support, as exemplified by the Ten Commandments

clear-cutting (p. 63) the cutting down of all trees on a given plot of land, regardless of age, health, or species

climate (p. 20) the long-term balance of temperature and precipitation that characteristically prevails in a particular region

cold war (p. 190) the contest that pitted the United States and Western Europe, espousing free-market capitalism and democracy, against the former USSR and its allies, promoting a centrally planned economy and a socialist state

cold war era (p. 46) the period from 1946 to the early 1990s, when the United States and its allies in Western Europe faced off against the Union of Soviet Socialist Republics and its allies in Eastern Europe

colonias (p. 132) see **favelas**

colonizing (or **mother**) **country** (p. 126) the country from which the people of a colony or former colony derive their origin

colony (p. 30) a (usually) distant land acquired by a more powerful country for economic gain

command economy (p. 243) an economy in which government bureaucrats plan, locate, and manage all production and distribution

communal conflict (p. 414) a euphemism for religion-based violence in South Asia

communism (p. 243) an ideology based largely on the writings of the German revolutionary Karl Marx, which calls on workers to unite to overthrow capitalists and establish an egalitarian society where workers share whatever they produce

Communist party (p. 243) the political party that ruled the former Russian Empire from 1917 to 1991

Confucianism (p. 458) a Chinese philosophy that teaches that the best organizational model for the state and society is a hierarchy based on the patriarchal family

contested space (p. 151) any area that several groups claim or want to use in different ways, such as the Amazon or Palestine

contiguous regions (p. 9) regions that lie next to or near each other

Corridor Five (p. 200) a planned and partially constructed major transportation roadway across southern Europe, stretching from Barcelona, Spain, to Kiev, Ukraine

crony capitalism (p. 533) a type of corruption in which politicians, bankers, and entrepreneurs, sometimes members of the same family, have close personal as well as business relationships

cultural diversity (p. 12) differences in ideas, values, technologies, and institutions among culture groups

cultural homogeneity (p. 11) uniformity of ideas, values, technologies, and institutions among culture groups

cultural identity (p. 11) a sense of personal affinity with a particular culture group

cultural marker (p. 11) a characteristic that helps to define a certain culture group

cultural pluralism (or **cultural complexity**) (p. 537) the cultural identity characteristic of a region where groups of people from many different backgrounds have lived together for a long time, yet have remained distinct

Cultural Revolution (p. 461) a series of highly politicized and destructive mass campaigns launched in 1966 to force the entire population to support the continuing revolution in China

culture (p. 10) everything we use to live on earth that is not directly part of our biological inheritance

culture group (p. 10) a group of people who share a set of beliefs, a way of life, a technology, and usually a place

currency devaluation (p. 359) the lowering of a currency's value relative to the U.S. dollar, the Japanese yen, the European Euro, or other currency of global trade

czar (p. 243) a ruler of the Russian Empire; derived from "Caesar," the title of the Roman emperors

death rate (p. 35) the ratio of total deaths to total population in a specified community

delta (p. 20) the triangular-shaped plain of sediment that forms where a river meets the sea

democratic institutions (p. 189) institutions—such as constitutions, elected parliaments, and impartial courts—through which the common people gained a formal role in the political life of the nation

demographic transition (p. 40) the change from high birth and death rates to low birth and death rates that usually accompanies a cluster of other changes, such as change from a subsistence to a cash economy, increasing education rates, and urbanization

demography (p. 34) the study of population patterns and changes

deposition (p. 20) the settling out of rock and soil particles from wind or water as it slows

desertification (p. 316) a set of ecological changes that convert nondesert lands into deserts

detritus (p. 520) dead organic material (such as plants and insects)

devolution (p. 219) the weakening of a formerly tightly unified state

Diaspora (p. 295) the dispersion of Jews around the globe after they were expelled from the eastern Mediterranean by the Roman Empire beginning in A.D. 73

diffusion (p. 26) the process by which agriculture and the domestication of animals spread around the world from a few places

digital divide (p. 73) the discrepancy in access to information technology between small, rural, and poor areas and large, wealthy cities; major government research laboratories; and universities

doi moi (p. 556) economic and bureaucratic restructuring in Vietnam

domestication (p. 26) the process of developing plants and animals through selective breeding to live with and be of use to humans

domestic spaces (p. 336) spaces in the home relegated to women

double day (p. 205) the longer workday of women with jobs outside the home who also work as family caretakers, housekeepers, and cooks at home

dowry (p. 417) a price paid by the family of a bride to the groom; the opposite of bride price

dry forests (p. 376) forests that lose their leaves during the dry season

dumping (p. 200) the cheap sale on the world market of overproduced commodities, which lowers global prices and hurts producers of these same commodities elsewhere in the world

early extractive phase (p. 135) a phase in Central and South American history, beginning with the Spanish conquest and lasting until the early twentieth century, characterized by colonial policies such as mercantilism that resulted in unequal trade

earthquake (p. 19) a catastrophic shaking of the landscape, often caused by the shifting and friction of tectonic plates

East African Community (EAC) (pp. 361–362) an organization formed by Kenya, Tanzania, and Uganda to promote economic links among the countries of East Africa

economic and technology development zones (ETDZs) (p. 470) zones in China with fewer restrictions on foreign business, established to encourage foreign investment and economic growth

Economic Community of West African States (ECOWAS) (p. 361) an organization of West African states working toward forming an economic union

economic core (p. 62) the dominant economic region within a larger region; specifically, southern Ontario and the north central part of the United States

economic diversification (p. 312) the expansion of an economy to include a wider array of economic activities

economic integration (p. 191) the free movement of people, goods, money, and ideas among countries

economic restructuring (p. 118) reorganization of an economy to encourage economic growth in markets free of government controls

economies of scale (p. 194) reductions in the unit costs of production that occur when goods or services are produced in large amounts, resulting in a rise in profits per unit

economy (p. 29) the forum where people make their living, including the spatial, social, and political aspects of how resources are recognized, extracted, exchanged, transformed, and reallocated

ecotourism (p. 152) nature-oriented vacations often taken in endangered and remote landscapes, usually by travelers from industrialized nations

El Niño (pp. 123, 575) periodic climate-altering changes, especially in the circulation of the Pacific Ocean, now understood to operate on a global scale

endemic (p. 575) belonging or restricted to a particular place

erosion (p. 20) the process by which fragmented rock and soil are moved over a distance, primarily by wind and water

ethnic group (pp. 10, 342) a group of people who share a set of beliefs, a way of life, a technology, and usually a geographic location

ethnicity (p. 342) the quality of belonging to a particular culture group

ethnocentrism (p. 342) the belief in the superiority of one's own cultural perspective

Euro, the (p. 195) the official currency of 12 of the 15 European Union countries, as of January 1, 1999

European Economic Community (EEC) (p. 194) an economic community created in 1958, when Belgium, Luxembourg, the Netherlands, France, Italy, and West Germany agreed to eliminate certain tariffs and promote mutual trade and cooperation in the interest of achieving stronger economic ties within Europe

European Union (EU) (pp. 182–183) a supranational institution including most of West, South, and North Europe, established to bring economic integration to member countries

exchange (p. 29) barter or trade for money, goods, or services

exclave (p. 226) a portion of a country separated from the main part and constituting an enclave in respect ot the surrounding territory

Export Processing Zones (EPZs) (or **free-trade zones**) (p. 139) specially created legal spaces or industrial parks within a country where, to attract foreign-owned factories, duties and taxes are not charged

extended family (p. 146) a family consisting of related individuals beyond the nuclear family of parents and children

external processes (geophysical) (p. 19) landform-shaping processes that originate at the surface of the earth, such as weathering, mass wasting, and erosion

extraction (p. 29) the acquisition of a material resource through mining, logging, agriculture, or other means

extractive resource (p. 29) a resource such as mineral ores, timber, or plants that must be mined from the earth's surface or grown from its soil

extraregional migration (p. 543) short-term or permanent migration to countries outside a region

evangelical Protestantism (p. 148) a Christian movement that focuses on personal salvation and empowerment of the individual through miraculous healing and transformation; some practitioners preach the "gospel of success" to the poor: that a life dedicated to Christ will result in prosperity of the body and soul

fall line (p. 78) a line of waterfalls and rapids that cannot be crossed by boats

fatwa (p. 417) a legal opinion based on an interpretation of Muslim law

favelas (p. 132) Brazilian urban slums and shantytowns built by the poor; called **colonias, barrios,** or **barriadas** in other countries

female genital mutilation (p. 372) the removal of the labia and the clitoris and often the stitching nearly shut of the vulva

Fertile Crescent (p. 294) an arc of lush, fertile land formed by the uplands of the Tigris and Euphrates river systems and the Zagros Mountains, where nomadic peoples began the earliest known agricultural communities in the world

feudalism (p. 187) a social system established by A.D. 900 in Europe, which created a class of professional fighting men, or knights, ready to defend the farmers, or serfs, who cultivated their protectors' land

floating population (p. 472) jobless or underemployed people who have left economically depressed rural areas for the cities and move around looking for work

floodplain (p. 20) the flat land around a river where sediment is deposited during flooding

food stability (p. 469) the ability of a state to supply a sufficient amount of basic food to the entire population consistently

forced migration (p. 544) the movement of people against their wishes

formal economy (p. 29) all aspects of the economy that take place in official channels

formal institutions (p. 10) associations such as official religious organizations; local, state, and national governments; nongovernmental organizations; and specific businesses and corporations

forward capital (pp. 176, 382) a city built to draw migrants and investment for economic development

fragile environment (p. 376) an area that contains barely enough water, soil nutrients, or other resources essential to meet the needs of plants and animals; human pressure in such environments may result in long-term or irreversible damage to plant and animal life

free trade (p. 31) the movement of goods and capital without government restrictions

free-trade zones (p. 139) see **Export Processing Zones**

frontal precipitation (p. 23) rainfall caused by the interaction of large air masses of different temperatures and densities

gender structure (p. 36) the proportion of males and females in each age group of a population

genetically modified organisms (GMOs) (p. 196) plants or animals whose genetic code has been modified to make them more attractive as commodities

genocide (p. 350) the deliberate destruction of an ethnic, racial, or political group

geopolitics (p. 46) the use of strategies by countries to ensure that their best interests are served

glasnost ("openness") (p. 244) an opening up of public discussion of social and economic problems, which occurred under Mikhail Gorbachev in the late 1980s

global economy (p. 27) the worldwide system in which goods, services, and labor are exchanged

globalization (p. 8) the growth of interregional and worldwide linkages and the changes they are bringing about

global scale (pp. 7, 9) the level of geography that encompasses the entire world as a single unified area

global warming (p. 45) the predicted warming of the earth's climate as atmospheric levels of greenhouse gases increase

Gondwana (p. 571) the great landmass that formed the southern part of the ancient supercontinent Pangaea

government subsidies (p. 74) amounts paid by the government to cover part of the production costs of some products and to help domestic producers to sell their goods for less than foreign competitors

grassroots rural economic development (p. 362) programs established to provide sustainable livelihoods in the countryside

Great Barrier Reef (p. 571) the longest coral reef in the world, located off the northeastern coast of Australia

Green (p. 209) name for environmentally conscious political parties

green revolution (p. 420) increases in food production brought about through the use of new seeds, fertilizers, mechanized equipment, irrigation, pesticides, and herbicides

gross domestic product (GDP) per capita (pp. 29, 32–33) the market value of all goods and services produced by workers and capital within a particular nation's borders and within a given year; dividing the value by the number of people in the country results in the per capita value

Group of Seven (G7) (p. 255) an organization of the most highly industrialized nations, now sometimes called Group of Eight (G8) when Russia is invited to meetings

growth poles (p. 471) zones of development whose success draws more investment and migration to a region

growth rate (p. 35) see **rate of natural increase**

growth triangles (p. 531) large transnational economic regions where firms attempt to find the best skills and resources for the best prices

guarantor (p. 474) in China, a third party who vouches for the honesty and reliability of one party in his or her dealings with another party

guest workers (p. 203) immigrants from outside Europe, often from former colonies, who came to Europe (often temporarily) to fill labor shortages; Europeans expected them to return home when no longer needed

Gulf War (1990–1991) (p. 313) one of the most massive and economically destructive wars that has ever taken place in the region, in which a military coalition of the United States and European and Arab countries quickly defeated Iraq after it invaded oil-rich Kuwait in 1990

hacienda (p. 135) a large agricultural estate in Middle or South America, more common in the past; usually not specialized by crop and not focused on market production

hajj (p. 296) the pilgrimage to the city of Mecca that all Muslims are encouraged to undertake at least once in a lifetime

Harijans (or Dalits or untouchables) (p. 415) members of a social group considered so lowly as to have no caste

hearth (p. 413) place of origin (of Hinduism)

hierarchical societies (p. 579) social structures in which groups are ranked according to power, prestige, and wealth, with layers of ruling elites at the top and undifferentiated commoners at the bottom

high islands (p. 572) volcanoes that rise above the sea into mountainous, rocky formations that contain a rich variety of environments

Hindustani (p. 403) during Mogul rule, the lingua franca (language of trade) of all of northern India and what is today Pakistan; the precursor of modern Urdu and Hindi

Hispanic (p. 58) a loose ethnic term that refers to all Spanish-speaking people from Latin America and Spain; equivalent to **Latino**

Horn of Africa (p. 344) the part of Africa that juts out from East Africa and wraps around the southern tip of the Arabian Peninsula

hub-and-spoke network (p. 71) the organization of air service in North America around hubs, or strategically located airports used as collection and transfer points for passengers and cargo traveling from one place to another

human geography (p. 5) the study of various aspects of human life that create the distinctive landscapes and regions of the world

human well-being (p. 10) the ability of people to obtain for themselves a healthy life in a place and community of their choosing

import quota (p. 31) a limit on the amount of a given item that may be imported into a country over a period of time

import substitution industrialization (ISI) (p. 136) a form of industrialization involving the use of public funds to set up factories to produce goods that previously had been imported

Incas (p. 124) Native American people who ruled the largest pre-Columbian state in the Americas, with a domain stretching from southern Colombia to northern Chile and Argentina

income disparity (p. 134) a dramatic gap in wealth and resources between the rich elite and the poor majority of a country or region

Indian diaspora (p. 402) the set of all people of South Asian origin living (and often born) abroad

Indus Valley civilization (or Harappa culture) (p. 401) the first substantial settled agricultural communities in South Asia, which appeared around 4500 years ago along the Indus River in modern-day Pakistan and along the Saraswati River in modern-day India

informal economy (p. 29) all aspects of the economy that take place outside official channels

informal economy (p. 140) all aspects of the economy that take place outside official channels

informal institutions (p. 10) ordinary or casual associations, such as the family or a community

information technology (IT) (p. 72) the part of the service sector that relies on the use of computers and the Internet to process and transport information, including banks, software companies, medical technology companies, and publishing houses

infrastructure (p. 194) basic facilities that a society needs to operate (transportation and communication systems, power and water, etc.)

institutions (p. 10) all of the associations, formal and informal, that help people get along together

internal processes (geophysical) (p. 19) processes, like plate tectonics, that originate deep beneath the surface of the earth

internal republics (p. 257) more than 30 ethnic enclaves in Russia, whose political status and relationship to the Russian state are continuously being renegotiated republic by republic

International Monetary Fund (IMF) (p. 47) a financial institution funded by the developed nations to help developing countries reorganize, formalize, and develop their economies

interstate highway system (p. 71) the federally subsidized network of highways in the United States

intertropical convergence zone (ITZC) (p. 342) a band of atmospheric currents circling the globe roughly around the equator; warm winds from both north and south converge at the ITCZ, pushing air upward and causing copious rainfall

intifada (p. 328) a prolonged Palestinian uprising against Israel

iron curtain (p. 190) a fortified border zone that separated Western Europe from Eastern Europe during the cold war

Islam (pp. 295, 413) a monotheistic religion considered an outgrowth of Judaism and Christianity; it emerged in the seventh century A.D. when, according to tradition, the archangel Gabriel revealed the tenets of the religion to the Prophet Muhammad

Islamic fundamentalism (p. 298) a grassroots movement to replace secular governments and civil laws with governments and laws guided by Islamic principles

Islamists (p. 299) fundamentalist Muslims who favor a religiously based state that incorporates conservative interpretations of the Qur'an and strictly enforced Islamic principles into the legal system

Israeli-Palestinian conflict (p. 314) a long-running conflict over the control of territory between Israel and Palestinians who live on lands now occupied by Israel

Jainism (p. 413) a faith tradition that is more than 2000 years old; Jains are found mainly in western India and large urban centers and are known for their strict vegetarianism

jati (p. 414) the subcaste into which a person is born, which largely defines the individual's experience for a lifetime

Judaism (p. 294) a monotheistic religion characterized by the belief in one God, Yaweh, a strong ethical code summarized in the Ten Commandments, and an enduring ethnic identity

kanban system (p. 466) an efficient Japanese manufacturing system in which suppliers are located in close proximity to the main factory where final assembly takes place

Kshatriyas (p. 415) members of the warrior and ruler caste

laterite (p. 344) a permanently hard surface left when minerals in tropical soils are leached away

Latino (p. 58) a loose ethnic term that refers to all Spanish-speaking people from Middle and South America and Spain; equivalent to **Hispanic**

leaching (p. 344) the washing out into groundwater of soil minerals and nutrients released into soil by decaying organic matter

liberation theology (p. 148) a movement within the Roman Catholic church that uses the teachings of Christ to encourage the poor to organize to change their own lives and the rich to promote social and economic reform

lingua franca (pp. 13, 374) a language used to communicate by people who don't speak one another's native languages

livestock ranch (p. 135) a farm that specializes in raising cattle and sheep

loess (pp. 58, 486) windblown dust that forms deep soils in North America, central Europe, and China

long-lot system (p. 95) a system of long, narrow plots of land stretching back from the edge of the St. Lawrence River, which gave French Canadian settlers access to resources extending inland from the river

lopsided spatial development (p. 132) the concentration of wealth and power in one place

machismo (p. 147) a set of values that defines manliness in Middle and South America

makatea (p. 572) raised coral platforms formed when atolls are uplifted

maquiladoras (pp. 110, 149, 161) foreign-owned factories, often located in Mexican towns just over the border from U.S. towns, that hire people at low wages to assemble manufactured goods that are then sent

elsewhere for sale; now also used for similar firms in other parts of Mexico and Middle America

marianismo (p. 147) a set of values based on the life of the Virgin Mary that defines the proper social roles for women in Middle and South America

marsupials (p. 575) mammals that give birth to their young at a very immature stage and then nurture them in a pouch equipped with nipples

material culture (pp. 10, 15) all the things, living or not, that humans use

megalopolis (p. 77) an area formed when several cities grow to the extent that their edges meet and coalesce

Melanesia (p. 578) New Guinea and the islands south of the equator and west of Tonga (the Solomon Islands, New Caledonia, Fiji, Vanuatu)

Melanesians (p. 577) a group of Australoids named for their relatively dark skin tones, a result of high levels of the protective pigment melanin; they settled throughout New Guinea and other nearby islands

mercantilism (pp. 126, 188) the policy by which the rulers of Spain and Portugal, and later of England and Holland, sought to increase the power and wealth of their realms by managing all aspects of production, transport, and trade in their colonies

Mercosur (p. 140) a free-trade zone created in 1991 that links the economies of Brazil, Argentina, Uruguay, and Paraguay to create a common market

mestizo (pp. 128, 146) a person of mixed European and Native American descent

metropolitan areas (p. 77) cities with a population of 50,000 or more and their surrounding suburbs

microcredit (p. 423) a program that makes very small loans available to poor entrepreneurs

Micronesia (pp. 577–578) the small islands lying east of the Philippines and north of the equator

migration (p. 131) the movement of people from one place to another

MIRAB (p. 589) an economy based on migration, remittance, aid, and bureaucracy

mixed agriculture (p. 386) the raising of a variety of crops and animals on a single farm, often to take advantage of several environmental niches

Moguls (p. 403) Muslim invaders of South Asia

Mongols (p. 242) a loose confederation of nomadic pastoral people centered in eastern Central Asia, who by the thirteenth century conquered an empire stretching from Europe to the Pacific

monotheistic (p. 294) pertaining to the belief that there is only one god

monotremes (p. 575) egg-laying mammals such as the duck-billed platypus and the spiny anteater

monsoon (pp. 23, 398) opposing winter and summer patterns of atmospheric and moisture movement between continents and oceans, in which warm, wet air coming in from the ocean brings copious rainfall during the summer months; in winter, cool, dry air moves from the continental interior toward the ocean

multiculturalism (p. 11) the state of relating to, reflecting, or being adapted to diverse cultures

multinational corporation (p. 30) a business organization that operates extraction, production, and/or distribution facilities in multiple countries

Muslims (p. 413) followers of Islam

nation (p. 46) a group of people who share a language, culture, political philosophy, and usually a territory

nationalism (p. 189) devotion to the interests or culture of a particular country or nation (cultural group); the idea that a group of people living in a specific territory and sharing cultural traits should be united in a single country to which they are loyal and obedient

nation-state (pp. 46, 189) a political unit, or country, formed by people who share a language, a culture, and a political philosophy

neocolonialism (pp. 136, 341) modern efforts by dominant countries to control economic and political affairs in other countries to further their own aims

noble savage (p. 579) a term coined by European Romanticists to describe what they termed the "primitive" peoples of the Pacific, who lived in distant places supposedly untouched by corrupting influences

nodes (p. 77) small regions with extremely dense populations

nomadic pastoralism (p. 509) a way of life and economy centered on the tending of grazing animals who are moved seasonally to gain access to the best grasses

nongovernmental organization (NGO) (p. 47) an association outside the formal institutions of government, in which individuals from widely differing backgrounds and locations share views and activism on political, economic, social, or environmental issues

nonpoint sources of pollution (p. 267) diffuse sources of environmental contamination, such as untreated automobile exhaust, raw sewage, and agricultural chemicals that drain from fields into the urban water supplies

North American Free Trade Agreement (NAFTA) (p. 140) a free-trade agreement created in 1994 that added Mexico to the 1989 agreement between the United States and Canada

nuclear family (p. 86) a family group consisting of a father and mother and their children

Ogallala aquifer (p. 91) the largest North American natural aquifer, which underlies the Great Plains

old-growth forests (p. 546) forests that have never been logged

oligarchies (p. 255) powerful elites that wield influence because of their enormous wealth

OPEC (Organization of Petroleum Exporting Countries) (p. 310) a cartel of oil-producing countries—including Algeria, Indonesia, Iran, Iraq, Kuwait, Libya, Nigeria, Oman, Qatar, Saudi Arabia, the United Arab Emirates, and Venezuela—that was established to regulate the production, and hence the price, of oil and natural gas

organic matter (p. 344) the remains of any living thing

orographic rainfall (p. 22) rainfall produced when a moving moist air mass encounters a mountain range, rises, cools, and releases condensed moisture that falls as rain

Pacific Rim (p. 66) all of the countries that border the Pacific Ocean on the west and east

Pacific Way (p. 585) the idea that Pacific Islanders have a regional identity and a way of handling conflicts peacefully, which grows out of their own particular social experience

Pancasila (p. 534) the Indonesian national philosophy based on tolerance, particularly in matters of religion; its precepts include belief in God and the observance of conformity, corporatism, consensus, and harmony

Pangaea hypothesis (pp. 18–19) the proposal that about 200 million years ago all continents were joined in a single vast continent, called Pangaea

Parsis (p. 413) a highly visible religious minority in India's western cities; they are descendants of Persian migrants who did not give up their traditional religion of Zoroastrianism when Iran became Muslim

perestroika (p. 244) a restructuring of the Soviet economic system in the late 1980s in an attempt to revitalize the economy

permafrost (p. 240) permanently frozen soil a few feet beneath the surface

permeable national borders (p. 110) borders subject to easy flow of people and goods

physical geography (p. 5) the study of the earth's physical processes to learn how they work, how they affect humans, and how they are affected by humans in return

pidgin (p. 585) a language used for trading made up of words borrowed from the several languages of people involved in trading relationships

plantation (p. 135) a large estate or farm on which a large group of resident laborers grow (and partially process) a single cash crop

plate tectonics (p. 19) a theory proposing that the earth's surface is composed of large plates that float on top of an underlying layer of molten rock; the movement and interaction of the plates create many of the large features of the earth's surface, particularly mountains

pluralistic state (p. 46) a country in which power is shared among groups, each defined by common language, ethnicity, culture, or other characteristics

pogroms (p. 327) episodes of persecution, ethnic cleansing, and sometimes massacre, especially conducted against European Jews

political ecologist (p. 41) a geographer who studies the interactions among development, human well-being, and the environment

polygamy (p. 307) the practice of having more than one wife at a time

polygyny (p. 372) the taking by a man of more than one wife at a time

Polynesia (p. 578) the numerous islands situated inside an irregular triangle formed by New Zealand, Hawaii, and Easter Island

population pyramid (p. 36) a graph that depicts age and gender structures of a country

populist movements (p. 148) popularly based efforts seeking relief for the poor

primate city (p. 131) a city that is vastly larger than all others in a country and in which economic and political activity is centered

privatization (p. 250) the sale of industries formerly owned and operated by the government to private companies or individuals

production (p. 29) a form of economic activity in which resources are transformed to produce new commodities or new bodies of knowledge

protectorate (p. 297) a relationship of partial control assumed by a European nation over a dependent country, while maintaining the trappings of local government

Protestant Reformation (p. 187) a European reform (or "protest") movement that challenged Catholic practices in the sixteenth century, resulting in the establishment of Protestant churches

public spaces (p. 336) social spaces such as the town square, shops, and markets, which are defined in contrast to the domestic (home and private) sphere

pull factors (p. 541) positive features of a place that attract people to move there

purchasing power parity (PPP) (p. 33) the amount of goods or services that U.S. $1 will purchase in a given country

purdah (pp. 410, 416) the practice of concealing women, especially during their reproductive years, from the eyes of nonfamily men

push factors (p. 541) negative features of the place where people are living that impel them to move elsewhere

qanats (p. 494) ancient underground conduits that carry groundwater for irrigation

Québecois (p. 58) French Canadian ethnic group or members of that group; also, all citizens of Québec, regardless of ethnicity

Qur'an (or Koran) (p. 296) the holy book of Islam, believed by Muslims to contain the words Allah revealed to Muhammad through the archangel Gabriel

racism (p. 17) the negative assessment of people, often those who look different, primarily on the basis of skin color and other physical features

rain shadow (p. 22) the dry side of a mountain range, facing away from the prevailing winds

rate of natural increase (growth rate) (p. 35) the rate of population growth measured as the excess of births over deaths per 1000 individuals per year, without regard for the effects of migration

region (p. 5) a unit of the earth's surface that contains distinct patterns of physical features or of human development

regional conflict (p. 426) a conflict created by the resistance of a regional ethnic or religious minority to the authority of a national or state government

regional geography (p. 5) the analysis of the geographic characteristics of particular places

regional self-sufficiency (p. 468) an economic policy that encouraged each region to develop independently in the hope of evening out the national distribution of production and income in communist China

regional trade bloc (p. 31) an association of neighboring countries that have agreed to lower trade barriers for one another

religious nationalism (p. 425) the belief that a certain religion is strongly connected to a particular territory and that adherents should have political power in that territory

remittances (p. 130) earnings sent home by immigrant workers

Renaissance (p. 187) a broad European cultural movement in the fourteenth through sixteenth centuries that drew inspiration from the Greek, Roman, and Islamic civilizations, marking the transition from medieval to modern times

resettlement schemes (p. 542) government plans to move large numbers of people from one part of a country to another to relieve urban congestion, dispense political dissidents, or accomplish other social purposes

resource (p. 29) anything that is recognized as useful, such as mineral ores, forest products, skills, or brainpower

resource base (p. 192) the selection of raw materials and human skills available in a region for domestic use and industrial development

responsibility systems (p. 468) economic reforms that gave the managers of Chinese state-owned enterprises the right and the responsibility to make their operations work efficiently

Ring of Fire (p. 19) the tectonic plate junctures around the edges of the Pacific Ocean, characterized by volcanoes and earthquakes

roaring forties (p. 573) powerful westerly air and ocean currents that speed around the far Southern Hemisphere virtually unimpeded by landmasses

Russian Federation (p. 257) Russia and its political subunits, which include 30 internal republics and more than 10 so-called autonomous regions

Sahel (p. 375) a band of arid grassland that runs east-west along the southern edge of the Sahara Desert

salinization (p. 311) damage to soil caused by the evaporation of water, which leaves behind salts and other minerals

Schengen Accord (pp. 201–202) an agreement signed in the 1990s by the EU and many of its neighbors that called for free movement across common borders

seclusion (p. 297) the requirement that a woman stay out of public view, a regional cultural practice that possibly predates Islam

secondary forests (p. 546) forests that grow back after the first cutting, often with decreased diversity of species

secularism (p. 13) a way of life informed by values that do not derive from any one religious tradition

secular states (p. 305) countries that have no state religion and in which religion has no direct influence on affairs of state or civil law

self-reliant development (p. 362) small-scale development schemes in rural areas that focus on developing local skills, creating local jobs, producing products or services for local consumption, and maintaining local control so that participants retain a sense of ownership

serfs (p. 243) persons legally bound to live and farm on land owned by the lord

service sector (p. 72) economic activity that amounts to doing services for others

sex tourism (p. 544) the sexual entertainment industry that services civilian men who arrive in Southeast Asia from around the globe to live out their fantasies during a few weeks of vacation

shari'a (pp. 304, 434) literally "the correct path"; Islamic religious law that guides daily life according to the principles of the Qur'an

sheiks (p. 324) patriarchal leaders of tribal groups on the Arabian Peninsula

shifting cultivation (pp. 124, 344) a productive system of agriculture in which small plots are cleared in forestlands, the dried brush is burned to release nutrients, and the clearings are planted with multiple species; each plot is used for only two or three years and then abandoned for many years of regrowth

Shi'ite (or Shi'a) (p. 305) the smaller of two major groups of Muslims with differing interpretations of shari'a; Shi'ites are found primarily in Iran

shogun (p. 462) a member of the military elite of feudal Japan

Sikhism (p. 413) a religion of South Asia that combines beliefs of Islam and Hinduism

Sikhs (p. 413) adherents of Sikhism

silt (p. 119) fine soil particles

Slavs (p. 242) a group of farmers who originated between the Dnieper and Vistula rivers, in modern-day Poland, Ukraine, and Belarus

slash-and-burn (shifting or swidden) cultivation (p. 528) an ancient form of multicrop gardening using small plots, often on lands that do not support more intensive types of agriculture

smog (p. 89) a combination of industrial air pollution and car exhaust (smoke + fog)

socialism (p. 244) a social system in which the production and distribution of goods are owned collectively, and political power is exercised by the whole community

social welfare (pp. 206–207) in Europe, elaborate tax-supported systems that serve all citizens in one way or another

Southern African Development Community (SADC) (pp. 361–362) an organization of 14 countries in Southern and East Africa working together for regional development and freer trade

spatial freedom (p. 307) the ability to move about without restrictions

special economic zones (SEZs) (p. 470) free-trade zones within China

steppes (p. 240) semiarid, grass-covered plains

Structural Adjustment Policies (SAPs) (pp. 31–32) requirements for economic reorganization toward less government involvement in industry, agriculture, and social services, sometimes made by the World Bank and IMF as conditions for giving loans to borrowing countries

structural adjustment programs (SAPs) (pp. 137, 358) policy changes designed to free up money for loan repayment to international banks

subcontinent (p. 397) term used to refer to the entire Indian peninsula, including southeastern Pakistan and Bangladesh

subduction (p. 119) the sliding of one lithospheric (tectonic) plate under another

subsidies (p. 200) monetary assistance granted by a government to an individual or group in support of an activity, such as farming or housing construction, that is viewed as being in the public interest

subsistence affluence (p. 589) the ability of Pacific Islanders who rely primarily on a home-grown diet, informal local economies, and remittances from overseas to maintain a safe and healthy lifestyle on relatively little formal income

subsistence lifestyle (p. 38) a way of life in which each family group produces its own food, clothing, and shelter

Sudras (p. 415) members of the caste of low-status laborers and artisans

Sunni (p. 304) the larger of two major groups of Muslims with differing interpretations of shari'a

sustainable agriculture (p. 41) farming that meets human needs without poisoning the environment or using up water and soil resources

sustainable development (p. 41) efforts to improve standards of living in ways that will not jeopardize those of future generations

syncretism (or fusion) (p. 370) the blending of elements of a new faith with elements of an indigenous religious heritage

taiga forests (pp. 104, 242) subarctic forests

Taliban (pp. 416, 434) an archconservative Islamist movement that gained control of the government of Afghanistan in the mid-1990s

tariff (p. 31) a tax imposed by a country on imported goods, usually intended to protect industries within that country

technology (p. 15) an integrated system of knowledge, skills, tools, and methods upon which a culture group's way of life is based

theocratic states (p. 305) countries that require all government leaders to subscribe to a state religion and all citizens to follow certain rules decreed by that religion

thermal inversion (p. 89) a warm mass of stagnant air that is temporarily trapped beneath heavy cooler air

thermal pollution (p. 266) the return of unnaturally hot water to the environment, as by industrial processes or hydro and nuclear power plants

tierra caliente (p. 122) low-lying "hot lands" in Middle and South America

tierra fria (p. 123) "cool lands" at relatively high elevations in Middle and South America

tierra helada (p. 123) very high elevation "frozen lands" in Middle and South America

tierra templada (p. 122) "temperate lands" with year-round springlike climates at moderate elevations in Middle and South America

transmigration (p. 542) the local term used for resettlement in Indonesia

transparent (policies or practices) (p. 474) open to public scrutiny

tsunami (p. 483) a large sea wave caused by an earthquake

tundra (pp. 104, 240) a treeless area between the ice cap and the tree line of arctic regions, which has a permanently frozen subsoil

underclass (p. 580) the lowest social stratum, composed of the disadvantaged

United Nations (UN) (p. 47) an assembly of 185 member states that sponsors programs and agencies that focus on scientific research, humanitarian aid, planning for development, fostering general health, and peacekeeping assistance

urban sprawl (p. 77) the encroachment of suburbs on agricultural land

Vaishyas (p. 415) members of the landowning farmer and merchant caste

volcano (p. 19) an area between plates or a weak point in the middle of a plate where gases and molten rock, called magma, can come to the earth's surfaces through fissures and holes in the plate

Wallacea (p. 518) a trench extending from Bali through the Makassar Strait and along the western side of the Philippines; within this zone there is some mixing of Asian and Australian flora and fauna

weather (p. 20) the short-term (day-to-day) expression of climate

weathering (p. 19) the process by which rocks are physically or chemically broken up

welfare state (p. 189) a social system in which the state accepts responsibility for the well-being of its citizens

wet rice cultivation (pp. 490, 528) a prolific type of rice production that requires the submersion of the plant roots in water for part of the growing season

World Bank (pp. 31, 47) a global lending institution that makes loans to countries that need money to pay for development projects

World Trade Organization (WTO) (pp. 31, 47, 472–473) a global institution whose stated mission is the lowering of trade barriers and the establishment of ground rules for international trade

yurt (or ger) (p. 494) a round, heavy felt tent stretched over collapsible willow frames used by nomadic herders in northwestern China and Mongolia

Zionists (p. 327) European Jews who worked to create a Jewish homeland (Zion) on lands once occupied by their ancestors in Palestine